AF773663

Proceedings of the Conference on

Grouting In Geotechnical Engineering

Edited by Wallace Hayward Baker

Sponsored by the Geotechnical Engineering Division
of the American Society of Civil Engineers

in cooperation with
ASCE/AIME Underground Technology Research Council
Louisiana Section, ASCE
New Orleans Branch, ASCE

Monteleone Hotel
New Orleans, Louisiana
February 10-12, 1982

AMERICAN SOCIETY OF CIVIL ENGINEERS FOUNDED 1852 ®

Published by the
American Society of Civil Engineers
345 East 47th Street
New York, New York 10017

The Society is not responsible for any statements made or opinions expressed in its publications.

Copyright ©1982 by the American Society of Civil Engineers,
All Rights Reserved.
Library of Congress Catalog Card Number 81-71798
ISBN 0-87262-295-9
Manufactured in the United States of America.

PREFACE

Since 1960, the Geotechnical Engineering Division has sponsored specialty conferences at two-year intervals. These highly successful meetings have covered a wide range of geotechnical subjects and have made significant contributions to engineering literature. Most of the specialty conferences have been organized by task committees appointed specifically and solely for such purpose. This, the first specialty conference to deal exclusively with grouting, has been organized by the Grouting Committee of the Geotechnical Engineering Division. Committee members involved during the past three years in the various tasks required include:

WALLACE HAYWARD BAKER, Co-Chairman
Hayward Baker Company, Odenton, MD

R. H. KAROL, Co-Chairman
Rutgers University, New Brunswick, NJ

PETER P. ABERLE
Bureau of Reclamation, Denver, CO

JOHN A. ALBRITTON
Corps of Engineers, Omaha, NE

RICHARD M. BERRY
Penetryn/BPR, Knoxville, TN

G. WAYNE CLOUGH
Stanford University, Stanford, CA

RAY COPE
Grouting Consultant, Lincoln, NE
(formerly Soil Conservation Service, ret.)

JAMES H. COULSON
Tennessee Valley Authority, Knoxville, TN

EDWARD D. GRAF
Pressure Grout Company, South San Francisco, CA

JEAN-YVES PEREZ
Woodward-Clyde Consultants, Chicago, IL

BIRGER SCHMIDT
Parsons, Brinckerhoff, Quade & Douglas, Inc.
San Francisco, CA

JAMES WARNER
Consulting Engineer, Mariposa, CA

JOSEPH P. WELSH
Hayward Baker Company, Odenton, MD

It has been almost two decades since the last publication of proceedings from a grouting conference. The 63 papers included in these proceedings come from 12 different countries and deal with the pertinent advances in grouting materials and technology that have occurred throughout the world over the past two decades.

A post-conference discussion volume will be published with distribution to registered attendees. In addition, the Grouting Committee welcomes communication from anyone with a contribution to make to the advancement of the grouting profession.

Wallace Hayward Baker
ASCE Grouting Conference Committee Co-Chairman

R. H. Karol
ASCE Grouting Conference Committee Co-Chairman
ASCE Grouting Committee, Chairman

CONTENTS

SESSION IV
DESIGN AND CONTROL FOR DAM GROUTING

SESSION V
CHEMICAL GROUTS MATERIALS

SESSION VI
BEHAVIOR OF CHEMICALLY GROUTED SOIL

SESSION VII
CHEMICAL GROUTING TECHNOLOGY AND APPLICATIONS

SESSION VIII
GROUTING FOR TUNNELS, SHAFTS, AND MINES

SESSION IX
ALTERNATIVE GROUTING TECHNOLOGY (A)

SESSION X
TESTING AND CONTROL FOR GROUTING

SESSION XI
ALTERNATIVE GROUTING TECHNOLOGY

SESSION XII
APPLICATIONS OF GROUTING TECHNOLOGY

CEMENT GROUTING FOR DAMS

By Adam Clive Houlsby, M.ASCE.

KEYNOTE PAPER

Grouting was once a mysterious operation shrouded in all sorts of mumbo jumbo. Fortunately this is no longer the case and we can nowadays carry out cement grouting as an engineered process in much the same way as other construction work is done in order to achieve specified standards of workmanship.

However, the techniques used to do this need a lot of understanding, much skill, and meticulous attention to detail. It is possible to explain their basics, give examples of their use and warn of what their improper use can produce, but, beyond that, it devolves on the grouting practitioner to apply his skill. This is where experience comes in! Grouting, more than most engineered processes, requires an intuitive perception of just what the liquid grout does as it flows through the open joints and cracks hidden away down there underground. How is this intuitive understanding gained? The only way is by experience - not just sitting in an office or shed either, but actually out amongst the grout holes working the valves oneself every so often, particularly when troublesome holes are encountered!

Lest those who yet lack sufficient experience feel discouraged, let them take heart now that in recent years the basic techniques have become developed to an advanced stage able to provide a stronger base than formerly for the gaining of further experience. This Keynote paper describes current philosophies on these basics, with an emphasis on the lesser understood aspects and on controversial issues. It seeks to provoke constructive thought and discussion leading, hopefully, to continuing improvements.

The paper is structured to deal firstly with some fundamental issues relating to the all-important decision as to whether grouting is really needed and, if so, how much of it. Equipment is then discussed, followed by injection techniques, grouting methods, pressures, treatment of moving rock, and drilling methods. The paper concludes with some thoughts about future progress in grouting technology.

SOME TYPICAL GROUTING ARRANGEMENTS

Foundation grouting at dams mostly takes the form of a curtain and this is for the purpose of influencing the amount of underseepage. Other forms of grouting include blanket grouting for strengthening of an area and grouting used in tunnel work.

Dealing with curtain grouting for the present, fig. 1 illustrates the usual range of positions of grout curtains for the main types of dam. At embankment dams with an earth core (or of the homogeneous type) a curtain, if used, should be located upstream of the core axis but not so far upstream that insufficient of the impervious fill foundation remains between it and the stored water influence. The shaded areas indicate the usual positions, ranging

---

Engineer Specialising in Grouting, and Senior Surveillance Engineer, Water Resources Commission, North Sydney, N.S.W. Australia.

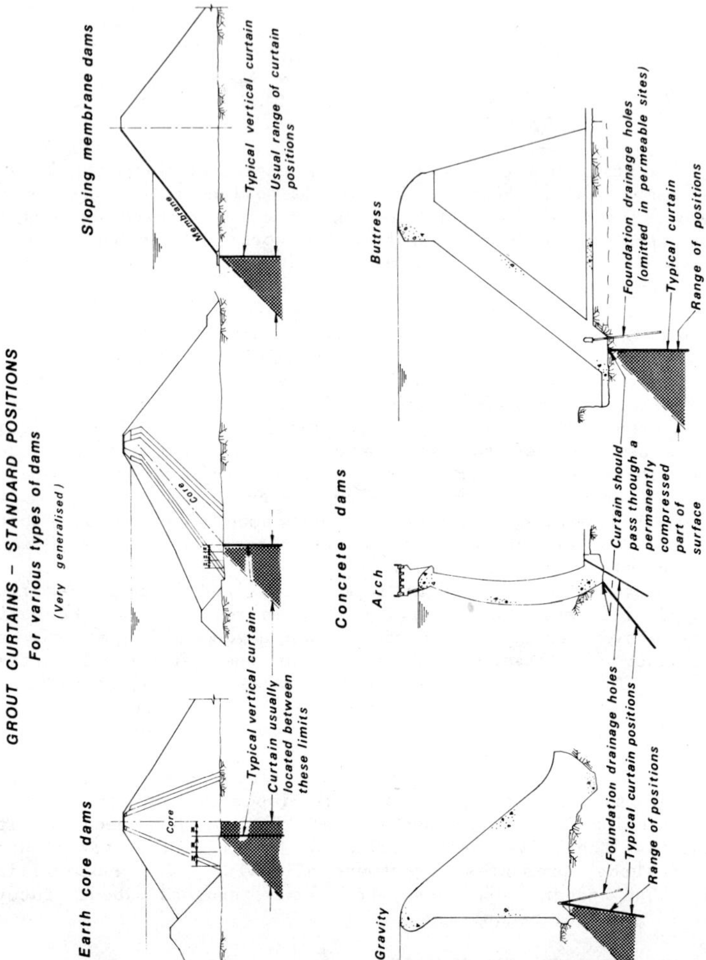

FIG 1 USUAL POSITIONS OF GROUT CURTAINS

from a vertical curtain at the axis to a steeply inclined one commencing at a third of the way between the axis and the upstream toe of the core.

Curtains at faced rockfill dams, commence from the upstream toe and range from a vertical to an inclined upstream orientation.

Concrete dams usually need foundation drainage holes as an adjunct to the grout curtain [or vice versa possibly (7)]. In gravity dams, both are commonly drilled from a gallery within the dam. Some arch dams likewise make use of a gallery for this purpose. The small arch dam shown on fig. 1, however, has the arrangement where the grouting is carried out from rock before concrete placement. At an arch dam, the curtain must pass through the part of the foundation which stays more or less compressed during flexure cycles of the dam.

Buttress dams likewise use arrangements with and without a gallery.

Fig. 2 gives a general impression of a basic layout for a grout curtain. There are many variations on the theme, however, and some jobs might not have much resemblance to this example.

Before proceeding to the question of whether a grout curtain is necessary and, if so, to what extent, it would be appropriate to outline relevant permeability and geological considerations.

THE LUGEON UNIT AND PERMEABILITY

This is rather a key unit in much grouting practice: it has been extensively used in Europe since its introduction by Lugeon in 1933(17). However, it has not had wide use in the United States and, therefore, a brief description might be of interest.

The Lugeon unit is a measure of permeability obtained from pump-in water tests. The basic definition: one Lugeon unit is a water take of 1 litre per metre of hole per minute at 10 bars pressure.

However, this pressure of 10 bars, which is 150 p.s.i., is usually too high for routine grouting work and much lower pressures are therefore used. A correction is applied to relate the lower pressures to that in the definition and is applied as follows:-

$$\text{Water Take (litres/metre/min.)} \times \frac{10\ \text{(bars)}}{\text{actual pressure (bars)}} = \text{Lugeon Value}$$

or in customary units

$$\text{Water Take (cu. ft./ft./min.)} \times 0.0107 \times \frac{150\ \text{(p.s.i.)}}{\text{actual pressure (p.s.i.)}} = \text{Lugeon Value}$$

To get a sense of proportion for the Lugeon unit, it might be noted that -

1 Lugeon = 1.3×10^{-5} cm/sec.

or 10 ft./year very approximately

1 Lugeon is the degree of permeability encountered in those nearly tight foundations which require almost no grouting at all.

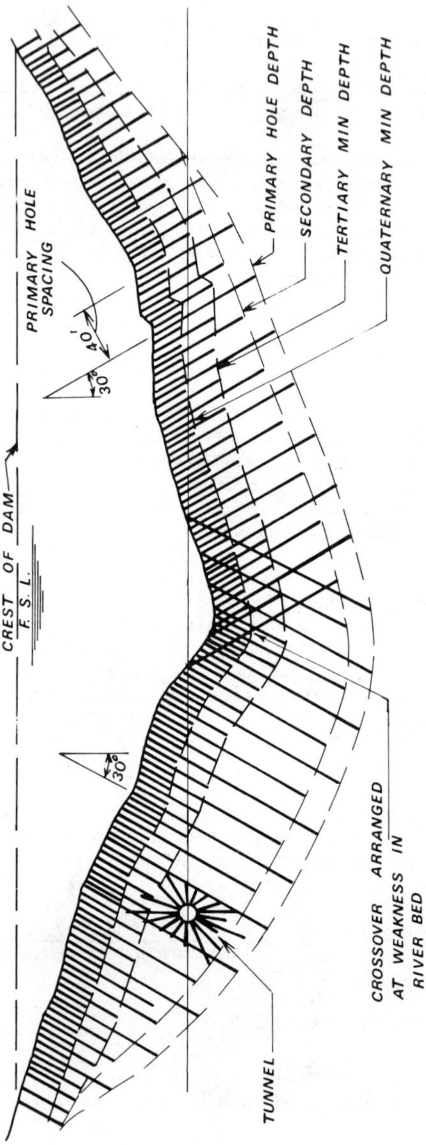

FIG 2 BASIC LAYOUT FOR A SIMPLE CURTAIN

<u>10 Lugeons</u> warrants grouting for most types of dams.

<u>100 Lugeons</u> is encountered in heavily jointed sites with relatively open joints. It is also met in sparsely cracked foundations where joints are very wide open

The Lugeon scale decreases in sensitivity as the values increase. The greatest sensitivity and importance are in the low values, 1-5. When values such as 50 are reached, an accuracy better than ±10 units is all that is warranted, and when 100 units are reached, ±30 units is the best accuracy needed. Although the scale has no upper limit, beyond 100 units values become meaningless, and for grouting work, can be conveniently designated as "greater than 100 Lugeons" without further elaboration.

Lugeons are quoted to the nearest whole number when less than about 30 Lugeons, (use of decimals implies an illusionary accuracy quite inconsistent with reality). Above that, quoting to the nearest 5 Lugeons is appropriate.

References (2, 6, 10, 24) deal further with Lugeon tests.

It is not absolutely necessary to use these units but it is helpful to do so in view of the units' international nature and appropriateness for grouting work. Alternatives are all rather clumsy. To go a step further and calculate water takes into "velocity" type permeability units such as ft./year or m/sec., may please the purists but retunds reality!

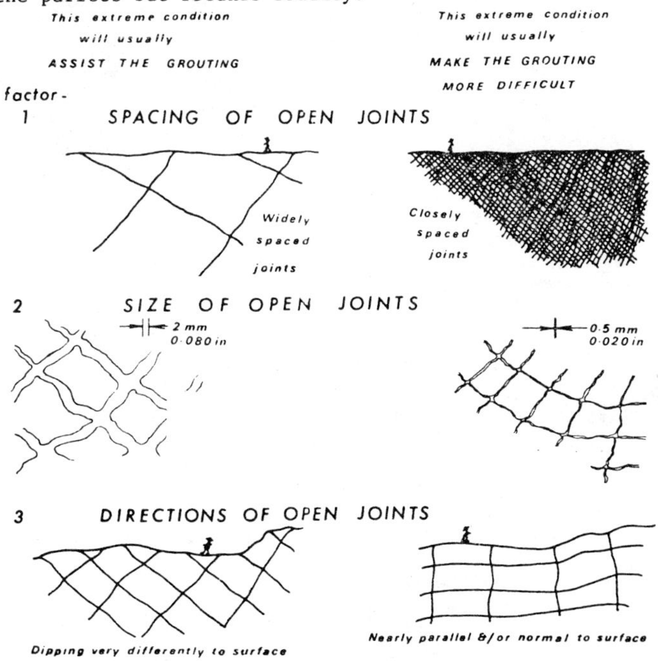

<u>FIG 3</u> EFFECTS OF GEOLOGY ON GROUTING

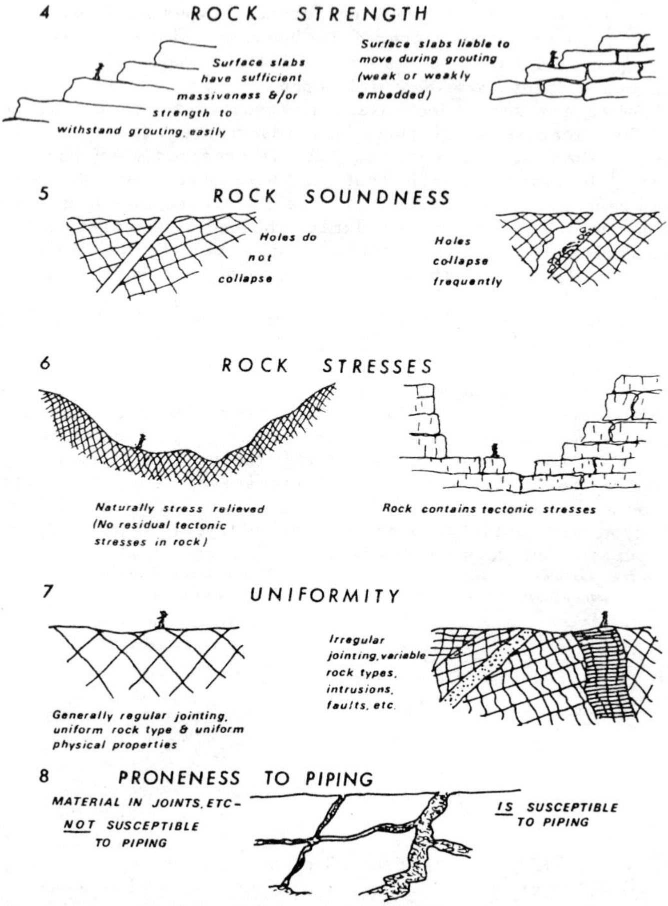

FIG 3 (CONTINUED)

GEOLOGICAL CONSIDERATIONS

This is not the place to expound on engineering geology in detail. The summary shown in fig. 3 draws attention, however, to aspects of geology which influence the grouting, either by making it easier or vice versa. These aspects are exaggerated into extremes on the figure to illustrate each point. Actual cases usually fall between the extremes. On some of the sketches a stick figure of a man is shown for scale.

Commenting on the factors of fig. 3:-

Factor 1 - Spacing of open joints ranges from widely spaced joints to closely spaced ones. In general, wide spacing of joints makes grouting easier. Close spacing can lead to many surface leaks, patchy grouting and other difficulties which can, of course, be overcome by using appropriate techniques.

Factor 2 - Joint openings wider than 0.080" (2 mm) assist the grout to penetrate easily although if they are very wide, such as a quarter of an inch (6 mm), the grout might penetrate so easily that proper tightening to refusal requires inhibition of excessive travel, by using multiple applications of grout at intervals sufficiently spaced that the grout has time to set between each. At the other extreme, joints tighter than about 0.020" (0.5 mm) will make penetration of the grout relatively difficult

Factor 3 - The direction of open joints has an influence on the orienting of the grout holes and on the possibility of rock movements during the grouting. Joints which dip between about 30° and 60° are easily intercepted by vertical holes and are less liable to permit rock movements than those of near-horizontal or near-vertical dip. These latter require inclined grouting holes which, although not really difficult, are not, however, quite as easy to drill as vertical holes are.

Factor 4 - Rock strength can assist the grouting if surface slabs are massive, tough and well anchored into bedrock. The other extreme of this is when the rock is weak with loose surface slabs or bands which move under the grouting pressure.

Factor 5 - Soundness of the rock is a factor which can be either helpful, where the rock is sufficiently sound that holes do not collapse, or troublesome if holes collapse frequently during or after drilling.

Factor 6 - Stress in rock is usually of tectonic origin and can be severe in massive strong rock. The presence of such stress can often be detected visually if cliff faces and outcrops show the characteristic near-vertical rebound or valley bulging jointing. This forms where abutment blocks part company with bedrock as the result of gradual propagation of these cracks to the stage where stress in the small area remaining attached becomes so intense that the strength of the rock is exceeded and sudden fracturing results. The strain energy released can frequently project the detaching rock away from the bedrock and leave these rebound joints open a matter of inches. The grouting practitioner needs to be able to recognise the presence of stressed rock because of its dominating effect on grouting procedures. The rather cumbersome word "tectonophysics" has been coined for study of stress in rock and it is the subject of much recent research, references (20, 22) are of interest in this regard. At the other extreme to high strength rock, is the soft, intensely cracked foundation of the type which has already stress relieved itself and will present no difficulties on this score during grouting.

Factor 7 - Uniformity of the foundation can be of assistance to the layout of the grout holes if everything is nice and regular so that grout holes can be evenly spaced and all can have the same inclination. Irregular foundations may require the holes to be at various inclinations and weak features may require especially intensive localised grouting.

Factor 8 - If material in joints is able to be removed by seepage, the grouting may need to be more intensive than otherwise to virtually eliminate seepage in the piping prone areas.

These factors are only the commonly met ones. There are many others and vigilance is necessary to find them before grouting starts so as to be able to design the grouting intelligently and appropriately.

WHEN IS GROUTING OF A FOUNDATION NEEDED?

The previous comments on geology relate mainly to considerations following a decision that grouting is indeed necessary. However, there are occasions when grouting can quite safely be omitted and likewise there are marginal cases where its need or its omission are equally hard to justify. In order to guide deliberation on such issues, fig. 4 [from (11)] shows a flow chart which commences with a question, "How valuable is water lost by leakage?". If water is precious or is sufficiently expensive to warrant the cost of intensive grouting, then grouting could be desirable if the foundation has a permeability generally in excess of 1, 2 or 3 Lugeons as indicated in the chart. Examples of extra-tight grouting, such as this, include pumped storages, particularly of the off-stream type, and storages in arid, water-short climates.

If water lost by seepage is of negligible value, which is, of course, the usual case, the flow chart of fig. 4 next leads to the question, "Does piping of foundation material need to be prevented?". A positive answer to this question could lead to the need for a fairly tight curtain such as 3 Lugeons. A negative answer, which again is the usual case, leads next to considerations concerning the type of dam. Embankment dams, including membrane faced embankments, require various degrees of tightness depending on the width of the core (if the embankment is zoned), and the number of rows in the grout curtain. Likewise, the requirements of concrete dams can vary as indicated. Additional considerations apply often, such as when there may be a special need to protect the underside of an earth core against erosion by seeping water.

Once the degree of tightness required of the grouting has been decided, the natural permeability of the foundation when compared with it gives an idea as to whether grouting is desirable or not. Sometimes there is no need to grout if the foundation is naturally tighter than the standard required of the grouting. At other sites, only part of it may be pervious enough to warrant grouting. The commonest case is when the whole foundation is too permeable naturally and requires grouting to tighten it up to the desired standard.

A fairly thorough sampling of the permeability of the site by the use of investigation holes is needed in advance of grouting design. By conducting pump-in tests at various pressures, preferably on the pattern of "low-medium-high-medium-low" pressures and then by interpreting the flow regime as indicated in references (2, 10), a fairly confident measure of permeabilities is obtainable. This multi-pressure test should not be confused with the water tests in the actual grout holes before grouting which are discussed later and are usually single pressure tests.

Fig. 4 is a guide only. It must not be regarded as a universal recipe. Every site has its own peculiarities and these

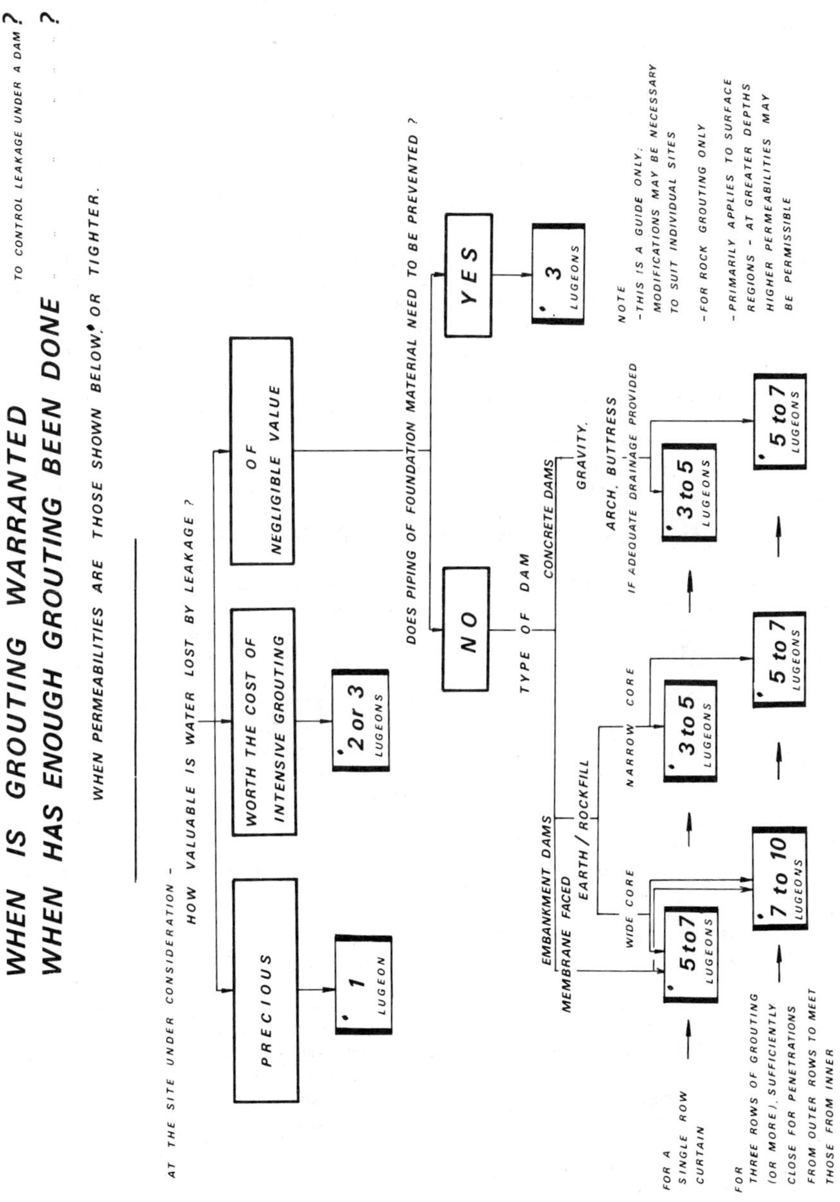

FIG 4 GUIDE FOR DECIDING WHEN GROUTING IS NEEDED, & IF SO, TO WHAT INTENSITY

must be taken individually into account particularly in relation to cracking frequencies and sizes, which may require evaluation by borehole T.V. or periscope in addition to permeability tests. Further, the pre-grouting water testing usually comes up with such a wide range of Lugeon values that sweeping generalisations can be necessary to produce a workable grouting design. Finesse is futile. Common sense is essential.

The original basis for fig. 4 came from Maurice Lugeon who, in 1933(17), originated his concept that grouting is necessary if the site permeability exceeds 1 Lugeon for dams over 100 ft. (30 metres) high, and 3 Lugeons for dams lower than this. His consulting practice dealt with concrete gravity and arch dams in the Alps and his criterion was intended for such dams. Subsequently, these standards have been eased as the result of experience and have been extended to other types of dams and to different foundations. Nonveiller (21), Arhippainen (2), Acker and Jones (1) and others have discussed the concept of basing a grout design on permeability, and Fergusson and Lancaster-Jones (8), in a comprehensive treatment of grouting design, indicated use of 2 Lugeons as the criterion for Dokan embankment dam in 1955 to 1960. Pratt, McMordie and Dundas (23) used 3 Lugeons for outer rows at W.A.C. Bennett Dam and 1 Lugeon for the centre row in the mid 1960's. The subsequent easing of criteria, as shown in fig. 4, and in particular those shown greater than 5 Lugeons, reflect the writer's experience on dams such as those documented in other papers of this Conference (13) taken in conjunction with his assessments of current published practice.

At some sites, the curtain may play only a temporary role until such time as silting proceeds sufficiently to clog most intake paths.

Fig 4 mainly considers curtain grouting. In weak or heavily loaded foundations, other forms of grouting, such as blanket grouting, may be necessary to strengthen the rock. The need for this is usually assessed by considering whether sufficient or poorly filled jointing exists to warrant filling with grout for the purpose of preventing undue settlement or dilation as the result of structural loads.

Tunnels and conduits may require special grouting treatment to provide a cut-off at some convenient point along their length. Blasting during excavation often loosens the rock surrounding or adjoining these, needing grouting to tighten it again to give adequate cut-off and structural strength.

In most situations requiring grouting at dams, cement is the most suitable type of grout. Alternatives such as clay grout and chemical grout do have occasional application, such as in alluvial foundations, and sometimes these are mixed with cement. However, they are not the subject of this present paper.

If cement grouting is decided upon, after considering these matters, together with any other special requirements of the site or dam, there can follow the need to select an appropriate grouting method (upstage, downstage, circuit etc.) together with deciding the hole positions, closure criteria, etc. These are dependent to some extent on the sort of equipment used and the injection techniques. Accordingly these are next outlined.

GROUTING EQUIPMENT

The development of grouting equipment has progressed from primitive, makeshift gear to fairly highly developed, purpose-made machinery. The greatest improvements have been in mixing and pumping.

The basic items are:-

(a) mixer - to mix the grout ingredients and then discharge to

(b) agitator - which keeps the grout stirred ready for the pump to take;

(c) pump - which draws from the agitator and pumps the grout around a circulation line;

(d) circulation line commencing at the pump and leading past the grout hole and then discharging unused grout back into the agitator;

(e) control fittings on the grout hole by which the injection rate and pressure can be controlled and the hole can be regularly bled of water and thin grout.

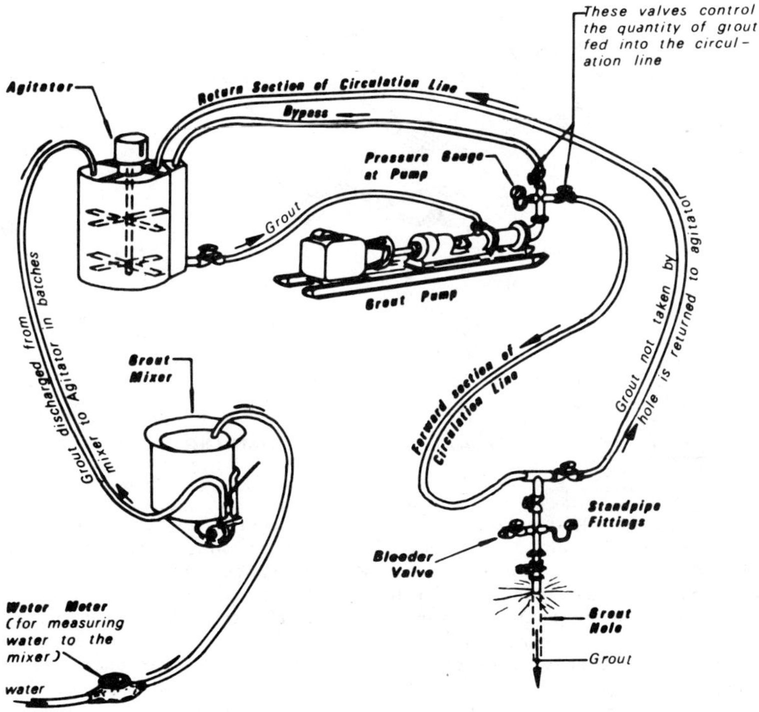

FIG 5 TYPICAL EQUIPMENT LAYOUT

Fig. 5 shows how these items are typically hooked together. On relatively small jobs, the individual items can be combined into a compact single unit of the type shown in Fig. 6. At the other extreme, on very large jobs, the machinery is sometimes installed in a central mixing and pumping station with circulation lines to various parts of the site (9); lines up to two miles return have been used successfully. Automated, computer controlled and monitored equipment is then feasible.

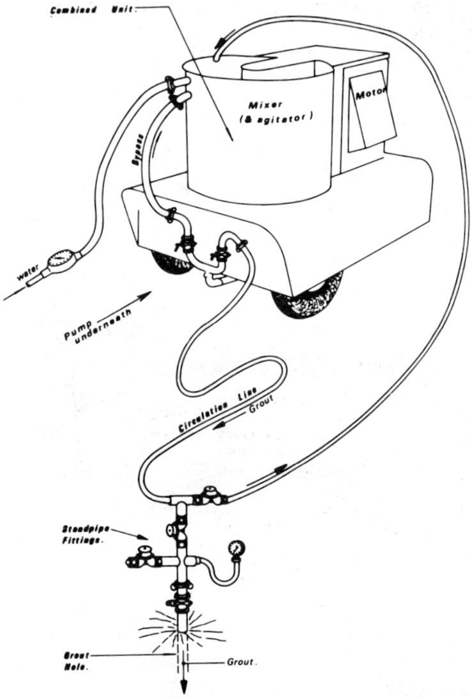

FIG 6 TYPICAL SMALL MIXING– PUMPING UNIT

Grout Mixers

The earliest mixers were simple stirring devices rotating inside a drum or tank. In some, the rotating shaft was vertical and in others it was horizontal. Rotation speed was usually less than 100 r.p.m. They resembled modern agitators.

The subsequent improvements on these were interesting cases of isolated inventiveness confined to particular countries and with very little interchange of ideas. For instance, the Colcrete high speed mixer was developed before World War II in England for the purpose of mixing the sand cement grout used in the Colcrete process of making intrusion grouted concrete (comparable to the Prepakt process used in the U.S.). The mixer was found to be so effective that it came into use for foundation grouting also.

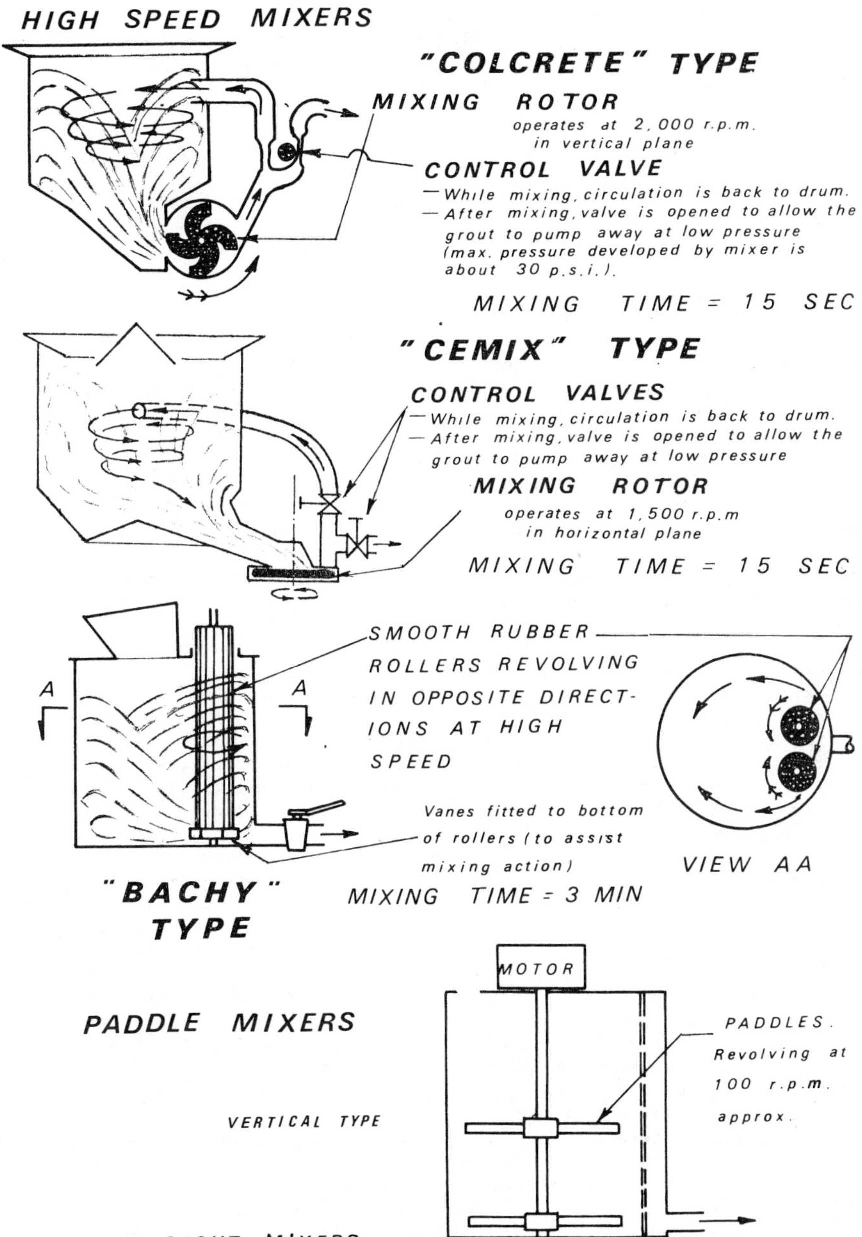

FIG 7 GROUT MIXERS

Meanwhile, in France the S.I.F. Bachy organisation developed its "high turbulence mixer" and in Sweden, Atlas Copco built its Cemix mixer. These three machines are all high speed mixers but operate on different principles as described shortly. The Colcrete mixer had very little use outside certain franchised English grouting contractors for many years. In 1955 a double drum Colcrete mixer was tried at Folsam Dam and then tested in the laboratories of the U.S. Corps of Engineers at Vicksburg (3). The Corps of Engineers' tests concluded in favour of the Colcrete mixer when compared with a paddle mixer but did not lead to widespread adoption of high speed mixing in general grouting practice in the United States, possibly because of the less favourable reports from the Folsam Dam tests. Kennedy has quoted them in ref. (15).

The Bachy mixer has found extensive use by French grouting contractors, although, like the Atlas Copco machine, it appears to have had little use in the U.S. Other manufacturers have produced similar machines and contractors have been known to make their own! The slow acceptance of high speed mixing in North America may have been influenced by the relatively high cost of the European machines compared with locally made paddle mixers and by a feeling that paddle mixers can be good enough if the mixing time is sufficiently long and if suspension agents are used to decrease the settlement rate of particles. Bussey of Harza Engineering Company was one American engineer whose experience led him to conclude differently, and in a paper prepared before his death in 1972 and published in ref. (5), he stated that "the high speed grout mixer is far superior to the old paddle mixers".

The basic differences between high speed and paddle mixers is the rotation speed and the type of mixing impeller: high speed mixers operate at 1500 to 2000 r.p.m. while paddle mixers usually rotate relatively slowly at about 100 r.p.m. or less. The mixing impeller in the Colcrete machine is like a turbine running in reverse. In the Bachy mixer it comprises rubber rollers and in the Atlas Copco Cemix machine it is a centrifugal pump. In paddle mixers the mixing element takes the form of a flat or curved paddle sweeping almost the full diameter of a drum. Fig. 7 shows the basics of each type. In essence, the paddle mixer uses a stirring action while the high speed mixer uses a violent shearing motion in conjunction with a vortex. The vortex is important. It works as a centrifugal separator feeding the impeller and ensures that uniform grout is obtained at the conclusion of mixing. Fig. 8 indicates the mixing process. It is very rapid. The actual mixing cycle takes a minimum of 15 seconds in a Colcrete mixer but is, in fact, longer because of the time taken to charge and discharge the drum. The total cycle time is typically half a minute. Paddle mixers take several minutes per batch and thus may require the use of more machines to feed a fast hole.

The high-speed mixer

(i) violently separates every cement grain from its neighbours; thus permitting,

(ii) thorough wetting of each grain. This chemically activates each, giving thorough hydration.

The resulting grout is commonly called "colloidal" but is not really. It is more correctly "semi-colloidal", and when set, is

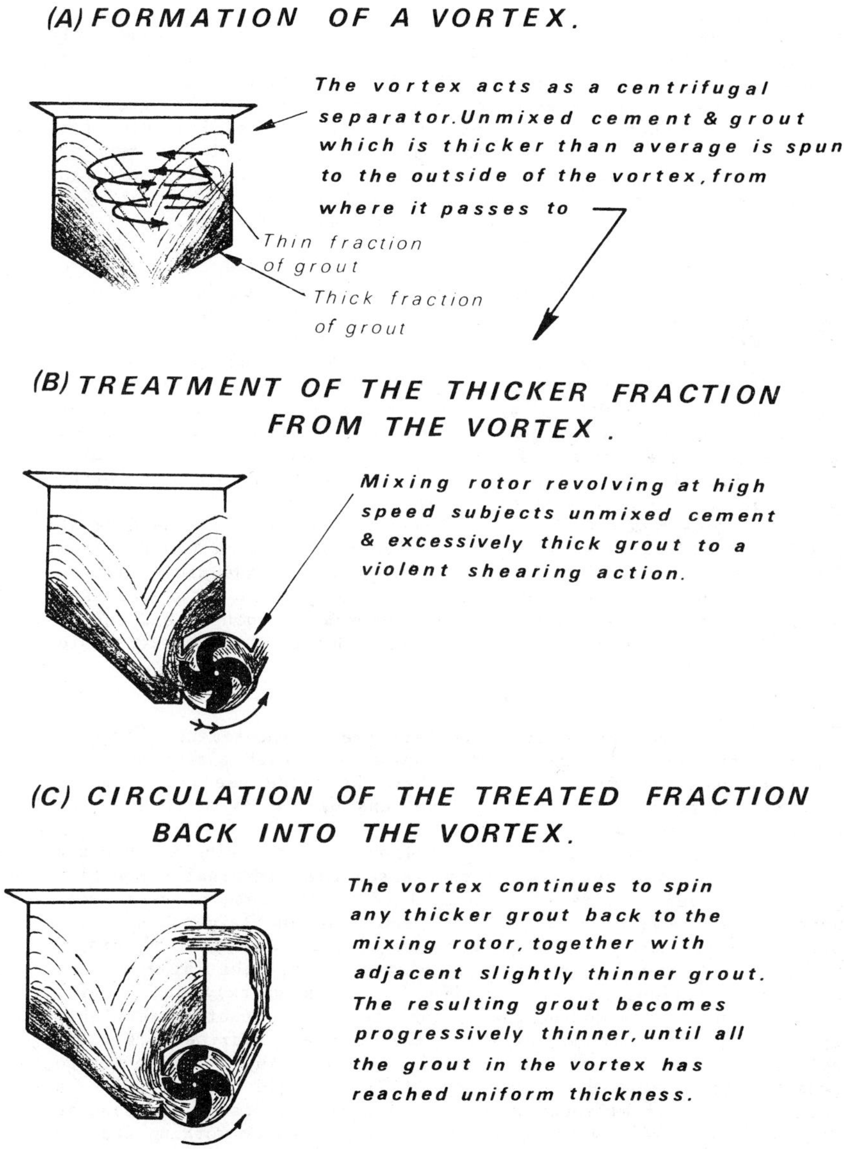

FIG 8 THE MIXING ACTION OF HIGH SPEED MIXERS

light in colour and density, and should be free of colour bands formed by differential segregation of the various components of cement. It can usually penetrate well due to the absence of flocs of cement grains and is a relatively stable grout.

The writer considers, along with may other grouting practitioners, that high speed mixing is an essential requisite for obtaining effective cement grouting in rock. There remain, however, applications where the paddle mixer is superior, such as compaction grouting as described by Warner (27) where the mix is so thick that instead of being stirred, it has to be chopped.

Agitator

After mixing, the grout is usually discharged to a holding tank or "agitator" so as to free the mixer for other work or to avoid the overheating which occurs with prolonged mixing, particularly at high speed. The agitator keeps the grout stirred ready for the pump to take and commonly has a size of about 3 ft. high and 2 ft. diameter, with a vertical spindle revolving at 100 r.p.m. approximately, having a couple of stirring arms, one of which sweeps the bottom of the drum as closely as practicable. Baffles fitted to the inside of the drum break up any vortex formation (which is undesirable here).

Measurements of grout volumes can be taken in the drum at regular time intervals by means of either a dipstick hanging from its rim or by use of projecting markings inside the drum surface. These measurements are used for finding the volume of grout taken in the grout hole. They are rough measurements though, and can be improved on if necessary by passing the grout through an electronic flow meter (28).

Grout Pumps

Like mixers, there have been great improvements in grout pumps since the early crude piston pumps. Improved piston pumps of various types and diaphragm pumps have had wide use but nowadays have been superseded in most applications by the Moyno (Mono) type of helical rotor pump.

The piston pumps were often of the single cylinder double-acting type operated by compressed air, internal combustion motors, hydraulic drive or even by hand in the case of very small units. Multi-cylinder pumps have been used on large jobs where high capacity and pressures were required. Problems on piston pumps come from their valves, piston wear, the slowness of cleaning, difficulties in clearing blockages quickly and general cumbersomeness when moving them around the job. Manufacturers have endeavoured to overcome these difficulties with various ingenious solutions but these have been easily bettered by the Moyno type except for compaction grouting (27). These pumps have a helical-shaped steel rotor revolving in a rubber stator as shown in Fig. 9. Both are shaped so as to provide a cavity working its way along the worm as the rotor turns. Grout or other abrasive fluids are displaced along within this cavity. There is no need for working parts such as valves. Wear is almost entirely confined to the rubber stator which can be quickly slipped out of the end of the pump and replaced when worn out. The units are light and easily moved around the job by manhandling if necessary. A standard unit develops about 125 to 150 p.s.i. and if pressures greater than this are required, units are hooked in series.

There has traditionally been a difference in approach on circulation line practices in Europe compared with the U.S.A. and this has led to differences in pump requirements. Much of European grouting uses, or did once use, non-return lines which connected the pump directly to the grout hole without continuing the line back to the agitator (and thus permitting circulation). This type of direct connection means that the pump speed and output have to be regulated to suit the demands of the grout hole. If the hole is taking grout fast, the pump has to be turning over rapidly and then later in an application, when the hole approaches refusal, the pump has to struggle to turn out a substantial pressure without much volumetric output. Consequently, pumps with hydraulic drive have been developed for these circumstances. An electric motor powers the hydraulic drive which in turn drives the pump, usually of a piston type. However, these are invariably cumbersome units and this confines them to use in central mixing stations virtually.

By contrast, when a circulation line is used, there is no need to match the pump output to the grout hole rate of take. Instead, the pump can turn over at a comfortable speed at all times. The surplus output is merely returned around the circulation line to the agitator. This permits mobile, direct drive pumps, such as the Moyno, and gives direct control of injection at the grout hole and is a very workable arrangement.

Even so, Moyno pumps have to be able to handle a range of pressures and outputs on any one site to suit different elevations of the various grout holes and therefore the drive motors must have a variable speed output such as is obtainable from compressed air or internal combustion motors. Direct coupling of constant speed electric motors, for instance, is unsuitable usually.

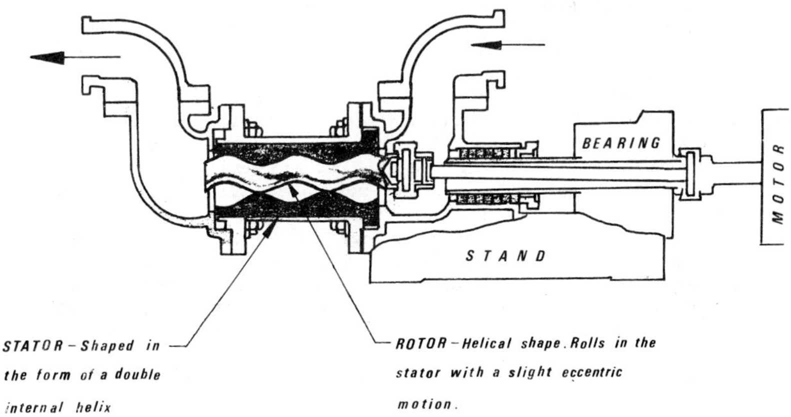

FIG 9 MOYNO (MONO) PUMP

This discussion so far has been about standard pumps for average size and for large jobs. On small jobs with low pressures and takes, equipping up with the standard machines may not be warranted. Small units having a high speed mixer and a Moyno type pump combined, are available such as the "Colmono" machine or of the type shown in Fig. 6. If required pressures are very low, and circulation lines are short, there may be no need at all for a pump if a mixer of the Colcrete or Cemix type is in use. The pumping action of the mixing unit on these develops 15 p.s.i. (1 bar) or more and can be used for pumping but not for more than a few minutes at a time.

On a very small scale, hand operated pumps are available of the piston type or of the diaphragm type operated by a long rocking lever.

Compressed air pans were once popular in tunnel work and are still occasionally used. These have a large chamber into which the grout is placed and is then extruded by the application of compressed air. A danger from these is that compressed air may travel with the grout into the foundation cracks and then build up pockets of air in the foundation cracks which may exclude grout and can build up pressure sufficiently to expel surrounding grout when application pressures are diminished.

Circulation Lines

As mentioned above, the use of a circulation line is not universal. Displacement grouting, whereby relatively high pressures are used to deliberately displace the rock to open cracks up and enhance penetration, is usually carried out with a non-circulation type of line. This means that the pump operator is in direct command of the injection. However, his remoteness from the grout hole coupled with the usual difficulties of communication on a noisy job, can lead to problems in promptly handling difficulties, if they occur, such as with movements and leaks.

These problems are diminished if a circulation line is used as shown in Fig. 5. All control is then directly at the hole and the pump can be unattended most of the time.

One of the constant hazards of grouting is sudden blockages in the lines. They can affect the quality of the work and can cause a lot of lost output. An effective way of minimising them is to keep the grout moving rapidly and this is easily done if the line is a circulating type. The non-circulating type can be prone to blockages, particularly at refusal when the whole line of stationary grout is more likely than not to block up and require washing out before the application can be properly finished off. The usual tendency with this layout is therefore to avoid thick grouts and to cease grouting before real refusal is reached. Both of these are considered by the writer to be inhibitive to good grouting.

The thixotropic nature of some grouts can mean that if travelling sluggishly around a circulation line, they can be more viscous than if moving vigorously. Less pumping effort can be needed for the vigorous movement than for the languid motion. To assist keeping the velocities high and hence vigorously moving the grout, the line should not be more than about 1" (25 mm) internal diameter.

3/4" (20 mm) lines work well if couplings do not project far inside them. The two mile line mentioned earlier was made of 1" I.D. galvanised pipe.

INJECTION TECHNIQUES

A generalised description of the rheological behaviour of grout during injection is:-

1. At the commencement of the injection, fresh grout is travelling fairly rapidly through the cracking system.
2. Then, when most of the easily reached cracks have become full with grout, the grout slows down and becomes sluggish. Thixotropic stiffening, together with settlement, set in and the injection pressure is no longer transmitted hydraulically outward from the hole. Instead, the pressure falls off rapidly near it because the semi-paste, which the grout has now become, offers considerable resistance to pressure transmission. The overall effect is to reduce the take gradually, and this gradualness comes about because the stiffening does not occur simultaneously in all of the cracks but instead occurs first in some of the finer ones and is followed later by stiffening in coarser cracks, and so on. If wide cracks carry un-stiffened, active grout past the ends of finer ones containing stiffened grout, there can occur at times a bursting of the active grout through the stiffened grout as the result of pressure built up in the active grout. Such build-ups can occur as the result of the changing pattern of flow paths accessible to the still-fluid grout and can occur even though the injection pressure in the hole remains constant. They do not of course, exceed the injection pressure.
3. Eventually the pressure is unable to push the outer fringe of the grout any further.
4. After that, the remainder of the application is limited to injecting fresh grout into passages and voids which have formed where bleedwater has risen up out of the grout injected earlier in the application.
5. Eventually refusal is reached when all bleed passages accessible from the hole are filled with fresh grout.

The resulting grout filling can have a "finger" shape at the limit of penetration and there can easily be pockets of bleedwater trapped in cracks, particularly those with little dip. There may be sufficient volume of trapped bleedwater to warrant giving a further application of grout 24 hours or more later.

The behaviour of grout during injection has been studied in both site and laboratory investigations. Amongst the better known work is that of the U.S. Corps of Engineers (26) in which preformed cracks between horizontal slabs were grouted with various mixes and under various pressures. The results are of particular interest, even though they were not fully explored in the official report. They emphasise the improbability of obtaining complete filling if excessive bleedwater is present and cannot escape, and they indicate the desirability of thick grouts such as w:c = 1:1 by volume or thicker for reasons of durability. This matter is taken up further at Ref. (12). Camberfort (6) suggests that these tests were not representative of site conditions because the concrete slabs were not permitted to deflect during the grouting. This raises the issue as to whether grouting should be performed at moderate pressures without disturbance of the foundation (penetration grouting) or whether high pressures should be used so as to force cracks to open appreciably during grouting (displacement grouting).

Displacement grouting makes use of multiple hydraulic fracturing and hydraulic propagation exerted through relatively thin mixes able to continue the transmission of injection pressures for an appreciable time after penetration has ceased - sufficiently long that much of the bleeding will have occurred - then, by turning off the injection pressure, the cracks close somewhat and squeeze the still-mobile bleedwater out. In some sites with fine cracking or almost nil cracking, displacement grouting has been successful where penetration grouting was not, but there could be doubt as to whether grouting was needed at all at such sites or, if so, whether cement grout was the most appropriate choice.

Displacement grouting is unsuitable for grouting of surface rock because of the probability of heave (6, 24) unless the surface is loaded with fill or concrete. At suitable depths, however, the method usually avoids production of surface heave because of the beneficial effect of steeply dipping cracks in limiting the propagation of near-horizontal fractures and joint widening.

It can be argued that displacement grouting ruptures natural bonding and therefore wounds the patient before healing him! If all fractures are fully filled, the "wounds" probably do not matter much but the method places a lot of reliance on the dispelling (and maybe prestressing) effect obtained when the displaced pieces of rock strain back into their former positions.

Penetration grouting avoids these problems by using natural cracking systems without any real attempts to distend them. Pressures are usually less than 2 p.s.i. per ft. of depth (0.3 bar per metre). Excess bleedwater escapes where it can, by ascending upward cracks. Of course there should be as little bleedwater as possible commensurate with reasonable penetration. Ref. (12) discusses this further. The majority of rock grouting throughout the world is done by the penetration method reflecting the opinion expressed by Morgenstern and Vaughan (19) that displacement grouting can present hazards of increased permeability and cost. At the risk of provoking objections from his European friends, the writer offers, for discussion, his opinion that adequate grouting can be obtained with the penetration method in nearly every foundation so there is rarely the need to resort to displacement grouting. This present paper is, of course, primarily concerned with penetration grouting.

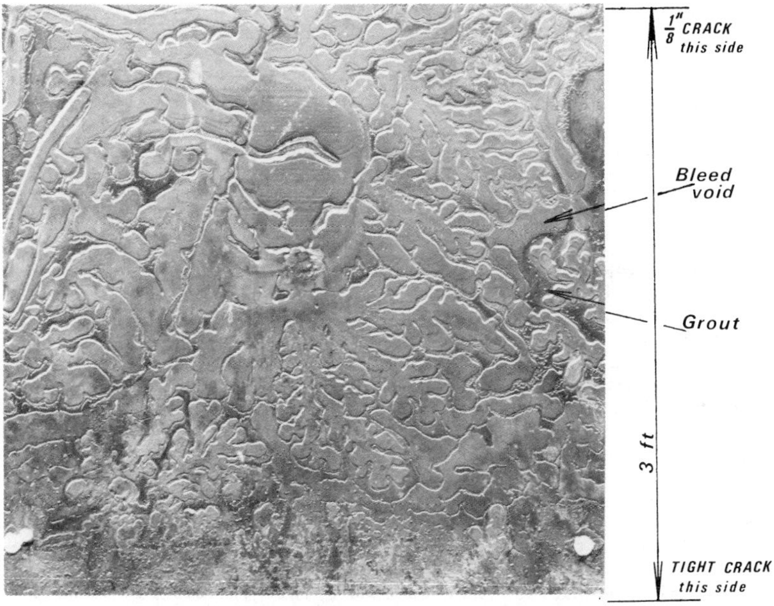

FIG 10 INCOMPLETE FILLING IN ARTIFICIAL HORIZONTAL CRACK

Returning now to the matter of grouting tests between slabs of concrete. Those of the U.S. Corps of Engineers were followed by other people, including the writer's organisation from which Fig. 10 and those shown in Ref. (12) come and are typical. Such tests, if performed between non-deflecting horizontal slabs with no opportunity for bleedwater to escape until it reaches the slab perimeter, almost always produce pockets and bleed paths interrupting the continuity of the grout fill. Hence the desirability of using the thickest grout practicable consistent with reasonable penetration.

However, the pessimism of such artificial testing is usually confounded by in-situ tests to some extent! This is because of the rareness of absolutely horizontal unvented cracks in nature. Even the gentlest slope is able to assist bleedwater to float out slightly and in steep cracks it has no trouble at all. In most field tests and inspections of actual grouting jobs, cracks are usually found to be completely full or nearly so, if suitably thick grout has been used with penetration methods. Simonds shows photos of typically satisfactory work (25).

The stability of the grout plays a role in obtaining effective grouting. If cement settles too rapidly after it enters the cracking system, penetration can be impaired. Klein and Polivka (16) and Jones (14) and Mayer (18) have reported on various additives for improving stability of which sodium bentonite in proportions less than a few per-cent of the cement weight has, in the past, had the most use. However, the need for additives has greatly diminished nowadays with the widespread adoption of high speed mixing, which usually imparts the requisite stability to the cement without needing assistance from any of these additional materials (18). As noted earlier, these mixers separate each cement grain and surround it with a water layer. Rarely is an additive necessary but if thought to be, polymers of the bleed-control type have become available, some of which have the additional property of lubricating the grout sufficiently to enable it to pass thrugh cracks finer than the usual minimum of about 0.015 in (0.3 mm). Improvement of these polymers is a suitable subject for future research.

Thixotropy plays a part in the whole penetration behaviour of grouts thicker than about w:c = 1:1 by volume and also in the later phases of injection of thinner grouts at the time when settlement produces thick grouts in lower parts of the cracks. Below the Bingham Yield Point, grout can be sludgy and unable to transmit pressures hydraulically to any extent, in contrast to the same grouts when moving rapidly. Advantage can be taken of thixotropic stiffening, aided by the onset of setting, to ensure grout stays in place after injection in a foundation against back pressure, or where groundwater is passing quickly through the cracking system. To this end, and even in quiescent foundations, good practice is to continue every grout injection to absolute refusal and then to hold pressure for a further 15 minutes afterwards. In a typical grout application taking two or three hours, the half hour or so taken to ensure full refusal and to hold it in this manner, is considered well worthwhile.

Fairly frequently, an unduly large quantity of bleedwater collects in the cracks in spite of the use of the thickest grout possible. This is especially the case where cracks are wide, such as 1/8th". Fig. 10, for instance, shows the effect in an exaggerated manner. The best remedy is to wait for a day or so and then give another application of grout. Second applications such as this can take quite surprising quantities of grout as the case histories of Ref. (13) show. If cracking or voids are more than about 1/4" open and extend for long distances such as 50 ft. (15 m) or more, the multiple application technique is a good answer, but instead of taking each application to refusal, it is terminated when it reaches some fairly large take, such as 100 bags of cement. Arbitrary curtailment such as this assists to confine the grouting to the region where it is really required instead of pushing it far and wide around the countryside. By persistent applications in this manner at daily intervals, refusal can usually be eventually obtained and a good, economical, though protracted job obtained.

Removal of bleedwater from grout holes and cracks in their immediate vicinity can be assisted by bleeding it off at frequent intervals (quarter hourly) through the bleeder or valve located in the control fittings mounted on the holes. A good test of the efficiency of bleeding comes the day after grouting when, on removal of surface valves, the holes should be found to be filled solidly right to the top unless, of course, packers were used down the hole.

An essential part of grouting operations is constant critical review carried out several times a day while the work is in progress. The reviewing should assess any need to change w:c ratios, pressures, techniques, etc. to improve the efficacy of the work. Some guidance for this is available in such publications as the Grouting Manual (28) where examples are given for detecting crack characteristics from grouting reports and hence for checking the suitability of the method in use. However, beyond such outline guidance, experience is very necessary for effective reviewing.

To help with these assessments, and also with those concerning when to cease closure of a curtain (mentioned shortly) a water test should be carried out in each stage before giving the first grout application to that stage. These water tests have the additional benefit of "rehearsing" the subsequent grouting operation and can help in w:c selection and in finding leaks, connections, possible rock movement problems and other matters needing remedial action before or during the grouting. The water test need only be brief, such as 15 minutes total time divided up into three successive tests of 5 minutes each. The comparative takes of these three tests indicate whether flow is steady or is causing washout or is filling blind voids or is unstable warranting a longer test. The pressure used need not be high, 15 p.s.i. (1 bar) maximum is sufficiently high to learn all that has to be discovered. The use of water test pressures equal to, or even nearly as high as, subsequent grouting pressures is unnecessary and can be downright hazardous because of the vastly greater penetrating ability of water in comparison with grouts.

Towards the completion of sections of a grout curtain, there arises the question of when to cease closure - should the grouting stop after completion of the tertiaries, or of the quaternaries, or of the quinaries, or of the sextaries . . . . ? The answer may be different for varying foundation conditions and for the various stages of depth.

The question has basic resemblances to those of "When is grouting needed?" and "How much is required?" discussed earlier and similar answers apply in general. However, those earlier questions were somewhat general whereas here the issues are more specific and require firmer resolution.

The salient consideration is the purpose of the grouting, i.e.

(1) A relatively tight curtain. If lost water is expensive or piping is to be prevented, or for some other reason the grouting has to be relatively tight, as indicated by the flow chart of fig. 4, then the criterion for cessation of closure should logically be based on the permeability of the completed work and should be in terms of the same Lugeon value as has been adopted for the standard of the grouting.

(2) A relatively permeable curtain. At the other extreme to a relatively tight curtain is the nominal one, which seeks to only reduce seepage to a tolerable amount and does this by closing off the significant cracks mainly and almost ignoring the

fine ones. Here the permeability of the work is not as important as getting worthwhile takes in deserving cracks.

Between these extremes lie most of the actual jobs. Which criterion is therefore applicable for the common cases? A criterion based on permeability, or one based on cement take? On this issue there has been some variation in practice and maybe some confused thinking over the years. A fairly common approach has been to set a closure cessation criterion in terms of cement take. Typical values are 0.25 cu. ft. of cement/ft. (0.25 litres of cement/cm) in the top stage, 0.33 in the second stage, 0.5 in the third and subsequent stages, and these or similar figures have been used for many dams. It is therefore interesting to note that, in the few cases where comparisons have been possible, curtains constructed to these criteria appear to have permeabilities in excess of 10 lugeons and are therefore rather "leaky" in comparison to those governed by permeability standards. This might be a surprise to the designers of some, who have rather tended to expect a moderately tight curtain from the "take" criteria. The performance of such curtains has however often been satisfactory and this, of course, provokes the thought that effective curtains may be more permeable than was once thought to be the case. Certainly the old concept that grouting produces an "impermeable barrier" holds water no longer!

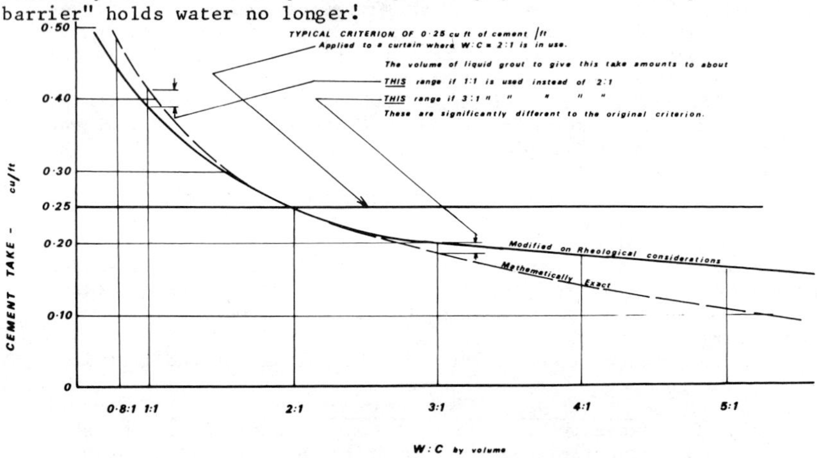

FIG 11 INCONSISTENCY OF A CLOSURE CESSATION CRITERION BASED ON CEMENT TAKE

Looking a little further into the "take" type of criteria from aspects of its constancy and reliability, there arise problems. These stem from its sensitivity to technique variabilities such as, which w:c ratio is in use, whether absolute refusal is practised, whether multiple applications are given, mix thickening practices and the pressures used. The effects of the variables can change the significance of the criterion by several hundred per cent. For instance, considering just the w:c effect, fig. 11 shows a cement take of 0.25 cu. ft./ft. at w:c = 2:1, relates this to its volume of liquid grout (12.5 cu. ft.) in a typical stage length of 20 ft., and then presumes that about this same volume would be taken at other w:c ratios in the same jointing system. Using 1:1 instead of 2:1, this volume of grout would give 0.42 cu. ft. of cement/ft. and at 3:1, 0.18. These are significantly different to the criterion at 2:1 and indicate the different degree of closure that curtains using these different w:c ratios would receive. If it is felt that this comparison is a little unfair in that penetrations with 1:1 would be less than 2:1, rheological modification based on experience and shown on the figure could cut the take back to say 0.38, and likewise an adjustment to the 3:1 figure to allow for easier loss of water could increase it to, say, 0.20. The message is, however, the same.

A further dominating factor is whether multiple applications are given. If significant takes are not followed with further applications much lower takes can result than the foundation can genuinely accept and hence a false assessment is obtained.

Hopefully, contractors are all honest! However, there may be a temptation sometimes to finish off a hole quickly for ulterior reasons and there are some difficult-to-detect ways of doing this by choking an application prematurely if the grout operator is so minded. The resulting falsely low take could give erroneous evaluation of need for further closure if this is based on considerations of take.

Many years experience of trying to use "take" type criteria and encountering these sorts of problems with them led the writer to a preference for the "permeability" type of criteria instead, whenever the curtain is required to be moderately to very tight. The logic of constructing such curtains to permeability standard is obvious. However, for relatively permeable curtains where the standard might be in excess of 10 Lugeons, the rough and ready but very simple "take" criteria are usually more practical than the "permeability" ones.

STAGE GROUTING METHODS

These can be chosen from:-

- Downstage without packer
- Downstage with packer
- Upstage
- Circuit grouting downstage

Fig. 12 shows these. Sometimes combinations of them are used.

For high standard grouting, the writer advocates downstage without packers. Although it requires repeated drill set-ups and is more expensive then upstage work, it provides proving of upper stages and if any weaknesses exist, it deals with them automatically. If any sensitive surfaces are present, they should be thoroughly tightened before grouting of lower stages proceeds.

The addition of packers to the downstage method is rarely warranted. It is occasionally done with the aim of protecting upper stages from the higher pressures used in stages below them but can give false security, particularly in foundations prone to surface movement where the presence of a packer interferes with the

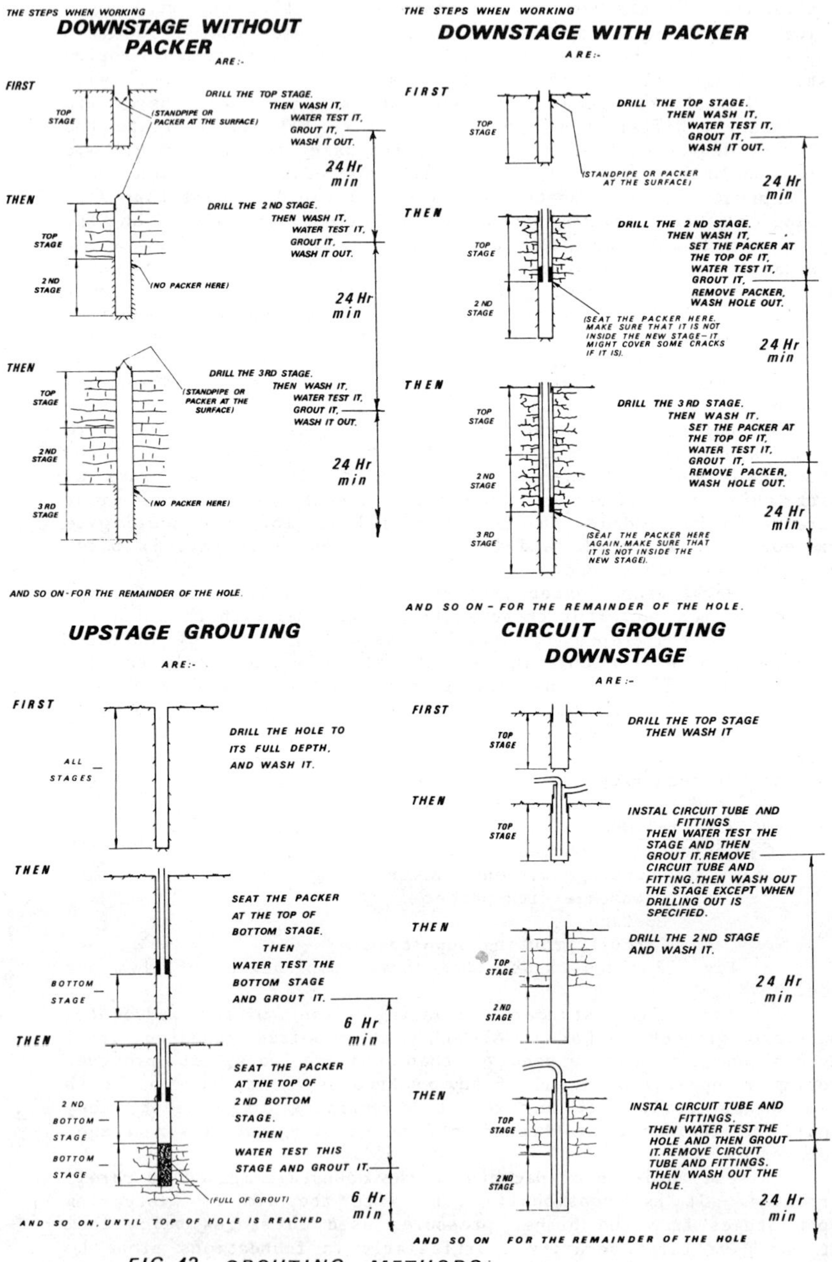

FIG 12 GROUTING METHODS

use of effective techniques for controlling movements caused by high pressure grout surfacing. Packers can become inadvertently grouted into place by grout bypassing them via cracks and, in any case, it is not possible to effectively hold pressure through a packer after refusal, nor is it possible to properly bleed the hole through one.

These packer problems also apply to the upstage method. It is certainly the cheapest but has practical limitations which make it, in general, less effective. In sites prone to rock movement, it is almost impossible to carry out grouting using upstage if grout persistently travels upwards at high pressures through as-yet ungrouted upper stages. In sites where holes collapse quickly, upstage grouting is, of course, impractical. Problems can arise when carrying out daily assessments and in deciding when to cease closure if the fastest upstage method is used where there is no time lapse between successive stages: no sooner has the grouting of one stage been completed to the extent that pressure dissipates, than the packer is lifted to the next stage and grouting commences again. No water test can be carried out on individual stages (except the bottom one) before grouting and hence the very useful information provided by such tests is not available. The impact of this can be reduced a little if a six hour pause is made between stages as Fig 12 shows, permitting a water test, although admittedly not a wise procedure on the wet grout just under it.

Circuit grouting is difficult, prone to blockages and very expensive, only warranting use in consistently caving country. The U.S. Corps of Engineers (4) used it at Folsom Dam; otherwise it seems to have had little general use.

GROUT PRESSURES

Allowable pressures increase as the depth of the stage increases. In penetration grouting, the rate of this increase is usually conditioned to the strength of the rock on the premise that the weight of overlying rock, plus some beam and cantilever strength, acts to hold the foundation against dislodgment by grout pressures. The well known rule of thumb of 1 p.s.i. per ft. depth is usually a happy compromise for average to weak rock. Pressures can be doubled for sound rock.

Fig. 13 has been adapted by the writer from earlier graphs of this type (4) with curves straightened out for simplicity and an upper limit of 150 p.s.i. (10 bars) applied on average conditions because of the rare need to exceed this pressure in dam grouting.

The metric rule of thumb of 1 bar per metre gives much greater pressures as discussed earlier for displacement grouting. Sabarly (24) has rather cannily suggested that the choice of grout pressures depends not so much on safe injection issues as on which system of units is in use - those who use metric units have traditionally used the metric rule of thumb while the "Anglo-Saxon" (his description) uses the p.s.i. per ft. depth rule. Hence the European penchant for displacement grouting?

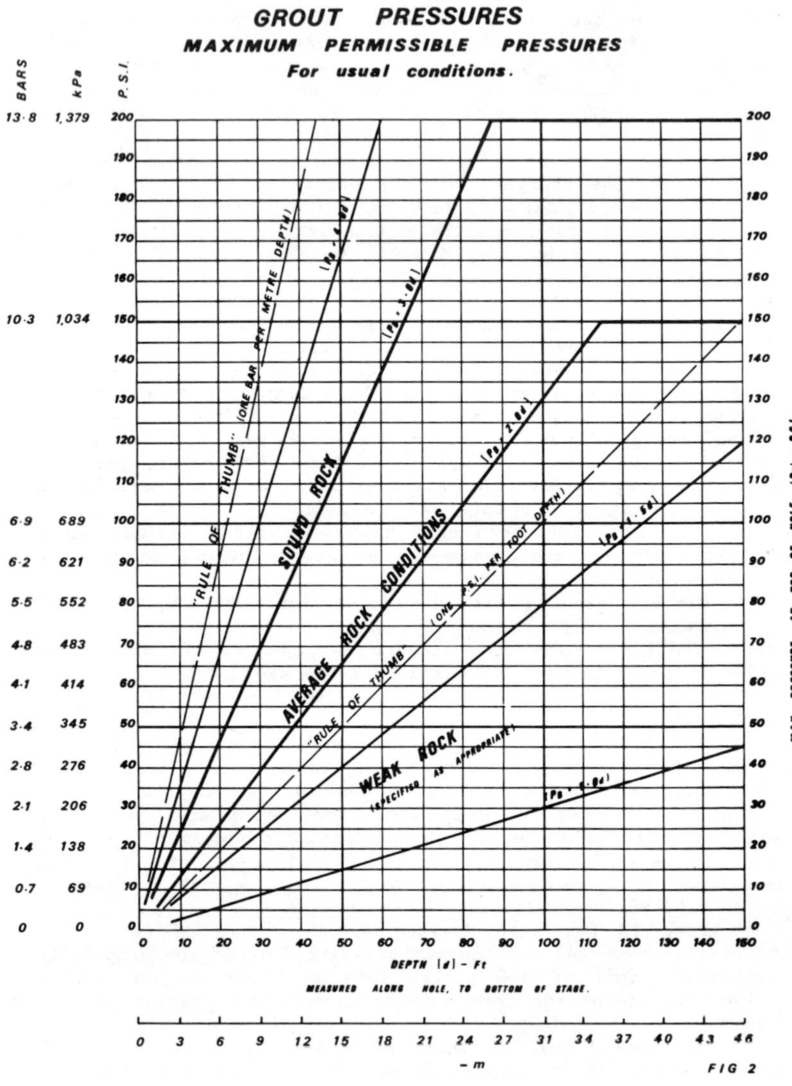

FIG 13 RECOMMENDED MAXIMUM GROUT PRESSURES

Good practice in penetration grouting is to slowly build up to these pressures. For instance, by using a relatively low tentative pressure for, say, the first 15 minutes and carefully watching for leaks, connections, rock movements etc., it is much easier to take appropriate remedial action than if full pressure is used straight off and maybe causes a runaway condition. At obviously trouble-free holes, this 15 minutes can be reduced to about 5 minutes. After the tentative period, the pressure should be gradually built up to reach full pressure no later than 30 minutes after commencement of the injection. The principle is to get the thickest possible grout into the cracks as early in the application as feasible so as to reach the limit of penetration while the grout is still able to move freely and transmit pressure effectively. The overall time available for this part of injections ranges from half an hour in the case of trouble free injection in fine cracks, to three hours or so for fast, high takes in wide cracks. Except when dealing with leaks, connections or rock movements, grout applications can generally be expected to reach refusal within about four hours of commencement.

SURFACE MOVEMENTS

A frequent problem in grouting is handling injections properly when surface movements occur as the result of the grouting. These movements mostly take the form of individual slabs deflecting slightly away from adjacent or underlying rock and thus permitting grout to escape, usually rather freely. The size of the moving slabs can range from tens of feet to a matter of inches, depending on the geology and degree of natural anchorage of the slabs.

A complication is when rock is tectonically stressed. When this happens, residual stresses in rock which are temporarily in equilibrium, become unbalanced by grout pressure and thus cause the rock to move in a very sensitive manner. Visual recognition of tectonically stressed rock was mentioned earlier (Fig. 3). If stressed rock is present, injection procedures have to be modified from those normally used, to be able to produce an effective grouting job without generating persistent rock movements and without using excessive quantities of cement. The general principle in grouting stressed rock is to control the amount of energy conveyed into the cracks via the grout. This energy can be loosely regarded as depending on both the rate of injection and the pressure. Note that it is not just a matter of restricting the injection pressure. The rate and hence the volume of grout injected is equally important (29). Very often in foundations which are actively relieving themselves of pressure during the gruting, the rocks creep apart during the time of grouting and for hours afterwards. To handle this, the best practice is to restrict the volume of grout entering the job to be only that quantity needed to fill initially open voids plus that subsequently needed to compensate for bleeding and leakage losses. This is all done at very low pressures and requires the use of a grout flow meter. Several applications are necessary to deal with overnight subsequent openings. In the hands of an experienced practitioner, tight, effective grouting can be eventually obtained but is very tricky and can cost possibly twice as much as grouting of unstressed rock.

Sometimes, but not always, bolting of surface slabs assists with grouting of stressed rock. However, in some of the heavily stressed sites such as canyons in massive rock, floor and wall rebound during grouting can be on such an extensive scale that the relatively puny efforts of even an extensive network of anchors does little more than stitch the surface rock together unavoidably permitting deepseated relief to occur.

Unstressed slabs of large dimensions can be effectively prepared for grouting by bolting but is not always necessary if injection techniques can be personally directed by an experienced person using a flow meter. The flow meter is not as essential when dealing with smaller unstressed slabs and in weak, broken country but the experience is! Ref. (28) describes the methods in essence. A surface coating of slush grouting or gunite is a suitable way to bind flaky, finely fragmented surfaces etc. and thus control against myriads of surface leaks.

DRILLING

Controversy about suitable drilling methods has persisted for years. The fundamental issue is to select economical drilling methods which do not clog cracks with cuttings. The choice lies between percussion drilling, rotary roller bit drilling and rotary diamond drilling. Percussion drilling can be a quarter to a third the cost of diamond drilling and this is a very significant cost advantage which has led to the use of percussion drilling on the majority of grouting projects. However, caution is needed - Firstly, the percussion drilling of grout holes must always use wet drilling, and cuttings must be thoroughly washed out at the completion of drilling by either a generous flow of water through the drilling rods while they are gradually withdrawn or, better, by use of special washout rods after withdrawal of the drilling string - Secondly, sites need to be tested to ascertain that cuttings can indeed be washed out effectively like this. These tests take the form of a few holes drilled by each favoured drilling method and a comparison of the cuttings and how well they can be washed out. If one of the drilling methods produces slimy, puggy, adhesive cuttings not able to be entrained by wash water, then that method is usuitable. Examination of the return wash water helps to assess the nature of the cuttings and can be assisted by borehole T.V. examination of the hole walls. Direct indications such as these are considered much more reliable than indirect ones such as conducting water tests or grout tests in the drill holes and comparing the takes: rarely can enough holes be afforded in such tests to provide a reliable statistical sampling.

Diamond drilling should not be looked upon as a panacea. Some sites have produced unacceptable cuttings with it but have yielded satisfactory, sharp cuttings with percussion drilling.

TUNNEL GROUTING

Mention was made earlier of grouting of rock from within tunnels. Concrete lined tunnels also require to have voids under the crown filled with grout. This is variously known as overbreak, back -pack, crown, backfill or contact grouting and is carried out through holes in the concrete lining. Voids are usually very long and therefore accessed by several grout holes. Hence, grout can be pumped into one hole and vented out others. Large steel lined tunnels can require a further grout operation if voids form under the liner due to shrinkage and bleeding of invert concrete. These voids are commonly of small extent and are accessed by single holes through the steel liner. Hence, in order to remove air and water from the voids to permit entry of grout, a vacuum ejector is used and grout is then released into the evacuated space.

SOME CONCLUDING COMMENTS

This paper has concentrated on controversial and on less well defined aspects of the subject. It has by no means covered everything relating to cement grouting at Dams nor has it said all that could be about those aspects mentioned. A large book would be needed for a thorough coverage of the subject. Hopefully, discussion during the conference will add to and enhance what has been said here, therefore a formal summary or conclusion is not appropriate at this juncture.

WHAT NEXT?

The vast amount of grouting experience gathered over the years in many parts of the world and by many separate individuals and organisations has tended to be unco-ordinated, scattered and documented to varying degrees of thoroughness. Some, indeed, has been deliberately kept confidential to firms or particular people in an effort to protect disclosure of the results of research or experience.

Believing that progress in grouting requires the gathering together of all this information, the writer has, over the last twenty years or so, assiduously done this as much as has been possible from published data and privately communicated matter. To this has been added his own experience. This present paper and others referred to (11, 12, 13), including the Grouting Manual (28), are the results to date. Further papers are envisaged. Other practitioners are encouraged to provide improvements on them wherever they have sound grounds for doing so.

However, the dominant role played by experience must not be forgotten. Each grouting site has its own peculiarities and grouting must never be relegated to a "recipe book" approach. Experience is the key to adapting the general principles to the individual idiosyncracies of each job. But the gaining of this experience takes much practice and can be difficult to attain unless one is engaged on a succession of large jobs and has initial training from other experienced people. Many dams have been grouted by the "learn as you go" process, unfortunately, and costs at them have been consequently usually very much greater than if supervisory staff had been sent away to train briefly elsewhere and the finished work may be technically inferior. Another problem comes from the transient nature of the grouting involvement of many design and site engineers. They are not given the several years' experience needed for amassing the fundamental feel for grouting before being put on to it.

This situation can never be completely remedied. It could be alleviated, however, by a far greater passing on of know-how from the competent, experienced practitioners amongst us. Grouting schools appear to be the best medium for this. There is an annual course in the United States but not elsewhere, as far as is known. Perhaps there is not the demand elsewhere at quite this frequency but nevertheless there should be courses available periodically in various part of the world, perhaps even with the same team of lecturers. There is a slight difference here in the teaching role of such courses in contrast to the sharing role of the various conferences which have been held.

This teaching would be enhanced by the inclusion of site demonstrations of grouting where half a dozen holes of various complexities are demonstration grouted. This could be very difficult and expensive to arrange on a regular basis, however.

And what is the future for research? The most pressing need is for thorough investigation of polymers for bleed prevention and penetration assistance. This could be financially profitable if patents are feasible on the results of the research.

In conclusion, we might remember the great strides made in grouting techniques already, and collectively do our part to continue them.

APPENDIX I - REFERENCES

1. ACKER, R.C. AND JONES, J.C., "Foundation and Abutment Treatment for Rockfill Dams", Journal of the Soil Mechanics and Foundations Division, ASCE, Vol. 98, No. SM10, Proc Paper 9303, Oct. 1972, p 995-1015.

2. ARHIPPAINEN, E., "Some Notes on the Design of Grouted Curtains on the Basis of Water Pressure Tests", 10th Congress on Large Dams, Montreal, 1970, Vol. II, p 137-145.

3. BENDINELLI, R.A., "Tests of Sanded Grout; Influence of Manufactured Sands and Admixtures on Pumpability and Evaluation of a Colcrete Mixer". U.S. Army Engineer Waterways Experiment Station, C.E., Vicksburg, Oct. 1958, Technical Memorandum No. 6-419, Report No. 4.

4. BURWELL, E.B., Jr., "Practice of the Corps of Engineers", Journal of the Soil Mechanics and Foundations Division", ASCE, No. SM1, Feb. 1958, p 1551-1 to 22.

5. BUSSEY, W.H., "Suspension Grouts and Their Injection", Bulletin of the Association of Engineering Geologists, Vol. X, No. 4, 1973, p 313-323.

6. CAMBEFORT, H., "The Principles and Applications of Grouting", Quarterly Journal of Engineering Geology, The Geological Society, London, England, Vol. 10, 1977, pp 57-95.
7. CASAGRANDE, A., "Control of Seepage Through Foundations and Abutments of Dams", Geotechnique, 11:3:161-181.
8. FERGUSSON, F.F., AND LANCASTER-JONES, P.F.F., "Testing the Efficiency of Grouting Operations at Dam Sites", 8th Congress on Large Dams, Edinburgh, 1964, Vol. 1, p 121-139.
9. GLOSSOP, R., Opening Address to "Grouts & Drilling Muds in Engineering Practice" Symposium, Institution of Civil Engineers, England, May 1963, Butterworths. p 1-3.
10. HOULSBY, A.C., "Routine Interpretation of the Lugeon Water Test", Quarterly Journal of Engineering Geology, The Geological Society, London, England, Vol. 9, 1976, p 303-313.
11. HOULSBY, A.C., "Engineering of Grout Curtains to Standards", Journal of the Geotechnical Engineering Division, ASCE, GT9, Sept.1977, p 953-970.
12. HOULSBY, A.C., "Optimum Water: Cement Ratios for Rock Grouting". This Conference.
13. HOULSBY, A.C., "A Digest of Typical Cement Grouting Takes". This Conference.
14. JONES, G.K., "Chemistry & Flow Properties of Bentonite Grouts", Symposium "Grouts & Drilling Muds in Engineering Practice", Institution of Civil Engineers, England, May 1963, Butterworths, p 22-28.
15. KENNEDY, T.B., "Symposium on Grouting: Research in Foundation Grouting with Cement", Journal of the Soil Mechanics & Foundation Division, ASCE, No. SM2, April 1961, p 55-81.
16. KLEIN, A. & POLIVKA, M., "The Use of Admixtures in Cement Grouts", Journal of the Soil Mechanics & Foundations Division, ASCE, No. SM1 Feb. 1958, p 1547-1 to 24.
17. LUGEON, M., "Barrages et Geologie". Dunod, Paris, 1933.
18. MAYER, A., "Modern Grouting Techniques", Symposium "Grouts & Drilling Muds in Engineering Practice", Institution of Civil Engineers, England, May 1963, Butterworth, p 7-9, and 58.
19. MORGENSTERN, N.R., & VAUGHAN, P.R., Discussion on Session 1, Symposium "Grouts & Drilling Muds in Engineering Practice", Institution of Civil Engineers", England, May 1963, Butterworths, p 61.
20. NICHOLS, T.C. Jr., "Rebound, its Nature & Effect on Engineering Works", Quarterly Journal of Engineering Geology, The Geological Society, London, England, Vol. 13, 1980, p 133-152.
21. NONVEILLER, E., "A Rational Approach to Design of Grout Curtains", Second International Congress on Rock Mechanics, Beograd, 1970, Vol. 3, p 217-222.
22. NUR, A: BARBER, D.W., "Tectonophysics". Theme II, Third International Congress on Rock Mechanics, Denver, 1974, Vol. III, p 53 et seq.

23. PRATT, H.K., & McMORDIE, R.C., & DUNDAS, R.M., "Foundations & Abutments - Bennett & Mica Dams", Journal of the Soil Mechanics & Foundation Division ASCE, Vol. 98, No. SM10, Proc. Paper 9290, Oct. 1972, p 1053-1072.

24. SABARLY, F., "Grouting & Drainage of Rock Foundations in Rock of Low Permeability", Geotechnique, 1968, p 229-249.

25. SIMONDS, A.W., "Cement & Clay Grouting of Foundations. Present Status of Pressure Grouting Foundations", Journal of the Soil Mechanics & Foundations Division, ASCE, No. SM1, Feb.1958, p 1544-1 to 11.

26. U.S. ARMY CORPS OF ENGINEERS, "Pressure Grouting Fine Fissures", Technical Report No. 6-437, Waterways Experiment Station, Vicksburg, Miss., October 1956.

27. WARNER, J., & BROWN, D.R., "Planning & Performing Compaction Grouting", Journal of the Geotechnical Engineering Division, ASCE, Vol. 100, No. GT6, June 1974, p 653-666.

28. WATER RESOURCES COMMISSION, N.S.W., "Grouting Manual", Third Edition, 1980, W.R.C., North Sydney, N.S.W. Australia, ISBN 0 7240 3786 1 (also available from University of Missouri-Rolla, Civil Engineering Dept.).

29. WONG, H.Y., & FARMER, I.W., "Hydrofracture Mechanisms in Rock During Pressure Grouting", Rock Mechanics, 5, 1973, Springer-Verlag, p 21-41.

DESIGN OF CEMENT BASED GROUTS

by G.S. Littlejohn, BSc(Eng),PhD,CEng.,FICE,MIStruct.E,FGS

ABSTRACT: Following a description of the main constituent materials of cement based grouts, the paper discusses major design characteristics such as bleeding, fluidity, setting, shrinkage, thermal properties, strength and durability, and provides predictive equations, together with test results.

1. INTRODUCTION

Cement grouts should be sufficiently fluid to allow efficient pumping and injection, and sufficiently stable to resist displacement and erosion after injection. Where a grout is used to waterproof zones of fractured rock or soil, fluidity, particle size and bleeding are important, whereas for grouted aggregate concrete, shrinkage, heat of hydration and strength may be the more dominant parameters in design. Cement grouts are formed basically from ordinary portland cement (Type I) and water. Occasionally other cements are used to obtain high early strength (Type III) low heat of reaction (Type IV) or resistance to chemical attack (Type V). Other solid materials such as fine sand, fly-ash or clay are added for economy or to obtain special grout characteristics. In this regard chemical admixtures designated according to their action as anti-bleed agents, fluidifiers, accelerators, retarders and expansion agents, may also be incorporated.

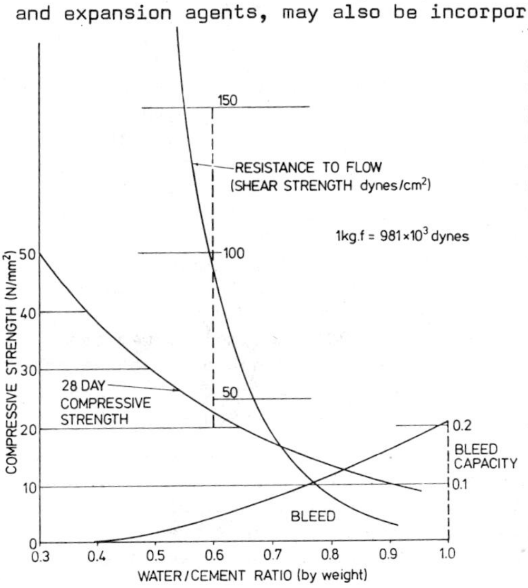

FIG. 1. - Effect of Water Content on Grout Properties

The principal variable affecting the properties of cement grouts is the water/cement ratio (w), the amount of water determining the rate of bleeding, subsequent plasticity and ultimate strength of the grout. The extent to which these, and also fluidity, are related to w of neat Type I cement grout is shown in Fig. 1. Excess of water causes bleeding, low strength, increased shrinkage and poor durability.

The purpose of this paper is to provide a simple guide to the main factors determining grout properties and thereby provide a logical basis for ascertaining the optimum mix design for a specific grouting application.

Technical Director,
Colcrete Ltd., Kent, England.

2. CONSTITUENT MATERIALS

Water As a general rule water which is suitable for drinking, except for the presence of bacteria, is suitable for cement grout formulation. On the other hand water containing sulphate (>0.1%), chloride (>0.5%), sugars, suspended matter such as algae, or high alkali content is technically dangerous particularly for high strength applications in the presence of steel e.g. duct grouting. If structural steel is not involved seawater can be employed e.g. underbase grouting of concrete platforms offshore. Satisfactory field experience of seawater concrete in fact dates back some forty years (1).

Cement The most common form of hydraulic cement is composed of the following compounds:

Tricalcium silicate ($3\ CaO.SiO_2$) - C_3S

Dicalcium silicate ($2\ CaO.SiO_2$) - C_2S

Tricalcium aluminate ($3\ CaO.Al_2O_3$) - C_3A

Tetracalcium aluminoferrite ($4\ CaO.Al_2O_3.Fe_2O_3$)- C_4AF

Gypsum is also included to slow down the set. Magnesium oxide, free lime and silica are present in minor quantities.

	C_3S	C_2S	C_3A	C_4AF	Others
Ordinary (ASTM-I)	45	27	11	10	7
High Early Strength (ASTM-III)	55	17	11	9	8
Low Heat (ASTM-IV)	30	46	5	13	6
Sulphate Resisting (ASTM-V)	45	35	4	10	6

TABLE 1. - Typical Percentage Compositions of Portland Cements

The main compounds behave quite differently and their relative proportions dictate cement properties (Table 1). C_3A virtually flash sets, great heat is evolved and some strength is attained at 1 day but no increase thereafter. C_4AF sets in minutes, with some strength development but less heat is evolved c.f. C_3A. C_3S sets in a few hours and attains almost full strength in 7 days. By contrast, C_2S sets very slowly with low heat but ultimately attains the strength of C_3S.

For most cements the maximum practical size (99% passing) ranges from 44 - 100 μ (Fig.2) and these particle sizes limit penetration of cement grout to soils with a permeability less than 5×10^{-4} m/sec, or fissures in rock of width less than 160 μ unless fracturing pressures are used.

Fillers Often the main purpose of fillers is to reduce the overall cost of the grout, without affecting significantly its physical properties. Certain fillers, however, give technical advantages, e.g. reduced bleeding improved fluidity or retardation. Three basic types are used; namely, pozzolans, sands and clays. Of these sands and clays are basically inert while pozzolans are reactive.

Clays with their capacity to absorb water and ability to form gel structures even at low concentrations as well as their small particle size are used to stabilise the cement, thus preventing its bleeding. The clay performs no important function in the final chemical set

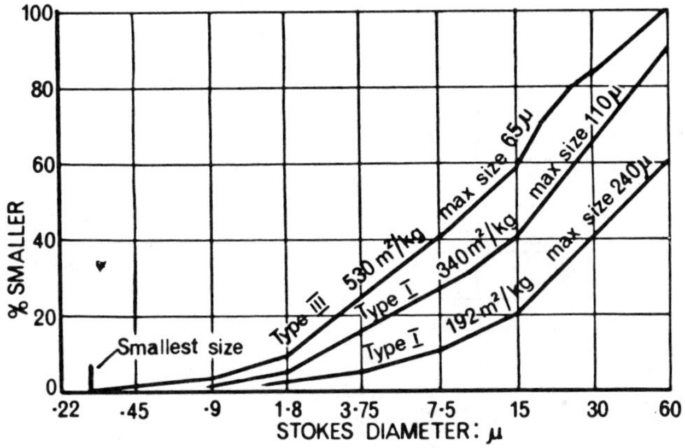

FIG. 2. - Size Analysis Curves for Cement

resulting from hydration of the cement, but development of strength is slow and there is no well-defined setting time. The special characteristics favourable to grout formulation are possessed most markedly by the sodium, and to a lesser extent by the calcium montmorillonites rather than by the kaolinites and illites which are mainly used as fillers.
In rock grouting and consolidation of soil where strength is required, the clay content is kept to a minimum by using bentonite (2-5% by wt. of water) for w = 1-3. For clay/cement grouts where high proportions of clay (50%) are employed, there is little demand for strength and the clay filler simply increases the volume yield per unit weight or cost of material.

Pozzolans, as silicates and aluminosilicates, are not themselves cementitious, but will react with free lime cement in the presence of water to form a cementitious compound. Naturally occurring pozzolans include finely ground shale, pumicite and diatomite. Flyash a combustion by-product of pulverised coal, and ground blast-furnace slag are examples of artificial pozzolans. Since flyash and slag are waste products they are normally used as cheap bulk fillers in low strength grouts for cavity filling eg consolidation of old coal workings. Precipitator flyash is preferred to lagoon ash since the latter is more variable. In this regard chemical composition is a useful indicator of variability but it is more usual to simply monitor grading, specifically maximum particle size and % passing the 45 μ sieve, moisture content and % combustibles. A maximum particle size of 0.5mm is recommended, and the amount retained at 45 μ can vary from 15 to 25%. Fineness is also related to pozzolanicity, thus finer ashes (less than 10% retained at 45 μ) react exothermically, contributing much earlier to strength than do coarser ashes (between 20 and 30% retained at 45 μ). These coarser ashes contribute little to early heat, but do not give such significant water reductions as the finer ashes. As a further quality control, there is good grouting experience where % loss on ignition $\leq$ 5%, and moisture content < 10%.

Fine sands can be added to neat cement/water suspensions to form an economical grout particularly where a high solids, low water grout with relatively high frictional shear strength is required. Sand is chosen as for concrete in relation to durability, shrinkage, and alkali reaction, and in general hard bulky crushed rock is preferred to flat

Admixture	Chemical	Optimum Dosage (% cement wt)	Remarks
Accelerator	Calcium Chloride	1-2	Accelerates set and hardening
	Sodium Silicate	0.5-3	Accelerates set
	Sodium Aluminate		
Retarder	Calcium Lignosulphonate	0.2-0.5	Also increases fluidity
	Tartaric Acid	0.1-0.5	
	Sugar	0.1-0.5	
Fluidifier	Calcium Lignosulphonate	0.2-0.3	
	Detergent	0.05	Entrains air
Air Entrainer	Vinsol Resin	0.1-0.2	Up to 10% of air entrained
Expander	Aluminium Powder	0.005-0.02	Up to 15% pre-set expansion
	Saturated Brine	30-60	Up to 1% post-set expansion
Anti-Bleed	Cellulose Ether	0.2-0.3 (for $w < 0.7$)	Equivalent to 0.5% of mixing water
	Aluminium Sulphate	Up to 20% (for $w < 5$)	Entrains air

TABLE 2. - Common Cement Admixtures

angular or flaky material which gives poor fluid handling properties. Evenly graded sands are preferred (5mm down to 75 μ) but for long pumping distances in excess of 300 metres the maximum size should ideally be reduced to 0.5mm and the maximum sand/cement ratio limited to 3 to maintain the particles in suspension and avoid segregation.

Admixtures can be added in relatively small quantities to modify grout properties (Table 2). Most commercial admixtures are compatible with Type I and III Portland cements, but many are incompatible with High Alumina and Supersulphated cements. Admixtures should not be regarded as a replacement for good grout practice nor be used indiscriminately. In general their suitability should be varified by trial mixes, and if two or more admixtures are proposed for a mix, the manufacturer of each should be consulted.

3. BLEEDING

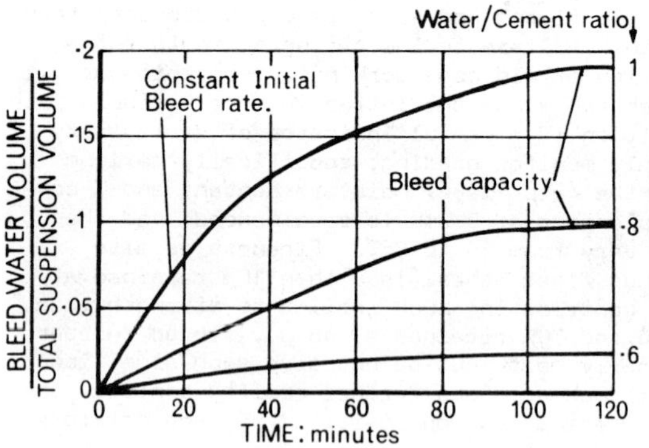

FIG. 3. - Rates of Bleed (Type I)

During mixing the cement particles are dispersed and suspended in water. Except in the case of a very dense paste, the resultant suspension is initially unstable and the cement particles settle under gravity. This bleeding mechanism is important in grout design particularly the bleeding rate and the final volume of bleeding (bleed capacity). The initial rate of bleeding (Fig. 3) can be estimated from D'Arcy's Law for $w \leq 1$ assuming acquiescent conditions.

$$q = \frac{9}{32} \frac{dm^2}{Y} \frac{Dc-Dw}{Dw} \frac{w^2}{(1 + 3w)} \quad \ldots\ldots 1.$$

where q = initial rate of bleed, dm = equivalent spherical particle diameter of cement in suspension, Y = kinematic viscosity of water, Dc = density of cement, Dw = density of water and w = $^{w}/c$ ratio by weight.

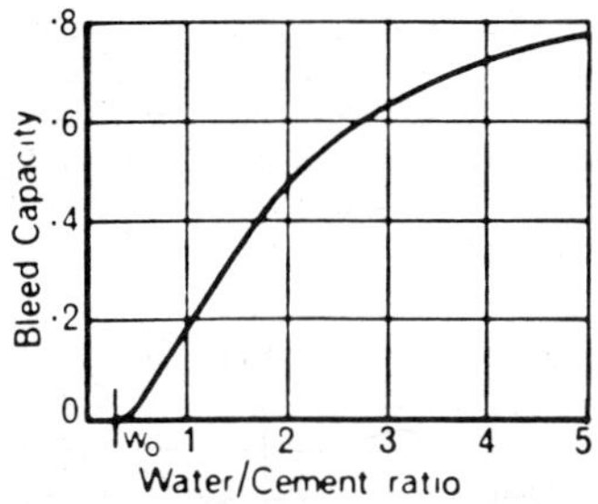

In less dense suspensions where particles are settling rapidly an analysis based on Stokes Law is more realistic. Fig. 4 shows results of bleed tests on high w grouts.

FIG. 4 - Effect of w on Bleed Capacity

4. FLOW PROPERTIES

Flow properties are affected principally by dynamic interparticle forces of attraction and repulsion and, in dense grouts, by dilatency of the moving particles. A dense grout can only be pumped easily when it contains sufficient fluid to prevent expansion of the particle matrix during shear and, generally speaking a wide well graded range of particles is preferred since the better the grading the lower the critical porosity at which the grout becomes pumpable. A reasonable percentage of fine particles is also desirable to increase the specific surface of the grout particles and thereby slow the separation of the liquid and solid phases.

Under conditions of laminar flow, cement grouts behave as Bingham fluids. The shear stress (T) necessary to cause the grout to flow at a constant rate of strain is given by

$$T = T_s + Y_p \frac{dS}{dt} \quad \ldots\ldots 2.$$

where T_s = initial shear strength,

Y_p = coefficient of plastic viscosity and

$\frac{dS}{dt}$ = rate of shear strain or velocity gradient.

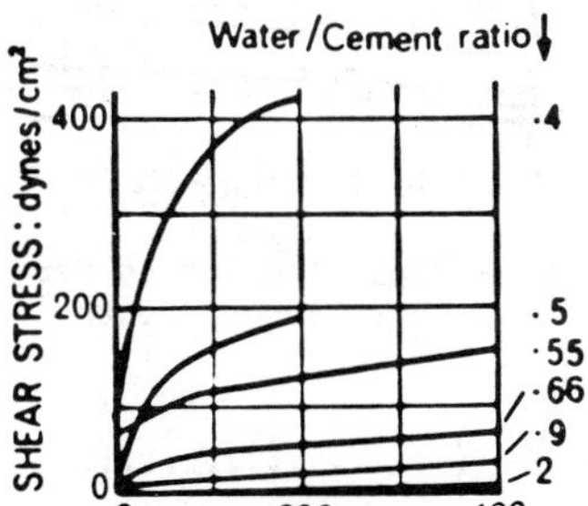

FIG. 5. - Flow Curves for Neat Cement Grouts

Typical flow curves in Fig. 5 indicate a rapid increase in viscosity and shear strength at w <0.9. The relationship between plastic viscosity and w may be expressed approximately in the form of the Arrhenius equation (2).

$$Y_p = Y_o e^{k/w} \quad \ldots\ldots 3.$$

where Yo = viscosity of water (0.01 poise) and K = constant (range = 1.6-2.2, and 1.8 assumed for Table 3)

Based on the work of ISH SHALOM & GREENBURG (3) and PAPADAKIS (4), a similar relationship can be established between shear strength and w.

$$Ts = To\ e^{k/w} \qquad \ldots\ldots\ldots 4.$$

where $To = 3.5$ dynes/cm^2 and $K = 2.1$

From equations 3 and 4, Table 3 has been compiled to relate the critical flow properties of a neat cement grout to its water/cement ratio. Bearing in mind that the most sophisticated chemical grouts eg resorcinol formaldehyde, have viscosities no lower than 0.015 poise, the traditional practice in fissure grouting of gradually thickening the cement grout starting from w = 20 hardly seems justified. As a starting point w = 5 would seem quite adequate from Table 3 and the volume of water for transport of cement is adequate with bleeding at 75% (Fig. 4).

w	Shear Strength Ts (dynes/cm^2)	Plastic Viscosity Yp (poise)
0.3	3840	4.03
0.4	670	0.90
0.5	230	0.37
0.6	120	0.20
0.7	70	0.13
1.0	29	0.06
2.0	10	0.025
5.0	5.3	0.014
10.0	4.3	0.012
20.0	3.9	0.011
Water	0	0.010

TABLE 3. - Shear Strength and Plastic Viscosity of Neat Cement Grouts

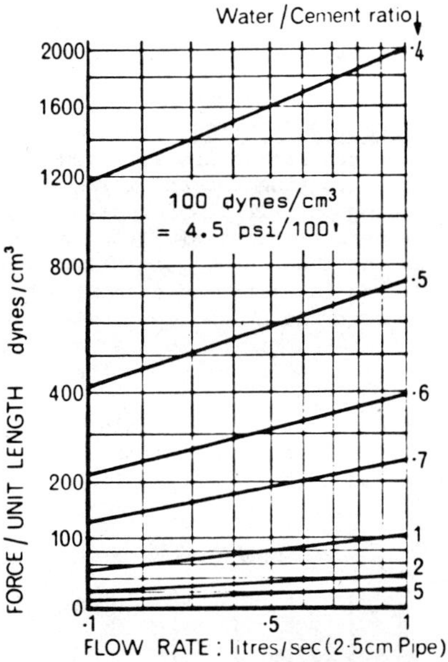

FIG. 6. - Pipe Flow Curves for Neat Cement Grouts

For laminar flow in pipes it may be shown that

$$V = \frac{D}{8}\frac{dS}{dt} \qquad \ldots\ldots\ldots 5.$$

and

$$\frac{dP}{dL} = \frac{4T}{D} \qquad \ldots\ldots\ldots 6.$$

where V = flow velocity, D = pipe diameter and dP = pressure increment/unit length of pipe dL.

Substituting for T and $\frac{dS}{dt}$ in equ. 2 gives

$$\frac{dp}{dL} = \frac{4}{D}\left(Ts + Yp\frac{32Q}{\pi D^3}\right) \qquad \ldots\ldots\ldots 7.$$

where Q = flowrate, $Ts = 3.5e^{2.1/w}$ dynes/cm^2 and $Yp = 0.01e^{1.8/w}$ poise.

This gives a series of laminar flow curves (Fig. 6) for estimating line pumping distances.

5. SETTING

The setting process has two stages, an initial stage in which the fluidity of the grout decreases to a level at which it is no longer pumpable and a second stage in which the set grout hardens and increases in strength. Generally speaking rates of setting and hardening are not related.

ISH SHALOM & GREENBURG (3) have indicated a near exponential increase in shear strength and viscosity with time (Fig. 7) which means that during the initial period following mixing these increases are small and do not effect pumping and grouting operations. For w = 0.35 an initial period of 1 hour is shown which extends to 2 and 3 hours for w = 0.45 and 0.55, respectively. These periods can be extended by continuous agitation of the grout. The rate of development of shear strength is also affected by age and particle size of the cement, which determine rate of hydration. This is low for coarse or old partly hydrated cement and high in new finely ground cement and Table 4 shows how initial Vicat set, equivalent to a shear strength of 170,000 dynes/cm^2 i.e., 2.5 p.s.i., is accelerated by fineness of cement.

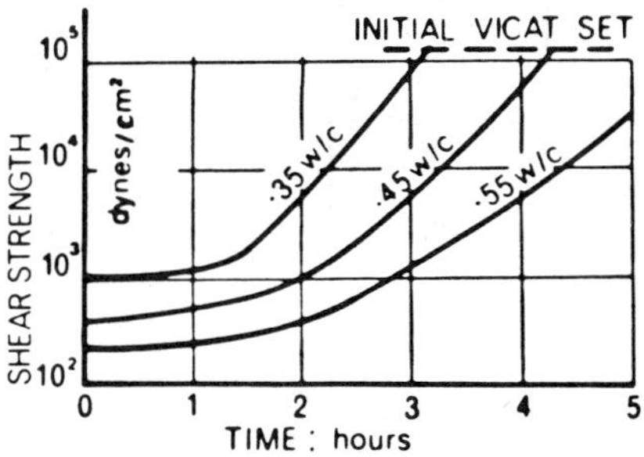

FIG. 7 - Setting times for Type I grouts (18°C)

Specific Surface Area (m^2/kg)	Initial Vicat Set (min)
1320	36
797	63
594	180
270	315

TABLE 4. - Effect of Cement Fineness on Initial Set (Type I - 18°C)

Rate of shear strength development can most readily be changed by the use of retarders or accelerators (Fig. 8). In the absence of test results calcium chloride reduces setting time by an amount roughly equal to 90 minutes per 1% added by weight of cement, but it is important to ensure that the $CaCl_2$ is evenly distributed by dissolution in the mixing water.

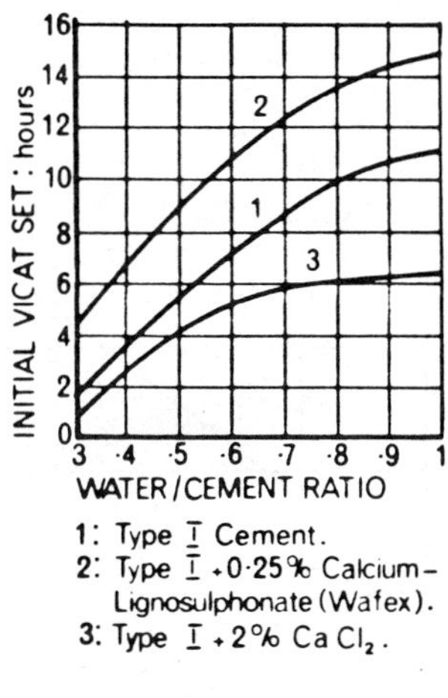

FIG. 8. - Effect of Retarder and Accelerator on Setting Time (18°C)

Although an excellent accelerator $CaCl_2$ has several important side effects. It may affect sulphate resistance, can corrode steel in contact with the grout and increases drying shrinkage. Where instantaneous "flash" setting is required High Alumina/Type I cement mixes may be appropriate, and Fig. 9 shows the relationship between setting time and proportion of Type I cement.

Retarders are normally used to reduce setting rates in hot conditions and when pumping over long distances. Sugar or tartaric acid in quantities of 0.05% by weight of cement can increase setting time by 100%, but trial mixes are recommended since results are variable.

Time (days)	Shrinkage (-%)
1	+ 0.08
2	+ 0.11
3	+ 0.11
7	- 0.72
8	- 0.94
10	- 1.30
14	- 2.37
17	- 3.24
21	- 4.19
28	- 5.14

TABLE 5. - Shrinkage of Neat Type I grout (w = 0.4 at 18°C and 70% relative humidity).

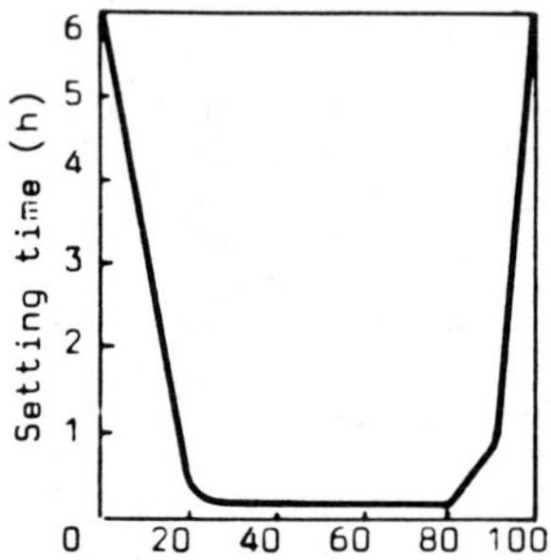

FIG. 9. - Setting Times for Type I High Alumina Mixes.

6. <u>SHRINKAGE</u>

Shrinkage of cement grout is related principally to the amount of water removed. Thus a moist cured grout remaining moist throughout its life, will not shrink and may in fact expand slightly with time. On the other hand, a cement grout dry cured or allowed to dry out after moist curing will shrink (Table 5). For rich large volume grouts autogenous shrinkage (self dessication) is known (5), but even for very rich mixes the maximum shrinkage recorded is only 60 microstrain, and under submerged conditions the mechanism tends to be self healing.

Shrinkage is not normally a serious problem in ground engineering where the environment is often damp or submerged. However, where shrinkage leading to formation of microfissures is likely to affect the permeability of a waterproofing grout, positive steps may be taken to counteract shrinkage, by introducing expanding agents into the grout (Table 2). Fig. 10 shows the effect of aluminium powder on unrestrained expansion during the pre set period.

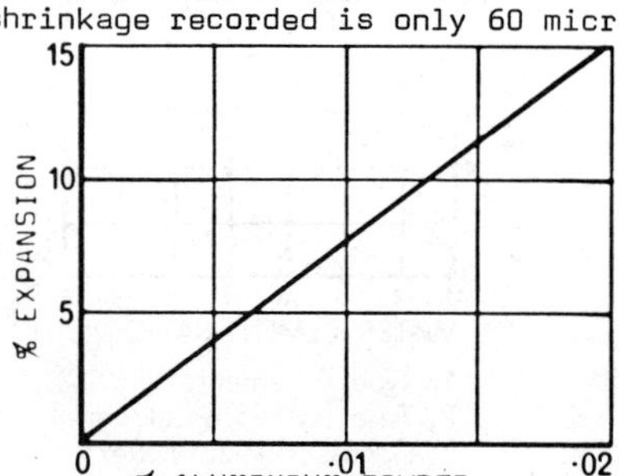

FIG. 10. - Expansion of Type I Grout

7. <u>THERMAL PROPERTIES</u>

Temperatures in a grout mass which are induced by heat of hydration are dictated largely by type and fineness of cement, cement content, placing temperature and insulation. Where thermal cracking is a major concern, low heat cements, low cement content, cool mixing water and cool constituent materials have been used with success to reduce the placement temperature and heat of hydration.

For thermal cracking field experience (6) indicates that cracks in pours up to 1.5m deep are initially formed by a temperature differential of 25 - 28°C. Provided that the mass is less than 1.5m thick the thermal problem is short term, the main temperature differentials dissipating within 7 to 10 days, the greatest temperature rises taking place in the first 3 days (7). In the long term there is a gradual cooling of the grout mass to ambient temperatures.

For mass grout or concrete pours the heat generated by the cement hydration process is controlled by using a low cement content, usually less than 300kg/m^3. It is often desirable to reduce heat even further substituting some of the cement with replacement material. In this regard ACI studies (8, 9) indicate that the early heat contribution of flyash may be conservatively estimated to be 5% to 35% of equivalent cement.

Key
Grade F concrete (138 kg/m^3 cement + 40 kg/m PFA)
Grade E (190+46)
Grade D (248+83)
Grade B (260+90)

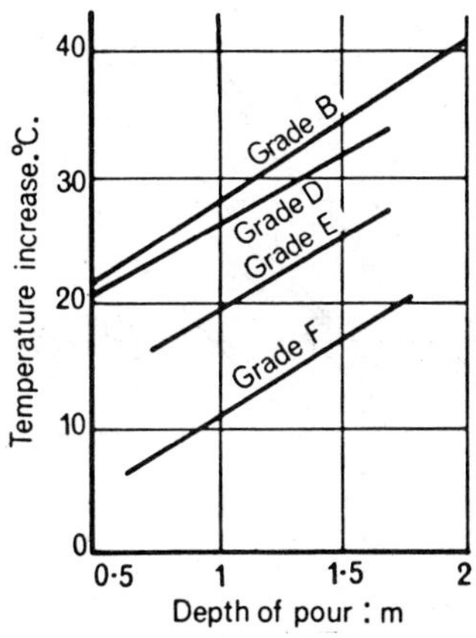

FIG. 11. - Relationship between maximum temperature increase and lift height: Cement content (6).

For Type I cement with flyash replacement Fig. 11 shows the maximum temperature rise which increases with pour depth. The results clearly show the temperature rise (per kg/m^3) of cement content decreases with decreasing cement content. There is also a distinct change in the amount of heat generated at a placing temperature of 16 - 18°C. Below this temperature the actual temperature rise is about 12°C per 100 kg/m^3 cement content (a common figure employed in design for Type I cement) whereas at 21°C say, the temperature rise increases to 16°C per 100 kg/m^3. In this connection for every 10°C increase in placing temperature the rate of heat generation is doubled.

Fineness of cement also has a considerable effect on rate of heat generation, although not total heat (10). When the specific surface is increased from 250^2/kg up to 350 and 450m^2/kg, the rate of heat evolution is increased by 45% and 80% respectively. The linear coefficient of thermal expansion of grout and grouted aggregate concrete ranges from about 5 to 15 microstrain per °C depending on richness of mix and aggregate quality. The lowest coefficients being related to lean grouts injected into preplaced flint or granite aggregate.

8. GROUT STRENGTH

The most important variables affecting grout strength are the original w, the pore space of the set grout and, in the case of times up to 28 days, the type of cement and presence of admixtures.

The dominant parameter is w and the unconfined compressive strength may be expressed in the form of Abrams' Law:

$$UCS = \frac{A}{B^{1.5w}} \qquad \ldots\ldots 8.$$

where UCS = unconfined compressive strength, A = strength constant of 14000 lb/in^2 and B = dimensionless constant depending on characteristic of cement at age of test. For Type I cement at 28 days, B = 5. Since full strength is only generated by complete hydration, Abrams' Law is best restricted to $w > 0.3$ and in grout subject to minimal bleed i.e. $w < 0.7$.

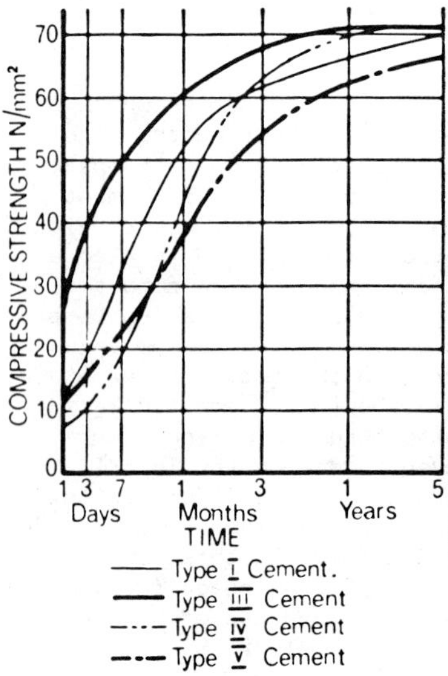

FIG. 12. - Gain in Strength of Set Grouts

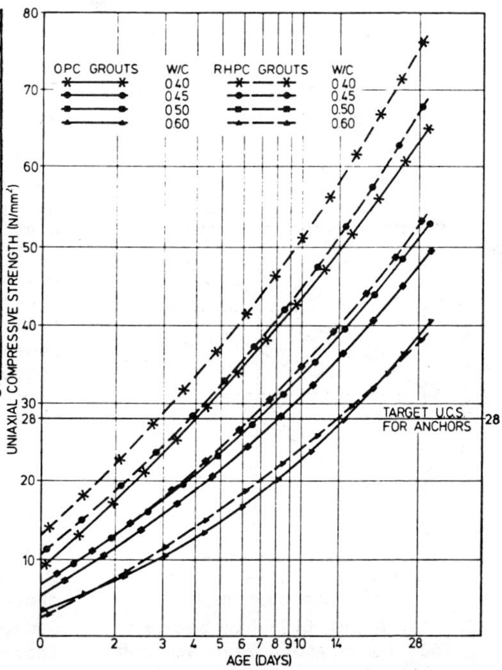

FIG. 13. - Effect of w on Strength Development (Types I & III)

Under optimum curing conditions and not subject to chemical attack a set grout will continue increasing in strength over a prolonged period (Fig. 12) and, irrespective of its short term strength gain, will attain a constant ultimate strength. There is a tendency for cements with a low rate of hardening to have a higher ultimate strength due to slow formation of denser gel during initial stages of setting. In general a Type I grout will have a set strength at 28 days equivalent to approximately 60-70% ultimate. This proportion may be obtained in 3 days with a Type III cement. The strength development curves of Types I and III cement are plotted in Fig. 13 to illustrate the influence of w but also to highlight that at $w \geqslant 0.6$ Type I cement will give a strength development equivalent to that using Type III.

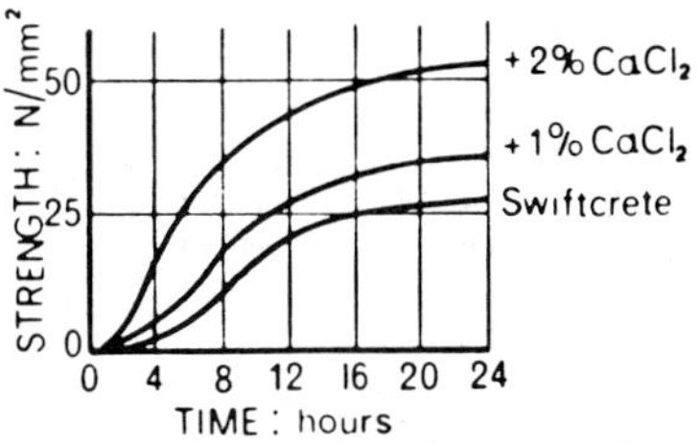

FIG. 14. - Strength Development of Finely Ground Type I Grout ($w = 0.45$).

Addition of fluidifiers permits a reduction of w of the grout at a given fluidity and causes a proportionate increase in set grout strength. Many fluidifiers also act as retarders, increasing the initial set and decreasing early strength. For early strength gain, tests show that $Ca\,C1_2$ gives 30% increased strength at 24 hours when added at 1% of cement weight, although the gain reduces to 10% at 28 days. Finely ground Type I cement ($830m^2/kg$) gives approximately the same strength increase and can be used safely in the presence of steel. The most rapid controlled acceleration is obtained by combining finely ground cement and $Ca\,C1_2$ (Fig. 14).

Porosity of Grouts A relationship between void volume and strength for brittle materials was originally developed by Feret (11) in the form:

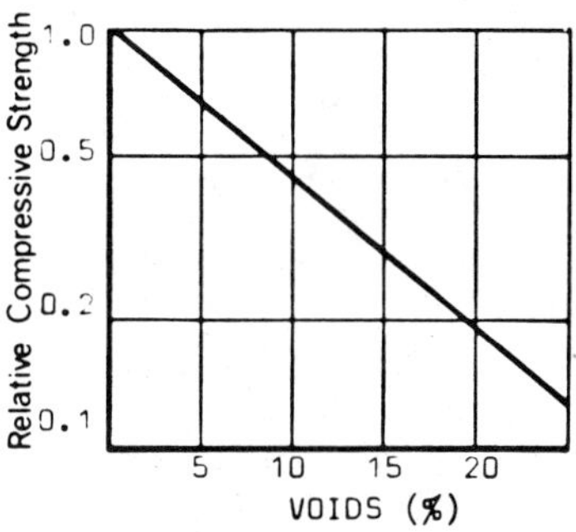

FIG. 15. - Effect of Gas Entrainment on Strength

$$UCS = K \frac{c}{c + e + a} \quad \ldots\ldots\ldots\ldots\ldots\ldots 9.$$

where K = constant and c e and a are the respective volumes of cement, water and air in the grout. In a grout containing little air, the porosity is determined by w, eg 6.5% and 3.5% for w of 0.55 and 0.4 respectively, thus relating strength to w. Where air or gas is released into the grout there is an additional effect on strength, eg 10% gas content can reduce strength by up to 50% (Fig. 15).

Modulus of Elasticity As for concrete an approximate relationship exists between compressive strength and static modulus for neat and normal sand/cement grouts giving in the latter case E values of $20\text{-}40kN/mm^2$ over the strength range $20\text{-}70N/mm^2$. The relative value of static to dynamic modulus is 60-80% for the strength range $20\text{-}45N/mm^2$. Grouts are not normally designed on the basis of E and where knowledge of this elastic property is important experimental determination is necessary using the desired materials and proportions. In this regard for most normal sand/cement grouts the higher the sand/cement ratio, the higher is the E value for a given strength, but E is sensitive to type of filler and may reduce considerably for lightweight materials.

9. DURABILITY

Cement grouts are durable under most normal conditions, but deterioration may be caused by abnormal environmental conditions, especially where there are deficiencies in grout quality, eg low density and high permeability. Of the adverse environmental conditions leading to grout deterioration the most common are chemical attack, notably through sulphates contained in ground water, and large scale temperature fluctuations.

The permeability of a mature cement grout is related, like its strength

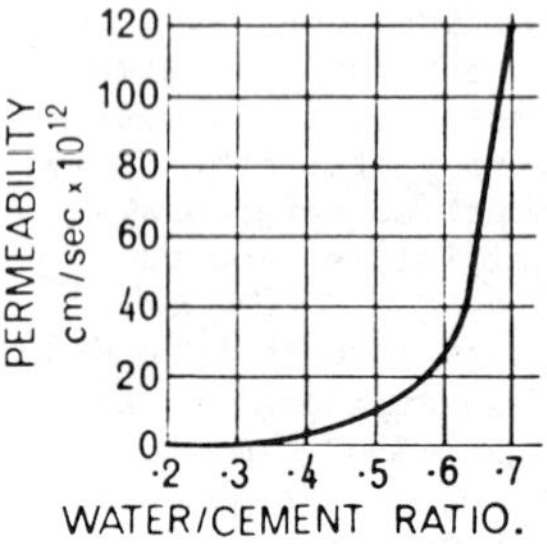

FIG. 16. – 28 day Permeability of Cement Grout

to its original w (Fig. 16). In a fresh or setting grout, permeability is related to the age of the grout (Table 6) and the subsequent degree of hydration which ultimately determines the extent of pores in the grout. Whilst there are no positive guidelines on grout design for durability, a minimum cement content is recommended to provide acceptable durability under the appropriate conditions of exposure as in conventional concrete technology e.g. 400Kg cement/m^3 for severe exposure to sea water, driving rain, alternate wetting and drying and to freezing whilst wet.

Age (Days)	Coefficient of Permeability (cm/sec)
Fresh	2×10^{-4}
5	4×10^{-8}
6	1×10^{-8}
8	4×10^{-9}
13	5×10^{-10}
24	1×10^{-10}
Ultimate (estimated)	6×10^{-11}

TABLE 6. – Permeability of a Type I Grout (w = 0.7)

Long and short term deterioration of cement grouts due to chemical attack may be caused by the presence of dissolved sulphates or acids in ground water, or by prolonged exposure to sea water. Since Type I cements have a poor resistance to chemical attack, this may be increased by use of blast furnace, sulphate resisting (for 0.5 – 1.0% total SO_3 in ground) or ultimately aluminous cements (for $\geq$ 2% total SO_3), but special cements should be considered primarily as an adjunct to high density and low permeability in increasing durability. Where grouts are subjected to frost attack, w should be less than 0.4 to reduce the capillary pore space and permeability of the grout. Alternatively, a rapid setting high heat of hydration cement might be appropriate, or entrainment of air if high water/cement ratios are considered inevitable.

10. QUALITY CONTROL

Variations in grout properties arise from three principal causes: (a) inadequate mixing, (b) variations in grout materials quantities and quality, and (c) apparent variations arising from the testing procedure. In order to obtain a satisfactory basis for grout mix design it is essential, prior to any contract, that methods of storage, batching, mixing and testing of materials be rigidly specified.

Mixing of cement grout leads to a sequence of exothermic reactions in four distinct stages, namely (a) initial highly exothermic reaction lasting 5–10 mins., (b) a dormant period of up to 2 hours with low rate of heat evolution, (c) an increasing rate of reaction leading to final set after 6+ hours and (d) a continuing decreasing rate of reaction after setting. During the dormant period, a cement grout should maintain a consistent physical state, when its properties can be measured. To achieve this state and at the same time avoid false sets, mixing for 5–10 mins. is normally required. Under most field applications this should be achieved by agitation during storage and pumping and placement after mixing.

Accuracy of measurement of grout properties is also an important factor in determining the variability of grout properties in the field. Some property measurements such as bleeding have been developed from laboratory tests, e.g. Powers Float Test and ASTM method, but bleed capacities $> 0.5\%$ can be easily detected in graduated wide, low containers. Laboratory measurements of grout fluidity in terms of shear strength and plastic viscosity are normally carried out with a rotating disc or coaxial cylinder viscometer. Two field instruments are the Colcrete flowmeter (which expresses fluidity in terms of horizontal slump - Fig. 17) and the Portland Cement Association cone (in terms of flow time - Fig. 18). Supplementary checks of w can be made on site by measuring the grout specific gravity using a Baroid mud balance. Hydrometers are not recommended since at low w errors of 25% may be introduced due to the thixotropy and particulate structure of the grouts. In most grouts the hydrogen ion concentration can reflect chemical contamination, thus pH is another useful parameter to monitor. For setting the Vicat apparatus is appropriate and for strength development characteristics 150mm or 75mm cubes can be crushed, the former invariably giving higher early strength (1-7 days) due to heat evolution but lower 28 day results due to sample size in relation to friction of platen faces.

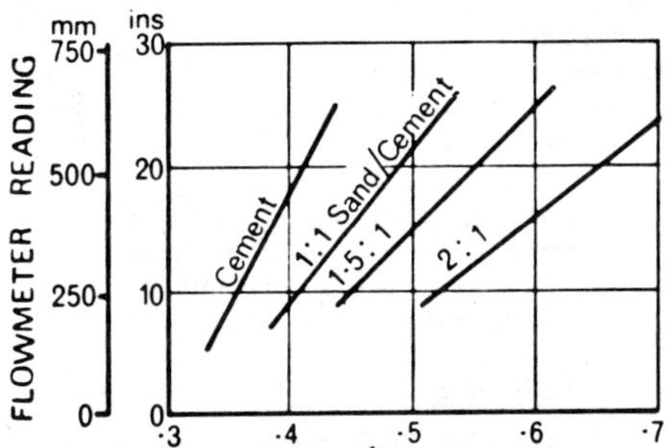

FIG. 17. - Colcrete Flowmeter Calibration

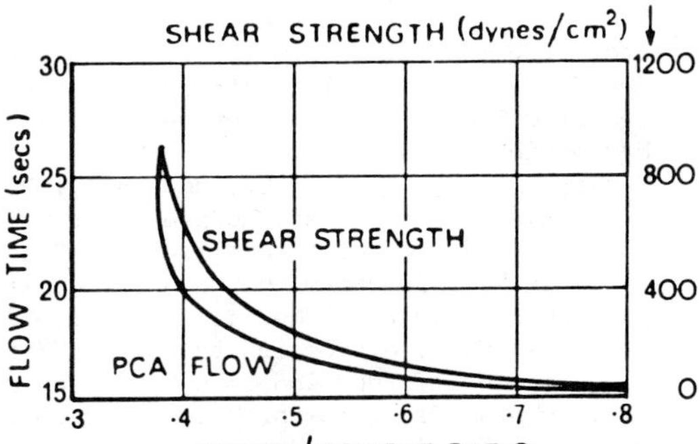

FIG. 18. - PCA Cone Calibration

As a principle in quality control, emphasis should be placed on those tests which permit the grout to be assessed prior to placement e.g. fluidity and density.

11. FINAL REMARKS

General guidance on factors determining grout properties has been provided for design purposes. Final designs should be confirmed, and adapted where necessary, through testing. Inevitable variations in the constituent materials and test conditions must be reflected in the quantitative data quoted and a coefficient of variation of at least 10% should be allowed. To obtain predictable and reproducible results will require:

a) obtaining cement, fillers and chemical admixtures from a reliable source,
b) storage of cementitious materials under dry and constant conditions,
c) accurate monitoring of moisture content of fillers.
d) use of fresh cement,
e) weigh batching of materials,
f) controlled water/cement ratio,
g) adequate rate and time of mixing,
h) early pumping and placement of grout after mixing, and
i) rigid supervision of all operations.

12. REFERENCES

1. OZA, H.P., "Concrete in Sea Water", Indian Concrete Journal, Feb., 1965, pp. 43.
2. Powers, T.C., "The Properties of Fresh Concrete", J. Wiley & Son, Inc., New York, 1968.
3. Ish-Shalom, M., & Greenberg, S.E., "The Rheology of Fresh Portland Cement Pastes", Proc. 4th Int. Symp. on Chem. of Cement Nat. Bur. Std. U.S. Monograph 43, Vol. 2, 1962, pp. 731-743.
4. Papadaxis, M., "Rheology of Cement Suspensions", Tech. Publ. No. 72. Centre d'Etudes et de Recherches de l'Industrie des Liants Hydraulic, Paris, 1955.
5. Troxell, G.E., Davis, H.E., and Kelly, J.W., "Composition and Properties of Concrete", McGraw-Hill, New York, 1968.
6. Dunstan, M.R.H., and Mitchell, P.B., "Results of a Thermocouple Study in Mass Concrete in the Upper Tamar Dam", Proc. Instn. Civ. Engrs., Vol. 60, No. 1, 1976, pp. 27-52.
7. Handcock, M.G., Discussion to Paper, "Shrinkage and Thermal Crackling in a Reinforced Concrete Retaining Wall", by E.P. Evans & B.P. Hughes, Proc. Instn. Civ. Engnrs., Vol. 38, Jan., 1968, pp. 111-125.
8. ACI, see "Features of Lednock Dam, Including the Use of Flyash", Proc. Instn. Civ. Engrs., Vol. 13, June, 1970, p. 179.
9. ACI. see "Mass Concrete for Dams and Other Massive Structures", J. Amer, Con. Inst., Vol. 67, April, 1972, pp. 273-309.
10. Forrester, J.A., "A Conduction Calorimeter for the Study of Cement Hydration", Cement Technology, Vol. 1, May/June, 1970, pp. 95-99.
11. Neville, A.M., "Properties of Concrete", Pitman, London, 1963.

TESTING FOR OPTIMUM PUMPABILITY OF CONCRETE

J.F. Best[1] and R.O. Lane[2]

A testing and evaluation program was conducted at TVA's SME Laboratory to determine significant parameters relative to pumping concrete and mortar mixtures and to develop a simple laboratory pump apparatus for quantitative comparisons. The effects of numerous variables on the pumping characteristics of concrete were investigated, including water-cement ratio; mortar volume; air content; slump; sand content; aggregate shape, size, and content; and the addition of fly ash. The study resulted in the design of a small pump test apparatus for pastes and mortars, and a slightly larger model for concrete mixtures. Both testers produced good correlation of results between lab and field tests utilizing full-scale concrete pumps.

Introduction

The conveyance of concrete by pressure pumping is recognized as an effective method of placing concrete, especially in inaccessible locations. The economy of pumping sometimes questioned for regular applications in concrete construction can be enhanced if mixtures providing adequate pumpability are properly proportioned in the laboratory, thus eliminating expensive trial and error in the field.

The need for a laboratory device that reliably correlates mix parameters to field conditions has resulted in several attempts by researchers to develop such an apparatus. Most efforts have been hampered by the inability to adequately simulate actual pumping resistance by using reduced scale equipment. If successful, however, a laboratory pumping device can assure greater reproducibility of a design mixture and improve pumpability of mixtures on site.

[1]Concrete Supervisor and [2]Head, Materials Engineering Lab., TVA, Knoxville, TN

Procedure

A program to develop a simple laboratory apparatus was initiated to quantitatively measure differences in the pumpability of plastic concrete. The criteria for the initial design included an apparatus to provide a reasonable simulation of pumping action in full-scale concrete pumps and that it be sensitive to significant changes in the consistency or proportions of the pump mixture.

Prototype models were developed to test cementitious mixtures in two phases:

1. Grouts or pastes, including water/cement blends and water/cement/fly ash mixtures
2. Concrete, incorporating various shapes, gradations, and maximum sizes of aggregates

Paste--Cement and cement/fly ash pastes were tested in a small laboratory apparatus shown in figure 1. This apparatus was designed with sufficient clearance between the plunger head and the Plexiglas tube to allow the plunger to pass through the grout rather than forcing the grout column out the top of the tube. A shear interface was thus created between the plunger, the grout, and the tube wall, with the shearing resistance being proportional to the grout consistency.

Grouts with water-cement ratios varying from 0.25 to 0.70 were tested, and the performance of the cement-water grout was compared to grouts containing 20, 40, and 60 percent fly ash by weight of cement. Maximum piston back pressure was used as the standard for comparison.

Concrete--The laboratory device developed to simulate pumping of concrete is shown in figure 2. Because of the coarse and fine aggregate in the concrete, no space could be tolerated between the

FIGURE 1

GROUT PUMPABILITY TEST APPARATUS

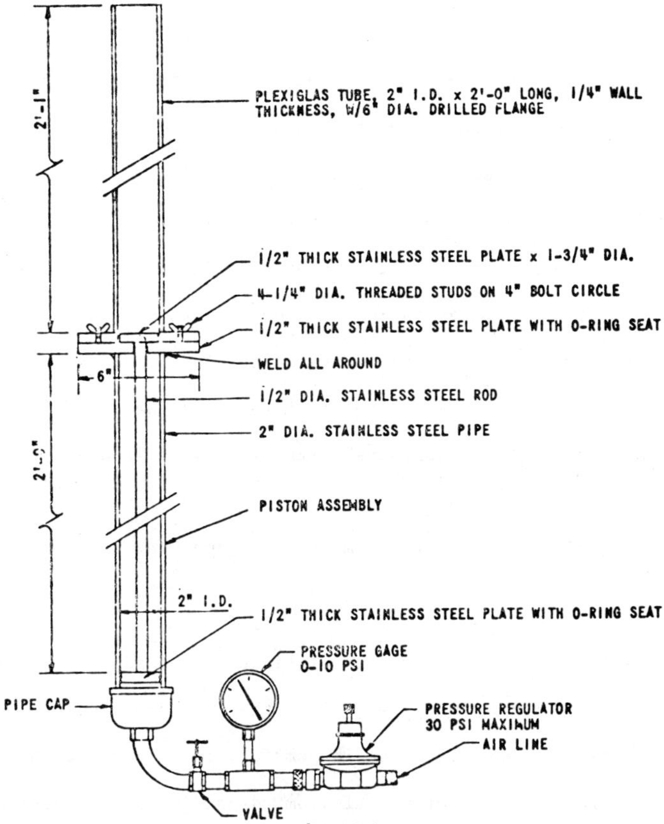

FIGURE 2

LABORATORY CONCRETE PUMPABILITY TESTER

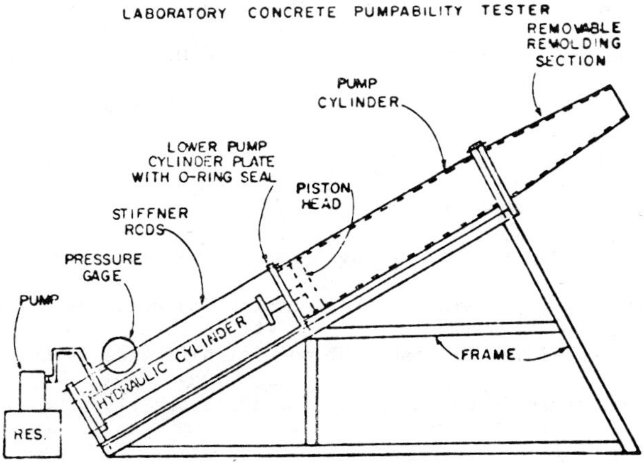

piston and cylinder wall. To simulate pumping resistance without an excessive length of pipe, a transition section was developed to provide a uniform pressure increase during pumping by simultaneously reducing and changing the shape of the cross-sectional pipe area. Experimentation showed the transition producing the most uniform pressure increase proved to be a circular to elliptical configuration. The transition section shown in figure 2 translates a 10-inch-diameter circular area to an ellipse having major and minor axis lengths of 12 and 7 inches, respectively, over an 18-inch distance. The resultant area reduction was approximately 17 percent. Advancing the ram at 1 inch per second, pressures were recorded for each test after uniform flow was obtained. This apparatus worked well with laboratory concrete mixtures and was sufficiently sensitive to provide good comparison of concrete pumpability with various mixture parameters.

Results

Paste--Figure 3 shows the relationship between the water-cement ratio versus pumping pressure for the laboratory pump. The curve reflects the sharp increase in pump pressure required when the water-cement ratio drops below a minimum, depending on the percentage of fly ash used. A minimum ratio is about 0.40 by weight of cement for the plain cement grout and 0.50, 0.55, and 0.60 for grouts containing 20, 40, and 60 percent fly ash. By selecting water contents at or near these minimum amounts, maximum grout strength may be achieved without significant loss in pumpability.

While the water-cement ratio is a meaningful strength-workability indicator for plain cement grouts, it is somewhat misleading for mixtures containing fly ash. Of more significance is the ratio of water to total cementitious material including the fly ash. This

FIGURE 3

LAB PUMPABILITY TESTING OF GROUT PASTES

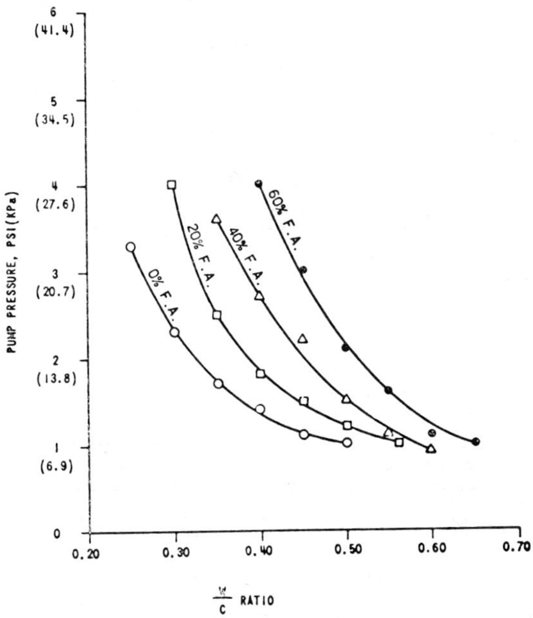

FIGURE 4

LAB PUMPABILITY TESTING OF GROUT PASTES

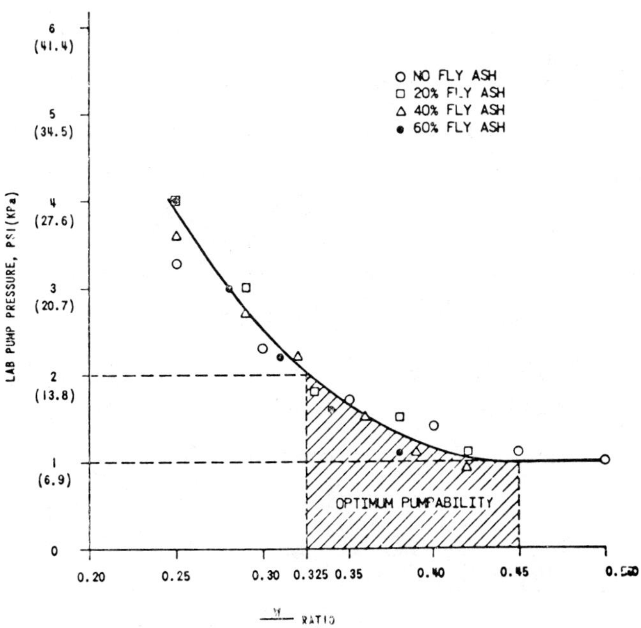

ratio, shown as W/C+P, is plotted versus pump pressure in figure 4. The use of the W/C+P ratio makes possible a single curve to define pump relationships for varying fly ash (P) contents.

Concrete--Figures 5 through 12 show the effects of several variables on the pumping pressure required for the concrete mixtures. Although pressure magnitudes varied depending on the varying test parameters, all curves show a marked pressure increase at slumps less than about 3 inches.

Conclusions

Results from the pump tests emphasized the fact that concrete pumpability is influenced not only by the physical characteristics of the concrete, but also by the relative proportions used in proportioning the mixture. The effects of slump, water content, mortar volume, fine and coarse aggregates, fly ash, and air content were evaluated independently, and their relative significance is discussed in the following summary.

Slump--The pumpability of concrete increases proportionally with slump, but the lower limit of slump for optimum pumping varies depending on the characteristics and proportions of the mixture ingredients. This is illustrated in figure 10 where the inflection point on the laboratory pressure curve occurs at a slump between 3 and 4 inches for the plain concrete mixtures and from 2 to 3 inches slump for the fly ash concrete. The inflection point on the pressure curves represents the approximate midpoint of the effective slump range at which concrete can be pumped without excessive pressure buildup. Concrete slumps approximately 1 inch on either side of this point defined the range of optimum pumpability for the field pump tests.

FIGURE 5

RELATIONSHIP OF PUMPING PRESSURE TO $\frac{W}{C+P}$ RATIO FOR CONCRETE MORTARS

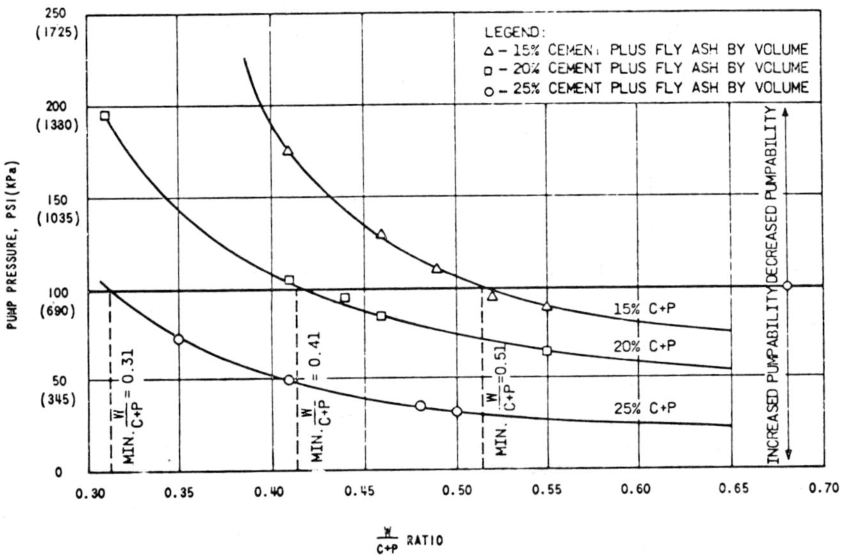

FIGURE 6

RELATIONSHIP OF PUMPING PRESSURE TO SAND VOLUME OF CONCRETE MORTARS

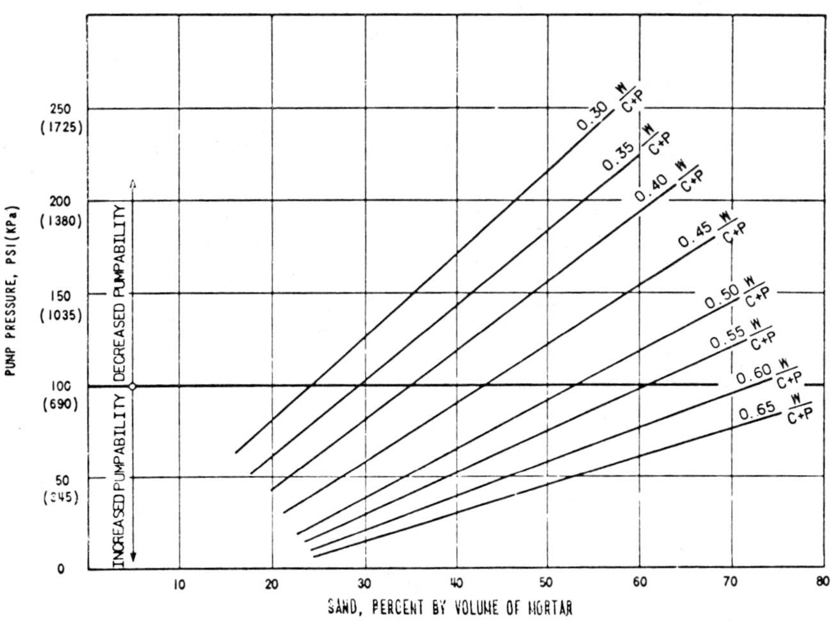

FIGURE 7

EFFECTS OF SLUMP ON PUMPABILITY

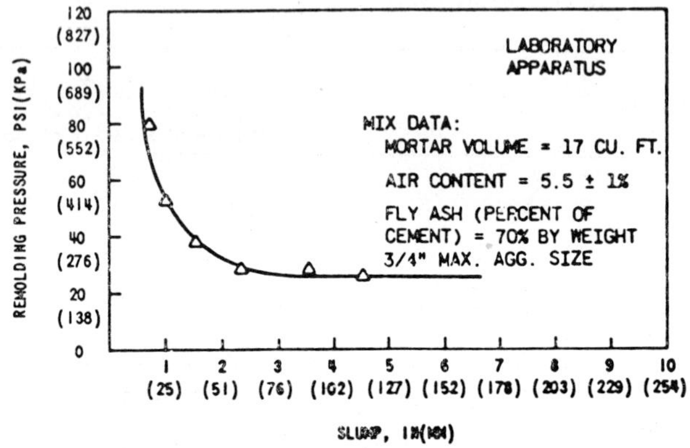

FIGURE 8

EFFECTS OF MORTAR VOLUME ON PUMPABILITY

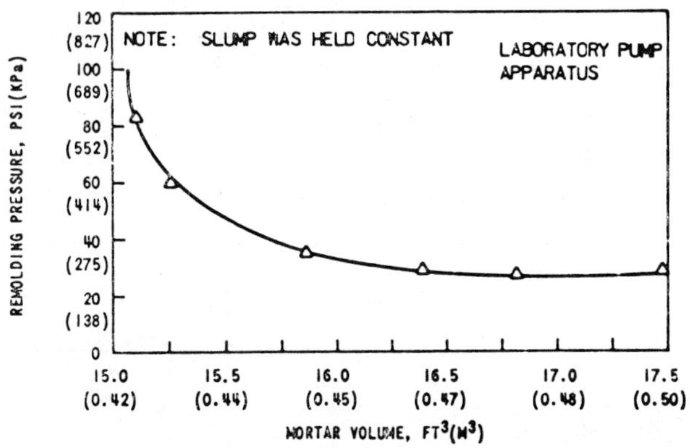

FIGURE 9

EFFECTS OF FLY ASH PROPORTIONS ON PUMPABILITY

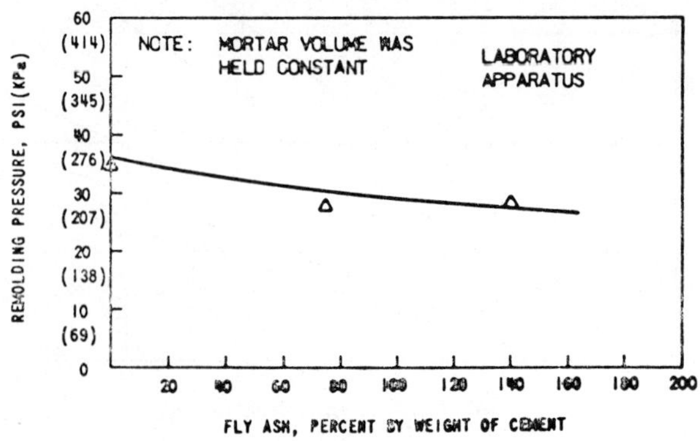

FIGURE 10

PUMPABILITY OF NON-FLY ASH VS. FLY ASH CONCRETE

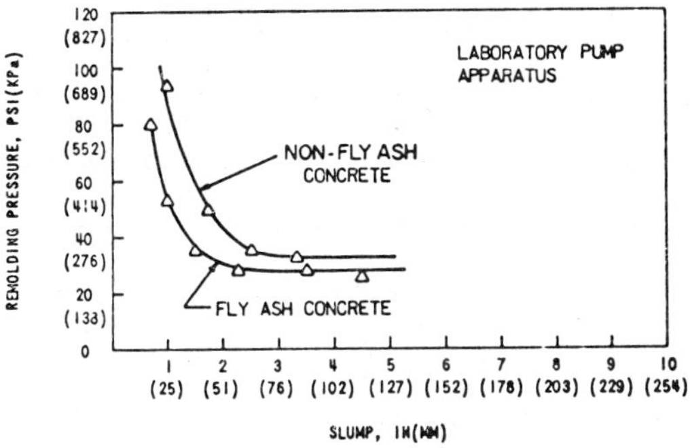

FIGURE 11

EFFECTS OF MAXIMUM SIZE AGGREGATE ON PUMPABILITY

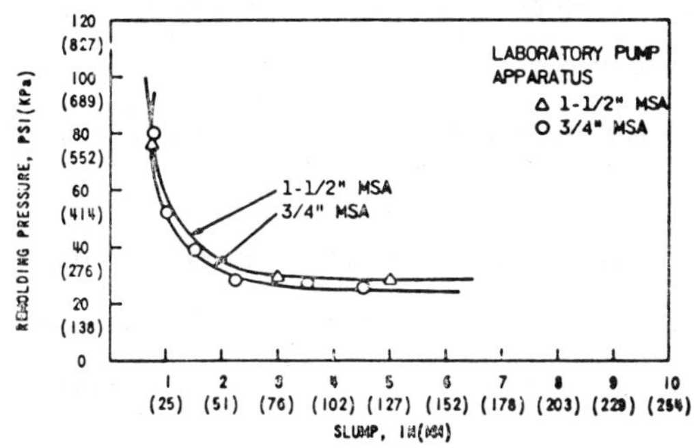

FIGURE 12

EFFECT OF AGGREGATE SHAPE ON PUMPABILITY

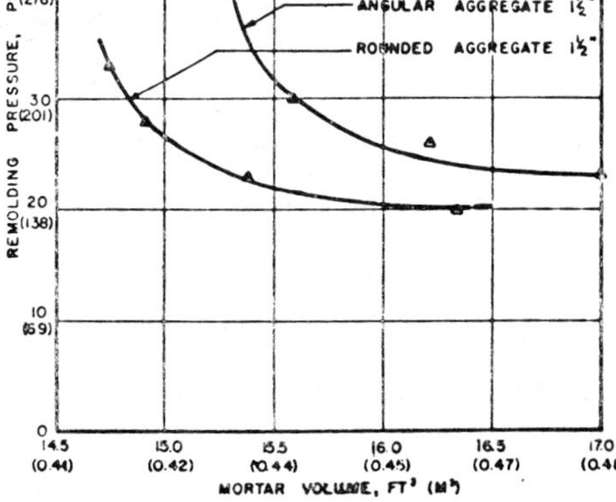

Water Content--The amount of water in pumped concrete should be the minimum necessary to produce fluidity adequate to transmit pressure without segregation or water migration. Too little water produces harsh mixtures which increase friction at the pipe-concrete interface, while too much water results in segregation of the mortar and paste from the aggregate during pumping. In addition, the strength of the placed concrete in the latter case is lowered.

Since the water content and slump of concrete are directly related, the curves comparing W/C or W/C+P ratios with pump pressures are similar in shape to the slump versus pressure curves. The inflection points for the W/C+P pump curves fall within the range of optimum slump previously discussed and also serve as a basis for estimating the correct amount of water for a trial pump mixture.

Mortar Volume--The relationship of mortar volume to the required pumping pressures for concrete is shown in figure 8. Although the curves shown can be used to select the optimum mortar volume for most 3/4-inch maximum size aggregate concrete mixtures, the actual pumping pressures may vary in magnitude if slumps and mortar proportions are different from those shown. For instance, figure 8 shows the relative difference in pumping pressures caused by varying the percentages of cement plus fly ash in the mortar. Significant changes in slump would cause a similar upward or downward shifting of the pressure curves.

For any pump mixture of given maximum aggregate size, a minimum mortar volume exists which will provide for adequate pumpability. Unlike slump and water content, however, there is no distinct upper limit for mortar volume at which pumping becomes more difficult. Practical maximum limits are imposed by cost considerations and the fact that excessive mortar volumes will increase shrinkage cracking and heat generation of the in-place concrete.

Fine and Coarse Aggregates--Several reports have dealt with the effects of aggregates on the pumpability of concrete, particularly with regard to optimum gradation limits and ratios of coarse to fine aggregate. The primary emphasis in this study was to determine the effects of aggregate shape, maximum size, and proportions on the pumpability of concrete.

The coarse aggregate shape has a significant effect on pumping pressures, particularly on mixtures with marginal mortar volumes. In a series run with 3/4-inch maximum size aggregates, mixtures containing an angular manufactured aggregate required about 15 percent greater pressures than rounded natural stone mixtures having 16.5 cubic feet of mortar. When the mortar volume was decreased to 15.5 cubic feet, the mixtures with manufactured stone required almost 40 percent greater pressure. See figure 12.

Maximum aggregate size had little effect on pumping due to the fact that mortar volumes were adjusted to provide additional mortar for the smaller sized aggregates. As long as sufficient mortar is coating all coarse aggregate particles, pumpability is not affected, assuming other variables, including pipe diameter, remain the same.

Fly Ash--The improvement in pumping characteristics due to fly ash in this study is typical of that reported by others. As with any finely divided mineral admixture increasing the effective size range, fly ash provides additional fines which reduce internal voids and improve workability. Additional benefits are derived from fly ash due to the rounded particle shapes, which further reduce internal friction and bleeding without increased water requirements associated with some of the fine-grained admixtures.

Air Content--The entrainment of 4 to 6 percent air in concrete increases pumpability by providing added plasticity, less bleeding, and subsequently less segregation due to water migration. Especially with crushed aggregates, too little air increases the line friction during pumping and makes startups after pumping delays more difficult. Too much air can decrease pumping efficiency by absorbing much of the pump stroke energy through compressibility.

Some loss of entrained air is inevitable due to the compression and shearing action of the plastic concrete during pumping. These losses can be minimized, however, by keeping the concrete slump levels as low as feasible to maintain uniform pumpability. Results of this study showed air loss to be more dependent upon initial concrete slump than the original air content. For a series of mixtures containing an air content of 5-1/2 ± 1/2 percent, the air loss after pumping averaged 8 percent for 1-inch slump concrete, 15 percent at a 3-inch slump, and 25 percent for the 5-inch slump mixtures. A second group of mixtures was pumped holding a constant slump of 3 ± 1/2 inches while varying the initial air content from 2 to 7 percent. Loss in air content after pumping these mixtures averaged 16 percent, regardless of the original air content.

Summary

The amount of data generated by this program has provided a substantial information base for determining the relative significance of the many variables associated with pumped concrete. The evaluation of the two laboratory pumps used in this study also indicates their reliability for predicting the relative performance of concrete pumped with piston-type pumps under field conditions. However, specific pump systems and layouts at the site require consideration prior to proportioning laboratory concrete mixtures for optimum pumpability.

CONVERSION FACTORS

U.S. CUSTOMARY TO SI UNITS

To convert British units of measurement to metric units:

Multiply	By	To Obtain
Inches	2.54	Millimeters
Feet	0.30480	Meters
Cubic feet	0.028317	Cubic meters
Pounds	0.45359	Kilograms
Pounds per square inch	6.89476	Kilopascals
Pounds per square inch	0.0068948	Megapascals
Pounds per cubic foot	1.60185	Kilograms per cubic meter

Effects of mixing on bentonite slurries and grouts

Dr. Stephan A. Jefferis[1]

ABSTRACT

Bentonite-cement mixes may be used to form hydraulic cutoffs by direct injection as grouts or by a variety of slurry trench techniques. It has been found that the properties of these materials are considerably influenced by the way in which they are mixed. For bentonites gel strength and viscosity increase and permeability of the gelled slurry decreases continuously with time from mixing. However, the data shows that poorly mixed slurries may never develop as good cutoff properties as better mixed slurries. For bentonite-cements good mixing reduces the bleeding of the fluid material and the permeability of the set material. However, it does increase the strength and brittleness of the mix though this can be offset by the addition of sand or spoil.

I INTRODUCTION

Bentonite slurries and bentonite-cement grouts can be used to form hydraulic cutoffs in the following ways:

(i) Excavation of a slurry trench under bentonite and: a) subsequent backfilling with soil to form a soil-bentonite cutoff or b) subsequent displacement of the bentonite by a bentonite-cement grout possibly containing graded soil/aggregate to form a "plastic concrete".

(ii) Excavation of a slurry trench under a bentonite-cement grout which is then left in the trench to set and harden.

(iii) Direct injection of a bentonite slurry or a bentonite-cement grout into a porous formation.

It has been found that the method of mixing the slurries or grouts has a considerable influence on their performance as cutoff materials. Therefore a detailed study of mixing was undertaken.

MIXING OF BENTONITE

Four different bentonites were used for this study of mixing. Three of these were of British origin and were natural calcium bentonites converted to the sodium form with sodium carbonate. The fourth bentonite was a natural sodium bentonite from Wyoming, U.S.A. The basic properties of the bentonites were as follows:

---

[1] Lecturer in Civil Engineering, King's College, University of London, London, England.

Bentonite number	Origin	Moisture content as supplied	Liquid limit
B1	U.K.	14.3%	>520%
B2	U.K.	21.7%	400-450%
B3	U.K.	9.9%	400-500%
B4	U.S.A.	17.4%	600-700%

Slurries of each of these bentonites were prepared in 3 litre batches using a 5cm diameter propeller mixer the speed of which could be varied between 2000 rpm and 10000 rpm. Initially 5% slurries (5 grams of bentonite as supplied per 100 grams water) of each of the bentonites were prepared at just two speeds, 2000 rpm and 10000 rpm with a constant mixing time of 5 minutes. These speeds were chosen as earlier tests had shown that 2000 rpm gave mixing similar to a rather poor site mixer and that 10000 rpm gave mixing comparable with good site mixing in a centrifugal pump type mixer (though not as good as could be obtained in a colloid mixer).

The development of the rheological properties of the slurries was monitored with a Fann viscometer. As shear in the viscometer could produce some additional mixing slurry samples were always discarded at the end of a test and not returned to the stock container. Figure 1a,b,c,d shows the apparent viscosity of each of the slurries as a function of the logarithm of time from the start of mixing (apparent viscosity has been plotted as the data is generally more precise than that for plastic viscosity or gel strength; problems in obtaining reliable gel data are discussed more fully in the section on gel strength development). It can be seen that for each of the slurries the apparent viscosity increases linearly with log (time from mixing). Furthermore for each of the slurries the data for the two mixing speeds follow separate lines which would not converge until very long times after mixing. The continuing development of viscosity with time as shown in Figure 1 was not expected. It was realised that the gel strength of an undisturbed slurry could show this type of behaviour. However, it was thought that viscosity, which is a property of a disturbed slurry, would tend to an equilibrium value within a few days of mixing and that this equilibrium value could be rather insensitive to mixing speed. However, the data suggests that there is no equilibrium value at least up to 6 months after mixing and very probably beyond 3 years (there is only one data point at 3 years and four at 6 months but all support the trend of increasing viscosity with time. Furthermore the data shows that at all times the apparent viscosity is dependent on the original mixing speed.

Remixing:- 24 hours after the original mixing samples of each of the slurries shown in Figure 1 were remixed at speeds of 2000 rpm and 10000 rpm. For clarity only the data relating to the 10000 rpm slurry remixed at 10000 rpm and the 2000 rpm slurry remixed at 2000 rpm are shown. It can be seen that remixing is a very effective at developing the early viscosity (and of course gel strength) though the effect of mixing speed is still significant. Actually remixing speed seems to become the dominant factor for when remixed at 10000 rpm both the 2000 rpm and 10000 rpm original mixes showed almost identical viscosities. Remixing at higher speeds than the original mix is of

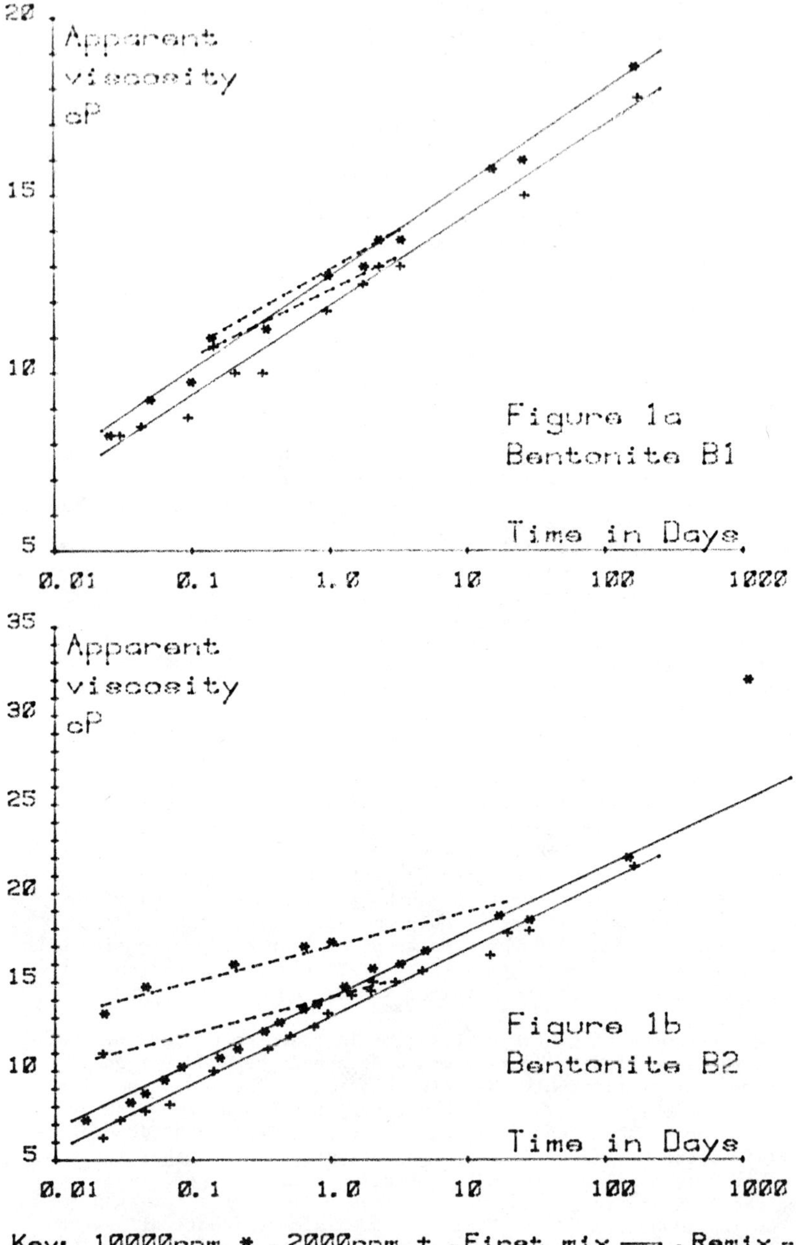

Key: 10000rpm * , 2000rpm + , First mix —— , Remix ---

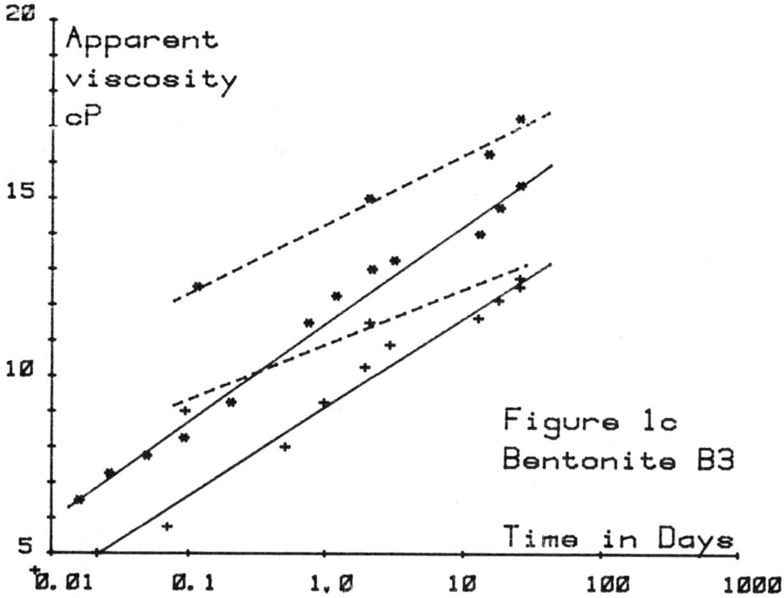

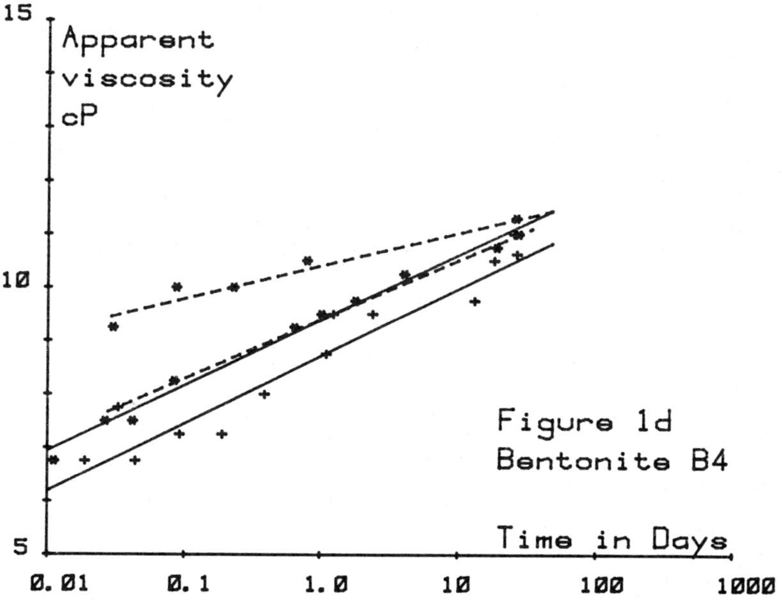

Key: 10000rpm * , 2000rpm + , First mix —— , Remix ---

little practical significance. For if such a mixer is available it should be used for the first mix as well as the remix as low speed original mixing may allow settlement of bentonite particles in the holding tanks.

Mixing speed and time:- As bentonite B3 showed greatest variation in viscosity with mixing speed it was used for a limited further study of mixing. Figure 2a shows the effect of mixing speed on the apparent viscosity (after 24 hours hydration) of 3.5%, 5% and 7% slurries of this bentonite and Figure 2b the effect of mixing time. For both these plots the data has been normalised by dividing the viscosity for any mixing speed or time by that for a slurry of the same concentration mixed at 5000 rpm for 5 minutes. From Figures 2a and 2b it can be seen that the effect of mixing speed increases with increasing slurry concentration but the effect of mixing time is rather similar at all three concentrations. In Figure 2a data has also been presented for mixing with a special high shear mix head. This had a rotor with two concentric circles of teeth and an intermeshing stator with three concentric circles of teeth. This head produces a very high level of mechanical shear and the data from it confirms that mix head design is probably more important in developing viscosity than simply mixing speed.

The sensitivity of a bentonite to mixing speed and time probably will be dependent on size of the original dry bentonite particles and concentration of cations in the resulting slurry (these cations may tend to inhibit dispersion, and thus this effect should be more noticeable at high slurry concentrations). Sensitivity to mixing speed and time maybe of rather little significance in itself provided that appropriate mixing procedures are used.

Development of gel strength:- In many situations (e.g. soil-bentonite cutoffs) the development of gel strength may be of more significance than the development of apparent viscosity. Unfortunately it is difficult to obtain sufficient and reliable gel strength data for long ageing times because:

(i) Gel strength is a function of the rate at which the gel is broken. High speeds of breaking lead to high gel strengths. This problem cannot be avoided by testing at constant rotational speed, for as the gel strength increases so does the shear modulus. Thus as the gel develops constant rate of straining will not correspond to a constant rate of stress application. This problem is exactly analogous to that encountered in the measurement of the compressive strength of concrete where constant rate of loading is necessary if good repeatability is to be obtained. Unfortunately although machines for testing concrete at constant loading rate are widely available few viscometers can be readily operated in this way.

(ii) For tests over ~24 hours anomalous very low results are often obtained. This may be due to some shrinkage of the gel (syneresis).

(iii) The sample must be prevented from drying out during the test.

(iv) As breaking the gel will distrub the sample timing must be restarted at the end of each test. Thus the rate at which data can be obtained will be very low.

(v) Gel strength will depend not only on time from last agitation but also the time from original mixing. Thus a considerable amount of data will be needed for a full understanding of the subject.

Some of the above problems can be avoided by using ultrasonic pulse velocity techniques similar to those used in the non destructive testing of concrete. A suitable apparatus for this is manufactured by Rank Brothers of Cambridge, England. Details of this can be found in Goodwin and Smith, 1974. The information obtained from the apparatus is basically the pulse velocity V for a shear wave travelling through the gel. At the frequencies used in the apparatus the shear modulus, G can be obtained from the following equation:

$$G = \rho V^2 \qquad \text{.. .. ..} \quad 1$$

where ρ is the density of the slurry. A major advantage of this system over rotational viscometry is that only a very low amplitude shear wave is propagated and thus the gel is not disturbed by the passage of the wave. Hence the pulse velocities can be measured as frequently as desired without having to start again at time zero after each determination. The disadvantage of the apparatus is that it gives data on shear modulus, G and not gel strength, τ and there does not appear to be a unique relationship between the two parameters (there is an exactly analogous problem in concrete testing). However, the following equation gives a reasonable fit to the data for all the slurries so far tested:

$$\tau = 10^{-3}\, G^{1.6} \qquad (\tau \text{ and } G \text{ in } N/m^2) \qquad \text{.. .. ..} \quad 2$$

Two slurries of bentonite B2 have been extensively studied by shearometry. Both slurries had a concentration of 5% and were mixed for 5 minutes one at 10000 rpm and the other at 2000 rpm. Immediately after mixing the slurries were poured into the test rig and the pulse velocity was then monitored over a period of 20 days. For both slurries the pulse V was found to be related to the time from mixing, t_m by an equation of the form (for each slurry the correlation coefficient was $>$ 0.996 for times in the range one hour to 20 days):

$$V = a \log t_m + b \qquad \text{.. .. ..} \quad 3$$

The constant, a was almost identical for both mixes but the constant, b was lower for the 2000 rpm mix. Thus just as for apparent viscosity the pulse velocity is correlated with log (time from mixing) and is lower for the low speed mix for all times (the pulse velocity plots would not appear to intersect until a time of $>$100 years).

Figure 3 has been plotted from the pulse velocity data using equations 1, 2 and 3. It shows the development of the shear modulus and the shear strength of the two mixes. After 20 days the 10000 rpm sample was stirred to break the gel and the subsequent recovery of shear modulus was followed. Data from this subsidiary test are also included in Figure 3. After stirring the pulse velocity no longer shows a simple relationship with solely log (time from mixing) or log (time after stirring). However, the data can be approximated by an equation of the following form:

$$V = a_s \log t_s + a_m \log t_m + b_s \qquad \text{.. .. .. ..} \quad 4$$

with $a_s < a_m$

and $a_s + a_m \simeq a$

$b_s \simeq b$

Where t_s is the time from stirring, a_s, and a_m are constants and a and b are the constants from equation 3. The first term in equation 4

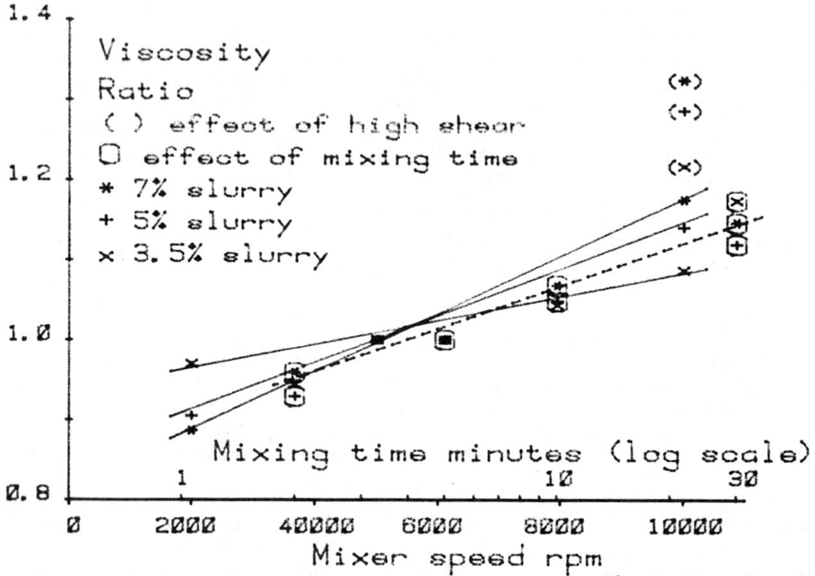

Figure 2a, b

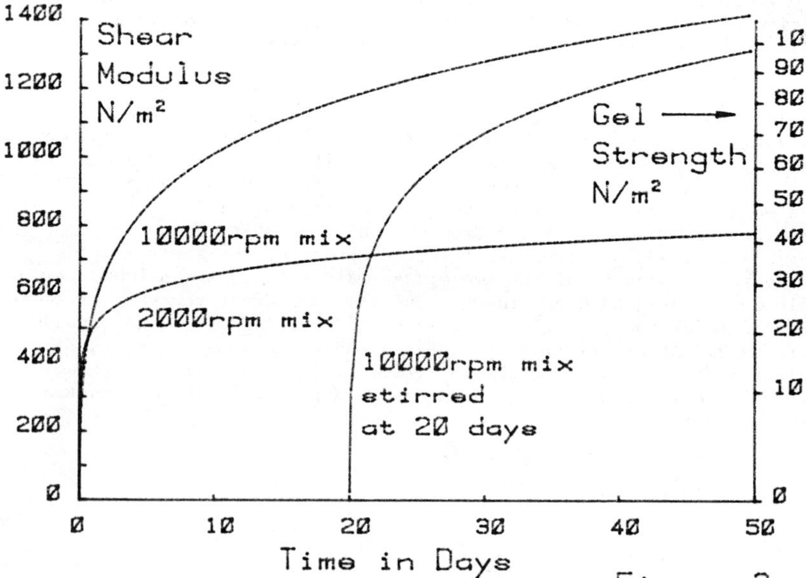

Figure 3

represents the development of pulse velocity soon after stirring. For much longer times ($\log t_s \simeq \log t_m$) the pulse velocity tends to that of an unstirred mix. Thus at long times from stirring the rate of increase of shear modulus and gel strength will be controlled by the time from original mixing, provided that any stirring is not so vigourous as to represent a new mixing.

The significance of the data presented in Figure 3 is that any extrapolation of gel strength based on data for gel times of hours rather than days (that is data that can be readily obtained by rotational viscometry) could lead to a very substantial overestimate of the long term gel strength. For example if the data either from first mixing or from stirring were extrapolated to one year the gel strength at that time would be only 8 times that at 24 hours from mixing. Whereas if the underlying trend were not known and the gel strength extrapolated from the early data a very much higher factor could be predicted. This will have significance in the prediction of long term behaviour of soil-bentonite cutoffs.

Shearometry is thus a very powerful tool for studying gel behaviour and more tests are in progress. Unfortunately, the present instrument cannot be used to study materials that set strongly as it would be impossible to remove them from the rig without damaging it. A modified instrument is currently being developed at King's College, so that grouts can be tested.

Reasons for the development of viscosity and gel strength with time:- The continuing development of viscosity and gel strength with time may be the result of many phenomena occurring within the slurry. However, just two will be considered here.

(i) Water may continue to penetrate between the layers of clay sheets that make up a clay particle. Thus the individual particles will swell with time (this is why remixing is so effective) and ultimately break up to form more but smaller particles. An increase in the number of particles will increase viscosity and gel strength. Obviously there is a limit to this process and that is when all the clay particles are broken down to individual clay sheets (if this is thermodynamically possible). However, this may take very long times and thus an apparent development of properties with log time may not be unreasonable. This process of swelling and separation will be referred to as delamination.

(ii) Orientation of the particles to form stronger interparticle bonds could explain the development of shear strength with time (and particularly the initial repid gain in gel strength that occurs when a slurry is agitated and then left quiescent). However, it is unlikely that this could explain the development of viscosity with time for such bonds would have to be fairly strong if they were not to be broken down by the shearing in a viscosity test. Samples of "old" (6 months) slurries sheared for more than one hour showed no significant drop in viscosity with time. Thus it seems likely that the increase in viscosity with time is mainly dependent on delamination or other phenomena such as continuing cation exchange. Further information on this is given in Stepkowska and Jefferis, (1981).

Significance of good mixing for bentonite slurries:- If the increase in the apparent viscosity of bentonite slurries with time and mixing quality is due to delamination or similar phenomena then the following effects will result:

(i) as delamination will lower the mean particle size of a bentonite slurry the permeability of a gelled slurry will decrease with time, mixing speed and mixing time. Thus for example, excluding factors such as aggressive ground water the permeability of a soil-bentonite cutoff should decrease with time. This effect has been observed in the laboratory though the data is as yet very limited. In one test the permeability dropped by 30% between the first and second months after mixing.

(ii) filtrate loss as measured in the standard American Petroleum Institute test rig will not necessarily decrease with time or mixing. The reason for this is that filtrate loss will depend on both the size of the particles forming the cake and its porosity. Particle size obviously decreases with delamination but gel strength increases. This increase in gel strength may lead to a cake of increased porosity as the applied pressure will not be able to compress it as much. In practice it seems that the effect of increase in gel strength may be dominant as increases of filtrate loss with mixing speed and time are often observed. Though the effect is not very great typically less than 5% for mixing speeds of 2000 rpm and 10000 rpm.

(iii) the increase of filtrate loss with mixing could have slight significance in soil-bentonite cutoffs where the filter cake that forms on the walls of the trench during excavation is intended to contribute to the cutoff or at least to provide an additional factor of safety (see D'Appolonia, 1980). For such cutoffs there could be a marginal advantage in adding a dispersing agent to reduce the gel strength of the slurry provided that it had no long term deleterious effects.

(iv) if the bentonite slurry is to be used to prepare a bentonite-cement grout then increased delamination will increase the strength and impermeability of the set grout and decrease the tendency of the fluid grout to bleed. Thus good mixing and long hydration times will be important in the preparation of bentonite slurries used in the preparation of bentonite-cement grouts.

(v) in general if any bentonite slurry must be used soon after mixing then high speed or prolonged mixing should be used.

MIXING OF BENTONITE-CEMENT GROUTS

There are three principal ways in which cement and bentonite may be mixed together:

Dry mixing:- The two dry materials are added together (possibly as a pre-blend) or closly following each other, to water in a mixer. If the materials are mixed in this way the calcium ions released during the early reactions between the cement and water will prevent the bentonite particles from dispersing. This type of mixing will only be suitable when relatively high strength grouts are required, that is fairly low water cement ratios. For if high water cement ratios were used severe bleeding would occur as the undispersed bentonite would not be able to control it. For these mixes the role of the bentonite is to improve the flow characteristics and the cohesiveness of the grout by providing additional fine material. The bentonite is unlikely to reduce the permeability of the grout as cement paste has an intrinsically very low permeability and any additional material is likely to cause an increase rather than a decrease. Grouts mixed in this way will be suitable as injection grouts but not for slurry trench work.

Dry cement addition to pre-hydrated bentonite:- In this type of mixing dry cement powder is added to pre-hydrated bentonite in a mixer. With this procedure soon after the first of the cement comes into contact with the bentonite the slurry stiffens and often becomes so stiff as to prevent circulation in the mixer. However, after a few minutes the mix becomes more fluid so that mixing can continue.

The rapid stiffening is caused by mutual flocculation of the negatively charged bentonite particles and positively charged cement particles. The fluidification probably results from complete covering of the cement particles by the finer clay particles. When this has occurred the positive charge of the cement particles is obscured and the strong floc structure breaks. However even by this stage the cement will have released calcium and other ions into solution. These will flocculate the bentonite (by promoting traditional edge to face bonding etc.) thus even after the fluidification the mix will be fairly viscous. A further factor promoting the fluidification is that some bentonite particles may aggregate (face to face bonding) as a result of the cations released by the cement. This will reduce the number of clay particles and thus reduce the floc strength.

When dry cement is added to bentonite it is unlikely that the particles will be fully dispersed (that is all grains completely separate) as they meet the bentonite. Thus bundles of cement grains will become coated with bentonite. Once formed such bundles are difficult to break down in the rather viscous mix unless prolonged or particularly high shear mixing is used. In practice such bundles often occur and can be seen as dark nodules if grouts prepared in this way are examined by hand lens. The presence of nodules will lower the effective concentration of cement in the mix and this could be regarded as a waste of cement. They also tend to increase the permeability of the mix but have the advantage of making it more plastic. These effects will be discussed more fully in the section on Significance of mixing procedure.

Pre-dispersion of the cement:- The cement may be mixed with water before it is added to the bentonite. A water cement ratio of about 0.5 gives a suitably fluid mix from which the cement does not settle too rapidly. The advantage of this mixing procedure is that the cement can be well dispersed before it is added to the bentonite. When it is added stiffening and subsequent fluidification occurs just as for dry cement addition (and for the same reasons). However, the mix is much more homogeneous and nodules of undispersed cement can seldom be found even using an optical microscope rather than a hand lens.

Mixing sequence:- The sequence of mixing the cement and bentonite can be important. If bentonite is added to dry cement or to a cement water slurry rather than cement to bentonite a severe flash set may occur and completely block the mixer. This is because when cement is added to bentonite the effect of the dilution prevents very severe stiffening.

Significance of mixing procedure:- The different mixing procedures outlined above have a very significant effect on the properties of the resulting grouts. As indicated in the section on dry mixing if the dry materials are added to water at the same time or closely following each other the role of the bentonite will be little more than that of an inert fine material. Unless rather high bentonite concentrations are used the properties of the grout will be dominated by the cement.

The mixing procedures using pre-hydrated bentonite are more suited to the weaker materials used in slurry trench cutoffs (either as excavation or replacement slurries). As already noted the addition of dry cement to bentonite results in the formation of cement nodules. When the resulting hardened material is subjected to stress the nodules act as local stress raising points so that these mixes tend to fail at lower average stresses than the more homogeneous mixes produced by pre-wetting the cement. The reduction in strength may be as much as 25% and of course could be due in part to the lower effective cement concentration in the dry cement mix. The nodules also tend to arrest the propagation of cracks. As a result the strain at peak stress for the dry cement mix may be substantially greater than that for the wet cement mix. Figure 4 shows typical stress-strain behaviour for two mixes of the same composition, one prepared with dry cement and one with pre-wet cement.

The effect of shear rate during the mixing of cement and bentonite is very similar to that of the two types of cement addition. Low shear mixes tend to be weaker and more plastic than high shear mixes. This is because there are always slight inhomogeneities in a mix even with pre-wet cement. Thus high shear mixing can produce a more homogeneous and thus a stronger and more brittle mix than low shear mixing.

From the foregoing it might appear that dry cement addition and low shear mixing are to be favoured as both will tend to produce weak mixes capable of withstanding large strains. For high strength is seldom required in slurry trench cutoffs and the ability to withstand large strains is often of great importance. However, it should be remembered that the more homogeneous a mix is the lower will be its permeability. Thus if low permeability is important wet cement addition and high shear mixing should be used. If permeability is not of prime importance the cheaper procedures of dry cement addition/low shear mixing may be appropriate.

The problem of brittleness can be easily overcome by adding sand to a mix. This provides sufficient inhomogeneity without significantly affecting the permability. Indeed in theory such addition should actually reduce the overall permeability as the sand grains are impermeable. However, in practice it normally produces a slight increase in permeability possible due to cracks at the sand/grout interface.

Sand addition will have an additional advantage for replacement slurries in that it will increase their density and thus ensure better displacement of the original excavation slurry. For self-hardening slurries sufficient spoil is normally picked up during the excavation phase that sand addition is not needed to control brittleness.

BLEEDING OF BENTONITE-CEMENT GROUTS

For bentonite-cement grouts the water/cement ratio could be typically in the range 3-11. At such water/cement ratios straight cement-water grouts would bleed very severely. Hence in bentonite-cement grouts part of the role of the bentonite is to prevent bleeding.

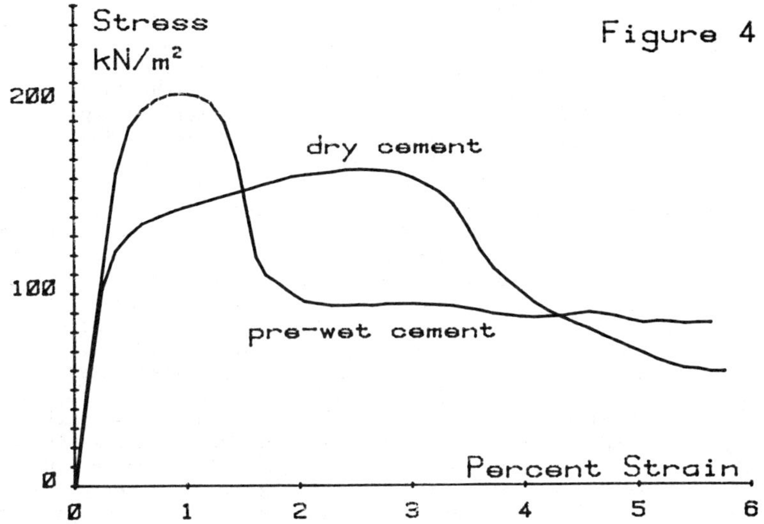

Mix: 250kg cement, 70kg bentonite to 890kg water

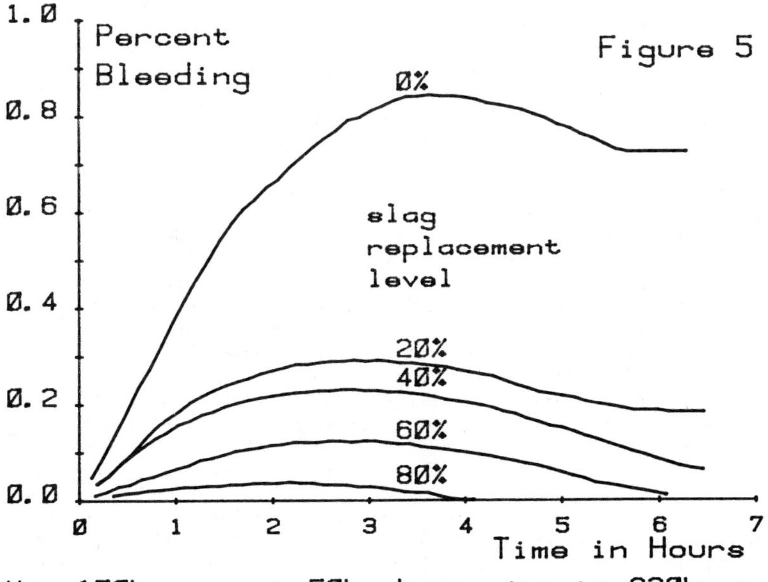

Mix: 150kg cement, 50kg bentonite to 930kg water

For it provides a floc structure which prevents excessive settlement of the cement particles until sufficient set has occurred to "lock them up". Thus the more developed the floc structure (that is more and finer particles) of the bentonite used to prepare the grout the lower will be the bleeding. High shear mixing and long hydration times should therefore reduce bleeding and this is observed in practice. However, sometimes sufficient control of bleeding cannot be obtained by just better mixing of the bentonite and it is necessary to add additional fine material. If such material is added in addition to the cement and bentonite the mix may become unacceptably viscous, particularly if the grout is to be used as an excavation slurry. Thus it may be necessary to add the additional fine material as a replacement for some of the cement and so if reasonable strengths are to be obtained it must have some cementing properties of its own. Materials which meet this requirement are ground granulated blast furnace slag and fly ash. Both of these are nominally waste materials and they may offer cost savings as well. Actually instead of adding slag to the cement, Portland blast furnace cement can be used if available. This is an interground mixture of slag and ordinary Portland cement. It will have the advantage of simplifying storage and mixing arrangements but the disadvantage that the fraction of slag in the mix cannot be varied at will. These materials can help to control bleeding as neither shows any rapid reaction with either plain water or bentonite slurry. In simple terms no reaction means no ions are released into solution and so they do not cause a bentonite slurry bleed as cement does. Also they tend to remain dispersed and so settle only very slowly.

In the presence of ordinary Portland cement both materials set as they are catalysed by/react with the lime released by the cement hydration. Thus they can contribute to the strength of a bentonite-cement grout. However, as they react rather slowly and only after some cement has reacted they will cause little bleeding in themselves (by the time they react the grout will probably have sufficient strength not to bleed). Furthermore they will reduce the bleeding caused by the cement as they will dilute it and possibly remove some of the lime produced by it.

In practice it has been found that slag can be used to replace cement at levels up to 95% though at high replacement levels the set slurries tend to be very strong and brittle, though they do have very low permeabilities (strength data is given in Jefferis, 1981). These properties probably stem from the highly homogeneous nature of such slurries for they are relatively fluid at the mixing stage and thus easily homogenised. Figure 5 shows the effect slag replacement on bleeding. It can be seen that even at 20% replacement slag has a very substantial effect (>60% reduction in peak bleeding). The slight decrease in bleeding observed at times greater than about 3 hours is probably due to swelling of the grout as it sets (a float attached to a very sensitive free aramature transducer was used to monitor bleeding and displacements as small as 10 microns could be recorded, see Guner, 1978).

Fly ash cannot be used at replacement levels greater than ~50% as at higher levels the set becomes very slow. However, it has the advantage of not producing a brittle mix possibly because it always contains some rather coarse material.

FILTRATE LOSS OF BENTONITE-CEMENT GROUTS

When a slurry trench is excavated under bentonite a filter cake forms on the walls of the trench and should prevent any significant fluid loss. With bentonite-cements a filter cake also forms but it is much more permeable and quite substantial fluid loss can occur (the slurry level in a trench could drop more than 0.5m overnight). Thus for both replacement and self hardening slurries a thick filter cake will build up on the walls of the trench. This may be an advantage, as when set, the cake should have a lower permeability than the set bulk material, though it may be rather strong and perhaps brittle.

Fluid loss will only cause concentration of the slurry in the cake. It should not effect the concentration of the bulk fluid unless the cake gets displaced and is redispersed in the bulk fluid. This may occur to some extent with self hardening slurries.

Although filtrate loss could be regarded as a somewhat similar phenomenon to bleeding it is much more difficult to control. High shear mixing has rather little effect probably for the same reasons as for pure bentonite slurries. Replacement of some of the cement by slag or fly ash will reduce the fluid loss but normally by less than 20% unless very high replacement levels are used. The reason for this is probably that cement, ground slag and fly ash all produce cakes of broadly similar fineness and thus porosity of the cake may be the controlling factor. However, as indicated earlier porosity is in part controlled by gel strength and slag or fly replacement has rather little effect on this. Thus fluid loss is not significantly effected by replacement.

Fluid loss can be controlled by special additives but this is seldom necessary as it does not seem to present any severe problems though, of course, it does waste slurry. Also a problem with such additives is that they are often bio-degradable and therefore must be checked to see that they have no long term deleterious effects on the cutoff.

CONCLUSIONS

The data presented on the effect of mixing on bentonite slurries show the somewhat unexpected result that apparent viscosity continually increases with time at least until 3 years after mixing. This may be explained by the continuing delamination of the clay particles. As a result of the delamination other properties of the slurry change with time and in particular the permeability of the gelled slurry decreases and the gel strength increases.

Shear wave velocity techniques have provided most useful information on the development of gel in slurries though much work remains to be done. The data show that gel strength is always influenced by time from mixing and that short term gel development data may substantially overestimate long term gel development.

The data on bentonite slurries enables some predictions to be made about the effect of mixing on bentonite-cement grouts. However, it seems that for these materials the dominant factor related to mixing is the dispersion of the cement.

Materials such as slag or fly ash can improve the fluid properties of grouts as they react more slowly than cement.

ACKNOWLEDGEMENT

The author gratefully acknowledges the advice and assistance so freely given by the late Professor Kevin Nash. This spanned more than twelve years and played a fundamental role in developing the author's research on bentonite slurries and grouts.

REFERENCES

D'Appolonia, David, J., "Soil-bentonite slurry trench cutoffs", Journal of the Geotechnical Engineering Division, A.S.C.E., Vol. 106, No. GT4, Proc. paper 15372, April 1980 pp 399-419.

Goodwin, J. W. and Smith, R. W. "Gels and gelling processes", Faraday Disc., Cehmical Society, 57, 129, (1974).

Guner, A., "Properties and behaviour of bentonite-cement slurries", Ph.D. Thesis, King's College, University of London, 1978.

Jefferis, S. A., "Bentonite-cement slurries for hydraulic cutoffs", Proc. Tenth International Conference on Soil Mechanics and Foundation Engineering, Stockholm, June 1981.

Stepkowska, E. T. and Jefferis, S. A., "Studies in microstructure of clay slurries", Seventh International Clay Conference, Bologna 1981.

CONVERSION FACTORS

Stress/shear modulus

$1 \text{ N/m}^2 = 0.102 \text{ kgf/m}^2 = 0.021 \text{ lbf/ft}^2$

Viscosity

$1 \text{ cP} = 10^{-3} \text{kg/ms} = 6.72 \times 10^{-4} \text{ lb/fts}$

ULTRA FINE GROUTING MATERIAL

Masao Shimoda*
Hiroshi Ohmori**

Abstract

The effectiveness of the ultra fine grouting material (hereafter, abbreviated as MC) is shown by the experimental results and practical applications.

MC belongs to the cement family materials and has an ultra fine grain diameter, that is, a mean grain size of 4 μm and the maximum grain size of approximately 10 μm. This property enables MC grout to penetrate into fine sand, where the coefficient of permeability is 10^{-3} - 10^{-4} cm/s.

1 Introduction

An ultra fine grouting material (hereafter, abbreviated as MC) has been developed with penetration ability comparable to many chemical grouts. This paper presents data from test programs and case histories to document its basic behavior and utility in practical applications.

2 Characteristics of MC

2.1 Toxicity

MC does not contain any organic or polluting element (see Table 1). Therefore, there is no contamination of underground water or soil.

Table 1 Chemical composition of MC(wt.%)

Ig.loss	SiO_2	Al_2O_3	Fe_2O_3	CaO	MgO	SO_3	Total
0.3	29.0	13.2	1.2	49.2	5.6	1.2	99.7

2.2 Penetrating power

The average particle size of MC is 4 microns. This fineness is the secret of the penetration comparable to chemical grout afforded by this suspension (see Table 2, Fig. 1 and Fig. 2).

* Chief Researcher, Central Research Laboratory, Onoda Cement Co., Ltd., JAPAN

** Researcher, Central Research Laboratory, Onoda Cement Co., Ltd. JAPAN

Table 2 Physical properties of MC

Appearance		Greyish white ultra fine particles
Specific gravity		3.00 ± 0.10
Unit weight (kg/ℓ) (Apparent bulk density)		1.00 ± 0.10
Fineness	Blaine specific area(cm^2/g)	about 8000
	50 percent grain size(μm)	about 4

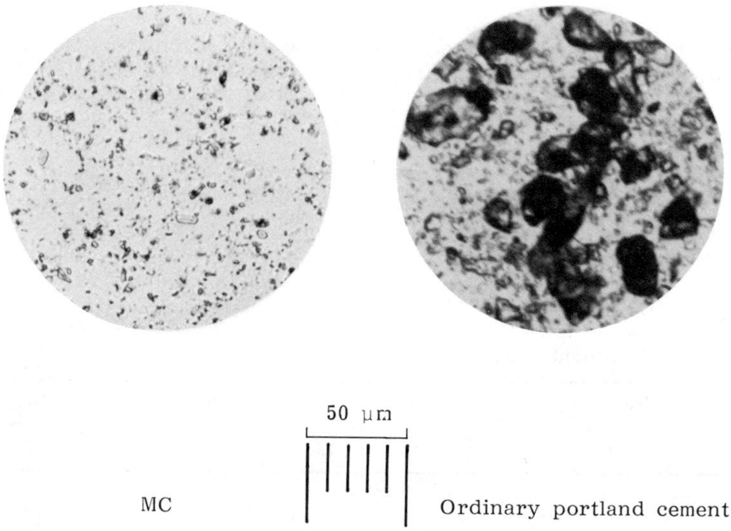

Fig.1 Optical micrograph

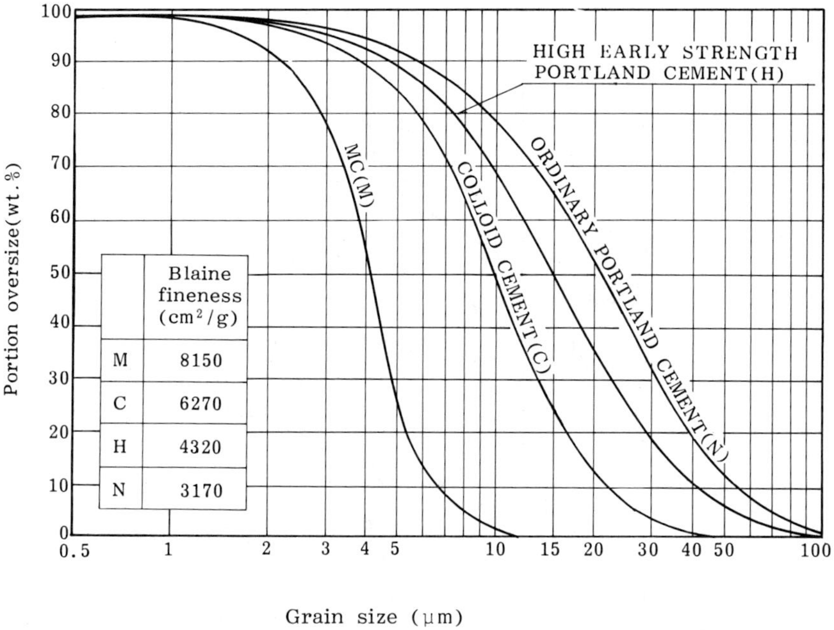

Fig. 2 Grain size distribution

2.3 Strength

The ultra fine grains of MC are chemically activated, so that curing provides rapid hardening to a high strength. This initial hardening is followed by further hardening during an extended period (see Fig. 3,4). Unconfined compressive strengths as high as 620 kgf/cm^2 are reached.

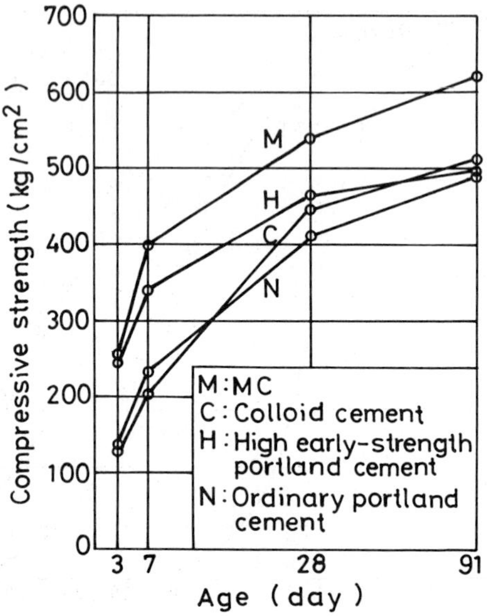

Fig. 3 Compressive strength of hardened mortar of MC and Cements (JIS R 5201-1977)

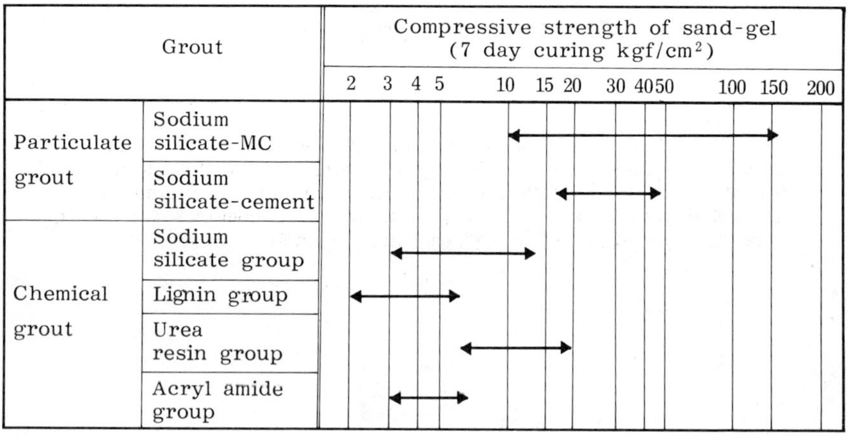

Fig.4 Sand-gel compressive strength of various grouts

2.4 Handling

With its strong dispersive quality, MC is resistant to separation and sedimentation, which provides protection against damage for the injection equipment, and prevents obstruction of pipework (see Fig. 5). No special grouting equipment or method is needed.

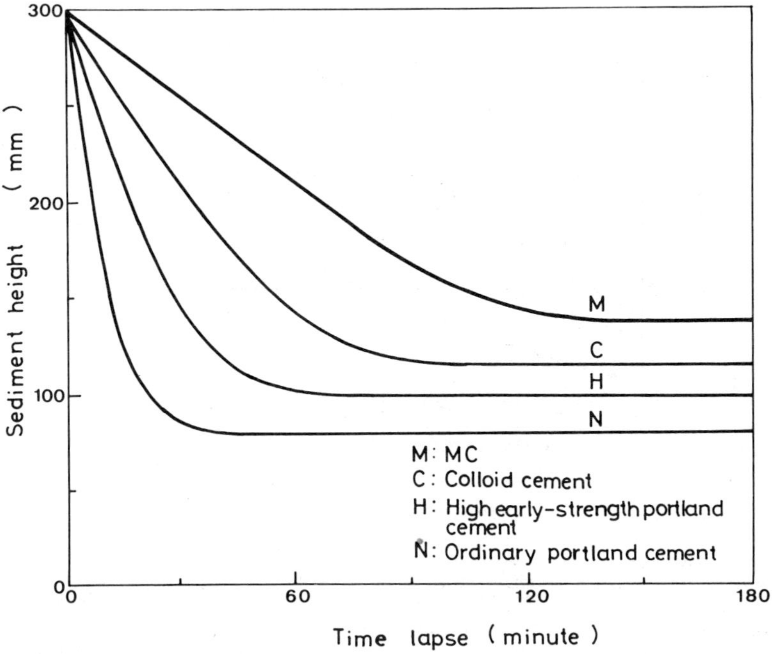

Fig. 5 Separation of MC milk and cements milk

2.5 Flexibility in respect of gel time
The gel time of MC can be regulated within a wide practical range of less than 20 seconds to several minutes by means of mixing with sodium silicate solution (see Fig. 6).

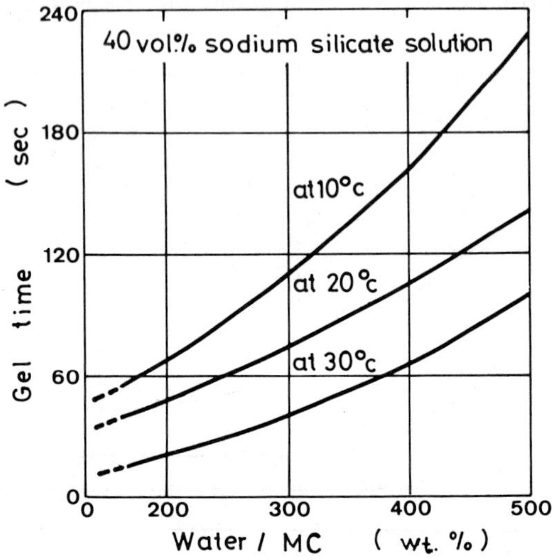

Fig. 6 Relation between Water/MC and gel time

3 Penetration of MC grout

Penetration ability of MC grout was measured using the equipment shown in Fig. 7. The grout 200 mℓ volume was dropped gently along the wall of the polyethylene bag filled with Tououra standard sand (see Table 3: density 1.58 g/cm^3). Penetration ability was judged from the final depth of penetration into sand mass. The results are shown in Table 4.

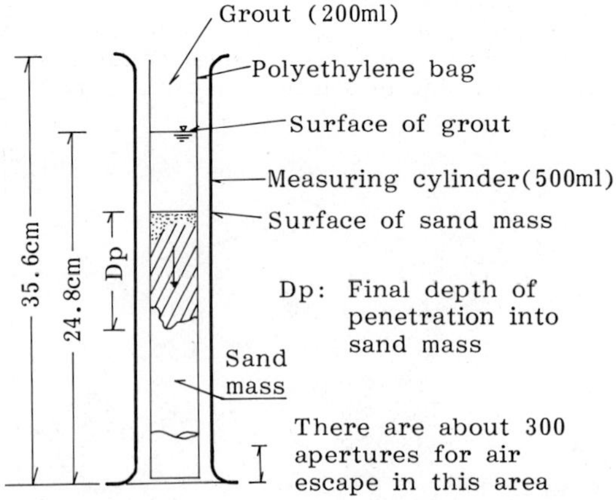

Fig.7 The equipment of the penetration test

Table 3 Physical properties of Toyoura standard sand

Item	Grain size: Grain coarser by weight(%) 297μ	210μ	149μ	105μ	Dmax (mm)	D_{60} (mm)	D_{10} (mm)	$Uc=D_{60}/D_{10}$
Toyoura standard sand	0.2	46.4	96.5	99.4	0.42	0.21	0.17	1.24

Dmax : Maximum grain size
D_{60} : 60 percent grain size
D_{10} : Effective grain size
U_c : Uniformity coefficient

Table 4 Results of the penetration test

Grout materials	Water/MC or cement (wt.%)	Dp (cm)	Compressive strength (σ_7, kgf/cm^2)
M C	200	10.6	32.7
	300	13.0	21.5
	400	14.9	9.2
Colloid cement	500	1.8	No penetration
Ordinary portland cement	500	0.5	No penetration

4 Standard mix proportion

MC may be used independently in slurry form (1-shot) or in mixture with sodium silicate (1.5 or 2-shot) for each purpose.
Representative mix proportion is shown in Table 5.

Table 5 Standard mix proportions

Item		M C No. 1	M C No. 2	M C No. 3
Application		1 shot	1.5 or 2 shot	
Gel time		4-8 hours	1-3 minutes	
Sand gel compressive strength (7 days curing)		70kgf/cm^2	10kgf/cm^2	35kgf/cm^2
Standard composition (per m^3)	M C	200kg	120kg	150kg
	MC admixture	2ℓ	1.2ℓ	1.5ℓ
	Sodium silicate No. 3*	–	200ℓ	200ℓ
	Water	931ℓ	759ℓ	749ℓ

* 40 vol.% $Na_2O \cdot 3SiO_2 \cdot aq$

5 Model experiment

Model experiments were made to prove the injection efficiency of MC grout under the practical injection pressure.

5.1 Equipment

The equipment used in this experiment consisted of a steel injection mold, two pumps and a steel injection pipe (see Fig.8). The two pressure hoses from the pumps met at the top of the injection pipe.

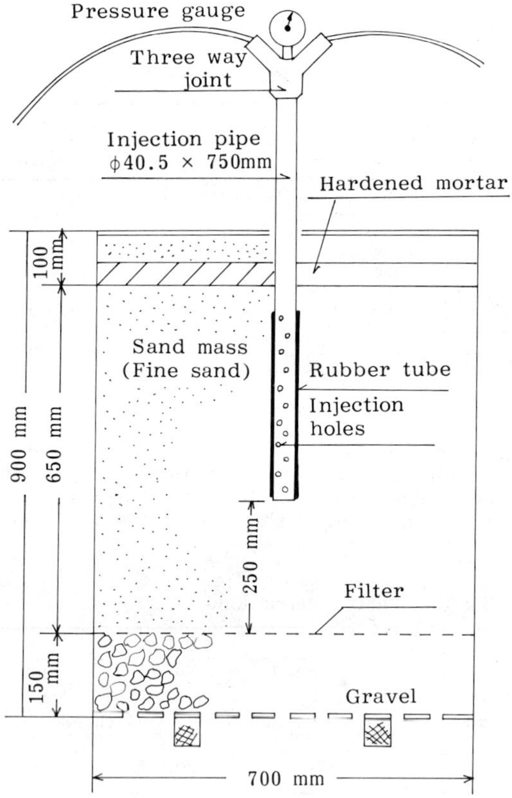

Fig.8 Cross-section of steel injection mold

5.2 Injection efficiency

Fig. 9 and 10 show cross-sections of the shapes of the stabilized sand masses made with MC grout and high early strength portland cement grout. As it was difficult to inject the latter into the sand mass, the injection efficiency was very poor. In comparison, it was very easy to inject MC into a sand mass having coefficient of permeability of 3.75×10^{-4} cm/s. The coefficient of permeability of the stabilized sand mass with MC grout was 1.00×10^{-6} cm/s(see Fig.10). Injection of the portland cement grout resulted in fracturing of the sand mass.

(a) MC (1-shot)
Water/MC = 200%
MC admixture:
MC x 1.5wt.%
Injection pressure:
0.2 - 0.4 kgf/cm^2
Volume of grout injected:
24.2 ℓ

(b) High early strength portland cement (1 - shot)
Water/Cement = 200%
MC admixture:
Cement x 2wt. %
Injection pressure:
0.5 - 0.7 kgf/cm^2
Volume of grout injected:
19.7 ℓ

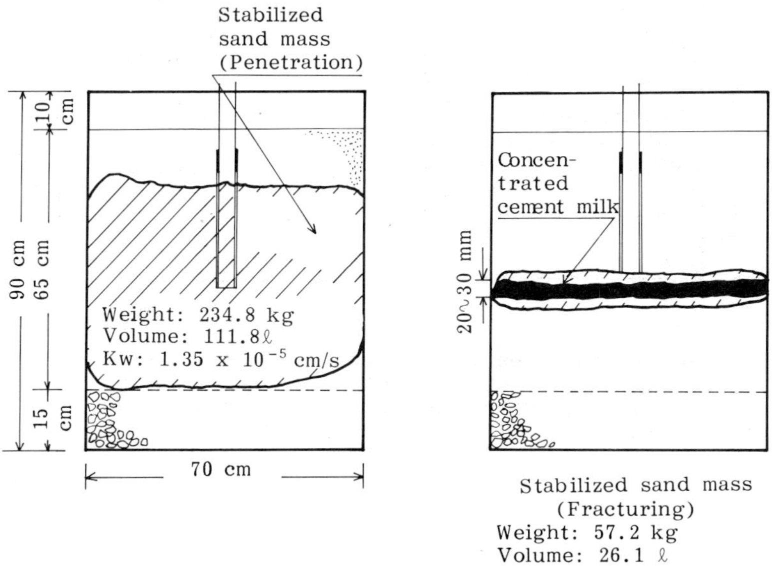

Fig. 9 Cross-sections of stabilized sand mass shapes made by injecting MC grout and high early strength portland cement grout with 1-shot method into sand masses.

(a) MC (1.5 - shot)
A: Water/MC = 300 %
MC admixture:
MC x 1.5wt.%
B: 40% sodium silicate
A/B = 1
Injection pressure:
0.7 kgf/cm^2
Volume of grout injected:
18.3 ℓ

(b) High early strength portland cement (1.5-shot)
A: Water/Cement = 350 %
MC admixture:
Cement x 2.0wt. %
B: 40% sodium silicate
A/B = 1
Injection pressure:
1.5 - 2.3 kgf/cm^2
Volume of grout injected:
9.8 ℓ

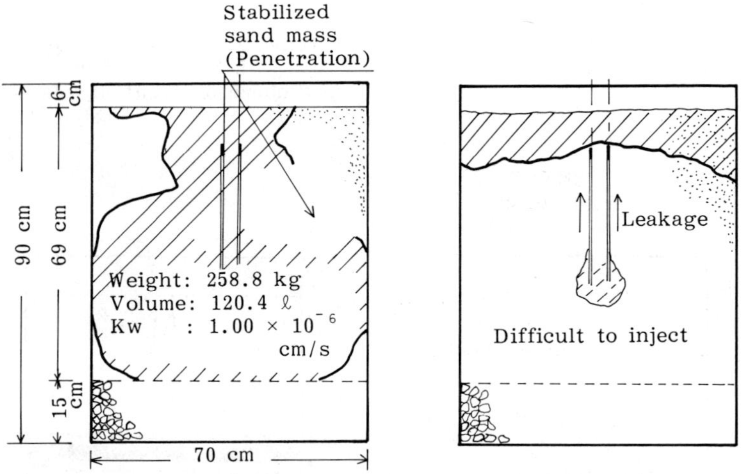

Fig.10 Cross-sections of stabilized sand mass shapes made by injecting MC grout and high early strength portland cement grout with 1.5-shot method into sand masses.

6. Case Histories

6.1 The Nakayama tunnel on the Joetsu Shinkansen

The Joetsu Shinkansen is a line which will connect to Tokyo with Niigata. Work is at present under way on the new line between Omiya and Niigata over a distance of 270 km. It involves tunneling in fractured rock to depths of 300 m. Water presents a major problem in the tunnel construction and extensive grouting is done to control the water. Table 6 presents the details concerning the project and the grouting application. Figure 11 and 12 show the tunnel cross-section and the grouting operations in progress.

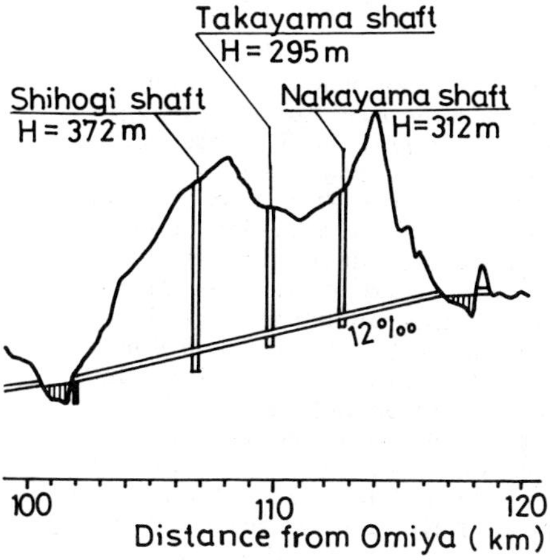

Fig. 11 Vertical section of the Nakayama tunnel

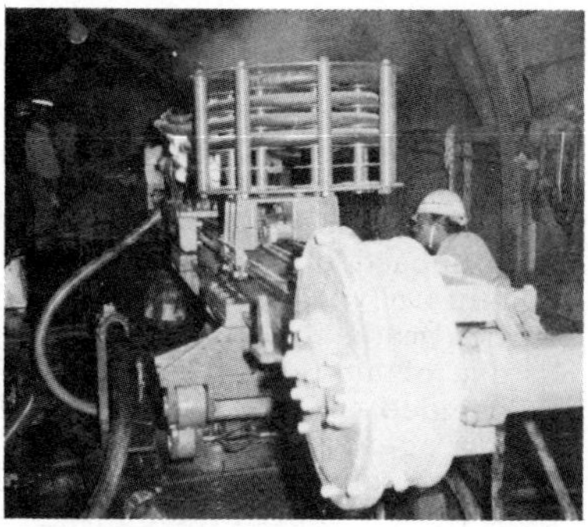

Fig. 12 Prevention works of ground subsidence in tunnel construction

Table 6 Injecting conditions

Location	Azuma county, Gunma prefecture
Purpose of Application	To seal off the high pressure underground water with 20 kgf/cm^2 and reinforce facing of tunnel
Type of Rock	Unconsolidified angular gravel tuff
Particulars of Grout	Mix proportion (1.5-shot) A: MC grout: water/MC = 200 (wt.%) MC admixture: MC x 2(wt.%) B: Sodium silicate (Mix ratio: A/B = 1) Gel time: 3-4 min. Homo-gel compressive strength: σ_7 = 50-60 kgf/cm^2
Results	The underground water was sealed off and the facing of the tunnel was reinforced.

6.2 Tunnel construction for the Kashiwazaki atomic power-plant

During the construction of a 550 m long road tunnel through hills of sandy soil facing the Sea of Japan, sand layers not foreseen from boring information created problems of insufficient bearing capacity provided by the ground on the lower part of the tunnel wall. This soil had poor adhesion, being composed of loose fine medium-grain sand deposited almost uniformly. The mode of tunneling adopted in the case in question made it impractical to overcome this difficulty by making use of an L-shaped section of tunnel wall: It was therefore decided to strengthen the ground through injection of hardening agent. The MC grout was injected on both sides of the tunnel along a length extending 114 m, with results that fully satisfied expectations in respect of penetration and reinforcement of bearing capacity. Figure 13 and Figure 14 show the entrance of the tunnel and the stabilized sand mass by injecting MC grout. Table 7 presents the details concerning the grouting application.

Fig. 13 Entrance of the tunnel under construction

Fig. 14 The stabilized sand mass made by injecting MC grout

Table 7 Injecting conditions

Location	Kariha county, Niigata prefecture							
Purpose of Application	To reinforce the foundation of the side walls of the tunnel							
Type of Soil	Nature	Specific gravity	Porosity	Moisture content (%)	Unit weight (kg/ℓ)	N-value	Coeff. of permeability (cm/s)	
	Sand	2.73	40.2	6.48	1.80	10~30	7.2×10^{-3}	
	Particle size distribution	Particle size (mm)	2.00	0.84	0.42	0.25	0.105	0.074
		Portion undersize (%)	99.9	99.1	80.3	30.7	6.2	5.9
Particulars of Grout	MC grout (1-shot) Water/MC = 300 - 600(wt.%) MC admixture: MC x 1.2(wt.%)							
Results	The bearing capacity of the foundation was improved.							

6.3 The Yamasa dam

The Yamasa dam is a gravity type concrete dam. It is founded on the eroded bedrock mainly consisting of granite, partially including faults. The Lugeon coefficient was 0.3 to 1.0 to inject MC grout for the consolidation grouting. Table 8 presents the details.

Fig.15 Side-view of the Yamasa dam under construction

Fig.16 Injection pipe and pressure gauge in place

Table 8 Injecting conditions

Location	Nogi county, Shimane Prefecture
Purpose of Application	To reinforce the foundation of the dam (Consolidation Grouting)
Type of Rock	Eroded granite including faults
Particulars of Grout	MC grout (1-shot) Water/MC = 1000 (wt.%) MC admixture: MC x 0.5(wt.%)
Results	The bearing capacity of the foundation was improved.

7 Conclusion

Recently cement family injection agents have found increasing acceptance over chemical (high polymer) injection agents which are inferior in strength and cause pollution in groundwater. Ordinary cement family grouting materials, however, can only penetrate into the coarse grained soil of which coefficient of permeability is more than approximately 10^{-1} cm/s.

It is against this backdrop that research was undertaken to develop a technology for producing ultra fine grouting material having permeability comparable to that of chemical injection agents.

The results of the experiments using the injection mold equipment show that MC grout can penetrate fine sand of which coefficient of permeability is 3.75×10^{-4} cm/s. In the practical application it can penetrate fine sand ground (coefficient of permeability 4.27×10^{-3} cm/s) and the bearing capacity increases from 22tf/m^2 to more than 160tf/m^2.

The compressive strength of the sand hardened with this material is confirmed to be 258 kgf/cm^2 and 540 kgf/cm^2, respectively, on the third and 28th day age.

MC may be used independently in slurry or in a mixture with sodium silicate. The former is suitable for permanent purposes, such as the construction of dam foundations and ground reinforcement of buildings, while the latter is more suitable for purposes such as the prevention of earth collapse and waterproofing in tunnels.

GROUTING OF THE KARSTIC CAVES WITH CLAY FILLINGS

Zhang Zuomei* Huo Pinshou*

Abstract

The present paper first briefly introduces the high pressure cement grouting works of karstic caves filled with soft clay at Wujiangdu Hydroelectric Project, then describes the main physical and mechanical properties of some specimens obtained from grouted zone in karstic formation and finally discusses the mechanism and technique of such grouting procedures based on the results of laboratory and in-situ investigations.

I. INTRODUCTION

The karstic caves are often filled with soft clay, which must be specially treated to prevent leakage and seepage failure when the karstic formation is used as a dam foundation. In the past a successful measure in many projects is to remove the clay fillings and provide a concrete diaphragm wall. However, it is very difficult and expensive to build such diaphragm in the case of the presence of a lot of deep caves.

The practice of Wujiangdu Hydroelectric Project in China has shown that a high pressure cement grouting technique can be applied to build a seepage resistant barrier in the karstic formation with clay fillings. This procedure is simpler and cheaper than the concrete diaphragm method; therefore, it is worthy of introducing to other projects with similar problems.

The clay fillings in the karstic caves are generally considered not groutable because of the large amount of clay fraction and the small void dimension. However, the successful application of cement grouting in Wujiangdu dam foundation reveals some aspects contradictory to the past practice. This fact shows that there must be some mechanism other than that of ordinary permeation grouting. This will be the main subject discussed in the present paper.

II. GENERAL FEATURES OF THE PROJECT

The Wujiangdu Hydroelectric Project is situated in the middle reach of the Wujiang River in the Guizhou plateaus. An arch-gravity

---

*Research Institute of Water Conservancy and Hydroelectric Power, Beijing, China.

dam of 165 m (540 ft.) in height was founded on a karstic limestone foundation. The strike of rock strata is nearly perpendicular to the river. The dip angle is between 60° and 70° towards the upstream.

The karstic erosion is quite remarkable at the dam site. There are a lot of karstic caves filled with soft clay and fine sand in the rock foundation, particularly in the left bank some big caves were discovered at a depth of about 250 m (820 ft.) beneath the river bed, as shown in Fig. 1.

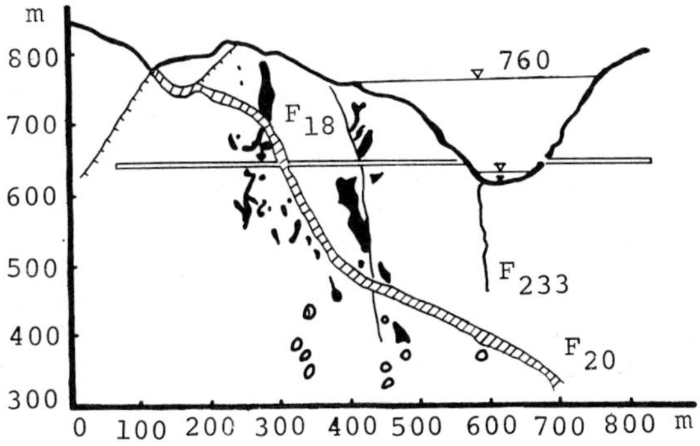

Fig. 1. Schematic diagram showing the caves in left bank.

●--Caves encountered in the process of drilling and grouting;

o--Caves found by geophysical exploration.

The clay fillings in the karstic caves in their natural state have a high water content near the liquid limit and a very low density (See Table 1). The ultimate hydraulic gradient (J_k) against piping or extruding varies with the diameter (D_s) of the test specimen subjected to seepage force, as shown in Fig. 2. Obviously, these soft materials are unsuitable for remaining in the dam foundation.

Table 1. Physical properties of soft clay.

Natural water content (%)	56.0
Natural dry density (g/cm^3)	1.10 (0.4 lb/in^3)
Clay fraction (%) (<0.005 mm)	54.0
Liquid limit (%)	57.5
Plastic limit (%)	32.4
Plasticity index (%)	25.1

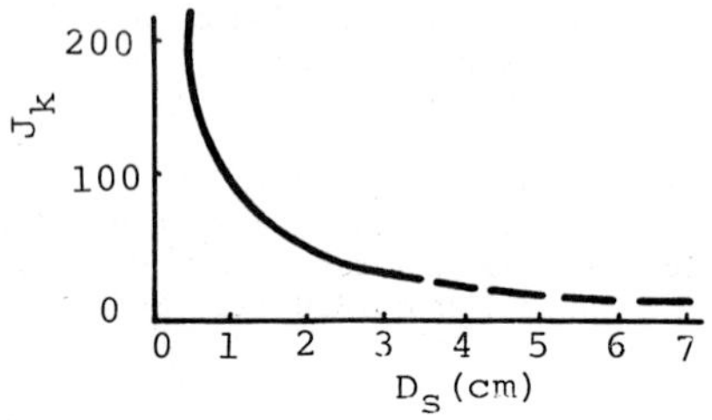

Fig. 2. Relationship between J_k and D_s.

In order to increase the resistance against high water pressure, the caves filled with soft clay were generally treated by means of cement grouting. The grout curtain was formed by two or three rows of boreholes with hole spacings varying between 1.5 and 2.5 m (4.9-8.2 ft.). The diameter of borehole was 53 mm (2.1 in.), and the holes were injected by the method of split spacing and descending stage grouting at a maximum pressure of 60 kg/cm^2 (852 lb/in^2). The material comprising the grouts used was ordinary Portland cement. The grout with a water cement ratio of about 8:1 by weight was initially injected, and this thin mix was gradually thickened to 1:0.8 and 1:0.5 according to the grouting conditions.

After grouting, the water-pressure tests were carried out in the curtain through boreholes, and the water acceptance was generally less than 1/2 Lugeon. When the reservoir was impounded to a level at 50 to 70 percent of reservoir head in 1970-1980, the uplift pressures and leakages measured by the piezometers installed downstream of the curtain were very small.

In order to further investigate the grouting effectiveness and injection mechanism of karstic caves filled with soft clay, some specimens were taken out from the grouted cave in the curtain line. These specimens were tested in laboratory to evaluate the structure and physical-mechanical properties. The main results are as follows.

III. STRUCTURE OF GROUTED SPECIMEN

All specimens were examined in detail, and the results showed that the clayey materials were cut by a lot of set cement grout networks existing on both longitudinal and transverse profiles, as shown in Fig. 3. The volumetric ratios of set cement grout to the total volume in each grouted specimen are given in Table 2. Fig. 3 and Table 2 show that the quantities of set cement grout in the specimens are rather high, but its destribution is non-uniform. It can also be seen from the specimens that the cement grouts neither penetrated into the voids nor mixed with the soils.

Table 2

Specimen number	2	3	4	5	Average
Volumetric ratio of set cement grout (%)	76	68	49	38	57

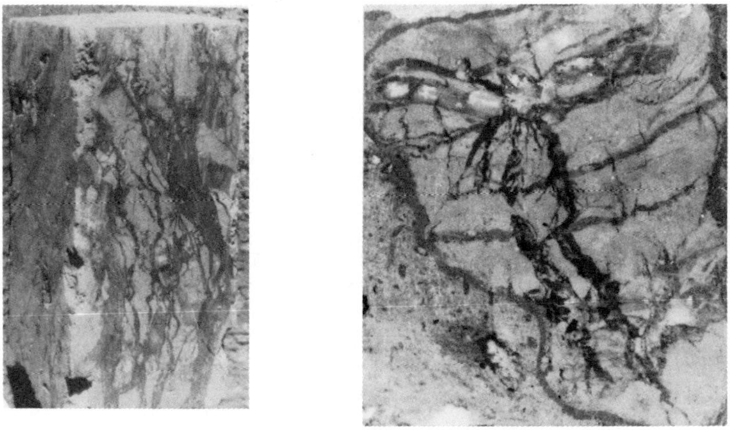

(a) (b)

(a) Longitudinal Profile (b) Transverse profile

Fig. 3. Photos showing the distribution of set grouts in the grouted specimens.

Besides the set cement grouts, a lot of white substances were found in the specimens. In order to make clear what they were, some soils in natural state and some from the grouting specimens had been tested by spectrum analysis. The test results showed that the calcium ion in clayey materials has obviously increased after grouting. These white substances can thus be considered as the calcium carbonate formed by grouting.

It has been indicated in Table 1 that the dry density of natural clay is very small (about 1.1 g/cm^3) (0.04 lb/in^3), but it has increased to some extent after grouting. The average value of six grouted samples is 1.26 g/cm^3 (0.045 lb/in^3) in the range from 1.13 to 1.44 g/cm^3 (0.041-0.052 lb/in^3).

IV. MECHANICAL PROPERTIES OF GROUTED SPECIMENS

Unconfined compression tests and shear tests (without application of normal stress) were carried out on grouted specimens with dimensions of 4 X 4 X 8 cm^3 (0.24 X 0.24 X 0.48 in^3). It will be seen from Table 3 that both the compressive strength and the shearing resistance of clay fillings considerably increased after grouting.

Table 3. The mechanical properties of grouted specimens.

No. of specimen	2	3	4	Average
Unconfined compressive strength, kg/cm^2	32.0 (455)	19.7 (280)	12.0 (170)	21.2 (302)*
Shearing resistance (zero normal stress), kg/cm^2	22.0 (313)	8.6 (123)	14.0 (199)	14.7 (209)*
Dynamic modulus of elasticity, 10^4 kg/cm^2	5.0 ($71X10^4$)	5.6 ($80X10^4$)	3.6 ($42X10^4$)	4.7 ($67X10^4$)*
Static modulus of elasticity, 10^4 kg/cm^2	3.6 ($42X10^4$)	2.0 ($28X10^4$)	1.9 ($27X10^4$)	2.5 ($35.5X10^4$)*

*The unit of values in the blanks is lb. per sq. in.

The deformability of grouted specimens was studied by the methods of sound wave test and unconfined compression test, and the results are expressed in terms of dynamic modulus and static modulus, see Table 3. The static modulus of natural clay did not exceed several tens of kg/cm^2. Accordingly, it was concluded that the deformability of clay fillings under load was considerably improved after grouting.

V. RESISTANCE TO SEEPAGE AND SLAKING

After grouting the permeability of clay fillings did not reduce obviously because their original coefficients of permeability were quite low (generally less than 10^{-6} cm/s). The main purpose of grouting in such condition was to increase the resistance of the clay against seepage failure caused by high water pressure.

The results of permeability tests on grouted specimens with diameters of 7 and 15 cm (2.8 and 6.0 in.) indicated that none of these specimens had shown any tendency of seepage failure including piping and extruding under a hydraulic gradient up to approximately 500. By comparing these results with the data shown in Fig. 2, it can be found that after grouting the resistance of clay fillings against water pressure increased considerably.

Permeability tests also showed that an obvious slope change had appeared in the hydraulic gradient (J) versus flow rate (q) curves as the water pressure exceeded a certain value (see Fig. 4). From the data obtained from three specimens, the hydraulic gradient at the turning point (i.e. so called yield gradient) is in the range of 200-250.

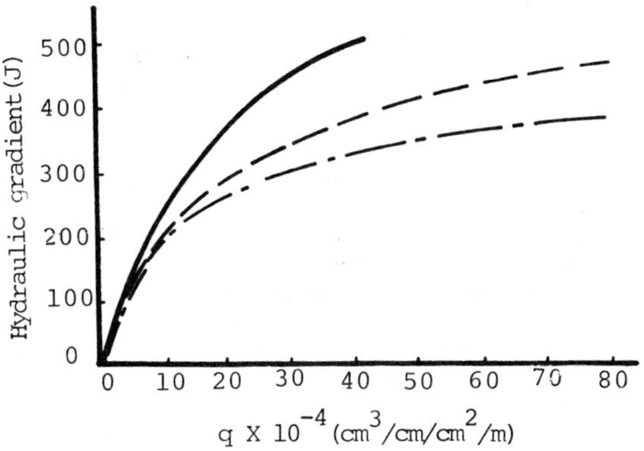

Fig. 4. Relationship between J and q.

Because the q values at the same gradient in the ascending and descending curves gradually approached each other as the gradient decreased (see Fig. 5) and because the amount of seepage water under a constant hydraulic gradient reduced with time (see Fig. 6), the yield gradient in the J-q curves indicated neither the failure nor the permanent deformation of the grouted specimens.

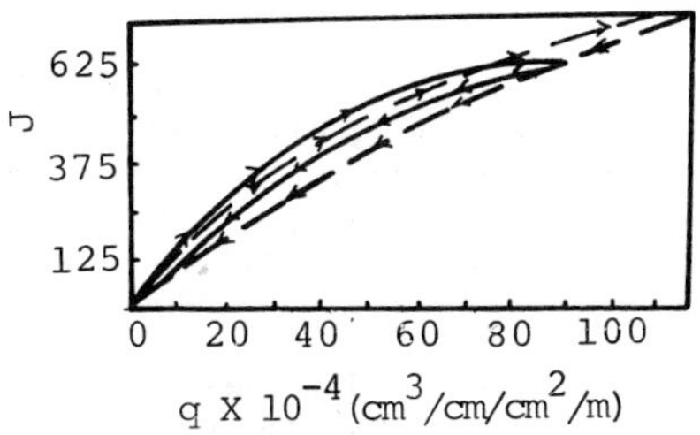

Fig. 5. Permeability test results
by increasing and decreasing pressure.

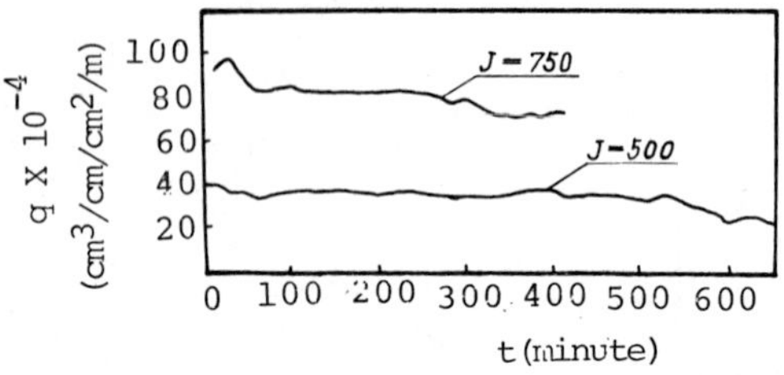

Fig. 6. Relationship between q and t.

Therefore, it was likely that the change in slope of the flow curve was caused by a temporary expansion of some voids or fissures of specimen under the action of high water pressure.

The grouted specimen also has a high resistance to slaking. The air-dried untreated clay thoroughly slaked after immersion into pure water for ten minutes. On the other hand, the grouted clay specimen air-dried for thirty days containing no set cement grout vein had kept its original shape at the end of immersing into water for a hundred days.

VI. GROUTING MECHANISM

Based on the test results mentioned above, it can be concluded that the clay fillings are subjected to four actions during injection under high grouting pressure.

Hydraulic Fracture

It is well known that hydraulic fracturing occurs when the fluid pressure in the voids or fissures significantly exceeds the initial stresses in the ground plus the tensile strength of the clay fillings. This is one of the grouting mechanisms employed to treat clay fillings which were practically impermeable to cement grout at Wujiangdu Project.

Because the holes were injected by descending stages and because a lot of major fissures and voids existed in the rock masses above the karstic caves, the cement grout would enter and fill up the top of the cave from various ways before the grouting of the cave itself or after completion of the first stage grouting in the cave. Fig. 7 illustrates a sketch of a cave with its top portion filled by cement grout.

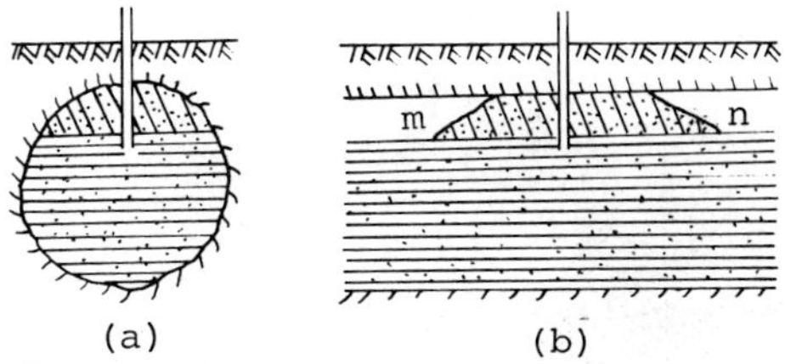

Fig. 7. Sketch of the cave with its top portion filled by grouts.

(a) Transverse section, (b) Longitudinal section.

--Natural clay, --Set cement grout.

According to the conditions shown in Fig. 7, it can be deduced that under the action of high grouting pressure hydraulic fracturing will take place in clay fillings in a radial direction from the wall of the borehole at first. Fractures then develop in a circumferential direction because of the existence of a pressure gradient among the first fracture lines. It is obvious that a part of the networks of set cement grouts, as shown in Fig. 3, was formed by hydraulic fracturing mentioned above.

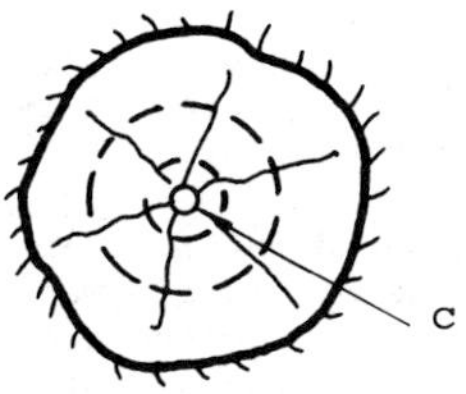

Fig. 8. Pattern of hydraulic fracture lines.

c--Injecting hole.

In addition, Fig. 9 shows a grouted sample taken out from karstic formation close to the boundary of the cave. From Fig. 9, it can be clearly seen that there is a layer of set cement grout (see "a") about 2 mm (0.08 in.) thick spreading around the boundary between clay-fillings and wall of cave. This is another evidence of the occurrence of circumferential fractures during grouting process.

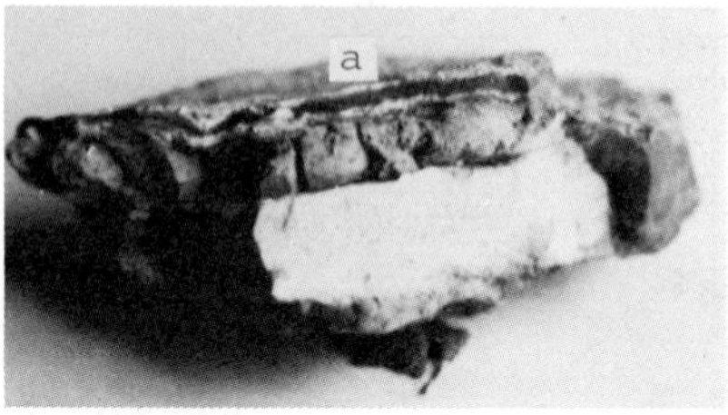

Fig. 9. Photo showing the circumferential fracture around the cave wall.

Extrusion

When the shearing stress in the clay fillings induced by grouting pressure exceeds the shearing resistance of the clay, some clay fillings will be extruded from the cave to other spaces. In the process of field grouting at Wujiangdu Project, it was often discovered that the soft clay fillings were oozed from the holes next to the injecting hole, or flowed out from the exposed rock fissures in the unlined grouting gallery. These phenomena served as good evidence of extrusion of the clay fillings.

According to the analysis of the available data and field observation, it is possible to deduce that there are two passages for extrusion of clay fillings from karstic cave.

The first passage is indicated in Fig. 7-b. Only a part of the top of the cave near the center line of curtain was filled up by cement grout. At a certain distance away from the center line of curtain in both upstream and downstream sides, the amount of cement grout gradually reduced and the original condition existed outside the points m,n. Accordingly, the space outside the points m,n became the exit of extruding of the clay fillings.

The second passage to extrude the clay fillings is the voids and fissures in the surrounding rock masses. In order to give a good illustration of this process, the cement consumptions in the rock mass surrounding the K_{104} cave during injection of the primary and secondary holes in the first row are quoted here (see Fig. 10). Fig. 10 shows that its cement consumption was considerably greater than that of the ordinary fissured rock showing the existence of big voids in the surrounding rock masses.

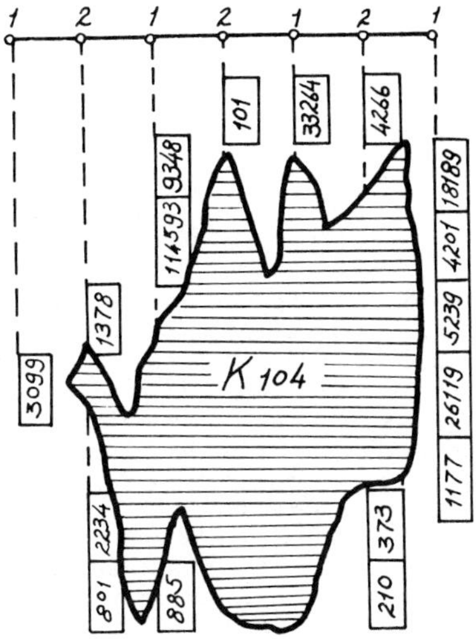

Fig. 10. Cement consumptions in the rock mass surrounding K_{104} cave.

Average cement consumptions per five-meter length along borehole (kg/m)

1--Primary holes, 2--Secondary holes.

Consolidation

The soft clay was consolidated under high grouting pressure. This process provided some spaces for cement grout to penetrate. The increase in dry density of clay fillings after grouting is an evidence of this effect.

The cement consumption caused by consolidation can be calculated by the following formula:

$$C = \frac{(1-V_c)\ (n_o-n_1)}{(1-n_o)} \times 100\% \qquad (1)$$

in which, C--volume of set cement grout injected due to consolidation (%);

V_c--total volume of set cement grout injected into soft clay fillings:

n_o--natural porosity of clay before grouting;

n_1--porosity of grouted clay.

According to the data listed in Tables 1 and 2 and the average dry density of the grouted clay fillings (1.26 g/cm^3) (0.045 lb/in^3), the value of C computed by expression (1) is 6.5%. It can be seen from this computation that the grout penetration due to consolidation is relatively small.

Chemical Hardening

It is well known that the hydrated lime will be liberated as a by-product when dicalcium silicate (C_2S), one of the cement minerals, reacts with water; and this hydrated lime can permeate into the voids of clay under high grouting pressure and react with the carbon dioxide (CO_2) existing in clay to form the white calcium carbonate. Furthermore, the calcium ion can enter into the clay fillings under the action of osmosis due to the difference in concentration and ion exchange. A large amount of calcium carbonate and exchangeable calcium ion plays the role of binder among soil particles and aggregates and tends to compress the double layer around the clay particles. The above mentioned effects lead to a great increase in the resistance of the clay fillings against seepage failure and slaking.

The mechanism of grouting the karstic caves filled with soft clay thus can be summarized as follows: The high grouting pressure leads to fracturing and extrusion of the clay fillings and simultaneously produces a series of cracks in the fillings. The cement grouts enter these cracks and set into hardened cement networks. This process is accompanied by the consolidation and chemical hardening of the clay fillings. Thus, the clay fillings in the karstic caves will be criss-crossed by networks of set cement grouts serving as the rigid skeleton. The clays enclosed within the networks become dense and hardened and are entirely different from those in natural state. The grouts will apply a rather high pre-stress to the soil and rock masses along their passages. The combination of set cement grout networks and enclosed hardened clay forms an anti-seepage barrier in the karstic formation with enough strength and stability to resist high water pressure.

VII. DISCUSSIONS ON SOME GROUTING TECHNIQUES

Caves filled by soft clay belong to distinct geological phenomenon, and the grouting mechanism in such a condition is essentially different as compared with the conditions in fissured rock and alluvium. The grouting technique also has a lot of different features because there is a strict causality between grouting mechanism and technique.

Because the fracture and extrusion actions perform a key function in a given case, the cement grout will begin to enter the clay only when the initial stress and the strength of clay are overcome by grouting pressure. As shown in Fig. 11, the grout take is practically equal to zero when the grouting pressure is smaller than a given value known as the yield pressure, which corresponds to the initial pressure to cause fracture and extrusion. This yield pressure (Py) is a fundamental parameter for fracturing grouting. Based on the practice at

Wujiangdu Project, the yield pressure often varied from 20 to 40 kg/cm^2 (284-568 lb/in^2).

It is evident that the maximum grouting pressure must be greater than the yield pressure, and the effective grouting pressure in a given case should be the difference of maximum grouting pressure and yield pressure.

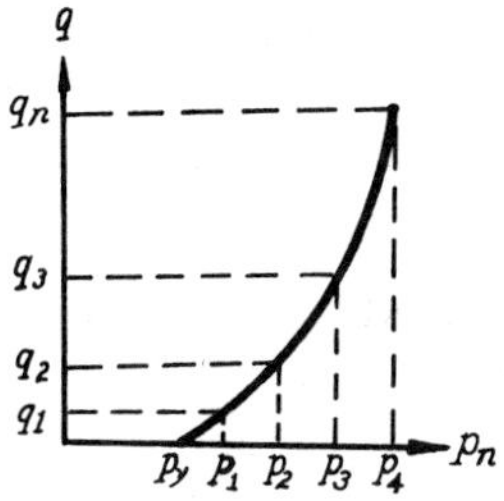

Fig. 11. Relationship between grout take and grouting pressures.

Because a yield gradient exists on the hydraulic gradient versus flow rate curve obtained from water pressure tests of grouted masses (see Fig. 4), it would be better to use this yield gradient as a basis for determining the allowable gradient of the grout curtain for the reason of safety.

As a principle of fracturing grouting, the injecting path must not be blocked by cement grouts from the beginning to the end of the grouting operation. By taking into account this factor, the conventional procedures of split spacing grouting, shown in Fig. 12-a, are not suitable, for the cement grouts injected from the boreholes of prior order are easy to block the injecting path of later order. Contrarily, the procedures indicated in Fig. 12-b will be able to reduce the block of injecting path and to increase the grout take in each hole.

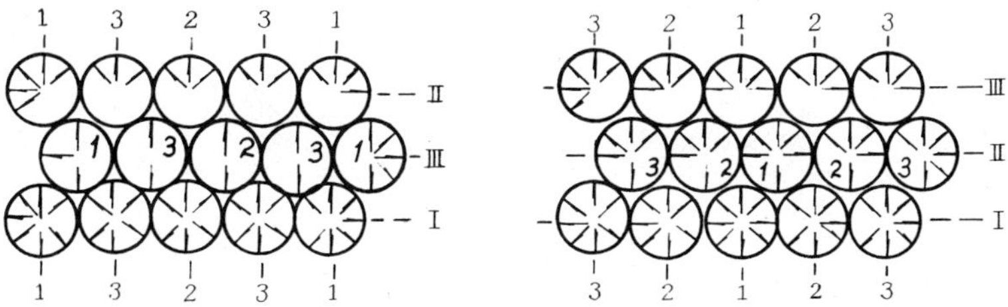

(a) (b)

Fig. 12. Two different orders of grouting works.

I, II, III--the order of lines

1, 2, 3--the order of holes

--extruding direction

In fissured rock, in order to meet the needs of grouting fine fissures, the cement slurry at any stage is always initially injected at a high water-cement ratio and then gradually thickened based on the rate of take. In grouting the caves filled with soft clay, the grout takes are independent of their concentration; therefore, it is unnecessary to follow the principle mentioned above. Besides, it is very difficult to extrude the excess water from the thin cement mix through soft clay because of the low permeability of the clay. This will produce a set cement vein with a low density and strength.

VIII. CONCLUSION

After grouting, the main physical and mechanical properties of soft clay filled in the karstic caves are improved to a varying extent, and the resistance to leakage and seepage failure increases considerably. It can be concluded that the karstic caves filled by the soft clay can be treated by means of cement grouting so as to form an effective water-tight curtain.

The laboratory and in-situ investigations have indicated that under high grouting pressure the clay fillings are subjected to four actions, i.e. fracture, extrusion, consolidation, and chemical hardening. Among them the actions of fracture and extrusion are fundamental.

The grouting mechanism of the clay fillings in the caves is essentially different as compared with that of the fissured rock and alluvium. Therefore, the grouting technique used should be chosen to meet the given conditions in order to obtain the satisfactory results.

MONITORING OF GROUT LEACHING AT THREE DAMS CURTAINS IN CRYSTALLINE ROCK FOUNDATIONS

by Miron B. Petrovsky[1], M. ASCE

ABSTRACT

Monitoring of leaching of cement grout curtains in crystalline rock foundations was conducted at three dams in the U.S.S.R. Monitoring of water chemistry and drainage flows was implemented during long-term investigations. The age of the grout curtains at the start of the investigations varied from 4 to 16 years. The investigations revealed that calcium concentration in seepage water passing through the older grout curtains (more than 12 years old) was relatively constant and low during the 9-year observation period. This result was affected by dissolution and removal of the main hydrated cement compounds from the seepage area when the leaching of CaO occurs by the diffusional mechanism of calcium conveyance from the deeper zones of grout to the seepage channels. Calcium concentration in the younger, 4-year old curtain was substantially higher indicating that a considerable amount of $Ca(OH)_2$ and hydrated cement compounds were maintained in the permeable areas. The study revealed that correlation between calcium concentration and drainage flow could reflect the degree of interconnection of permeable channels in the monitored crystalline rock foundations. Linear relationships between log CaO and log Q were used to compute the amount of lime leaching from the grout curtains. Rate of lime leaching depended mostly on seepage duration or age of the grout curtains: the younger curtain was leached more vigorously than the older curtains. The origin of flow in a drainage system was studied at one dam to identify the foundation flow through the grout curtain and determine the representative amount of leached lime.

INTRODUCTION

Leaching of cement grout curtains by foundation seepage is a continuous process acting throughout the life of the dam. Gradual leaching results in deterioration of the curtains. This deterioration can lead to an increase in the permeability of the grout curtains, a rise in the foundation flow, an increase on the uplift pressure, and subsequent decrease of stability of the dams. Monitoring of the leaching process is one of the means determining grout curtain condition, and hence, dam safety.

---

[1]Geotechnical Engineer
International Engineering Company, Inc., 777 Post Rd., Darien, Connecticut 06820

The ability of a grout curtain to withstand leaching depends on many factors including curtain design concepts (depth, width, permeability), technology of foundation treatment (equipment, injection pressure, hole spacing, type of cement, grout mix), natural conditions surrounding a curtain (foundation geology, hydrogeology, water chemistry), age of a curtain before reservoir filling, etc. Leaching of cement grout and subsequent deterioration of the dam grout curtains by seepage water has been noted in many publications (e.g., 1,2,4,5). Chemical analyses of the seepage water coupled with measurements of the quantity of drainage flow passing through dam foundations have been used to monitor and evaluate the condition of the curtains during the dam operation period (2,6,7). This paper presents the results of long-term investigations of the durability of conventional grout curtains in crystalline rock foundations conducted at three concrete gravity dams in the U.S.S.R. The purpose of the investigations was to monitor curtain behavior by determining the magnitude and intensity of the leaching.

DETERIORATION OF CEMENT GROUT CURTAINS BY LEACHING

Leaching corrosion of cement grout is caused by the dissolution and removal of hardened cement compounds by seepage water (1,5). The main component of hydrated cement is CaO (lime) which can comprise 60 to 65 percent of the total cement volume in the form of free $Ca(OH)_2$ (calcium hydroxide) and hydrated calcium compounds such as silicates, aluminates, and alumino-ferrites. These hydrated compounds remain stable in calcium hydroxide solution at concentration about 1,300 mg CaO/l and a pH of 12.0. Leaching of $Ca(OH)_2$ is accompanied by the hydrolysis of the hydrated cement compounds and the release of lime to the seepage water. The reduction of CaO content in the cement grout leads to the gradual decomposition of the cement binder and deterioration of the grout. The removal of 25 percent of the CaO from grout can reduce its strength as much as 50 percent.

In permeable cement grout, three stages of leaching can be schematically identified. The first stage consists of rapid dissolution of the free calcium hydroxide. During this stage, which occurs at the onset of the leaching process, the concentration of calcium in the seepage water reaches its highest point. The second stage occurs when lime is liberated from the hydrated calcium compounds by the hydrolytic action of the seepage water. During this stage calcium concentration in the seepage water diminish. The third stage of leaching is characterized by dissolution of $Ca(OH)_2$ and decomposition of grout hydrated compounds in the seepage area. The main source of calcium in seepage water at this stage is diffusion of calcium from the deeper zones of the grout. During this stage, the concentration of calcium in seepage water is low and remains relatively constant.

DAM DESCRIPTIONS AND FOUNDATION TREATMENT

DAM 1, a 36-m-high and 513-m-long structure on the Onda River (9), was completed in 1956 (Fig. 1). It is founded on moderately fractured granite-gneiss bedrock. The vertical 6-m-deep, single row

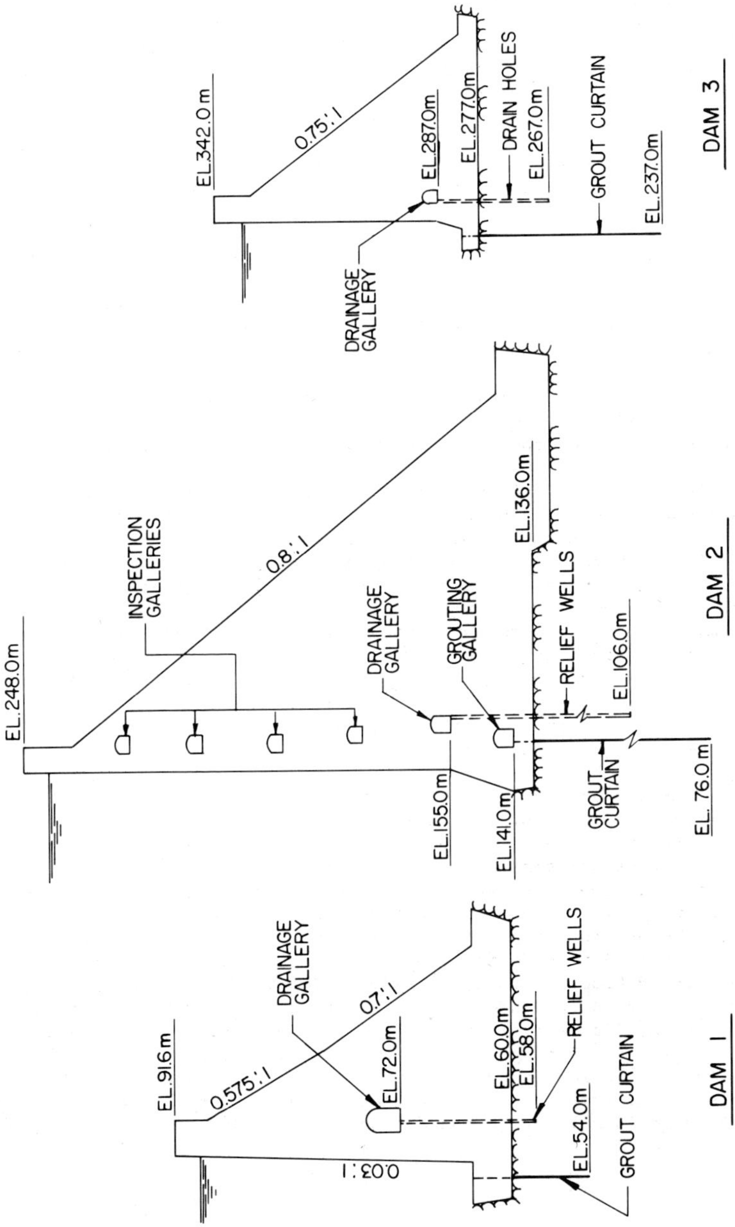

Fig. 1 - Typical Sections of Investigated Dams

grout curtain was constructed using the split spacing method with 1.5 m spacing between the boreholes. A 1-m-deep upper portion of the dam foundation was consolidated by blanket grouting. The neat cement grout had a maximum water-cement ratio by volume of 5:1 and was injected under pressures of 200 to 500 kN/m^2. The total quantity of injected cement was 51 metric tons with an average grout take of 12 kg/linear m. The compound composition of the portland cement used for construction of this grout curtain is shown in Table 1.

Table 1

Compound Composition of Grout Curtain Cements

	Compound Composition (%)						
Dam	C_3S	C_2S	C_3A	C_4AF	MgO	CaO	Free Ignition Loss
1	51.32	25.10	6.60	13.19	1.76	0.28	0.24
2	48.00	26.20	7.00	14.70	1.18	0.22	0.31
3	48.80	23.20	10.30	13.70	2.80	0.31	0.70

The 10-cm-diameter drain holes located 3 m downstream from the grout curtain were drilled to a depth of 1 to 2 m into the foundation from the drainage gallery of the dam at a spacing of 4 m on centers.

DAM 2, a 124-m-high and 1,072-m-long structure on the Yenisei River (8), was constructed in 1970 (Fig. 1). The dam is founded on fractured granites with zones of tectonic crushing ranging in thickness from several cm to 3 m. A 60-m-deep vertical grout curtain, which was enlarged in the contact zone with adjoining grouting, was installed using grouting pressures of 2,000 to 3,000 kN/m^2 and a grouting mix with a water-cement ratio of from 5:1 to 1:1. The 11-cm-diameter drain holes were spaced 3 to 5 m on centers and were sunk 20 to 30 m into the foundation to reduce uplift pressure. A total of 25,000 metric tons of cement (Table 1) was injected to construct the grout curtain, with an average grout take of 120 kg/linear m.

DAM 3, a 65-m-high and 391-m-long structure on the Irtish River (10), was completed in 1952 (Fig. 1). It is founded on gabbro with fractured zones and several inclined and subvertical faults. The grout and drainage curtains and the blanket grouting were installed to increase watertightness and strength of the foundation and to reduce uplift pressure. The 40-m-deep grout curtain consists of a single row of vertical boreholes spaced at 0.5 to 1.4 m. Injection was performed in stages of 5 to 6 m in depth extending from the top down. Grout pressures ranged from 300 to 1,500 kN/m^2. The 5-m-deep blanket grouting was constructed under the entire base of the dam. The total amount of cement (Table 1) injected into the foundation was 12,000 metric tons with an average grout take of 64 kg/linear m. The drainage curtain is located 8 to 10 m downstream of the grout curtain and consists of 15-cm-diameter boreholes spaced at 3 to 5 m on centers extending 10 to 15 m into the foundation.

METHODS OF MONITORING AND ANALYSIS

The investigation at the three dams consisted of a series of measurements taken over varying periods of time. Flows, water levels, and chemical concentration were measured at the reservoirs and at the drain holes. The reservoir levels were recorded daily. Flows and water levels in the drains were measured 3 times a month (Dam 1) and 1 to 2 times a month (Dams 2 and 3). Water samples for chemical analysis were collected 2 to 4 times a year from the reservoirs, 3 to 4 times a year from the Dam 1 drains, and 2 times a year from the Dam 3 drains. One hydrochemical survey of the drainage system was conducted at Dam 2. At this dam, electrical conductivity measurements of seepage water were also employed. The Van Dorn-type sampler was used to obtain water samples from specific depths. Table 2 presents mean values of chemical concentration in the water at each reservoir. The samples were collected over a five to fifteen year period.

Table 2

Water Chemical Composition in Dam Reservoirs (mg/l)

Dam	pH	Hardness as $CaCO_3$	Ca^+	Mg^{2+}	Na^+	HCO_3^-	SO_4^{2-}	Cl^-	CO_2 free
1	6.84	12	3.0	0.9	2.8	18.1	5.65	2.4	9.83
2	8.06	56	17.3	4.0	1.5	73.2	4.0	2.6	-
3	8.64	74.5	21.82	4.88	16.81	82.35	24.95	2.69	5.29

In addition to the parameters indicated in Table 2, the concentrations of potassium, iron, carbonate, and hydroxide were also determined. At Dam 1 thermal and electrical resistivity methods were used for continuous logging of the drains to detect water stratification and leakage zones. A sensing thermister probe calibrated to 0.01°C was used for thermal logging.

The calcium concentration was the main indication of the rate and extent of leaching during the investigation periods. This concentration was compared to the drain flow which can characterize seepage velocity and, thus, length of interaction time of water with cement grout. The relationship between flow and calcium concentration was defined by methods of regression analysis. Influence of seepage duration on Ca^{2+} concentration (Dams 1 and 2) was evaluated by dividing of the whole observation period into separate intervals and by mutual comparison of data (Ca^{2+} and Q) in the intervals.

Transformation of calcium concentration into lime removed from the grout curtains during a monthly period was computed taking into account the drain flow and calcium quantity entering the foundation from the reservoir: $CaO = a\,(Ca_d^{2+} - Ca_r^{2+})\,Qt$, where a = conversion factor from Ca to CaO (a = 1.4); Ca_d^{2+} = calcium concentration in water flowing from drains; Ca_r^{2+} = calcium concentration in the reservoir water; Q = flow rate from drains; t = number of seconds in a month. Leaching of CaO from the crystalline rock surrounding the grout curtains is assumed to be negligible. The average amount of CaO in igneous rock is about 5% (3). Then a relationship of log CaO to log Q was determined and further computations of leached lime were performed for each discharging drain using the derived equation for the dam being considered.

INVESTIGATION AT DAM 1

A long-term investigation at Dam 1 was conducted from 1967 to 1975. The grout curtain was 12 years old at the beginning of the monitoring program. The drains in the dam extended into the foundation only about 10 percent of their total length (Fig. 1). As a result, seepage flows into the drains may have originated from the foundation and permeable zones within the concrete dam. Evaluation of the quantity of the foundation flow affecting the magnitude of leaching was an important element of the investigation.

Figures 2 and 3 show the major types of foundation drains at the dam and typical flow hydrographs recorded in the drains. The type 1 drains are fed from the dam foundation and possess a uniform distribution of water parameters along the entire drain length. Due to the narrow range of pool level fluctuations (0.7 m to 1.4 m), the water level in or flow from these drains was relatively constant during a year (Fig. 3). Permeability of the dam foundation, evaluated by pressure and pumping tests, was found to be approximately 10^{-4} cm/sec to 10^{-2} cm/sec.

The type 2 drains were determined to be clogged. The drain water in the concrete length had a comparatively constant distribution of chemical parameters and electrical resistivity and had a sharp increase of these parameters in the rock foundation area. The drain water level varied insignificantly over the observation period. The permeability coefficients of the concrete-rock area intersected by the drains, estimated by pumping-out and falling-head tests, in the drains, were on the order of 10^{-6} cm/sec and lower, which corresponds to the quality of sound concrete. Relief wells clogged by mud, fine sand, microorganisms, and carbonate have been encountered at many concrete dams (6,7,9).

Water flows into the type 3 drains through discontinuities in the concrete, resulting in distinct stratification of the water's chemical and physical parameters along the drain length. Permeability of these zones was usually 1 to 2 orders of magnitude higher than the foundation permeability in the drain area. Flow hydrographs (level or discharge) in these drains had clearly recurrent patterns that reflected seasonal changes in the concrete permeability under the action of air and reservoir water temperature fluctuations (Fig. 3).

This study allow to identify the drains with evident flow from the dam foundation (which comprised about 40 percent of the total number of the relief wells) which were used in computation of lime leaching from the curtain.

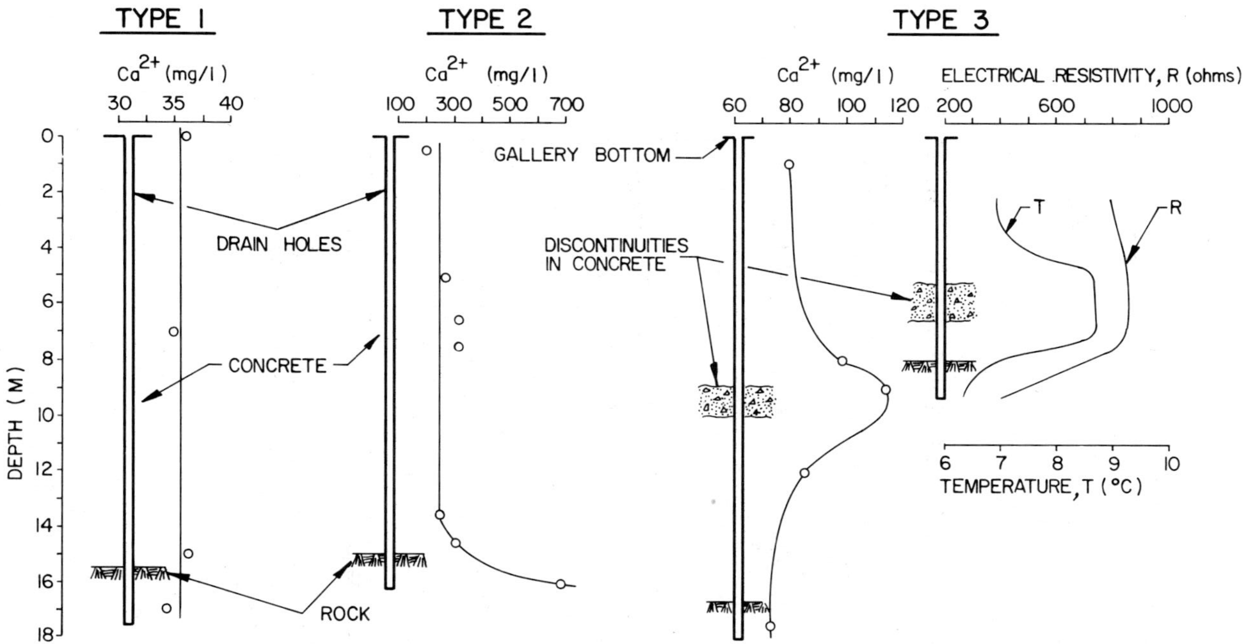

Fig. 2 — Dam 1. Patterns of Distribution of Calcium Content, Temperature, and Electrical Resistivity in Water along Drain Hole Depth: Type 1 — Flow from Foundation; Type 2 — Clogged Drain; Type 3 — Flow through Discontinuities in Concrete.

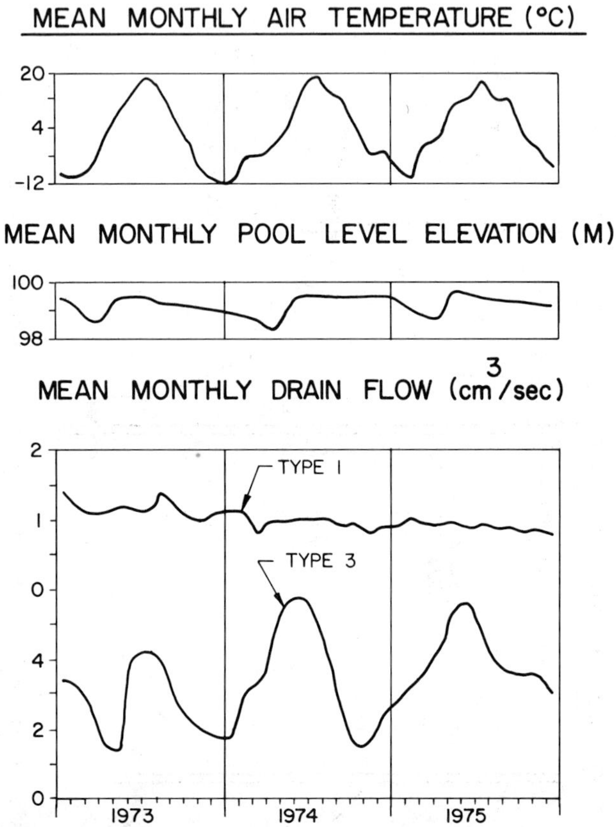

Fig. 3 - Dam 1. Flow from Typical Drain Holes: Type 1 - Flow from Foundation; Type 3 - Flow through Discontinuties in Concrete

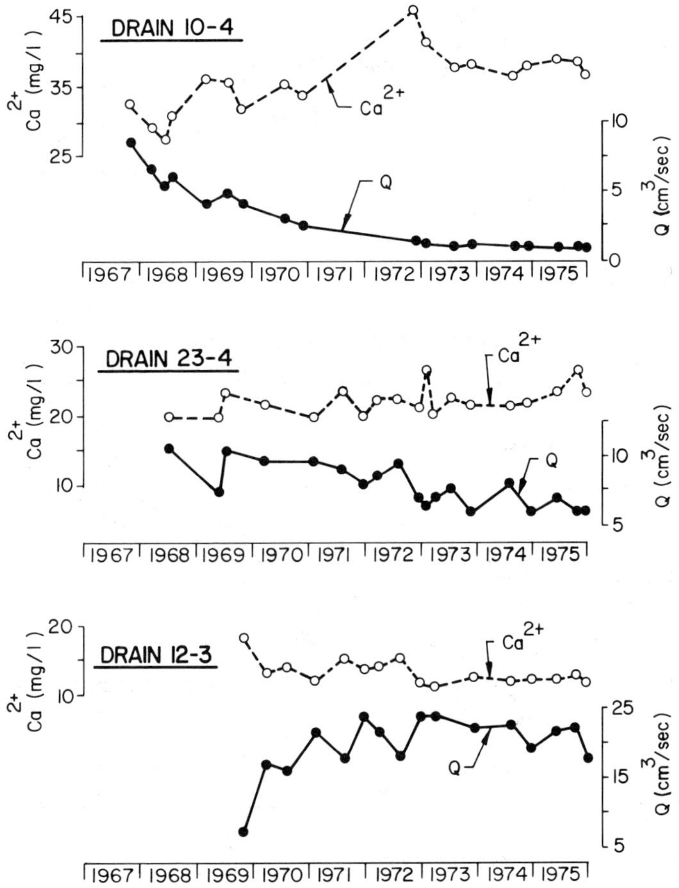

Fig. 4 - Dam 1. Typical Variations of Calcium Content and Flow in Foundation Drains

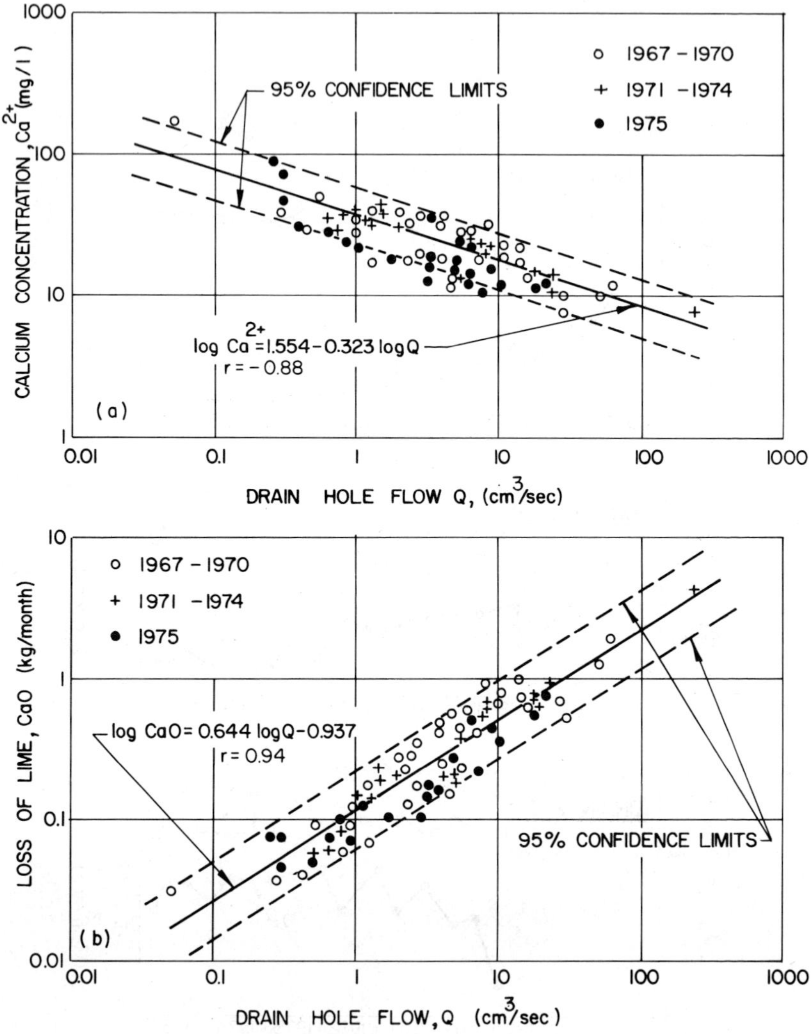

Fig. 5 - Dam 1. Relationship of Calcium Concentration (a) and Loss of Lime (b) to Drain Hole Flow

The typical Ca^{2+} variations in water flowing from three foundation drains are shown on Figure 4 for periods of 7 to 9 years. Based on the graphs, the calcium concentration in the seepage water did not diminish. There is an evident correlation between Ca^{2+} concentration and amount of flow, Q: the smaller flow rate conforms to the greater calcium concentration (drain 10-4) and the greater flow rate corresponds to the smaller calcium concentration (drain 12-3). Influence of seepage duration on Ca^{2+} concentration was evaluated by clustering all data (131 values of Q and Ca^{2+}) into three groups: periods of 1967-1970 (42 values), 1971-1974 (47 values), and 1975 (42 values) (Fig. 5a). The results of the study indicated that during the 9-year period calcium concentration did not change significantly. Variations in the statistics between groups were negligible and permitted consideration of the original data as one uniform group and computation of the final regression equation (Fig. 5a). The high correlation between calcium concentration and flow rates could be an indication of isolated permeable areas in the dam foundation. Numerous pumping tests conducted in the drain holes did not reveal, in most cases, hydraulic interference between the neighboring foundation drains. The relationship between log CaO and log Q was very close as shown on Figure 5b.

The average annual foundation flow and total annual amount of CaO removed from the Dam 1 grout curtain and tapped by the drains are presented in Table 3.

Table 3

Average Annual Drainage Flow and Loss of Lime From Dam Grout Curtains

Years	1968	1969	1970	1971	1972	1973	1974	1975
				Dam 1				
Flow(l/sec)	0.134	0.146	0.164	0.173	0.18	0.158	0.153	0.154
CaO(kg)	195	200	207	218	208	189	204	173
				Dam 2				
Flow(l/sec)					10.45	9.66	7.9	7.2
CaO(kg)					8,976	8,295	6,786	6,240
				Dam 3				
Flow(l/sec)	0.435	0.395	0.35	0.37	0.41	0.43		
CaO(kg)	286	248	256	272	264	273		

INVESTIGATION AT DAM 2

The hydrochemical survey of seepage water in the foundation of Dam 2 was conducted in 1975 on 140 flowing relief wells. At that time the age of the grout curtain was 4 years. The study of the observation data, plotted on Figure 6a, did not reveal a dependence of log Ca^{2+} from log Q in the foundation drains of Dam 2. The average Ca^{2+} value was about 46 percent higher than the calcium content in the drain waters of Dam 1. Relatively constant calcium concentrations in most of the drains might be explained by the numerous interconnected seepage channels in the foundation intersecting the deep relief wells averaging the calcium concentration along the grout curtains. This was confirmed by hydraulic interference between the drains located on

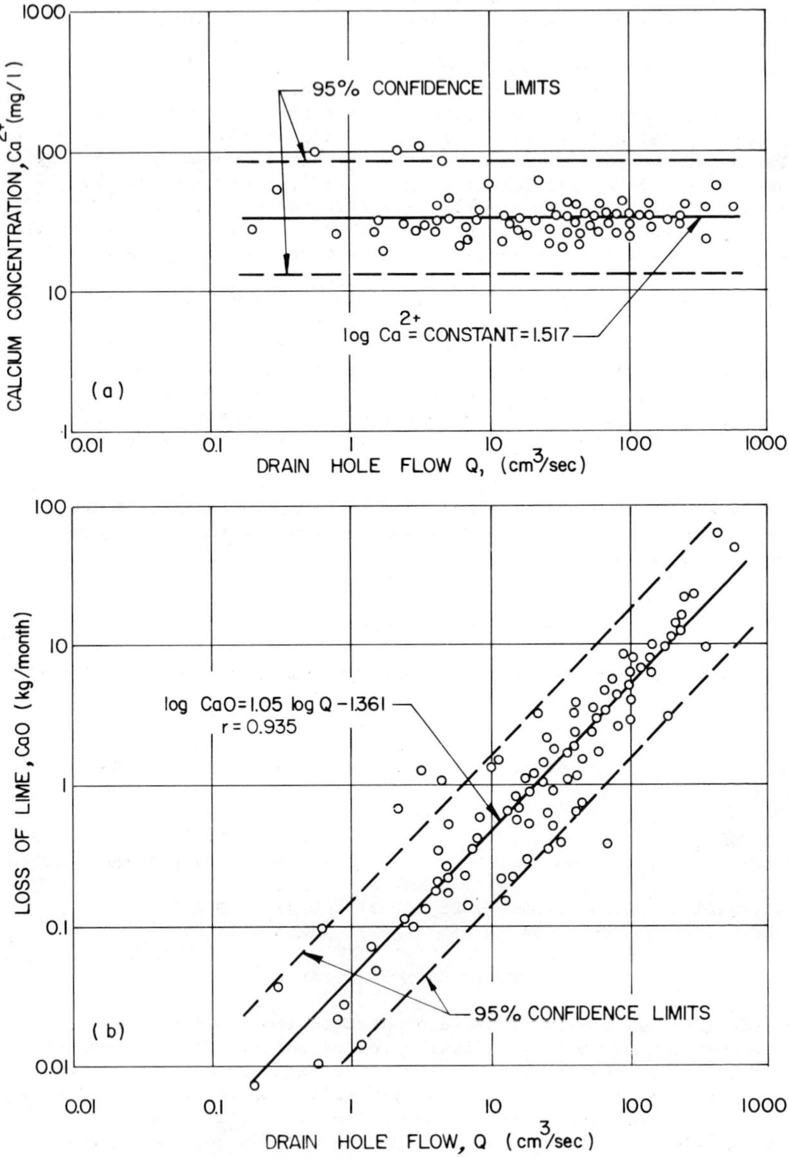

Fig. 6 — Dam 2. Relationship of Calcium Concentration (a) and Loss of Lime (b) to Drain Hole Flow a distance of 25 m to 35 m. The relationship between log CaO and log Q in Dam 2 (Fig. 6b) is characterized by a high correlation. Using the flow records, this permitted computation of the lime quantity leached out from the grout curtain for each discharging foundation drain and for the entire drainage system (Table 3). The amounts of lime leached for the period 1972 to 1974 were estimated assuming that a calcium concentration for this period of record approximated the concentration determined during 1975.

INVESTIGATION AT DAM 3

The investigation of the dam was conducted over the 1968-1973 period when the grout curtain was 16 years old. Calcium concentration for the entire observation period was relatively constant and low corresponding in some extent to the general results obtained at the Dam 1 investigation.

Lime leaching from the Dam 3 grout curtain was computed employing the regression equation $\log CaO = 0.563 \log Q - 1.117$, which was derived from field data (number of readings was 153, $r = 0.78$). The total loss of lime from the grout curtain for the 6-year observation interval is shown in Table 3.

ANALYSIS OF RESULTS

The monitoring programs described were performed at dams having grout curtains ranging in age from 4 to 16 years. Under these conditions second and third stage leaching is expected.

The concentration of calcium in the seepage water remained constant at each drain during the monitoring program, but could vary between drains along the length of the curtains. Generally, the variation can be attributed to the nature of permeability of the grouted rock. In isolated seepage channels variations in calcium concentration are experienced with varying flow rates when slower flows result in high concentration on the other hand. Interconnection of permeable channels in the foundation rock may yield averaging of concentration along the seepage front.

Figure 7 represents the leaching of lime from the three grout curtains. From these diagrams, intensity of CaO leaching can be characterized by the slope of the straight lines to the horizontal axis or the value of the coefficient "a" in the derived equations: $\log CaO = a \log Q - b$. The highest rate of CaO leaching was observed at Dam 2, which differed substantially from the rates at Dams 1 and 3. The main cause in the difference of the leaching intensities is related to the age of the grout curtains or the length of seepage duration at the time of the investigations: 4 years for Dam 2 and 12 to 16 years for Dams 1 and 3. The younger grout curtain with a relatively short period of seepage can contain a considerable amount of unleached calcium hydroxide in the permeable areas. Hydrolysis of the other cement components occurs during this stage. This causes relatively high average calcium content in seepage flows and more intense grout leaching (Dam 2). In the older grout curtains with the long periods of seepage (Dams 1 and 3), dissolution of $Ca(OH)_2$ and

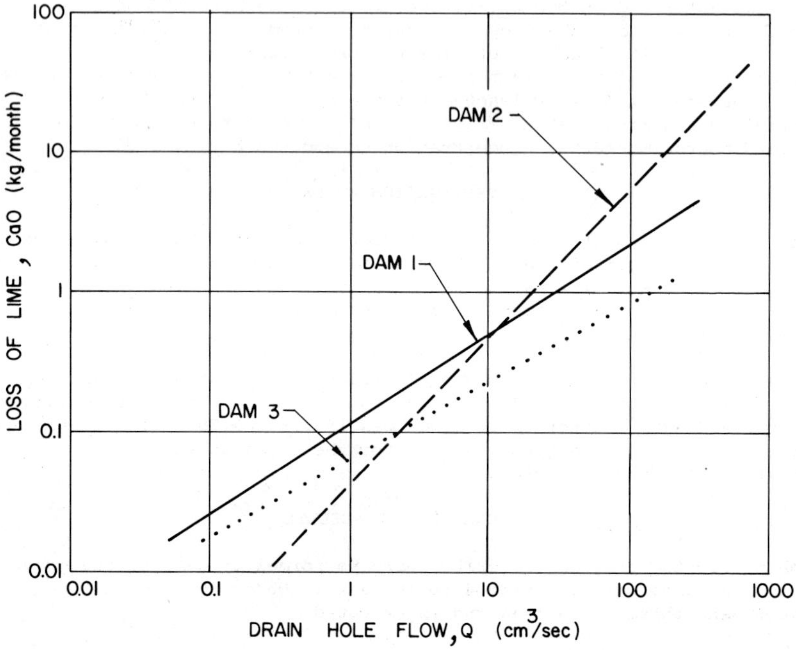

Fig. 7 — Leaching Intensity for Dam Grout Curtains

hydrolysis of hydrated compounds on surfaces and adjacent voids has nearly ceased over the entire length of the seepage channels. The corrosion process has moved into the deeper zones of the cement grout. However, this process is slowed by products of cement hydrolysis which block the capillaries feeding the main seepage channels. At this stage, the leaching of CaO occurs mainly by the diffusional mechanism of calcium conveyance from the sound zones of cement grout to the seepage channels. As a result, calcium concentration in seepage water remains relatively constant over the long period of time. Accordingly, the leaching intensity is decreased and the quantity of lime leached corresponds to the amount of seepage flow.

CONCLUSIONS

At two dams with grout curtains more than 12 years old in crystalline rock foundations, calcium concentrations in seepage water have been shown to remain nearly constant and relatively low over a nine year observation period, indicating the curtains to be in the last stage of the leaching process. Because the concentrations were found to be near constant, the frequency of water chemical analysis of seepage flows can be reduced.

Analysis of seepage flows at a dam with a 4-year old grout showed that calcium concentration was much higher, indicating that the grout curtain is in the intermediate stage of the leaching process. Decreases in calcium concentration in seepage flows can be expected. The correlation between calcium concentration and drainage flow could reflect the degree of interconnection of permeable channels drained by the relief wells in the crystalline rock foundations.

Linear relationship between log CaO and log Q can be used to compute the amount of lime leaching from a grout curtain. At the dams studied, the amount of leached lime was relatively negligible with respect to the total volume of hydrated lime in the grout curtains. However, taking into account the selective nature of seepage in crystalline rock foundations, leaching action can be concentrated in local areas, thereby weakening the grout curtains.

Intensity of lime leaching depends largely on seepage duration or age of grout curtains. Corrosion of the cement grouts was more vigorous in the younger grout curtain having been exposed to seepage for a shorter period of time than for the older curtains. Combined monitoring of the foundation seepage and the grout curtain leaching can be a means effective and efficient gauging of foundation condition resulting in adequate interpretation of the leakage and uplift pressure variations during the dam operation period..

In old and deteriorated concrete dams where discontinuities in concrete conduct water from the reservoir, and intersection of the vertical foundation drains is possible, the origin of flow tapped by the relief wells should be determined. This would permit identification of the foundation flow and allow quantification of the representative amount of CaO leaching from the grout curtains.

References

1. Biczok, I. "Concrete Corrosion and Concrete Protection", Eighth Editon, Academiai Kiado, Budapest, 1972.

2. Gebhart, L. R., "Foundation Seepage Control Options for Existing Dams", Inspection, Maintenance and Rehabilitation of Old Dams, ASCE Proceedings, 1974, pp. 660-676.

3. Gilluly, J., Waters, A. C., and Woodford, A. O., "Principles of Geology", Third Edition, W. H. Freeman & Co., 1968.

4. Houlsby, A. C., "Engineering of Grout Curtains to Standards", Journal of Geotechnical Engineering Division, ASCE, Vol. 103, No. GT9, Proc. Paper 13214, September, 1977, pp. 953-970.

5. Kind, V. V., "Cement and Concrete Corrosion in Hydraulic Structures", Moscow, Gosenergoizdat, 1955 (in Russian).

6. Ley, J. E., "Foundation of Existing Dams-Seepage Control", Inspection, Maintenance and Rehabilitation of Old Dams, ASCE Proceedings, 1974, pp. 584-608.

7. Lundin, L., "A Review of TVA's Problems and Treatment of Foundations of Existing Dams", Inspection, Maintenance and Rehabilitation of Old Dams, ASCE Proceedings, 1974, pp. 641-659.

8. Petrovsky, M. B. and Sergeeva, N. S., "Monitoring Seepage Flows through the Body and Foundation of the Concrete Dam at the Krasnoyarsk Hydroelectric Plant", Hydrotechnical Construction, No. 12, December, 1976, pp. 1210-1217.

9. Petrovsky, M. B., Sergeeva, N. S., and Maslova, N. P., "Hydrochemical and Microbiological Studies at the Onda Dam", VNIIG Proceedings, Vol. 91, Leningrad, 1975, pp. 11-17 (in Russian).

10. Zadvorny, G. M. and Petrovsky, M. B., "Condition of the Concrete Gravity Ust-Kamenogorsk Dam", Hydrotechnical Construction, No. 9, September, 1967, pp. 34-37.

GROUTING PLANT & EQUIPMENT

By
A.W. Gourlay* & C.S. Carson+

ABSTRACT The paper concentrates on the development of plant and equipment over the last 25 years and describes the range and types of mixing, batching, pumping, recording and ancillary equipment available today. Since grouting applications are so varied and range from bulk infilling of mine workings with large volumes of low strength cement/flyash grout to sophisticated Tube à Manchette injection programmes in alluvial deposits using complex chemical grout systems, perforce the equipment required also varies widely.

The advantages of the colloidal mixing principle are explained for cement based systems and the point is made that the choice of grout pumping equipment is often dictated more by subjective considerations and personal prejudice rather than by documented evidence of the suitability of any one type of pump for any particular application.

The recent development of multi-hole injection units and electronic recording systems is outlined and the paper lists and describes the main items of ancillary equipment in common use.

INTRODUCTION

Grouting techniques have been practised for centuries, but the last twenty-five years have seen a rapid improvement in the understanding of the physical and rheological properties of the various systems, and application techniques have become more scientific and progressively more sophisticated. The development of plant and equipment has had to keep pace with these advances in technique and with the consequent demands from specialist contractors for higher outputs, longer pumping distances and higher pressure together with the attendant requirements for greater control in terms of batching materials, regulating flow and monitoring grout injection.

In the early 1930s grouting practice, although advancing rapidly, still tended to rely on cement based systems and the early emphasis was therefore placed on designing plant and equipment for handling cement based grouts. This led to the production of high speed, high shear colloidal mixers and the first prototype units were developed in England in 1935 with production units being manufactured by the end of 1937. At this time pumping equipment tended to fall into two categories; low pressure screw type units capable of reasonably large outputs and high pressure ram pumps giving relatively low outputs. Ancillary equipment was very basic and monitoring systems virtually unknown.

*Managing Director Colcrete Group
+Equipment Development
Manager Colcrete Group
Strood, ENGLAND

The limitations of cement as a grouting material were also well recognised by then in that the larger particles in OPC prevent penetration of fissures in rock finer than 160μ and likewise restrict the permeation of alluvial deposits with a permeability less than $5x10^{-4}$m/sec. The subsequent development of essentially non-particulate chemical grout systems which were designed to fill the finer fissures and smaller pores in the ground did not require high shear mixing and simple mixing tanks with rotating paddles proved perfectly adequate. However, pumping and monitoring equipment then had to be designed to allow much greater control of the rate and pressure of injection. As a result sophisticated central batching stations are now available for dispensing clay/cement or chemical grout and typically these stations include individual regulation and recording of flow and pressure for each sleeve in a Tube à Manchette injection programme. Often the stations are containerised for ease of transport and to facilitate rapid site installation. Again multi hole injection systems are now available which allow volume and pressure limits to be set and recorded for any individual hole or sleeve.

The introduction in more recent years of epoxy and polyester resins for certain grouting applications has demanded further plant development in mix/metering to prevent wastage of these very expensive materials.

With the advent of hydraulics, the range of drive options for mixers and pumps has been increased and various prime movers are available from diesel and petrol engines to electric or compressed air motors.

The range of grouting plant and equipment currently available and the applications for which it has been designed can best be discussed under the following headings:-

- Mixers
- Weigh Batching Equipment
- Pumps
- Combined Mixer/Pump Units
- Central Mixing/Pumping/Monitoring Grout Stations
- Ancillary Equipment

MIXERS

The main types are:- colloidal, paddle and jet.

Colloidal

These mixers are based on the Colloidal Mill principle of high speed shearing of the cement in the water to remove any air attached to the particles of cement and to ensure thorough wetting. The early work was carried out by J.S. Morgan in England and the original machines had an output of around $3m^3$/hour. Today the larger, hydraulically driven mixers have outputs up to $16m^3$/hour (Fig 1) and prototype machines have been developed for outputs of $25m^3$/hour. Colloidal mixers are undoubtedly the most efficient means of hydrating cement and given a minimum specific surface for a typical OPC of $225m^2$/kg, the difficulty in achieving thorough wetting of the particles can be gauged by appreciating that for one 50kg bag the area that has to be wetted is $11,250m^2$ (equivalent to two football pitches).

Fig 1. Semi Automatic Single Drum Colloidal Mixer

Output - Up to $16m^3$/hour depending on mix
Batch capacity - 680 litres
Power required - 60 hp

Colloidal mixers can produce pumpable neat cement mixes down to 0.36 water: cement ratio without the use of admixtures and where admixtures are used for various purposes they can be intimately incorporated in the mix by the high speed mixing action.

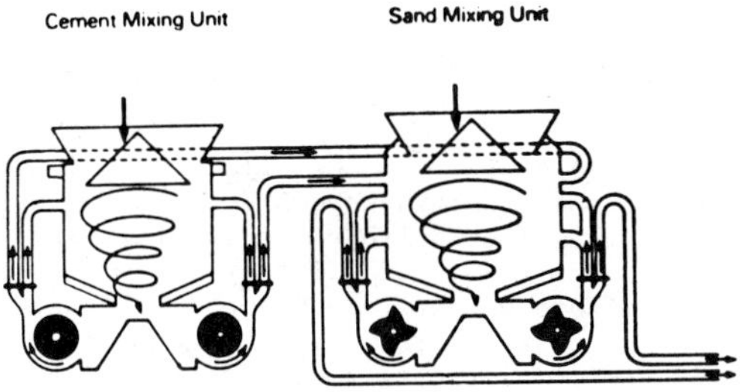

Fig 2. Flow Diagram of Double Drum Colloidal Mixer with Vertically Mounted Impellers

The mixers work on the principle of rotating an impeller at high speed (normally 2000 rpm) within the narrow slot of a steel casting to achieve a high shearing action and the vortex created in the cylindrical tank as the grout is recirculated through tangential inlets ensures rapid assimilation of the constituent materials. Different impeller shapes are used when sand or other fillers have to be incorporated in the cement slurry. Single and double drum versions of the mixer are available and in the double drum type (Fig 2) the cement slurry is mixed in the first drum using a plain steel impeller before being transferred to the second drum where the sand or other filler is added, the mixing impeller in this case being of a special shape. While some manufacturers mount the impeller in the vertical plane (Figs 1 and 2) others produce mixers with Horizontally Mounted Impellers (Fig 3). The only practical difference is that greater sand: cement ratios can be accommodated with the vertical configuration (up to 4:1 sand:cement by weight).

Fig 3. Single Drum Colloidal Mixer
Output - Up to $3m^3$/hour depending on mix
Batch capacity - 175 litres
Power required - 10 hp

Gravity discharge roller mixers are also available where the colloidal mixing action is produced by rotation of a drum or roller. These mixers are generally used for producing either heavy density grout using, for example, barytes or powdered iron as part replacement of the sand or other filler or for lightweight grout using a foaming agent. Grout densities in the range $5.35Mg/m^3$ to $0.32Mg/m^3$ can be produced by these roller mixers.

A large range of colloidal mixers is now available with outputs from $1m^3$/hour up to $16m^3$/hour. Power units can be diesel, petrol, air or electric and drive may be direct, through V belts or hydraulic.

Paddle

These are simple mixing tanks with a rotating paddle or series of paddles and the grout is thrown against baffles attached to the side of the tank. Whilst quite effective for mixing cement slurries down to water:cement ratios of 0.5 they are slower in operation than colloidal mixers and are discharged by gravity which means they have to be set up on a platform discharging into a pump receiving hopper at a lower level. Alternatively, paddle mixers can be coupled direct to a hand pump but each mix then has to be discharged completely before a new one can be made and this is impractical except for very small scale work. Paddle mixers can incorporate small amounts of sand, but when used with flyash this tends to float on the water and "ball-up". Most chemical grout systems can be accommodated by paddle mixers but special resin lined or stainless steel tanks are required for corrosive chemicals. Air, electric or hydraulic motors can be used to drive the paddle shaft and tanks are available in a variety of sizes.

Jet

The high pressure jet-type mixer revolutionised oil well cementing practice when it was first introduced nearly 50 years ago although it has never found the same acceptance in Civil Engineering practice. The mixing principle is based on forcing a stream of water through a jet, into a mixing bowl thus creating a vacuum which pulls in dry cement from a hopper located directly above the mixing bowl. As the cement enters the jet stream of water it is mixed in the turbulent flow that is created in the discharge pipe. Very high outputs (up to $100m^3$/hour) can be achieved and the units are simple and reliable.

WEIGH BATCHING EQUIPMENT

Weigh batching of the individual constituent components is essential for high quality work involving cement based grouts. While this can be achieved by off-site preblending with delivery to site in bags or in bulk for use with silos this can prove to be expensive when the site is remote from the preblending plant. Cement is normally batched on site by conventional silo, although a special hopper with a large orifice is required to feed the larger output mixers. Dry flyash can also be weigh batched by conventional silo. Moistened flyash and sand can be batched by purpose built equipment with outputs geared to the capacity of the grout mixer. Most systems rely on Dali type scraper units feeding hoppers fitted with load cells, the hoppers being elevated by a variety of means from winch and rail to hydraulic ram. Larger grout batching stations are fed from overhead hoppers supplied in turn by conveyor systems which can be fitted with vibrating screens. Particular care has to be taken in the design of units for handling bulk flyash since it typically contains 10-12% moisture as delivered to site which creates difficulties with flow and leads to bridging or arching in hoppers. Large batching plants have been built for specific contracts or for long term use and these are available in fully automatic or semi automatic form with push button operation (Fig 4). These plants are either electric or electro/hydraulic in operation and include water pumps, water meters, admixture dispensers, cement silos with screw feeds and sand feeding and weigh batching arrangements. The largest known batching plant manufactured to date was supplied in 6 modular units and had a total output of $100m^3$ of grout per hour (Fig 5).

Fig 4. Grout Weigh Batching Plant
Output - $50m^3$/hour
Mixer units - 4 No. DD16 Semi Automatic Colloidal Mixers
Cement Handling - Silo and Screw Feed
Sand Handling - Dali scraper and elevating hopper

Fig 5. Grout Weigh Batching Plant
Output - $100m^3$/hour
Mixer units - 6 No. SD24 Semi Automatic Colloidal Mixers
Cement Handling - Silo and Screw Feed
Flyash Handling - Belt Conveyor with Time Switch

PUMPS

There is considerable variety in the pumping principle adopted for grout pumps depending on the duty required of the pump. The main types are:- helical screw, piston, ram and diaphragm.

Helical Screw

The principle of operation involves rotation of a single helical chrome plated rotor within a double internal helix, resilient stator (usually of rubber, elastomer or polymer material). The rotor is of constant circular cross section, the centres of sections forming a helix which is eccentric to the rotor axis. The pitch of the stator is twice that of the rotor and the two engage in such a fashion that the rotor section traverses the stator aperture. The rotor maintains, up to certain limiting pressures, a positive seal along the length of the stator, and this seal progresses continuously through the pump. The rotor units can either be mounted horizontally or vertically but vertical mounting ensures a consistent flow of grout to the rotor and evens out wear on the rotor and stator. The pumps have no valves and have a characteristic pulse free flow.

Fig 6. Helical Screw Pump with Vertical Mounting of Rotor
Output - Up to $4m^3$/hour
Maxm. Pressure - $1000kN/m^2$
Power required - 6 hp

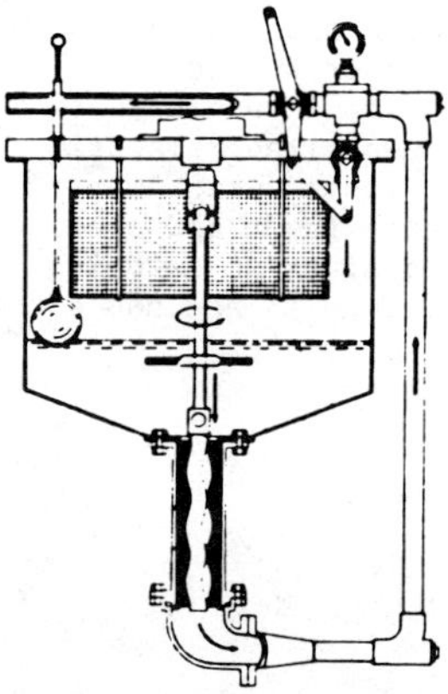

Fig 7. Flow Diagram of Helical Screw Pump with vertically Mounted Rotor

Helical screw pumps are available in single or multiple stage versions, depending on the operating pressure required. While the theoretical pressure potential is in excess of $15,000kN/m^2$ with multi stage units the higher pressure versions require excessive rotor lengths. Normally these pumps are used in double stage form with a consequent pressure limitation of $1000kN/m^2$. The outputs of helical screw pumps are a function of the size of rotor and stator and outputs up to $20m^3$/hour can be achieved from standard units. For applications demanding very high outputs individual pump units with a capacity of up to $50m^3$/hour can be supplied. Sanded grouts up to 4:1 by weight sand:cement can be handled with the larger helical screw pumps but wear tends to be high and slower rotation speeds are then used to combat wear.

Piston (Figs 8, 9, 10 and 11)

Fig 8. Hydraulic Piston Pump
Output - up to $2m^3$/hour or $6m^3$/hour depending on cylinder size.
Maxm. pressure - $10000kN/m^2$ or $3300kN/m^2$ depending on cylinder size
Power required - 10 hp

Fig 9. Hydraulic Piston Pump
Output - up to $2.6m^3$/hour or $8.2m^3$/hour
Maxm. pressure - $7300kN/m^2$ or $3000kN/m^2$ depending on cylinder size
Power required - 7.5 hp

These pumps comprise a driving and a liquid end. The driving end is invariably a double acting reciprocating piston operated by air or hydraulically, although diesel drive via a cam can also be employed to produce the reciprocating motion. The connecting rod couples the driving end to the liquid end piston (Fig 12). As the piston moves one stroke suction is created on the retreating side of the piston, thus inducing grout to flow into one pump chamber whilst, on the other side of the piston, grout is expelled through the other chamber. On the subsequent stroke the reverse happens. Non return valves at each inlet and outlet prevent interaction between the sides of the piston making the output per cycle sensibly constant.

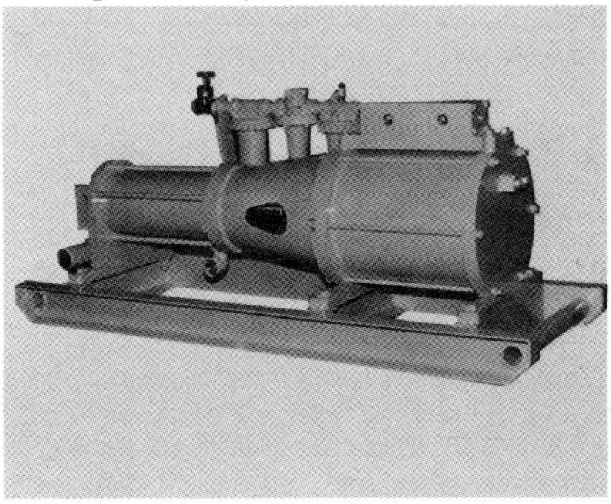

Fig 10. Pneumatic Piston Pump
Output - up to $10m^3$/hour
Maxm. pressure - $4800kN/m^2$
Power required - 17 hp

The valves fitted to the pumping chambers are either of the ball or disc type, the former being gravity returned and the latter spring loaded. These valves are critical to the effective operation of the pump and the seatings need regular inspection to maintain full pump efficiency and to prevent excessive wear and minimise blockages. Generally, ball valves give much less trouble and most piston pumps are fitted with this type of valve.

The output per stroke is slightly different, this being due to the effective sweeping area of the piston being greater on one stroke (full piston area) than the other (full piston area less rod area). However, as the output of reciprocating piston pumps is sensibly constant they can be considered as positive displacement pumps. As the reciprocating piston comes to a stop at the end of each stroke the pressure output is of a pulsating nature. The simplex type of piston pump (i.e. one cylinder only) is not favoured because this pulsating output characteristic is exaggerated and consequently duplex or triplex versions are commonly used.

Fig 11. Proportioning Pump - Arranged for injecting Sodium Silicate/Sodium Aluminate solutions in the ratio of 2:1. Ratios up to 10:1 can be dispensed for other chemical grouts

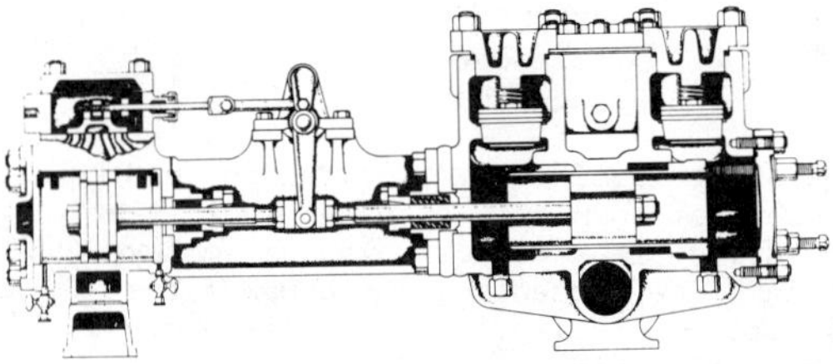

Fig 12. Cross Section of Typical Piston Pump

A feature of the multi-cylinder pump is an interlinked and phased piston arrangement such that when one piston is at the end of its stroke another is in the middle of its stroke. Consequently, the output pressure characteristics of duplex and triplex pumps have less pronounced pulsations (Fig 13). For some applications an accumulator is used to "smooth out" the output still further. Different pressure and output ranges can be achieved from the same pump by sleeving the liquid end cylinders and fitting smaller pistons.

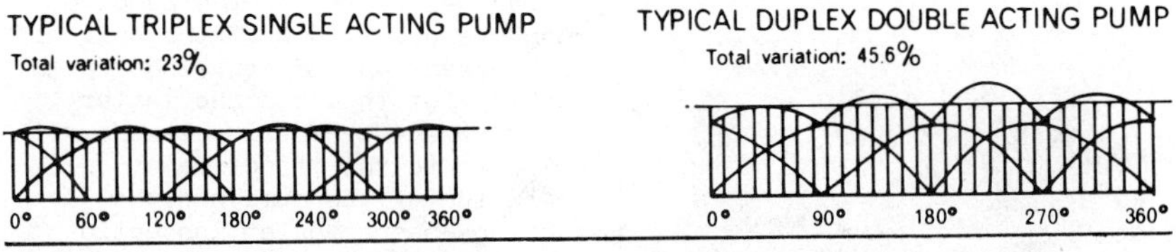

Fig 13. Discharge Variations of Triplex and Duplex Piston Pumps

Higher pressures can then be generated but at the expense of output volume. The output of large triplex piston pumps can be as high as $100m^3$/hour with pressure potential when sleeved of up to $100,000kN/m^2$. Those parts of the pumps in contact with the grout are often chrome plated or hardened to give extended life, but relatively high wear is to be expected with sanded grout. Sand:cement ratios of 3:1 can be accommodated.

Ram (Figs 14 and 15)

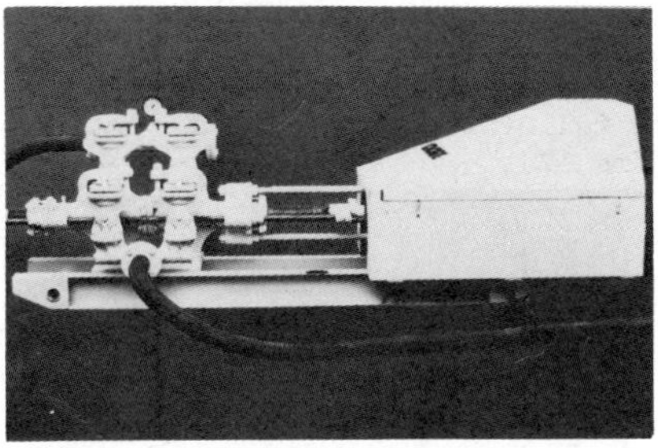

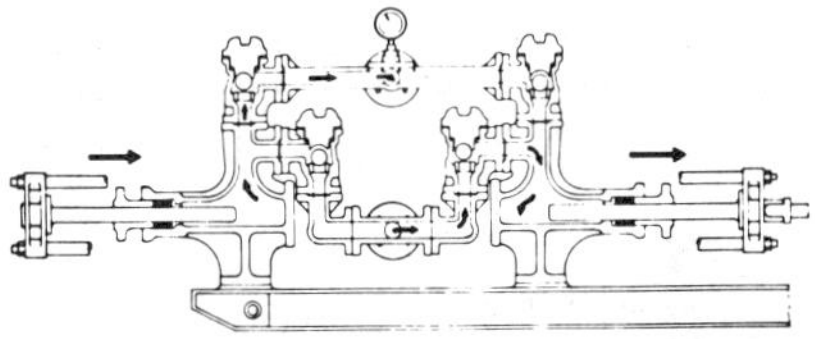

Fig 14. Hydraulic Ram Pump
Output - Up to 5.8m^3/hour
Maxm. Pressure - 5,800kN/m^2
Power required - 20 hp

Fig 15. Flow Diagram of Ram Pump

Ram pumps use a similar reciprocating principle to Piston pumps, but the ram is a loose fit in the cylinder and consequently the full volume of the cylinder is not displaced by each stroke. Each cylinder operates on suction during one direction of travel of the ram and on discharge on the reverse stroke. Invariably, such pumps are double acting to give a relatively "smooth" output but characteristically they still provide a pulsating flow. They are extremely robust in construction and easy to maintain which accounts for their popularity with grouting contractors. Pressure/output characteristics are similar to Piston pumps and again non return ball valves ensure positive displacement output per stroke. Ram pumps in common use give flow outputs up to 10m^3/hour with pressures up to 10,000kN/m^2. 3:1 sand cement grouts can be handled but as with all grout pumps the higher the sand content, the greater the wear on those parts of the pump in contact with the grout.

Diaphragm (Fig 16)

The activation of diaphragm units is reciprocating, normally by means of a piston although the drive can be rotary via a crank shaft. Each diaphragm is single acting (i.e. output on actuating stroke, suction on return stroke) but invariably in the larger pumps, there are 2 opposed diaphragms actuated alternately to give a double acting output. As in Piston and Ram pumps non return valves are used on the inlet and outlet chambers. The flow and pressure characteristics of diaphragm pumps are similar to Ram pumps although maximum pressure is normally limited to 1,200kN/m^2 with air operation and 5,500kN/m^2 when hydraulically driven. Outputs of up to 10m^3/hour are claimed. These pumps will handle 3:1 sand cement grouts and while heavily sanded mixes cause rapid wear on the diaphragms these can quickly be changed.

General Observations on Grout Pumps

Hydraulically driven pumps have an excellent characteristic in that injection pressures can be limited by regulating the hydraulic fluid pressure instead of having to rely on pressure relief valves fitted to the pump discharge or injection manifold. They also maintain excellent volume/pressure relationships, producing full theoretical output for some 80-85% of the pressure potential (Fig 17).

Fig 16. Double Diaphragm Pump
Output - Up to 8m^3/hour
Maxm. Pressure - 5,500kN/m^2
Power required - 7.5 hp

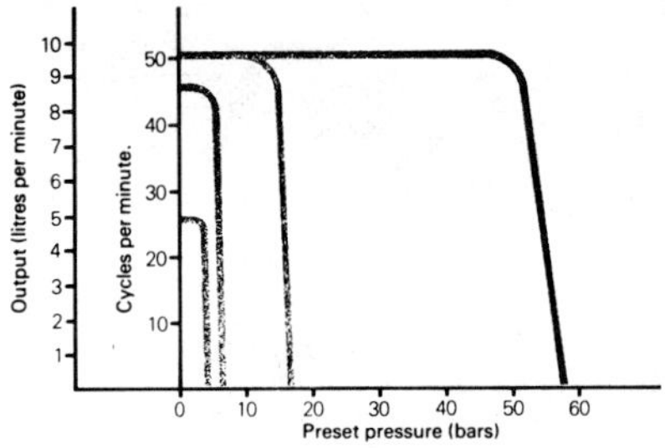

Fig 17. Output of Hydraulic Ram Pump at Various Preset Stalling Pressures

For applications involving the injection of fissured rock such as curtain grouting on dams or contact grouting programmes in tunnelling there is a distinct preference in British and European practice for ram type pumps. It is believed that the pulsating flow characteristics previously referred to give better results by allowing a longer period of injection in fine fissures before eventual refusal. It is interesting to note that American practice for these same applications recommends the use of helical screw, continuous flow pumps and since no conclusive evidence is ever presented either way this difference of opinion shows how subjective is the choice of the right grout pump for any particular application.

It is important to have the hopper capacity of the pump (or the agitating tank in the case of those pumps working on suction inlets) related to the batch output of the mixer being used to feed it, and care should always be taken in matching mixer/pump outputs. Most grout pumps will handle chemical grout systems but as already noted for mixing tanks, agitating tanks or pump hoppers may have to be resin lined to combat the more corrosive chemical grouts. The pump working parts may also have to be manufactured from stainless steel or other corrosion resistant material. For most chemical grouts proportioning pumps are recommended to give the necessary degree of control over the constituent elements of the grout system. Piston pumps are often used for two component systems with one cylinder suitably sleeved to give the required ratio between the two chemicals (Fig 11). Proportioning is of paramount importance when short gel times are required and should be specified for all but the smallest contracts.

Batch type mixing and pumping of epoxy and polyester resin systems is not recommended due to the high wastage that can occur with the expensive resin material and more sophisticated mix/metering equipment should be used to allow the resin and hardener elements to be pumped separately to the injection nozzle or to special mixing heads.

Various drive options are available for the different types of pump. Where diesel or petrol prime movers are used special variable ratio or stepless gear boxes have to be fitted to the pump drive to allow the output to be varied or alternatively a hydraulic pump can be used for the final drive to the pump unit.

COMBINED MIXER/PUMP UNITS

These units are designed for one man operation with the mixer, pump and agitating tank all mounted on the same chassis. Various combinations are available and a typical example is illustrated in Fig 18.

Fig 18. Electro Hydraulic Mixer/ Pump Unit
Output - Up to $4.8m^3$/hour
Maxm. Pressure - 4,500kN/m^2
Power required - 28 hp

Central Mixing/Pumping/Monitoring Grout Stations

For most large alluvial grouting contracts where for example it is required to reduce the permeability to $1x10^{-6}$m/sec it would be common to inject clay/cement grout to fill the larger pores followed by a secondary injection phase with either a silicate, chrome lignin or resin based chemical system. Such work can be carried out from a central batching station with grout of the appropriate type being fed from a ring main to individual pump units for injection and recording (Fig 19). Alternatively, more sophisticated stations can be used comprising mixing units, a series of central "slave pumps" feeding individually controllable injection cylinders and multi-hole or multi-sleeve units which automatically record the volume and pressure for each type of grout for each hole or sleeve being injected. Fully equipped Containerised units are available (Figs 20 and 21) which only require the connection of a supply of electricity and water to provide a complete mixing/batching/pumping/monitoring station capable of being controlled by one man.

The choice of whether to adopt the more labour intensive ring main and individual pump approach to the sophisticated centralised multi-hole injection station is dictated primarily by the size or duration of the contract and ultimately by the cost per m^3 of grout injected. Only the larger International grouting contractors can afford the capital investment for multi-hole units and so their use tends to be confined to alluvial injection programmes on major dams or to complex work where the Consulting Engineer recognises that the right equipment and experience should be employed and goes out to selective tender or negotiates with one firm in whom he has confidence.

Fig 19. Pumping Station for Clay/Cement Grout at Asprokremmos Dam, Cyprus

Fig 20. Containerised Mixing/Pumping/Monitoring Station showing slave cylinders and presetting switches and controls

Fig 21. Arrangement of pressure gauges and recording units used in Containerised Pumping Station

ANCILLARY EQUIPMENT

The following items of equipment tend to be purpose made for particular applications, and as with mixers and pumps care must always be taken to choose equipment sufficiently robust for the duty intended.

Water Meters

These are essential for high quality work and can be fitted direct to mixer inlets. Presetting switches are available and for Batching Plants "fail-safe" systems can be incorporated to prevent material loading if the preset water volume has not been fed into the mixer.

Mix/Volume Counters

Digital units can be fitted to the mixer water meter to record the number of batches, or flow recording units can be supplied for the pump to record output. For critical Tube à Manchette injection programmes in fissured rock or alluvium it is most important to monitor the "take" per sleeve.

Admixture Dispensers

Normally dry powder admixtures are packaged by the supplier in doses suitable for individual or multiple 50kg charges of cement but liquid admixtures such as plasticisers are supplied in bulk and dispensers have to be used. These range from simple hand operated piston/plunger pumps to push button dispenser units with built in "fail-safe" operation to prevent under or over dosage.

"Water in Sand" Estimators

For high quality work where careful control of water:cement and cement: sand ratios is necessary either the Gammon Morgan or Speedy systems can be used to determine the moisture content of the sand being fed into the mixers.

Flowmeters

As an approximate check on the grout viscosity of cement based systems as mixed, a Flowmeter or Flow Cone can be used. The Flowmeter measures the horizontal slump and the Flow Cone the time taken to pass a specific quantity through a standard orifice. Flowmeters are used to give an indication of pumpability rather than to register precise flow properties. A more accurate indication of grout viscosity can be obtained by using a Viscometer.

Pressure Relief Valves

When it is necessary to limit the injection pressure either to prevent uplift or control hydrofracture, pressure relief valves should be fitted to the injection banjo. The spring loaded diaphragm type must be properly maintained and regularly washed out to ensure satisfactory performance. It should be noted that as mentioned in the Pump Section hydraulically operated pumps can be set to stall at specific pressures by pre setting the hydraulic operating pressure. Special gauges are available to correlate hydraulic operating pressure and the corresponding grout injection pressure.

Continuous Recording Pressure Gauge Units

For critical Tube à Manchette injection operations or where it is necessary to record the injection sequence continuous recording pressure gauge units are recommended. Various pressure scales are available and the larger units operate over a period of 24 hours before the graph has to be changed. 8 hour scales are however preferred for most work.

Grout Recorder Sets

These units are designed to register and record flow and pressure. The flow circuit instrumentation comprises a magnetic flow meter and amplifier and the output signal is sent to the recording unit and to two impulse counters for summation. One counter is fitted with a zero adjustment. The pressure circuit consists of a transmitter with pressure unit and capillary tubing. The signal is displayed on an indicator and automatically recorded by pen on a graph of suitable scale. There is a free passage through the system of 20mm and since the measurements are made electronically there are no mechanical parts that can be contaminated by cement or chemicals. Viscosity and grout composition have no influence on the accuracy which is claimed to be ± 2% although at the extreme ends of the injection rate range accuracy could be ± 5%. Units are available to monitor and record single injection lines (Fig 22) but multi-hole units can also be supplied for 6 or 12 injection points (Fig 23).

Fig 22. Individual Hole Grout Recorder Set

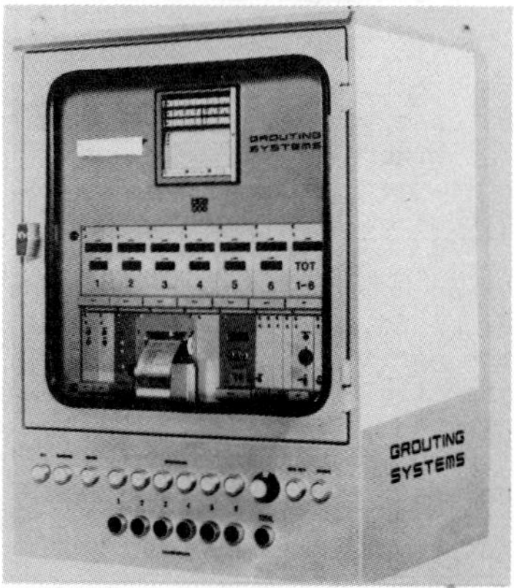

Fig 23. Multi Hole Grout Recorder Unit

Grout Level Indicators

Grout level indicators are used to monitor grout travel and can be used underwater or in the dry for such as grouted aggregate concrete projects or to check top grout level in ground anchorages when such control is necessary. Various different principles of operation can be used including electrical conductivity, temperature and photo electric cells.

Water Pressure Test Sets

Specially built units include pressure regulator and pressure gauge, volume recorder, filter and recirculation valve. They are used in conjunction with packers for permeability testing in rock and alluvium.

Grout Injection Banjos

Grout injection banjos comprise a top hole packer unit, pressure gauge, inlet and recirculation valves and either bypass valves or pressure relief valves to control pressure.

Top Hole Packers

These packers are normally hand operated mechanical units in which a rubber sleeve is expanded to fit inside a borehole or a standpipe to allow holes to be injected and sealed off under pressure. Standard units are available for 25, 38, 50, 75 and 100mm holes.

In-hole Packers

In-hole packers can be mechanically activated but more normally are inflated by compressed air for normal duty or hydraulically where high pressure injection is being carried out. Different lengths can be supplied depending on the pressures to be generated. Standard diameters range from 50mm to 150mm. Care should be taken to have the largest possible internal diameter of grout tube through the packer to prevent unnecessary restriction of grout flow and to minimise risk of blockage.

Double In-hole Packers

These units are used for permeability testing of isolated stages of a borehole.

While certain sizes and types of the above items of ancillary equipment can be regarded as normal stock items for the bigger grouting equipment manufacturers, special units often have to be made to accommodate specific project requirements. Given the very wide range of grouting applications and the correspondingly wide demands with respect to mix, output, pressure, depth and size of hole, it should be appreciated that there are very few ancillary units that can be called standard and a good manufacturer is always prepared to purpose build items to suit particular requirements.

REMEDIAL DRILLING AND GROUTING OF TWO ROCKFILL DAMS

Robert C. Fox [1]
and
Martin C. Jones [2]

ABSTRACT: Remedial drilling and grouting, utilizing a cement/bentonite/silicate grout, was used to seal off seepage through two rockfill dams in Eastern Canada. The dams are two of several which form a reservoir for power generation. The work was carried out with the reservoir in operation over two consecutive summer/fall periods. The cores of the dams were constructed of glacial till and, in both the drilling and grouting operations, great care was used to prevent hydro-fracturing of the embankment. The procedures, which were developed in the field between the Engineer and the Contractor, are described in detail. Information on initial and final seepage flows, number and depth of holes, grout takes and costs of the operations are given. This data indicates that there were isolated zones within the core where loss of material occurred prior to grouting. The reduction in seepage flows, sinkhole activitiy or surges of muddy water through the dam, indicates that the drilling and grouting programs were successful.

INTRODUCTION:

A year following impoundment of water, abnormal seepage developed in two rockfill dams on a large hydro-electric project in Eastern Canada. Sinkholes appeared on the crests and surges of seepage water occurred, sometimes carrying fines. This indicated that there had been a loss of finer size material from within the core zone of the embankment. The cause of migration of finer material was not known, nor was it known whether the seepage was through the embankment itself or the foundation rock or, possibly, a combination of both.

Both dams are rockfill, founded on metamorphic bedrock, and are similar in design. The core zones are composed of glacial till which is well graded, silty sand containing 9 to 20% passing the 200 sieve, with gravel and occasional boulders up to 12 inches (304.8 mm) in diameter, and compacted to an average density of 95% of the Standard Proctor Test. Each core is 12 feet (3.65 m) wide at the crest, and maximum dam heights are 67 and 86 feet (20.44 and 26.21 m). The outer shells of the dams are rock and transitions of processed sand and gravel are provided between the cores and shell zones, which are between 8 and 16 feet (2.44

---

[1] Senior Geologist, Bechtel Civil and Minerals Inc., San Francisco, CA

[2] Vice-President Ground Engineering, The Cementation Company (Canada) Limited, Brampton, Ontario.

and 4.88 m) wide. A section through Dam 'A' is shown in Figure 1, page 12.

Initially, a blanket of glacial till was placed against the upstream face of Dam 'A'. This blanket was not completely effective and at this juncture several other methods, such as a slurry trench through the core were considered. A grouting program was adopted to determine the locations of seepage and correct it. Reduction of seepage was desired to ensure the integrity of the structure, although the actual water loss from the reservoir was insignificant. Normal soils drilling and sampling was not acceptable because removal of any material from the core of the embankment, without simultaneous replacement, could cause further distress to the structure. In addition, should drilling cause problems, grouting equipment would be required immediately.

The first drilling and grouting program was at Dam 'A' as it was considered less risky due to the presence of the additional blanket. This program was designed to be experimental, so that while investigation and drilling and grouting to reduce seepage was carried out on Dam 'A', the program was also developing methods and procedures for grouting Dam 'B' during the following year.

The programs called for both dams to be drilled and grouted with the reservoir drawn down 8 feet (2.44 m) to the lowest level possible for power generation. This reduced the buoyancy of the dams, thus increasing the earth pressures within the dam core and, therefore, reduced the risk of hydro-fracturing of the dam during drilling and grouting operations.

Further precautions were taken to safeguard the dams during the grouting operation by channelling the seepage water through weirs which were fitted with electronic alarms which would sound in the powerhouse control room should seepage flows increase appreciably. As a further precaution large earthmoving equipment and operators were kept on standby on the dam during the grouting operations, in case an emergency should occur.

Due to the experimental and investigatory nature of the drilling and grouting program, procedures were based on guidelines, rather than rigid specifications, so that flexibility could be maintained in the field. Thus, the Owner and Engineer laid particular emphasis on obtaining a competent grouting contractor, with qualified personnel, who could work together with them to produce these guidelines.

The contract was based on labour costs and equipment rentals, and all materials, transportation and support facilities were provided by the Owner. Once the general drilling and grouting guidelines had been developed, the day-to-day field direction became the responsibility of the Engineer, with one inspector directing each drill and grout operation.

DRILLING PROCEDURES:

(a) Casing Through the Core of the Dam:

Drilling through the core of the dam was performed using a modified

OD72 system developed by Atlas Copco. This system uses a 4.06 inch (103.19 mm) outside diameter heavy wall casing equipped with a carbide insert ring bit. A percussion bit 2.75 inches (69.85 mm) in diameter, mounted on 1.50 inch (38.10 mm) diameter R38 drill rods is normally used inside this casing and the casing and percussion bit are advanced together or separately, using separate top rotation and percussion drives. The percussion and rotation were provided by two Atlas Copco BBE 57 air driven rock drills on Dam 'A'. On Dam 'B' one of these machines was modified to a RoPerc A55 so that hydraulic rotation was used with air percussion. These drills were on Atlas Copco ROC 600 and ROC 601 crawler mounted air tracks with heavy duty 22 foot (6.71 m) masts for Dam 'A'. For Dam 'B' a MK IV Acker all hydraulic crawler mounted drill carriage was used instead of the Atlas Copco ROC 601.

In normal usage, the OD72 system uses air, water or drilling mud for the drilling medium and this returns to the surface around the outside of the casing. This was not acceptable in this work because any blockage of the passage to the surface could have resulted in undesirably high pressures being produced within the embankment core. To prevent this, the casing and ring bit were advanced using no circulation medium except in the bedrock. The core material was forced into the casing during this operation and formed a plug. In most cases, the casing could be driven and rotated to bedrock without cleaning out this plug. In a few cases, it was necessary to partially clear out the plug by using the inner rods and bit within the casing by the method described below.

Once the ring bit was seated into the bedrock, the casing was cleaned out using flushing air through the drill rods and the ports in the inside cross bit which carried the cuttings inside the casing to the surface. The cuttings were finally ejected through the normal air inlet to the casing.

Both vertical and inclined cased holes were advanced to bedrock using these methods, during the work on both Dams 'A' and 'B', to a maximum vertical depth of 85 feet (25.93 m), and only a very few of these holes had to be abandoned because of inability to reach bedrock.

(b) <u>Drilling in the Bedrock Foundation of the Dam:</u>

An air driven Boyles VGR diamond drill, mounted on an A-frame was used on Dam 'A', and a skid mounted, diesel-driven Longyear 34 on Dam 'B', for core drilling the bedrock. Initially, holes were advanced in bedrock with BW rods and BXL core barrels. Excessive vibration of the BW rods within the OD72 casing produced highly fractured, broken core which did not reflect the true nature of the bedrock. Subsequently, AW casing was placed within the OD72 casing and AW rods and AXL core barrels were used. This produced a representative core, although smaller than originally required. Coring was only carried out in the initial holes, or when necessary to check an area of bedrock. Other holes in rock were advanced by standard percussion drilling methods; however, it was felt that the most useful information, as to the character of the rock, was obtained by water testing.

During operations on Dam 'A', it was necessary to enclose the core drill and grout plants because of cold weather. It was not practical to enclose the air tracks, although some shelter was provided for the driller and helpers, and production was adversely affected towards the latter part of the contract because of these conditions.

GROUTING PROCEDURES:

It was not known, prior to grouting, what type of grout would be required to seal off the seepage in the dams, but a hard rigid mass or column of grout was not desirable. Grout plants and materials were provided for cement, bentonite, silicate grouts and combinations of these.

Each of the two grout plants were fully enclosed because winter conditions were expected and each was operated by electric power provided by the Owner.

Electric powered, high speed colloidal mixers were used, together with air operated holding tanks which transferred to electric powered 3L8 Moyno grout pumps for injection in the grout hole.

After some initial experimentation, it was found that a cement/bentonite/silicate grout, of the following typical proportions, was best suited for the grouting of the core of the dam.

Cement:	Type 30
Bentonite:	200# mesh, 20% by weight of Cement
Sodium Silicate:	Type O or N solution, 5% by weight of Cement
Water/Cement Ratio by Volume:	8:1 to 2:1 (varied)

The mix for grouting in rock had the following proportions:

Cement:	Type 30
Bentonite:	200# mesh, 3.4% by weight of Cement
Water/Cement Ratio by Volume:	6.3:1

Generally, the water/cement ratio was varied, based on the characteristics of the stage of the hole being grouted but, usually, holes either accepted grout freely or refused. In most cases, refusal was achieved with the thickest mix but, in two cases, sand was added to achieve refusal.

All holes in rock were water tested so that seepage in the bedrock could be determined. A packer was set in the rock, just below the bottom of the casing, and the rock was tested with water at a pressure of 0.5 psi per foot (11.31 $kN/m^2/m$) of vertical depth to the packer. The takes were less than 1 Imp. gal/min (0.0047 m^3/min) in all cases, and there was no need to carry out testing of shorter intervals in the rock. Once water testing of each hole was completed, the bedrock was grouted with cement/bentonite at the same pressure used during testing. Due to the low grout takes it was not necessary to reduce the water/cement ratio in the rock.

Once the grouting of the bedrock in each hole had been completed, the casing was filled, using the mix stated above. A funnel was attached to the top of the casing and the grout level was, generally, about 2 feet (0.61 m) above the top of the dam. In this way no pressure, above the static head, could be applied to the core of the dam. The casing was first retracted a short distance above the bedrock and grout allowed to flow until refusal was achieved, which was defined as a take of less than 1 Imp. gallon (0.0047 m^3) of grout in 2 minutes. The casing was then slowly retracted, about 10 feet (3.05 m), dictated by the length between the casing joints, and grouting continued in the same manner until the hole was filled.

DRILLING AND GROUTING OF DAM 'A':

Dam 'A' had a crest length of 1,355 feet (413.28 m) between Stations 9+25 and 22+80 and a maximum height of 67 feet (20.44 m). Two areas for drilling and grouting were selected on this dam, where sinkholes on the crest had been observed. A third area was added when a test hole accepted appreciable grout.

(a) South Section: Stations 20+05 to 22+05:

Initially, holes were drilled vertically on 20 foot (6.1 m) centres to bedrock at a depth of 28 and 63 feet (8.54 and 19.22 m). These holes were grouted with particular attention being given to the interface between the embankment and foundation. Practically no grout was injected at this interface. Consequently, holes were core drilled between the first holes, nominally 30 feet (9.1 m) into bedrock, and were grouted. Later, because of low grout takes, lack of useful information obtained from the core and the need to expedite the work, the core drilling was replaced by percussion drilling.

In this section, 39 holes were drilled in the core of the dam. Only 12 stages, of 10 feet (3.05 m) each in 7 holes, accepted an appreciable grout take ranging from 25 to 172 sacks of cement, and only one of these was near the interface with the foundation bed rock. When intermediate split spaced holes were drilled, immediately adjacent to those holes which did accept appreciably, they reached refusal rapidly.

Two angled holes were also drilled to bedrock, and then cored beneath the area of the highest grout take, to intersect near vertical jointing in this area, and these holes were found to be tight.

(b) North Section: Stations 16+50 to 17+55:

This section was grouted in a similar manner to the South Section, with final spacing at 10 feet (3.05 m) and depth in the core between 33 and 72 feet (10.07 and 21.96 m). The holes that were advanced into bedrock were drilled with percussion methods and extended 30 ft. (9.15 m) into the rock. Holes accepting appreciable grout were closed out by the split spacing method.

(c) Centre Section: Stations 18+35 to 19+05:

This section was not scheduled to be grouted originally. A hole was drilled at Station 18+95 to record temperatures within the core at various depths, and it was decided to grout this with chemical grout to determine grout penetration in a supposedly tight area. When the bottom 14 feet (4.27 m) of this hole readily accepted 1,800 Imp. gallons (8.45 m^3) of silicate grout, it was apparent that grouting would also be required in this section of the dam. Grouting of the hole continued, with cement/bentonite/silicate, and 113 sacks of cement were placed.

Check holes were drilled and grouted 10 feet (3.05 m) either side of this hole and the procedure repeated until the end holes in the sequence were tight.

(d) Seepage Flows:

Readings were taken prior to, and during, the period of the drilling and grouting on Dam 'A' at a weir placed downstream of the three areas of seepage. The readings, prior to grouting, indicated flows of between 240 and 290 Imp. gal/min (1.31 and 1.58 m^3/min), which had dropped to 30 to 35 Imp. gal/min (0.16 to 0.19 m^3/min) by the completion of the grouting.

(e) Program and Costs:

This program was started in the fall and planned on a six day week, 10 hours per day basis. The week was extended to a 7 day basis, with extended shifts while grouting, as it was necessary to raise the reservoir level to avoid formation of unfavourable ice conditions in the reservoir during winter.

Fifty-four days were required, from the start of drilling to completion of grouting, on Dam 'A'. The Contractor's personnel included one superintendent, 6 skilled operators and 5 labourers, while the Engineer maintained a site manager, a technical advisor and 3 inspectors.

The total cost of the program for Dam 'A' was about $637,900. Forty percent was direct costs for the Contractor's supply of labour and plant; thirty percent was the Owner's costs for site preparation and maintenance, supplies, transportation and site accommodation for the crews; and thirty percent was for design and engineering costs, including field inspection and management.

DRILLING AND GROUTING OF DAM 'B':

Dam 'B', located adjacent to the intake structure, was grouted the following year. This dam is 6,065 feet (1849.83 m) long, between Stations 9+50 and 70+15, with a maximum height of 86 feet (24.21 m). The dam and foundation are similar to Dam 'A' except shear zones, and zones of closely spaced joints, were mapped in the bedrock during construction. Dam 'B' also did not have a glacial till blanket upstream. Substantial seepage had been observed previously, at four locations along the dam, and weirs were installed to measure the surface flows. Seepage occurred near Station 46+55, where there is an outlet conduit which was used for diversion during construction, and near Station 39+00 where there is a steep bedrock face nearly normal to the dam axis. Appreciable seepage also occurred near Station 23+02 that could not be related to any known foundation feature, and near Station 12+25, adjacent to the intake structure, where significant shear zones were mapped. Prior to grouting it was not known if seepage flows were related to these foundation features.

The drilling procedure was similar to that used at Dam 'A', except the ROC 600 was replaced by an Acker MK IV all hydraulic, tracked drill fitted with an A55 RoPerc hydraulic rotary and BBE 57-01 air powered percussion drill. In some cases, refusal was reached before bedrock was encountered. Refusal was reached when no rotation or penetration was possible due to the denseness of the material and/or friction on the walls of the casing. In these cases, the holes were terminated and grouted. The casings were cleaned by drilling and flushing, using grout for circulation. The grout column, inside the casing, was maintained at the top of the dam at all times to create a positive head within the casing. The holes were then grouted, using the same procedure as at Dam 'A'.

(a) Conduit Area: Stations 45+93 to 47+62

Grouting started in the Conduit Area. Initially, holes were drilled some distance away from the conduit, to familiarize the crew with the equipment and procedures. Care was taken not to penetrate the conduit with any drill hole.

The first hole, located 77 feet (23.49 m) north of the conduit, accepted 280 sacks of cement at 20 to 35 feet (6.10 to 10.68 m) depth; bedrock was over 80 feet (24.40 m) deep. Small quantities of grout were injected in the next five holes. The seventh hole, drilled directly over the top of the conduit, accepted 745 sacks of cement and 246 cubic feet (6.89 m^3) of sand from 0 to 35 feet (0 to 10.68 m) above the conduit. Another hole, located 5 feet (1.53 m) north of the conduit, accepted 84 sacks. The only other appreciable grout take was 15 sacks in the bedrock immediately south of the conduit.

Seepage measured over the weir in this area, with full reservoir head, was reduced from 450 to 575 Imp. gal/min (2.45 to 3.14 m^3/min) to 42 to 46 Imp. gal/min (0.23 to 0.25 m^3/min). A tabulation of the reduction in flows over the weirs at Dam 'B' is presented in Table 1.

FIGURE 1: A TYPICAL SECTION THROUGH DAM 'A'

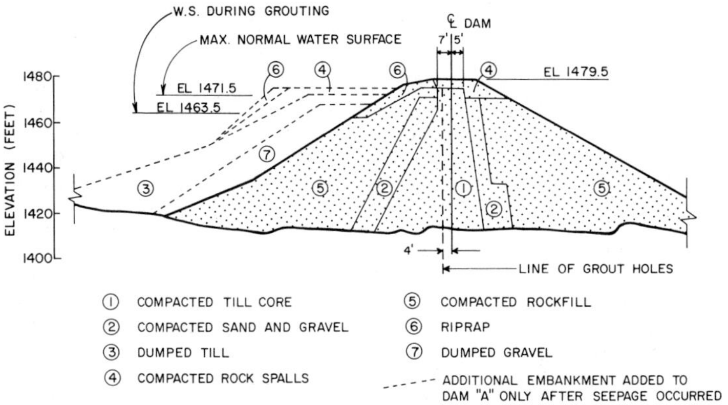

Dam 'B' was similar except for the additional embankment. (Conversion: 1 foot = 0.305 metres).

TABLE 1: SEEPAGE FLOWS DAM 'B'

		FOREBAY 8 FEET BELOW NORMAL OPERATING LEVEL		FOREBAY AT NORMAL OPERATING LEVEL		
AREA	WEIR	PRIOR TO GROUTING	AFTER GROUTING	PRIOR TO GROUTING	AFTER GROUTING	REDUCTION IN FLOW
Station 14+00	14+00	760 (4.15)	540 (2.95)	1030 - 1160 (5.62-6.33)	760 - 790 (4.15-4.31)	26 - 31%
12+25	12+25	630 (3.44)	400 (2.18)	*	690 (3.76)	-
Station 23+02	23+02	240 (1.31)	10 (0.05)	300 - 390 (1.64-2.13)	10 - 12 (0.05-0.06)	97%
Station 39+00	38+50	300 (1.64)	16 (0.09)	290 - 375 (1.58-2.05)	20 (0.11)	96 - 97%
Conduit Area	46+42	495 (2.70)	35 (0.19)	450 - 575 (2.45-3.14)	42 - 46 (0.23-0.25)	91 - 92%

* The weir, at Station 12+25, was not functioning prior to grouting.

Listed above are the seepage flows prior to, and after, grouting the four main seepage areas at Dam 'B'. Figures are in Imperial gals/min. with cubic meters per minute in brackets.

(b) Station 39+00 Area: Between Stations 38+21 and 39+65

The second area to be grouted was at Station 39+00. Eighteen grout holes in this area were inclined 60 degrees from horizontal, parallel to the dam axis, to intersect the steeply dipping face in the bedrock. The first hole penetrated the rock near the base of the steep face, at a vertical depth of 81 feet (24.71 m). All circulation was lost 10 feet (3.05 m) into rock. The hole was grouted and accepted 110 sacks of cement in the rock. An adjacent hole, located 10 feet (3.05 m) away and penetrating rock near the top of the steep face at a vertical depth of 68 feet (21.0 m), also lost circulation. Ninety-eight sacks were placed in the bedrock, and 230 sacks were placed at the foundation interface and 501 sacks with 250 ft^3 (7.00 m^3) of sand, were placed at a vertical depth of 40 to 57 feet (12.20 to 17.39 m) in this hole. Two straddle holes, on either side, took 45 and 37 sacks in the bedrock. The remainder of the holes were tight. Seepage was reduced from between 290 to 375 Imp. gal/min (1.58 to 2.04 m^3/min) to 20 Imp. gal/min (0.11 m^3/min).

(c) Station 12+25 Area: Between Stations 9+71 and 15+00

The third area grouted was near Station 12+25, adjacent to the intake structure. Seepage was greatest in this area; however, it was believed most seepage was through the rock. Sixty-five vertical holes were drilled, generally spaced 10 feet (3.5 m) apart, 529 feet (161.21 m) along the crest. The depth to bedrock varied between 30 and 58 feet (9.15 to 17.69 m) and holes penetrated 20 feet (6.1 m) into rock. In the rock six widely spaced holes accepted between 11 to 62 sacks of cement and only one of these had an appreciable effect on seepage flow. Three holes accepted between 11 to 44 sacks at the foundation interface and five widely spaced holes accepted 12, 72, 114, 117 and 167 sacks at various elevations in the embankment. The remainder of the holes were tight.

Two weirs had been established in this area. The station 14+00 weir was located some distance downstream from the dam and collected the seepage water from the entire station 12+25 area, including water from the 12+25 weir. The Station 12+25 weir was near the downstream toe and collected most, but not all, of the seepage water from this area. The weir at Station 12+25 was not functioning prior to the lowering of the reservoir. Flows over a weir, at Station 14+00, were 1,030 to 1,160 Imp. gal/min (5.62 to 6.33 m^3/min) prior to grouting, and 760 to 790 Imp. gal/min (4.14 to 4.31 m^3/min) after grouting. With the reservoir drawn down, flow over the 12+25 weir dropped from 630 to 400 Imp. gal/min (3.45 to 2.18 m^3/min).

(d) Station 23+02 Area: Between Stations 21+80 and 22+60

This area was designated by the location of the weir and no grouting was actually required near Station 23+02. In the Station 23+02 area, there was no significant foundation feature and no indication as to where the seepage was occurring. The limits were assumed to be the high points in the foundation, about 1,000 feet (305.0 m) apart. A streaming potential survey was performed, to more accurately define the seepage area. This survey measures the natural electrical potential between two electrodes in contact with the ground. Significant anomalies were found along a 150 foot (45.75 m) length of the crest. Grouting was, therefore, started in this area.

Twelve vertical holes, totalling 778 lineal feet (237.29 m), were grouted. Two holes, spaced 10 feet (3.05 m) apart, accepted a total of 926 sacks. The remaining holes accepted only 11 sacks. Holes were drilled 20 feet (6.10 m) deep into rock and no grout was placed in the rock. Flows, over the weir in this area, dropped from between 300 to 390 Imp. gal/min (1.64 to 2.13 m^3/min) to between 10 to 12 Imp. gal/min (0.05 to 0.06 m^3/min).

(e) Program and Costs:

The initial grouting program consisted of only two sections; the Conduit Area and Station 39+00 Area. It was estimated that this program would require 78 working shifts. Work was performed on a 10 hours per day, 6 days per week basis, with extended shifts as dictated by the grouting operation. The first two sections were completed after working only 35 days, mainly due to reduction in the number of holes required. The program was extended to the two other sections, and the entire program required 63 working days. The Contractor's personnel consisted of one superintendent, 5 skilled operators and 5 labourers. The Engineer maintained a site manager, who was also the technical advisor, an assistant site manager and 3 inspectors.

Total cost of the program at Dam 'B' was about $799,600. Fifty-one percent was direct Contractor's cost, for the supply of labour and plant; twenty-six percent was direct Owner's costs for site preparation, maintenance, supplies, transportation and site accommodation for the crews; twenty-three percent was for design and engineering costs including field inspection and management. This cost was 20 percent below the Engineer's estimate and was due, mainly, to completing the work ahead of schedule.

OBSERVATIONS ON DRILLING, GROUTING AND SEEPAGE DATA FOR DAMS 'A' and 'B'

Tables 2 and 4, on pages 13 and 15, on Drilling and Grouting Data for Dams 'A' and 'B', are based on overall figures for the work and contain the quantities of drilling in the embankment and rock foundation, the number of holes, the number of stages and the total grout takes in each area. The average grout take per stage does not give a true guide to the variation of grout take in each hole nor, indeed, across each area in the two dams.

Tables 3 and 5, on pages 14 and 16, on Drilling and Grouting Data for Dams 'A' and 'B', are based on selected data. Only holes that had a grout take, in any stage, of 10 sacks or more have been recorded and the average grout take per stage then computed. This gives a clearer picture of the variation of grout take in each area.

TABLE 2: SUMMARY OF DRILLING AND GROUTING DATA FOR DAM 'A' USING OVERALL DATA

SECTION	FOOTAGE DRILLED	NO OF HOLES	STAGES	TOTAL GROUT TAKE	AVERAGE GROUT TAKE PER STAGE
SOUTH: 20+05 - 22+05 200 feet long (61.00 m)					
28 - 63 feet deep (8.54 - 19.22 m)	Core 2,007 ft (612.14 m)	39	200 x 10 feet (3.05 m)	842 sacks	4.2 sacks/stage
30 - 38 feet deep (9.15 - 11.59 m)	Rock 425 ft (129.64 m)	13	9 x 33 ft av (10.07 m)	18 sacks	2.0 sacks/stage
NORTH: 16+50 - 17+55 105 feet long (32.05 m)					
33 - 72 feet deep (10.07 - 21.96 m)	Core 2,584 ft (788.12 m)	43	258 x 10 feet (3.05 m)	810 sacks	3.1 sacks/stage
30 - 37 feet deep (9.15 - 11.28 m)	Rock 296 ft (90.28 m)	9	9 x 33 ft av (10.07 m)	8 sacks	0.9 sacks/stage
CENTRE: 18+35 - 19+55 120 feet long (36.60 m)					
62 - 72 feet deep (18.91 - 21.96 m)	Core 676 ft (206.18 m)	10	68 x 10 feet (3.05 m)	637 sacks	9.4 sacks/stage
	Rock --	--	--	--	--

TABLE 3: SUMMARY OF DRILLING AND GROUTING DATA FOR DAM 'A' USING SELECTED DATA

SECTION		NO OF HOLES	STAGES OF 10 SACKS OR MORE	TOTAL GROUT TAKE	AVERAGE GROUT TAKE PER STAGE
SOUTH: 20+05 - 22+05 200 feet long (61.00 m)					
28 - 63 ft deep (8.54 - 19.22 m)	Core	7	12	784 sacks	65.3 sacks/stage
30 - 38 ft deep (9.15 - 11.59 m)	Rock	1	1	12 sacks	12.0 sacks/stage
NORTH: 16+50 - 17+55 105 feet long (32.03 m)					
33 - 72 ft deep (10.07 - 21.96 m)	Core	11	15	728 sacks	48.5 sacks/stage
30 - 37 ft deep (9.15 - 11.29 m)	Rock	--	--	--	--
CENTRE: 18+35 - 19+55 120 feet long (36.60 m)					
62 - 71 ft deep (18.91 - 21.66 m)	Core	6	11	601 sacks	54.6 sacks/stage
	Rock	--	--	--	--

TABLE 4: SUMMARY OF DRILLING AND GROUTING DATA FOR DAM 'B' USING OVERALL DATA

SECTION	FOOTAGE DRILLED	NO OF HOLES	STAGES	TOTAL GROUT TAKE	AVERAGE GROUT TAKE PER STAGE
CONDUIT AREA: 45+93 - 47+62 169 feet long (51.55 m)					
36 - 85 feet deep (10.98 - 25.93 m)	Core 1,780 ft (542.80 m)	28	178 x 10 feet (3.05 m)	1,434 sacks	8.1 sacks/stage
34 - 40 feet deep (10.37 - 12.20 m)	Rock 480 ft (146.40 m)	14	14 x 34 ft av (10.37 m)	19 sacks	1.4 sacks/stage
STATION 39+00 AREA: 38+21 - 39+65 144 feet long (43.92 m)					
40 - 130 feet deep (12.20 - 39.65 m)	Core 1,307 ft (398.65 m)	18	131 x 10 feet (3.05 m)	1,033 sacks	7.9 sacks/stage
10 - 40 feet deep (3.05 - 12.20 m)	Rock 389 ft (118.65 m)	14	14 x 28 ft av (8.54 m)	296 sacks	21.1 sacks/stage
STATION 12+25 AREA: 9+71 - 15+00 529 feet long (161.35 m)					
21 - 55 feet deep (6.51 - 16.78 m)	Core 2,708 ft (825.94 m)	65	271 x 10 feet (3.05 m)	653 sacks	2.4 sacks/stage
20 - 33 feet deep (6.10 - 10.07 m)	Rock 1,209 ft (368.75 m)	59	59 x 20 ft av (6.10 m)	217 sacks	3.7 sacks/stage
STATION 23+02 AREA: 21+80 - 22+60 80 feet long (24.40 m)					
41 - 46 feet deep (12.51 - 14.03 m)	Core 538 ft (164.09 m)	12	54 x 10 feet (3.05 m)	937 sacks	17.4 sacks/stage
20 feet (6.10 m)	Rock 240 ft (73.20 m)	12	12 x 20 feet (6.10 m)	--	--

TABLE 5: SUMMARY OF DRILLING AND GROUTING DATA FOR DAM 'B' USING SELECTED DATA

SECTION		NO OF HOLES	STAGES OF 10 SACKS OR MORE	TOTAL GROUT TAKE	AVERAGE GROUT TAKE PER STAGE
CONDUIT AREA: 45+93 - 47+62 169 feet long (51.55 m)					
36 - 85 ft deep (10.98 - 25.93 m)	Core	3	7	1,357 sacks	193.9 sacks/stage
34 - 40 ft deep (10.37 - 12.20 m)	Rock	1	1	15 sacks	15.0 sacks/stage
STATION 39+00 AREA: 38+21 - 39+65 144 feet long (43.92 m)					
40 - 103 ft deep (12.20 - 31.42 m)	Core	1	3	981 sacks	327.0 sacks/stage
10 - 40 ft deep (3.05 - 12.20 m)	Rock	3	3	290 sacks	96.7 sacks/stage
STATION 12+25 AREA: 9+71 - 15+00 529 feet long (161.35 m)					
21 - 55 ft deep (6.41 - 16.78 m)	Core	9	9	575 sacks	63.9 sacks/stage
20 - 33 ft deep (6.10 - 10.07 m)	Rock	5	5	185 sacks	37.0 sacks/stage
STATION 23+02 AREA: 21+80 - 22+60 80 feet long (24.40 m)					
41 - 46 ft deep (12.51 - 14.03 m)	Core	2	4	926 sacks	231.5 sacks/stage
20 feet (6.10 m)	Rock	--	--	--	--

In Dam 'A', the average grout take per stage (which took 10 sacks or over in the core) is somewhat similar between areas; ranging from 65 sacks per stage in the South Section, to 48 sacks per stage in the North Section, to 55 sacks per stage in the Centre Section. The maximum take per stage was 172 sacks in the South Section, 157 sacks in the North Section and 87 sacks in the Centre Section. Grout take in the rock was not significant.

In contrast, on Dam 'B' the average take per stage (which took 10 sacks or more) varied in each area from 327 in the Station 39+00 Area, to 64 in the Station 12+25 Area, with 231 in the Station 23+02 Area and 194 in the Conduit Area. Maximum takes ranged from 905 to 167 sacks per stage. Also, there were significant takes in the rock, varying from 15 sacks per stage to 97 sacks per stage.

It is suspected that much of the apparent take in the core escaped into the shell of the dam. Appreciable grout was observed flowing from the downstream toe during the grouting of some holes where high takes occurred; however, the data presented does indicate that grout takes varied considerably in different areas of the two dams and at different levels in the same hole. This indicates that, although finer material was deficient in the core, the areas were not necessarily interconnected and were of local extent only. This is also indicated by the very small percentage of holes which actually took significant grout and the fact that holes, either side of significant takes, were tight. Furthermore, there does not seem to be any correlation between levels in holes that took grout.

Table 1, on page 12 shows that in each area seepage was reduced considerably; in Dam 'A' by, approximately, 88% and on Dam 'B', in three areas, by over 90%. In only one area, Station 12+25 Area, was the reduction of a lower order and this was between 26% and 31%. The flows in this area were much higher prior to grouting, and it is thought that the majority of the seepage was due to water flowing through the rock. This was not considered a hazard to the dam. Grouting was carried out only to ensure that seepage was not through the core and to seal off leakage in the core and interface of the rock foundation.

POST GROUTING HISTORY:

The grouting program was completed in the late summer of 1979. The flows at both dams remained fairly constant until February, 1981. At that time, seepage at Dam 'A' increased to a maximum of about 290 Imp. gal/min (1.23 m^3/min) and some fines were observed in the flow. A month later, the seepage had decreased to 140 to 150 Imp. gal/min (0.66 - 0.71 m^3/min) and no fines were observed. The flows have remained fairly constant since that time. It is believed, based on geophysical data, that this seepage was around and not through the grouted areas. At Dam 'B' there has been a slight increase in seepage in the Station 39+00 Area and 15 to 20 Imp. gal/min (0.06 to 0.09 m^3/min) increase in the Station 12+25 Area. Seepage in the other areas has remained nearly constant.

The Owner maintains a surveillance program of their dams, and both Dams are inspected at least once a week. When changes in seepage rates occur at any of the dams, it is inspected several times a day. Currently, both dams are inspected once a day.

CONCLUSIONS:

These two drilling and grouting programs were successful in restoring, at least partially, portions of the core zones where high seepage flows had occurred, as evidenced by the substantial reduction in seepage flows in all grouted areas. This work was carried out under minimum operating reservoir head, and required a good understanding of grouting principles and embankment design, and close co-operation between the Engineer, Contractor and Owner during the design and field operations.

The manner in which the grout take varied, in the core, indicated that there were zones which were deficient in finer material prior to grouting, even though a considerable part of the core did not accept grout. The cause of the deficiency was not investigated, and is not within the scope of this paper.

The amount of grout, retained in the core zone of the dam, is not known and could not be estimated. In general, the majority of stages of each hole accepted only enough grout to fill the holes, indicating that the core material would not accept a particulate grout which was to be expected. In some cases, grout flowed from the downstream toe of the dam, indicating that grout penetrated from the line of grout holes to the downstream edge of the core zone.

The grouting, performed on this project, demonstrates that the till core of rockfill dams, of this type, can be successfully drilled and grouted, to reduce seepage flows under minimum reservoir operating water levels. Great care must be taken, however, to minimize any possible distress to the structure while the drilling and grouting operation is taking place, and in the design of grout mixes which are compatible with the structure.

SPECIAL GROUTING AT TARBELA DAM PROJECT

John Lowe, III,[1] F. ASCE and T.C. Sandford,[2] M. ASCE

Abstract

Two types of special grouting were performed at the Tarbela Dam Project which are different from the curtain grouting and consolidation grouting usually used in connection with hydro projects. Subsurface exploratory grouting with grout containing alluvial sand was carried out to determine whether the size of voids in openwork gravel, indicated to exist in the deep alluvial foundation of the Main Embankment Dam, was sufficiently large to permit ingress of fine sand sizes from adjacent sand beds or from adjacent sand choked cobble gravel. The paper describes the grouting technique used and presents results of the exploratory grout program. The second special type of grouting consists of the grouting of long water passageways of relatively small cross-sectional area in karstic limestone containing erodible materials. The grout was designed to travel as far as possible along such encountered passageways rather then to create a narrow curtain. The paper describes this grouting, which was carried out in the foundation of the Auxiliary Spillway during the 80/81 construction season.

Introduction

Tarbela Dam Project is a major irrigation storage and hydroelectric project on the Indus River in northern Pakistan. The major structures (Figure 1) include the Main Embankment Dam and two auxiliary dams, two spillways, five tunnels and a powerhouse (Lovell et al, 1972). The structures are among the largest in the world with the Main Embankment Dam being the world's largest dam. Between the left and right rock abutments, the Main Embankment Dam rests on alluvium of variable thickness with a maximum depth of 756 ft. (230 m) to rock at the right hand side of the valley (Figure 2). To minimize the underseepage losses and foundation gradients, an upstream impervious blanket which ties into the main dam inclined core was placed on the alluvium.

Various kinds of grouting were carried out for the project: curtain, consolidation, exploratory, and recently grouting of long karstic passageways. Single and in places triple line grout curtains

[1]Partner, Tippetts-Abbett-McCarthy-Stratton, New York, NY
[2]Former Senior Geotechnical Engineer, Tippetts-Abbett-McCarthy-Stratton, New York, NY

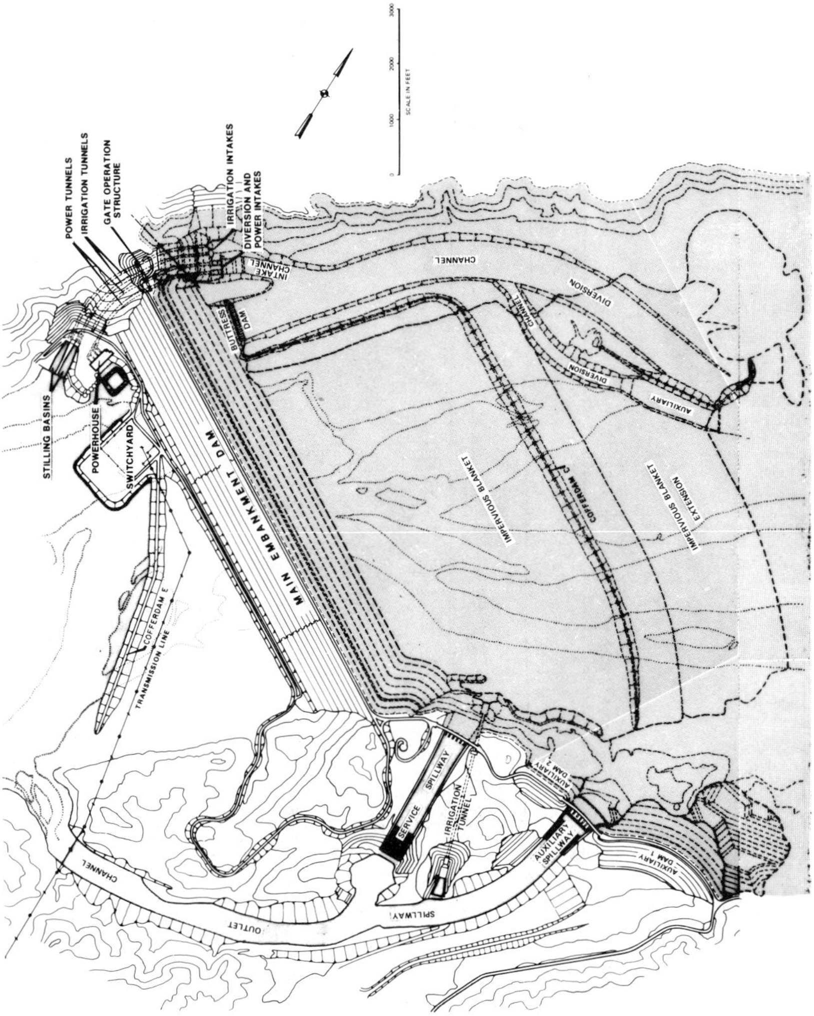

FIGURE I GENERAL PROJECT PLAN

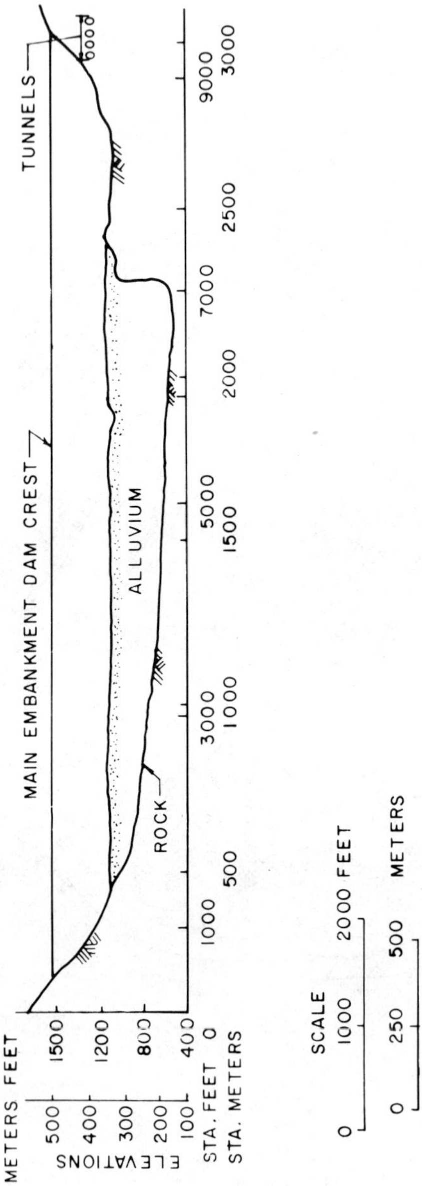

FIGURE 2 SECTION AT MAIN EMBANKMENT DAM

were installed in the relatively poor abutment rock at all core contacts for the Main Embankment Dam, Auxiliary Dam 1 and Auxiliary Dam 2 to fill possible preferential flow paths to a depth equal to half the net head. Grout curtains backed by drainage curtains were constructed beneath the two spillways and into the left abutment and far into the right abutment to protect the stability of downstream excavation slopes and structures by limiting piezometric pressures. Upstream of the Auxiliary Spillway a triple line grout curtain was installed in rock and tied into an upstream blanket overlying the rock and into the ends of the grout curtain of the adjacent Auxiliary Dam 1 and Auxiliary Dam 2. This triple line curtain was designed to lengthen the seepage path in the rock formation underneath the Auxiliary Spillway and thus reduce the gradient in beds of fine materials within this rock. Consolidation grouting was done under the Auxiliary Spillway, Service Spillway, Buttress Dam and around all the tunnels to minimize the possiblility of differential movement under loading. Exploratory grouting in the foundation alluvium was conducted to ascertain the properties of the alluvium to accept fine alluvial sand. Grouting of long water passageways of relatively small cross-sectional area in karstic limestone is presently being done at the Auxiliary Spillway approach area. An approximate total of 980,000 ft (229,000 m) of grout holes were drilled and 1,500,000 cubic feet (42,000 m^3) of grout were injected for the grouting at Tarbela Dam excluding the present limestone grouting at the Auxiliary Spillway. This paper describes the exploratory grouting in the alluvium and the grouting of passageways in the limestone.

Subsurface Exploratory Grouting

Grouting, whether impervious curtain or consolidation type, normally combines exploration with the grouting operations. The technique of split spacing which depends upon the results of grout takes in adjacent holes or modification of grout mixes during production utilizes the knowledge of the subsurface gained during the prior holes to modify the later operations. Except for this use of grouting results, grouting in the past has rarely been used strictly as a subsurface exploratory tool. Subsurface exploration tools have developed to primarily obtain estimates of compressibility, strength and permeability by means of classification, field tests and sampling for laboratory testing. Grouting will not improve our knowledge of these properties obtained from other exploration methods, but for the Tarbela Dam foundation another property of the soil, i.e. the capacity of voids in cobble-gravel to accept alluvial sand, required definition for which grouting could provide some fundamental information.

A variety of subsurface exploration methods were employed to investigate the alluvial material filling the Indus Valley at the site (Figure 3). The alluvium is composed of essentially two components: fine sand; and cobble-gravel. The cobble-gravel component consists of hard granite and quartzite particles which are not only difficult to drill through, but which make undisturbed sampling impossible. Drive samples and undisturbed samples could be obtained only in sand beds and lenses having no gravel or cobble particles.

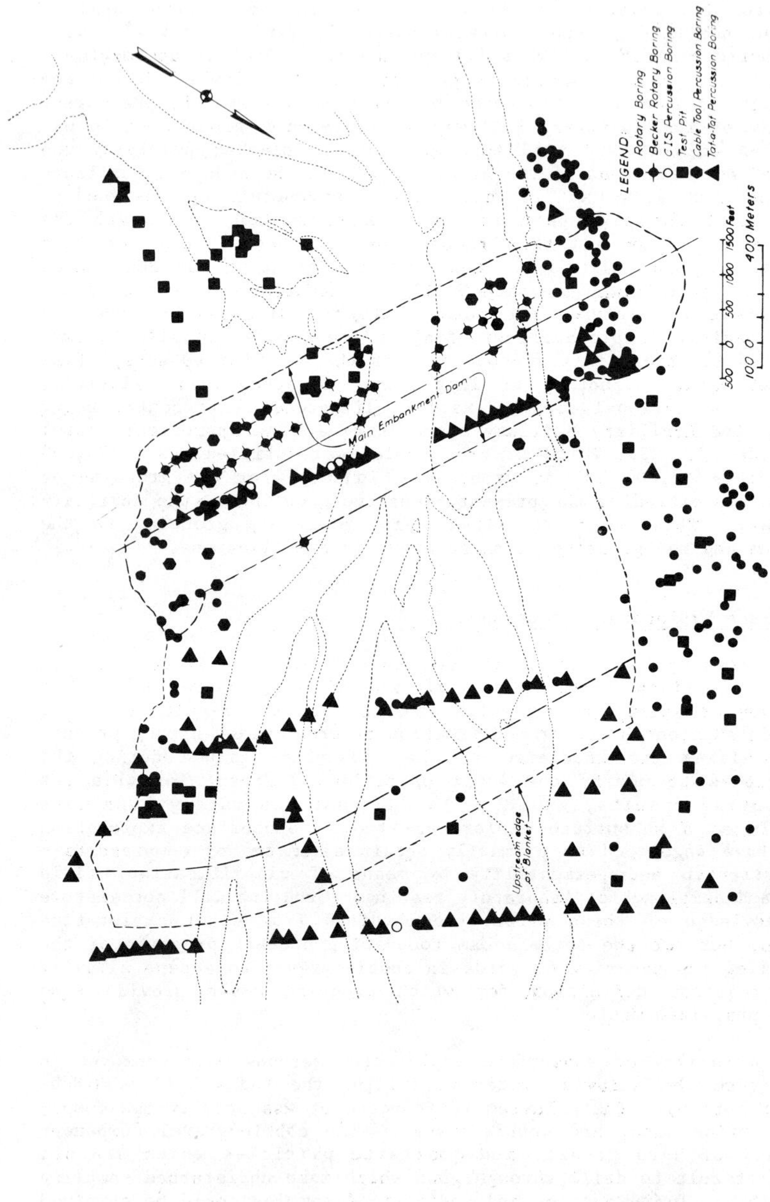

FIGURE 3 PRECONSTRUCTION SUBSURFACE EXPLORATIONS

The first subsurface explorations during the design stage were of three types: borings made with cable tool well drilling rigs, a type of wash boring termed tat-a-tat, and test pits. Subsequently, CIS (Circulation Inverse Sondage) borings and a few rotary borings were made using a tricone bit where gravel and cobbles were present and using a Denison double-tube core barrel to obtain samples from sandbeds. At the end of the program, a few Becker double-wall casing drive borings were made and shallow test pits were constructed into the alluvium. A description of each is given by Lowe (1978).

All overburden materials encountered in the borings were alluvial. Tests of samples from the shallow test pits in the valley bottom showed gradation limits of the alluvium as given on Figure 4. Gradations for three materials are shown: for openwork cobble gravel, cobble gravel choked with fine sand, and fine sand. The D_{15}* size of the cobble-gravel is 50 to 100 times larger than the D_{85}** size of the fine sand. This ratio of D_{15}-filter/D_{85}-protected is thus many times the ratio of 4 to 5 required in filter design and indicates that the fine sand can readily pass through the voids of the coarser cobble gravel. Drilling mud containing fine sand was lost during both CIS and rotary borings.

During construction, additional rotary borings and Becker type borings were conducted as shown on Figure 5 to further define subsurface conditions underneath the dam and the upstream blanket. An 88 ft. (26.8 m) deep freeze shaft, shallow test trench, thermistor studies and piezometric information gave further information concerning the nature and patterns of deposition of the alluvial formation. As a result of these additional explorations and observations made during the construction period, it was evident that exceedingly deep alluvium with zones of openwork existed at the dam site. Differences of opinion existed as to whether the openwork indicated by such losses occurred in limited pockets or extended over long distances.

If continuous openwork extended from upstream of the upstream end of the impervious blanket to downstream of the downstream toe of the Main Embankment Dam then water would flow from the bottom of the reservoir through the sand choked cobble-gravel and sand overlying the openwork cobble-gravel and into the openwork under high gradients. Serious internal erosion thus could possibly occur in the choked alluvium if the openwork beds did not have a natural filter surrounding them. Migration of sand from the choked alluvium would convert choked cobble-gravel to openwork and in effect cause the openwork layer to extend upstream and upward, possibly until it reached the bottom of the reservoir. If a sand layer or sand lens were encountered as the openwork extended upstream, a void space would be left as the sand migrated into the openwork, and collapse of the void space would occur. Then, either possible subsidence, or sinkhole formation caused by loss of sand layers into the openwork could cause detrimental cracking and loss of integrity to the impervious blanket of the Main Embankment Dam. Thus exploratory grouting

---

*D_{15} - diameter of particle for which 15% by weight is finer
**D_{85} - diameter of particle for which 85% by weight is finer

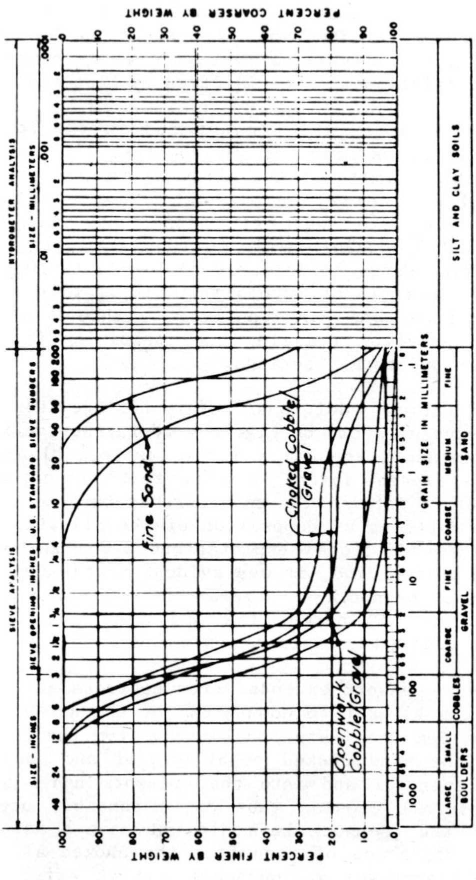

FIGURE 4 GRADATIONS OF INDUS VALLEY ALLUVIUM

FIGURE 5 SUBSURFACE EXPLORATIONS DURING CONSTRUCTION

was undertaken to check the concern that the openwork would accept alluvial sand and to see if it could be demonstrated that the openwork zones were not continuous.

To investigate whether the openwork occurred in pockets or extensive layers, a program of exploration using sanded grout was carried out in 105 of the 605 above-mentioned Becker borings. When the drill hole had penetrated to its maximum depth, a packer was set in the inner casing to block the inner casing itself as well as the ports from the annular space between casings. The grout mix was generally 1.0 cement: 0.20 bentonite: 5.0 sand: 11.0 water by weight. Local bentonite* was used although it was less thixotropic than bentonite normally used in drilling muds and grouting. The pressure used for grout penetration was created primarily by the difference in unit weight between the grout and the water below ground water table in the alluvium, and the short column of grout above ground water table, i.e. about 20 ft (6.1 m). Maximum gauge injection pressures were as follows:

Below 60 ft	(>18m)	depth, 30 psi (207 kPa)
30-60 ft	(9-18m)	depth, 20 psi (138 kPa)
10-30 ft	(3-9m)	depth, 10 psi (69 kPa)
0-10 ft	(0-3m)	depth, Gravity pressure

If no grout takes occurred, the casing was withdrawn slowly by jacking. As soon as grout take commenced, jacking was stopped until grout take ceased, i.e. until more or less refusal: a grout take of 0.5 cu ft/min (14 l/min) or less; then the casing was raised again. Grout takes in openwork varied from 800 to 2500 cu ft of solids/ft (75 to 230 m^3/m). The bottom of an openwork layer was determined quite accurately since until the casing rose above the bottom of the openwork, grout would not be taken. The top of the openwork layer could be determined only approximately as the grout tended to flow upward and fill the upper portion of the entire openwork before the bottom of the casing reached that portion. Nevertheless, valuable information was obtained on the location of openwork.

Exploratory grouting of openwork was carried out along three profiles and at several isolated locations as shown on Figure 6. Profiles of grout takes are shown on Figure 7. The intent was general mapping of openwork conditions. No attempt was made to create a grout curtain.

The exploratory grouting program indicated that wide zones of openwork which would accept alluvial sand existed on the right and left sides of the valley (Figure 8) and occurred generally at the 80 to 120 ft (24 to 37 m) depth and also at depths of 180 ft to 250 ft (55 to 76m). Because alluvial features are expected to be more extensive in an upstream/downstream direction and since all explorations consistently showed openwork on the right and left sides of the

*Liquid limit equal to 141: plasticity index equal to 105. All was finer than .074 mm but approximately 30% was larger than .002 mm, i.e. within the silt size range.

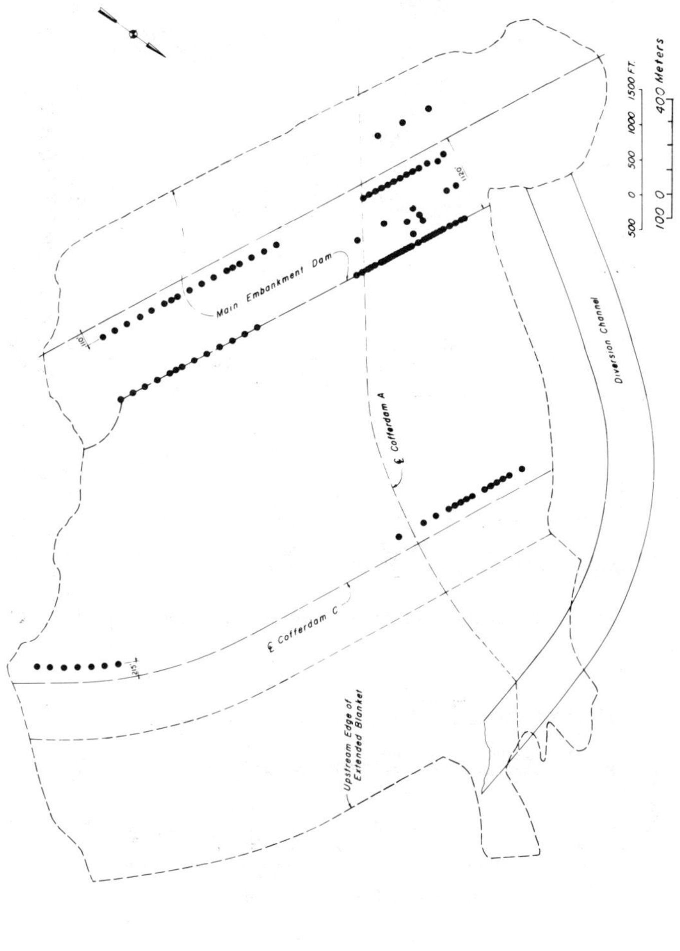

FIGURE 6 LOCATION OF EXPLORATORY GROUTING HOLES

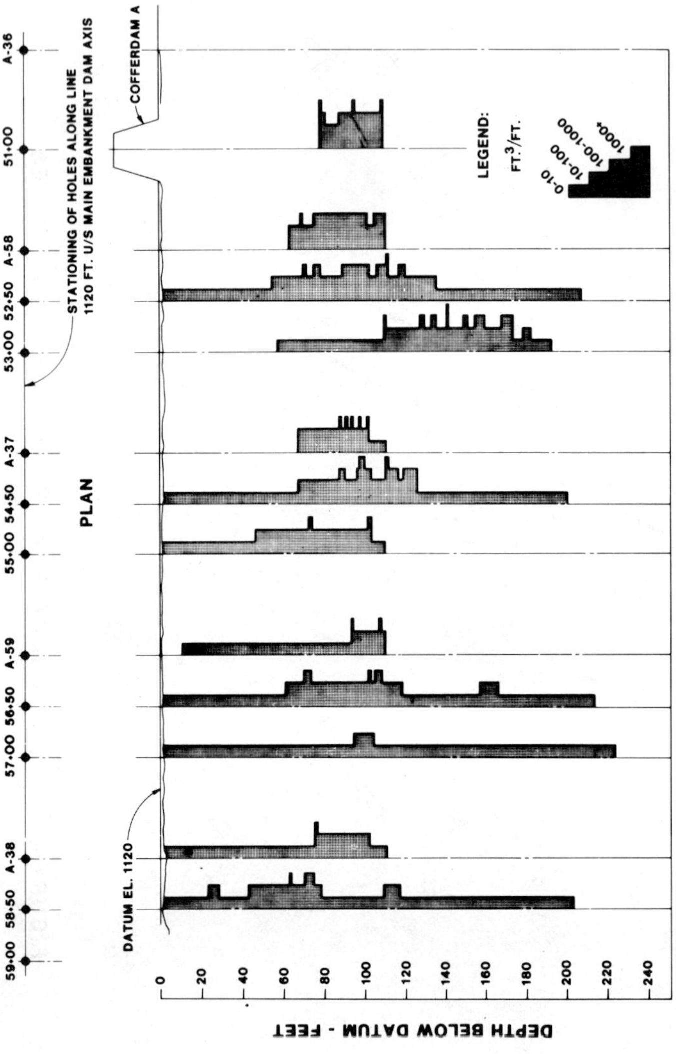

FIGURE 7 GROUT TAKE PROFILES - EXPLORATORY GROUT HOLES ALONG UPSTREAM TOE OF MAIN EMBANKMENT DAM

valley, the continuity of high permeability openwork zones on both sides of the valley was indicated. This was confirmed by a large scale pump-in test with piezometric monitoring which showed responses up to 2500 ft (762 m) away in openwork zones. The openwork acted like a confined aquafer.

As a result of the indicated presence of continuous openwork zones which would accept alluvial sand, the original upstream blanket was lengthened and the "observational" method was adopted for monitoring the behavior of the foundation and blanket during impounding.

Grouting Passageways in Limestone

At the Auxiliary Spillway, weathered dolomitic limestone and limestone exists interbedded with phyllite and beds consisting of cohesionless marly silt; gouge is also present on faults. At the Auxiliary Spillway, the beds dip 30° to 75° in a downstream direction and strike 20°-40° with the upstream-downstream direction. Located within the rock are bedding and fold joints, tension cracks from folding and shear faults which allow water movement across bedding as well as along the bedding within the rock. The rock is dolomitic limestone and some karstic features of relatively small cross section have developed along bedding and possibly along cross bedding faults. Some of these small karstic features are open and others are partially filled with silt. A plan view of the Auxiliary Spillway is given on Figure 9 and a cross-section on Figure 10.

To insure the stability of the spillway ogee and chute structures, an extensive "V" shaped drainage curtain was installed around these structures and a 225 ft. (68 m) long upstream impervious blanket with a triple line grout curtain was provided. Some movement of fines has occurred into the drainage system in recent years although the drains were lined with slotted polyvinylchloride (PVC) pipe having a nominal slot width of .02 in (0.5 mm). The fine material has proven to be of a uniform size with approximately 85% passing the No. 200 sieve (.074mm). Practically no coarse material is present in the erodible material to form a natural filter outside the PVC pipe. The drains have been systematically replaced with special double PVC tubes having a sand filter between the tubes. Each tube is slotted PVC with .02 in (0.5 mm) wide slots, and the interlayer sand gradation has been designed to filter the above mentioned fines. The filtered drains are spaced 10 ft. apart to minimize the possibility that preferred seepage paths might pass between the drain holes and create excessive piezometric pressures downstream of the drainage curtain. The erodible materials are so fine that they are ungroutable with portland cement grout. An initially intact narrow, 15 ft (5m) wide, grout curtain in these materials is subject to high gradients under which beds or faults containing the above mentioned ungroutable fines would pipe into voids downstream of the curtain thus perforating the grout curtain and invalidating its integrity. Thus a program of grouting was undertaken to fill preferred paths and voids in a wide zone upstream of the drainage system to reduce the possibility of any preferred paths existing or developing from the reservoir through the drainage curtain.

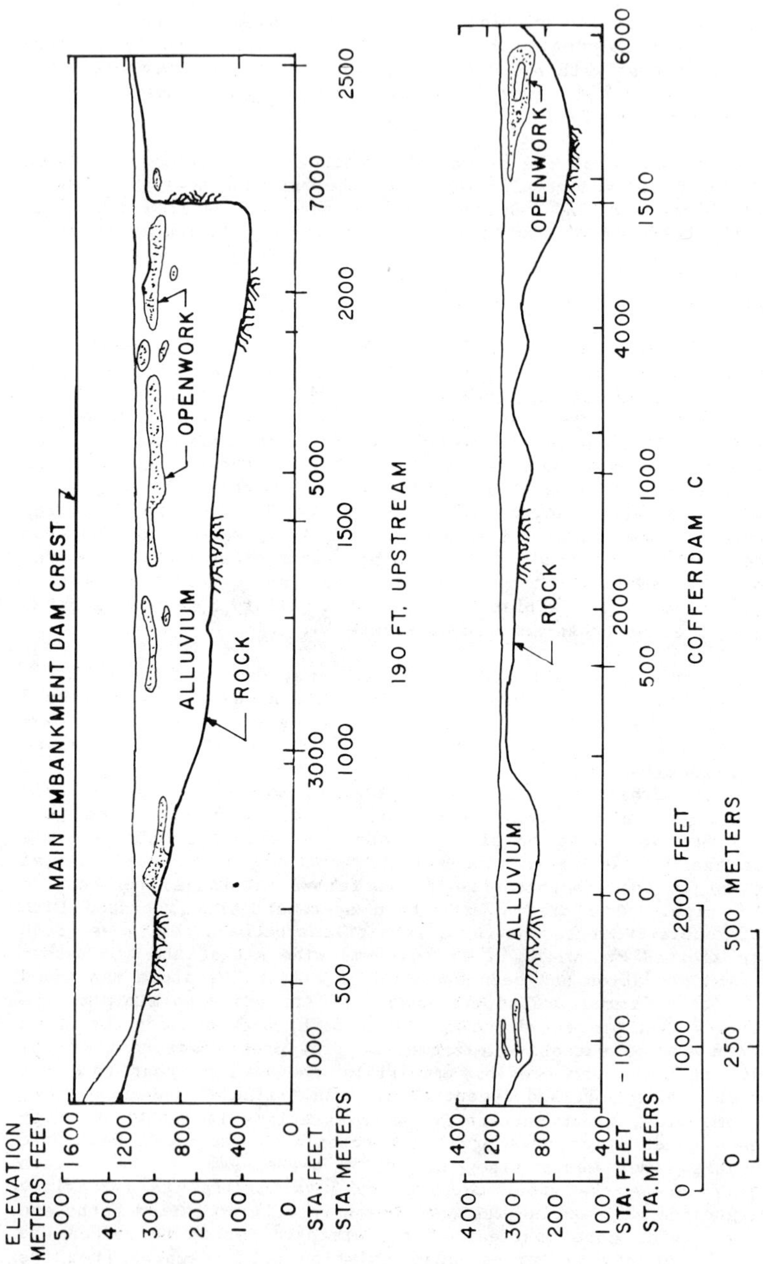

FIGURE 8 OPENWORK IN ALLUVIAL FOUNDATION

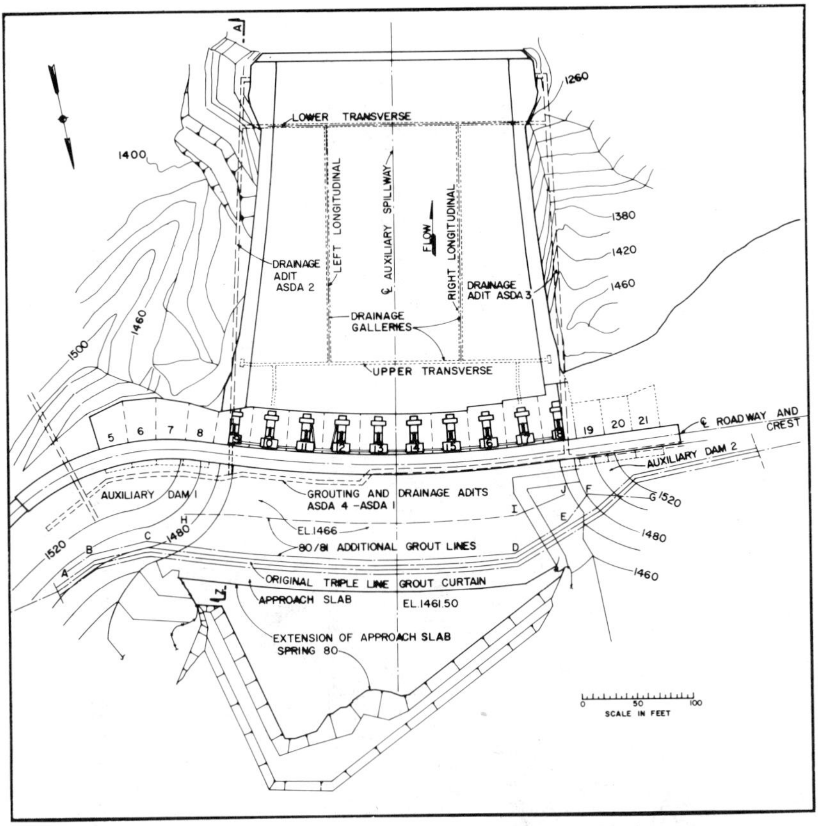

FIGURE 9 AUXILIARY SPILLWAY PLAN

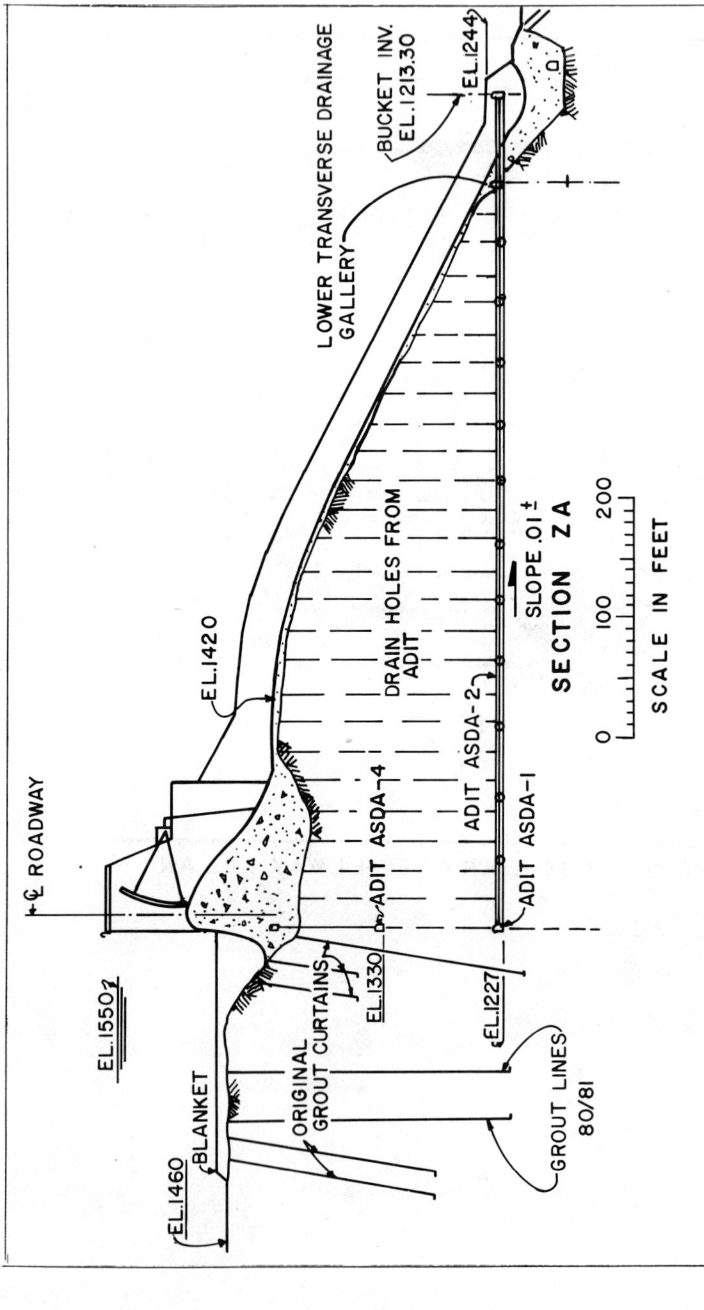

FIG. 10. AUXILIARY SPILLWAY SECTION

To grout long water passageways of relatively small cross-sectional area, it was desired to have a low viscosity grout so that the distance of travel would be maximized but at the same time to have a stable grout so that the properties would not be lost during the time of travel. A grout with sufficient strength so that the packer could be seated when a hole was redrilled for the next stage was also needed. Accordingly, utilizing locally available materials a series of test mixes with different water-cement ratios and percentages of bentonite were prepared in the grout plant. The bentonite had a liquid limit of 436 and a plasticity index of 379 with 99% finer than .074 mm and 80% finer than .002 mm.

The Marsh flow cone results for test mixes of neat cement grout and cement grout with 5 different levels of bentonite are summarized in Figure 11. The Marsh flow cone indicated that the grout must have a water-cement ratio essentially higher than 0.9 except for neat cement grout and cement plus 1.0% bentonite (by weight of cement) to maintain a low viscosity. When the water-cement ratio increases, however, then the grout becomes less stable as measured by sedimentation tests. Figure 12 shows the sedimentation (bleed) measured in a 12 in (300 mm) high by 3 in (75 mm) diameter cylinder to the time of grout set which ranged from 1 to 5 hours depending upon the mix. For a sedimentation of less than 10%, the number of acceptable mixes were further reduced to those having a water-cement ratio of less than 1.4 and bentonite percentages greater then 1.5%. The 7 day cube strengths (Table 1) show that the strengths for 4 and 6% bentonite at a 1.36 water-cement ratio were equal to or less than 300 psi (2.1 mPa). In the previous season, packer seating problems after redrilling previously grouted stages had occurred with a grout with approximately this same level of strength. Deere (1980) notes that at higher bentonite levels (near 6%) squeezing out of bentonite from the grout sometimes occur. Thus these mixes with higher water-cement ratios and higher bentonite percentages were eliminated from consideration. The selection had thus narrowed to a water-cement ratio near 1.0 and a bentonite percentage ranging from 1.5 to 6.0%. Since high quality bentonite was expensive relative to cement and a lower percentage produced higher strengths, the grout with 1.5% bentonite was selected for grouting of the long water passageways. Gelimeter tests indicated that little change occurred for approximately 2 hours which thus appeared to be adequate time for sufficient penetration.

TABLE 1

CUBE STRENGTH, (7 DAYS)

Bentonite, % (By Weight of Cement)	Water-Cement Ratio 0.93		1.36	
	psi	(mPa)	psi	(mPa)
0	1524	(10.5)	978	(6.7)
1	1956	(13.5)	631	(4.3)
2.5	2050	(14.1)	684	(4.7)
4	728	(5.0)	306	(2.1)
6	866	(6.0)	251	(1.7)

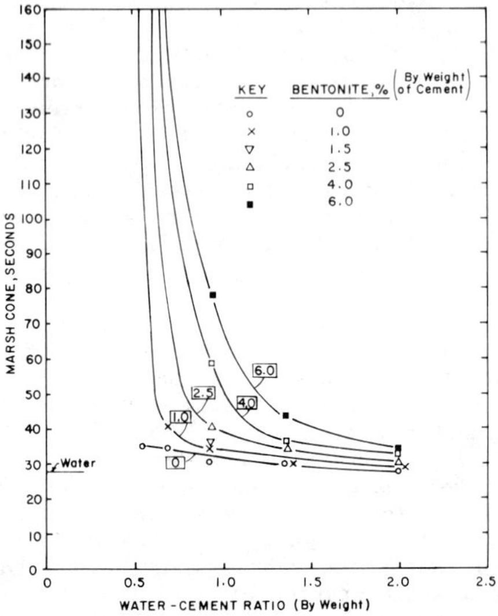

FIGURE 11. MARSH CONE RESULTS FOR TEST MIXES

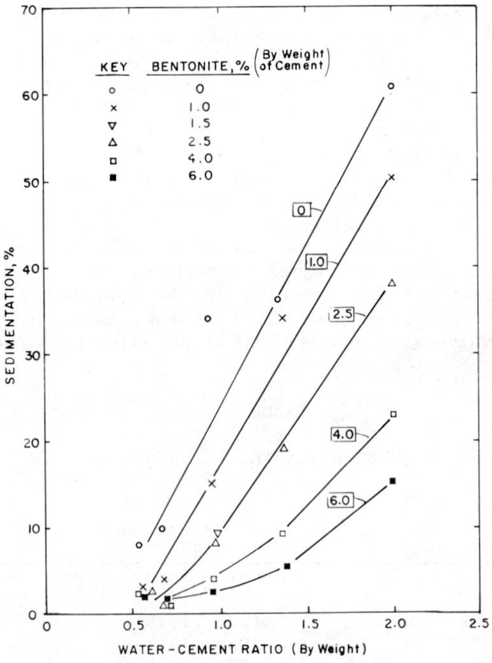

FIGURE 12. SEDIMENTATION RESULTS FOR TEST MIXES

In contrast to typical grouting procedures where the mix is varied depending upon the takes, the qualities required of this grout resulted in the selection of a single grout mix from those tested. Since it was desired to have the grout travel as far as possible, thick mixes were not needed for blocking off passages. The only foreseeable mix change was the possible addition of inert filler, i.e. silt, as an economical extender of the grout. Any grout mix with silt must meet similar requirements for stability, viscosity and strength as without silt.

Since the rock to be grouted was of generally poor quality, the method of descending stage grouting was utilized. Mechanical packers were set at the top of the stage in order to allow progressively increasing pressures as the hole was deepened. However, this packer could not be placed below a depth of 105 ft (32 m) and thus for stages below this depth, the packer remained at this constant depth. The grout holes were drilled utilizing rotary and plug bits since the average depth of hole of 260 ft (79 m) and the maximum depth of 385 ft (117 m) were beyond the depth capability of the on site percussion equipment. When rock sampling was desired, then the holes were drilled with diamond core bits. Stage lengths were limited to 20 ft (6 m) for less than 100 ft (31 m) depth, 30 ft (9 m) to 225 ft (69 m) depth and 40 ft (12 m) for more than 225 ft (69 m) depth. Following drilling of the stage, the hole was washed until there was no further removal of fine material or for twenty minutes if there was water loss. Gage pressures at the top of the hole for water tests and grouting were limited to 1.0 psi (6.9 kPa) per foot of drilled hole except for the top two stages where the maximum gage pressure was stipulated to be 10 psi (69 kPa) and 30 psi (207 kPa) respectively. Water tests were conducted in all stages prior to grouting and provided information on subsurface conditions giving a check on whether the grout operations were functioning properly. Grout was mixed in an 8 cu ft (0.2 m^3) capacity mixer tank which was connected to a 17 cu ft (0.5 m^3) capacity agitator tank which force fed the grout into the pump suction by oriented paddles. Grout was pumped by Wirth double acting piston type pumps with pressures monitored at the pump and at the top of the hole. A single grout mix had production and control advantages as well as serving as a more reliable indicator of subsurface conditions. If grout mixes are changed during grouting, it is difficult to evaluate whether changes in grout takes are due to changes in subsurface conditions or due to changes in the grout mix. Grouting was conducted to refusal which was defined as a grout take per stage of less than 20 liters in a 10 minute period. Between 3 and 6 hours after completion of grouting for a stage, redrilling for the succeeding stage was stipulated to be done. This minimum time for redrilling corresponded approximately to the set time of the selected grout as measured by the Rodio Gelimeter.

At the Auxiliary Spillway, the initial hole layout was a single line transverse to the direction of water flow and located at a distance of approximately 200 ft (61 m) upstream of the drainage curtain. The holes were split spaced starting at a 40 ft (12 m) spacing and working down to 5 ft without a closure criteria since the initial concern was to hit practically all karstic passages with this line of closely spaced holes. It was anticipated that the passages would be preferential in an upstream-downstream direction since this was approximately the strike direction of the bedrock and also more or less the direction of maximum water gradient tending to create new passageways. Small 1 ft drain sinkholes had occurred in the previous two seasons approximately 250 ft (76 m) upstream of the drainage curtain. Water injected in these sinkholes appeared at the drainage curtain.

During grouting, grout appeared in one of the few drains that had not had a filter replacement and thus the grout was capable of traveling the entire 200 ft (61 m) to near the drainage curtain. Some substantial takes, up to 70 cubic feet per foot (6.4 m^3/m) for a stage, indicated that significant travel was being accomplished. Despite the anticipated preferential direction of the passages, there was a substantial drop in grout take as the hole spacing was reduced by split spacing (Figure 13). This indicated that most passageways were connected in a lateral direction possibly by joints across the bedding. Since very few large takes occurred with the final spacing, not all the holes for the last split spacing were put down along the initial line. Instead, a second line placed 100 ft (31 m) downstream of the first line was started. Actually the number of holes required for quarternaries is the same as the sum of primary, secondary and tertiary holes. The grout takes in this line were of the same order of magnitude as the initial line indicating that the grout travel had not penetrated en masse as far as 100 ft (31 m). These second line takes indicated that, in general, the existing karstic voids did not form long continuous preferred paths in the upstream-downstream direction. Thus more than one line was needed in order to fill the voids in a wide zone.

Conclusions

Grouting as a subsurface exploratory tool was utilized at the Tarbela Dam Project to define locations and continuity of openwork gravel in the alluvial foundation. The acceptance by the openwork of alluvial sand in the injected grout indicated that alluvial sand from adjacent beds could be piped into the openwork beds. The design of the dam was then modified to account for this situation.

To fill existing preferred paths and forestall development of new preferred paths in a dolomitic limestone containing erodible materials, a grouted zone of significant width was created. A fluid grout consisting of cement with bentonite having low sedimentation and a long enough time to initial set was selected so that the grout would travel a maximum distance in the passageways. Despite the grout selection, the initial results indicated that the upstream/downstream trending passageways were not continuous enough for complete grouting from one line and therefore additional lines were necessary.

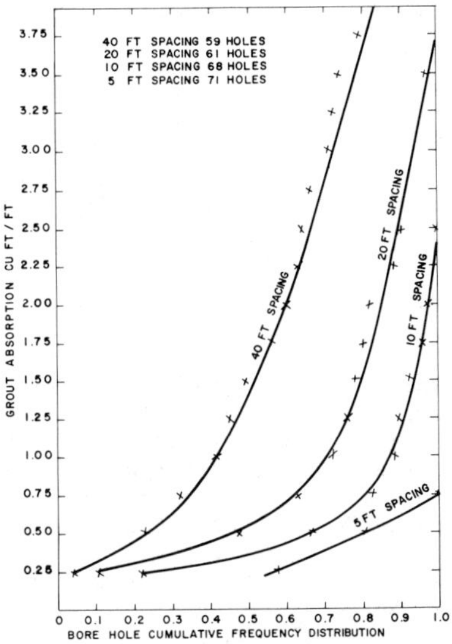

FIG.13 EFFECT OF SPLIT SPACINGS ON GROUT ABSORPTION

Acknowledgements

The owner and operator of the Tarbela Dam Project is Pakistan Water and Power Development Authority (WAPDA). During the period that the above described grouting was carried out at the Auxiliary Spillway, Major General Fazli Raziq was Chairman; Mr. Gulzar Ahmed, General Manager (Dams); and Mr. Asif Kazi, General Manager and Project Director at site. Design and supervision of construction was carried out by Tippetts-Abbett-McCarthy-Stratton (TAMS) Engineers, Architects and Planners, New York and Pakistan with Dr. Paul C. Chao, Project Manager, New York and Mr. Lyle H. Hixenbaugh, Chief Resident Engineer, Pakistan. Harza Engineering Co., WAPDA's general consultants for Indus Basin projects, and Sir Alexander Gibb and Partners, advisors to the International Bank of Reconstruction and Development, reviewed and contributed to the designs. Colrok were subcontractors for drilling and grouting to the general contractor, Indus River Contractors, a joint venture headed by Impregilo, Milan, Italy. The exploratory grouting was carried out in 1970 under the direction of different staff in Pakistan.

References

1. Deere, D.U. " Some Recent Experiences in Curtain Grouting for High Dams" Presented at Joint Meeting of ASCE and Assoc. of Eng. Geologists, Denver Section, September 8, 1980.

2. Lovell, L. A., Lowe, J. and Binger, W. V. "Tarbela Dam Construction Reaches Half-Way Mark" Water Power, September-October, 1972.

3. Lowe III, J. "Foundation Design - Tarbela Dam" Fourth Nabor Carillo Lecture, Presented at the Mexican Society for Soil Mechanics, X National Meeting, Yucatan, Mexico, November 18, 1978.

A CUT OFF IN ROCK AND ALLUVIUM AT ASPROKREMMOS DAM

By L.A. BELL[1] B.Sc.,M.Sc.,M.I.C.E.

1. Introduction

Asprokremmos Dam is being constructed on the Xeropotamos River for the Government of Cyprus as part of the $80 million Paphos Irrigation Project. The aim of the project is to irrigate about 5,000 ha of net irrigable area lying in the south western coastal plain of Cyprus on both sides of the town of Paphos. The water requirements for the irrigation of this area are estimated to be 36 million cubic metres per year, and will be provided by the Xeropotamos flow (22 million cubic metres), regulated at Asprokremmos, and the alluvial aquifers in the main river beds East of Paphos.

The dam is situated nine miles East of Paphos and is of the zoned earth fill type with central clay core and gravel shoulders. The embankment will be 50m high with a crest length of 600m, and total volume of approximately 2 million cubic metres. The dam will impound 53 million cubic metres of water compared with a total capacity of all the reservoirs presently in use in Cyprus of 65 million cubic metres.

Geologically the site is underlain by a sedimentary sequence of Eocene and Miocene age, known locally as the Lefkara and Paknos Formations. The lower part of the sequence, the Eocene Lefkara Formation consists of alternating bands of white chalks and marly chalks with chert bands or nodules succeeded by massive chalk bands. This in turn is overlain transitionally by an alternating sequence of yellowish white chalk, marly chalk and silty marl, the latter being associated with thin black shale bands.

Superficial deposits of late Pleistocene to recent age include terrace deposits up to 20m thick and river alluvium up to 30m thick. It was the depth of the alluvium infilling the river bed which presented the principal problem on the site, and required a cut-off some 180m long and 28m deep.

The Main Contractor for the dam construction is a Joint Venture of two Cypriot-based firms, Joannou and Paraskevaides Limited and Medcon Construction Limited. The specialist Geotechnical sub-contract was let in April, 1978, to Colcrete Limited of Wetherby, England, at an original contract value of STG£1.5 million ($3.3 million). The geotechnical work included the construction of a 28m deep concrete diaphragm wall cut-off, multiple row blanket and curtain grouting in rock and alluvium, pressure relief and drainage wells, tunnel grouting and rock anchors for the spillway slab and chute.

[1] Contracts Engineer, Colcrete Limited, Strood, ENGLAND

The construction of the diaphragm wall was sub-let to ICOS (G.B.) Ltd. The ICOS sub-contract excluded the provision of concrete, steel cages and trench spoil removal.

Due to problems encountered with the diaphragm wall, additional alluvial grouting was ordered by the Engineer to augment the original cut-off. The cost of this additional grouting, together with an increased volume of alluvial grouting and the effects of inflation, increased the final value of Colcrete's sub-contract to almost STG£2.7 million ($6 million). The geotechnical sub-contract was completed in February, 1981.

2. Diaphragm Wall Cut-Off

The Engineer's design required a permanent cut-off in the alluvial deposits which would give a residual permeability of 1 x 10^{-6}cm/sec and four alternative proposals were put forward at tender stage.

Alternative (A) - A single plastic concrete diaphragm 1.0m wide with full depth alluvial grouting upstream and downstream and blanket alluvial grouting below base of clay core.

Alternative (B) - A double bentonite/cement diaphragm, both 0.8m wide and spaced 3m apart, with full depth alluvial grouting between diaphragms and blanket alluvial grouting below base of clay core.

Alternative (C) - A concrete diaphragm wall 0.8m wide with blanket alluvial grout below base of clay core.

Alternative (D) - A bentonite slurry trench 3.0m wide.

In all alternatives a single row of holes was included to form a grout curtain in rock underlying the alluvium, directly beneath the cut-off wall.

Alternative (C) proved to be the cheapest and was finally adopted.

Work on the diaphragm wall started in September, 1978, and was completed in February, 1979. In all, a total of 38 panels, 4.7m wide ranging from 6.5 to 29m deep (representing 3182m^2 of wall), were constructed in 20 weeks. The total quantity of concrete placed was 3358.5m^3, which is equivalent to 32% overpour on the theoretical volume.

Throughout their operations, ICOS used a 7.5 ton cable operated clamshell grab mounted on an NCK-Rapier 605 crane chassis to excavate and clean the cut-off trench. During excavation a 6% bentonite slurry, with a density of less than 1.15g/ml., was introduced into each panel to support the side walls.

Towards the completion of the diaphragm wall construction, the Engineer decided to carry out some investigation on the base of the wall to determine the quality of the contact between the concrete and the underlying rock. Sediment was discovered at the interface, and the Engineer considered that this would leave a high permeability seam through which reservoir water might subsequently escape. As a result of these early findings, the Engineer ordered that a core sample be taken through the base of each panel to determine the extent of the problem. Out of 37 results obtained, 7 holes showed that there was good contact between rock and concrete, 7 holes had sediment between 1-10cm in thickness, 14 holes had sediment 11-20cm in thickness, 4 holes had sediment depths of 21-40cm, and 5 holes had sediment over 40 cm in depth.

Based on these results, the Engineer rejected part of the diaphragm wall as in his opinion it did not conform to the Specification, and instructed the following additional/remedial grouting to be carried out in order to rectify the deficiencies which he considered to exist (Figure 1):-

(i) Zone D - a 4m x 4m thrust block at the downstream toe of the diaphragm wall to prevent lateral movement where it had not been keyed into the rock.

(ii) Zone C - a 6m x 6m impermeable (1×10^{-4}cm/sec) block at the upstream toe of the wall to prevent seepage through the concrete/ rock interface.

(iii) Zone B - a 3m wide zone up the upstream face of the wall to connect Zone C with the original blanket treatment (Zone A) which would also act as additional insurance against seepage through panel joints (again, 1×10^{-4}cm/sec. required).

As the intention was to improve the shear strength of Zone D, the Engineer decided to grout this zone with a neat cement grout using the hydrofracture system.

Zones B and C had to be grouted to the same residual permeability as the original Zone A (1×10^{-4}cm/sec), and therefore the same layout of holes was adopted for these three zones based on the principle of permeation.

All additional/remedial works were completed successfully and the Engineer has now accepted the diaphragm wall cut-off.

3. Alluvial Grouting

The methods employed by Colcrete for the revised alluvial grouting are typical of current European practice and an outline is given below of the techniques used together with details of the acceptance criteria and water test programme.

3.1 Tube-à-Manchette

The technique of using tube-à-manchette for alluvial grouting has been well described elsewhere (Glossop and Ischy, 1962), but a brief description of the system used will be given here for completeness.

The tubes-à-manchette comprised of a upvc perforated tube of 38mm bore which was placed into a borehole 96mm diameter and sealed into the ground with a relatively weak bentonite/cement sleeve grout of 100 parts water:50 parts cement:10 parts bentonite by weight.

Along the outside length of the plastic tube, rubber sleeves were fitted covering perforations, normally sets of 12 no. holes each hole of 6mm diameter, spaced at 0.5m intervals of depth.

Injections were made through a double packer system inserted into the tube and positioned opposite a sleeve, rupturing the sleeve grout to enter the ground at the selected depth (Figure 2). The rubber sleeve acts as a one-way valve allowing grout to pass into the ground but preventing its re-entry by closing onto the tube once the injection pressure has dropped.

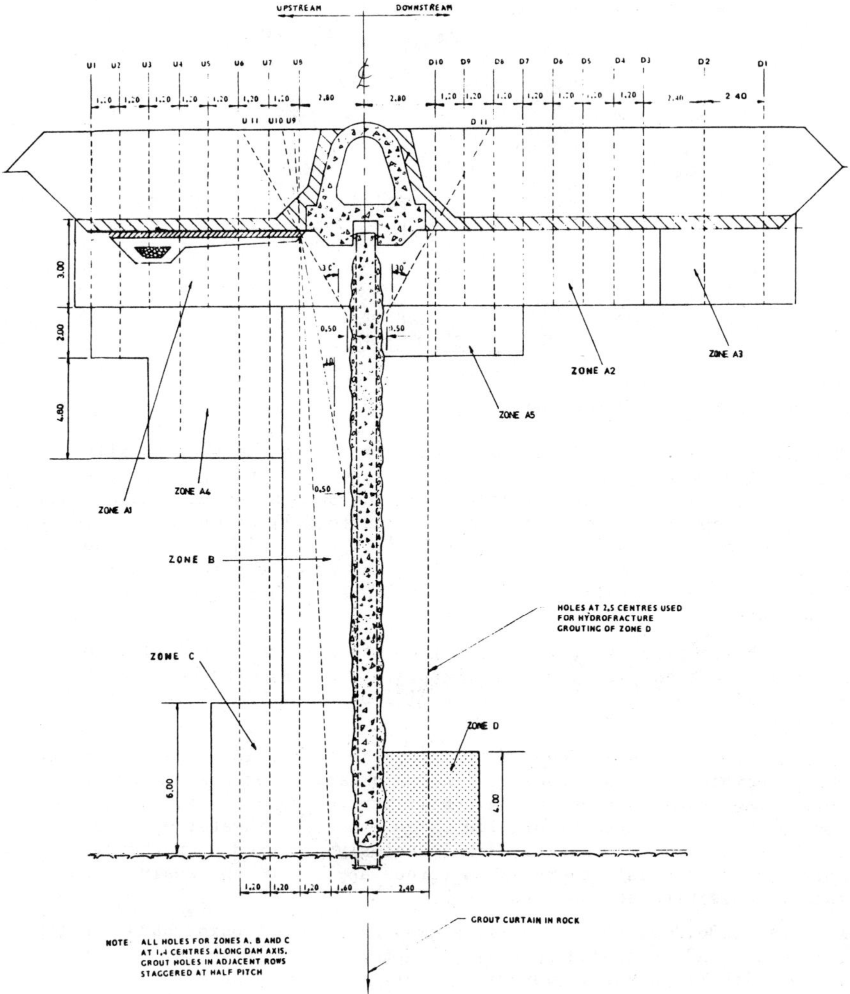

Figure 1. - Revised Alluvial Grouting

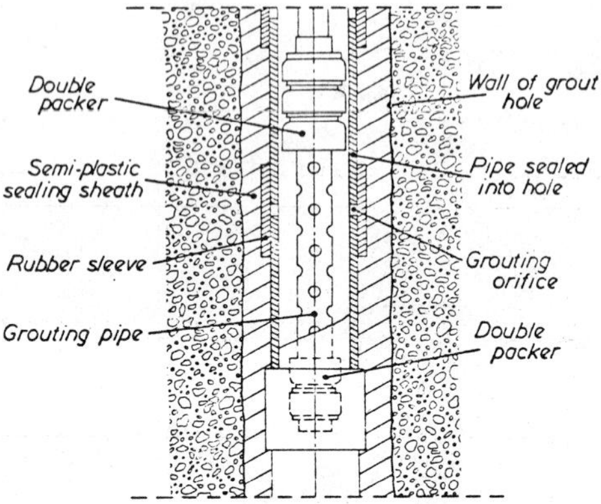

Figure 2. Sleeved Grout Tube

The great advantage of the tube-à-manchette system is that it allows grouts of differing properties to be injected into the ground in any sequence and at any order of depth from the same borehole. As a result, low-cost suspension grouts can be injected to fill major voids prior to using the more expensive chemical solution grouts with lower viscosity. By keeping records of grout acceptances and water testing as work proceeds, it is relatively easy to detect where further injections are required.

The manchette pipes were installed from the surface of the clay core after it had been placed to a thickness of 4m. All the holes were drilled with hydraulic rotary/percussive drilling rigs using the ODEX 76 system of casing.

The ODEX system (overburden drilling by the eccentric method) works on the principle of under reaming. The main feature of the method is the swivelling reamer bit which, by a half turn on and relative to the pilot bit, enables a hole to be drilled to an oversize diameter with respect to the casing. The casing is simultaneously hammered down the oversize hole, and as a result there is no torque applied to the casing, and casing shoes/bits are not required.

A total of 36,000m of ODEX drilling was carried out using Hands England Drillmaster A36 rigs and specially adapted Hymac hydraulic excavators fitted with Krupp HB101 hydraulic rotary/percussive hammers.

Many difficulties were encountered whilst drilling to the full depth of the alluvium, and very high casing wear was experienced, particularly in zones containing a high percentage of cobbles and boulders. Also, as the drill rigs themselves were unable to extract casing which had been driven to a depth in excess of 12-15m, special extraction jacks had to be used.

3.2 Choice of Grout System

Before commencing any alluvial grouting project, it is essential that full consideration be given to all available grouting systems in order to choose the most appropriate for a particular application.

From a practical point of view grouts are most simply divided into two broad groups as particulate grouts (suspensions) and non-particulate grouts (solutions). This division is important in grouting, since the presence of solid particles in a grout immediately places an absolute limit on the type of ground which the grout can effectively permeate, i.e. the soil pores must be sufficiently large for the solid particles to pass through.

Chemical grouts are solutions which act as true or Newtonian liquids, i.e. the rate of shear is proportional to the tangential stress applied to it. On the other hand, particulate grouts normally act as Bingham fluids, having a measureable shear strength which must be overcome before flow will commence, the rate of shear then being proportional to the tangential stress.

Obviously Newtonian fluids can be defined by a single measure of viscosity. However, Bingham fluids require a number of measurements for various rates of shear, which can be conveniently carried out by means of the Fann rotational viscometer (Littlejohn, 1982).

As the maximum size of grout particles limits the type of soil which may be treated, analysis of borehole grading curves allows an assessment to be made of the groutability of the deposit. The effective size (D_{10}) and uniformity coefficient (U) provide an approximate description of a material. By using these regularly, it is possible to assess subjectively the type of material in question.

It has been found in practice that, for continuous flow without blocking the capillaries, the diameter of the solid particles must not be more than about one-tenth of the D_{10} size of the soil.

Another way of expressing the relation between the particle size of the grout and the grain size of the soil to be grouted is sometimes indicated in terms of a groutability ratio specified by Karol (1960) as follows:-

$$\text{groutability ratio} = \frac{(D_{15}\ \text{soil})}{(D_{85}\ \text{grout})} > 25$$

One problem not yet resolved is the maximum allowable grout particle size for a particular pore size distrubtion, since penetration is not through a single opening, but through a long series of alternate widening and narrow pores radiating into the soil.

For soils that are not uniformly graded, an estimate of the condition for blockage of a particulate grout can be obtained by applying the Kozeny relationship (Raffle and Greenwood, 1961),

$$a = \left[\frac{8NK}{\gamma_w g n}\right]^{\frac{1}{2}}$$

to find the diameter of 2a of the average pore passage, where N is the viscosity, k the permeability, γ_w the density of water and n is the porosity of the soil. For instance, if the average pore diameter is taken to equal in size a 100μm particle of cement, positive blockage can be expected on this basis for a soil permeability of about $10^{2.5}$m/s.

A 'rule of thumb' generally accepted, states that the soil pore should be three times the grout particle diameter. This will allow the grout to flow in the pore with little likelihood of bridging occurring.

Experimental work has been carried out to define the limiting particle size distribution of soils which are capable of being injected by a suspension of a given particle size, or a solution of a given viscosity. However, results are inconsistent, as the porometry is often the deciding factor. Karol (1960), Cambefort (1977), both agree that the permeability of the soil is, therefore, a much more accurate criterion upon which to base the choice of grout system.

Apart from the limits of injectability, many other important factors have to be considered, such as:- required performance (e.g. impermeabilisation or improved strength, or both), stability, permanence, viscosity, setting time, toxicity and cost.

In order to satisfy the requirements for blanket grouting at Asprokremmos and to treat economically the highly permeable zones, it was recommended that primary grouting should be carried out using a clay/cement grout. However, due to particle size of cement based grouts, the practical limit for grout permeation is in soils with a k value greater than 5×10^{-2}cm/sec. The addition of bentonite to cement grout helps to keep the cement particles in suspension and thus prevent premature blockage of the pores. By using bentonite which has been fully hydrated for at least 24 hours, the quantity of bentonite required to hold the cement in suspension, and hence minimise bleed, is reduced. A series of tests were carried out on site to produce the most satisfactory grout mix with regard to yield and suitability for this stage of the work. A mix of 2:1 water/cement ratio with 2½% bentonite by weight of water was finally adopted.

Before it was possible to select a proposed hole layout it was necessary to choose a chemical grout with properties acceptable to the Engineer. In relation to gel permeability, permanence and grout viscosity, several grout systems could be considered, namely chrome Lignin/dichromate (e.g. TDM and Sumisoil), resorcinol formaldehydes (e.g. MQ14), acrylamides (e.g. AM-9), phenol formaldehydes (e.g. MQ4) and silicate based grouts. When considering potential hazards from the materials, however, the choice narrows further, since the acrylamide AM-9 is neurotoxic and the chrome lignin/dichromate systems, whilst less toxic, are dermatitic. The phenol formaldehydes by comparison are non-toxic, and tests to date on resorcinol formaldehydes show that there is little danger if the materials are ingested. Silicate based systems are also non-toxic and have been used successfully on other dams and are less expensive than any of the other chemical grouts mentioned above. It was on this basis that sodium aluminate/sodium silicate grout was proposed to the Engineer.

Extensive tests were carried out on a sodium aluminate/sodium silicate grout which had been successfully used at Mattmark Dam in Switzerland (Einstein and Schnitter; 1970), to establish the suitability of this type of grout for Asprokremmos Dam. The mix tested comprised of:-

Solution (1) Sodium aluminate 16 parts/wt.
Water 320 parts/wt.
(i.e. 4.8% Sodium aluminate).

Solution (2) Sodium silicate (M75) 522 parts/wt.
Water 284 parts/wt.

and from the results obtained, the following conclusions were drawn:-

The sodium aluminate/sodium silicate mix tested had an initial viscosity of 3.6cP and gel time of approximately 33 minutes at 20°C when tested under laboratory conditions. The gel time of this mix could be extended to approximately 80 minutes by reducing the sodium aluminate content of solution (1). A reduction in sodium aluminate content, however, resulted in a weaker gel and at concentrations less than 3.6% in solution (1), the gel strength was judged to be too weak to be of any value where "permanence" is required.

The gel time could also be extended by increasing the total water content of the mix. However, for gel times greater than 60 minutes, the degree of dilution required is excessive (see Graph 3).

The viscosity of a sodium aluminate/sodium silicate grout solution increased gradually with time and a mix with an initial viscosity of 3.6cP and gel times 73 minutes at 20°C remains below 5.0cP for 50% of its life and below 10cP for 75% of its liquid life.

The sodium aluminate/sodium silicate reaction is temperature sensitive and a rise of approximately 8°C will halve the gelation time.

Graphical representation of the results obtained are shown in graphs 1 to 6.

Further tests carried out by an independent body, as directed by the Engineer, also showed that the chemical grout proposed was satisfactory with regard to permanence and that there was very little likelihood of any leached material acting as a dispersive agent on the clay core.

3.3 Design of Grout Hole Layout

When designing a grout system, whether in homogeneous or heterogeneous deposits, the aim of permeation grouting is to inject the grout evenly and comprehensively into the soil porespace, thus forming a series of near-symmetrical inter-connecting spheres or cylinders around the injection points. These injection points should be spaced so that the area can be treated, as required, from the minimum number of points.

Realistic expressions relating to the rheological characteristics of grouts and their capacity to permeate soil have been developed by Raffle and Greenwood (1961). They have shown that for injection from a spherical injection source of radius r, the resistance to flow may be thought of as that due to a sphere of grout which has penetrated to a radius R, with water on the outside reaching to infinity.

If the grout has reached a radius R at the time t, Raffle and Greenwood show that the volume flow rate Q is related to the hydraulic head H, at the source radius r by the following expression:-

$$H = \frac{Q}{4\pi k}\left[N\left|\frac{1}{r} - \frac{1}{R}\right| + \frac{1}{R}\right] \qquad (1)$$

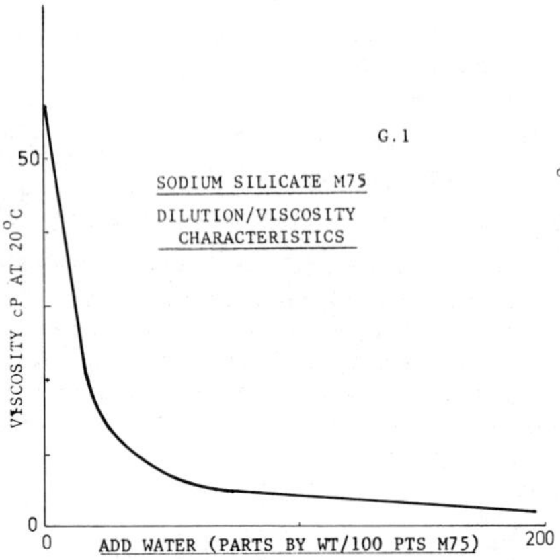
G.1
SODIUM SILICATE M75
DILUTION/VISCOSITY CHARACTERISTICS
VISCOSITY cP AT 20°C
50
0
0
200
ADD WATER (PARTS BY WT/100 PTS M75)

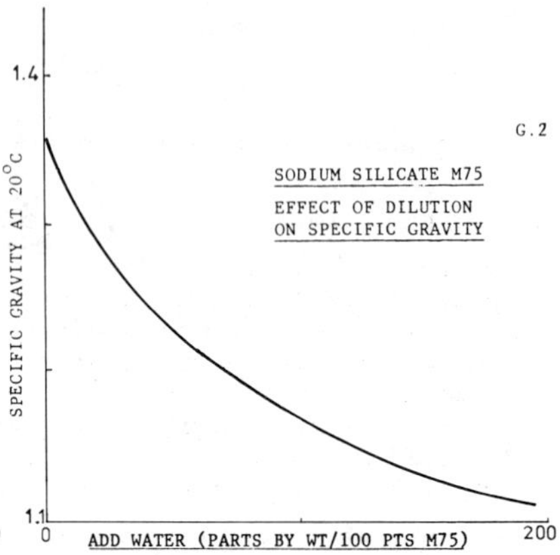
G.2
SODIUM SILICATE M75
EFFECT OF DILUTION ON SPECIFIC GRAVITY
SPECIFIC GRAVITY AT 20°C
1.4
1.1
0
200
ADD WATER (PARTS BY WT/100 PTS M75)

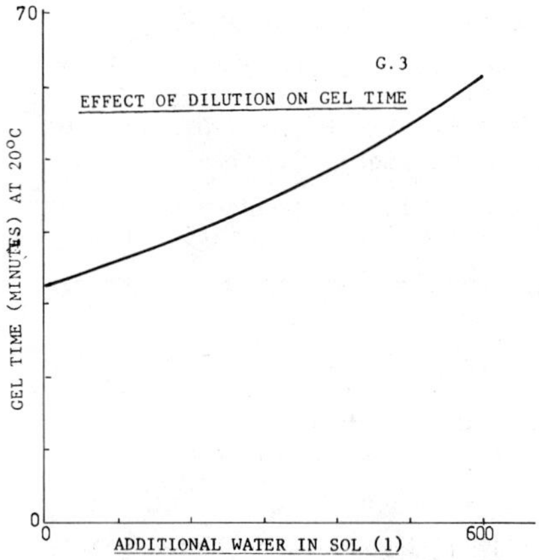
G.3
EFFECT OF DILUTION ON GEL TIME
GEL TIME (MINUTES) AT 20°C
70
0
0
600
ADDITIONAL WATER IN SOL (1)

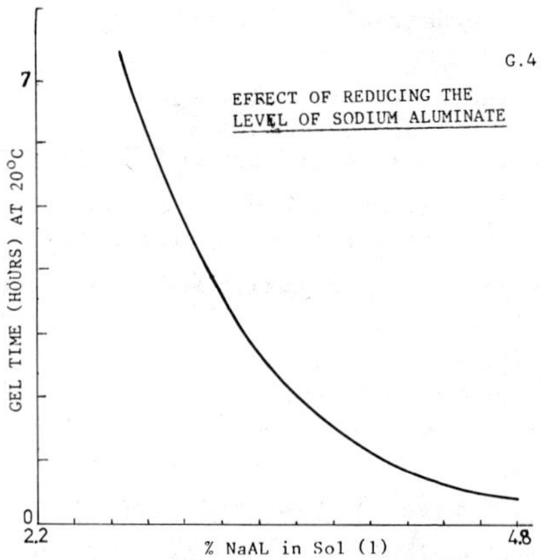
G.4
EFFECT OF REDUCING THE LEVEL OF SODIUM ALUMINATE
GEL TIME (HOURS) AT 20°C
7
0
2.2
4.8
% NaAL in Sol (1)

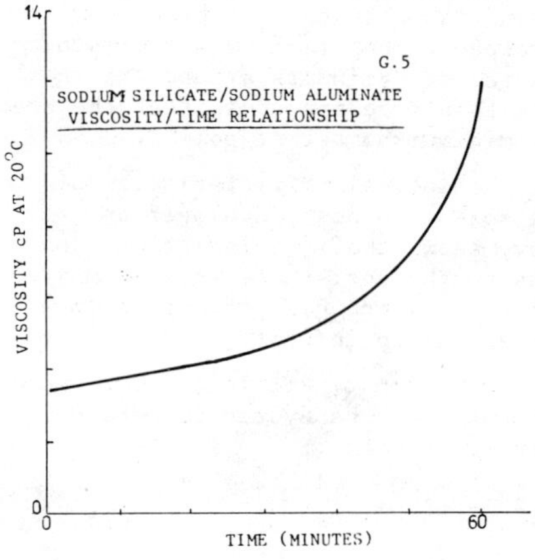
G.5
SODIUM SILICATE/SODIUM ALUMINATE
VISCOSITY/TIME RELATIONSHIP
VISCOSITY cP AT 20°C
14
0
0
60
TIME (MINUTES)

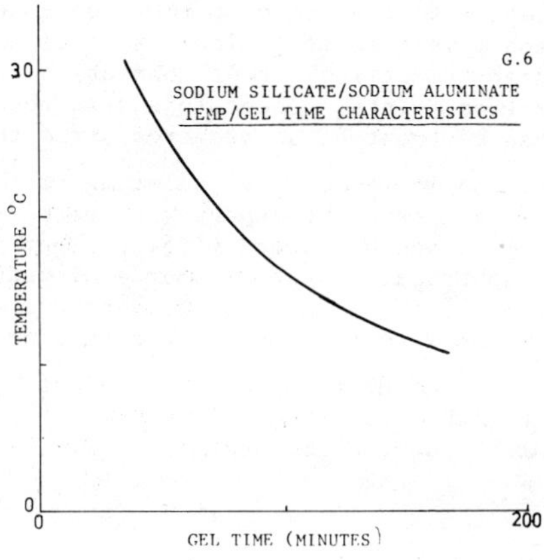
G.6
SODIUM SILICATE/SODIUM ALUMINATE
TEMP/GEL TIME CHARACTERISTICS
TEMPERATURE °C
30
0
0
200
GEL TIME (MINUTES)

By re-arranging, the ratio of source driving pressure to flow rate (hydraulic resistance) becomes:-

$$\frac{H}{Q} = \frac{1}{4\pi kr}\left[N - (N-1)\frac{r}{R}\right] \quad (2)$$

Ignoring the second component of the main bracket, this can be reduced to:-

$$\frac{H}{Q} = \frac{N}{4\pi kr} \quad (3)$$

For a cylindrical injection stage length L and diameter d, r can be replaced (approximately) by $\frac{1}{2}\sqrt{Ld}$. Thus, substituting for r in equation (3) gives the simplified working formula for grout flow rate for a cylindrical source:-

$$Q = \frac{3.77\ kH\sqrt{Ld}}{N} \times 10^5 \text{ litres/min.} \quad (4)$$

Where Q is flow rate (litres/min), k is permeability (m/sec), H is hydraulic head (m), N is viscosity of grout (cP), L is length of cell (m) and d is diameter of cell (m).

Raffle (et al. 1961) also state that the time taken by grout to fill out to radius R, in the ground of porosity n, is given by:-

$$t = \frac{nr^2}{KH}\left[\frac{N}{3}\left[\frac{R^3}{r^3} - 1\right] - \frac{N-1}{2}\left[\frac{R^2}{r^2} - 1\right]\right] \quad (5)$$

Again, ignoring the second component of the main bracket in equation (5) this relation simplifies to Maag's (1938) theory of alluvial injection:-

$$t = \frac{Nn}{3kHr}\left(R^3 - r^3\right) \quad (6)$$

which tends to err on the conservative side (i.e. increased values of t). Also bearing in mind the small value of r, and relatively small influence of the term in brackets, the equation may be simplified further:-

$$t = \frac{NnR^3}{3kHr} \quad (7)$$

substituting again for $r = \frac{1}{2}\sqrt{Ld}$, the equation becomes:-

$$t = \frac{NnR^3}{90kH\sqrt{Ld}} \text{ mins.} \quad (8)$$

where t is pumping time or gel time (mins), R is radial distance of grout penetration (m), N is viscosity of grout (cP), n is porosity of soil expressed as a fraction, k is permeability (m/sec), H is hydraulic head (m), L is length of cell (m) and d is diameter of cell (m).

These relationships are based on several simplified assumptions and obviously an alluvial grouting programme has to be flexible to take into consideration the variability of the deposits. They do, however, provide a good first approximation for the designer to work from.

Based on the water test data, the application of Maag's formula, together with the Engineer's requirement of a residual permeability of 1×10^{-4} cm/sec. after grouting, calculations showed that a 1.4m equilateral spacing of holes should be planned. This was achieved by a primary grid 2.8m (parallel to dam axis) x 2.4m (normal to dam axis), followed by secondaries at intermediate points on the main lines parallel to the dam axis to bring the centres down to 1.4m. Thereafter, tertiaries at 1.4m centres along the dam axis were constructed midway between the existing major lines. The terms primary, secondary and tertiary are used here simply to indicate the sequence of injection.

3.4 Proof Grouting

Before hole spacing, injection sequences, injection pressures and grout mixes could be finalised, a proof grouting exercise was carried out in an area selected for extensive investigation and treatment. 46 No. holes were drilled in an area approximately 10m x 11m, bounded on one side by the diaphragm wall, in the 1.4m equilateral triangular pattern described earlier. Sleeve permeability tests were carried out at successive stages of grouting to assess the efficiency of grout treatment in relation to the quantity and type of grout injected. These tests indicated that the hole layout proposed should be adopted for the production grouting.

3.5 Production Grouting

The grouting of Zone D was carried out from the base of the key trench in advance of the other alluvial grouting to coincide with construction of the cap gallery, in order to assist the Main Contractor's programme.

As Zone D did not have any residual permeability requirements, the Engineer instructed the use of neat cement grout (1:1 w/c ratio) via a single line of holes 2m from the downstream face of the diaphragm wall and at 2.5m centres across the valley. The Engineer required 2-3 tonnes of cement to be injected per metre depth over the bottom 4m of each hole regardless of pressure requirements. In all, 60 holes were drilled over a total length of 1353m. 440 tonnes of cement were injected over an injected length of 240m.

In some instances, extremely high pressures of up to 50 bars were required to achieve the injection quantities specified by the Engineer, and on several occasions grout appeared on the surface, sometimes from depths as great as 26m.

The layout of production grout holes for zones A, B and C remained the same as that previously recommended and used in the proof grouting. Additional rows of holes (D11, U11 and U10) were also introduced by the Engineer to ensure complete treatment to the recess just below the cap gallery (Figure 1).

The production mix formulation for the clay/cement grout was a 2:1 w/c ratio with 2½% hydrated bentonite by weight of water. This produced a grout with a compressive strength of 1.5N/mm^2 at 7 days, and a yield of 2.32m^3 per tonne of dry cement.

As holes were spaced at 1.4m centres across the valley, in order to pick up lenses of high permeability, alternate sleeves i.e. 1m centres, were injected with clay/cement in all grout holes creating a 1m x 1.4m rectangular grid in elevation. Grouting was carried out in ascending stages in the order primary (2.8m or 2.4m centres), secondary (reducing 2.8m centres to 1.4m), and tertiary (giving a final 1.4m triangular grid). At the start of the treatment all sleeves on all holes on outer lines to each zone were treated with clay/cement in order to limit the future grout flow.

For clay/cement grout an injection pressure of up to 0.45 bars/metre depth of overburden was employed together with an upper volume limit of 700 litres per sleeve. Where the maximum sleeve consumption was reached at an injection pressure below the limiting value, the adjacent sleeves (0.5m distance on either side) were also injected with clay/cement up to the pressure or volume thresholds specified above. Injection was carried out using a recirculation system and ceased when grout flow was less than 8 litres per minute.

All the clay/cement grout materials were mixed in a Colcrete colloidal SD10 mixer which has an output of up to 4.5m^3/hour and a batch capacity of 280 litres. The grout was then transferred to agitation tanks and pumped by pneumatic or electric/hydraulic Colcrete Evans pumps, each with a rated output up to 5.8m^3/hour.

The production mix formulation generally used for the chemical grout was:-

522 parts (by weight)	Sodium Silicate M75
16 parts (by weight)	Sodium Aluminate
604 parts (by weight)	Water

to give gel times in the range of 30-60 minutes. Variations of the above proportions were allowed to cater for changes in air and ground temperatures in order to maintain sensible control of the gel times.

Bearing in mind the increased shear strength and reduction in permeability after clay/cement grouting, an increase in injection pressure was permitted to facilitate chemical treatment. An injection pressure of 0.75 bars/metre depth was allowed initially. To confirm whether these higher injection pressures were safe, hydrofracture tests through existing sleeves were carried out at various chainages and depths. Surface levels were also monitored by the Engineer during grouting to determine the presence of heave.

An upper limiting quantity of 1250 litres of chemical grout was applied to each alternate ungrouted sleeve. Again, where alternate sleeves consumed the limiting quantity without achieving the maximum pressure, the adjacent sleeves were also injected with chemical up to the same pressure or volume thresholds. Changes in both the limiting pressures and the volumes were introduced as work proceeded to cater for variations in ground conditions. In all instances, however, injection was stopped when grout flow fell below 6 litres per minute.

Whilst the mixing and injection of clay/cement grouts is relatively simple and straightforward, the mixing and injection of 'one-shot' silicate grouts needs very careful control. The sodium aluminate/ sodium silicate reaction is very sensitive to variations in sodium aluminate content and it was essential that metering operations were carried out to an accuracy $\pm$ 5%. To cope with these problems, a semi-automatic dispensing and batching plant was designed. From the batch plant the grout was transferred to chemical grout proportioning pumps situated in the field. This meant that the constituent solutions were not mixed until they reach a high velocity mixing head at the top of the grout hole thus permitting far greater control over gel times and hence penetration.

Table 1 shows the total volumes of cement and chemical grout injected into zones with a residual permeability requirement of 1 x 10^{-4} cm/sec, compared with the theoretical volume treated.

Zone	Theoretical Volume Treated (m^3)	Volume of Cement Grout Injected (m^3)	Ratio of Cement Grout Volume to Treated Vol.	Volume of Chemical Grout Injected (m^3)	Ratio of Chemical Grout Volume to Treated Vol.	Ratio of Total Grout Volume to treated Volume
A1	6,900	787	0.114	1,068	0.155	0.269
A2	6,900	276	0.04	858	0.125	0.165
B	4,515	754	0.167	148	0.033	0.183
C	5,499	594	0.11	166	0.03	0.14
TOTAL	23,715	2,411	0.102	2,240	0.094	0.196

Table 1. Ratio of Grout Volumes to Treated Volumes

3.6 Water Testing

In order to test the efficiency of the grouting as work proceeded and to establish whether the required residual permeability had been achieved, a comprehensive water testing programme was established.

This was based on a series of permeability tests carried out through existing sleeved tubes at selected locations and horizons as instructed by the Engineer. Generally, however, tests were always carried out on interior lines in any grouted zone chosen, in order to provide at least one outer barrier as a restraint to flow. Both simple and multiple tests were carried out, but in all cases the pressure was always kept below the chemical grout injection pressure.

The following target specification was set by the Engineer for acceptance of the alluvial grouting:-

$\leqslant 1 \times 10^{-4}$ cm/sec	65%
$> 1 \times 10^{-4} \leqslant 5 \times 10^{-4}$ cm/sec.	25%
$> 5 \times 10^{-4} \leqslant 10 \times 10^{-4}$ cm/sec.	10%

The water test results actually obtained are shown in Table 2.

Zone	No. of Sleeves Tested	% of Holes Tested	$\leqslant 1\times10^{-4}$ cm/s		$>1\times10^{-4} \leqslant 5\times10^{-4}$ cm/s		$>5\times10^{-4} \leqslant 10\times10^{-4}$ cm/s		$>10\times10^{-4}$ cm/s	
			No.	%	No.	%	No.	%	No.	%
A2	171	6.0	129	75.4	21	12.3	15	8.8	6	3.5
A1	109	5.1	82	75.3	13	11.9	9	8.3	5	4.6
B	82	21.0	65	79.2	11	13.4	6	7.4	-	-
C	100	12.4	84	84.0	10	10.0	5	5.0	1	1.0
Overall U/S	291	8.6	231	79.4	34	11.7	20	6.9	6	2.0
Overall U/S & D/S	462	11.0	360	77.9	55	11.9	35	7.6	12	6.6

Table 2. Summary of Water Test Results in Alluvium (Note: U/S - upstream D/S - downstream)

It can be seen from these results that the final permeability of the grouted alluvium is not uniform, as Kw depends almost entirely upon the proportion of non injected voids which must always remain e.g. in silt and sandy-silt lenses.

As a final check on the permeability the Engineer requested additional tests to be carried out during the drilling for the installation of piezometers. In this instance, a 1.0m perforated AW drill rod was driven into the alluvium approximately 1 to 1.5m beyond the end of the drill casing. Water was then pumped through the rods at various increments of pressure and the flow recorded. By previously testing and calibrating the perforated rod in a material of known permeability it was possible to calculate the corresponding permeability of the alluvium. Results from these tests compared favourably with the sleeve test results.

4. Other Geotechnical Activities

Although the alluvial grouting and the diaphragm wall formed the greater part of the geotechnical sub-contract, considerable rock grouting was also carried out in order to complete the cut-off beneath the diaphragm wall and to form wing curtains in the abutments.

Rock grouting started on the left abutment once the Main Contractor had stripped off the overburden and formed the key trench in the rock. The design required three rows of grout holes, the centre row on the dam axis for curtain grouting and the other two located 4m on either side for blanket grouting.

The depth of the grout holes were calculated from the following formulae:-

Curtain holes to abutments	$\frac{1}{2}$H + 15m
Curtain Holes beneath Diaphragm Wall	$\frac{1}{2}$H + 5m
Blanket Holes	1/3H + 10m

where H is the height of top water level above rock head.

In general, the holes on the abutments were drilled at 30° to the vertical, with primary holes at 6m centres and secondaries midway between. Beneath the diaphragm wall, the holes were drilled vertically with primaries at 4.7m centres.

The majority of rock grouting was carried out using the descending stage method. However, in places the abutments were quite steep (up to 30° from the vertical), and as grout consumption was very low the Engineer allowed ascending stage grouting to alleviate the problem of continually setting-up the drill on the slopes.

All holes were drilled with rotary percussive drills using 50mm diameter cross-bits and high pressure water flush. After drilling, each stage was flushed out, and a simple 10 minute (0.25 bar/m) water test carried out prior to grouting.

Grouting started with a 6:1 water:cement (by weight) mix which was progressively thickened as necessary up to 1:1 mix and injected at a pressure of up to 0.35 bar/m. Re-circulatory grouting was used and grouting ceased at the limiting pressure or when grout consumption fell below 2 litres/minute, averaged over a 10 minute period.

Special multiple pressure water test control holes were drilled before and after grouting to determine the efficiency of the grouting. These holes were cored (NW 61mm core dia.) at 45° or 30° to the vertical across the curtain line.

Water tests were undertaken over each 5m stage in an unbroken sequence at varying pressures in the ratio of 1:2:3:2:1 (1 = 0.1 bar/m depth) with each of these five pressures held for 10 minutes and the flow recorded. These results were then tabulated and the corresponding Lugeon value calculated.

In all, 20,000m of drilling was carried out and 420 tonnes of cement injected in the rock grouting programme.

In addition to the rock grouting, other geotechnical requirements were 2310m of rock anchors/bolts, cavity and pressure grouting to a 315m long, 4m diameter diversion tunnel and 2,000m of drainage holes and pressure relief wells.

Acknowledgements

The Author is grateful to Mr. Lytras, Director of the Water Development Department, Ministry of Agriculture and Natural Resources, Cyprus, for permission to publish this paper. He also acknowledges the assistance received from Mr. J. Fleming, Partner of Sir M. MacDonald & Partners, Dr. P. Creighton, Resident Engineer, and Mr. K. Antoniades, Project Manager for the Joint Venture, Joannou and Paraskevaides Limited and Medcon Construction Limited.

References

Cambefort, H. (1977) "The Principles and Applications of Grouting", Q.Jl.Engng.Geol.Vol.10, 57-95.

Einstein H.H. and Schnitter, M. (1970), "Selection of Chemical Grout for Mattmark Dam". J.Soil Mech. and Found.Divn, SM6, A.S.C.E., Vol. 96.

Glossop, R. and Ischy, E. (1962) "An Introduction to Alluvial Grouting", Proc. I.C.E., Vol. 21, 449-474.

Karol, R.H. (1960) "Soils and Soils Engineering", Prentice-Hall.

Littlejohn, G.S. (1982)"Design of Cement Based Grouts", A.S.C.E. Geotechnical Engineering Speciality Conference, New Orleans.

Maag, E. (1938) "Veber die Verfestigung und Dichtung des baugrundes (injektionen)", Course on soil mechanics of the Zurich Technical School.

Raffle,J.F. and Greenwood, D.A. (1961) "Relations between the rheological characteristics of grouts and their capacity to permeate soil", Proc.5th Int.Conf.Soil Mech. and Found.Engng., S. 789-794.

REMEDIAL GROUTING AT HILTON FALLS DAM, CANADA

By Michael G. Jefferies[1] and John B. Davis[2]

ABSTRACT

A remedial cement grouting program was employed to reduce seepage through the east abutment of Hilton Falls dam, near Milton, Ontario. Constructed in 1972-3, the dam provided satisfactory service until 1978. In June 1978, dirty seepage was observed flowing from the downstream face of the dam and from the east abutment. A rapid lowering of the reservoir was carried out, and an investigation started into the cause of the problem. Four boreholes put down in the east abutment area generally confirmed the stratigraphy encountered during the initial site investigation. A grout curtain in the east abutment area was recommended as an appropriate remedial measure.

Grouting was carried out during the winter of 1979/80. The remedial program involved the drilling of 45 holes and injection of 24,400 gallons of relatively thick grout. The reservoir was subsequently refilled, during which time piezometric conditions in the dam were frequently monitored. The remedial grouting completely stopped seepage from the downstream face of the dam, substantially reduced seepage from the east abutment, and substantially reduced the piezometric pressures in the dam and foundation downstream of the grout curtain.

Piezometric pressures in the dam have continued to be monitored and the dam has now given satisfactory service for more than one year.

THE SITE AND GEOLOGY

The site of Hilton Falls dam (for location, see Figure 1) is a 'U' shaped valley, cut 100 ft. deep into the Niagara Escarpment by a glacial or pre-glacial predecessor to the present drainage system. The valley walls have been

---

[1]Senior Civil Engr., Nova Corporation, Calgary, Alberta (formerly with Golder Associates, Toronto, Ontario)

[2]Principal, Golder Associates, Toronto, Ontario.

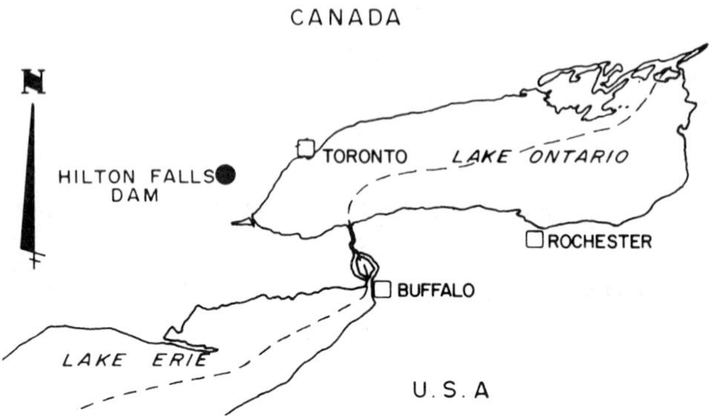

FIG. I SITE LOCATION

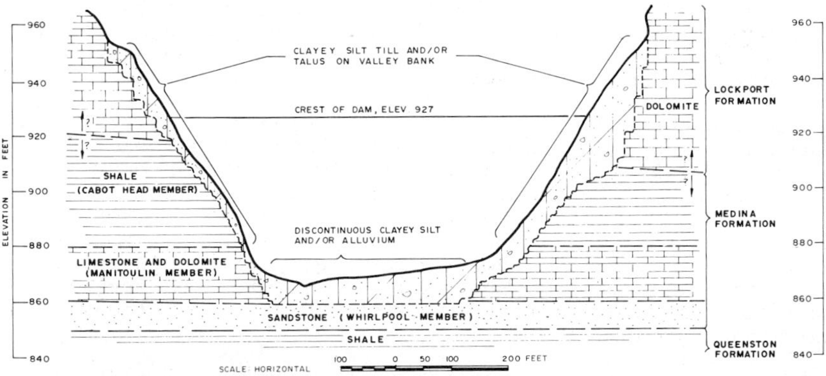

FIG. 2 SECTION SHOWING SIMPLIFIED LITHOLOGY ACROSS THE SITE

'plastered' with shale till and/or talus. However, it is not possible on the basis of borehole samples to differentiate between glacially transported soil and what may be classified as residual soil derived by weathering of the underlying shale bedrock. The valley floor is underlain by thin and discontinuous deposits of glacial till and alluvium.

Bedrock at the site comprises, in order of increasing geologic age, Lockport Formation, Medina Formation and Queenstone Shale. The Lockport Formation is a dolomitic limestone which forms the cap rock of the Escarpment, and extends down to an elevation some 10 to 20 ft. below the crest of the dam. The Medina Formation comprises interbedded shales, sandstones and dolomite, and is generally fractured. The Medina is usually 50 to 60 ft. thick and forms the lower portion of thc valley. Finally, the Queenstone Shale, which extends to considerable depth, consists of fairly sound to sound reddish brown shale although the upper few feet are weathered and altered.

A generalized geologic section of the valley is shown on Figure 2.

THE DAM

The dam, constructed in 1971-73, is a 60 ft. high earthfill embankment (crest elevation 927) with a culvert type spillway. The upstream face slopes at 3 horizontal to 1 vertical, and the downstream face slopes at 2 horizontal to 1 vertical, with a berm at about elevation 900. A sloping clay core is incorporated in the upstream portion of the dam together with a graded filter system. A plan of the dam is shown on Figure 3 and a section on Figure 4.

During site preparation and excavation of the cut-off trench in late 1971, the rock forming the valley floor (Whirlpool sandstone, see Figure 2) was found to be quite pervious, and curtain grouting was carried out across the valley floor with some primary groutholes extending into the abutment. Based largely on economic considerations, complete grouting of each abutment was not carried out as it was considered that the 'till' would act as an upstream blanket.

POST-CONSTRUCTION SEEPAGE

Seepage from bedrock was observed for a considerable distance downstream of the dam along the east valley wall during first filling of the reservoir in 1973. This seepage was anticipated during the design phase and was neither a constraint on the operation of the dam nor a shortfall in performance. Little change in the seepage pattern appears to have occurred until June, 1978.

Early on June 1, 1978, when the reservoir was at about elevation 914, seepage of "dirty" water was observed on the downstream face of the dam. The main seepage was concentrated in an area about 90 ft. east of the spillway and between elevations 872 and 877 where small cavities were observed in the dam face (see Figure 3). The entire face of the dam over a distance of about 100 ft. east of the spillway was damp or spongy below approximately elevation 875.

An emergency lowering of the reservoir was initiated during the morning of June 1, 1978, because of the volume

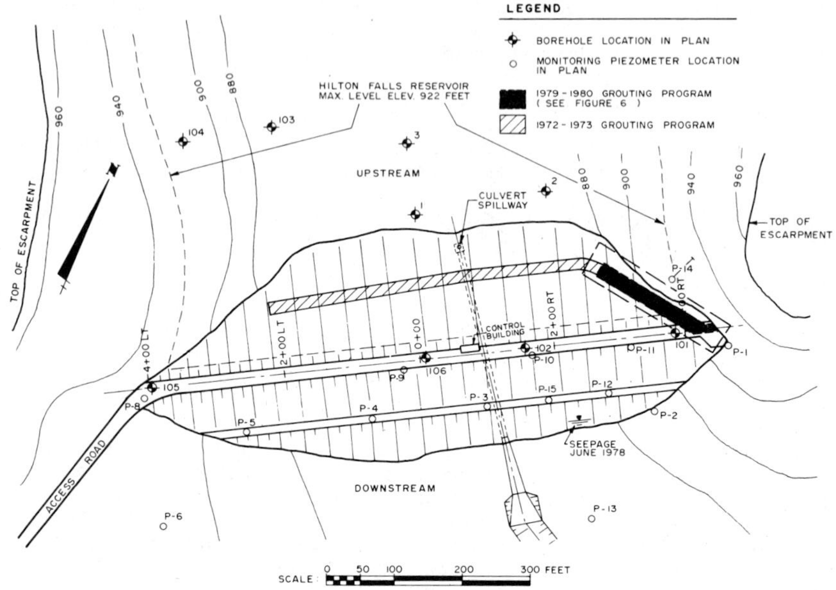

FIG. 3 PLAN OF DAM

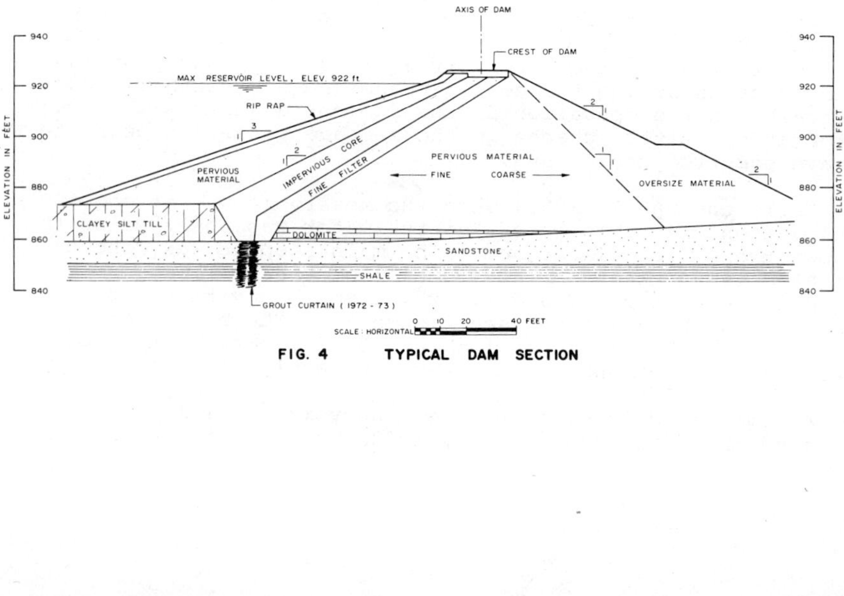

FIG. 4 TYPICAL DAM SECTION

of the seepage and the fact that the seepage now contained fines. In addition, a toe stabilizing berm was constructed extending from the east abutment to the spillway outfall.

THE INVESTIGATION

The seepage conditions in and around the east abutment were investigated during June and July 1978, by drilling boreholes, carrying out Lugeon pressure packer tests and installing piezometers.

The results of the additional borings in the east abutment area generally confirmed the stratigraphy disclosed during the pre-design investigation. The pressure packer test results were not significantly different from those carried out during the pre-design investigation, and are illustrated on Figure 5. Monitoring of water levels in the new and original (1973) piezometer installations indicated that the existing grout curtain was continuing to function but that a steep hydraulic gradient existed in the east abutment area in both the bedrock at depth (dolomite and sandstone) and the contact between either the dam fill or native till and the bedrock. It appeared likely therefore that seepage was "spilling" around the end of the dam and exiting from the bedrock into the downstream portion of the dam fill.

Examination of the dam operating records indicated that seepage appeared to only become evident when the reservoir rose above about elevation 900 ft. One of the recommendations from the seepage investigation was therefore to continue operating the dam subject to frequent piezometric monitoring and inspection of the seepage zone (the dam by now had a stabilizing berm in place, complete with filter system channelled into a 'V' notch weir). The dam was operated in this condition for over one year and much valuable piezometric data accumulated. This data forms the 'before' part of the 'before and after grouting' data presented later in this paper.

PROPOSED REMEDIAL GROUTING AND CONTACT

The results of the seepage investigation suggested that extension of the existing grout curtain into the east abutment was the most feasible and certain way of preventing excessive seepage through the east abutment area. Guidelines by Houlsby (3) for grouting of bedrock to control seepage beneath a narrow core earth/rockfill dam by a multiple row grout curtain, would imply a target permeability of 5 to 7 Lugeons on completion of grouting. By definition, one Lugeon unit is equal to a water take of one litre per metre per minute at a pressure of 10 bars.

The designed remedial grouting was a mixture of a three row curtain and a closure grouted single row curtain, the

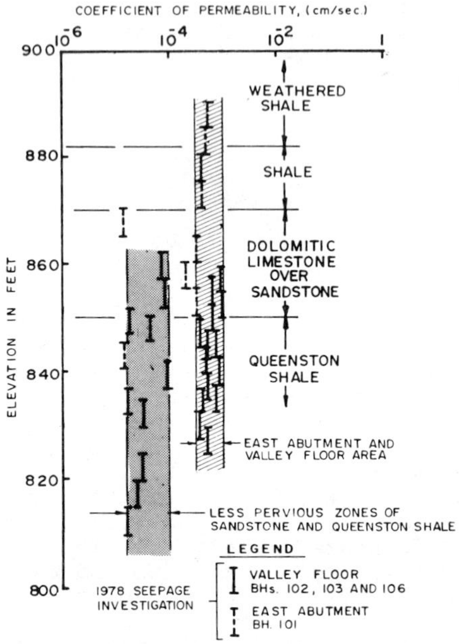

FIG. 5 PRESSURE PACKER TEST RESULTS (1978)

grouting being carried out along the contact of the dam core and original ground, see Figure 6. The reason for this choice was that it was not possible to dewater the reservoir completely. Grouting had therefore to be carried out in the presence of a significant differential head (about 15 ft.). The grout specified was a water/cement/ bentonite mixture, to be blended with a high speed mixer. The primary hole spacing was set at 20 ft. based on the 1972-3 grouting experience at the site. The method of drilling grout holes was not specified, other than a minimum hole diameter (2 inch) and water flush. Grouting of individual holes was to be by the upstage technique with a Lugeon water pressure test of each stage of each primary hole prior to grout injection.

The contract was called on a unit basis with a cost item for each of the contractor's significant identified costs, generally following the concept of method related charges (Barnes, 1). No attempt was made by the owner to shed risks associated with the subsurface conditions onto the contractor. The contractor allowed the resident engineer considerable freedom of action and indeed imposed on him the duty of technical direction of the grouting. Quantities were estimated using the data from the 1972-3 grouting at the site.

Five bids for the work were received, all of them unqualified. The bid range was from $105,000 to $179,000 with three bids less than $115,000, the latter figure being the project estimate.

INITIAL GROUTING OPERATIONS

Grouting operations started on December 6, 1979, some 18 months after the seepage problem was first observed.

Grouting was carried out in four phases, designated primary, secondary, tertiary and quaternary. The associated groutholes were designated A, B, C and D series respectively and the complete layout of groutholes is shown on Figure 6.

The groutholes were drilled using casing through the dam fill and triconing uncased holes in bedrock. Drill-water return was lost shortly after penetrating bedrock in most holes and rock slabs layered with soft clay and/or voids were encountered in the upper bedrock. Below this highly weathered shale/clay layer, the rock was very heavily fractured and rock wedges generally slid into the primary holes after the drill rods were removed. Immediately re-drilling the hole often did not prevent the hole remaining obstructed when the drill rods were removed a second time. This tendency caused considerable difficulty in getting packers in and out of the holes and prevented the vast majority of the scheduled water tests in the primary groutholes from being satisfactorily completed. Only two water tests were successful in the primary phase; both were in the same borehole A1, which is where the new grout curtain intersected the old, (see Figure 7). The results indicated permeabilities of about 2×10^{-3} cm/sec. - marginally greater than encountered during either the initial site investigation or the seepage investigation (Figure 5). However, these two results were obtained in a borehole with little fracturing or caving.

It was decided, in light of the difficulty with the Lugeon water tests, to carry out the primary grouting in essentially two stages. First, the hole would be filled with grout using a tremie pipe, subsequently seating a packer in the casing and pressure grouting from the casing. Second, the casing would be extracted from its seat in bedrock and grout pressure/flow maintained so as to inject grout in the bedrock/overburden or bedrock/core contact. Grouting was initially carried out using a 2:1 water/cement mix with 5% bentonite as a stabilizing agent (water/cement ratios are expressed by weight, while bentonite is given as percent by weight of cement). It became obvious very rapidly that

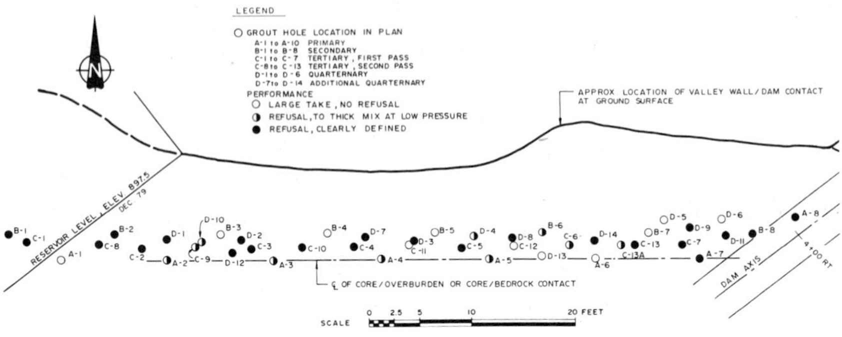

FIG. 6 GROUT HOLE LOCATION PLAN

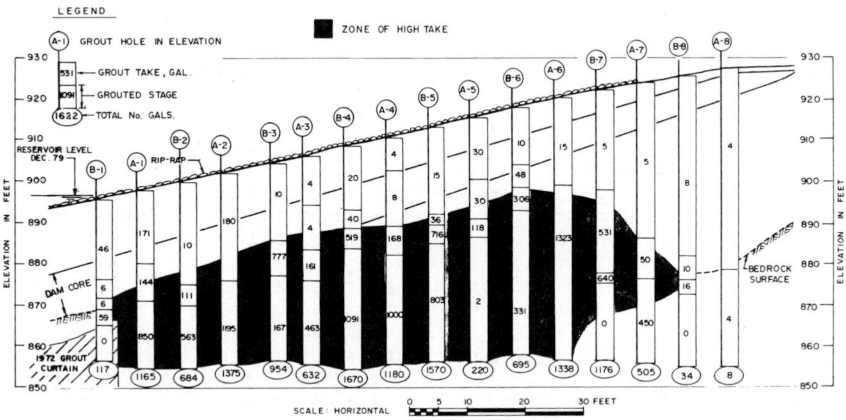

FIG. 7 PRIMARY AND SECONDARY HOLES IN ELEVATION

this grout was too thin, typically flowing into the grout-holes in large quantities (600+ gallons) with minimal injection pressure required. The ease of injecting grout was inferred to mean that fissures were relatively open. Accordingly, it was judged important to use a no-bleed grout so as to avoid significant secondary fissures being created as the grout set. The grout sequence adopted for each stage of the primary holes was to inject 500 gallons of 2:1:10%, 500 gallons of 1:1:5% and then 0.75:1:5% as required to refusal, the latter grout having the consistency of toothpaste. However, if pressure build up or flow reduction occurred with a particular grout, then injection of this grout was continued until refusal without changing to a thicker mix.

The volume of 500 gallons was calculated using the fracture porosity observed in rock exposures, the average stage length and a desired radius of penetration - the predicted volume then being near doubled to obtain 500 gallons. Grout injection pressure was allowed to build up until 20 psi was reached, 20 psi being treated as an upper limit. The grout flow rate/pressure relationship was continuously monitored and plotted so that hydrofracture could be immediately detected and the grout pressure reduced.

The pattern of primary grout take is illustrated on Figure 7. It can be seen that the actual east abutment (holes A8 to A10 inclusive) was relatively tight. The grout takes in the remainder of the holes indicates that the bedrock was extremely pervious as was the overburden/bedrock contact. Both were substantially more pervious than the seepage investigation program had suggested.

SECONDARY GROUTING

The secondary groutholes were located approximately 10 ft. upstream of the primary line of groutholes.

Injection of grout was carried out using the same techniques and pressures as outlined for the primary holes. Grout takes were essentially as large as the primary takes (Figure 7). Lugeon water tests were attempted but were unsuccessful because of the continuing problem of rock wedges sliding into the groutholes.

The flow from the toe stabilizing berm decreased significantly during the course of secondary grouting. The flow of water had been about 5.5 gal/min prior to the start of grouting and remained unchanged during the primary grouting. However, by the time 50% of the secondary holes had been grouted, the flow had reduced to just 0.4 gal/min and, by the time secondary grouting had been completed, the flow had entirely stopped.

TERTIARY GROUTING

The required nominal tertiary spacing was 10 ft. and tertiary grouting was therefore carried out in two passes, working up-slope in each case. The tertiary holes were located in plan midway between the lines of primary and secondary groutholes (Figure 6).

The problem of rock wedges sliding into the groutholes was found to be much reduced and it was possible to carry out successful Lugeon tests. The results of these tests

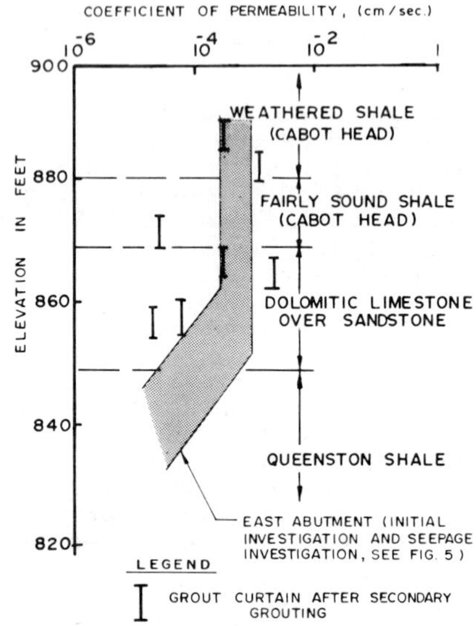

FIG. 8 PERMEABILITIES MEASURED AFTER SECONDARY GROUTING

are compared on Figure 8 with the data from the preceeding investigations. The comparison indicated that the permeability of the bedrock had been reduced to less than the surrounding ground with the exception of a zone in the weathered Cabot Head Shale and a zone within the upper dolomitic limestone.

It was decided to continue with the essentially one stage injection technique, because the water test data indicated that there were essentially only two pervious zones, each of limited thickness and each having very similar permeabilities (2×10^{-3} cm/sec. or 200 Lugeons). Grout for the tertiary injection was typically 1:1 water/cement for the holes with Lugeon values of 50 or more and 2:1 for less pervious holes. It was not necessary to reduce the water/cement ratio of grout mixes to achieve refusal although in some holes the bentonite content was progressively increased from 10% to 20%.

The pattern of grout take for the tertiary holes is shown on Figure 9. The total tertiary take was approximately 40%

of either primary or secondary takes. The take pattern indicated however, that two pervious 'holes' remained in the grout curtain, adjacent to grouthole C9 and between C11 and C13.

It is interesting to note that the tertiary grouting indicated a one-for-one relationship between Lugeon value (or permeability) during water tests, and initial grout flow rate for 2:1:10% to 2:1:15% grout (typical variation less than ± 10% for the same injection pressure). This correlation was used together with the water test data itself to conclude that the lower bedrock was relatively tight. This conclusion was checked with a further two water tests after completion of the tertiary grouting, which suggested that the lower bedrock in the grout curtain could be characterized by an upper bound Lugeon value of about 6.

QUATERNARY HOLES

Six quaternary holes were planned to close the two 'windows' in the tertiary grout line. Subsequently, the large grout takes encountered in some of these six holes indicated that yet more quaternary holes were required. Eventually, a total of fourteen holes were drilled, the holes being as close as 2.5 ft. centre to centre. The quaternary holes were drilled only 5 to 7 ft. into bedrock below the Cabot Head Shale.

The pattern of quaternary grout takes is illustrated on Figure 9. Some of the quaternary takes were large, of the same order as the primary holes. However, the holes with the large takes (D5, D6 and D13) were located at the edge of the three rows of groutholes (Figure 6). The response of these holes is, therefore, with hindsight, reasonable. It follows that the grout curtain is only a few feet thick at most. The remainder of the quaternary holes averaged 2/3 of the tertiary grout take, and the grout/water flow rate correlation suggest the pervious zones could be characterized by 20 to 30 Lugeons prior to the quaternary grouting, although a few holes were substantially tighter than this, 10 Lugeons or less.

PIEZOMETRIC RESPONSE BEFORE GROUTING

There are 14 piezometers installed in 7 boreholes in the east abutment area (Figure 3).

The reservoir elevation upstream of the dam fluctuates as dictated by operational requirements, producing a corresponding fluctuation in the piezometric elevations within the the dam. A clearer understanding of the piezometric behaviour can be obtained by plotting piezometric elevation against reservoir elevation. A plot for the upper piezometer at monitoring location P11 is shown on Figure 11.

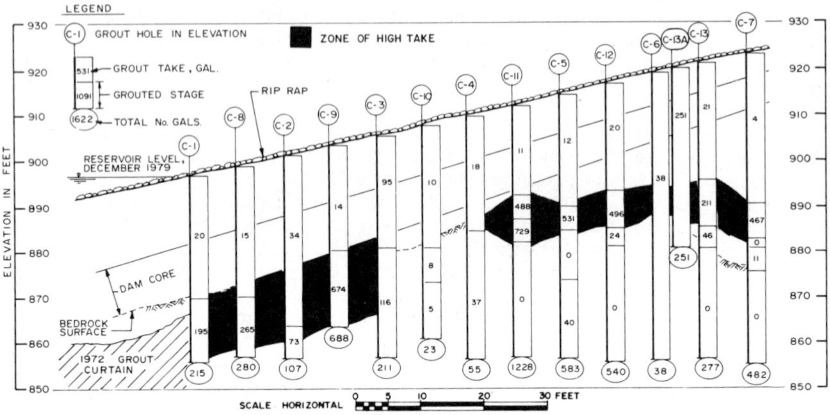

FIG. 9 TERTIARY HOLES IN ELEVATION

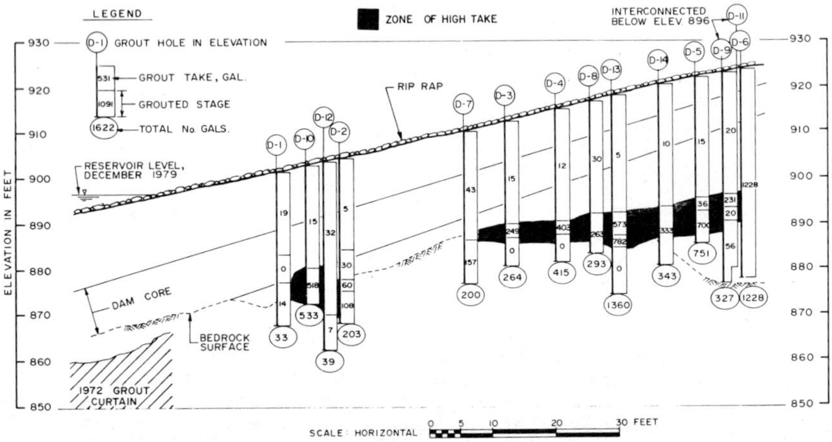

FIG. 10 QUARTERNARY HOLES IN ELEVATION

This piezometer measures the piezometric elevation in the dam fill at the fill/overburden contact approximately midway between the dam core and the downstream toe. Figure 11 illustrates that a 1 ft. rise in reservoir elevation produced nearly a 1 ft. rise in piezometric elevation at this piezometer prior to extension of the grout curtain, (on average, the response was 95% of the reservoir change). Further, the piezometer indicated virtually reservoir elevation.

The other 13 piezometers in the east abutment area all exhibited similar excessively large responses to reservoir elevation.

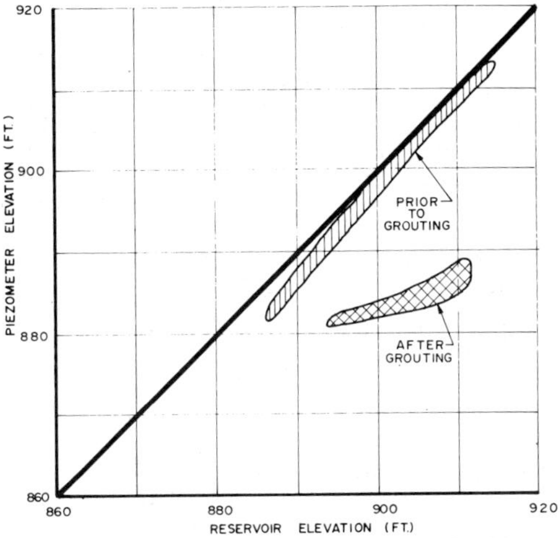

FIG. 11 P11 UPPER PIEZOMETER

PIEZOMETRIC RESPONSE AFTER GROUTING

One month was allowed to elapse following completion of the grouting before increasing the reservoir elevation. The piezometric response of the upper piezometer at location P11 following completion of grouting is shown in detail on Figure 11. This piezometer is the closest downstream to the curtain, and provides the clearest measure of the performance of the grout curtain. The remaining east abutment area piezometers all indicated a similar favourable response, the magnitude of the change in response becoming less with distance from the grout curtain.

The piezometric response measured in the upper piezometer at P11 during first filling of the reservoir following completion of grouting was approximately 50% of the maximum differential head across the dam. However, an inspection of the piezometric/reservoir elevation plot, Figure 11, indicates that the apparent datum hydraulic head for this piezometer is no longer the tailwater elevation (859 ft. approx.) but rather an elevation of about 875 ft. The piezometric response in terms of the 875 ft. datum is approximately 38% of the differential head. This can be contrasted with a response of 95% of the differential head prior to grouting. It is interesting to note that piezometric elevation measured at the piezometer started declining early in February, 1980, and declined steadily until February 25, 1980. This corresponds to the tertiary and quaternary phases of grouting. The primary and secondary grouting had no effect on the piezometric elevation at this piezometer.

Cedergren (2) has published an analysis of the line of seepage in an earthfill dam of similar geometry to Hilton Falls for various thicknesses and permeabilities of grout curtains. The post-grouting behaviour of the upper piezometer at P11 approximately corresponds to Cedergren's analysis for a grout zone which is 10% of the foundation permeability, and in which the ratio of the width of the grout zone to the depth of the reservoir is 0.2. In addition, the flow through the V-notch weir is now less than 0.1 gal/min after grouting (and frequently zero) compared to more than 5 gal/min before, a reduction of some 98 percent or more.

It was noted earlier than, on completion of grouting, the curtain probably could be characterized by an upper bound of 6 Lugeons at depth with zones of 10 Lugeons close to the bedrock/overburden contact. The available data on the bedrock permeability away from the remedial grout curtain (Figure 5) shows permeabilities in the more pervious strata ranging from 4 x 10^{-4} cm/sec to 1 x 10^{-3} cm/sec (about 40 to 100 Lugeons). The likely permeability contrast between the grout curtain and surrounding bedrock is therefore about 10. This corresponds well with the contrast ratio deduced by working back from Cedergren's analysis.

CONCLUSIONS

1. A grout curtain was successfully constructed to curtail the seepage through the east abutment of Hilton Falls Dam.

2. Data collected during the grouting and the subsequent performance of the dam provide evidence confirming the adequacy of the target permeabilities presented by Houlsby (3), at least in connection with narrow core earth fill dams.

3. The piezometric data collected before and after grouting clearly demonstrates that a grout curtain can substantially change the piezometric pressure regime within a dam and its foundation. At Hilton Falls, a permeability contrast of about one order of magnitude was established between the surrounding bedrock and the grout curtain. An analysis presented by Cedergren (2) fitted the data well.

ACKNOWLEDGEMENTS

The writers wish to acknowledge the permission of the owner of Hilton Falls dam, the Halton Region Conservation Authority, to publish the data contained in this paper.

Further, the writers gratefully acknowledge the assistance of the Authority's staff, and in particular, that of Mr. M.E. Stephen, General Manager and Mr. Carey Moore, P. Eng., Hydrology Engineer. Finally, we would like to thank all our colleagues in Golder Associates who were involved with the project and preparation of this paper.

APPENDIX - REFERENCES

(1) Barnes, M.; "Measurement in Contract Control", Institution of Civil Engineers, 1977.

(2) Cedergren, H.; in "Embankment Dam Engineering, Casagrande Volume", editors Hirschfield & Poulos, publ. Wiley, 1973.

(3) Houlsby, A.; "Engineering of Grout Curtains to Standards", J. Geotechnical Eng. Div., ASCE, Volume 103, GT9, 1977.

FOUNDATION GROUTING OF THE LG-2 MAIN DAM LA GRANDE COMPLEX, QUEBEC

BY Z. AZIZ, M. ASCE* and J. LEVAY **

ABSTRACT

Located on the La Grande River in the James Bay Region of northern Quebec, the main dam at LG-2 is a 552 foot (168,2 m) high and 9,272 foot (2 826,1 m) long earth rockfill structure. The dam rests on bedrock except locally in the riverbed where some alluvial material was left under the downstream shell.

Taking into consideration the abundance of rebound fractures and open joints near surface and the relative tightness of the foundation bedrock at depth, a program of shallow blanket grouting and a single line grout curtain was designed. The remote and harsh winter conditions of the Canadian North were also taken into account by making provisions in the specifications to monitor the freeze and thaw cycles of the bedrock and to grout the foundation, only when it was frost free. This particular aspect posed problems to the planners and the builders but was overcome by careful scheduling.

The project was commissioned in late 1979 and so far, all indications are that grouting has effectively sealed the foundation.

INTRODUCTION

La Grande Complex (Figure 1), now under construction in the James Bay region of north western Quebec, Canada, is one of the world's major hydroelectric undertakings. The Complex is being developed and managed by Société d'énergie de la Baie James, a wholly owned subsidiary of Hydro-Quebec.

Located 650 miles (1046 km) north of Montreal, the La Grande Complex comprises a series of projects (Figure 2) that develop, as a first phase, the hydroelectric potential (Figure 3) of the lower reach of the La Grande River which has a net head of 1235 ft (376,4 m). The river flows in an east-west direction for 535 miles (856 km). In order to increase the potential, parts of the adjacent drainage basins of the Caniapiscau and Eastmain rivers are diverted into the La Grande basin [1]. The overall development includes nine powerhouses having an annual generating potential of 13 750 MW.

---

* Chief Geology Division, James Bay Projects, ABBDL-TECSULT

** Deputy Head of the Department of Geology and Soil Mechanics Société d'énergie de la Baie James, Montreal, Quebec, CANADA

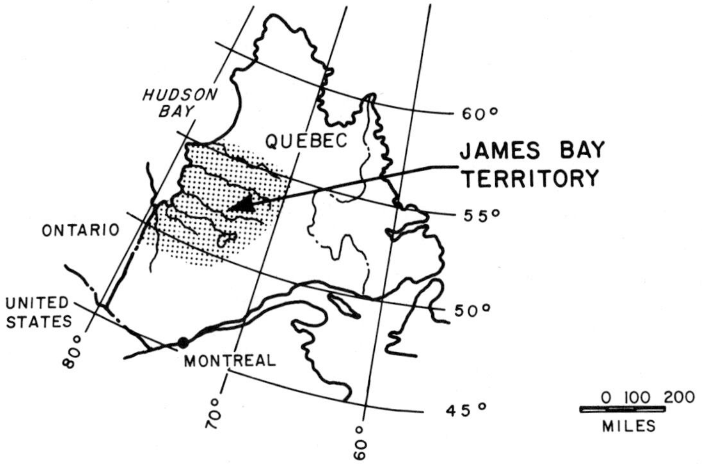

LOCATION PLAN (FIG. 1)

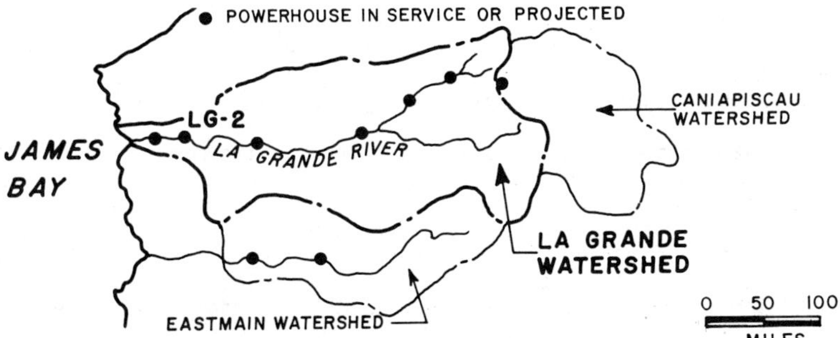

LA GRANDE COMPLEX (FIG. 2)

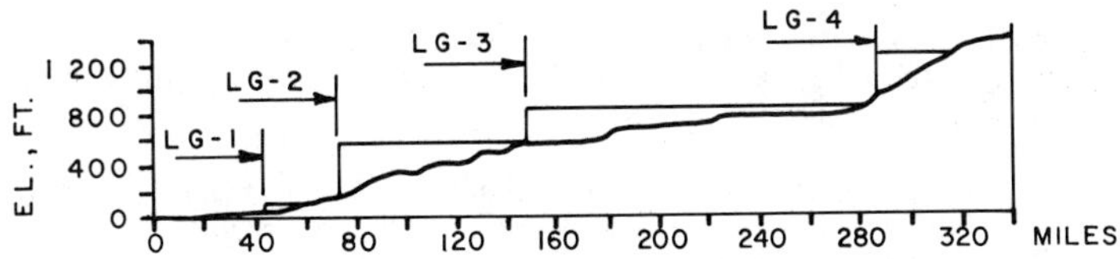

LA GRANDE RIVER-LONGITUDINAL PROFILE (FIG. 3)

The first phase of the construction involves three generating stations: LG-2, LG-3 and LG-4, along with their reservoirs and diversion works. Construction started in 1972 and, when completed in 1985, will generate 10 282 MW. First commercial power has been on line from LG-2, one of the world's largest underground powerhouses [2], since October 1979.

LA GRANDE 2 (LG-2) PROJECT

LG-2 (Figure 4) is situated 75 miles (121 km) upstream of the mouth of the river. The main dam [3] is of the earth rockfill type, 552 ft (168,2 m) high involving 30,333,500 yd^3 (23 193 x 10^6 m^3) of fill. The reservoir is enclosed by 31 dykes with a total crest length of 98,097 ft (29 900 m) and a fill volume of 30,083,448 yd^3 (23 x 10^6 m^3). The reservoir occupies an area of 1772 square miles (4589 km^2) and has an active storage of 25,322 x 10^6 yd^3 (19 360 x 10^6 m^3). The powerhouse is underground with 16 units and has a generating capacity of 5280 MW. The spillway is of the open channel cascade type [4] and can handle a flood of 540,000 cfs (15 282 m^3/s). The diversion was achieved with two 66 ft (20 m) diameter unlined tunnels with an average length of 2600 ft (792,5 m) and the height of the upstream and downstream cofferdams was respectively 180 and 70 ft (55 and 21 m).

CLIMATE

The climate of the region is of the continental type (i.e. four principal seasons with strong interseasonal fluctuations) and marked by extreme temperatures. The mean annual precipitation in the La Grande basin is about 30 inches (76 cm), 40% of which falls as snow.

The temperatures range between the lows in the minus fifties and the highs in the eighties (°F) as shown on table 1; the mean annual air temperature is 23°F (-5°C). The weather starts turning cold in September or early October and begins to warm up in late May, thus making the winters very long and cutting the construction period down to about five months a year. A typical set of freezing indices from various stations in the region for the years 1972 to 1980 is shown on figure 5.

PERMAFROST AND FREEZE-THAW MEASUREMENTS

The complex area is categorized in the "Southern Fringe of Permafrost" region by Brown [5] and in the "Scattered Discontinuous Permafrost" zone by Johnston [6] (Figure 6).

Aware of the problem and the consequences of grouting in a frozen rock mass which has a temperature below 32°F (0°C) or might have capillary or mineralization water at or below 32°F (0°C), a programme of freeze and thaw measurements was undertaken.

Observations were carried out using methylene blue frost indicators [7], copper constantan thermocouples [8] and resistance thermometers grouted in holes drilled in the bedrock. The depth of the holes varied between 20 and 30 ft (6 and 9 m) and their locations were chosen so that favourable (south facing slopes, deep snow cover, low wind

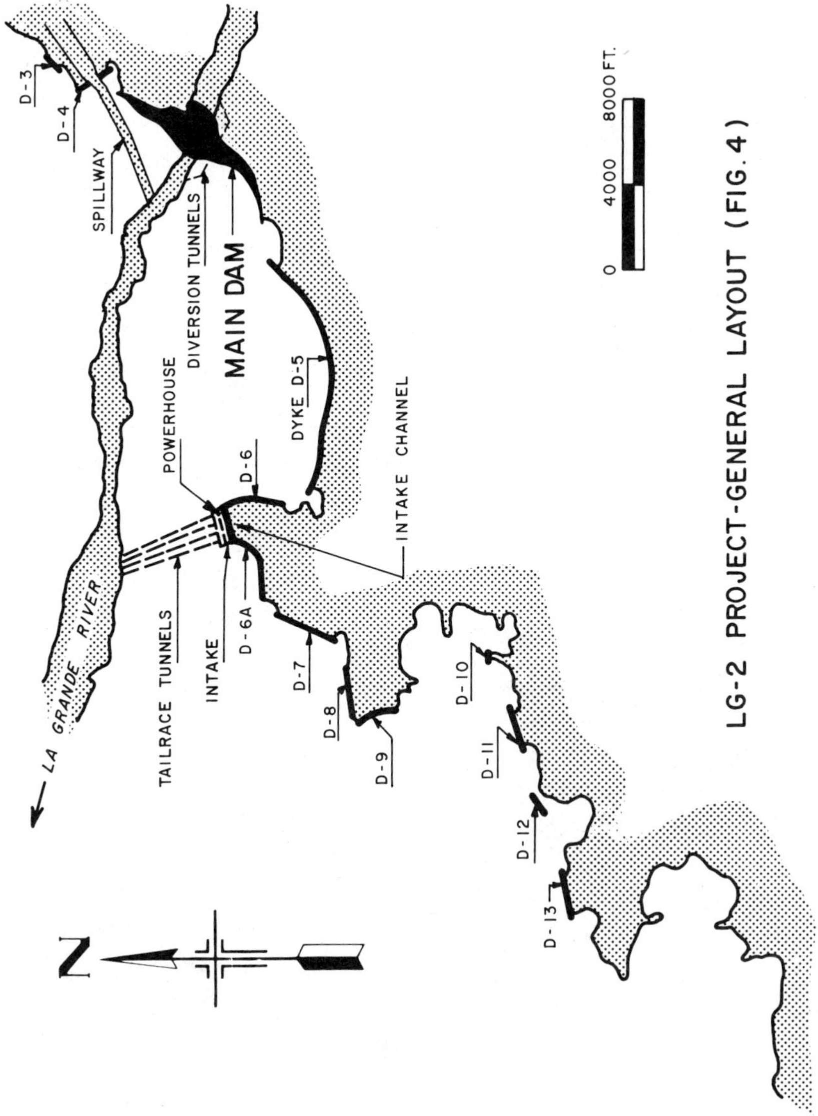

LG-2 PROJECT-GENERAL LAYOUT (FIG. 4)

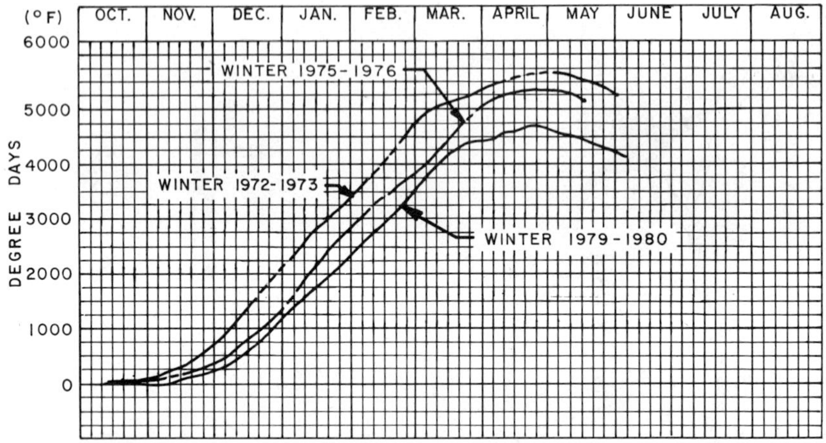

JAMES BAY TERRITORY FREEZING INDICES (FIG. 5)

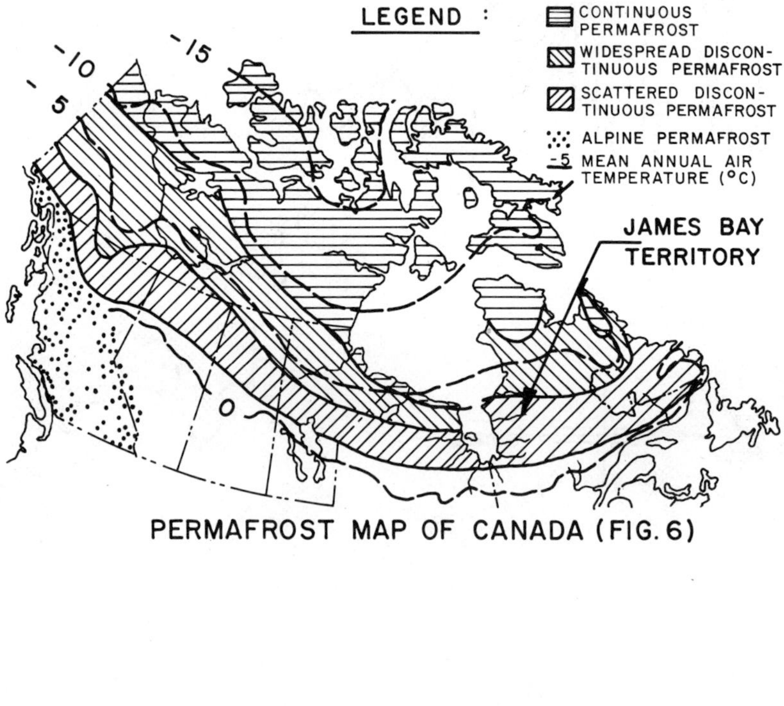

PERMAFROST MAP OF CANADA (FIG. 6)

TABLE 1

MONTHLY VARIATION IN TEMPERATURE *

MONTH	AVERAGE OF MAXIMUM DAILY TEMPERATURE		AVERAGE OF MINIMUM DAILY TEMPERATURE		AVERAGE TEMPERATURE		MAXIMUM TEMPERATURE		MINIMUM TEMPERATURE	
	°F	°C	°F	°C	°F	°C	°F	°C	°F	°C
January	1.3	-17,1	-19.5	-28,6	- 9.1	-22,8	39	3,9	-57	-49,4
February	7.0	-13,9	-15.2	-26,2	- 4.1	-20,1	38	3,3	-55	-48,3
March	18.5	- 7,5	- 4.9	-20,5	6,8	-14,0	47	8,3	-49	-45,0
April	32.2	0,1	11.3	-11,5	21.8	- 5,7	56	13,3	-31	-35,0
May	44.3	6,8	26.9	- 2,8	35.6	2,0	79	26,1	- 8	-22,2
June	58.7	14,8	40.4	4,7	49.6	9,8	90	32,2	20	- 6,7
July	64.8	18,2	48.6	9,2	56.7	13,7	88	31,1	33	0,6
August	61.4	16,3	46.6	8,1	54.0	12,2	84	28,9	33	0,6
September	50.9	10,5	38.5	3,6	44.7	7,1	79	26,1	19	- 7,2
October	38.7	3,7	27.4	- 2,6	33.1	0,6	66	18,9	1	-17,2
November	23.8	- 4,6	11.1	-11,6	17.5	- 8,1	47	8,3	-25	-31,7
December	7.5	-13,6	- 9.5	-23,1	- 1.0	-18,3	44	6,7	-52	-46,7
Annual	34.1	1,2	16.8	- 8,4	25.5	- 3,6	90	32,2	-57	-49,4

* Average of 29 years

factor, etc.) and unfavourable (north facing slopes, lack of snow cover, high wind factor, etc.) locations could be monitored. Simultaneously, a similar programme was established to measure the freeze and thaw in the glacial till deposits which were to be used in the embankment construction [9].

The results of this monitoring showed that the rock started freezing around October and thawed completely only between mid-June and mid-August, depending on the location and other factors. The depth of frost penetration varied between 22 and 26 ft (6,7 and 7,9 m) (Figures 7 and 8). It was also noted that as the rock thawed, thawing took place from the surface downwards with very little heat transfer and upward thawing. This observation was further reinforced by installing resistance thermometers in a few 100 ft (30,5 m) deep holes which show that the warming trend with depth is very slow and temperatures close to freezing linger on up to a depth of 45 ft (14 m).

These studies, which were undertaken at the design stage, indicated that there was no permafrost at the site.

GENERAL GEOLOGY

The Complex lies within the Canadian Shield, a glaciated peneplain, developed on Precambrian basement rocks of igneous and metamorphic origin. The LG-2 site is underlain by granitic rock ranging in texture from massive to gneissic. The bedrock is generally of excellent quality and practically unaltered. Faults and shear zones in various orientations are present, but usually are narrow and reconsolidated. The bedrock is dissected by a set of vertical east-west and north-south joints, both of which are tight.

Successive phases of Pleistocene glaciation culminating in the last advance of the Laurentian ice sheet at the end of Wisconsin period, stripped the old land surface to fresh hard surfaces and left deposits of glacial and proglacial deposits that cover most of the region. Till is widespread in the form of ground moraine, locally including some drumlin forms and provides an essential source of borrow material for the core in the dam and dyke construction. Glacio-fluvial deposits are abundant and are used as filter and transition materials.

The thickness of ice in the area is estimated to have been around 7500 ft (2300 m) during glaciation and the associated post-glacial isostatic rebound has caused extensive near-horizontal fractures in the rock. These rebound fractures are parallel to the surface and are more predominant in the top 30 ft (9 m) or so, but can be found at depth also. The fractures are often filled with silt and sand and near the surface tend to be open due to consecutive freeze and thaw cycles.

The area was the site of marine and lacustrine deposition during the recessional phase of glaciation and the marine deposits in the form of silty clays are extremely sensitive so that they cause various engineering problems.

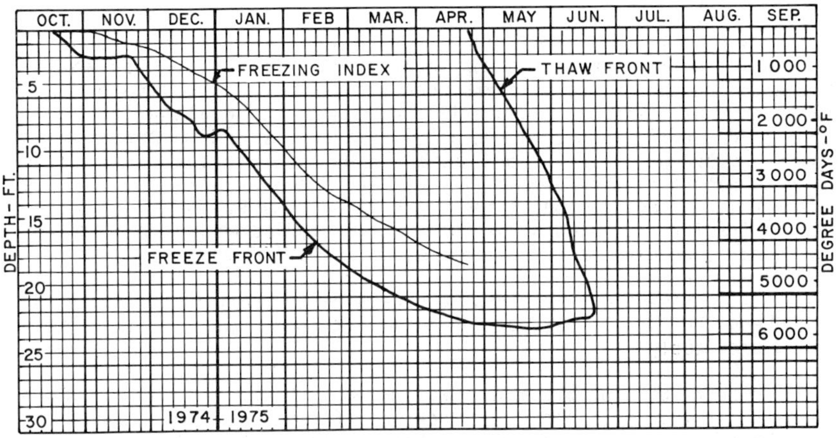

FREEZE & THAW CYCLE
NORTH BANK (FIG. 7)

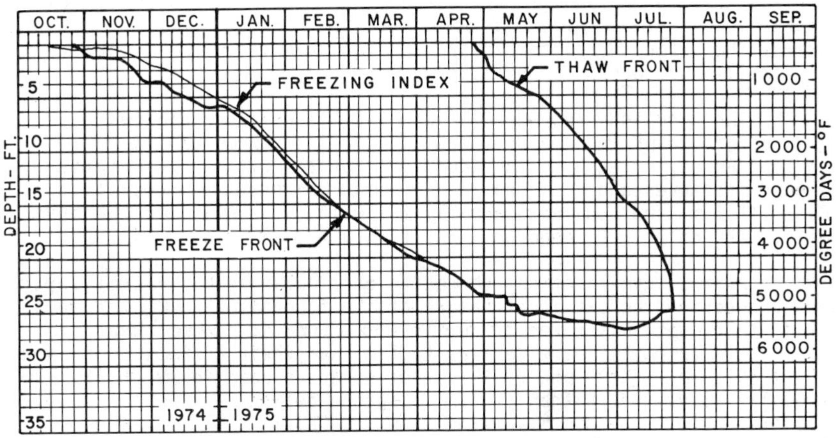

FREEZE & THAW CYCLE
SOUTH BANK (FIG. 8)

SITE INVESTIGATION

A site programme using the refraction seismic method and calibration and verification of the seismic results with diamond drill holes was the first investigation carried out. This helped establish the depth and the nature of overburden materials and thus the bedrock topography. NX calibre diamond drill holes were done along the abutments to determine the bedrock quality. Water absorption tests using double packers were carried out at 10 ft (3 m) intervals in all the holes to determine the permeability of the rock. Of particular interest was the determination of the depth to which previously mentioned rebound fractures existed. Borehole camera studies were done in selected holes to acquire more knowledge about these fractures.

In conjunction with this, a programme to monitor the freeze and thaw phenomenon in the bedrock and glacial till was undertaken as described previously.

The results of diamond drilling and borehole camera studies demonstrated that the foundation rock was of excellent quality and that the rebound joints frequency diminished significantly with depth.

DESIGN OF GROUTING PROGRAMME

As the thickness of overburden at the dam site was nominal, the dam cross-section (Figure 9) was based on a total excavation scheme. The typical section consists of a morainic core slightly inclined towards the upstream, protected by granular transitions and supported by rockfill shoulders.

The grouting programme (Figure 10) was divided into two parts; blanket grouting and curtain grouting. The depth of the blanket grouting holes, related to the occurrence of the previously mentioned rebound fractures and based on the exploration results, was fixed at 40 ft (12,2 m) and reduced gradually to 30 ft (9,1 m) along the abutments. The holes were done on a 20 ft (6,1 m) staggered grid.

As the investigations showed no particular weak zones at depth and considering the excellent nature of the foundation rock, the maximum curtain depth was fixed at about one third the water head. The holes were spaced at 20 ft centers (6,1 m) and their maximum depth was 170 ft (51,8 m). The depth of these holes was reduced to 30 ft (9,1 m) gradually along the abutments. Split spaced between these holes were shallower holes, the depth varying between 120 ft (36,6 m) and 30 ft (9,1 m).

All grouting was performed using the split spacing method. The grouting pressures measured at the hole collar were 1 psi/ft of rock cover (22,6 kPa/m) above 50 ft (15,2 m) depth in all curtain and blanket grouting holes, while below this depth the pressures were 1.5 psi/ft (34 kPa/m). Refusal criteria was fixed as the rate of absorption of cement in a section of 10 ft (3 m) being inferior to 0.10 ft^3 (0,0028 m^3) in 10 minutes using a 5:1 mix. The specifications were modeled after Elston [10].

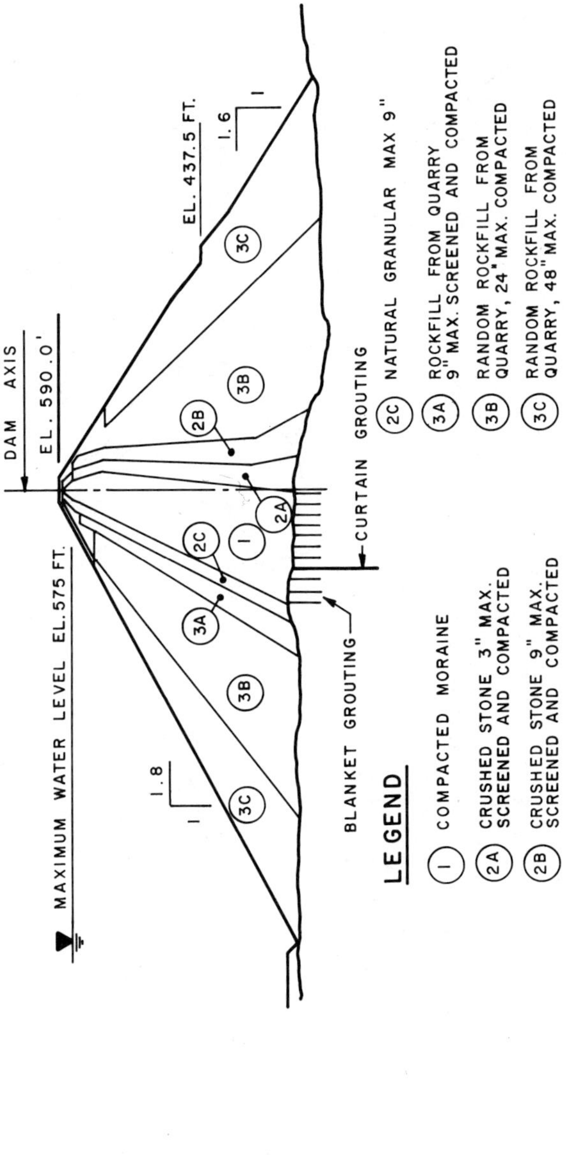

MAIN DAM-TYPICAL SECTION (FIG. 9)

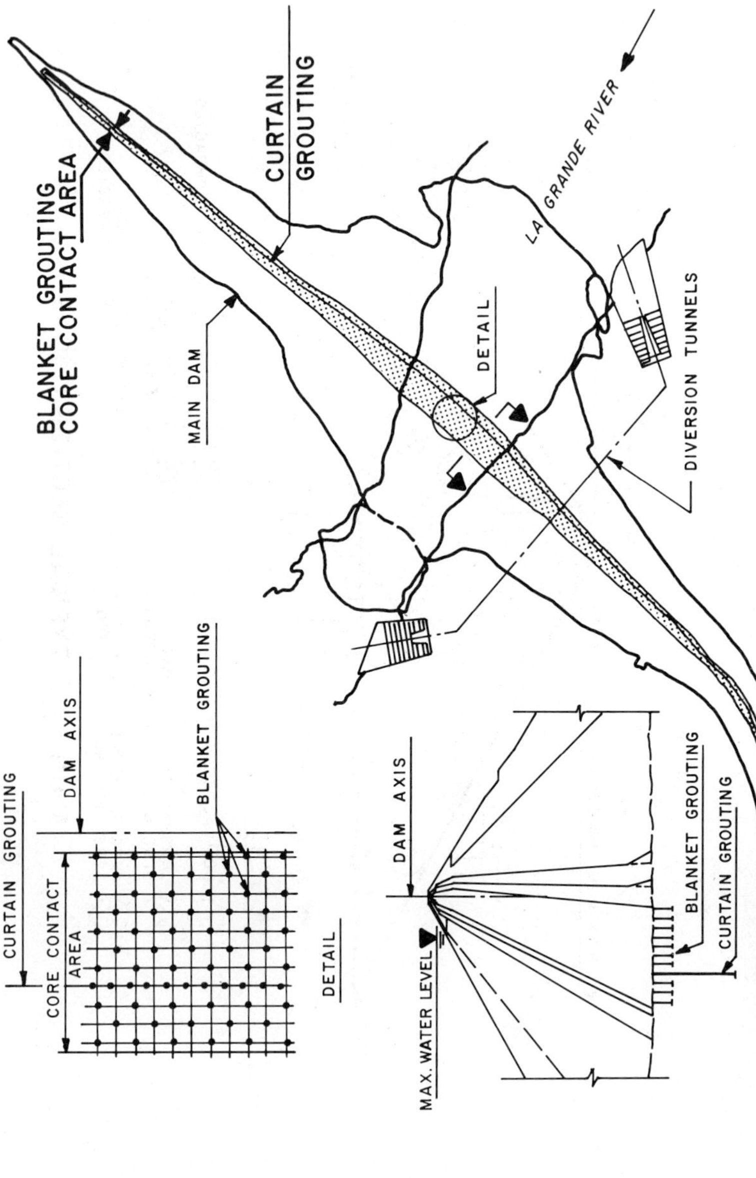

GROUTING (FIG. 10)

Realizing the futileness of grouting the rock which had a temperature inferior to 32°F (0°C), the specifications required that the contractor plan his grouting operations in such a way that during these operations the rock was never frozen and that the joints and cracks in the rock contained no frozen material. He was, however, allowed to grout the unfrozen rock through frozen layers of overburden or rock. The contractors were provided with the results of freeze and thaw data at the time of bidding to make them aware of the importance of the problem and its impact on overall construction schedule.

In instances where frozen rock was encountered, the contractor was required to thaw out the rock before grouting by circulating hot water or steam. In cold weather the contractor was required to keep the materials and water warm before grouting and to protect the grouted area against freezing for 24 hours after grouting.

EXECUTION OF THE GROUTING PROGRAMME

The contractor started his full grouting operations in 1975 and completed the work by June 1978. Five percussion drills with wash water were used to produce 2.5 in. (6,3 cm) holes and an equal number of grout plants were employed. Table 2 gives a summary of yearly grouting operations. The absorptions were very low, but normal for the type of rock grouted. The absorption for blanket grouting was generally higher than for the curtain grouting, as expected, because of the rebound fractures and near surface jointing. The average absorption per linear foot for blanket and curtain grouting combined was 0.19 ft^3 (0,008 m^3/m). In areas where risk of uplift existed, particularly in areas where rebound fractures were present, uplift gauges were installed to monitor any tendency to jack-up the foundation.

TABLE 2
SUMMARY OF YEARLY GROUTING OPERATIONS

	1975	1976	1977	1978	Cumulative
Blanket (ft)	42,460	25,474	49,810	1,632	119,376
(m)	12 942	7 764	15 182	497	36 386
Curtain (ft)	16,654	19,106	44,056	6,255	86,071
(m)	5 076	5 824	13 428	1 907	26 236
Cement absorption					
(pi^3)	5,472	3,353	25,592	4,922	39,339
(m^3)	153	94	717	138	1 102

The key to the success of the grouting programme was that during the short grouting period, the contractor would advance his grouting operations sufficiently ahead of the embankment construction, so that, the following year he could start placing the fill at the beginning of the summer while waiting for the rock to thaw out in the next area to be grouted. At the end of a grouting season, methylene blue frost indicators were set routinely in the area to be grouted the following year. In between the frost indicators, temperature measurements were made using resistance thermometers lowered at regular depths in an alcohol filled tube lowered in the hole.

No patches of permafrost were encountered. However, frozen rock was found in one area on the south bank where a construction road was built on the foundation on top of unthawed rock at the beginning of one summer and removed in the following summer, thus exposing the rock to two successive seasons of freezing, without an intermediate thaw. The rock in this area was effectively thawed by circulating hot water and was grouted only after positive observations with the help of resistance thermometers.

GROUTING CRITERIA ESTABLISHED FOR THE COMPLEX

LG-2 was the first site on the Complex to be grouted. The following assumptions made at the design stage were confirmed by the grouting results:

- the rock was of excellent quality, absorptions were low and high absorptions were sporadic in the case of curtain grouting;
- relatively high absorptions occurred in blanket grouting holes because of the presence of rebound fractures;
- there was no permafrost in the region and only the top 25 to 30 ft (7,6 to 9,1 m) of rock was affected by seasonal freeze and thaw.

SEBJ was able to establish a set of grouting criteria for the rest of the sites based on this information, as shown on Table 3. These criteria have been successfully used on the other sites in the Complex.

TABLE 3
RESUME OF GROUTING CRITERIA FOR DYKES AND DAMS [11]

WATER HEAD	CURTAIN GROUTING			BLANKET GROUTING		
H ft (m)	NUMBER OF LINES	DEPTH ft (m)	SPACING ft c/c (m)	NUMBER OF LINES	DEPTH ft (m)	SPACING ft c/c (m)
0 to 25 (0 to 8)	1	25 (8)	40 (12)	--	--	--
25 to 75 (8 to 25)	1	25 (8)	20 (6)	--	--	--
75 to 200 (25 to 60)	1	H/3	20 (6)	2	25 (8)	20 (6)
> 200 (> 60)	1	H/3	20 (6)	Core width	25 (8)	20 (6)

CONCLUSIONS

The successful completion of the grouting programme and performance of the LG-2 dam in the far north demonstrates that, with careful planning and diligent work, projects can be built in remote areas under very difficult conditions so that hydro resources of the region can be developed.

ACKNOWLEDGEMENTS

The writers are indebted to the Société d'énergie de la Baie James for permission to publish this paper. The Committee of Engineering Experts composed of Messrs W.K. Chadwick, J.W. Cooke, D.D. Campbell, R.B. Peck, T.A.L. Paton, P. Larochelle and G. Turenne has reviewed the design and construction on the Complex. Special thanks is given to its Chairman, Mr. Wallace Chadwick for his interest in obtaining results on frost penetration, to Dr. D.D. Campbell the geological member of the Committee for his support throughout the work and finally to Richard Hardy, geological specialist at LG-2 site, for his conscientious monitoring of freeze and thaw in rock.

REFERENCES

[1] Amyot, P., Laliberté, C., Larocque, G.S. - "Meeting Quebec's Power Needs for the Eighties"
Water Power & Dam Construction, Pt.1, July 1976, pp. 27-30, pt. 2, Aug. 1976, pp. 17-23

[2] Murphy, D.K., Levay, J., Rancourt, M. - "The LG-2 Underground Powerhouse"
Rapid Excav. and Tunnelling Conf., Las Vegas 1976, pp. 515-533

[3] Larocque, G.S. & Pigeon, Y. - "Aménagement La Grande 2, Barrage principal et digues du bief amont"
Réunion d'automne de la section Énergie hydraulique de l'Association canadienne de l'Électricité et l'assemblée générale annuelle de CANCOLD, Montréal du 10 au 13 septembre 1978

[4] Levay, J. & Aziz, Z. - "Siting and Excavation of the La Grande 2 Spillway Channel"
Fall meeting of the Canadian Electrical Association, Hydraulic Power Section and Annual General Meeting of CANCOLD, Montreal, September 1978

[5] Brown, R.J.E. - "Permafrost in Canada. Its Influence on Northern Development"
University of Toronto Press, Toronto and Buffalo, 1973

[6] Johnston, G.H. - "Permafrost. Engineering Design and Construction" - John Wiley & Sons, Toronto, New York, 1981

[7] Mackay, R.J. - "A Frost Tube for the Determination of Freezing in the Active Layer above Permafrost"
Canadian Geotech. Journ. 10, pp. 392-396, 1973

[8] Johnston, G.H. - "Instructions for the Fabrication of Thermocouple Cables for Measuring Ground Temperatures"
Technical Paper No. 157 of the Division of Building Research, National Research Council of Canada, Ottawa, September 1973

[9] Paré, J.J., Lavallée, J.G., Rosenberg, P. - "Frost Penetration Studies in Glacial Till on the James Bay Hydroelectric Complex"
Canadian Geotech. Journ. Vol. 15, pp. 473, 1978

[10] Elston, P.J. - "Cement and Clay Grouting of Foundations: Suggested Specifications for Pressure Grouting"
Journ. of the Soil Mechanics and Found. Div. Proc. Am. Soc. of Civ. Eng., pp.1548-1530, Feb. 1958

[11] Paré, J.-J., Larocque, G.S. and Schneeberger, C.E. - "Design of Earth and Rockfill Dams for La Grande", Water Power and Dam Construction, Vol. 30, No. 1, I.P.C. Electronics Press Ltd, London, England, Jan. 1978, pp. 38-45

ELECTRIC LAKE DAM GROUTING

By Jagan N. Gosain[1], M. ASCE and John R. Kunz[2]

INTRODUCTION

Electric Lake Dam, located on Huntington Creek in Emery County, Utah, is owned and operated by Utah Power and Light Company of Salt Lake City, Utah. This 228 ft. (69.5 m) high zoned earth-fill embankment retains a reservoir providing approximately 30,000 acre feet ($37x10^6 m^3$) storage, primarily for use as cooling water for the coal fired Huntington generating plant. Construction of the dam started in April, 1972 and was completed by December, 1973.

During the first filling season, the spring of 1974, seepage developed at the downstream end and increased when the reservoir rose during 1975. Increased seepage and washing of fines resulted in a sinkhole in the dam crest, in 1975.

The remedial grouting program was carried out during 1974, 1975, 1976 and 1980 working seasons. An extensive monitoring program was initiated during 1975 and 1976. The instrumentation consisted of 58-open stand pipe observation wells, 15 pneumatic piezometers both upstream and downstream of the impervious core, in the fill and foundation rock and 21 settlement monuments.

DAM DESIGN

The dam is zoned earthfill structure 228 feet (69.5 m) high with a crest length of 920 feet (281 m) and crest elevation 8586 above MSL. The upstream slope is 2.4H to 1V to a 6 foot (1.8m) berm at elevation 8525 and 2.25H to 1V from berm to the crest. The downstream slope is 2.3 H to 1V steepened to 2H to 1V in the top 40 feet (12.2 m). The embankment consists of five zones as shown on Figure 1.

The Morning Glory spillway and the discharge conduit are located near the right abutment. Typical Section also is shown on Figure 1.

---

1. Deputy Chief Field Engineer, Tunnel and Reservoir Plan, Harza Engineering Co., Chicago.

2. Manager Civil Engineering, Utah Power & Light Co.

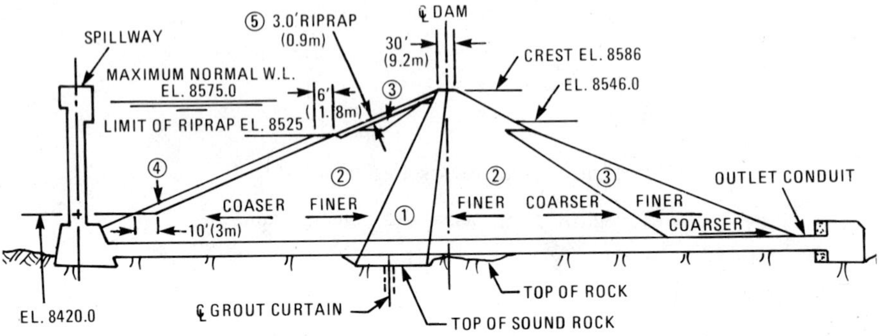

FIGURE 1. TYPICAL SECTION

Zone 1	Impervious material
Zone 2	Sandy silty gravel and silty sands
Zone 3	Sandy gravel with a maximum o 15% passing the #200 sieve.
Zone 4	Required oversize-large gravel, cobbles, boulders and sandstone fragments.
Zone 5	Riprap moderately hard, massive sandstone.

FOUNDATION CONDITION

a. Area Geology. The rock exposed at the damsite and in the reservoir are part of the Blackhawk Formation of Upper Cretaceous Age. They consist predominantly of blocky, fractured, moderately-resistant sandstone and easily-weathered lenticular shale interbeds. Soil and slopewash mask the bedrock in most of the area. The Blackhawk is the principal coal-producing formation of the Wasatch Plateau and the coal deposits are usually associated with shale beds. The beds in the vicinity of the project area are essentially flat lying. In general, they dip less than 10° to the west. Gentle localized folds are imposed on the regional dip. North-south trending block faulting is common in the Wasatch Plateau. No faults were mapped proximate to the dam or reservoir site.

b. Site Geology. Soil and loose slopewash, consisting of blocks of sandstone in a silt and sand matrix covers most of the slopes at the damsite. It had an average thickness of 5 feet (1.5 m) on the left abutment and on the right abutment it averaged about 8 feet (2.4 m) in thickness and extended in a few places to a maximum of 15 feet (4.5 m). Stream deposits consist generally of 4 to 9 feet (1.2-2.7 m) of soft organic silts, sands and coal mine tailings overlying 9 to 15 feet (2.7-4.5m) of dense sand and gravel.

The sandstone is in beds which range in thickness from several inches up to several feet. It contains dark, occasionally carbonaceous shale interbeds from 5 to 50 feet (1.5-15 m) thick. Occasionally the sandstone grades into the shale over a distance of 1 to 2 feet (0.3-0.6 m). Two sets of nearly vertical joints intersect the foundation rock. Many of these joints are open and quite permeable.
The major joint set strikes about northeast; a secondary, less prominent set, strikes to the northwest. The water pressure tests conducted indicated that the sandstone is quite open down to 60 to 100 feet (18-30m) below the ground surface and becomes fairly tight in the more shaley zones and massive gray sandstone below those depths.

The reservoir area is almost completely blanketed with slopewash and terrace deposits. There are numerous springs which discharge into Huntington Creek in the reservoir area. Some of these exit from the pervious sandstone or where the sandstone unit has become saturated with groundwater. Other springs appear to exit from perched water tables on shale beds high on the abutments. One spring exits near the downstream toe on the left abutment.

FOUNDATION TREATMENT

A single row grout curtain along with two rows of holes for consolidation grouting spaced 7.5 feet (2.3 m) on either side of the grout curtain was provided. The grout curtain was provided along the center line of the core as shown in Fig. 2.

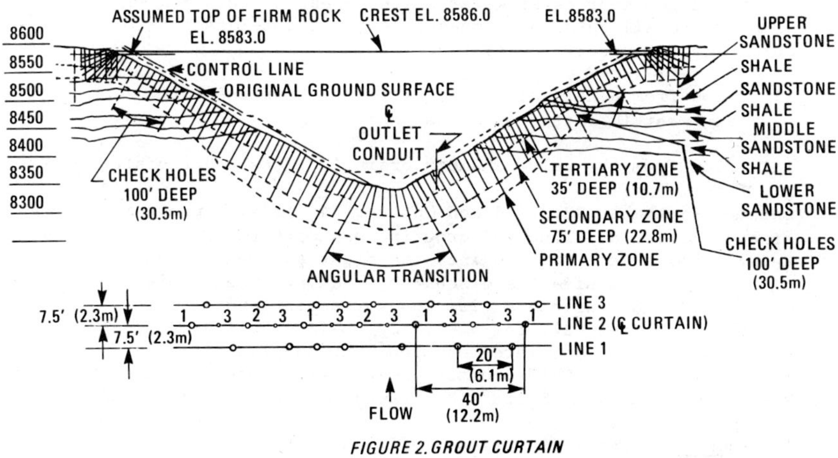

FIGURE 2. GROUT CURTAIN

Many deep open joints were discovered running parallel to the river in the dam foundation. These joints, which were probably the result of gravity slump and stress relief, were open to widths ranging from one to twelve inches (25-300 mm) and were up to 30 feet (9 m) or more in depth. It was necessary to fill these jounts in the vicinity of the core in order to prevent water from flowing through them and bypassing the grout curtain. Many were filled with sand cement grout by gravity from the surface either by pumping through a hose or by dumping directly from a transit mix truck. Others were filled during the consolidation grouting and curtain grouting operations. A total of 52,336 sacks (2,228,571 Kg) of cement were pumped, to provide 93,750 square feet (8720 m^2) of grout curtain or about 0.56 bags per square foot (256 Kg/m^2) of curtain constructed. The total amount of solids, cement, sand, fly ash and bentonite were about 7.5 times the original estimated 7000 sacks (298000 Kg). About 12,100 lineal feet (3690 m) of Ax (48 mm) grout holes were drilled which in nearly twice the estimated quantity of 6,000 lineal feet (1830 m). Nearly 4.3 bags of solids were pumped per foot (600 Kg/m) of drilled length.

RESERVOIR FILLING

a. 1974 Runoff. On June 24, 1974, the reservoir rose to a maximum elevation of 8,511 or 64 (19.5 m) feet below spillway crest, after which it was lowered steadily for irrigation releases.

On July 26, 1974, with the reservoir at El. 8,506, the total seepage was measured at approximately 5.5 cfs (155 l/sec). This seepage was measured at a weir across the stream approximately 320 feet (98 m) downstream from the toe of the dam. Of this 5.5 cfs (155 l/sec), approximately 1.0 cfs (28 l/sec) was coming from the left abutment with the remainder coming from the right abutment and stream banks. There are two sources of water which appear in this measurement and all subsequent measurements of total seepage which are not reservoir related.

i) flow from a natural spring at the base of the left abutment (0.3 cfs - 0.5 cfs) and

ii) flow from the valve chamber drain (0.1 cfs - 0.2 cfs).

Between September 23, 1974 and October 24, 1974, a program of exploration drilling and grouting was performed. Several holes were drilled into the right abutment foundation rock from inside the discharge conduit. These holes were drilled at various angles and varied in length from 7 feet to 50 feet (2.1 to 15 m). Approximately 520 sacks of cement (22,140 Kg) were injected into 352 lineal feet (107 m) of holes. This grouting reduced the total seepage from approximately 5.2 cfs (14 71/sec) to 4.0 cfs (113 l/sec).

b. 1975 Runoff. The 1975 spring runoff added approximately 15,000 acre-feet (18.5x$10^6 m^3$) of water to the Electric Lake reservoir. The reservoir crested at El. 8,555, 20 feet (6 m) below the spillway crest, on July 8 to 13, 1975

The seepage from the reservoir increased with increasing reervoir elevation as shown on Figure 3. The total seepage increased at a linear rate from about 4.0 cfs (1131/sec) at reservoir elevation 8,498 to 5.2 cfs (14 71/sec) at El. 8,519. When the reservoir rose above El. 8,520, the seepage began increasing at an increasing rate. The peak rate of seepage was 11.7 cfs (331 l/sec) with the reservoir at El. 8,555.

The left abutment seepage increased linearly from about 1.0 cfs (28 l/sec) with the reservoir at El. 8,498 to 2.1 cfs (59 l/sec) with the reservoir at El. 8,555, which indicated a washing out and enlargement of the seepage network on the left abutment.

The flow for the right abutment weir located on top of the stilling basin increased from 0 cfs with reservoir Elevation 8,498 to approximately 0.3 cfs (8 l/sec) at Elevation 8,542. At the same time the total seepage increased from 2.0 cfs (56 l/sec) to 6.8 cfs (192 l/sec). Most of this increase was observed coming from a 6 inch (150 mm) wide vertical joint in the rock on the right side of the discharge channel just downstream from the stilling basin. The seepage water was cloudy with the water from the vertical joint more turbid than that coming over the weir. As the reservoir Elevation increased from 8,542 to 8,555, the right abutment weir discharge increased from 0.3 cfs (8 l/sec) to approximately 2.5 cfs (70 l/sec). The total seepage increased from 6.8 cfs (192 l/sec) to 11.7 cfs (331 l/sec). The right abutment discharge remained cloudy.

The seepage points are shown on Figure 4.

SINKHOLE

A sinkhole, as shown on photo 1, was found on the dam crest in the afternoon of August 14, 1975. The sinkhole was centered around dam station 7+25, about 5 feet downstream of the dam centerline. It was approximately 6 feet (1.8 m) in diameter at top and about 25 feet (7.6 m) deep. The cave-in belled to an irregular elliptical shape at the bottom, approximately 25 feet (7.6 m) wide parallel to the dam axis and approximately 20 feet (6 m) wide normal to the axis. There was about 3 feet (0.9m) of water standing in the sinkhole. There was no noticeable change in either the quantity or turbidity of the seepage at the downstream end.

The following morning, the sinkhole was filled with approximately 150 cubic yards (115 m^3) of pea gravel and the day following that another 10 cubic yards (8 m^3) of pea gravel was added to fill in approximately 3 feet (0.9 m) of settlement which had occurred. Settlement of the sinkhole pea gravel after that date was minor.

Immediately after the discovery of the sinkhole, the reservoir was drawn-down at an average rate of 0.8 foot (250 m) per day for the next 37 days. The initial rate of draw-down was about 1.0 foot (300 mm) per day.

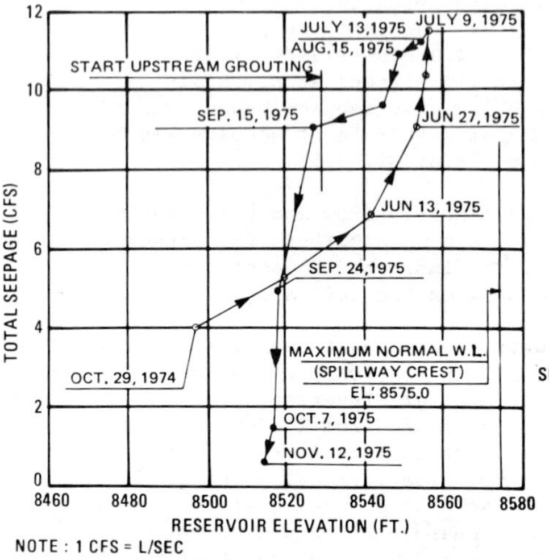

FIGURE 3. SEEPAGE HISTORY

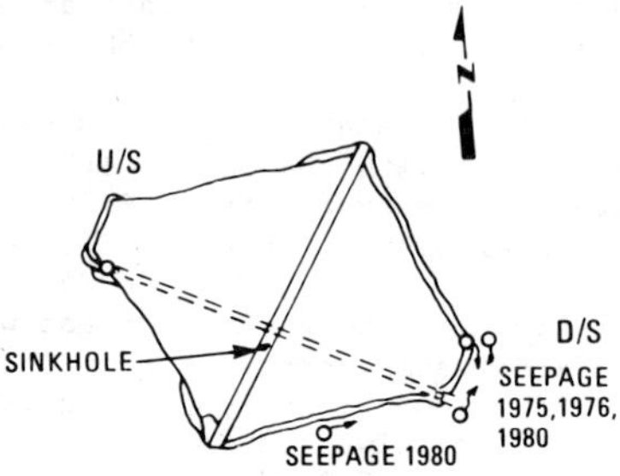

FIGURE 4. OBSERVED SEEPAGE

PHOTO 1. OBSERVED SINKHOLE ON DAM CREST

EXPLORATION FOR SINKHOLE

An exploration program was started to determine the extent of the sinkhole and the amount of disturbed embankment material. Wash borings with standard penetration test samples at 10 feet (3.05 m) vertical intervals were drilled 30 feet (9.15 m) on each side of the sinkhole and 30 feet (9.15 m) directly downstream from the sinkhole from a berm cut into the downstream slope of the dam. One hole was drilled through the center of the sinkhole. Subsequently, other holes, varying in distance from a few feet to 105 feet (32 m) from the center of the sinkhole, were drilled in a similar manner.

During drilling of the hole in center of the sinkhole, the right abutment weir discharge became very muddy. Dye introduced in another hole located about 30 feet (9 m) downstream of the center of the sinkhole at a depth of 110 feet (33.5 m) showed up at the right weir in 35 minutes. The inferred limits of the sinkhole are shown on Figures 5 and 6.

Water level readings in the exploration holes showed a pronounced hydraulic gradient in the middle sandstone unit toward the base of the sinkhole, indicating that this rock was providing the leakage water carrying fines from the sinkhole area. This was confirmed by introducing dye in holes which penetrate into the middle sandstone. Dye from these holes was observed in the right weir seepage approximately 25 minutes after to 45 minutes being introduced.

The exploration holes provided detailed information on the foundation conditions under and adjacent to the sinkhole. Water level readings in these holes showed the existence of perched water levels onthe various major shale units. The foundation beneath the sinkhole consists of two sandstone beds separated by a thin (7 to 12 feet thick) layer of shale. The lower sandstone consists of a very thick series of sandstone beds, some of which are separated by thin shale beds and others are themselves somewhat shaley. The other sandstone bed under the sinkhole is approximately 25 feet (7.6 m) thick and referred to as the middle sandstone. Higher up the abutment to the right of the sinkhole area, the sandstone is overlain by a very thick shale unit.

TREATMENT PROGRAM, 1975

It was apparent as the reservoir was rising that supplemental grouting of the right abutment would be necessary. Mobilization of grouting crews and equipment was started in early July and the first hole was grouted on July 22, 1975. The initial grouting effort was directed at the upper massive sandstone unit above El. 8,500. It was felt that the large increase in the seepage rate indicated a "window" in the grout curtain in this uppermost sandstone unit which had not previously been subjected to reservoir head. Seven holes had been drilled and grouted in this sandstone unit when the existence of a sinkhole on the dam crest was found on August 14, 1975. At this time, the grouting had not significantly affected the quantity or turbidity of the seepage.

After discovery of the sinkhole, the following program of treatment was formulated:

1. Reduce the quantity and velocity of the seepage water in the foundation by tightening the grout curtain directly upstream of the sinkhole. The holes were to be drilled upstream of the core to prevent hydraulic fracturing of the core by the drilling fluid and to avoid the possibility of increasing the seepage.

2. After controlling the foundation seepage, grout the top 15 (4.5 m) to 20 (6 m) feet of foundation rock under the sinkhole to prevent further movement of embankment fines into the foundation in this area.

3. Grout the gravel and stones in the lower 20 feet (6 m) of the sinkhole to consolidate this material and prevent the movement of embankment fines into or through it.

4. Investigate the condition of the impervious core adjacent to the sinkhole to determine the amount of penetration, if any, of the sinkhole into the core.

5. Consolidate, if necessary, the sinkhole material by grouting or other means to provide horizontal support to prevent adverse deflection of the core under full reservoir head. However, avoid grouting a stiff strong column through the dam which would have a tendency to cause cracking in the surrounding materials.

   a. Drilling

   The grout holes were drilled from the dam crest and inclined to intercept near-vertical relief joints upstream of the core, as shown on Figure 5. NX (89 mm) and BX (73 m) wireline casing were used to case the holes to rock. The casings were drilled approximately 2 feet (600 mm) into rock and cemented. All of the grout holes were drilled with BX (60 mm) core barrels. In most of the holes an initial stage varying from 15 feet (4.6 m) to 30 feet (9.2 m) was drilled in rock and grouted before extending the holes. Grout seams from the original construction grouting program were found at various elevations in all of the holes which were cored. These seams were dense and in good contact with the adjacent rock.

b. Grouting

The first hole was grouted on September 9, 1975. When coring the rock in this hole, dye was injected and appeared in the downstream seepage water in 65 minutes. The hole accepted water at 2 cubic feet (56 l) per minute with no gage pressure and grouting was started with a 5:1, water-cement grout. Cement appeared downstream in about an hour and the flow over the right abutment weir began to decrease. The water level in the observation well located in the center of the sinkhole dropped from 149.0 feet (45.4 m) to 153.8 feet (46.9 m). After grouting this stage of the first hole, the flow over the right abutment weir had decreased from 1.4 cfs (39.6 l/sec) to 1.1 cfs (31 l/sec) and the discharge was clear. The hole was refused with a 3:1 grout mix and 5 psi (0.35 Kg/cm^2) gage pressure. This procedure was repeated for the other grout holes located about 20 feet (6 m) on either side of this hole.

In view of the success of the right abutment grouting, this procedure was repeated on the left abutment in the same sandstone formation.

The total grout take for the upstream grouting on the right side was 9468 sacks (403,166 Kg) of cement. The total take for the upstream grouting on the left side was 4465 sacks (190,120 Kg) of cement.

The total seepage measured at the downstream gaging station with a current meter was 0.7 cfs (19.8 l/sec) on November 12, 1975. This includes the left abutment natural spring and the valve chamber drainage. Reservoir Elevation on November 12, 1975, was approximately 8516.

c. Grouting the Sinkhole

The top 15 feet (4.5 m) to 20 feet (6.1 m) of foundation rock under the sinkhole and fili was grouted from 6 holes DH-4, DH-6, SH-13 DH-14, DH-15 and DH-16 as shown on Figure 6. Four holes were drilled to intersect the foundation rock approximately 15 feet (4.5 m) from the center of the sinkhole. The four perimeter holes were drilled and grouted one at a time. The grouting was done by gravity pressure alone, starting with a 4:1 mix by volume but thickening the mix quickly to 3:1, 2:1 and finally 1:1 for all of the holes to confine the grout to the sinkhole area and prevent sealing the existing downstream drainage. After grouting the 4 perimeter holes, hole in the center of sinkhole would not accept a 5:1 grout with 5 psi (0.35 Kg/cm^2) gage pressure.

During grouting, the casing was raised in 10 foot (3 m) increments while attempting to keep the casing full of a 1:1 grout. Using this procedure, 381 sacks (16223 Kg) of cement were injected into the fill material and rock at the bottom of the sinkhole.

After grouting of the fill materials in the bottom of the sinkhole, three exploration holes DH-17, DH-18, and DH-19 were drilled into the core of the dam, as shown on Figure 6, recovering core in two of these holes. The holes were started using NX wireline (89 mm) casing with a bottom discharge bit and a bentonite mud for drilling fluid. The core recovery was good and showed tight, well compacted core material.

All three holes were grouted by raising the casing and injecting grout by gravity. Mixes were initially 5:1 but were thickened to 1:1 as the holes accepted grout and a total of 1607 sacks (68,426 Kg) of cement were injected.

TREATMENT PROGRAM, 1976

The remedial program formulated in 1975 included installation of piezometers and grouting of the body of the sinkhole which, because of the press of events and short construction season, were scheduled for 1976.

The management of Utah Power & Light Company appointed an independent Board of Consultants to review the safety of Electric Lake Dam and consult on the continuing evaluation of the remedial work. The Board of Consultants consisted of Mr. J. Barry Cooke, Mr. James L. Sherard and Mr. Thomas F. Thompson.

A program of exploration and installation of additional observation wells and settlement monuments was started in April 1976, prior to the beginning of the 1976 runoff.

The remedial program was expanded to include the following:

1. Grout fans in the upper sandstone on both abutments, located 5 feet (1.5 m) downstream from the existing main grout curtain.

2. A series of inclined grout holes on both the abutments similar to those constructed in 1975 to inject grout into the existing seepage network and to seal off the rock under dam core.

3. Grouting of the right abutment foundation rock through three fans from the discharge conduit, with one fan under the upstream half of the core, one fan under the downstream of the core and one fan under the sinkhole.

4. Installation of additional instrumentation in the embankment, foundation and abutments to define the various piezometric surfaces and to measure surface movements of the dam, particularly near the sinkhole.

The fan grouting was executed using the split spacing technique. On the left abutment 2,332 sacks (99,296 Kg) of cement were used for 3,588 lineal feet (1,094 m) of AX (48 mm) holes. The right abutment fan took 370 sacks (15,755 Kg) of cement for 1,775 lineal feet (541 m) of AX holes.

Similar technique as used for 1975 was adopted for both left and right abutment grouting. The total take for left abutment was 8,732 sacks (371,810 Kg) of cement in 1,347 feet (411 m) of hole, and on the right abutment 6,754 sacks (287,585 Kg) of cement was injected in 1,569 feet (478 m) of hole.

CONDUIT FAN GROUTING

Three fans were drilled and grouted through the conduit, as shown on Figure 7. All three of the fans are located in the lower sandstone unit. Primary holes are located 3-feet (0.9 m) apart in plan along the conduit and are spread 10° apart vertically. Secondary and tertiary holes were located using split spacing techniques.

Holes were drilled with AX (48 mm) plug bit and the flow of water from the hole and the static pressure were measured at 25-feet (7.6 m) intervals, or whenever an increase in the flow of water was encountered. Initially, the grouting was done in stages from the conduit with the packer set in the concrete. To save time and cost of redrilling, cup packers were used successfully in the later stages. The grouting pressures used were 25 psi (1.76 Kg/cm^2) in excess of the static water head measured at the conduit, plus the elevation head of the grout for the stage being grouted. Whenever the water from the holes was found to be carrying fines, the drilling was immediately stopped and the hole was grouted, irrespective of the length of the hole.

Ten holes varying in length from 105 (32 m) to 160 feet, (48.8 m) were drilled and grouted in Fan A. A total of 1.037 bogs (44,072 Kg) of cement were injected in 1,469 ft. (448 m) of hole. Fan B took 157 bags (6685 Kg) of cement in 1,170 feet of hole, while Fan C, which was directly below the sinkhole, accepted 556 bags (23,674 Kg) of cement in 493 feet (150.4 m) of hole.

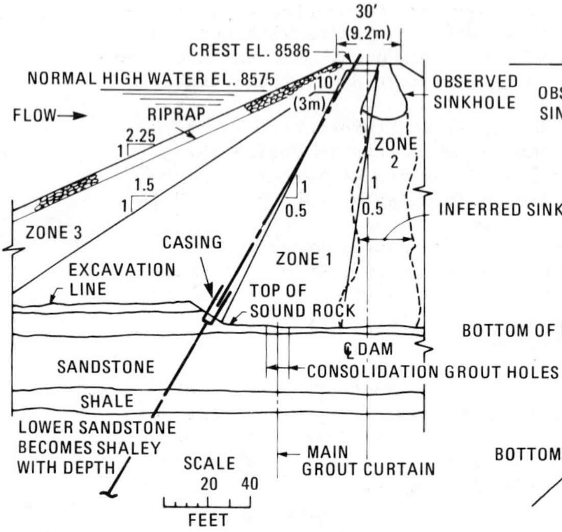

FIGURE 5. TYPICAL GROUT HOLE

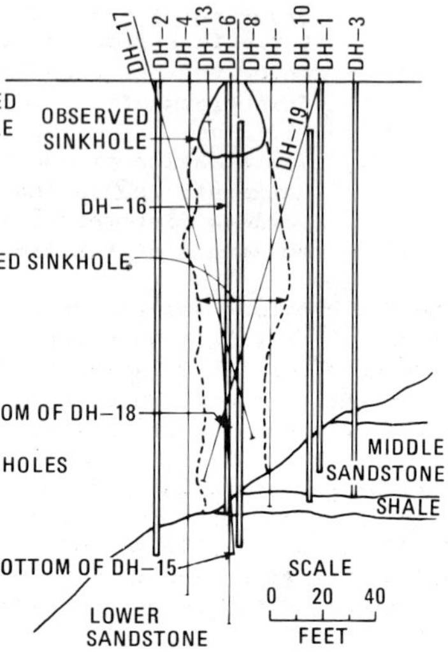

FIGURE 6. SINKHOLE GROUTING

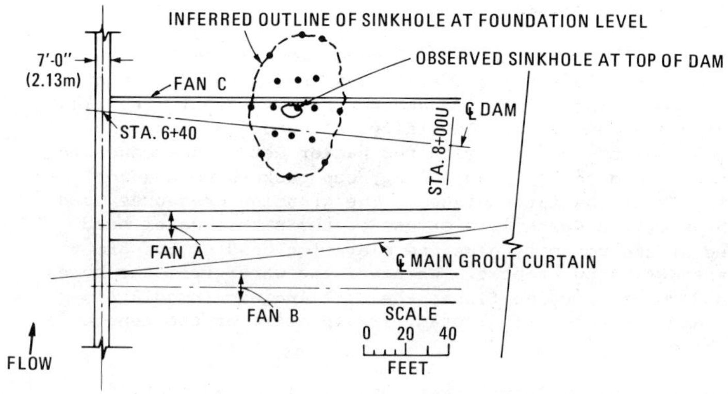

FIGURE 7. CONDUIT FANS AND SINKHOLE GROUTING

GROUTING THE SINKHOLE

The exploration programs in 1975 and 1976 for the purpose of determining the extent of sinkhole and disturbed embankment material had demonstrated that it was generally confined to zone 2 material. The sinkhole is believed to be generally elliptical in shape in plan, about 60-feet (18.3 m) wide normal to the dam axis and about 45 feet (13.7 m) wide parallel to the axis at the bottom.

The 1976 grouting program for the sinkhole consisted of (1) grouting the foundation rock directly under the sinkhole (2) grouting the fill-rock contact and (3) grouting the body of the sinkhole.

1) The foundation rock under the sinkhole was grouted by drilling 17 holes, as shown on Figure 7, from top of the dam at various angles to provide coverage of the foundation rock at the bottom of the sinkhole. These holes penetratec 30 (9.1 m) to 45 feet (13.7 m) into rock.

2) The bottom of the sinkhole, varying from 2.5-feet (0.75 m) to 10-feet (3 m) deep, was grouted using 1:1 neat cement grout under low pressures.

3) Grouting of the body of sinkhole was accomplished by injecing grout through the casings of the holes drilled into rock as the casings were withdrawn upward.

A plastic soil-bentonite-cement grout was used for the sinkhole. One batch of this mix consisted of:

Water	22.6/169	cu. ft./lbs
Cement	3/282	sacks/lbs
Bentonite	84.6	lbs.
Sand	100	lbs.
Silt	100	lbs.

Note: 1 cu. ft = 28.3 liters, 1 lb. = 0.453 Kg.

This soil-cement grout showed an unconfined compressive strength of approximately 1 ton/ft^2 (0.97 Kg/cm^2) at an age of 7 days.

This mix was gravity fed while the casing was slowly withdrawn at the rate of 1 to 2 feet per minute. A total of 152 batches of this mix (480 cubic feet of solids) were used.

In 1976 the total seepage at the downstream end measured after grouting was 0.26 cfs (7.4 l/sec) and reservoir rose to maximum elevation of 8535. In 1977 the reservoir level was only 8524 with no visible change in the quantity of seepage, which remained clear. Runoff during 1978 brought the reservoir level to 8561.5, 6.5 feet (1.98 m) higher than the previous high of 8555 during 1975. There was a small increase in leakage over the 1977 levels but remained clear.

TREATMENT PROGRAM, 1980

During the 1979 runoff, the reservoir peaked at an elevation of 8568.25, 6.75 feet (2 m) below the normal maximum elevation 8575. This represents about 97% of the maximum normal head, the highest level, since reservoir filling was started in 1974. The seepage as measured at various downstream monitoring points increased. This trend was noted at the left abutment weir, at cracks through the rock at the right abutment an in the total seepage. The seepage at the left abutment weir is shown on Figure 8. An examination of the plot indicates the possibility of an open area in the upper sandstone unit above elevation 8564. A remedial program was prepared, to be undertaken during the 1980 working season for the long-term stability of the dam. The program consisted of groutng the upper sandstone unit with about half the proposed holes going through the middle sandstone.

On June 1, 1980, as work was beginning on the 1980 grouting program, a leak appeared on the right abutment at elevation 8522. The flow was estimated at 3.5 - 4.0 cfs (99 to 112, l/sec) and issued from the talus-covered abutment slope immediately adjacent to the embankment slope. This leakage developed as the reservoir reached elevation 8573.5, 5.25 ft. (1.60m) above the 1979 peak elevation of 8568.25 ft. The leakage was timbid for about one day, but then cleared up.

The 1980 grouting program was modified to concentrate efforts on this newly developed leak.

The program adopted consisted of:

1. Right abutment grouting to reduce the newly developed leakage, and to reduce leakage exiting from the weir located on top of the stilling basin and leakage through the jointed rock downstream of the stilling basin.

2. Left abutment grouting to reduce the small amount of leakage at approximate elevation 8520, and the downstream toe of the dam, which was noted at 2.04 cfs (57 l/sec) on June 15, 1980.

3. Extension and reinforcement of existing fans on both left and right abutments.

The procedure for drilling and grouting was similar to the one used during the 1975, 1976 grouting program.

During drilling for the newly developed leakage, it became evident that the flow path for the leakage was concentrated in several steeply inclined, open joints, in a 20 foot (6.1 m) wide zone of the Upper Sandstone Unit. Some open joints were found directly at the rock/fill contact.

The flow path was very direct, as evidenced by the 8 minute elapsed time for dye injected upstream through one of the grout holes, to appear at the newly developed leakage.

A total of 23 grout holes were drilled and grouted to treat this leakage.

The 1980 remedial grouting program was undertaken during the highest reservoir level, peaking at elevation 8576.5.

The newly developed leakage on the right abutment was completely stopped by the end of the grouting program, and the following reductions were also achieved:

	1980 Leakage Reduction
Left Abutment, El. 8375 ±	2.24 cfs to 0.49 cfs
Left Abutment, El. 8520 ±	7.5 gpm to 0 gpm
Right Abutment, El. 8376 Weir + rock leakage	1.0 cfs to 0.02 cfs

Note: 1 cu. ft. = 28.3 liters, 1 gallon = 3.78 liters

The treatment program extended from June 1, 1980, to October 13, 1980. During this period, 101 holes were drilled for a total of 14,792 ft (4511 m) ; 25,383 sacks (1,080,808 Kg) of cement were injected; and approximately 100 gallons (378 l) of chemical grout were injected.

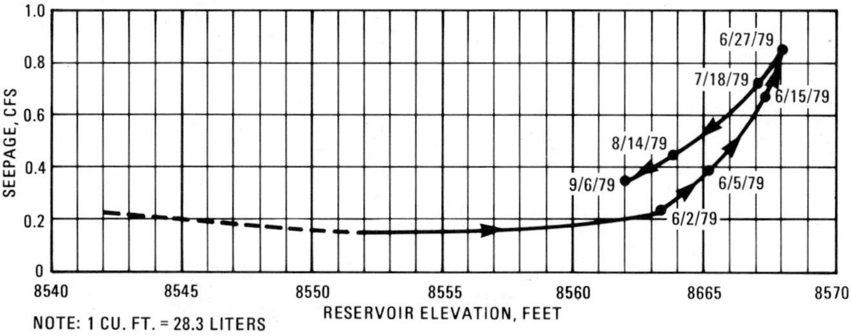

FIGURE 8. LEFT ABUTMENT LEAKAGE DURING 1979, RUNOFF

CONCLUSION

The remedial grouting program has reduced the seepage from 11.7 cfs during 1975, to nearly 0.5 cfs. The sinkhole has been consolidated with soil-bentonite-cement grout. A total of 67,317 sacks (2,866,358 Kg) of cement were used for 30,700 lineal feet (9363 m) of holes in the foundation, fill and the sinkhole.

The surveillance of the dam is being continued. The quality and quantity of seepage water, piezometers and settlement monuments are being monitored and data plotted. The piezometers upstream of the core, in the abutments and in the downstream shell have defined the water surface. This water surface has stabilized at a relatively low level and is not expected to present a hazard to the stability of the structure.

The surface monument records do not indicate any further movement of the embankment in the vicinity of the sinkhole. There are no apparent cracks on the surface of the dam or other evidence of distress in the dam. In the event that any increase in the quantity or turbidity of seepage is noted, the remedial program will be undertaken.

GROUTING OF THE FOUNDATION OF DARTMOUTH DAM

by Brian Raymond Cole[1], M. ASCE

KEYWORDS : Dam construction, Dam design; Dams (rockfill); foundations; Grouting; Permeability; Seepage.

ABSTRACT : Dartmouth Dam a 180 m high embankment dam was completed in 1978. It is located in north-eastern Victoria, Australia and is an earth and rockfill type with a wide central core. The grouting pattern consisted of blanket grouting to 5 m depth over most of the core foundation and a three row grout curtain ranging in depth from 25 to 60 m. Conventional cement grouting techniques were proposed. The conditions encountered necessitated a reassessment of the grouting techniques to be used and resulted in changes in the layout of holes and modifications to the techniques. In one location chemical grouting was required to satisfactorily grout out the area. A borehole TV camera was used to study the effectiveness of the grouting.

INTRODUCTION

Dartmouth Dam a 590 ft (180 m) high embankment dam was completed in 1978 after a construction period of about three years. The dam is located on the Mitta Mitta River in North Eastern Victoria and will impound a reservoir of 2,240,000 acre feet (four million megalitres) capacity.

Aspects of the design and construction of the project have been covered elsewhere, e.g. Michels and Dickson (7). The embankment consisted of earth and rockfill. A wide central core (of approximately seventy per cent of hydraulic head) flanked by two filter zones was provided. The cross section of the dam and the layout of the dam and appurtenant works are shown in Fig. 1. The foundation consisted of granitic gneiss with minor intrusions of mainly acidic dykes. The foundation varied from fresh to completely weathered and the permeability varied from nil to 60 Lugeons (600 ft/year).

GROUTING

General - The grouting pattern consisted of eight lines of blank grouting and a grout curtain comprising three rows of grouting as shown in Fig. 2. The foundation grouting operations included 143,360 ft (43 670 m) of drilling, 3130 drill set ups and 25260 cft (713 m^3) of cement was injected.

[1]*Geotechnical Engineer, John Connell Group, Melbourne, formerly Designing Engineer, State Rivers and Water Supply Commission, Victoria, Australia.*

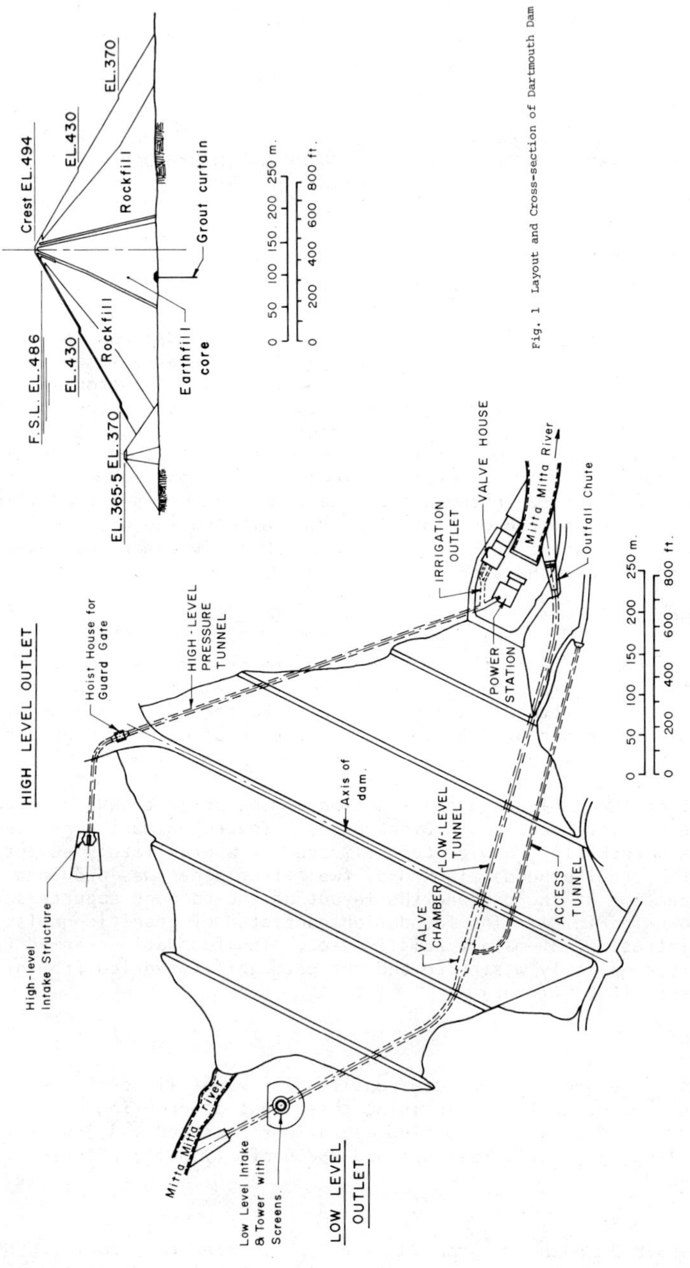

Fig. 1 Layout and Cross-section of Dartmouth Dam

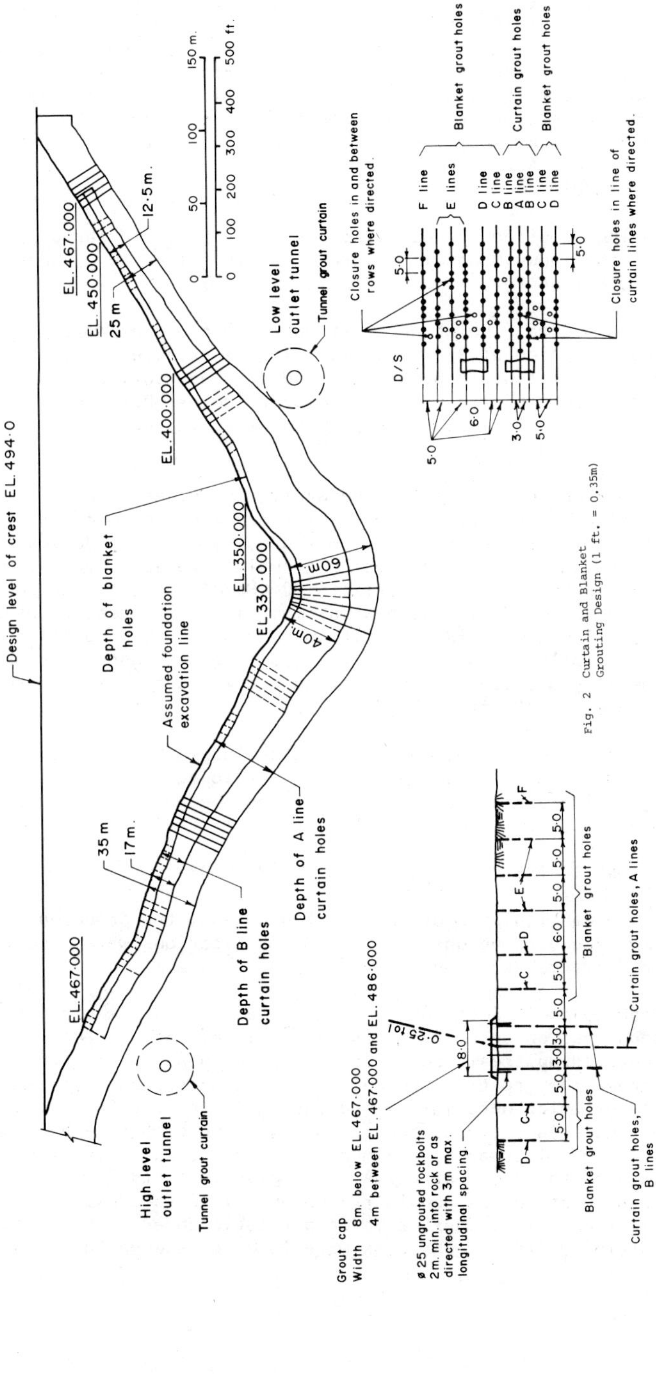

Fig. 2 Curtain and Blanket Grouting Design (1 ft. = 0.35m)

Layout - The blanket grouting pattern provided for generally 16.5 ft (5 m) deep holes at 16.5 ft (5 m) centres, supplemented by closure holes at reduced spacing and also holes of increased depth to intersect particular geological features. The curtain consisted of three rows of holes identified as the A, B1 and B2 lines. A continuous reinforced concrete grout cap was constructed on, and anchored into, the foundation rock along the line of the grout curtain.

GROUTING METHODS

General - The grouting methods used generally followed the procedures adopted by the Snowy Mountains Hydro-Electric Authority as described by Wallace and Hilton (9).

Methods - The curtain holes were set out on a closure pattern with primary holes spaced at 33 ft (10 m). It was expected that primary and secondary holes would be drilled and grouted routinely and tertiary holes (and quaternary etc.) used as closure holes where required. Prior to grouting all curtain holes were water tested with a packer placed at 30 ft (9 m depth).

Split space grouting was used in general. The upper stage of curtain holes was grouted as part of the blanket grouting sequence. The remaining length of the curtain hole was grouted using predominantly the upstage grouting method. The design grouting pressures varied over a range from the conventional 1 psi/ft (20 kPa/m) depth to the packer, to 2 psi/ft (40 kPa/m) depth, with a maximum allowable pressure of 200 psi (1400 kPa). The grout mixes specified ranged from 5:1 to 1:1 by volume. Initially a standard for the grouting was based on cement take, i.e., generally a closure hole was required where the take on the last hole exceeded 0.3 cft/ft (1 bag/m).

The procedures can be considered as conforming with conventional grouting practice. However, as the grouting developed these procedures and the design philosophies were re-examined on the basis of the conditions encountered on site.

MODIFICATIONS TO THE GROUTING DESIGN AND METHODS

General - In particular areas of the foundation, the proposed design or methods were found to be unsuitable and modifications were made to achieve the desired standards. These are described below.

Major Shear Zones in Both Abutments at About El 350 - several features were encountered on both abutments near this elevation. Many of these were orientated upstream-downstream and were open. During the grouting of this area, the grouting procedures were assessed and modifications made, many of which were maintained during later grouting. The modifications included reducing the maximum spacing of A line holes to 8.25 ft (2.5 m), increasing maximum grouting pressures by fifty per cent and increasing the depth of the A line holes to fifty five per cent of reservoir head at shears considered potential leakage paths. At El 365 on the right abutment, where a major permeable shear zone crossed the grout curtain additional curtains were located between the A and B lines.

Major Feature on Left Abutment at El 420 - At about El 420 on the left abutment, a major shear zone striking parallel to the river and varying from 0.4 ins (10 mm) of clay to 3.3 ft (1 m) of closely jointed rock and clay was exposed in the upstream and downstream rock zone foundations. The drilling of this zone was characterised by water losses and drills encountering soft rock and cavities. Water testing of the zone indicated that its permeability was in excess of 50 Lugeons. During grouting a maximum hole spacing of 8.25 ft (2.5 m) was adopted on the A and B lines.

It was decided to provide a continuous curtain of grout in the shear as far back into the abutment as was practical. This was achieved by deepening the foundation curtain so that it extended to a point opposite where the high level tunnel grout curtain commenced to intersect the major shear zone. The horizontal distance of 60 ft (18 m) separated the two curtains at this point in the plane of the shear. The gap between the two curtains was grouted from the foundation by drilling four number 150 ft (45 m) deep grout holes along the line connecting the two curtains. A total of 4,500 bags of cement was injected over the 100 m the shear zone was intersected.

Treatment of the Bench Area at El 420 on the Right Abutment - Initial abutment excavation on the right abutment revealed very deep weathering extending from above the dam crest down to a steeply dipping sheared zone at El 420. As the depth of weathering was at first unknown, final excavation of the bench was delayed until the grouting was within 16.5 ft (5 m) of the base of the weathering trough.

The final excavation resulted in an almost horizontal bench across the core width at about El 420 which was about 80 ft (25 m) below natural surface. However, up to 33 ft (10 m) of the excavated material had to be removed by blasting because of the existence of massive blocks of fresh rock which were underlain by completely weathered seams. These were due mainly to the weathering of joints opened by stress or creep.

When excavation reached the El 420 level it was acknowledged that further seams probably existed below the level, but as the foundation treatment had been completed to near this elevation it was decided that further excavation would cease and the foundation would be treated by grouting.

During blanket grouting of the area some of the lines of blanket grouting were extended, holes were increased in depth and a large number of closure holes were used to attempt to counter the adverse geologic conditions.

The curtain holes between El 425 and 422 intersected a weathered seam approximately 3.3 ft (1 m) in thickness, 100 ft (30 m) long and at a depth of 25 to 40 ft (8 to 12 m) below the grout cap. During the grouting it became apparent that there were going to be problems in grouting the seam as high water takes were recorded, but only low cement takes resulted. The grouting procedures were varied in an attempt to improve grout take. For instance the pressures were increased by fifty per cent above the design pressures and a 5:1 mix was retained to improve grout penetration. As well the holes were grouted to complete refusal. However, none of these measures appeared effective in increasing grout takes.

A check core hole was drilled to investigate the effectiveness of the grouting. The core from the hole include material from a highly weathered seam which contained little evidence of grout. The seam was found to have a permeability of 20 to 30 Lugeons when water tested. Further investigation revealed that a seam 3.3 to 5 ft (1 to 1.5 m) thick containing weathered rock and some clay did exist below the level of the bench. Such a major continuous seam would present a major leakage problem.

Additional curtain grouting was undertaken. This consisted of further holes drilled on the A line to reduce the spacing to 2.5 ft (0.75 m). A curtain line between the A line and the B1 line was drilled with an average spacing of 3.3 ft (1 m) and these holes were grouted with pressures increased one hundred per cent above the design pressures. The results of this additional grouting was to inject a small amount of grout with no appreciable reduction in permeability.

It was apparent that the grouting standard achieved was insufficient. Two approaches were contemplated to overcome the problem. These were flushing and chemical grouting.

The experience of Kariba Dam Lane (6), suggested that flushing out of the material in the seam and its replacement with cement would be a satisfactory means of treating the seam. Trial flushing was attempted and appeared to be successful in providing connections between holes which would allow "cells" to be grouted. There were however doubts as to how much of the weathered material would be removed and therefore this method of treatment was not proceeded with.

Two chemical grouts were considered, sodium silicates and AM-9. A grouting consultant was contacted and after inspecting the site proposed grouting with sodium silicates using a "tube a manchette", a technique often used for alluvial grouting. The techniques proposed called for long gel times and low injection rates which would have required about five weeks grouting. At this stage clay placing operations were being affected because they were approaching the El 420 level. In view of the time involved in the silicate grouting and some doubts raised about the long term stability of the grout, this proposal was rejected.

The use of AM-9has been widely covered in the literature, refer Karol (5). At the time of this work it was still available in Australia. Although it has since been withdrawn from the market, the techniques adopted apply to chemical grouts in general.

It is often suggested that the use of short gel times less than one minute are most appropriate for grouting with AM-9. However, these require sophisticated pumping equipment to allow for mixing at the point of injection of the grout with the catalyst. After advice from a grouting consultant, Mr. P. Ennis of Cementation Australia, it was decided that it was more practical to aim at a longer gel time and therefore use cruder equipment. The method adopted was to use a mix with an AM-9 concentration of ten percent and aim at a gel time at 10 minutes. This required an injection rate of 2.5 to 3.3 gpm (12 to 15 L/min.).

The technique used was to drill two lines of holes at 3.3 ft (1 m) intervals, 3.3 ft (1 m) apart. Into these was injected using normal packers a certain amount of grout calculated as equal to the volume of a cylinder around the hole assuming a twenty five percent voids ratio. Then a line of holes was drilled between these two rows and grout pumped in until refusal. Following this, three holes were drilled and water tested. The results indicated negligible permeability. In all 3100 gals (14 000 litres) of grout were injected and the area was grouted out over 32 working hours.

USE OF A BORE HOLE TV CAMERA

Introduction - the specification provided for the drilling of diamond drill holes into the foundation to enable cores of the rock to be extracted to enable the degree of penetration of the cement grout to be observed. The interpretation of these was difficult because when a core was examined it was difficult to determine whether joints were open or closed in situ. It was therefore decided to use a bore hole TV camera to study the sides of the core hole and therefore assist in interpretation of the core.

Bore Hole in The Left Abutment - The bore camera revealed that the bore hole intersected several grouted seams. This was apparent from the core. The objective of the use of the camera to examine the major shear at El 420 was achieved. Where the hole intersected the shear, the core consisted of fragments of rock and grout. The camera pictures showed the shear had been well grouted with no open fissures.

Bore Hole in the Right Abutment - When the borehole camera was used, all of the core holes on the right abutment had been grouted so a percussion hole was drilled. The pictures in the percussion hole confirmed that poor material existed for the full length of the hole. Several seams of completely weathered material were apparent. There was little or no evidence of grout which was consistent with grouting results near the hole. There were no signs of smearing on the walls of the hole which was thought to have been possible in percussion drilling of such material, instead there were several instances of caving of the hole which was indicative of the silty nature of the material.

Use of the Camera - The camera was shown to be useful in interpreting the effectiveness of grouting. The limit of the thickness of a joint which was discernible with the camera was about 1 mm for a horizontal joint. Therefore there are joints which would be filled with cement grout which would not appear on television pictures.

FOUNDATION PERMEABILITY

In Fig. 3, the results of the water tests have been plotted. These show that the average permeability of the foundation was about 10 Lugeons and it ranged up to about 20 Lugeons. The average permeability, prior to final grouting, was about 1 to 2 Lugeons with a maximum of up to 10 Lugeons.

These results must be considered in the context of the long stages used in the water testing. As a result, the permeabilities of particular features would tend to be underestimated. This was found when particular features were tested using shorter stage lengths and permeabilities in excess of 50 Lugeons were recorded.

Therefore it seems reasonable to suggest that the foundation was grouted to at least a permeability of 10 Lugeons, and generally better than 5 Lugeons.

GROUT TAKES

The grout takes recorded are shown in Fig. 4. The takes for the blanket grouting were substantial which probably reflects the incidence of near surface or relief joints.

The curtain grouting involved the grouting of 969 holes. Some 98,000 ft (29 800 m) were drilled and 15,032 bags of cement injected. The takes were high about El 420 on the left abutment and between El 395 and El 420 on the right abutment. These high takes could be related to particular features in the abutment. The results also show the influence of the B line on the subsequent grout takes for the A line.

COMMENTS ON GROUTING METHODS

Hole Spacing - The hole spacing adopted initially was 33 ft (10 m) spacing for primary holes. The split spacing method used meant that hole spacing was 16.5 ft (5 m) for secondary holes and 8.25 (2.5 m) for tertiary holes. These spacings were based on the practice adopted on other damsites having similar foundations. At that time much of the grouting had ceased at secondary grouting. The potential for grout to travel was considered, but the general spacing was maintained. Later on as mentioned previously the grouting was continued routinely to tertiary holes so that the maximum hole spacing was 8.25 ft (2.5 m) which appeared to be appropriate.

The degree of "interference" between the grout holes can be studied by comparing grout takes (and water takes) for various grout holes. Some examples are shown in Table 1 for an area on the left abutment. These indicate that there was "interference" between the grout holes and that a 33 ft (10 m) primary spacing was suitable.

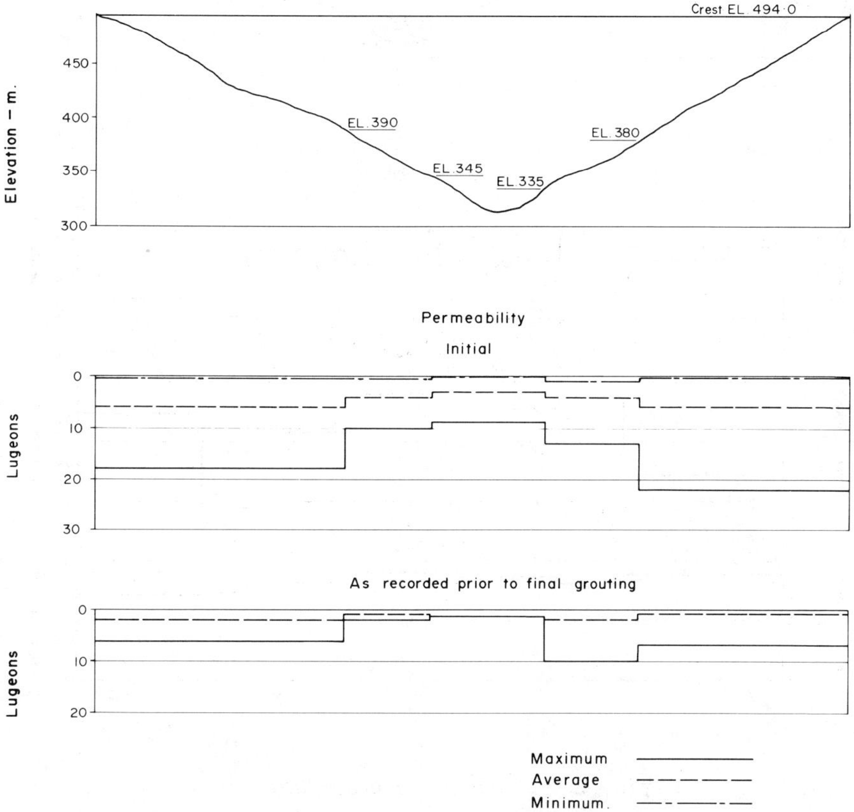

FIG. 3 - Foundation Permeabilities

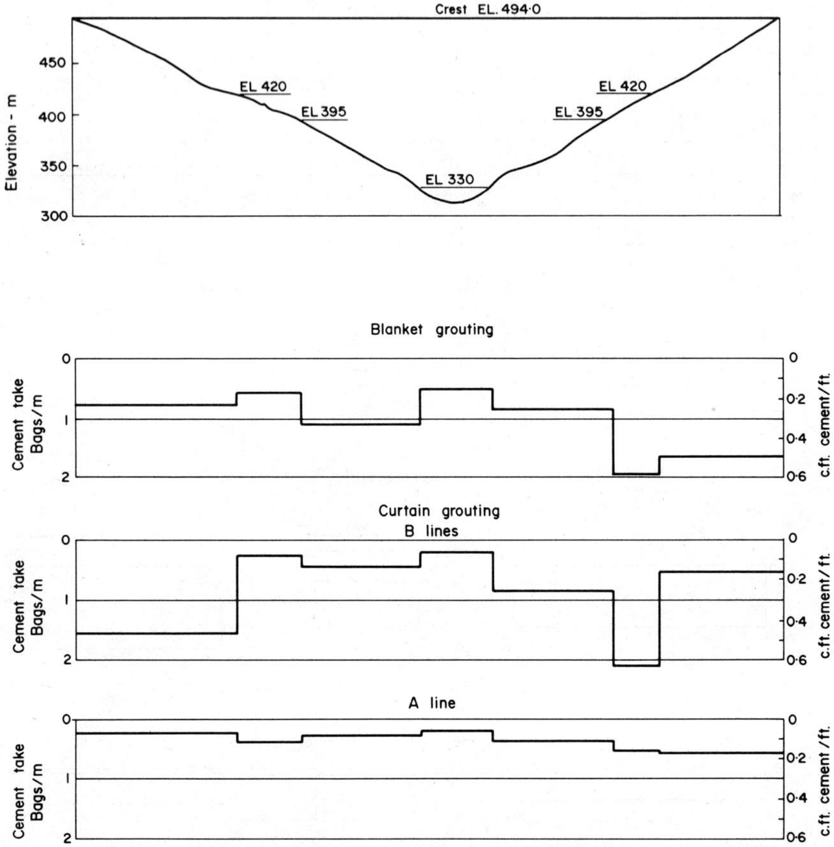

FIG. 4 - Analysis of Grout Takes

TABLE 1 - Comparative Grout Takes (bags/m)

Hole Designation	B1 Line	B2 Line	A Line
Primary	0.38	0.27	0.37
Secondary	0.23	0.26	0.17
Closure	0.11	0.09	0.22

Water Testing - The procedure adopted for water testing was to place a packer at 30 ft (9 m) and test the full depth of the hole at a pressure of 40 psi (300 kPa). The water test results were used mainly to give some indication as to the starting mix to be used for the grouting. The above procedure was used for each hole.

Misleading assessments of the foundation permeability can result from the use of such procedures, as has been pointed out by DeMello (2). However, economy was a major constraint on the water testing.

Grouting Pressures - The design pressures proposed ranged from 5 psi to 20 psi (35 kPa to 1400 kPa). These pressures are equivalent to the pressures recommended for sound rock (Task Committee on Cement Grouting, 8), which would appear reasonable given the geology of the damsite. During the grouting operations situations arose in which the water takes were high and yet grout could not be injected into the foundation in any appreciable quantities. The natural tendency in these cases is to increase the pressures (and possibly thin the grout). Increases up to fifty per cent above those proposed in the design were permitted generally and up to one hundred per cent in certain locations. However, this was associated with close inspection of the foundation both visually and using uplift gauges to ensure that uplift did not occur. The use of displacement grouting techniques or "Claquage", refer to Cambefort (1) was considered but not adopted. The locations in which increased pressure was used were probably situations where neat cement grout was not appropriate and chemical grout or cement grout with additives should have been used.

Standard of Grouting - The concept of a standard of grouting has been formalised in recent years by Houlsby (4). For the Dartmouth grouting, the standard was expressed in terms of a grout take per foot (metre) of hole. The figure adopted for much of the grouting was 0.3 cft/ft (1 bag/m).

As the grouting operations continued, the inappropriateness of such a standard became apparent particularly where the grout takes and water takes showed great disparity. By the time the grouting was completed the degree of grouting was considered in terms of a water take and grout take. The standard of grouting was never formalised in terms of a particular water take, but the objective would have been to aim at something less than 5 Lugeons.

Grout Mixes - The grout mixes ranged between 5:1 to 1:1. There has been some criticism of water cement ratios in excess of 3:1 (Houlsby, 3) because of the low durability of this grout. The average grout mix used for the Dartmouth foundation was probably between 2:1 and 3:1 as on holes where takes were large the mix would be thickened or a 3:1 starting mix would be used. The mixes used were therefore considered appropriate as much of the grout would have been 3:1 or thicker, but there was still provision for thinner mixes where it was thought necessary to ensure better penetration into finer features.

Grouting Techniques - The Specification called for grouting to be undertaken using cement grouting exclusively. Problems arose in areas of the foundation where, as has been mentioned, cement grouting was found to be inappropriate and chemical grouting had to be resorted to. The problem with this philosophy was that when problems arose, the tendency was to persevere with the methods allowed in the Specification and as a result, the chemical grouting had to be adopted as a "fire fighting" technique at a very late stage.

PERFORMANCE

The performance of the grouting has been monitored using weirs constructed at the downstream toe and at the downstream portals of the tunnels. The interpretation of the flow readings is complicated by rainfall and groundwater effects. As well, there is some uncertainty as to whether the weirs are collecting all the seepage.

From a study of the readings it is possible to discern a trend toward increasing seepage with rise in reservoir level.

To date a rise in the reservoir level to El 445 has resulted in an increase in seepage of about 160 gpm (750 L/min.).

CONCLUSIONS

The grouting of the foundation of Dartmouth Dam involved conventional cement grouting techniques. The grouting design features a conventional blanket grouting pattern and a three-row grout curtain.

The features encountered in the grouting necessitated that changes be made to the layout of holes and the techniques used. For instance hole spacing, hole depth, grouting pressures and closure requirements were modified during the grouting. As well, blanket areas were extended and additional curtain lines introduced. Finally, chemical grouting was used in our area of the abutment where cement grouting was found to be inappropriate.

The experience of Dartmouth indicates that although a detailed grouting design may be evolved and procedures set out, these may have to be modified to suit the conditions as a "feel" for the foundation is developed.

ACKNOWLEDGEMENT

In the preparation of the paper reference was made to internal reports prepared by R. Keen, Executive Engineer and P. Wood, Geologist, both formerly of State Rivers and Water Supply Commission.

APPENDIX - REFERENCES

1. Cambefort, H., "The Principles and Applications of Grouting", *Quart. Journ. of Eng. Geol.*, Vol. 10, 1977, pp 57-95.

2. DeMello, V.F., "Reflections on Design Decisions of Practical Significance to Embankment Dams", *Geotechnique,* Vol. 27, No. 3, 1977, pp 279-355.

3. Houlsby, A.C., "Foundation Grouting for Dams", *ANCOLD Bulletin,* 1977, No. 47, pp 5-12; No. 48, pp 13-23.

4. Houlsby, A.C., "Engineering of Grout Curtains to Standards", *Proc. ASCE,* Vol. 103, GT9, 1977, pp 953-970.

5. Karol, R.H., "Chemical Grouting Technology", *Proc. ASCE,* Vol. 94, SM1, 1968, pp 175-204.

6. Lane, R.G.T., "The Jetting and Grouting of Fissured Quartzite at Kariba", *Grouts and Drilling Muds in Engineering Practice,* Butterworths, London, 1963.

7. Michels, V.S. and Dickson, R.S., "Dam - abutment and core-filter interfaces, Dartmouth Dam Australia", *Trans. 13th Inter. Cong. Large Dams,* New Delhi, 1979, Vol. 1, pp 563-590.

8. Task Committee on Cement Grouting, "Cement Grouting, Progress Report, *Proc. ASCE,* Vol. 88, No. SM2, 1962, pp 49-97.

9. Wallace, B.J. and Hilton, J.I., "Foundation Practices for Talbingo Dam", Australia, *Proc. ASCE,* Vol. 98, No. SM10, 1972, pp 1080-1098.

CHICOASEN PROJECT GROUTING AND DRAINAGE SYSTEM

by Armando Rábago[1]

ABSTRACT

The Chicoasén Project is part of the hydroelectric development of the Grijalva River in the southern region of Mexico. The zoned embankment dam is 859-ft (262-m) high with an impervious core of compacted clay; it is the highest dam in the country and one of the largest in the world in its type.

The main embankment is located in a narrow canyon with abrupt walls carved through stratified and massive limestone formations from Mid to Upper Cretaceous. Four sets of fractures govern seepage conditions in the rock mass.

A comprehensive program of grouting and drainage was planned and implemented to protect the structures against the heavy loads imposed by the reservoir and to reduce seepage to a minimum. The design criteria took into account the local topographical and geological features, the degree of fracturing, the behavior of the rock, the absorption of water and grout as determined from Lugeon and injectability tests, and the location of unstable zones and of the structures to be protected. About 3 miles (5 km) of galleries were excavated to carry out the curtain grouting and the drainage system. The former covers from ground surface to a depth of 393 ft (120 m) below the river bottom, extending about 1639 ft (500 m) normally to the abutments. The total length of boreholes was 144 miles (232 km) including the short holes drilled to connect the galleries with the grout curtain. Blanket grouting was performed from the contact of the foundation rock with the core, penetrating an average length of 33 ft (10 m) in the basal rock; the total drilled length was 84 miles (135 km). The drainage system comprised a total length of borings of 15.6 miles (25 km). Stable and unstable suspensions of water, cement and bentonite were injected.

The overall grout takes recorded at the curtain and blanket grouting were of about 19 lb/ft (29 kg/m) and 12 lb/ft (18 kg/m) respectively, and the average permeability was reduced from 1 to 41 UL before the treatment, to 1 to 6 UL after it. Total seepage observed during the first months after filling of the reservoir was in the order of 11 cfs (300 ℓ/s) with the water at maximun level; the efficiency of the grout curtain can thus be considered satisfactory as confirmed by the piezometric curves obtained at the damsite.

INTRODUCTION

This paper describes in brief the most relevant aspects of

[1] Field Rock Treatment Engineer at Chicoasén Project during construction, Federal Commission of Electricity, Mexico

the grouting and drainage operations performed at the Chicoasén damsite; the development of seepage through the grout curtain and through the rock mass are also commented upon for the period comprising the first months after reservoir impounding.

GENERAL DESCRIPTION OF THE PROJECT

The Chicoasén Hydroelectric Project is located on the Grijalva River in the southern state of Chiapas, Mexico, 13.7 miles (22 km) from the State Capital of Tuxtla Gutiérrez. It was built from 1975 to 1980 by the Federal Commission of Electricity to develop an installed capacity of 2400 MW for peak operation delivered to the national interconnected network.

The project is a part of the hydroelectric development scheme of the Grijalva River comprising also Malpaso, La Angostura and Peñitas dams, this latter under construction at present. The system will provide an installed capacity of 5000 MW while regulating an average flow of 28 240 cfs (800 m^3/s).

The main structures comprising the project are the following (Fig. 1): (1) An earth and rockfill embankment dam 859-ft (262-m) high with an impervious core of compacted clay, complemented by a 180-ft (55-m) high dike of the same type, founded upon an old river bed on the upper part of the right bank; the total volume of materials including the dike is of about 19 x 10^6 cu yd (14.5 x 10^6 m^3); (2) a spillway structure formed by an approach channel and three tunnels 49-ft (15-m) diameter, having a combined discharge capacity of 529 500 cfs (15 000 m^3/s); (3) the power plant that includes an intake channel leading to eight penstock tunnels 20 ft (6 m) in diameter with an slope of 50 degrees and a hydraulic head of 590 ft (180 m), an underground powerhouse with a portal cross-section 141-ft (43-m) high with eight Francis turbines of 300 MW each, an underground transformers gallery, two surge chambers for five and three units, respectively, and three outlet tunnels; (4) two diversion tunnels with a capacity of 158 850 cfs (4500 m^3/s); and (5) an auxiliary middle outlet tunnel at El. 320* with a capacity of 17 650 cfs (500 m^3/s) for control during first filling of the reservoir and a safety provision during subsequent operation.

The dam will impound a reservoir of 1.36 x 10^6 acre-ft (1680 x 10^6 m^3) at maximum elevation (El. 395).

GEOLOGY OF THE DAMSITE

The damsite is located in a highly seismic area of the country and lies within the physiographic province "Altos de Chiapas" formed by mountain ridges and wide valleys of sedi-

* Elevations are expressed in meters above mean sea level (1 m = 3.3 ft)

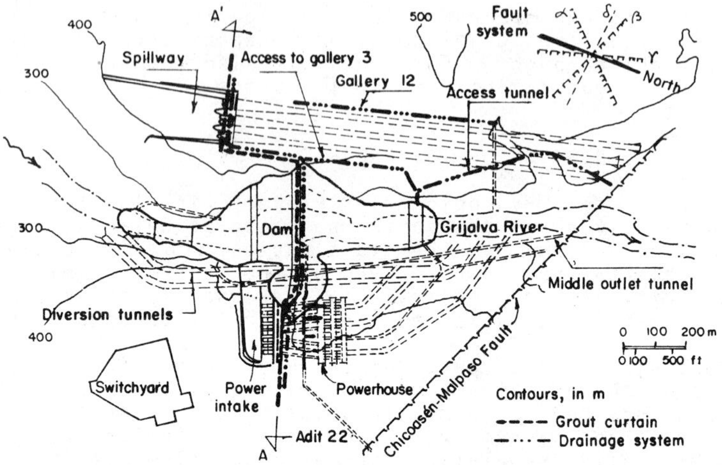

FIG. 1.- Plan view of Chicoasén project including grout curtain and drainage system

mentary rocks from the Cretaceous and Tertiary, covered by recent alluvial and piroclastic deposits from the Quaternary along the valleys. The damsite proper is a narrow canyon formed by nearly vertical walls up to 656 ft (200 m) in height cut through limestone rock; the gorge intersects the eastern side of an anticline the axis of which strikes at an angle N 80° W, and is situated between two regional faults, namely, the inverse fault Muñiz approximately 3 miles (5 km) upstream of the dam axis and the normal fault Chicoasén-Malpaso, less than 1639 ft (500 m) from the dam. Three lithological units have been identified at the site (Fig. 2): (1) Unit U1 formed by limestone interstratified by layers of clay shales; (2) unit U2 is a massive limestone concordant with U1; and (3) unit U3 of limestone interbedded with thin layers of dark clay. An outcrop of this unit can be seen at river level (El. 200) and it constitutes the foundation of the impervious core. These formations from the Middle and Upper Cretaceous are covered by clay shales from the Eocene.

This lithological sequence within the canyon generally dips from the left bank to the right bank. At the damsite, the strikes vary from NE to NW with angles ranging from 8 to 24 degrees at the left bank and from 24 to 36 degrees at the right abutment.

Four sets of main subvertical faults of tectonic origin cross the lithological units, and their average strike is shown in Fig. 1. These discontinuities are generally present without fill, having an irregular trajectory and with evidences in the scarps of horizontal displacements with a slight vertical component. Those fractures with the widest openings correspond to systems α and γ, the latter induced by the rock

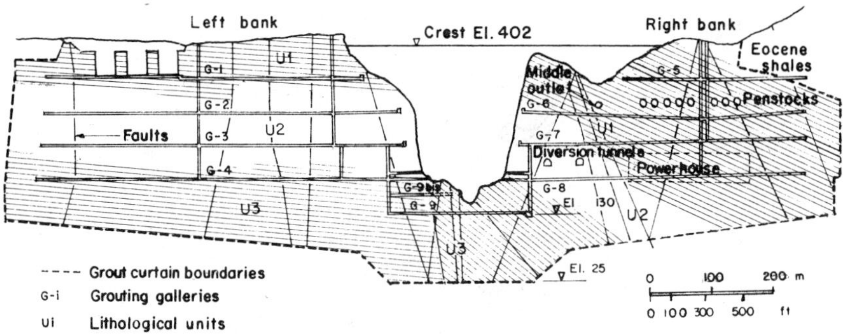

FIG. 2.- Cross-secion A-A' of Fig. 1 showing the geology of the site and grout curtain boundaries

decompression in the proximity of the canyon; this feature is more conspicuous at the left bank due to the orientation of the strata. Units U1 and U3 are the most affected by fracturing particularly the one close to the surface (U1). Microfissures observed in the rock blocks are closed or have a reduced width, and they have traces of calcite. Spacing between fissures is not greater than 6 in. (15 cm). Karsticity observed at the damsite is related to fracturing, bedding planes and microfissuring; unit U1 at the left bank shows the greatest effect at the surface. Karsticity is little developed with depth and cavities (apparently discontinuous) do not exceed 5 cm in width.

GEOHYDROLOGY

From piezometric measurements undertaken at the left bank during one year following the start of construction, a clear trend of seepage following the path of the α faults (Fig. 3) was observed at the upper part of the rock mass, which apparently deviates further on through a sinkhole (near the left wall of the spillway channel) toward fault systems α and β. The most important flow lines of seepage at the river bottom were through the open faults α and γ.

The phreatic level reached El. 400 at a distance of 1.25 miles (2 km) from the left abutment whereas in the right bank the same elevation was observed at only 0.6 miles (1 km) from the canyon.

Two springs were identified, one of them above the river, upstream of the dam, and the other at the left bank near the intersection of the spillway outlet closest to the river and the path of the Chicoasén-Malpaso Fault, probably associated with a fault of the α family. Flow discharges were not measured.

The continuity of karsticity could not be traced although faults through which the main seepage paths occur show karsticity in parts of their lengths.

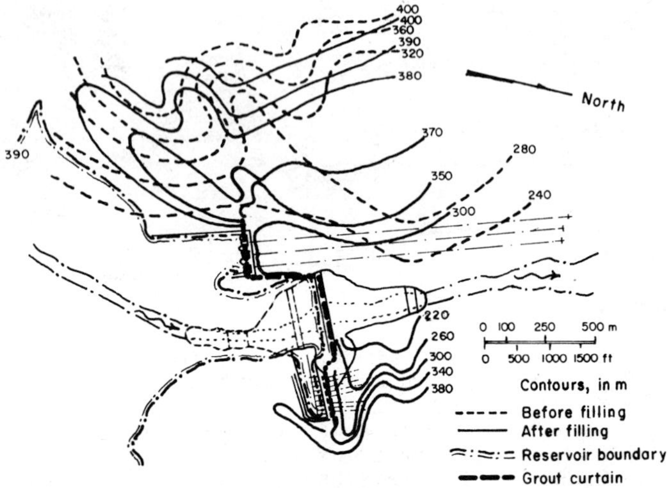

FIG. 3.- Piezometry before and after first filling

PERMEABILITY

Permeability of the rock formations is governed by the existing fracturing. The sound rock is little pervious and has a porosity of only 2 percent and a coefficient of air permeability of 4×10^{-4} fps (12×10^{-3} cm/s).

For each of the lithological units, the average permeability as determined from Lugeon tests prior to treatment of the rock, varied from 3 to 9 UL* for unit U1 (210 stages were tested), from 1 to 12 UL in U2 (509 stages tested) and from 5 to 41 UL for unit U3 (364 stages tested). Very seldom high or very high values (in excess of 100 UL occasionally) were encountered.

ROCK TREATMENT

Following the preliminary geological investigations and taking into consideration the heavy hydrostatic loads to be imposed by the reservoir, it was decided to improve the properties of the rock mass by means of a grout treatment oriented towards stopping the seepage paths, avoiding excessive losses due to seepage and reducing or eliminating the unfavorable effects caused by seepage forces and by the hydrostatic thrust acting within the geological formations and against the structures excavated through rock.

* 1 Lugeon unit (UL) = 0.01 cu ft/ft/min (1 ℓ/m/min) for 142 psi (0.98 MPa) of pressure.

The treatments that were carried out included the following: (1) Blanket grouting of the rock underlying the impervious clay core, to depths of about 33 ft (10 m) in the average, necessary to protect the foundation zone of the embankment against internal erosion (piping) and to reduce the rock deformability at the base and the abutments; (2) curtain grouting of variable thickness performed from the surface and from underground galleries for waterproofing and protection of the underground structures; and (3) a drainage curtain on the downstream side of the grout curtain, to dissipate the hydrostatic forces applied to the appurtenant structures excavated through rock, and to surface rock blocks. Additional treatment comprised blanket grouting in the area of gates to the spillway channel and at the approach channel to the intake works; consolidation and contact grouting at the interfaces concrete-rock of lined sections and of plugs; backpack and consolidation grouting around the steel penstock pipes; auxiliary curtain grouting at the cofferdams and grout treatment of faults, voids and structurally weak zones. Table 1 shows total lengths

TABLE 1.- General data of main activities involved in the grout treatment and drainage system

Activity		Drill length (meters)	Injected cement (tons)	Average grout takes (kg/m)
Blanket grouting in embankment	S	127359	2409	19
	G	7833	52	7
Blanket grouting in spillway	S	12665	233	18
Blanket grouting in power intake	S	3295	77	23
Grout curtain ‡	S	49863	2528	51
	G	182330	4144	23
Drainage system		24654	274	161 *
Short drainage holes		19310	26	21 **
Concrete-rock contact		50000	1283	26
Consolidation grouting		31400	481	15

Notes: 1 m = 3.3 ft S = From surface
1 ton = 2204 lb G = From galleries
1 kg/m = 0.67 lb/ft ‡ Includes connecting holes and reinjected holes
* 1705 m injected after first filling
** 1221 m injected after first filling

of boreholes drilled and the cement consumption, as well as the average grout takes per meter recorded at each treatment zone.

For the blanket grouting a value of 2 UL was set for the final water absorption and between 2 to 10 UL for the curtain, depending on the location of the treated zone with respect to the dam and appurtenant structures.

The following general treatment procedure was specified: Rows or grids of holes for split spacing grouting; grouting through ascending stages 16 ft (5 m) each; simple mechanical packers; a water-cement-bentonite mix for grouting; and Lugeon-type tests before and after the treatment. The spacing of the holes was established upon the results achieved from groutability tests, their orientation depending on the local geological conditions; the refusal pressures were defined in terms of such geological conditions and the response of the rock as observed during Lugeon tests.

Drilling. Pneumatic percussion track-drills were used for the semihorizontal or slightly inclined zones of the blanket grouting, and light pneumatic percussion drills when steep slopes were treated from scaffolding. The hole diameter was 3 1/4" (82.6 mm) in all cases.

For 15 percent of the curtain grouting treatment, neumatic percussion equipment was used and rotary drilling was employed for the rest, being electrically driven in galleries and with diesel engines at the surface. Most of the holes drilled with percussion were 3 5/8" (92 mm) diameter, whereas the rotary holes drilled were 1 7/8" (47.6 mm) diameter.

In 10 percent of the drainage curtain holes rotary drilling was used and in the rest the same type of percussion equipment employed for curtain grouting. The hole diameter was 3" (76 mm).

Grout mixes. Three groups of grout mixes were used for both the blanket and curtain grouting and in most of the other treatment procedures. Their design and control was the responsibility of the field laboratory. Table 2 shows their main physical and mechanical properties.

The first two groups of mixes are essentially unstable suspensions with over 5 percent of free water, whereas the third one corresponds to stable mixes. All of the grout mixes were formulated on the basis of using water, cement and bentonite, with the water/cement ratios (W/C) expressed in terms of volume, considering a unit weight of the cement equal to 98.3 pcf (1.575 g/cm^3). The bentonite content is expressed in percentage by weight of cement. The bentonite slurry was prepared in advance by adding 10 parts of water and one part of bentonite and leaving the mix curing for at least 12 hours. Three different commercial brands of bentonite and five of Portland cement, types I to III, were made available to the project. The finest type of cement (type III with a Blaine specific surface greater than 2000 sq ft/lb (4000 cm^2/g)) was recommended for the grout blanket and the two other types for the grout curtain and the special treatment. In those zones

TABLE 2.- Properties of grout mixes

Grout mix group	Water - Cement ratio (W/C)	Bentonite content (%)	Fluidity (sec Marsh)	Density (g/cm^3)	Sedimentation (% of solids in suspension)	Unconfined compressive strength (at 28 days) (MPa)	Ratio ℓ/kg *
1	6	10	32	1.17	75	0.7	4.3
	5	10	38	1.21	89	1.0	3.6
	4	10	41	1.17	95	1.5	3.0
2	6	2	28	1.15	50	3.0	4.3
	5	2	29	1.19	53	3.5	3.6
	4	2	30	1.22	65	4.3	3.0
	3	2	33	1.31	80	5.9	2.3
	2	2	34	1.44	88	8.7	1.7
	1	2	—	1.70	96	13.7	1.0
3	1.67	4.8	37	1.30	93	2.2	2.0
	1.25	2.8	37	1.40	96	3.8	1.6
	1.10	1.1	34	1.50	97	8.6	1.2

Notes: The grout mixes included here were prepared with type III cement (Blaine specific surface = 2000 sq ft/lb (4000 cm^2/g)) and bentonite with liquid limit of 340 and plasticity index of 303

*Liters of grout mix per kilogram of dry cement

1 g/cm^3 = 62.3 pcf   1 MPa = 145 psi

1 ℓ/kg = 0.016 cu ft/lb

of very high takes, water-cement-bentonite-sand grouts were injected, supplemented in some cases by sodium silicate or by retarding admixtures. Where large volumes of seepage water were encountered (*e.g.* over 0.7 cfs or 20 ℓ/s), in addition to the above mentioned grout mixtures, organic products were also used such as wheat, barley, sawdust, etc. and inorganic fillers such as cellophane sheeting, to seal the voids.

Switching from mixes of group 1 to those of group 2 was made with the aim of improving the resistance to washing out, and from group 2 to 3 to essentially maintain the resistance, enhance the penetrability and in addition have the certainty that the volume of voids grouted with the stable mixes was filled with a percentage of solids greater than that obtained with unstable mixes.

Grouting was performed starting with the most fluid mixes of either group, and changing to the thicker ones depended upon the rate of absorption of the hole. Due to the generally low grout takes, the most common mixes injected were the unstable mixes with W/C = 6 and the stable mix of group 3 with W/C = 1.67.

Mixers and grouting equipment. Central plants for mixing and delivering of grout mixes were installed to supply the reception and injection plants placed in the neighborhood of the grout holes. In some cases central plants were also used for grout injection; these plants consisted of high-speed mixers (1800 rpm) to prepare the bentonite slurry and the mixes; storage tanks for the bentonite slurry; calibrated containers for proportioning water and bentonite with manual operation; low-speed mixers (70 rpm) for the agitation of mixes; and Moyno pumps of large capacity and outflow to apply pressures of 256 psi (1.8 MPa) necessary to deliver the grout mixes. The injection plants carried only low-speed mixers and pumps of the Clivio type with double piston and low discharge, used for applying refusal pressures of 853 to 1706 psi (5.9-11.8 MPa). For the blanket grouting Moyno pumps were also employed for injection by adding return pipes even when stable mixes were used, because they lacked a valve to regulate the discharge.

The central plants were subsequently modified to operate semiautomatically thus achieving an increased efficiency in the delivery of mixes to the different work sites.

The control of the refusal pressures was made by means of Bourdon gages installed in manifolds having an outlet for the return pipe used for injecting unstable mixes, and also a pressure relief valve. For intercommunication between any of the plants and the grouting personnel, wireless telephones were available.

CUT-OFF WALLS IN COFFERDAMS

Grout treatment at the damsite was started by constructing cut-off walls through the alluvium underlying both cofferdams in order to reduce seepage towards the foundation zone of the impervious core and to facilitate the excavation operations, the rock treatment and placement of materials in the embankment. Both plastic cut-off walls 2-ft (0.60-m) thick were placed to a depth limited by the size of particles larger than 2 ft (0.60 m); it was supplemented by a grout curtain executed 6.6 ft (2 m) above the lower end of the plastic cut-off wall and ending 16 ft (5 m) below the contact between the alluvial deposits and the rock. The cut-off walls were constructed by excavating overlapped trenches 6-ft (1.8-m) long by 2-ft (0.6-m) wide and then pouring a slurry made of water, cement, bentonite, sand and setting retarder; the curtain grouting was performed using *tubes à manchette* with grout mixes composed of water, cement and bentonite in the first sequence of injection and a gel made of sodium silicate and aluminate during the second sequence.

The average area of the upstream and downstream cut-off walls was 20 400 and 31 200 sq ft (1900 and 2900 m^2), with a maximum depth of 216 and 184 ft (66 and 56 m), and a width of 148 and 354 ft (45 and 108 m), respectively. The downstream cut-off wall was left 148 ft (45 m) short of the left abutment and this distance was covered with a grout treatment. Regardless of these walls, the volume of water pumped from the excavation between cofferdams with a differential head of 60 m was rather high, in the order of 11 to 14 cfs (300-400 ℓ/s), mainly due to water seeping through the abutments.

BLANKET GROUTING

Grouting operations for the blanket and curtain treatment of the rock were started simultaneously at the contact with the base of the impervious core, once the cofferdams cut-off walls and the excavation between El. 190 (33 ft or 10 m below the river level) and El. 140 (river bottom) had been completed. It was necessary during the excavation to remove large blocks of rocks and alluvial deposits from a very narrow gorge; the blocks were the result of rock slides from the canyon walls. Removal of these fragments gave way to unstable conditions at the left bank of the excavated narrow gorge, due to the unfavorable attitude of the unit U3 strata at this particular site.

To perform the blanket grouting six treatment zones were defined in terms of their topographical setting, as shown in Fig. 4. The semihorizontal zones A, C and E were treated with split spacing grouting arranged in a grid pattern that closed down in some areas to quaternary holes; primary holes were spaced at every 20 ft (6 m) and in the final sequence 5-ft (1.5-m) spacing was used.

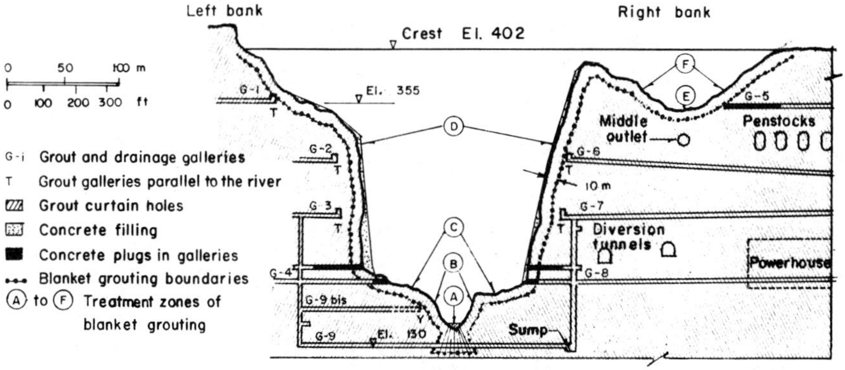

FIG. 4.- Blanket grouting and drainage galleries along the dam axis

At the zones with steep slopes (B, D and F) the treatment was made with a vertical spacing of primary rows of 20 ft (6 m) and closing with secondary rows at every 10 ft (3 m); tertiary holes spaced at every 5 ft (1.5 m) were only drilled in zone B. Each row was carried out with split spacing grouting with primary and secondary holes spaced horizontally at 20 and 10 ft (6 and 3 m) respectively. Additional boreholes were drilled at the sides of holes corresponding to the last sequence which showed grout takes in excess of 33 lb/ft (50 kg/m). Fig 5 shows the treatment zones A, B and C at the bottom gorge of the canyon (see Fig. 4).

FIG. 5.- Blanket grouting treatment zones A, B, and C at the bottom gorge of the canyon, as seen downstream

The blanket grouting covers an area of 882 000 sq ft (82 000 m^2) with an effective depth in rock of 33 ft (10 m), with holes drilled down to 49 ft (15 m) inclined 45°, except at zone A where the depth varied progressively from 33 to 59 ft (10-18 m), with holes 49 to 121 ft (15-37 m) long drilled in rock with the same inclination; the blanket was to be connected with the curtain grouting below gallery 9 at El. 123.

The most difficult conditions during the blanket grouting were encountered at the bottom gorge (zones A and B) due to the instability of the blocks of the left bank and to water seeping through the fault sets α and γ, towards which most of the holes were oriented. In this zone the blocks were anchored and drained and a concrete slab 16-ft (5-m) thick in the average was poured on top of the bottom rock. Both the unstable blocks and the slab were instrumented during the treatment to detect any displacement induced by the grout pressure. Artesian water was observed when the first holes were drilled, with pressures up to 10 psi (0.07 MPa), and turbidity could be noticed in the water due to the presence of soil fines; there were frequent intercommunications between holes which decreased with time, as well as artesianism as the treatment progressed.

Refusal pressures prescribed initially for zone A were very low (3 psi or 0.02 MPa) and they were progressively increased to 7 and 28 psi (0.05 and 0.2 MPa) at the uppermost and lowest grout stages, respectively, taking advantage of the concrete slab and of the first grouted holes. Subsequently, when the blanket was deepened and once the quaternary holes in the zone 33-ft (10-m) thick had been completed, pressures increased from 28 psi (0.2 MPa) at the shallowest stage to 256 psi (1.76 MPa) at the deepest stage (105 to 121 ft or 32-37 m). Grout mixes containing 10 percent of bentonite (group 1) were applied to 50 percent of the holes and this proportioning was reduced to 2 percent (group 2) for the rest; the permeability was thus lowered to 6 UL. At zone B the procedure applied was similar to the above and at the left bank pressures were increased up to a maximum pressure of 21 psi (0.15 MPa) once the anchoring of the blocks had been performed. Grout mixes with 10 percent bentonite were used in approximately 30 percent of the holes drilled in this zone, and in only 16 percent at zone C, with no further application whatsoever in zone D. In this latter zone, mixes of group 3 were used starting at El. 270 approximately.

Refusal pressure ranges, initial and final average permeabilities and measured grout takes, together with the different groups of mixes employed in each zone, are presented in Table 3. It should be pointed out that during the Lugeon - type permeability tests carried out in the blanket grouting, pressures applied did not reach 142 psi (1 MPa) and hence the results were extrapolated as a function of the measured flow *versus* pressure graphs plotted for each test.

In spite of the fact that the average grout takes were low and that the initial permeabilities were greatly reduced, certain doubts rose regarding the actual success of the blanket grout; as a result, an additional treatment was undertaken, monitoring and adding new holes drilled from cross-adits excavated at the ends of galleries 1, 2, 3, 6 and 7 parallel to the river, as well as from gallery 9 bis at El. 160 where two adits in a Y shape were excavated at the zone of unstable blocks (Fig. 4); holes were also drilled from gallery 9 towards the upper rock beds. The confinement obtained from the weight of the partially built embankment was used to apply higher pressures than those originally recommended at the surface. The treatment implied the need of extreme precautions to avoid the injection of large volumes of grout within the impervious core, thus eliminating the

TABLE 3.- Average values obtained in blanket grouting

Zone (see Fig 4)	Bank	Lithological unit	Refusal pressure range (MPa) S	Refusal pressure range (MPa) D	Grout mix group	Initial permeability (UL)	Final permeability (UL)	Grout takes (kg/m)
A	river bottom	U3	0.02-0.20	0.02-0.60	1 and 2	41	6	29
			1.8*					
B	L	U3	0.02-0.05	0.05-0.15	1 and 2	16	1	18
	R		0.05-0.10	0.15-0.20		10	5	18
C	L	U3	0.02-0.10	0.02-0.30	1 and 2	5	4	12
	R		0.10	0.60		7	2	16
D	L	U3 U2 U1	0.10-0.20	0.30-0.80	2 and 3	10 (U2)	6(U2), 2(U3)	22
	R		0.10-0.20	0.60-0.80		12 (U2)	2(U2), 5(U3)	15
E	dike foundation	U1	0.20	0.80	3	—	—	18
F	dike L	U1	0.20	0.80	3	—	—	22
	dike R		0.20	0.80		—	—	17

Notes: 1 MPa = 145 psi; 1 kg/m = 0.67 lb/ft; 1 m = 3.3 ft; L=left bank R=right bank; S=shallow stage (0 to 5 m); D=deep stage (10 to 15 m); *from 32 to 37 m

risk of hydraulic fracturing that could lead to weak zones susceptible to piping. Allowable grout and refusal pressures were decided upon from the hydraulic pressures required to cause fracturing of the clay core by using clear water and grout. All of these pressures were measured with test holes by injecting water and grout mix with W/C = 1.67 of group 3; the density of the grout mix and the height of the free surface of the impervious core were taken into account during the grouting operation. The presence of concrete walls constructed in most part of the contact area with the core, made it possible in several instances to avoid punching the core and to increase the refusal pressures.

General criteria adopted for the grouting operations, contemplated a refusal pressure for the stage closest to the impervious core, equal to half of the core weight on top of the treated zone plus 14 psi (0.1 MPa), up to a maximum of 142 psi (1 MPa), limiting the amount of grout injection to 53 gal. (200 ℓ). The refusal pressure was increased in 28-psi (0.2-MPa) increments in each of the other two upper stages located within the 33-ft (10-m) thickness of the blanket. Those stages outside the blanket grout were injected with grout and mortar. The holes were arranged in a grid pattern 40 ft (12 m) on the side, distributed through the whole area covered by the blanket and the average grout takes did not exceed 7 lb/ft (10 kg/m), with a weighted average of 4.7 lb/ft (7 kg/m), due to the original treatment performed from the surface; closing of fissures due to the stresses induced by the embankment may have also influenced these results.

CURTAIN GROUTING

The curtain grouting was performed from galleries excavated at different depths and also from the surface. Fig. 2 shows the geometry of the profile developed along the galleries and the lithological units encountered. The treatment covered an area of about 4.2 x 10^6 sq ft (392 000 m^2). The galleries are located along a vertical plane save in the zone of the approach channel to the spillway where the axes of the deepest galleries are displaced towards upstream in a stepped fashion. The average vertical spacing between galleries was 180 ft (55 m); galleries 4, 8, 9, 9 bis and 5 were entirely lined whereas the others were only partially lined along the badly fractured stretches. The excavation was made with a portal cross-section 7.2-ft (2.2-m) wide and 10-ft (3-m) high with a drainage ditch in the downstream side.

The geometry of the grout curtain was defined after taking into account the topography of the damsite, the layout of the structures where protection was required, the observed piezometric levels and the variation of the permeability of the rock mass.

Grouting was carried out by means of rows of holes inclined towards upstream. For the treatment of zones A, B and C of the blanket grouting, five rows of holes were drilled down to 16 ft (5 m) below gallery 9, arriving at a grout width of about 23 ft (7 m); within the rock mass proper, the number of rows drilled from the galleries was decreased progressively from three to one, leaving either one or two rows at the end of the galleries. The hole spacing was decreased at each row from 79 to 5 ft (24 to 1.5 m), the smallest value being used in highly fractured zones. The proposed number of rows was based on the local geological features. The farthest downstream line was performed first following the extreme upstream row and then the intermediate and/or central rows.

Each row consisted of primary and secondary holes as a minimum except for the central row where an unlimited number of holes was carried out until grout takes less than 67 lb/ft (100 kg/m) were achieved.

The average drilled length varied from 198 to 262 ft (60-80 m), except in those holes made from zones near the cliff in galleries 4 and 8 and in those drilled from gallery 9, where the length ranged from 393 to 443 ft (120-135 m). Boreholes oriented towards upstream and with a variable direction that were performed to intersect any of the fault sets α, β and γ found at the site, were drilled from the floor of the galleries or from the surface, at a distance of 33 ft (10 m) from the lower gallery. The space left between the gallery and the grout curtain was covered with three interconnecting holes 49-ft (15-m) long spaced at every 16 ft (5 m) in the south wall of the galleries; simultaneously, short holes were made for consolidation grouting and drainage and for contact grouting in the lined tunnel sections.

Grout mixes are similar to those used for the blanket grouting. Holes drilled from zones A, B and C at the beginning of the treatment were injected with mixes of groups 1 and 2; on the other hand, grout mixes of group 3 were employed in most of the holes advanced from galleries.

To define the refusal pressure for each zone to be treated, several factors were considered such as the presence of potentially unstable blocks at the surface, the density of local fracturing and the behavior of the rock as observed from Lugeon tests. The treatment of zones A and B from the surface involved very low grout pressures, applying 0.87 psi per feet of depth (0.02 PMa/m); at zone C pressures were increased to 2.6 psi/ft (0.06 MPa/m) at the left bank and to 3.5 psi/ft (0.08 MPa/m) at the right abutment. With these pressures leaks were observed at the left part of gallery 9 (gallery 9 bis had not yet been excavated) and therefore the depth of the curtain was limited to El. 140 in this zone; it was subsequently reinjected from gallery 4 taking advantage of the buttressing effect of the embankment and of the lining of gallery 9. When grouting from the galleries, the pressures increased at a rate of 4.3 psi/ft (0.1 MPa/m); however, it was feared that below a depth of 164 ft (50 m) the applied pressure would exceed the range of elastic yielding of the rock, and the pressure was thus limited to 711 psi (5 MPa) for depths in excess of 164 ft (50 m). This pressure was applied at a distance of 33 ft (10 m) from the lower galleries without observing neither grout leakage nor fracturing or block thrust. This enabled the increase of grout pressures in the first four stages of each hole drilled from the lower gallery to 71, 142, 427 and 711 psi (0.5, 1, 3 and 5 MPa), thus obtaining a more homogeneous treatment: in this way differences in the refusal pressures between the holes drilled from the upper gallery and those advanced from the lower gallery were eliminated. In highly fractured zones and in open joints the pressures were limited to 427 psi (3 MPa). For the holes 49-ft (15-m) long connecting with the grout curtain, pressures decreased from 427 to 71 psi (3 to 0.5 MPa) when advancing from the deep to the shallow stages.

In addition to the grout curtain, other injection operations were carried out such as radial holes from the end of galleries 4 and 8 closest to the canyon to check and strengthen the treatment performed from the surface; additional curtains from the surface (zone D) between Els. 300 and 365 and from cross-adits to intercept the blanket grouting; and finally, window treatment around penstocks from galleries 5 and 6.

Generally speaking the average maximum grout takes were recorded at the left bank and in unit U1, followed by unit U2 and to a lesser degree in U3. The row with the greatest consumption corresponded to the farthest downstream line, with 14.7 to 55 lb/ft (22 to 82 kg/m) in all the galleries except during grouting from surface in the left bank where it was about 155 lb/ft (232 kg/m) that was reduced to 39 lb/ft

(59 kg/m) by means of additional holes. Grout takes greater than 335 to 670 lb/ft (500-1000 kg/m) were very seldom recorded.

The average final permeability resulting from the complete treatment was determined from about 800 tests carried out in all the lithological units and it did not exceed 3 UL. However, larger values were obtained at localized zones, with permeabilities as high as 42 UL. These zones were grouted again but no further Lugeon tests were subsequently performed.

DRAINAGE SYSTEM

To build the drainage curtain, holes 3 in. (75 mm) diameter were drilled from galleries, in the zone downstream of the grout curtain, with a spacing varying from 20 to 115 ft (6-35 m) depending on the closeness to structures. All holes were advanced from the roof of the galleries (ascending drains) and only in galleries 4, 8 and 9 holes were also drilled downwards to a depth limited to 164 ft (50 m). The average length of the ascending drains was 197 ft (60 m). The drainage curtain covers approximately 70 percent of the grout curtain area.

The drainage system also comprised holes made in gallery 12, access facilities to gallery 3, an auxiliary tunnel for construction of the powerhouse, a drainage gallery for the powerhouse and adit 22 (Fig. 1); the aim of these activities was to relieve the water pressures from the spillway tunnels, the jointed surface blocks, the penstocks and the powerhouse. The drainage system as a whole covers an area of about 4.4×10^6 sq ft (410 000 m^2).

All the boreholes in galleries 4, 8 and 9 were adapted to function as open-type piezometers by placing a mouthpiece with a stopper, valve and manometer in order to measure periodically hydraulic pressures; in addition, 17 holes were drilled in galleries of both banks to operate also as individual piezometers downstream of the drainage curtain and close to the cliffs.

In all galleries gauging weirs were placed to measure seepage coming from short relief wells in lined stretches of galleries, flow through fissures in unlined zones and from the long drains of the curtain itself. The amount of seepage water collected in galleries 4, 8, 9 bis and 9, which represents approximately 60 percent of the combined flow in all the galleries, is pumped from gallery 9 to the surface, whereas seepage from the rest of the galleries is diverted to the surface.

The total volume of water recorded in galleries during the first months of the first filling of the reservoir, are presented in the logarithmic plot of flow *versus* pool level of Fig. 6.

The amount of water collected in the whole drainage system with the reservoir at El. 391 (13 ft or 4 m below the maximum water level) reached a value of 10 cfs (290 ℓ/s) and peak

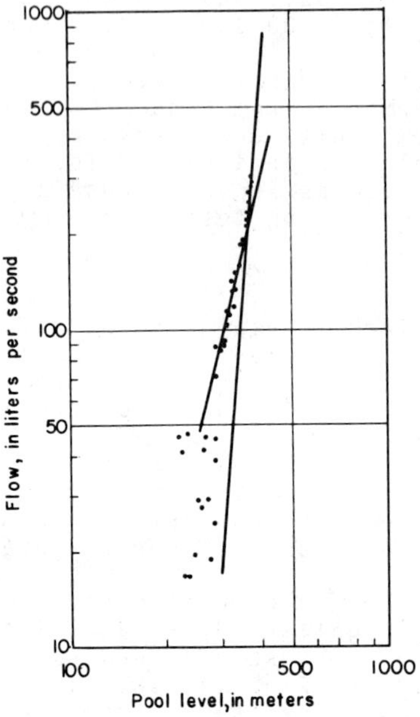

FIG. 6.- Flow in galleries *vs.* pool level during the first filling

values of about 1.8 cfs (50 ℓ/s) were recorded in galleries 3, 4 and 8. Seepage is not concentrated and its distribution is quite regular within most of the drainage system. Only in 16 holes out of the total of 467, flows between 0.14 and 0.35 cfs (4-10 ℓ/s) were gauged and in seven of them it exceeded 0.35 cfs (10 ℓ/s). The drainage system takes care of all seepage, and no springs have been observed at the abutments.

The ratio between the seepage flow and the area covered by the drainage system is of 3.3×10^{-6} cfs/sq ft ($0.8\ cm^3/s/m^2$) which represents a very low value.

PIEZOMETRIC CONDITIONS AFTER FIRST FILLING

The elevation of the phreatic surface in both river banks after the first reservoir impoundment could be recorded by means of 39 piezometers installed in the left bank and 13 in the right abutment. Piezometric changes were more noticeable during the first four months after filling was started and equilibrium was achieved when the reservoir reached El. 390 approximately, with only slight variations thereupon.

Fig. 3 shows also the distribution of the contours of equal piezometric elevation one month after completion of reservoir filling. It can be observed that in the left bank there is a tendency of the water to flow along the α set close to the grout curtain. The hydraulic head loss through the grout curtain observed in holes close to its end is of only 115 ft (35 m); however from readings taken at the piezometers in galleries, an abrupt change of about 262 ft (80 m) can be observed in the zone of the spillway tunnels immediately downstream of the curtain.

A clear trend of the underground seepage cannot be defined at the right bank. In one of the piezometers of this zone located upstream of the grout curtain, the level rose from El. 227 to El. 367 during the first filling, whereas in other piezometers placed in the downstream zone, the water level remained practically constant, a fact that confirms that the overall efficiency of the waterproofing and drainage operations has been successful to date.

CONCLUSIONS

The permeability, grout takes and seepage quantities observed at the rock banks of the Chicoasén damsite before, during and after the comprehensive grout treatment described above, show that extreme seepage conditions may develop locally through the rock; although the real behavior of the grouting and drainage systems has been quite satisfactory, as demonstrated by measurements and records taken one year after the river was closed and following 9 months of reservoir operation under practically its maximum elevation, periodical monitoring of the water seepage should be continued, measuring carefully the amount and velocity of seepage through faults and in springs, and the pressures and characteristics of the flow regime in both abutments, in order to be able to control any unexpected situation that might develop and also to evaluate the actual efficiency of the system with time.

ACKNOWLEDGEMENT

The author is indebted to the Federal Commission of Electricity (CFE) for permission to publish data presented and to Raúl Esquivel for reviewing and translating the manuscript. Messrs. A. Pautre and F. Sabarly acted as consultants to CFE for the grout treatment and drainage system. Field operations and quality control were carried out directly by CFE through its SUTRAROC and quality control branches, respectively.

CEMENT GROUTING PRACTICES U. S. ARMY CORPS OF ENGINEERS

by
John A. Albritton*

ABSTRACT

Cement grouting, as applied to water retention structures, is one of the most frequently used methods of foundation treatment within the U. S. Army Corps of Engineers. Current Corps practices are discussed. These practices are based on both experience and research. Much of the information presented will also be published in a joint Air Force/Army Technical Manual "Grouting Technology" which is presently under preparation.

INTRODUCTION

Cement grouting is the most widely used method within the Corps of Engineers of reducing rock mass permeability of dam foundations and increasing the strength of foundation materials. When properly designed and constructed, effective results are achieved. Grouting in Corps of Engineers work is performed as design requirements of permanent construction, post construction remedial treatment, or an increment of expedient construction or repair. This discussion deals primarily with cement grouting as applied to earth and rockfill dams, concrete dams, tunnels, and shafts.

In its broadest sense, grouting can be defined as injecting chemical grouts or suspended solids into voids or compactable material to reduce permeability; to consolidate, compact, strengthen or stabilize the host medium; to lift, move, and/or support structures; or to provide contact between structures or between structures and soil or rock.

Grouting has not totally achieved the status of a science but in some respects remains an art--primarily due to the large number of variables and unknowns involved. Grouting practices vary widely within the Corps and even on the same job. Good results have been attained by widely differing procedures and methods. Consequently, Corps practices are not highly standardized.

An important requirement for a good grouting job is field control by a geologist or geotechnical engineer who understands the geology and subsurface conditions, and also has grouting experience on similar jobs.

---

*Engineering Geologist, Division Geologist, U.S. Army Engineer Division, Missouri River, Omaha, Nebraska

Because of physical and procedural limitations common in grouting, the Corps of Engineers does not rely on grouting as a single line of defense but integrates it with other design features which act together to reliably achieve their purpose. One important design consideration is the possible requirement to regrout segments of a grout curtain at some future date. The curtain should be designed so that it is accessible with a full pool. Present practice by many designers is to avoid designing grout curtains under the upstream toe of embankment dams, but rather to design the embankment for the grout curtain beneath a central impervious core. This allows access from the crest of the dam for possible remedial grouting. Internal access may be provided by tunnels and galleries in the original design. These features may add to the effectiveness of seepage and uplift control by providing for foundation drainage in addition to grouting.

INVESTIGATIONS

Project investigations for impoundments are designed to assess the need for grouting and to provide complete information for design of the grouting program. The need for grouting is not based on average permeabilities or average foundation conditions but is assessed by detailed project investigations. Major considerations include leakage potential, areal and structural geology, in-situ stress conditions, geohydrology, geochemistry, and compatability of in-situ and grouting materials. The presence and characteristics of anomalous conditions are ascertained and appropriate treatment planned. Grout takes, mixes, procedures, and pressures are best determined or estimated by conducting a grout test program at the site.

Numerous case histories have time and again demonstrated the necessity of thorough geological exploration prior to grouting and day-by-day assessment during grouting. Until recent years, geological investigations for design of grouting treatment have more often than not been limited by economic considerations or a failure to recognize their importance. Discovery of solution cavities during construction of major projects such as the Gathright Dam in Virginia and the Clarence Cannon Dam in Missouri have led to extensive investigations and design changes during construction which resulted in massive cost overruns. Such experiences have mandated more thorough investigation for design.

Investigations for grouting may encompass any geological or geotechnical method normally used for regional and site investigations. The investigations are in sufficient detail to eliminate major surprises. The full depth and lateral extent of the grout program is explored.

Test grouting is accomplished when feasible. In addition, regional groundwater and geological studies are made of the reservoir area. Piezometers are installed upstream and downstream of the proposed grout curtain in order to obtain baseline data prior to construction. In addition, weirs are installed and periodic measurements made of downstream springs and seepages prior to and after construction. Water well measurements and records are obtained prior to construction.

TYPES OF TREATMENT

Cement grouting done by the Corps of Engineers normally falls into one of seven categories: curtain grouting, area grouting (sometimes referred to as consolidation grouting or blanket grouting), tunnel grouting, cavity filling, backfilling boreholes, contact grouting, and specialized applications. Dental treatment and concrete backfill, mortaring of joints, broom or slush grouting and positive cutoffs are frequently associated with and integrated with the grout curtain or area.

CURTAIN GROUTING is drilling and grouting one or more lines of grout holes in the foundation, abutments, or reservoir rim to produce a barrier to seepage. The Corps normally uses a single line beneath concrete dams. The grout holes are generally vertical or angled in an upstream direction and drain holes are installed downstream of the grout curtain to reduce uplift pressures. Grout curtains beneath the impervious core of embankment dams usually consist of multiple-line curtains at least in the upper zone. The curtain grouting is generally done before any fill is placed. Consequently, low pressures are necessary in the upper zone. Grouting through fill is sometimes necessary for remedial or deferred grouting. Special precautions are necessary to avoid fracturing or eroding the fill.

A single line is generally used for rim or upland grouting. Specifications are flexible enough to add additional lines of grout holes at any location or depth as determined necessary in the field.

AREA GROUTING normally consists of grouting a shallow zone in a particular area utilizing grout holes arranged in a pattern or grid. Its purpose is to make the area more impermeable or to consolidate fractured and jointed rock. It has been used at many Corps projects and is performed most often in foundations for the impervious core of embankment dams or for concrete structures. Deeper area grouting is sometimes done to grout specific geologic conditions such as fault zones or to consolidate subsurface materials at shaft or buried structure locations. Area grouting near the surface is with low or gravity pressures; however, where deeper zones are grouted, higher pressures are used.

TUNNEL GROUTING. Grouting for tunnels may be for backpacking tunnel liners, consolidation of material surrounding the bore, seepage control, contact grouting, or grouting a portion of the grout curtain. Preexcavation grouting may be required for consolidation and water control. Where grouting is accomplished after tunnel excavation, imbedded pipes or formed holes are provided through the liner for grout or drain holes and venting. Holes may also be drilled through the liner, if necessary. Pressure grouting for backpacking behind cast-in-place concrete liners is not done until 7 days after placement of the liner. In bad ground where precast concrete or steel ring liners are used, grouting for lagging and backpacking is accomplished as quickly as possible after ring placement. A sanded mix is required for grouting of tunnel liners. Injection begins at the bottom and is moved up as grouting proceeds. The final step is contact grouting with neat cement grout at the crown after the backpacking has been completed.

CAVITY FILLING. Cavity filling is one of the least standardized types of grouting. The effectiveness of grouting clay-filled cavities is questionable; however, air- or water-filled cavities or large open joints can be grouted successfully with cement grout. The extent of a cavity is unknown after penetration by just one grout hole. More exploration or drilling may be necessary before treatment is completed.

When a cavity is encountered in drilling, the hole is normally stage grouted. A packer is set near the top of the cavity and the cavity is grouted. Frequently, a sanded mix is required in order to complete the grouting of the cavity.

Intermittent grouting may be necessary. Intermittent grouting is the process of injecting grout (usually a thick sanded mix) in the hole and then waiting several hours before injecting more grout. Several waiting periods may be necessary. In intermittent grouting, the last batch of thick, mortar grout injected into the hole before the waiting period is followed by injecting water into the hole to keep the hole open. After the waiting period, grouting is resumed with neat cement grout before going back to the mortar mix. The amount of grout to be injected during each period is normally a predetermined limit. In practice, the limit varies from about 300 sacks to over 1,000 sacks per injection period. A limit is also placed on the maximum amount of grout to be injected into a single hole.

When refusal is reached, it is assumed that grout has filled at least the portion of the cavity penetrated by the grout hole. Additional grout holes are then drilled and grouted until the desired results are achieved. If pressures fail to build up or the cavity is obviously too large to grout in this manner, grouting should continue with the curtain beyond the cavity. Additional exploration, consultation, evaluation, and design of treatment can then take place without delaying the contractor. These measures may call for specialized grouting procedures or materials such as foaming agents, positive cutoff diaphragm or formed concrete wall, additional excavation, or some other solution.

BACKFILLING BOREHOLES and grout holes is an important part of the grouting program. These holes may become wells under the reservoir head, and if not properly grouted, could contribute to seepage and piping. In the rock foundation, holes are backfilled with grout having a water-cement ratio of about 0.7:1 with 4 percent bentonite. A minimum 1-inch diameter (25mm) delivery line with a steel section at the end is extended to the bottom of the hole after the line is completely full of grout. Grout is then pumped until it flows from the hole; then the delivery line is slowly withdrawn while pumping continues. If settlement of grout occurs, the holes are topped off (rebackfilled) before placement of fill.

Where holes are backfilled in embankments or fill, the tremie method is also used; however, the backfill mix is designed to be more plastic. It is designed to have the approximate consistency of the fill.

CONTACT GROUTING is injecting neat cement grout at the contact of a structure with a vertical or high angle surface, or at the contact of backpack grout with the crown of a tunnel. Its purpose is to fill the void at the contact that results from shrinkage. Contact grouting may be done either through header pipes installed for that purpose or through drill holes. The installations or grout holes are thoroughly washed before grouting. Grouting pressures vary but should be 25 psi (172kPa) or higher where the higher pressures can be used safely.

SPECIALIZED APPLICATIONS. There are many specialized applications of grouting by the Corps of Engineers. They are frequently the subject of special reports and will not be discussed herein.

GROUTING METHODS

GENERAL. Corps of Engineers grouting methods have become less standardized in recent years. The current Corps guide specification only prescribes stage grouting, but current practice is to use the method which best meets specific job requirements. It is not unusual to utilize several different methods at the same project. Before discussing grouting methods individually, it will be beneficial to consider drilling, washing, and pressure testing methods.

DRILLING METHODS. The actual drilling methods are normally left up to the contractor subject to specified restrictions and Corps approval. Coring is allowed but not required. A minimum size is normally specified for grout holes. Many jobs have been successfully grouted with grout holes as small as 1-1/2 inches (38mm) in diameter. Larger diameters are frequently required for deep, inclined grout holes. The main reason for this is stiffness of the drill rods resulting in a straighter bore. Drift can be substantial at depths greater than 200 feet (61m). Another reason for larger holes is related to injection pressures. It has been shown that the distribution of grouting pressures in the formation is affected by borehole size. Smaller size holes require greater pressures to achieve the same results as less pressure in larger holes. Some Corps projects have specified minimum size grout holes as NX rather than the smaller EX holes sometimes used for grout holes.

Orientation of inclined grout holes is based on the geology. Where practical, a statistical study of rock joints is made, the trends identified, and angles and direction of inclination of grout holes based on the results. Grout holes are oriented to intersect all joint trends. They are angled either in an upstream direction or parallel to the dam axis. In the abutments, the holes are also angled with a component into the abutment rather than toward the valley. Where low angle features are the object of the grouting, vertical or high angle holes are used. Conversely, where high angle or vertical features are the object of the grouting, holes are on flatter angles. Stress release or sheet joints are considered even if not identified as statistical trends. Horizontal grout holes are considered for abutments and integrated into the grout curtain when conditions indicate they will add to grouting effectiveness.

Normally, it is Corps practice to require water as a drilling fluid rather than air. In some instances, however, water causes problems such as hole caving, erosion, or binding of the bit, and in these cases, air or some other agent is required for best results. Air may also be used to evacuate cuttings when water circulation cannot be achieved.

Rotary drilling methods are specified for many grouting applications, but recent years have seen greater acceptance of percussion drilled grout holes utilizing water. Percussion drilling in fine materials such as shale or percussion drilling with air can cause premature plugging of fine fractures and joints. The clay slurry or dry rock flour, if forced into fine fractures, or if not completely cleaned from the hole, can either plug fractures or mix with the initial grout causing a reduction in the water-cement ratio and instant plugging of the fractures.

Blind drilling (i.e., drilling after lost circulation occurs) is not allowed for grout holes where it is avoidable. It is considered better to stop drilling and stage grout the hole when lost circulation occurs. Some blind drilling is allowed to avoid excessive staging. For instance, if the hole is stage grouted, and upon resumption of drilling, lost circulation occurs immediately, then blind drilling is allowed for a few feet before stage grouting again.

In drilling grout holes, care must be exercised not to overpressurize the formation. A maximum pump pressure is not normally specified for drilling because water circulation must be with sufficient velocity to remove the drill cuttings, and the low pressures required for grouting some zones would be unduly restrictive as a maximum pump pressure for drilling. Adequate annular space between the hole and the drill rods is required in order to prevent hole blockage and pressure buildup. In addition, the drilling rate sometimes needs to be controlled to allow better evacuation of the drill cuttings.

WASHING. Upon completion of drilling, the grout hole is washed or flushed until all drill cuttings and turbidity are removed. This is a separate operation from pressure washing which is performed with the pressure testing rather than drilling equipment.

PRESSURE WASHING AND PRESSURE TESTING are required immediately before pressure grouting operations are begun for the hole. Pressures used for pressure washing and testing never exceed the maximum allowable grouting pressures. Where connections exist between grout holes, pressure washing is accomplished in each of the holes for best results. Washable materials are ejected from holes hydraulically connected to the hole being washed. Washing continues as long as clay or washable materials are being removed from an interconnected hole or a surface leak, or as long as the rate of water injection increases at a given pressure.

After pressure washing, pressure testing with clean water is accomplished. Holes in which pressure cannot be built up (i.e., open holes) or in which substantial water takes are experienced, are tested by moving the packer, or with straddle packers, in order to isolate the take zone and determine the grouting procedures to be used. In packer grouting, this is particularly important so that packer depths can be determined which will most effectively grout the hole. Additional packer settings to those at the top of each predetermined grout zone are often required based on high water takes, or artesian pressures. Conversely, sometimes predetermined packer settings are eliminated if water takes for a particular zone are negligible.

Grouting methods used by the Corps are usually a combination of the methods outlined in the following discussion. Field adjustments are often necessary to meet actual conditions encountered.

THE SPLIT-SPACED METHOD OF DRILLING AND GROUTING is a procedure whereby primary holes are drilled and grouted on a maximum predetermined spacing, then secondary holes are drilled and grouted at locations midway between primary holes. Each subsequent order of holes are at locations which split the spacing of the previous order of holes. Sometimes secondary or even tertiary holes are mandatory with subsequent split-spacing based on grout takes. At other times, only the primary holes are mandatory and the split-spacing is based entirely on grout takes. Primary grout holes are rarely spaced closer than 10 feet (3m) apart but the maximum spacing may be expanded considerably until the holes are more properly called exploratory grout holes. Split-spaced drilling and grouting is usually combined with other selected methods of grouting. Grout take criteria for split-spacing grout holes varies from job to job.

STAGE GROUTING. Stage grouting involves cycles of drilling, washing, pressure testing, grouting, and hole clean out, beginning with the primary holes in the shallowest stage of the first zone of a given section. The depths of the stages are governed by foundation conditions. Before the injection of grout in any hole is started, the nearest two holes in advance of the hole to be grouted are drilled and the nearest hole is thoroughly washed. If pressure tests indicate a tight hole, grouting of that hole may be omitted for that stage and the hole left open for drilling and grouting of the next stage. After a hole is grouted, the grout within the hole is removed by washing or jetting before it has set sufficiently to require redrilling.

After all first stage grouting of the primary holes in the section is completed, a waiting period of 24 hours is required. Then the primary holes are deepened to their second stage of depth and grouted at higher pressures. The process of successively drilling primary holes to additional depths and grouting at higher pressures is continued until all primary holes in the section have been grouted to the depth of the first zone. Secondary holes are then drilled and grouted using the same procedures as for primary holes. The process of reducing the grout-hole interval by the split-spacing method and grouting by stages is continued until grouting of the section is complete to the full depth of the first zone. Using the same grout holes, successively lower zones are drilled and grouted in like manner until the full depth of grouting treatment is reached. Other sections are drilled and grouted in the same manner until the grout program is complete.

Usually, grout is injected at the top of the holes for each stage using higher pressures as deeper stages are grouted. Where bedrock conditions cause concern, surface leaks develop, or there exists a danger of lifting the foundation, packers may be used at the bottom of the first or deeper stage in order to safely use higher pressures.

STOP GROUTING (or packer grouting) is a method of drilling and grouting whereby each hole is drilled to full depth and grouted through an expansion plug or packer set at successively shallower stops. An exception to the requirement of drilling the hole to final depth is made when lost circulation or artesian conditions are encountered. In either of these cases, the hole is grouted prior to drilling to final depth. The depths of the stops are at the top of the zone being grouted, or lower, depending on foundation conditions.

For a given section, primary holes are drilled to full depth. Before grout injection is started, the nearest two advance holes are drilled; the hole to be grouted washed and pressure tested; and the adjacent hole thoroughly washed. If communication exists between holes, packers are used in holes other than the one being grouted to shut off the leak. The packer is normally left in place after each injection until the pressure on the newly placed grout has dropped to or below the maximum pressure contemplated for the next higher stop. The packer is then moved to the next higher stop and grout placed at lower pressure. This procedure is repeated until grouting of the hole is completed.

After the primary holes for a given section are grouted, secondary and succeeding order holes, as determined by the split-spacing method, are drilled and grouted in like manner until the section is completely grouted. Other sections are drilled and grouted in similar manner until the grouting program is completed.

CIRCUIT GROUTING is a method of grouting where the supply line and return line are connected to the hole by means of a special header that permits the injection pipe to be extended to the bottom of the grout hole.

This method can be used in combination with stage grouting procedures or with the single stage grouting method. The grout slurry is made to circulate from the sump or pump to the bottom of the hole, thence up the annular space between the injection pipe and the wall of the hole to the header, thence through the return line from the header to the sump. Pressure is controlled by a valve on the return line. The method is well suited for caving or sloughing holes, since the injection tubing can be washed or jetted through the debris effectively. Much of the infill material from caving or sloughing is removed from the hole by the rising grout column. A 100-mesh screen is placed between the return line and the sump to remove materials picked up by the circulating grout column.

SINGLE STAGE GROUTING METHOD. The single stage grouting method is used for grouting shallow grout holes and sometimes for other applications as conditions warrant. It consists basically of drilling the hole full depth and grouting with one connection. Where conditions permit its use, it can be effective and economical. It is sometimes used in conjunction with the circuit grouting method for grouting fairly deep holes, but can also be combined with injection through a normal circulating grout header, a tremie pipe, or a funnel at the top of the grout hole.

SERIES GROUTING METHOD. The series grouting method is a variation of stage grouting whereby each successively deeper zone is grouted by means of a nearby drilled hole to eliminate the need for washing grout out of the hole before drilling deeper. This method is rarely used in current Corps practice. It is the most expensive of the methods described and has a much greater danger of lifting and damaging the foundation.

TREMIE METHOD. The tremie method of grouting consists of placing a grout filled tube or pipe to the bottom of the hole and pumping until grout flows from the hole and then slowly withdrawing the injection pipe while pumping continues. It is used in foundation grouting programs mainly for backfilling drill holes but may also be combined with single stage grouting.

GROUT MIXES

Most of the Corps' foundation grouting is done with grout composed of portland cement, bentonite, and water. The addition of a small percentage (2 percent - 4 percent) of sodium bentonite produces beneficial results. Settlement is almost eliminated without significant reduction in strength or increases in setting time. Fluidizing agents may be added to reduce the viscosity of very thin grouts or for sanded mixes. Generally, the fewer the additives in a mix, the better. Additives are included in the mix design when needed for a specific purpose. Other than bentonite, additives may be used to serve as fillers, accelerators, retarders, lubricants, nonshrinkage agents, stabilizing agents, admixtures to increase or decrease specific gravity, or to increase strength of the grout. Where specialized mixes containing additives are contemplated, it is standard practice to design and test trial mixes in the laboratory prior to preparing the specifications.

In most Corps of Engineers foundation grouting, where the mix consists of cement, bentonite, and water, the water-cement ratio is expressed in terms of cubic feet of water per bag of cement. For example: A 2:1 grout consists of 2 cubic feet (.056m^3) of water and one sack (94lb=42.6kg) of cement. Bentonite is expressed in terms of pounds of bentonite per bag of cement, i.e., for a mix with 2 percent bentonite, 2 pounds (.9kg) bentonite are used per bag of cement.

Recent experience and case histories have shown that thin grouts are more effective than they were previously thought to be. Rather than starting with a water-cement ratio of 2:1 or 3:1 as has been done at many jobs in the past, the philosophy of starting with a thin mix, such as 5:1 or 6:1, is gaining more acceptance. Additional costs are negligible and benefits can be substantial. For instance, a 13 cubic foot (.36m^3) batch of 6:1 grout only has two sacks of cement. It has been injected into holes which plugged quickly when thickened to a 5:1 mix. Even where 1:1 grout is required before refusal, the thin grout injected first may grout fine fractures which would have been plugged immediately with thicker grouts. Instead of just grouting the larger features with thick grout, by starting with thin grout, the fine fractures are also treated.

GROUTING PRESSURES

The Corps of Engineers is conservative in regards to allowable grouting pressures. A common rule of thumb in determining maximum allowable pressures at the collar of the hole is that the pressure in pounds per square inch at any elevation should not exceed the depth in feet in rock (22.6kPa/m) plus 1/2 the depth in feet of overburden materials over the rock (11.3kPa/m). This rule was derived considering the weight of materials over the zone being grouted. Other factors affecting the maximum safe grouting pressure include rock strength and physical characteristics, consistency of the grout, tightness of the hole, geology, in-situ stresses, and hydrologic conditions. The rule of thumb is generally conservative. Higher pressures can safely be used in many cases.

Where grouting is done in the upper zone from the surface, gravity grouting is recommended. An overflow standpipe is frequently used to keep from overpressurizing the foundation. Pressure relief valves have also been used but are not as reliable. Another method of limiting pressure is to inject grout into a funnel at the top of the grout hole, nipple, or standpipe. If a circulating header is used without some sort of pressure relief, the risk of damaging the foundation is frequently present, as even small pressures of less than 10 psi (68.9kPa) can cause damage. It is very difficult to control the header so that pressures do not reach dangerous levels when grouting from the surface.

GROUTING EQUIPMENT

Many types of grouting equipment are available and have been used successfully by the Corps of Engineers. The main requirements are: suitable equipment of a capacity adequate for the job which mixes grout to a uniform consistency; continuing to agitate the grout to prevent settlement; and a pump capable of close pressure control to constantly circulate the grout through a system of lines with a header for injecting grout into the hole as desired. Suitable packers, gages, valves, and accessories are required for the job.

The grout mixer and sump should be of the same capacity. Where high takes are anticipated, two mixers may be arranged to discharge into the same sump. Both the mixer and sump should continuously agitate the grout until it is either injected or wasted. High speed colloidal grout mixers are superior to standard slow speed mechanical mixers. They produce grout of greater uniformity with better penetrability and pumpability. Cement clusters are separated and the individual particles are often broken and rounded to a significant degree in these mixers making it possible to grout tighter fractures than with standard mixers. Colloidal mixers are required for mixing and hydrating bentonite. Bentonite is mixed in a separate mixer and fully hydrated before being introduced into the grout mixer. Water is metered to the mixers and the meter should be calibrated in cubic feet and be large enough for easy reading. A direct reading totalizer is usually required.

Various types of pumps are available for grouting. The pump should be specified with selection based on the individual job requirements. For the bulk of Corps of Engineers work, either piston pumps or progressing cavity pumps are used. Many prefer the progressing cavity (or double helical screw-type) pumps such as Robbins & Meyer "Moyno." They can reliably pump grout slurries with or without sand or gravel up to 7/8 inch (22mm) size under constant pressure without surging. They will operate at working pressures up to several hundred psi. If piston pumps are used, they should be double acting, duplex slush pumps of the long stroke type. Either type can be powered by air, gasoline, or electricity. A standby pump is required on all Corps jobs of significant scope.

Most Corps jobs require a circulating grout header. Pressure gages are required at the hole and also at the pump. The pressure range of the gages is based on the pressures contemplated for the program. For instance, a 200 psi (1378kPa) gage is not appropriate where a maximum pressure of 50 psi (345kPa) is anticipated. Where a possibility of doing serious damage by too much pressure exists, duplicate gages should be required.

Flexible hoses are generally used for grout lines. They should be capable of withstanding the maximum grouting pressure with an ample safety margin. Grout lines are normally a minimum of 1-1/2 inches (38mm) inside diameter. Pipe for pressure lines, fittings, header, nipples, etc., should be of standard black steel or better, capable of withstanding maximum grouting pressures, with a minimum inside diameter of 1-1/2 inches (38mm).

Valves for the grout line are used to control the flow of grout. They should be hand operated, quick acting valves capable of withstanding the maximum grouting pressure and capable of accurately controlling pressure and rate of injection.

The packers specified should be capable of sealing the grout hole at any elevation and of withstanding the maximum grouting pressures. In packer grouting, it is not unusual to grout a packer in the hole. For this reason, the contractor is required to have an adequate number of extra packers on the job in order to avoid unnecessary delays.

For most jobs, grouting equipment is arranged so as to provide a continuous circulation of grout throughout the system and to permit accurate pressure control regardless of how small the take. The equipment and lines are prevented from becoming fouled by periodically flushing the system with water and by the constant circulation of grout.

GROUTING TECHNIQUES

Grouting techniques depend on the job, policy, objective, geology, contractor, field personnel, and individual judgment and preference. Procedures subject to field technique include selection and adjustment of mixes, changing grouting pressure, flushing the holes and washing the pump system during grouting, use of delays, intermittent grouting, and treatment of surface leaks.

Regardless of how well conceived and designed the grouting program, its success depends upon the field techniques used and upon good judgment by field personnel. Grouting techniques for Corps of Engineers jobs are not subject to contractor quality control but are directed by the Corps field personnel. For this reason, it is mandatory that an experienced geologist or geotechnical engineer is in charge of the grouting program and that he is provided with an adequate staff.

For winter grouting, all grout is maintained at temperatures above 50°F (10°C) until injected. The temperatures of mixing water range from 50°F to 100°F (10°C to 37.8°C) when added to the grout mixer. Storage of grouting materials is at temperatures above freezing. In addition, the grouted rock should be no colder than 40°F (4.4°C) when grout is injected and for a period of 5 days thereafter. It is normally left up to the contractor as to how to accomplish the above. Insulation, heated enclosures, and water heaters are frequently necessary. It is not unusual for the contractor to schedule grouting in nonwinter months in order to avoid winter operations. In extremely hot weather, grout and grouting materials are protected from direct sunlight. It is desirable to maintain the grout at temperatures at or below 90°F (32.2°C). The higher temperatures increase water demand and consequently shrinkage, as well as accelerate the setting time of the grout which decreases the working time.

Thin initial mixes are usually recommended, especially if the hole is dry or pressure tests indicate slow or small takes. Some fractured formations such as meta volcanics and fractured limestone and dolomite are quite pervious but refuse 3:1 grout very quickly and accept considerable 6:1 or 5:1 grout. This justifies beginning with a thin grout even in pervious conditions. If the hole accepts a few batches of the starting mix without pressure buildup, thicker mixes are required; however, if the pressure builds up, grouting may continue with the same mix until refusal.

Mixes are usually thickened by batching a new mix in the mixer and discharging it into the sump after most of the thinner grout has been injected. If immediate thickening is required, the hole is shut off temporarily and cement is added to the sump. Mixing is accomplished by agitation of the sump and circulation through the pump and lines.

If the hole accepts a few more batches of the new thickened mix without pressure buildup, the next thicker mix is used. The process of thickening the grout continues until the pressure builds up. Then injection continues, and the rate of injection into the hole is slowly cut back when the pressure tends to rise until the hole refuses to take grout at the maximum pressure. It should be noted that as the water-cement ratio decreases, each integer change requires more cement. For example, going from 2:1 to 1:1 requires 67 percent more cement. Therefore, mixes of intermediate consistency are used after reaching a 3:1 or 2:1 consistency. When sudden refusal and pressure buildup is experienced, premature plugging may have occurred. If the hole is still taking a small amount of grout, water should be pumped into it to reopen it if possible. After the water injection, a thinner grout mix may be required. If the hole is plugged, a new hole may be required. Other causes of sudden refusal include a blocked line, packer or hole, a collapsed hole, or the voids may be full.

During grouting when pressures begin to rise, they are controlled so that they are raised slowly in increments until the desired injection pressure is reached.

It is recommended that the pump system be flushed with water at intervals during grouting with thick mixes (i.e., 2:1 or thicker). A few cubic feet of water is also injected into the hole when this is done.

A maximum pumping rate is specified for injecting grout in order to restrain grout travel within reasonable limits and have better control of the job. Three ft^3/min (85L/min) is considered a reasonable maximum pumping rate for most foundation grouting.

When pressures cannot be built up using the thickest mixes allowed, or when it is desirable to prevent grout from spreading too far, delays may be used. They may last from a few minutes to several hours. The amount of grout injected per delay is controlled to fulfill the intended purpose. If the delays are very long and thick grout is being used, the hole and pump system should be flushed before each delay. Also, the contractor's efforts are allowed to be directed elsewhere during the delay. If the delays are short and the contractor is required to standby, provisions are made in the contract for payment for standby time. In intermittent grouting, previously described for cavity filling, delays of several hours are required. Intermediate shorter delays during a single injection period may be required in order to restrict grout travel.

Surveillance of the area is made frequently during grouting to check for surface leaks and collect monitoring data from other holes, springs, piezometers, wells, and seeps. Records are kept of any discolorations, changes in flow or changes in water levels. Leaks are controlled, if necessary, by dikes or calking with materials such as oakum, wood wedges, burlap, or other materials. If the leaks are serious, an accelerator may be added to the ponded grout within a diked area and a delay used to allow the grout to set. If the leak cannot be stopped, grouting may be continued at reduced pressure and a thicker mix used.

Up-to-date, detailed records are kept of drilling, pressure testing and grouting operations for technical evaluation as well as for determining payment quantities. In addition, geologic sections and profiles are kept up to date with drilling, testing, and grouting data, and records are made of monitoring data in order to evaluate the grouting program. This information is included in the foundation report for future reference.

Evaluation of grouting effectiveness must be constant and continuous during the program. It is a joint effort between engineering and construction personnel. If problems develop, reaction must be expeditious. Flexibility is maintained for making changes and improvements as the program progresses. Design changes of other project features are sometimes made based on knowledge of foundation conditions gained during grouting.

ESTIMATES AND PAYMENT

Because of uncertainties involved in grouting, the requirement for maximum flexibility to meet field conditions, and the exploratory nature of grout programs, accurate estimates of quantities are extremely difficult. Even though the best estimates possible are made for bidding purposes, the Corps reserves the right to increase or to eliminate any part of the drilling and grouting program without changing unit prices. This is done by subdividing the bid items for drilling and grouting and only guaranteeing the work required by the first subitems which may be either an estimated quantity or mobilization and demobilization.

Grout and exploratory hole drilling are paid for on the basis of the linear feet of holes actually drilled and may include the cost of washing, pressure testing, pressure washing, and removal of grout from holes for deeper stage drilling. Sometimes pressure testing and washing are separate pay items and are paid for by the hour. Should the contractor be directed to leave grout in a hole until it takes a hard set and redrilling is required, he is paid at one-half his bid price for grout hole drilling. Also, if he is directed to leave packers in the hole until they are grouted in, the packers are paid for at a unit bid price per each for packers.

Cement, sand, and mineral filler are paid for on the basis of the number of cubic feet of each injected into the holes. Other additives are paid for by the pound or cubic foot as appropriate.

Grout injection (or placement) is paid for on the basis of either the number of cubic feet of solids injected or by the hours.

SUMMARY

The most common and accepted cement grouting practices of the Corps of Engineers have been discussed. While these are the most often used practices, others are used and nonstandardized practices may be adopted to meet specific objectives or requirements. Because the Corps of Engineers is a world-wide construction agency, grouting practices may be applied in any geologic environment and under various hydrologic, geochemical, and geophysical conditions. Current Corps practices, properly applied, have proven effective under a variety of conditions and environments. Improvements have been and will continue to be made as indicated by new experience, developments, and research.

ACKNOWLEDGMENTS

The writer is grateful to Ray Cope for his critical review, and to members of the U.S. Army Corps of Engineers Grouting Work Group for their contributions through extended discussions of much of the content of this manuscript. The Work Group members are: Paul Fisher, Wayne McIntosh, John Boa, Ralph Bendinelli, Lawson Jackson, John Albritton, Todd Riddle, Jim Erwin, and Dan Parillo.

CEMENT-BENTONITE GROUTING FOR DAMS

by Don U. Deere,[1] M. ASCE

ABSTRACT

Two simple field tests may be used to control the mix design of a cement-bentonite grout. One is the measurement of the slurry density which may be conveniently done using the mud balance, a device used in the drilling mud industry. The slurry density gives a check on the cement content. The second test is the measurement of the viscosity by means of the Marsh funnel, also a device of the oil drilling mud industry. This test is sensitive to the amount of bentonite for a given water-cement ratio. With these two rapid field tests it becomes possible to check routinely on the accuracy of the mix proportions, as was recently done on three major dam projects. Correlations are also given of laboratory values of sedimentation of cement grains which indicate that moderately thick mixes with a small percentage of bentonite may be preferred. Current design practices of grout-curtain layout and specifications are reviewed. The specifications form the link between the design as intended and the construction of the grout curtain.

INTRODUCTION

Curtain grouting constitutes an important part of the foundation treatment for all high dams. The purpose of the curtain grouting is to form a zone of low permeability to some given depth below a specified portion of the upstream half of the dam. Under reservoir conditions this zone forms a partial barrier to seepage flow below the dam.

Seepage losses are reduced by the presence of the grout curtain. The uplift pressure and piping potential also are more easily controlled by appropriate downstream drainage provisions. During the drilling of the grout holes some additional information may be obtained regarding the dam foundation geology. Adverse geologic features, if encountered, may be treated by more extensive grouting or by other remedial measures.

The grouting process during construction involves three essential phases: drilling the grout holes, mixing the grout ingredients, and pumping the grout slurry into the jointed rock mass through the drillholes. A variety of methods exists for performing each of these phases as well as differing opinions as to which methods are preferable. The design engineers must study the site conditions and on the basis of that information, their

[1]Don U. Deere & Andrew H. Merritt, Inc. and Adjunct Professor, University of Florida, Gainesville, Florida.

own experience, and precedent decide on the appropriate methods to be used. Specifications are then prepared for these methods.

The design process also entails the layout of the grout curtain in terms of hole depth, size, and spacing; single-row vs. multiple-row; type of grout slurry; maximum grouting pressures to be employed; sequence of drilling and grouting; nipple grouting vs. packer grouting; and others. The specifications and design drawings will then form the link between the design as intended and the construction of the grout curtain. Field control verifies that the specified materials and methods are employed as well as noting problems that may require the attention of the designers and possible design modifications.

This paper treats both design aspects and field control of grouting. Certain recommended field control procedures are emphasized.

DESIGN OF THE GROUT CURTAIN

The design of a grout curtain is not a rigorous procedure but rather an almost arbitrary selection of a geometric layout of extent of curtain, layout of hole depths, and spacings, and number of grout lines. These are selected, however, only after consideration has been given to the size of the river, the dam type, the hydraulic head, the details of the site geology, the potential for leakage and piping, and the consequences of such.

Extent of Curtain

The lateral extent of the grout curtain at the dam crest is usually a minimum of about 150 ft. (46 m). Where critical topographic situations occur or where it is desired to increase the length of the seepage path because of leakage or piping potential, the grout curtain may be extended several hundred feet into the abutment. It may be desirable to use an upper grouting gallery for doing the work rather than drilling from the surface, particularly if the surface topography rises abruptly so that the drillholes would be very deep.

The depth of the grout curtain ideally should be a function both of the hydraulic head and the rock quality. This is often not the case in the river bottom where the head is the greatest, presumably requiring the deepest grouting, but where the rock quality may be the best. For this reason, recommended practice is to augment the geologic and permeability data obtained during the design exploration phase with the initial grout holes made during construction.

These initial holes are referred to as the <u>exploratory grout holes</u>. Commonly they would be located on about 130-ft (40-m) spacing along the line of the grout curtain. They would be taken to a depth of perhaps 2/3 H to H, where H is the hydraulic head above the point in question, but with a minimum depth of 150 ft (46 m). The holes would be cored holes of NX size with Lugeon type water-pressure tests in about 15-ft (4.6-m) stages. Packer-grouting in 15-ft (4.6-m) ascending stages would be done following drilling. On the basis of the results obtained both from these exploratory grout holes and from the previous design borings, appropriate depths for the primary holes would be selected for each reach of the grout curtain. The selected depth would normally be within the range of 1/3 H to 2/3 H with a minimum depth of 130 ft (40 m), all of which should be indicated in the specifications.

For some geologic and topographic situations it may be desired to carry the grout curtain much deeper into the abutments. In such cases, two or three grouting galleries may be excavated in each abutment to depths of 300 ft (91 m) or greater. Drainage holes may also be drilled in a downstream direction from these galleries if desired.

Progressive Grouting and Single-Line vs. Multiple-Line

Most grout programs are set up as a progressive type program where a basic pattern of grout holes is specified with certain minimum requirements. For instance, the exploratory grout holes on 130-ft (40-m) centers, as described above, and the primary holes on say 32.5-ft (10-m) centers would be required to the depth intervals specified. Secondary holes on split spacing would also be required, but perhaps of slightly less depth. Split-spaced tertiary holes might or might not be required, depending on the grout absorptions and the specified requirements.

In some cases the specifications state that tertiary holes will be executed depending on the grout "takes" in the various intervals of the secondary holes as directed by the resident engineer or grouting supervisor. An alternate method is to give guidance in the specifications so as to require tertiary holes on either side of a secondary hole that has one or more intervals of "moderate" take (33.5-67 lb cement/ft; 50-100 kg cement/m). The specifications may also allow greater limiting absorptions at depth but may be more restrictive at shallow depths. Table 1 gives as a function of depth some suggested upper limits of grout absorptions in intervals of the secondary holes (above which tertiary grout holes would be required).

TABLE 1. Suggested Upper Limits of Grout Absorptions for Secondary Holes as Function of Depth

Depth Interval in feet (meters) (1)	Grout Absorption in lb cement/foot of hole (kg cement/meter of hole) (2)
0-33	17
(0-10)	(25)
33-66	23
(10-20)	(35)
66-98	33.5
(20-30)	(50)
98+	67
(30+)	(100)

In applying the criteria one should consider whether only one grouting interval or several are involved, and also whether a fault or shear zone is suspected which might deserve especially close grouting.

Where the tertiary holes also fail to meet the criteria of Table 1, quarternary holes at split spacing would normally be specified. The depths would correspond to those of the zones showing larger than the limiting takes. An alternate method is recommended by Houslby (6) for defining closure in which a target value of Lugeon permeability, such as 3 or 5, is set for different depths and for various geologic conditions and dam types.

The previous considerations apply to a single-line curtain. Upon occasion the designers will specify a multiple-line curtain, particularly when it is desired to have a wide curtain to reduce hydraulic gradients and piping potential (6). Another condition meriting a 3-line curtain is that of permeable rock with open fractures where it is desired to have tight closure but to limit the grout absorption. This may be done by grouting first a downstream line, then an upstream line, and finally a central line. The downstream line is grouted first so that the grout of the second line is preferentially forced upstream.

Closure is not attempted on the two outer lines but the primary, secondary, and selected tertiary holes will be grouted. At any grouthole interval having a grout absorption greater than 1 ton of cement, grouting will usually be temporarily halted and started again after 24 hours, possibly using fine sand or sodium silicate admixture to decrease the distance of travel. The central row will be grouted to refusal using tertiary holes and quarternary and quinary holes if needed.

A combination of a single-line and multiple-line curtain may be designed. The single-line curtain is initially grouted but in any zone where large absorptions are experienced in the secondary holes, or where the rock conditions are poor, a 3-line curtain may be established locally. This combined procedure has been used successfully on several occasions within the writer's experience.

DESIGN OF THE GROUT SLURRY MIX

An ideal mix should be a uniform, stable mix that has a viscosity sufficiently low to be easily pumped into rock cracks but not so low as to travel long distances without appreciable pressure drop. For most applications the slurry should set up in a number of hours to a strong, impermeable solid with permanent cohesive bonds. The cost should also be reasonable. Cement slurries with or without bentonite or other admixtures have been used to satisfy these requirements over the years. A variety of proportions of the ingredients has been employed and there is no general agreement as to the best proportions. Burwell (2) has pointed out the influences that have retarded standardization.

Water-Cement Proportions and Slurry Density

It has been common practice to start the grouting with a thin mix such as 6:1, water:cement (W:C) by volume. The unit weight (density) of the mix would be only about 1.15 gm/cm^3. It is a watery mix and is unstable as the cement grains settle out when agitation of the mix ceases. In practice after a certain amount of this very thin mix has been pumped in with little or no increase in pressure, the mix would progressively be thickened to 4:1, 3:1, 2:1, and 1:1. The density of the thickest mix commonly used, 1:1 by volume, is about 1.68 gm/cm^3.

The Corps of Engineers' practice is to use the thicker mixes (2):

> ...Field experience and laboratory research have led the Corps to believe that grouts of high water-cement ratio are unnecessary and undesirable, primarily because of the poor quality of the end product but also for economic reasons. Therefore, the maximum water-cement ratio of neat cement grout permitted under the Corps' specifications is 3 or 4 and the bulk of the Corps' foundation grouting is done at ratios of 1 or less. . .It is the practice of the Corps to express the water-cement ratio in terms of cubic feet of water per bag of cement...

In the U.S. a bag of cement has a bulked volume of dry cement of 1 cu ft. Therefore, the above mentioned 3 or 4 water-cement ratio would be 3:1 and 4:1, W:C, by volume.

Some design organizations use weight rather than volume proportions. A sack of cement weighs 94 lbs. and having a volume of 1 cu ft, the density of the dry, bulked cement is therefore 94 lb/ft^3, or 1.5 gm/cm^3. The 6:1, W:C, ratio by volume becomes 4:1 by weight; the 1:1 by volume becomes 0.67 by weight, etc. Since both methods of expressing the proportions are in common use, it is necessary to clearly identify the method being specified.

Yet another procedure is followed in parts of Europe (F. Sabarly, personal communication). The weight ratio is used but as the inverse C:W. This method has appeal in that the values when expressed as decimals normally range from about 0.25 to 1.25, with the higher values corresponding to higher cement contents and to higher densities. Some common mix proportions are shown in the three manners in Table 2.

In order to visualize better these relationships a plot is given in Figure 1. This figure shows the curved-line relation between W:C by volume and C:W by weight. The W:C by weight can be obtained by taking the inverse of the C:W values given along the abscissa.

Also of interest is the plot of slurry density or unit weight vs. the water-cement proportions (Fig. 2). The full line is for a neat cement slurry and the dashed line for a cement slurry with 8 percent bentonite (in terms of weight of the cement). It can be

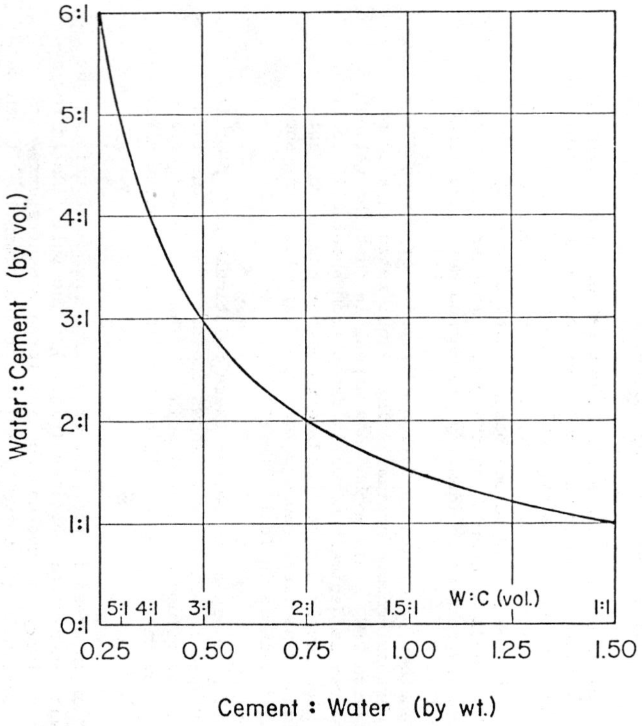

Fig. 1 - RELATION OF WATER-CEMENT RATIO BY VOLUME AND CEMENT-WATER RATIO BY WEIGHT (After Burgin, Ref. 1)

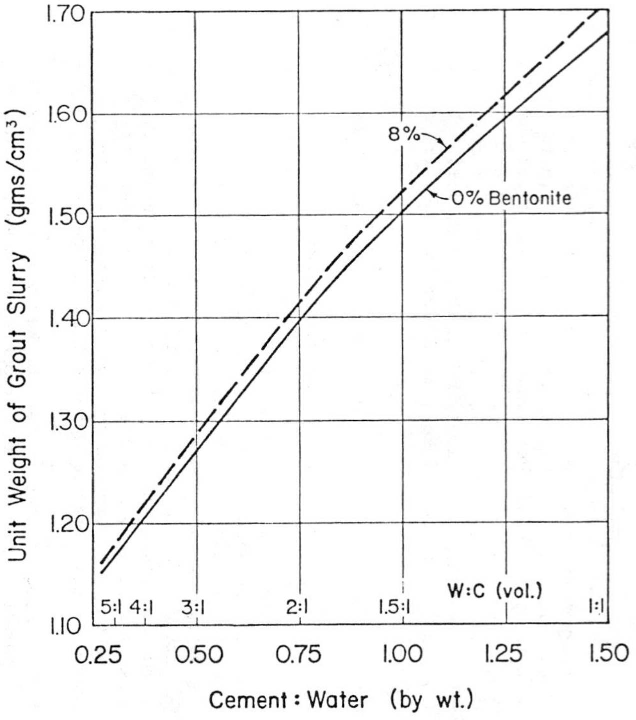

Fig. 2 - EXPERIMENTAL RELATION OF UNIT WEIGHT OF GROUT SLURRY AND CEMENT-WATER RATIO BY WEIGHT (After Burgin, Ref. 1)

TABLE 2. Relationships of Different Mix Proportions of Cement and Water in Grout Slurries.

Water:Cement by volume (1)	Water:Cement by weight (2)	Cement:Water by weight (3)	
6:1	4:1	1:4	(0.25)
4:1	2.67:1	1:2.67	(0.37)
3:1	2:1	1:2	(0.50)
2:1	1.33:1	1:1.33	(0.75)
1.5:1	1:1	1:1	(1.00)
1:1	0.67:1	1:0.67	(1.50)

seen that the unit weight of the slurry increases markedly with an increase in cement content but only slightly with an increase in bentonite content from zero to 8 percent. The relationships shown in this figure suggest that the slurry density could be a good index property to use in the field to check and control the mix proportions, particularly, the cement content.

The density of the slurry may be readily measured in a mud balance, a portable instrument sufficiently rugged for field use. The cup is filled with the slurry, the attached beam is balanced, and the density is read directly from the scale (American Petroleum Institute, Dallas, Texas; RP13B, Standard Procedure for Testing Drilling Fluids, Section 1, Density-mud Weight).

Sedimentation of Cement Grains

A grout slurry is essentially an unstable water-cement suspension. In a very thin mix such as 6:1, W:C by volume, there may be as much as 60 percent sedimentation of the cement grains in a 2-hr. period. A sedimentation test commonly used is carried out as follows. A liter of grout slurry is placed in a standard 1000-ml graduated cylinder. At the end of 2 hours the volume of clear liquid that has formed at the top of the cylinder due to sedimentation of the cement grains is noted. This volume expressed as a percentage of the total volume gives the percent sedimentation.

The thicker the mix, the less is the sedimentation as shown in Figure 3. For example, using the bottom curve which is a water-cement slurry without bentonite, it is noted that a 3:1 volume mix has a sedimentation of 53%, a 2:1 mix 26%, and a 1:1 mix only 5%.

From European practice (F. Sabarly, personal communication) comes the concept of "stable mixes" or "unstable mixes." An unstable mix is commonly taken as one with more than 5% sedimentation. Most cement-water slurries are therefore unstable mixes with the

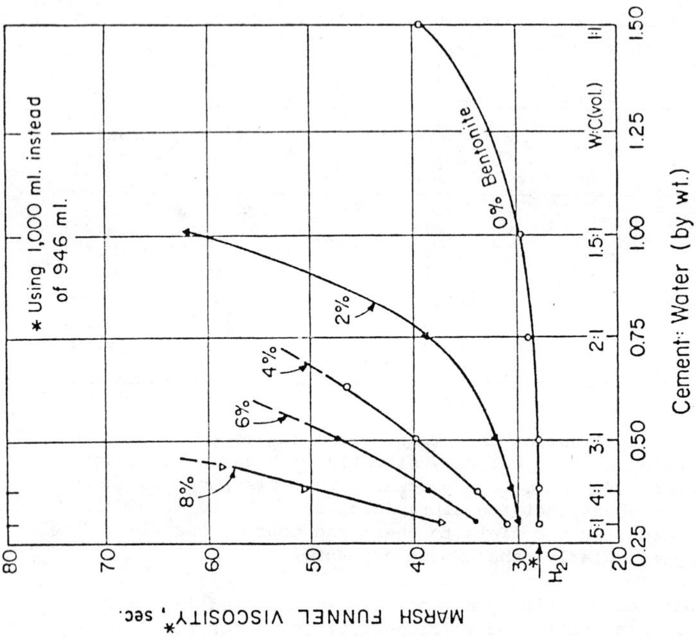

Fig. 4 - EFFECTS OF W:C RATIO AND PERCENT BENTONITE ON MARSH FUNNEL VISCOSITY (After Burgin, Ref. 1)

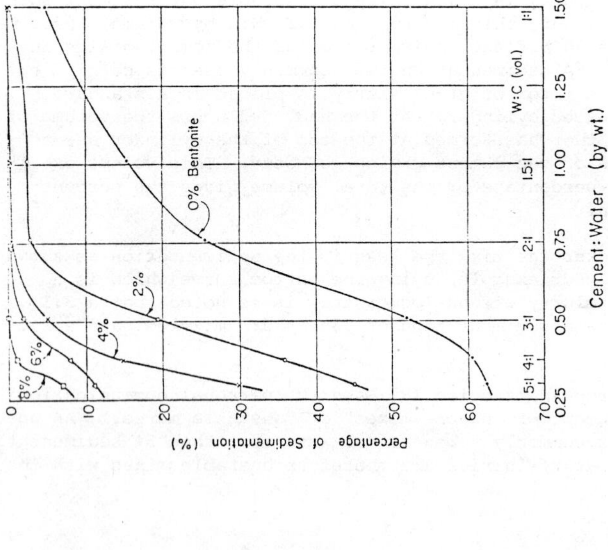

Fig. 3 - EXPERIMENTAL RELATION OF PERCENTAGE OF SEDIMENTATION AND CEMENT-WATER RATIO BY WEIGHT (After Burgin, Ref. 1)

exception of the moderately thick 1:1 volume mix as noted above. The low sedimentation value may also be achieved by adding to the thin mixes a small percentage of activated bentonite (liquid limit of 400%). As can be seen from Figure 3, a 3:1 volume mix requires 4% of bentonite and a 2:1 mix only 2% of bentonite to limit the sedimentation to 5% or less. The figure indicates that the 1:1 volume mix would require no bentonite; however, this would depend on the fineness of the cement.

Stable mixes, then, may be obtained by the addition of small percentages of bentonite. A more uniform, stable mix will result which will maintain its characteristics during pumping and crack penetration. The effect of the bentonite on the viscosity of the slurry and on the compressive strength of the hardened grout must also be considered.

Viscosity of Grout Slurries

Viscosity may be readily determined in the laboratory by a concentric cylinder-type viscometer in which various rotation speeds are used. Both viscosity in centipoises and the yield-point stress can be obtained. These instruments are not commonly used in the field and are not available in many laboratories. Therefore, some form of funnel viscosity is usually preferred for giving information about fluidity of the grout slurry.

Although funnels of a variety of shapes and sizes have been used, the author prefers the standardized Marsh funnel which was developed for testing drilling muds in the oil drilling industry (American Petroleum Institute, RP 13B, Section 2, Viscosity and Gel Strength). The funnel cone has a diameter of 6 in (152.4 mm) and a height of 12 in (304.8 mm). The bottom orifice outlet has an inside diameter of 3/16 in (4.76 mm) and a length of 2 in (50.8 mm). The top of the cone has a screen with 1/16 in (1.59 mm) opening for screening out lumps of poorly mixed slurry.

The slurry is poured into the funnel until it touches the screen, a volume of around 1.6 quarts (1500 ml). The time is recorded for 1 quart (946 ml) to pass through the funnel into a measuring container. The result to the nearest second is the Marsh funnel viscosity. For clear water at 21°C (±3°C) the Marsh funnel viscosity is 26 seconds (±0.5 sec). For grout slurries the values range from 26-28 sec for the thin slurries to more than 60 sec for the thicker ones.

The test frequently is done incorrectly. A common error is to record the time for passing 1 liter (1000 ml) instead of 1 quart (946 ml). This error is small, about 2 sec. A more serious error is to start with only 1 liter of slurry and to allow it to drain away completely. This procedure causes a significant increase in the time required and should not be used. Also, for consistent results the orifice length and diameter must be as specified.

In Figure 4 are shown the effects of the W:C ratio and the percent of bentonite on the Marsh funnel viscosity. It is noted that 1000 ml rather than 946 ml (1 quart) of slurry were used in this study, giving about 2-second greater Marsh funnel viscosity values than for the standard conditions. The bottom curve of Figure 4 is for a neat cement slurry, i.e., with no bentonite. It is noted that for 5:1, 4:1, and 3:1 volume mixtures, the viscosity is essentially the same as for water. Even the 1.5:1 mix has a Marsh funnel viscosity of only 30 sec., and the 1:1, 39 sec. The next curve above shows the striking influence of an addition of only 2% bentonite. The 2:1 mix for example goes from 29 sec for zero bentonite to 39 sec for 2% bentonite. The upper curves indicate that higher percentages of bentonite, i.e., 4%-8%, would only be appropriate for the thin mixes.

It is thus apparent that bentonite does increase appreciably the funnel viscosity of a cement slurry. Small amounts of bentonite appear to be preferable, sufficient to reduce sedimentation and bleeding but not so great as to impare significantly the pumpability and penetrability.

The question is often raised if funnel viscosity values from funnels of different sizes, shapes, and capacities can be correlated with the Marsh funnel values. The Corps of Engineers, for instance, has used a larger funnel in some of its research at the U.S. Army Engineers Waterways Experiment Station, Vicksburg, Mississippi. The volume of the funnel was 1.82 quarts (1725 ml) and orifices of either 0.5 in (12.77 mm) internal diameter and 1.5 in (38.1 mm) length or of 0.375 in (9.52 mm) internal diameter and 6 in (152.4 mm) length were used (8). In Brazil this funnel was accepted but with a modified orifice of intermediate length (3.93 in; 100 mm) and with a slightly smaller diameter (0.315 in; 8.0 mm). The funnel cone and upper cylindrical portion have a diameter of 7 in (177.8 mm) and an overall height of 10.75 in (273 mm) as does the Corps of Engineers funnel; the funnel is filled with 1.82 quarts (1725 ml) and the time for 1.06 quarts (1 liter; 1000 ml) of fluid to flow out is recorded (15).

The relationships of the funnel viscosities determined by the two types of funnels are shown on Figure 5. Two of the curves were developed at the Tucurui Project Concrete Laboratory, Eletronorte, Brazil, and one at the Itaipu Binational Project (Brazil-Paraguay) by the Division of Geology, Rock Mechanics, and Instrumentation. The relationships are essentially linear up to a Marsh viscosity value of 40 to 50 sec above which the curves steepen. These upper portions correspond to the thicker slurries with greater density and viscosity which apparently decrease the flow rate proportionately more in the Marsh funnel with the smaller diameter orifice than in the Brazil funnel. The difference in position of the Tucurui curves and the Itaipu curve may be a result of the different cements used but may be caused by slight differences in dimensions of the Brazil funnels which are shop-made as needed. In the writer's opinion the Marsh funnel is preferred in that it has greater sensitivity and its use is standardized.

Another question that has frequently been raised is whether the Marsh funnel, or any other funnel, gives apparent viscosity

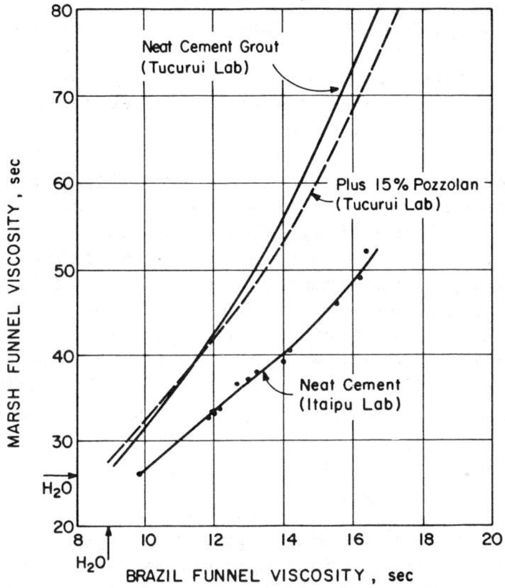

Fig. 5 - RELATION OF MARSH AND BRAZIL (8 mm) FUNNEL VISCOSITIES

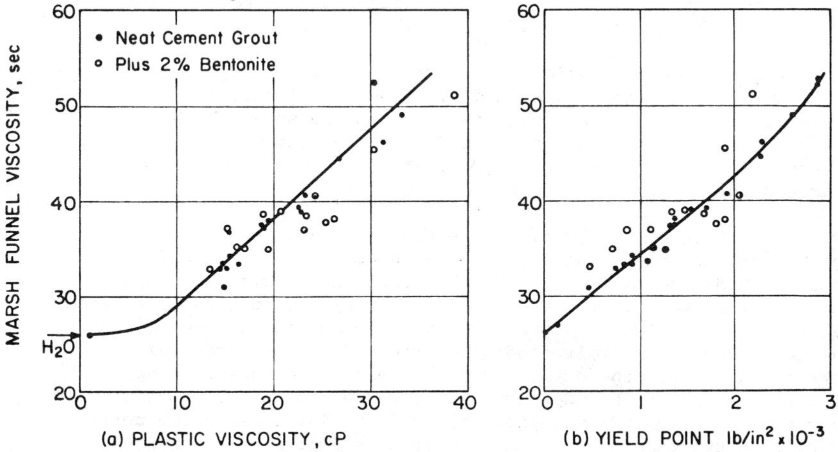

Fig. 6 - RELATION OF MARSH FUNNEL VISCOSITY AND (a) PLASTIC VISCOSITY AND (b) YIELD POINT FOR NEAT CEMENT AND CEMENT-BENTONITE GROUTS

values that have any relationship with the true viscosity of the slurry. Figure 6 indicates that there is a good relationship, nearly straight-line in the range of interest of 35-50 sec, between the viscosity in centipoises determined from a co-axial cylinder viscometer and the Marsh funnel viscosity (after test results from the Itaipu Binational laboratory).

The results of the viscometer tests indicate that the grout slurries are Bingham viscous fluids with well-defined yield point stresses and plastic viscosities, i.e., viscosity of flow beyond the yield point. Figures 6a and 6b show the relationship of these values and the Marsh funnel viscosity for several grout slurries. The curves are drawn on the basis of the neat cement grout (filled circles). The grout slurries with 2% bentonite (open circles) also follow the same trend but with more scatter. These points were obtained from the field control records of the Itaipu Binational project grouting (all with 1:1, W:C by weight, and 2% bentonite). The funnel viscosities were obtained with the Brazilian funnel and were then transformed to Marsh funnel values by correlation.

The foregoing discussion would appear to support the idea that the Marsh funnel viscosity is an important measure of the true viscous nature of grout slurries; as the Marsh funnel reading increases, so do the yield point stress and plastic viscosity of the grout (Figs. 5 and 6). Also, for a given W:C ratio the Marsh funnel viscosity increases rapidly with an increase in bentonite content (Fig. 4). Thus, the Marsh funnel viscosity is an important index property for quality control of field mixes, particularly of the bentonite content.

<u>Required Strength of Hardened Grouts</u>

Two problems are associated with the question of the required strength of hardened grouts. One concerns the strength that is needed to resist the forces exerted by a high dam and its reservoir. The second is the actual strength that is achieved when the grout hardens in the rock fractures.

With respect to the first question, the main reason for a grout curtain under a high dam is to reduce the underseepage below the dam so that uplift can be controlled. With the construction of the grout curtain a partial barrier to flow is created as the rock fractures intersected by the grout curtain become injected. Under reservoir conditions the underflow is reduced but the change in seepage paths resulting from the grouting brings the reservoir head farther downstream under the dam as the grout barrier restricts the underflow. Thus, there is an increase in the uplift pressures upstream of the grout curtain as well as an increased hydraulic gradient across the zone of the grout curtain. With a downstream drainage curtain there becomes even a higher hydraulic gradient.

The grout in the fractures must be strong and well bonded to the walls of the rock fractures so that it can resist the high hydraulic gradients without eroding and piping. Certainly, a grout with the strength of a soil would not be sufficient; nor, would a grout that was porous and friable. In addition to being mechanically resistant, it must also resist long-term dissolution by the leaching of calcium from the grout mortar by the action of the percolating water. Thus, the hardened grout should be dense and of low permeability. The grout must be fairly strong to have these desirable properties. A strength in the 700-1400 lbs/in^2 range (5-10 MPa; 50-100 kg/cm^2) is the minimum desirable in the author's opinion.

The second problem is difficult as well. It is not easy to find hardened grout samples in an injected rock mass of sufficient size for samplings and testing. An alternative is to study strengths of grout samples prepared in the laboratory, although there are drawbacks to this approach as well.

In the laboratory one is confronted with the problem of sedimentation of the cement grains from thin to medium grouts which are unstable. For instance, a thin mix such as 6:1, W:C by volume without bentonite will sediment by about 50-60% so that the sample of hardened grout will correspond to a W:C ratio of 2:1 to 3:1. If the same mix were prepared with bentonite, much less sedimentation would occur and the grout would harden with a higher void ratio, higher water:cement ratio, and a lower density. The strength would be correspondingly lower.

Burgin (1) investigated the laboratory strength of hardened grout samples. For the thinner grout mixes the bentonite lowered the strength by 50-75%. For the thicker mixes the strength was reduced by only 25% or so, and it made little difference if the bentonite content were 2% or 8%. The samples with 1.5:1, W:C by volume, regardless of bentonite content gave 28-day strengths around 90 kg/cm^2 (1,280 lb/in^2; 9 MPa). This is equivalent to a weak concrete and in the author's opinion is the minimum strength to be achieved.

During the grouting process several actions take place which cannot be directly observed but which can be visualized. One of these is sedimentation of the cement grains when the flow velocity decreases. Such decreases would occur in a dead-end passage, in the bottom of drill holes where the hole is full of grout but the grout take is higher up in the hole[2], and in those cases where a rock fracture narrows and the flow eventually stops. An unstable grout would sediment leaving a large void space above.

Another phenomenon believed to occur is the squeezing out of water from the grout by seepage pressure as a filter cake forms in front of a narrow opening. This would decrease the water-cement ratio and would yield locally a denser, stronger grout. However, the process is random and cannot be foretold as it depends on the pattern of cracks and intersections and their widths.

---

[2]Minear (11) reports such an experience where each hour 15 ft (5 m) of sedimented cement accumulated in the bottom of the hole.

Selection of Design Mix for Grout Slurry

On the basis of the previous discussion it is possible to select the water-cement ratio and the bentonite content so as to obtain a stable pumpable mix of uniform character with a minimum of sedimentation and an acceptable minimum 28-day compressive strength. A neat cement grout is subject to excessive sedimentation unless it has a W:C ratio by volume of 1:1 or thicker. This is the range of mix preferred by the Corps of Engineers (2). Houlsby (6) also notes its considerable use in Australia.

It is also possible to obtain a stable mix by adding 4%-6% of bentonite to a thinner mix, such as 6:1 or 5:1, W:C by volume. These could be acceptable mixes but the final hardened grout would be of quite low strength unless seepage pressures densified the grout (which would not be uniformly done). An undesirable mix is a thin mix such as the above but with only 1% or 2% of added bentonite. There is not enough bentonite to reduce the sedimentation and any water squeezed out tends to take the bentonite with it, depositing it in surrounding rock fractures. The author has observed this in the field when test shafts and adits were excavated into a previously grouted area to observe the end product.

In consideration of these observations and in attempting to achieve the desired results, the author prefers a moderately thick grout with only a small percentage of bentonite (such as 1.5:1, W:C by volume, and 2% of bentonite). The exact proportions should be determined by laboratory testing, determining slurry density, Marsh funnel viscosity, sedimentation, time of setting, and compressive strength at 7, 14, and 28 days.

This series of tests should be made by the design engineers during the design period using locally available products and then repeated by the field control grout engineers at the field laboratory using the contractor's proposed cement and bentonite. A worthwhile by-product of this procedure is that both the design engineers and field control grout engineers become familiar with the effects of varying W:C ratios and bentonite contents as well as the use of the mud balance, the Marsh funnel, and the sedimentation test. The mixes to be tested could include the 3:1, 2:1, 1.5:1, and 1:1, W:C by volume with bentonite contents of 0%, 1%, 2%, and 3%. The author has been involved in such programs on a number of major hydroelectric projects over the past few years and can vouch for their effectiveness.

Additional Specifications

Extent of Grout Curtain. - The lateral and vertical extent of the grout curtain and whether single- or multiple-lines will be shown on the drawings. The results of the exploratory grout holes (depth and spacing would be specified) will allow field judgements to be made regarding depths of secondary and tertiary holes, etc. Minimum acceptable depths and maximum spacing should be specified, however.

Closure Criteria. - Normally, the specifications will require either a maximum grout take value for the secondary or tertiary holes, or a Lugeon permeability value as proposed by Houlsby (6).

Type of Drill. - Many of the former specifications in the U.S. required the holes to be drilled by rotary means. Now, drilling is usually permitted by rotary-percussion (including down-the-hole hammers) wherever the conditions so permit, i.e., slimes are not formed or where the bit does not plug. Rotary-percussion drilling is at least five times faster and cheaper than rotary drilling and is practiced in most countries where conditions allow.

Mixing Equipment. - High-speed, colloidal type mixers with 1200-1500 minimum r.p.m. are normally required with appropriate agitators. Where a central mixing plant is used and grout is furnished to a distant area for injection, an additional agitator would normally be required.

Grout Pumps. - In Europe and in much of South America, special positive-displacement grout pumps are used in which the pressure may be selected and maintained by means of an automatic by-pass of the hydraulic fluid at the pump. These pumps provide control of the grouting pressure without the necessity of a hand-controlled valve at the top of the hole.

In the U.S. and several other areas the Moyno slurry transport pump is preferred because of its ruggedness and its smooth and rapid delivery. For efficient control of pumping rate and pressure it is necessary to have a multi-speed transmission (personal communication, A. Sabarly). In addition, a grout by-pass valve is needed with hand control at the head of the hole.

Grout Circulation Line. - U.S. practice is to use a grout circulation line to return to the agitator the grout which is not absorbed by the hole. This practice has developed because of two aspects of traditional U.S. grouting: one, the use of an unstable grout which has excessive sedimentation and which requires constant circulation; and, two, the use of the by-pass valve to control injection pressures.

The author favors European practice which controls sedimentation by specifying a stable mix and which controls pressure by means of a special grout pump (such as the Clivio of Italian make). No circulation line is required for returning the grout to the agitator although an emergency by-pass valve and pressure gage are still used at the top of the hole.

In addition to the inconvenience of providing and maintaining a return line, there is also the problem of the circulating grout heating up by friction and by sunshine as it recirculates various times. Measurements made at one project showed a 9°F (5°C) temperature rise for each circulation round-trip, with temperatures of the slurry reaching 113°F (45°C). There was fear, and some evidence, of flash setting during circulation with a lack of final setting. On another project in Brazil-Paraguay where a return line was used, a serpentine cooling coil was used to cool the return grout to 77°F (25°C) before allowing it to flow into the agitator (personal communication, A. Barbi).

Grouting Pressure. - There is much controversy about the magnitude of grouting pressures that should be specified (2,3,4,5,6, 7,9,10,12,14). The U.S. engineers have often been accused of using pressures too low (such as a maximum of 1 lb/in^2 per ft of depth) and the Europeans of using pressures too high (such as 1 kg/cm^2 per meter of depth) which is about four times higher. In reality, many U.S. organizations used higher pressures (2,5,9,13) and many Europeans use lower pressures than these quoted rules-of-thumb (personal communication, B. Gilg, G. Lombardi, A. Pautre, F. Sabarly). The author believes that the lower of these quoted pressures should be used in the upper part of the rock mass where rock jacking could occur (particularly where rock joints parallel the surface, such as horizontal bedding or joints below a horizontal surface, or inclined relief joints parallel to an inclined valley wall). With depth much higher pressures are justified.

One reason that European engineers are confident with higher pressures is that they require grout flow meters and pressure meters of the recording types so that any hydrofracturing can be immediately identified and permanently recorded. At any sudden drop in pressure or increase in flow, the pumping pressure is immediately lowered. This restricts the fracturing and lifting process (personal communication, A. Pautre). The high pressures attributed to European grouting practice are the final "refusal pressures)." The main grouting is carried out at lower pressures. Like U.S. practice, the main objective is to provide a steady intake of grout slurry without choking off the permeable channels. As grout take declines, the pressure is slowly raised and held, ultimately arriving at the specified final refusal pressure.

On several recent major jobs totaling more than 300 miles (500 km) of grout holes the grouting has proceeded from beginning to end with only one grout mix, or at the most, two. These have been moderately thick, stable mixes with 2%-3% bentonite. The author agrees with the philosophy that it is the size of the cement particles that determines whether a fine fissure will be grouted or not. As long as the hole is not choked off by rapid increases in flow or pressure, a thick mix will penetrate the crack but at a rate proportional to its viscosity. Therefore, in a given period of time a stable grout of 20 cP viscosity (such as 1.5:1, W:C by volume, with 2% bentonite) will penetrate only one fourth as far as a grout of 5 cP viscosity (such as 3:1, W:C by volume with 0%-1% bentonite). To arrive at the same distance, the thicker grout would have to be grouted for 4 times as long a period of time or at 4 times the pressure, or with a suitable combination of both. Actually, the higher viscosity is of value in many cases: (1) in reducing the distance of travel in a wide, open fracture, and (2) in allowing a pressure drop to actualize as the grout moves away from the hole, thus reducing the chance of hydrofracturing and surface uplift.

Type of Cement. - Type 1 cement is often the only type available. For underground work, Type 2 is preferred as it is moderately sulphate resistant, a favorable property where acid waters from sulphide oxidation are often present. The finer the grind the better as less bentonite is required to achieve a stable mix. Also, the percent retained on the 200 and 325 meshes should be less than about 2% and 5%, respectively, to limit sedimentation and choking off of fine fissures. On some jobs, special grinds at the cement mill have been made every month or so to furnish the fine cement for grouting. At a project in Brazil several years ago, fine cement was obtained from the dust collector at the cement mill for use in the grouting (personal communication, M. Ruiz).

It can be seen that because of the many variables of the cement properties (and the bentonite properties as well) it is really a design necessity to have laboratory tests on the proposed materials and mixes. Even the time required before washing out a grout hole depends on the time of setting of the grout, a property that should be determined by the design laboratory tests and then checked during field operations.

FIELD CONTROL

The field control of a grouting program will vary in scope considerably because of the differences in the specifications that one finds from job to job. On one hand, the specifications may be vague with much of the field control left up to field decisions. This approach essentially leaves much of the design as well as the field control to the field grouting supervisor. The merit of this approach is that the field grouting supervisor may be more knowledgeable than the designers in some cases. The drawback to this approach is that the designers are turning over to the field some of their design responsibility. Often, it is not known what organization will ultimately be given the field supervision. It may not be the design firm, nor even the owner agency's staff. The grouting supervisor cannot be identified in advance and his own particular philosophy about grouting mixes, pressures, and closure criteria are usually not subject to prior screening.

The second approach is to prepare very detailed specifications which gives clear guidance to the field about grout curtain layout, design of the grout slurry, mixing the ingredients, drilling the grout holes, and pumping the slurry down the holes. The design drawings and the specifications should clearly show the grout curtain layout, the details of the progressive grouting system, and the acceptable closure criteria. This approach is to be preferred in the author's opinion and increasingly is being used.

Field Control of Design Slurry

Current practice is trending toward the use of only 2 or 3 different mixes for a job: a starting mix, a production mix, and a special mix for large absorptions. Perhaps, 99% of the grouting would be done with the production mix, such as 1.5:1, W:C by volume, with 2% bentonite. The starting mix might be a little thinner, and the special mix a little thicker, perhaps with fine sand or sodium silicate admixture.

Ideally, the mixes will have been previously tested in the laboratory during the design period and the specifications will provide for field laboratory tests using the cement and bentonite proposed by the contractor for use in the actual grouting. Tests should be conducted on each of the mixes for sedimentation, slurry density, Marsh funnel viscosity, Vicat needle setting time, and unconfined compressive strength at various ages.

The values obtained will serve as guides during the grouting operations. It is recommended that grout samples be obtained hourly from the agitator or grout by-pass valve at the head of the hole and that three tests be done: temperature of slurry, density of the slurry using the mud balance, and Marsh funnel viscosity. These tests can be carried out on a table located at the grout plant and only take minutes. The values serve as control of the ingredients (quality and mix proportions) and of the mixing procedure. The bentonite should be pre-mixed with about 14%-15% of water and aged for at least 2 hours before adding to the grout slurry.

<u>Additional Duties of the Grouting Supervisor</u>

By the use of a small number of mixes and routine field sampling and testing by his inspectors of the grout consistency, the grouting supervisor is free to concentrate on the other aspects of the specifications and field operations. The routine ones include checking on the equipment, drilling depths, hole layout, grouting pressures, etc. The more specialized items are the method of drilling, the method of grouting, and closure criteria.

Rotary-percussion drilling is often specified as it is faster and cheaper. However, in shales or other altered or clayey rocks there may be problems because of sliming of the fines into a clay slurry or with balling up of the bit. The supervisor would then have to recommend alternative rotary drilling and should discuss the resulting cost and schedule problems with the design engineers.

Full-depth nipple grouting and descending-stage grouting (with or without packer) are not used as much as formerly (9). Many projects specify packer grouting in ascending stages because of its rapidity and the advantages of applying the higher pressures at depth where rock uplift and surface leakage are minimized. However, field conditions are met where the hole cannot be drilled full-depth because of caving hole, water-loss, or water-gain conditions. The grouting supervisor again must decide on alternative procedures and discuss them with the designers regarding adequacy, cost, and schedule.

In the progressive grouting system certain field judgements must be made as to the need for additional holes or lines, even in those cases where fairly detailed specifications exist. Again, the grouting supervisor and design engineers will be jointly involved in interpreting the specifications and evaluating the results.

Summary Diagrams

Experience has indicated that various types of summary diagrams are helpful in evaluating the grouting results during the course of the grouting program. These allow comparisons to be made of one abutment vs. the other, one geologic area vs. another, secondary holes vs. primary, etc. The writer (3) and various of his colleagues and clients have used for several years the following terminology (Table 3) in order to allow comparisons to be made and to facilitate communication.

TABLE 3. - Terminology of Grout Absorptions

Grout Absorption kg cement/m of hole (1)	Descriptive Terminology of Absorption (2)
0-25	Low
26-50	Moderately Low
51-100	Moderate
101-200	Moderately High
201-400	High
Greater than 400	Very High

Note: These categories have been used mostly in countries with metric units. Multiply by 0.67 to obtain lb cement/ft of hole.

Figure 7 illustrates another type of diagram that is helpful in illustrating the decrease in grout absorption with increasing number of holes or phases. Again, this type of diagram allows comparisons to be made for different areas and zones within a site as well as from one project site to another. For multiple-line grouting an average hole spacing is obtained by dividing a given length along the line by the total number of holes in all the lines within the given distance.

Frequency diagrams (Fig. 8) are also useful in showing how the number of holes of very low absorption increase and the number of holes of high to very high absorptions decrease as the grouting progresses. The average grout absorption for each phase may also be indicated. A trend curve may be drawn for each phase; as can be noted from Figure 9 the shapes of the curves are quite characteristic.

While the summary diagrams are useful, each hole is an entity in itself and careful control of the drilling, washing, and grouting of

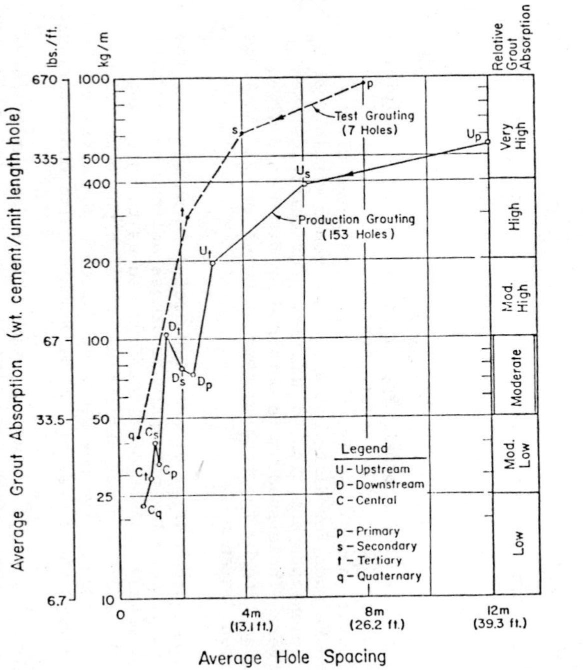

Fig. 7 - RELATION OF GROUT ABSORPTION AND HOLE SPACING AT VARIOUS PHASES OF TEST GROUTING AND PRODUCTION GROUTING

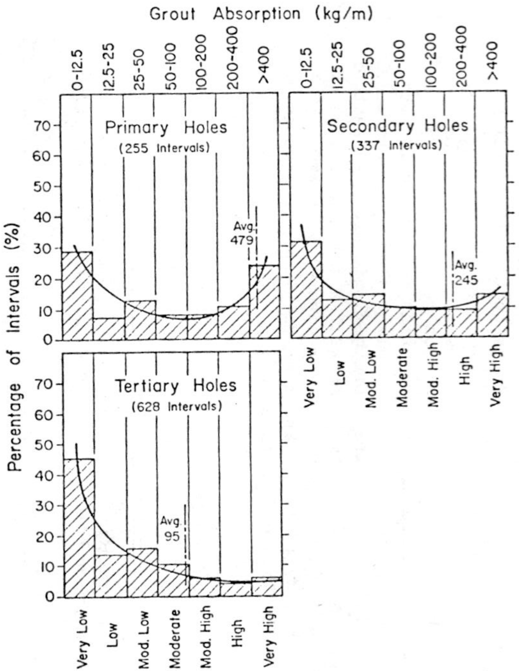

Fig. 8 - FREQUENCY DIAGRAMS OF GROUT ABSORPTIONS FOR VARIOUS PHASES OF GROUTING

every hole is important (6). Only when good design and explicit specifications are put together with careful field control by experienced field supervision will an optimum grout curtain be achieved, and only then if ample dialogue ensues among the designers, the grouting supervisor, and the contract management team.

APPENDIX.-REFERENCES

1. Burgin, C. R., "Investigation of the Physical Properties of Cement-Bentonite Grouts for Improvement of Dam Foundations", thesis presented to the University of Florida in 1979 in partial fulfillment of the requirements for the degree of Master of Science.

2. Burwell, E. B., "Cement and Clay Grouting of Foundations: Practice of the Corps of Engineers", Journal of the Soil Mechanics and Foundations Division, ASCE, Vol 84, No. SM 1, Proc. Paper 1551, Feb. 1958, pp. 1-21.

3. Deere, D. U., "Dams on Rock Foundations-Some Design Questions", Rock Engineering for Foundations and Slopes, Proceedings of Specialty Conference of Geotechnical Engineering Division, ASCE, Boulder, Colorado, Vol. 2, 1976, pp. 55-86.

4. Elston, J. P., "Cement Grouting", Journal of the Soil Mechanics and Foundations Division, ASCE, Vol. 88, No. SM 2, Proc. Paper 3098, Apr. 1962, pp. 49-98.

5. Hilf, J. W., and Deere, D. U., "Summary-Panel Discussion of Dam Foundations", Rock Engineering for Foundations and Slopes, Proceedings of Specialty Conference of Geotechnical Engineering Division, ASCE, Boulder, Colorado, Vol. 2, 1976, pp. 147-156.

6. Houlsby, A. C., "Engineering of Grout Curtains to Standards", Journal of the Geotechnical Engineering Division, ASCE, Vol. 103, No. GT 9, Proc. Paper 13214, Sept. 1977, pp. 953-970.

7. Jones, J. C., "Rock Foundations and Abutments for Dams", Rock Engineering for Foundations and Slopes, Proceedings of Specialty Conference of Geotechnical Engineering Division, ASCE, Boulder, Colorado, Vol. 2, 1976, pp. 91-100.

8. Kennedy, T. B., "Pressure Grouting Fine Fissures", Journal of the Soil Mechanics and Foundations Division, ASCE, Vol 84, No. SM 3, Proc. Paper 1731, Aug. 1958, pp. 1-36.

9. Lippold, F. H., "Cement and Clay Grouting of Foundations: Pressure Grouting with Packers", Journal of the Soil Mechanics and Foundations Division, ASCE, Vol. 84, No. SM 1, Proc. Paper 1549, Feb. 1958, pp. 1-10.

10. Mayer, A., "Modern Grouting Techniques", Grouts and Drilling Muds in Engineering Practice, Butterworths, London, 1963, pp. 7-9.

APPENDIX.-REFERENCES (continued)

11. Minear, V. L., "General Aspects of Cement Grouting of Rock", Journal of the Soil Mechanics and Foundations Division, ASCE, Vol. 83, No. SM 1, Proc. Paper 1145, Jan. 1957, pp. 1-11.

12. Morgenstern, N. R., and Vaughan, P. R., "Some Observations on Allowable Grouting Pressures", Grouts and Drilling Muds in Engineering Practice, Butterworths, London, 1963, pp. 36-42.

13. Simonds, A. W., Lippold, F. H., and Keim, R. E., "Treatment of Foundations for Large Dams by Grouting Methods", Transactions, American Society of Civil Engineers, Vol. 116, Paper No. 2439, 1951, pp. 548-567.

14. Szalay, K., "Dams on Rock Foundations-Some Design Questions", Rock Engineering for Foundations and Slopes, Proceedings of Specialty Conference of Geotechnical Engineering Division, ASCE, Boulder, Colorado, Vol. 2, 1976, pp. 114-124.

15. Tatamiya, C. T., et al, "Proposta de Adocao de uma Sistematica de Ensaios de Caracterizacao de Cimento Usadas em Injecoes", Proceedings, 8th Seminario Nacional de Grandes Barragens, Sao Paulo, Brasil, 1972.

ASPECTS OF ROCK GROUTING PRACTICE ON BRITISH DAMS

D.A. BRUCE*

Summary

As a means of illustrating certain aspects of contemporary British cement grouting practice in rock, three recent English dam grouting projects are reviewed. The first, a large concrete buttress dam, involved over 94km of drilling and 3500 tonnes of grouting materials. Details regarding selection of grouting methods, curtain geometry, mix design, pressures and plant are provided, together with guides to analysis and assessment. On this project, the extremely poor nature of the bedrock in one particular area necessitated additional grouting and the conduct and effectiveness of this work are also described. Data are cited from the second example, Kielder Dam, to highlight the relevance of special Lugeon water testing programmes as a means of independently assessing the effectiveness of curtain grouting. The third project, Grimwith Dam, provides a good example of the application of tube-à-manchette grouting - a system developed when the originally foreseen method of descending stage proved inappropriate in the variable drift deposits encountered in a section of the curtain near the original stream course.

Key Words

Dam Grouting: Design: Drilling and Grouting Plant: Stage Grouting: Lugeon Water Testing: Tube-à-manchette.

1. Introduction

Particularly in North America there has been a tendency to regard dam grouting as a fully developed and established science. For example, Polatty (1974) reviewed his experience in the subject and concluded there had been very little change in Specifications over 30 years: "And, this is good". More recently, however, grouting engineers have become involved with complex remedial domestic works and with large schemes on new overseas projects, often with foreign grouting specialist contractors. Much has been made of the traditional differences between U.S. and European grouting procedures, in terms of techniques and equipment, and the relative merits of the different approaches have often been examined (e.g. at Asilomar, 1974). However, the Author's experience supports Rigny's statement at that Conference, that the most important contrast is now contractual as opposed to technical. For instance, it is now common on overseas dams for the Engineer to adopt an "open mind" approach by, for example, permitting the Grouting Contractor to select optimum methods, often on the basis of short test programmes.

Relatively few papers have been published in the last decade illustrating current British cement grouting practice in rock. American engineers may, therefore, be unaware of current methods and developments. Since space does not permit a comprehensive survey in this contribution, the more important, or potentially controversial, aspects are highlighted from works completed recently by my Company

*Technical Manager, Colcrete Limited, Wetherby, England.

on three large water supply dams in Britain (Figure 1).

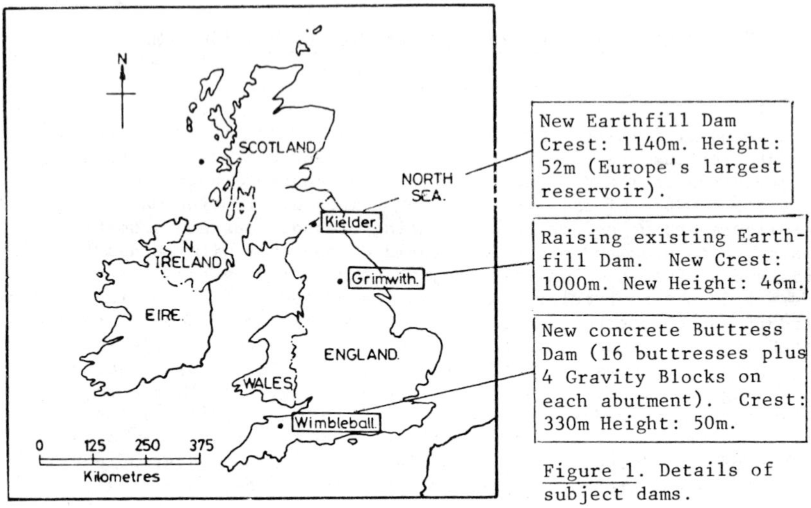

Figure 1. Details of subject dams.

The first, Wimbleball Dam, Somerset, is essentially a straightforward statement of blanket (consolidation) and curtain grouting design, construction and assessment, although the importance of post-completion monitoring is underlined clearly. This project involved over 94km of drilling and 3500 tonnes of injected materials.* At Kielder Dam, Northumberland, a full scale Lugeon water test programme was conducted to determine the effectiveness of the curtain grouting and to investigate the relationships between water test values and levels of grout acceptance. The potential of these techniques was fully realised at Grimwith Dam where they permitted potential problems with a supposedly routine grout curtain to be highlighted at an early stage. As a consequence, an alternative grouting technology, tube-à-manchette as opposed to descending stage, was employed most successfully.

2. Wimbleball Dam

2.1 Geology: The dam is located in a deep v-shaped valley incised into Upper Devonian sediments, mainly hard sandstones, siltstones and mudstones. Dips vary from 45° on the North Bank to 14° on the South. From site investigation studies, including seismic refraction, deep sub-vertical shatter and crush zones were inferred, striking oblique to the line of the dam and becoming more frequent towards the South Bank. Although extreme vertical and lateral lithological variations were noted, a common feature across the site was the closely fractured nature of the bedrock.

2.2 Design: This very fissured nature of the rock mass, with its attendant potential for hole instability, dictated the use of descending stage grouting methods for the curtain. The relative merits of descending and ascending stage procedures were described recently by Bruce and George (1981), but it may be noted that very few British dam sites have enjoyed foundation rock sufficiently hard, massive and competent to permit the latter method. In addition, the shallower

*SI units are now invariably used in British practice. Relevant conversion factors are provided at the end of the paper for easy reference.

blanket (consolidation) grouting by its nature automatically favoured descending stages. The blanket grouting preceded the curtain in each area, to a maximum depth of 15m. The 3m stage length which was selected typifies British practice, although allowance is frequently made to increase this in late phases of grouting when grout consumptions may be expected to be lower. Hole diameter for rock fissure grouting has no significant influence on the efficiency of the injection within the normal range of diameters (40 to 100mm), and commonly, 50mm is regarded as the best practical and economic compromise.

Mix designs were finalised during early injection tests and ranged from w/c (OPC) ratios of 5 (apparent viscosity 1.4 centipoise) to 0.5 (37 centipoise), both by weight. Occasionally, Specifications require starting mixes as thin as 20:1 but rarely in practice do they exceed 8:1. In cases where stages were still accepting the thickest mix freely, after a total cement injection of 1000kg, fine sand was added up to a 1:1 ratio by weight to cement. Even then, however, final grouting of the stage was accomplished by reverting to the neat cement mix to ensure the fullest treatment of the finer fissures (down to 160μ) incapable of being filled with sand particles (Littlejohn, 1975). Stages which did not come up to pressure after a total material injection of 3000kg were abandoned for 24 hours prior to the resumption of the neat cement grouting sequence as before.

Refusal pressures were carefully designed to prevent the possibility of any structural uplift occurring during injection. Since most of the grouting was conducted early in the construction of each buttress, maximum pressures (as measured at the top-hole packer) were restricted to the effective theoretical overburden pressure at the top of the stage, to an overall maximum of 3.5 bar. The work of Houlsby (1977) suggests that this approach at Wimbleball may seem unduly conservative (Figure 2), but wherever there is a possibility that horizontally bedded strata exist at shallow depths, particular consideration and caution must always be exercised to avoid structural movements (e.g. Weyerman, 1977).

Since clay infilling of joints was recognised from investigatory boreholes as being neither extensive, nor significant from a geomechanical viewpoint, no major programme of clay expulsion by air and water flushing was specified. This decision again typifies current British practice, the tenent being that such attempts merely create a "pipe" between the boreholes in question, whilst not removing significant amounts of joint infill material.

Since unstable grout mixes were used (i.e. those with a very high bleed capacity due to their high w/c ratios) consumptions were recorded in terms of kg/m or kg/m^2. Regarding the former measure, Deere's classification (1976) is gaining acceptance in Britain (Table 1).

Grouting was conducted using "Split Spacing" (closure) procedures, each successive phase being at least two stages in arrears of the preceding one. Examination of absolute and relative phase takes formed the basis of determining the intensity of grouting: with the former, Deere's recommended (1976) "Low" average consumption was the target for the last phase, whilst Reduction Ratios (e.g. the ratio of average Secondary to Primary takes) in the range 25-75% (Cope, 1978) were sought.

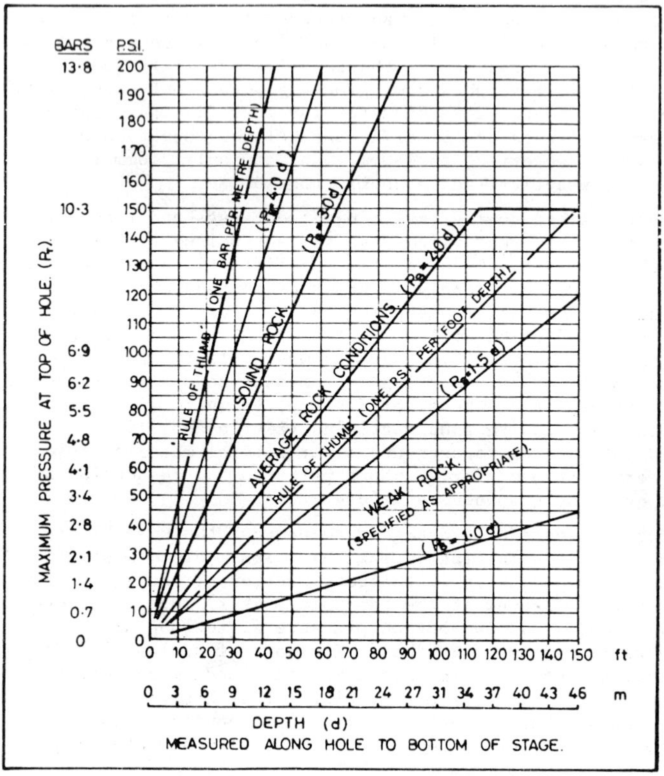

Figure 2.

Typical permissable Grout Pressures for various foundation conditions (Pressures shown are at surface; P_B denotes pressure at bottom of stage, and d is the depth to there). (After Houlsby, 1977).

Average Lineal Grout Consumption (kg/m)	Designation
400+	Very High
200-400	High
100-200	Moderately High
50 -100	Moderate
25 - 50	Moderately Low
12½- 25	Low
0 - 12½	Very Low

Table 1. Proposed grout consumption classification (after Deere, 1976)

Conclusions on the "tightness" of the rock at any point were confirmed with simple stage water tests. In general, a target permeability equivalent to 3 Lugeons was required (1 Lugeon: 1 litre/m/min at a pressure of 10 bar).

The dam curtain was formed in a vertical plane descending continuously from concrete scarcements on the upstream side of the dam. Locally, holes were angled at 15° to encourage intersection of subvertical fissures or shear zones. It is noteworthy that inclined drilling is more expensive than vertical drilling and should be specified only on sound geotechnical or practical grounds.

The dam curtain was extended laterally for 270m as the North Wing (orientated 40° downstream for technical and financial reasons) and 318m as the South Wing (extending out along the dam centre line).

A single row of closely spaced holes was designed since (i) large amounts of erodible joint fill material were neither foreseen nor recorded, (ii) no phase of chemical treatment was envisaged to achieve the target permeability, (iii) there was no evidence to suggest that the variably orientated fissures would be more successfully intercepted by a multiple row curtain than by an intensive single row. In addition, there was limited access on the narrow dam scarcements and this would have restricted markedly the potential "width" of a multirow curtain, bearing in mind the extent of lateral grout travel believed to occur (Champion, 1961).

Prior experience in similar rock suggested that final interhole spacings in the range 0.75 to 0.90m would be necessary to provide the target degree of curtain grout and water-tightness. In order that the progress could be monitored over three phases (i.e. Primary, Secondary and Tertiary) a Primary spacing of 3m was selected. As noted in Section 2.4, this estimate was in general well-founded, although the extremely poor rock encountered under some southern buttresses required Quaternary treatment to certain depths giving locally a final interhole spacing of 0.375m.

In general, the curtain depth was to be equal to at least 75% of the maximum depth of the reservoir when full and in detail this was varied depending on drilling, grouting and water take observations. This rational selection of appropriate curtain depths is obviously facilitated in practice by the use of descending stage techniques. Rarely are curtains extended to depths in excess of the storage head.

2.3 Plant: Grout hole drilling was conducted exclusively by the rotary percussive method with water flush. A variety of air- and hydraulic-powered wheel and track-mounted rigs were used depending on access and depth considerations (to 110m). Air flushed rotary or down-the-hole drilling was permitted only for Relief Wells, Piezometers or standpipe holes. This widespread use of rotary percussion was formerly one of the major contrasts between American and European practice (e.g. Gebhart, 1976), but increasing amounts of experimental data (e.g. Doty 1970) have led to a more equivocal view in both camps. Pure rotary drilling is two to five times more expensive than rotary percussion (Deere, 1976) and so merits consideration in only exceptional circumstances in Britain. For example, a certain number of Primary curtain holes may be cored to provide additional geotechnical data. It is considered that flush characteristics are often more critical in terms of subsequent hole "groutability" than the actual drilling mode.

Grouting featured the use of diesel driven Colcrete colloidal mixers (Figure 3) linked via storage tanks to Colcrete Evans Pumps, operating on reciprocating ram principles (Gourlay and Carson, 1982). Rock grouting is considered by European engineers to require a pulsating pressure flow to inhibit premature blockages arising from the temporary build-up of coarse grout particles in the fissures ("pressure-filtration"). The hydraulic Evans pumps also permitted the maximum stage grouting pressures to be pre-set, so avoiding the possibility of accidentally exerting excess pressure in any stage - an extremely valuable feature in all dam grouting applications.

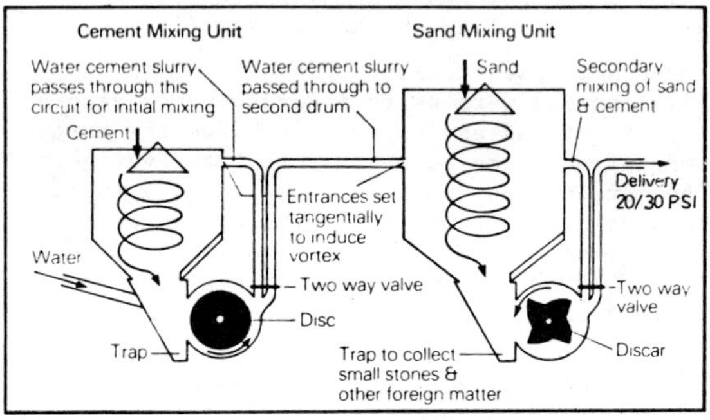

Figure 3.

Operating principle of Double Drum colloidal mixer.

2.4 Grouting Analysis: The blanket treatment involved over 22km of drilling, 4450 stages, 742 tonnes of cement and 65 tonnes of sand. Overall lineal consumptions per buttress ranged from 19kg/m in the North, to over 165kg/m under South Gravity Block 1 - over three times the overall average blanket consumption. This evidence of marked rock mass deterioration southwards, suggested by site investigation, was further confirmed by an increase in holes per buttress, stages per hole and the use of sand, in that direction. However, despite this great inhomogeneity - from "Low" to "Moderately High" characteristics - special instrumentation of the dam during and after impounding confirmed the satisfactory response of the structure.

Regarding the dam curtain, the salient figures were 20km of drilling, 5888 stages, 597 tonnes of cement and 7 tonnes of sand. The pattern of lineal grout consumptions was similar to that of the blanket: 19kg/m rising southwards to over 92kg/m under South Gravity Block 4. When the corresponding average areal takes per buttress were reviewed, the deterioration was even more acutely marked: 22kg/m^2 to 279kg/m^2. In addition, it is significant that Quaternaries were required only from Buttress 13 south, i.e. for four buttresses and the four South Gravity Blocks.

The efficiency of the dam curtain in operation was confirmed by data provided by Casagrande Piezometers, Relief Wells and standpipes. The piezometers registered no excess head under the structure, whilst the Wells gave low flows (up to 2 litres/sec) except in the southern section (to 10 litres/sec). The standpipes allowed the phreatic surface along and across the curtain to be monitored.

Data from the North and South Wings confirmed their geotechnical continuity with their contiguous dam curtain sections and on the South Wing the highest individual stage takes ("Very High") of the whole programme were recorded. The overall areal consumption of 24kg/m^2 on the North Wing contrasted with the figure of 82kg/m^2 on the South Wing.

The pattern and scale of grout takes, coupled with the Relief Well flows strongly suggested that more intensive additional grouting be done on the South side, and this need was underlined with the appearance of a series of Springs some 200m downstream on that Bank. Their flow varied with reservoir level and extrapolation showed that it could reach 114 litres/sec. at full reservoir level. Chemical analyses suggested that 23 litres/sec. of this was a run-off "constant", whilst data from the standpipes confirmed the likelihood of a major seepage path in the vicinity of the southern end of the dam. This quantity of flow together with its potentially detrimental effects on the stability of the valley side was unacceptable, and a further phase of grouting was agreed, extending under approximately 100m of the Southern section of the dam.

It was reasoned that the high Well and Spring flows were occurring through fissures which were locally orthogonal to the line of the curtain, and had, therefore, "combed" through it, escaping full treatment during the original grouting. In such circumstances, a multirow solution is considered preferable and the additional grouting was conducted as a second parallel row of holes, 1.5m upstream, with a Primary spacing of 3m. In an effort to further inrease the zone of influence of the curtain holes, grouting pressures were raised by 25%. Flows and piezometric levels were monitored closely throughout the work, to optimise its conduct in technical and economic terms. In total, a further 6.5km of grout hole drilling, 1887 stages, 482 tonnes of cement and 95 tonnes of sand were involved. Reduction ratios of 50% dropped the Tertiary average lineal take to 28kg/m whilst 62% of the Tertiary stage takes were in the "Low" to "Very Low" range. The overall areal take averaged $70kg/m^2$. The effectiveness of this treatment may be examined after three important intervals (Figure 4). During the first phase, major open "pockets" were located and treated, mainly in the upper sections of the curtain, and the average flow reduction was 3.7 litres/sec/working week. Over the same period, Well flow dropped by 50%. In Phase 2, the curtain was lengthened by 30m and deepened by several stages but with an average reduction of 1.1 litres/sec/week, whilst Phase 3 consisted of local "weakspot" Tertiary grouting (average flow reduction 0.9 litres/sec/week). Thus, by the end of the treatment, through flows had been dropped by a factor of three to 30 litres/sec, Well flows by the same ratio, and the piezometric contours showed a markedly increased gradient across the curtain line. By analysing the Phase 3 grout takes, it was concluded that this residual flow was occurring through small fissures (i.e. less than 160μ) or in deeper fissures more distant from the dam. To have further reduced it would have required either a different grouting material or more extensive conventional treatment. On balance, neither alternative was considered cost effective, and the additional grouting was discontinued at this point.

3. Kielder Dam

3.1 Geology: The North Hillside of the valley comprised gently dipping Lower Carboniferous sediments, mainly sandstones, but with mudstones, shales and thin coal beds. The whole mass was closely jointed and faulted and expected to be highly permeable. The overall dam design featured a clay core linked to an upstream clay key by a clay blanket, to minimise underseepage. However, the high permeability of the North Hillside, in addition to allowing high reservoir leakage, could have threatened the integrity of the clay by piping, and a grout curtain was therefore planned in that area.

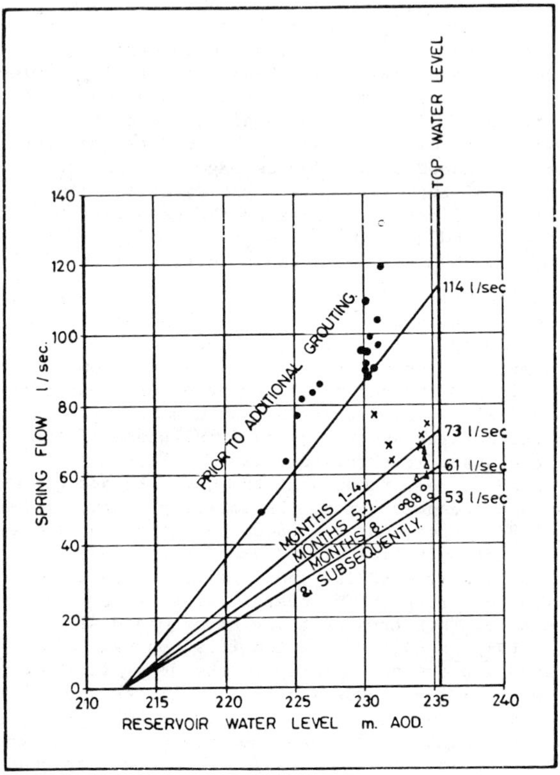

Figure 4.

Relationship between Reservoir Level and Interpreted Spring Flow, illustrating effect of grouting at different periods.

3.2 Design: The orientation of the major discontinuities, often gouge filled, led to the selection of a multirow curtain. Two Primary rows, 3m apart, with the same interhole spacing, extended over a length of 150m. It was foreseen that an intermediate row of Secondary holes would be required with any Tertiary treatment also being on this central line. The holes were vertical, 50mm in diameter and from three to ten 3m descending stages into rock with the option to deepen if necessary to terminate in relatively impermeable mudstone. Neat cement grout mixes with w/c ratios in the range 14:1 to 0.7:1 by weight were specified, with provision for sand to be added in the case of especially large takes. Early tests reduced the range to 7 to 0.49, to be thickened in ten gradual steps as necessary. A minimum of 3m of clay had to be placed before grouting, both to provide a reaction when treating the upper layers and to permit grouting of the clay-rock contact. Maximum grouting pressures were to be as high as was consistent with minimum surface displacement or splitting of the rock. These were verified by special test holes into which water was pumped at successively higher pressures and the pressure-flow relationship observed. No hydrofracture was noted despite pressures equivalent to over two times theoretical total overburden pressure being used, and grouting pressures were set conservatively at overburden pressure plus 0.4 bar (as measured at the top of the hole). It may be emphasised that here British practice is decidedly in contrast to the hydrofracture ("claquage"), high pressure grouting methods favoured by certain other European countries.

This project was remarkable in that the permeability of the rock mass before and after grouting was to be gauged in independent water test holes, following the methods of execution and interpretation proposed by Houlsby (1976). The general philosophy of the dam was "controlled underseepage", and, bearing in mind the presence of the clay blanket, it was held that the curtain need not be engineered to extremely "tight" standards (as at Wimbleball). A general target of around 10L was considered acceptable. Equally, it was not anticipated therefore that successive phases of grouting would be conducted until Deere's "Low" figure for final phases would be met. The general precept was that although specific procedures had been detailed, early results would be assessed with a view to amending or optimising the execution of the work.

3.3 Execution and Results: Standard rotary-percussive drilling plant and colloidal mixers with ram pumps were again used. Only where the clay thickness exceeded 10m was casing required to rockhead to permit the insertion of the plastic standpipe, and for this the ODEX eccentric overburden drilling method was adopted (Figure 5). This system had been used in Britain for the first time on a large production scale (viz. 75km of drilling) only recently before in conjunction with a mine infill project in South Wales (Patey, 1977), and again proved extremely effective in this more restricted application.

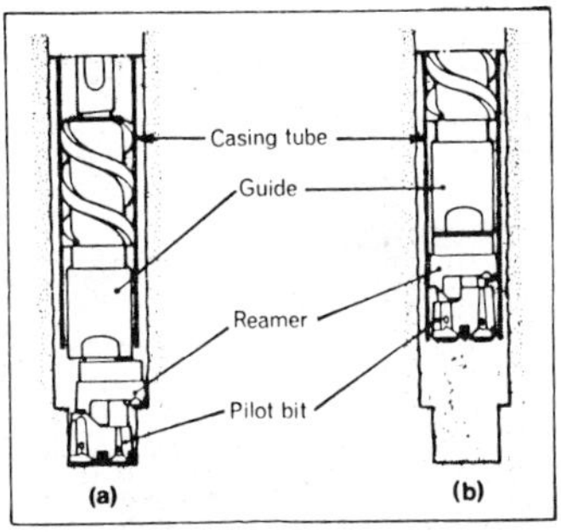

Figure 5. Major features of ODEX drilling system.

Overall, 175 grout holes were drilled, involving 4530m of rock drilling, 1365 stages and over 306 tonnes of cement. Full Secondary treatments were carried out in especially fractured areas where average Primary hole takes over 120kg/m were recorded, although their "Moderately Low" nature permitted 6m Secondary stage lengths to be used. A few Tertiary holes were also installed locally. Overall, the Primary holes averaged 85kg/m and others 44kg/m. The average areal take was $78kg/m^2$.

3.4 Lugeon Testing: An independent programme of Lugeon water test holes was planned to assess the effects of the grouting, in terms of the permeability of the curtain zone. Eight vertical holes were involved, on the centre line of the curtain, comprising four pre-grouting (37 stages) and four post-grouting (38 stages). Each hole was drilled and tested in 3m descending stages, each isolated by a pneumatic down-the-hole packer. Each stage was tested at five pressures in the sequence "a b c b a" with the flow over a 10 minute period at each pressure being recorded. In this case, the maximum pressure (c) on the deepest stage was 10 bar, with the lowest steps (a & b) being 0.4 and 0.7 times this maximum respectively. An equivalent Lugeon value was calculated for each test pressure, and then a representative value selected for that stage with regard to Houlsby's chart (Figure 6) based on inferred type of flow regime.

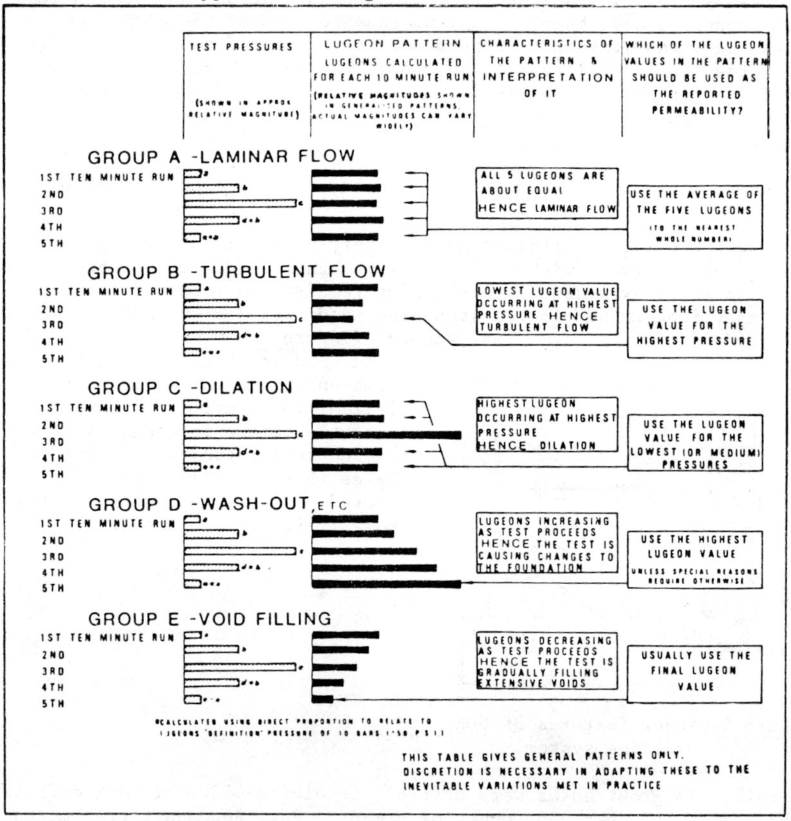

Figure 6. The interpretation of Lugeon Test Results (after Houlsby, 1976).

Regarding the pre-grouting tests, a wide scatter of values was obtained ranging from impermeable (shale/coal) to over 100L in especially fractured sandstone. Although 49% were 5L or less, 35% were above 20L, and these stages, like the very tight ones, tended to congregate in distinct pockets. The post-grouting values did range to a 28L maximum, but 74% were 5L or less, and barely 5% were above 20L, confirming the effectiveness of the grouting as executed.

Houlsby (1976) notes that geological factors such as wall roughness, frequency, orientation and geometry, strongly influence flow through fissures. One consequence is a considerable variation in results from even neighbouring holes in apparently uniform areas. On these grounds alone, close correlation may not be expected between water test results and earlier or subsequent grout consumptions. Detailed examination of the test hole and adjacent grout hole data does, however, lead to the following conclusions for cementitious grouts:-

(i) Even when water test holes are as close as 1.5m to grout holes, close correlation of grout and water test patterns may not be noted.

(ii) A low water take will indicate a subsequent low grout consumption but a high water take will not necessarily be followed by a high grout consumption.

(iii) The relative magnitude of Lugeon values is no certain guide to the magnitude (relative or absolute) of the subsequent grout takes.

(iv) When an area is grouted to refusal as indicated by cement consumptions, it may still yield moderate to high Lugeon values.

The explanation for these conclusions is related to the particulate nature of cement grouts. It may be assumed that the minimum fissure width which can be reliably treated with OPC (Type 1) is 160μ (Littlejohn, 1975). A 3m stage with one such fissure, will give a Lugeon value of 10L. Conversely, a stage with five 100μ fissures will also give a value of 10L. Thus, although the first case will be groutable, the second will not, and so a post-grouting Lugeon test near the first stage will show tight ground, whereas one near the second will still show 10L. It is significant that of the 31 post-grouting stages of permeability 10L or less, 71% exhibited the "Laminar" flow characteristic, compatible with "fine" fissures.

These tests confirmed, therefore, that only the fine fissures remained untreated in the zone of the curtain, and that the great majority of the wider fissures encountered had been effectively sealed. They also highlighted the importance of considering grout and water test data together when attempting to analyse the effects of such grouting projects.

4. Grimwith Dam

4.1 Geology: With the incorporation of the existing 23m high earth-fill dam into the upstream shoulder of the new structure, a grout curtain was to be formed in the permeable strata present beneath the base of the clay core on the eastern side of the valley. The clay core was to be carried only to limited depths below existing ground level. Geological conditions beneath the western side of the valley were such as to render a grout curtain unnecessary there. On the eastern side of the valley the bedrock comprised alternating Carboniferous sandstones (or gritstones) and shales (or mudstones) with occasional interbedded coal seams which had been worked many years previously. This sequence overlay alternating limestones and shales (or mudstones). The strata dipped generally in an upstream direction at a small angle, and the sandstones in particular were extensively fissured and jointed. The solid rocks were overlain by a variable thickness of superficial deposits, thinning from the western, or "stream", end of the curtain (CH.800), to virtually zero thickness on the higher ground of the eastern valley side (beyond CH.900). The purpose of the curtain was to control possible seepage through the Upper Millstone Grit sandstones (estimated mass permeability 10^{-5}m/s), present in the west abutment of the dam, and it was anticipated that it would descend to terminate in relatively impermeable shale (or mudstone) at a maximum depth of 45m.

4.2 Design: A single row of 45 Primary and 45 Secondary vertical holes was planned originally, giving a total curtain length of 250m. Again, descending stage methods were adopted, reflecting the expected nature of the rock, and standard cement grout mixes and pressures were employed. As at Kielder, the overall design of the curtain did not require it to have an extremely high degree of efficiency or relative impermeability since some clay blanketing was also to be placed within the reservoir basin. A general target figure of about 10L was again considered adequate.

4.3 Results: Using conventional rotary percussive drilling rigs, and colloidal grouting plant, an intensive first phase was executed. The "Very High" Primary and Secondary takes required the installation of a further 81 Tertiary holes, located (due to the detection of clay in major fissures) in a second row, 0.75m upstream of the first. Some of these Tertiaries were locally inclined at 30° to ensure intersection of suspected subvertical fissures (as at Wimbleball). Overall, 5013m of grout holes were drilled, and 1597 tonnes of material injected at an average of 317kg/m or 225kg/m^2.

Although satisfactory reduction ratios of 52-78% had been recorded, the Tertiary average was still 255kg/m (vertical holes) or 133kg/m (inclined holes). In addition, surface breakout of grout was noted, even at lowest pressures, and there were instances of stages collapsing before grouting could commence. Simple stage water tests also pointed to erratic and high Lugeon values after Secondary grouting. Eighteen full scale vertical Lugeon test holes were then installed, at an average spacing of 15m along the line of the grouting, and their results precipitated the decision to conduct a second major phase of treatment, off-set 1m upstream of the original.

These holes, plus evidence from further core holes, confirmed the viability of descending stage grouting in the eastern 160m of the curtain and the subsequent re-treatment of that area proceeded with the same methods as before. After that phase, post-grouting Lugeon values were considerably below 10L. However, the same sources indicated that the western 90m of the original curtain did in fact comprise a major geological complexity, possibly in the form of a large sandstone and shale slump, intimately immersed in the clayey overburden and resting on massive limestone. Due to this area's lithological heterogeneity, its remnant high and variable permeabilty, and the inherent problems in relation to hole stability, descending stage grouting was reckoned not to be the most appropriate system. An alternative technology in the form of tube-à-manchette* was agreed. Since this obviously constituted a radical change in method, it was decided to test its efficacy in the most westerly (and poorest) area, extending for a length of

*Also known as "Sleeved Pipe" e.g. Rigny (1974).

approximately 60m. The section between there and the major "descending stage" length was to proceed simultaneously with further descending stages, but with the effect carefully assessed by Lugeon test holes. There was thus the option of selecting tube-à-manchette for this middle section also.

4.4 Tube-à-Manchette: This system features the use of a plastic tube (Figure 7), with groups of holes drilled at regular intervals along its length, protected by rubber sleeves (manchettes). This tube is fixed into the ground with a "sleeve" grout - a weak clay-rich mix. Normally the ground is too unstable to support an "open hole", and has to be cased when drilling to permit tube installation. This casing is withdrawn as the sleeve grout is added. After a suitable period, grouting of the ground is then conducted via a double packer lowered through the tube to isolate each sleeve location in turn. The grout forces the rubber sleeve open, flows through the sleeve grout and then exits into the surrounding ground. Upon cessation of grouting, the sleeve recloses, and prevents ingress of injected grout. The system permits sleeves to be reinjected if required, and with a variety of materials.

During the first phase of (descending stage) grouting, the western area had 16 holes each with three to five 3m stages commencing from 9-15m below the surface (i.e. at interpreted rock head). Overall, the average lineal take was 359kg/m. Post-grouting Lugeon tests confirmed unstable holes of high permeability, with the great majority of test stages giving values above 20L. In addition, analysis of the test characteristics indicated "Wash out" phenomena (Houlsby, Class "D", Figure 6).

A line of 40 vertical tube-à-manchette holes was then located 1m upstream of the original, with an interhole spacing of 1.5m. Sleeves were located on the 48mm o.d. tubes at 50mm vertical intervals, and the assemblies installed in ODEX cased holes of 84mm diameter. A sleeve grout of composition (by weight) Bentonite:OPC:Water = 1:10:20 was used at an average consumption of 19kg/m of tube annulus. This mix gives a 28 day strength in the region of $4N/mm^2$. Alternative holes were designated Primary and Secondary, and the sleeves in each likewise. This permitted the progress of the grouting to be assessed over four initial phases of sleeve injections.

After initial experimentation, the maximum amount injected through each sleeve was limited to five batches (i.e. 250kg OPC, plus 5% bentonite by weight of water), provided the limiting flow pressure, for that depth of sleeve, had not been reached already. This refusal pressure was taken to be the same as in the descending stage grouting (i.e. approximately overburden pressure). Most sleeves required a higher initial pressure (up to 10 bar) to open them and fissure the sleeve grout, although some sleeves, owing to the "tightness" of the adjacent ground, scarcely accepted any grout at all. The grouting of each sleeve commenced with a mix of w/c ratio (by weight) 4:1, thickening to 2:1 as injections proceeded. Those sleeves which had accepted the original five batch injections were later selected for reinjection, to refusal (up to three further injections for some sleeves). 26% of the original 1013 sleeves were injected following the initial phase of treatment.

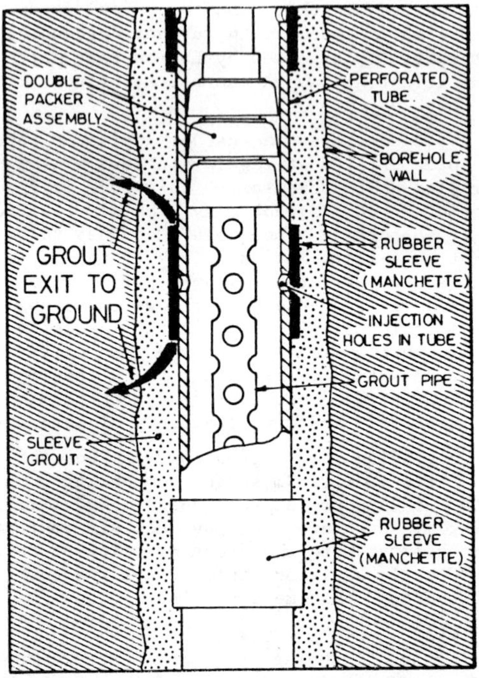

Figure 7.

Operating principle of tube-a-manchette grouting system.

Overall, 142 tonnes of cement, 14 tonnes of bentonite and 1356 sleeve injections were involved (average 114kg/injection). Regarding the four steps of the initial phase of sleeve injections:-

(i) Average sleeve takes decreased progressively (223 to 73kg) and satisfactorily (reduction ratios 63-76%).

(ii) The percentage of "maximum batch" sleeves decreased (42 to 10%), and those which were completely "tight" increased (16 to 35%).

(iii) The rate of grout acceptance decreased from about 590kg/hour to 310kg/hour.

(iv) The cement:bentonite ratio (by weight) decreased steadily from 12.4 to 7.1 mirroring the fact that fewer sleeves took the full five batches (ending in cement-rich mixes) as treatment proceeded.

After all sleeves had been subsequently grouted to refusal, Lugeon water test holes showed results in stark contrast to the pre-grouting values: a maximum value of 6L with half 1L or less. In addition, no hole stability problems were recorded during the "open hole" Lugeon testing, further underlining the effectiveness of the grouting method.

Meanwhile, in the adjacent area, 58 additional holes had been grouted in three phases of descending stage work, interspersed with Lugeon testing. Although a slight overall improvement in permeability and hole stability was achieved, particularly by the last phase of inclined holes, test stages of up to 38L were still encountered. These descending stage treatments had involved 747m of drilling and over 130 tonnes of materials (average consumption 174kg/m). These data, plus the notable success of tube-à-manchette in the first area, led to the adoption of the alternative technology in this area also. Using identical methods and procedures, post-grouting water tests on twelve stages showed eight of 1L or less, three of 2L and one of 4L. No collapse of holes was encountered.

5. Summary and Conclusions

Principally due to the quality of the bedrock, British dam grouting projects nearly always feature descending stage grouting. Stage length may vary depending on economic and technical considerations, but is usually 3m. Grout holes are drilled rotary percussive, with water flush, typically 50mm diameter, and depart from the vertical only for sound geotechnical reasons. Cement grouting is conducted with mixes most commonly in the range 8:1 to 0.5:1 (w/c by weight), although inert fillers may be added in the case of large takes, and small proportions of bentonite may be incorporated depending on the fissure characteristics. Refusal pressures, as measured at the top hole packers, are generally conservative by European standards, although trials are often conducted to optimise grouting procedures. "Split Spacing" techniques are employed, with due regard to lineal and areal consumptions, reduction ratios and final phase average consumptions. High speed grout mixers linked to fluctuating pressure ram pumps predominate. Grouting effectiveness is generally reflected in Relief Well flows, Piezometer readings, and simple stage water tests. More recently, the worth of independent Lugeon test hole schemes has been clearly indicated. Where the ground is too unstable even for descending stage work, tube-à-manchette grouting has proved extremely effective, usually linked with the ODEX method of poor ground drilling.

Acknowledgements

At Wimbleball Dam, the Client was the Wessex & South West Water Authorities, the Engineer Rofe, Kennard & Lapworth, and the Main Contractor Bovis Civil Engineering Limited. At Kielder, the Client was the Northumbrian Water Authority, the Engineer Babtie, Shaw & Morton, and the Main Contractor, Balfour, Beatty Fairclough Kielder Joint Venture Limited. At Grimwith, the Client was the Yorkshire Water Authority, the Engineer Rofe, Kennard & Lapworth, the Main Contractor Sir Alfred McAlpine & Son (Northern) Limited. The Author gratefully acknowledges permission to refer to these projects and adds that the conclusions presented in this paper are not necessarily those of the above-mentioned bodies.

References

Bruce, D.A. & George, C.R.F. (1981) Rock Grouting at Wimbleball Dam, Somerset. Geotechnique - In preparation.

Cope, F.W. (1978) Dam Grouting. Cement & Concrete Association Seminar, Fulmer Grange, Wexham Springs, England, January.

Champion, S., (1961) Correspondence to: Geotechnique, 11 (4) Dec., Page 365.

Deere, D.U. (1976) Dams on Rock Foundations - Some Design Questions in; Rock Engineering for Foundations and Slopes Conference II, Boulder, Colorado, August. pp55-86.
Doty, J.W. (1970) Standards in Evaluating Percussion and Rotary Drill Methods in Foundation Grouting. Kansas City, District Corps of Engineers, U.S. Army. pp28.
Gebhart, L.R., (1976) Foundation Grouting. Bur.Rec., Division of General Research, Concrete & Structural Branch. pp15.
Gourlay, A.W. & Carson, C.S. (1982) Grouting Plant and Equipment. This Conference - In print.
Houlsby, A.C. (1976) Routine Interpretation of the Lugeon Water Test. QJEG 9 (4), pp303-313.
Houlsby, A.C. (1977) Engineering of Grout Curtains to Standards. Jour. Geot. Divn., ASCE, 103(9), pp953-970.
Littlejohn, G.S. (1975) Acceptable Water Flows for Rock Anchor Grouting. Ground Engineering, 8(2), pp46-48.
Patey, D.R. (1977) Grouting Old Mine Workings at Merthyr. Ground Engineering, 10(8), pp24-27.
Polatty, J.M. (1974) Comments on U.S. Grouting Practices in Foundations for Dams Conference, Asilomar, California, March 17-21. pp47-55.
Rigny, P., (1974) European vs U.S. Grouting Practices in Foundations for Dams Conference, Asilomar, California, March 17-21. pp37.45.
Weyerman, W.J. (1977) Rock Conditions Improved through Pressure Grouting. Inst. Eng. Res. Foundation Kollbrunner - Rodio, Zurich, April. pp202.

Measure	S.I.	British or Conventional Metric Unit
Length	1mm 1m 1km	0.0394in. 3.281ft. 0.621 mile
Mass	1kg 1 tonne (≡1Mg)	2.205lb 0.984 ton
Volume	1 litre	0.220 U.K. Gal.
Pressure	$1N/mm^2$	$0.145x10^3 lbf/in^2$ $10.20 kgf/cm^2$

ALSO

1. To approximate w/c ratios by weight to w/c ratios by volume, multiply the former by 1.5.
2. 1 bar is approximately 14.5 psi. ($0.1N/mm^2$)
3. 1 micron (μ) = 10^{-3}mm.
4. 1 kg/m ≡ 0.672lb/ft.
5. 1 kg/m^2 ≡ 0.205lb/ft^2.

OPTIMUM WATER : CEMENT RATIOS FOR ROCK GROUTING

By Adam Clive Houlsby, M. ASCE.

INTRODUCTION

The production of a durable, satisfactory cement grout job in a rock foundation is, to a large extent, dependent on the water : cement ratios used.

Water is necessary to give the grout mobility, but if in excess, some can separate from the grout and may be trapped, leading to the formation of interruptions to the continuity of the cement filling, such as pockets, bubbles and bleed paths. The possible penalty for incompletely filled cracks is that subsequently seepage may pass along any continuous paths available and this presents the hazard of gradual leaching of grout from the paths. Cases are known where grout curtains have been completely leached away; some of these are described here.

The aim is, therefore, to use the minimum amount of water able to produce a sound filling within a reasonable radius of penetration from the grout holes. This paper has the purpose of indicating optimum w:c ratios for various situations. It is based on thousands of actual grout applications; details of some of these are shown.

The subject of w:c ratios has always been a controversial issue amongst grouting practitioners. The responses on the subject to (3) led the writer to present this paper.

THE DURABILITY OF GROUTING

Whenever flowing groundwater can pass through a grouted area after the grouting is complete, there arises the possibility of removal or weakening of the grout. If the water is aggressive, loss of most of the effectiveness of the grouting can even occur within a year or so if "watery" grouts were used; innocuous water might take many years in the same circumstances.

To appreciate how this can happen, Figures 1 and 2 show the types of bleed paths and pockets which can result from use of excessive water in the grout. These paths and pockets permit the passage of groundwater past grout surfaces which have a very weak bonding and can be easily attached chemically, and eventually can be "eaten" away. Loss of the surfaces permits attack deeper into the grout and the process can then continue until most of the grout is gone. In this type of attack the cementitious materials are removed in solution in the water and if this water emerges from the rock in a suitably quiescent environment, the cementitious component is liable to come out of solution and produce calcite formations resembling those in limestone caves. Figure 3 shows an example of this. The stalagtites, shawls, and other "limestone" formations in this tunnel emerge from pink granite and can only have come from the grout curtain which is located in a vertical plane 3 ft. upstream of this tunnel, and parallel to it. Over the past 30 years, the transported calcium has been gradually accumulating and is now estimated to approximate the initial volume of the placed grout. This grouting used w:c ratios ranging from 5.5:1 to 30:1 (by volume); the average was 12:1.

Engineer Specialising in Grouting, and Senior Surveillance Engineer, Water Resources Commission, North Sydney, N.S.W. Australia.

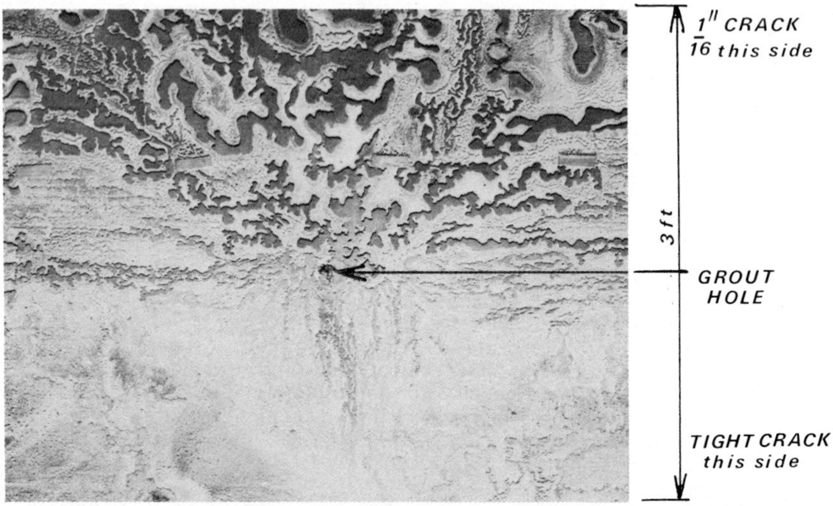

FIG 1 BLEED PATHS & POCKETS MEANDERING AMONGST SOLID FILLING IN ARTIFICIAL CRACK BETWEEN TWO HORIZONTAL SLABS OF CONCRETE. PADDLE MIXER, PISTON PUMP. REF (5)

W:C = 3:1

FIG 2 AS FOR FIG 1 EXCEPT THAT HIGH SPEED MIXER, MOYNO (MONO) PUMP USED. REF (5)

W:C = 3:1

The surface of watery grout films is akin to the weak, milky laitance formed on concrete which was too wet and was excessively vibrated. Laitance such as this on concrete is not usually durable against weathering; likewise weak surfaces on grout are liable to be impermanent. (4) gives a laboratory description of such surfaces obtained in grouting tests.

How then can durable grouting be obtained? It can be obtained by ensuring that negligible bleed water is trapped in the grout film. This is fairly easily achieved in vertical and near-vertical cracks, where density differences cause bleed water to float out of its own volition, but can be very difficult to achieve in near-horizontal cracks, particularly if they are not vented by numerous upward cracks.

FIG 3 CALCITE LEACHED FROM NEARBY GROUT CURTAIN. W:C= 5·5:1 TO 30:1. Average = 12:1

Some years ago, it was thought that moderate pressures were able to expel excess water from near-horizontal cracks almost immediately after injection and before thixotropic stiffening can set up resistance to migration of this water. However, research (4) (5) has subsequently shown that this does not happen, generally. This particular research did not extend to the high pressure range but fundamentals indicate that acceleration in water separation from grout freshly placed in a near-horizontal crack is unlikely to result from the existence of high pressure along the crack. This pressure does not necessarily pass through the denser grout; it is more likely to be transmitted near the roof of the crack through the less dense fluids which rise to there. Because bleeding of excess water is a gravity process primarily, it is very slow, commonly taking in excess of an hour for initial bleeding to get under way in temperate climates, and even longer in cold climates. Therefore, prolonged grout applications can be necessary if the high pressure "displacement" techniques of holding the crack un-naturally open are used, i.e. waiting for the bleeding and then expelling the products.

This paper relates to w:c ratios used with moderate rather than high pressures. Therefore, the grouts are relatively thick so as to introduce the minimum of water into the job.

When relating the foregoing considerations about durability, to grouting engineered to achieve designed standards of permeability in a curtain (3), it is important to appreciate that durability requires that grout fillings be solid, but that the various standards should be obtained largely by varying the radius of open cracking thus filled solidly from each grout hole. This is perhaps a difficult concept to grasp; it is perhaps best visualised by regarding each grout hole as the centre of a zone where groutable cracks should be solidly filled (this achieves the durability aspect), but the amount of ungrouted open cracking between these grouted zones is a measure of the standard of the grouting: i.e. for the relatively/ tight standard of say, 3 Lugeons (the Lugeon unit of permeability in grouting is explained at (3)), sufficient grout holes are placed to leave virtually no ungrouted cracks, whereas for a rather "leaky" curtain having a standard of 10 Lugeons, there could, in some foundations, be ungrouted gaps of about the same width as the solidly grouted area around the holes. This visualisation is a convenient generalisation for explanatory purposes, but its practical limitations in relation to the vaguarities of sites should not be forgotten.

The distinction should be noted between this approach to obtaining varying standards, and the cheaper alternative whereby low standard work can be produced by reducing the intensity of grout filling in the cracks around each hole. This happens (perhaps unintentionally) when relatively watery grouts are used to obtain extended penetrations, grout applications are concluded before firm refusal is reached, there is omission of further applications of grout following substantial takes, some rapid forms of upstage grouting are used, there is omission of regular bleeding from the top of the hole, and/or by use of ineffective mixers. Because the lesser standards produced in these ways do not necessarily result in solid filling of cracks, durability is suspect. Figure 4 shows two examples resulting from most or all of these. They were sketched from borehole periscope observations of the walls of vertical holes drilled after grouting for checking purposes. Note the voids where trapped bleedwater collected during the grouting and were unable to escape upwards. In both cases, the grouting was done with 5:1 grout. Figure 4(a) was observed at Blowering Dam in 1972 in an area where the grouting is disintegrating. It had been carried out in 1967-68. Some parts of this curtain are now completely disintegrated. Figure 4(b) was typical of wide near-vertical open joints at Copperlode Falls Dam. At the time of the observation water was continuously spurting out of these voids. Further grouting using 2:1 grout and improved techniques was subsequently carried out and has proved effective.

TERMINOLOGY

Before proceeding, it is necessary to define the terminology used for describing water:cement ratios. This paper, in accordance with customary U.S. practice, denotes them by volume in terms of loose volume of cement,

i.e. w:c = 3:1 is a grout made of 3 volumes of water to one of cement. This would be typically batched as 3 cu. ft. of water to 1 bag of cement.

The standard 94 lb bag of cement is traditionally assumed to have a loose volume of 1 cubic foot. This is a reasonable assumption and is eminently practical.

The philosophy behind the preference for volume batching is discussed in (2) together with the reasoning behind the recommended usage of litres of cement per centimetre for the metric equivalent of cubic feet of cement per foot. One of the practical reasons for this is the coincidence that both of these units give the same numbers for the take i.e. 0.50 cu. ft. of cement per ft. is actually equal to 0.50 litres of cement per cm.

Precision with specifying w:c ratios is not warranted. Recognisable differences in viscosity and rheology occur for the following ratios; use of any ratio intermediate between them is not realistic:-

0.5:1 (a viscosity like paint),
0.6:1, 0.8:1, 1:1, 2:1, 3:1, 4:1, 5:1,
7.1 10:1.

Only grouts made from Portland cement without any additives at all are dealt with in this paper. The use of Bentonite, sodium silicate or other additives to improve stability and penetration is not considered necessary in the writer's experience and, in any case, modern polymers appear to offer improvements over these older materials.

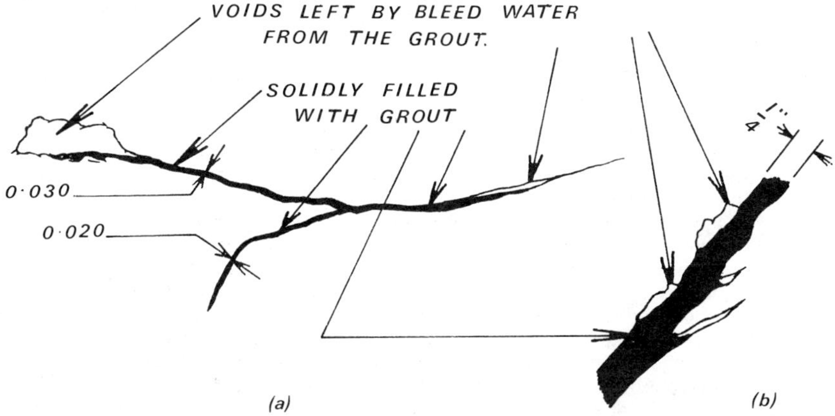

FIG 4 TYPICAL INCOMPLETE GROUT FILMS DUE TO USE OF EXCESSIVELY THIN GROUT, & UNSATISFACTORY INJECTION TECHNIQUES. W : C = 5:1 WAS USED IN BOTH CASES.

CASE HISTORIES

The following selection is regarded as representative of many thousands of holes grouted in the writer's practice and in the practice of others. The selection shows dam sites only but the principles apply substantially to other types of sites.

Grouting suffers from the inherent difficulty that the completed work can rarely be actually seen. Evaluation of its success (or otherwise) is therefore dependent on indirect and perhaps inconclusive performance monitoring of various types. With this in mind the following case histories have been grouped as far as possible as "successful" operations or as "not successful as was intended". Note the inability to give a definite "success" or "failure" verdict to grouting, usually.

In general the "successful" jobs were grouted with 3:1 grout or thicker. The others used 5:1 or thinner. This is not to say that all jobs grouted with 5:1 or thinner are, therefore, suspect, but is presented as a discerned distinction between the "successful" and the "unsuccessful" curtains in the limited number of cases known to the writer. Other practitioners may be able to instance further cases to widen the scope of the comparison.

"SUCCESSFUL" CASES

(A) COPETON DAM

This is a central earth core fill dam with a maximum height of 325 ft. (100 m) and a length of 5000 ft. (152 m).

The foundation is mainly coarse grained massive biotite granite with an inclusion of a small mass of fine grained granite. The rock is in tough, near-horizontal sheets and is tectonically stressed. Main joints are relatively widely spaced and are of considerable extent. Surface joints exhibit effects of stress relief and are frequently wide open. From grouting aspects, the foundation could be regarded as an extreme of which the example shown at (D) is the other extreme. The extreme illustrated by Copeton Dam is that of a foundation where grouting is relatively easy, is obviously needed, and has cracks so wide and extensive that they provide adequate penetration with no difficulty. In fact, penetration had to be curbed in some of the long horizontal cracks which would otherwise have conveyed grout to areas remote from the curtain. However, there were in addition fine, short joints which the grouting also had to cater for.

A single row grout curtain was constructed by stage grouting, using split spacing methods, with widely spaced primary holes then successively halving the distance between holes until eventually the desired standard of permeability was achieved in accordance with the principles of (3).

Figure 5 summarises the grouting. It shows a cross section of the dam and curtain to scale and gives data plotted on ordinates for each sequence of closure. The first ordinate deals with primary holes, the second with secondary holes and so on, and on each is the data relating to starting mixes and then to finishing mixes. This distinction between mixes is necesary because frequently the starting mix is thickened during a grout application and in such cases the starting mix is not really indicative of most of the application.

Most applications started with 2:1 grout but 1:1, 0.8.1 and even 0.6:1 were used where water testing immediately prior to grouting indicated the presence of cracks wide enought to accept these mixes readily and the geological investigation indicated absence of fine cracks which would be clogged by such a thick mix. 2:1 was the thinnest grout used.

In primary holes, 2:1 starting mix was used in 165 cases, 1:1 was used six times, 0.8:1 thirteen times and 0.6:1 twice. In secondary holes 97 cases were commenced with 2:1, four with 1:1, and two with 0.8:1. Then as closure progressed, the thicker mixes were needed less, until at the quinary stage (5 ft. or 1.5 m, spacing between holes), 475 of them were started with 2:1, twenty with 1:1 and none with thicker mixes. The total number of cases commenced at the various mixes is listed at Table 1.

The finishing mix on all these applications was either the same as the starting mix or was thicker. Some applications commenced with 2:1 and were found to require progressive thickening right through to 0.6:1 but the majority of thickening was through less of the range than this. As indicated on Figure 5, 114 cases in primary holes were finished with 2:1 (these had not been thickened during the application), 21 with 1:1, 33 with 0.8:1 and 18 with 0.6:1. Subsequent sequences were in similar proportion. Table 1 also gives in brackets the number of cases finished at the various mixes.

The case numbers quoted are for curtain grouting only, and include holes where surface leaks or connection occurred. They also include any multiple applications given at 24 hour intervals due to the take on the first, or on an earlier application, being sufficiently great to warrant further tightening grouting.

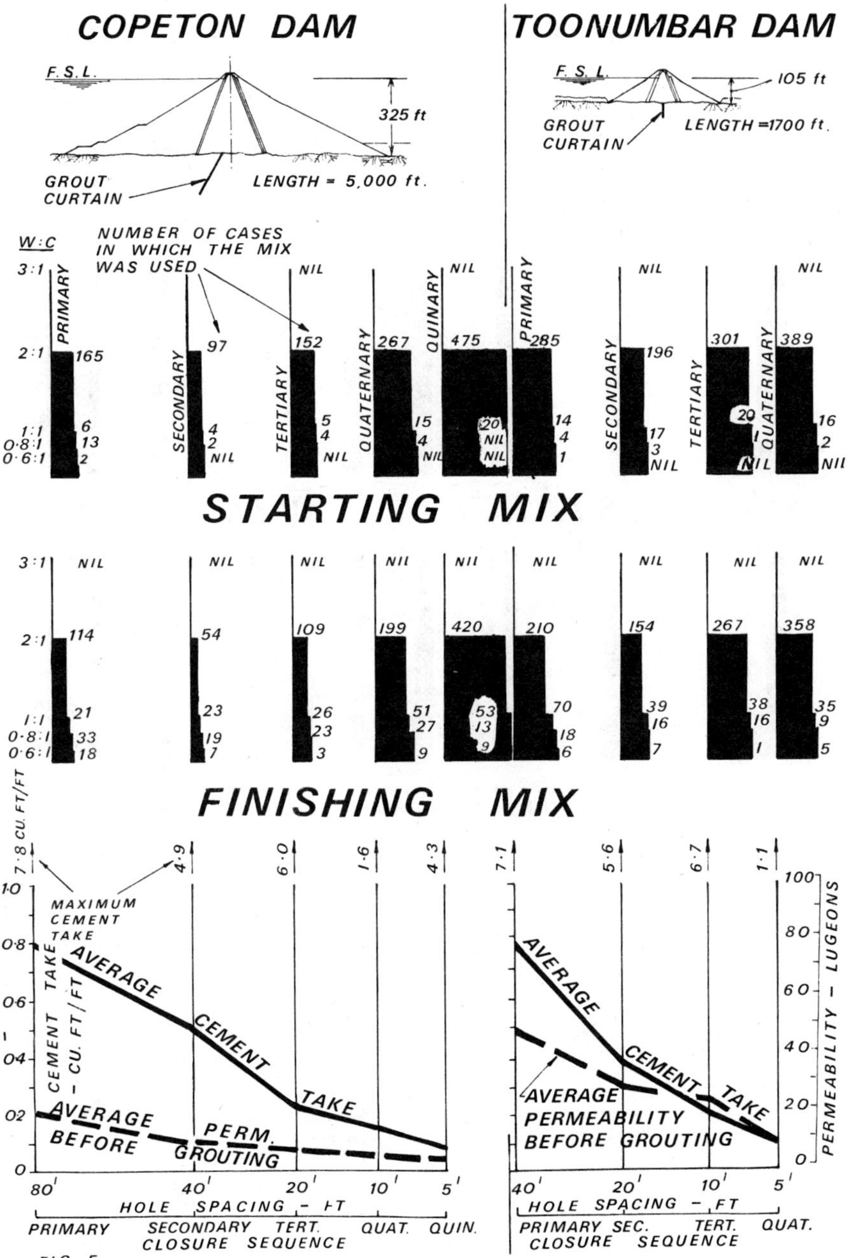
COPETON DAM
F.S.L.
325 ft
GROUT CURTAIN
LENGTH = 5,000 ft.
TOONUMBAR DAM
F.S.L.
105 ft
GROUT CURTAIN
LENGTH =1700 ft.
W:C
NUMBER OF CASES IN WHICH THE MIX WAS USED
PRIMARY
SECONDARY
TERTIARY
QUATERNARY
QUINARY
STARTING MIX
FINISHING MIX
MAXIMUM CEMENT TAKE
7·8 CU. FT/FT
CEMENT TAKE – CU. FT/FT
AVERAGE CEMENT TAKE
AVERAGE PERM. BEFORE GROUTING
AVERAGE PERMEABILITY BEFORE GROUTING
PERMEABILITY – LUGEONS
HOLE SPACING – FT
PRIMARY SECONDARY TERT. QUAT. QUIN.
PRIMARY SEC. TERT. QUAT.
CLOSURE SEQUENCE

FIG 5

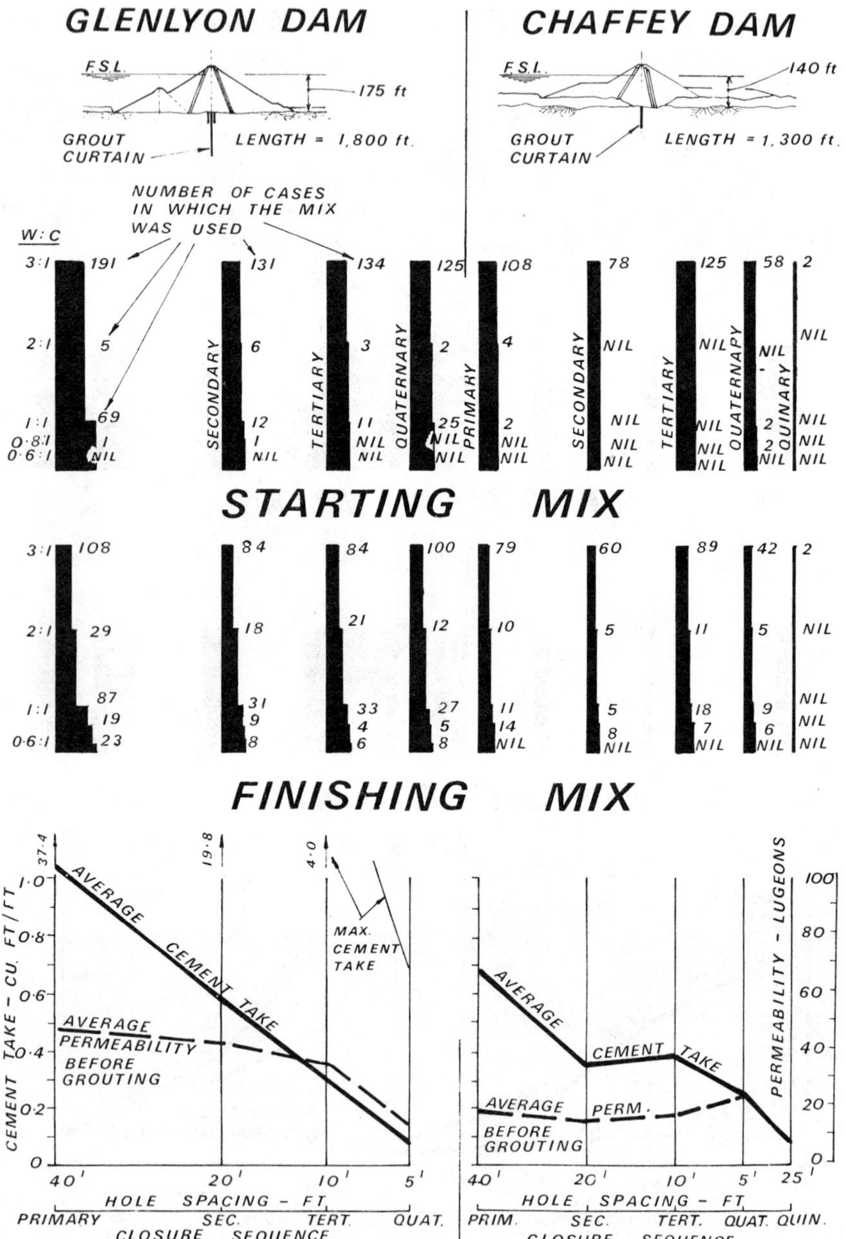
GLENLYON DAM
CHAFFEY DAM
F.S.L.
175 ft
140 ft
GROUT CURTAIN
LENGTH = 1,800 ft.
LENGTH = 1,300 ft.
NUMBER OF CASES IN WHICH THE MIX WAS USED
W:C
3:1
2:1
1:1
0·8:1
0·6:1
SECONDARY
TERTIARY
QUATERNARY
PRIMARY
QUINARY
NIL
STARTING MIX
FINISHING MIX
AVERAGE CEMENT TAKE
AVERAGE PERMEABILITY BEFORE GROUTING
MAX. CEMENT TAKE
AVERAGE PERM. BEFORE GROUTING
CEMENT TAKE - CU. FT/FT
PERMEABILITY - LUGEONS
HOLE SPACING - FT.
PRIMARY
SEC.
TERT.
QUAT.
PRIM.
QUIN.
CLOSURE SEQUENCE

FIG 6

At the foot of Figure 5 is a summary graph showing cement takes at the various closure sequences. The average take on primary holes was 0.79 cu. ft./ft. (= litre/cm. See Appendix II for comments on metric relationships) with a maximum of 7.8 cu. ft./ft. (= litre/cm). This reduced to 0.51 on secondaries with a maximum of 4.9. Subsequent closure showed progressive reductions although the maximum for quinary holes (4.3) showed a higher figure than earlier (1.6) because a number of these holes picked up isolated short joints. This event corroborated the need for closure to 5 ft. in order to produce a thorough curtain to the design standard.

The graph on Figure 5 also shows the average permeability before grouting, at each closure sequence. Before any grouting of primary holes at all, the average permeability was 21 Lugeons. The grouting of primary holes reduced this to 10 Lugeons before secondary grouting. This was reduced to 9 Lugeons by the secondaries, 8 by the tertiaries, 6 by the quaternaries and to an estimated 5 by the quinaries.

The penetration of grout from primary holes at 80 ft. (24 m) spacing exceeded 40 ft. (12 m) in a sufficient number of cases to appreciably reduce the demand in secondary holes.

The grout was mixed at high speed (greater than 1,500 r.p.m.) and was pumped with Mono (Moyno) pumps. Grouting techniques were generally in accordance with (6); some of these are later discussed.

The dam has been storing water since 1974. It shows negligible leakage, typically 0.07 c.f.s. (2.2 l/sec) for full storage conditions, and pore pressures are satisfactory.

TABLE 1

W:C RATIO BY VOLUME	NUMBER OF CASES COMMENCED (FINISHED) WITH THIS W:C RATIO:-			
	DAM			
	COPETON	TOONUMBAR	GLENLYON	CHAFFEY
(1)	(2)	(3)	(4)	(5)
3:1	NIL	NIL	581(376)	371(272)
2:1	1156(896)	1171(989)	16(80)	4(31)
1:1	50(174)	67(182)	117(178)	4(43)
0.8:1	23(115)	10(59)	2(37)	2(35)
0.6:1	2(46)	1(19)	0(45)	NIL

(B) TOONUMBAR DAM

This resembles Copeton Dam in cross section approximately but is only 105 ft. (32 m) high and 1700 ft. (518 m) long. On Figure 5 it is drawn to the same scale as Copeton Dam.

The foundation comprises various beds of fine, medium or coarse grained sandstone separated by thin lenses of fine conglomerate shale breccia or shale and laminite. The general strike is normal to the axis of the dam: dip is about 10°.

The rock is relatively weak, particularly in the laminated horizons. Cross bedding dips in variable directions at up to 15° from the true bedding. The beds tend to fracture along cross bedding causing local high angle discontinuity planes. The beds have been fractured by many near-vertical joints and piping of weak material from seams had to be considered when designing the grouting.

From considerations of ease of grouting, the foundation could be regarded as intermediate between examples (A) and (D).

It has a single row curtain constructed by closure methods and the effect of its relatively tight curtain on seepage pressures is discussed at (3), which at its Figure 5 shows pressure profiles through the dam and indicates why a tight curtain of 2 Lugeons is necessary.

The grouting data is presented here on Figure 5 in the same way as for Copeton Dam. The grouting was all at 2:1 or thicker.

285 of the primary holes were grouted with 2:1 as the starting mix and 14 with 1:1. The joints were too fine to require thicker starting mixes than this apart from 4 cases of 0.8:1 and one case of 0.6:1. This pattern was similar in subsequent sequences. Thickening during grouting resulted in finishing mixes in primary holes of 1:1 in 70 cases, 0.8:1 in 18 and 0.6:1 in six. Table 1 gives total number of cases for each of the starting and finishing mixes.

The graphs at the foot of Figure 5 show the reductions in cement takes and permeability which occurred as a result of closure. The average cement take in primaries of 0.79 cu. ft./ft. (= litres/cm) indicated a ready degree of acceptance of the 2:1 grout in the relatively fine jointing.

Grouting equipment and methods were the same as for Copeton Dam. The dam has been storing water since 1971 and is behaving satisfactorily.

(C) GLENLYON DAM

This is another central core, earth fill dam. Dimensions are shown on Figure 6, where its cross section is drawn to the same scales as those of Figure 5.

Indurated fine grained claystones form the foundation, with thin quartzite bands and frequent calcite and quartz veins. Secondary structural features cause discontinuities such as Breccia zones, shearing and multi-directional jointing.

Overall, the foundation could be regarded as fairly weak structurally. It is intensely jointed in near-vertical and near horizontal patterns, some of which are associated with narrow shear zones. Preliminary grouting was carried out to tighten some exceptionally wide cracks. After that its groutability was about the same as example (B).

3:1 was the thinnest mix used here. The details are on Figure 6 and Tables 1. These include the Preliminary grouting. Closure produced the unusual situation where the plot of take-reduction is almost a straight line. Maximum cement takes on primary and secondary holes were unusually high at 37.4 and 19.8 cu. ft./ft. (= litre/cm).

The pattern of permeability reductions during the closure showed that, on average, tertiary sequence was the first which produced substantial change. Tertiary holes brought the spacing down to 10 ft.; penetration was of the order of 5 ft. therefore the subsequent (quaternary) holes encountered lower permeabilities than earlier holes had. Even so, the permeability prior to quaternary grouting was 15 Lugeons average and therefore the quaternary grouting was essential to bring the permeability down to the desired standard.

Equipment and methods were the same as for Copeton and Toonumbar Dams. The dam is behaving satisfactorily. It has been storing water since about 1977, but has not yet (February 81) completely filled.

(D) CHAFFEY DAM

This dam is included because it illustrates a somewhat extreme situation where penetrations were untypically short because of fine cracking.

As Figure 6 indicates, closure produced little effect until after tertiary sequence was reached, but because the aim of this particular grouting was somewhat unusual in that it did not have the purpose of producing a low permeability curtain but had the purpose of dealing with significant cracks only, very little grouting was carried on beyond the tertiary sequence, which was at 10 ft. (3 m) spacing. The foundation had an average permeability of 19 lugeons before grouting of primary holes. The increase to 24 lugeons measured in quaternary holes before grouting, as shown on Figure 6, does not result from hydraulic fracturing during previous sequences but is merely an example of the vicissitudes produced by the statistical averaging of grout takes. Averaging is never a wise approach considering the wide ranges experienced in grout takes, but is used here of necessity for illustrative purposes. Quaternary and quinary holes were used only where takes in tertiaries warranted them. The overall permeability on completion of the grouting is, of course, rather variable in such a situation. It appears to be of the order of 15 lugeons which is much the same as before grouting and raises the issues as to why do the grouting at all! The purpose was, as indicated above, to deal with discrete significant open joints. The criterion for continuation of closure sequences was based on cement takes rather than on permeability and was set at 0.5 cubic feet of cement per foot.

The geology of the foundation can be simplified into bars of tough, almost tight, vertically dipping massive jasper, interspersed with areas of weak siltstone and chert. The strike of beds approximates the axis direction. Seepage emerges through deep alluvial beds without known hazards to the stability of the dam and seepage losses of several c.f.s. are tolerable.

The range of penetration from many of the grout holes possibly averaged only a foot or two, hence in the sections where closure ceased at 10 ft. spacing there conceivably remain ungrouted cracks, possibly wide enough to warrant grouting but not accessed from the grout holes. A closer spacing of final holes would be necessary to thoroughly grout all groutable cracks.

This dam is listed here amongst the "successful" cases, but at the time of writing (February 1981), this verdict is a presumption because a sustained drought has meant that the dam has accumulated very little storage. The relatively high permeability of the curtain permits core pore pressures upstream of it to exhibit unusual behaviour in which they drain to the foundation there and are equal to about tailwater head rather than storage pool head.

From the grout technology point of view this foundation is of the type which might suggest that use of very thin mixes would have been better than those actually used, or to some European practitioners it might suggest the use of displacement high pressure techniques to burst open long lengths of cracks and then gain wide access from few holes. When deciding the w:c. ratio to be used, the durability of the completed work was regarded as paramount, and hence 3:1 and thicker mixes were used for reasons described in this paper. Realistic improvements in penetration would have required the use of mixes of about 10:1 or thinner; the use of these is regarded as a wasted effort in terms of long term efficacy. The displacement method is regarded as somewhat hazardous, disrupting as it does the natural bonding in near tight rock (which would not have warranted grouting if undisturbed) and opening up, in this site, numerous near vertical cracks along the axis, most of which would leak grout embarrassingly to the surface. Hence the grouting was carried out by normal methods, using 3:1 or thicker.

(E) WARRAGAMBA DAM

This is a concrete gravity dam, the highest in the southern hemisphere with a height of 450 ft (137 m) and a crest length of 1152 ft. (351 m). The foundation is horizontally bedded sandstone mainly, exhibiting near-vertical and horizontal cracking with openings varying widely from almost tight to several inches. The larger openings were due to tectonic stress relief manifesting itself in rebound of the valley floor and block separation of the canyon walls before and during grouting.

3:1 or thicker was used for all grouting. This was a most extensive grouting operation comprising several curtains and including shallow blanket grouting of the whole foundation. The total length of holes was 1,117,000 ft. (340 000 m) and the total cement injected was 293,850 cu. ft. (8 320 000 litres). The average take was 0.26 cu. ft. of cement/ft (= litre/cm).

The grouting was carried out between 1953 and 1960 using paddle mixers and reciprocating pumps of various types, although towards the end of the job, the deficiencies of this equipment led to experiments with better equipment such as subsequently has been generally adopted elsewhere. An unusually great number of grout holes was necessary in a number of locations in order to compensate for these equipment deficiencies.

The dam has been storing water for 20 years approximately and is performing satisfactorily.

CASES WHERE GROUTING WAS NOT AS SUCCESSFUL AS WAS INTENDED

(A) BURRINJUCK DAM

This is an old concrete gravity dam which leaked so much through the concrete that it was grouted over the period 1936 to 1948. The grouting comprised a 3 row curtain drilled from the crest through the dam and into the foundation. The old concrete was of poor quality and cracks in it took most of the grout. In the initial curtain row, the average cement take was 0.18 cu. ft./ft. (= litres/cm) with a maximum of 8.1 (in concrete). The later two rows took less.

Water:cement ratios were 5.5:1 to 30:1 with an average of 12:1.

As illustrated at Figure 3, the grout curtain in the rock is being gradually leached away by seepage which then emerges in the tunnels and deposits the leached calcium in the form of stalagtites and shawls. This grouting appears to have been non-permanent because of the use of excessively thin grout. However, remedial work is not necessary in this case because the foundation was really too tight to need grouting in the first place, and because the tunnels provide adequate drainage without assistance from grouting.

(B) KEEPIT DAM

This is a modern concrete gravity dam with conventional single row grout curtain and drainage hole arrangements. The foundation is finely jointed conglomerate and tuff and the curtain grouting was commenced with thin grouts, such as 8:1. However, it was found that this grouting was ineffective in reducing uplift pressures, probably partly because seepage passing through the foundation at the time of grouting washed much of the grout away before initial set occurred. Tests indicated that initial set for the 8:1 grout could take several days, and even then the resulting product was very weak.

A limited test in situ using 0.8:1 in new holes alongside those using 8:1 average, was able to inject the same small quantities of cement in the cracks as had been injected with 8:1. The initial setting time of 0.8:1 is only a matter of hours and assisted as it is by thixotropy, it resisted removal by seepage.

(C) HUME DAM

This is another concrete gravity dam and was constructed initially between 1919 and 1936. By 1954, some concern for uplift pressures then existing, together with modifications for increasing the storage, prompted attempts to construct a grout curtain. The foundation is granite with occasional shears and jointed areas.

A conventional curtain was attempted at various locations along the dam using 20:1 grout. This mix was adopted because, at the time, it was believed by those concerned that grout thicker than this would not penetrate in the very fine joints. A small amount of bentonite was used in the grout for lubrication purposes.

However, after attempts in several short areas it soon became apparent that the grouting was producing no effect on uplift pressures. Some foundation drainage holes were then drilled as an alternative and were so successful in reducing pressures that further grouting was discontinued. Instead, curtains of drainage holes were drilled.

In retrospect, it seems probable that a mix of about 3:1 would have been successful. However, curtain grouting is obviously superfluous at this site, in general, in view of the efficacy of drainage.

(D) BLOWERING DAM

Some parts of the curtain at this dam have disintegrated. The grouting used 5:1 mix mostly and techniques were not the best. It should be noted, however, that some parts of the curtain remain effective as illustrated at Fig. 4 of (3) and discussed there for a particular section which has 10 lugeons permeability and is, therefore, relatively leaky. Although this is a greater permeability than the design intended, it is producing an almost ideal pattern of pore pressure relationships and this pattern satisfactorily repeats itself during each cycle of storage filling and depletion.

SELECTION OF OPTIMUM W:C RATIOS

With the lessons of these case histories, and others, in mind, we turn now to their application to the selection process when determining optimum mixes for future jobs.

The selection should be based on a combination of experience and of review during the progress of the grouting. Case histories assist with experience, but must always be adapted to the variations needed to satisfy each fresh job.

The process of deciding what w:c ratios to use for each grouting application involves -

(a) selection of starting mix based on information obtained from water testing carried out in the grout hole, taken in conjunction with geological information such as the size of joint openings, their orientation etc. and experience from ealier holes, then

(b) once the grouting has commenced, selection of successively thicker mixes depending on how well the hole has accepted the grout.

A brief description of this selection process follows:-

Selection of Starting Mix

In the majority of sites, the writer's experience has indicated that 2:1 is usually suitable for first trials. Obvious exceptions to this are where -

(i) the cracks are relatively fine, such as 0.030 in (0.75 mm) or finer .... start with 3:1 rather than 2:1

(ii) the cracks are fairly wide, such as 0.50 to 0.100 in (1.25 to 2.5 mm) .... start with 1:1 rather than 2:1

Surface leaks or connections are another reason for starting with 1:1 or thicker.

Note that realistically, the choice of mixes is limited to 3:1, 2:1, 1:1 or thicker, as mentioned earlier. It is considered unwisely pedantic to use intermediate ratios such as 2.5:1 or 1.5:1.

Selection of Thickening Mixes

After grouting of an application commences, it is necessary to decide whether the mix should be thickened or not. If the hole has accepted the starting mix slowly and it is apparent that thickening would choke it off prematurely, then the starting mix should be retained until refusal is reached.

If, however, the hole is taking grout at an appreciable rate, thickening might be carried out to the next thickening step . Further thickening after this may be warranted through several steps to quite a thick mix if large cracks are indicated.

By taking measurements of the grount taken in regular intervals, such as every 15 minutes, and immediately comparing the amount taken with the preceding amount, the strategy for the following 15 minute run can be quickly decided and then implemented. This strategy relates to possible thickening of the mix and changes in the applied pressure. Presuming that departure from straightforward grouting is not occasioned by a need to deal with leaks etc., the manner in which the 15 minute takes can indicate appropriate strategy, is based on a desired situation where the hole takes a little less in each successive run until the take gradually ceases. If, instead of this happening, there are -

(a) equal takes for protracted periods and these are about equal to the initial 15 minute take and continue for, say, three more runs without diminishing, a thicker mix will probably be accepted readily and would produce a better quality (and hence more durable) grouting job. If thickening of the mix is not carried out, a patchy job can be expected, taking much longer and costing more than if it were properly thickened.

(b) Sudden reductions in take, where the application has proceeded happily through the initial run or two with a reasonable take (in terms of its permeability) and then, as a result of a subsequent thickening, suddenly slows up or even stops. The thickening has choked off the hole and was inadvisable. Note, however, that after an hour or two of grouting, sudden take up is frequently experienced without any thickening of the grout as refusal approaches, and is a satisfactory occurrence at this juncture.

(c) Sudden increases in take, when the opposite to (b) occurs. The cause is probably rock movement. Treatment must be prompt and appropriate (unless the displacement method is in use) and, although outside the scope of this paper, is, in brief, a matter of thickening the mix, reducing the pressure and restraining the amount of energy conveyed into the grout hole.

To achieve the ideal gradual reduction of takes at each succeeding run, the feel of the hole should be gauged, and thickening then carried out only as prudent to avoid an excess of (a), (b) and (c) and to be within the general guidance of the considerations discussed.

The setting of the cement, once it proceeds sufficiently to increase shear strength and viscosity, come to appreciably dominate the injection process and finally dominates all other factors. Ideally, the bulk of the cement should be injected within an hour of commencement, in average crack sizes.

The ideal procedure is to commence grouting with a starting mix which is sufficiently thin to penetrate a few feet into the finest groutable cracks anticipated to be cut by the grout hole. Where wider cracks than this comprise more than about 30% of the anticipated crack population, it is necessary to thicken the mix after sufficient of the starting mix has been injected to fill the finer cracks, otherwise too much water will be injected and this excess water can produce gaps in the grout filling if trapped. Further thickening beyond this is warranted if relatively wide cracks are also present.

CONCLUSIONS

(1) w:c = 2:1 by volume is regarded by the writer as the optimum basic mix. It has been successfully used in thousands of cases.
(2) This basic mix should be varied to suit site conditions, such as using 3:1 in fine cracking, or thicker mixes in wide open joints.
(3) 5:1 mix and mixes thinner than this, have been known to produce non-durable grouting.
(4) Non-durable grouting can result from incomplete grout filling such as can be caused by trapped bleed pockets and meanders. These can be minimised by using the thickest grout possible consistent with reasonable penetration.
(5) After commencement of a grout application, benefit can be obtained from thickening the mix during the course of the injection.

ACKNOWLEDGMENTS

Particular thanks are due to Mr. Jim Warner (Consulting Engineer) Mr. Ray Cope (National Grout Specialist, U.S. Soil Conservation Service) and members of the U.S. Water and Power Resources Service (former U.S.B.R.) for their helpful comments on drafts of the paper. Appreciation is also expressed to the Water Resources Commission, N.S.W., Australia, for much of the data used.

APPENDIX I - REFERENCES

1. Cambefort, H., "The Principles and Applications of Grouting", Quarterly Journal of Engineering Geology, Vol. 10, The Geological Society, London, England, Vol. 10, 1977, pp 57-95.

2. Houlsby, A.C., "Towards Appropriate Metric Units for Grouting", Ground Engineering, Vol. 10, No. 5, July 1977 Foundation Publications, Brentwood, Essex, England, p. 39-42

3. Houlsby, A.C., "Engineering of Grout Curtains to Standards", Journal of the Geotechnical Engineering Division, ASCE, Vol. 103, No. GT9, Proc. Paper 13214, Sept. 1977, pp 953-970.

4. U.S. Army Corps of Engineers, "Pressure Grouting Fine Fissures", Technical Report No. 6-437, Waterways Experiment Station, Vicksburg, Miss., October 1956.

5. Water Conservation and Irrigation Commission, "Grouting Tests". Unpublished 1962. N.S.W. Australia.

6. Water Resources Commission, N.S.W., "Grouting Manual", Third Edition, 1980, W.R.C., North Sydney, N.S.W. Australia, ISBN 0 7240 3786 1 (Also available from University of Missouri-Rolla, Civil Engineering Dept.).

GROUTING AT CLARENCE CANNON DAM: A NEW APPROACH

by

Michael J. Klosterman, Michael M. Easterly, and Nels G. Jahren 1/

ABSTRACT

The discovery of large-scale solution features during excavation for Clarence Cannon Dam, Missouri, resulted in a redesign of the grouting program. This new approach to the design took into account not only these newly discovered geologic features but also recent litigation regarding "standard" grout specifications, possible lack of grouting experience by Government or Contractor personnel, and results of a recent grout testing program conducted at the Meramec Park damsite in Missouri. This input produced a unique set of specifications in which a detailed approach to grouting was taken.

A computer system was developed in which daily grouting data could be entered into an on-site computer terminal which tabulates quantities, calculates and compares grouting variables among holes, and prepares and graphs parameters which reflect grouting effectiveness.

Readvertising the grouting under new specifications produced a bid significantly lower than the price negotiated under the old contract. There were no major problems with the new specifications. The results indicated that approximately 15 percent of the holes intercept rock defects which take a considerable amount of grout. The remaining holes take no grout whatsoever. Pressure test results in exploratory holes indicated that a positive cut-off was achieved.

INTRODUCTION

Rapidly changing construction philosophy and technology coupled with a long project completion time resulted in a series of mid-contract design changes at the Clarence Cannon Dam in northeastern Missouri. Located on the Salt River, 63 miles upstream from the Mississippi River, the 2,000-foot long, 138-foot high combination concrete and earth fill dam will impound a 30-mile long, 38,000-acre lake (Figure 1). The reservoir is intended to provide

1/ Engineering Geologists, U.S. Army Engineer District, St. Louis.

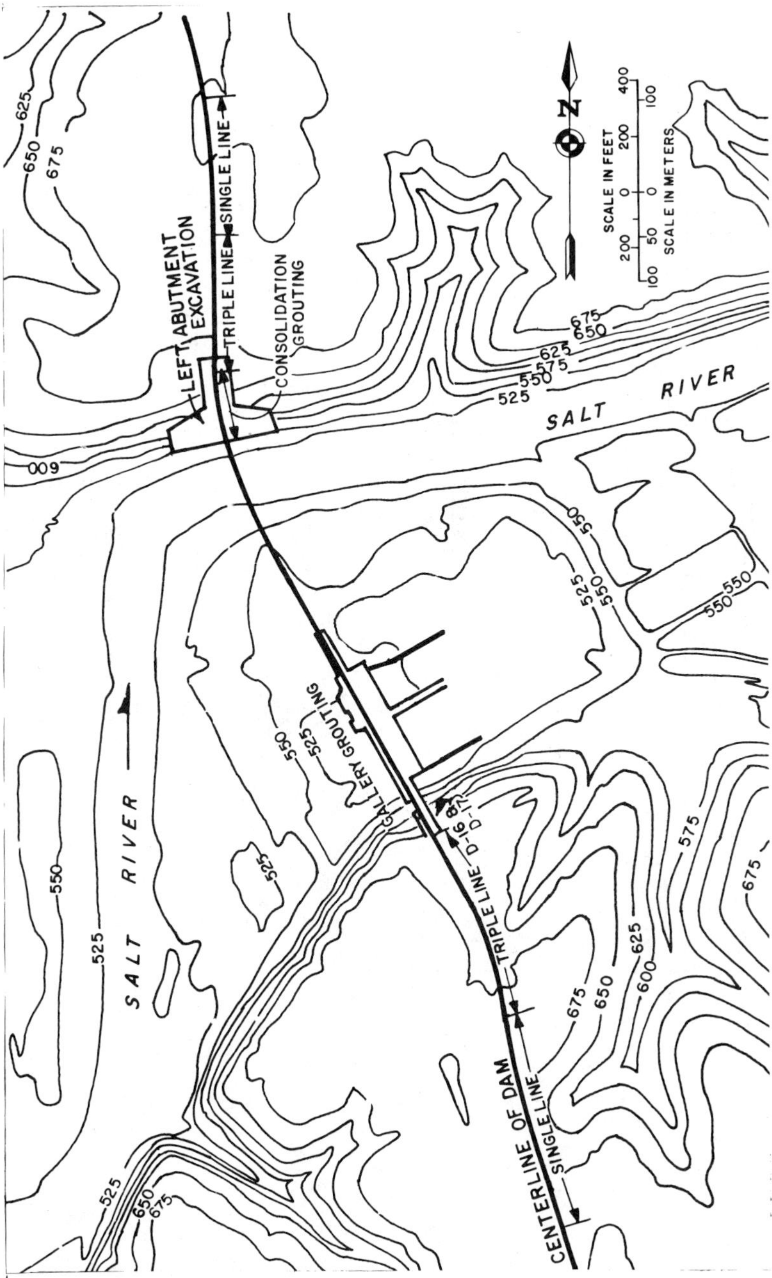

Figure 1. Site Plan. Areas of curtain grouting indicated.

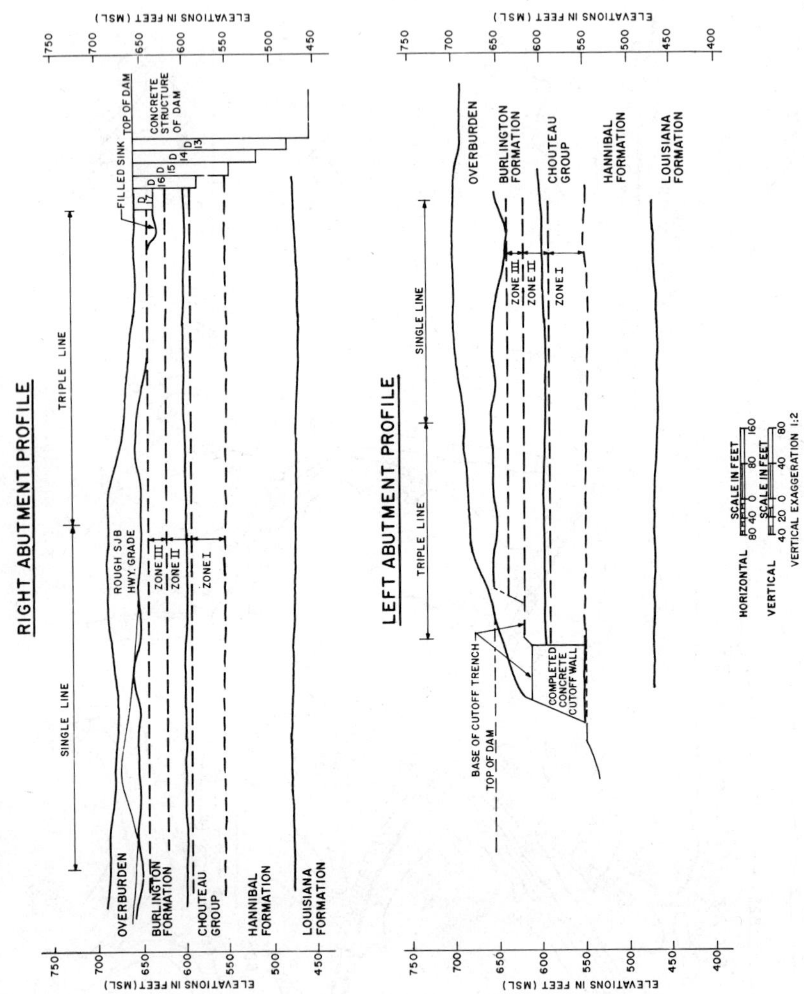

Figure 2. Site Profile. Solid lines mark formation contacts; dashed lines grout stages (zones).

flood control, hydropower, recreation, and wildlife conservation benefits. A small dam 10 miles downstream will provide pumpback storage. Construction began in 1970 and is currently 90 percent complete, with final closure expected in fall, 1982.

FOUNDATION CONDITIONS

The Cannon dam site is located in the Disected Till Plains Section of the Central Lowlands Physiographic Province. The local geology consists of essentially flat-lying lower Mississippian sedimentary formations (Figure 2). The uppermost unit in the abutments is the Burlington limestone formation, which contains a silty, cherty upper member and 20-foot thick, pure, massive lower member. The Chouteau Group underlies the Burlington and is composed of thinly bedded limestone with silt and shale content increasing toward a gradational basal contact with the Hannibal Formation. The Hannibal Formation is a siltstone whose air-slaking properties have resulted in strict requirements for exposure time during construction. The lowest unit exposed at the site is the Mississippian-Devonian Louisiana limestone, which occurs only in a buried channel under the Salt River and in the excavations for the right abutment concrete monoliths.

The uplands bedrock surface is typified by general pinnacle weathering and occasional paleo-sinks filled with Pennsylvanian residuum. Relief jointing of the abutments parallel to the valley slopes is common, both in the shale and the limestones. High-angle joints in both the relief and regional sets increase in depth toward the valley slopes where they occasionally extend through the limestones into the shale, and may be solutioned and filled with residual clay.

No major foundation defects were encountered in the pre-design exploration or in the right abutment excavations which were commenced in 1973. Solution-widened and clay-filled joints encountered on the excavation surfaces were dentally excavated and backfilled with concrete, and it was assumed that any similar openings occurring below the structure would be filled during grouting.

Excavation began on the left abutment in 1977 and solutioning there was found to be much more intense. Two major vertical cavities were uncovered, each 10-30 feet wide and 30-50 feet deep, transecting the cutoff trench. Additional exploration delineated four more cavities, as well as upstream-downstream extensions of the first two (Figure 3). All of the cavities appeared to be formed by phreatic solutioning of joints subparallel to the abutment slope, with well-developed vertical pipes occurring through the highly soluble lower Burlington and upper Chouteau limestones.

After the left abutment cavities were thoroughly investigated, the main dam construction contract was modified to provide for additional treatment. The cutoff trench was deepened to 40-50 feet, and a 12-foot wide, 60-foot deep concrete cutoff wall was installed from the base of the cutoff trench to the top of the Hannibal Formation. Consolidation grouting in the cutoff trench, a triple line grout curtain landward of the cutoff wall, and a sand blanket under the downstream embankment were also added, and the main grouting program was redesigned.

DETAILED SPECIFICATIONS

The grouting work originally was to be performed as part of the plans and specifications for the main dam and spillway awarded in early 1973. The grouting work, which was subcontracted, was begun in the fall of 1977. The grouting specifications, which followed the Corps of Engineers' guide specification, left many technical considerations open to the interpretation of the Contracting Officer. This type of loose specification is successful when mutual respect and trust exists between the Contractor and the Government, when the anticipated foundation conditions and quantity of work are met, and when both parties are adequately staffed with experienced grouting personnel. At the Cannon project, there were significant disagreements over the specifications to the detriment of the quality of the work. The Government decided to limit to a minimum the grouting done under the initial contract and readvertise the work under a separate contract. New specifications were prepared to eliminate the problems of the previous contract. The new specifications also took several other factors into account. One major factor the new specifications had to take into account was the discovery in 1977 of large scale solution features in the left abutment. These features, which have been previously discussed, injected conservatism into the grouting procedures developed for the new specifications. A second factor was a 1978 court decision involving measurement and payment of grouting at Corps of Engineer dams on the Fresno River, California. This decision required a restructuring of the traditional sub item method of presenting grouting quantities, and placed increased importance on developing an accurate Government estimate. A final influencing factor was results obtained during a 1976 grouting test conducted by the St. Louis District in carbonate bedrock. This test involved pregrouting and postgrouting pumping tests from water wells completely surrounded by a 50-foot square grout curtain (Figure 4). This test provided quantitative data on seepage through a grout curtain as well as the effectiveness of various mixes, pressures, and procedures (Klosterman, 1981).

After considering the above factors and the original contract problems, the St. Louis District decided to prepare a more precise specification. Detailed procedures were developed based on the grouting experiences of St. Louis District and other Corps of Engineers' personnel, grouting consultants, and the literature. This experience was adapted to the foundation conditions known to exist at the Cannon damsite. This specification enumerated in detail the exact procedures to be used for all phases of grouting including the following: (1) equipment operation, (2) sequence of drilling and grouting, (3) pressure testing, (4) raising and seating the packer, (5) initial hook up, mix and pressure, (6) injection of

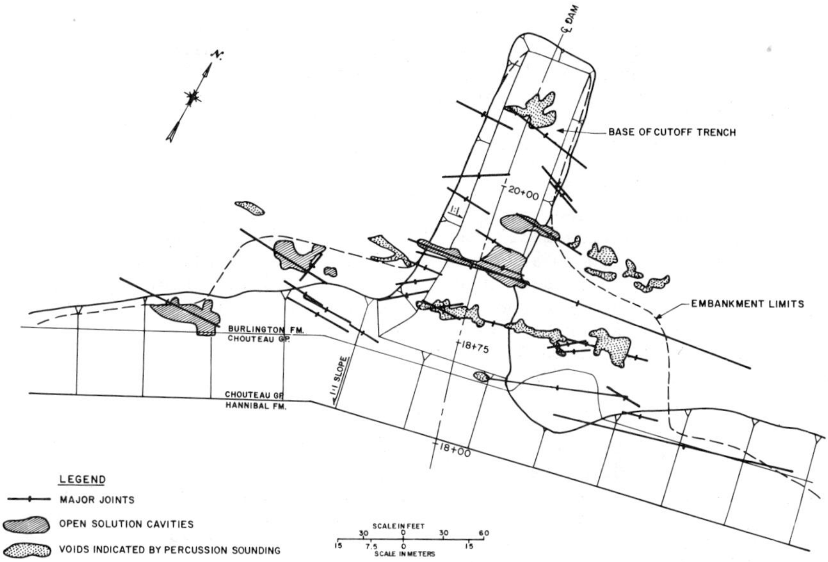

Figure 3. Plan View of Solution Features.

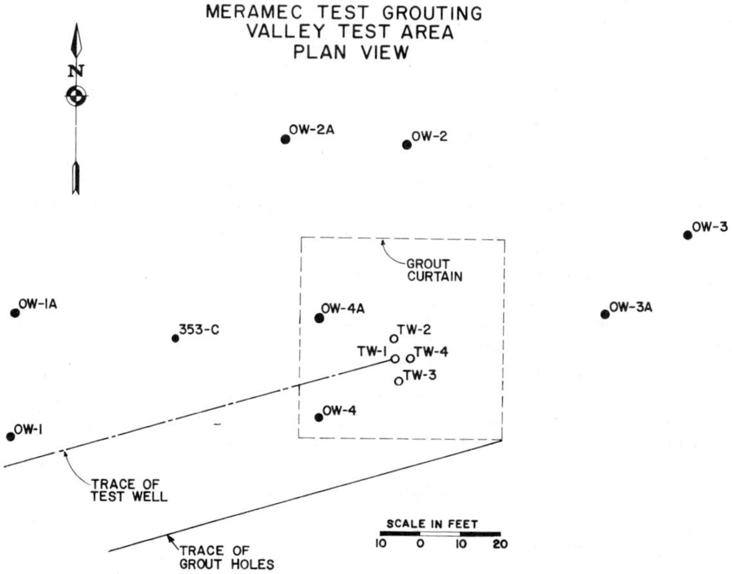

Figure 4. Test Grouted Square. Meramec Park Dam. Test well (TW). Observation well (OW).

grout, (7) when and how to change mixes and pressures, (8) maximum and minimum quantities of grout to be injected of any specific mix, (9) when to add sand and other admixtures, (10) intermittent grouting, (11) unusual conditions, (12) reaching refusal. Just as experienced individuals may treat the same foundation condition slightly differently, these specifications may not employ the exact procedures everyone would use in each case. However, the chance of a serious mistake occurring while following this specification is extremely low.

A detailed specification has several additional advantages over a more general specification. The Contractor is able to make a lower bid since he will know in advance what will be acceptable procedure. Future claims will be reduced since areas of possible contention between the Government and the Contractor have been reduced. A full staff of highly experienced grouting personnel is not needed, since the exact procedures to be followed are enumerated in the specifications.

The test of any specification is the results achieved under the contract. One positive result noticed immediately was that three of the four contract bids were lower than the Government estimate. Another positive result was that the St. Louis District has not found it necessary to hire field personnel with grouting experience. Finally, there were no modifications required or significant claims filed. Elston (1958) has argued that the very nature of groutable features makes it impossible to formulate procedures in advance to cover all potential problems. He states that "to establish pressures, water/cement ratios, and mix changes in advance is to defeat the purpose, compound the confusion, and smother the intent of the work." The St. Louis District, Corps of Engineers did not share this philosophy and felt that detailed grouting specifications could be developed to effectively grout all anticipated foundation conditions - and still remain easy to understand and follow. To insure this, three guidelines were followed in preparing the specifications: (1) Obtain significant subsurface information; (2) Knowing the foundation exploration results, assume the worst conditions when detailing procedures; (3) Have some one with grouting experience and a thorough knowledge of the foundation conditions prepare the specifications. It is our opinion that this specification has overcome Mr. Elston's objections. During the initial six months of grouting, the St. Louis District retained a geotechnical firm to evaluate the effectiveness of the specifications. Their report stated, "the detailed specifications were very beneficial . . . had a definite advantage over a performance specification . . . grouting was accomplished with less difficulty [than with a standard specification]."

GROUTING OPERATION

Equipment and Materials.

Grout holes were drilled through overburden and the upper Burlington chert beds by percussion with air circulation, and then by rotary methods and water circulation to final depth. The predominant piece of equipment used for drilling grout holes in rock was an air-powered rotary drill on air tracks. Stepped plug bits were found to have the highest efficiency in drilling the required NX-size hole in limestone. Use of a down hole hammer was allowed for penetration of the upper Burlington chert beds.

Each grout plant consisted of a mixer, an agitator tank, and a progressive cavity pump. Although a paddle mixer was allowed by the specifications, the contractor elected to use a mixer equipped with a high-speed pump. The agitator tank was equipped with rotating paddles. Each entire plant was air or hydraulic-powered. Pneumatic packers were used, with 1-1/2-inch grout lines, except in consolidation grouting where mechanical packers were employed.

Grout mixtures consisted chiefly of portland cement mixed with water from a supply well in the Salt River alluvium. Bentonite (2 percent by weight) was added to mixes of water-cement ratios of 5:1 or higher. Fluidifier was added in the same proportions to mixes of 1:1 or thicker. Sanded grouts were composed of natural sand, cement, water, and 1 percent fluidifier.

Procedure.

The grout curtain was divided into sections 120 feet in length, which were grouted as separate units. The basic curtain grouting strategy was to drill and grout holes of the same order in alternating sections before proceeding to higher order holes. Drilling was not permitted in any section while grouting was in progress. First, primary holes were drilled on 40-foot centers and grouted in odd-numbered sections, then primary holes were drilled and grouted in the even-numbered sections. Upon completion of the primary grouting, secondary holes were drilled and grouted between the primary holes, first in odd-numbered sections, then in even-numbered sections, and so on. This process was continued until all tertiary holes were completed, resulting in a grouted hole spacing of 10 feet. Further split-spacing was done as necessary, based on the grout takes of adjacent holes. In the triple line portions of the curtain, the downstream line was grouted first, the upstream line second, and the center line last. After completion of an area, exploratory holes were drilled and pressure tested, and additional grouting was done based on these results.

Grouting of each separate hole was done by the upstage method, however, holes were grouted in sets of 3. The Burlington and Chouteau limestones were divided into 3 stages, based on lithology and known weathering characteristics. First, a packer was set at the bottom stage in each of the 3 holes, and each hole was water pressure tested. The holes were then grouted, beginning with the hole with the highest water take. If water take was less than 1 cubic foot in 10 minutes in any stage of a hole, the packer was raised to the next higher stage and pressure testing continued.

The water-cement ratio of the grout mix used to start grouting in any stage of a hole was determined by the water take, and varied from 6:1 for less than 4 cfm. to 3:1 for 7 cfm. and above. The maximum allowable total grouting pressure used was 1.5 psi/ft. of depth for the lower stage, and 1 psi/ft. of depth for the upper 2 stages. If 3/4 of the maximum allowable pressure could not be built up, the mix was gradually thickened. If maximum pressure could be maintained, the mix was not changed, unless a sudden decrease in grout take was experienced, in which case the mix was thinned immediately. Once the grout take decreased to less than 1 cubic foot per 10 minutes, grouting of the stage was considered complete.

COMPUTER ANALYSIS

The evaluation of grouting effectiveness involves the statistical manipulation and analysis of a large volume of semiquantitative data: cement take, grout take, water take, injection pressure, rate of injection, etc. In order to determine the need and location of additional split spaced holes as well as the effective completion of the grout curtain in any area, this evaluation must be accomplished while grouting is underway. This places a burden of time and accuracy on the geologist directing the grouting operation. A computer program was developed by the St. Louis District to remove this burden and allow the project geologist to spend his time controlling the grouting operation in the field.

The system was designed so that grouting data could be entered into an on-site terminal by the grouting inspector at the completion of his shift. The standard grout form was designed as the computer form and the data was entered and manipulated in an interactive graphics format using the English language. A hard copy print out was available of any graphics display. No knowledge of computers was required to use the system. The Boeing CDC time sharing system was used in conjunction with the Tektronix 4014 graphics terminal.

Following input of data from the standard grout form, the computer tabulates the quantities of all the materials (cement, fluidifier, bentonite, sand, etc.) used in grouting in each stage of each hole. This information can be used directly to determine pay quantities. The computer calculates injection time, average and maximum total pressures, and the number and injection sequence of each type of grout mix in each stage of each hole. This information can be used to determine if fracturing of the foundation or premature plugging of a grout hole may have occurred. The computer also calculates and visually displays a bar graph showing cement take and normalized water take per linear foot of each stage of each hole. The computer print-out presented in Figure 5 shows data for all primary, secondary, and tertiary holes which were pressure grouted in zone 1 between stations 2500 and 3000. Figure 5 gives detailed data only for tertiary holes; however, the operator can select to display all data on any order hole in any zone. This information can be used to determine split spacing requirements and evaluate grouting effectiveness. A plan and profile view of grout hole locations is also displayed by the computer.

RESULTS

A total of 91,776 linear feet of grout holes were drilled for the abutment and uplands grout curtain; this was within 8 percent of the Government estimate. A total of 48,332 sacks of cement were injected; this was within 12 percent of the estimate. The average unit take for the entire grout curtain was 0.53 sacks per foot. Only 15 percent of the grout holes required pressure grouting using the grouting criteria of 0.1 cfm during water pressure testing. The average unit take for the grouted holes was 3.2 sacks per foot.

Left Abutment Triple Line.

The average unit take for the left abutment triple line was 0.63 sacks per foot, with separate figures of 1.25, 0.08, and 0.01 for the downstream, upstream, and center lines, respectively (Figure 6). Final closure of the downstream line required 50 splits (fourth through sixth order holes) whereas only six were drilled in the upstream line and none in the center line. These numbers indicate that almost all voids were filled during grouting of the upstream line. Upstream line averages were 2.27, 0.50, and 0.15 sacks per foot for stages 1, 2 and 3 respectively. Normally this would indicate increasing permeability with depth, however, due to the occurrence of high angle relief jointing in the abutment, it is suspected that upper stages were grouted from below by communication. The use of higher grouting pressures in the lower stages could also have contributed to this phenomenon. The lack of uniform decreases in unit take with increasing order indicates that the grouted voids were localized and discontinuous. The take of 5.0 sacks per foot for secondary holes in the downstream line compared to 1.93 sacks per foot for primary holes suggests that the primary spacing of 40 feet was more conservative than necessary, and that 20 feet would have sufficed. Twelve exploratory holes were drilled in this reach, with an average pressure testing water take of 0.0006 cfm per foot.

Left Abutment Single Line.

The average unit take for the left abutment single line was 0.59 sacks per foot, indicating a landward improvement in foundation conditions when compared to the left abutment downstream line take of 1.25 sacks per foot. Similar to the triple line, unit takes increased with depth, with figures of 0.94, 0.53, and 0.03 for stages 1, 2, and 3, suggesting vertical communication between stages. Decreasing unit take with increasing order was more evident than in the triple line, although tertiary holes (0.80 sks./ft.) took more than secondary holes (0.53 sks./ft.). A relatively large number of fourth and higher order holes was required for closure of the line. Twenty-one of the 64 fourth through eighth order holes were drilled and grouted between stations 24+90 and 25+50, where a seismic anomaly had been detected during exploration. Twelve exploratory holes were drilled in this reach, with an average pressure testing water take of 0.0008 cfm per foot.

Right Abutment Single Line.

The average unit take for the right abutment single line was 0.88 sacks per foot. A total of 100 splits was required to obtain closure. Unit takes were 1.46, 0.49, 0.56 sacks per foot for stages 1, 2, and 3, respectively. Eighteen exploratory holes were drilled in this reach with an average water take of 0.0013 cfm per foot.

Right Abutment Triple Line.

The average unit take for the right abutment triple line was 0.29 sacks per foot, with separate figures of 0.68, 0.05, and 0.04 for the downstream, upstream, and centerline, respectively. Final closure of the downstream line required 53 splits (fourth and fifth order holes); 13 fourth and fifth order splits were required in the upstream line; but only 6 were needed in the centerline. This indicates that most voids were filled during grouting of the downstream line. Downstream line averages were 1.26, 0.25, and 0.17 sacks per foot for zones 1, 2, and 3, respectively. This decrease with depth, similar to the left abutment, suggests vertical communication. The take of 1.6 sacks per foot for secondary holes in the downstream line indicates that many large voids were not affected by grouting of the primary holes. Thirteen exploratory holes were drilled in this reach with an average water take of 0.0014 cfm per foot.

Geologic Interpretation.

Only fifteen percent of the holes drilled accepted grout; however, these few holes had a high grout take. This suggests that the features taking grout are large scale, open, and spaced, on the average, about 75 feet apart. Increased grout take with depth indicates these features extend into the lower Chouteau Limestone. Grout or water communication occurred in less than ten percent of the primary and secondary holes and less than five percent of the remaining holes. This suggests that: (1) The large scale features are transverse to the grout curtain; and (2) they are normally isolated features, only rarely interconnected by small scale features which are themselves transverse to the grout curtain. Previous surface and subsurface exploration indicates that these large scale features are probably high angle relief joints which have been further enlarged by solution activity. Areas of high grout take showed a high degree of correlation with anomalies defined by in-hole seismic techniques. The usefulness of this geophysical method has been demonstrated by Hempen (1977).

EVALUATION

The success of the grouting program will not ultimately be known until several years after reservoir filling, now scheduled to begin in the spring of 1983. However, several qualitative criteria do indicate that grouting has been successful. These qualitative indicators have been shown to provide an accurate means to assess

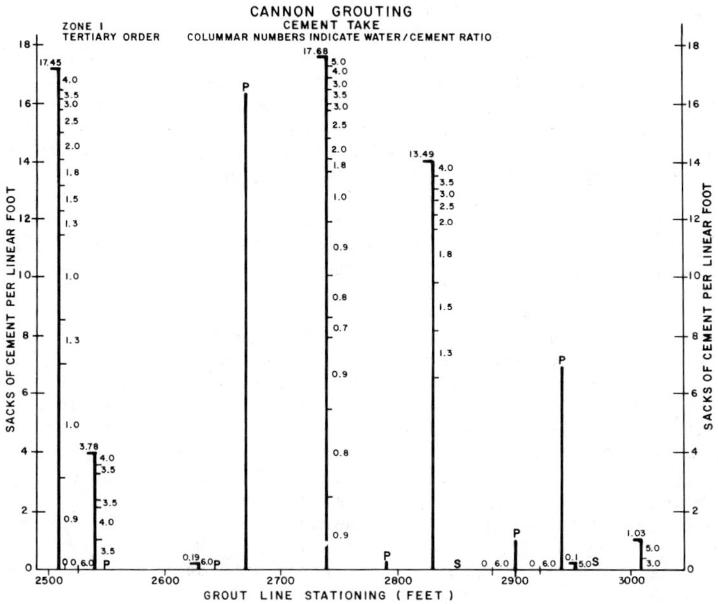

Figure 5. Computer Graphics Display. Primary (P), secondary (S), tertiary holes. Compares grout take (sacks/ft.) and amount of each mix (w/c) injected for split spaced holes in interval requested.

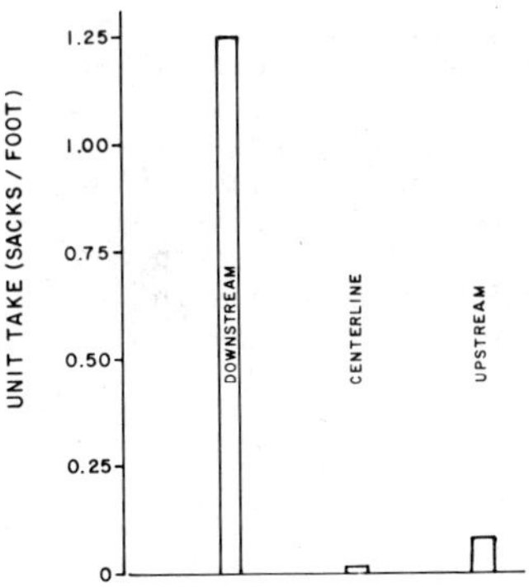

Figure 6. Grout Take in Multiline Curtain.

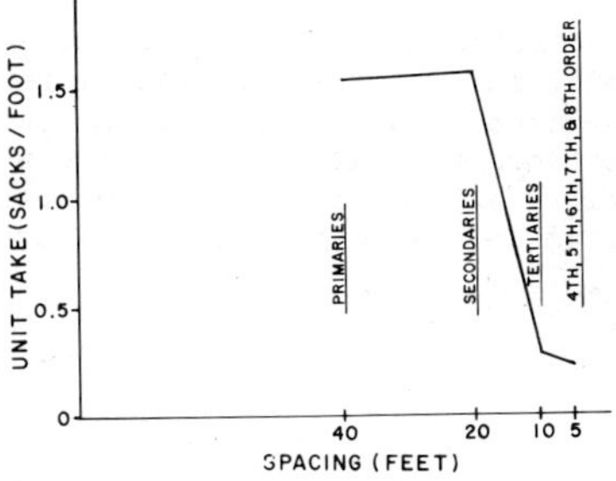

Figure 7. Grout Take in Split Spaced Holes.

grouting effectiveness (Klosterman, 1981). One criteria is the degree of reduction in grout take between primary/secondary and higher split-spaced holes. This reduction was 87 percent, 92 percent, and 71 percent in the left abutment triple line, right abutment triple lines, and left abutment single line, respectively (Figure 7). Klosterman (1981) indicated 65 percent was needed to create an impermeable barrier; Fergusson and Lancaster-Jones (1964) stated 20 percent - 80 percent was a satisfactory range.

A second criteria by which to assess grouting effectiveness is reduction of grout take between lines in a multiline curtain. This reduction was 99 percent between the initial and final lines in the left abutment.

Another traditional method of assessing grouting effectiveness is water loss during pressure tests in completed sections of grout curtain. This water loss is normalized for volume (liters), time (minutes), hole length (meters), and pressure (atmospheres) and expressed as Lugeon Units or coefficient of water loss. An average Lugeon value of 0.1 has been obtained for the 55 exploratory holes tested to date. No individual hole has exceeded a value of 0.50. DeMello and daCruz (1959) state that a value of 2.0 Lugeons indicates no additional grouting is needed.

No unanticipated problems arose either with regard to the foundation or the specifications. This is confirmed by the following: (1) No changes were made, either formally through modification or informally in the field, to the grouting procedures as detailed in the specifications; (2) the work remained on schedule and was completed within the allotted time; (3) the actual quantities for drilling and grouting were within 8 percent (below) and 12 percent (below), respectively, of the estimated quantities.

Many factors contributed to the success of this program; good management, sufficient subsurface information, computer analysis, a good contractor, etc. However, the performance of the detailed specifications indicates that this new philosophical approach can successfully eliminate built-in Owner-Contractor conflicts in grouting as well as other technical contracts.

REFERENCES

DeMello, V. F. B. and daCruz, P. T. (1959), "Some Quantitative Investigations on Curtain Grouting in Rock Foundations of Earth Dams," Proceedings on the First Pan American Conference on Soil Mechanics and Foundation Engineering, Volume 2, pp. 699-720.

Elston, J. P. (1958), "Cement and Clay Grouting of Foundations: Suggested Specifications for Pressure Grouting," Journal of Soil Mechanics and Foundation Division, ASCE, Volume 84, No. SMI, 30p.

Ferguson, F. F. and Lancaster-Jones, P. F. F. (1964), "Testing the Efficiency of Grouting Operations at Damsites," Proceedings of the Eighth International Conference on Large Dams, Edinburgh, Volume 1, pp. 121-140.

Hempen, G. L. (1977), "Detection of Solution Features in Karst Terrain," Third Annual American Geophysical Union Midwest Meeting (Abstract, No. H3).

Klosterman, M. J., Jahren, N. G., and Wolff, T. F. (1982), "Quantitative Relationship of Grouting to the Reduction of Seepage Through Rock Foundations," Association of Engineering Geologist Bulletin, Volume 19, No. 1.

THE STATE OF GROUTING IN THE 1980's

Dr CLAUDE CARON
Consulting Engineer

Abstract

Grouting a soil to insure its imperviousness and/or its consolidation is a process known since the beginning of the 19th century. But it was only at the beginning of this century that the first serious developments took place.

Except for very open ground conditions, injections were carried out at high pressure and fracturing and uplift of the ground frequently occured. Examples have been cited where uplift of several inches were measured. Of course, under such conditions, it was impossible to carry out large injection works at urban sites.

In the 1960's developments in chemical grouting techniques provided injectors an adequate tool to resolve the specific problems of urban sites i.e. to be able to grout without fracturing and uplift.

In view of the wide performance of this new range of grouts it became possible to think of grouting as a normal construction process. As a result, grouting is now applied to works of increasing size and importance. These new possibilities oblige the injector to optimise all grouting techniques in order to devise a competitive solution.

The fundamental progress made in the 1960's opened new horizons in grouting techniques. These improvements related to new types of drilling and grouting methods (more rapid, more rational and consequently more economical) and new types of fluid grouts coming mainly from the field of organic chemistry (hard gels : silicate + organic hardener, aqueous resins : acrylamide, phenoplasts and aminoplasts). With these new grouts whose viscosity is as low as water, there were literally no soils impossible to grout.

At the International Symposium of Soil Mechanics (London 1963), describing the state of the art in grouting, twenty years ago, these new techniques (methods and grouts) were

---

President of GEATEC Consultant
Val de la Torse B2 - 13100 Aix-en-Provence FRANCE

fully dealt with and around the same period, at a conference of the Technical Institute of Building and Public Works, reverting again to the quotation from Matthew : "The Sage built his house and the winds and tides could not touch it because it was built on rock" (Book 6 - Verse 25), I was able to add : "but the Sage can build his house too, on any ground if he calls in the services of an injector".

The main accomplishments of recent years (for example 150,000 m^3 of soil treated by chemical grouting only for the construction of the "Regional Express Railway"in PARIS) have confirmed this judgement and grouting has now outgrown the restricted and rather mysterious range of applications known only to a few specialists.

Does this mean that this well established method is now fixed and codified and that it is sufficient for each new site to repeat what was done with success 10 or 20 years ago ? It would be a grave error for the future of grouting to take such an easy attitude.

The important progress made in grouting in the past twenty years has led to both an advantage and a disadvantage :

- the advantage is that the cases where grouting can be applied have been greatly extended.

- the disadvantage is that the grouting contractor is now in competition not only with other grouting specialists but also with general contractors. For example, a cutoff can be conceived either by grouting or by sheetpile curtain, a tunnel can be excavated after protective grouting has been carried out or by using shield techniques, and so on... These competitive possibilities oblige the contractor to develop a very sound mastery of all grouting techniques in order to devise a competitive solution. Reverting again to the idea that the injector now knows how to transform and strengthen every type of soil and consequently the Sage can build on any ground, I added in a recent conference at the "Institute of Building and Public Works" : "but the Sage wants also economy".

This optimisation with regard to results and cost of the injection can and must relate to :

. the means of drilling and injection
. the best choice of grout
. the techniques of injection

I THE MEANS OF DRILLING AND INJECTION

I.1 DRILLING

Concerning the drilling, the grouting contractor now disposes of more powerful drilling machines adapted to every kind of soil (loose sand, rock and so on ...). These drilling machines are generally fitted with a borehole parameter recorder.

Since many years, the systematic use of drilling muds resulted in reduced drilling time. And for such soils, the recent use of jet grouting (from Japan) constitutes an elegant solution in several cases (for example formation of an injected watertight bottom). With this process, drilling by jetting water and grouting are successively carried out with the same equipment. Drilling is performed under pressure and after switching from the bottom jet to the horizontal jet, grouting is realised radially by turning out the tool with a slow rotary movement (see the following figure).

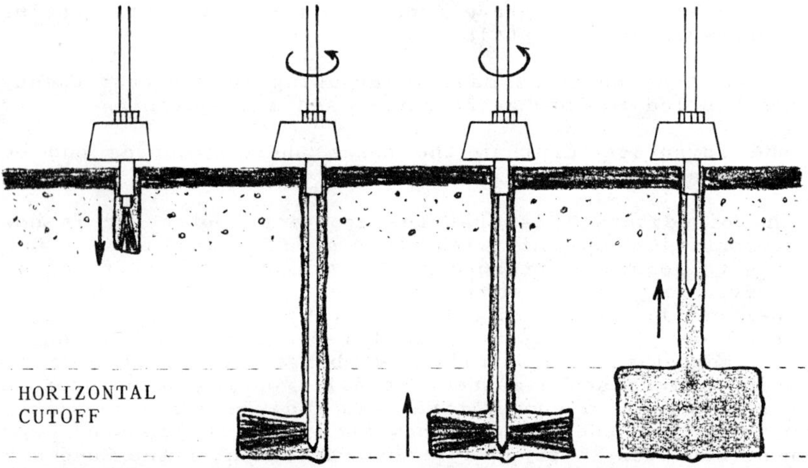

Fig. 1 - **JET GROUTING METHOD**

I.2 GROUTING

For the preparation and injection of grout, these two operations are more and more being systematically carried out with automatic plants which assure the following functions :

- storing and diluting grout components
- proportionning, mixing, metering and pumping grout
- monitoring grout pumping pressure and flow-rate

This automation assures at the same time a reduction in the preparation time and accurate dosing.

This precision became essential with the extensive introduction of chemical grouts : an error of 20 to 30 % in the composition of a cement grout is not too important but a much lesser error in the case of a chemical grout cannot be tolerated.

Precision in the dosage and measurement of quantities of injected materials are obtained by proportional pump or by tanks fitted with electrodes for the liquid constituents and by automatic balances for the powdered materials.

II BEST CHOICE OF GROUT

According to the soil and the desired result, the grout is selected in order to :

penetrate the soil effectively, and assure the required degree of watertightness and consolidation.

II.1 PENETRATION OF THE GROUT IN THE SOIL

Grout penetrates the soil in three ways : by fracturing, by filling the soil voids (impregnation), and by fracturing (this third way has hardly been described in the past).

II.11 Penetration by fracturing

Penetration by fracturing is only used in exceptional cases : when a building must be raised in order to make it level or when a watertight bottom needs to be created. When grouting by fracturing, cement-based grouts are generally used.

II.12 Penetration by impregnation

When grouting by impregnation, the choice of grout is more vast, because the grout, in order to penetrate the soil voids, must be adapted to the size of the voids.

a) In fissured-coherent soils, under the effect of grouting pressure, the opening of the fissures allows penetration of cement-based grouts which are injectable up to 5 LUGEON units ($K = 5.10^{-7}$ m/sec). So it is only for exceptional cases that it is necessary to use non-granular grouts (chemical grouts).

b) On the other hand, in granular soils there is a filtration of cement-based grout as soon as the soil is to fine. Even when adding colloïdal clay to cement, it is difficult to inject, with these granular grouts, soils having a permeability inferior to 10^{-3} m/sec. It is then necessary to pass to non-granular grouts whose viscosity will be much lower as the ground becomes more and more fine.

Using the criteria of grainsize distribution, permeability,

specific surface of the soil, Terzaghi, Riedel, Glossop, Skempton, Askalohoff, Cambefort, Miki gave successively the types of grout appropriate to each kind of granular soil :

- binghamian sand granular grouts (clay-cement or bentonite cement, for coarse soils characterized by :

$d_{10} > 0.5mm \quad K > 10^{-3} m/sec \quad S > 10 \ cm^{-1}$

- colloïdal solutions or prepolymers (silica or lignochrome gels, tannins, organic or mineral colloids, polyurethane..) for medium grained soils characterized by :

$0.02 < d_{10} < 0.5mm, \quad 10^{-3} > K > 10^{-5} \ m/sec, \quad 100 > S > 1000 \ cm^{-1}$

- pure solutions (acrylamide, phenoplast, aminoplast) for very fine soils characterized by :

$d_{10} < 0.02mm \quad K < 10^{-5} \ m/sec \quad S < 1000 \ cm^{-1}$

The limit between cement-based grouts and chemical is clear : when the soil is too fine the cement grouting can only proceed by fracturing. From experience at many sites, we have observed the accuracy of this limit. We can cite, in particular, a systematic examination of the results at Lock and Dam 26, Mississipi River : as certain horzons have a K of 0.05×10^{-3} m/sec, close to the theoretical limit of bentonite-cement grouts, the injection of this grout was consequently attempted, and it was observed, upon excavation, that grout had penetrated only by fracturing.

On the other hand, the dividing line between gels and resins is not so clear because the only difference between these two chemical grouts derives from a difference in viscosity which can be compensated by adjusting the pumping rate. It is therefore possible to treat fine sand with $K = 10^{-6}$ M/sec with silica gels on the condition of grouting at a pumping speed less than 100l/hour.

II.13 Penetrating by fracturing-impregnation

If, in this kind of soil, the grouting of silica gels is attempted at higher pumping rates (consequently by increasing the pressure), the grout does not penetrate the soil by a real impregnation but by both fracturing and impregnation. Under the effect of the grouting pressure, a series of disc-shaped fractures are created around the borehole, and chemical grout penetrates into the soil from these fractures (fig.2).

By applying the LEFRANC-MANDEL's laws, it can be shown that high pumping-rates (0.5 m^3/h) can be achieved with two or three discs of 50 to 60cm diameter, per pass of 33cm.

If the grouting is simply to achieve watertightness, this type of treatment is sufficient. But if the grouting aims

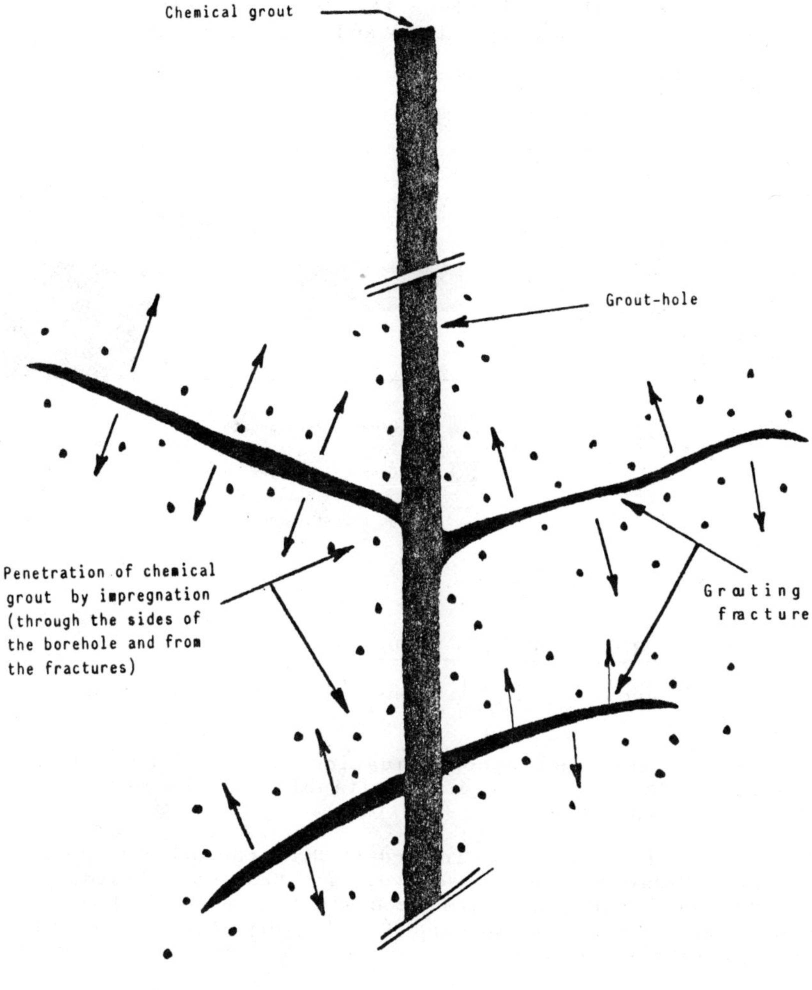

Fig. 2 GROUTING BY FRACTURING-IMPREGNATION

at a consolidation, fracturing with gel constitutes a weak sliding plane in the soil mass and the use of an aqueous resin would be preferable.

The following table summarises the influence of both parameters : type of grouting method and type of ground.

Type of ground / Type of grouting method	Coherent fissured soils		Loose soils	
	Large and medium fissures $K > 5.10^{-7}$ m/sec	Very fine fissures $K < 5.10^{-7}$ m/sec	Coarse and medium $K > 10^{-3}$ m/sec	Fine $K < 10^{-3}$ m/sec
by fracturing			Cement-based grout	Cement-based grout
by impregnation	Cement-based grout	Chemical grout	Cement-based grout	Chemical grout
by fracturing-impregnation				Chemical grout

Cement-based grout

Chemical grout

No case for grouting

II.2 OBTAINING THE DESIRED RESULT

The aim of grouting is to consolidate and make the soil impermeable.

It is exceptional that a consolidation without impermeabilising is sought. This case is justifiable but too rare to be developped in this text.

The essential condition, in order that the watertightness and consolidation be effective, is that all voids are filled with grout and this necessitates a good choice of grout, especially with regard to its ability to penetrate the soil (see chapter II.1).

When this condition is carried out, the imperviousness and, if necessary, the consolidation is reached, but if it is necessary, too, that the result will remain effective despite the passage of time.

II.21 Imperviousness

In circulating through the injected soil, groundwater can

dissolve the soluble elements of grout and can also cause certain chemical or physicochemical changes to the grouts.

These two effects lead to a very variable loss of imperviousness from one grout to another. Each type of grout indeed had a specific behaviour when subject to leaching.

Besides this parameter concerning the nature of grout, external factors such as the nature of water, granulometrical composition of soil...intervene.

In the face of this complex situation, it is difficult to give general rules and it is prudent to make a specific study for each case taking into account individual criteria of ground and water.

We can indicate, by way of orientation :

. the ageing behaviour of cement-based grouts and aqueous resins is generally good.

. the behaviour of silica-gel is questionable : in addition of the syneresis phenomenum (contraction of the gel makes the treated soil more permeable), the inverse reaction gel→solution can take place.

It is certain that the stability is greater if the content of hardener is higher. It is likely that the chemical nature of the hardener has an influence but this is a subject that has not yet been systematically studied.

II.22 Consolidation

As for the watertightness, the ageing behaviour is considered good for cement-based grout an aqueous resins and questionable for silica-gels. The destabilizing of a part of the silica gel by leaching leads to a loss of strength. But, above all, the clear influence of the creeping of silica-gels is placed in evidence.

All the parameters of the loss of strength of silica-gel by creeping have not yet been basically studied. This is nevertheless a problemn of great practical interest because it is not use to design a grout that will impart a short term strength of 150 psi to the soil, if, by creeping, failure occurs at a load of 50 psi within one or two months. The NUREMBERG incident is a vivid remember that in choosing a consolidation gel, contractor has to take into account the loss of strength under the effect of time.

According to my own experience the ageing behaviour of silica-gels is highly variable, the tendancies being :

. the finer the injected ground, the better the ageing behaviour
. the higher the content of hardener, the better the ageing

behaviour.

At present, strengths are always determined from rapid standard crushing tests. For a silica gel, this standardisation is erratic : for a given mix there is, when we operate classically, with a crushing time of a few minutes, a relation $R_c = f(D)$ such as AB (fig. 3 hereunder) but in equipping the compression apparatus in order to have a crushing time of several weeks, we measure values very different from those given officially (as shown in curve AC).

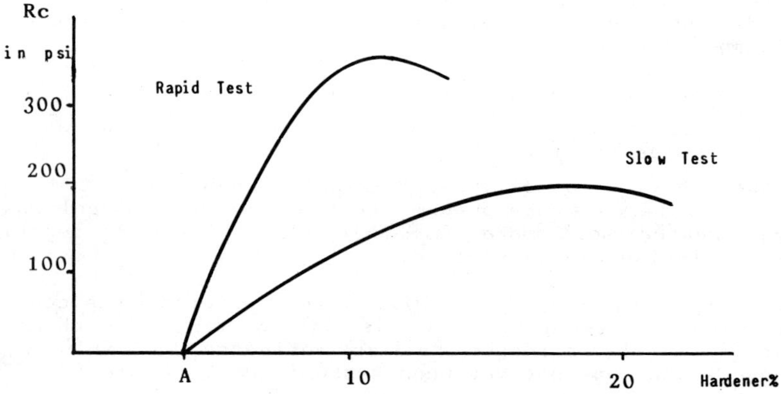

Fig. 3 EFFECT OF STRAIN RATE

We see that :

. on the one hand, these values, more credible because obtained by slow testing, are systematically lower than those given by a rapid test.

. on the other hand, the maximum compressive strength of the curve $R = f(D)$ occurs for a higher dosage of hardener.

All the strength formulas of soils injected with gels should be calculated again in order to take into account the influence of creep and the tendancy would be to increase the dosage of hardener. But this practice often leads, with classical hardeners, to shorter gelling times.

III GROUTING TECHNIQUES

In order that the grouting result be good, it is evident that the injector has to choose the successive grouts most adapted to the problem (viscosity, setting time, strength) but it is also necessary to know how to distribute the grout in the soil. For this, it is necessary to make a correct choice of :

. the grouthole equipment
. the distance between boreholes
. the length of injection passes
. the number of grouting phases
. the grouting pressure and the pumping rate.

It is in making the best choice of the values of these 5 parameters that the injector will arrive at an excellent result without cost overuns. We have, for each parameter, arrived at empirical relationships that allow the orientation of this best choice according to the nature of the soil and the aim sought which will, without risk, result in the best quality/price ratio. The first parameter (grouthole equipment) will be discussed because it is impossible to treat all five in this text.

Since several decades two classical methods have been used to achieve this equipment : the open bottom pipe method and the sleeve pipe method. These two methods are sufficiently well known that they will not be described here.

The most universally adopted method in tight formations (such as fissured rock) is grouting from the bottom of an open grout pipe. On the other hand for loose formations (sands and gravels) in the U.S and in Japan, grouting from the bottom of an open pipe is the most accepted method, in Europe sleeve pipe grouting is almost universally employed.

The fact that neither of the two methods has prevailed in all countries is because both have a series of advantages and disadvantages as summarised in the following table.

Criteria	Open bottom hole method	Sleeve pipe method
a Possibility of separating the two site operations drilling and grouting.	no	yes
b Ability to inject different grouts successively in the same borehole.	no	yes
c Risk of resurgence along the length of the borehole.	yes	no
d Ability to adjust the length of the injection pass.	yes	no
e Ability to probe the ground during the injection.	yes	illusory

A thorough analysis of these five criteria should permit a functional choice and lead to the development of new methods having the disadvantages of both the traditional equipment without the disadvantages.

a The possibility of separating the drilling and grouting operations

Using the open bottom pipe method for loose formations it is evident that grouting must follow drilling. So, the drilling machine is immobilised during the whole time of grouting.

On the contrary inthe sleeve pipe method, all the boreholes are completed, sleeve pipes are sealed into the ground and they are injected afterwards. This is considered a decided practical advantage : by separating the drilling from the grouting, the contractor can rapidly free his drilling equipment.

b The ability to inject several different grouts in the same groutholes

Only the sleeve grouting method offers this possibility which is particularly interesting in very heterogeneous soils.

c The risk of resurgence along the grouthole

In the sleeve pipe method, the sleeve grout sets along the entire length of borehole and thereby prevents the resurgence of grout.

Statistically, the absence of resurgence is more difficult to achieve with the open bottom pipe method. The normal method consists in utilising a packer. In coherent ground (fissured rock) the packer can be set securely against the ground and the result are good, but il loose soils, the packer setting is less secure, even with long packers.

According to a method currently used in Japan, the absence of resurgence along the grouthole is automatically obtained by using a grout having a short setting time. Effectively, the results are good, but this method has two disadvantages: it results in fracturing and the radius of action is smaller than when using classical methods.

d The ability to adjust the length of injection pass

In the open bottom pipe method the length of the pass to be injected can be chosen in accordance with the soil conditions. This choice is impossible in sleeve pipe grouting: the length of the injection pass corresponds to the fixed spacing between sleeves (generally 33cm).

The grouting rate Q without causing fracturing is given by the equation :

$$Q = \frac{2 \pi L P K}{\eta \ln \frac{L}{r}}$$

K is fixed by the nature of the soil, r (the borehole radius) is pratically constant, η (the viscosity of the grout) is fixed by the permeability of the soil to be treated.

So, the injector has only two parameters, P and L (pressure and length of grouting pass), to use to reach an injection rate compatible with an economic operation (i.e. at least 150 l/h). Consequently, if L is constant (which is the case with sleeve pipe grouting), the contractor can vary only the pressure and this is highly undesirable. It is therefore better to limit P and adjust L according to the permeability.

e The ability to probe the ground during the injection

An examination of the pressure-discharge curve is extremely interesting, because the progress of treatment can be followed and zones which would require further treatment can be identified.

When using the open bottom pipe method, this continuous examination is obtained automatically without ambiguity and so the injector has a good knowledge of the ground conditions and is able to direct the grouting by judicious selection of the characteristics of the grout (viscosity, setting time) and choosing the best injection method.

This type of fruitful examination is rather illusory when applied to readings taken during sleeve pipe grouting, since the pressure measured is affected by loss of head in passing the fractures in the sleeve grout.

From this analysis of the characteristics of general methods available at present for equiping groutholes, it appears that neither of the two classical methods can be inversally applied : each has advantages as well as disadvantages. In addition, the use of a grout with a short setting time to resolve the problem of resurgencies, brings other disadvantages. It is therefore necessary to have complete mastery of the classical methods in order to be able to apply to each particular case the solution that is best adapted to the soil. It appears, too, that there is room for new types of grouthole equipment. We have, at the moment, in the stage of study method of equipment which would have none of the disadvantages cited above.

CONCLUSION

Until recent times, grouting has been only used when traditional methods were not feasible. In these interventions, often of a catastrophic nature, there were no question of searching for optimisation of costs.

But, in recent times, grouting is more and more being used as a construction method and not merely for repairing. In each practical case of underground construction, the cost analysis of each of the competitive methods often includes the grouting solution. So the grouting contractor has to find more and more competitive solutions.

The total cost of grouting is the sum of three factors : drilling (1), grout materials (2), injection (3). To arrive at a competitive project, the grouting contractor must concentrate on each of these factors, for example: maximise the spacing between groutholes (factor 1), inject without wastage (factor 2), and at the maximum pumping speed compatible with the ground conditions (factor 3).

This optimisation is not always easy to achieve because there are interactions between the three above-mentioned factors. For example, by increasing the speed of injection (to gain on factor 3), the ground may be fractured with consequent grout loss (increasing factor 2) and so on.

The thing that is difficult in the work of the injector is that he has to deal with soils with changing characteristics. In order to achieve a better result, he must therefore continuously adapted himself to these changes. Presently, the means at the diposition of the injector are fortunately varied and the best solution can often be found.

In order to realize this review of points where progress was made in recent years or is yet to be made, we relied on our european experience and on the survey of works carried out in countries such as United-States and Japan.

This confrontation which is the purpose of this congress, must lead to new ways of improving the grouting progress. I would draw particular attention to :

. effective means to analize the ground permeability during the drilling and the grouting

. new type of groutholes equipments

. better knowledge of the ageing behaviour of grouts.

CHEMICAL GROUTS
AND THEIR PROPERTIES

R. H. Karol[1]
Member, ASCE

For many years, the term "chemical grout" was synonymous with sodium silicate and the Joosten process. However, sodium silicate is the basis for many chemical grout formulations in addition to the silicate-chloride two-shot injection. Further, the last three decades have witnessed the introduction and use of many materials totally unrelated to the silicates. Today, commercial chemical grouts cover a wide range of materials, properties and costs, and give the grouter the opportunity to select a grout for specific job requirements.

The mechanical properties of any grout that are important factors in its selection for a specific job include mechanical permanence, penetrability, and strength. Similarly, the chemical properties include chemical permanence, gel time control, sensitivity, and toxicity, and the economic factors include availability and cost.

a. Permanence

All grouts which contain water not chemically bound to the grout particles are subject to mechanical deterioration if subjected to alternate freeze-thaw and/or wet-dry cycles. The rate at which such deterioration occurs varies with the amount of free water available in the grout, as well as with the degree of drying or freezing. Laboratory tests on small sand samples grouted with a gel having a high free water content indicated that significant deterioration began only after the completion of four to six full cycles. The test conditions were probably more severe than field condition would be, since total drying seldom occurs close to the water table, and freezing in temperate zones seldom extends more than 5 to 10 feet below the ground surface. Nonetheless, mechanical deterioration of grouts can occur under certain environmental conditions.

Chemical deterioration of grouts can occur if the grouts react with the soil or groundwater to form soluble reaction products, if the grout itself is soluble in groundwater, or if the reaction products which form the grout are inherently unstable. Generally, materials with any of those unfavorable characteristics would not be proposed or used as a grout. However, there may be locations in which unusual concentrations of strong reactants are

[1]Professor of Civil Engineering
Rutgers University
New Brunswick, N.J.

present in groundwater, and these may have deleterous effects on grouts which are otherwise considered permanent.

b. Penetrability

The comparative ability of grouts to penetrate a formation is mainly a function of their relative viscosities. Although there is truth to the contention that lubricating fluids pass through pipes with less friction loss than non-lubricating fluids, the difference is small, and cannot make up for significant viscosity differences. The effectiveness of pumping silicates into a formation to increase the take of the following cement grout is more probably due to fracturing the formation than to making the voids more "slippery." In general, viscosity alone should be used as a comparative guide to penetrability of chemical grouts.

Penetrability can be related to permeability. A conservative criterion is that grouts with viscosities less than 2 cp (such as acrylamide-based materials) can usually be pumped without trouble into soils with permeabilities as low as 10^{-4} cm/sec. At 5 cp, grouts (such as chrome lignins and phenoplasts) may be limited to soils with permeabilities higher than 10^{-3} cm/sec. At 10 cp, grouts (such as some silicate-based formulations) may not penetrate soils below 10^{-2} cm/sec. All grouts may have trouble penetrating soils when the silt fraction exceeds 20% of the total. (This is based on the assumption that the soil is not fractured by the grouting pressure.)

c. Strength

Chemical grouts (except for epoxies and polyesters and some polyurethanes) have little strength compared to cement, and the actual strength of a solid mass of chemical grout is of academic interest only. However, the strength of soil formations stabilized with chemical grout is of very practical interest.

Grouts which completely displace the fluid in the soil pores form a continuous but open and non-uniform latticework which binds the grains together. In so doing, the grout increases the resistance to relative motion between soil grains, and thus adds shear strength to the soil mass. Grouts which form as fingers and lenses within a soil mass also add shear strength, depending upon the degree to which the lenses and fingers interlock.

Grouting with chemicals is often done to add strength to formations. Other major applications of chemical grouts are to prevent anticipated groundwater problems, or alleviate existing ones. Most applications, regardless of purpose, result in placing grout below the water table, and drying of the grouted mass will never occur. Thus, the most significant strength factor is the "wet" strength: the strength of the stabilized soil formed and remaining immersed in a saturated formation.

In contrast, much of the literature devoted to strength data reports "dry" strength: strength of stabilized soils in which some or all of the contained water has been removed by air or oven drying. When water held by a grout is lost through desiccation, the grout matrix shrinks. Shrinkage of the grout between soil grains sets up forces analogous to capillary tensile forces, but often much stronger. This increases the resistance to relative motion between grains, and adds shear strength to the soil mass. Thus, desiccated grouted soils will show strengths higher than "wet" strengths, often by as much as ten times. "Dry" values of strength are meaningless for grouts which are placed and remain under the water table.

The strength of a fully permeated grouted soil depends upon the specific grout used, but also upon other factors. Chief among these are the density, the average grain size and grain size distribution of the soil. (2) In general, strength increases with increasing density and with decreasing effective grain size. Well graded soils give higher strengths than narrowly graded soils with the same effective grain size. (23) Because of these variables, general discussion of the anticipated strength of grouted soils should present ranges of values rather than a specific number.

Cohesive soils almost always have permeabilities too low for effective grouting. Thus, virtually all soil grouting is in granular soils. Grouting a granular soil has the effect of adding a small amount of cohesion while the friction angle stays the same, or decreases slightly. Thus, at small soil loads, the effect of grouting is to give a significant increase in strength, but at large loads the increase due to grouting is insignificant. Each specific job must be analyzed separately to determine if grouting can indeed provide a significant strength increase.

Almost all strength values reported in the literature for grouted soils are the results of unconfined compression tests (samples loaded in compression along one axis only). However, some of those tests are on cube-shaped samples, and many others are on samples whose height (long axis) is less than twice the minimum cross section dimension. All such tests give higher than true strength values, because there is not sufficient height for a failure plane to develop without intersecting the loading blocks. It is also probable that (since no standards exist) many tests are run at strain rates too rapid for determining true static strength. Those tests also give results higher than true strength values.

Unconfined compression tests measure the strength of the test sample at zero lateral pressures. For in-situ soils, this occurs only very close to ground surface. Most grouting work places the grout below the ground surface, and even at shallow depths significant lateral pressures develop. Therefore, the triaxial test is a better replication in the laboratory of actual field stress conditions. Whenever it is important to know the actual strength of a specific grouted formation, triaxial tests are preferred. The unconfined compression test is still very useful for comparative purposes, for example checking the effects of additives to a grout, or comparing different grouts.

Although adequate testing has not been performed on all grouts, unless contrary evidence can be produced, it should be assumed that chemically grouted soils will be subject to creep. (2,10) Stress controlled unconfined compression tests on acrylamide and lignin based grouts have shown that sustained loads well below the unconfined compressive strength can cause creep leading to failure in hours, days, or even months, depending upon the ratio of applied load to unconfined compressive strength. (11)

It is possible to delineate a "creep endurance limit," below which failure will not occur regardless of the load duration. Limited data suggests that the endurance limit is about 25% of the unconfined compressive strength for the materials tested, when applied to field conditions where stabilized soil masses are unsupported on one face, as in open cuts, tunnels and shafts. For field conditions where lateral support exists for the grouted mass, triaxial tests are more suitable. Limited data indicates that, using "at rest" lateral pressures, the triaxial creep endurances limit approaches half of the unconfined compressive strength. These data, while approximate and in need of further study, indicate the values of the safety factor to be used when grouted soil strength is a design consideration. (2,10,11)

d. Gel Time Control

The ability to change and control the setting time of a grout can be an important factor in the successful completion of field projects. Excepting the Joosten Process, there is always a time lag from the mixing of the chemical components to the formation of a gel. If all other factors are held constant, this time lag (referred to as the gel time or induction period) is a function of the concentration of activator, inhibitor and catalyst in the formulation. With most grouts, the gel time can be changed by varying the concentration of one or more of those three components. For some grouts, a wide range of gel times can be obtained and accurately repeated. For others, only a narrow range of gel time is possible, and repeatability is poor.

Some chemical grouts, after catalyzation, maintain a constant viscosity and at the completion of the induction period, turn from liquid to solid almost instantaneously. Other grouts maintain viscosity for much less than the total induction period. Still others increase in viscosity from the moment of catalysis to gel formation. For these latter two categories, it may not be possible to pump grout into a formation during the full induction period.

e. Sensitivity

For any given grout formulation, the gel time is a function of temperature. Many other factors may affect the gel time when a grout is used in the field. Primary among these is dilution with groundwater which (in addition to changing the solution temperature) will dilute the concentration of activator, inhibitor and catalyst. Excessive dilution may bring the grout itself below a concentration where gel will form at all. Groundwater may also carry dissolved salts which affect the gel time either by chemical activity or by change in pH. Gel times may also be affected by contact with undissolved solids, such as the materials used in pumps, valves and piping. The degree to which a grout responds to such factors is termed its sensitivity. Materials which are highly sensitive generally cause difficulty in handling in the field.

f. Toxicity

Health and occupational hazards have come under increasing scruitiny in the last decade, resulting in several instances of banning of grout components by Federal Agencies, both in the U.S. and abroad. Among the products used in the past, and still in use, are ones classed as neurotoxic,carcinogenic, toxic, corrosive, highly irritating, etc.

All chemical grouts should be handled with care in the field, with safety and cleanliness equal to, or better than, the manufacturer's recommendations. Under such conditions, occupational hazards are reduced to acceptable limits.

g. Economic Factors

Cost is a most important factor in selecting the construction method best suited to solving a specific field problem. When grouting is compared with other methods such as wellpointing or slurry-trenching, chances are that the cost comparisons will be realistic. When comparing two different grouts, however, the mistake is often made of comparing raw material costs rather than job costs. Chemical grouts are available commercially at prices ranging from half to ten dollars per gallon. However, the ease of placing and the effectiveness of these materials vary, and the cost of placement is almost always a major part of the in-place cost. Thus, while raw material costs may vary over a 20 to 1 ratio, in-place costs generally range from 3 to 1 to about equal. Selecting the most suitable material for a specific job will generally overcome the possible lower material cost of other products.

Commercial Chemical Grouts

If the goals of research to develop a new chemical grout were to be listed, they would probably state that the materials should be:

a. a powder readily soluble in water (this eliminates the expense of transporting a solvent, and water is the least expensive solvent).
b. inexpensive, and derived from chemicals in abundant supply
c. stable at all anticipated storage conditions
d. non toxic
e. non corrosive
f. not explosive, and the grout solution should be
g. of low viscosity, preferably that of water
h. stable under all normal temperatures
i. non-toxic, non-corrosive, non-explosive
j. catalysed with common, inexpensive chemicals, meeting the criteria of h and i
k. insensitive to salts normally found in groundwater
l. of stable pH on the positive side (so that it may be used in conjunction with cement)
m. readily controlled for varying gel times
n. able to withstand appreciable dilution with groundwater, and the gel should be
o. permanent
p. unaffected by chemicals normally found in groundwater
q. non-toxic, non-corrosive, non-explosive
r. high strength

Of course, no such material exists. However, every criterion listed can be found in one or more commercially available materials. It is important, therefore, to determine which grout properties are critical to a specific project, in order to have a sound basis for selecting a grout.

If a count were made of all the chemical grout formulations which have passed through patent offices in the past thirty years, it would reach into the hundreds. Although most of those formulations have commerical possibilities, very few have been developed commercially. The majority have been eliminated by the normal industrial selection process. In the U.S., from about 1970 onward, two specific materials (silicate and acrylamides) shared a market percentage estimated at between 75% and 90%, with the remaining portion of the market divided among only six other products.

It should be expected that new products will appear on the market from time to time. In fact, three have appeared within the past year: AC-400, Injectite 80, and a modified polyurethane. Materials in current use will not be readily displaced unless the new products have significant advantages. Thus, it is reasonable to expect that in the future only a limited number of materials will have significant commercial use, and that these will behave much the same as those in current use, except for improvements in such areas as toxicity, control and cost.

The most widely used chemical grouts are aqueous solutions, and these are divided into the following chemical families:

a. Sodium Silicate Formulations
b. Acrylamides
c. Lignosulfites
d. Phenoplasts
e. Aminoplasts

a. Sodium Silicate Formulations

Gels can be formed from many silica derivatives. For example, fluorosilicates can be precipitated by hydroxides, and silicon esters precipitate in the presence of an alkaline solution. Organic derivatives (such as methyl silicate) can be gelled by polybasic acids (such as citric acid) in various intervals of time after mixing. The silicate derivatives form the largest single group of related grouting materials. However, the alkali silicates, and in particular sodium silicate, is the only one used to any extent for chemical grouting.

Sodium silicate, $nSiO_2 \bullet Na_2O$, is commercially available as an aqueous (colloidal) solution. The silica/alkali ratio, n, is important in that ranges of 3 to 4 yield gels with adhesive properties, particularly suitable for grouting.

When sodium silicate solution and a concentrated solution of appropriate salt are mixed, the reaction forming a gel is virtually instantaneous. The earliest successful field process is credited to Joosten, and consisted of injecting a concentrated solution of sodium silicate at about 38° Bé into the ground immediately followed by an injection of Calcium Chloride solution at about 35° Bé. The reaction in the ground is expressed

$$SiO_2 \quad Na_2O + CaCl_2 + H_2O \longrightarrow Ca(OH)_2 + SiO_2 \quad 2NaCl$$

The grouting sequence results in an unstable interface between the two solutions, and if the silicate solution has not been moved away (by groundwater or gravitational forces), it is penetrated by thin fingers and lenses of chloride solution. The resulting gel contains calcium hydroxide, silica, and sodium chloride. Because the reaction is so rapid not all of the solutions can reach contact, but the unstable interface generally insures that sufficient contact occurs to provide a continuous gel network through stabilized soil.

The Joosten process, properly used, results in a strong gel which can give unconfined compressive strengths above 500 psi. However, the utility of the process is limited by the high viscosity of the solutions, and the need for many closely spaced grout holes. Also, the nature of the process prohibits complete reaction of the two liquids. Research and development was for many years aimed at a silicate based grout with gel time control, to eliminate the inherent disadvantages of the two-shot system. As those research efforts met with success, the Joosten system was phased out as a construction tool. Today, it is virtually a method of the past.

Silica is a weak acid, and sodium silicate is, therefore, basic. Silicate will be precipitated as a gel by neutralization. Thus, a dilute sodium silicate solution, mixed with certain acids or acid salts, will form a gel after a time interval related to chemical concentrations. The best known combination is the sodium silicate - sodium bicarbonate mixture. The key word is "dilute," and in essence a reaction that would be instantaneous in concentrated solutions is delayed by diluting the sodium silicate with water, and also limiting the amount of reactant by dilution with water. The dilution yields a grout of very low viscosity, but also of very low strength. (In all silicate based grouts, the strength and viscosity are functions of the sodium silicate content, and are reciprocally related.) Control of gel time is obviously not precise. Such grouts are considered temporary, although they do have application for stabilization of tunnel faces, where structural support in the form of a liner is added within days or weeks.

The search for a single shot permanent silicate grout with high strength eventually led to the use of organic compounds. These react with the water-silicate mixture to form an acid or an acid salt, which causes precipitation of the silica. Setting time is controlled by the rate of acid formation and therefore by the quantity of organic compound. Since the sodium silicate need not be diluted, strong gels can be formed.

In the late 50's, a French patent proposed the use of ethyl acetate with a detergent, and a U.S. patent proposed the use of formamide. Refinements of these two patents have formed the basis for most of the silicate formulations used throughout the world at present.

Since 1960 (until several years ago), silicate using formamide as a reactant, and one of several salts as accelerators was the most widely used formaulation in the United States. Gel time was controlled by varying either the reactant or the accelerator, or both. Formamide, however, was recently labeled as a possible carcinogenic, and its use is dwindling.

Within the past several years, considerable field work has been done using sodium silicate solution and organic reactants without accelerators. Much of this relates to vehicular tunnels in the Washington, D.C area and is well documented. Portions of this work are discussed elsewhere, in these proceedings.

In Europe, most of the silicate grouts use organic reactants only. One manufacturer described his product as a "mixture of open chain compound diacid esters."

The strength that a silicate gel imparts to a stabilized soil is primarily a function of silicate content. For consolidation of strata where laboratory UC values of 100 psi or more are desired silicate contents required are such that the solution viscosity is in the 10 cp and higher range. For waterproofing of soils where less strength is needed, solution viscosity will be in the 3 to 10 cp range. (These are initial viscosities. During the first three quarters of the induction period, the viscosity approximately doubles, for most silicate formulations.)

Short term strength in one study has shown that final strength of the pure gel is inversely proportional to setting time. (10) Some of this relationship must carry over to grouted soil, and perhaps less to long term strength. Since setting time is related to catalyst concentration, a relationship is thus established that strength increases with increasing catalyst concentration. Strength will also vary with different catalysts, all other factors being equal.

The strengths obtained in laboratory UC tests of grouted soils also depend upon the curing conditions and the testing rate. Higher strain rates give higher strength.

Dry cure of grouted soils gives numbers much higher than wet cure, and tend to be misleading. Wet cure of a small stabilized mass in a large quantity of water may also be misleading, because of possible dissolution (discussed later).

Although most of the silicate formulations are considered permanent materials, the end product is subject to two phenomena which often tend to cause doubt about permanence. A newly made silica gel will, upon standing, exude water and shrink. This phenomenon is called syneresis, and occurs at a decreasing rate. The total water loss is related to the gel properties, generally decreasing with increasing silicate content and shorter setting times. Syneresis also takes place in the voids of stabilized soil masses. In a soil whose voids are completely filled with new gel, the shrinkage accompanying syneresis results in a residual permeability after several weeks. In coarse grained soils, this may partially negate the initial effectiveness of water shut-off. As soils become progessively finer, the practical effects of syneresis become smaller. For fine sands, the effects are generally considered negligible.

Stabilized sand cylinders stored under water for a period of time show loss of strength which varies from negligible to total, depending upon the grout chemistry. (10) The dissolution of the gel is caused by the unreacted portion of the soda, which reverses the process which formed the polysilisic acid. High soda contents generally result from low reactant concentrations associated with long setting times. High enough soda content to completely dissolve the silica gel is generally due to error or improper use of materials. However, most gels will contain some soda, and therefore gels made in the field under water will always show some loss of strength. This is a phenomenon limited in extent to the contact zone between grouted soil and groundwater, and for the normal scale of ground injection may not have significant effect on the overall project. In contrast, on a laboratory scale, the phenomenon may readily encompass the entire grouted sample.

Sodium silicate solution is generally considered totally non-toxic and non-corrosive (although it will etch glass). This material is considered free of health hazards and environmental effects. However, sodium salts may be exuded from silicate gels and these might be classed as environmental hazards in special circumstances. Some of the organics used for reactants may have toxic, corrosive and/or environmental effects. Manufacturers recommendations for handling those products should be followed.

b. Acrylamide Grouts

The first of the new chemical grouts, AM-9, discovered in 1951, was a mixture of organic monomers which could be polymerized at ambient temperatures, with setting time a direct function of catalyst percentage. The solution had a viscosity and a density close to those of water, and both properties remained virtually constant during the induction period. The change from liquid to solid was almost instantaneous for the shorter gel times. The introduction of this product commercially in 1953 was the start of a new era in chemical grouting technology. Over the years, over 1000 pages of technical literature related to it were published. Much of that data is applicable to other grouting materials.

Acrylamide-based materials come closest in terms of performance to meeting the specifications for an ideal grout. They penetrate more readily, maintain constant viscosity during the induction period, have better gel time control, and adequate strength for most applications. They are, however, more costly per pound (or per gallon) than silicates, and acrylamide is neurotoxic.

Until early 1978, three acrylamide grouts and one acrylamide-based grout were commercially available. At that time, the manufacture of the U.S. product AM-9 was stopped.

The market gap left by the withdrawal of AM-9 was quickly filled. By early 1979, a Japanese product Nitto SS and a French product, Rocagil BT were both available in the United States. (Rocagil BT was later withdrawn.) Nitto SS has essentially the same toxicity level as AM-9 (it was banned in Japan in 1974, following a careless application near a well which lead to several cases of acrylamide poisoning). It is marketed as a dry powder. Rocagil BT contains methylolacrylamide, and its toxicity is about half that of Nitto SS. Rocagil BT is marketed as a 40% solution in water. There is also Rocagil 1295 grout (a mixture of acrylamide and methylolacrylamide) not currently marketed in the United States, as well as a discontinued Japanese material, Sumisoil, which is in essence a copy of AM-9. Two products formerly marketed in the United States, Q-Seal and PWG, were actually distributors' trade names for AM-9. Another grout, Terragel, a mixture of acrylate and acrylamide monomers, was marketed in 1979, and withdrawn a year later.

All of the acrylamide based grouts have similar characteristics and properties, and are treated together in the following discussion.

The acrylamide-based grouts consist of a mixture of two organic monomers, acrylamide (or methylolacrylamide, methacrylamide, etc.) which is about 95% of the mixture and will polyerize into long molecular chains, and 5% cross linking agent such as methylene-bis-acrylamide, which binds the acrylamide chains together. The stiffness of the grout can be varied by changing the 95-5 ratio. If 97-3 is used, a sticky, very elastic transparent gel of low strength is obtained. If 90-10 is used, a harder, stiffer, opaque white gel is obtained. All commercial products are sold as a mixture of fixed ratio, and the user must purchase the components separately if he wishes to modify the ratio.

Grout solutions up to 20% solids have viscosities well under 2 cp. Such solutions, when properly catalysed, will, after a length of time which depends upon catalyst concentration, change almost instantly to a solid irreversible gel.

The minimum concentration of grout from which a gel will form is temperature dependent, ranging around 3%. Percent monomer selected for field use must take into account the possibility of dilution with groundwater below the values at which gels will form. Most field work is done at about 10%. When pumping into flowing water, concentrations as high as 15% to 20% may be used. When large grout volumes are placed at short gel times 7% to 8% may be used, and in special cases it may be reasonable to work at 5%.

Gels which form from grouts normally used for field work contain 8 to 12% solids. These solids, in the gel, are in the form of long molecular chains, randomly cross-linked to each other. The rest of the grout, 88 to 92% by weight, consists of water molecules mechanically trapped in the "brush-heap" structure of the gel. If submersed in water or under saturated soils, or placed in an environment of 100% relative humidity, the water in the gel remains in place, and the gel undergoes no volume change. However, gel placed in an arid environment will lose water, and can shrink to the volume occupied by its solids, (about 10% of the original gel volume). Mechanical pressure can also force water from the gel. Hydraulic pressure, on the other hand, will force water into the gel, and cause it to expand. When the forces causing volume change are removed, in the presence of a water source, the gel will return to its original volume. Thus, gels can reabsorb the 90% water they have lost, and expand to fill the voids in the formation in which they were placed. Shrinkage cracks which may have formed in the gel will re-seal but not heal, so the permeability of the formation may be somewhat greater and the strength somewhat less than prior to dessication.

Ions can migrate through the water in the gel. Soluble salts which are added to a grout solution for various reasons may be expected to leach out in time. Similarly, gels made with pure water can be expected to absorb salts from a saline environment.

Acrylamide grouts at ambient temperatures are catalysed with a two component redox system. One part, the initiator or catalyst, can be a peroxide or a persalt. Ammonium persulfate (AP), is most commonly used. The second part, the accelerator or activator, is an organic such as nitrilotrispropionamide (NTP), dimethylaminopropionitrile (DMAPN, currently labeled a possible carcinogenic), and triethanolamine (TEA). All three have disadvantages. DMAPN, a liquid, is best from a control point of view, but is considered a health hazard. NTP, a powder, has limited solubility in water, particularly at low temperatures. TEA, a liquid, is somewhat metal sensitive. At the present time, virtually all U.S. applications use TEA. There are also materials which act as inhibitors, and can be used reliably to control gel time. Potassium ferricyanide, KFe, is most often used.

Gel time is independent of monomer concentration, but is directly dependent upon temperature and concentration of catalyst, activator and inhibitor. Although the gel time can be changed by varying any one component, there is an optimum ratio between catalyst and activator for the most effective use of each. Short gel times, down to 5 and 10 seconds are obtained by omitting the KFe and increasing both catalyst and activator. For work at temperatures above 100°F, AP is generally used without an activator. For very long gel time (12 hours, for example) AP is used with KFe. Whenever an activator is not used, a buffer such as disodium phosphate is required to maintain the pH of the solution around 8.

Sunlight can cause catalysis of grout solutions to which an activator has been added. Such solutions should be kept covered in daylight. Entrained air from overmixing can reduce catalyst concentration and prolong gel times.

A catalysed grout solution stays at constant viscosity until just before gelation. At that time, the temperature begins to rise and the grout gels. For a 20 minute gel time, the time from start of gelation to completion is less than a minute, in the laboratory. The reaction is exothermic, and the exotherm contributes to the speed with which gelation is completed. When grouting soils, the temperature rise is much less than for a solid gel, and in the ground the change from liquid to solid is not as rapid as in samples of pure grout.

The gel time of a grout solution can also be affected by chemicals dissolved in formation groundwater or by contact with the formation itself. Sodium and calcium chlorides, in particular, tend to shorten gel times, and in fact are sometimes diliberately added for this purpose when working at low temperatures. The effects of groundwater on gel time can be cancelled by using groundwater to mix the grout solution.

Grouted soils have short term unconfined compressive strengths ranging from 50 to 200 psi. Strengths increase with increasing soil density, decreasing soil grain size, and increasing range of grain sizes. Creep endurance limits range from 1/4 to 1/2 of the U.C. values. (11)

Acrylamide gels are considered permanent. The gel is unaffected by exposure to chemicals, except for very strong acids and bases - materials not naturally found in soils. Tests on samples stored under saturated sand for 10 years showed no loss in strength, and many field jobs have shown adequate performance for times reaching well into the second decade.

Gels are, however, subject to mechanical deterioration when exposed to alternating drying and/or freezing cycles. These will eventually break the gel into smaller pieces and reduce both the strength and imperviousness of the treated formation. Additives such as antifreeze, glycerine and calcium chloride may be used to counter the effects of freezing and drying.

Acrylamide in solution or powder form is neurotoxic. The gel is non-toxic. The type and degree of health hazard is well documented, and can be readily overcome by consistent use of simple safety measures.

c. Lignosulfonate Grouts

Lignosulfonates are the waste liquor by-products of the wood processing industries (such as paper mills). Since the product is not controlled, it is widely variable in content, depending upon the specific source of trees and the particular wood processing. Not only will the liquor vary from country to country, but also with different mills closely located to each other, and even with the same mill at different times of the year. Lignosulfonate solutions used for grouting purposes range from the raw liquor trucked directly from the mill to the job site to dried, pre-catalysed powders with other additives to assist in gel quality and quality control.

Lignosulfonate grouts always consist of lignosulfonates and a hexavalent chromium compound, hence the name "chromelignins." Most generally, calcium lignosulfonate is used with sodium dichromate. In an acid environment, the chromium ion changes its valence from plus 6 to plus 3, oxidizing the lignosulfonate to produce a gel. If the lignosulfonate is of itself sufficiently acid (pH of 6 or below), no other additives are needed. If the pH is above 6, acids or acid salts are generally added to control pH.

Although the composition of the product is variable, in a unit weight of solids approximately three quarters will be lignosulfonates, one fifth reducing sugars and the rest ash. Concentration of solids for field use vary between 200 to 600 grams per liter. Initial solution viscosities for these concentrations vary between 3 to 8 cps.

The viscosity of any specific grout mix increases at an increasing rate from the instant of catalysis to the formation of a gel.

The setting time of a catalysed grout solution varies with the concentration of solids, decreasing as the percent solids increase. Setting times also vary with the dichromate content, decreasing as the dichromate increases.

The strength of soils stabilized with lignosulfonates is of the same order of magnitude but somewhat less than that of acrylamide grouts. In common with other grouting materials, the lignosulfonates show higher strengths in finer materials, and are subject to creep and consolidation phenomena. The creep endurance limit is of the order of 1/4 to 1/2 of the U.C. value.

The strength or lignosulfonate gels increases dramatically (tripling or quadrupling) as the solids content varies from 300 to 500 grams per liter. (24)

The strength also increases with increasing dichromate content, although this may be partially due to shorter gel times (which will also generally increase the gel strength). A dramatic increase in gel strength is obtained by lowering the pH, which promotes quicker and more complete chemical reactions. All of these affects are carried over in subdued form to stabilized soil samples.

Gelled soils under the water table which are not subject to alternate wet-dry or freeze-thaw cycles have good stability. Under these conditions, the lignosulfonates are considered permanent materials. However, they respond very quickly to wet-dry and freeze-thaw conditions, and should not be used under those conditions for permanent installations.

Since lignosulfonates are by-products of other processes, they are relatively inexpensive and can compete on a cost basis with any other grouting material. The major drawback lies in the toxicity of the dichromate salt, present as a catalyst in all lignosulfonate grouts. Hexavalent chromium is highly toxic. It is reduced to the non-toxic trivalent form during the reaction but the reduction is not necessarily complete, particularly at higher range of pH and when using long gel times. Thus, gels which should be innocuous may leach toxic materials into the surrounding environment.

d. Phenoplast Grouts

Phenoplast resins are polycondensates resulting from the reaction of a phenol on an aldehyde. They set under heat over a wide pH range at elevated temperatures. At ambient temperatures, the reaction for most phenols requires an acid medium. (Most soils are neutral or slightly basic, and a grout requiring acid conditions is undesirable.)

There are several materials which will react at ambient temperatures. One of these (the only one generally used for grouting) is resorcinal, and it is most commonly reacted with formaldehyde. A catalyst is required, whose main function is to control pH. Sodium hydroxide is commonly used.

The theoretical proportion for complete reaction is two molecules of resorcinal with three molecules of formaldehyde. Optimum mechanical properties are obtained with that ratio, for any given amount of dilution. The only control of setting time is the dilution of the grout components. Setting time varies greatly with solution pH, being shortest for any given grout concentration at a pH slightly above 9.

Since mechanical properties of the gel are optimum at the shortest setting time for any given grout concentration, this fixes the amount of catalyst. (Since the catalyst functions primarily as a pH control, many different soluble materials including hydroxides, carbonates and phosphates can be used. These will give different setting times, equal to or longer than those obtained with sodium hydroxide.) Thus, the three ingredients of the grout always have fixed proportions to each other, and the only field variable is the dilution with water.

The initial viscosity of resorcinal formaldehyde grouts ranges from 1.5 to 3 cp for concentrations normally used for field work, and like the acrylamide grouts, the viscosity remains constant at those low levels until gelation starts. Again, like the acrylamides, the change from liquid to solid is almost instantaneous.

The strength of soils stabilized with resorcinal formaldehyde grouts is directly proportional to the resin content. Values are comparable to the high concentration silicates. The phenoplasts in general appear to be less sensitive to rate of testing strain than other grouts, and their creep endurance limits appear to be a higher percentage of their UC values. Resistance to wet-dry cycles is poor, however, and can lead to complete disintergration. Except for grouts placed in a dehydrating environment, the phenoplasts are considered permanent materials.

Phenoplasts always contain a phenol, a formaldehyde and an alkaline base. All three components are health hazards and potential environmental pollutants. Resorcinal is a phenol, and although not as hazardous as some phenols, it is still toxic and caustic. Formaldehyde is considered a dangerous material, and at low atomspheric concentrations can lead to chronic respiratory ailments. Sodium hydroxide is, of course, well known as a caustic material.

Phenoplast gels, if proportioned properly, will not leave an unreacted excess of either of the major components. Thus, only the catalyst could possibly leach out to cause environmental pollution. Gels from properly proportioned constituents are generally inert (i.e. non-toxic and non-caustic).

Because of the fixed relationship of ingredients, strong gels are always associated with short gel times, and weak gels with long gel times. This property of the grout is undesirable. Waterproofing operations, for example, do not need strong grouts but very often do need short gel times. One method of solving this problem is to combine the phenoplast system with another grout, such as a silicate or a chrome-lignin, and control the gel times with the latter materials. These act as vehicles to move the phenoplast into the formation and then hold it in place over the time it takes for gelation.

There is also another problem, that of obtaining a longer gel time with a high grout concentration, to give high strength. This problem cannot be solved by using another grout, as a vehicle. However, it could be solved by the use of inhibitors, if they could be developed, and the long gel time problem could be solved by the use of accelerators. Chemical companies recognized the problem, and in the middle 60's began looking for ways to produce a commercial product with all the advantages of the resorcinal formaldehyde system, and without the disadvantages.

Several commercial products are now available. One of these, Rocagil, is described as consisting of partially sulfonated tannin (a natural polyphenol), which is reacted with formaldehyde in an alkaline environment. This product is similar in action and properties to the resorcinal-formaldehydes, except for a higher initial viscosity (5 to 10 cp).

Another product, Geoseal, is described in the patent as consisting of a vegetable tannin extract which can be from mimosa, formaldehyde or paraformaldehyde, sodium carbonate, metaborate or perborate, water soluble inorganic salts of sodium, calcium magnesium or aluminum, hydroxyethyl cellulose (a thickening agent) and bentonite.

With all of these possible additives the end product is quite different from the original phenol-formaldehyde concept. While Geoseal does succeed in providing gel time control without dilution of the phenol, and can be used at short gel times, the viscosity of the initial solution is dependent upon concentration, varying from 2 to 10 cp for field use. The viscosity increases continuously from catylization to gel, and the final gel is considerably weaker than the resorcinal-formaldehyde formulation.

A third product, Terranier, could be grouped with polyphenols or lignosulfonates, since it contains both a polyphenol base and a dichromate salt. Its properties are similar to those of the chrome-lignins.

e. Aminoplast Grouts

Aminoplasts are grouts in which the major ingredients are urea and formaldeyhde. The formation of a resin from these two materials required heat, and initial suggestions for use, (as with the phenoplasts) were in the oil industry. Also paralleling the phenoplasts, the aminoplasts require an acid enviornment to complete the reaction. However, there are no exceptions to this requirement, as there are for phenoplasts. Thus, all urea-formaldehyde grouts will set up only under acid conditions, a distinct limit to the utility of these grouts. They should be used only when it is known that ground and groundwater pH is below 7. Such conditions exist in coal mines, and may exist elsewhere.

Urea solutions have very low viscosities, similar to the acrylamides and phenoplasts. The reaction with formaldehyde, in addition to requiring elevated temperatures, is rapid and difficult to control. However, there are intermediate stages in the reaction, when the urea is still soluble in water . Such materials, called precondensates or prepolymers, are readily available from industry, since urea-formaldehydes are used in large quantities as adhesives. Of course, prepolymers are more viscous than the initial urea solution, but products are made which permit the final grout to be used at viscosities in the 10 to 20 cp range. The tradeoff in viscosity is made to obtain a product easy to handle, with good gel time control at ambient temperatures.

The prepolymer is a material in which the reaction has been suspended, either by inhibitors, pH control, or both. The reaction continues once the pH is brought down. Catalysis is therefore done with an acid or an acid salt.

Soils stabilized with urea-formaldehyde have strengths comparable to the phenoplasts, and like those materials, are less sensitive to testing strain rate than other chemical grouts. Little data is available, but it is probable that aminoplasts break down comparatively quickly under cyclic wet-dry and freeze-thaw conditions. Creep endurance limit is probably a relatively high percentage of the UC. Except as noted above, the resins have good stability and are considered permanent.

Grouts using urea solutions are toxic and corrosive due to the formaldehyde and acid (catalyst) content. Processes using a prepolymer contain much less free formaldehyde than those using a monomer, and therefore present much less of a hazard.

The gel, or resin formed at the conclusion of the reaction is inert, but it generally contains small amounts of unfixed formaldehyde. In an enclosed space such as a mine drift, this can become a nusiance and a hazard.

Several commercial products are available. One of the grouts in the Rocagil series is an aminoplast advertised specifically for use in coal mines.

Herculox has been used in the U.S. for several decades. It is a proprietary material and specific engineering data is lacking, but appears to be a pre-polymer.

Diarock, marketed internationally for about a decade beginning in the mid 60's appears (from the manufacturer's literature) to be basic urea formaldehyde resin (not a pre-polymer).

Cyanaloc 62 is a pre-polymer marketed as a concentrated liquid which is diluted with water for field use. Sodium bisulfate is the catalyst normally used. For very short gel times, or for shorter times at low temperatures, sodium chloride can be added. The gel time is actually the time required for formation of a soft gel (with consistency similar to the acrylamides). Over the next few hours to a day or more, depending upon the gel time, the gel cures to a much harder and stiffer consistency.

f. Water Reactive Materials

Materials which gel or polymerize upon contact with water offer obvious possibilities for use as grouts. Materials which can be foamed also offer possibilities.

Many foams were investigated in the late 1960's for their potential as soil grouts. (25) The study concluded that on formation methods alone polyethelene, polyvinyl, CCA, urea formaldehyde and syntatic foams can be eliminated from consideration, and that of the four other major types of foam (silicone, phenolic, epoxy and polyurethanes) the polyurethanes excel, having the best mechanical properties and the widest range of conditions of formation.

Polyurethanes are produced by reacting a polyisocanate with a polyol, or with other chemicals such as polyethers, polyesters and glycols, which have hydroxyl groups.

Catalysts, generally tertiary amines and tin salts, may be used to control the reaction rate, and synchronize the gelation and foaming reactions.

Surface active agents are used to control bubble size. The foam structure itself is produced by a blowing agent, reacting chemically to produce carbon dioxide.

The urethane linkage is not the only one occuring in the polymer. The isocyanate groups also react with carboxylic groups, hydrogen and nitrogen ions, and water. Thus, water plays a role in both the formation of a polymer and the reaction which produces the foam.

Under closely controlled conditions, a manufacturing process, for example, all of the chemical elements may be mixed simultaneously. For grouting purposes, use of a "prepolymer" is advantageous. The prepolymer is formed by adding a hydroxyl-containing compound to an excess of polyisocyanate. The prepolymer thus formed contains attached isocyanate groups, which can later be foamed and reacted with water and more hydroxyl groups.

Two commercial products are available as water reactive grouts. One of these, TACSS, was introduced in Japan in 1967. TACSS offers a choice of a number of different prepolymers, with viscosities ranging from 22 to 300 cp. The relatively high viscosity precludes the treatment of fine grained soils.

A second polyurethane grout (CR250) is marketed primarily for sewer sealing applications. It is applied internally through a packer inside the sewer, and acts as a gasket in the spigot.

Reports from PRC indicate that low viscosity polyurethane grouts are currently in use for grouting soils and fissured rock.

The isocyanates and other components of the grout are health hazards. As with most grouts, consistent safety procedures must be observed.

g. New Products

Late in 1980, AC-400 was commercially introduced. This is an acrylate based product, which is cross linked the same way as the acrylamide grouts, and uses the same catalyst system. The product is not neurotoxic, and its general level of toxicity is claimed to be about one percent that of the acrylamides.

Early in 1981, Injectite 80 was commercially introduced. It is based on an acrylamide pre-polymer, thus eliminating the toxicity problems, at the expense of viscosity. Reactions and end product are exactly like the acrylamides.

Also in 1981, a modified "CR250" polyurethane grout, intended for grouting soils, was placed on the market.

SUMMARY

Formulations based on sodium silicate are the most widely used chemical grouts. Formulations based on acrylamide are next, in order of usage. Silicates offer (relatively) low cost and high strength, with a minimum of environmental hazards. These materials are also (relatively) high viscosity, and under some conditions may suffer deleterious syneresis and dissolution. Low viscosity silicate solutions can be used, at the expense of strength, gel time control and possibly permanence.

Acrylamide grouts offer low viscosity and excellent gel time control, but are high in cost, low in strength and pose handling problems.

Research and development continues, in an attempt to combine the best properties of the silicates and acrylamides. Over the years, these have led to the marketing of commerical products classified (chemically) as phenoplasts, aminoplasts and lignosulfonates. The ideal product, however, is still unavailable.

REFERENCES

1. Adamovich, A.N., and V.K. Baushev, "Methods of Improvement of the Properties of Fissured Rock Masses by Injecting Synethetic Furan Resins," Proceedings of the International Society on Rock Mechanics, 2nd Congress, Belgrade, Vol. 3, Theme 6, no. 2, 1970, p. 137-141.

2. American Cyanamid Company, Am-9 Chemical Grout Field Manual, Engineering Chemicals Research Center, Princeton, N.J., 1976.

3. - - - - -, Cyanaloc 62 Chemical Grout, Explosives and Mining Chemicals Department, Wayne, N.J., 1962.

4. Badappanavar, M.G., "Chemical Grouting in Pervious and Rock Foundations," Journal of the Institution of Engineers (India), Civil Engineering Division, vol. 54, no. C13, 1974, p. 110-115.

5. Baron, J., "Silica Grouts," Draft Translation, U.S. Army Cold Regions Research and Engineering Laboratory, Hanover, N.H., no. CRREL-TL-552, vol. 31, no. 12, 1974, p. 2579-2601.

6. Bishop, A.W., M.F. Kennard, and P.R. Vaughan, Grouts and Drilling Muds in Engineering Practice, A.D.M. Penman, London, 1963.

7. Bowen, Robert, Grouting in Engineering Practice, John Wiley & Sons, New York, 1975.

8. Caron, C., New Horizons in Construction Materials, Enro Publishing Company, Lehigh Valley, Pa., vol. 1, 1976, p. 237-247.

9. - - - - -, "Special Grouts: Repair, Compaction, Sealing," Annales de l' Institut Tecynique de Batiment et des Travaux Publics, Enro Publishing Company, Lehigh Valley, Pa., no. 261, 1969, p. 1307-1318.

10. Chemical Grouts for Soils, Report No. FHWA RD-77-50 and 51, Vol. 1 and Vol 2), NTIS, Virginia, June, 1977.

11. Creep Tests, American Cyanamid Company, Soils Engineering Research Center, Princeton, N.J., November, 1959.

12. Drake, J.E., Salt for Short Gel Time and Low Temperature, American Cyanamid Company, Wayne, N.J., 1962.

13. Egorov, I.V., "Deep Consolidation of Ground by Carbamide Resin," Soil Mechanics and Foundation Engineering, no. 5, 1965, p.261-265.

14. Ellis, R.L., Epoxy Grouts and Their Various Uses in the Pulp and Paper Industry, Tappi Engineering Conference, Toronto, Onto., Vol. 2, 1975, p. 125-128.

15. Karol, R.H., Aging Tests AM-9 Chemical Grout, American Cyanamid Company, Wayne, N.J., 1963.

16. Karol, R.H., "Chemical Grouting Technology," Journal of the Soil Mechanics and Foundations Division, American Society of Civil Engineers, vol. 94, no. SM1, Paper 5748, 1968, p. 175-204.

17. - - - -, Chemical Grouting, Marcel Dekker, New York City, N.Y., 1982.

18. Lamberton, B.A., Terranier Chemical Grout, American Society of Civil Engineers, Structural Engineering Conference--Preprint 517, 1967.

19. Petrasovits, G., "Tests on Chemical Grouting and Waterproofing by Using Aminoplast Resins," Proceedings, International Conference on Soil Mechanics and Foundation Engineering, Budapest, p.429-440, 1963.

20. Plaisted, A., "Application of Chemical Grouts to Ground Consolidation," Ground Engineering, Vol. 7, no. 4, 1974, p. 42-44.

21. "Principles of Grouting Selection," Civil Engineering and Public Works Review, vol. 66, no. 783, 1971, p. 1137.

22. Skeist, I., Epoxy Resins, Reinhold Publishing Corporation, New York, N.Y. 1958.

23. Schiffman, Robert L. et al, Mechanical Behavior of Chemically Treated Soils, Lehigh University, Bethlehem, Pa., 1957.

24. Terra Firma Chemical Grouts, Intrusion Prepakt Company, Cleveland Ohio, August 5, 1963.

25. Vinson, Ted S. and James K. Mitchell, "Polyurethane Foamed Plastics in Soil Grouting," Proceedings of the American Society of Civil Engineers, Journal of the Soil Mechanics and Foundations Divison, Vol. 98, No. SM6, Proc. Paper 8947, June, 1972, p. 579-602.

A NEW GROUTING MATERIAL OF NON-ALKALINE SILICA SOL

Masaru Hoshiya[1]

Shunsuke Shimada[2]

Hitaka Kanematu[3]

Kenji Kayahara[4]

Abstract

Injection engineering in Japan has progressed rapidly during the past few years, specifically with regard to efficiency, environmental safety, and economy of money and resources.

One of the latest innovations has brought out a new grouting material based on non-alkaline silica sol which is obtained from water-glass with its alkali removed.

Also, a special device has been developed which carries out the production of silica sol and adjusts it to be used for grouting automatically and continuously at a construction site.

The grouting material maintains the pH-value of the underground water of the injected ground within the neutral range and causes no problem to underground water. Further, it has higher strength and penetrability than water-glass grouts with organic reactants.

In this paper, the principle of gelation and the results of laboratory tests of this material are described, and examples of practical applications are presented.

Preface

The ground which presents most problems in construction are weak alluvium deposits in urban areas and the highly fractured rock in mountain tunnels. In alluvium, one has to insure a homogeneous

---

[1]Professor, Musashi Institute of Technology, Tokyo, Japan.

[2]Kyokado Engineering Co., Ltd., Tokyo, Japan.

[3]Executive Director, Sanshin Kensetsu Kogyo Co., Ltd., Tokyo, Japan.

[4]Kyokado Engineering Co., Ltd., Tokyo, Japan.

solidification to avoid local collapses. The major problem in addition to proper homogeneity in grouting is to avoid environmental contamination. Here we want to prevent injected grout from leaking outside grout zones and to avoid changing the pH value and BOD of underground water within the prescribed ranges. Accordingly, we should develop injection techniques that enable permeation grouting while preventing leakage and also enable the use of injection materials which show gelation behavior suitable to the injection techniques and have high safety in view of environment preservation.

In order to resolve these problems, a non-alkaline grout of silica sol has been developed. In addition, a unique manufacturing plant to automatically produce the silica sol and injection method effectively taking advantage of the behavior of gelation of the silica sol is proposed. These technologies are now being put into practical use in Japan. This report will give the summary of the technologies.

General Characteristics of New Grout

Conventional water-glass grout gels in the alkaline region, but this is avoided with non-alkaline silica sol grout using only inorganic materials. Normally when water-glass is gradually added to an aqueous solution of an acid reactant, the pH of the solution is changed from acid to weak acid to neutral and its gel time becomes faster. (Fig. 1) The non-alkaline silica sol pH is regulated to a predetermined region to obtain a formulation which is the non-alkaline silica sol grout used in the injection works.

The non-alkaline silica sol grout has the following general features necessary for a grout:

(1) The gelation, as shown in Fig. 1, occurs in the form of a convexed curve indicating that the solidification of the sol takes place over a gel time ranging quick setting of several seconds to several hours. (Fig. 2)

(2) The pH of the underground water in the ground being grouted can be kept within practically neutral region (Fig. 3), and also no problems with BOD and COD arise.

(3) No unreacted water-glass is left behind because SiO_2 present in the grout mixture is completely embraced into gelation. For this reason, its gel strength is high and also its durability is excellent even if the concentration of water-glass is low. In addition, the non-alkaline silica sol grout is excellent in penetrability because of the low viscosity of the grout mixture. (Fig. 4)

(4) The economy of the grout mixture is excellent because the grout can be obtained by using small amounts of inorganic materials.

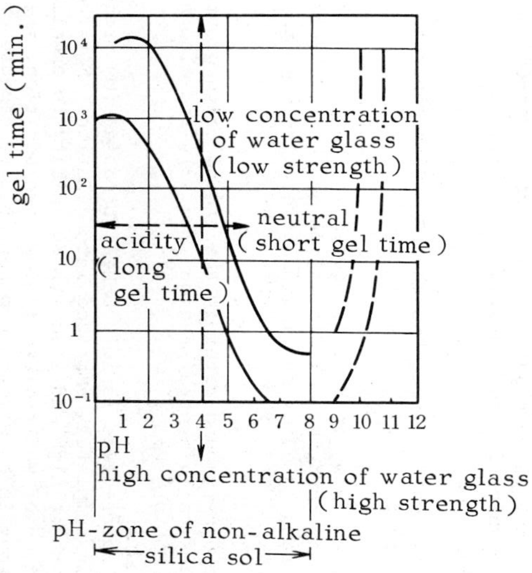

Fig. 1 pH range of non-alkaline silica sol grout and the general tendencies of the concentration of the water glass, the gel time and the strength.

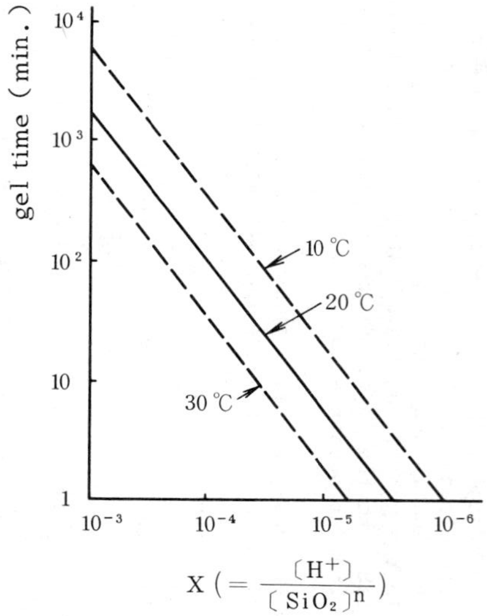

Fig. 2 Gel time of non-alkaline silica sol (n = mol ratio)

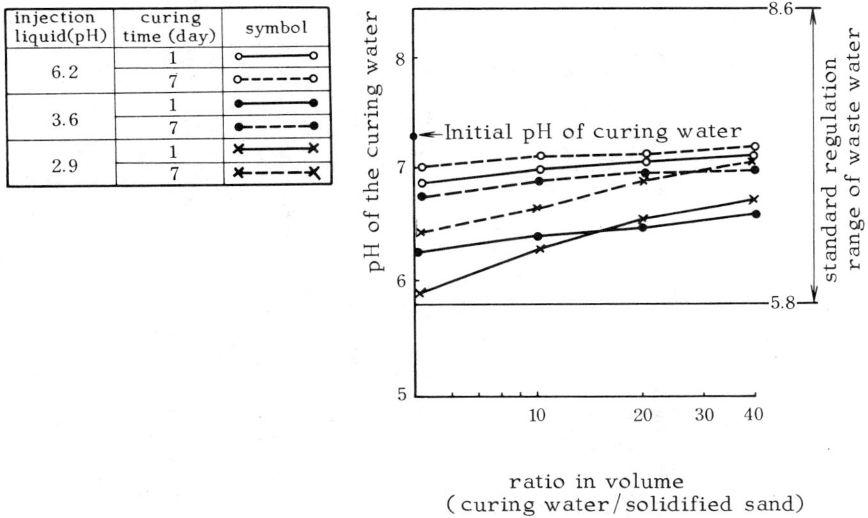

Fig. 3 Relation between the pH-value of cured water and the ratio of volume (no replacement of cured water)

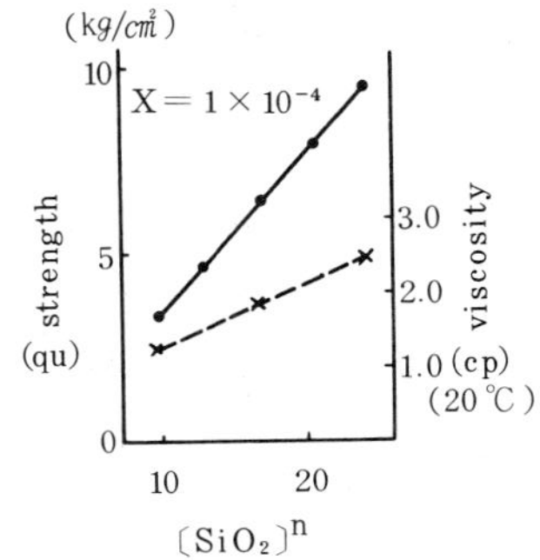

Fig. 4

Uniaxial compression strength of solidified standard sand (continuous line) and initial viscosity (broken line)

$$X = \frac{[H^+]}{[SiO_2]^n}$$

$n = 3.3$

Injection Techniques

The following three grouting techniques are used:

(1) A short gel time method with a double pipe

(2) A long gel time method with tube a manchette

(3) A chemical packer injection method with a double pipe, (CPI method)

By selecting the appropriate one from these three injection methods according to the purpose of injection and the conditions of ground, the grouting work of unsolidified ground could achieve most economical grouting results.

(1) A short gel time method with a double pipe

A. Principle

A double pipe rod provided with a special tip grout monitor whose tip is incorporated with a mixing chamber is used.

After drilling with the double pipe, the boring water is switched into chemical grouts, where a silica sol as the solution A and accelerator as the solution B are each sent through respective piping systems, mixed together in the grout monitor provided at the tip, and then injected into the ground. The grout is called "Clean-Rock." (Fig. 5, Photo. 1).

The gel time can be up to 5 minutes, but about 5 to 10 seconds is used in most cases.

B. Example--Chemical grouting work in a shield tunnelling work for laying a telephone cable in Yokohama city.

In this work a shield tunnel was being constructed of 3,658 mm diameter in sand. At a certain point the face collapsed, and underground water flowed out the lower part of the shield making further excavation impossible. It was decided to strengthen the ground by grouting, and the silica sol grout was chosen. A double pipe rod injection technique for the 17 m long sand layer area was selected. When a grout mixture of the liquid A and liquid B, having a gel time of 5 seconds and pH of 5 to 7, was injected into the sand layer, the resulting strength of the solidified sand was 3 to 4 kg/cm^2. The face was stabilized and further excavation was carried out safely.

A summary of the work is as follows:

Number of borings:	206
Total length of borings:	1,790 m
Total quantity of Clean-Rock injected:	302,367 liters

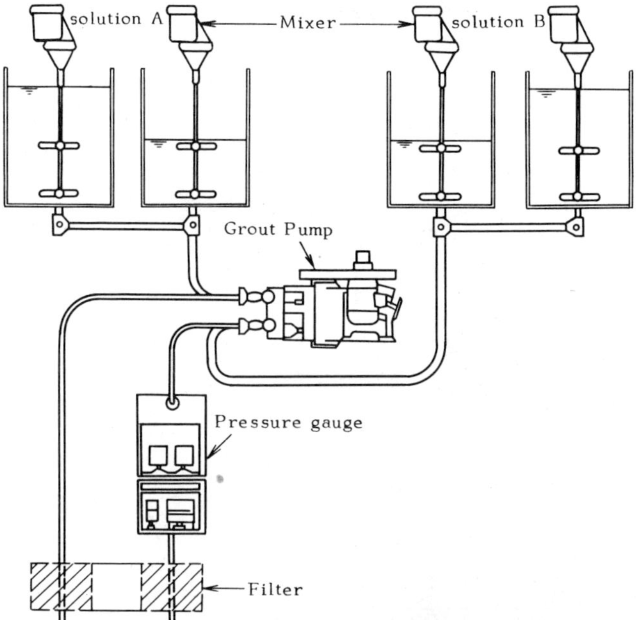

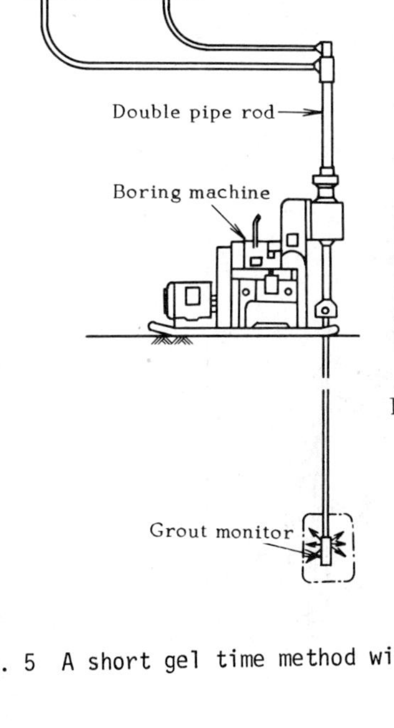

Photo. 1 Plant for Clean-Rock

Fig. 5 A short gel time method with Clean-Rock

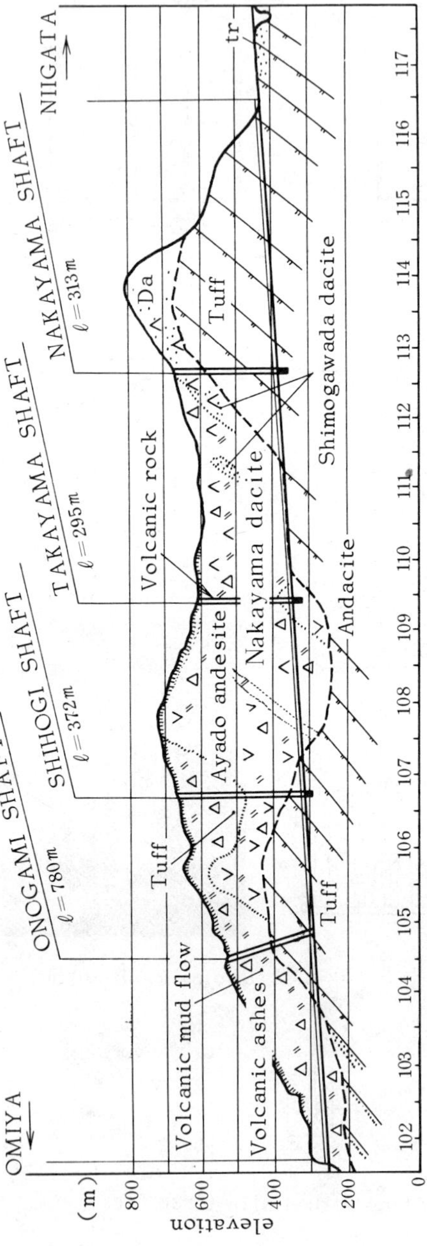

Fig. 6 Nakayama Tunnel of the Joetsu-Shinkansen Railway

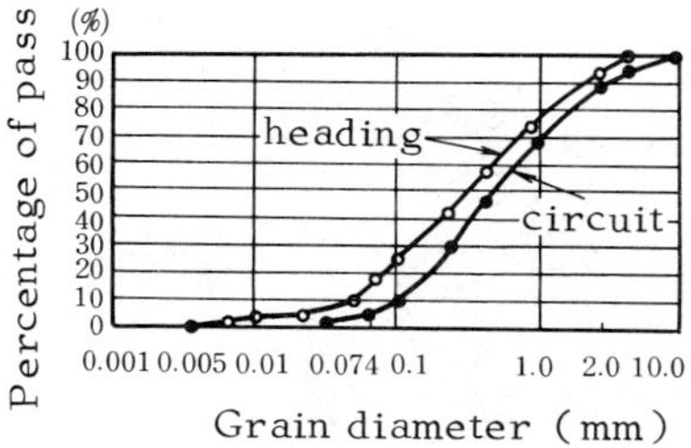

Fig. 7 Grain size curve of the grouted ground

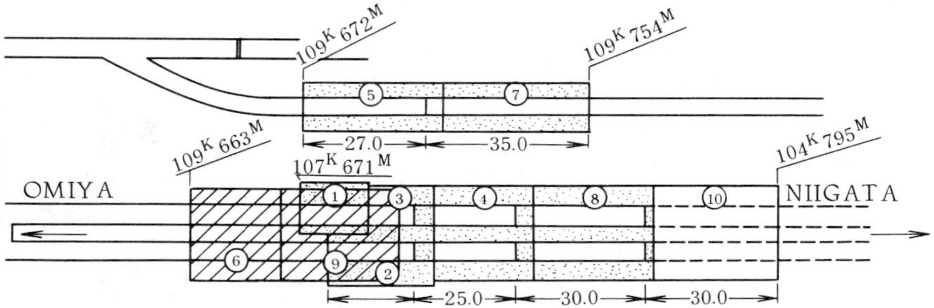

Fig. 8 Plain view of the Joetsu-Shinkansen Nakayama Tunnel work (Takayama section) showing the injection location by the Soletanche Method using silicalizer

(1) collapsed portion in the left heading

(2) the right heading, first injection

(3) the left heading, the first injection

(4) the both headings (right and left), for the second injection

(5) circuit heading, for the first injection

(6) main upper-half heading, for the first injection

(7) circuit heading, for the second injection

(8) the both headings (right and left), for the third injection

(9) main upper-half heading, for the second injection

(10) the both side headings, for the fourth injection

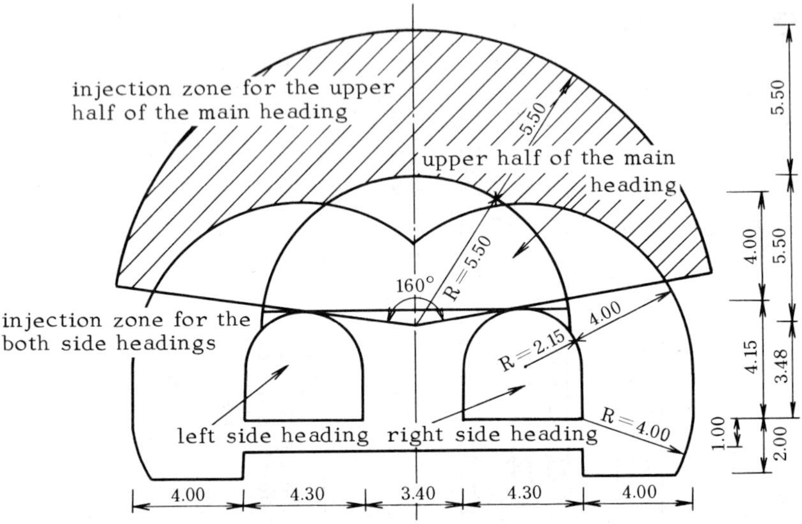

Fig. 9 Improved zone by Soletanche Injection Method with Silicalizer

Photo. 2 Control board of the plant which produces silicalizer

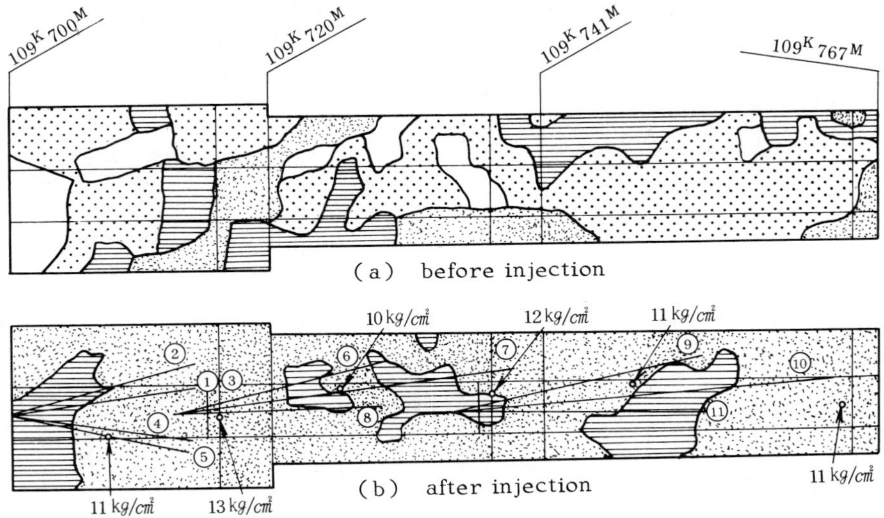

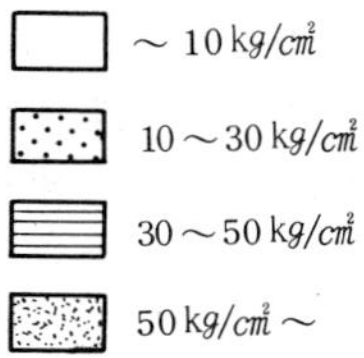

Fig. 10 Cross-sectional figure for penetration pressure distribution corresponding to penetration rate (10 l/min) before and after injection by penetration test using Tube a manchette

The lines shown radially show the positions of check boring.

In all cases, water spout was not observed.

The uniaxial compression strength of samples taken after injection is shown at the parts with o marks.

(2) A long gel time method with tube a manchette

A. Principle

In Europe, technological development of grouting was attained by the invention of tube a manchette with one-solution water-glass grout using an organic reactant having a long gel time and capable of giving a high strength. In our work, instead of the non-solution type water-glass grout using an organic reactant, a one-solution type silica sol grout "Silicalizer," having a long gel time and capable of giving a high strength, is used.

B. Example

In the Nakayama tunnel of the Joetsu Shinkansen Line (Fig. 6), the construction of Takayama area was started in 1972; and the excavation was plagued by various problems due to a large quantity of water. The excavation of a heading on Niigata side, from Takayama Shaft, depth of 295 m, was carried out through andesite strata, adopting a water extraction method; but upon reaching Station 109.7 km, excavation became impossible due to a cave-in of ground.

The ground is composed of unsolidified volcanic ejecta under artesian pressure (Fig. 7), and the rock is hard to permeate because the coefficient of permeability is 10-3 - 10-4 cm/s. However, in this material once a water path is formed, soil and sand can flow at once with high pressured underground water; and cave-ins occur. Therefore, the penetration property of injection materials became especially important regarding the stability of the ground.

As the result of investigations into the conditions of ground and the conditions of construction, the Soletanche injection method using tube a manchette with silica sol grout "Silicalizer" (one solution type non-alkaline silica sol grout with long gel time) was selected. First, a trial injection and excavation were carried out at the cave-in site. This showed that the excavation could be carried out in entirely in a dry state.

Subsequently, as shown in Fig. 8 and Fig. 9, grouting was applied to left and right side wall headings and a circuit heading.

Owing to the problem of workability in the heading, a system for continuously producing Silicalizer with gel time of 1.5 - 2 hours was developed in which the silica sol is produced with a plant outside a pit and transported through a pipeline to a plant inside. (Photo. 2) This plant enables the automatic continuous production of the silica sol having been adjusted to the prescribed pH region. Since it never involves manual operation directly, labor is saved and quality control is enhanced. The success of this system enabled the application of Silicalizer to this site.

A summary of the key project data is as follows:

Number of borings:	972
Total length of borings:	18,517.5 m
Total quantity of cement grout:	1,251.68 m^3
Total quantity of Silicalizer:	6,259,420 liters

The state of solidification of the heading faces where the injection was carried out is shown in Photo. 3; it was very solid. Therefore, the excavation was carried out by blasting, but there was no problem of water and cave-ins at all. The results of the uniaxial compression test of the specimens taken by excavation showed 4.5 kg/cm^2 in loose cave-in section, and 9-16 kg/cm^2 in undisturbed ground.

Fig. 10 shows the distribution of penetration pressure before and after the injection corresponding to the penetration rate of 10 l/min in the longitudinal cross section at tunnel center line.

From this figure, it can be seen that the unsolidified ground was irregularly constituted; but after the injection, water tightness and strength were improved remarkably.

(3) Chemical packer injection method with a double pipe

A. Principle

The chemical packer injection method with a double pipe (CPI method) was developed with the intention of attaining complete permeation grouting. A compound grouting in which formation of quick setting chemical packer by short gel time grout and permeation by long gel time chemical grout are combined through double pipe rod by one continuous process can be applied.

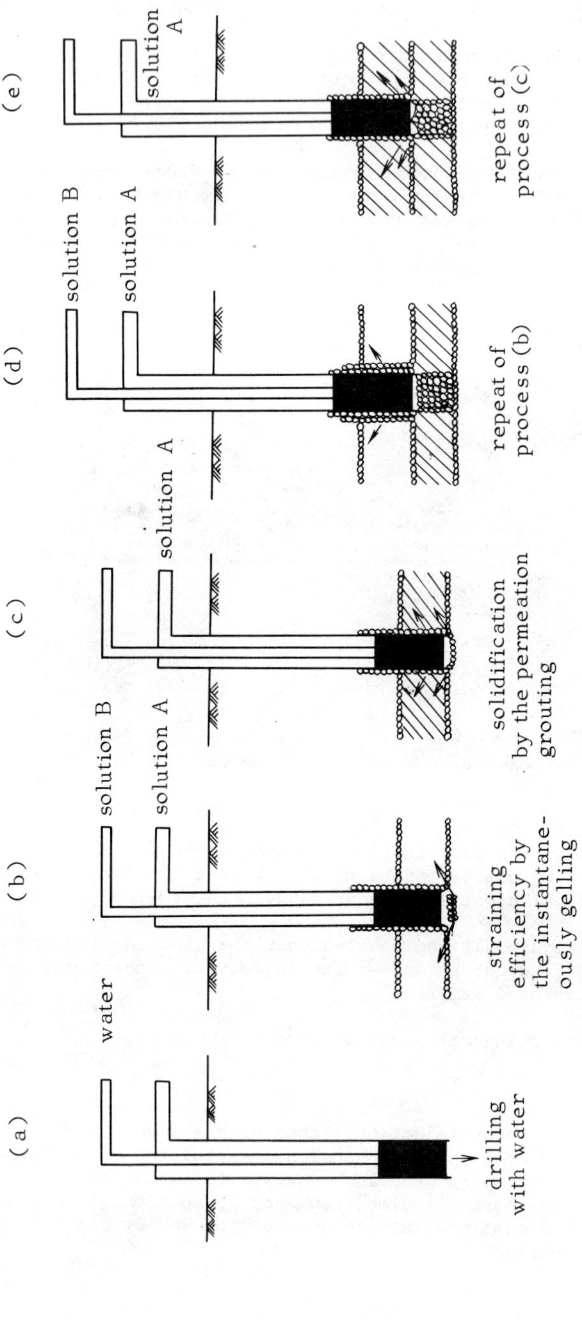

Fig. 11 Injection order of chemical packer injection method (CPI method)

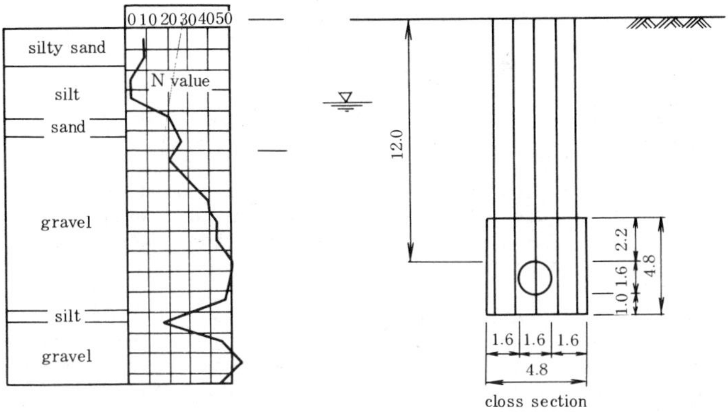

Fig. 12 Improved zone by chemical packer injection method with silicalizer

Photo. 4 State of facing after injection by chemical packer injection method

M. Hoshiya

Fig. 11 shows that the short gel time grout (gel time of several seconds) acts as a packer by solidification of the voids and coarse particle layers around the grouting pipe, and the long gel time grout serves for permeation grouting.

B. Example--Chemical grouting work in a pipe jacking project at the right bank of the river Katsura in Kyoto.

Conditions of ground: The ground consists of a dense sand layer and a pebble layer having a permeability of 1 to 1 x 10^{-2} cm/sec. and a high water level of - 4.0 m below ground surface; the ground is in contact with the river Katsura.

Grouting: Silica sol grout for permeation grout of a gel time of more than 30 minutes was intermittently combined with an accelerating agent at the end portion of a double pipe in order to form a quick-setting grout of gel time of several seconds.

Fig. 12 gives the cross-sectional view of injection work area.

The results of the uniaxial compression test of the specimens taken from the solidified ground showed 3 to 4 kg/cm^2. A summary of project data is as follows:

Number of borings:	1,500
Total length of borings:	71,000 m
Total amount of grout injected:	2,470,000 liters
Total length of executed grouting work:	500 m

The work was conducted safely and with no groundwater problems. (Photo. 4)

Conclusion

It can be said that the techniques presented herein lead the following conclusions:

(1) The new grout does not use directly petroleum products and high solidifiability can be obtained by small amounts of inorganic materials.

(2) Non-alkaline silica sol grout can attain long gel time as well as high strength by use of only inorganic materials.

(3) The non-alkaline silica sol grout solves the problem of water qualilty concerning pH and BOD.

(4) The development of a compact automatic plant of manufacturing silica sol made it possible to the practical use of the non-alkaline silica sol grouting.

(5) Development of injection procedures utilizing the gelation characteristics of non-alkaline silica sol made it possible to apply compound injection by simple operations.

Acknowledgment

In writing this paper, we were assisted by Japan Railway Construction Public Corporation, Raito Kogyo Co., Ltd., Sanshin Kensetsu Kogyo Co., Ltd., and Kyokado Engineering Co., Ltd. We express gratitude to the above people.

References

(1) T. SASAO, T. SUGA, S. TUCHIYA, and S. SHIMADA, Development and Applications of Injection Method at the Takayama Construction Area in Nakayama Tunnel, the Joetsu Shinkansen Line. Journal of the Japan Society of Civil Engineers. Sept., 1980.

(2) S. SHIMADA, K. NISHIDA, and T. MURASE (1978), Characteristics of Non-Alkali Silica Sol Grout. Collection of Synopsis, 33rd Annual Lecture Meeting, the Japan Society of Civil Engineers.

(3) S. SHIMADA, K. KAYAHARA (1979). Behavior of pH of Grout of Suspension and Solution Type of Silica Sol. Collection of Papers, 15th Report Presentation Meeting, Japan Society of Soil Engineers, pp. 1673-1676.

(4) S. SHIMADA, and K. KAYAHARA (1980), Technical System of Compound Injection Method by Double-Tube Rod and Instantly Consolidated Packer System. Collection of Synopsis, 35th Annual Lecture Meeting, The Japan Society of Civil Engineers.

Injectite®-80 Polyacrylamide Grout

Richard M. Berry, ASCE Member[1]

Abstract

A new polyacrylamide grouting system was commercially introduced to the sewer grouting industry in the Fall of 1980. The system uses a radically different catalyst system. All the components of the system are low toxicity, biodegradable organics or inorganic salts. Initial evaluation of the material shows it in all respects equal to and in some respects superior to the mono-acrylamide system used for the past twenty years.

This paper reviews early data on the system, field performance in water shut off, use in waterstop applications in dams, locks and other concrete and masonry structures, and presents initial performance data in soil grouting applications.

Properties and performance are compared to similar applications of the mono-acrylamide system.

Background and Chemistry

Injectite -80 is the trade name for a new polyacrylamide chemical grout system[2]. It has many similarities to the older widely known mono-acrylamide product (AM-9) now produced in Japan. It differs in its pumping characteristics, its mixability with groundwater, its long term stability, and its low toxicity.

Injectite® is a trademark of BPR Corporation

(1) Technical Director, Penetryn Restoration, A Division of BPR Corporation, Knoxville, TN

(2) U.S. Patent No. 4,199,625

First, the similarities: Injectite-80 grout (I-80) has the ability to form a relatively stable gel in water down to concentrations of about 5-6% polymer, 94-5% water. This is about the same threshold as AM-9. AM-9 is acrylamide monomer crosslinked with about 3% methylene-bis-acrylamide using a complex catalyst initiator system of ammonium persulfate and an amine. I-80 is a low molecular weight linear polymer of acrylamide crosslinked directly with itself using a totally different catalyst system consisting of active chlorine; bleach, in a word. In both systems, the polymer has the ability to entrap large quantities of water. The gels have a considerable memory. That is, they will shrink somewhat when removed from a wet environment, but will re-swell to approximately their original shape when re-wet. The chemistry in the two systems is shown below in Figure 1.

Figure 1
Chemical Structure

Injectite-80

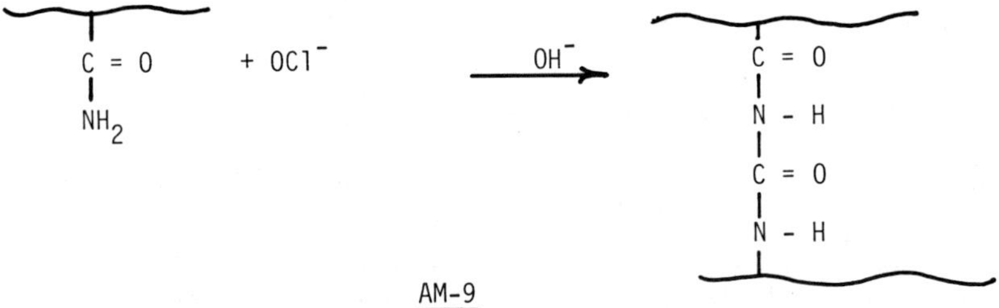

AM-9

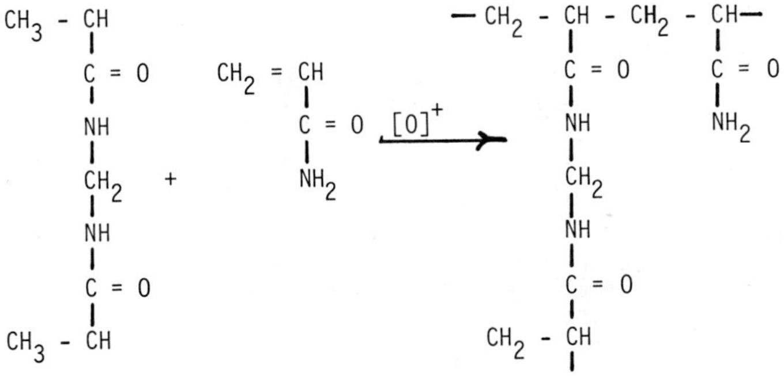

Although serving the same general function, I-80 was developed in response to toxicity in the AM-9 acrylamide monomer system. Acrylamide itself is toxic, and some of the catalyst materials have similar problems. The most toxic part of the I-80 system is a hypohalite initiator, normally calcium hypochlorite, a common swimming pool algicide.

However, the I-80 polymer is formed through a different mechanism, and this probably accounts for some differences. While the AM-9 system directly crosslinks acrylamide molecules with bis-acrylamide, the I-80 system most likely is a direct carbon to carbon linkage of the polyacrylamide molecules. A very small amount of glyoxal is used to effect a rapid gel [0.1%], but the stable chemical bond is formed by the hypochlorite activated crosslink, which takes longer. Comparison of 3 minute and 60 minute strengths [yield stress of 10% gel] shows the early strength of AM-9 higher, but the ultimate strength never as great, as shown below:

	Gel Time	3 Min.		60 Min.	
	Sec.	kN/m^2	psi	kN/m^2	psi
I-80	20	179	26	990	144
AM-9	20	516	75	790	115

Wet/dry cycling in tests shows I-80 to be somewhat less fragile than AM-9. On shrinking as a result of drying, the material is somewhat more leathery and it cracks less. When re-wet, it swells back to its original shape without breaking, exhibiting a physical "integrity" better than AM-9. Some data on chemical stability is shown below:

Effect of Chemicals on Grouts

	I-80	AM-9
Hydrochloric Acid 2%	none	none
Sulfuric Acid 2%	slight swelling	swelling
Sodium Hydroxide 2%	swelling	swelling
Sodium Chloride 15%	slight swelling	slight swelling
Laundry Detergent 1%	swelling	swelling
Sodium Sulfite 5%	slight swelling	slight swelling
Toluene	none	none
Heptane	none	none
Isopropyl Alcohol	shrinkage	shrinkage
Ethylene Glycol	slight shrinkage	shrinkage

The I-80 polymer is a solution in water, normally shipped as a 20% concentration. That solution has a standard viscosity of around 300 centipoise (cps). When diluted to 10% concentration at the mixing head, the viscosity reduces to about 50 cps at the injection point. If a 6% gel is acceptable, the viscosity can be as low as 17 cps. This viscosity effect is important to soil grouting and water stop processes, both pro and con. The AM-9 system has substantially the same viscosity as water. The I-80 uncured gel will mix with water but reluctantly, as maple syrup mixes with water. It tends to displace water in soil and around structures rather than to mix with it. In general, some 20-40% less material is used than with AM-9. The material remains more concentrated around the point of injection. Soil samples from test work confirm that logic in that the grout is often nearly pure gel adjacent to the point of injection, and a very rich concentration in sand or soil close in. By contrast, AM-9 injection shows greater penetration through the soil, but a much more ragged gel front out at a distance, often tailing off to a soft, weak, less stable material at the outer fringes.

The dilution effect in water of the ca. 2 centipoise viscosity AM-9 is obvious. In running or moving water, the 10-50 centipoise I-80 provides a considerable advantage. In order not to wash away, operations using AM-9 compensate by increasing the polymer concentration, often up to 20%, so dilution can be tolerated without going below the 6% polymer level. With I-80 this is not as necessary. A viscosity vs. concentration curve is shown in Figure 2.

Figure 2
I-80 Viscosity

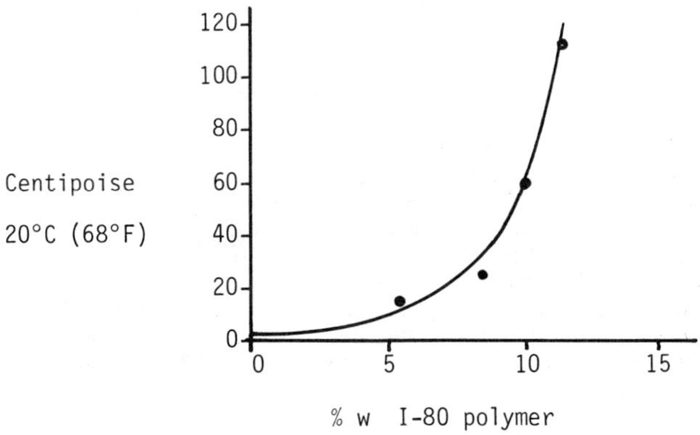

The intent of the grouting is a good starting point for any discussion of choice of product. In the case of I-80 and AM-9, one of the major points to consider is whether the area to be treated needs saturation with the material or whether some displacement of water and/or soil makes for a better job. Theoretical arguments provide very little in the way of guidance. It is the author's present opinion that developing a high polymer concentration close in to the zone with the problem is the best solution in the great majority of cases where water shut off or curtaining is the objective. Although pumping pressures have to be higher with the more viscous I-80 system, and longer gel times may be required if deep penetration is desired, the lesser quantity of grout used and the greater assuredness of the system are major compensating factors. The worry about water dilution to below the threshold concentration, about 5-6% polymer, to make a "good" grout is significantly reduced.

Stability of the I-80 system is great. The polymer generally fits the chemical category of a nylon, inert to most things. This is also true of a well cured AM-9 gel. Both are inert to most things contained in groundwater, seawater, and municipal and industrial waste. The chemicals which affect I-80 grout most are strong concentrations of hydrating and dehydrating agents. Strong caustic (say 20%) causes significant swelling and the alcohols and glycols draw water out and cause shrinkage. The I-80 grout appears stable in moving water. It does not exhibit the slow deterioration, presumably by hydrolysis, shown by polyurethanes, nor the solution deterioration apparent in the silicates when subjected to moving water.

Applications

The primary application of I-80 grout so far has been in water shut off, preventing the leakage of water through structures buried in the soil, along waterways, or sometimes within the structure. Perhaps this is the most significant application of chemical grout.

I-80 has been used to seal major leakage in a seawall in Florida in 1980. The anchor bolts had rusted and eroded and large quantities of water were entering a concrete foundation some ten feet behind the structure. Holes were drilled in the structure and I-80 grout forced into the holes to intercept the water at its source rather than to try to seal up the structure, as ingress was spread out. The viscous, non-mixing characteristics of the grout were a significant advantage, and grout penetrated all the way to the bolt holes in the seawall.

At the Watts Bar lock in the Tennessee River, a lock with an 85 foot lift, water was leaking in for almost the full length of a vertical construction joint giving a particular problem to small boats using the lock. Water pressure near the bottom was nearly 40 psi. The project was carried out in April of 1980 while the lock was down for other maintenance. Plastic ports were installed at two foot vertical intervals through holes drilled directly into the construction joint. The holes were drilled to a six foot depth, although only the first two feet were through concrete, in an attempt to saturate the area. Injection was begun at the bottom. Using a 75 second gel time, each port was pumped to refusal pressure or until grout appeared at the port above. Some 3/4 of the ports exhibited connection to the next port while the rest went to refusal pressure. The cure with I-80 is somewhat less abrupt than with AM-9. From the initial set at 75 seconds, the material slowly thickens but remains pumpable for about 50% longer time, or about 110 seconds. The operators say it is more "forgiving." It does telegraph something about what is happening, rather than exhibiting the abrupt cutoff characteristic of AM-9. The injection was done on a hot spring day, and another characteristic of the I-80 system became evident. The gel time is less affected by temperature than AM-9, and control problems are less where the temperature varies during the day as shown in the data below:

Temperature		Gel Time, Sec.	
°C	°F	I-80	AM-9
0	32	100	225
20	68	20	20
40	104	16	6

A unique application took place at the Passaic Valley Sewage Commission wastewater treatment plant, under construction in New Jersey at the time. Four inch electrical conduits were buried in the floors of the buildings often running through the expansion joints. The floor of the main electric gallery in the clarifier building is some 25 feet below mean high tide. Many of the conduits leaked, some badly. The conduit turned up and surfaced at control centers and pull boxes, all of which had couplings above the floor. When not plugged, the conduit leaked into the boxes and onto the floor. If cable caps were inserted, the couplings leaked. The solution adopted was to go ahead and pull the cables in the wet conduit, and then seal the conduit, full if possible, with a chemical gel to cut off the water at its source. Work on the project began in July, 1981. The worst leak flowed at about 12 gallons/minute from a 180 foot line. Gel times were set to try to attain complete filling of the line, and for the most part, this succeeded. In some of the wettest cases, conduits were pumped from both ends as leakage out the conduits far exceeded the calculated amount of grout the line could take. In a few of the conduits, the second end was filled as much as possible, allowing the heavier than water grout to sink and overflow standing water from the conduit. Over 200 conduits were sealed in this manner.

Only one has leaked and that is a persistent wet one around the outside of the conduit; none have leaked inside. AM-9 was used to seal some of the conduits, but failure to attain a seal in those with large amounts of running water caused a switch to I-80. We attribute the failure of AM-9 to dilution and mixing in fast flowing water. Telephone cable conduit was initially sealed by New Jersey Bell with a foamable polyurethane grout, but the incidence of failure was high. The worst of the leakage has now been re-sealed with I-80.

A water cutoff project was undertaken at a water pumping station in late 1980 in Florida, where construction had begun on an expansion project. Water from a nearby river had begun to flow into the excavation. Both AM-9 and I-80 were used on the project, pumping the material directly into the soil, largely a mixture of clay and rock. Nearly equal areas (footages) were pumped with each product. Both were successful in stopping the leakage, which was flowing at over 500 gallons/minute. Grout consumption of AM-9 was 50% greater than I-80, with 1,120 gallons of AM-9 used compared to 825 gallons of I-80. The sections grouted were kept as nearly as possible equal, and equal 10% polymer concentrations were used. The cost of I-80 is about 10% greater than AM-9 for equal grout concentrations. The applied cost of I-80 has been significantly lower in most cases because of the smaller quantity of grout used.

Operational Aspects

In order to keep the pumping pressures low in the viscous polymer side (300 centipoises for a 20% solution), we have adopted a 5/1 polymer to catalyst ratio. This is attained with a positive displacement multiple cylinder pump running off a single air motor. A typical mix is shown below:

Typical Components and Formula
36 gallon mix

Polymer Tank		
18 gallons	I-80 polymer	(20% concentration)
14 ounces	glyoxal	(40% concentration)
12 gallons	water	
Catalyst Tank		
1.5 lb.	calcium hypochlorite	(commercial 65% swimming pool grade)
0.5-4.0 lb.	trisodium phosphate	(shop detergent grade)
6 gallons	water	

The catalyst solution is highly corrosive to steel. Dilute chlorine solutions used as catalyst are about 10 times the concentration as those encountered in a swimming pool. Everything on the catalyst side must be a high alloy or plastic. Maintaining

Table 1

Safety Data Comparisons

Polyacrylamide System	LD50 mg/kg	Flash Point	Storage & Handling Precautions
Injectite-80 Grout	5,000	None	32-110°F for 1 year storage 150°F shorter term
40% Glyoxal (1%)	5,000	None	Glasses, rubber or vinyl gloves
Trisodium Phosphate (1%)	4,500	None	Glasses, rubber or vinyl gloves
Calcium Hypochlorite (1%)	850	None	Glasses, rubber or vinyl gloves Gives off chlorine when heated Shipped as yellow label
Acrylamide System			
Acrylamide Grout (AM-9)	150	None	Accumulative neuro-toxin
Triethanolamine (10%)	5,000	None	Eye Irritant
Ammonium Persulfate (10%)	820	None	Skin Irritant
Urethane System			
CR202 Urethane Grout or CR250 Urethane Grout	5,000	20°F	Store under 80°F, shipped as red label, flammable liquid
Toluene Diisocyanate (2%)	5,000	270°F	Organic vapor mask Respiratory irritant TLV 0.02ppm
Acetone (used for cleanup)	5,000	0°F	Storage below 80°F Red label shipment, highly flammable
CR205 Cleaning Compound	Caustic	58°F	Flammable, may cause caustic burns, poisonous. Red label shipment
Comparative Substances			
Ethyl Alcohol	5,000		
Table Salt	4,000		
Aspirin	1,050		
DDT Insecticide	100		

the proper amount of chlorine in the system is important to end up with a permanently stable polymer. Therefore, the correct ratio of polymer to catalyst is important, more so than with AM-9. The polymer is stable under storage conditions, and is good for 3 or more days even after the glyoxal is mixed in, even at elevated temperature. The chlorine solution is unstable and must be mixed each day. However, the chemicals involved, calcium hypochlorite and trisodium phosphate, are inexpensive and used in small quantities.

The Injectite-80 system was originally developed in response to a need for a safe sewer grout. It is safe and easy to handle. The base I-80 polymer has a toxicity about equal to ethyl alcohol. The worst chemical in the system is calcium hypochlorite, dry bleach, widely used as a swimming pool chemical. Some safety and toxicity data on I-80 is compared with AM-9 and polyurethane grout in Table 1. The polymer makes a concrete surface slippery when spilled on it. About the worst complaint of the operators is that it bleaches their clothes.

Conclusion

Injectite-80 polyacrylamide grout is a new safe-to-use chemical grout with wide potential in soil stabilization and waterstop operations. It is especially useful in applications where dilution by running water or large quantities of groundwater constitute a problem.

References

1) Minsk, L.M., et al. U.S. Patent 2,486,192 (1949)
2) Meinel, K., German Patent 940,734 (1956)
3) Wacker-Chemie, British Patent 771,478 (1957)
4) General Mills, British Patent 865,755 (1961)
5) Cassella Farbwerke Mainkur, British Patent 891,268 (1962)
6) Gentile, A.J., U.S. Patent 3,114,651 (1963)
7) Suen, T.J., et al., U.S. Patent 3,168,500 (1965)
8) Davis, C., Jr. German Offen Patent 2,231,810 (1973)
9) Bracke, W.J.I., German Offen Patent 2,108,811 (1972)
10) Coscia, A.T., U.S. Patent 3,556,932 (1971)
11) American Cyanamid Company, AM-9 Chemical Grout (1965)
12) Pilny, R.J., et al., U.S. Patent 4,199,625 (1980)
13) Adams, R., et al., Organic Reactions, Wiley & Sons, NY (1946) pp 267-306
14) Flory, P.J., Principles of Polymer Chemistry, Cornell University Press, Ithaca, NY, 1953, pp. 317-398

POLYURETHANE GROUTING IN HYDRAULIC ENGINEERING

Liu Jiacai* **

Wang Baochang* Cui Wengguang*

Guo Yuhua* Cheng Hesheng*

Abstract

The paper describes the principles of the polyurethane grouting, the mechanisms of the chemical reactions of grouts on water as well as the main properties of the grouts and its hardened polymers. Some formulations of the grouts are given for various applications.

The injecting process control is based on the theoretical relations between the grouting parameters. The grouting technology has been applied to hydraulic engineering for various purposes.

Polyurethane grout is a high molecular grouting material developed recently. It is primarily produced by mixing the polyol with the excessive polyisocyanate to form a low prepolymeric compound, containing some free (OCN) groups. The compound produced is so called prepolymer. The grout used for injection is composed of prepolymer,

---

*Research Institute of Water Conservancy and Hydroelectric Power, Beijing, China.

**Member of Committee of Geotechnical Engineering, Chinese Society of Hydraulic Engineering plasticizer, diluter, surfactant and catalyzer.

The mechanism of reactions among the isocyanate, polyol and other components is rather complicated. According to the general theory of polymerization of high molecular compound, it can be expressed as follows:

1. The reaction between the excessive isocyanate and polyol yields prepolyurethane.

$$\begin{array}{l} OCN\text{-}R\text{-}NCO{+}HO\text{-}R'\text{-}OH \longrightarrow \\ OCN\text{-}R\text{-}NH\text{-}\underset{\underset{O}{\|}}{C}\text{-}O\text{-}R'\text{-}O\text{-}\underset{\underset{O}{\|}}{C}\text{-}NH \end{array} \qquad (1)$$

2. The reaction of polyisocyanate with water is accompanied by the formation of CO_2 and urea derivatives.

$$\text{-}R\text{-}NCO{+}H_2O{+}OCN\text{-}R\text{-} \longrightarrow \text{-}R\text{-}NH\text{-}\underset{\underset{O}{\|}}{C}\text{-}NH\text{-}R\text{-} \quad {+}CO_2\uparrow \qquad (2)$$

3. The reaction of polyisocyanate with ureido causes the development of molecular links and high molecular formation.

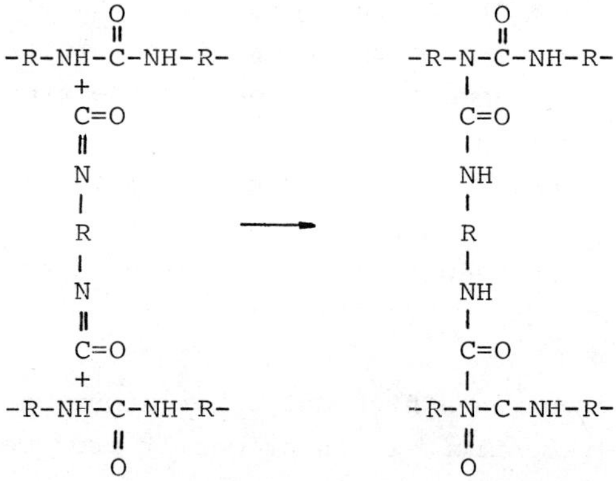

(3)

The reactions given above occur because of the existence of the free (OCN) groups in the grout, which can react with the compounds containing active hydrogen atom (H), such as hydroxy, water, amino and ureido.

The essence of such reactions is that the hydrogen atoms (H) move to link up with the nitrogen atoms (N) of the polyisocyanate and form high molecular polymer having rather high strength and low permeability.

The main factors influencing the properties of the polyurethane grouts and its hardened polymer can be listed as follows: the number of hydroxy and molecular weight of polyol, the kind of polyisocyanate, the percentage content of isocyanato (OCN) groups in grouts, the kind and amount of additives and catalyzer, etc. The more the number of the hydroxy, the higher the strength of the hardened polymer. In the same series, the larger the molecular weight of the polyol, the lower the strength of the polymer. The more the free (OCN) groups contained in the grout, the more the activity, and the more the amount of CO_2 liberated. When the percentage content of the free (OCN) groups exceeds the optimum value, the strength of hardened polymer will decrease. The optimum value should be found out by means of tests. The higher the environmental pressure, the denser the hardened polymer.

The viscosity of the prepolymer is rather high, therefore, it is necessary to be diluted by adding some one of the following: acetone, xylene, ethylacelate or dichloroethane. Among them acetone is the best. Besides these, plasticizers such as dibutylphthalate also serve the same role as a dilution agent. The effects of dilution are illustrated in Figs 1 and 2.

For the purpose of reducing the surface tension of the grout and improving its properties, stabilizers and surfactants are applied by an amount of 0.5-1.7% of the prepolymer (by wt.).

Tertiaryamine reagents, such as triethylamine, triethanolamine and triethylenediamine, can be used as the catalyzers for accelerating the polyisocyanate/polyol reaction. The catalyzing action of triethylamine is more powerful than that of triethanolamine, therefore the former is often used in cases of short setting and the latter in cases of long setting. Fig.3 shows the effects of these two catalyzers.

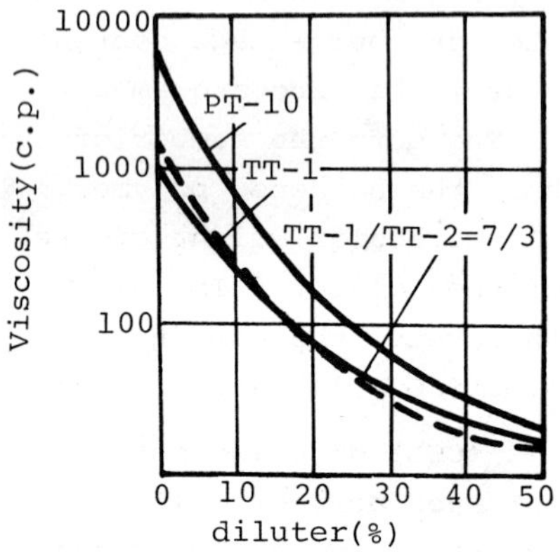

Fig.1. The relation between viscosity and the percentage content of diluting agent.

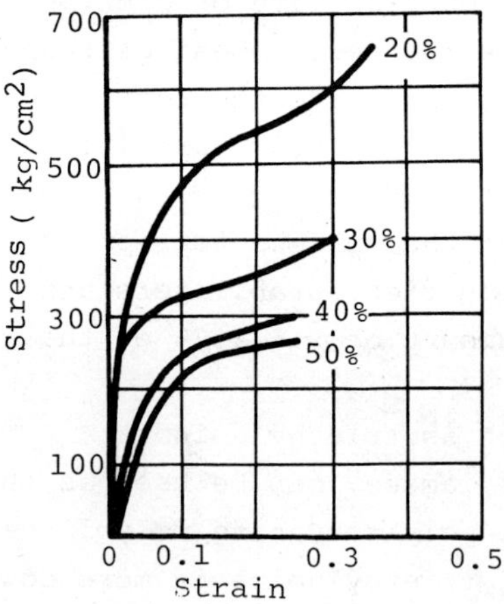

Fig.2. Stress-strain curves of the hardened grouts containing different amounts of diluting agent.

Laboratory and in-situ tests have led to the development of several formulations of polyurethane grouts for application to hydraulic engineering (Table 1)

Problems.

The percentage contents of the free isocyanato (OCN) groups in these formulations are suited to application in hydraulic engineering. The Sk-1 grout contains quite a number of the aromatic groups, when fresh, it has a low viscosity and, when hardened, it has a rather high strength. The Sk-3 and Sk-4 grouts have high adherent strength and low modulus of elasticity, so that it can stiffly adhere to the surface of the concrete cracks and accommodate to the temperature change. The triethanolamine is an active catalyzer which also takes part in the reaction with the polyisocyanate.

Table 1. Formulations of Grouts

Component	Composition (by wt.) \ Formulation	Sk-1	Sk-3	Sk-4
Prepolymer	PT-10	100		
	TT-1/TM-1		100	
	TT-1 / TP-2			100
Plasticizer	Dibutylphthalate	10	10	10
Diluter	Acetone	10	10	10
Catalyzer	Triethanolamine	0.5-2.0		
	Triethylamine		0.2-0.8	0.2-0.8
Surfactant		1.0	1.0	1.0
Water		Theoretical Value	Theoretical Value	Theoretical Value
Field of application		Consolidation and sealing the minor fissures, faults and weak seams	Sealing the cracks against flowing water	

The corresponding properties of the polyurethane grouts and its hardened polymers are listed in Table 2.

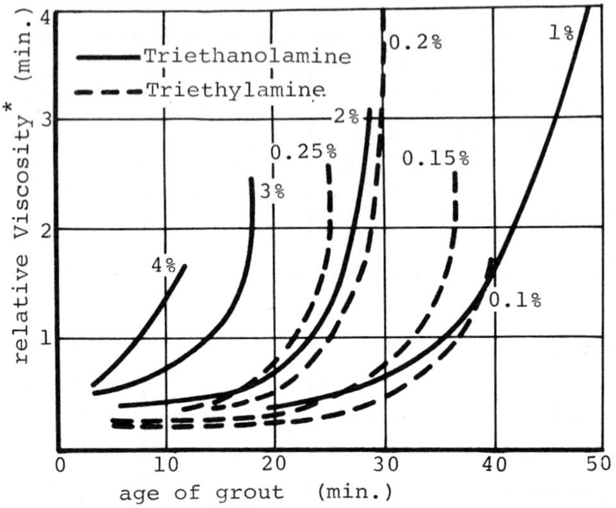

Fig. 3. The relation between the relative viscosity* and the age of grout with different percentage contents of catalyzer

Polyurethane grouting is characterized by the following processes. When th grout is introduced into the water bearing fissures or cracks, the reaction of polyisocyanate with water forms urea derivatives and simultaneously liberates CO_2. At the same time, the grouts transform from single phase (liquid phase) system into two-phase (liquid and gas) system. The pressure of the system itself rises and the grouts have a tendency to spread toward the zone with lower pressure. Thanks to this phenomenon, known as "secondary spreading", the spreading radius of the injected gouts will be greatly increased. As the viscosity of CO_2 is extremely low, the penetrability of the system is promoted.

---

*The relative viscosity of the grout is defined by the falling time that a steel ball, 4^{mm}(o.157") in diameter, needs to travel 50 cm (19.68") in the grout, in min.

Table 2 The properties of the polyurethane grouts and its hardened polymer

Item	Formulation / Property index	Sk-1	Sk-3	Sk-4
Grout	Free(OCN) (%)	21.2	18.4	18.3
	Specific gravity	1.12	1.14	1.15
	Viscosity at 20°C (C.P.)	19.1	161.9	166.6
Hardened* solid of grout	Compressive strength (kg / cm^2)	240 (3406)	200 (2839)	230 (3264)
	Adherent strength with concrete (kg / cm^2)	16.6 (236)	18.3 (259.7)	28.3 (402)
	Water resisting pressure (kg / cm^2) (length of sample=6cm)	15 (213)	10 (142)	10 (142)
Hardened mixture of grout with standard sand**	Compressive strength (yielding value) (kg / cm^2)	140 (1987)	100 (1419)	99.8 (1416)
	Modulus of elasticity (kg / cm^2)	9250 (1.3×10^5)	2870 (4×10^4)	2962 (4.2×10^4)

* Samples prepared under confined condition

** Samples prepared under unconfined condition

*** The unit of values in the blanks is lb. per sq. in.

It can penetrate into the fissures with a width of 0.01-0.03mm. (3.9×10^{-4}-1.2×10^{-3} in.) The setting time of the grouts can be regulated in a wide range according to the needs of injection.

As described above, the properties of polyurethane grouts are different from those of other chemical grouts, and a corresponding grouting technology must be adopted in order to make the grouting works successful. It can be concluded as follows:

The contact of grout with water in the pipelines for grouting will cause it plug up. A single direction valve should be placed at the lower end of the pipe to keep away from water.

As soom as the grouts are injected into the water-bearing fissures, the reaction between the polyisocyanate and water begins. The viscosity of the grouts increases as the reaction is going on. In general, the injecting process should be performed as quick as possible by using a high allowable pressure at a low viscosity.

The setting time of the grouts must meet the time needed for injection, corresponding to the designed spreading radius. It can be determined by formula (4):

$$T = \frac{1.02\text{X}10^{-7}\,\eta\,(R^2-r^2)\ \ln\frac{R}{r}}{(P-P_o)\ b^2} \qquad (4)$$

Where: T —— Time needed for injection, min;

η —— Viscosity of grouts, c.p;

R —— Spreading radius of grouts, cm;

r —— Radius of the grouting hole, cm;

P —— Pressure in the grouting hole, kg/cm^2;

P_o —— Hydrostatic pressure of the groundwater in the water bearing fissure, kg/cm^2;

b —— Thickness of the fissure, mm.

The secondary spreading action of the polyurethane grouts is favourable to the grouting work, because it can increase the spreading radius. The tests performed in the laboratory proved that the secondary spreading radius is nearly equal to 50% of the primary one.

For the promotion of the secondary spreading, the packer should not be relaxed until the process of the secondary spreading is over.

As noted, the polyurethane material has properties suitable for hydraulic engineering. It has been used to construct permanent impervious curtains, to consolidate and seal the minor fissures, faults and weak seams in the rock masses, and to stop the flowing water in the cracks in water retaining works. The following examples describe the application of the polyurethane grouting.

Fengtan and Chencun Dams

An example of the application of polyurethane grouting to treating the faults and weak seams in the rock foundation is the Fengtan dam, situated on the Youshui River. It is a hollow gravity arch concrete dam with a height of 111 m and a storage capacity of $1.55 \times 10^9 m^3$ ($5.4 \times 10^{10} ft^3$). The capacity of the power plant is 400 MW. The dam foundation is composed of sandstone, quartz-sandstone and feldspathic quartz sandstone interbedded with thin slate and arenaceous slate. There exist a number of faults, weak strata and seams (Fig4a, 4b). They were considered as the main defects needed for foundation treatment. Although the faults and weak seams are stable in the natural state, their stability under high water head due to the filling of reservoir is doubtful because of the piping of the clayey materials. The dam would be unsafe under high uplift pressure due to the increase of permeability. Therefore, grouting was used to raise the resistance of the faults and weak seams against high water pressure. For these purposes, the grouting materials used must be of rather high strength and low viscosity, and the polyurethane grout of formulation Sk-1 was adopted.

After extensive laboratory and field investigations, a grout curtain (Figs. 4 a.b.c) consisting of three rows of boreholes was designed. The holes of the two side rows were grouted with cement and those of the middle row were grouted with polyurethane. The spacing was 1.5 m (4.9ft) between rows and 1.5 m (4.9ft) between boreholes. The depth of grout curtain is 48 m (157.4ft).

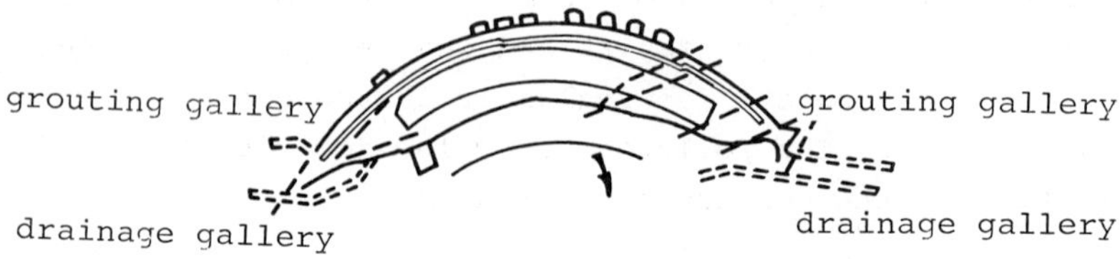

Fig. 4.a Plan of Fengtan dam

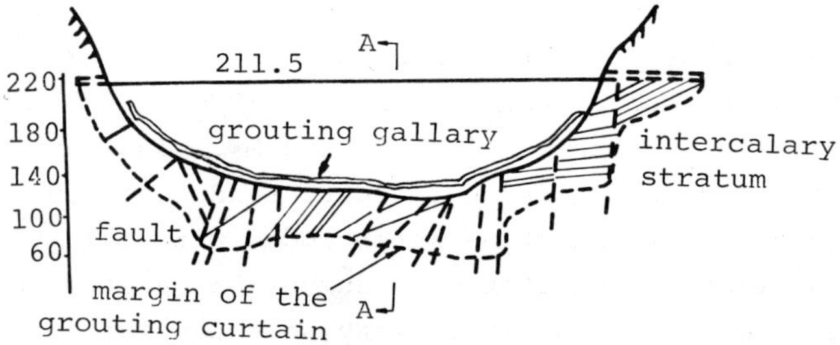

Fig. 4.b Section passing through the axis of the grouting gallery

The polyurethane grouting did not begin until the cement grouting in the neighbouring zone was completed. The injection pressure used was up to 30 kg/cm^2(425.8 lb/sq. in.)

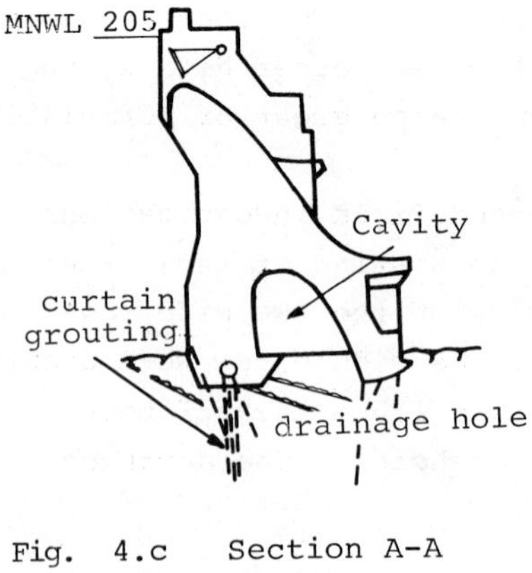

Fig. 4.c Section A-A

Expression (4) for calculating the necessary injecting time was used as the criterion for controlling the injection process.

The samples obtained from the inspection holes show that the hardened polyurethane stiffly adhered to the surface of the fissures in the faults and the weak seams (Fig. 5).

Fig. 5 The cores obtained from the inspection hole drilled in grouted zone

After the formation of the grout curtain, the observed uplift pressure at the position of the downstream drainage holes near the crutain was equal to 0.01 H-0.03 H, where H is the difference of upstream and downstream water levels.

A set of in-situ tests of polyurethane grouting was carried out in the fault F11 in the tunnel in left abutment of Chencun Dam for the same purpose as in the case of Fengtan Dam. The depth of the grouting hole was about 5 m (16.4ft) and the spacing between holes was nearly 1 m (3.28ft). The grouting pressure was 10-15 kg/cm^2 (142-213 lb/sq. in.). The samples obtained from the grouted zone of the fault (Fig. 6) showed that all breccias and mylonites in the fault were completely consolidated. The water pressure test in the grouted zone showed that the ultimate pressure of seepage failure had increased from 0.5 kg/cm^2 (7.1 lb/sq.in.) before grouting to 8 kg/cm^2 (113.6 lb/sq.in.) after grouting (Fig. 7).

Fig. 6 The grouted fault breccia and mylonite obtained from field testing zone

These results show that the polyurethane grouting of faults and weak seams is effective.

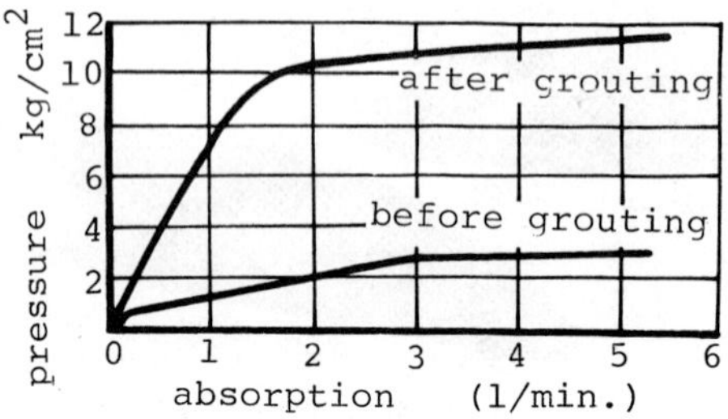

Fig. 7 The relation between the injecting pressure and the water absorption

Laopoling Dam

This is an example of grouting fine sand with polyurethane. The Laopoling concrete gravity dam was founded on sand and gravel deposit. A cast-in-situ concrete diaphragm wall was built as the anti-seepage barrier. A grout curtain was provided near to both abutments owing to impracticability of building such concrete diaphragm in these areas. Because cement grout could not penetrate into the fine sand deposit of dam foundation, polyurethane grouting was accepted.

In order to verify the injectability of the sand deposit and to inspect the properties of the grouted sand, the grouting tests were carried out in the laboratory and in-situ. The test results showed that the permeability of the grouted sand greatly decreased to that equivalent to concrete.

The strength of the grouted sand was so high (140 kg/cm^2) (1988 lb./sq.in.) that it was possible to build a thin grout curtain capable of resisting a hydraulic gradient of 38. Fig. 8 shows the grouted sand using polyuretane.

The grout curtain is composed of one row of boreholes spaced 0.5-0.7 m (1.6-2.3ft) apart and extended into the bedrock. The thickness of the curtain is nearly 0.7m(2.3ft).

The overlapping length of the grout curtain with the concrete wall is 1.5m(4.92ft) (Fig. 9). The grouting pressure was 3.5 kg/cm^2(28.4-71 lb/sq.in.). The grout take per meter length of the hole was equal to 27 liters.

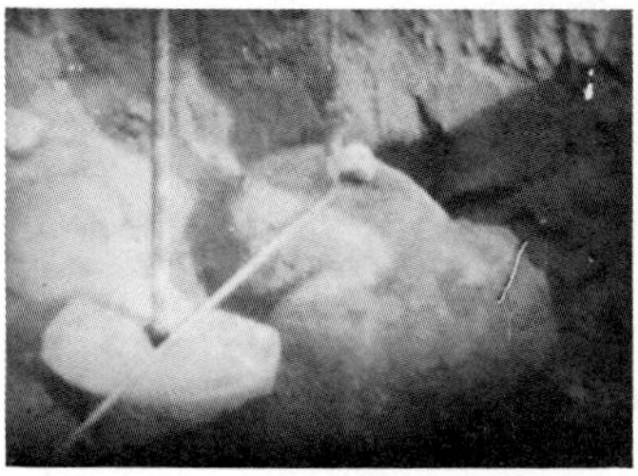

Fig. 8 The grouted sand in the testing pit at Laopoling

Because both the necessary amount of the polyurethane materials used per unit seepage control area of the grout curtain and the needed thickness of polyurethane grout curtain were less than those of cement, the total cost of polyurethane grout curtain was estimated to be 36% of that of the cement grout curtain in spite of its higher unit price.

Under the combined effect of the concrete diaphragm and polyurethane grout curtain, the ground water level downstream of the barrier is greatly lowered.

Zhuwo Dam

This is an example of stopping flowing water in concrete cracks. The Zhuwo concrete gravity dam, 33 m (108.2ft) high, comprises 7 monoliths. Among them the middle 5 are the spillway. In the ninth year after the first filling of reservoir, transverse vertical cracks appeared right in the middle of each concrete monolith and water leaked through these cracks (Figs. 10 and 11).

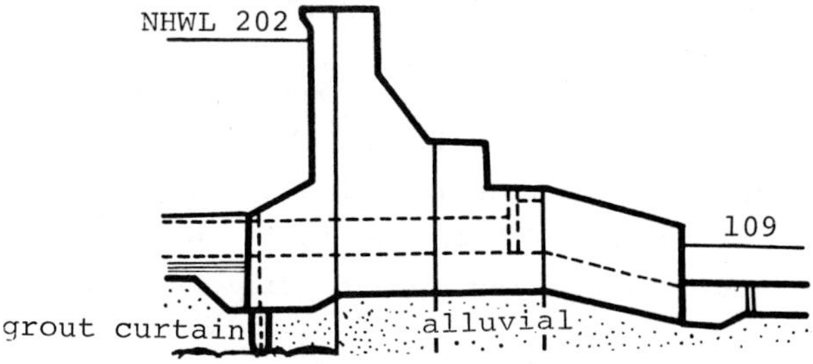

Fig. 9 Laopoling dam

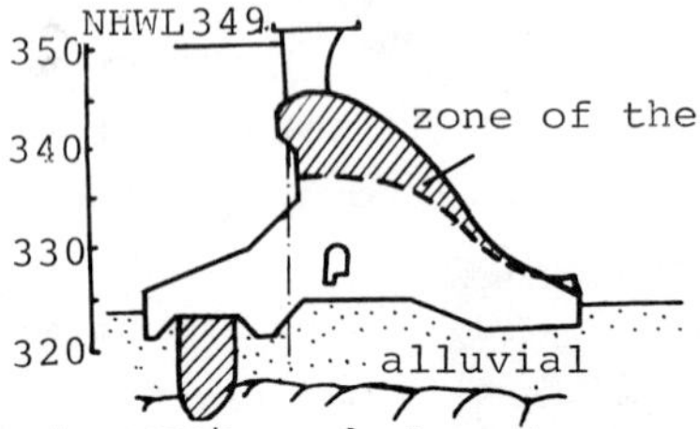

Fig. 10 Zhuwo dam

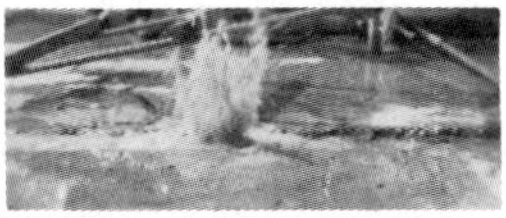

Fig. 11.a Leakage from the crack

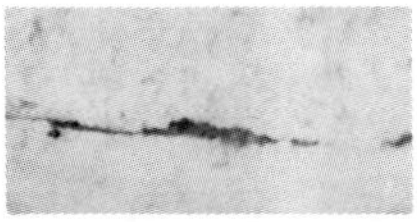

Fig. 11.b The leakage disappeared after grouting

In order to seal the cracks and to stop the water leakage without lowering the water level in the reservoir it was necessary to have a grouting material which could be effectively used in flowing water and could accommodate the seasonal change of the width of the concrete cracks. The polyurethane material was well suited for these purposes.

The grout holes were drilled with inclination from the surface of the spillway crest down to intersect the cracks at a given elevation. The Sk-3 and Sk-4 formulation of grouts were used.

The grouting work was carried out under the condition of maximum storage water level. As the grout was introduced into the cracks, the leakage immediately stopped (Fig.11). Through several cysles of high and low temperature over the past five years, the sealing effect still has been maintained.

Reference

Liu Jiacai, Principle and Technology of Polyurethane Grouting, Journal of Hydraulic Engineering, No. 1, 1980 (in Chinese).

Performance Characteristics of Acrylate Polymer Grout

William J. Clarke[1]

Abstract

Acrylate polymer chemical grout (AC-400 grout) is a new monomer system grout designed specifically for ground water control soil grouting, seepage control in concrete and infiltration control in sewer networks. The grout was introduced in 1980 to replace acrylamide (AM-9) grout which was removed from the market in 1978 because of its toxicity problems.

The new grout is a mixture of acrylate monomers that are polymerized with an oxidation/reduction catalyst system and crosslinked with a small amount of methylenebisacrylamide. The grout has approximately one one-hundredth the toxicity exposure of acrylamide grout, operates in existing grouting equipment and uses current triethanolamine and ammonium persulfate catalysts.

AC-400 grout comes in liquid form (40% solids), exhibits low viscosity (2 cps), low permeability (5×10^{-9} cm/sec) and controllable gel times for soil stabilization and underground water control applications. The performance characteristics of the new acrylate grout are compared to those of acrylamide and silicate grouts.

Background

Acrylamide grout, introduced by American Cyanamid in 1955, came close to meeting the goal of a perfect solution grout with accurate gel time for underground water control. Extensive use of the grout was disrupted in 1978 when the grout was effectively removed from the market when American Cyanamid stopped manufacturing AM-9 grout.

Because the USEPA Construction Grants Sewer Infiltration/Inflow (I/I) program was based on acrylamide grout, its removal from the market resulted in a search for an exact replacement grout. To accelerate the development of new grouts and provide guidelines for testing new grouts, EPA initiated grant R-806567-01 for "Assessment of Sewer Sealants" in 1979.

AC-400 grout was developed as an exact replacement for acrylamide grout using existing grouting equipment, the same catalyst system and having very low toxicity and controllable gel time. The grout consists of a mixture of acrylate monomers that are catalyzed with triethanolamine and ammonium persulfate and crosslinked with methylenebisacrylamide (not neurotoxic). The grout contains no acrylamide monomer, the toxic component of acrylamide grout.

(1) President, Geochemical Corporation, Ridgewood, NJ

The new grout formulation resulted from a research program that started with sodium acrylate and ended with magnesium and calcium acrylate monomers because of the stability of their gels in water. The acrylate monomers are highly stable in water, have excellent temperature stability and form gels that meet requirements for soil stabilization, underground water control and meet the criteria outline in the "Assessment of Sewer Sealants".

AC-400 grout has been used to stabilize the crown of a subway tunnel in Cambridge, seal coreholes in Leominster, Massachusetts, seal water leaks in the Washington, D.C. metro and seal sewer joint infiltration in various localities. The grout's low viscosity (2 cps), low permeability (5 x 10^{-9} cm/sec) and controllable gel time allows use in the following areas of chemical grouting:

Soil Stabilization

Seepage control in concrete

Infiltration control in sewer networks

AC-400 grout must be used judiciously in soil stabilization application because its cost is higher than silicate grout and equal to acrylamide grout. The grout may be used in place of silicate grout where low viscosity, fast gel time and low permeability are required for underground water control. The grout also may be used in place of acrylamide grout in seepage and infiltration control applications where the toxicity of acrylamide grout presents a problem.

Properties of Concentrated Grout

The new grout consists of a mixture of acrylate monomers in water solution with the following properties:

Appearance	Straw yellow color
Density	9.8 lbs/gal (1.18 gm/ml)
Stability	Good
Percent Solids	38 - 42%
pH	6.5 - 7.5

The grout has good storage stability but should be kept above its freezing point -4°F (-20°C) and below 100°F (38°C). It cannot be stored in contact with steel.

Properties of Grouting Solution

AC-400 grouting solution contains 10 percent monomer solids and sufficient accelerator (TEA) and initiator (AP) to give the required gel time. The properties of a typical grouting solution follow:

Viscosity	2 cps (20°C, 50 rpm, Brookfield)
Density	8.6 lbs/gal (1.04 gm/ml)
pH	7.5 - 9.5
Stability	3 days catalyzed

Grouting solutions containing only accelerator TEA are stable for 3 days before initiation with AP for grouting use. However, additional amounts of TEA and AP may be required to give very short gel times (30 seconds or less).

Description of Reaction

The grout is a mixture of acrylate monomers and a small amount of methylenebisacrylamide (MBA) crosslinking monomer that produce flexible gels from water solution when properly catalyzed. The gel is formed by the free radical polymerization of the monomers in a two component, equal volume system as follows:

Monomers, accelerator (TEA) and water
Initiator (AP) and water

First, the accelerator (TEA) is added to the monomer system and then the initiator (AP) is added from the initiator system, thereby, starting the induction period or gel time. After the induction period, heat of polymerization evolves and long, flexible, crosslinked polymer chains are formed.

Properties of Acrylate gels

The long, flexible, crosslinked polymer chains of acrylate gel at 10 percent solids in water have these properties:

Appearance	White, flexible gel
Solubility	Insoluble in water, kerosene, gasoline. The gel will swell slightly in the presence of water.
Permeability	5×10^{-9} cm/sec
Stability	Under conditions that allow water to evaporate, gel will gradually dehydrate. In underwater or at 100 percent relative humidity (conditions underground), the gel will maintain its volume for years. The gel is resistant to attack by bacteria, fungi and the dilute chemicals that may be found underground.
Longevity	Because of the permanence of acrylate polymers in soils and the similarity between acrylamide monomer and acrylate monomer polymerizations, it is believed that AC-400 gel will remain unaffected for many years in soils.

Catalysts

Triethanolamine (TEA) may be used to activate the acrylate monomer system. TEA is available in 85 and 99 percent concentrations with freezing points at 60°F and 70°F respectively. A new, low freezing point grade of 85 percent TEA is available that freezes at 41°F.

Ammonium persulfate (AP) may be used to initiate the polymerization of the acrylate monomer system. When equal volumes of activated grout and AP solutions are mixed, the induction period or gel time begins.

Typical Formulations

The following typical formulation of AC-400 grout, TEA and AP will give approximately 30 seconds gel time at 20°C:

	Weight Percent	Pounds	Gallons
AC-400 grout (40%)	25.00	125.0	12.7
Triethanolamine (85%)	1.0	5.0	0.5
Water	24.0	120.0	14.4
Ammonium Persulfate	1.0	5.0	5.0 (Lbs.)
Water	49.0	245.0	29.4
	100.00	500.0	60.0

The formulation is calculated on a weight percent and volume basis to accommodate existing 500 pound or 60 gallon capacity Cues grouting equipment. Gel time may be changed by varying the amount of TEA and AP. TEA should be increased incrementally with AP to avoid low pH from excess AP. Longer gel times may be obtained by adding small amounts of potassium ferricyanide (KFe).

Unconfined Compressive Strength (UCS)

Acrylate and acrylamide grouts were tested for UCS using ASTM D-21.66 for low strength chemical grouts. A standard filler of coarse sand was used to make 3.77 cm diameter by 9 cm long samples giving the Stress Strain curves in Fig. l. The sand was a coarse 20 - 30 Ottowa sand and the strain rate used was 0.02 inch/inch/minute.

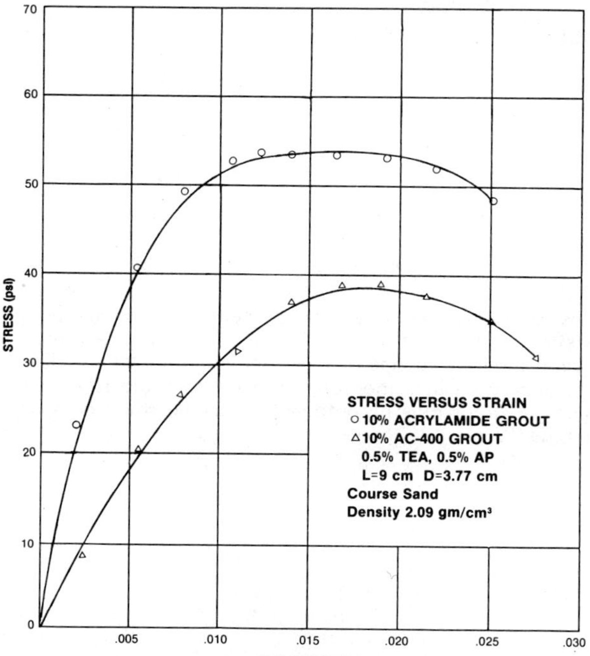

FIG. 1 STRESS VERSUS STRAIN FOR CHEMICAL GROUTS

The UCS testing of acrylamide grout gave maximum strength at 53 psi while the acrylate grout curve peaked at 38 psi. This indicates that acrylate and acrylamide grouts may be used for underground water control.

Toxicology

AC-400 grout has an LD_{50} of approximately 5000 mg of concentrate per kg body weight of white, albino rats. This is the highest level of oral dosing that is normally administered in the LD_{50} test. The grout should not present a problem in the following areas of toxicology:

1. Because of the low single dose toxicity, ground water contamination of residual monomer should not present a problem.

2. Low eye and skin irritation properties present minimal hazards to operators. However, rubber gloves, goggles and waterproof shoes are required for handling the grout and catalysts.

3. General environmental hazard is low because of the biodegradability of acrylate monomers.

Product Reaction Characteristics

AC-400 grout may be modified for field application by addition of Ethylene Glycol (EG) for shrinkage control and freeze stability and Diatomaceous Earth (DE) for increased strength as follows:

(Weight %)	Standard	EG Admix	DE Admix
AC-400 grout (40%)	25.0	25.0	25.0
Ethylene Glycol (EG)			5.0
Diatomaceous Earth (DE)			5.0
Triethanolamine (TEA)	1.0	1.0	1.0
Water	24.0	19.0	19.0
Ammonium Persulfate (AP)	1.0	1.0	1.0
Water	49.0	49.0	49.0
	100.0	100.0	100.0

The standard grouting mix should be used for sealing sewer joints and soil stabilization where low viscosity is required. The Standard, EG and DE Admixes will give approximately 30 seconds gel time at 20°C using one percent TEA and AP catalysts.

Ethylene Glycol (EG) may be added to the grout for soil stabilization at freezing ambient temperatures or for sealing the top of manholes that may be subject to heat loads during the summer. Diatomaceous Earth (DE) may be added to the grout for improved compressive strength in special applications for grouting crushed stone bedded sewer joints. A good suspending grade of Diatomaceous Earth (Johns Manville Celite 266) should be used to avoid settling problems in the grouting equipment.

Gel Time Control

AC-400 grout gel time may be controlled by varying the amount of accelerator (TEA) and initiator (AP). The relation between gel time and temperature is linear between 10°C and 20°C as shown in Fig. 2 and 2A.

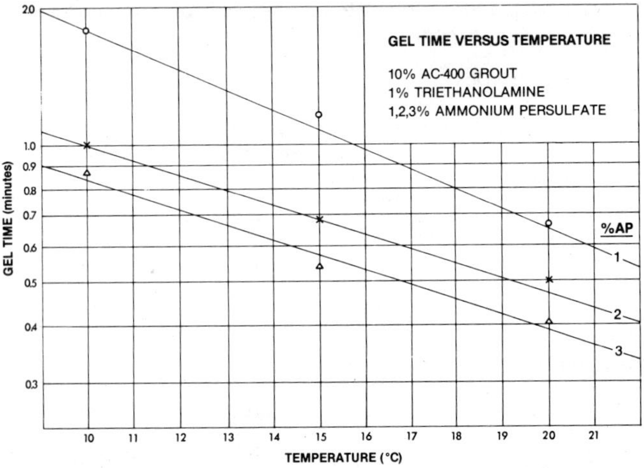

FIG. 2 GEL TIME VERSUS TEMPERATURE FOR 10% AC-400 GROUT

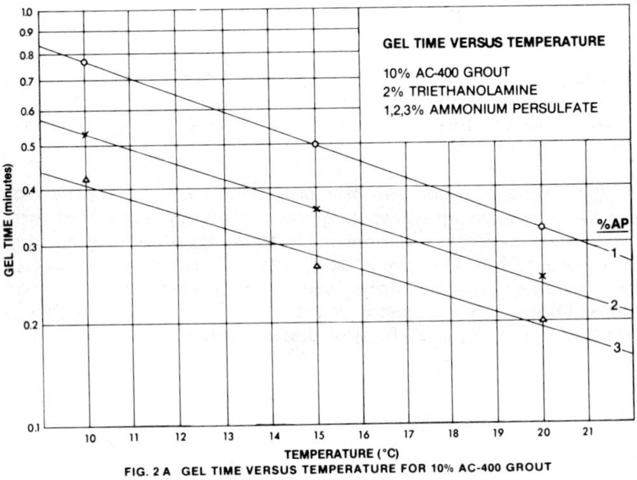

FIG. 2 A GEL TIME VERSUS TEMPERATURE FOR 10% AC-400 GROUT

Longer gel times may be obtained by adding small amounts of potassium ferricyanide (KFe) as shown in Fig. 3.

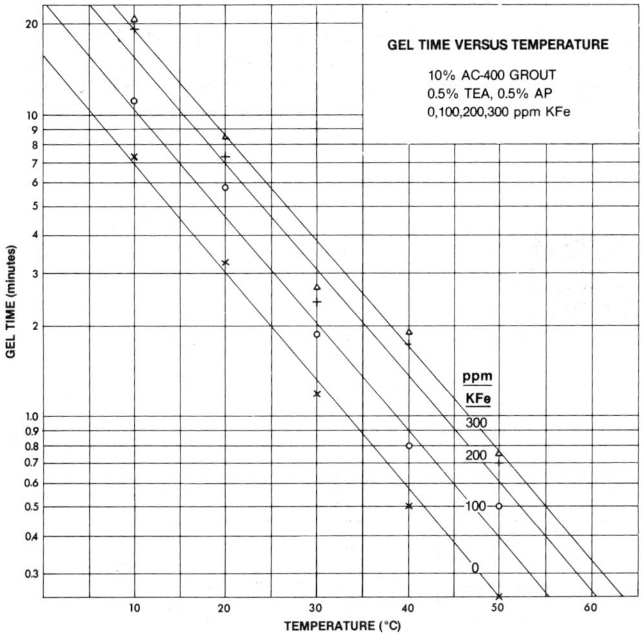

FIG. 3 GEL TIME VERSUS TEMPERATURE FOR 10% AC-400 GROUT

For longer gel time, the TEA and AP levels may be reduced to 0.4 percent and the KFe level increased to 600 ppm. Smaller amounts of TEA and AP or larger amounts of KFe will give erratic results with the AC-400 grout.

Viscosity Versus Gel Time

Where continued agitation of silicate grouts can prolong complete gelation indefinitely, acrylate and acrylamide grouts have abrupt gelation characteristics even when agitated. Viscosity versus gel time experiments were carried out at Northwestern University using a Brookfield Viscometer (Model RTV) with a UL adaptor. As shown in Fig. 4, silicate grout does not have a definite gel time but increases slowly in viscosity. These properties indicate that acrylate and acrylamide grouts may be used for underground water control where fast viscosity development is required.

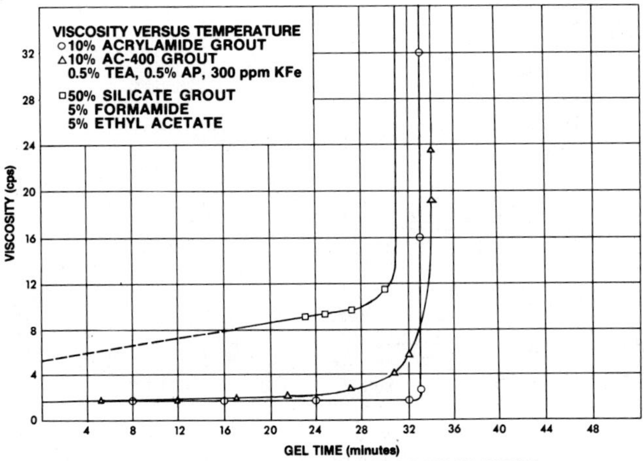

FIG. 4 GEL TIME VERSUS VISCOSITY FOR CHEMICAL GROUTS

Solubility and Chemical Reaction

AC-400 grout is essentially insoluble in most dilute chemicals that may be found in sewer systems. The grout is unaffected by 10 percent solutions of alcohols, ketones, hydrocarbons, acids and metal salts in 16 week tests as follows:

Chemical	Volume Change	Comments
Heptane	+5%	No deterioration
Methyl Ethyl Ketone	None	" "
Propanol	None	" "
Carbon Tetrachloride	+3%	" "
Toluene	+3%	" "
Sulfuric Acid	+5%	" "
Potassium Chlorate	None	" "
Acetic Acid	+6%	Gel turned clear, No deterioration
Potassium Hydroxide	+10%	"
Hydrochloric Acid	+4%	"

Flammability

AC-400 grout consists of 40 percent acrylate monomers in 60 percent water solution. Therefore, the grout is not flammable or explosive and not subject to DOT regulations.

Acid/Base Reaction

Gel time of AC-400 grout changes slightly with pH change of the grouting solution. The grout may operate in soil with pH ranges between 6 and 8.5 at 20°C as follows:

pH (1% AP & TEA)	Gel Time (Seconds)
6.0	43
6.5	37
7.0	35
7.5	32
8.0	30
8.5	28

The soils around sewer joints do not exhibit high ranges of pH as the joints are constantly purged with ground water, which tends to be in the alkaline range.

Permeability

Permeability tests run at Northwestern University in fine, medium and coarse sand demonstrated AC-400 gel to be essentially impermeable (5×10^{-9} cm/sec). Barely measurable flow was obtained from acrylate and acrylamide grouted specimens (less than 1 cc per day) in 6 months continuous testing under about 3 kg/cm^2 pressure.

Freeze/Thaw Cycling

Five 100 gram samples of 10 percent acrylamide gel and 10 percent AC-400 gel were cycled from 20°C to -10°C in 7 day periods through 5 cycles. Weights of the samples after each cycle follow:

Acrylamide Gel

Days	7	14	21	28	35	Net (grams)
Sample						
1	-0.1	-0.3	-0.3	0	+0.1	-0.6
2		0	-0.1	0	0	-0.1
3			0	-0.1	-0.2	-0.3
4				-0.1	-0.4	-0.5
5					-0.2	-0.2

AC-400 Gel

Days	7	14	21	28	35	Net (grams)
Sample						
1	0	+0.2	-0.2	-0.3	-0.1	-0.4
2		-0.8	-0.2	-0.1	-0.2	-1.3
3			-0.7	-0.7	-0.0	-1.4
4				-0.9	-0.1	-1.0
5					-0.3	-0.3

The acrylamide gel samples lost some weight during freeze/thaw cycling. This weight loss is believed to be the result of the low relative humidity during the freeze cycle. The AC-400 gel also lost weight but did not crystalize during the freeze cycle. The gel maintained its properties and did not crystalize during five freeze/thaw cycles.

Wet/Dry Cycling

Sets of four 100 gram samples of 10 percent acrylamide gel and 10 percent AC-400 gel were cycled at 20°C through wet (water) and 60 percent and 100 percent Relative Humidity in 7 days cycles for 6 weeks with weight gain (+) or loss (-) as follows:

60% Relative Humidity

Day	7	14	21	28	35	42	Net
Cycle	Wet	Dry	Wet	Dry	Wet	Dry	
Acrylamide	+6.2	-1.2	+1.1	-0.7	+2.6	-1.3	+6.7
AC-400	+4.0	-1.2	+1.1	-0.7	+1.3	-0.7	+3.8

100% Relative Humidity

Day	7	14	21	28	35	42	Net
Cycle	Wet	Dry	Wet	Dry	Wet	Dry	
Acrylamide	+2.6	-0.4	+0.5	-0.3	+1.4	-0.2	+3.6
AC-400	+2.7	-0.4	+1.0	-0.5	+2.4	-0.2	+5.0

AC-400 grout operates very much like acrylamide grout in that weight lost during dry cycling is gained during wet cycling. Both grouts showed a net weight gain during 6 weeks of wet/dry cycling. The Relative Humidity in the soil around sewer joints very rarely goes below 90 percent, therefore, both the AC-400 and acrylamide gels should operate satisfactorily in soils during wet/dry cycling.

Pot Life

Catalyzed AC-400 grout was tested for pot life at 20°C for five days as follows:

Day	1	2	3	4	5
Gel Time	30"	38"	1'1"	1'23"	1'47"

A maximum of 3 days pot life is recommended for AC-400 grout catalyzed with TEA. Before 3 days, the gel time of the catalyzed grout may be decreased to 30 seconds by adding equal amounts of TEA and AP.

Grouting Crown of Subway Tunnel, Cambridge, Massachusetts

As shown in Figures 5 and 6, the crown of a subway tunnel that entered a 30 foot diameter air shaft at 60 feet depth was grouted with AC-400 chemical grout. Because of the damaged soil and water conditions around the air shaft, the crown of the subway tunnel was grouted to a depth of 20 feet from each side of the air shaft.

The crown of the subway tunnel, containing water-bearing layers of silty sand and gravel, was stabilized with the new grout. Eighteen grout pipes on approximately 2 feet centers were drilled in the pattern shown in Fig. 6.

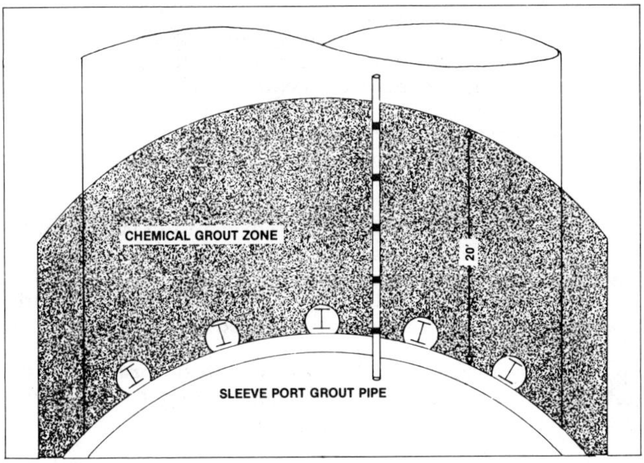

FIG. 5 AC-400 GROUT PATTERN - PLAN VIEW - SUBWAY TUNNEL

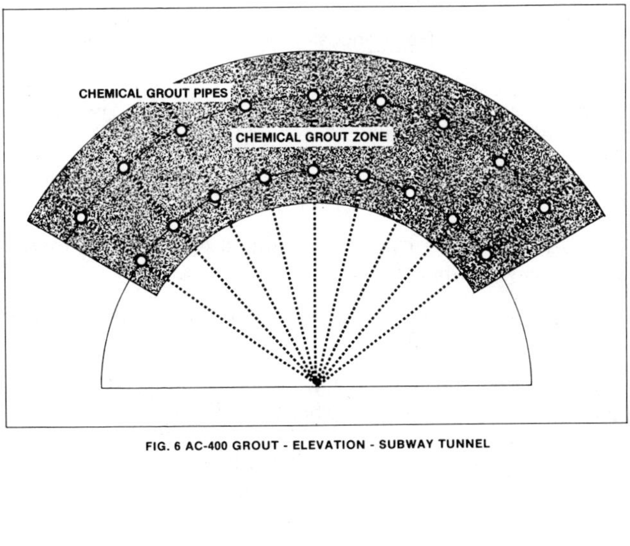

FIG. 6 AC-400 GROUT - ELEVATION - SUBWAY TUNNEL

The grout pipes consisted of 1.5" diameter PVC pipe with rubber sleeve ports on 18" centers which acted as one-way valves. Using a packer in the PVC pipes, the grout was injected in a first stage pattern to a predetermined volume based on the estimated zone of treatment. In a second stage grouting in the same area, injection was continued until an increase of grouting pressure was observed, indicating complete grouting of the affected zone.

The AC-400 grout was injected with an automatic proportioning grout plant that metered the volume and flow rates automatically and provided for continuous variation of grout proportions and pumping rates. The two components of the chemical grout system were metered through separate hoses to the grout site and joined at a mixing "Y" where the chemicals combined to give a gel time of 5 minutes at 30°F. The mixed grout passed to a grouting manifold fitted with valves, flow meters and pressure gauges, where the grout was distributed to as many as eight grout pipes simultaneously. The manifold permits the independent control of different grout flow rates and pressures to each grout pipe. The combination of AC-400 chemical grout and the continuous mixing and pumping system permitted the use of long pumping times and relatively short gel time for maximum efficiency injection water-control grouting.

The volume of AC-400 grout was precalculated from the total volume of the soil to be grouted. Using 18 grout pipes on approximately 2 feet centers and 20 feet deep, the grouting volume for each pipe calculated to approximately 80 cubic feet. Based on an estimated 30 percent void volume in the grouting zone, each pipe was injected with approximately 180 gallons of grout.

The grout was pumped at pressures of 10 to 20 psi with flow rates of 1 to 2 gpm to each grout port. Approximately 4,000 gallons of grouting solution were pumped in two days.

The subway tunnel was successfully bored through the bottom of the air shaft with the grout in the crown of the tunnel controlling water and stabilizing soil during the tunneling operation.

References

1. "Manufacturers Test Report", Low Toxicity AC-400 Grout, 1980, Geochemical Corporation, Ridgewood, NJ 07450.

2. Product Bulletin and Specification, AC-400 Grout, Geochemical Corporation.

3. "All about Cyanamid AM-9 Chemical Grout" (EN-4), 10/75, American Cyanamid Company, Wayne, NJ 07470.

4. "Chemical Grouting" Lecture Notes, 1981, R. H. Karol, Rutgers University, New Brunswick, NJ 08900.

5. "Assessment of Sewer Sealants", 1980, Grant No. R806567-01 USEPA, Municipal Environmental Research Laboratory, Edison, NJ 08817.

6. AM-9 Chemical Grout Field Manual (EN13A), 1/76, American Cyanamid Company, Wayne, NJ 07450.

PROPERTIES OF CHEMICALLY GROUTED SAND AT LOCKS AND DAM NO. 26

Richard R. Davidson, A.M., ASCE and Jean-Yves Perez, M., ASCE*

ABSTRACT

A full-scale field chemical grouting test program was performed to study the effectiveness of grouting the foundation alluvial sand to rehabilitate the existing Locks and Dam No 26. Changes in soil properties such as in situ stress, density, elastic modulus, shear strength, creep and permeability caused by grouting are discussed. Grouted soil properties were evaluated from the results of in situ testing and laboratory testing of lab grouted and undisturbed samples. The information presented allows comparisons of the properties of soils grouted with the four different grouting methods, eight grout types and two grout-hole spacings studied.

INTRODUCTION

A foundation investigation and test program at the existing Locks and Dam No. 26 was performed for the St Louis District of the Corps of Engineers to investigate the feasibility, safety and reliability of various rehabilitation techniques. Two basic avenues of rehabilitation were addressed:

(1) improvement of the foundation soil characteristics by chemical grouting; and
(2) improvement of the stability of the existing structures by construction of additional support elements (driven or drilled-in piles and anchors).

One of key objectives of the chemical grouting test program was to assess changes in soil properties caused by injection of chemical grout. Soil properties such as stress conditions, water content, total unit weight, elastic deformation modulus, shear strength, time dependent properties and permeability were characterized by field in situ testing and laboratory testing of undisturbed samples. Changes in soil properties were evaluated by comparisons of similar in situ and laboratory measurements made before and after grouting. This paper discusses the characterization of grouted soil properties at Locks and Dam No. 26. A comprehensive report of the results and interpretation of the test program was prepared by Woodward-Clyde Consultants (1979).

DESCRIPTION OF TEST PROGRAM

The chemical grouting test program involved testing four different grouting methods, eight grout types, and two grout-hole spacings. A total of 467,500 ℓ (123,500 gal) of grout was injected through 74 grout holes to grout approximately 1100 m^3 (1440 yd^3) of alluvial sand. The test program was unique in that the properties of the grouted soil were not only characterized by sampling and in situ testing performed from the surface, but the grouted soil was actually fully excavated and mapped.

*Woodward-Clyde Consultants, Chicago, Illinois

Grouting Activities

Grout injection was done by a joint venture of Raymond International Builders, Inc and Soletanche and Rodio, Inc. A modular grouting plant was used to mix and pump grout. Grout was pumped through open-bottom pipes and through sleeve-pipes into the grout hole configuration shown in Fig. 1. Fourteen different grouting test subareas were used to study the effects of grout types, grouting methods and grout-hole spacings. Table 1 presents the types of low strength and high strength grouts used, and the appropriate grouting method and grout-hole spacing for each subarea. Of the eight grout types, seven were sodium silicate and one was cement-bentonite. The percentage in the designation of the grout type refers to the concentration of sodium silicate by volume in the grout mix. The various reactants used in the grout included: sodium aluminate, formamide, calcium chloride and di-methyl-succinate (R600).

The pumping pressure was kept at less than 23 kPa per meter (1 lb/in^2 per foot) of soil above the bottom of open-bottom pipes in accordance with conventional grouting practice for open-bottom pipe grouting (grouting method O_1) in the United States. Grout was pumped at an average rate of 800 ℓ/h (3.5 gal/min) through each open-bottom pipe. The average grout take was approximately 25 percent of the volume of soil to be grouted on the basis of the contractor's (Raymond) usual practice in this type of alluvial sand. By making this selection, it was implictly acknowledged that the soil would not be completely saturated with grout because its porosity is approximately 35 percent.

Three grouting methods were tested using sleeve-pipes. In the S_1 method, grout was pumped using a pressure limiting criterion of 23 kPa per meter (1 lb/in^2 per foot) of overburden. In the S_2 method, the intended volume of grout was injected in one stage using a pumping rate less than 85 percent of the rate that would induce hydraulic fracturing of the soil. This pumping rate was usually between 300 ℓ/h and 450 ℓ/h (1.3 gal/min and 2 gal/min). The volume of soil to be injected was selected as being equivalent to approximately 45 percent of the volume of soil to be grouted, consistent with the contractor's (Soletanche and Rodio) experience. In the S_3 method, the intended volume of grout was injected in several separate stages using the same rate of pumping as in the S_2 method. The grout take for the third procedure averaged 53 percent of the volume of soil to be injected.

Borehole Sampling and In Situ Testing

Borehole samples were taken and in situ tests were performed before and after grouting. Prior to grouting, one standard penetration test boring, two static cone penetration test soundings, two pressuremeter borings and one permeability test boring were performed at the test site at the locations shown in Fig. 1. After grouting, forty five borings were drilled from the original ground surface and five borings were drilled after the first 6.5 m (21 ft) of overburden was excavated.

After grouting each type of in situ testing and sampling boring was performed in each subarea to facilitate comparisons between grout types and between grouted and ungrouted soil. In situ testing performed in these borings included continuous standard penetration, continuous static cone penetration, Menard pressuremeter and falling head permeability tests. Undisturbed borehole samples were obtained

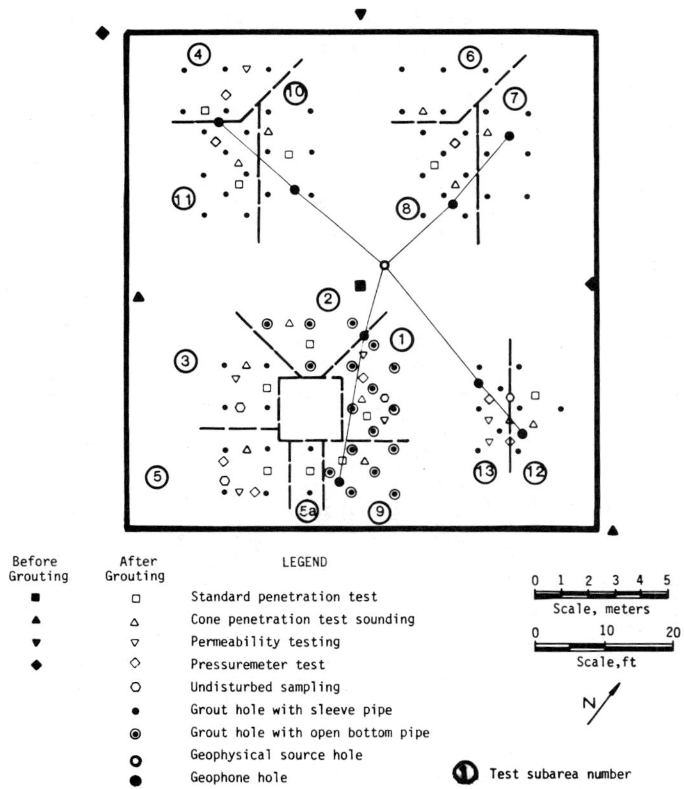

Fig. 1 LAYOUT OF GROUT HOLES, TEST SUBAREAS, AND BORINGS PERFORMED BEFORE AND AFTER GROUTING

	GROUT TYPE	GROUTING METHOD	GROUT HOLE SPACING m (ft)	MAXIMUM GROUT PRESSURE OR RATE OF PUMPING	TEST SUBAREA NO.
LOW-STRENGTH	35% SIROC 142	O_1	1.3 (4.2)	23 kPa/m (1 lb/in.^{2}/ft)	1
			1.8 (6)		2
		S_2		85% of hydraulic fracturing rate of pumping	4
	25% Silicate / Aluminate	S_1		23 kPa/m (1 lb/in.^{2}/ft)	5A
		S_3		85% of hydraulic fracturing rate of pumping	6
	Some Cement-Bentonite and 25% Silicate / Aluminate				7
			1.3 (4.2)		8
	28% Silicate / R600	S_2	1.8 (6)		3
HIGH-STRENGTH	45% SIROC 132	S_2	1.8 (6)	85% of hydraulic fracturing rate of pumping	5
	55% SIROC 142 / 132	O_1	1.3 (4.2)	23 kPa/m (1 lb/in.^{2}/ft)	9
		S_2	1.8 (6)	85% of hydraulic fracturing rate of pumping	10
			1.3 (4.2)		11
	46% Silicate / R600	S_3	1.8 (6)		12
			1.3 (4.2)		13

Table 1 COMBINATION OF TEST VARIABLES

using a Pitcher piston tube sampler. Crosshole shear wave velocity measurements were made before and after grouting in four arrays of geophysical boreholes shown in Fig. 1.

Excavation Sampling and In Situ Testing

After grouting, the test area was dewatered, and then excavated carefully to investigate the limits and extent of the grouted soil mass. As the excavation proceeded, fifty undisturbed samples were taken from exposed grouted soil masses for laboratory testing. Twenty two samples were carved as approximately 25-cm (10-in.) cubes. In soil grouted with high-strength grout, block sampling was difficult and time consuming. After some attempts at block sample carving, a small electric diamond core drill was successfully used. Twenty-eight 11-cm (4.5-in.) dia core samples were taken in hard grouted soil in groups of four 25-cm (10-in.) long cores.

Seventeen plate load tests were made at various elevations on grouted and ungrouted sand by applying an incremental load to a 35-cm (13.5-in.)-dia steel plate placed on top of a leveled soil surface. Twenty one in-place density tests were made in the grouted and ungrouted soil during excavation. The in-place unit weight was measured using the water balloon method (Washington Densometer).

Laboratory Testing

Laboratory testing was performed on undisturbed samples of ungrouted sand and included: water content measurements; in situ, maximum and minimum unit weight determinations; grain size analyses; and consolidated drained and undrained triaxial compression tests. Water content and unit weight determinations, unconfined compression tests, unconsolidated undrained triaxial tests and falling head permeability tests were performed on undisturbed grouted samples recovered in boreholes.

In addition to index property tests, short term unconfined compression and consolidated drained triaxial compression, long term (creep) unconfined compression and consolidated drained triaxial compression, sonic velocity, and constant head permeability tests were performed on excavation block and core samples. Samples of sand grouted with low-strength 25% silicate/aluminate grout could be trimmed using standard soil procedures. Samples of sand grouted with high-strength 55% Siroc 132/142 and 46% silicate/R600 grouts required coring from the block. For unconfined compression test specimens, saturation was achieved by soaking in water; however, this could not be done for samples grouted with 25% silicate/aluminate because these samples disintegrated during soaking. Triaxial compression specimens were saturated by back pressure in the testing cell.

Laboratory testing was also performed on reconstituted sand samples grouted in the laboratory with 35% Siroc 142. The sand samples were compacted to field density and water content conditions, injected with grout under field grouting pressures, and submerged cured under field confining pressures.

SUBSURFACE CONDITIONS

The sand strata grouted during the test program comprised the upper 7 m (20 ft) of the recent alluvium deposits of the Mississippi River valley. The recent alluvium sand was overlain by approximately 7 m of silty clay and silty sand flood plain deposits and underlain by coarser grained glacial outwash deposits. The alluvial sand was generally fine- to medium-grained and contained on the average no more than 5 percent fines (that is, 5 percent by weight of soil particles passed through a No. 200 US sieve with 75- μm square openings). The sand was medium dense and cohesionless with an angle of internal friction of 36 to 40°. In situ stress measurements indicated that the sand was normally consolidated with a K_o = 0.4 to 0.5. The coefficient of permeability in the zone that was grouted ranged from 5×10^{-3} cm/s to 3×10^{-2} cm/s. During excavation of the test area, abundant concentrations of carbonaceous material (wood, charcoal, and lignite) ranging in size from silt to large tree trunks, were observed. The alluvial sand was always cross-bedded.

GROUTED SOIL PROPERTIES

Properties of grouted soil were interpreted from the results of the in situ and laboratory tests performed after grouting. Comparisons with results of testing performed before grouting are discussed subsequently. A summary of grouted soil properties for the various subareas is presented in Table 2.

Stresses

In situ horizontal stresses were inferred from pressuremeter test results. In all cases, the in situ horizontal stresses measured after grouting were higher than those measured before grouting, even in poorly grouted or ungrouted soil which was surrounded by grouted soil within a subarea, regardless of the grouting method used. K_o values increased from 0.45 measured before grouting to between 1 and 3 measured after grouting. The ratio of the after-grouting in situ horizontal stress to the average peak grouting pressure in nearby grout holes ranged from 0.5 to 1.0 and averaged 0.77. Therefore, it appears that the pressure applied to the soil during injection had not completely dissipated at the time the pressuremeter tests were made which was up to more than one month after completion of grouting. This phenomenon could be of importance when designing foundation elements such as retaining walls which would be influenced by lateral stresses.

Density

The unit weight of the grouted soil was determined by three different methods: laboratory measurements on undisturbed Pitcher samples; in-place measurements with Washington Densometer during excavation; and laboratory measurements on block and core samples. The water content of the grouted soil was 3 to 6 percent less than the water content before grouting. The dry unit weight of the grouted soil measured in place averaged 16 kN/m^3 to .47 kN/m^3 (1 to 3 lb/ft^3) more than the dry unit weight of the ungrouted soil.

Elastic Deformation Modulus

The elastic deformation modulus of grouted soil was calculated from the results of pressuremeter, plate load, crosshole shear wave velocity, and laboratory unconfined and consolidated drained triaxial compression tests. Average modulus values for each test subarea and each type of test are presented in Table 2. Modulus values from these tests varied widely because of the different conditions associated with each type of test. These differences include principal stress

							Index Properties		Stress	Stiffness							Shear Strength					
										Unconfined Compression		CID Triaxial Compression					CID Triaxial Compression					
Subarea Number	GROUT TYPE	Grouting Method[8]	Grout-Hole Spacing, ft	Standard Penetration Resistance, bl/ft N	Static Cone Penetration Resistance t/ft^2 q_c	Shear Wave Velocity (Intact Mass), ft/s V_s	In-Situ Dry Density, lb/ft^3 γ_d	Water Content, % ε_n	Coefficient of Earth Pressure at Rest K_o	Initial Tangent Modulus, t/ft^2 E_t	Secant Modulus at Failure, t/ft^2 E_{sf}	Initial Tangent Modulus, t/ft^2 E_t	Secant Modulus at Failure, t/ft^2 E_{sf}	Plate Load Test Modulus, t/ft^2 E_{plt}	Pressuremeter Modulus, t/ft^2 E_{pmt}	Cross Hole Shear Wave Velocity Modulus, t/ft^2 E_{csv}	Cohesion, t/ft^2 C	Friction Angle° ϕ	Unconfined Compressive Strength, t/ft^2 q_u	Ultimate Stress Plate Load Test, t/ft^2 q_{u_p}	Pressuremeter Limit Pressure, t/ft^2 P_L	Coefficient of Permeability, cm/s K
U	Ungrouted	-	-	20	125	600	105	22	.45	-	-	500	150	144	130	1,800	0	39.5	-	5	25	2×10^{-3}
6	25% Silicate/Aluminate	S_3	6	-	275	-	106.3	20	-	-	-	-	-	534	-	-	-	-	-	8	-	-
7	25% Silicate/Aluminate	S_3	6	-	280	-	110	17	-	-	-	-	-	311	-	-	-	-	-	6.4	-	-
8	25% Silicate/Aluminate	S_3	4.2	22	280	1,100	112.4	17	2.3	58	39	801	145	246	119	4,000	.34	35	.32	8	59	1.5×10^{-4}
5a	25% Silicate/Aluminate	S_1	6	20	-	-	-	-	-	-	-	-	-	-	-	-	-	-	-	-	-	-
3	28% Silicate/R600	S_2	6	60	300	-	103.9	17	-	90[1]	55[1]	-	-	1,613	-	-	-	-	.29[1]	11	-	1.2×10^{-5} [4] 1×10^{-4} [5]
1	35% SIROC 142	0_1	4.2	45	320	1,190	110.3	19	1.3	1,231	757	902	155	338	151	3,500	.36 .70[2]	35.5 39.5[2]	4.0	10	27	2×10^{-5} [4] 2×10^{-4} [5]
2	35% SIROC 142	0_1	6	42	250	-	110.2	19	-	625	471	-	-	1,815	-	-	-	-	1.91	12	-	-
4	35% SIROC 142	S_2	6	100+	450	-	111.7	17	1.6	12,000	8,000	-	-	1,218	459	-	-	-	18.38	>14	69	1.2×10^{-4}
5	45% SIROC 132	S_2	6	100+	500+	-	-	-	3	-	-	-	-	8,233	1,576	-	-	-	-	>17	100+	8×10^{-5}
12	46% Silicate/R600	S_3	6	100+	500+	-	-	-	-	-	-	-	-	911	-	-	-	-	-	>18	-	-
13	46% Silicate/R600	S_3	4.2	100+	500+	2,110	105[3]	1.8	3.3	3,006	2,106	2,146	880	962	860	27,000	3.5	39.5	11.88	>14.5	100	1×10^{-4}
10	55% SIROC 142/132	S_2	6	100	425	-	-	0	-	-	-	-	-	9,290	-	-	-	-	-	>15.8	-	-
11	55% SIROC 142/132	S_2	4.2	100+	500+	3,547	104[3]	20	2.5	5,192	4,910	6,667	3,673	12,098	1,327	16,000	5.82	39	15.60	>14	100+	2×10^{-4}
9	55% SIROC	0_1	4.2	90	375	-	-	-	-	-	-	-	-	5,464	-	-	-	-	-	>16	-	-

General Note

Unless otherwise noted, laboratory tests were made on excavation block and core samples.

Notes:

(1) From undisturbed borehole samples

(2) Reconstituted sample grouted in the laboratory

(3) From excavation block sample

(4) Laboratory permeability test

(5) Borehole permeability test

Table 2 SUMMARY OF GROUTED SOIL PROPERTIES

direction, strain rate, stress duration, strain amplitude, drainage conditions and many others. However, all of the testing indicated that chemical grouting increased soil stiffness as the concentration of silicate in the grout increased.

An elastic deformation modulus was calculated from the pseudo-elastic portion of the pressuremeter curves. These results indicated a slight increase in modulus for subareas 1 and 8, and significant improvement for subareas 4, 5, 11 and 13. The same 35% siroc 142 grout was used for subareas 1 and 4 but the difference in modulus may have been a result of the grouting method because subarea 1 used open bottom pipes and subarea 4 used sleeve pipes. Although the increase for subarea 13 was substantial, it was less than the other high-strength grouted areas.

An elastic deformation modulus was calculated from the initial portion of the plate load test pressure-settlement curves. The plate load test results indicated a small increase (2-4 times) in modulus for subareas 1, 6, 7 and 8, a moderate increase (6-10 times) for subareas 2, 3, 4, 12 and 13, and a significant increase (60 times) for subareas 5, 9, 10 and 11. Although the ungrouted plate load test modulus agreed well with the pressuremeter derived modulus, only the tests from subareas 12 and 13 agreed with the after grouting pressuremeter test modulus from the same subareas. Moduli from the other test areas in some cases significantly exceeded the pressuremeter modulus. Generally modulus values increased with increasing percentage of silicate and decreasing grout hole spacing. Interestingly the modulus from subareas 12 and 13 was significantly less than the modulus measured in other high-strength grouted sand. These results may be explained by the creep-prone behavior of the R600 grout which becomes a factor with the relatively long stress duration of the plate load test, see the subsequent section on Time Dependent Properties.

Elastic modulus values were calculated from laboratory unconfined and consolidated drained triaxial compression tests on undisturbed field and laboratory reconstituted grouted samples. Initial tangent modulus and secant modulus at the peak stress (failure) were obtained from stress-strain curves. The influence of silicate content on modulus is illustrated in Fig. 2(A). Modulus values increased with silicate content as expected, which is in agreement with recent work by Clough et al (1979). The excavation block and core samples, however, were significantly stiffer than the laboratory reconstituted or borehole samples. For grouts having a silicate content greater than 25 percent, unconfined compression moduli were greater than consolidated drained triaxial moduli for tests made on block samples. Generally, Table 2 indicates that the moduli measured in situ with pressuremeter and plate load tests compare favorably with triaxial secant moduli of failure measured on block samples taken from the same subareas.

From the limited range of strain rates studied, it appears that modulus values increased with faster strain rates, although considerable scatter was found at strain rate of 0.5 percent per minute. This is consistent with the trend observed by Clough et al (1979). Koenzen (1978) showed that the effect of strain rate on modulus diminishes with decreasing percentage of silicate. The influence of confining pressure on modulus is illustrated in Fig. 2(B). Modulus values increased with confining pressure. Clough et al (1979) showed similar trends, with the rate of increase of modulus with confining pressure being greater for higher silicate content. Modulus values did not appear to increase with curing time, contrary to the findings of Clough et al (1979).

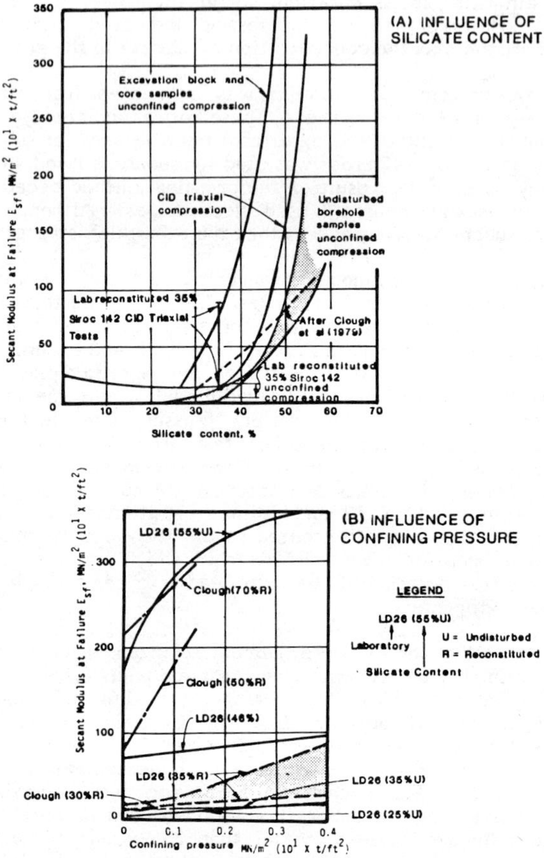

Fig. 2 INFLUENCE OF VARIOUS FACTORS ON LABORATORY ELASTIC DEFORMATION MODULUS

Small strain level elastic deformation moduli were also calculated from crosshole shear wave velocity measurements. The increase in shear wave modulus was not proportional only to the percentage of silicate content in the grout because the increases for subareas 7-8 and 12-13 exceeded subareas 1-9 and 10-11, respectively. Shear wave-derived moduli were probably more significantly influenced by the extent of grout penetration. Subareas 7-8 and 12-13 were more completely grouted due to multiple stage sleeve pipe grouting than the single stage open pipe and sleeve pipe grouting in subareas 1-9 and 10-11, respectively. These modulus values are representative of the bulk of the entire grouted and ungrouted soil mass.

Shear Strength

The shear strength of grouted soil was found to be the result of a frictional component ϕ mobilized between sand grains and a cohesion component c induced by the grout. Shear strength of ungrouted and grouted sand was assessed from results of unconfined compression tests on undisturbed borehole samples of grouted sand, excavation block and core samples; and on reconstituted sand samples grouted in the laboratory; from results of triaxial compression tests on reconstituted sand samples ungrouted and grouted in the laboratory, and on excavation block and core samples.

Triaxial Test Results

The ϕ and c components of shear strength were calculated using p-q diagrams, where each Mohr circle is represented by one point. Such a p-q diagram is shown in Fig. 3. All triaxial test results analyzed are summarized in this figure. The curve drawn through points on this figure is the K_f-line. The slope α of the K_f-line and its orgin intercept a are related to ϕ and c as follows:

$$\sin \phi = \tan \alpha \quad \text{and} \quad c = \frac{a}{\cos \phi}$$

Results from tests on reconstituted sand samples grouted in the laboratory (35% Siroc 142), and on undisturbed samples from Subarea 11 (55% Siroc 132/142) and Subarea 13 (46% Silicate/R600) are in agreement with results reported by other investigators (Clough et al, 1979): injection of silicate grouts did not significantly change the angle of internal friction but increased the cohesion. The increase in cohesion generally increased with the silicate content of the grout. Results from Subareas 1 and 8, however, show a smaller increase in cohesion, and a decrease in friction. In the case of low-strength grouts (Subarea 8), it may be postulated that the grout tended to lubricate the sand grains, thus reducing the internal friction. In the case of medium-strength grouts (Subarea 1), the lower friction angle may have resulted from sampling disturbance.

Unconfined Test Results

The unconfined compressive strength is useful as an index to shear strength. Undisturbed borehole samples, excavation block and core samples, and laboratory reconstituted samples were used in unconfined compression testing. Results from specimens prepared in the laboratory fall within a relatively narrow range, while results from field samples are much more scattered. This is explained by the variability of the natural soil structure and non-uniform in situ grout penetration. Average results for 35% Siroc 142 from tests on laboratory reconstituted samples, borehole samples, and excavation block samples are plotted in Fig. 4. The block samples had the highest unconfined compressive strength and showed a brittle response. The borehole samples had the lowest strength and were the least stiff, indicative of the highest degree of disturbance.

The influence of silicate content on unconfined stress-strain characteristics is illustrated in Fig. 5. The unconfined compressive strength increased with silicate content, as expected. Data scatter for field samples is attributed to sampling disturbance, but the general trend is in agreement with findings reported by Clough et al (1979) for laboratory reconstituted samples. The axial strain at failure decreased as the silicate content increased (that is, the response became more brittle and the modulus larger as the silicate content increased).

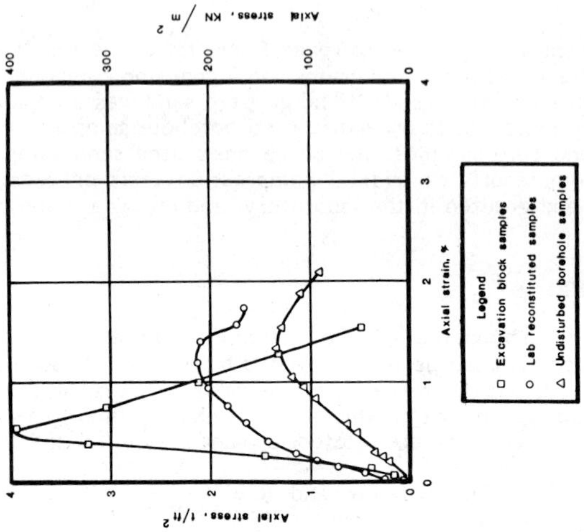

Fig. 4 AVERAGE STRESS-STRAIN CURVES FROM UNCONFINED COMPRESSION TESTS ON FIELD AND LABORATORY SAMPLES GROUTED WITH 35% SIROC 142

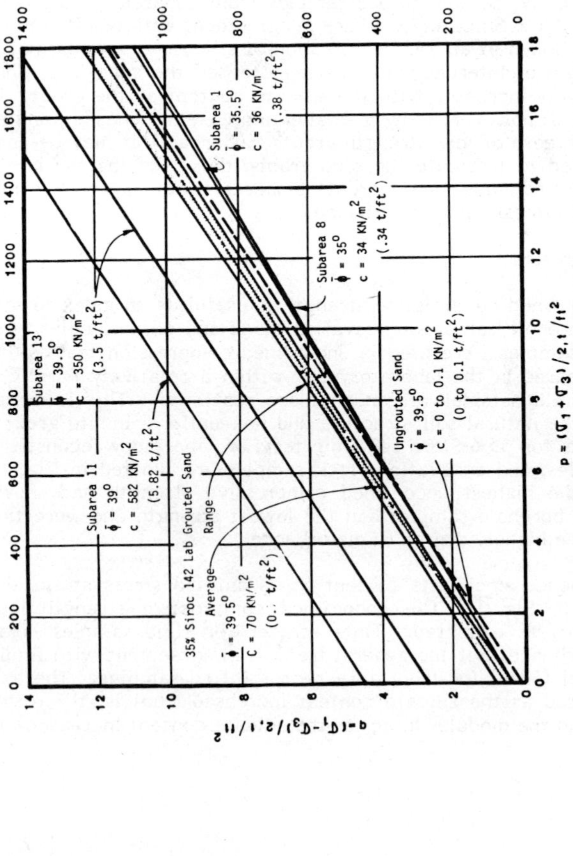

Fig. 3 SHEAR STRENGTH FROM TRIAXIAL TESTS ON UNGROUTED AND GROUTED SAND SAMPLES

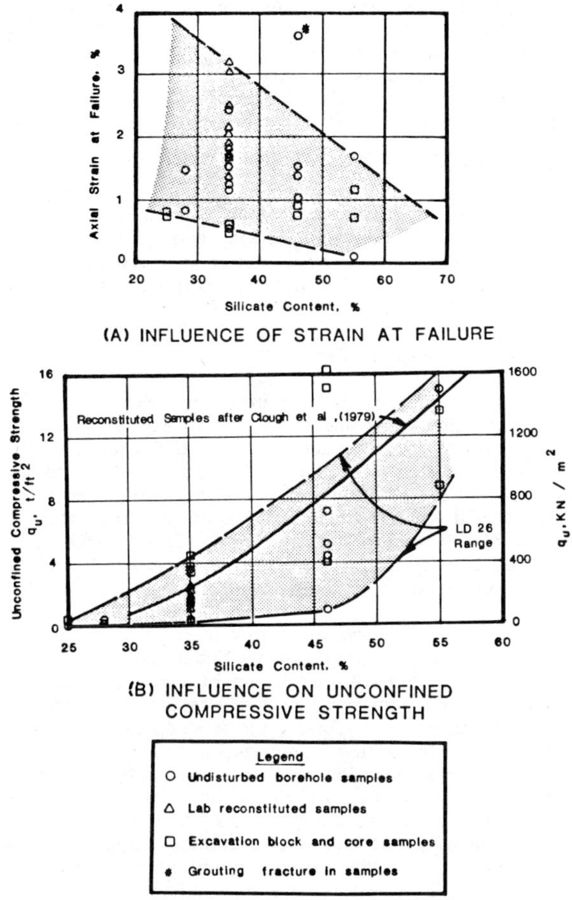

Fig. 5 INFLUENCE OF GROUT SILICATE CONTENT ON UNCONFINED STRESS-STRAIN CHARACTERISTICS

The unconfined compressive strength increased as the strain rate increased. This is in agreement with results reported by Clough et al (1979). Koenzen (1978) found that for 70% silicate grouts, the strength decreased by 45 percent when strain rate was decreased by three orders of magnitude. From those results, it is concluded that conventional laboratory loading rates (that is, about 0.5 percent per minute) may yield strength results up to 50 percent greater than slower loading rates that are more likely to occur in actual field loading.

The influence of curing time on unconfined and triaxial stress-strain characteristics was also evaluated. No significant strength and axial strain at failure change were noted as curing time increased from 2 days to 100 days, contrary to results reported by Clough et al (1979). The disparity can be explained by the different curing environments used in WCC and Stanford laboratories.

Clough allowed the grouted samples to stand in air for 24 hours before placing them in a humid room. Locks and Dam No. 26 samples were placed in a humid room immediately after grouting, then submerged in groundwater in a curing tank pressurized to the average hydrostatic pressure expected at midpoint of the grouted zone. The later curing procedure was more representative of the expected field curing conditions.

Time-Dependent Properties

Laboratory, pressuremeter, and plate load tests indicated that grouted soil behavior under a sustained or slowly applied load was different than under a fast loading rate. The increase in bearing capacity and stiffness due to chemical grouting must be permanent to be a suitable rehabilitation scheme for Locks and Dam No. 26.

Long-Term Unconfined Compression Tests

Results of long-term unconfined compression tests on reconstituted sand samples grouted in the laboratory are presented in Fig. 6. For constant stress ratios CSR* ranging from 15 to 50 percent, none of the tests approached failure or creep rupture. Creep rates, that is the ratio of change in strain to change in time, were almost identical at CSR of 50 and 37 percent. Similar tests were performed by Clough et al (1979) and Koenzen (1978). Results of these tests are compared with Locks and Dam No. 26 test results in Fig. 6. The greater tendency towards creep rupture reported by Clough and Koenzen may be attributed to the greater silicate content of the grouts used by these investigators.

Long-term unconfined compression tests were also performed on excavation block and core samples (Fig. 7). Samples from Subareas 8 and 11 did not undergo creep rupture. Samples from Subareas 1, 3, and 13, however, failed at CSR as low as 30 to 35 percent. This is inconsistent with the results obtained on reconstituted samples grouted in the laboratory. Again, block samples from Subareas 1 and 2 showed a brittle response in unconfined compression characterized by a small creep strain until the strain or stress duration reached a point where rapid, brittle failure developed. The block sample from Subarea 13 showed a steady increase in strain up to failure. The creep behavior of soil grouted with 46% silicate/R600 appears to be inherent to this type of grout and not related to sampling disturbance. The block sample from Subarea 8 (25% silicate/aluminate) showed a tendency for stabilization of creep strain at about 300 minutes, indicative of strain-hardening.

Long-Term Triaxial Compression Tests

Long-term triaxial compression tests on reconstituted sand samples grouted in the laboratory were performed and no creep rupture was observed under CSR of 40, 59 and 79 percent. Koenzen (1978) reported creep rupture for CSR greater than 65 percent for similar tests on samples injected with grout containing 70 percent silicate. From these results it can be concluded that confinement significantly decreased creep tendency of grouted soil, and that creep tendency may have decreased when silicate content was reduced.

---

* CSR: Constant stress ratio = ratio of sustained axial deviator stress to unconfined compressive strength or peak triaxial deviator stress measured at strain rate of 0.15 percent per minute (or at some other typical short term test strain rate)

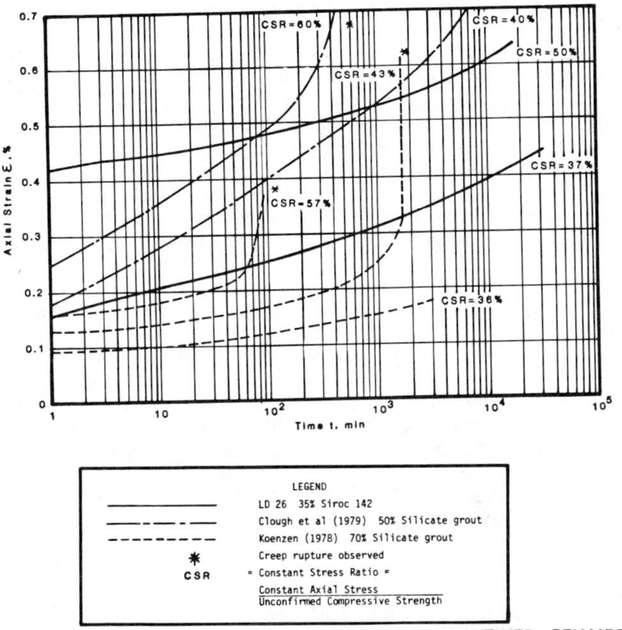

Fig. 6 COMPARISON BETWEEN LONG-TERM UNCONFINED BEHAVIOR OF LAB GROUTED SAND USING 35, 50 AND 70% SILICATE GROUTS

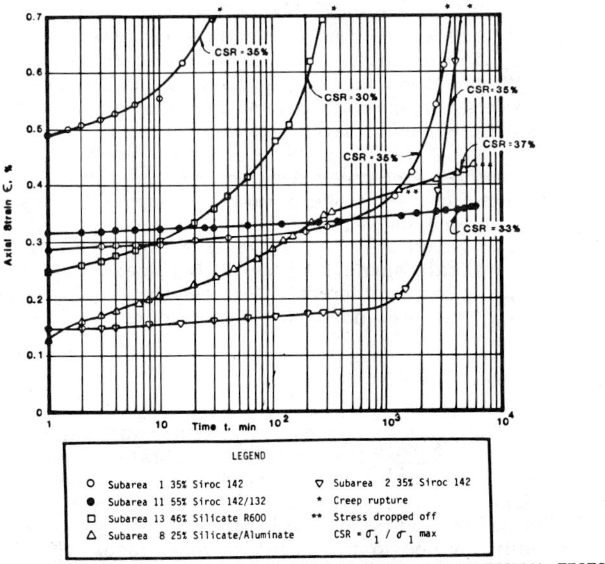

Fig. 7 RESULTS OF LONG-TERM UNCONFINED COMPRESSION TESTS ON EXCAVATED GROUTED SAND SAMPLES

Results of long-term triaxial compression tests on excavation block and core samples from Subareas 1, 8, 11, and 13 are presented in Fig. 8. For reference, results of similar tests on ungrouted sand samples obtained from the excavation and reconstituted in the laboratory are also plotted in that figure. No creep rupture was observed at CSR of as much as 38 percent for all grouted samples, and for one sample from Subarea 8 at a CSR of 68 percent. At CSR values of as much as 38 percent, creep rates for samples from Subareas 1, 8, and 11 did not differ significantly from the results obtained with the ungrouted sample. The sample from Subarea 13 (46% silicate/R600), however, showed a much larger creep rate than the other samples and a tendency towards creep rupture. The samples from subarea 8 all showed a slight tendency for strain hardening between 10^2 and 10^3 minutes.

The stress path followed during triaxial compression more realistically models typical field loading conditions than unconfined compression (except for unsupported excavations and tunnels). It is recommended that triaxial creep tests be performed instead of unconfined creep tests to assess long-term behavior of grouted soil. Lack of confinement in laboratory testing may indicate a tendency for larger creep deformations than would occur under actual in situ stress conditions. In triaxial compression, creep did not appear to be significant up to CSR values of about 60 percent for all grouts tested, except for 46% silicate/R600. On the basis of these results, lower allowable load levels should be used in design for R600 grouts than for Siroc grouts.

Field Creep Tests

The results of long-term plate load tests (see Woodward-Clyde Consultants, 1979) indicated that the creep rate of grouted sand was larger than ungrouted sand at 50 and 100 percent of the ultimate plate bearing capacity. In Subareas 5, 10, and 11 (55% Siroc 132/142), the maximum applied stress never reached 50 percent of failure, and creep rates for these subareas were less than 0.007 percent per minute. However, in Subareas 12 and 13 (46% silicate/R600), which exhibited short-term strength similar to that of 55% Siroc 132/142, the observed creep rate was at least two orders of magnitude faster. Low- strength 25% silicate/aluminate also had a noticeable creep deformation tendency.

Results of pressuremeter creep tests are presented in Fig. 9 as radial strain vs time during which constant probe pressure was applied. Results of pressuremeter creep tests showed trends similar to those found in triaxial creep tests. Small creep strain occurred up to a CSR level of 50 percent in subareas grouted with Siroc and 25% silicate/aluminate grouts. But again, sand grouted with 46% silicate/R600 had a much faster creep rate. Tendency for strain hardening at a CSR of 50 percent was also noted for sand grouted with low-strength 25% silicate/aluminate.

Permeability

The reduction of permeability due to grouting was assessed from results of borehole falling-head permeability tests and laboratory permeability tests on borehole undisturbed samples, excavation block and core sample, and reconstituted sand samples grouted in the laboratory. A summary of the results is presented in Fig. 10, together with results obtained before grouting. Generally, grouting reduced permeability by one to two orders of magnitude. The reduction did not

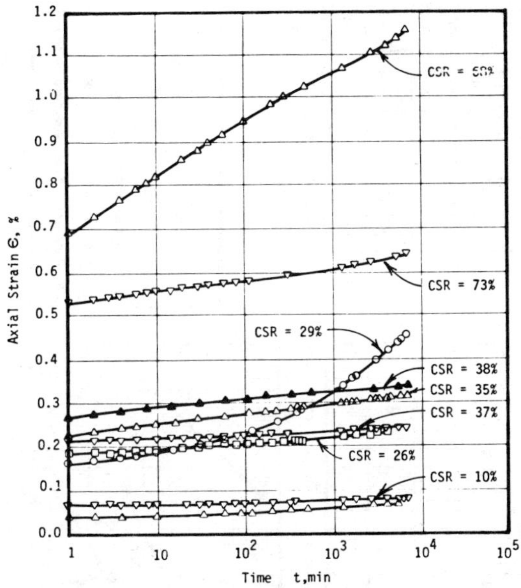

Fig. 8 RESULTS OF LONG-TERM TRIAXIAL TESTS ON GROUTED AND UNGROUTED SAND SAMPLES

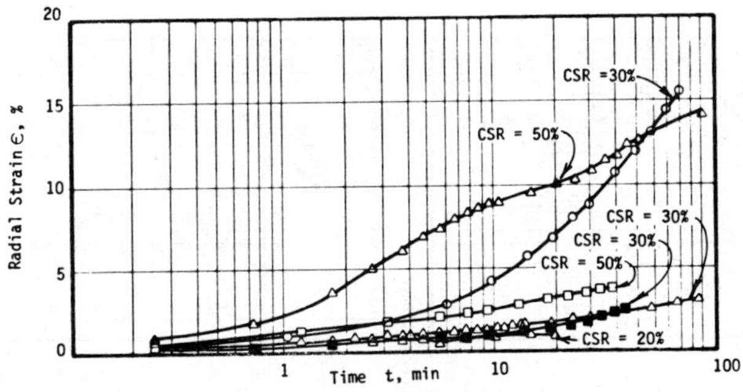

Fig. 9 RESULTS OF LONG-TERM PRESSUREMETER TESTS

LEGEND

- △ Subarea 8 25% Silicate / Aluminate
- ▲ Subarea 1 35% Siroc 142
- □ Subarea 11 55% Siroc 142
- ■ Subarea 5 45% Siroc 142
- ○ Subarea 13 46% Silicate / R600
- ▽ Ungrouted Soil

CSR = $(\sigma_1 - \sigma_3) / (\sigma_1 - \sigma_3)$ max.

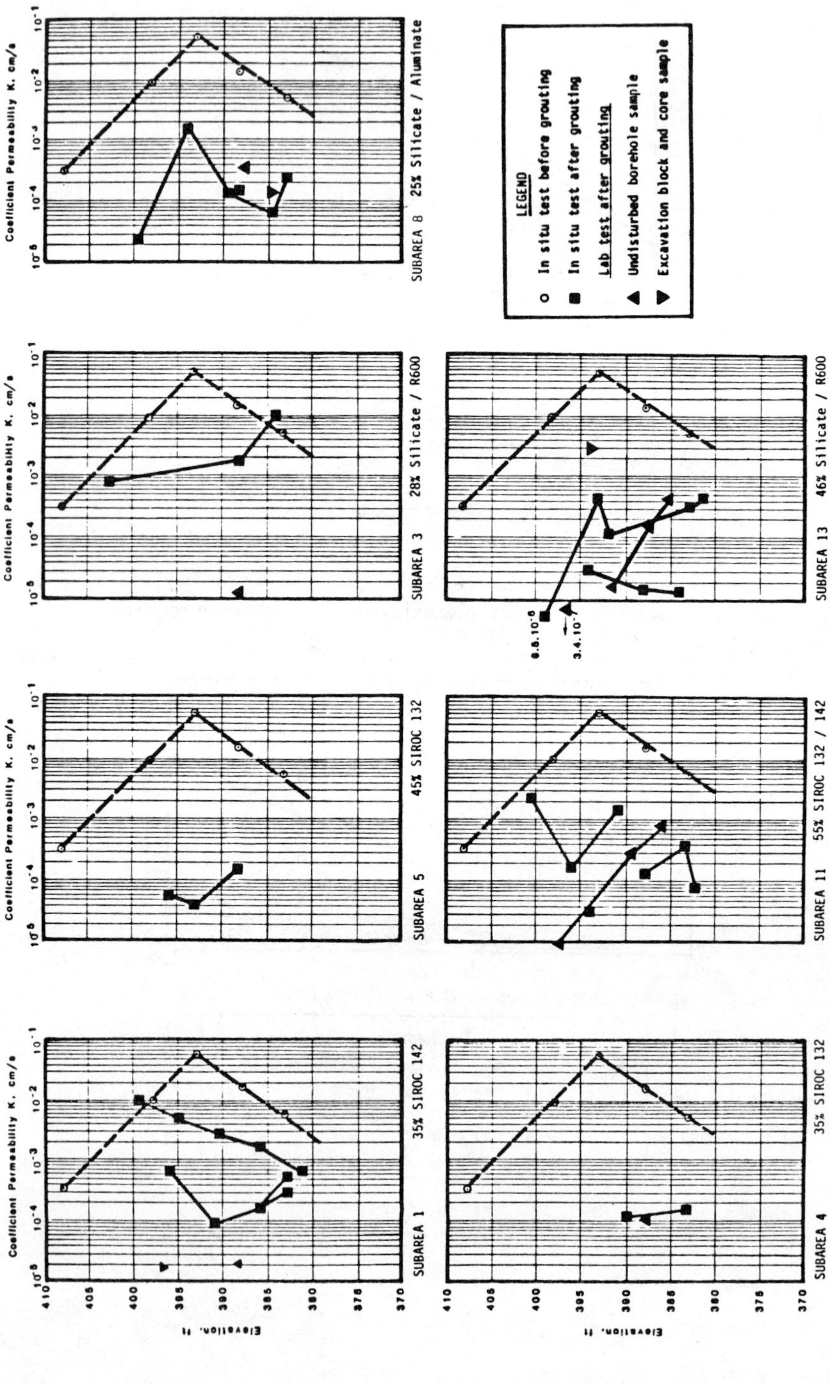

Fig. 10 FIELD AND LABORATORY FALLING HEAD PERMEABILITY TEST RESULTS

appear to be a function of the grout type; rather, it was found to be a function of the grouting method. Comparison between mapping observations and field permeability test results clearly showed that zones in which permeability was not decreased were ungrouted. Ungrouted zones surrounded by well-grouted zones, however, showed a marked decrease in permeability.

Permeability measured on reconstituted sand samples grouted in the laboratory averaged 4.8×10^{-7} cm/s; that is, two to three orders of magnitude less than field and laboratory tests on sand grouted in situ. The difference is explained by natural soil discontinuities and borehole disturbance which was not a factor for the laboratory prepared specimens.

CONCLUSIONS

The test program at Locks and Dam No. 26 provided a unique opportunity to assess the index and engineering properties of chemically grouted sand by a wide variety of in situ and laboratory tests and measurements. In situ horizontal stresses increased by two to three times due to grouting. In-place dry unit weight of grouted soil averaged .16 kN/m^3 to .47 kN/m^3 (1 to 3 lb/ft^3) more than before grouting. Strength and deformation modulus values were generally increased manyfold by grouting; the actual increases depended on the strength of the grout injected. The strength increase was due only to an increase in cohesion, the angle of internal friction remaining practically unchanged or decreasing slightly. Time dependent properties of grouted soil were tested. Significant creep was observed in some grouted zones, although tendency to creep failure was not generally pronounced. Siroc-type grouts exhibited less creep than grouts containing aluminate reactant. Grouts containing R600 reactant were the most creep-prone. Grouted soil was found to be much less permeable than ungrouted soil. Generally, grouting decreased the coefficient of permeability by one to two orders of magnitude. The reduction in permeability did not appear to be a function of grout type, but rather it was found to be a function of grouting method (that is, of grout penetration).

ACKNOWLEDGEMENTS

The test program was designed and implemented by Woodward-Clyde Consultants under contract No. DACW43-78-C-0005 with the Corps of Engineers, St Louis District. The technical input of Bruce H. Moore and Edward C. Demsky of the District was instrumental in the development of the program. Jin Kim was the task leader for WCC. WCC retained a joint venture of Raymond International and Soletanche and Rodio, Inc to perform the grouting. Marcel P. Rigny and Joseph P. Welsh directed the joint venture team.

REFERENCES

Clough, G. W., Kuck, W. M., and Kasali, G. (1979) "Behavior of Silicate Stabilized Sands" JGED, ASCE, 105, GTI, Jan

Koenzen, J. P. (1978) "Time-Dependent Stress-Strain Analysis of Chemically Stabilised Building Elements" Symposium on Soil Reinforcing and Stabilising Techniques in Engineering Practice, New South Wales, Australia

Woodward-Clyde Consultants (1979) "Results and Interpretation of Chemical Grouting Test Program Existing Locks and Dam No. 26" Phase IV Report II, Prepared for US Army Corps of Engineers, St Louis District, 15 July (Available from NTIS)

CREEP BEHAVIOR OF SILICATE-GROUTED SAND

BY

Roy H. Borden,[1] M.ASCE, Raymond J. Krizek,[2] M.ASCE, and Wallace H. Baker,[3] M.ASCE

ABSTRACT

The macroscopic time-dependent behavior of sands injected with sodium silicate base grouts, using a formamide-ethyl acetate reagent, is investigated as a function of the silicate content. Approximately twenty-five unconfined compression tests, thirty unconfined creep tests, and seven triaxial creep tests were conducted on laboratory injected dense sand (Ottawa 20-30) specimens using three different silicate content grouts, chosen to span the range of concentrations generally used for structural grouting. The stress-strain-time behavior of silicate stabilized sands subjected to sustained mechanical stress is described by a simple three parameter phenomenological model. The creep strength is explained by utilizing the results of unconfined compression tests and the theoretical framework of the Mohr-Coulomb failure criterion. The consideration of time-dependent and time-independent components of strength gives rise to a strength parameter which relates the applied stress level to the time-to-failure and describes both unconfined and confined creep behavior.

INTRODUCTION

Chemical grouting has been used as a soil stabilization technique in Europe for many years and is now finding significant application in the United States in underground construction projects. The design of grouted soil support systems is still considered an art practiced by a small number of highly specialized experts utilizing emperical rules developed over years of extensive personal experience. In the past two decades experimental studies have been reported in the literature which document the mechanical properties of soils stabilized with a number of different chemical grout formulations. Due to the variety of grout types and possible combinations of components within each chemical system the collection of data relating to any one grout mixture tested under comparable conditions is sparce. In order to rely on the strength imparted to a soil through the injection of chemical grouts it is necessary that one know the relationship between the strength as measured in conventional laboratory tests (a strength measured in minutes or hours) and that which will be realized over a longer period in a field application. It is this factor which will

[1]Assistant Professor, Department of Civil Engineering, North Carolina State University, Raleigh, North Carolina.

[2]Professor and Chairman, Department of Civil Engineering, The Technological Institute, Northwestern University, Evanston, Illinois.

[3]President, Hayward Baker Company, Odenton, Maryland.

allow for the more economical future use of chemical grouts for soil stabilization purposes.

The major objectives of the investigation reported herein were to explain the stress-strain-time and strength-time behavior of sodium silicate grouted sands based on an investigation of the time-dependent behavior of laboratory prepared specimens subjected to sustained mechanical stress (creep). The study was designed to utilize several silicate content grouts which span the range generally used in engineering practice for obtaining moderate to very strong gels, while maintaining the reagents used to promote gelation in the grout system constant to provide a uniform basis for comparison. A combination of unconfined compression tests, unconfined creep tests and triaxial creep tests were conducted on sand specimens injected with each of the three grouts, with a total of approximately seventy specimens tested.

Based on the results of the creep tests, and correlation with short term laboratory strength tests, the capability of a single stress-strain-time function to explain the behavior of the range of silicate content grouts tested under various confining stresses and subjected to a range of creep stresses is investigated. The development of a relationship capable of predicting the length of time over which a grouted soil element, subjected to a known confining stress, could sustain a given creep stress without failure is also addressed.

BACKGROUND

The application of chemical grouting to geotechnical problems related to foundations, tunnels, bridge piers, dams, mine shafts, and other types of structures led to the documentation of numerous case histories (1). More recently, a comprehensive performance record of the use of chemical grouting in tunneling for the Washington Metro construction was reported in the literature (5). Silicate grouts have even been used to reduce the liquefaction potential of loose sands in one documented case (8). In conjunction with the development of grouting technology, numerous laboratory studies have been performed to characterize the mechanical properties of both the gel and soils which have been stabilized and, as yet, no simple relationship between the two has been established. The use of chemical grouts for the strengthening of soils is largely done on an empirical basis and the composite material is in its early stages of understanding as an engineered material by the geotechnical profession. The questions which have generally been addressed in experimental studies refer to (1) the difference between the response of similar density specimens injected with different grouts, (2) the effect of the grout on the frictional response of a granular soil, (3) the magnitude of the cohesion imparted to a previously cohesionless soil, (4) the effects of strain rate or applied stress rate on the measured strength and stiffness, (5) the effects of specimen preparation techniques and curing conditions on the above factors, and (6) the behavior of specimens subjected to sustained shear stress (creep), with emphasis on establishing the long term creep strength. Although the literature contains sometimes conflicting conclusions, certain findings have repeatedly been reported; these are:

1. The unconfined compressive strength of similar density specimens, subjected to a fixed rate of strain or applied stress, increases

with the silicate content of the grout mix, and increases to some limiting value as the amount of a given type of reagent used to cause gelation is increased for a given concentration of silicate (6, 9, 12, 14, 17, 18).

2. The presence of a silicate grout does not alter the frictional response of a granular medium and the relationship between void ratio or density and the internal angle of friction is the same for grouted sands as ungrouted sands (6, 7, 16).
3. The magnitude of the cohesion intercept in confined compression increases with the strength of the grout (6), in accordance with the silicate and reagent concentrations as explained in statement number one. For a 50% silicate content grout the cohesion intercept was increased from 2.3 kg/cm^2 for low relative density, uniform medium sand specimens to 3 kg/cm^2 for densely prepared specimens of the same sand (7) showing a small density effect. The cohesion also increases with the fineness or specific surface of the sand (6, 16, 17).
4. The strength and stiffness of silicate grouted sands are very much influenced by the rate of shear stress application or strain rate. Although no studies have been reported in the literature that show the development of Mohr-Coulomb failure envelopes for different rates of strain, virtually all investigations have found decreases in unconfined strength with decreasing strain rates (4, 6, 10, 12, 19). In one particular study (12) the unconfined strength (and consequently the cohesion intercept by virtue of the applicability of the Mohr-Coulomb failure criterion) was shown to decrease at different rates for different grout components. Although a quantitative explanation was not advanced, the following generalizations were made: (1) the influence of variations of the strain rate is greater when a smaller percentage of the potential strength of the silicate is activated due to a low concentration of reactant and (2) the sensitivity to variations in the strain rate decreases as the silicate concentration decreases.
5. The effects of specimen preparation and curing conditions have been shown to exert a substantial influence on the strength and stiffness of silicate grouted sands. However, it is in this area that probably the greatest inconsistencies between investigators are found and due to the lack of comparable detail in all studies a reconciliation of these differences is difficult. For example, it has been suggested that the strength of soils injected with silicate gels can be assumed to have reached a maximum in two days (17) and this finding has been substantiated for one grout mix in another investigation (19). However, one study has shown this time to be in excess of seven days and possibly as long as 200 days (6) while still another has found the maximum strength obtained in four days with no significant variation over 200 days (4). It is clear from these differences that the specification of a silicate grout does not adequately describe the curing time at which a near steady state strength will be reached; one must know the type and concentration of the reagent used to gel the silicate. A comprehensive study of this problem has not been reported in the literature and at this time one must require that the system of components being employed for any

grout mix be studied to determine the relationship between curing time and resultant strength.

As for the influence of the curing environment, those investigators performing comparative studies have found that with reference to specimens cured in sealed molds and stored in humid rooms (moist-cured) the strength of specimens cured exposed to the atmosphere are 50% stronger (4) while those cured submerged under water were found to have strengths ranging from 50% of the moist-cured specimens (6) to 20% of the strength (4), to as little as 10% or 20% for several silicate grouts (18). It is clear from the above comparisons that the behavior of the silicate grout chemical system is a complicated one which has not yet been adequately characterized.

6. The response of silicate grouted soils to sustained shear stresses (creep) is characterized by an increase in the time at which rupture is observed as the creep stress is decreased from the maximum stress level reached in a short term test (6, 10, 18, 19). The results of these studies provide insight and a general characterization of the creep phenomenon for silicate grouted soils. However, in only one study reported herein (19) has it been shown that the silicate grout mix used had achieved its near maximum strength before the start of the creep tests. This creates the problem of evaluating the response of a material that is gaining strength during the test. For specimens which failed in a relatively short period of time (less than one day, for example) this is probably not significant. However, for specimens which were subjected to creep stresses for longer periods of time, the actual ratio of the applied stress to the "strength" of the specimen is not known. Only one study on the behavior of silicate grouted sands subjected to confined creep has been reported (19) and indicates that the creep potential of 35% silicate specimens was significantly reduced by the presence of a confining stress of 2 kg/cm^2. No creep rupture was observed under compressive stress levels, defined as the ratio of applied shear stress, $(\sigma_1-\sigma_3)$, to the triaxial strength at failure under a similar confining stress, $(\sigma_1,-\sigma_3)_f$, of 40%, 59% and 79%. For similar tests on specimens injected with a 70% silicate grout it was reported that rupture occurred for stress levels greater than 65%. From this it was concluded that the creep tendency of silicate grouted soils may increase as the silicate content is increased.

The available literature provides a framework in which to view the creep behavior of silicate stabilized soils, but as was noted, the relationship between the compressive stress level and the time to failure has not been generalized for the range of grouts typically used in practice. The literature suggests that the presence of a confining stress significantly decreases creep potential with the magnitude of this decrease, possibly a function of the strength of the grout.

EXPERIMENTAL PROGRAM

The basic objectives of this experimental investigation were (a) to determine the creep strength of sand stabilized with three different sodium silicate grouts, (b) to evaluate the time-deformation response of silicate grouted sands under sustained stress, and (c) to develop a consistent theory for explaining the creep strength of silicate stabilized soils. Accordingly, constant stress level unconfined creep tests were performed on sand specimens stabilized with one of three silicate grouts (chosen to represent the range of strengths generally used for the strengthening of soil masses). For each of the grouts studied, a number of conventional unconfined compression tests (failure achieved in minutes) were performed to serve as a control and to develop a reference strength. Drained triaxial creep tests were also performed to extend the findings of the unconfined creep tests to develop a more general explanation of the stress-strain-time and time-dependent strength of silicate grouted soils.

Dense sand specimens (γ_d = 18 kN/m^3, $D_r \approx$ 95%) were chosen due to their ease of duplication. The three grout mixes each contained 10% by volume of a reagent composed of equal parts ethyl acetate and formamide, and either 40%, 50% or 60% by volume of sodium silicate. In each mix the remaining component was tap water to create a total volume of 100%.

Unconfined creep tests were performed on moist-cured specimens containing one of the three grouts over a range of creep stresses ranging from approximately 30% to 80% of the short term strength as determined from unconfined compression tests. Submerged-cured specimens, stabilized with one of the grouts were also tested, but only in the more important range of lower stresses (between 25% and 40%) to evaluate the influence of the curing environment on the creep strength. Triaxial creep tests under a range of confining stresses were performed on several specimens from each grout series.

Materials Tested

To provide a uniform basis for comparison, Ottawa 20-30 sand was used exclusively in the study. The maximum and minimum void ratios are 0.768 and 0.462, the specific gravity is 2.67, and the grain size distribution of the sand is considered uniform with a nominal diameter of 0.59mm. The three silicate grouts used were composed of all commercially available components. The sodium silicate used is produced by the Philadelphia Quartz Company, Philadelphia, Pennsylvania, carries the trade name N-grade silicate and has a silica to soda ash weight ratio (SiO_2/Na_2O) of 3.22. The silicate is composed of 28.7% SiO_2 and 8.9% Na_2O, exhibits a density of 13.8kN/m^3 and a viscosity of 180 centipoise at 20^oC and has a pH of 11.3. The ethyl acetate (99.5% purity) and formamide (97% purity) components are laboratory grade and were supplied by the Aldrich Chemical Company of Milwaukee, Wisconsin.

Specimen Preparation Technique

Specimens were prepared using a plexiglass mold, 38mm in inside diameter and 368mm long, which was split longitudinally into three pieces. This mold allowed the formation of four specimens in one grouting operation, and the ability to remove them with minimal disturbance. Circular pieces of nylon screen (38mm in diameter) were positioned in the mold at the predetermined ends of each specimen, thus eliminating the necessity of cutting the grouted sand to the desired specimen length.

The grout injection system consisted of a mixing tank, the specimen mold, a regulated air pressure source, and associated connecting hoses. The mixing tank was initially filled with water and, under an air pressure of 7 kN/m^2, at least 10 specimen void volumes of water were flowed through each specimen to enhance saturation. The desired grout solution was then made by pouring the required quantities of water, ethyl acetate, and formamide into the tank and mixing with a hand stirrer for one minute. After adding the premeasured quantity of silicate the grout solution was stirred for an additional three minutes, at which time the tank was closed and pressurized to approximately 40 kN/m^2. Injection was continued under a constant pressure until at least $0.0008m^3$ (800 cc; equivalent to two void volumes of the sand in the mold) of the $0.001m^3$ (1000 cc) was used. The mold remained connected to the injection system for approximately one day, when it was dismantled for specimen storage.

Curing Time and Environment

One of the major concerns in producing specimens to be used to conduct long term tests of any kind is to ensure that the specimens do not change material properties during the period of the test. In this experimental investigation the major concern about curing was that the grout had completely gelled, as evidenced by the unconfined strength of similar specimens, prior to the initiation of the creep tests. For the grouts used in this study the intermediate silicate content mix (50% by volume) was investigated to determine the increase in strength as a function of time. The results indicated that the maximum strength of specimens cured in a sealed environment and stored in a 98% relative humidity chamber at room temperature (≈20°C) is achieved in less than four days. Recognizing that the higher silicate content specimens would take longer to cure and that some of the creep tests would probably be conducted over a period of several months a conservative curing time of one month was adopted.

The majority of specimens were cured in a moist condition, that is, they were removed from their molds, sealed in plastic wrap, and stored in a controlled humidity room. However, due to the variability in the findings reported in the literature as to the influence of curing specimens under water, this mode of curing was also investigated. Because it was felt that the total submersion of a specimen in a container of water was an unrealistic model of the in situ conditions below the ground water table, each group of four specimens placed in a plastic container (100mm x 100mm x 100mm) surrounded with sand to provide lateral confinement, and then submerged.

Upon removal from the humid room for testing, the weight of each specimen was determined and indicated that no loss of weight occurred during storage. A sulfur-cement capping compound was used to provide flat parallel ends for those specimens to be tested in either unconfined compression or unconfined creep. Specimens that were to be tested over a period exceeding five minutes were enclosed in standard triaxial membranes and coated with grease.

Testing Techniques

The testing conducted in this experimental investigation included unconfined compression tests and unconfined and triaxial creep tests.

Unconfined compression tests were performed at a strain rate of approximately 0.5%/min. using a controlled deformation rate Geonor testing machine. Unconfined creep tests were conducted using a standard dead load consolidation loading frame. This frame incorporates a lever system that enable multipliers of 10 and/or 40 to be used to achieve the desired load. Only simple height adjustments of the equipment were necessary to accomodate the approximately 16mm (3 inch) long specimens. Prior to the application of the desired load a small seating load, approximately 2% of the test load, was applied to the loading frame to ensure complete contact between the swivel loading head and the specimen end cap and to minimize the possibility of impact during the application of the full test load.

Triaxial creep tests were conducted utilizing standard laboratory triaxial equipment which incorporates a hanger system for the application of dead load (conventionally used for incremental load controlled shear tests or for anisotropic consolidation). The procedure for conducting the triaxial creep tests was as follows: (1) The specimen to be tested was mounted in the triaxial cell and the desired isotropic confining stress was applied. Consolidation was generally observed to result in the immediate discharge of some fluid from the system into the burettes with all volume changes completed within 10 minutes. An additional 10 minutes was allowed before the application of the design creep stress; (2) The desired stress difference, $\sigma_1 - \sigma_3$, was applied in increments of 140 kN/m^2 or less, and in all cases the time between the successive application of loads was no longer than 5 seconds; and, (3) The deformation of the specimen was recorded at successive time increments and the time of failure was recorded when appropriate.

RESULTS AND INTERPRETATION

In this test program a series of four specimens was obtained from each sand column, which had been injected with a single batch of grout. At least one unconfined compression test was performed on a specimen from each series and in numerous cases a second test. In ten instances unconfined creep tests were discontinued before failure occurred and these specimens were immediately tested in unconfined compression. The results of all unconfined tests are presented in Figure 1. The major emphasis in this investigation is on the behavior of moist-cured specimens stabilized with grouts containing 40%, 50% and 60% by volume of sodium silicate. Figure 1 shows the increase in unconfined strength (determined at a strain rate of approximately 0.5%/min.) as a function of silicate content, with the average strength of these specimens being approximately 300 kN/m^2, 900 kN/m^2, and 2500 kN/m^2, respectively. Also shown are the results of tests conducted on submerged cured specimens, and only for the 60% silicate content grout is any significant difference in strength observed, where the average strength is about 1600 kN/m^2, or slightly more than 60% of the strength of the moist-cured specimens. The data also indicate that the short term unconfined strength of silicate grouted soils is not influenced in any consistent way by the previous application of a sustained compressive stress, in some instances as high as 40% to 50% of the unconfined strength for a period of a week or more.

Within each series of specimens (the four specimens formed at the same time) the variation in unconfined strength was generally less than

15%, regardless of whether the specimens were tested "fresh" or after a period of sustained stress. However, in several instances the difference in average strength for a series and the average of all other specimens of that silicate content, was greater than 40%. The reason for these differences is not apparent, as a consistent procedure for specimen preparation was adopted and all specimens were prepared by the same technician. When interpreting average parameters of a large number of specimens these infrequent variations would not pose a serious problem, but when referencing an applied creep stress to an assumed short term strength this would introduce significant error. For this reason the stress levels reported for the creep tests (defined as the ratio of applied stress to the short term strength) are referenced to

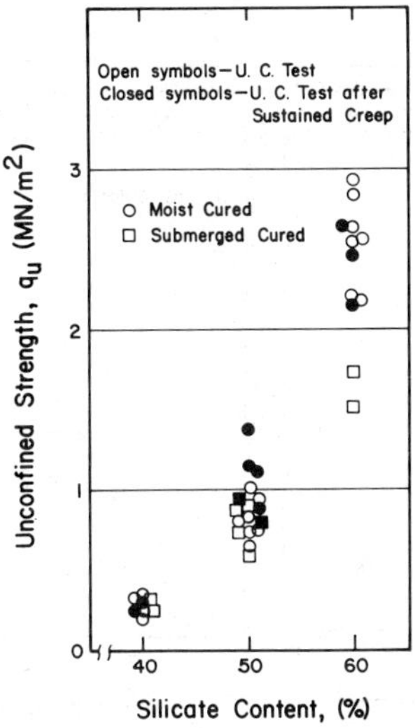

Figure 1. Unconfined Strength versus Silicate Content.

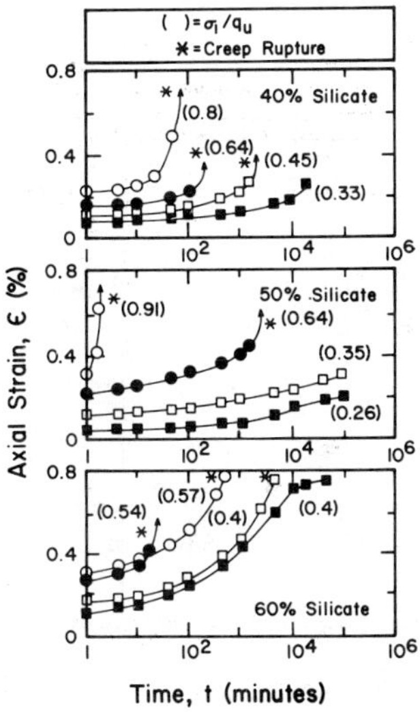

Figure 2. Typical Axial Strain versus Log-Time Response.

the average strength of the unconfined tests conducted on specimens from the same series.

Typical strain-log time behavior for specimens stabilized with each of the three grouts and subjected to nominal stress levels ranging from 30% to 80% of the short-term strength, and tested in unconfined creep, is shown in Figure 2. Generally, it can be said that the initial strain, the rate of strain increase, and the tendency for creep rupture increases with increasing stress level. Figure 3 shows the relationship between the compressive stress level and the length of time over which the stress was supported before rupture or the discontinuance of the test. These data show the general trend toward increasing times to failure as the mobilized percentage of the short term strength is decreased. There does not appear to be a significant difference in the response of the three grouts, indicating that the time-dependent strength of silicate grouted soils can be formulated in terms of the ratio between the applied shear stress and the short term strength with no loss in accuracy. General conclusions that can be drawn from Figure 3 include:

1. The highest compressive stress level for which failure did not occur in one week was approximately 50%, and this was observed for both 40% and 50% silicate content specimens which had been moist-cured.

2. The moist-cured specimens subjected to 40% compressive stress levels or less sustained their stresses without rupture for periods ranging from one week to two months in all but one case (7 of 8).

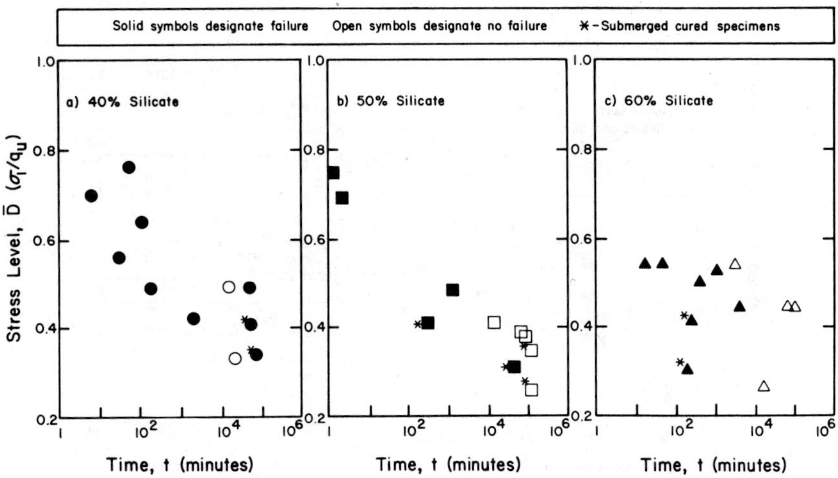

Figure 3. Stress Level versus the Logarithm of Time for Unconfined Creep on (a) 40%, (b) 50%, and (c) 60% Silicate Content Specimens.

3. At stress levels below 40% the submerged cured specimens showed a larger amount of scatter in their response. The two 40% silicate specimens failed in slightly more than one month, a time which supports an increasing time to failure function with no minimum stress level to prevent rupture, that is, no "fundamental" strength. Of the 50% silicate content specimens three of four sustained their stresses in excess of three weeks. Two of these three (both from the same series) did not fail in over three months. Finally, of the two 60% silicate specimens both failed unexpectedly in a short period of time.

One possible explanation for early failure of the 60% silicate specimens can be qualitatively advanced based on the reported findings of Luong et al (12). Their data on a number of silicate content grout mixes showed that the effect of insufficient reagent on the strength of silicate grouted specimens was greatest for higher silicate content mixes and seen most clearly at lower strain rates. Although for all grouts used in the study mixing was intended to be adequate to uniformly disperse all components, it is possible that this goal was not achieved. This could also be the reason for the difference in the average unconfined strength between series (that is, those specimens injected with different batches of grout). It is also possible that some of the reagent was leached out of the specimen during the submersed curing period with the effect being magnified in the 60% silicate content grout due to the already reduced ratio of reagent to silicate in comparison to the lower silicate content grouts. From their findings it also follows that, even if this lower value of strength is used to normalize the applied creep stress to calculate the stress level used to present the data, the specimen response will not be consistent due to the further magnification of the effects of a weaker gel at the lower strain rates existing during creep.

Time-Dependent Strength

The length of time over which an unconfined grouted soil element can sustain a given percentage of its potential short term strength is shown to be reasonably well defined as a decreasing linear function between stress level and the logarithm of time. As the majority of unconfined creep tests in this study were conducted on moist cured specimens they were chosen as the basis for developing this relationship. Figure 4a shows the relationship between the stress level and time to failure for all unconfined creep tests on moist-cured specimens and the curve chosen to represent the length of time over which one may expect a stress level to be sustained can be expressed as:

$$t_f = 10^{(10-14\,\bar{D})} , \qquad (1)$$

where t_f represents the predicted failure time in minutes and $\bar{D}$ the applied stress level. The applicability of the expression to submerged-cured specimens is shown in Figure 4b where the curve resulting from Equation (1) is plotted in conjunction with the actual data. For the 40% and 50% silicate content specimens the agreement generally is good; however, less satisfactory correlation is observed for the 60% silicate specimens. It appears that not only are the submerged cured 60% silicate specimens weaker in the short term compression tests (approximately 60% as strong as the moist-cured specimens) but the response under sustained stress is poorer, resulting in shorter failure time than predicted. Unfortunately, because all four specimens of this series were submerged-cured it is not possible to tell if the observed weakness is due to the curing method which caused reduction of the reagent concentration by dispersion, or inadequate mixing of the reagent with the silicate; both of which, according to Luong et al (12), could have resulted in poorer long term response. Figure 4c presents a composite of the creep data reported in the literature, plotted in conjunction with the relationship developed above, and the relationship between the predicted failure time and the reported data is seen to be very good.

Figure 5 is a composite presentation of the results of confined creep tests conducted on the range of silicate content specimens. The stress-strain response during loading of the 40% silicate content specimens is shown in Figure 5a, and reflects an increasing initial stiffness as a function of confining stress. Figure 5b shows typical strain versus the logarithm of time data for each of the three silicate contents and Figure 5c presents the data in the form of stress level versus time to failure, or discontinuance of the test. From Figure 5c it can be seen that the long term strength of a grouted soil element subjected to a confining stress can not be simply expressed as a function of the applied stress level, where the stress level is defined as the ratio of the applied stress to the short term triaxial strength.

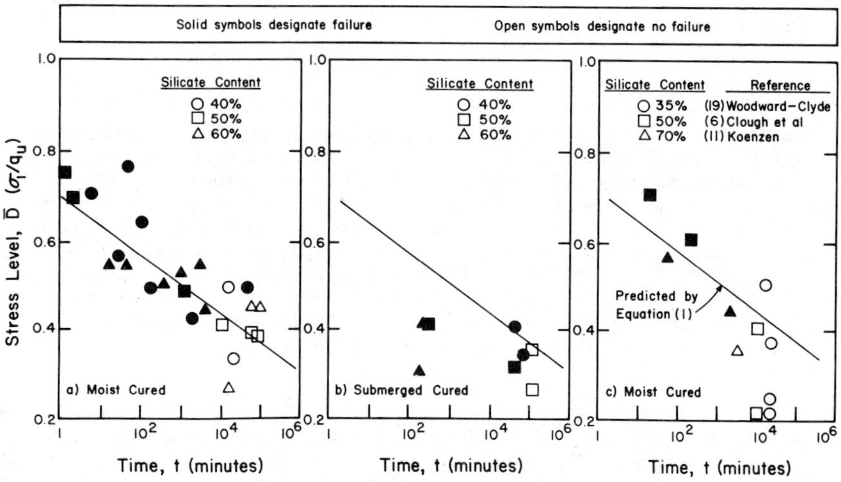

Figure 4. Relationship between Stress Level and the Logarithm of Time for Unconfined Creep.

The failure strength of soils that are generally not considered to exhibit time dependent properties can be represented by the Mohr-Coulomb failure criterion without reference to the strain rate at which the strength test is conducted. The strength formulated in terms of the principal stress difference, $\sigma_1-\sigma_3$, (with total stresses assumed to be equal to effective stresses, representing drained conditions) can be expressed as

$$\sigma_1-\sigma_3 = \sigma_3\left[\frac{1+\sin\phi}{1-\sin\phi}-1\right]+c\left[\frac{2\cos\phi}{1-\sin\phi}\right], \qquad (2)$$

where ϕ is the internal angle of friction (a measure of frictional resistance) and c the cohesion intercept (generally thought of as arising from some type of attractive and/or adhesive forces between particles). In general, however, the value of c which is determined from a given series of strength tests on specimens having the same density (and therefore representing essentially the same frictional response), and tested at different confining stresses, is not independent of the rate of strain at which the tests are conducted. This is especially true for silicate grouted soils.

Although no reported data have been found in the literature where Mohr-Coulomb failure envelopes have been developed at various rates of strain, the tests of Koenzen (11) conducted on 70% silicate stabilized

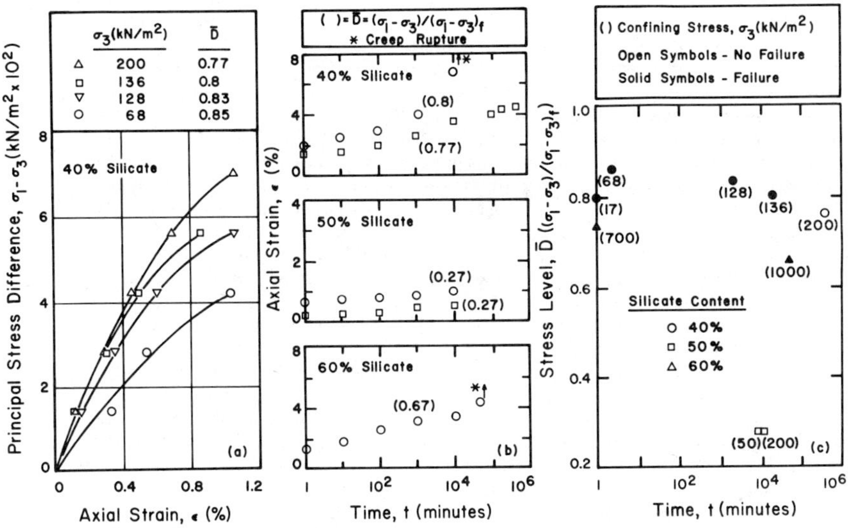

Figure 5. Composite of Confined Creep Data: (a) Stress-Strain Response; (b) Axial Strain-Log Time Response; and, (c) Stress Level-Log Time Response.

specimens at one confining pressure and various rates of strain, combined with the results of the previously mentioned unconfined compression tests, suggests that the observed cohesion will decrease as the strain rate decreases. The strength of the grouted specimen can therefor be regarded as being composed of a time independent component, resulting from the frictional resistance of the sand grains and related to the confining stress σ_3, by the expression

$$(\sigma_1-\sigma_3)_{f\ \mathrm{sand}} = \sigma_3\left[\frac{1+\sin\phi}{1-\sin\phi}-1\right] \quad , \qquad (3)$$

and a time-dependent component due to the presence of the grout, which is the second term in Equation (2) and equal to the unconfined strength. Clearly, as the ratio of the confining stress to the unconfined strength or cohesion intercept increases the strength contribution from the cohesive nature of the soil that is required to prevent failure decreases, for any given applied stress level. It also follows that for any confining stress to unconfined strength ratio there exists an applied stress level below which failure will not occur, as the frictional strength of the soil will not have been exceeded. The practical significance of this relationship between the (σ_3/q_u) ratio and the strength components is that if a series of triaxial creep tests are performed on similar density grouted specimens stabilized with different strength grouts and the confining stress is maintained constant for all tests, although they may all be subjected to the same stress level (i.e. stressed to the same percentage of their triaxial strength) the specimen with the strongest grout will have a larger portion of its time dependent component mobilized.

In order that a general capability to predict the response of a grouted soil element may be based on a knowledge of the applied shear stress level, some method must be employed to take into consideration the influence of confining stress on the relationship between the time-dependent and time-independent components of strength. One relationship which can be used and is proposed herein, is the ratio between the applied stress in excess of the frictional strength of the specimen (which would exist as a result of the confining stress without consideration of the presence of the grout), to the unconfined strength of the specimen, which essentially arises due to the grout (It can be shown that the unconfined strength is an accurate measure of the cohesion within $\pm$ 10% of a medium density sand state without reference to density).

Figure 6 shows the confined creep data developed in this study plotted as the modified stress level versus the logarithm of time. Also presented is the curve described by Equation 1 and the agreement is generally quite good. An attempt was made to show the applicability of the expression developed in this study to confined creep tests reported in the literature. However, for virtually every test reported (19), the relationship between the applied shear stress and the confining stress was such that the frictional strength of the sand was not exceeded, and therefore, failure would not have been expected, and in no cases was failure observed.

Stress-Strain-Time Behavior

In addition to determining the overall stability of a soil structure system, which requires the need to know the relationship between applied stresses and material strength, there also exists a need to estimate the deformations that are likely to result in and around a grouted soil mass due to the imposition of stress variations. In order to predict these deformations, stress-strain-time functions must be established for each of the materials within the domain. The results of the unconfined and confined creep tests provide substantial data on which to develop such a relationship for silicate stabilized soils. Although a microscopic model of grout-soil interaction is not possible at this time due to the very limited knowledge about the character of silica gels, the literature does offer numerous phenomenological models proposed for the description of both natural and frozen soils.

In order to characterize the nature of the creep response of the grouted soils in this study, the incremental strain rate was calculated for each specimen. The relationship between the logarithm of strain rate and the logarithm of time was generally found to be linear, at least over the first three or four log cycles of time, and the strain rate at a given point in time after the initiation of creep was shown to be an increasing function of the applied stress. When these data are plotted as the logarithm of strain rate versus stress level the relationship can be approximated by a linear function. One of the

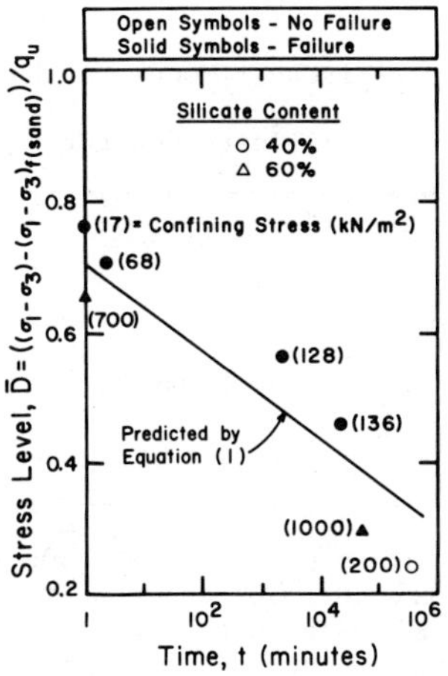

Figure 6. Modified Stress Level versus Time for Confined Creep; Comparison of Predicted and Observed Response.

simplest of the phenomenological creep models, and one capable of representing the type of behavior displayed by the silicate grouted soils, is that proposed by Singh and Mitchell (15). The relationship takes the form

$$\dot{\varepsilon} = Ae^{\alpha D}\left(\frac{t_1}{t}\right)^m , \qquad (4)$$

where $\dot{\varepsilon}$ = the strain rate at any time, t; A = a constant that determines the order of magnitude of the initial creep rate; α = the value of the slope of the linear portion of the log-strain rate versus major principal stress difference plot; D = the value of the stress; and, m = the absolute value of the slope of the straight line on the log-strain rate versus log-time plot.

In an attempt to generalize the results of this study for the range of silicate content grouts investigated, Equation (4) was formulated in terms of the stress level, $\overline{D}$ (that is, the ratio of the applied stress difference to the short term strength, D/D max), yielding the following expression

$$\dot{\varepsilon} = Ae^{\overline{\alpha}\,\overline{D}}\left(\frac{t_1}{t}\right)^m, \qquad (5)$$

where $\overline{\alpha}$ = αD max, the dimensionless function, is appropriate. A tabulation of the values of m, A, and $\overline{\alpha}$, as a function of the silicate content in the grout mix, is given in Table 1 for the unconfined creep tests. Integration of Equation 5 and evaluation of the initial strains

Table 1. Parameters for Singh-Mitchell Model for Unconfined Creep.

Parameter	Silicate Content		
	40%	50%	60%
m	0.77	0.75	0.72
A	1.5×10^{-5}	2.0×1.0^{-5}	5.0×10^{-5}
$\overline{\alpha}$	1.5	1.5	1.5

from the average stiffness values of 110 MN/m^2, 300 MN/m^2, and 600 MN/m^2, determined from the 40%, 50% and 60% silicate contents specimens, respectively, results in an expression capable of predicting the time-dependent strain. A comparison of the predicted and actual strains for specimens which did not fail within a week is shown in Figure 7a.

The values of m for the confined creep tests were also determined and are significantly higher than those of the unconfined specimens, with an average of 0.91, versus the previously reported averages of 0.72 to 0.78. Even the specimens tested at a fairly low confining stress (50 kN/m^2) exhibited a greater decrease in strain rate with time (larger m value) than did the unconfined specimens. In addition, there is no discernible trend with respect to either the consolidation stress or the ratio of the confining stress to the unconfined strength of the grouted specimens, a relationship which was previously shown to be an important factor in explaining the time-dependent strength behavior. It may be significant that the confined creep tests were initiated some six months after the specimens were prepared (in comparison to one month for the unconfined specimens), and although the unconfined strength had been shown not to change over this period of time, it is possible that some other chemical factor in the gel caused the resistance to creep to increase. Data have shown (4) an indication of increasing stiffness in grouted specimens tested in unconfined compression beyond the time at which the strength appears to cease increasing. A final factor which may be responsible for the difference in response between the unconfined and confined specimens is the magnitude of the initial strains under application of the creep stress. The stress-strain response of the 40% silicate specimens during the incremental application of the creep stress (Figure 5a) shows the strains achieved in this short period of time are on the order of 5 to 10 times those observed in the unconfined tests and may have resulted in a more rapid mobilization of the frictional resistance of the sand structure.

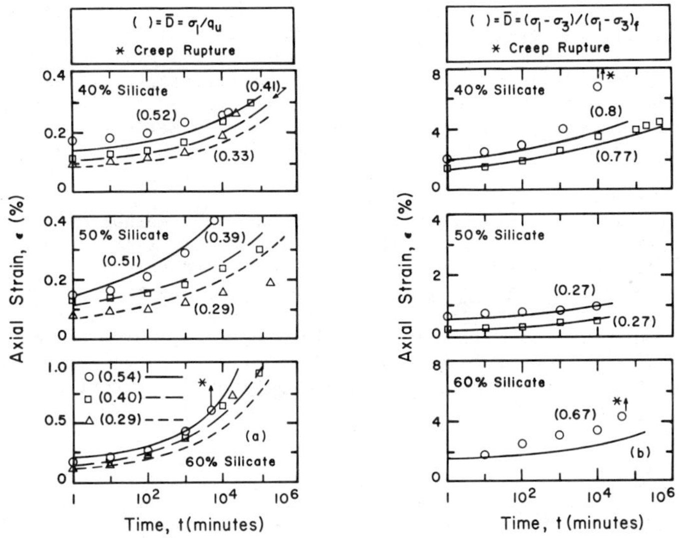

Figure 7. Comparison of Actual and Predicted Strain-Time Response for (a) Unconfined and (b) Confined Creep.

Because sufficient data was not accumulated at various stress levels for specimens injected with each of the three grouts in the confined tests, a meaningful relationship between the logarithm of the strain rate and the applied stress level at the reference time of one minute (that is, A and $\overline{\alpha}$ for t - 1 minute) could not be determined for each silicate content. A composite of the data that was available for all silicate contents was evaluated, and the relationship between stress level and log strain rate was represented by the slope, $\overline{\alpha}$, equal to 3.5. This value is higher than that determined for the unconfined creep tests (1.5) and may be influenced by the generally higher stress levels investigated in confined creep and the larger strains necessary to mobilize sufficient frictional resistance. Further data are needed to evaluate this relationship. The value of the parameter A was found to be 8.0×10^{-5} min^{-1}. Utilizing the values for m, A and $\overline{\alpha}$ which have been determined for the confined specimens and using the known values of strain at one minute, a comparison between observed and predicted response for specimens that did not fail in a week or more is shown in Figure 7b.

Engineering Relevance

The practical significance of the work reported herein lies partially in that an explanation of the time-dependent strength of a wide range of silicate content grouts has been presented. It has been shown that the strength of both unconfined and confined specimens can be viewed as arising from a time-dependent value of cohesion, superimposed on the frictional strength arising from the nature of the soil and the magnitude of the confining stress. In other words, when a silicate grouted soil element is subjected to shear stress, the resistance offered by the presence of the grout is a rate dependent function; at very high strain rates or rates of stress application, a high resistance will be observed, while at very low strain rates a much smaller resistance will be ovserved.

For example, if grouting is being considered as a method of stabilization to allow for a near vertical cut to be made in a previously cohesionless soil, stability analyses may be performed to verify the safety of the proposed design (It should be recognized that deformations may also need to be estimated if they are deemed critical). In the performance of a stability analysis (for example, using the method of slices) if one chooses to represent the strength of the soil along the assumed failure plane by the Mohr-Coulomb failure criterion, such that

$$\tau = c + \overline{\sigma}_n \tan \phi , \qquad (6)$$

where τ = the shear strength; c = the cohesion intercept; $\overline{\sigma}_n$ = the effective normal stress on the plane; and, ϕ = the internal angle of friction; the results of this work show that the time-dependent nature of the silicate grouted soil can be represented by the proper choice of the value of cohesion. If one adopts the design methodology that representative values for c and ϕ for each of the soils within the problem domain will be estimated and the required zone of stabilized material will be the design variable, the following procedure might be used.

If the applied stress variation as the result of excavation is to act for several weeks or less, it appears reasonable to assume, based on Figures 4 and 6 that a stress level of 40% or less would be supported without failure. If one chose to use a 50% silicate content grout (which yielded an unconfined strength of about 900 kN/m^2 for dense medium sand specimens) one can determine that the cohesion intercept, for $\phi = 40^o$, is slightly more than 200 kN/m^2. Utilizing 40% of this value as the appropriate stress level to be sustained for several weeks will result in an assumed cohesion of 80 kN/m^2, and this becomes the appropriate value for use in Equation (6) for the grouted soil. Stability calculations can then be performed to determine critical modes of failure and the required extent of grouting to achieve the desired factor of safety. If the strength imparted due to grouting is not sufficient or requires a larger area to be stabilized than is considered practical, one option available is to increase the strength of the grout. Utilizing the 60% silicate content grout would result in a stabilized soil with a higher unconfined strength (care must be taken to maintain fairly constant silicate to reagent concentrations). As before, the frictional characteristics of the soil can be used to estimate the cohesion intercept from the short term test, and by application of the stress level factor of 40% an appropriate cohesion value can be determined. Based on the 60% grout used in this study, the cohesion value appropriate for use in Equation (6) would be 230 kN/m^2.

CONCLUSIONS

This study was undertaken to determine the macroscopic time-dependent behavior of sands injected with sodium silicate base grouts, as a function of the silicate content. To achieve this goal, approximately twenty-five unconfined compression tests, thirty unconfined creep tests, and seven triaxial creep tests were conducted on laboratory injected dense sand specimens using three different silicate content grouts, chosen to span the range of concentrations generally used for structural grouting. Based on the findings of this experimental investigation, and a synthesis of these data with the existing literature, the following conclusions may be advanced:

1. The unconfined compressive strength of sodium silicate grouted sands increases as the silicate content increases and the gain in strength appears to be more exponential than linear (at least within the range of 40% to 60% by volume).
2. The curing environment does appear to have some effect on the strength of silicate grouted sands. With respect to the strength of moist-cured specimens, submerged-cured specimens of the 60% silicate content grout were weaker in unconfined compression and did not behave consistently with the bulk of experimental evidence when viewed from the perspective of long term strengths. It is important to note that the ratio of reagent to silicate content was not held constant during this test program (the absolute percentage of reagent was constant) and extrapolations from the observed behavior should not be made to the properties of higher silicate content grouts in general, but rather, these results should be viewed as indicative of a need for further research.
3. The relationship between the applied stress level and the time-to-failure (creep rupture) is well described by a linear relationship between the stress level and the logarithm of time. This relationship is valid regardless of the silicate content, although the 60% silicate content specimens that were submerged-cured failed much earlier than predicted by the general trend of the overall data.
4. The triaxial creep strength of silicate grouted specimens is well explained by utilizing the unconfined strength of similar specimens and the theoretical framework of the Mohr-Coulomb failure criterion. The consideration of time-dependent and time-independent components of strength gives rise to a strength parameter which relates the applied stress level to the time-to-failure and in the limit of zero

confining stress, duplicates the expression used for explaining unconfined creep behavior.

5. The stress-strain-time behavior of silicate stabilized sands subjected to unconfined creep appears adequately described by a simple three parameter phenomenological model. The creep potential, as evidenced by the slope of the log strain rate-log time relationship, was seen to increase slightly with silicate content, although an argument could be made for considering a single value for all unconfined tests reported. The stress-strain-time behavior of confined creep specimens was characterized by lower creep potential than was observed for unconfined creep. This response was thought to be either a function of a longer period of curing prior to testing or to the rapid mobilization of frictional resistance due to the larger strains induced during initial loading.

APPENDIX I - REFERENCES

1. ASCE Committee on Grouting, (1966), "Bibliography on Chemical Grouting, Third Progress Report Committee on Grouting," Journal of the Soil Mechanics and Foundations Division, American Society of Civil Engineers, Vol. 92, No. SM6, November, pp. 39-66.
2. Caron, C., (1963), "The Development of Grouts for Injection of Fine Sands," Proceedings of the Symposium on Grouts and Drilling Muds in Engineering Practice, Butterworths, London, May, pp. 136-141.
3. Christopher, B. R., Borden, R. H., Benlayf, M., and Krizek, R. J. (1977), "Triaxial Compression Tests on Large Size Specimens of Grouted Sands," Technical Report No. HB-1, Northwestern University, Evanston, Illinois, pp. 1-23.
4. Christopher, B. R., Atmatzidis, D. K., Borden, R. H., Krizek, R. J. (1979), "Evaluation of Specimen Preparation and Testing Procedures for Chemically Stabilized Granular Materials," Technical Report No. HB-5, Northwestern University, Evanston, Illinois, 112 pp.
5. Clough, G. W., Baker, W. H., Tan, D. Y., and Kuck, W. M. (1977), "Effect of Chemical Solidification of Soils Around a Shallow Tunnel," American Society of Civil Engineers, Preprint No. 3006, San Francisco, California, October, 29 pp.
6. Clough, G. W., Kuck, W. M. and Kasali, G. (1979), "Silicate-Stabilized Sands," Journal of the Geotechnical Engineering Division, American Society of Civil Engineers, Vol. 105, No. GT1, January, pp. 65-82.
7. Diefenthal, D. C., Borden, R. H., Baker, W. H., and Krizek, R. J. (1979), "Strength and Stiffness of Silicate Grouted Sand with Different Stress Histories," Geotechnical Testing Journal, Vol. 2, No. 4, December, pp. 200-205.
8. Graf, E. D., and Zacher, E. G. (1977), "Chemical Grouting of Sand for Earthquake Resistance," American Society of Civil Engineers, Preprint No. 3007, San Francisco, California, October, 19 pp.
9. Herndon, J., and Lenehan, T. (1976), Grouting in Soils, a State of the Art Report, Haliburton Services, Duncan, Oklahoma, pp. 20-22, 106-112.
10. Koenzen, J. P. (1977), "Time-Dependent Stress-Strain Behavior of Silicate-Grouted Sand," Journal of the Geotechnical Engineering Division, American Society of Civil Engineers, Vol. 103, No. GT8, August, pp. 903-908.

11. Koenzen, J. P. (1978), "Time-Dependent Stress-Strain Analysis of Chemically Stabilized Building Elements," Sumposium on Soil Reinforcing and Stabilizing Techniques in Engineering Practice, New Southwater Wales, Australia.
12. Luong, M. P. Gandais, M., and Allemand, P. (1978), Time-Dependent Stress-Strain Behavior of Silicate-Grouted Sand, "Journal of the Geotechnical Engineering Division, American Society of Civil Engineers, Vol. 104, No. GT6, June, pp. 770-772.
13. Mitchell, J. K. (1976), "Stress-Strain-Time Functions and Rheological Models," Fundamentals of Soil Behavior, John Wiley and Sons, Inc., New York, New York, pp. 327-333.
14. Rhone Progil (1975), "Hardener 600 Series," Rhone Progil Brochure, C10B, France.
15. Singh, A. and Mitchell, J. K. (1968), "General Stress-Strain-Time Function for Soils," Journal of the Soil Mechanics and Foundations Division, American Society of Civil Engineers, Vol. 94, No. SM1, pp. 21-46.
16. Skipp, B. and Renner, L. (1963), "The Improvement of the Mechanical Properties of Sand," Proceedings of the Symposium on Grouts and Drilling Muds in Engineering Practice, Butterworths, London, May, pp. 29-33.
17. Tallard, G. and Caron, C. (1977), "Chemical Grouts for Soils," Report Number FHWA-RD-77-51, Federal Highway Administration, Department of Transportation, Washington, D. C.
18. Warner, J. (1972), "Strength Properties of Chemically Solidified Soils," Journal of the Soil Mechanics and Foundations Division, American Society of Civil Engineers, Vol. 98, No. SM11, November, pp. 1163-1185.
19. Woodward-Clyde Consultants (1979), "Results and Interpretations of Chemical Grouting Test Program," Locks and Dam 26, Mississippi River, Alton, Illinois, Phase IV Report, Volume 11, May 213 pp.

APPENDIX II - NOTATION

The following symbols are used in this paper:

A = Strain at unit time at zero stress difference;
D, $\overline{D}$ = Stress and stress level;
D_r = Relative density;
c = Cohesion intercept;
m = Slope of log strain rate, log time curve;
q_u = Unconfined compressive strength;
t, t_f = Time and time-to-failure;
α = Slope of log strain rate versus stress intensity;
$\overline{\alpha}$ = Slope of log strain rate versus stress level;
$\varepsilon, \dot{\varepsilon}$ = Axial strain and axial strain rate;
ϕ = Friction angle;
$\overline{\sigma}_n$ = Normal effective stress;
σ_1 = Axial stress;
σ_3 = Confining stress;
$(\overline{\sigma}_1 - \sigma_3)_f$ = Principal stress difference at failure;
$(\sigma_1 - \sigma_3)_{f\ sand}$ = Principal stress difference (strength) associated with confining stress; and
τ = Shear strength.

LONG TERM AGING EFFECTS ON CHEMICALLY STABILIZED SOIL

By Thomas E. Graf,[1] G. Wayne Clough,[2] M. ASCE, and

James Warner,[3] F. ASCE

Abstract

The results of a testing program conducted on samples of chemically stabilized soil samples aged for nine to eleven years is presented. Comparisons of behavior are made relative to that observed for the same materials tested at ages of a few days to two years. The findings show that the extended aging process generally increased the strength of the stabilized soils, and reduced their susceptibility to creep or creep rupture under sustained loading. Significant strength reductions were observed if the samples were subjected to extended immersion under water. The observed trends in behavior are explained in terms of formation of additional bonds during aging between the sand particles and the available unneutralized silicate of the grout mix.

Introduction

Grouting solutions to civil engineering problems are becoming more popular alternatives. As a result, significant research efforts have been directed towards defining engineering characteristics of chemically stabilized soils. Recent investigations have presented information as to the effect of many parameters such as mix design, curing environment, confining pressure and rate of loading (1,2,3,4). However, because of the limited time frame of most studies, the effects of extended aging have not been evaluated. This is of importance to any engineer considering grouting as a solution for permanent applications. These would include grouting to improve bearing capacity or reduce liquefaction potential and grouting of dam foundations.

In this paper, test results are reported for chemically stabilized samples aged up to 11 years. The work represents an extension of that presented by Warner (4) in 1972, who carried out an extensive test program on samples treated by a variety of grout mixes aged up to two years. His results indicated that the factors significantly affecting strength included curing environment, rate of drying, method and rate of loading, and the chemical grout system itself. Warner's work also showed that the strength of most of the stabilized sands increased rapidly within the first several months, and only gradually increased

---

[1] Soils Engineer, Woodward-Clyde Consultants, Oakland, California
[2] Professor of Civil Engineering, Stanford University, Stanford, California
[3] Consulting Engineer, Mariposa, California

thereafter. Following the original testing, the remainder of Warner's samples were left in their molds and stored. Subsequently, in 1979, the senior two authors obtained the samples and renewed the testing program. This involved investigating the integrity of the samples, performing fast and slow load tests on both dry and rewetted samples and analyzing thin sections to determine the nature of the grout bonds.

Materials and Sample Preparation

Samples used for testing were prepared by the Warner Construction Company in the late 1960's and early 1970's. The procedure involved first premixing the grout and pouring it into a cardboard mold. Next, the sand was added and manually pressed into the mold. After gelation, the samples were stored for curing. Those used for this investigation were kept in the molds with the top open, and no attempt was made to control moisture loss from the specimens. Further details concerning sample preparation are given by Warner (4)

A range of grout mixes were used to stabilize the sands, including 15 silicate, one acrylamide and two polyphenol solutions (see Table 1). The grouts were mixed with two different, but reasonably

TABLE 1 - GROUT MIXES

OF SAMPLES TESTED IN THIS INVESTIGATION

	Grout Type	Base	Reactant	Modifier	Water	Gel Time
Silicate-Bicarbonate	BE	200 ml	200 ml	--	200 ml	*
	DO	350 ml	350 ml	--	350 ml	2-1/2 hrs.
	DP	350 ml	525 ml	--	350 ml	45 min.
Glyoxal	BB**	190 l	1.1 kg	19 l	171 l	*
	BA	190 l	2.4 kg	19 l	171 l	20 min.
	AJ	190 l	3.2 kg	11.4 l	125 l	25 min.
	AO	170 l	2.9 kg	15.2 l	171 l	18 min.
Modified Earthfirm	A2	420 ml	6.72 gm	21 ml	259 ml	3 hrs.
	A3	420 ml	6.72 gm	14 ml	266 ml	*
	A4	420 ml	6.72 gm	7 ml	273 ml	*
	X1	420 ml	6.72 gm	35 ml	245 ml	1 hr.
	X3	420 ml	6.72 gm	14 ml	266 ml	5 hrs.
	X5	350 ml	6.72 gm	14 ml	336 ml	1.25 hrs.
Siroc	S1	140 ml	49 ml	70	441 ml	45 min.
	S2	350 ml	63 ml	70	217 ml	25 min.
Acrylamide	BD***	1000 ml (10%)	10 ml	4 ml	w/base	20 min.
Terranier	TI	900 ml (10%)	10 gm	--	w/base	<1 hr.
	AG	160 gm	?	--	640 ml	--

*Very long gel times not recorded
**4.5 lb casting plaster
***4 ml Potassium Ferrocyanide

similar, commercially processed, uniform sands. The sands, designated SN-20 and S-8-100, have D_{50} values of 0.33 mm and 0.47 mm and uniformity coefficients of 2.6 and 2.7, respectively. Most of the samples were made using the SN-20 sand.

In order to prepare a specimen for testing, the first step involved is stripping the cardboard mold and inspecting it for defects. Some of the specimens disintegrated during removal of the mold, and in a few cases the top or bottom surfaces displayed poor or inconsistent gelation. If the samples were in a very poor condition, they were not used for load testing. Those that maintained sufficient integrity were capped using molten sulfur. In order to enhance the efficiency of this process and ensure parallel ends, a special capping device was developed. After capping, the specimen was either immediately tested or submersed in water for wet testing.

Testing Program

The loading of the specimens was carried out using conventional unconfined test procedures. In most cases, the stress was monotonically increased until failure was reached using loading rates from 0.2 to 30 psi/sec (1.4 - 210 kN/m^2 per sec). The majority of these tests were performed with the faster rate so as to conform to the original procedures used by Warner. The slower rate was used to investigate the effect of speed of load application, which has been shown to significantly influence behavior of grouted soils in some cases. As a general practice, each test was repeated at least once, and in some cases as many as five times to insure that representative results were obtained. In a few tests a constant stress was applied to the samples at levels of 50, 75, 90 and 100 percent of the average ultimate strength determined from the increasing load tests. The constant stress procedure was used to determine if the soils would fail at reduced levels under sustained loading.

Most of the samples were tested in their original state as they existed upon removing them from the molds. However, in a number of cases samples were immersed in water, and kept submerged for 21 days in order to evaluate the effects of a drastic change in environment. It was felt desirable to determine if the prolonged drying during the ten-year plus aging period might lead to property changes when the samples were submerged.

Thin sections of the stabilized soils were made by injecting an inert epoxy resin into the void space, allowing it to harden, and cutting thin slices of the soil. The slices were then ground and polished, and set onto a clear specimen glass. In this form, the soil grains and the interstitial materials could be examined by a high power microscope. Using normal and polarized light, the soil grains and grout cements were defined. The grout could be distinguished clearly relative to the epoxy injected to make the thin sections since the epoxy shows up as a consistent battleship grey color under polarized light which does not change regardless of the angle of the projected light.

Typical Sample Characteristics

The first indication of the strength of the samples was obtained by the character they exhibited upon stripping away the cardboard molds. There was a direct correlation between the level of the strengths reported by Warner (4) and the stability of the exposed samples. After two years aging, the weakest samples from the original study were the low silicate mixes of SIROC and Terranier, and these gave the most trouble upon opening of the molds. All of the 20% silicate SIROC samples crumbled completely, while three of seven of the Terranier samples broke before testing. The 50% silicate specimens of SIROC also required careful handling and the tops and bottoms often were not consistently intact. Since Warner (4) reported similar problems in his original work, it seems unlikely that the aging process contributed significantly to the deterioration of these samples.

In contrast to the foregoing problems, sodium bicarbonate, modified earthfirm and glyoxal specimens all demonstrated excellent integrity during removal of the molds. As will be shown subsequently, these grout mixes yielded the highest strengths in the load tests.

Stress-Strain Response

Typical stress-strain curves for unconfined compression testing are given in Fig. 1. The curves all show an initial concave portion,

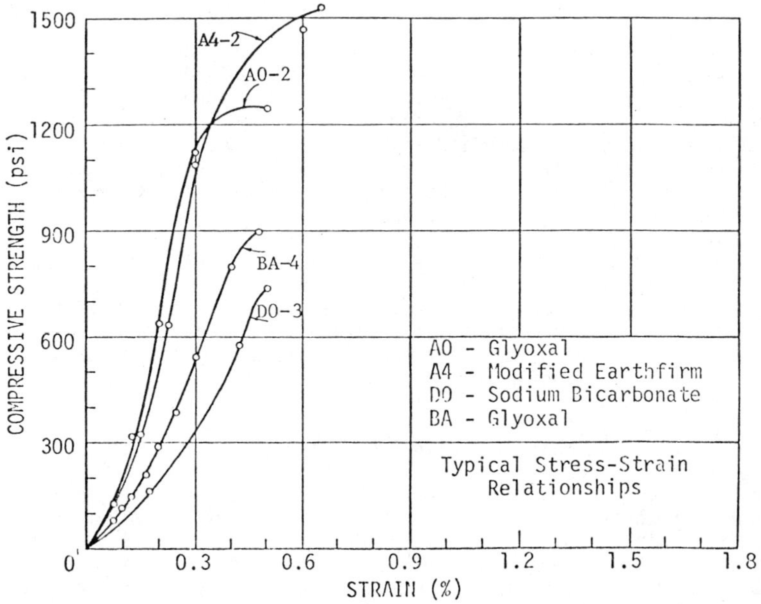

Figure 1. Typical Stress-Strain Curves in Unconfined Compression. Note: 1 psi = 6.9 kN/m^2

just before failure. Failure was generally sudden with a complete breakage of the sample. Both shear and splitting failure modes occurred as demonstrated in the photographs in Fig. 2. Axial strains to failure were generally small, ranging from 0.4% to 0.8% for the weaker ones which were stabilized by SIROC and Terranier.

(a) Tensile Splitting (b) Shear Failure

Figure 2. Samples Exhibiting Tensile and Shear Splitting Failure Modes.

Thin Sections

Photographs showing areas of two thin sections magnified 125 times are given in Fig. 3. The soil grains are distinguished as the particles with sharp, clean boundaries. The silicate-based grout is the amorphous mass located in the center of the photographs. By rotating the thin section under polarized light, the sand particles change colors dramatically, reflecting the light passing through the crystalline structure. The grout on the other hand shows little change in color since the crystalline structure is not as well developed as the individual soil grains.

It is clear from the photographs that not all the void space is occupied by grout. Where it is present, it has attached itself to the particles at certain prominent points and not along all the boundaries. The separations between the grout and particle boundaries may have existed from the beginning, or could have developed as a result of the long curing process.

Of course, the photographs in Fig. 3 are taken from only a small area of each thin section. By scanning the entire thin section a larger viewpoint is developed, although this cannot be shown in a photograph because of magnification problems. Careful examination of all of the thin sections showed that only selected areas of the void space are filled with grout. No direct evidence could be found that the grout had undergone significant shrinkage away from the particles since, as shown in Fig. 3, contact exists at least over a certain portion of the interparticle boundaries.

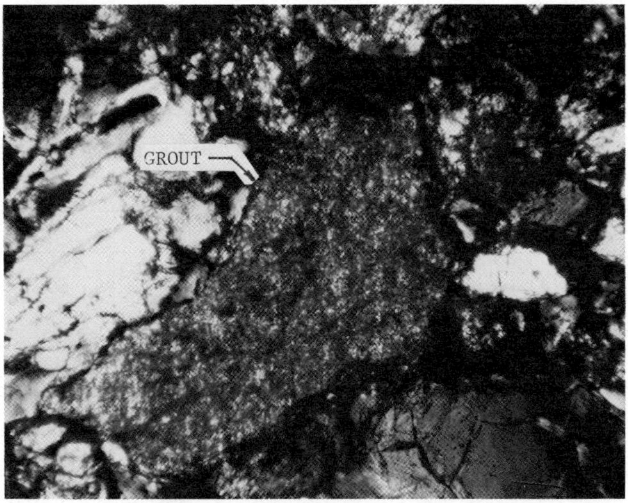

(a) Bicarbonate Mix

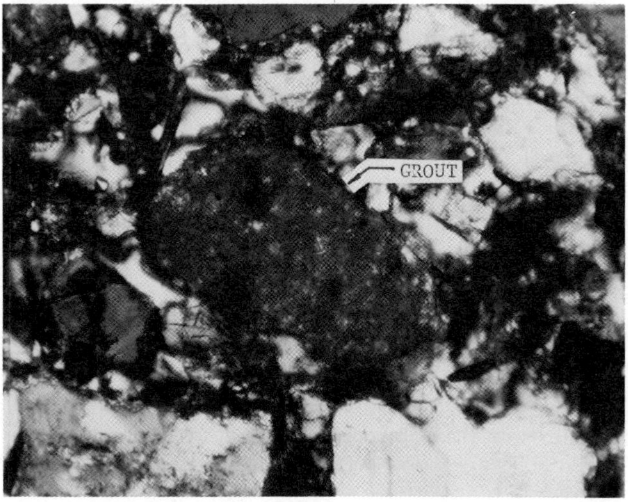

(b) Glyoxal Mix

Figure 3. Photomicrographs (125X) of Thin Sections: (a) Bicarbonate Mix; (b) Glyoxal Mix.

Variation of Strength with Aging-Rapid Tests

The unconfined tests performed using the rapid loading rate yielded consistent results. Table 2 presents the average strengths from all of these tests along with the ranges where multiple tests were performed on similar specimens. Typically the strength values from the repeated tests were within 15% of each other.

TABLE 2

UNCONFINED COMPRESSIVE STRENGTHS

				Present Study							Previous Study				
				Dry Cure			Submerged 21 Days				Dry Cure Strength (psi)				
	Grout Mix	Sand Type	Age (yrs)	No. of Tests	Average Strength (psi)	Strain to Failure	No. of Tests	Average Strength (psi)	Strain to Failure	% Strength Loss	10 Days	30 Days	90 Days	1 Year	2 Year
Sodium Silicate Base	Sodium Bicarbonate														
	BE-33% Sil	SN-20	10	3	2102(±)220	.5%	2	318(±)31	.875%	85%	Too weak to test . . .				772
	DP-28% Sil	SN-20	9	5	503(±)80	.4%	2	167(±)40	.6%	67%					463
	DO-33% Sil	SN-20	9	3	721(±)16	.45%	2	294(±)6	1%	59%					690
	Glyoxal														
	BA-50% Sil	SN-20	10	4	1313(±)176	.77%	2	716(±)48	.69%	45%	115	165	742		1185
	BB-50% Sil	SN-20	10	3	1306(±)175	.5%							658		
	AJ-60% Sil	SN-20	10	3	1316(±)191	.55%	2	621(±)48	1%	53%					1415
	AO-50% Sil	SN-20	10	3	1210(±)32	.57%					64	142			949
	Modified Earthfirm														
	X1-60% Sil	SN-20	11	2	525(±)16	.375%	2	334(±)16	.7%	46%	352		340	350	
	X3-60% Sil	SB-100	11	2	1305(±)0	.45%	2	637(±)32	.69%	51%	157		652	956	
	X5-50% Sil	SB-100	11	5	742(±)127	.58%					162		500	733	
	A2-60% Sil	SN-20	11	2	1130(±)430	.625%	2	493(±)15	.625%	56%	180		625	661	685
	A3-60% Sil	SN-20	11	3	1264(±)111	.625%					187		925	935	
	A4-60% Sil	SN-20	11	3	1762(±)287	.7%					55		1145	1103	
	Siroc														
	S1-20% Sil	SN-20	11	3	Samples too weak to test			-	-	-	19	Too weak			
	S2-50% Sil	SN-20	11	2	239(±)0	1-2%					227		225	225	177
Acrylamide	AM-9														
	BD-10% AM-9	SN-20	10	1	1242	2%	1	-	-	-			1140	1230	-
Polyphenol	Terranier														
	AG-20%Resin	SN-20	10	2	17(±)1	1.4%		-	-	-	-	46	71	63	-
	TL-10%Resin	SN-20	11	2	18(±).5	1.5%								46	

Note: 1 psi = 6.9 kN/m^2;

Figure 4 presents a plot of the average strengths of the various grouted samples versus age. The points up to 2 years age for this plot were obtained from the original work of Warner (4), while those at 9 to 11 years were obtained from the present work. The samples stabilized with grout types other than Terranier or SIROC show an increase in strength with extended aging. The trends indicated by Warner's data at 2 years age are continued, as the strength increases were gradual over the last ten years. Typically, the most dramatic strength gains were made in the first 120 days of the life of a specimen. The only exception to this rule were the sodium bicarbonate specimens which were too weak to test up to one-year aging period, but which picked up strength rapidly thereafter.

The SIROC samples which could be tested (those with 50% silicate base) showed essentially a constant strength with extended aging. Only the very weak Terranier specimens yielded a decrease in strength with time. Thus, the general picture derived from the results is that the strengths of the stabilized soils were maintained at, or above, the

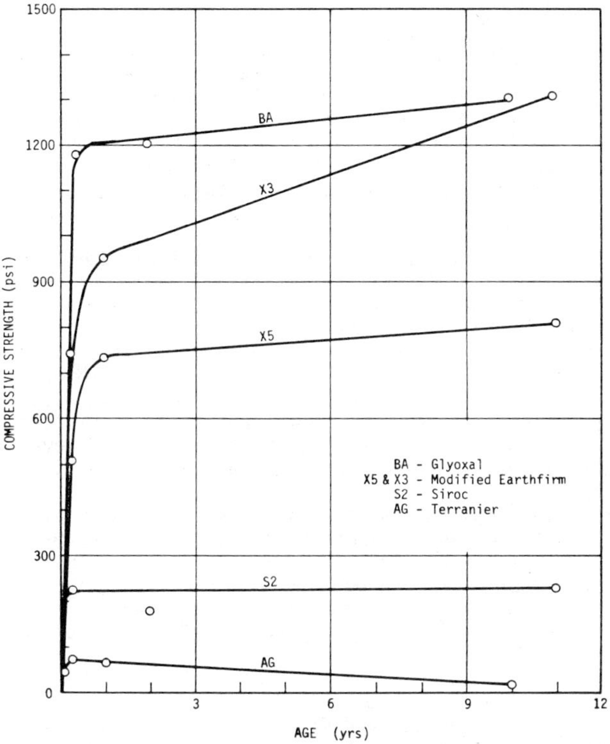

LONG-TERM STRENGTH GAIN

Figure 4. Variation of Average Unconfined Strength with Age of Sample (Dry Cure, 30 psi/sec Loading)

levels determined at the one to two-year age period. The durability of the chemical treatment appears to be excellent over long periods of time.

Effect of Submergence on Strength- and Deformability-Rapid Tests

Sixteen samples of the silicate based groups were submerged for 21 days and then loaded in unconfined compression at a load rate of 30 psi/sec (210 kN/m^2 per sec). The results, given in Table 2, show that the strengths upon rewetting were reduced in every case from 45 to 85%. Strain to failure was also affected by submergence, showing an increase in almost all cases. On the average, the strain to failure was increased by 150%.

Qualitatively, the loss of strength of the stabilized soils is consistent with observations made by Warner (4) in his original tests. He subjected samples to alternative wet and dry cycles and observed strength loss for most of the grout mixes. The exception was Modified Earthfirm which maintained its strength; however, in the present study this grout mix also showed a decrease with wetting.

Effects of Rate of Loading

It has been generally observed that chemically stabilized soils show decreasing strengths as rate of loading decreases (2,3). Also, many grouted soils have been shown to creep and fail under sustained load levels below those defined in rapid load tests. In particular, Warner (4) was one of the first to observe these phenomena in his original tests. He referred to the term "fundamental strength" to define the strength below which sustained loading would not cause failure.

As noted earlier, in this investigation differing load rates were used in the monotonically increasing load tests, and a series of tests were performed using a sustained constant load to investigate the effects of these procedures on the aged specimens. The constant load tests were conducted on eight specimens, two treated by Modified Earthfirm and six by Sodium Bicarbonate. Initially a load of about 50% of the average unconfined strength from the rapid tests was placed on the sample. This was held for up to 2 hours, and if no significant creep was observed, theload was then increased to the 70 to 80% level. The procedure was then repeated, increasing the load in increments.

The results of these tests show a very different behavior from those reported on "young" grouted specimens. First, in the monotonically increasing load tests variations in load rate from 0.2 to 30 psi/sec (1.4 - 210 kN/m^2 per sec) were not found to significantly influence the strength values of any of the grout mixes. Any differences in strengths which did occur were within the normal scatter of the results obtained for any one load rate.

Second, in the sustained load tests, little, if any, creep occurred, and failure was induced at load levels generally within the expected values from the rapid tests. Of the eight samples tested, six failed at sustained loads from 90 to 125% of the average rapid load test result. Two of the samples failed with loading increments applied at 55 to 85% of the rapid load strength. The response of the samples to the sustained loading is shown in Fig. 5, a plot of stress vs. strain. The lack of any significant creep strains is apparent at all load levels as the deformations primarily occur immediately upon load application. These results suggest that the extended dry curing process considerably reduces the time-dependent response of a chemically stabilized soil.

Analysis of Results

A number of trends stand out from the test results:

1) Except for the weak Terranier samples, the strength of the stabilized soils subjected to only a dry or moist environment held constant or increased under the extended aging period.

2) Submergence of the specimens for 21 days resulted in a substantial decrease in strength in all cases.

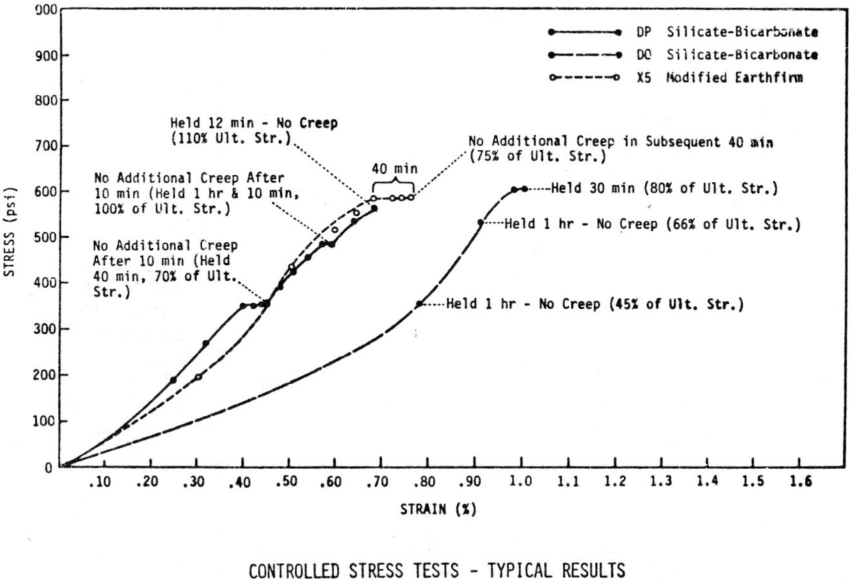

Figure 5. Sustained Load Test Results

3) Time-dependent response of the "old" sample was minimal compared to that observed for comparable "young" samples.

On the basis of these trends, it may be inferred that the extended aging led to formation of additional bonds between the groud and the sand, bonds which yield a higher dry soil strength, which are not susceptible to creep effects, but which are damaged by exposure to water. An insight to the nature of these bonds is found by examining the relationship between the soil response and the quantitites of silicate and the additives which cause the gelling process. The best data for this purpose are available for the Modified Earthfirm samples which show that the smaller the amount of modifier used in the mix, the higher the ultimate strength obtained in the dry state (see Fig. 6). Similar, but less well defined trends occur for other grout mixes. These trends suggest that after aging the more free, or unneutralized, silicate available (more free silicate is present if less modifier is used), the stronger the sample is in a dry state. It can be hypothesized that over the aging period, glassy bonds were formed between the free silicate and the sand particles. This hypothesis can be further supported by the response of the samples to wetting, since the glassy bonds cannot be maintained in water, thus explaining, in part, the strength loss observed upon submergence.

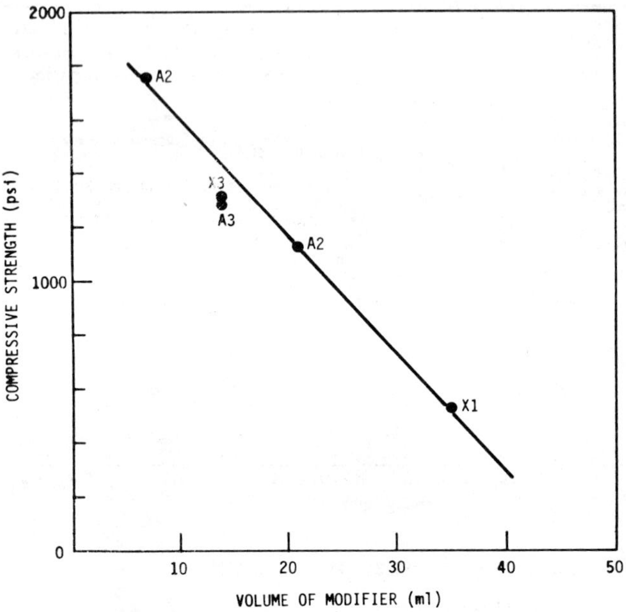

Figure 6. Variation of Unconfined Strength With Amount of Modifier in Grout Mix.

MODIFIED EARTHFIRM

Base = 420 ml
Reactant = 6.72 gm
Water +
Modifier = 280 ml

Conclusions

The investigation involved testing of samples of chemically stabilized soil which had been aged for nine to eleven years. During the aging process, the samples were kept in cardboard molds, open at the top, and no attempt was made to prevent moisture loss. Major conclusions of the study are as follows:

1) Except for the very weakest samples, the strengths of the stabilized soils tested with no environmental change, showed no change or a modest increase over those measured after one to two years aging.

2) Submergence of the samples for 21 days resulted in a significant strength decrease from 45 to 85%.

3) Effects of changes in rate of loading or sustained loading on the "old" specimen were minimal as opposed to the strong sensitivity of "young" specimens to these parameters.

4) The long-term increase in strength of the stabilized soils appears to be directly related to the amount of unneutralized silicate

available; the higher the final strength, the larger the quantity of unneutralized silicate present.

5) Bonds between the grout and the sand particles are created and maintained mainly at the most prominent portions of the sand particles. Additional glassy bonds are apparently formed with the unneutralized silicate with aging. These bonds are broken down if the soil is immersed in water.

Acknowledgements

The writers would like to thank Mr. E. D. Graf for his encouragement and cooperation during the course of the study. Appreciation is also expressed to Mr. Roy Wright for his help in assessing the reasons for the data trends. The work was in part financially supported by funding from the contract DOT-TSC-1726 between Stanford University and the U.S. Department of Transportation.

References

1. Cambefort, P. H., Injection des Soils," Vol. I and II, Eyroller, Editeur, Paris, 1967.

2. Clough, G. W., Kück, W. M. and Kasali, G., "Silicate Stabilized Sands," Journal of the Geotechnical Division, ASCE, Vol. 105, No. GT1, January, 1979, pp. 65-82.

3. Koenzen, J. P., "Rheologische Eigenschaften Silikat-Injizierter Korngeruste," Ph.D. Dissertation, Universität Karlsruhe, 1975.

4. Warner, J., "Strength Properties of Chemically Solidified Soils," Proceedings, Journal of the Soil Mechanics and Foundations Division, ASCE, Vol. 98, No. SM11, November, 1972, pp. 1163-1186.

EFFECTIVE STRESS-STRAIN-STRENGTH BEHAVIOR OF SILICATE-GROUTED SAND

by

Raymond J. Krizek[1], M. Adel Benltayf[2], and Dimitrios K. Atmatzidis[3]

ABSTRACT

The effective stress concept was used to interpret the results of drained and undrained triaxial compression tests on grouted sand specimens (Ottawa 20-30 sand in both a loose and dense state grouted with a sodium silicate grout consisting of 50% sodium silicate, 40% water, 5% formamide, and 5% ethyl acetate by volume). The grout was injected into the specimens under at-rest conditions or under a confining stress. When cured for six or more days before testing, no appreciable differences were observed in the stress-strain, volume change, pore pressure, and shear strength behavior. All undrained tests resulted in cavitation of the pore water, regardless of the applied back-pressure. The Mohr-Coulomb criterion was employed to analyze the data in terms of cohesion and friction components, both of which were found to vary as different functions of axial strain. The cohesive component of the shearing resistance reached a maximum at low levels of axial strain (between 0.5% and 1%) and then decreased rapidly to a value equal to that measured at failure. The frictional component of the shearing resistance increased gradually with increasing axial strain, approaching a maximum which is approximately the same as that of ungrouted sand. Hence, the relative contributions of cohesion and friction to the shearing resistance of a grouted sand change considerably as deformations take place, with cohesion being more significant at very small strains and low confining stresses and friction governing the response at larger strains and higher confining stresses.

INTRODUCTION

The continued and expanded use of chemical permeation grouting to stabilize in situ soils depends in large part on our ability to (a) perform appropriate laboratory control tests, (b) interpret and eval-

[1] Raymond J. Krizek, M.ASCE, Professor and Chairman, Department of Civil Engineering, Northwestern University, Evanston, Illinois.

[2] M. Adel Benltayf, Department of Civil Engineering, College of Engineering, University of El-Fatah, Tripoli, Libya.

[3] Dimitrios K. Atmatzidis, Professor of Soil Mechanics, School of Engineering, University of Patras, Patras, Greece.

uate correctly the results obtained therefrom, and (c) apply this knowledge to predict field behavior within acceptable limits of reliability. The design of structural grouting projects has been mostly empirical, and the mechanical properties of a grouted mass have usually been evaluated from the results of unconfined compression tests. Although it is generally accepted that, of the conventional laboratory tests, the triaxial compression test best simulates field conditions whose load application is multi-dimensional, very limited and inconclusive evidence exists regarding methods for applying the effective stress principle to analyze the mechanical behavior of grouted soils. In this study drained and undrained triaxial compression tests were conducted on specimens prepared with one type of sand and one type of sodium silicate grout, and the effective stress concept was used to interpret the results of all tests. The stress-strain, volume change, pore pressure, and shear strength behavior of grouted sand specimens was compared to that of ungrouted sand. Shear strength was interpreted in terms of cohesion and friction components, each of which was found to develop as a different function of axial strain.

BACKGROUND

Numerous investigations have been conducted to evaluate the mechanical behavior of grouted soils. Among the variables investigated are the grain size, grain shape, gradation, relative density and mineralogy of the soil component, the type and amount of reagent and silicate content of the grout mix, the specimen preparation and grouting techniques, the curing time and curing environment of the specimen, and the deformation or loading rate of the test (Schiffman and Wilson, 1958; Karol, 1968; Warner, 1972; Rhone-Progil, 1973; Tallard and Caron, 1977; Christopher, 1979). For the most part these investigators used the results of unconfined compression test to make qualitative comparisons between the effects of various parameters. However, such tests can not be considered representative of the in situ behavior of grouted soils where a three-dimensional state of stress exists. In a few cases the triaxial compression test has been used to measure the shear strength parameters of grouted soils, but very limited information is available on the stress-strain behavior, including volume change during drained shear and pore pressure response during undrained shear (Skipp and Renner, 1963; Farmer, 1974; Koenzen, 1978; Clough et al, 1979; Diefenthal et al, 1979; and Woodward-Clyde Consultants, 1979). Most information in the literature indicates that the Mohr-Coulomb criterion is valid for silicate grouted sands and that grouting increases the strength of sands by imparting a cohesive component while having a negligible effect on the frictional component. However, the evidence is inconclusive with respect to the applicability of effective stress concepts to the interpretation of data from triaxial compression tests. For example, Skipp and Renner (1963) performed both drained and undrained tests on similarly grouted specimens, and they reported that the values obtained for the effective cohesion and angle of friction were dependent on the type of test. However, it is questionable whether the pore pressures measured during undrained shear were truly indicative of the actual pressures inside the specimens. No other comparisons of results from drained and undrained triaxial compression tests were found in the literature.

EXPERIMENTAL PROCEDURES

In order to minimize, if not eliminate, variations in the mechanical properites of grouted specimens due to variations in the properties of their constituents, all specimens were prepared using one type of sand (Ottawa 20-30) and one type of grout 5% hydrated sodium silicate, 40% water, 5% ethyl acetate, and 50% formamide). Grouted specimens with a diameter of 70 mm and a length of 170 mm were produced using a specially constructed mold. To provide the specimens with smooth flat ends so that point loading could be avoided during testing, porous stones were incorporated at the ends during specimen preparation; the lower porous stone also helped to distribute the grout over the entire base of the specimen during the grouting process. The porous stones (prepared from a mixture of 36 parts Ottawa 20-30 sand and one part epoxy resin) had a diameter of 70 mm and a thickness of about 6 mm, and numerous holes were made in the stones to enhance unimpeded drainage.

Dense specimens were prepared by placing the sand in the mold in three layers of approximately equal thickness and vibrating each layer horizontally for three minutes (one minute at each of three points located 120° apart). To prepare loose specimens the sand was pluviated into the mold through a glass tube with a 5 mm tip diameter. The relative densities obtained ranged from 84% to 91% for dense specimens and from 36% to 41% for loose specimens. Each specimen was saturated by permeating through it a volume of distilled water equal to about five times the void volume of the specimen. A constant pressure of 2.5 psi and 1.5 psi were used to grout dense and loose sand specimens, respectively. The grout was injected through the bottom of the mold, displacing the water in the voids of the specimen as it moved upward and exited through the top of the mold. The amount of injected grout (maximum pumping time of about 15 minutes) was about double the void volume of the specimen.

Approximately 12 to 24 hours after grouting, the specimens were removed from their molds, the holes in the porous stones were cleaned, and the specimens were sealed in plastic bags and stored in a high humidity environment. Specimens were allowed to cure for six days to one year before testing. A few specimens were grouted directly inside a triaxial cell under a confining pressure according to procedures described by Diefenthal et al (1979). These specimens were allowed to cure inside the triaxial cell for six days to simulate field conditions, and they were then tested in triaxial compression.

A total of eleven consolidated-drained and twenty-two consolidated-undrained triaxial compression tests were conducted. The effective confining stress ranged from 1.44 kg/cm^2 to 21.60 kg/cm^2 for the drained tests and from 1.44 kg/cm^2 to 14.40 kg/cm^2 for the undrained tests. The back-pressure applied during the undrained tests varied from zero to 14.40 kg/cm^2. The axial load was applied at a rate of 100 pounds per minute through a Southwark-Emery Universal Testing Machine. The cell pressure and the back-pressure were obtained from a compressed air tank, and both were controlled by pressure regulators and measured by electrical transducers.

To minimize errors in the measurement of pore pressures during testing, the following precautions were taken: (a) electrical transducers were used; (b) brass pipes with the shortest possible length were used to connect the transducers with the triaxial cell; (c) all connections were made using no-volume-change valves; (d) four O-rings were used to provide an appropriate seal at the cap and base blocks of the triaxial cell; (e) two 0.3 mm (0.012 inch) thick membranes, with silicone grease in between, were used to enclose each specimen; and (f) specimens with a diameter of 70 mm (2.8 inches) were tested. According to Poulos (1964) and Wissa (1969), these precautions should reduce errors in pore pressure measurement substantially. However, for pore pressure measurements to be realistic, the specimen must be saturated with water.

As proposed by Skempton (1954), saturation is checked by measuring the change in pore pressure, Δu, induced by a change in the all-around total stress, $\Delta\sigma$, or cell pressure increment. The ratio $\Delta u/\Delta\sigma$ is called the pore pressure parameter, B, and should yield values approaching unity for most soils if saturation is complete. According to Wissa (1969), the value of B could be less than unity for saturated cemented soils and dense sands under a high confining stress. Wissa (1969) suggests that the value of B for stiff soils should be determined for several increments of back-pressure under a constant effective confining stress; saturation is indicated when this value remains constant (not necessarily equal to unity) and independent of the applied back-pressure.

The foregoing procedure was followed to evaluate the degree of saturation of the grouted sand specimens tested in this program, and typical results are shown in Figure 1a. It was observed that B increased continuously as the back-pressure was increased up to a value of about 6 kg/cm^2, after which it remained constant. If air were present in the pore fluid, increasing pore pressures would result in a reduction of the air volume, a decrease in the compressibility of the pore fluid, and an increase in B. Since B remained constant after several back-pressure increments, the amount of air in the pore fluid and test system can be considered negligible, and B values lower than unity should be attributed to cementation effects or high effective confining stresses. The typical results shown in Figure 1b indicate that increasing effective confining stresses resulted in decreasing values for B for both ungrouted and grouted sand specimens. Furthermore, gelled grout provides some "cementing" effect between the sand grains and this could reduce the compressibility of the solid skelton and consequently the value of B.

The foregoing observations indicate that (a) the grouted sand specimens can readily transmit pore pressures through the fluid which saturates their interconnected voids; (b) the saturation of specimens can be confirmed by appropriately checking the value of the B parameter; and (c) pore pressure changes during testing can be monitored. Accordingly, the results of drained and undrained triaxial compression tests may be interpreted in terms of effective stresses to quantify the strength characteristics of grouted sand.

RESULTS AND DISCUSSION

The results of all triaxial compression tests, including those of Diefenthal (1978), are summarized in Table 1. Due to limitations of the compression machine, reliable stress-strain data could not be obtained after the maximum principal stress difference had been reached.

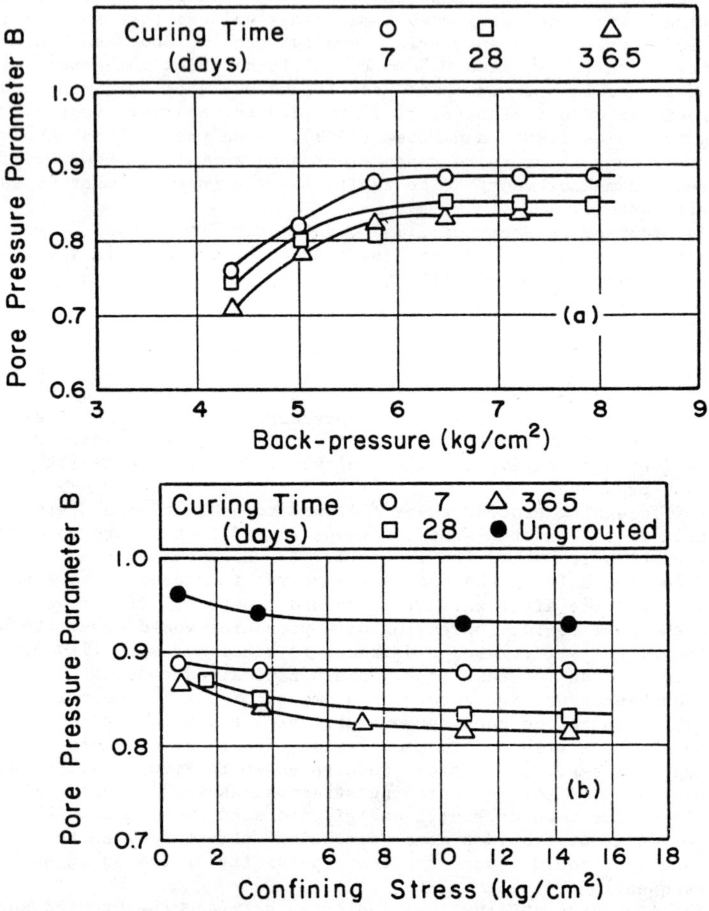

Figure 1. Pore Pressure Parameter as a Function of Back-pressure and Confining Stress

Table 1. Summary of Test Data

Sample Number	Relative Density (%)	Curing Time (days)	Effective Confining Stress (kg/cm^2)	Back Pressure (kg/cm^2)	Initial Tangent Modulus (kg/cm^2)	At Failure (maximum principal stress difference): Minor Principal Effective Stress (kg/cm^2)	Principal Stress Difference (kg/cm^2)	Axial Strain (%)	Volumetric Strain (%)	Effective Angle of Internal Friction (degrees)	Effective Cohesion (kg/cm^2)
34-CD	87	Sand	3.6	0	2600	3.6	12.4	9.9	3.2	37	0
35-CD	90		14.4		4200	14.4	45.2	3.3	1.3		
UL-2	40	Sand	3.6	0						32	0
UL-3	40		14.4								
40-CD	84	6	3.6	0	1800	3.6	21.8	5.1	2.5	36.5	2.7
41-CD	87		10.8		4600	10.8	44.6	5.3	2.0		
42-CD	86		14.4		5300	14.4	53.4	5.5	1.8		
43-CD	88		21.6		7000	21.6	78.0	5.7	1.8		
45-CD	91	7	14.4	0	7100	14.4	58.6	7.0	4.3		
46-CD	91	28	3.6	0	4870	5.6	19.7	6.4	6.7	35.6	3.0
47-CD	84		14.4		7000	14.4	56.8	6.4	4.1		
50-CD	86	365	3.6	0	4300	3.6	21.0	6.0	5.2	37	3.0
51-CD	84		14.4		7400	14.4	59.6	5.8	3.2		
FL-1	35	6	0							32	2.5
FL-2	48		3.6								
FL-3	36		14.4								
36-CU	91	Sand	3.6	7.2	2000	10.8	29.7	5.7	0	36	0
37-CU	90		14.4		4500	21.5	59.7	3.7			
38-CU	37	Sand	3.6	7.2	1100	10.0	18.9	10.2	0	30	0
39-CU	36		14.4		3500	21.5	39.9	7.3			
52-CU	90	6	3.6	7.2	2100	11.4	47.4	5.2	0	36	3.0
53-CU	87		14.4		6600	22.0	75.9	5.8			
55-CU	89	7	3.6	7.2	3100	11.5	44.7	5.4	0	36.5	2.8
56-CU	87		14.4		6500	22.3	77.8	6.7			
57-CU	84	28	1.4	7.2	1800	9.3	37.5	6.0	0	36	2.7
58-CU	87		3.6		3300	11.5	44.9	5.3			
60-CU	87		14.4		6500	22.5	75.0	5.3			
66-CU	83	365	3.6	7.2	3500	11.5	49.0	6.5	0	36.5	3.0
61-CU	84	28	14.4	0	6500	15.5	57.7	5.2	0	36	2.8
60-CU	87			7.2	6500	22.5	75.0	5.3			
62-CU	85			10.8	6500	25.8	89.7	5.8			
64-CU	84	183	1.4	10.8	2000	12.8	52.8	5.7	0	37	3.0
65-CU	84			14.8	2100	16.7	64.6	6.4			
67-CU	41	7	3.6	7.2	4400	11.8	33.9	10.0	0	31	2.6
68-CU	41		14.4		4800	22.0	57.6	9.0			
69-CU	41	28	3.6	7.2	4400	11.9	35.7	8.3	0	31	2.5
70-CU	41		14.4		4800	22.3	59.5	8.0			
71-CU	40	365	3.6	7.2	4400	11.4	35.3	7.5	0	32	2.3

CD: Consolidated-Drained Test
CU: Consolidated-Undrained Test

UL: Ungrouted Loose Specimen (Diefenthal, 1978)
FL: Field-simulated Loose Specimen (Diefenthal, 1978)

Therefore, failure was assumed to initiate at this point and the results presented in Table 1 reflect this assumption. Typical curves of principal stress difference, volumetric strain, and pore pressure change as a function of axial strain for drained and undrained triaxial compression tests on dense and loose grouted sand are shown in Figure 2.

In general, specimens subjected to drained shear exhibited a linear stress-strain curve until at least half of the maximum principal stress difference was reached; however, as indicated in Table 1, grouted specimens had higher initial tangent moduli than ungrouted specimens. This behavior was not evident during undrained shear. Furthermore, the stress-strain curves of loose grouted specimens in undrained shear were not continuously smooth and exhibited some yielding at rather low stress levels. As shown typically in Figure 2d, increasing the shear stress resulted in a high, nearly constant, modulus for the low-strain stage of loading (up to about 0.5% axial strain), at which point there was a significant increase in the axial strain for a small change in the principal stress difference. The shape of these curves may reflect the behavior of the grout in the soil voids. In all likelihood, the initial stress-strain behavior of the grouted specimens is dominated by the properties of the grout and the bonding of soil grains by grout, but, as the shear stresses increase, the grout yields and the bonds between soil grains break; as stresses continue to increase, the frictional resistance of the sand probably dominates the behavior.

The volume change behavior of all grouted specimens (both loose and dense) was very similar to that of ungrouted dense sand; there was an initial reduction in volume at low strain levels and then a continuous dilation until the maximum principal stress difference (assumed failure) was reached. This behavior, although not expected for loose sand specimens, can be attributed to the effects of the grout in the voids; apparently the sand particles were restricted from moving and rearranging in a denser condition and the grouted soil was forced to dilate.

Consistent with the volume change behavior during drained shear, the pore pressure response of both loose and dense grouted sand specimens in undrained shear was very similar to that of ungrouted dense sand. As shown typically in Figure 2d, pore pressure increased during the early stages of shear and reached a maximum for axial strains of about 0.2% to 0.5%. Subsequently, the pore pressures decreased continuously until a negative pore pressure of about -0.6 kg/cm^2 (-8 psi) was reached, at which point cavitation of the pore water occurred.

Shear Strength

Effective stresses were used to plot Mohr failure circles and obtain Mohr-Coulomb failure envelopes (shown typically in Figure 3) and the resulting values for the effective angle of internal friction and the effective cohesion are summarized in Table 1. Ungrouted dense specimens (relative density from 87% to 91%) exhibited an angle of internal friction of 37° and 36° in drained and undrained shear, respectively, and ungrouted loose specimens (relative density from 36% to 40%) exhibited comparable angles of 32° and 30°, respectively. The cohesion of the grouted specimens (all ungrouted specimens had no effective cohesion) ranged from 2.3 kg/cm^2 to 3.0 kg/cm^2, with the higher values generally associated with high relative density specimens.

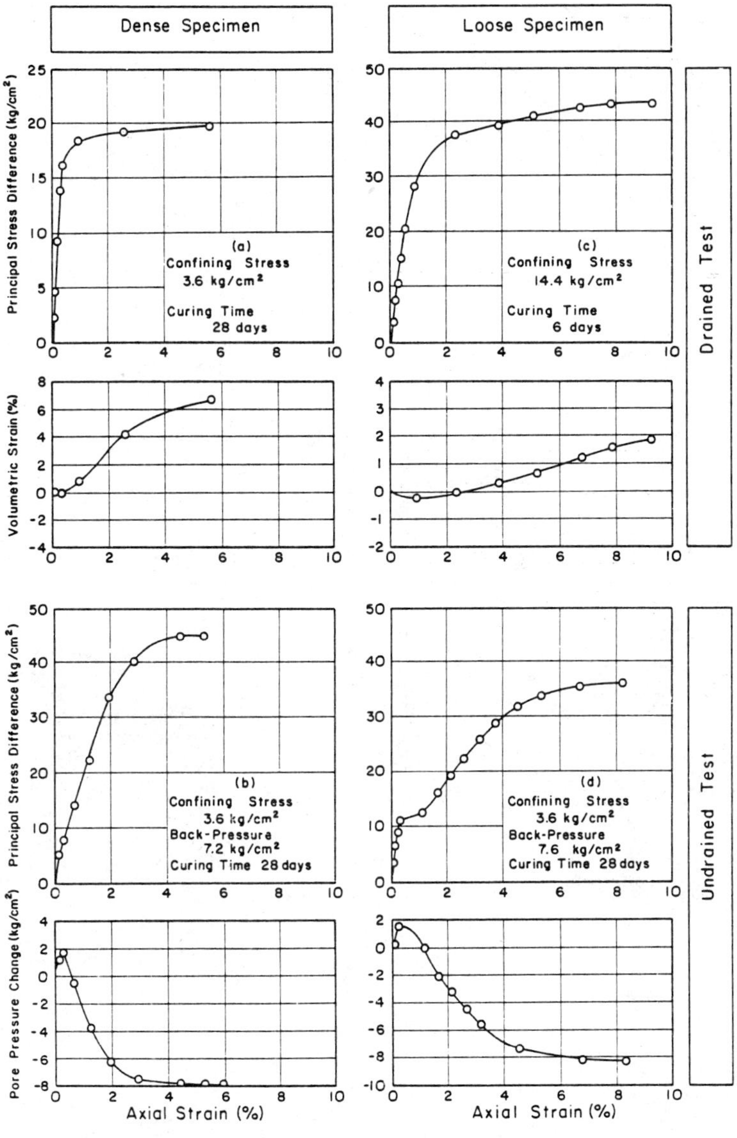

Figure 2. Typical Response Curves for Grouted Specimens Subjected to Triaxial Compression Tests

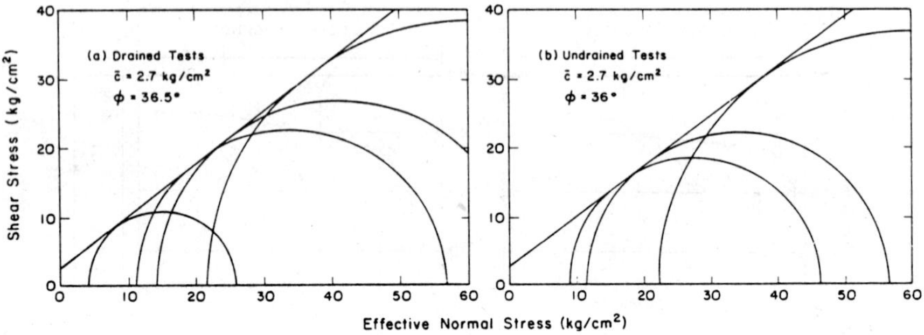

Figure 3. Mohr-Coulomb Failure Envelopes in Terms of Effective Stresses

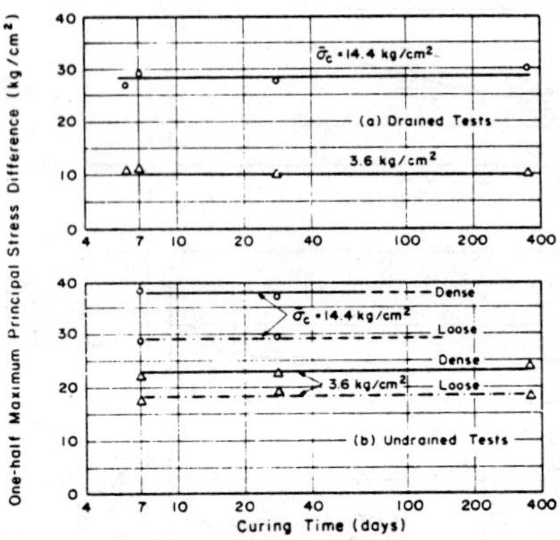

Figure 4. Shear Strength as a Function of Curing Time

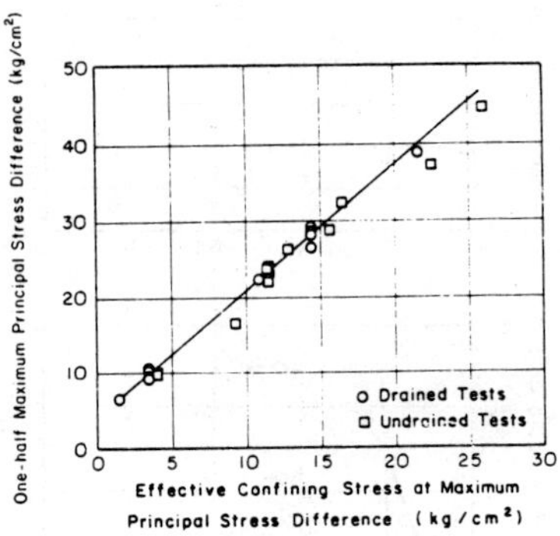

Figure 5. Shear Strength as a Function of Effective Confining Stress

Grouting did not alter the effective angle of internal friction of Ottawa 20-30 sand. The results of the consolidated-drained triaxial compression tests are in agreement with those reported by Diefenthal et al (1979), Clough et al (1979), and Woodward-Clyde Consultants (1979).

The shear strength (in terms of maximum principal stress difference) of dense and loose grouted specimens is plotted in Figure 4 as a function of curing time. When the initial effective confining stress, back-pressure, and drainage conditions were held constant, the shear strength did not vary with curing times ranging from six days to one year, but, when these parameters were varied, the measured values of the shear strength were substantially affected, as shown by the maximum principal stress differences given in Table 1. The shear strength of both grouted and ungrouted specimens subjected to either drained or undrained tests increased with increasing effective confining stress. However, for specimens with the same relative density and initial effective confining stress, undrained tests yielded higher strengths than drained tests; this difference was found to depend on the value of the back-pressure applied during the undrained tests, higher back-pressures yielding higher shear strengths.

The observed differences in shear strength between specimens tested in drained and undrained shear can be explained in terms of variations of the effective confining stresses which result from the testing procedures. During drained tests the effective confining stresses did not vary, because pore pressures were allowed to dissipate; therefore, the effective confining stresses at failure were the same as those applied at the beginning of the tests. During undrained shear the initial pore pressures in the grouted specimens were equal to the applied back-pressure; however, as the axial loads was increased, the specimens tended to dilate, and, since the volume of each specimen was held constant, pore pressure changes occurred. For tests conducted in this program the pore pressure in each specimen reached a negative value of about 0.6 kg/cm^2, at which point cavitation of the pore water developed. Accordingly, the pore pressures during each test were reduced by an amount equal to the applied back-pressure plus the value of the final negative pore pressure, and the effective confining stresses were increased by the same amount.

The variation in the shear strength of dense grouted specimens tested in drained and undrained triaxial compression is shown in Figure 5 as a function of the effective confining stress at failure (point of maximum principal stress difference). The resulting linear relationship between shear strength and effective confining stress at failure is independent of the drainage condition (type of triaxial test), back-pressure, and initial effective confining stress. This observation confirms that, when interpreted in terms of effective stresses, the results of both drained or undrained tests on grouted specimens describe equally well the shear strength behavior of the material at failure.

Development of Cohesion and Friction

If Mohr circle plots in terms of effective stresses are made for specific axial strain levels lower than the axial strain at failure, determination can be made of the development of shearing resistance with strain and the effective shear strength parameters, $\bar{c}$ and $\bar{\phi}$, can be evaluated as a function of axial strain (Schmertmann and Osterberg, 1960). For reasons of convenience and simplicity, $\bar{p}$-$\bar{q}$ diagrams were used for the analysis instead of Mohr-Couloub failure envelopes. Although the tests were constant-rate-of-load tests and not constant-rate-of-deformation tests, the strain rate was essentially constant up to axial strains of about 2%. Hence, axial strain contours ($\bar{p}$-$\bar{q}$ diagrams for specific levels of axial strain) were drawn only for axial strains of 2% or less.

As explained by Rowe (1962) the resistance of densely packed cohesionless materials during drained shear may be attributed to mineral-to-mineral friction, geometric interference between particles, and work done during dilation. Mineral-to-mineral friction is related to the coefficient of sliding of the grains on a smooth surface of the same material; particle interlocking depends on the packing and shape of the grains; and shearing resistance due to dilation can be determined from the volume change versus axial strain relationship. Under such conditions the maximum particle interlocking occurs at relatively small strains, where the volume of the specimen is minimum; upon further straining the interlocking decreases because the specimen volume increases, and it reaches a minimum when the residual strength has been attained and the specimen volume is maximum.

During undrained triaxial tests, the specimen volume remains constant (up to the point where cavitation of the pore water occurs). Since dilatancy does not occur in this case, Rowe (1962) speculated that the shear resistance consists only of the frictional and particle interlocking components. In this investigation values of the effective angle of internal friction obtained from drained triaxial tests are slightly higher than those obtained from undrained triaxial tests, and this difference may well be due to the effect of dilation during drained tests.

Figure 6 shows axial strain contours drawn on $\bar{p}$-$\bar{q}$ diagrams for the ungrouted sand. Based on the assumption that the strain contours are straight lines, no cohesion intercept is indicated at any strain level (that is, the behavior of the material is purely frictional). The strain-dependent increase of frictional resistance, as represented by values of the effective angle of internal friction, is illustrated in Figure 7; these data show that $\bar{\phi}$ increases rapidly with increasing axial strain up to an axial strain of about 0.5%, after which it continues to increase at a much lower rate and reaches a maximum at the point of failure.

Typical variations of the effective cohesion and effective angle of internal friction as a function of axial strain for dense and loose grouted sand specimens tested under drained and undrained conditions are shown in Figure 8. For specimens tested under drained conditions, the low-strain shear resistance consisted of a rapidly increasing cohesive component and a moderately increasing frictional component. At an axial strain of about 0.5% for the dense specimens and about 1% for the loose specimens, the cohesion component of the shear resistance reached a maximum and started to decrease, while the frictional resistance continued to increase. For axial strains larger than about 2%, the resistance due to cohesion attained a more-or-less constant value and any increases in shear resistance were due to increases in frictional resistance. The development of frictional and cohesive resistance during undrained shear follows the same trend as during drained

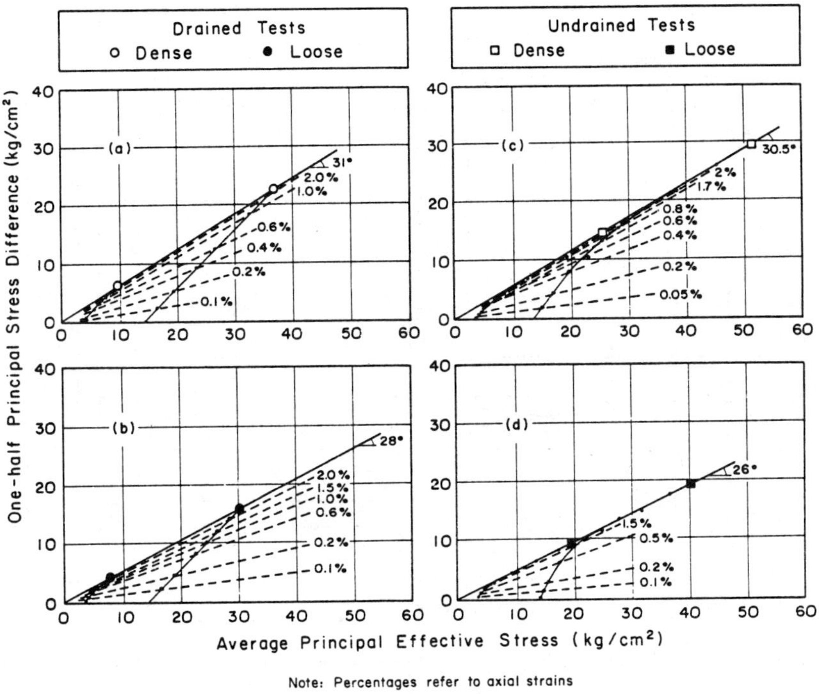

Figure 6. Axial Strain Contours for Ungrouted Sand

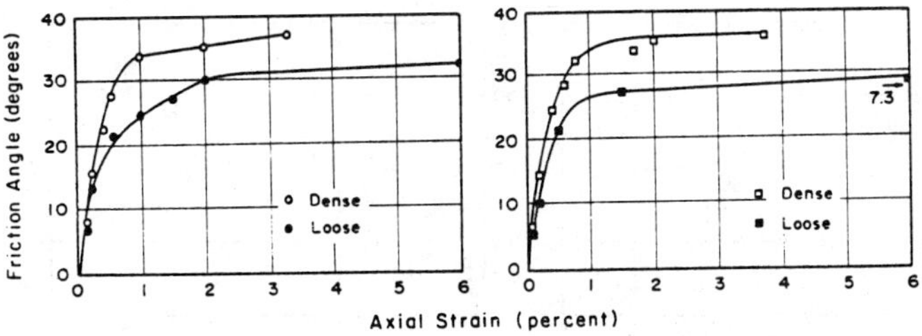

Figure 7. Variation of Friction Angle with Axial Strain for Ungrouted Sand

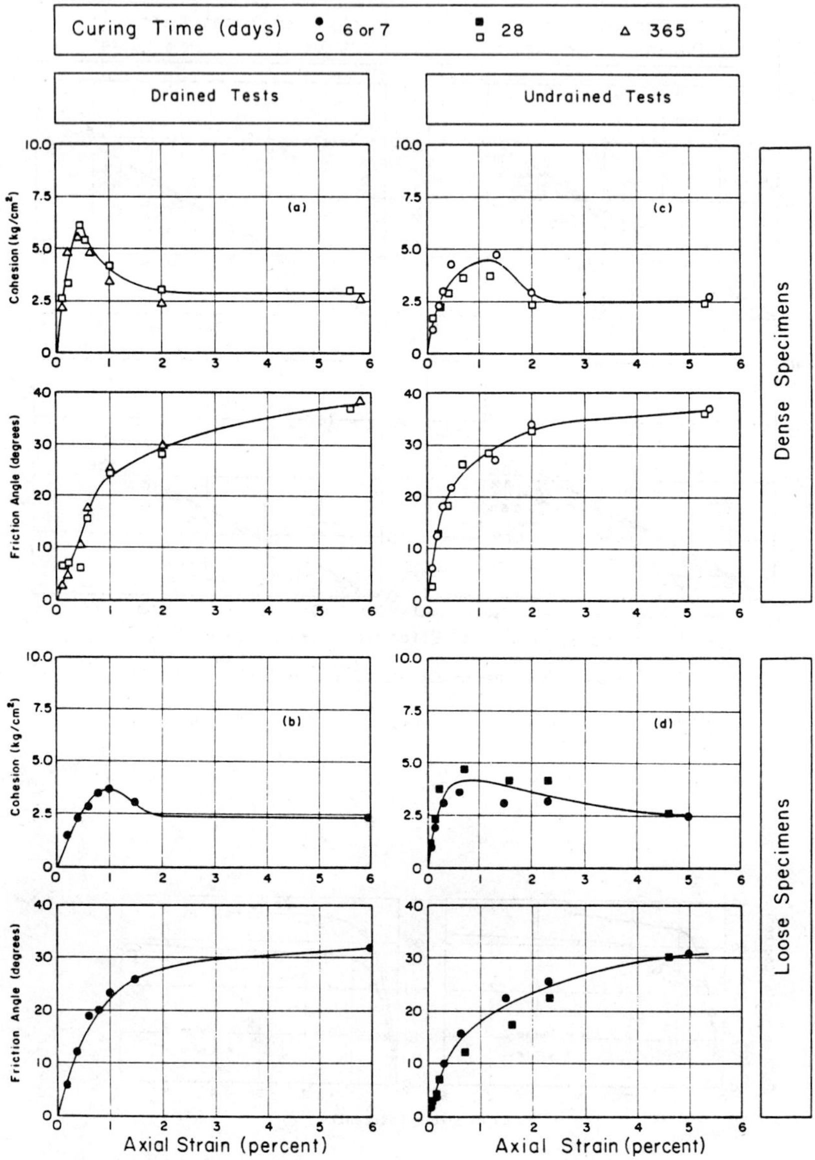

Figure 8. Variation of Cohesion and Friction Angle with Axial Strain for Grouted Sand

shear. The cohesive resistance increased with increasing axial strain, reached a maximum value at an axial strain ranging from 0.6% to 1.2%, and decreased upon further straining. The frictional resistance increased rapidly with increasing axial strain up to strains of about 1%, after which it continued to increase at a constantly decreasing rate until a maximum was attained at the point of maximum principal stress difference.

CONCLUSIONS

Within the scope and limitations of the foregoing experimental observations, interpretations, and discussions, the following conclusions can be advanced:

1. The results of drained and undrained triaxial compression tests can be interpreted in terms of effective stresses to describe the mechanical properties and behavior of grouted sand specimens; in particular, the Mohr-Coulomb failure criterion is valid for grouted sands and the same failure envelope in terms of effective stresses is obtained for both conditions.
2. Grouting has no identifiable influence on the angle of internal friction of the sand, but it effectively increases the strength of the sand by imparting a cohesion.
3. The cohesion and friction components of the shear resistance of grouted sand are a function of the axial strain, and their sum is maximum at the point of failure. The shear resistance due to cohesion increases rapidly with strain and reaches a maximum at low strain levels; it then decreases rapidly and stabilizes at a value which is equal to that measured at failure. The shear resistance due to friction develops gradually with increasing strain and reaches its maximum at failure; trends are similar for both loose and dense specimens.
4. Grouted sand specimens with both high and low relative density manifest volume change and pore pressure response characteristics which are very similar to those of dense sand.
5. Although grouted sand specimens may be saturated, measured values for the pore pressure parameter B are usually less than unity, probably due to the high stiffness of the specimens.
6. Cavitation of pore water may develop in both dense and loose grouted specimens tested in undrained shear.

Since the results of triaxial compression tests on grouted sand can be affected significantly by the procedures employed during testing, the following cautions and recommendations are advanced to improve the quality and reliability of the data obtained:

1. All specimens should be cured for at least six or seven days prior to testing to minimize the effects of curing time on their mechanical behavior.
2. When attempting to saturate specimens, values on the order of 0.8 or 0.9 for the pore pressure parameter B should not be considered as indicative of less than full saturation; if these values remain constant for a number of successive back-pressure and cell-pressure increments, the specimen can be considered fully saturated.
3. Care must be taken to provide full drainage for specimens

tested in drained shear, especially if the specimens are grouted in the cell prior to testing (thereby clogging the stone with grout unless special precautions are taken to render it porous again). Also, correct volume change meaurements can be made during drained shear only if the loading rate is sufficiently slow to allow for full drainage and complete dissipation of pore water pressures. Similarly, correct pore pressure measurements can be made during undrained shear only if the specimens are fully saturated, the amount of air in the pore water is minimized, and the volume changes of the connecting lines and the pore pressure measuring device are negligible.

4. Pore water cavitation during undrained shear may be avoided if very high back-pressures are applied.

ACKNOWLEDGEMENT

The partial support of the National Science Foundation under Grant CEE-8025163 is gratefully acknowledged.

APPENDIX I - REFERENCES

1. Christopher, B. (1979), "Evaluation of Specimen Preparation and Testing Procedures for Chemically Stabilized Granular Materials," Masters Thesis, Department of Civil Engineering, Northwestern University, Evanston, Illinois.
2. Clough, W., Kuck, W., and Kassali, G. (1979), "Silicate-Stabilized Sands," Journal of the Geotechnical Engineering Division, American Society of Civil Engineers, Volume 105, Number GT1, Paper Number 14335, pp. 65-82.
3. Diefenthal, D. (1978), "Shear Strength of Sand Grouted under Confining Stress," Master Thesis, Department of Civil Engineering, Northwestern University, Evanston, Illinois.
4. Diefenthal, D., Borden, R., Baker, W., and Krizek, R. (1979), "Strength and Stiffness of Silicate Grouted Sand with Different Stress Histories," Geotechnical Testing Journal, American Society for Testing and Materials, Volume 2, Number 4, pp. 200-205.
5. Farmer, I., (1974), "Undrained Strengths of Chemically Grouted Soils," Proceedings of the Second International Congress of the International Association of Engineering Geology, pp. 1-6.
6. Karol, R. (1968), "Chemical Grouting Technology," Journal of the Soil Mechanics and Foundations Division, American Society of Civil Engineers, Volume 94, Number SM1, Paper Number 5748, pp. 175-204.
7. Koenzen, J. (1978), "Time-Dependent Stress-Strain Analysis of Chemically Stabilized Building Elements," Symposium on Soil Reinforcing and Stabilizing Techniques in Engineering Practice, University of New South Wales.
8. Poulos, S. (1964), "Report on Control of Leakage in the Triaxial Test,"Harvard Soil Mechanics Series Number 71, Cambridge, Mass.
9. Rhone-Progil (1973), "Hardener 600 Series," Rhone-Progil Brochure, France.
10. Rowe, P. (1962), "The Stress Dilatancy Relation for Static Equilibrium of Assembly of Particles in Contact, "Proceedings of the Royal Society of London, Series A, Volume 269, pp. 500-527.

11. Schiffman, R. and Wilson, C., (1958), "The Mechanical Behavior of Chemically Treated Granular Soils," Proceedings of the American Society for Testing and Materials, Volume 58, pp. 1218-1244.
12. Skempton, A. (1954), "The Pore Pressure Coefficients A and B," Geotechnique, Volume 4, Number 4, pp. 143-147.
13. Schmertmann, J. and Osterberg, J. (1960), "An Experimental Study of the Development of Cohesion and Friction with Axial Strain in Saturated Cohesive Soils," Conference on Shear Strength of Cohesive Soils, American Society of Civil Engineering, Boulder, Colorado, pp. 643-694.
14. Skipp, B. and Renner, L. (1963), "The Improvement of the Mechanical Properties of Sand," Grouts and Drilling Muds in Engineering Practice, Symposium organized by the British National Society of the International Society of Soil Mechanics and Foundation Engineering at the Institute of Civil Engineers, London, England, pp. 29-33.
15. Tallard, G. and Caron, C. (1977), "Chemical Grouts for Soils," Report Number FHWA-RD-77-50/51, Volumes 1 and 2, Federal Highway Administration, Department of Transportation, Washington, D. C.
16. Warner, J. (1972), "Strength Properties of Chemically Solidified Soils," Journal of the Soil Mechanics and Foundations Division, American Society of Civil Engineers, Volume 98, Numer SM11, pp. 1163-1185.
17. Wissa, A. (1969), "Pore Pressure Measurement in Saturated Stiff Soils," Journal of the Soil Mechanics and Foundations Division, American Society of Civil Engineers, Volume 95, Number SM4, Paper Number 6670, pp. 1063-1073.
18. Woodward-Clyde Consultants (1979), "Results and Interpretation of Chemical Grouting Test Program, Existing Locks and Dam No. 26 Mississippi River, Alton, Illinois," Phase IV Report, Volumes 2 and 3, Woodward-Clyde Consultants, Chicago, Illinois.

MECHANICAL BEHAVIOUR OF SILICATE-GROUTED SOILS

by Birgit U. Stetzler

INTRODUCTION

In foundation engineering it is often necessary to improve the load-bearing properties of the subsoil.

Consolidation grouting with cement or clay-cement suspensions or chemical solutions enable the engineer to solve foundation problems economically and thereby to achieve a high technical standard. When the subsoil consists of fine-grained particles grouting is not effective because the cement or clay-cement particles are filtered out before complete penetration.

Therefore, in such cases, it is necessary to use chemical solutions of a viscosity appropriate to the soil conditions so as to obtain a ready penetration into the voids between the particles. The majority of the conventional chemical grout systems are based on silicate solutions which, after sufficient penetration, ensure proper bonding of the grain particles within a specific gel time. The new material obtained after grouting consists of an amalgamation of soil particles and silicates.

The mechanical behaviour of the material is determined by the viscous and time-dependent stress-strain properties of the hardened silicates and the time-independent character-

[1] Research Eng. of Civil Eng. Univ. of Karlsruhe, 7500 Karlsruhe, Germany

istics of the soil structure. This results in obvious creep deformations which have to be ascertained in the chemically grouted soils under constant load.

So far, the time dependent load-bearing and deformation characteristics of chemically grouted subsoil have usually not been considered in structural analysis of foundations and undercuttings. Research is now proceeding towards a proper evaluation of the strain-stress behaviour and the establishment of practical criteria for calculations in soil mechanics.

THE NATURE OF THE PROBLEM

Although silicate solutions show favourable flow properties, their complex load-bearing and deformation characteristics are a real handicap, even at the end of the specific gel time, unless they are carefully investigated. Matured silicates still retain some fluidity, a high viscosity respectively. Beyond that, the load-bearing properties depend on strain rate, lateral pressure, and loading time. Compression tests at rapid strain rates and/or stress rates produce parameters which exceed eligible bearing properties. This has been ignored so far in test procedures and structural calculations of chemically treated soils.

The experience gained from the general practice clearly reveals the shortcomings of the conventional determination of parameters. Extensive creep tests and strain-controlled slow-compression tests have to be made to ascertain the values permitting to form an idea of the bearing capacities of treated soils.

The time needed for such tests may be unacceptable for the requirements of a construction site. Shorter tests producing the same results would be ideal, but do not exist. Therefore, it is indispensable to proceed to both short-time and long-time tests to establish the general relationship that exists between the individual stress-strain characteristics of the chemically grouted soils.

To do this, an extensive test program was established and carried out for a particular silicate solution and a particular sand quality.

On the supposition that
- stress level $(\sigma_1 - \sigma_3)$
- loading time t
- strain rate $\dot{\varepsilon}_1$
- water content w

are the main factors determining the behaviour of grouted soils, only these are varied in the test program.

PREPARATION OF SPECIMENS

The specimens used in the test program are formed by means of perspex cylinders 60 cm high, 9 cm diameter. Top and bottom of these cylinders are each closed with a metal plate with rubber joint and central hole (1 cm diameter). They serve as inlet and outlet for water and grout.

A layer of gravel, 3 cm thick, grain size 1 - 2,5 mm, forms a filter at the bottom of the system. Then follows the proper sample, dry sand layer, grain size 0.2 - 1.0 mm. This sand is slowly introduced through a funnel to obtain a uniform density of D = 0.56 corresponding to a dry unit weight of γ_d = 15.9 kN/m^3 and a porosity of n = 0.4. This sand layer is topped by another layer of gravel, 3 cm thick, which fills up the cylinder. The air is removed from the voids by water slowly introduced through the bottom hole, moving through the cylinder and freely flowing from the top. Subsequently, the water is displaced by chemical grout injected at a constant pressure of 0.7 bar until approximately 85 % of the voids are filled with grout. The chemical grout used to produce the specimens is made up as follows:

"Monodur":

66,2 % of weight Na-Waterglas 37/40 KA γ = 13,6 - 13,8kN/m^3
27,5 % SiO_2
8,3 % Na_2O

6,9 % of weight formic acid
1,0 % of weight acetic acid
25,9 % of weight water

During the injection the grout is mixed by a mixer to ensure homogeneity of the material. The grouted specimens are taken from the perspex cylinders one day after grouting. After removal of the gravel layers at top and bottom, the remaining block of material is cut into two identical cylinders, each ~ 18 cm long. These two cylinders are kept in water at a temperature of 20°C for at least 33 days. After this period they are ready for the tests.

TESTING SYSTEM

Two series of compression tests are carried out. The top and bottom plates used are rough and unlubricated. The specimens are appropriately trimmed to produce regulare cylinders,with height twice their diameter.

a) Uniaxial stress-controlled compression tests:
A constant load is instantly applied to the specimen through weights, levers, a piston - well lubricated to prevent friction - and the top plate. During the test the deformation is measured outside the cell to an accuracy of 1/100 mm (fig. 1). A rubber sleeve preserves the water content of the specimens even for a period of more than 100 days. Drainage of the specimens is ensured during the tests by porewater pipes

each in top and bottom plate.

b) Uniaxial and triaxial strain-controlled compression tests: The specimens are deformed at constant strain rate, the water content being again preserved by a rubber sleeve (fig. 2). During the whole tests specimens are drained as explained before. Load and deformation are measured outside the cell, and the circumferential variations are measured by three rings equally spaced over the height of the specimens. The data obtained permit to determine the change of volume (ε_v), strain (ε_1) and deviatoric stress ($\sigma_1 - \sigma_3$) during the tests. For triaxial tests water introduced into the cell is designed to transmit the lateral pressure to the specimen. The further procedure is as before.

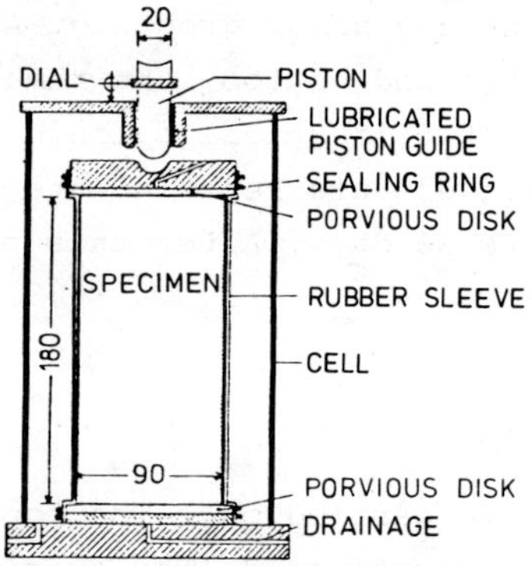

FIG. 1 TESTING SYSTEM FOR STRESS-CONTROLLED CYLINDER COMPRESSION TESTS

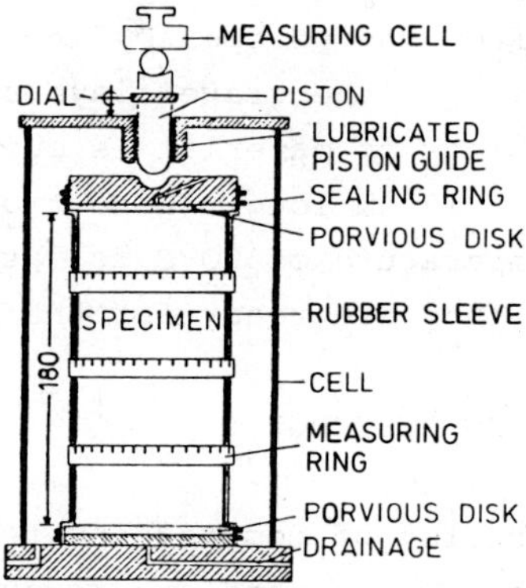

FIG. 2 TESTING SYSTEM FOR STRAIN-CONTROLLED CYLINDER COMPRESSION TESTS

RESULTS AND INTERPRETATION OF CREEP TESTS

EMPIRICAL CREEP EQUATION

Fig.3 shows idealized creep curves. They represent the results of creep tests where specimens are deformed under constant load in a uniaxial state of stress during a specific loading period. In the case of undercritical loads, the deformation characterstics of chemically grouted soils are those of fig. 3, curve σ_a. They show a high strain rate just after the application of the load and subsequently a continous decrease as a function of time.

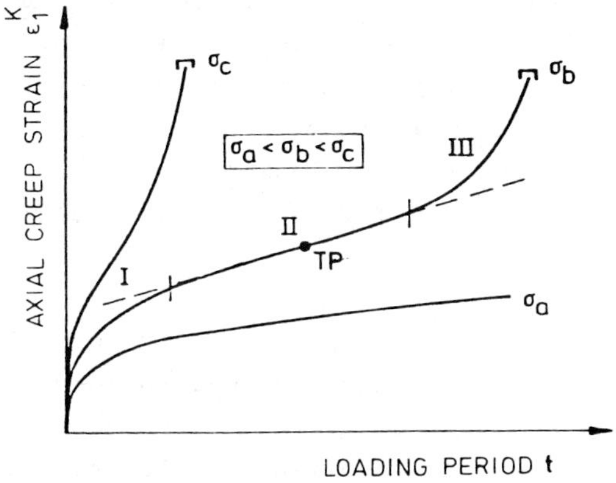

FIG. 3 IDEALIZED CREEPCURVES

In the case of overcritical loads, after a period of diminishing creep deformation, an increasing strain rate results in failure of specimen (fig. 3 σ_b and σ_c).

So far creep curves used to be divided into three stages (Koenzen 1975)

1. decreasing strain rate,
2. invariable strain rate,
3. increasing strain rate, resulting in failure of specimen.

Tests with both frozen (Eckardt 1979; Vyalov 1963) and grouted soils show that the constant strain rate of stage 2 is only a linear indication of the actual deformation curve. Stability of specimens can be expected only at decreasing strain rates. As against the conventional strain-stress concepts according to which the creep behaviour splits into three stages, the new empirical creep equation (Stetzler, 1981) eliminates the elastic time-independent deformation and considers exclusively the time-dependent deformation occuring during the tests.

The stability of grouted materials, particularly in the case of long-time loads depends only on the creep deformation. One of the test series concerned is presented in this paper. Fig. 4a shows the well-known non-linear relationship that exists between the axial creep strain and the loading period. The measured and calculated values are plotted in standardized form.

Long-time creep curves representing tests of more than 100 days duration and plotted in double-logarithmic diagrams (fig. 4a and 4c) clearly show that tests where age, temperature and water content of specimens remain unchanged can be described by a double-logarithmic creep equation,

$$\varepsilon_1^K = a \cdot \sigma_{1,norm}^n \cdot t_{norm}^m \quad \ldots\ldots (1)$$

until the limiting creep strain $\varepsilon_{1,G}^K$ is reached. For the symbols used, see appendix II.

Creeping in this case depends only on the axial load and the loading period. This equation permits to formulate a design criterion for each case of practical application with due consideration to the period during which the load is applied to the treated soil.

The limiting creep strain $\varepsilon_{1,G}^K$ can be seen from the curves in the double-logarithmic diagram to the extent as the creep properties follow the empirical creep equation and do not show a faster creep than that prescribed by the derivation of equation (1) which results in:

$$\dot{\varepsilon}_1 = \frac{d\varepsilon_1^K}{dt} = a \cdot m \cdot \sigma_{1,norm}^n \cdot t_{norm}^{m-1} \quad \ldots\ldots (2)$$

This limiting creep strain ($\varepsilon_{1,G}^K$) is lower than the creep

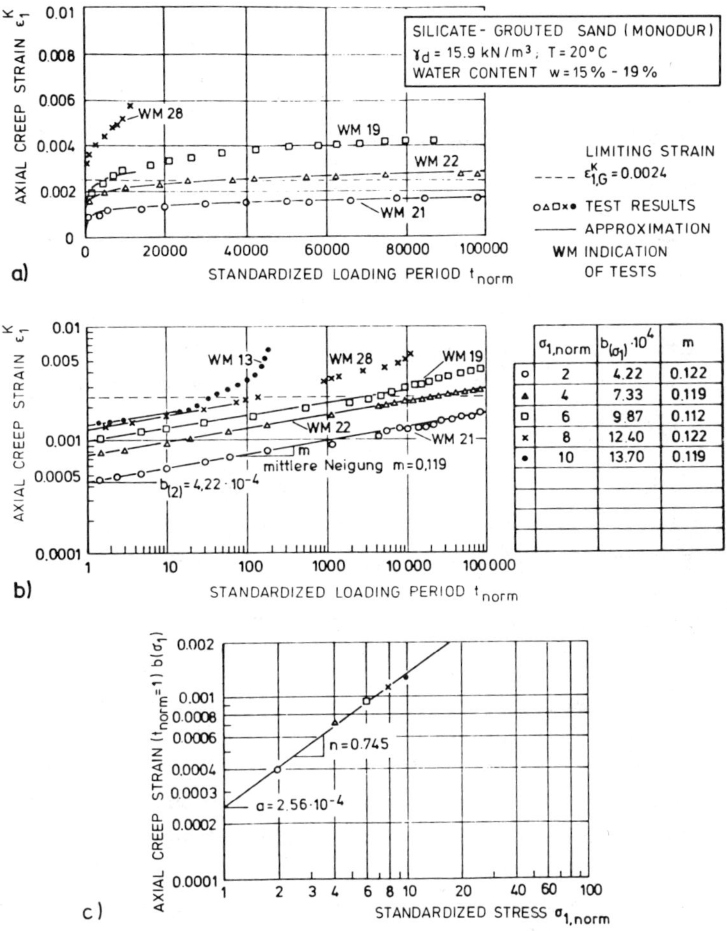

	$\sigma_{1,norm}$	$b_{(\sigma_1)} \cdot 10^4$	m
○	2	4.22	0.122
▲	4	7.33	0.119
□	6	9.87	0.112
×	8	12.40	0.122
•	10	13.70	0.119

FIG. 4 UNIAXIAL CREEP TESTS

strain at the turning-point (TP) (fig. 3), a fact which provides an additional safety factor.

The limiting stress,chemically treated soil can bear for a specific loading period can be calculated on transformation of equation (1)

$$\sigma_{1,G} = \frac{\varepsilon^{K}_{1,G}}{a \cdot t^{m}_{norm}} \qquad \text{(3)}$$

This means that the limiting stress decreases as the loading period increases.

The recommended creep equation can serve to describe the strain-stress time relationship of the material provided the creep period is long enough. For each loading period it supplies safe limiting stresses on the supposition of the previous definition of the limiting strain and the ageing process in the grouted soil.

The following table shows the parameters ascertained by the tests (fig. 4). It supplies for a loading period of 30 years the expected limiting stresses $\sigma_{1,G}$ with the aid of formula (3).

w (%)	$a \cdot 10^4$	n	m	$\varepsilon^{K}_{1,G}$ -	$\sigma_{1,G}$ (30 years) (MN/m^2)
15-19	2,56	0,745	0,119	0,0024	0,15

Equations (1), (2) and (3) describe fully the creep properties of silicate-grouted soils under constant uniaxial load.

Where for a special grouting problem the parameters a, n, and m have been determined by long-time creep tests as described, the bearing capacity of in-situ stabilized soils can be found by creep tests within a few days. Thus, it is possible to ensure that the proposed load is safely taken by the stabilized soil.

STRAIN-CONTROLLED COMPRESSION TESTS

To obtain parameters for conventional stability analysis it

is necessary to proceed to strain-controlled compression tests on drained cylindrical specimens. In such tests, the specimens are deformed at constant strain rate ($\dot{\varepsilon}_1$) up to final strains of about ε_1 = 10 % to 16 %. Fig. 5 shows a group of idealized stress-strain curves and the development of the volumetric strains observed in such tests under varying cell pressure and equal strain rates.

Two stress peaks are characteristic for the mechanical behaviour of silicate grouted soils in triaxial strain-controlled tests. After a steep rise of the deviatoric stress the first peak occurs already at very small strains. Under uniaxial stress conditions this first peak results in immediate failure of the specimen. On the other hand, specimens tested in the triaxial state of stress "recover" again and after a temporary decrease in "strength" reach a second peak. It is only at this point that failure definitively occurs.

Under failure condition, a slip joint develops at an angle of approximately 67° while simultaneously the deviatoric stress is reduced to a constant residual stress dependent on the lateral pressure. There is a sliding motion of one of the split specimen over the other.

Fig. 6A and B shows the stress-strain diagrams of four groups of tests carried out at varying cell pressures each group with a different but constant strain rate, in order to take account of the time-dependent behaviour of silicate-grouted soil. The tests show that the "intensity" of the two peak stresses depend on both strain rate and lateral pressure, and that the first peak disappears with an increasing lateral pressure and/or with decreasing strain rate. These facts show that the first peak can be attributed to the viscous properties of the chemical grout in the voids between the grains. A clear first peak shows that

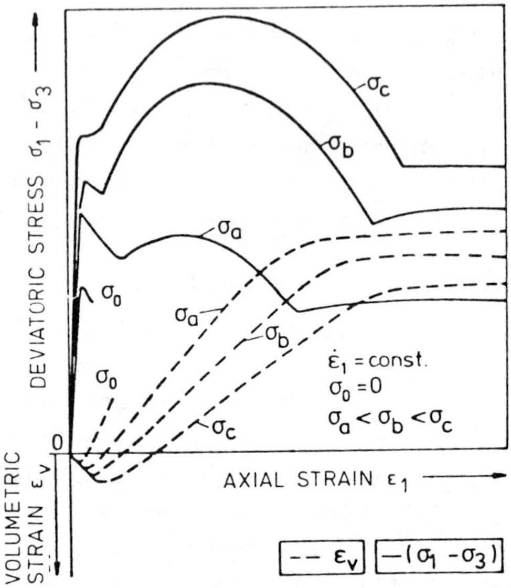

FIG. 5 IDEALIZED STRESS-STRAIN AND VOLUMETRIC STRAIN CURVES OF SILICATE-GROUTED SOILS

the load is carried mainly by the viscous grout. At greater deformations, the load is shifted from the grout to the whole system formed by grout and grains until the definite failure at the second peak.

Tests carried out with identical cell pressure but $\dot{\varepsilon}_1$ = 2 %/min and $\dot{\varepsilon}_1$ = 0,004 %/min respectively show that the ultimate strength diminishes by approximately one half.

The second-peak values can be used to draw a Mohr's rupture diagram (fig. 7). Each group of tests with equal strain rates gives one shear line. This diagram permits to determine the cohesion (c^V) and the friction angle (ϕ_s^V) as a function of the strain rate.

The cohesion (c^V) decreases with a decreasing strain rate ($\dot{\varepsilon}_1$) but not right down to the value c = 0 of the untreated dry sand used to prepare the specimens (fig. 7).

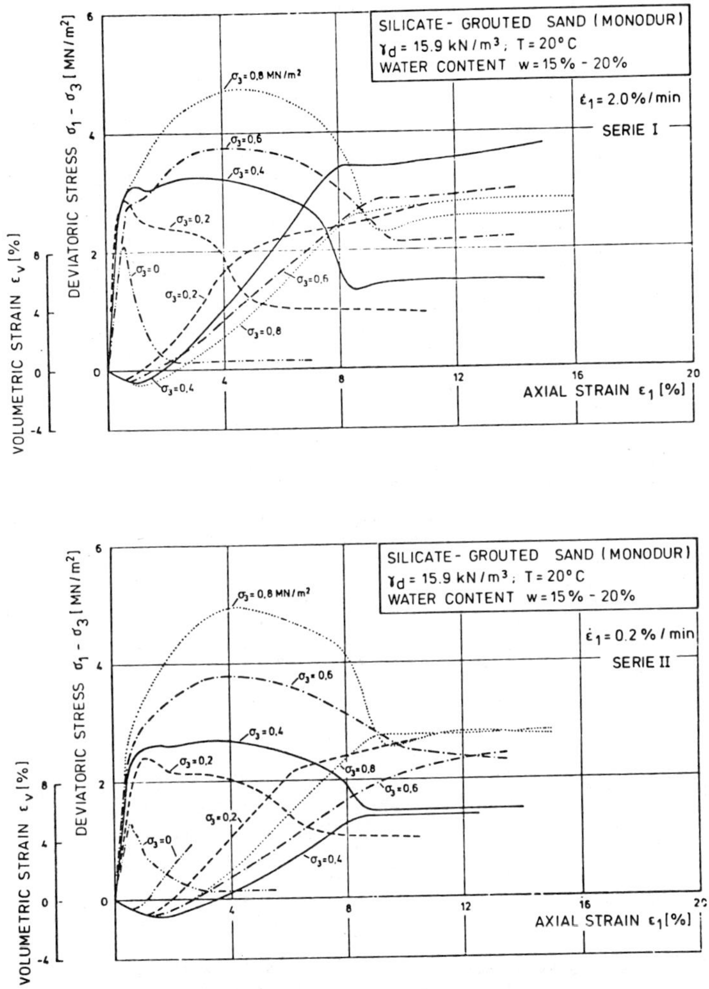

FIG. 6A STRESS-STRAIN PROPERTIES AND VOLUMETRIC STRAIN OF SILICATE GROUTED SAND AT VARIED EXTERNAL PRESSURE

The friction angle (ϕ_s^v) decreases when the strain rate is reduced. It may diminish to values comparable to those of untreated dry sand provided the strain rate is slow enough.

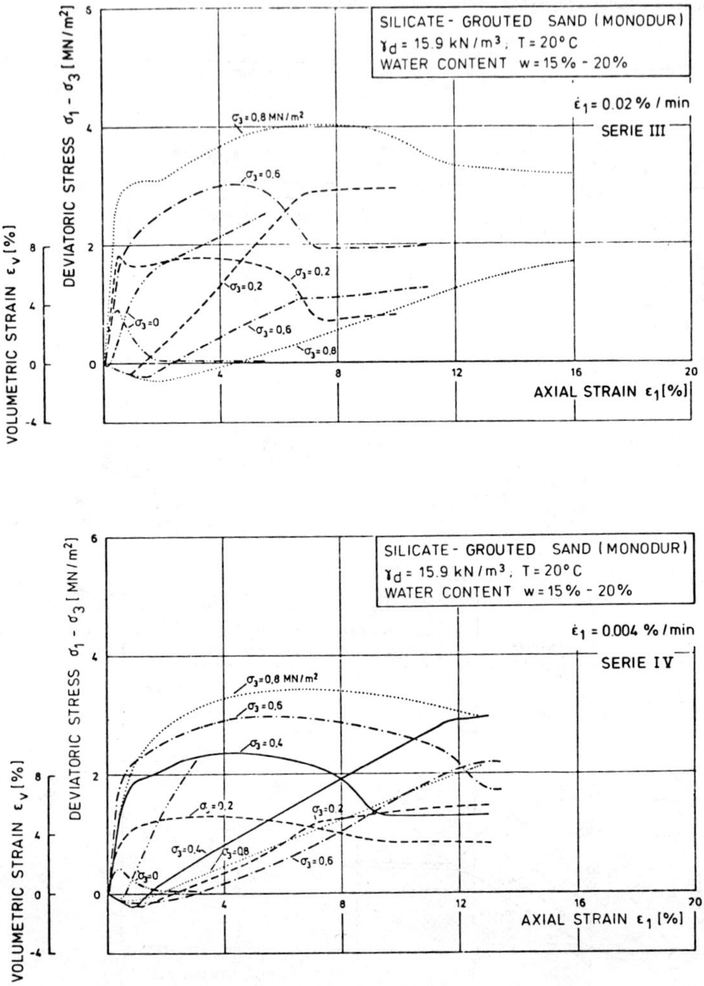

FIG. 6B STRESS-STRAIN PROPERTIES AND VOLUMETRIC STRAIN OF SILICATE GROUTED SAND AT VARIED EXTERNAL PRESSURE

Consequently the cohesion (c^V) ascertained in the foregoing tests determines the efficiency of chemical grouting and can be calculated as follows:

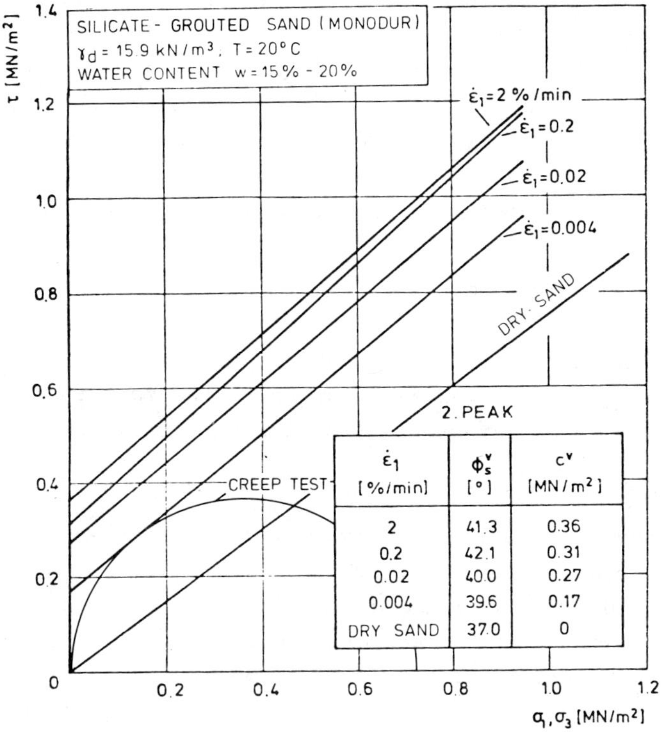

FIG. 7 FRICTION ANGLE AND COHESION OF SILICATE-GROUTED SAND AND UNTREATED DRY SAND

$$c^v = \frac{1}{2\cos\phi}((\sigma_1 - \sigma_3) - (\sigma_1 + \sigma_3)\sin\phi) \quad \ldots\ldots\ldots\ldots\ldots\ldots (4)$$

the friction angle being that of the untreated soil.

An extensive test program carried out at the Karlsruhe University shows that the uniaxial strain-controlled compression tests do not fit the Mohr's rupture diagram (fig. 7) because of the scattering of the first-peak stress values.

On the other hand the ultimate stress ($\sigma_{1,Br}$) determined by the uniaxial creep tests (fig. 4) amounts to 0,6 MN/m^2 <

$\sigma_{1,Br} < 0{,}8\ MN/m^2$ and is plotted in the form of a Mohr's circle in fig. 7. It fits the line of rupture determined by the slow triaxial tests at a strain rate of $\dot{\varepsilon}_1 = 0{,}004$ %/min. Thus the cohesion may also be determined by uniaxial creep tests and equation (4) in the following simplified form:

$$c^V = \frac{\sigma_{1,Br}}{2\cos\phi}(1 - \sin\phi) \qquad (5)$$

SUMMARY AND CONCLUSIONS

So far the estimation of the mechanical properties of "chemically grouted soils" was handicapped by the fact that the considerable creep characteristics of such materials were not known.

In this paper, theoretical and experimental investigations of ultimate bearing capacities and creep properties have been presented. The creep theory here developped is the result of and proved by the aforementioned long-time creep tests. A design criterion is suggested, established with the aid of the empirical creep equation (1) which specifically includes the loading period in each case of practical application.

Beyond that, conventional design parameters are determined in strain-controlled compression tests on drained cylindrical specimens, with due consideration of the time dependency of the grout by varying the relevant strain rate ($\dot{\varepsilon}_1$).

Two stress peaks in the stress-strain diagram are characteristic features of grouted soils in the triaxial state of stresses. The peaks occur owing to the fact that the grout remains viscous even after gel formation and that the sand grains are speed-independent so that both materials behave differently at a given stress and strain rate. The testing method used permits to find cohesion decreasing to the specific limit of a given grouted soil when the strain rate moves towards zero. The cohesion, dependent on the loading period of a special foundation and the friction angle of the untreated soil can be used for conventional calculations.

The relation between stress-controlled and strain-controlled compression tests has been established by Mohr's rupture criterion. The above tests have been performed to obtain parameters permitting to find better solutions to practical design problems. They can help to increase the accuracy of estimation of the bearing properties of time-dependent grouted soils.

APPENDIX I - REFERENCES

Eckardt, H. (1979): Tragverhalten gefrorener Erdkörper. Veröffentlichungen des Institutes für Bodenmechanik und Felsmechanik, Universität Karlsruhe, Heft 81.

Koenzen, J.P. (1975): Rheologische Eigenschaften silikatinjizierter Korngerüste. Veröffentlichungen des Institutes für Bodenmechanik und Felsmechanik, Universität Karlsruhe, Heft 64.

Stetzler, B. (1981): Retardationsverhalten von silikatgelverfestigtem Mittelsand im einaxialen Druckversuch. Veröffentlichungen des Institutes für Bodenmechanik und Felsmechanik, Heft 87.

Vyalov, S. (1963): Rheology of Frozen Soils. Proc. NAS NRC, Int. Permafrost Conference, Purdue Univ. Lafayette, Indiana.

APPENDIX II - NOTATION

The following symbols are used in this paper:

a	=	coefficient of uniaxial creep
c^v	=	cohesion
D (cm)	=	diameter of specimen
H, h_o (cm)	=	height of specimen before test
Δh (cm)	=	axial deformation
n	=	coefficient of uniaxial creep
m	=	coefficient of uniaxial creep
t (min)	=	loading period
t_{norm}	=	$\frac{t}{t_{ref}}$ = standardized loading period
t_{ref}	=	chosen reference time (here: 1 min)
w (%)	=	water content
TP	=	turning-point
γ_d (kN/m^3)	=	unit weight of dry soil
ε_1	=	axial strain
ε_1^K	=	axial creep-strain
$\varepsilon_{1,G}^K$	=	$\frac{\Delta h}{h_o}$ = limiting axial creep-strain
ε_v (%)	=	volumetric strain
$\dot{\varepsilon}_1$ (%/min)	=	strain rate
σ_1 (MN/m^2)	=	axial stress
σ_3 (MN/m^2)	=	lateral pressure
$\sigma_{1,Br}$ (MN/m^2)	=	ultimate creep stress
$\sigma_{1,norm}$	=	$\frac{\sigma_1}{\sigma_{ref}}$ = standardized axial stress
$\sigma_{1,ref}$	=	chosen reference stress (here: 0,1 MN/m^2)
$\sigma_{1,G}$ (MN/m^2)	=	limiting axial stress
τ (MN/m^2)	=	shear stress
ϕ_s^v (^{o})	=	angle of internal friction

PLANNING AND PERFORMING STRUCTURAL CHEMICAL GROUTING

By Wallace Hayward Baker,[1] Member ASCE

ABSTRACT

Structural chemical grouting has developed into a rational geotechnical construction process for in-place modification of sandy soils. Planning and technical performance concepts are presented which were developed on numerous large chemical grouting projects for underground rapid transit contruction. The discussion is from the designer's point of view and focuses on the objectives of structural chemical grouting, important geotechnical parameters, the design process itself, injection planning details, and monitoring and evaluation of the work.

INTRODUCTION

Structural chemical grouting is the injection under pressure of granular soils with a gellable fluid grout that hardens in-place to produce a sandstone-like material. Structural chemical grouting is done to increase the strength and stiffness of sands. This paper summarizes the basic concepts involved in the planning and technical performance phases of structural chemical grouting, primarily from the designer's point of view. The ideas presented herein have been developed and tested over numerous chemical grouting projects, including over a dozen major applications for excavation of rapid transit tunnels and stations (Clough, Baker and Mensah-Dwuman, 1979; Ziegler and Wirth, 1982), and have been refined during research sponsored by the Federal Highway Administration (Baker, 1982). The chemistry of available grouts and the mechanics of grout injection systems are treated in FHWA sponsored reports by Tallard and Caron (1977), and Herndon and Lenahan (1976).

The steps involved in a chemical grouting program are: (1) the establishment of specific objectives for the grouting program, (2) definition of the geotechnical conditions requiring and controlling the treatment, (3) development of an appropriate grouting program design and matching specifications, (4) the planning and execution of a detailed injection work plan, and (5) monitoring and evaluation of the grouting program.

---

[1] President, Hayward Baker Company, Odenton, Maryland

GROUTING OBJECTIVES

The term "structural chemical grouting" is applied where the objective of the grouting is to improve the strength and/or rigidity of the groutable soils to prevent ground collapse, reduce otherwise unacceptable ground movement during construction, improve bearing capacity, etc. The term "waterproof grouting" is used to describe chemical grouting aimed primarily at stopping the flow of water, which otherwise would provoke ground movements or the flow of unacceptably large amounts of water into a construction area, or both. Although many grouts can be considered to be permanent, i.e., have a service life in excess of 20 years under normal conditions, most structural chemical grouting is required for only a few days to several months.

The design objective is often to give the non-cohesive ground (no strength under unconfined conditions) sufficient cohesion so as to prevent the beginning of collapses or soil "runs" into excavations, tunnels or shafts. Chemical grout underpinning is another application of structural chemical grouting, wherein granular foundation support soils are strengthened so as to permit excavation adjacent to footings. In these cases, the soil strength lost by the reduction in confining stresses is replaced by the cohesion imparted to the soil by the grout. Figure 1 shows schematically several different structural grouting applications.

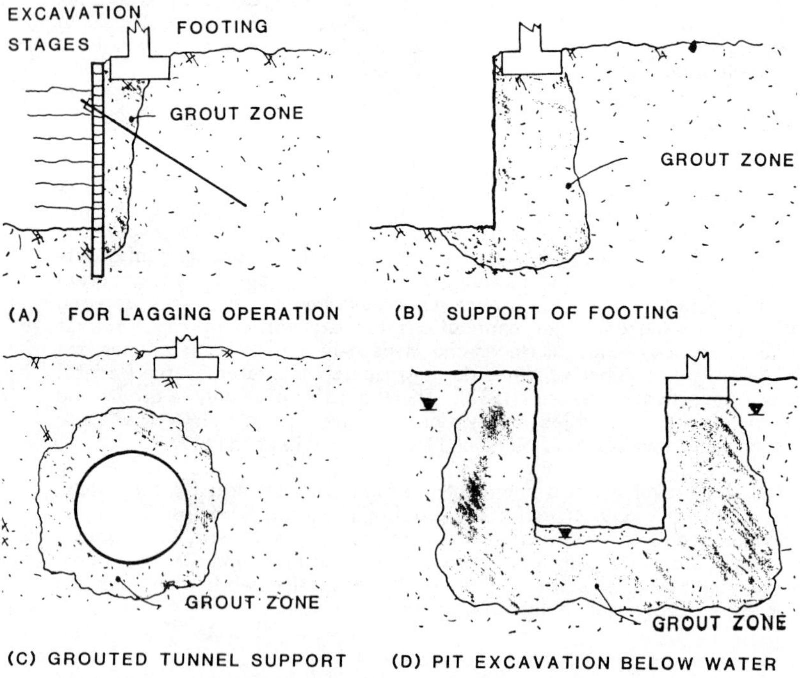

FIGURE 1 - STRUCTURAL CHEMICAL GROUTING APPLICATIONS

Most construction contracts are focused on the legal, commercial and liability aspects of the client-contractor-engineer relationships, and not on the technical intent of the construction project itself. This leads to such language as "...the contractor shall perform chemical grouting to protect the indicated structures from any damages due to adjacent excavation work." This language tends to hide the designer's intent, which may have been only to provide a temporary structural support system to prevent structural and major cosmetic damage to the building, and not reduce settlements to "zero". In this case, settlements of up to 15 mm may be anticipated by the designer, but not by those reading the specifications. In such cases, the designer should state his goals in engineering terms, such as, "...the contractor shall limit vertical settlements of footings of adjacent structures to 15 mm. The chemical grouting scheme shown on the drawings is shown as a typical plan for accomplishing this purpose. The contractor shall verify that his injection plan will accomplish the stated building settlement reduction purpose".

Compare the varying consequences of failing to meet the grouting objective in the following three situations:

1. Chemical grouting underpinning to reduce settlements in a two-story building above and adjacent to a subway tunnel constructed 7 meters below in granular dense ground which experienced settlements up to 3 cm in a previous section of the tunnel.

2. A large diameter tunnel in soft ground passing 3.5 meters immediately below a corner column of a 5 story apartment house with a basement.

3. A subway tunnel passing 1.5 meters below an old, active 1.5 meter diameter sewer main that cannot be bypassed.

In the first example, chemical grouting is to reduce commercial liability due to cosmetic damage to a structure and associated commercial activity disruption. In the second example, major structural damage could occur if the chemical grouting were not entirely successful, but not so quickly that loss of life would result. In the third example, the consequences of the chemical grouting program failing to achieve the objective to protect the sewer could be catastrophic.

The early establishment of clear, quantitative objectives to be achieved by a chemical grouting program is a basic prerequisite to good design and satisfactory economical performance.

IMPORTANT GEOTECHNICAL CONSIDERATIONS

Since chemical grouting is typically used to solve the special construction problem rather than as a part of a global construction process, its use is often related to geological anomolies or special conditions. This requires additional subsurface information not usually obtained in a conventional soils investigation. It may be necessary to perform additional borings with semi-continuous soil sampling, more detailed soil classification of the samples, and additional special laboratory testing to carefully define the special geotechnical situation to be solved by the grouting process. In the least case, a separate interpretation of all available subsurface information should be made.

Groutability

One of the fundamental questions to be asked when grouting is considered is whether the ground involved is groutable. A "groutable" soil is one which will, under practical pressure limitations, accept injection of a given chemical grout at a sufficient flow rate to make the project economically feasible. The permeability of groutable sands may vary as much as 3 or 4 orders of magnitude, from 1 cm/sec for clean sand to 10^{-4} cm/sec for silty sand containing about 20% silts and clays (passing a No. 200 sieve).

Realistic injection rates per grout port experienced in practice range from 2 to 20 liters/min, but sometimes may be as low as 1 liter/min or as high as 40 liters/min. Injection rates higher than 40 liters/min become hard to control, and may suggest that cement grouting could be used. Injection rates slower than 1 liter/min become impractical, since the volume of grout placed per day at less than this rate, even with a multiple hole injection system, is very low. In addition, very long gel times are required to obtain adequate flow times within the soil for practical grout port spacings.

The injection flow rate increases about directly proportionally to increase in the injection pressure, up to the point when uncontrolled ground fracturing occurs. When fracturing of ungrouted soil extends more than a short distance beyond the point of injection, the adjacent ground will probably not be impregnated properly with grout. The effect is loss of control of the grouting process and erratic results. Fracturing of grouted soils is another phenomenon and is considered necessary following primary injection for full treatment of adjacent ungrouted zones.

Hydraulic fracturing for a given injection pressure is largely controlled by local effective stresses, strengths and permeabilities. Thus, in a real (heterogeneous) soil, hydrofracturing may start at a point of low effective stress and/or low permeability and propagate throughout the zone until it reaches a region of higher effective stress/strength and permeability, thus blocking unlimited extension of the fracture. The minor principal stress will determine the pressure at which hydraulic fracturing will occur. Initial hydrofracturing occurs along generally vertical planes. Hydrofracturing of previously grouted soils can occur at pressures as low as one-half the overburden pressure, or may not occur until pressures reach three or four times the overburden (Woodward-Clyde, 1979). Thus, pressure control is not a reliable way of controlling hydrofracturing in grouted soils.

Initial soil permeability is thus the primary guide to establishing the groutability of a soil mass. Soils having permeabilities in the range of 10^{-1} cm/sec to 10^{-3} cm/sec are easily groutable. Soils showing permeabilities in the range of 10^{-3} cm/sec to 10^{-4} cm/sec are moderately groutable. When the permeability is from 10^{-4} to 10^{-5} cm/sec, the soil is usually only marginally groutable and may be ungroutable from a practical point of view. Soils with permeabilities above 10^{-5} cm/sec are considered ungroutable. One rule-of-thumb suggests that soils with less than 12% fines are easily groutable, soils with 2 to 20% fines may be only moderately groutable, and soils with 20 to 25% fines are only marginally groutable. Figure 2 shows typical ranges of groutable soils.

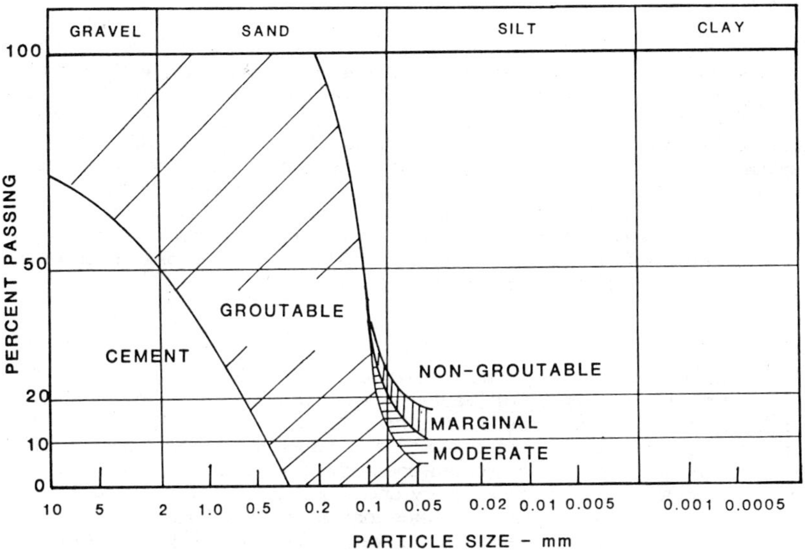

FIGURE 2 - GRAIN SIZE RANGES FOR CHEMICALLY GROUTABLE SOILS

Groundwater

Chemical grouting can be performed in pervious soils either above or below the groundwater surface with about equally successful results, provided both the chemical and hydraulic effects of the groundwater are taken into account. Samples of the local groundwater should be tested to check for compatibility with the chemical grouts to be used. For example, groundwater with high pH can be very destructive to sodium silicate based grouts, preventing initial gellation and/or encouraging grout degradation with time. Low pH groundwater conditions can accelerate setting of sodium silicate grouts while preventing the setting of acrylamide or acrylate grouts. The presence of organic materials in the ground or groundwater can also have a dramatic effect on the gel times and quality of chemical grouts. It is also important to establish the directions and rates of groundwater flow, to distinguish between perched water and groundwater, to establish the presence of any artesian pressures, and to estimate the possible effects the injection program will have on the groundwater levels.

Porosity

In order to calculate the volume of chemical grout needed to impregnate a given soil volume, it is necessary to have a fairly accurate estimate of the porosity of the groutable soils. Typical groutable soils have porosities of 0.25 to 0.45. For a porosity of 0.35, some 350 liters of chemical grout will be required for every cubic meter of soil treated (2.62 gallons/cubic foot). Because a major cost of chemical grouting is the cost of grout chemicals, the porosity has important cost consequences. A correct porosity estimate is necessary to predict the point at which additional chemical grouting will start to cause heave. Since more precise data are not usually available, crude estimates of soil porosity are often obtained from published correlations with Standard Penetration Test "N" values.

Environmental History

To define subsurface ground conditions as accurately as possible, it is important to obtain a history of previous construction activities in the area. The presence of old shafts, wells, cisterns, etc., can provide preferred grout migration locations away from the grout zone, rather than into adjacent, less pervious soils intended to be grouted under the original plan. Drillers and grouting technicans should record every anomoly encountered in the drilling and grouting operations. Such anomolies include a sudden drop in the drill rods, sudden increases or decreases in the ease of drilling, sudden fluctuations in grouting pressures (especially after grouting has been proceeding at a particular grout port for some time), and inconsistencies in development of injection pressure with flow rate. These anomolies should be explained and their significance evaluated before conducting grouting.

PROGRAM DESIGN

Structural grouting design usually involves the solution of a soil-structure interaction problem. The structural behavior of the sandstone-like grouted mass is the basis of the design. Selection of the chemical grout will be partly based on the required grouted-soil properties, which in turn will influence the shape and size of the treatment zone. All this is considered in a performance prediction analysis for proper grouting program design.

Chemical Grouts

Since their first introduction in the early 1950's, numerous "single-shot" chemical grouts have been developed which use catalysts or reactants that provide a time-controlled, delayed gellation long enough to permit premixing and injection of the mixed liquid grout into the desired subsoil zone prior to solidification. In terms of total grout volumes employed in geotechnical chemical grouting, it is estimated that sodium silicate-based chemical grouts are used for over 90% of the cases. The other minor grouts for geotechnical grouting include acrylamide grouts, acrylate grouts, urea-formaldehyde grouts, polyureathane grouts, and resin grouts. A detailed study of chemical grouts was performed by Tallard and Caron (1977) under FHWA sponsorship. This two-volume report represents important reference materials for chemical grouting designers.

Sodium silicate, once called waterglass, is a heavy, syrupy liquid having a pH of 11. Upon reduction of pH, as by acidification or saponification, a gel of silicon dioxide and hydroxides is precipitated. Modern silicate grouts can be prepared with reactants that permit gel times from a minute to over an hour. SIROC, a once widely used silicate grout, involves formamide and dissolved salts as reactants. GELOC-4, TERRASET and HARDENER 600 are now widely used non-toxic sodium silicate grouts which use mixtures of organic esters as reactants.

The properties that affect a grout's injectability, i.e., the ease with which it flows through a porous material under pressure, are not completely understood. Grout viscosity, a measure of shear resistance to shear strain rate, is the main grout property that relates to the injectability of grout in a given soil formation. Surface tension is another important property that has had little study in relation to its effect upon injectability. Neat sodium silicate has a viscosity of over one hundred centipoise, but upon the addition of water, the viscosity is drastically reduced. Structural grouts typically have 40 to 60 percent by volume of sodium silicate, and display viscosities between four and ten centipoise, enabling them to be injected into sand and slightly silty sand, but not silt or clay. Typical waterproofing chemical grout viscosities range from 1.2 centipoise for the acrylamide and acrylate grout systems to about 3 centipoise for dilute (25%) sodium silicate grouts. ASTM Standard D 4016 was recently established for testing grout viscosity.

Sodium silicate grouts tend to increase in viscosity gradually over time, whereas the acrylate and acrylamide grouts typically remain at about their initial viscosity over most of the time until gellation, when a dramatic viscosity increase occurs. Figure 3 shows the typical variation in viscosity versus time for silicate, acrylamide and acrylate grouts when left undisturbed. It is well established, however, that sodium silicate grouts, and to a lesser extent the acrylamide and acrylate grouts, have their gel times dramatically increased by continued agitation. Where silicate grout experiences continued agitation because of turbulent flow through the soil, actual gel times in the ground may be much longer than those taken in cup samples.

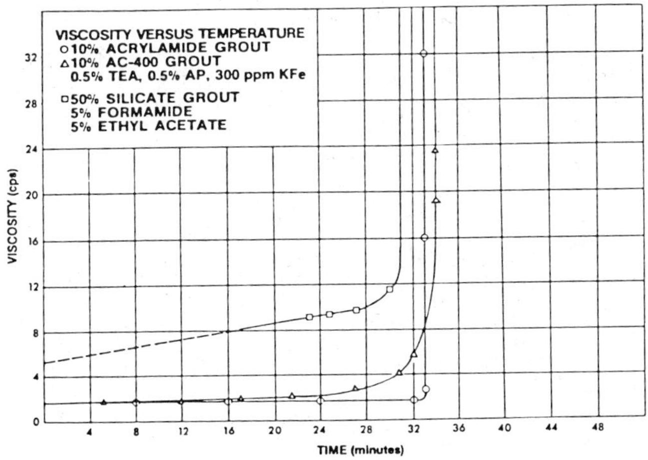

FIGURE 3 - GROUT VISCOSITY VERSUS TIME

Properties of Grouted Soils

Much progress has been made in recent years in defining the parameters that govern the stress-strain and strength behavior of silicate-stabilized sands (Warner, 1972; Koezen, 1977; Diefenthal et al, 1978; Clough, Kuck and Kasali, 1979; Tan and Clough, 1980; and Borden, 1980).

It is helpful to view chemically grouted soils as "composite materials", consisting of a granular soil mass and a cohesive void filler that acts as a glue; the filler (chemical grout) is added to the granular soils by injection as a liquid. The action of a sodium silicate grout in the voids of a soil can be conceptually described as that of a "glue", which bonds soil particles together, and a "filler", which reduces the void volume of the soil. As the grout gels and cures, it undergoes a volume reduction which is directly related to the phenomenon of syneresis (pure silica gel contracts and expels free water). At the same time, the mechanical properties of the grouted sand vary continuously with time.

Grouted granular soils should be tested under appropriate states of confinement and with proper consideration given to time-dependent effects in order to reflect a realistic representation of their field behavior. The recently proposed ASTM procedure for preparing chemically grouted laboratory sand specimens consists basically of placing the sand in a split mold at the desired density and injecting the sand with grout. Care should be taken to minimize disturbance of the grouted specimens; in particular, jacking the specimens out of their molds and trimming their ends should be avoided.

Unconfined and triaxial compression tests on silicate grouted sands indicate that the strength but not the stiffness are dependent on relative density, denser specimens being stronger. Furthermore, strength and stiffness increase with increasing curing time, but this effect is more pronounced during the early stages of curing. The axial strains at failure decrease with increasing curing time and approach an approximately constant value of less than 1% after about one week of curing, regardless of relative density.

Grouted specimens have an initial coefficient of permeability which is about three to six orders of magnitude lower than that of the ungrouted sand. However, due to syneresis the grout shrinks with time and interconnected passages develop within the mass of the grouted soil, thereby increasing the permeability. This makes it possible to conduct triaxial compression tests under either drained or undrained conditions.

The Mohr-Coulomb failure criterion has been found to describe reasonably well the strength behavior of grouted specimens subjected to drained or undrained triaxial compression. The Mohr-Coulomb failure envelopes obtained on the basis of effective stresses at failure yield the same effective cohesion and effective angle of internal friction, regardless of drainage conditions during testing. In general, the magnitude of the cohesion intercept obtained for loose grouted specimens is somewhat smaller than that of dense grouted specimens. This could be attributed to the smaller number of grouted (bonded) grain contacts associated with loose specimens. Grouting has little effect on the angle of internal friction of the sands. These effects can be seen in Figure 4.

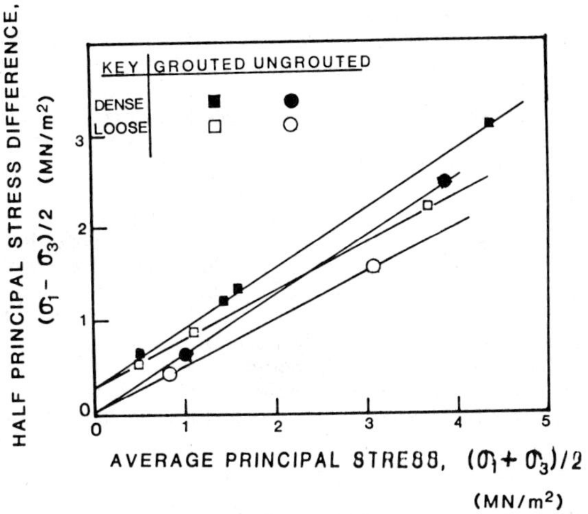

FIGURE 4–MOHR–COLOMB ENVELOPES FOR GROUTED AND UNGROUTED SANDS

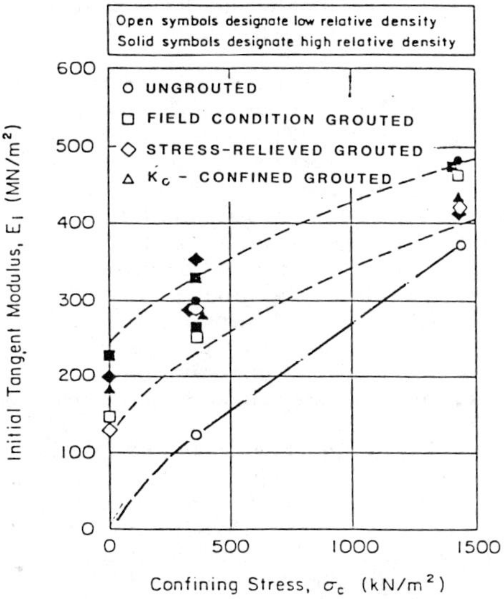

FIGURE 5 – TANGENT MODULUS VERSUS CONFINING STRESS

Figure 5 shows the relationship between the confining stress and the initial tangent modulus of grouted specimens. Grouted specimens (regardless of the relative density) tested under higher confining stresses exhibit a stiffness approximately equal to that of ungrouted sand in a dense state. One explanation for the observed behavior is that, when a specimen is subjected to confined compression, the gelled grout in the soil voids has sufficient shear strength and flow resistance to initially inhibit reorientation and collapse of the soil fabric.

Grouted specimens manifest a time-dependent response to sustained loading (creep). The limited information available indicates that silicate grouted specimens subjected to unconfined compression creep loading could fail after being subjected for only a few days to a load equal to 30 to 50% of their rapid-loading unconfined compressive strength. Confined specimens are much less creep dependent due to the dominant effect of the sand. This characteristic of grouted sands is of special importance in field situations where a grouted mass is required to sustain loads for several months.

Performance Prediction Methods

The rational design of structural chemical grouting requires some prediction of the project performance, including evaluation of soil stress and strains. This requires: (1) a useful model of the construction environment and sequence; (2) a stress-strain-strength model of the grouted and ungrouted soils; (3) an understanding of the structural behavior of the chemically stabilized mass in resisting loads and deformations; and (4) validation of the approach by comparison with case histories.

Consider the frequent case of chemical underpinning of a conventional footing to permit adjacent excavation below the footing. If the excavation is to be made with no lateral bracing (Figure 1-b), the grouted mass must provide total vertical support to the footing and lateral support to the retained soils. Due to the reduced creep strength of chemically grouted soils, it is important to know how long the area will be left exposed before final lateral support is provided. On the other hand, if the footing is located within an area that will be laterally supported by conventional soldier piles and lagging (Figure 1-a), the role of the grouted mass may be reduced to providing limited vertical footing support near the cut face and to prevent any loss of ground during the actual lagging process. Long term strength would not be a factor. It is apparent that for this case the role of the chemically grouted mass is much different than in the former example.

When chemical grouting is used to reduce settlements for tunnels in sandy soils (Figure 1-c), the specific tunneling procedures used are important. Face control, use or nonuse of a shield, size of overcutting bars, and tail shield thickness, among other things, all effect the maximum potential for ground movement. Tail void grouting or liner jacking procedures also are important in determining the standup time required of the grouted zones prior to the performance of backpack grouting. If a portion of the tunnel is below the groundwater table and no dewatering has been provided, the chemically grouted mass will be expected to provide water control as well as structural ground support.

Tan and Clough (1980) modified non-linear stress-strain soil models to describe silicate grouted sands, and employed these models for finite element analysis of the chemically stabilized soft-ground tunnel problem. For the case of a circular tunnel opening in chemically grouted sand, they showed that the grouted zone acts primarily as a compression ring, absorbing most of the stress changes caused by the excavation. Using these models in an appropriately configured finite-element network analysis, the effects of grout thickness, grout strength, tunnel depth, groundwater, and ungrouted zones were studied. Using the results of numerous different case analyses, simplified design charts were developed that permit rapid initial selection of grout strength and thickness to control soft ground tunnel subsidence using silicate-grouted masses.

In many practical cases, the presence of ungroutable layers within the idealized grout zone may have important effects on the behavior of the grouted system. Obviously, the properties of the ungroutable layers are not changed or improved by chemical grouting. When grouting to prevent excavation collapse and retain water behind and a below a grouted pit, special consideration must be given large hydraulic pressures acting against the impervious grouted mass.

INJECTION PROCESS PLANNING AND PREPARATION

Planning the injection program requires a fundamental understanding of how the liquid grout flows in the porous soil from the injection point. With this understanding, the planning aspects for the actual injection portion of a chemical grouting program involve the following four major steps: (1) definition of the geometry (shape and size) of the grouted zone; (2) estimation of the total liquid grout volume required for injection of the grouted zone; (3) definition of the grout pipe layout, including pipe location and spacing and pipe installation scheme; and (4) establishment of an injection staging and sequencing procedure, including an indication of the order in which the various pipes and grout ports will be injected and the partitioning of the estimated grout volumes among the individual grout ports.

Injection Concepts

Fluid chemical grout injected into pervious granular soil under pressure will permeate the soil following the paths of least hydraulic resistance. This means that grouted three-dimensional masses form radially around each injection point. Their shapes are generally irregular spheroids or ellipsoids (called grout bulbs) for the first stage of grouting. Where a very pervious channel is intercepted, a much larger proportion of grout will be found in the pervious channel. Moving water will also displace the grouted mass in the direction of flow, as a function of groundwater flow rate versus injection rate times the gel time. The most important factors affecting the location of the grouted zones are the locations and arrangements of injection points, which in turn is a function of the grout pipe location and type. Grouting sequence, i.e., the order in which the various grout points is injected, can be important.

The final grouted soil mass is produced by the logical geometrical arrangement of small, contiguous grouted masses. Second and third stages of grouting are often performed to fill in the small ungrouted zones between the original grout bulbs. Gel time, grout viscosity, pumping pressure, water flow, and soil anisotrophy also affect the eventual grout distribution.

The injection of chemical grouts into granular soils for either structural grouting or waterproofing purposes is generally expected to follow the basic principles of Darcian flow through porous media where the flow rate, v, for a given soil and permeant, is considered to be directly proportional to the pressure gradient, i, such that $v = ki$.

The proportionality constant, k, is termed the coefficient of engineering permeability. Darcy's Law has generally been found valid for laminar flow of water through sands. It has been suggested (Huang, Borden and Krizek, 1979) that for silicate grout flow conditions, the engineering permeability of the medium can be related to the effective grain size, D_{10}, the soil porosity, n, the viscosity, u, and unit weight, w, of the grout by the equation

$$k = CnD_{10}^2 w/u$$

where C is a constant, probably related to the shape of the sand particles. For fine or silty sands, chemical grout flow is expected to be generally laminar, except perhaps very near the grout pipe at high injection pressures. Where Darcy's law and the above equation are approximately valid, the following general phenomena will be observed for chemical grouting in permeable soils:

1. Grout flows generally radially away from the source of injection pressure. Thus no preferential direction of flow is possible and grout "spheres" tend to be developed from point injection sources.

2. Injection rate is directly proportional to injection pressure.

3. Injection rate varies directly with apparent soil permeability.

4. Apparent permeability varies inversely with the grout viscosity.

5. Apparent permeability varies directly with the soil porosity.

6. Apparent permeability varies directly with the square of effective grain size.

Grout Zone Geometry

The idealized shape and size of the chemically grouted zone is established in relation to the associated construction and excavation details of the project and, for each trail grout zone geometry, an estimate of the grouted soil-structure interaction response is required.

In the case of chemical grouting to reduce settlements due to construction of large-diameter (subway) tunnels, for example, an initial estimate of the desired thickness of the grouted soil around the tunnel is needed. For chemical underpinning of shallow footings adjacent to an open-cut excavation, the idealized cross sectional shape and depth of the underpinning pedestal is needed. When planning for a groundwater cutoff floor, the thickness and extent of the floor must be estimated.

In Figure 6-a, a typical idealized cross section of the required grouted zone for a subway tunnel is shown, as frequently appears in construction specifications. In Figure 6-b, the actual grouted cross section is shown, as dictated by the location of a typical grout pipe array placed from the ground surface. The difference between the two is due to the fact that the grout ports are located linearly along the grout pipes, and it is not possible to truncate a grout sphere that invades the ground beyond the idealized cross section. Designers must recognize that the consequence of the necessary actual grouted cross sectional shape as compared to the idealized shape is to increase the grout quantities.

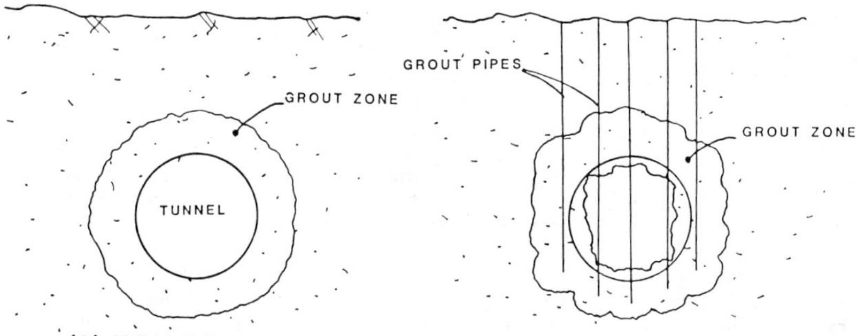

(A) IDEALIZED GROUT ZONE (B) ACTUAL GROUT ZONE

FIGURE 6–IDEALIZED AND ACTUAL GROUT ZONES

Estimation of Liquid Grout Volumes

The volume of liquid grout required to chemically solidify a given volume of granular soil is generally calculated based on the assumption that the grout must fill almost all of the soil voids, that the grout flows radially away from the injection points leaving no ungrouted zones in its path, and that some grout will penetrate beyond the idealized flat faces of the grout zone due to the generally spherical shape of the grout bulbs. This is represented in the equation

$$\text{Liquid Grout Volume} = V_z(n \times F)(1 + L)$$

where V_z is the total volume of the treatment zone, n is the soil porosity, F is the void filling factor, and L is the loss factor for grout placed outside the treatment zone.

The porosity of granular soils may vary from 25% for very dense, well-graded silty sands to almost 50% for loose, uniform sands. The actual porosity depends upon the grain size distribution and the relative density of the deposit.

In practice, it has been found that typical sodium silicate-based grouts fill from a minimum of about 85% of the void volume up to nearly 100% of the void volume, depending on various factors. Generally, the lower percentage of void filling occurs for well-graded sands with fines, where only a single stage or limited second stage of grouting is performed under low pressures. Some correlation seems to exist between injection pressure gradients and the percentage of void filling.

The grout loss factor may vary from about 5% to 15%, depending upon the shape of the grouting zone, the frequency of injection points per unit volume, and the presence of highly porous layers in the groutable soils.

Grout Pipe Layout

Spacing. Grout pipe spacing is arranged to provide for primary and secondary injection locations, with these locations generally falling at the midpoint from each other. Grout pipes spaced too closely to each other will result in excessive drilling costs. Grout pipes spaced too far apart will require extremely long pumping times to place the necessary volumes of grout at each location, with loss of control of the grouting operation as the location of the grout front becomes unknown. In practice, spacing between grout pipes varies from a minimum of about 0.5 m to a maximum of 2.5 m, with most grouting projects having pipe spacings on the order of 0.8 to 1.5 m.

The simplest grout pipe array occurs when pipes can all be placed parallel to each other. In many cases, however, construction constraints require grout pipes to be placed in a fan array. This results in considerable additional drilling and pipe placement requirements since near the fan apex, pipes are too close together for all to be used effectively. Figure 7 shows different grout pipe layouts required by specific site conditions.

Grout Pipe Installation. Next to the proper location of the grout injection point, one important factor affecting the successful distribution of grout in the intended grout zone is the proper mechanical grout pipe installation. The grouting engineer anticipates that grout under pressure will flow evenly in all radial directions away from the grout injection point, filling the voids in the porous granular media. Actually, the liquid grout under pressure will flow along the path of least resistance, and any open channel left by the grout pipe installation procedure will permit the grout to flow away from the intended injection zone. Thus, success requires that grout pipe installation procedures permit no unexpected flow channels to remain within the borehole.

Many simple grouting projects have been successfully completed by using the expedient grout pipe installation method of temporarily sealing an open-ended pipe with a drive point, driving the pipe to the desired grouting depth, removing the point, and injecting grout while withdrawing the grout pipe. As the grout pipe is withdrawn, it is assumed that all of the grout is entering the ground at the elevation of the grout pipe tip. After the pipe has been withdrawn some distance during the injection process leaving an open channel in the ground, this likely is not the case. Control of the injection location has been lost especially if a more porous strata is encountered and the grout can travel to it along the space created by withdrawal of the pipe.

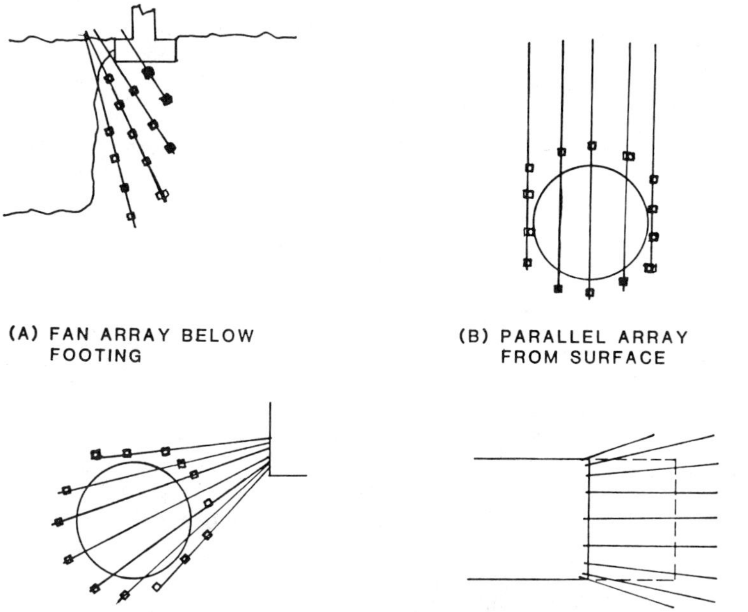

(A) FAN ARRAY BELOW FOOTING

(B) PARALLEL ARRAY FROM SURFACE

(C) FAN ARRAY FROM PIT

(D) FAN ARRAY FROM SUCCESSIVE HEADINGS

FIGURE 7 - GROUT PIPE LAYOUT PLANS

These problems can be partly resolved by the use of needle pipes, long tapered drive pipes with grout holes located near the pipe tip and which are driven successively downward as grouting progresses. The needle grout pipe is advanced downward through continual driving rather than withdrawn, and can provide effective control over the actual injection location. A common drawback of all of these pipe-driving procedures is that during second stage and third stage grouting, the ground has become considerably stronger and the grout pipes may actually encounter refusal in the previously grouted zones through which they must penetrate.

All of the above problems are effectively resolved by the use of sleeve-port grout pipes, sometimes known by the French term, "tube-a-manchette". This system consists of a three to six centimeter diameter pipe that has small, periodically spaced grout holes drilled through the pipe wall at the preferred grouting locations. The grout holes are in turn covered by snugly fitting rubber sleeves, which act as one-way check valves when grout is injected one port at a time from the inside of the pipe out into the adjacent soil. The entire rubber sleeved grout pipe system is installed in an oversized borehole, and the annular space between the sleeve-port pipe and the borehole wall is then sealed with a brittle mortar to prevent movement of grout along the grout pipe. This sheath of mortar is fractured when the sleeve is expanded during grouting from the inside of the pipe. Grouting is done through an internal double packer pushed ahead of a separate, small diameter, internal grout pipe. The packer is then used to locate the sleeve-ports one by one, and to inject the desired quantity of grout at each sleeve-port location. The sleeve-ports can be injected in any sequence or may be reinjected as desired. In Figure 8, a diagram of the sleeve-port system is shown.

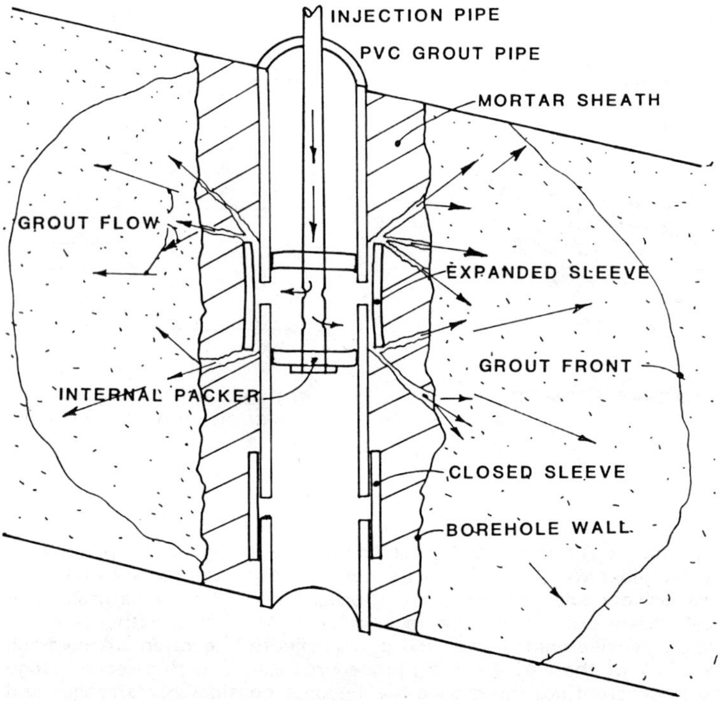

FIGURE 8 - SLEEVE-PORT GROUT PIPE

The resulting excellent control that is possible over grout injection locations using the sleeve-port system generally justifies the additional expense and effort required in their installation. Actual project economies usually result, because the sleeve-ports can be regrouted without subsequent pipe installation cost and a primary-secondary grouting effect can be obtained in a single grout pipe by alternating sleeve-ports during the two stages of grouting. Initial high injection pressures may be necessary for a few brief moments to crack the mortar sheath around the sleeve pipe and initiate the flow of grout into the ground. The exact grouting pressures acting in the ground cannot be measured at the grout pipe header, because of the grout pressure dissipated in keeping open the sleeve-port. This "cracking pressure" can be estimated by plotting the flow rate versus pressure curve for the hole during initial stages of grouting and tracing the flat portion of the curve back to zero flow. The pressure intercept at zero flow then approximates the pressure dissipated within the sleeve-port. In Figure 9, a view of a grouting face with the sleeve-port pipe installed and multiple ports being grouted simultaneously is shown.

FIGURE 9 - SLEEVE-PORT GROUT PIPES AT GROUTING FACE

Injection Staging and Sequencing

Injection staging refers to separate episodes of grouting done at different times in the same zone. Injection sequencing refers to the order in which grouting proceeds from one zone to another.

Primary stage chemical grouting refers to the initial volume of grout placed in essentially ungrouted zones where the grout bulbs either are not in contact or overlap only over a small proportion of their boundary surfaces. Secondary stage grouting refers to the next grouting episode following primary grouting in the same zone. Tertiary, quaternary, etc., stages refer to the third, fourth, etc., subsequent grouting episodes in the same zone.

If the initial injection spheres were to exactly touch, they would occupy 52.4% of the theoretical cubes in which they are contained. Primary injection is usually carried out so that adjacent grout sphere radii would overlap each other about 10% beyond the initial midpoint radius. This represents about 70% of their cube volume. Thus, it is a typical practice to inject between 65% to 75% of the planned total liquid grout volume in the primary injection ports, and to inject the remaining 25% to 35% of the total grout volume in secondary and tertiary grout ports. This volume distribution is depicted in Figure 10. Where 70% of the theoretical required grout volume is injected in the primary stages of an outside row of grout pipes, up to 15% of the grout will necessarily be placed outside the intended grout zone.

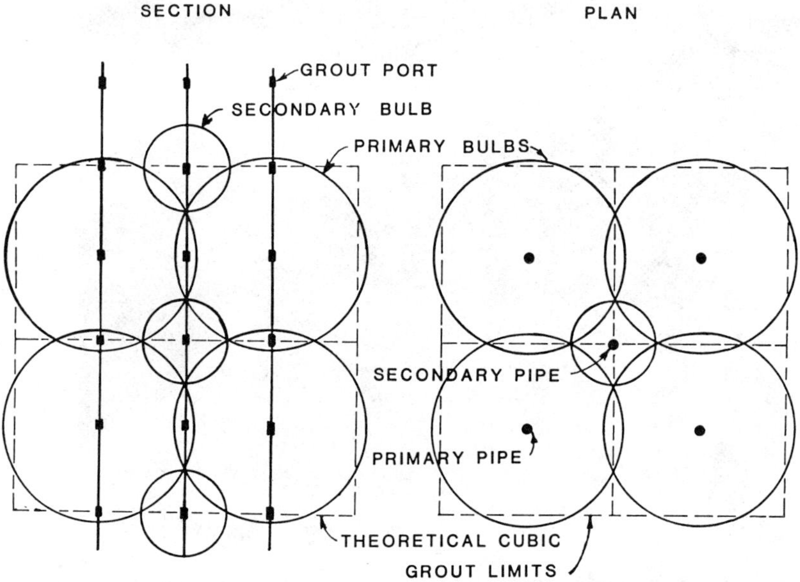

FIGURE 10 - GROUT DISTRIBUTION AT PRIMARY AND SECONDARY GROUT PORTS

After primary grouting large ungrouted spaces still exist in the intended grout zone. Secondary grouting is usually intended to inject all but the final 5% to 10% of the anticipated grout volumes. After secondary grouting, a very large proportion of the intended grout mass is fully impregnated with hardened grout. Tertiary grouting then necessarily requires careful selection of grout pipe ports so as to be near any ungrouted zones. During secondary and tertiary grouting, it is usually necessary to provoke localized fracturing of previously grouted ground, to permit migration of the liquid grout from the injection point to nearby ungrouted regions. It should be noted that for structural grouting where no waterproofing effect is needed, complete tertiary grouting may neither be necessary nor cost effective. Tertiary grouting is often used as a form of check injection or in-situ pump-in test, to verify that the major proportion of a zone has been effectively treated.

Anticipated primary, secondary and tertiary grout volumes are noted on the grout port layout drawing. Actual grout takes are later noted for each grout port on the same drawing, so that major differences between planned and actual grout injection volumes can be analysed.

In grouting above the water table, injection sequencing does not have the same importance it does when grouting below the water table where trapped water can prevent the migration of grout to ungrouted zones. Injection below the groundwater table must be done in such a way that escape locations are always provided for groundwater to be pushed or "herded" ahead of the advancing grout front, so that water is not trapped within the final solidified mass.

MONITORING AND EVALUATION

The best rational design is eventually subjected to the reality of field injection practice. Effective construction management of the actual injection process will assure that:

1. Grout pipes are placed properly.

2. Grout components are properly formulated and mixed.

3. Planned grout volumes are accurately injected at specified grout ports, in a logical sequence, and with acceptable flow rates and pressures.

4. Injection process data are recorded and used as feedback to infer grout is where and in the condition expected.

5. Final quality assurance testing is provided after completion of the injection work.

Grouting Systems

Grouting systems may be distinguished by the particular combinations employed of grout mixing methods (batch vs. continuous) and grout injection methods (open pipe vs. grout ports). Because control and evaluation depend upon the particular grouting method used, these factors should be considered when establishing the quality control and quality assurance plans.

Chemical grouting by continuous mixing uses metering or proportioning pumps and totalizing meters for grout components. Continuous mixing systems permit better control over the injection process since short gel times can be used. Typical gel times used with batch systems are 45 to 90 minutes, whereas gel times used with the continuous mixing systems are usually 10 to 25 minutes. Combining the continuous mixing system with short gel time thus avoids the formation of large pools of ungelled grout in the ground. Further argument for short gel time grouting was provided by Karol (1968), who conducted laboratory tests in samples subjected to lateral flow of groundwater. Tests using short gel times produced balls of stabilized soil around the injection point, but long gel time grout was diluted and washed away before it could gel. In this respect, consideration should be given to the tendency of grout gel times to be delayed during agitation due to high injection flow rates.

Injection Measurements

Volume. Accurate volume measurements of individual grout components are required to confirm proper grout mix proportions and calculate total grout volumes for pay items. For both batch and continuous mixing type grout plants, positive displacement meters should be used for this purpose. Conventional water meters or waffle-plate meters cannot provide required accuracy for the variable flow rates and viscosities involved. Meters should be provided with filters and be protected from overpressuring.

Pressure. Where continuous electronic recording is not done, pressure measurements are commonly made by bourdon tube pressure gages. Indicated pressures are periodically noted by hand on a data sheet. Bourdon gages will not register below about the lowest 5% of their range. Thus, high pressure gages must not be used to measure low grouting pressures. Positive gage protection against plugging by gelled grout and against overpressuring during grout port cracking must always be provided.

Flow Rate. Grout flow rates at individual grout ports must be measured in an easily interpreted form. This preferably involves a direct reading device such as a flow-column type meter, a mechanical turbine, or a positive displacement meter. Back calculation of flow rate based on the rate of liquid drop in chemical tanks is not considered accurate enough for good control. Flow rate measuring devices must be constructed so as not to have dead spots where grout gel can be trapped, to later break loose and jam flow paths.

Continuous Monitoring. Injection pressures and flow rates should be continuously monitored. The grouting technican should not have major duties other than monitoring the injection process itself. Pressure and flow rate records are made for each grout port whenever injection pressure changes of more than 0.25 kg/sq cm or flow rate changes of more than one liter per minute occur, or at least every 10 minutes. For simultaneous multiple hole grouting, special automatic recording equipment may be indicated (Mueller, 1982).

Data Recording and Evaluation

The accurate measurement of grout flow rates, injection pressures and total grout volume with time per injection location are fundamental quality control requirements of any chemical grouting project. In conjunction with the known geometrical array of grout port locations, these data are used to infer the location and behavior of the grout underground. Grouting flow rates, pressures and volumes are used in real time by the grouting technican to decide whether to reduce or increase flow and pressure at any given moment and to decide when to end injection altogether at a given port.

Specifications usually require that a detailed grout monitoring program be submitted by the grouting contractor for approval by the Engineer. The detailed methods for measuring grout flow rates, pressures and volumes and recording these items are usually left up to the contractor.

Graphical Grout Take Log. An important adjunct to field record keeping is the use of a graphical grout take log, showing grout volume injected at each injection point. Upon a cross-sectional display of injection pipes and ports, a graphical representation of actual injected grout volume is drawn at each injection point. Linear, logarithmic, or circular representations of grout volumes have been used. Color coding can be used on the graphical grout take log to highlight areas of serious over and under grout takes, making it an excellent tool for contractor and inspector alike for visualizing project progress. By referring to such a graphical log as the job proceeds, variations in injection conditions can be observed and any unusual conditions spot-lighted. A graphical grout take log using linear scale grout take representations is shown in Figure 11. Grout volumes are drawn at each grout port proportional to the volume injected. In this example, it can be immediately seen that grouting of pipe 5 has not been completed and that the upper primary grout port at pipe 7 was undergrouted.

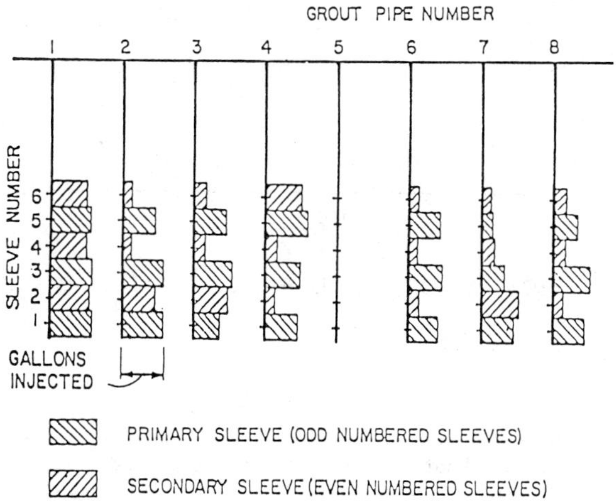

FIGURE 11 - GROUT TAKE LOG

Secondary-Tertiary Grout Port Test. An important tool used to test the adequacy of grouting is the secondary-tertiary grout port test. Grouting is traditionally conducted in stages, with primary, secondary, and sometimes higher level injection points.

If the secondary stage injection points produce a rise in pressure and reduction in flow as the projected volumes are reached, this is an indication of nearing complete grout saturation and confidence in the primary grouting is increased. Figure 12 shows in a typical injection record for secondary grout ports. For the curve displaying a strongly decaying flow rate pressure ratio with time, injection was terminated when the designed secondary grout volume had been injected, with good confidence in the effectiveness of primary grouting in this particular example. If the flow rate pressure ratio response characteristic of refusal does not occur during the second stage, a third stage of grouting may be advisable to check on the completeness of the second stage. This saturation response is not typical of a single row grout pipe array, or grout ports on the perimeter of the grout zone.

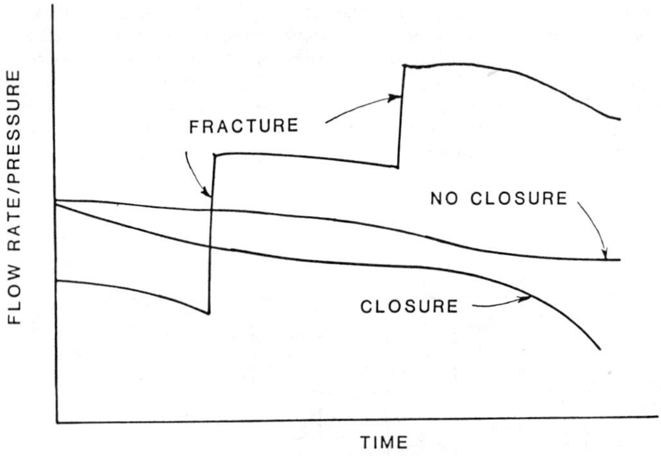

FIGURE 12-TYPICAL FLOW RATE-PRESSURE RATIO CURVES

Hydraulic fracture is indicated by large increases in flow rate with only small pressure increases. That is, the flow rate pressure ratio curve is concave upward with time. Though relatively easy to diagnose during injection into impermeable materials, hydraulic fracture is more difficult to identify in the permeable soils in which grouting takes place.

Acoustic emission monitoring, a relatively new geotechnical monitoring tool (Huck, Waller, and Fowler, 1982), may be used to detect structural distress in geotechnical materials. In grouting, it may be used to detect hydraulic fracturing and therefore allow for control of this phenomenon. Indications of fracturing are bursts of microseismic noises heard by the system, denoted by increased acoustic emission count rates.

Quality Assurance

Quality assurance methods include those systems of measurement and documentation that are useful in proving that the specified project objectives were accomplished. Unfortunately, one cannot "see" the grouted ground.

Conventional Testing. Attempts are frequently made to evaluate chemical grouting effectiveness using conventional site exploration tools. Tools that may be applied include the standard penetration test, borehole pressuremeter, undisturbed sampling, cone penetration resistance and the excavation of test pits. These systems are useful under some conditions, but under other conditions, they are ineffective or even misleading.

The Standard Penetration Test (SPT) is a favorite site exploration tool in spite of its crudeness. The primary advantage of the SPT appears to be the wide familiarity geotechnical engineers have with its use. Unfortunately, it is a dynamic test and grouted soil is an easily shattered brittle material. These factors combine to increase the already large variability of the SPT when it is applied to grouted soil. While it is usually apparent in the blow counts that there has been an increase in soil strength after grouting, the increase in blow count is not commensurate with strength increase.

Occasional attempts are made to obtain undisturbed samples by core drilling. Shelby tube or split-spoon sampling is out of the question in grouted soil. Rotary drilling with a core barrel is sometimes successful. The sample may be destroyed when small gravel particles or broken pieces of grouted sand work their way into the core barrel and abrade the sides of the sample, usually breaking it in flexure. Sample erosion by drilling water is also a problem, partially solved by drilling with air flushing. Even if a sample is recovered intact, it is of questionable value because of the rough handling it undergoes during the coring process. It is difficult to hand trim an undisturbed specimen from a block sample in the lab, much less to drill one in the field. Undisturbed samples can, however, be obtained by trimming block samples recovered at tunnel headings or from test pits.

The most effective traditional grout evaluation method is the excavation of test pits. One can then enter the grouted zone and recover undisturbed samples, conduct plate bearing, CBR or wall reaction tests in-situ, and generally evaluate the grouted soil by personal inspection. If it is difficult to detect the grouted soil either by odor or color, an acid/base indicator such as phenothalene can be sprayed on the soil to detect the presence of high pH silicate grouts.

Geophysical Testing. In addition to conventional site exploration tests, selected geophysical tests are now being used to evaluate grouted soils. These include borehole radar and cross hole acoustic velocity (Huck, Waller, and Fowler, 1982), to determine the grout location and condition underground.

Borehole radar profiling involves the transmission of microwaves through the ground from one borehole to another. Silicate grouted sand becomes opaque to the microwave transmission, such that loss of signal is interpreted as a sign of good grout penetration. Recent development of small-diameter (35 mm) down-hole transmitters and antennas now permit economical use of plastic sleeve-port grout pipes as monitoring holes. This eliminates the need for expensive special instrumentation holes and greatly reduces the cost of radar profiling of grouted masses. Another needed development is data processing equipment that will allow on-site interpretation of the before- and after-grouting radar logs by non-specialist technical personnel.

Acoustic velocity profiling involves the transmission of shear waves from one borehole to another. Medium strength sodium silicate grouts may double or quadruple the acoustic velocity of the impregnated soil. Thus this testing procedure permits inference of the quality of grouting work.

CONCLUSIONS

Recent improvements in the engineering practice of structural chemical grouting has provided increased confidence in its ability to control settlements and prevent excavation failure.

Designers can expect significantly improve the success of structural chemical grouting by defining their grouting program objectives and focusing on the geotechnical conditions controlling the process. New developments in analyzing the structural behavior of grouted soil masses permit numerical performance predictions prior to construction for more realistic designs and better project economy.

Detailed planning of injection locations, grout volumes staging, and injection sequencing can provide a specific work program against which subsequent quality control management can best be viewed.

Finally, new geophysical methods of evaluating chemically grouted soils, utilizing subsurface radar and cross-hole acoustic velocity profiling, can provide for quality assurance testing that contributes to increased confidence in this method of ground modification.

REFERENCES

Baker, W.H., "Engineering Practice of Chemical Grouting", Vol. III of Improved Design and Control of Chemical Grouting, Report No. FHWA/RD-82-038, for U.S. Dept. of Transportation, February 1982.

Borden, R.H., "Strength of Silicate Grouted Sands", Ph.D. Thesis, Northwestern University, Evanston, Illinois, August 1980.

Clough, G.W., Kuck, W.M., and Kasali, G., "Silicate Stabilized Sands" ASCE Journal of Geotechnical Engineering, Vol. 105, No. GT 1, pp. 65-82, January 1979.

Clough, G.W., Baker, W.H., and Mensah-Dwumah, F., "Ground Control for Soft Ground Tunnels Using Chemical Stabilization - A Case History Review", Proceedings, 1979 Rapid Excavation and Tunnelling Conference, Vol. 2, June 1979.

Diefenthal, D.C., Borden, R.H., Baker, W.H., and Krizek, R.J., "Strength and Stiffness of Silicate Grouted Sand with Different Stress Histories", ASTM Geotechnical Testing Journal, Vol. 2, No. 4, pp. 200-205, December 1979.

Herndon, J. and Lenahan, T. Grouting in Soils, Report No. FHWA-RD-76-26-27, Vol. I and II, for U.S. Dept. of Transportation, June 1976.

Huang, A., Borden, R.H., and Krizek, R.J., "Non-Darcian Flow of Viscous Permeants Under High Gradients", Technical Report No. 4 to Hayward Baker Company, September 1979.

REFERENCES (Cont'd)

Huck, P.J., Waller, M.J., and Fowler, J., "Construction Control", Vol. I of Improved Design and Control of Chemical Grouting, Report No. FHWA/RD-82-036, for U.S. Dept. of Transportation, February 1982.

Karol, R.H., "Chemical Grouting Technology", ASCE Journal of Soil Mechanics and Foundation Division, Vol. 94, No. SM1, pp. 175-204, January 1963.

Koezen, J.P., "Time Dependent Stress-Strain Behavior of Silicate- Grouted Sand", ASCE Journal of Geotechnical Engineering Division, Vol. 103, No. GT8 pp. 903-908, August 1977.

Krizek, R.J., and Baker, W.H., "Materials Description Concepts", Vol. II of Improved Design and Control of Chemical Grouting, Report No. FHWA/RD-82-037, for U.S. Dept. of Transportation, February 1982.

Mueller, R.E., "Multiple Hole Grouting Method", ASCE Specialty Conference on Grouting in Geotechnical Engineering, New Orleans, LA, February 1972.

Tallard, G.R., and Caron, C., Chemical Grouts For Soils, Vol. I and II, Report No. FHWA-RD-77-60, for U.S. Dept. of Transportation Contract No. FHWA-FH-11-8826, June 1977.

Warner, J., "Strength Properties of Chemically Solidified Soils", ASCE Journal of Soil Mechanics and Foundation Division, Vol. 98, No. SM11, pp. 1163-1185, November 1972.

Woodward-Clyde Consultants, Results and Interpretation of Chemical Grouting Test Program, Existing Locks and Dam No. 26, Mississippi River, Alton, Illinois, Final Report for Department of the Army, St. Louis District, Corps of Engineers, Contract No. DACM43-78-C-0005, June 1979.

Ziegler, E.J., and Wirth, J.L., "Soil Stabilization by Grouting on Baltimore Subway", ASCE Specialty Conference on Grouting in Geotechnical Engineering, New Orleans, LA, February 1982.

INJECTION AND DISTRIBUTION OF SILICATE GROUT IN SAND

by

Talat A. Bader[1]

Raymond J. Krizek[2], M. ASCE

ABSTRACT

The mechanism by which injected silicate grout permeates a soil mass is due in large part to the complex interaction of both chemical and physical factors that are not well understood. In an effort to explain certain observed field phenomena and associated hypotheses, an extensive laboratory model study was undertaken to relate the effects of grout composition and flow configuration to the injection pressure, distribution of pressure in the soil mass, and flow rate. Three different experimental models, each with a different geometry, were studied; the basic difference between these models is the rate at which the grout front advances as the grout is injected at a constant flow rate. The magnitude and distribution of the pore fluid pressures developed during the grout injection process were measured, and the results indicate a rapid decrease in pore fluid pressure with distance from the injection point. The strength and modulus of the grouted soil mass showed a tendency to decrease with distance from the injection point, but the data are not completely consistent in this trend. Upon injection, the grout appears to flow progressively into the soil, with the fresher grout displacing the older grout toward the exterior of the grouted mass. After the grout is pumped into the soil for about five or six times the design gel time, there is limited evidence to suggest that the older grout farther from the injection pipe develops a sufficiently high viscosity and/or adherence to the sand grains to allow the fresher grout to pass through preferred flow channels, but there was no conclusive evidence of soil fracturing due to grout gelation in any of the tests.

INTRODUCTION

The increasing use of silicate-base grouts to stabilize permeable soils has directed considerable attention to our need to understand more fully the mechanism by which the grout flows into the soil

[1]Assistant Professor, University of Petroleum and Minerals, Dhahran, Saudi Arabia.

[2]Professor and Chairman, Department of Civil Engineering, Northwestern University, Evanston, Illinois, U.S.A.

mass and its resulting distribution in the treated soil zone. Although a few studies have been undertaken to evaluate this phenomenon, relatively little is known with any degree of certainty. For example, based on the results of a laboratory study using AM-9 grout, Karol (1968) indicated the existence of a "break-through phenomenon" or flow inversion when pumping times significantly exceed grout gel times. Alternatively, Baker (1979, Personal Communication) performed similar injection studies using a silicate chemical grout and found that the laboratory tests gave no indication of a "break-through phenomenon" whereas the field tests suggested such an occurrence, as detected by acoustic emission techniques and rapid decreases in the pumping pressure.

In addition, it has been observed on numerous occasions in the field that pumping could be continued for a period much longer than the design gel time of the grout; there are three possible explanations for this situation. The first is simply that the grout does not gel as it permeates through the voids of the soil; reasons to support this explanation are (a) turbulence occurs as the grout passes through the voids of the soil and this agitation prevents the grout from gelling, (b) dilution takes place along the interface zone between the grout front and the ambient groundwater and this alters the concentration of reagent and associated gel time of the grout, and (c) reagent is adsorbed by the sand grains, thereby reducing the concentration of reagent in the grout mix and increasing its gel time. The second is the possible existence of a flow inversion or "break-through phenomenon", whereby grout pumped into the ground later in the operation fractures the zone of earlier grout that has gelled and propagates to points external to the heretofore grouted soil mass; if this is the case, the rate of advance of the grout front is believed to affect the response, because gelation must take place at the leading edge in order to develop the potential for a fracture and subsequent "break-through" of fresh grout as pumping continues. The third explanation is due to the geological features of the grouted soil zone; these include heterogeneity, stratification, anisotropy, and flowing groundwater.

OBJECTIVES

The objectives of this work are (a) to investigate whether or not a flow inversion or "break-through phenomenon" occurs as the external shell of grout begins to gel during continuous pumping; (b) to understand more fully the mechanisms by which silicate grouts permeate cohesionless soils; (c) to assess the magnitude and distribution of fluid pressures that are developed during the grout injection process; (d) to examine the strength and modulus of a grouted soil mass as a function of distance from the point of injection; and (e) to develop a consistent explanation for the flow of silicate grouts in a soil mass and the resultant distribution of the grout.

SCOPE OF STUDY

The experimental program consists of grout injection studies using three different models, each with a different geometry. The basic difference between the three models is the rate of advance of the grout front as the grout is injected at a constant front rate. In the linear model the advance of the grout front is directly proportional to the pumping rate, regardless of the position of the grout front relative to the injection point. In the cylindrical model the advance of the grout front is directly proportional to the inverse of the instantaneous radial distance to the grout front. This means that, as the grout front moves outward from the injection point, its rate of advance decreases and perhaps reaches a point where the grout at the front is in a quasi-static condition, thereby enhancing the opportunity to gel. Once the grout front begins to gel (if indeed it does), a noticeable increase in the injection pressure will result; driven by this pressure, the fresh grout will crack through the gelled grout in the outer shell and the injection pressure will decrease. In the spherical model the grout front advances at a rate that is directly proportional to the reciprocal of the square of the instantaneous radial distance to the grout front. The same physical sequence described for the cylindrical model will apply for the spherical model, except that the tendancy for a "breakthrough phenomenon" is enhanced.

MATERIALS TESTED

The grout used in this test program consists of hydrated sodium silicate ($Na_2SiO_3 \cdot nH_2O$), water (H_2O), and organic reactants formamide ($HCONH_2$) and ethyl acetate ($CH_3COOC_2H_5$). The reactants were combined with tap water to form a homogeneous solution, which was then blended with the sodium silicate and agitated mechanically to insure a homogeneous mixture. The percentage of each component was varied to comply with needs dictated by the different phases of this study.

The sand employed was a commercially available Torpedo sand purchased from Materials Service Company in Evanston, Illinois. Torpedo sand is a well graded gray sand with traces of medium to fine gravel and very fine sand; its grain size distribution yields C_u equal to five. The specific gravity of a representative fraction of Torpedo sand was determined to be 2.69, and its coefficient of permeability at a dry density of 119 lb/ft^3 is about 1.4×10^{-2} cm/sec; use of the Hazen formula yields a coefficient of permeability of about 3.2×10^{-2} cm/sec. The maximum and minimum dry densities were found to be 132 lb/ft^3 and 107 lb/ft^3, respectively. In general, the sand was placed at a dry density of about 123 lb/ft^3 for most of the tests.

EXPERIMENTAL SYSTEM

Two constant-displacement Milroyal double diaphragm pumps, which provide for the accurate metering and safe transfer of corrosive viscous chemicals, were used to pump the silicate and reagent. There is no mechanical connection between the diaphragms and the pump driving mechanism; both diaphragms are hydraulically balanced during pump operation. The flow rate of each pump was controlled manually to within ± 1% accuracy by adjusting the length of the plunger stroke; if necessary, this could be done while the pump was operating. Each pump is driven by a 1½ horsepower motor that is capable of pumping against 130 psi, and the maximum output capacity is 132 gallons per hour.

Two types of mixing devices were used to blend the grout components immediately prior to entering the soil mass. The first was a 2½ inch diameter, 7-inch long pipe with 1-inch diameter inlet and outlet connections; enclosed in the pipe was a randomly coiled mass of wire to enhance mixing by inducing turbulent flow. The second device was simply a magnetic stirrer inside a 1½-inch diameter, 3-inch long plexiglas chamber. A few tests with the first device indicated that the degree of mixing actually achieved was very small, and the magnetic stirrer was used in the majority of the tests in this experimental program; this stirrer was very effective and no separation between the reagent and the sodium silicate was observed in the mixed grout.

A medium-scale grouting plant, depicted in Figure 1, was assembled in the laboratory to undertake this test program. Three elevated tanks were used to store the sodium silicate (Tank 1), reagent (Tank 2), and water (which was used only to flush the system after each grouting process was finished). The reagent mix for a given test (gel time) was proportioned and mixed in a large container and pumped into Tank 2 by means of a small pump. Then, this reagent mix was combined with an equal amount of sodium silicate from Tank 1 to form the grout. During the grouting process, valves 1, 3, and 4 were open and valves 2 and 5 were closed; during the flushing process valves 1 and 3 were closed and the others were open. The storage tanks were connected to the pumps by 2-inch diameter steel pipe. Pump A was used only for sodium silicate, and pump B was used only for the reagent mix.

At the outflow end of the pumps the piping was reduced to 1-inch diameter, and a 25-psi Swagelok back-pressure valve was installed in the discharge line near each pump (a) to insure that a sufficient discharge pressure was maintained for proper pump metering action and (b) to minimize the tendency for siphoning due to reduction of the discharge piping. The discharge pressure for each pump was monitored with standard Marsh pressure gages.

Valves 6 and 7 are three-way valves that enable flow to be directed to the mixing chamber after each individual pump has been calibrated for the design flow rate. Joining the mixing chamber and valve 8 was either of two different sizes of tubing; the linear model had a 0.375-inch diameter, 40-inch long flexible tube, whereas both the two-dimensional and three-dimensional models had a 0.75-inch diameter, 26-inch long flexible tube. Valve 8 is a three-way valve that allows the grout to be sampled at the injection point.

In some cases two chambers with different-colored reagent mixes were connected to the pipe which supplied the reagent mix to pump B. The reagent mix in these chambers consisted of the same constituents, except for the addition of a coloring agent, as that in the main supply tank. During a particular test a colored reagent mix was forced into the supply line by use of a small overpressure (about 3 psi) and manipultation of the appropriate valves.

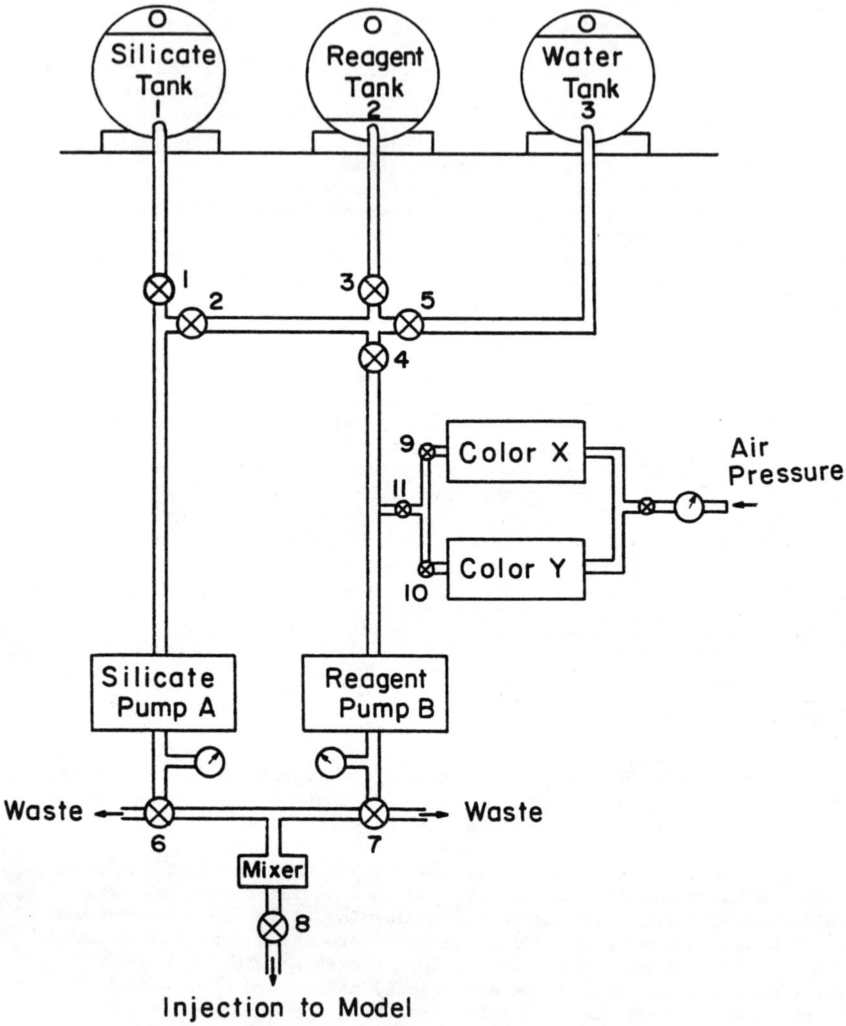

Figure 1. Schematic Diagram of Grouting Plant

Prior to starting a test, valves 1 and 3 were closed. However, once the grouting process began, valves 2 and 5 were closed and valves 1 and 3 were opened. Regardless of whether colored reagent was to be used at some point during the test, valves 9, 10, and 11 were closed at this time. Three-way valves 6 and 7 were adjusted to direct the flow of each component (sodium silicate and reagent mix) to graduated cylinders; this "wasted" grout was used to ascertain system stability and flow rate prior to initiating a test. The stroke length of each pump was adjusted, if necessary, to produce the design flow rate. Once the proper flow rates were achieved, three-way valves 6 and 7 were adjusted to direct the flow into the mixing chamber and the tubing between the mixing chamber and the three-way valve 8. At this point in the test valve 8 was adjusted to waste the grout until system stability was verified, at which point valve 8 was changed to direct the flow of grout into the model and the test was started.

During the process of injection, pressure gages located at various points in each model were read every few minutes, and grout samples were collected from different sampling ports along the model. Prior to collecting each sample (about 30 cc in volume), the sampling port was cleansed of any grout that entered during the previous sampling operation. Each sample was allowed to stand in an at-rest condition, and the gel time was recorded. The advance of the grout front was visible only in the tests on dry sand. When the colored reagent mixes were introduced, valves 9 and 11 or valves 10 and 11 were opened, as appropriate. The grouting process was terminated when (a) either of the pump pressures reached 130 psi, (b) the grout components in the supply tanks were depleted, or (c) the test was determined to have progressed sufficiently long to yield the information desired.

LINEAR MODEL

The primary purpose of the test series involving a linear model is to investigate the flow of grout through a soil for the case where the rate of advance of the grout front is linearly related to the flow. This was accomplished by conducting one pilot test (flow direction was horizontal) and a series of four other tests (flow direction was vertical), two of which were performed on dry sand and two on water-saturated sand. The pilot test served only to provide guidance for conducting the principal series of tests, and no data are reported for this test.

Description of Model

The sand was placed in a plexiglas tube 6½ inches in diameter and 72 inches long. Pairs of pressure gages and sampling ports 90° apart were located at 6, 18, 30, 39, and 60 inches from the injection point. All pressure gages were protected from direct contact with the grout to avoid corrosion and clogging. A screen was placed inside the sampling port connection to prevent the piping of sand during the process of sampling. End caps were held in place with four tie bars, and the first 18 inches of the tube was reinforced around the circumference with aluminum mesh to withstand the high pressures expected near the injection end.

Prior to placing the sand in the tube, a thin layer of petroleum jelly was applied to the inside wall to inhibit channelized flow along the sand-tube interface. At the bottom of each column was a bed of filter gravel 3 inches thick. After the gravel was leveled, a 20-30 mesh screen was placed over it to prevent migration of small sand particles into the gravel while the column was filled and compacted. A funnel and long hose with an inside diameter of 0.75 inches was used to place the sand in the column; this procedure was used to minimize the separation of fine and coarse sand particles that would otherwise occur when sand is poured from heights of 5 or 6 feet. The funnel was raised slowly as the sand was deposited in a cone-like fashion, and the soil surface was gently leveled and lightly mixed to enhance a more uniform particle distribution. Ten blows of a small wooden hammer were then applied to the wall of the column at each of three points spaced 120° apart. Successive layers of sand were placed in this manner, and the average density of each layer was determined until the column was almost filled. The top layer consisted of a 3-inch deep bed of gravel. Once the top cap was positioned, the end caps were held together longitudinally with four tie bars. To saturate the sand tap water was introduced into the bottom of the column under approximately two feet of head; saturation was achieved in 2 to 3 hours.

Tests on Dry Sand

Two tests were performed on dry sand to eliminate any effects of dilution on the gel time of the grout. In each case the tube was filled with compacted sand at an average dry density of 122.8 lb/ft^3. The grout needed to fill the voids was calculated to be about 2.5 gallons, and a flow rate of 6.6 gallons per hour was selected to fill the voids in about 25 minutes. For the first test (Test L-1) the grout mix unfortunately had an extremely short gel time of about 30 seconds, instead of the intended 5 minutes; this was attributed to a large change in temperature. Nevertheless, pumping was continued and the grout front advanced about 7 inches in the first 3 minutes and then about 2 more inches in the next 2 minutes. When the injection pressure reached 125 psi after about 5 minutes, the pumps were stopped and the test was terminated. Although this test did not proceed as planned, it is significant that pumping continued for a period about 10 times longer than the gel time.

The gel time of the grout used in the second test on dry sand (Test L-2) was about 9 or 10 minutes. Although the chosen flow rate of 6.6 gallons per hour should have filled the voids in about 25 minutes, the actual advance of the grout front was somewhat faster. Once pumping started, grout samples taken at the injection point exhibited gel times consistent with the design gel time. In general, grout samples taken just as the grout front passed a given sampling port had a slightly longer gel time than those samples taken later at the same port; also, the gel times were longer than those of grout samples taken before the grout entered the soil (counting the time required for the grout to move from the injection point to the sampling point). This behavior of the gel time might be due to the adsorption of some ethyl acetate by the sand particles, which will increase the gel time of the grout; after this adsorption is complete, more of the ethyl acetate will serve to enhance the reaction and the gel time will decrease to its design value. The last important observation from this test is that, although the gel time of the grout was about 9 minutes, pumping was able to continue for 34 minutes. In general, the grout samples from all ports gelled about 7 minutes after they were taken, as long as pumping was continued; however, once pumping was stopped, the grout gelled quickly.

Tests on Saturated Sand

Two tests were performed on saturated sand; in one case (Test L-3) the grout had a gel time of 10 minutes and in the other case (Test L-4) it had a gel time of about 5 or 6 minutes. Typical results for Test L-4 are shown in Figure 2. The dry density of the sand in this test was 126.3 lb/ft^3, and a pumping rate of 7.3 gallons per hour was chosen to fill the sand voids in about 22 minutes.

During the injection phase grout samples were taken as shown in Figure 2a, and the gel time of each was recorded. During the injection process red color was introduced after 20 minutes of pumping, and the dots next to each gel time indicate that the color was present in the grout sample taken from that particular sampling port at that particular time. If a line were drawn through the points indicating the first arrival of the color at each sampling port, it would nearly parallel that which represents the advance of the grout front. Figure 2b shows the time-dependent nature of the pressure distribution along the length of the tube, as well as at the pump and injection point. While pumping at a constant rate, a pattern of gradually increasing pressure with time was measured at the injection point, as illustrated in Figure 2c. Pumping was terminated when the pressure at the pump reached 135 psi. Although the gel time was about 6 minutes, pumping continued for 43 minutes, which is about 7 times longer than the design gel time.

Interpretation of Results

The experimental results from the tests on the linear model indicate that silicate-base grout can be injected as long as pumping is continued, regardless of the gel time of the grout (as long as it doesn't gel prior to being injected). Apparently the pumping action, along with the permeation of the grout through the pores of the soil, produces a continuous agitation that prevents gelation. In general, the gel time of grout samples taken from various points along the length of the column was somewhat shorter than the design gel time, not counting the time required for the injected grout to permeate to the sampling port.

The pressure distribution along the injection tube was observed to decrease almost linearly between the injection point and the point to which the grout front had advanced at some particular time. The injection pressure was observed to increase gradually in all tests, except the test with a 30-second gel time. The increase in injection pressure during pumping is due to (a) the fact that, while the sodium silicate permeates and coats the sand particles and

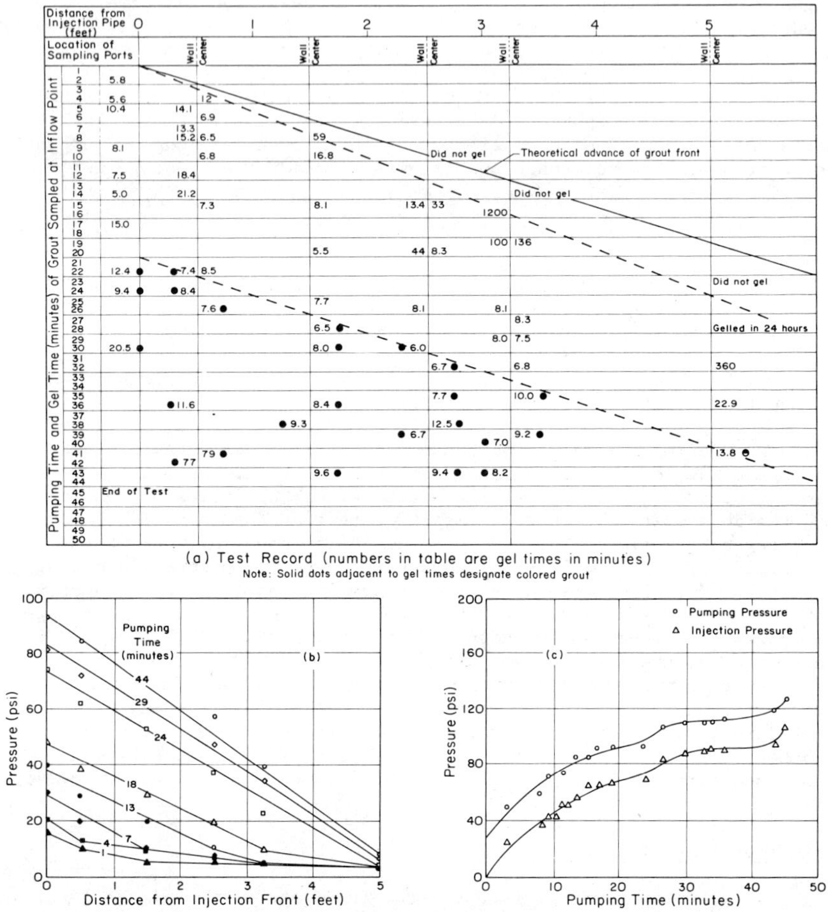

Figure 2. Results from Typical Test on Linear Model

replaces the water, it coats the sand particles with a thin film that causes a decrease in void volume and (b) the time-dependent change in the viscosity of the grout. No significant difference was noted in the pressure distributions in the tests on saturated and dry sand.

The water in the voids of a saturated sand apparently causes the grout in the vicinity of the front to be diluted. Samples taken a few minutes after the grout front had passed a given sampling port yielded a gel time that was higher than the design gel time. However, once grout replaced all of the water in the sand voids, the gel time of the grout stabilized to a value very close to the design gel time. This initial increase and subsequent decrease in gel time might also be due in part to the adsorption of some ethyl acetate by the sand particles; once this adsorption was complete, the gel time of the grout decreased to the design value or less.

CYLINDRICAL MODEL

The primary aim of the test series involving the cylindrical model was to investigate the flow of grout through a soil for the case where the rate of advance of the grout front is proportional to the inverse of the instantaneous radial distance to the grout front. Special attention was directed toward investigating the variation, if any, of the strength and modulus of the grouted sand with radial distance from the line of injection; this was accomplished by trimming specimens from the grouted mass and testing them in unconfined compression. Included in this test series is one pilot test and three other tests on saturated sand confined in a segment of a cylinder. Since these tests were performed prior to the tests on both the linear model and the spherical model, they served the auxiliary purpose of providing an appreciation for the problems involved in preparing and controlling grout mixes designed to have a specific gel time; as a consequence, adequate provisions were not taken in this test series to exercise definitive control over the monitoring of grout gel times, as was done for the tests involving the other two models.

Physical Model

The physical model consisted of a $22\frac{1}{2}^{\circ}$ wedge-shaped section of a radial flow configuration; the thickness was 2 feet and the length was 5 feet. The radius was chosen to simulate the range of grout permeation from a typical grout pipe. The model was designed so that both sides and the end could be separated from the top and bottom after the grouting process was completed. Pressure gages were installed along one of the radial boundaries at distances of 1, 2, 3, 4, 6, 9, 12, 18, 24, 30, 36, 42, 48, and 52 inches from the line of injection; all pressure measurements were taken along the inside surfaces because data from the pilot test indicated no difference in the readings taken at the interface and in the interior of the sand mass. A porous stone was located at the end opposite the injection pipe to serve as a reservoir and allow a constant hydraulic head to be maintained at this boundary; it also provided drainage during the grout injection process. The grout injection pipe (0.75 inches in diameter) was positioned so that its axis coincided with the axis of revolution of the cylindrical sector. Thirteen 0.25-inch diameter holes were spaced 1.5 inches center-to-center along the length of the injection pipe to simulate a line source for the grout; the first hole was located 0.75 inches from the bottom, and the top hole was located 7 inches below the sand surface to reduce the potential for soil heave during injection.

Test Program

Sand was compacted in 2-inch lifts; after each lift the surface was gently leveled and lightly mixed to insure uniform particle distribution. The densification of each layer was produced by tamping it with about 75 evenly distributed blows of a standard-Proctor hammer (2-inch diameter face; 5.5 lb weight) dropping one foot. The compacted sand in all cases had an average dry density of about 120 lb/ft^3. In the pilot test and the first test of the basic series (Test C-1) a layer of styrofoam was placed on the sand surface and compressed slightly as the cover was installed; this proved to be unsatisfactory in Test C-1 (although it worked in the pilot test) because the chemical action of the grout dissolved the styrofoam. This may have been due to the fact that the grout used in Test C-1 (as well as Tests C-2 and C-3) was mixed immediately prior to injection, whereas the grout used in the pilot test was batch-mixed, and the ethyl acetate in the fresher grout was apparently more aggressive to the styrofoam. In Tests C-2 and C-3 the surface of the sand was covered with a 0.75-inch layer of bentonite slurry (consisting of 12% bentonite and 88% water) prior to placing the cover. Saturation was achieved in about 5 to 6 hours by introducing tap water through the porous stone under a small head.

After the model was filled with sand and saturated with water, a permeability test with water was conducted and the pressure was measured as a function of the distance from the injection pipe. The resulting pressure distributions agree with those calculated from classical theory based on radial flow from a well in a confined aquifer and the permeability values agree with those obtained from both the linear model and constant head permeability tests.

After grouting was completed, the wedge-shaped mass was allowed to cure for 5 days before the model was disassembled; then, the grouted mass was sectioned with a chain saw and specimens were trimmed from different locations. All specimens were capped and tested on a stress-controlled Southwark Emery press with a capacity of 60 tons. Loading was applied at a rate of 100 pounds per minute until failure, and axial deformations were measured by a strain gage with an indicator dial sensitivity of 0.001 inches.

Test C-2 was conducted with a grout mix having a gel time of about 24 minutes. The grout needed to fill the voids was calculated to be about 18 gallons, and a flow rate of 24 gallons per hour was selected to fill the voids in about 45 minutes (approximately twice the gel time of the grout). During the grout injection, time-dependent variations in the pressure distributions with distance from the injection pipe were measured every 2 minutes, and the resulting distributions are illustrated in Figure 3. Two observations in the pressure distributions are worthy of note. The first is that the pressure in the first 9 inches

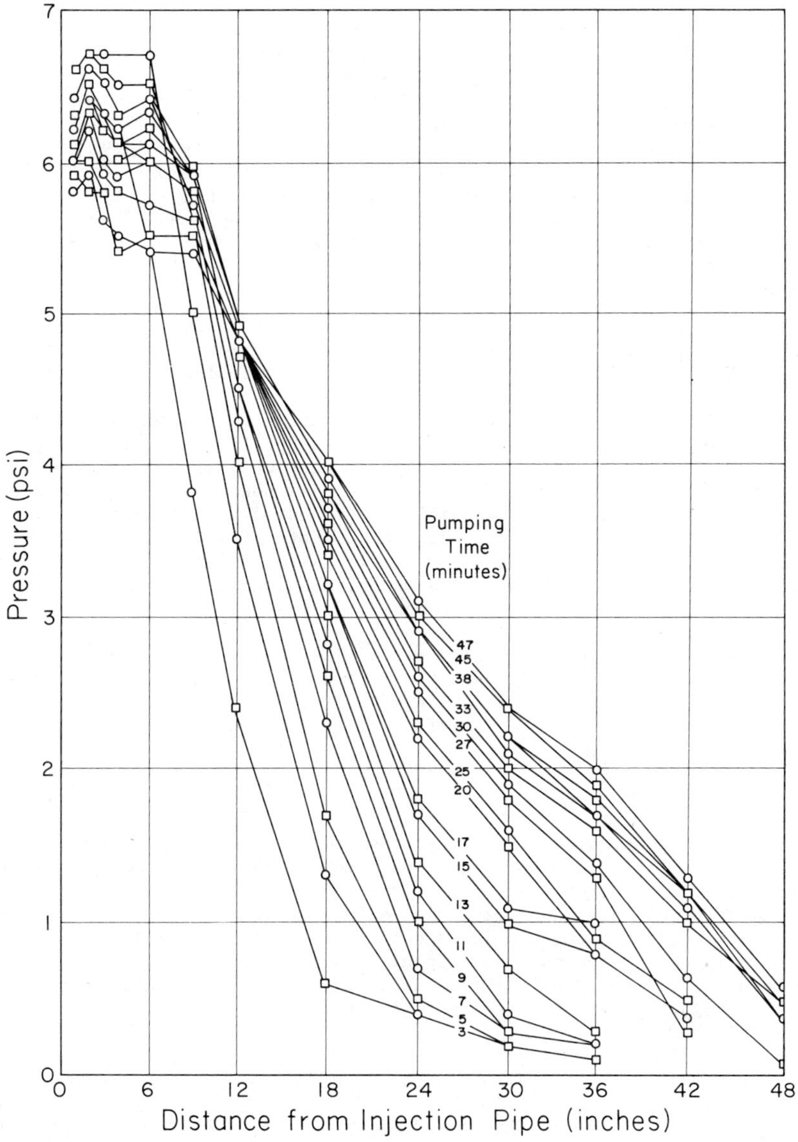

Figure 3. Results from Typical Test on Cylindrical Model

was more or less constant at any given time and dropped about 1 psi during the test; after disassembling the model, a large void with emanating fractures was discovered near the injection pipe, and this was clearly responsible for the observed condition. The second observation is that, in the region from 9 inches to the end of the model, the pressures increased with time, reaching a nearly steady-state condition after about 39 minutes of pumping. This time-dependent pressure increase is attributed to the increase in the viscosity of the grout in the sand voids. Based on the flow rate chosen and pumping time, the grout should have filled all of the voids in the sand and it apparently did. It is important to note that, although the gel time of the grout was about 24 minutes, pumping continued for 47 minutes (about twice as long as the grout gel time).

Test C-3 was conducted with a grout mix having a gel time of about 10 minutes. Once pumping was started, a gradual decrease of about 2 psi was noted in the injection pressure during the first 10 minutes, and this was followed by an extremely slow increase of about 0.4 psi during the following 20 minutes. After 30 minutes of pumping, a red coloring agent was introduced, and the color appeared in the effluent 3 minutes later; this indicated that channelized flow was occurring and the test was terminated. A few hours after the test was complete, the model was disassembled to reveal a vein of pure grout having a width of about 1.5 inches and a depth of about 0.25 inches near the injection pipe, becoming smaller with distance from the injection pipe. The grout front had apparently advanced about 1.5 feet from the injection pipe before the channelized flow developed. However, once the channel was formed, virtually all of the grout flowed through the channel rather than the sand mass. The results of this test suggest that some gelling had occurred in the sand mass, thereby increasing the flow resistance and causing the fresh grout to follow a preferred flow path along the interface of the sand with the face of the model.

Interpretation of Results

The experimental results from this series of tests indicate that a "break-through" phenomenon may occur during the process of grout injection, but the evidence is not clear. This is due in large part to the development of channelized flow along the interface boundaries between the sand and the model, especially for grouts with a short gel time. In many respects this channelization and the "break-through" mechanism are caused by similar factors, but the presence of one can not necessarily be taken as proof of the existence of the other.

Figure 4a summarizes the strength results for all of the specimens tested and indicates that there is a decrease in the strength of the grouted material with increasing distance from the injection pipe. One possible reason for this trend might be that the grouted sand near the injection pipe has retained more of the sodium silicate as films coating the sand particles, thereby effectively increasing the concentration of silicate in this region and decreasing it at points farther away. A second possible explanation is that the water in the soil voids dilutes the grout mix; this dilution would be more extensive closer to the grout front and the strength of the grouted material

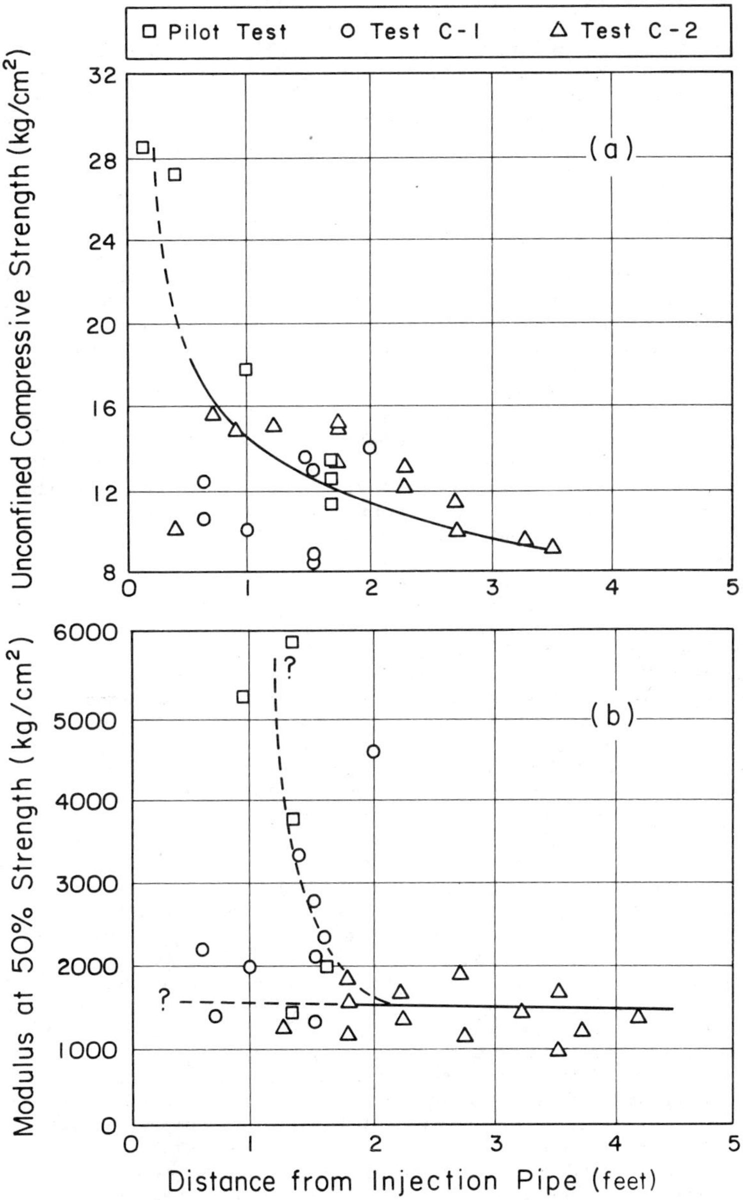

Figure 4. Variation of Strength and Modulus with Distance from Injection Pipe

would be correspondingly lower. Figure 4b suggests that the modulus also tends to decrease with distance from the injection pipe, but there is substantial scatter in the data and a definitive relationship is difficult to identify. Preliminary pumping tests (prior to injection of grout) with various mixtures of water and sodium silicate (no reagent added) indicated that the pressure decreased logarithmically with distance from the injection pipe once a steady-state condition had been reached.

SPHERICAL MODEL

The primary aim of the test series (consisting of Test S-1 on dry sand and Test S-2 on saturated sand) involving the spherical model is to investigate the flow of grout through a soil for the case where the grout front advances at a rate that is directly proportional to the reciprocal of the square of the instantaneous radial distance to the grout front. The physical model consists basically of a two-foot cubical box that was designed to allow the top and all four sides to be readily removed after the grouting process was completed. The grout was injected into the center of the sand mass through four 0.25-inch diameter holes evenly spaced in a horizontal plane at the end of a 0.5-inch inside diameter pipe. Sand was glued on the outside surface of the embedded section of the injection pipe to minimize the possibility of channelization along this boundary. The pressure was measured only at the injection point.

Sand was compacted in three-inch lifts; after each lift the soil surface was leveled and scarified to enhance uniform particle distribution and proper mixing with the subsequent layer. Densification of each layer was accomplished by tamping it with about 35 evenly distributed blows of a Standard Proctor hammer (2-inch diameter face; 5.5-lb weight) dropping one foot. After the model was one-half filled with sand, the injection pipe was positioned in the center and the process of placing and compacting the sand was continued until the model was filled, after which the cover was placed on the model. In both tests the model was filled with compacted sand at an average density of about 116 lb/ft^3.

Saturation (in the case of Test S-2) was achieved in about 4 to 5 hours by introducing tap water from four 0.25-inch diameter tubes placed at each corner of the model. After saturation was completed, grout was pumped at a flow rate of 6 gallons per hour to form approximately a 2-foot diameter grouted mass in about an hour. The injection pressure was recorded as a function of time, and grout samples were taken every two minutes during the injection process and checked for gel time. The reagent tank (Figure 1) in this operation was replaced with three independently valved 3-gallon reagent tanks, each containing a different color reagent. These new tanks were located at a lower level and pressurized to 6 psi to compensate for the head difference achieved by the elevation of the original tank; this pressurization is required to insure an accurate flow rate to the pump. After grouting was completed, the more-or-less spherical grouted mass was cut into halves and the location of various colors of grout was observed. Finally, 1.4-inch diameter specimens were trimmed from the grouted mass at locations relatively close to the injection pipe and near the outermost boundary of the grout bulb.

Test on Dry Sand

Test S-1 was performed on dry sand to eliminate any effect of dilution on the gel time of the grout. Once pumping was initiated, the injection pressure, as shown in Figure 5, increased slowly to about 5.6 psi in the first 12 minutes; however, this was followed by a rather abrupt decrease to about 5.2 psi. Since the gel time for the grout was about 7½ minutes (based on numerous samples taken during the injection process), this drop (although admittedly very small) suggests that fresh grout may have "broken through" a gelled front. However, the sharp increase shortly thereafter tends to discredit this possibility. Subsequently, the injection pressure decreased gradually during the next 8 minutes, reaching a plateau of 3.5 psi for the following 14 minutes. This drop in pressure (2.5 psi) strongly suggests that new channels have been formed in the gelled grout front. Next, a slight increase of about 0.3 psi was observed, and this could be due to a build-up in the viscosity of the grout or the formation of gelled grout at the grout front.

Test S-1 was terminated after 56 minutes of pumping. Approximately 3 hours after grout injection, the model was disassembled and the loose (ungrouted) sand surrounding the grouted mass was removed. The resulting grouted mass was an irregular sphere with a very rough surface. A close observation of the external surface (Figure 5) suggests that the grout front did not move as a progressively increasing spherical surface, but rather the grout appeared to move sequentially through the easiest path, forming small bulbs in different areas and a resulting irregular surface. Based on the flow rate chosen and the pumping time, the grout should have filled all of the voids in the sand to form a sphere about 22 inches in diameter, and this was indeed found to be approximately the average diameter of the grouted mass.

Subsequently the grouted mass was cut into halves and the color sequence was observed; the resulting pattern is sketched in Figure 5. The sequence of the three different colored grout solutions that were used verifies that, as the pumping began, the grout moved radially (spherically) from the openings in the injection pipe, and the fresher grout apparently displaced the older grout farther from the injection pipe. This mechanism is believed to prevail because the first color injected was observed at the outer boundary of the sphere. It also appears that some of the grout that was injected earlier in the pumping process adhered to the sand particles and allowed the fresher grout to pass; evidence of this situation is primarily the presence of a combination of mixed colors in certain areas shown in Figure 5.

When the spherical mass was cut along other planes, it was observed that a thick (0.7 inches) vein of fresher grout was present in the older grout. However, since there was no significant increase in the injection pressures (less than 0.3 psi), it is believed that the fresher grout developed this vein before the older grout had gelled. Once this vein had developed, the older grout surrounding it had a

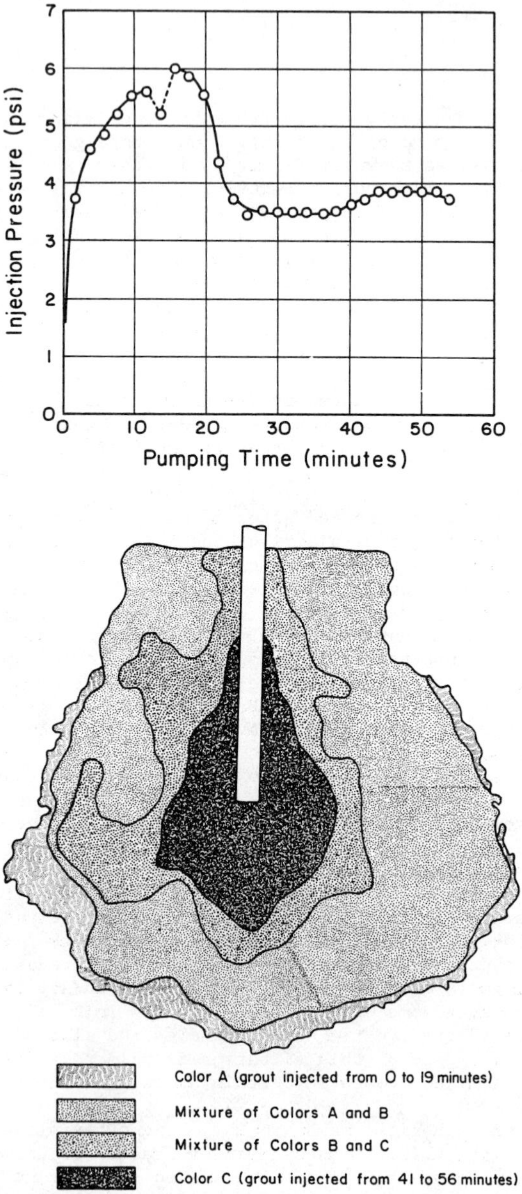

Figure 5. Results from Test on Dry Sand in Spherical Model

chance to gel. It is significant to note that the observed veins consisted only of the grout which was introduced during the last 16 minutes (40 minutes after the onset of pumping); by this time pumping had continued for at least 4.5 times longer than the grout gel time.

Test on Saturated Sand

Test S-2 was performed in a similar manner on saturated sand using a grout mix with a design gel time of about 12 to 18 minutes. As illustrated in Figure 6, the injection pressure remained essentially constant during the first 30 minutes. Based on samples taken during the first 26 minutes of pumping, the actual average gel time was about 22 minutes. Although some degree of gelation may have occurred, the injection pressure showed no signs of a "break-through".

This test was terminated after 60 minutes of pumping. Approximately 24 hours after the test was terminated, the model was disassembled and the ungrouted sand surrounding the grouted mass was removed; the resulting grouted mass was a relatively smooth sphere. The pressure of pure water apparently contributes toward the formation of a smooth relatively perfect sphere by lubricating the sand particles and allowing more uniform flow. Based on the pumping rate used and the duration of pumping, a 2-foot diameter sphere should have been formed, and this was essentially the observed result. Subsequently, the sphere was cut into halves and the color sequence is shown in Figure 6; based on the sequence of the colors injected, it appears that the newer grout displaced the older grout radially because the grout that was injected first is found at the outer boundary of the sphere. Cutting the sphere through other planes did not reveal any clear channelization or veins, such as found in Test S-1.

Interpretation of Results

Based on the experimental results from these two tests with the spherical model, it appears that, although pure grout samples will gel if left unmixed, the process of flow through a soil, caused by continuous pumping, prevents gelation. Perhaps the grout in the voids of the soil does increase in viscosity, but it apparently does not gel sufficiently to develop a "hardened outer shell" at the grout front. Tests showed that grout can be pumped as long as flow is continued, regardless of the gel time of the grout. Furthermore, during the early stages of injection the flow of grout through the soil mass seems to be consistent with classical flow concepts whereby the grout flows radially (spherically) from the opening in the injection pipes.

Since the grout front advances at a rate that is directly proportional to the reciprocal of the square of the instantaneous radial distance to the grout front when the pumping rate is held constant, it is believed that some of the older grout farther from the injection pipe develops a higher viscosity and adheres to the sand particles, thus allowing the fresher grout to pass. This mechanism will continue

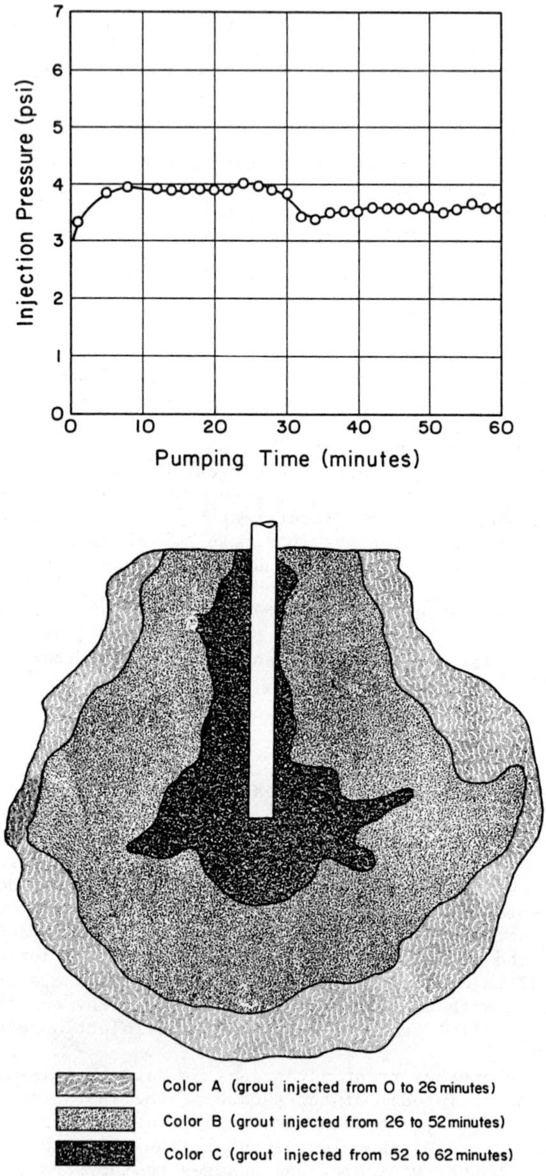

Figure 6. Results from Test on Saturated Sand in Spherical Model

to prevail until the grout front reaches a certain radial distance from the injection pipe, after which the fresher grout will develop veins. Such a mechanism is probably most applicable to grout injection into dry sand where the leading edge of the grout front has more chance to gel. When injecting grout into a saturated soil, the leading front will be continuously diluted and thereby hindered from gelling.

About 30 minutes after the test on the dry sand was terminated, it was observed that the surface of grouted mass had already gelled and exhibited considerable strength, whereas the surface of the grouted saturated sand was very soft and showed no signs of gelling for about 2 hours. The injection pressure during the grouting process in both tests showed no significant drop that could be attributed to signs of a "break-through" phenomenon; however, rather small gradual increases or decreases in injection pressure might be related to changes in viscosity or possibly the development of a sequence of preferential flow paths.

ANALYSIS AND ENGINEERING SIGNIFICANCE

All the foregoing results are synthesized and interpreted within the context of the limited available data on field behavior to deduce guidelines and observations of engineering significance regarding the injection and resulting benefits of chemical grouting in engineering practice.

Break-through Phenomenon

It has been observed in the field that silicate grout can usually be pumped in the ground for a period much longer than the design gel time, and, if the pumping rate is held constant, there is generally a build-up of pumping pressure. In addition, Karol (1968) observed in the laboratory that, when AM-9 grout was injected into a sand mass, some of the fresher grout that was injected later in the grouting operation flowed to regions beyond that already occupied by the older gelled grout that was injected earlier. This flow inversion indicated the existence of some type of "break-through" mechanism. If this is indeed the case, the rate of advance of the grout front will likely affect the response, because gelation must take place at the leading edge of the grout front in order to develop the potential for a fracture and the resultant "break-through" of the fresh grout as pumping continues. Accordingly, three different experimental models, each with a different geometry and associated relationship between the constant pumping rate and the rate of advance of the grout front, were constructed and tested to investigate the relationship among the injection variables and whether or not a flow inversion or "break-through" does indeed occur as the external shell of grout begins to gel during continuous pumping.

Results from the linear model, where the rate of advance of the grout front is directly proportional to the pumping rate, indicate that grout can be pumped for periods of time much longer than the gel time of the grout. Based on data obtained from grout samples taken from various points along the length of the column during grout injection into a dry soil, an initial increase and subsequent decrease in the gel time was noted; this is felt to be due in part to the initial adsorption of some ethyl acetate by the sand particles; once this adsorption was complete, the gel time of the grout decreased to approximately the design gel time or shorter. Grouting into a saturated medium showed that the water present in the soil voids apparently caused the grout front to be diluted. Samples taken a few minutes after the grout front had passed a given sampling port yielded a gel time that was higher than the design gel time. However, once the grout replaced all of the water in the sand voids, the gel time of the grout stabilized to a value very close to the design gel time or slightly shorter. A progressive flow mechanism of silicate grout in the linear model was observed; the newly injected grout pushed the older grout forward, and no signs of gelation or "break-through" were observed, although pumping was continued for at least seven times longer than the design gel time.

Results from the cylindrical model, in which the advance of the grout front is directly proportional to the inverse of the instantaneous radial distance to the grout front, suggests that a "break-through" phenomenon may occur during the process of grout injection, but the evidence is not clear. This is due in large part to the development of channelized flow along the interface boundaries between the sand and the model, especially for grout with a short gel time.

Results from the spherical model, in which the grout front advances at a rate that is directly proportional to the reciprocal of the square of the instantaneous radial distance to the grout front, suggest that the grout was displaced radially from the point of injection and the leading surface of the grouted mass formed a sphere. This was determined by the sequential injection of three different-colored grout solutions and then cutting open the grouted mass to ascertain the actual locations of the colored grouts. Observations indicate that there was a definite progressive flow of the grout in this model. The newly injected grout rather obviously displaced the older grout and pushed it to the outer boundary of the grouted sphere. Data from all three models show that pumping was never stopped due to the gelation of grout in the voids of the soil during the injection process, although rather short gel times were used in some cases. Pumping was continued in some tests for six times longer than the grout gel time without any sign of gelation in the soil voids.

Grout Flow Mechanism

Based on (a) the measured gel times of grout samples taken immediately prior to injection, (b) the relative locations of sequential injections of colored grout, and (c) data and observations obtained from the models tested, it is believed that a progressive flow mechanism governs the advance of the grout; namely, the newer grout displaces the older grout and gelation does not occur at the outer boundary of the grouted mass, even though the rate of advance may decrease as pumping is continued. However, as injection progresses in a given cylindrical or spherical situation, the rate of advance of the grout front and the associated agitation of the grout in the soil voids decrease as the radial distance of the grout front increases. This allows the grout to increase in viscosity, but it does not gel, and, due to this increase in viscosity, the newer grout will seek preferred flow channels through the more viscous older grout. Once these channels are formed, most of the newer grout will flow through them, allowing the older grout, (which is essentially not moving) to gel. These channels do not appear to start developing until the pumping time exceeds several times the grout gel time.

Injection Pressure and Pressure Distribution

The injection pressure is an important parameter in a typical grouting operation. In the United States one commonly quoted rule-of-thumb limits the injection pressure to 1 psi per foot of overburden depth; British practice suggests that 2 psi per foot of depth is more reasonable with possible changes based on experience at each site; and French practice does not specify any injection pressure-overburden depth relationship, but sets limits (based on field experience) on the flow rate. Baker, (1979, Personal Communication) has found that, based on injection of grout into the center of an uncovered cubic-meter sample (that is, the over-burden was about 1½ feet), an injection pressure of 12 to 16 psi did not cause heave. Although injection pressures up to 140 psi were measured inside the linear model, no indication of any hydraulic fractures were found in the specimens that were cut open.

The pressure distribution was measured as a function of the distance from the injection point in both the linear model and the cylindrical model. Based on results from the linear model, the pressure was found to decrease more-or-less linearly between the injection point and the point to which the grout front had advanced at some particular time. The injection pressure, as well as the pressure along the length of the model, increased during pumping; this increase is believed to be due to (a) the fact that the grout coats the sand particles with a thin film that causes a decrease in void volume and (b) the time-dependent change in the viscosity of the grout.

When several different silicate-water mixtures (no reagent added), each with a different viscosity, were used as permeants in some preliminary tests with the cylindrical model, the pressure was found to decrease logarithmically with distance from the injection pipe once a steady-state flow condition was reached. As the viscosity of the permeant increased, the head loss with distance from the injection pipe became more rapid. However, the limited data for grout injection are too inconsistent to either verify or refute the logarithmic nature of the pressure variation with distance. Furthermore, no pressure distribution data were measured in the tests with the spherical model.

Data from an instrumented field project (Baker, 1980, Personal Communication) showed that the piezometric pressure 1.25 feet from the injection pipe was much lower than the injection pressure (less than one-half at most and usually less than one-quarter). At 2.75 feet the greatest piezometric pressure was about one-quarter of the injection pressure. In the next few feet the pressure was nearly constant at about 10% to 15% of the injection pressure. Comparing field data with results from Test C-2 shows a general agreement in the pressure distribution as a function of distance from the injection pipe; in the laboratory test drops of 65% and 35% of the injection pressure were measured at distances of 1.25 feet and 2.75 feet from the injection pipe. A large part of the reduction in the field test probably occurs within the mortor jacket, which is not present in the laboratory tests.

Strength Distribution

Based on data from unconfined compression tests performed on grouted specimens trimmed from the cylindrical model at different distances from the injection pipe, the strength apparently decreases with increasing radius. This decrease is attributed to the dilution that takes place in the grout as the front advances and the incomplete replacement of the water in the soil voids. The greater strength near the injection pipe may be due to a higher concentration of silicate on the sand particles, because this region was subjected to many throughputs of grout as the front advanced. The modulus appears to decrease with distance from the injection pipe, but there is substantial scatter in the data and a definite relationship is difficult to identify.

Unconfined compression tests were performed on specimens (1.5 inches in diameter) trimmed from the grouted masses in the spherical model tests. These specimens were taken from two locations -- one about 1½ inches from the injection point and the other near the grout front about 8 or 10 inches from the injection point. These results showed an increase in strength with distance from the injection point; this is opposite to the trend found in the cylindrical model tests. The weaker strength of the samples near the injection pipe may be due to either a loosening of the soil near the injection pipe or the effect of the coloring agents on the strength of the grout (this effect was not studied, but it is not expected to be significant).

Based on data from unconfined compression tests on specimens of a field-grouted Maryland sand and tested after 1½ years (Spino, 1981, Personal Communication), some differences in strength were noted between samples trimmed horizontally and vertically; horizontally trimmed samples showed a slight decrease in strength with increasing distance from the injection pipe, but vertically trimmed samples showed an increase in strength. Although the results obtained in this study suggest that the strength decreases with increasing distance from the injection pipe, this was not the case for tests on field-grouted samples; however, variations in the results obtained from the tests on field specimens could be due to geological factors.

CONCLUSIONS

Based on the analysis and interpretation of the data obtained in this study, the following conclusions can be advanced:

1. Upon injection, the grout appears to flow progressively into the soil, with the fresher grout displacing the older grout toward the exterior of the grouted mass, even for grout with a short (5 to 10 minutes gel time.

2. Silicate grout can be pumped into a dry sand for at least five to six times longer than the design gel time before viscosity increases begin to be manifested; the apparent dilution that takes place at the advancing grout front in saturated soils increases the available pumping time even further.

3. When pumping grout with a short gel time into a soil mass for long periods of time (more than five or six times the gel time) the agitation as the grout flows through the soil voids will decrease as the radial distance of the grout front increases, and this allows the older grout to become viscous; when this occurs, the newer grout may develop prefereed flow channels through the region where more viscous grout is present, thereby allowing the older grout to gel while the newer grout is bypassing it.

4. No clear evidence of a "break-through" phenomenon was found, when interpreted in the sense that fresher grout breaks through an external shell formed by older grout that has gelled in place. This is in contrast to some evidence of preferred flow paths being sought by newer grout through zones of more viscous grout which subsequently gels in place.

5. The linear model showed that the injection pressure decreased almost linearly between the injection point and the point to which the grout front has advanced at some particular time.

6. The cylindrical model showed that the pressure decreased logarithmically with distance from the injection pipe for silicate-water mixtures, the head loss with distance being much more rapid for the more viscous permeants, but pressure data during grout injection are inconsistent.

7. General agreement in the pressure distribution as a function of distance from the injection pipe was noticed from results obtained from the cylindrical model and field data. Both results showed a sharp decrease (50% to 60%) in the injection pressure at 1.25 feet, and the pressures in the next few feet were about 10% to 15% of the injection pressure.

8. Time-dependent increases in injection pressure may be attributed to (a) coating the sand particles with sodium silicate and thereby decreasing the void volume and (b) changes in the viscosity of the grout.

9. Data from laboratory tests with a cylindrical flow geometry showed a very definite decrease in the unconfined compressive strength with increasing distance from the injection pipe; this can be explained by (a) increased dilution closer to the grout front and (b) a higher concentration of silicate closer to the injection pipe. In contrast, results from the spherical model and a field test showed a tendency for strength to increase with distance, but these data are not as conclusive as those from the laboratory cylindrical model.

10. The modulus is believed to decrease with distance from the injection pipe, but substantial scatter in the data precludes the establishment of a clear relationship.

REFERENCE

Karol, R. (1968), "Chemical Grouting Technology," Journal of the Soil Mechanics and Foundations Division, American Society of Civil Engineers, Volume 94, Number SM1, Paper Number 5748, pp. 175-204.

SEEPAGE CONTROL WITH CHEMICAL GROUT

R. H. Karol[1]
Member, ASCE

Abstract

Grouting often has, as its primary purpose, the replacement of the existing fluids in the formation voids by a different material (the grout) which will modify the formation properties. When grouting for seepage control, the formation voids must be substantially to completely filled, and the grout must be an impervious material with sufficient strength to resist extrusion by the existing or anticipated hydraulic head.

Seepage control is often remedial, rather than preventive, and as such is concerned with a small number of seepage channels. Field techniques involve the use of sophisticated equipment which can exploit grout properties of accurate gel time control and short gel times. The sequential steps in the technique are dictated by specific field conditions. Curtain grouting, by contrast, can be completely pre-planned.

Seepage Control With Chemical Grout

Although every grouting job in construction may be different and unique, most jobs have one or both of two purposes: 1) increasing the formation strength, and 2) modifying the flow of groundwater through the formation.

Increase in strength can be for the purpose of increasing bearing capacity, increasing resistance to soil failure along slopes or against retaining walls, reducing formation and surface settlements, and limiting movement of soil particles and masses. All of these purposes are accomplished by filling the formation openings partly or completely with a material which increases the resistance of the formation particles to relative movements, but which is not necessarily impervious. Modifying groundwater flow is accomplished by filling the formation openings with an impervious material, that has sufficient strength to stay in those openings when subjected to the normal local groundwater pressure.

[1]Professor of Civil Engineering
Rutgers University
New Brunswick, N. J.

Grouting techniques for increased strength differ from groundwater control grouting. The presentation which follows is related specifically to grouting for groundwater control.

Water problems during construction are generally associated with the inflow of sufficient water to retard or close down construction operations. While water inflow in small amounts (generally termed seepage) may be tolerable during a construction phase, it sometimes becomes intolerable during the operational phase of the completed structure. Seepage may also begin after construction is completed, because the elements of the structure modify ground water conditions, and/or faulty construction, or minor structural cracks caused by shrinkage or foundation settlements.

Water flow can also occur in large volumes, particularly at great depths below the surface, or when the flow zone is very large. In deep shafts, for example, single drill holes with flows of hundreds of gallons per minute are not uncommon, and in the vicinity of dams, exposed pervious strata can, if not treated, drain the entire reservoir.

Typical seepage problems include infiltration of water through fissures in rock (such as in tunnels, shafts and mine slopes or drifts), through cracks and construction joints in concrete, through pervious zones in stratified granular deposits, and through porous concrete and defective waterproofing membranes.

The term "seepage" is difficult to define quantitatively. The ASCE Glossary of Terms states "the flow of small quantities of water through soil, rock or concrete." This definition, of course, depends upon the interpretation of the word "small." Five gpm of water entering the bottom of a deep shaft is a small amount. The same quantity entering a domestic basement is a large amount. Seepage, then, is better defined in terms of the procedures used to eliminate it, rather than by job or quantity of water involved. The term "seepage" is generally understood to refer to an existing condition rather than an incipient one. Thus, seepage control may not require complete grouting of a formation. Seepage is generally associated with a relatively limited number of flow channels, sometimes only one, and is more often remedial than preventive.

Many construction techniques are in use for groundwater and seepage control. These include drainage, sump pumps, caulking, surface coatings, well pointing, compressed air and freezing, as well as grouting.

"Grout" is defined (ASCE Glossary) as "a material injected into a soil or rock formation to change the physical characteristics of the formation." Grouts can be divided into two categories, the suspended-solids grouts, such as cement and bentonite, and those materials which are true solutions, or chemical grouts. Physically, in order for a suspended-solids grout to penetrate a formation, the particles must be smaller than the void size. Practically, in order to penetrate a formation at reasonable pressures and rates (without blinding the formation), the maximum particle size must not exceed one-third to one-tenth the size of the voids. The penetrability of the chemical grouts, on the other hand, is limited primarily by the solution viscosity, with consideration of reasonable flow at practical pressures. The discussion which follows is applicable specifically to chemical grouts.

A wide variety of materials is available as shown in Table 1.

Table 1

Chemical Grouts

	Corrosivity or Toxicity	Viscosity	Strength
SILICATES			
Joosten Process	low	high	high
Siroc	medium	medium	medium-high
Silicate-Bicarbonate	low	medium	low
LIGNOSULFITES			
Terra Firma	high	medium	low
Blox-all	high	medium	low
PHENOPLASTS			
Terranier	medium	medium	low
Geoseal	medium	medium	low
AMINOPLASTS			
Herculox	medium	medium	high
Cyanaloc	medium	medium	high
ACRYLAMIDES			
AV-100	high	low	low
Rocagel BT	high	low	low
Nitto-SS	high	low	low
POLYACRYLAMIDE			
Injectite 80	low	high	low
ACRYLATE			
AC-400	low	low	low
POLYURETHANE			
CR-250	high	high	high
CR-260	(scheduled for marketing in 1982)		

The material most suitable to a particular job is dependent primarily on the site conditions. Although the costs of grouting materials varies widely, the cost per unit volume of grout in place shows much less cost variation. Where the cost of materials is a small portion of the total job cost (generally true in seepage problems) overall economy is best served by selecting the grout whose properties are most suitable, with little consideration of material costs.

Since "seepage" is defined in terms of low flow rates and volumes, small seepage channels are implied. For such conditions, low viscosity materials are to be preferred. For all conditions, accurate control of setting times is necessary, whether this is done by mechanical equipment or by special field procedures. Viscosity and gel-time control are the two most important factors in grout selection.

The importance of viscosity in a grouting process can be inferred from Poiseuille's Law, valid for the case of a liquid flowing slowly through a long tube of small diameter:

$$V = \frac{\pi p r^4 t}{8L\eta}, \text{ where}$$

V is the volume of liquid moving a length L in time t
t is the time for V
p is the pressure differential between tube ends
r is the tube diameter
L is the tube length
η is the viscosity of the liquid

In this equation all factors except "r" are to the first power. Thus, the following conclusions can be drawn, as quantitatively applicable to a grouting operation:

1. For any given pressure, the rate at which a formation will accept a chemical grout will vary inversely with the grout viscosity. (Viscosity is the major factor in determining the penetrability of a grout. The relative lubricity of the material has a small effect on penetrability, and surfactants may be added to improve penetrability to a small degree.)

2. Acceptance rate and pumping pressure are directly proportional.

3. Acceptance rate is directly proportional to the fourth power of the average void size (and therefore for the average grain size, for granular deposit comparisons).

These extensions are subject to the consideration that the initial data are referenced to a "liquid flowing slowly through a long tube of small diameter." None of these conditions may exist in an actual grouting operation. The conclusions, therefore, should be used qualitatively.

Pumping equipment for seepage control work must be reliable and accurate. Two or more variable volume pumps must be combined with other components to permit the operator to exercise volume control at constant (gel) setting times, and gel time control at constant volume, by mechanical means.

A wide variety of commercial equipment is available for placing holes in soft and hard materials. These range from small hand-held electric drills through trailer mounted drill rigs. Full details on drilling equipment are available from manufacturers and distributors.

When grouting through a drill hole it is necessary to seal the hole exposure to prevent loss of grout. In uncased holes through rock or concrete, packers are used for this purpose. Mechanical and pneumatic packers are commercially available in EX rod and larger.

Double packers, generally known as pressure testers, can be used for isolating any desired length of drill hole for more precise grout placement, or for checking the results of grouting.

Surface packers are specially made devices useful in treating cracks and construction joints in concrete. They consist of long plates or channels with sealed ends, with an "O" ring or sponge rubber gasket around the periphery. In use, they are located directly over the crack or joint, and held in place by bracing from an opposite wall. Grout can thus be pumped into the source of seepage.

The procedures currently in use for controlling seepage by chemical grouting were developed several decades ago by field and laboratory experimentation. Much of the early field work was done in Canadian mines.

Deep mines often have severe seepage problems. Typically, many thousands of feet of drifts and raises are required to provide access from the main shafts to the ore bodies. The drifts will intersect numerous minor fault zones, as well as major ones. Some of the cracks and fissures may conduct surface water from rivers, lakes and precipitation into the drifts. Water entering the drifts must be collected and pumped out of the mine. As the ore zones are developed, total seepage increases. Eventually, it becomes more economical to shut off the seepage rather than endure inefficient working conditions and pay continuously increasing pumping costs.

Figure 1 is a geologic map of a portion of the 250 level in a Canadian mine, showing the cracks and faults as plotted by the mine geologists. A number of grout holes were drilled into fault zones identified on the geologic map. It was anticipated that grouting through these holes would fill the fault zones with grout and thus eliminate seepage by blocking the supply of water. In the mine, grouting through these holes proved totally ineffective in reducing the seepage. This unsuccessful field work made it evident that the procedures for handling seepage problems are only as effective as the ability to define and locate the flow channels. It also demonstrated that in many cases the complexity of seepage through a fissured rock mass will virtually preclude the effective pre-planning of a grouting operation.

New field techniques were needed in order to be able to control seepage efficiently and economically, in mines and in other applications. The development of these techniques hinged upon the availability of grouts with special properties--particularly low viscosity, excellent gel time control and the ability to set while in motion.

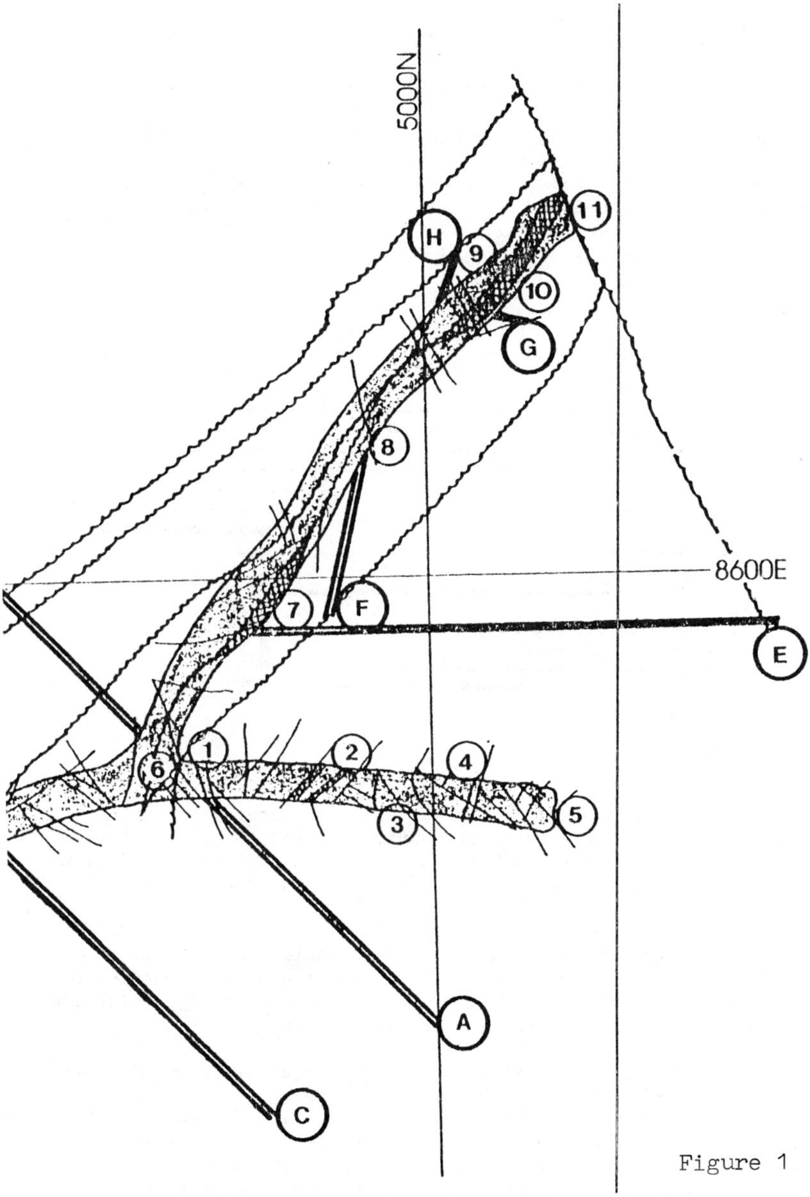

Figure 1

Commercial products with such properties were unavailable until the early 1950's. Thus, as is normally the case, the technology and the materials evolved almost simultaneously, and the total process could not have evolved sooner.

Following the initial unsuccessful mine grouting, a series of laboratory experiments were set up to simulate field seepage conditions. Lucite tubing was used to represent seepage channels and drill holes. In the model shown in Figure 2, BCDE is 3/8" ID, AC and FG are 3/16" ID, lines 1, 5 and 6 are 1/8" ID, and the rest are 1/16"ID. ACDE represents a fault zone, BD represents a drill hole. Lines 1 through 13 represent seepage channels leading from the fault zone to the mine drift.

Figure 2.

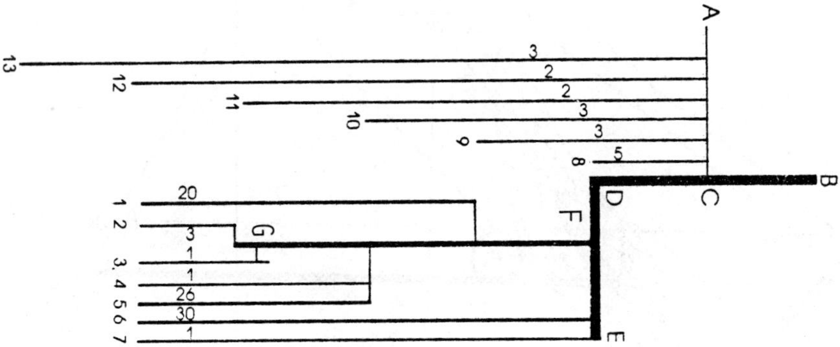

In one experiment, point B was shut and water was pumped through point A at a constant rate until equilibrium was reached. When the volume pumped through point A was changed, and equilibrium again established, the percent of total flow going through each line varied as shown in Table 2.

Other experiments corroborated the conclusion that can be drawn from this one, that percent flow was not a direct function of total flow, and that the variation was much higher in pipes with low percentage flows, than in those with high percentages.

In other series of experiments, dyed water was pumped in at both points A and B to study the paths and mixing of grout and groundwater. After equilibrium had been established with a static head of red-colored water at A causing seepage in each of the 13 pipes, blue-colored water was pumped in at B. At a pumping pressure less than the static head at A, no flow from B occurred. As the pressure at B was increased to a value above that at A, flow

Table 2

2 Minute Flow Through Model

Pipe No.	cc	%	cc	%
1	305	20-	169	20+
2	43	3-	22	3-
3	21	1+	8	1-
4	15	1	9	1+
5	395	26-	232	28+
6	455	30+	210	26-
7	12	1-	5	1-
8	78	5	44	5+
9	48	3	32	4-
10	46	3	37	4+
11	35	2+	20	2+
12	34	2+	7	1-
13	39	3-	33	4
TOTAL FLOW	1526		828	

in the BC direction started. Mixing occurred at point C, with all the fluid from B (blue) moving in the CD direction. As the flow from B increased, the color in line CD became predominantly blue and equilibrium could be established with lines 1 through 7 flowing blue, and lines 8 through 13 flowing red. This condition persisted over a considerable pressure increase at B. Only when the volume entering at B became very large compared to that at A, did any of the blue fluid begin to enter the pipes 8 through 13. It was found by experimentation that if the larger openings such as 1, 5 and 6 were plugged, much less flow was required at B to cause blue fluid to move into pipes 8 through 13.

A number of different models were used in the experiments partially described above. After working with any given model under varying conditions, sufficient data were available so that the seepage system could be sealed with a pre-planned procedure. The required data for any system could be summarized by the following descriptions:

1) the magnitude of pumping pressure to fill all the seepage channels, with groundwater dilution held to negligible proportions.

2) the time lapse from the start of pumping until the pumped fluid reaches the end of each seepage channel.

3) the volume of pumped fluid required to fill all the channels at the necessary pumping pressure.

In the field, such data can only be obtained by a pumping test. It cannot be obtained by visual examination even for very simple seepage systems, because the addition of external pumping pressures, which will be required for grouting, changes the characteristics of the seepage system.

Lab and field work both indicate that when pressure conditions within a seepage system are changed, the very small leaks are much more affected than the large ones. Thus it would be expected that a field pumping test will reveal leaks which were not flowing under normal static head conditions.

The laboratory studies showed that at some point within a seepage system, the external fluid pumped mixes with the internal fluid. In field work, this means grout would mix with groundwater. At small pumping volumes it should be expected that somewhere within the seepage system, grout will be diluted with groundwater to the extent that it will not gel. As the pumping volume is increased, the detrimental dilution will decrease. At either small or large pumping volumes, the grout tends to flow toward and into that portion of the seepage system farthest from the source of groundwater supply.

The laboratory studies led to development of procedures refined by field experimentation, which work well in stopping seepage in fractured rock, and which are also useful in treating fractured and porous concrete. The technique consists of first drilling a short hole that will intersect water bearing fissures and cracks. Holes are drilled in the simplest fashion, often by jack hammer. Dry holes are generally ineffective, and for this reason are abandoned. Wet holes (holes that strike water) are generally useful, and are tested as soon as completed.

The test should be done with the grout plant itself, in order to insure that the pumping pressure and volume used in the test can be duplicated when pumping grout. Both pressure and volume used in the test should be kept well below the pump capacity, so that unforseen conditions while pumping grout will not lead to early refusal because of pump limitations. If the grout to be used has a viscosity close to that of water, water can be used for the test. If the grout viscosity is significantly higher than water, additives should be used to raise the water viscosity. It is necessary to use additives which do not produce liquid lubricities much different from that of the grout.

The test consists of pumping water colored with dye (and if necessary, with additives to increase the viscosity) through a packer placed at the collar of the wet hole just drilled. Pumping pressure should be kept within 1/2 to 3/4 of the pump capacity, and remain relatively constant during the test. If the volume (pumping rate) is less than 1 gpm, it may be best to abandon the hole. Specific job experience dictates such decisions.

Dye concentration must be carefully controlled. Ideally, the concentration of dye should be such that if the dyed water is diluted with formation water to an extent that would preclude a similarly diluted grout from gelling, the dye should no longer be visually identifiable. Field use of dye concentrations which are identifiable when diluted more than 100% to 500% will identify seepage channels which often cannot be shut off from the hole tested (because of excessive grout dilution).

When the test begins, the adjacent wall area is carefully watched for evidence of dye. When dye is first seen at any point,the time lapse since the start of pumping is noted, as well as the pumping pressure and rate. Tests may be stopped when dye appears at one point, or may be continued until a number of different locations show dye. (Generally, the points where dye appears are in areas which are already wet or flowing. This is the assumed condition in the discussion which follows. If dye appears only in areas which were dry prior to the dye test, the hole should be abandoned.) For each point, time, pressure and rate are recorded.

Every time dye appears, this indicates that an open seepage channel exists between the packer and the point where dye appears. The exact location of the channel within the rock mass is not known, but if the drill hole being tested made water, then the established channel must reach into the water source somewhere. Therefore, the hole is worth grouting.

The time recorded for the appearance of dye is the maximum gel time that can be used to seal that particular channel. (If longer gel times are used, the grout will run out of the leak before it sets.) If the time is short, of the order of 5 or 10 seconds, this may be too short a gel time to handle readily. The effective time can be lengthened by lowering the pumping rate. This is a necessary step when attempting to seal a number of zones from one hole, and one of the zones has a very short return time.

Metering pumps are highly desirable for dealing with seepage problems, for both the dye tests and the actual grouting. Once a hole has been drilled, tested and it has been determined that grouting is in order, the pump suction lines may be switched directly from the dyed water tanks to the grout tanks. Dye may also be used in the grout, a different color than that used for dye testing.

When pumping begins, the pumping volume should be brought as quickly as possible to that used during the tests, and the grout itself should have a gel time of about three-quarters the previously recorded return time. The pumping pressure is monitored to make sure it doesn't exceed the allowable, but otherwise no attempt need be made at this stage to keep the pressure at dye test values. The leak is watched closely. It should begin to seal at about the recorded return time. If this does not happen, and dye (of the color used in the grout) does not appear at the leak, then dilution beyond the ability to gel has occurred. (This would normally mean too high a dye concentration was used in the dye tests.) If dye does appear, but the leak doesn't seal, then dilution of the grout has extended the gel time beyond the return time. This may be compensated for by shortening the gel time (easy to do with metering equipment, but very difficult with equal volume equipment) or by decreasing the pumping rate (easy to do with either kind of equipment).

As the leak begins to seal, the pumping pressure will rise, particularly if a single channel is being treated. If the rise is rapid or reaches high enough values, it will blow out the seal just made and re-open the leak. Therefore, it is important, as the leak begins to seal, to keep the pressure from rising by reducing the pumping rate. It is desirable at this time to continue pumping, and if possible to place additional grout, since the grout now being pumped is most probably going directly toward or into the source of the seepage.

Once gel begins forming in the seepage channel, the entire seepage net is altered, and all previously gathered seepage data may become totally unreliable. For this reason, holes should be drilled, tested and grouted one at a time. For holes which feed more than one leak, considerable change in return times will occur, once sealing of one leak begins. For such holes, sealing of additional leaks becomes a trial and error proposition, with a good chance of blowing out earlier seals, with the higher pressures which may be needed for other leaks.

The process described can be readily used to seal one leak or zone at a time, and if the total number of leaking zones is small, the technique is economical for complete seepage control. However, if the number of leaks are many, the method becomes very time consuming, and it becomes necessary to shut off leaks by shutting off the source of seepage. Grout pumped through each hole after all the external leaks have closed is often very effective for this purpose.

After a drill hole has been grouted as part of a seepage control program, it is possible for gel to be extruded from the drill hole, and water flow may begin again.

There are two extreme results at the grout hole itself: 1) complete shut-off of every leak feeding the hole, and 2) failure to shut off any leak feeding the hole. Under the first condition, gel would not be extruded from the hole. Under the second condition, gel would usually extrude from the hole. A fact which is generally overlooked is that at every degree of success varying between the extreme limits listed, gel will normally extrude. Thus, the extrusion of gel should not be considered a failure of the grouting operation, but rather an indication that the degree of water shut-off is less than 100%.

The pattern of fissures intercepted by a drill hole cannot be assessed accurately. Since the fissures will never be exactly the same size, there will be differences in the extent of grout penetration into the fissures. Since there are many fissures treated with a limited amount of grout, the chances are that penetration into one or more fissures may not be great enough to prevent extrusion of gel when the pumping pressure dissipates (or is removed). In time, this could lead to extrusion of gel from the drill hole along its length between the fissure and the face, even though most of the fissures have been effectively sealed. Regrouting after extrusion has occurred will generally seal the remaining open fissures.

If fissure size varies from very fine to large, the first treatment may not penetrate the very fine fissures due to insufficient pressure, and may not penetrate the largest sufficiently to prevent extrusion. In such cases, two or more regrouting efforts will be needed for an effective treatment.

If gel extrudes immediately upon removal of the packer, this indicates some relatively large channels are unfilled. If there is considerable time lapse prior to gel extrusion, the unsealed fissures are the finer ones.

The effectiveness of a drill hole treatment can best be assessed by flow measurements from the hole at equilibrium conditions before and after grouting.

Local groundwater elevation is related to the quantity of seepage into zones from which drainage can occur. As grouting closes drainage paths, groundwater elevation will rise, increasing the head on the grouted zones. (The increase may not be immediate.) Thus, the higher pressures (particularly if they reach values greater than the grouting pressure previously used) may push grout out of seepage zones, and re-open seepage channels. One method of reducing this possibility is to pump some grout into each hole at the close of treatment at pressures higher than the future anticipated groundwater head.

When it is necessary to exclude groundwater from flowing into a specific location through all of many seepage channels, this is generally done (if grout is to be used), by constructing a cut-off wall or grout curtain. Curtains are generally thought of as impermeable membranes or sheets constructed vertically, as for dams, shafts and deep excavations for buildings, etc. However, the construction may be horizontal or sloped, as required for tunnels and other underground conduits. In contrast to seepage problem procedures, a grout curtain can be engineered from start to finish, and carried out with minimum deviation from initial plans. Grout curtains are discussed elsewhere in these proceedings.

SUMMARY

Control of seepage through a finite number of channels is accomplished by making a portion of each seepage channel part of an artificial seepage system which has been deliberately made amenable to filling with grout. In the field this is done by placing a drill hole to intersect the seepage channel a short distance from its outlet. Dye tests are then performed to determine the flow characteristics of the new circuit consisting of drill hole and channel. The data gathered is then used to determine the pressure, volume and gel time limitations for pumping grout.

SOIL STABILIZATION BY GROUTING ON BALTIMORE SUBWAY

By Edward J. Zeigler[1], M. ASCE, and John L. Wirth[2], A.M. ASCE

ABSTRACT: Soil grouting procedures were extensively used during construction of the Northwest Line of the Baltimore Region Rapid Transit System to successfully minimize the impact of underground construction on adjacent structures. Chemical grouting of soil served to stabilize soil masses adjacent to both cut-and-cover excavations and driven tunnels. Loss of ground beneath the spread footing foundations for five important commercial buildings was minimized during excavation of an adjacent slurry trench by stabilizing the dense sand and gravel with a sodium silicate grout. Soil masses surrounding two railroad tunnels were likewise chemically stabilized prior to excavation of twin transit tunnels beneath the existing tunnels. Compaction grouting techniques were employed to fill voids and recompact the soil above the crown of twin soft ground tunnels. In this unique application, grout injection was synchronized with progress of the shield-driven tunnels so that ground settlements were prevented from progressing upward to the foundations of adjacent buildings. Field data concerning grouting procedures are summarized and instrumentation data are used to indicate the extent of ground and structure movements.

INTRODUCTION

Construction of Section A of the Northwest Line of the Baltimore Region Rapid Transit System required tunneling and the cut-and-cover excavation of underground stations in dense sands and gravels of the Coastal Plain in the central city area of Baltimore. Underlying the sands and gravels were the residual materials derived from weathering of the parent crystalline metamorphic bedrock. These residual materials were encountered in the deeper portions of several of the cut-and-cover excavations and in short lengths of the tunnels. In general, the sands and gravels are sufficiently dense to provide adequate spread footing support at shallow depths for even the most heavily loaded structures. Excavation below the foundations, however, presented the potential for loss of ground, subsidence, or even loss of bearing for spread footing foundations. While structural underpinning may have been the traditional approach to preserving the integrity of the existing structures, soil stabilization by grouting was chosen as an economical construction expediency. Both chemical and compaction grouting were employed in a variety of applications, including

[1]Associate, Rummel, Klepper & Kahl, Consulting Engineers, Baltimore, Maryland

[2]Senior Geotechnical Engineer, Rummel, Klepper & Kahl, Consulting Engineers, Baltimore, Maryland

stabilization of foundation soils adjacent to cut-and-cover excavations and of soil masses above twin subway tunnels excavated below existing railroad tunnels, conduits, or surface structures. Grouting patterns, pressures, and takes, and the data from monitoring structural settlements indicate the scope and effectiveness of the grouting program. The results should be useful for future projects involving soil stabilization of granular soils in an urban environment.

PROJECT DESCRIPTION

The sands and gravels encountered on the Baltimore Subway represent the Cretaceous deposits of the Coastal Plain Physiographic Province, overlying the residual materials and rock of the Piedmont Physiographic Province outcropping to the northwest. The granular soils were encountered to thicknesses of as much as 80 feet (24 m), thinning toward their surface contact with the Piedmont formations approximately 2.5 miles (4 km) from the central city terminus of the project alignment. The northwest limits of the Coastal Plain sediments and the portion of the subway alignment discussed in this paper are shown in Figure 1.

The Charles Center Station is the terminus of the Northwest Line in the heart of Baltimore's central business district. The 60-foot (18-m) wide station structure extends virtually from building face to building face along Baltimore Street, a principal east-west street. Construction of slurry walls as part of the temporary support of excavation system required that the footings of the major buildings adjacent to the subway structure be exposed, or even trimmed by as much as 20 percent of their bearing area. Stabilization of the cohesionless soils below the footings to minimize loss of ground during slurry trench excavation was accomplished by chemical grouting.

Proceeding west from Charles Center Station, the subway tunnels pass approximately 5 feet (1.5 m) below the invert of a 90-year-old brick railroad tunnel. Minimizing the displacement of the granular soils surrounding the tunnel was essential in preserving the structural integrity of the structure. Chemical grouting from the ground surface and from the tunnel invert was performed to stabilize the soil mass.

The effect of ground subsidence on surface structures over shield-driven soft-ground tunnels on the Bolton Hill Tunnels contract was considered to be a significant problem. The solution to minimizing structural settlements was to consolidate the granular soils over the tunnels before the subsidence had progressed to the surface. Timely injections of low-slump soil-cement grout under high pressure above the tunnel crown after the tunneling shield had passed resulted in re-compacting the soil before ground settlement could progress upward to building foundations.

Compaction grouting, combined with chemical grouting, was employed in lieu of underpinning for a 100-year-old stone church adjacent to a 70-foot (21-m) deep access shaft at the Bolton Hill Station. The granular soil was recompacted to a depth of 45 feet (14 m) below the foundation, minimizing the potential for settlement of the foundations and consequent damage to the structure. At the same location, the sands and gravels at the portal of the shield-driven tunnel adjacent to the church were also chemically grouted to control loss of ground prior to the application of compressed air.

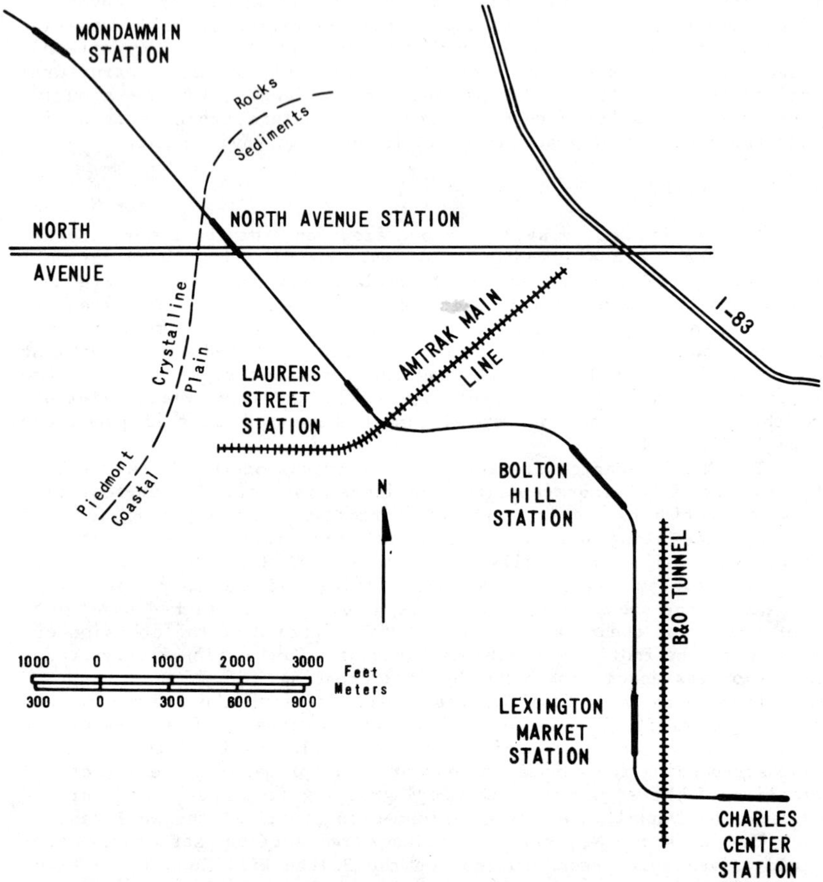

Fig. 1.-- Central City Portion of the Northwest Line of the Baltimore Region Rapid Transit System.

The main line of the Amtrak Northeast Corridor is carried by a 108-year-old brick-lined tunnel crossing the subway alignment just southeast of the Laurens Street Station. Since the subway tunnels were to be excavated to within 13 feet (4 m) of the Amtrak tunnel invert and it was absolutely necessary to maintain Amtrak service, the granular soil mass surrounding the tunnel was chemically grouted.

At Laurens Street Station, vent shafts were to be excavated immediately adjacent to 80-year-old row houses in fragile structural condition. Prior to excavation, the granular soils were chemically grouted, and loss of ground below the buildings was negligible.

SOIL STABILIZATION BY CHEMICAL GROUTING

Support of Surface Structures.-- Construction of the Charles Center Station required the opening of an excavation 60 feet (18 m) deep in a principal east-west street in the central business district of Baltimore. In lieu of underpinning the major buildings adjacent to the cut-and-cover excavation, an excavation support system sufficiently rigid to minimize lateral deflections and consequent settlement was installed. This system included a 24- to 30-inch (0.6- to 0.75-m) thick cast-in-place concrete slurry wall, reinforced with embedded steel H piles, with internal bracing concentrated at the locations of lateral surcharge loads caused by spread footing bearing pressures. It was estimated that this system would limit structural settlements during excavation of the cut-and-cover station to amounts that could have been anticipated had the originally conceived structural underpinning been employed.

Slurry trench excavation involved exposing, and in some cases, partially removing the footings of five important commercial buildings along Baltimore Street. These buildings most affected were the Union Trust Bank, the 34-story Maryland National Bank, the 15-story W. R. Grace Building, the 12-story B & O Railroad Building, and the 17-story Lord Baltimore Hotel, as shown in Figure 2. The bentonite slurry used as trench backfill during excavation was not considered to be solely able to prevent loss of ground caused by the action of the excavator bucket or by sloughing of the granular soils subject to bearing pressures estimated to be as high as 12 KSF (580 kPa). Therefore, prior to slurry trench excavation, the soil masses adjacent to the Station walls to depths of 10 to 15 feet (3 to 4.5 m) below the footings of the five buildings were stabilized by chemical grouting. The desired effect of the grouting was to impart to the sands and gravel a cohesion of 50 to 100 psi (350 to 700 kPa).

The construction sequence first involved excavating to the tops of the spread footings so that grout pipes could be installed through

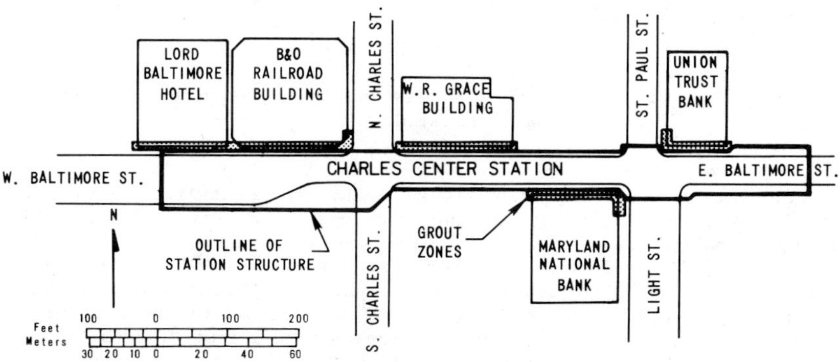

Fig. 2.-- Plan View of Charles Center Station Showing Chemical Grout Zones at Major Buildings.

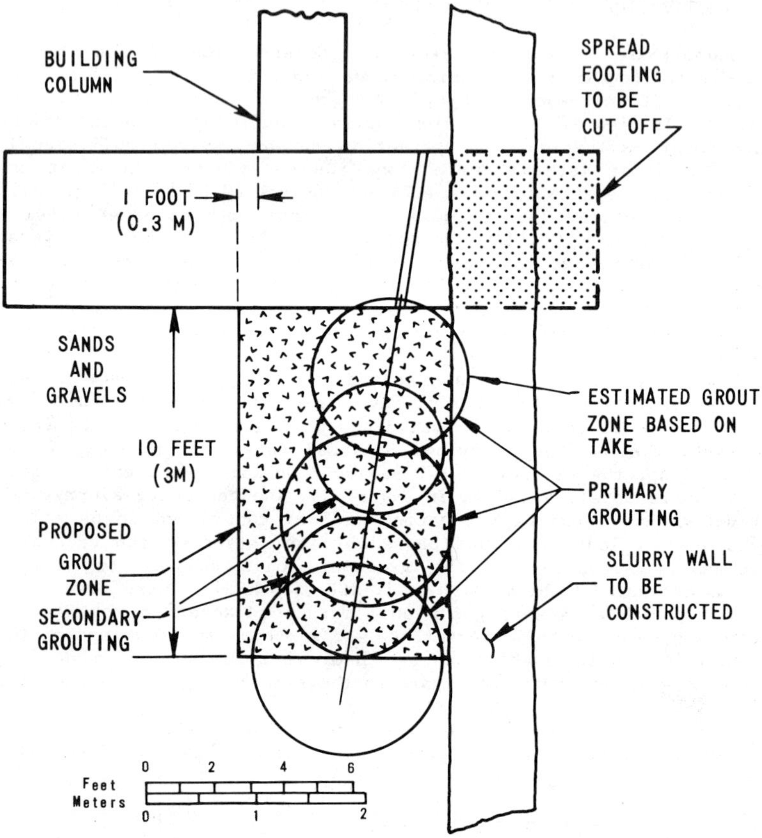

Fig. 3.-- Cross Section of Spread Footing at Maryland National Bank Showing Chemical Grout Zone

the footings, except at Union Trust Bank, where utility ducts and manholes prohibited access to the footings. At Union Trust, all drilling and grouting operations were performed from street level. Otherwise, 3-inch (75-mm) diameter holes were drilled through and adjacent to the building footings to a minimum of 10 feet (3 m) (15 feet (4.5 m) at the W.R. Grace Building) below the bottoms of the footings, as shown in Figure 3. The holes were spaced 4 to 5 feet (1.2 to 1.5 m) along the lengths of the buildings adjacent to the Station, except at Union Trust Bank where the grouting effort was more concentrated at the individual footings. Also at Union Trust, grouting was performed to a depth of 40 feet (12 m) below the footings at the southwest corner of the building where soldier piles and slurry wall could not be installed due to utility obstructions.

The grout holes were generally inclined under the buildings to provide as large a grout zone as practical. Sodium silicate (Geloc-3) grout was injected through 1.5-inch (38-mm) diameter PVC pipe under a pressure of 35 psi (240 kPa) at 1.5- to 2.0-foot (0.45- to 0.6-m) vertical intervals. Primary and secondary phases of grouting were effected by alternating grout intervals, beginning at the bottom of the hole, with considerably less grout "take" measured during secondary grouting. Figure 3 schematically shows the estimated grouted volume based on the measured take at each interval. At completion of the grouting, the footings were cut off as necessary and the excavated volume above the footings was backfilled to street level so that the cast-in-place concrete slurry wall could be installed from street level.

Grout takes were estimated based on approximately 25 percent of the soil volume being groutable voids. The total grout take for the five buildings was approximately 79 000 gallons (300 kl), or about 78 percent of the estimated volume of 101 000 gallons (383 kl). Expressed as a percentage of the estimated takes, the actual pumped volumes ranged from 72 percent (B&O Railroad Building) to 109 percent (Union Trust Bank), indicating that, in general, the porosity of the granular soils was slightly less than anticipated.

Total settlement of the five major buildings, as measured from survey reference points, was negligible at the completion of slurry trench excavation.

Chemical soil stabilization was also employed to control ground movements adjacent to the 80-foot (24-m) deep cut-and-cover excavation required for construction of the Laurens Street Station. Vent shafts protruding 4 feet (1.2 m) beyond the side walls of the station proper involved excavating cohesionless soils to within 2 feet (0.6 m) of and as much as 40 feet (12 m) below stone and rubble foundations for 80-year-old two-story row houses. To minimize settlement and consequent structural distress of the buildings, the sands and gravels were chemically grouted for the lesser of the depth of the shafts or the depth to the non-groutable residual soils. Grout holes, 3.5 inches (89 mm) in diameter, were drilled along the lengths of the shafts, spaced about 4 feet (1.2 m) and inclined beneath the buildings. The holes were fitted with 1.5-inch (38-mm) diameter PVC pipe with grout ports at 2.5-foot (0.75-m) vertical intervals, through which sodium silicate (Geloc-3) grout was injected into the soils at pressures of as much as 100 psi (690 kPa) to provide an 8-foot (2.5-m) wide grout zone below the building foundations. Settlement of the survey-monitored building reference points during excavation of the shafts was less than 0.02± feet (6 mm), well within the tolerable limits for structural deflections.

Support of Existing Tunnels.-- During the project design phase, it was recognized that the 80-year-old Baltimore and Ohio Railroad Howard Street Tunnel was in a deteriorated structural condition. The subway alignment was to pass beneath the railroad tunnel and was to be excavated as 18-foot (5.5-m) diameter twin tunnels, spaced 33 feet (10 m) center to center, shield-driven under compressed air. Any significant soil movements due to excavating the subway tunnels less than 7 feet (2.1 m) below the brick railroad tunnel would result in deformation of the arch and walls and the potential for structural failure. Therefore, structural reinforcement of the railroad tunnel and the stabilization of the granular soils surrounding the tunnel by chemical grouting were required by the subway tunneling contract.

The soil stabilization program included grouting from street level and from within the B & O Railroad Tunnel. Approximately 170 3-inch (76-mm) diameter holes were drilled from the street surface to either the subway tunnel crown level or two feet (0.6 m) into the fine-grained residual soils underlying the sands and gravels, whichever was deeper. The surface holes were arranged in two staggered rows on each side of the tunnel, spaced 5 feet (1.5 m) transversely and longitudinally, and 1.5-inch (38-mm) diameter PVC pipe with grout ports spaced 2.5 feet (0.75 m) vertically was inserted into the holes. Grout holes drilled from within the tunnel were 2 inches (51 mm) in diameter and fitted with 1.25-inch (32-mm) diameter PVC pipe. The tunnel holes were arranged in grout "fans" as shown in Figure 4, with 43 "fans" spaced 5 feet (1.5 m) longitudinally along the railroad tunnel and were

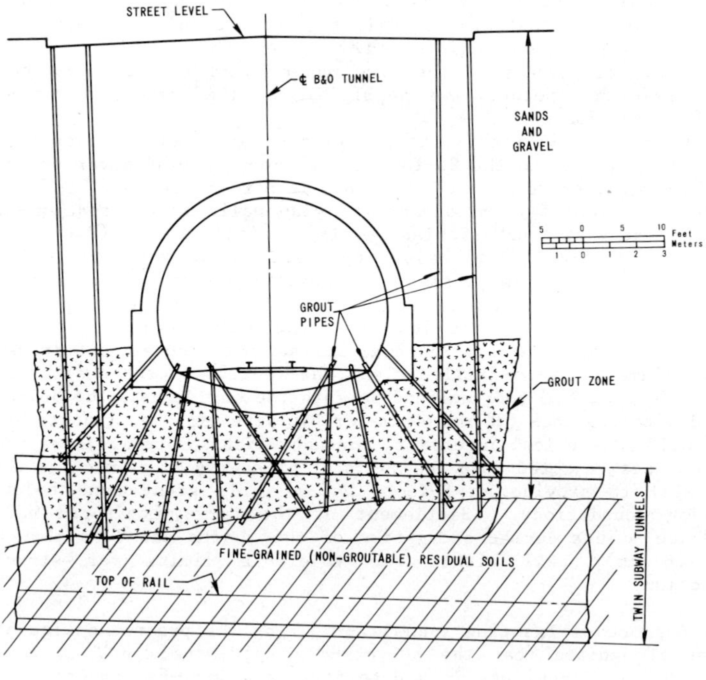

Fig. 4.-- Cross Section of B & O Railroad Tunnel Showing Chemical Grout Zone and Grout Pipe Arrangement

likewise drilled to crown level or into the residual soils.

Based on an estimate of 33 percent of the soil volume to be groutable voids, the total volume of sodium silicate (Geloc-3) grout calculated by the grouting contractor to effectively stabilize the specified grout zone was 340 000 gallons (1290 kl). In fact, approximately 354 000 gallons (1340 kl) of grout were pumped into the sands and gravels, at pressures of as much as 100 psi (690 kPa). Grout flows were generally less than 2 gallons per minute (7.5 l/min.), and the gel times were correspondingly adjusted to as high as 45 minutes. During the initial stages of tunnel grouting, the high grouting pressures and low flow rates combined to produce a heave of the survey-monitored rails of as much as 2 inches (50 mm). Reducing the maximum grout pressure to 80 psi (550 kPa) and terminating grouting when flows were reduced to less than 0.5 gallons per minute (1.9 l/min.) prevented further heave of the tunnel invert and rails.

Unfortunately, settlement of the rails during subway tunnel excavation was not monitored. However, settlement of the street surface over the B & O tunnel was measured to be only on the order of 1/8 of an inch (3 mm). Considering there was no apparent structural distress in the tunnel liner or visually observed settlement of the rails, the twin subway tunnels were successfully excavated beneath the brick railroad tunnel without adverse impact.

A second and even more critical situation existed in tunneling beneath the Amtrak Wilson Street Tunnel. This tunnel carries the main rail line between New York and Washington, perpendicularly crossing the subway alignment approximately 200 feet (60 m) southeast of the Laurens Street Station (see Figure 1). The tunnel was originally constructed in 1873 by the Baltimore and Potomac Railroad with stone-faced walls backed by concrete and capped with a brick arch roof. The need for additional clearance to accommodate newer trains resulted in an alteration of the structure in 1916, which alteration consisted of severing the concrete invert slab from the side walls, deepening the side walls, and constructing a lower invert slab which is free of the walls. Settlement of the walls since 1916 and the minimal clearance provided by the existing structure for modern trains made it essential that excavation of the twin subway tunnels with tunnel crown only 13 feet below the Amtrak tunnel be carried out in a manner which would preclude settlement of the arch relative to the invert slab. Therefore, chemical stabilization of the granular soils adjacent to, as well as below, the railroad tunnel was specified to minimize movements of the soil within a broadly defined zone of influence.

The soil stabilization program was accomplished in primary and secondary phases by both surface and tunnel grouting. The surface, or street level, grouting program included primary grout holes in a 10-foot (3-m) square grid pattern spaced throughout the entire area of the zone of influence, which was specified as being 150 feet (46 m) long (parallel to the subway tunnels) and 70 feet (21 m) wide (parallel to the Amtrak tunnel). The secondary grout hole pattern also consisted of a 10-foot (3-m) square grid, offset 5 feet (1.5 m) in each direction from the primary pattern. Approximately 142 grout holes of 3.5-inch (89-mm) diameter were drilled from the surface to the non-groutable fine-grained residual soils encountered about 5 feet (1.5 m) below the level of the Amtrak tunnel invert. Each hole was fitted with 1.5-inch

(38-mm) diameter PVC pipe, with grout ports spaced 2.5 feet (0.75 m) vertically from the Amtrak tunnel springline to the top of the residual soil.

The tunnel grouting program consisted of seven primary grout "fans" spaced 10 feet (3 m) longitudinally along the Amtrak tunnel, with six secondary "fans" midway between the primary fans. The orientations of the primary and secondary grout holes is shown in Figure 5. Forty-six holes were drilled through the tunnel invert to the residual soils, approximately 5 feet (1.5 m) below the invert of the Amtrak Tunnel.

Pressures as high as 80 psi (550 kPa) were used to inject the sodium silicate (Geloc-3) grout into the dense granular materials, with gel times set at 15 to 60 minutes. The Contract Specifications required that the grouted soil achieve a minimum unconfined compressive strength of 100 psi (690 kPa) prior to subway tunnel excavation. Four 4-inch (100-mm) borings were drilled into the stabilized zone of soil, and the grouted soil was sampled in a relatively undisturbed manner by air-coring. Two samples from each of the four borings were tested in unconfined compression, with test values ranging from 100 to 1000 psi (690 to 6900 kPa). These exceedingly high test values appear to have resulted in part from the drying of the samples during the sampling process.

It was estimated that the primary and secondary phases of surface grouting would require 259 000 gallons (980 kl) and 88 000 gallons (330 kl), respectively, of the Geloc-3 grout to be injected into 70 primary and 72 secondary holes. The actual takes were 189 000 gallons (715 kl), or 73 percent of the estimated volume, for the primary phase, and

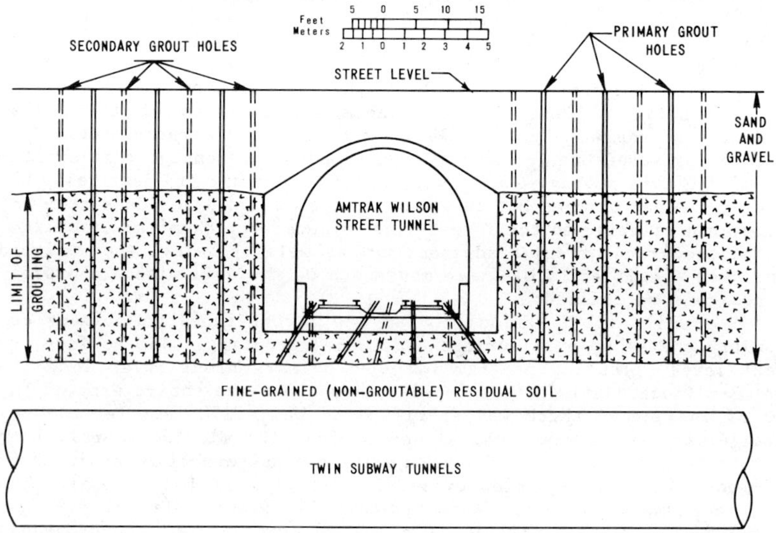

Fig. 5.-- Cross Section of Amtrak Wilson Street Tunnel Showing Chemical Grout Zone and Grout Pipe Arrangement

61 600 gallons (233 kl), or 71 percent of the estimated volume, for the secondary phase. It was determined that the grout takes were substantially less than anticipated due to lenses of non-groutable fine-grained soils within the predominantly granular soil mass, as encountered by the four borings drilled to sample the grouted soil, rather than due to inadequate grout penetration into the soil.

Excavation of the twin horseshoe-shaped subway tunnels was accomplished by the top-heading-and-bench method in mixed-face conditions. Temporary support of the excavation was provided by steel liner plates, rib sets, and invert struts. A comprehensive program of settlement monitoring was established to measure the influence of subway tunnel excavation on the Amtrak tunnel and the soil below the Amtrak tunnel invert, the ground surface, and the subsurface soils within the zone of influence of the Amtrak tunnel. It was anticipated that settlements would be larger in the immediate vicinity of the subway tunnels and at structures due to their concentrated loads. This was found to be the case, with the most settlement being observed at the Amtrak tunnel invert over the subway tunnels, with the edges of the invert slab measured to have settled as much as 0.10± feet (30 mm). Settlement of the soil below the invert slab over the subway tunnels was slightly less, averaging approximately 0.08± feet (25 mm). The walls of the tunnel, however, settled only an average of 0.03± feet (10 mm). Surface and subsurface settlement over the Amtrak tunnel was on the order of 0.05± feet (15 mm). Considering that the purpose of the soil stabilization was to prevent settlement of the Amtrak tunnel walls relative to the invert slab, the results of settlement monitoring indicate that the grouting accomplished the stated objective.

SOIL STABILIZATION BY COMPACTION GROUTING

Compaction Requirements.-- Compaction grouting was used extensively on the Bolton Hill Tunnels contract as an alternative to conventional underpinning of buildings over or adjacent to twin soft-ground tunnels. The Bolton Hill Tunnels contract included construction of more than one mile (1.7 km) of twin trnasit tunnels to be shield-driven under compressed air, with segmented steel liner rings as permanent support. Structures over and adjacent to the twin tunnels were designated as Type 1 structures if the structures were to be supported by compaction grouting. Type 1 structures were indicated on the Contract Drawings by building number, and the contractor was required to submit a lump sum price for compaction grouting of the soil beneath each of the 40 Type 1 buildings. Compaction grouting was specified to be performed beneath 10 of the buildings with the remaining 30 buildings to be compaction grouted only if the contractor was so directed by the Engineer.

In general, the structures involved were three- and four-story brick commercial and residential buildings, estimated to range in age from 50 to 100 years. There were, however, several more modern buildings including a six-story reinforced concrete parking garage. Foundation levels were generally on the order of 40 feet (12 m) above tunnel crown.

Compaction grouting was defined in the specifications as a grouting technique using the injection of a highly viscous soil-cement grout to form a grout bulb. The grout was to be extruded through pipes installed to the necessary depths into the soil mass, causing a bulb to form which would physically displace the soil particles adjacent to the bulb. Controlled pressure exerted by the grout bulb was to be used to lift the soil mass and structure, thereby accomplishing a re-leveling of the previously settled structure. It was required that when the settlement of any portion of a Type 1 structure reached a maximum of 1/4 inch (6 mm), the contractor was to proceed immediately with compaction grout lifting of the structure.

Contractor's Alternate Proposal.-- Following award of the contract, the grouting subcontractor (Hayward Baker Company of Odenton, Maryland) proposed a different and new approach to stabilization of the granular soils beneath the Type 1 buildings. The alternate compaction grouting program was intended to recompact the soils loosened by tunneling operations immediately following passage of the shield rather than waiting for the settlements to reach the building foundation and then attempt to raise the foundations back to their original elevations. The proposed procedure was expected to be able to significantly reduce or eliminate building settlement and to greatly reduce the horizontal movements that usually occur at the ground surface.

The Contractor's proposal indicated that the grout pipes would be installed from the ground surface at least 30 days before tunneling operations would arrive at a subject building. Compaction grouting was planned to begin sometime after the front of the shield passed below the grout pipe and before the tail of the shield passed by, although this timing was subsequently modified based on field experience. Grouting would occur before settlements were observed on the surface and would be continued until a specific grout volume had been pumped, or until a slight rise in surface elevation occurred, or until an excessive grout pressure developed, or until the tunnel passed near another grout pipe, depending on the judgement of the grouting engineer.

To monitor subsurface soil movements through a non-grouted area and an area grouted with the proposed new dynamic compaction grouting procedure, an expanded subsurface instrumentation program was implemented. Two test sections were established, each consisting of six combined inclinometer/settlement columns with a considerable number of shallow surface settlement points and building reference points. This instrumentation program was monitored before, during, and after passage of the tunnel beneath the test sections and the data were summarized in a report prepared by the Hayward Baker Company (1).

Execution of the Work.-- Following approval of the Contractor's alternate proposal, compaction grouting was used to recompact the soil beneath 39 of the 40 Type 1 buildings so designated in the contract. In the case of the fortieth building, the City Temple of Baltimore, a large stone church over 100 years old, a combination of chemical soil stabilization and compaction grouting was employed as described in a subsequent section of this paper.

Grout holes were installed by percussion drilling a 3-inch (75-mm) diameter steel casing to the bottom of the planned grout bulb elevation. Just prior to, or at the start of grouting in a particular pipe, the pipe was pulled out approximately five feet (1.5 m) in order to be able to start the injection process. In some cases it was necessary to blow out about one cubic foot (30 l) of soil to form an initial void space so that grout would start moving into the ground. Two staggered rows of grout pipes on 10-foot (3-m) centers were placed over each tunnel, with the pipe tip in each row located about five feet (1.5 m) off the tunnel centerline and about 10 feet (3 m) above the tunnel crown, as shown in Figure 6.

The grout was a stiff, viscous mortar composed of a well-graded sandy soil having 10 to 25 percent fines passing the No. 200 sieve, two bags of Portland Cement per cubic yard of grout material, and enough water to obtain the desired consistency of about two inches (50 mm) of slump in the concrete cone slump test. The grout plant consisted of a heavy duty pugmill-type double-bin mixer and a high-pressure double-piston pump capable of pumping up to 2.5 cubic feet (75 l) per minute at pressures up to 650 psi (4500 kPa).

Injection of grout in any particular grout hole was performed just after the tail of the shield had passed the location of the grout pipe. In general, grout injection occurred after the tail of the shield was 5 to 10 feet (1.5 to 3 m) beyond the grout pipe; however, it was found that no significant surface settlement occurred if the tail of the shield was as far as 15 to 20 feet (4.5 to 6 m) beyond the grout pipe. Grout pressures were limited to a maximum of 400 psi (2800 kPa) after it was found that pressures of 500 psi (3500 kPa) resulted in one instance in enough deflection of the steel tunnel lines to cause the lines to become ironbound in the tail of the shield.

The dynamic compaction grouting procedure utilized on this

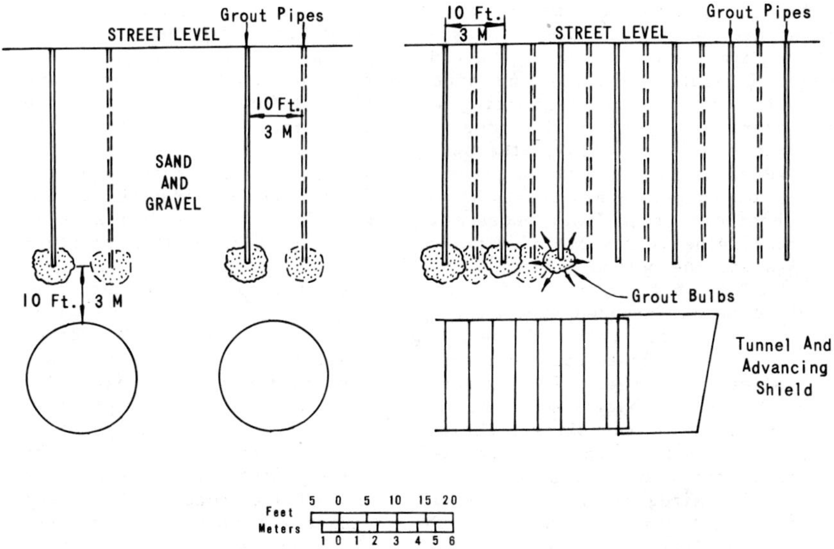

Fig. 6.-- Cross Sections of Twin Soft-Ground Tunnels Showing Compaction Grouting Sequence and Arrangement

contract was considered to be highly successful. In general, surface settlements over the tunnels were less than 1/4 inch (6 mm) where compaction grouting had been performed. In contrast, surface settlements in ungrouted areas ranged up to 1.5 inches (40 mm).

Further verification of the success of the procedure is indicated by the fact that damage caused by settlement of the 39 buildings which had their foundation soils compaction grouted was virtually non-existent. There was, however, one instance of heave of a concrete basement floor caused by excessive grout pressure in a bulb only 20 feet (6 m) below the basement floor.

SOIL STABILIZATION BY CHEMICAL AND COMPACTION GROUTING

During construction of the Bolton Hill Tunnels, much concern was expressed for the City Temple of Baltimore. This historic structure, constructed of stone masonry, is more than 100 years old and has a steeple well over 100 feet (30 m) in height. The building was located only 23 feet (7 m) from the 70-foot (21-m) deep north tunnels access shaft, with a support of excavation system consisting of soldier piles and timber lagging. The steeple portion of the church, however, was not adjacent to the access shaft, but was ahead on line and adjacent to the outbound tunnel. Specific information concerning the foundations of the church was not available, but it was known that shallow spread footings had been used. Following practice of the time, it was believed that the footings were constructed of rubble masonry.

Instrumentation used to monitor the church consisted of reference points for survey measurement of settlement and interior and exterior reference points for measurement of horizontal spreading of the structure through the use of a tape extensometer instrument.

Careful monitoring of the church during excavation of the access shaft indicated virtually no settlement or horizontal movement, and therefore no compaction grouting was performed for the portion of the structure adjacent to the access shaft. There was, however, much concern for the portion of the structure, including the steeple, which was adjacent to the outbound tunnel. Based upon experience gained at the south tunnel access shaft, it was believed that cutting the lagging and beginning tunneling, necessarily under free air, would result in several inches (100± mm) of settlement of the portion of the structure closest to the outbound tunnel.

Since settlement of this magnitude was unacceptable, a scheme was devised which included chemically grouting the soil around the upper periphery of the outbound tunnel prior to tunneling and compaction grouting the loosened soil over the tunnel as tunneling was in progress. As indicated in Figure 7, eight grout holes were drilled horizontally from the access shaft around the upper periphery of the tunnel prior to the beginning of tunneling. The holes were drilled for a length of 30 feet (9 m) and sodium silicate (Geloc-3) grout was injected into the granular soil, forming a collar of chemically stabilized soil. This collar minimized the loss of ground associated with cutting the timber lagging of the access shaft and the beginning of free-air tunneling. Compaction grouting through grout pipes installed from the street surface to approximately 10 feet (3 m) above tunnel crown was accomplished using the dynamic approach previously developed on this project. As a result of the grouting procedures utilized for this

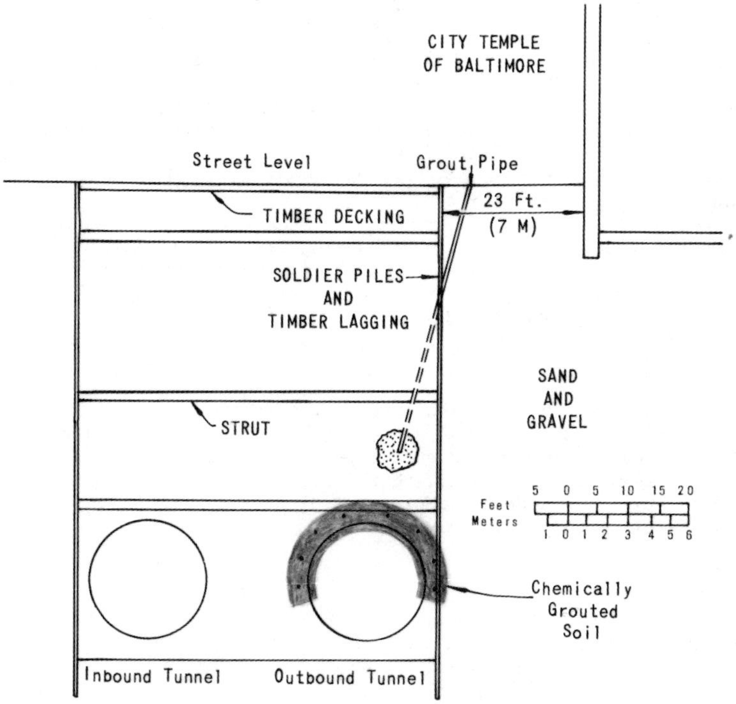

Fig. 7.-- Cross Section of North Tunnels Access Shaft at City Temple of Baltimore Showing Chemical and Compaction Grouting

structure, tunneling of both the outbound and inbound tunnels was successfully completed without measureable settlement and horizontal movement of this historic church. The only restoration required was the repair of holes caused by the insertion of reference points on the exterior and interior of the building.

SUMMARY AND CONCLUSIONS

Field observations lead to the conclusion that soil stabilization by chemical and compaction grouting procedures successfully minimized the impact of underground subway construction on adjacent structures in Baltimore, Maryland. The soils which were stabilized consisted of the dense Cretaceous sands and gravels of the Coastal Plain.

A sodium silicate grout was employed to impart cohesion to the sands and gravels which comprise the foundation material for spread footings beneath important commercial buildings as tall as 34 stories. The chemical grouting was performed prior to the excavation for cast-in-place concrete slurry wall. A similar chemical grouting procedure stabilized the soil surrounding two old but critically important railroad tunnels prior to the excaation of twin subway tunnels beneath the railroad tunnels. Sodium silicate grouting was also used at several other locations on the Baltimore subway to stabilize granular soils, thereby minimizing the loss of ground associated with tunneling and cut-and-cover excavation procedures.

Compaction grouting, using a viscous soil-cement grout, recompacted the soil above the crown of twin shield-driven tunnels. Grout injection was synchronized with tunnel shield advance so that ground setlement was prevented from progressing upward to the foundations of adjacent buildings.

ACKNOWLEDGEMENTS

The authors wish to express appreciation to the Mass Transit Administration of the State of Maryland for granting permission to use the data acquired during construction of the Baltimore Region Rapid Transit System. In particular, thanks are given to Mr. L. A. Kimball, Administrator, Mr. Robert J. Murray, Project Manager, and Mr. Frank A. Hoppe, Director of Construction, for their cooperation.

REFERENCES

1. The Hayward Baker Company, "Ground Deformations at Compaction Grouted Tunnel Section, Building No. 115, North Eutaw and Biddle Streets", June, 1978 (unpublished).

CHEMICAL GROUTING FOR CONSTRUCTION OF 8 FOOT DIAMETER TUNNEL THROUGH "LITTLE ITALY" BALTIMORE, MARYLAND

BY

Manu Patel[1], S. Murray Miller[2], Kwok Leung Li[3], Howard Waugh[4], Joseph P. Welsh[5]

Abstract

In conjunction with the installation of a new pumping station to upgrade the City of Baltimore's sanitary sewerage system, an 8-foot diameter force main was required beneath Fawn Street in the historic "Little Italy" section of Baltimore. A tunnel which could be installed with minimal affect on the numerous restaurants, delivery and patron traffic and the four blocks of 3-story row homes was mandated by the City Fathers to the Public Works Department and their consultants. The ground water was over the crown of the proposed sewer and the soil varied from granular fill to organic clays. After exploring all known methods of ground water and settlement control, it was decided to place a 3-foot annular ring of chemical grout around the tunnel utilizing the sleeve-pipe method of grouting. To determine the effect of the tunneling operations on the streets and adjacent buildings, an extensive instrumentation program was developed consisting of deep and shallow settlement points, inclinometers, tilt plates, tape extensometers and water observation wells. Since the entire mile-long installation was instrumented while only 1,330 linear feet was grouted, the differences in the response of the instrumentation between the grouted and ungrouted section can be analyzed, compared and discussed together with the design and installation of the annular ring of grout.

Introduction

A section known as Little Italy, located near the Inner Harbor, is one of the cultural and ethnic centers of Baltimore City. Many of the residences date from the early days of the City and were subsequently occupied by Italian immigrants and their descendants. In addition to the historic significance of the area, many popular Italian

---

[1]Facilities Section Chief, City of Baltimore, Baltimore, Maryland
[2]Chief Geotechnical Engineer, Greiner Engineering Sciences, Inc.
[3]Geotechnical Engineer, Greiner Engineering Sciences, Inc.
[4]President, Faisant & Associates, Baltimore, Maryland
[5]Vice President, Hayward Baker Company, Odenton, Maryland

restaurants, bakeries and groceries have been established by business people throughout the neighborhood.

This project developed as a replacement for twin cast iron 42-inch force mains which were originally installed through Little Italy in 1911. These mains have suffered numerous faiures in recent years and show signs of heavy erosion of their inverts. They carry sewage 5,221 feet from the Eastern Avenue Pumping Station to discharge into a gravity main. Approximately 25 percent of the City sanitary sewage drains to this pumping station and since there is no storage capacity, uninterrupted flow to the gravity main is required to prevent sewage overflow into the nearby Inner Harbor.

A feasibility study was undertaken in 1976 for a new force main which would be the primary discharge outlet for the pumping station. The study indicated that the required design peak capacity of 100 mgd could be obtained with a single 60-inch diameter main. Seven different alignments were investigated by City engineers taking into consideration various factors such as length, traffic volumes and density of existing utiities. It was determined from the study that the most cost effective route would not parallel the existing mains but would take an entirely different route as shown on Figure 1. The first leg after leaving the pumping station runs approximately 1,100 feet under Fawn Street, a narrow, busy street in Little Italy lined on both sides by row houses used as private residences, restaurants and businesses.

Following the geotechnical investigation, cost analyses indicated that tunneling would be approximately equal in cost to an open cut installation. However, tunneling would have the advantage of minimizing interruption to businesses and annoyance to the community.

Little was known about the foundations of the buildings along Fawn Street. Probably they were constructed on rubble foundations only slightly lower than their basements due to the presence of ground water which is controlled by tidal action of the adjacent harbor. The street is 30 feet wide and due to limitations imposed by existing utilities, the center of the proposed tunnel had to be located very close to the east curbline. The depth to the crown of the tunnel varies from 6 to 14 feet.

In order to minimize property and utility damage along the Fawn Street leg, a soil stabilization program was incorporated into the contract requirements. The method selected required that the three feet of soil surrounding the proposed 96-inch tunnel bore be injected with a chemical grout designed to strengthen the material and prevent or at least reduce the inflow of ground water with attendant loss of fines.

In addition, an extensive instrumentation program was designed and established as a contract requirement. The purpose of this program was intended to serve primarily as an early warning system for detecting incipient damage to adjacent structures in time to take preventive action. Secondarily, it was intended, in the event damage did occur, to provide baseline data as well as ongoing data to determine

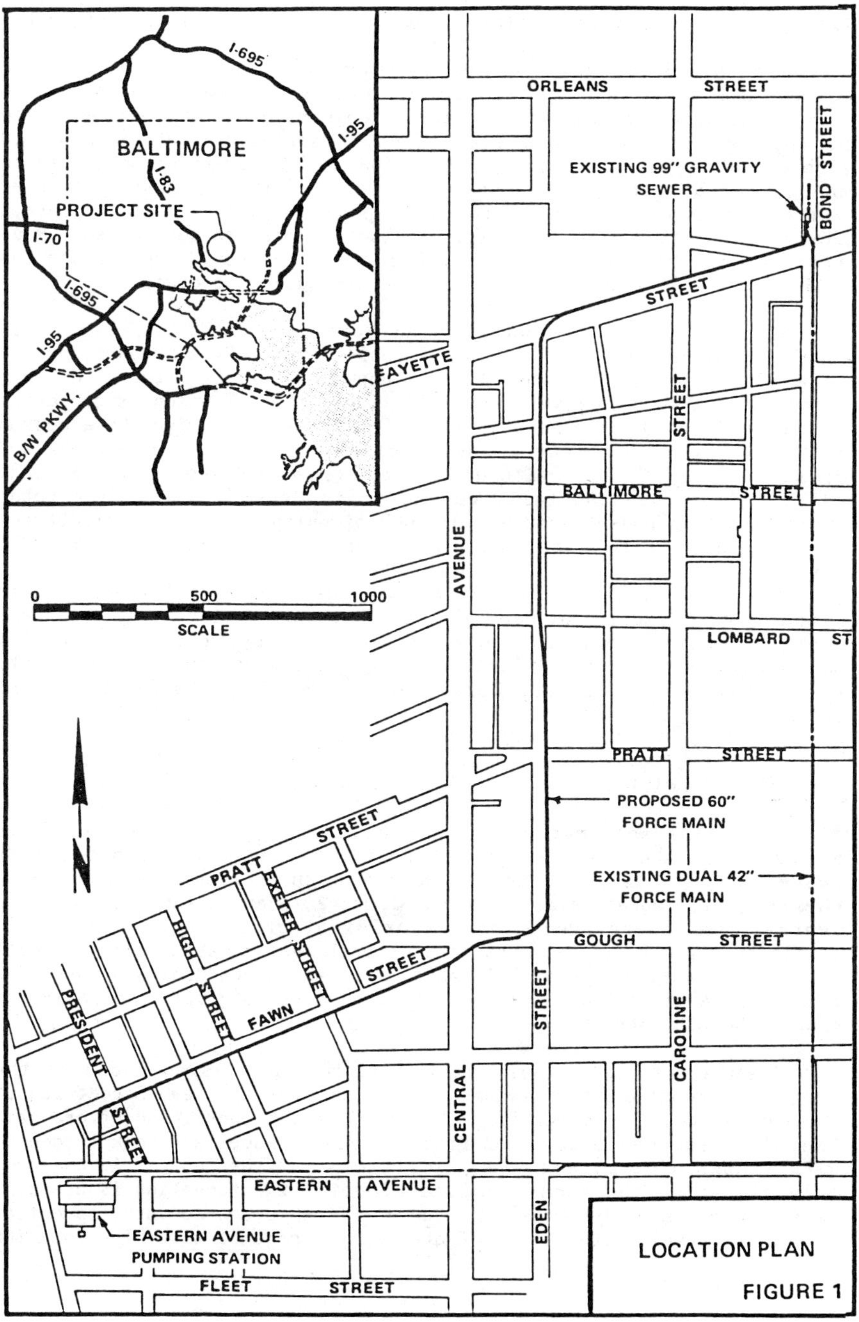

LOCATION PLAN

FIGURE 1

the extent and cause of the damage. It was also hoped that analysis of the data would provide useful insights for design of future projects.

The soil stabilization and instrumentation also were found to be most useful in a public relations program which the engineers undertook to allay the fears and counter the objections of the residents and businessmen of the Little Italy area. A series of meetings was held with community organizations at which the entire construction procedure was outlined together with an explanation of the safeguards and precautions which had been included as part of the contract to protect their interests. These safeguards included a preconstruction survey of each structure along the route, insurance to provide for damage, as well as the soil stabilization and instrumentation programs.

The New Eastern Avenue Pumping Station Force Main project as finally designed consists of 4,355 linear feet of 60-inch concrete pipe in a 99-inch ring and timber lagged tunnel, 243 linear feet of 60-inch force main in an open cut trench, 188 linear feet of which is supported by augered-in-place concrete piles. The tunnel was driven by a custom designed Lovat full face tunnel machine with hydraulically operated closure doors. Construction of the new force main was started in June 1979, and was substantially completed in December 1981. The total project cost is budgeted at $10,082,000 of which $7,561,500 will be reimbursed to the City by the U.S. EPA under the Clean Water Act of 1972.

The construction contract was awarded to Driscoll Construction Company of Philadelphia, Pennsylvania. The construction inspection and geotechnical instrumentation were performed by Greiner Engineering Sciences, Inc. Faisant Associates, Inc., acted as General Consultants to the City of Baltimore, Department of Public Works for the entire project.

Subsurface Conditions

Shoreline and drainage maps published by the Maryland Geologic Survey indicate the Fawn Street portion of the tunnel project to traverse soils filled into the waters of the Baltimore Harbor beyond the original shoreline as shown on Figure 2, whereas the soils east of Central Avenue are predominantly original ground covered by a thin mantle of fill. It is apparent that Central Avenue was a major drainage way in the early days of Baltimore City, which now drains by use of a 5'-6"x16'-6" brick and concrete arch buried beneath fill and paving to create the city street.

A total of 55 borings drilled during the design phase of the project throughout the alignment confirms that the Fawn Street leg of the tunnel would be driven through fill material consisting of gray and tan, fine to coarse sand and fine to medium gravel and rock fragments with lenses of silt and clay, while east of the Central Avenue arch culvert, which the tunnel passes beneath, the tunneling is predominantly through native brown and tan wet fine to coarse sand, some fine to medium gravel, with lenses of gray clayey organic silt. The location

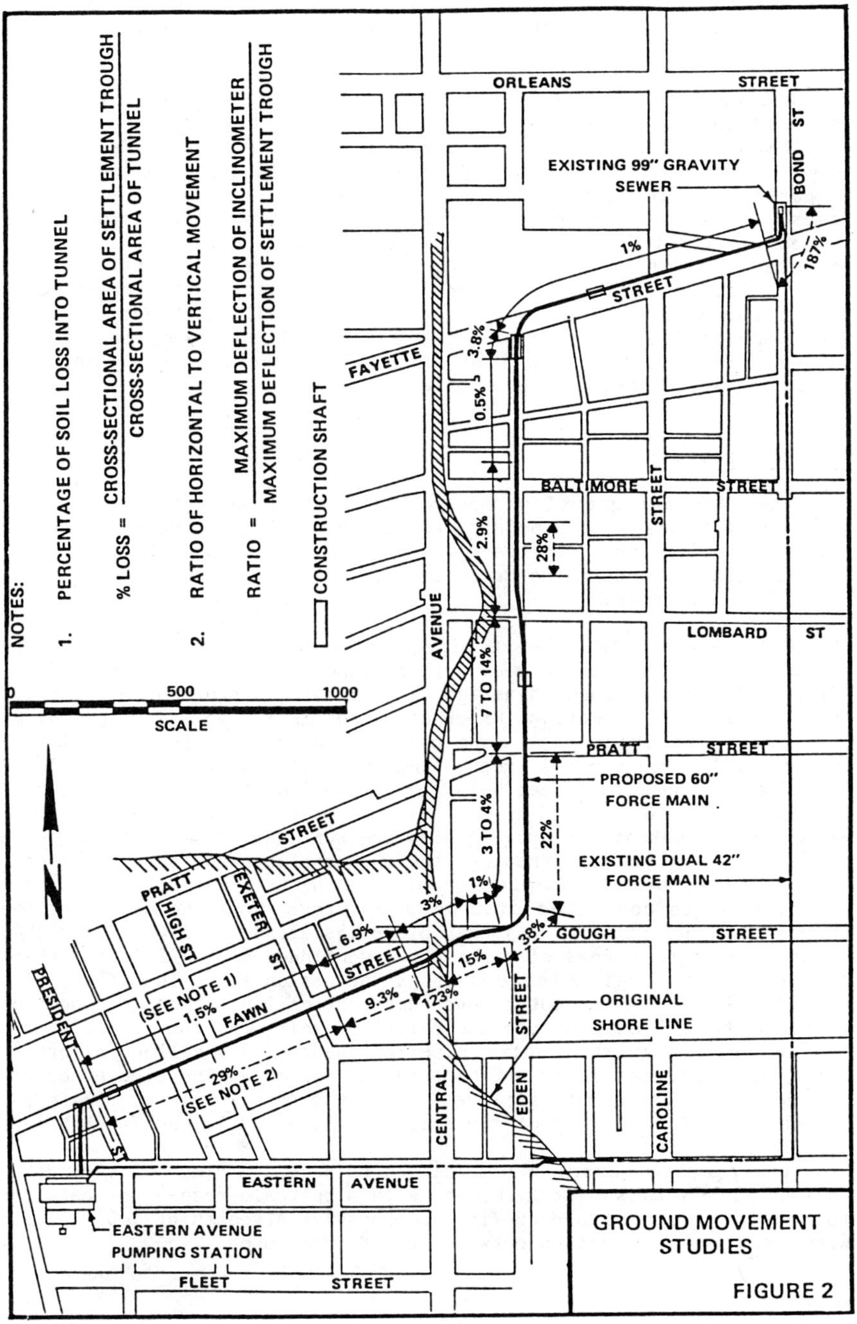

GROUND MOVEMENT STUDIES

FIGURE 2

of the water table varied from a few feet below the tunnel crown along Fawn Street to 5 to 8 feet above the crown throughout the remainder of the project.

A study of the geotechnical properties of the soils along the proposed tunnel alignment with respect to dewatering, grouting and settlement characteristics was performed by the Robert B. Balter Company, Geotechnical Consultants. A laboratory testing program consisting of particle size distributions, consolidation tests, permeability tests, unconfined compression tests, and direct shear tests were performed on selected disturbed and undisturbed samples. Based upon these test results, the Geotechnical Consultant prepared a report including a chart listing both the drainage and groutability characteristics of the soils. Approximately 70 percent of the soils within the project site were indicated to be groutable and with the exception of some lenses of fines, nearly all of the filled materials along the Fawn Street leg were considered groutable.

Chemical Grouting

The chemical grouting phase of the project was performed by Hayward Baker Company, a subcontractor to Driscoll Construction Company. First, an Atlas Copco air track was used to drill 4-inch holes through the asphalt and underlying brick and carefully probe to 6 feet for unidentified utilities.

Drilling for the sleeve pipes was then performed by an Acker holegater and two Mobile B-61 drills using hollow shaft augers. These drills offered the added advantage of retrieving disturbed soil samples for visual identification. Grout pipes consisting of 1 1/4 inch diameter polyvinyl chloride sleeve pipes with sleeved ports at 1 foot centers, which would permit regrouting, were placed by augering a 3 1/2 inch diameter hole to 3 feet below the elevation of the proposed tunnel invert. Prior to inserting the sleeve pipes, the hole was held open with cement bentonite slurry. A high pressure water pump with a double packer was utilized to fracture the rubber sleeve and the cement bentonite slurry before pumping grout. The pattern established for the grouting pipes consisted of a line of pipes down the center of the proposed tunnel with two adjacent pipes installed 4 1/2 feet on either side. These three rows of grout pipes continued on 4 1/2 feet centers through Fawn Street. Figure 3 shows the design injection volume with the circles on the two outer rows representing 150 gallons of grout or approximately 7 gallons per cubic feet of soil with the circle at the top and bottom of the center pipe representing 225 gallons of grout. Chemical grout was injected through the column at the sleeve nearest the center of the grout circle and if that sleeve would not take grout due to the impermeability of the soil, the adjacent sleeves were tried.

The chemical grout was specified to have a minimum 35 percent by volume of liquid sodium silicate with a suitable reagent or catalyst to provide an in situ unconfined compressive strength of 100 ± 25 psi with reaction time ranging between 10 and 60 minutes.

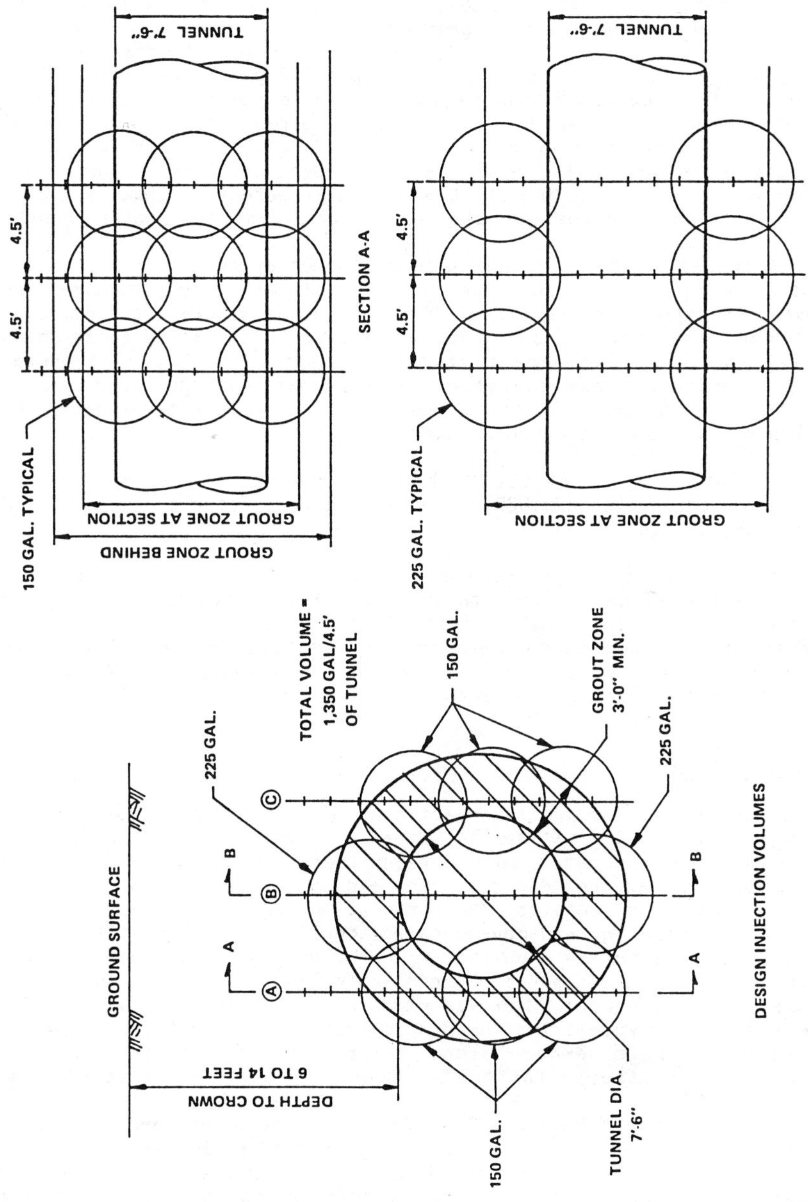
TUNNEL 7'-6"
4.5'
4.5'
150 GAL. TYPICAL
GROUT ZONE AT SECTION
GROUT ZONE BEHIND
SECTION A-A
TUNNEL 7'-6"
4.5'
4.5'
225 GAL. TYPICAL
GROUT ZONE AT SECTION
SECTION B-B
GROUND SURFACE
DEPTH TO CROWN
6 TO 14 FEET
225 GAL.
TOTAL VOLUME =
1,350 GAL/4.5'
OF TUNNEL
150 GAL.
GROUT ZONE
3'-0" MIN.
225 GAL.
A
B
C
A
B
A
B
150 GAL.
TUNNEL DIA.
7'-6"
DESIGN INJECTION VOLUMES

FIGURE 3

The chemical grout used in the project was GELOC which consisted of 35 percent sodium silicate grade 40 with a Baume of 41.6°, 5 to 10 percent of an organic solvent, calcium chloride, and the remainder water. The chemicals were pumped from a 5,700 gallon compartmentized tanker using a series of four parallel operated pumps all mechanically linked with variable speed transmissions to provide positive control for both gel time and flow rate which varied between 0.5 and 2 gallons per minute for any individual sleeve. Pumping pressure at the header varied between 10 and 80 psi. Grout samples were taken hourly during pumping and checked for grout quality, gel time and neat strength.

Field permeability tests were performed every 50 feet along the proposed tunnel before and after grouting. In all cases, the pressure required to inject a given volume of water in the sleeve after grouting was higher than the pressure used initially. Soil samples cored using air from 5 bore holes were tested for grout content and strength. Results indicated that the strength exceeded that specified by over three hundred percent. This is believed to result from allowing the samples to air cure rather than water cure.

The total volume of GELOC chemical grout injected was 321,974 gallons compared with 290,000 gallons estimated. Grout pipes installed totaled 16,884 linear feet.

One of the critical tests of the grouting technique occurred where the tunnel passed beneath the old brick and concrete arch storm sewer at Central Avenue. Although the invert of the arch was only 2'-6" from the top of the tunneling machine and flash summer storms had caused the sewer to flow full at the time of passage, no movement was noted as the tunnel passed beneath.

Instrumentation

In developing the instrumentation program, the experience acquired during the tunneling of the Baltimore subway was utilized and, where applicable, incorporated into the project. The instrumentation as finally established included both deep and shallow settlement points to monitor settlement profiles transverse to the alignment, inclinometers to detect lateral ground movements adjacent to buildings, tilt plates and tape extensometers mounted on buildings to detect significant building movements, and water observation wells to monitor ground water levels. The instrumentation was most intensive in the critical Fawn Street area. A typical instrumentation layout in Fawn Street is shown on Figure 4a and elsewhere along the project alignment on Figure 4b. The placement criteria for the various instruments are as follows:

Settlement Points - Shallow settlement points were placed to approximately 1 foot below the depth of pavement or sidewalk while deep settlement points were driven to approximately 3 feet above the crown of tunnel. A cross section of settlement points transverse to the tunnel alignment was generally formed by 5 to 8 shallow settlement points with one near the centerline of tunnel and a deep settlement

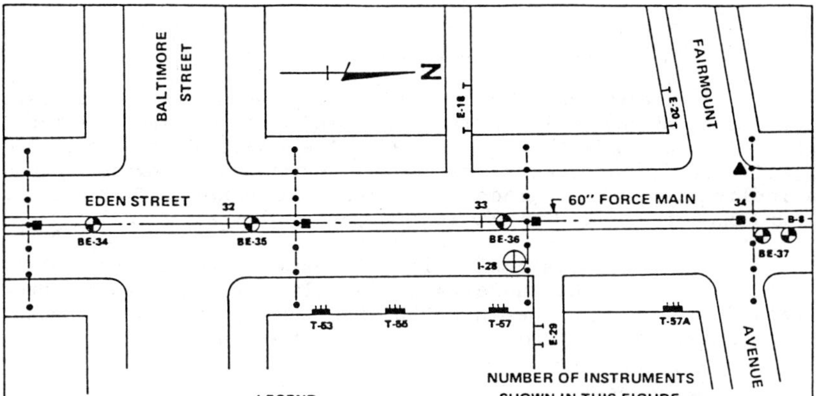

LEGEND		NUMBER OF INSTRUMENTS SHOWN IN THIS FIGURE
•	SHALLOW SETTLEMENT POINT	4 SECTIONS
■	DEEP SETTLEMENT POINT	4
⊕	INCLINOMETER	1
▲	PIEZOMETER	1
	TILTPLATE	4
⊣	TAPE EXTENSOMETER POINTS	3 PAIRS
	TEST BORING	5

FIGURE 4a – INSTRUMENTATION IN UNGROUTED AREA

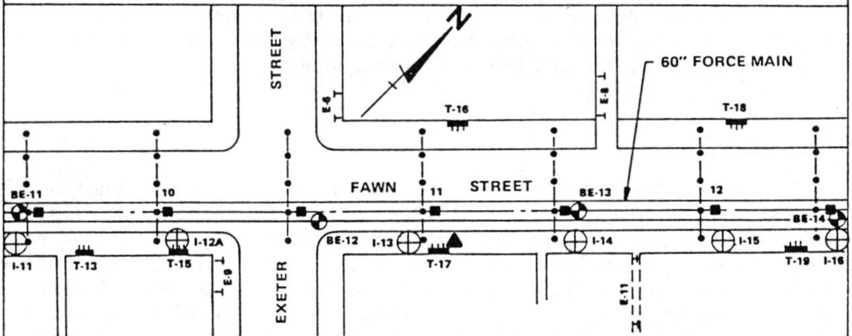

LEGEND		NUMBER OF INSTRUMENTS SHOWN IN THIS FIGURE
•	SHALLOW SETTLEMENT POINT	7 SECTIONS
■	DEEP SETTLEMENT POINT	7
⊕	INCLINOMETER	6
▲	PIEZOMETER	1
	TILTPLATE	6
⊣	TAPE EXTENSOMETER POINTS	4 PAIRS
	TEST BORING	5

FIGURE 4b – INSTRUMENTATION IN GROUTED AREA (LITTLE ITALY)

point adjacent to it. The longitudinal spacing of these cross sections was approximately 50 feet along Fawn Street and approximately 100 feet throughout the remainder of the project.

Inclinometers - Inclinometer casings were generally installed on the side of the tunnel nearest to an adjacent building. The casings extended to approximately 10 feet below the invert of the tunnel and a digitilt indicator sensor was used to measure the progressive change in inclination. The casings were spaced at about 50 foot intervals along Fawn Street and over 100 feet apart in other areas.

Water Observation Wells - Observation wells were installed along the side of the tunnel with the filter screen located approximately 10 feet below the invert of the tunnel. They were spaced at approximately 200 feet centers along Fawn Street and 300 feet in other areas.

Tape Extensometer Reference Points (TEP) - Most of these points were installed approximately 4 1/2 feet above street level. Change in distance between pairs of points placed perpendicular to the tunnel alignment was measured by a Tape Extensometer. There was no spacing pattern, as the points could be installed only by permission of a building's owner.

Tiltplates - Tiltplates were affixed by epoxy glue to building walls parallel to the tunnel alignment and located approximately 6 feet above street level. A tiltmeter placed against the tiltplate was used to measure angular movement of these buildings. As in the case of the TEP, tiltplates could only be installed by permission of the building's owner, hence, no spacing was prescribed.

Data Collection

To ensure the consistency of the data collected, inclinometer casings and tiltplates were installed in known positions at relatively remote distances from the project area. These dummy set ups were used as standards for periodic checking of the readout units. Also, to ensure continuity of readings, a second readout unit for each instrument, with the exception of the survey levels used in settlement point readings, were used periodically to obtain auxillary data in case the primary unit was damaged or needed repair.

Readings were scheduled as follows:

Instruments within a distance of approximately 100 feet in advance of the tunnel shield and approximately 150 feet behind the tunnel shield were read daily. Those between approximately 150 feet and 850 feet behind the shield were read weekly and those beyond 850 feet behind the shield were read monthly throughout the construction period. During the warranty period which extends 10 months beyond the completion of tunneling, readings were reduced to a bimonthly schedule. This overall schedule was based on a tunnel advance rate of approximately 30 feet per day. Basic readings were taken once during the day and additional readings were made when the level of distress indicated by instrumentation approached or had exceeded the established limits. When tunneling activities were reduced due to equipment repair or other incidents for more than two weeks, weekly readings were reduced accordingly until tunneling was resumed.

Establishment of Movement Criteria

The movement criteria established for this project were based on the "threshold of architectural damage"[1] to structures and corresponds to an angular or lateral distortion of the ground surface as indicated below:

Settlement Points - 1 inch settlement at centerline of tunnel.

Inclinometer - slope of 1:1000 measured to the point of maximum deflection.

Water Observation Well - 1 foot drop in water level.

Tape Extensometer Point - tensile strain of 1:1000 of distance measured between two points.

Tilt Plate - rotation of 1:1000.

These criteria serve as an alarm for corrective measures (compaction grouting) if deemed necessary or precaution to prevent the adjacent buildings from suffering structural damage if this level of distress was exceeded.

Findings

Ground loss due to tunneling was estimated from the settlement troughs which developed over the tunnel. The trough is believed to reflect the soil displaced into the tail void, effect of the timber lagging expansion, the effect of grouting and consolidation of the adjacent soil mass. The ratio of unit ground loss (unit volume of settlement trough) to the unit volume of soil excavated ranged from 0.5 to 6.9 percent as shown on Figure 2 for different segments of the project.

It is apparent that throughout the major portion of the grouted fill materials along Fawn Street as well as the natural soil materials along Fayette Street, the settlements were nominal and within the normal range of one percent ±[1], resulting from good tunneling practice. Between Stations 12+50 to 14+00, within the grouted area, a high loss of ground ratio (6.9 percent) was experienced. This was associated with a portion of the tunnel alignment that, although grouted, was sheeted and shored to construct a shaft for realigning the tunneling machine. To compound the poor results at this location, the tunneling machine actually stove in the back wall of the shaft displacing the rear sheeting as a result of an over-exuberant tunneling crew. Other areas of moderately high ratios occurring along Eden Street, in natural soil formations, are directly attributable to the breaking of water and sewer utility lines.

The ratio of maximum horizontal movement as indicated by inclinometers and the maximum vertical surface movements are shown in Figure 2. Through the majority of the area, this ratio is approximately 30 percent(2) which is considered normal for tunnel construction. Also, unlike the ground loss due to tunneling, the ratio of maximum horizontal movement to vertical movements appears to be independent of soil formation or whether the soil has been grouted or not. However, near the shaft locations, the ratios were much higher and ratios above 120 percent were recorded. If cut-and-cover construction techniques had been employed which are similar to the construction method used at shaft locations, it is apparent that far greater movements would have been experienced throughout the tunnel alignment.

Both vertical and lateral ground movements were less in the grouted area as compared to those in the ungrouted area as shown in typical data plots on Figures 5a, 5b, 6a, 6b, 7a and 7b. In the grouted area, some of the inclinometers indicated movements away from the proposed tunnel after chemical grout was injected as shown by curve "1" in Figure 7b. Some settlement points also seemed to indicate a very slight upheaval in the order of .006 feet. However, this effect diminished rapidly as the chemical grout set up.

Water levels in all functioning observation wells dropped as the tunneling machine approached regardless of whether the area was grouted or not as indicated in Figures 8a and 8b. The water level remained essentially at the elevation of the tunnel invert until the line pipes had been placed and grouted.

Conclusion

Through good workmanship, the use of a closed-face tunneling machine and chemical grout a successful and safe tunneling operation was accomplished. Environmental impacts on the surrounding neighborhood were kept to a minimum.

By use of chemical grout in filled materials, the surface settlements and disturbance to adjacent buildings were limited to similar small movements experienced in natural soil formations. No utility breaks occurred in the grouted zone while a number of utility failures were experienced in the ungrouted section.

For a tunnel project of this magnitude spacing of instruments at 50 feet to 100 feet in sensitive areas and 100 feet plus in less sensitive areas is adequate.

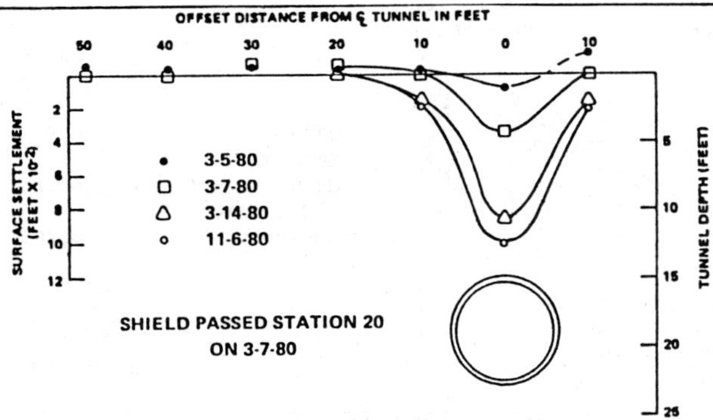

FIGURE 5a – SETTLEMENT PROFILE AT STATION 20+00
(UNGROUTED AREA)

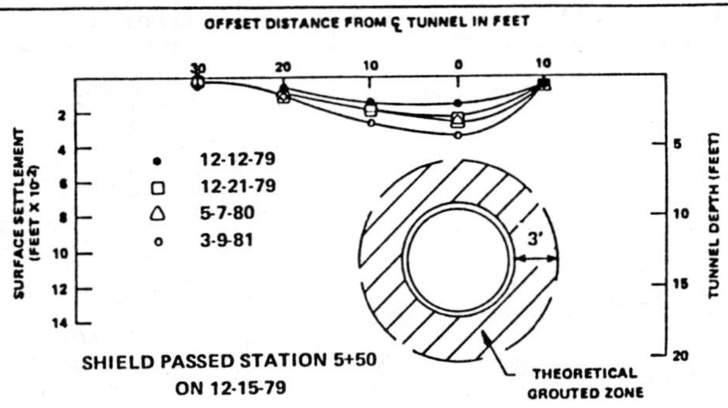

FIGURE 6a – SETTLEMENT PROFILE AT STATION 5+50
(GROUTED AREA)

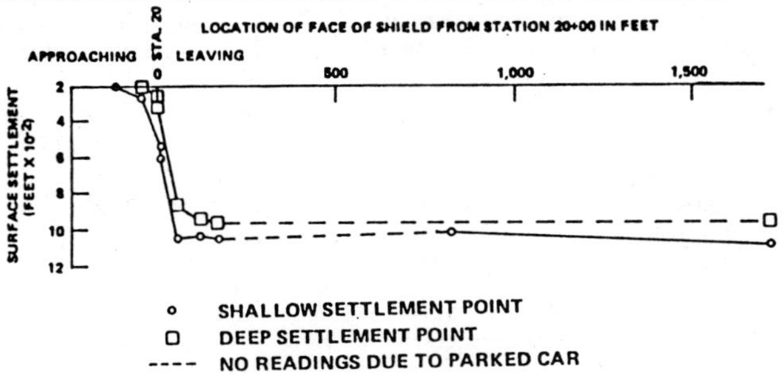

FIGURE 5b – ℄ SETTLEMENT AT STATION 20+00
(UNGROUTED AREA)

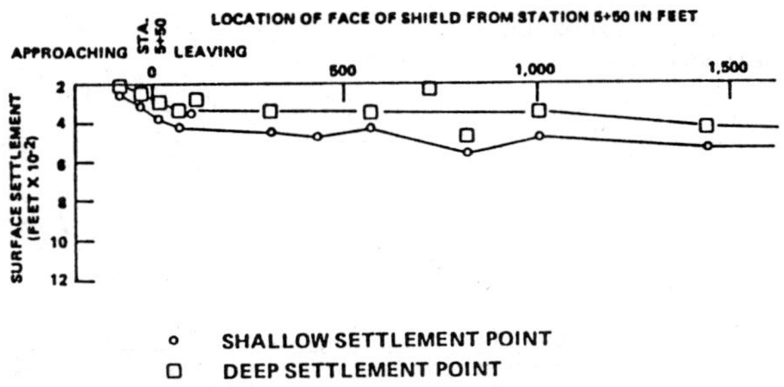

FIGURE 6b – ℄ SETTLEMENT AT STATION 5+50
(GROUTED AREA)

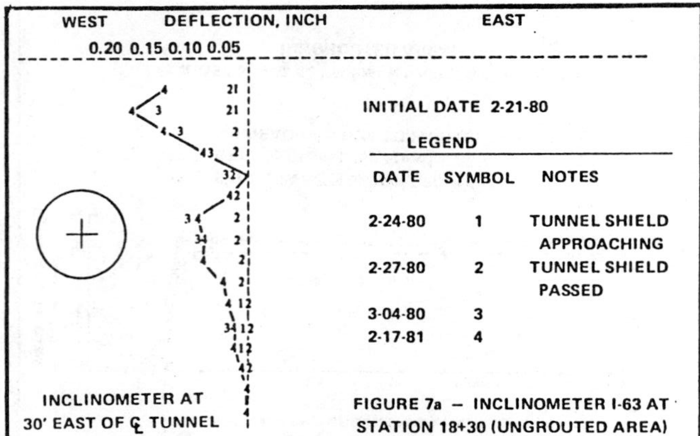

FIGURE 7a – INCLINOMETER I-63 AT STATION 18+30 (UNGROUTED AREA)

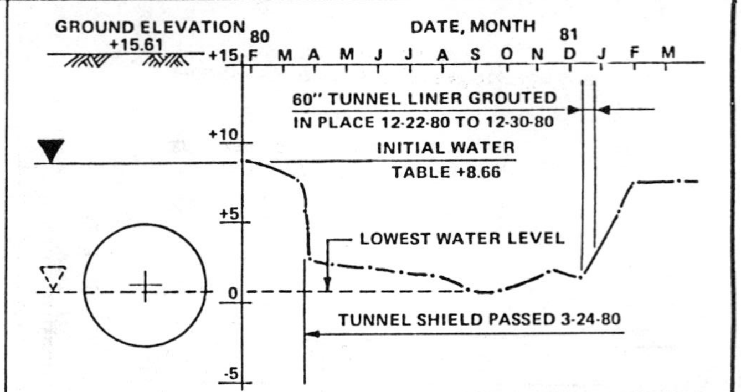

FIGURE 8a – WATER TABLE ELEVATION VS. TIME AT STATION 24+50 (UNGROUTED AREA)

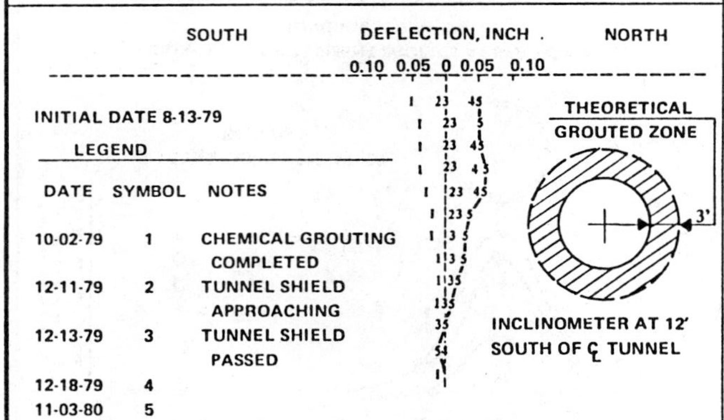

FIGURE 7b – INCLINOMETER I-5 AT STATION 6+00 (GROUTED AREA)

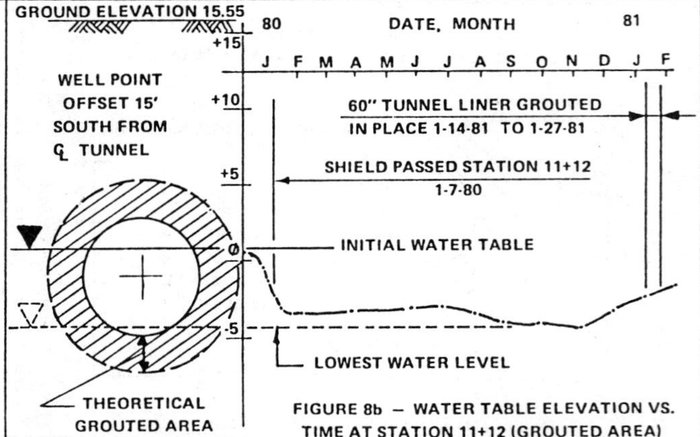

FIGURE 8b – WATER TABLE ELEVATION VS. TIME AT STATION 11+12 (GROUTED AREA)

If there are budget limitations emphasis should be placed on surface and deep settlement points, inclinometers, and observation wells rather than tilt plates and tape extensometer points.

When properly planned and implemented, instrumentation and chemical grouting serve well as part of a policy to insure the safety of buildings and protect the Owner from claims.

Acknowledgments

The authors express their thanks to Francis W. Kuchta, Director of Public Works of Baltimore City, and Theodore M. Andriotis, Head, Bureau of Water & Waste Water, for permission to present and publish the data collected on this project. Special thanks are also extended to Nelson A. Luthy, Chief - Division of Waste Water Engineering, for his comments and guidance during the design of the project.

REFERENCES

1. Cording, E. J. et al. "Methods for Geotechnical Observations and Instrumentation in Tunneling," The National Science Foundation UILU-ENG 75 2022, 1975, pp.97-100.

2. Schmidt, B. et al. "Construction Monitoring of Soft Ground Rapid Transit Tunnels - Volume I: A Definition of Needs and Potential Developments," U.S. Department of Transportation, Report PB 241 536, 1974, pp. 2-7 to 2-9.

3. O'Rourke, T. D. et al. "The Ground Movements Related to Braced Excavation and Their Influence on Adjacent Buildings," U.S. Department of Transportation, Report No. DOT-TST 76T-23, 1976, pp. 4-1 to 4-29.

GROUTING TECHNIQUES USED IN DEEP SOUTH AFRICAN MINES

Hans K.O. Dietz, Pr.Eng., F.E.A.S.A.

ABSTRACT : This paper deals with grouting techniques typical of Mines operating in depths between 500 and 4000m below surface. Very high sealing pressures are applied and the transport of all grouting material is effected through pipe lines. Unusual ground conditions require unusual grouting methods which in turn necessitate stringent safety precautions.

A Mine is quite naturally an extensive drainage system in itself but there are other reasons which make grouting necessary, such as thickness and nature of overburden and of overlying strata, nature and situation of the natural ground water table or for instance fossile water trapped under high pressure in greater depth, but also the need to combat extensive moisture in refrigerated areas or sealing off of methane gas yielding fissures.

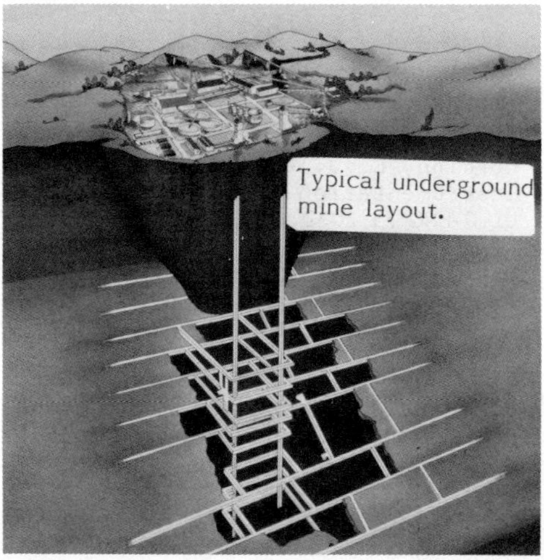

All Mines which have to combat water must answer the question : Draining of the water and pumping it out or sealing off potential water ingress?

Sealing may be expensive but is a once and forever operation.

Draining and pumping is an ongoing activity the cost of which depends mainly, but not exclusively on :

Life span of Mine;

Inflow being constant or not;

Author: H.K.O. DIETZ, General Manager, Rodio (South Africa) (Pty) Ltd.

Depths from which pumping is carried out;

Rising cost of energy;

Maintenance expenses for pump system.

For safety reasons, it is simply not possible to allow the discharge of water in big quantities and under considerable pressure into underground workings.

Consequently, there are many cases where grouting ahead of development is imperative and cannot be questioned.

GROUTING PROCEDURES.

General:

The really deep Mines in the country are Gold Mines. Presently the most shallow "reefs" being mined are gold bearing quartzites of 10-50cm thickness which are found approximately 600m below surface.

At the other extreme the deepest point of development has passed the 4000m mark.

For practical reasons emphasis is laid on preventative grouting while the available pumping installations have to cope with the normal seepage and provide standby capacity for emergency cases.

There are some basically different grouting procedures :

1. Cover grouting ahead of the excavation with the purpose of creating a dry zone through which the gallery or face is then advanced.
2. Ring or fan grouting against flowing water which becomes necessary when for whatever reason water runs in from the exposed surface at an intolerable rate. This is always an emergency operation and unavoidably delays the development.
3. Stope cover grouting to prevent cracks from opening and yielding water when the hanging wall of a mined out area subsides.
4. Pregrouting in connection with shaft sinking with the purpose to seal water in overburden (occasionally also layers of granulated rock fragments), weathered rocks and water bearing fissures in the underlying bed rock. At the same time broken, weathered and unstable ground shall be consolidated so as to create a water tight solid zone through which the shaft can be excavated without the hazard of water or crumbling walls.
5. Packgrouting which involves pumping of large quantites of mortar in order to create a concrete core in the supporting timber packs of worked out stopes.

   Remark : The grouts which are necessary for the foregoing operations are all based on cement.
6. There is also the need for sealing off very fine cracks by injecting chemicals to prevent the yield of gas or of seeping water.

7. Recently some Mines have started to build underground compressed air storage chambers by plugging no longer required haulages and/or driveways. There plugs are constructed basically of mortar intrusion concrete whereby the coarse aggregates are taken from the waste of the nearest development area, the mortar being prepared and intruded by the available grouting installations. Inside the chamber also very fine cracks must be sealed to prevent the air from escaping.

EXECUTION OF GROUTING WORK

Typical features of grouting in deep South African Mines are :

Extremely large quantities of solid injection material are sometimes necessary in dolomitic formations.

Necessity to work simultaneously at various places on different levels far from each other and far from surface.

Transporting large quantities of solid grouting material which when done by means of conventional hoisting and lowering plant would be very involved and deprive the mine of valuable transport capacity.

Very high water counter pressure in the rock which needs to be overcome (up to 15 MPa static pressure and even more for fossil water contained in compressed pockets).

Different mix compositions and injection pressures necessary at the various working places.

All ongoing major grouting work must be supervised from one control station.

General procedure for cover grouting.

Preconsiderations.

Consult overall geology
Make use of observations and experience in neighbouring mines.

Practical Execution.

Exploratory drilling is done ahead of the excavation.

Rotation drilling is executed to a length of ± 100m, usually in AX size (=48mm) from bays cut some distance behind the development face. Bays are alternately situated on either side of the gallery approximately 70m apart, thus holes overlap ± 30m.

Safety devices must be installed in case water under high pressure is struck. If water is struck the rate of flow and pressure are estimated.

The exploratory holes are used as access for grouting.

In order to reduce the need for redrilling of a cement filled portion of the hole a packer is installed and connected to the grouting pump.

After grouting, the hole is reopened and drilling is continued through the grouted zone to the full length.

These two illustrations hereunder show the rigid anchoring of drilling machines which may intersect water under high pressure.

DIAMOND DRILL SAFETY RIG

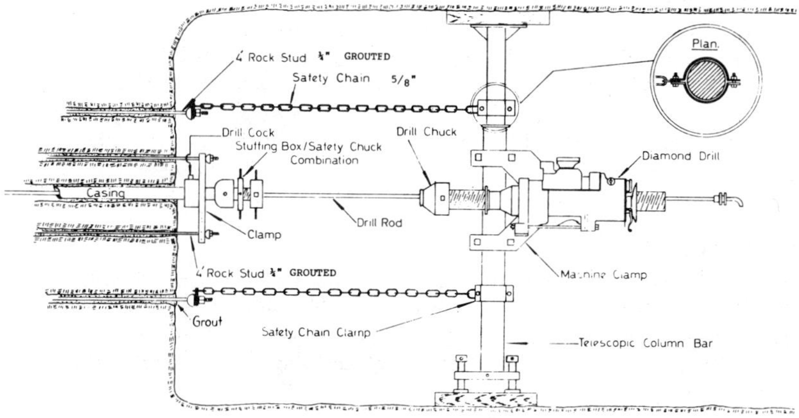

PERCUSSION DRILL SAFETY RIG.

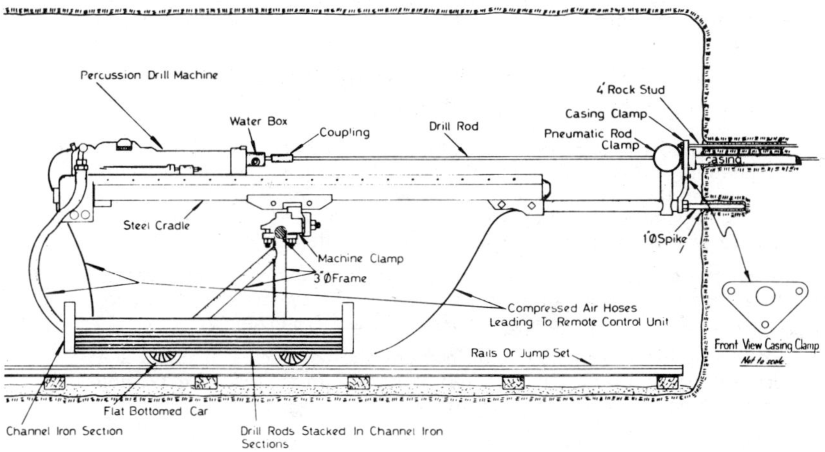

Ring or fan grouting.

The most probable procedure is :

Install short pipes, fitted with a cock on the exposed end at the spots where maximum seepage occurs.

Patch water discharging fissures between pipes with adequate mortar thereby directing the water flow through the pipes keeping the cocks open.

Inject through one pipe until grout emerges from the adjacent one. Close discharging cock and continue grouting until pressure builds up. Now close the cock at point of connection and disconnect groutline.

Apply the same procedure to another pair or system of pipes and continue until no pipe yields water when its cock is opened.

This method is easier described than successfully executed.

Stope cover grouting.

Fans of holes more or less parallel to the plane of the hanging walls are drilled from specially excavated bays before stoping is commenced.

Depending on the extent of the unavoidable rock movement after excavating the reef, a "post grouting" applying the same procedure may become necessary.

Pregrouting of shafts.

A pregrouting campaign consists generally of drilling holes of a suitable size within or close to the perimeter of the future shaft and of injecting a suitable grout whenever unstable ground or water bearing fissures are encountered in order to fill the voids and/or fissures properly.

Two basically different methods are presently being applied.

Working from the bottom of excavation.

This method foresees grouting through holes which are drilled by means of pneumatic machines from the shaft bottom, either within the section of the shaft or from specially excavated bays. Such holes the number of which may vary according to the local conditions, are generally drilled to a depth of 30-50m and then injected.

When the grouting is completed, the shafts are further excavated to a depth of about 6m above the bottom of the treated zone, in order to leave a cover against water inrush from below.

From this newly reached shaft bottom another series of holes are drilled and grouted. The shaft is thus sunk in stages. Excavation and grouting being executed alternately.

Precementation from surface.

When the grouting work is done from surface, before the commencement of shaft sinking, the cover holes are laid out near or along the perimeter of the shaft and can be drilled down to any required depth well before the actual shaft skinking is started. This eliminates almost all interruptions and delays arising from ground water and broken rock during the actual shaft sinking an operation burdened by a very high fixed overhead cost.

Drilling is usually done by the rotation method in stages of approximately 30m followed by water pressure testing, which decides whether grouting is necessary or not. For executing the test, a packer is inserted which will remain in position for the subsequent grouting operation.

As no hole can be maintained absolutely straight over long stretches, it must be surveyed from time to time and if necessary deflected in order to bring it back into proper direction.

Grouting by means of a packer, which reaches deep into the hole has the advantage that much greater pressures can be applied than is possible when working from the bottom of the shaft. In this way, the spread of grout can be increased considerably.

Filling of grout packs used as stope support.

For this work a fairly stiff cement mortar is required, however, the same equipment used for processing and transporting of grout is used. Although no pressure at the point of discharge is required, high pressure pumps must push the mixture over long distances through small diameter pipe ranges and thereby overcome high frictional pressure drops. The mortar must travel very fast at velocities of more than 2m/sec in order to avoid blockages arising from segregation.

Sealing fine cracks which discharge water (refrigeration area), gas or air from compressed air storage chambers.

Conventional grout is not suitable.
Small quantities and very low viscosity are required.
Normally two component chemicals are used which are mixed on the spot and injected by manually operated grout pumps.

OPERATIONAL PARAMETERS FOR GROUT INJECTION.

The main objective of pressure grouting in mining is to seal off water which otherwise would enter the workings and not consolidation or improvement of the physical properties of the surrounding rock.

For this purpose it is not always necessary to mobilise the maximum mechanical strength of the cement.

The hardened grout must however withstand erosion. Any grout contains an enormous excess of water required as a transport medium which must eventually drain away.

The following operational particulars must also be considered :

Distance of and time required for transporting grout to the spot.

Pressure under which injection must take place.

Frictional losses in pipe range.

Commencement of setting and time required for hardening of injection material.

Time available for the operation.

One paramount factor must always be kept in mind : The grout must be the cheapest mixture just good enough to achieve the objective.

The actual injection process consists of :

Removing of drilling equipment.

Insertion and setting of a packer (obturator) against high water pressures.

Connecting of injection pump to the packer.

Pumping under high pressures until limit pressure is reached.

Recording of quantities pumped, taking into account the contents of the pipe range.

Monitoring of injection progress and (telephone) communication with supervisor in central plant on surface.

Emptying and flushing of pipe ranges after injection.

DOSING AND MIXING OF GROUT AND TRANSPORTING IT TO THE INJECTION FACE.

In wet Mines more than 1000 metric tonnes of solid grouting material may be required per month. Transporting such quantities in bulk or in pockets down the shaft and underground to the working places is

- very labour intensive
- occupies a lot of the otherwise urgently required underground transport capacity
- results in breakage and wastage

It has been established that up to 50% of conventionally transported grouting materials have been wasted and virtually gone down the drain.

Between 1958 and 1960 Rodio (South Africa) (Pty) Limited, introduced a conveying system through pipe ranges which has now been commonly adopted. All grouting materials such as cement, bentonite, filler, sand and additives, are pumped to the working places from surface. In one case the absolute distance was more than 2,300m including a vertical drop of 1,850m down the shaft.

Particulars of the pipe range system.

One inch diameter pipes are used.

The velocity of grout must be high in order to prevent decantation and blocking.

There is one pipe range per shaft with a conveyor pump attached to it in the central plant through which the basic grout is transported to a relay station down in the underground workings.

Frictional losses must be taken into account.

Abrasion of the pipes is highest in the bends.

The required and permissible time of conveyance is a factor which governs many of the beforestated features.

Basic principles of handling grouts.

A fully mechanised and semi automatic central dosing and mixing station on surface prepares a concentrated basic mixture (Water: Solid ratio by weight = 0,6 to 1,0).

If required, only one solid component plus water can be handled.

The optimum viscosity of the grout must be established.

The basic mix is pumped down to the underground relay station where the grout is diluted to requirement and transferred to the receiving tanks of the actual injection pumps.

Each injection pump is connected to an injection hole or other outlet by means of high pressure pipes.

The distance between injector and rock face can be more than 300m and a further 100m may have to be travelled in the hole before the spot is reached which actually requires treatment.

Features of the centralised mixing plant.

Designed to dose and mix at any ratio within 2% accuracy four components (three solids plus 1 water) with a capacity of 2 to 3 metric tonnes of solids per hour. Measuring is always automatic thus eliminating human errors.

As required the plant can run fully automatic or can be operated manually in order to suit varying conditions.

GROUTING STATION: OUTSIDE VIEW

DESCRIPTION AND TECHNICAL FEATURES OF MACHINERY AND EQUIPMENT EMPLOYED.

The dosing and mixing station consists of- Storage silos for bulk delivery of

Cement-Flyash-Slime-Bentonite

Batch measuring scales for

Cement-Flyash or Slime-Bentonite

Mixing tanks with high frequency centrifugal type mixing pumps handling the dosed components.

Transfer or storage tanks fitted with a rotating paddle into which the mixers transfer the grout after processing.

Transport plunger pumps which draw from the storage tank and convey the grout to the underground relay station through a one inch diameter pipe range.

An underground relay station consists in general of a receiving tank with a rotating paddle and two or three double drum storage tanks into which the received grout is transferred by means of a centrifugal pump. Attached thereto is a water supply line with a water meter which allows a controlled dilution of the basic mix if required.

As the relay station is usually situated close to an underground shaft station, the prime energy for mixers and transfer pumps is usually electric power.

However, the actual injection pumps which draw the grout from the storage tanks in the station and force it into the ground are propelled by compressed air.

The pneumatic piston motor usually has a 14"Ø (350mm) cylinder and drives via the piston rod the reciprocating actual plunger injection pump. Details of this unit as shown on the illustration hereunder.

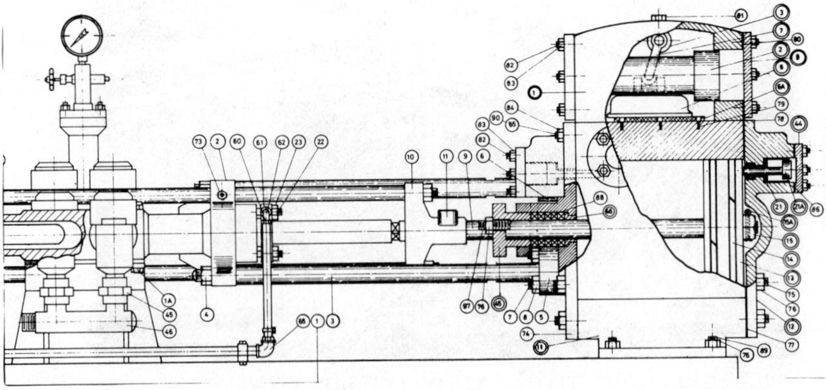

Underground grouting is characterized by very high sealing pressures. Grouting is mainly done in connection with developments. These are usually far away from permanent installations which could be affected by rock movements. Thus the high grouting pressures which are required can be applied without undue risk.

The pressure which the injector pump must generate and which has to be taken by the pipe ranges must amongst others overcome

the water counter pressure in the hole generated by the static head or the compression force in case of fossil water pockets;

the frictional losses in the ground when travelling through fissures and cracks;

the frictional losses in the pipe range which are always comparatively high as the aim is to inject the greatest possible quantity of solid injection material and hence a grout of the highest possible viscosity.

These facts can bring the injection pressure to 35MPa ($\pm$ 5000 psi). An occasional widening of the fissures cannot be excluded. It is, however, accepted as unavoidable.

Although the pipe ranges leading from the injector to the hole must stand the highest pressures all pipe ranges are assembled of the same type of drawn steam pipes which have successfully been tested at 43 MPa ($\pm$ 6000 psi).

The assembling and maintenance however, requires special care and attention.

The pipes must be joined by special high pressure couplings with tee-piece unions at regular distances. Excessive stress on the pipe range resulting from its own weight (for example, down the shaft) must be avoided.

The bends for change of direction must have a radius of not less than 1,25m (=4ft). Suspended pipes must be fixed to the rock at short distances so as to prevent swinging and breaking under the strokes of the pumps.

From time to time the pipe ranges must be pressure tested and if necessary partially or in full replaced when their strength has deteriorated as a result of abrasion.

Oxidations and corrosion from outside however, affect the lifespan of the pipes more than inside abrasion.

The free flow in the pipe ranges must be checked continuously. Plugging must be avoided at all costs.

From the moment a range is assembled it must be kept full of water when not in use. Water must also be pumped before conveying grout to make certain the range is free from obstructions. After completion of a grouting operation, the range must be flushed.

A long range holds a considerable quantity of grout. This must be taken into account when completing a given job. Flushing a pipe full of grout into the mine's drainage system must be avoided under all circumstances.

Each end of a pipe range which comes from an injection pump is connected to the hole by a three metre (10') 3/4" Ø high pressure hose attached to a high pressure cock and a discharge device provided with a pressure gauge. The whole system is safely chained to the rock.

In the places where men are working all piping and hosing must be extra covered as grout escaping through leakages can inflict painful wounds.

Whenever possible packers are inserted into a grout hole in order to reduce the need for redrilling after completion of the injection work.

In slim diamond drilled holes which usually have smooth walls in sound rock the simple leather cup packer is found best suited. Twelve cup packers are used up to a distance from the mouth of ± 200m standing up to injection pressures of 36 MPa (5100 psi). When, however, a hole discharges great quantities of water placing a packer may become impossible.

INJECTION MATERIALS

Selecting the technically best and economically cheapest material requires a lot of consideration, studying and testing.

The hereto attached schedules and graphs give particulars of the various characteristics of grouting materials and their behavious under differing conditions. Assessing of grouts follows in broad lines the rules of concrete technology

With reference to the mentioned annexures, this paper restricts itself to mentioning only the most important features to which attention must be paid.

Water should be taken from the nearest source provided it is suitable. Mine water, however, is not always appropriate.

Binders are usually not judged by their ultimate mechanical strength but by their fineness of grain size and most of all by their cost. The ultimate aim of the grouting operation in a mine is sealing, not stabilisation. Normally Portland- or premixed Blast Furnace Cement is used, sometimes also Portland Cement plus slag Ash is blended on site.

When greater quantities of grout are required fillers are added to the binders in order to reduce cost. Such fillers are usually fine sand, mine (tailings) slime, (if not chemically polluted) or quartz flour.

Chemical additives are used to control the viscosity of the grout, to prevent it from settling and thus plugging the pipe range but still ensuring sedimentation once inside the rock fissure.

The mostly used additives are

Bentonites, Retarding agents, or Cement-Plasticisers

the following three pages give operational data which are self-explanatory.

COST ELEMENTS OF GROUTING OPERATIONS

When assessing the operational costs of ongoing grouting one realises that the biggest single factor is the transport of the solid materials to the rock face.

Considering the enormous quantities to be transported the high initial costs for setting up a surface plant with substations and long pipe ranges are nevertheless a very economic proposition.

RODIO (SA) (PTY) LIMITED Grouting-Operational Data

CERTAIN CHARACTERISTICS OF SOLIDS

Binders	Sp. Mass	Blaine Cm^2/gr	Liquid Limit	Plastic Limit	Chemical Compos-tion % CaO	SiO	MGO
Portland Cement	3 113	3020	-	-	64,7	21,8	1,7
PBFC	3 010	3520	-	-	50,4	26,8	8,1
Slag	2 902	3520	-	-	30,8	32,4	19,9
Fillers							
Ground Quartz	2 610-2770	4540	-	-			
Slime	2 647	1920	-	-			
Additives							
Bentonite (Wyoming)	2 420	-	554	38,4			
S.A. Bentonite	-	-	526	42,8			
Bentonite Clay							
(Natural)	-	-	101,5	-			
(Activated)	-	-	134,0	-			

VISCOSITY OF BENTONITE

	Marsh viscosity (seconds) for Bentonite plus water, (ratio by weight) 1%	5%
Wyoming Bentonite (natural)	13,6	17,5
South African Bentonite (activated)	13,8	19,2
Bentonite Clay (natural)	13,1	13,2
Benotnite Clay (activated)	13,3	13,8

The Marsh viscosity of a bentonite suspension is accepted as a measure of efficiency in respect of sedimentation and volumetric index, but not as a measure of pumpability.

VOLUMETRIC INDEX OF GROUTS

Bentonite increases the amount of water permanently retained, thus improves the volumetric index.

Portland Cement	Slime	Bentonite	W/S Ratio
100	0	0	0,27
100	0	2	0,365
75	25	2	0,36
50	50	2	0,35

RODIO (SA) (PTY) LIMITED — Grouting-Operational Data

COMPRESSIVE STRENGTH OF BINDERS (Comparison)

Binders	7 Days	28 Days
Portland Cement	100,00%	100,00%
Portland Blast Furnace Cement	63,10-71,88	84,20-85,90%
Portland Cement + Slag :		
ratio 75/25	83,00%	101,00%
ratio 80/50	65,80-74,2%	81,30-88,50%

INERT FILLERS

Grain Size : Below 0,60mm, the grains tend to floculate, making it difficult to establish a grain-size curve for the finer portion. Slime is somewhat coarser than the Portland Cement with which it is mixed.

SIEVE CURVE

Of Mine tailings (slime) as compared to ordinary Portland cement.

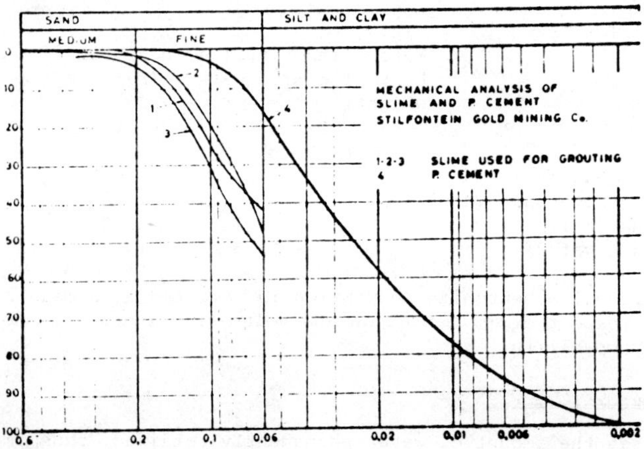

Chemical Composition

SiO_2	89,3%	96,5%	98,74%
Al_2O_3	6,48)		
Fe_2O_3	1,2%)	2,2%	1,13%
SO_3	1,8%		
Trace Elements	1,3%	1,3%	0,13%
	100,00%	100,00%	100,00%

RODIO (SA) (PTY) LIMITED — Grouting-Operational Data

MECHANICAL STRENGTH OF HARDENED GROUTS

RATION OF SOLIDS				WATER	SPECIF.	COMPR.STRENGTH AFTER	
PC	PBFC	SLAG	SLIME	%	MASS	7 DAYS	28 DAYS
						%	%
100	-	-	0	35,7	2,08	100,00	100,00
75	-	-	25	33,7	2,02	65,4	78,1
50	-	-	50	31,8	1,97	34,3	37,1
25	-	-	75	29,8	1,93	9,8	14,8
-	100	-	0	33,6	1,06	71,8	84,2
-	75	-	25	32,1	1,96	40,8	62,9
-	50	-	50	30,7	1,94	19,7	35,3
-	25	-	75	29,2	1,91	7,9	14,4
50	-	50	0	33,4	2,07	74,2	81,3
32,5	-	32,5	25	32,0	2,00	62,1	68,3
25,0	-	25,0	50	30,6	1,97	38,8	48,5
12,5	-	12,5	75	29,2	1,88	9,9	19,2

PC + Slime. The strength at 7 days bears witness to the retarding effect of the impurities of the slime.

PBFC + Slime. PBFC appears to be less sensitive than PC to the effect of the impurities of the slime, when in a proportion above 50%.

SETTING TIME OF GROUTS

PORTL. CEM.	SLIME	BEGINNING	END
100	0	3 hrs 50 min	6 hrs 10 min
90	10	4 hrs	6 hrs 20 min
80	20	4 hrs 10 min	6 hrs 40 min
70	30	4 hrs 20 min	7 hrs
60	40	4 hrs 35 min	7 hrs 30 min
50	50	5 hrs	8 hrs

These times are within the limits prescribed by the British Standards: not less than 30 minutes, but no more than 10 hours.

The beginning of the setting time shows the time available for mixing, transporting and injecting the grout into the rock fissures. The end shows the delay before the mine work (blasting) may sefely be started.

Note : P.C. = Portl. Cement; P.B.F.C. = Portl, Blast Furnance Cement

When big quantities for filling big fissures or cavities are required the cost per unit injected decreases because

- cheap filler is added to the grout;
- the pumping velocity is much greater than normal.

To compare the cost of all grouting work with the overall working cost of a mine does not really say much because the question whether to execute grouting or not depends in the first instance on the geological circumstances and is not so much a matter of discretion.

Experience has shown that with greater depth of the workings and with increasing age of the mine the necessity for grouting decreases significantly. The author recalls cases where the mine was virtually rendered dry and the specialist contractor has grouted himself out of business.

CONCLUSION.

Finally, I like to give some figures on the dimensions and quantities indicative for a Mine which has employed the services of the author's company. The Harmony Gold Mine in the O.F.S. which is the second largest of its kind in the world,

employs	approximately 32 000 officials and workmen
excavates] and hoists]	in excess of 700 000 tons of rock per month (± 265 000m³ of rock in situ)
circulates	1'650m³ air per sec or in other words 5,400 tons air per one ton rock excavated
stopes	ore along a face front of almost 14 500m.
installs	besides rockbolts approximately 400 rock anchors of between 500 and 1000 KN retention force per month.
injects	between 500 and 900 tons of solid grouting material per month.

Gloss : The South African Mining Industry and its engineers are renowned for their high standards and achievement. Grouting in deep Mines, however, and related developments always was and still is a matter of civil engineers.

ACKNOWLEDGEMENTS The author is indebted for allowing to publish operational and/or research data related to the subject of this paper to:

General Mining - Union Corporation Ltd. Johannesburg.

Rand Mines Limited, Johannesburg.

H.L.E. Du Bois, Pr.Eng., Consulting Engineer, Johannesburg.

CEMENTING STEEL LININGS IN A DEEP BLIND-BORED SHAFT

By Ray W. Amstutz[1] and James E. Hawkins[2]

Abstract

The U.S. Bureau of Mines contracted in 1977 for the boring of a 120-inch diameter hole to a depth of about 2400 feet. The shaft is located in Rio Blanco County in the central area of the Piceance Creek Basin in northwestern Colorado. Purpose was to reach the deep, thick oil shales and to conduct experimental mining tests.

A large, oil-field-type drilling rig was used. Drilling mud was designed and used to stabilize the hole walls, retard dissolution of water soluble minerals, prevent the inflow of water and/or gases, provide cooling of the bit and to transport the cuttings to the surface. After all boring was completed, sections of the 96-inch ID casing were welded into a continuous closed liner and floated into the hole.

Standard hydraulic oil field cementing techniques were used to pump the slurry down multiple strings of small tubing placed in the annulus. The annulus was filled from the bottom to the surface in 9 stages. Considerations included in the cement design were mud properties, casing collapse strength, rock conditions, formation dissolution characteristics and the viscosity, density, strength and setting time of the cement. The mix was designed for maximum bond to the steel pipe and also for best attainable bond to the various rock types.

Cost effectiveness of the various operations and procedures are included in the report. Considerations are given to possible improvements including (1) use of thin-wall open-ended casing, (2) providing a larger annulus area for easier pipe emplacement and cementing operations, and (3) improved drilling muds to prevent hole irregularities and enlargement.

Total cement pumped during the 9-stage operation was 62,000 sacks yielding 83,000 cubic feet of slurry. Operation was completed in 9 working days during a calendar period of 17 days.

---

[1]Williams Brothers Engineering Company, Tulsa, Oklahoma
[2]U.S. Bureau of Mines, Denver, Colorado

Introduction

In 1974, the U.S. Bureau of Mines began an extensive research program directed primarily toward providing a sound technical approach to accessing and mining the deep, thick saline oil shale deposits in the Piceance Creek Basin of northwestern Colorado. All prior mining efforts had been conducted in the shallower, non-saline facies of the Green River oil shale formation. This present effort was the first attempt to construct and develop a mining operation in the deeper, richer shales. Several unknowns, among which were the nature and extent of the aquifers above the mining horizon and the behavior of the oil shale beds containing bedded and disseminated saline minerals, made this pioneer effort important to the future development of these deeper horizons.

These deeper shale horizons are extremely important because they contain the bulk of the oil shale resource in the Piceance Creek Basin.

After a careful engineering assessment, the decision was made to bore the shaft rather than sink it by conventional methods. The primary factor in this decision was the presence of two separate high-pressure aquifers above the proposed mining zone and the lack of knowledge concerning their hydrologic characteristics. Freezing or grouting was considered but the economics of each could not be determined with the required degree of certainty. Drilling, on the other hand, could be quantified because the drilling mud could be tailored to cover all known variations in permeability, porosity and pressure present at the site. Additionally, the drilling mud could be maintained, by the addition of chemicals, in a state that would minimize or prevent the dissolution of the saline minerals known to be present in the mining horizons. It now is known that the calculations in this regard were too optimistic as explained in the cementing section.

Drilling Operations

Engineering and planning for the project began in 1976 and the actual drilling operations on March 25, 1977. The schedule required that drilling operations be performed 24 hours per day, seven days per week.

A surface hole, 140 inches in diameter was drilled to a depth of 195 feet below the collar. This hole was lined with 122-inch ID steel casing to a depth of 190 feet. The bit size was reduced to 120 inches and drilling resumed.

Drilling the 120-inch hole proceeded smoothly until a depth of 845 feet was reached. At this depth, two of the roller cutters from the bit were lost in the hole. Retrieval was achieved after seven days of "fishing". Prior to and after this time, numerous bolts were lost from the drilling assembly, but they were considered to be only minor incidents at that time.

After recovery of the two cutters, cutting was resumed and proceeded to a depth of 1781 feet. The drill pipe was pulled from the hole to change bits and, in so doing, the entire bit body and cutter assembly were lost in the hole. It was an unfortunate point in the operation for this event. Both high-pressure aquifers had been penetrated and all of the bedded halite (rock salt) had been drilled. The base of the deepest bedded salt was at a depth of 1777 feet. The bit body lodged in the hole at a depth of 1245 feet, thus effectively sealing off the bottom of the hole from circulating fluids until the fishing job was completed 60 days later. This was to have unfortunate consequences as detailed later.

Failure analysis of the two-inch bolts used to secure the bit cutter plate to the bit body indicated that the bolts had failed because of a combination of over-torquing and the corrosivity of the saline drilling fluid. Further investigation revealed that the bolts were manufactured from steel that had a very low fatigue life in this severe environment. All bolts in the drilling assembly were replaced. The new bolts were of a more fatigue-resistant material, had finer threads and were secured by lock-nuts. A program was adopted of changing all bolts prior to expiration of their fatigue life and insuring that all bolts were carefully torqued to their proper limits. No additional serious bolt failures were encountered.

Future shaft drilling assemblies, particularly those penetrating saline rock sections and/or using saline drilling fluids, should be carefully fitted with proper fasteners and the rig crews should be trained and supervised to insure proper installation techniques.

After recovery of the bit, drilling proceeded smoothly to the scheduled total depth of 2371 feet.

The custom designed and fabricated casing was run in the hole in 40-foot lengths. Each length was field-welded in the vertical position on the drilling rig floor, x-rayed and then lowered into the hole. The bottom of the first joint was sealed with an hemi-spherical steel head, thus providing the required buoyancy. It was necessary to float the casing into the hole in this manner because its total weight, in air, exceeded the weight capacity of the derrick by a factor of about five. The total length of the casing set in the well was 2352 feet.

The fact that the casing had to be designed to resist the external hydrostatic pressure of the mud column necessitated using thick-walled casing at the bottom. The casing wall was graduated in thickness from two inches at the bottom to three-fourths of an inch at the shaft collar. In addition, heavy steel reinforcing rings were welded to the exterior of the casing for further reinforcement. These rings ranged from 3 5/8 x 8 inches to 3 x 6 inches in cross section. They were spaced two feet apart at the bottom of the string and ten feet apart near the surface. The total weight in air of this mass of steel exceeded five million pounds. Some possible economics in casing design are included under Cost Effectiveness and Possibilities for Improvements.

Cementing the 144-Inch Corrugated Metal Pipe (CMP)

The first cementing job on the shaft project was the cementing of twenty-one feet of 144-inch diameter CMP into the shaft collar. The hole was blasted prior to constructing the collar pad. The CMP was lowered into the hole with a crane. The rock at the bottom of the blasted hole was badly fractured. When 1500 sacks of neat cement slurry with three percent calcium chloride was pumped into the annulus around the CMP, the level of the cement would not rise above the fractures at a depth 9½ feet below the surface. When this cement had set, the remaining annulus was filled from the surface with 85 cubic yards of lean (1500 psi) concrete.

Cementing the Steel 122-Inch ID Surface Casing

The second cementing operation consisted of pumping neat cement into the annulus between the 140-inch diameter surface hole and the 122-inch surface casing.

The ten joints of surface casing, each 20 feet long, 5/8-inch thick with an ID of 122-inches were run into the hole open-ended and landed at a depth of 190 feet below the pad.

Theoretical calculations, based on the setting depth, wall thickness of pipe and hydrostatic pressure of cement outside the pipe during cement emplacement had shown that the cement had to be added in three stages to prevent casing collapse.

The first stage, from 195 feet to 185 feet was calculated to require 1054 cubic feet. Actual requirement was 1150 cubic feet, 9.1 percent over the theoretical requirement. Neat, Class G cement slurry with three percent calcium chloride was used for this first stage. Emplacement was accomplished by pumping the slurry down 2 7/8-inch steel tubing using standard oil field hydraulic cementing pump trucks. For this first stage, the tubing was placed inside the casing so the cement would displace the drilling mud equally up the annulus and up the inside of the casing. It was allowed to set 12 hours.

The second and third stages were pumped into place with the 2 7/8-inch tubing in the annulus. The second stage consisted of 2164 cubic feet of Class G cement with two percent of pre-hydrated bentonite and two percent calcium chloride. This volume was 5.2 percent over the calculated theoretical requirement. Yield was 1.15 cubic feet per sack of cement.

After waiting 12 hours for stage two to set, stage three was started using the same mix as was used in stage two. Required volume for stage three to fill the casing annulus to the surface was 2450 cubic feet or 3.8 percent more than the calculated theoretical volume.

Cementing the 96-Inch Steel Casing

The 96-inch steel casing was cemented from the total depth of 2371 feet to the surface in nine stages as shown in Figure 1.

The cement was pumped through three strings of 1.9-inch tubing and one string of 2 7/8-inch tubing down the outside of the 96-inch steel casing to the bottom of the shaft. The tubing strings were manifolded into a common flow line at the surface. Three oil field cementing trucks were used to mix and pump the cement. As each stage was pumped, the rise of the cement was monitored by means of a nuclear logging tool. Each time the top of the cement rose 90 feet the tubing was withdrawn a like amount to reduce friction in the tubing.

Before each stage was pumped in, water was added to the inside of the casing to a pre-determined level to partially offset the external cement pressure.

Stages four, five and six required a total of 21,060 cubic feet (176 percent) more than the theoretical volume of 11,939 cubic feet. These stages occurred between the depths of 1253 feet down to 1792 feet and correspond to the section of hole sealed off for 60 days while the bit was being fished out of the hole. The pH of the mud last taken prior to this incident was 9.1 and the salt saturation was only 20 percent. Had the mud been fully salt saturated, this dissolution would not have occurred and the excess cement would not have been required.

Post-Cementing Observations and Conclusions

Following completion of the shaft, two mining levels were opened. This permitted some visual observation and evaluation of the actual cement in place and permitted limited testing of the cement.

The first penetration of the casing was a 1/4-inch test hole at Level 2 (1659-1669 feet). This hole penetrated a void in the cement behind the casing and a large quantity of entrapped drilling fluid under a pressure of 610 psi was encountered. After draining the hole for two days, the flow stopped. Further drilling indicated a grout thickness of seven feet at that point. Additional drilling mud was drained and methane was vented to the surface through a 4-inch pipe. The project was classified by MSHA as a gassy mine. Appropriate safety procedures were immediately implemented.

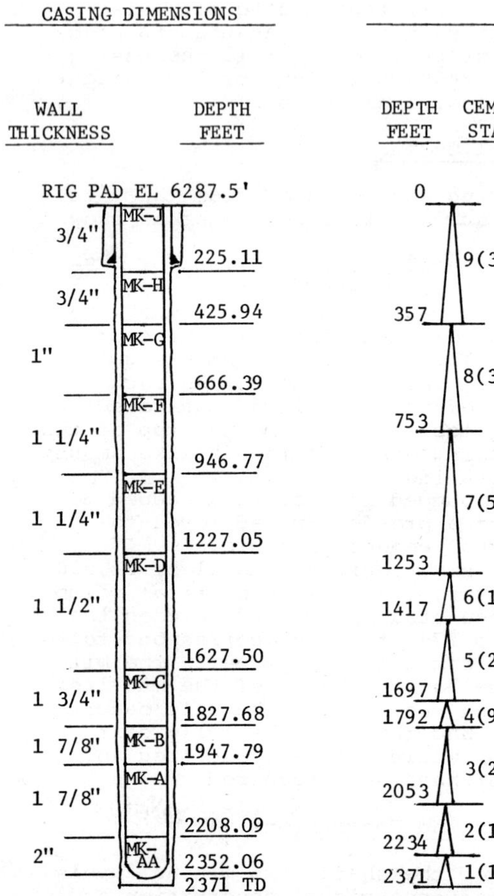

CEMENT STAGE	THEOR. HOLE VOL. FT^3	ACTUAL SLURRY VOL. FT^3	ACTUAL AS % OF THEOR.
9(357')	9,843	10,524	104
8(396')	9,990	10,102	101
7(500')	11,945	16,542	138
6(164')	3,660	5,048	138
5(280')	6,211	17,838	287
4(95')	2,068	10,113	489
3(261')	5,068	5,587	110
2(181')	3,870	3,514	91
1(137')	3,844	3,844	100

STAGE 1 THRU 7 - CLASS "H" SATURATED
STAGE 8 & 9 - CLASS "H" W/2% PREHYDRATED BENTONITE PLUS 2% $CaCl_2$

TOTAL NET THEOR. HOLE VOL = 57,041 FT^3
TOTAL (BY STAGE) = 83,122 FT^3 SLURRY (146% OF THEORETICAL)

TOTAL (BY WGT TICKETS) = 62,374 SKS
TOTAL (BY STAGE) = 63,125 SKS

FIGURE 1

Decision was made by the contractor and the U.S. Bureau of Mines to seal off upper formation water by providing a grout curtain above Level 2. The first grout hole penetrated 60 feet and drained an estimated 45,000 gallons water and considerable methane. New pockets of methane were encountered with each few inches of drilling through the oil shale. An estimated 7,000 cubic feet of voids were opened during the exploration drilling of Level 2.

The grout make-up included the following additions to each sack of cement: 1.9 pounds of bentonite, 20 pounds of sodium chloride, 0.25 pounds of defoamer and 6.5 gallons of water with a yield of 1.48 cubic feet per sack. Salt addition was required to prevent dissolution of halite. Red dye was added to permit future identification.

A total of 335 sacks of cement was pumped into the voids at Level 2 filling a volume of about 5300 cubic feet (75 percent of the estimated voids). This operation provided the desired seal from water and methane.

An opening was cut into the casing and a mining drift was begun. The drift revealed the following: (1) cement bond to the casing was excellent, (2) the casing was tight against one side of the drilled hole, (3) numerous voids in the cement extending above and below the drift appeared to have been caused by channeling of the cement as it was pumped in, (4) small spaces below the stiffener rings welded to the exterior of the casing trapped drilling mud instead of filling with cement because of the square shoulders on the rings and (5) the cement had numerous closed vertical fractures which appear to be cooling cracks caused by the heat-sink effect of the thick-walled steel casing. These cracks were not visible in situ but when an attempt was made to cut a diamond core upward in the cement parallel to the casing it was impossible to obtain a whole core. When the core barrel was emptied, the stress relief caused the cement core to break along the fractures and only small flat-sided pieces of cement were recovered.

It is concluded that: (1) a drilled shaft should either be absolutely straight or it should be sufficiently larger than the casing that deviations will not permit the casing to lodge against the shaft wall, (2) tubing strings used to pump cement behind the casing should be numerous and closely-spaced to prevent channeling of the cement, and (3) stiffener rings should have a chamfered edge to prevent entrapment of fluids.

The cooling cracks probably cannot be completely prevented but the design mix, casing thickness, temperature of fluid inside the casing and other factors can be tailored to minimize these effects now that this condition is known to exist.

Some pieces of the cement core cut from the 2300-foot level in the shaft were subjected to physical testing in the Bureau of Mines Laboratory at Denver. Results are as shown on Table 1.

The colored grout pumped by the grouting contractor was detected as low as Level 5 (2069 to 2079 feet) proving that the channeling was extensive, extending as a minimum from cement stage five into stage two. The fact that the cement sealed the aquifer overlying the mining horizon indicates that the channeling stopped once the greatly over-sized hole through the salt section had been cemented.

Cost Effectiveness and Possibilities for Improvements

The most cost effective improvement would be to use thin-walled, open-ended casing instead of the thick-walled closed-end casing. When the closed pipe is used, costs escalate rapidly.

The use of a thin-walled, open-ended casing would allow major savings in the cost of the casing, its transportation and handling and would permit the use of a much lighter, cheaper drilling rig. Thin, open casing would require a larger hole to insure that there was no possibility of collapsing it due to bumping the hole wall. This larger hole would require more cement and the casing would have to be accurately placed to prevent collapsing. However, since the casing serves no useful purpose other than as a form for the cement, the thickness of the pipe wall is immaterial after the cement is placed and cured. The economy resulting from thinner open pipe would more than offset the incremental increases necessitated by larger hole diameter, care in handling and extra costs of cementing. Another area for cost improvement is in design of the external reinforcing, or stiffener, rings to eliminate the square shoulders and the consequential entrapment of fluid in these spaces.

Improvements in the drilling mud program could be made, particularly in geological sections including water-soluble minerals, to prevent large solution cavities such as those found in this shaft. The 60-day wait during fishing resulted in dissolution of large chambers in the salt beds and nahcolite beds. Despite the tremendous quantities of cement pumped into these zones during cementing of stages four and five, there were still large quantities of grout required to completely seal these zones when mining commenced. At least 15,000 sacks of cement could have been saved in these two areas if a fully salt saturated drilling mud had been in place when the bit was lost in the hole. Additional benefits would include reduced risk to personnel and reduced expenses encountered later during the mining phase.

Table 1

Test Results, Core Samples of Cement, 2300-ft Level

	Sample Number 1	2				3	4
Length, inches	5.313	5.314				2.435	2.690
Diameter, inches	2.100	1.975				2.110	2.050
Specific gravity	1.517	1.582				1.567	1.595
Lateral pressure, psi	0	0	100	200	300	0	0
Compressive strength, psi*	549	718	783	849	1061	801	879
Comp. strength ρc_{1-1}, psi	634	835	910	986	1233	825	928
Young's mod. x 10^6 @ 50% ultimate stress							
Secant	.185	.290	.209	.183	.170	.220	.286
Tangent	.147	.228	.194	.167	.147	.268	.221
Poisson's ratio							
Secant	.16	.10	-	-	-	-	-
Tangent	.13	.08	-	-	-	-	-
Shear strength, psi	-	-	340	-	-	-	-
Tan ϕ, (Coef. of Int. Fric.)	-	-	.150 (8.52°)	-	-	-	-

*Uncorrected

**Corrected using the formula: $C_s = C_1 \left[0.778 + 0.222\left(\frac{d}{h}\right)\right]$

Where C_s = compressive strength for $\left(\frac{d}{h}\right)$ not equal to 1.

C_1 = compressive strength for $\left(\frac{d}{h}\right)$ equal to 1.

d = diameter of specimum

h = height of specimum

A two-volume history of this precedent-setting project is on file in all Bureau of Mines library facilities as Open File Report number 79-009(1)(2). A report on the subsequent mining phase was prepared by Williams Brothers Engineering Company for the Bureau of Mines under the title of "Equipping a Shaft and Mining Bulk Samples of Oil Shale, Nahcolite and Dawsonite".

SQUEEZE GROUTING UNSTABLE GROUND IN DEEP TUNNELS

By David A Greenwood[1] and M Thomas Hutchinson[2]

ABSTRACT

The technique is described of squeeze grouting for stabilising finely fractured or pulverised rocks, soils and mud runs encountered in tunnelling. These materials cannot be permeated with grouts, and must be consolidated. Empirical practice usually involves injections of chemical followed by cementitious grouts under high pressure. A hypothesis is proposed for estimating injection hole spacing and injection time in relation to ground and grout properties to allow better process design for efficient ground stabilisation.

GROUND STABILISATION BY SQUEEZING OR GROUT WEDGING

Squeeze grouting techniques are adopted in tunnelling where ground to be traversed must first be stabilised to permit excavation, but cannot be permeated by grouts. Stabilisation is achieved by consolidating ground ahead of excavation to increase its shear strength, the consolidation stresses being imposed by grout injections under high pressure. On excavation the consolidated ground is held tight by wedges of set grout. Quantities of grout injected are sometimes very small, but intergranular pressures are increased sufficiently to lock the ground in place. Porosity is reduced only a little according to volume of set grout.

The principle of squeeze grouting is similar to that of compaction grouting, differing in that whilst compaction treatment induces localised consolidating stresses centred on the injection point, squeezing encompasses a larger ground volume stressed by fissures of grout penetrating by hydrofracture in soil-like material or along paths pre-determined by rock fractures. The mortar-like grouts associated (3, 9) with compaction processes could not be injected into finely fractured rock mosaics nor into over-consolidated pulverised rock flour; for these especially, a significant sized pressure "bulb" can only be achieved by extending the radius with more penetrating grouts and technique.

MECHANISM OF SQUEEZE GROUTING

When injecting grouts into silt sized materials, too fine for permeation flow, two extreme types of behaviour can occur. These are illustrated by the results of laboratory experiments in Figures 1 and 2.

[1]New Projects Director, Cementation Specialist Holdings Limited.
[2]Director, Cementation Research Limited, Rickmansworth, ENGLAND.

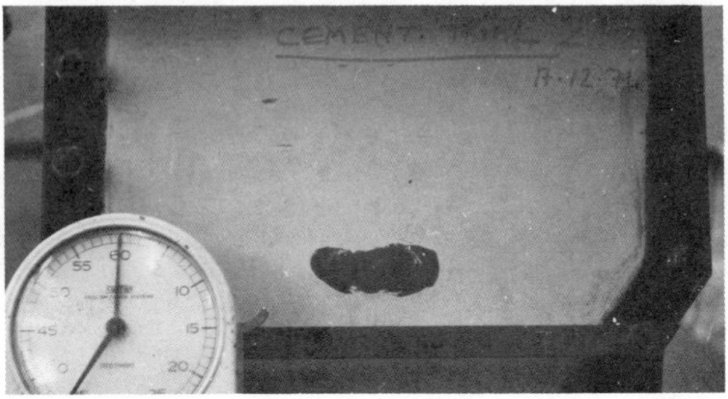

FIGURE 1 CEMENT GROUT INJECTION BALL

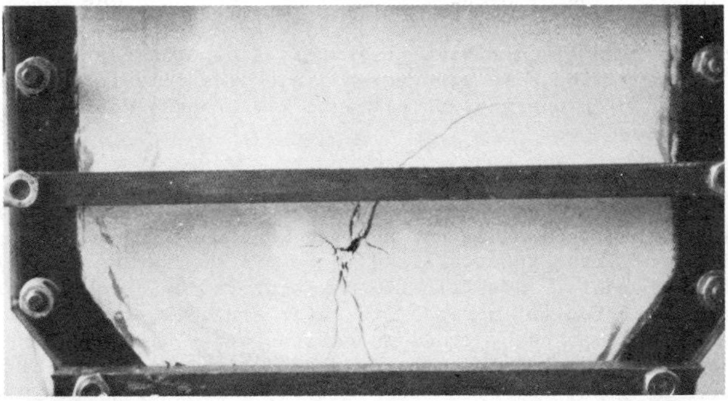

FIGURE 2 CHEMICAL GROUT INJECTION FISSURE PATTERN

FIGURE 3 CEMENT/ANTI-BLEED INJECTION FISSURE PATTERN

At one extreme, the grout remains round the injection point as a ball (compaction grouting) and at the other it fissures the ground and forms a network of grout filled cracks (squeezing). In the former case, high pressures are concentrated around the ball and consolidation is therefore restricted in extent. In the latter case, consolidation takes place around high pressures in fissures and is much more extensive; if the fissuring is too extensive, however, grout can escape entirely from the consolidation zone and control of the process will be lost.

The essence of squeeze grouting is therefore controlled fissuring of the ground combined with the exertion of pressures above those occuring naturally in the soil until sufficient strength is gained for stability. To achieve this it is necessary to control the grout properties in two basic respects - bleed and rheological properties.

On completion of consolidation the grout must set to lock in enhanced effective stresses. Cementitious grouts are usually essential for this.

Grout Rheology

Fluids used for grouting can be classified into two types:

a) Newtonian fluids, characterised by their viscosities. Water and chemical grouts before gelation are of this type.

b) Fluids characterised by both viscosity and yield stress. These approximate to materials called Bingham bodies and they do not flow until the shear stress applied to them is greater than the yield stress. Examples are cement and bentonite based grouts.

Flow of grout along a fissure is restricted as regards rate by viscosity, and extent by yield stress; a point is reached where the force exerted on the fluid by the injection pressure equals the yield stress acting on the walls of the fissure and flow ceases. In practice, if the grout has initiated its own fissure pattern in a silty or clayey soil, the continuing injection of grout at pressures necessarily above the hydrofracture pressure will serve only to widen the fissure; thus the higher the yield stress, the thicker and shorter the induced fissure.

If the grout has no yield stress, the viscous pressure loss due to flow will at some stage result in insufficient pressure being available at the end of the fissure to propagate it further. As before, continued grout injection will thicken the fissure until the pressure loss (which is a function of increasing crack width) drops sufficiently for further hydrofracture to take place. Thus for a given injected volume, the more viscous the grout the thicker and shorter the fissure.

It follows that, in principle, control of the extent of a fissure pattern induced by hydrofracture can be exercised by manipulation of grout rheological properties.

Grout Bleed

Cement based grouts are suspensions of solids in water and, as such, if they are pressurised against a particulate soil, where the cement particles are greater than about one tenth of the soil particle size, a filter cake will form; water is squeezed out from the grout into the soil and the grout thickens considerably.

This is most undesirable for squeeze grouting, for two reasons:

a) The rapid thickening results in a very high yield stress, which often prevents fissuring of the soil to any extent; a grout ball forms around the injection point (cf compaction grouting).

b) The bleed water forced into the soil increases rather than decreases its water content; this extra water must also be squeezed from the formation for consolidation to occur. Often as in over-consolidated ground, the amount of water to be squeezed out to increase effective stresses significantly is only a few percent of the total volume, so that bleed water may cause considerable slowing down of the process of consolidation.

Control of bleed by the use of anti-bleed admixtures is thus essential to the use of cement grouts for squeeze grouting. Even when the inclusion of such an admixture results in a more viscous grout, it will enhance its fissuring capability considerably. This is shown by laboratory experiment as in Figure (3), which should be compared with Figure (1), where no anti-bleed admixture was used.

Consolidation Theory

To design a treatment schedule for squeeze grouting it is necessary to understand the manner in which consolidation occurs, so that the spacing of injection holes, injection pressures, and treatment time can be related to the properties of the material being treated. The theory is a general one but as detailed below assumes consolidation of a uniform silty or clayey material, such as would be found in pulverised weak rocks or in milonite filled chimneys in fault or pressure zones; or a mud run following accidental loss of ground.

The extent of the material is assumed to be greater than the range of influence of the consolidation process. In practice, it is frequently impossible to know the extent of the soil to be treated, and the properties of the material may not be uniform nor well characterised, so that the theory is best used in conjunction with experience gained from initial trial injections to help to choose optimum grout properties for controlling the extent of the hydrofracture pattern.

A model of a typical hydrofracture pattern is sketched in Figure 4. In the model we assume:

a) A "radius of influence" of the injected grout as defined by the limits of the hydrofracture pattern. To this volume we ascribe a radius, a, of the equivalent sphere upon which the mathematical treatment is based.

b) The grout itself does not permeate the soil, but raises the total stress of the soil within the "sphere of influence" to the hydrostatic grouting pressure.

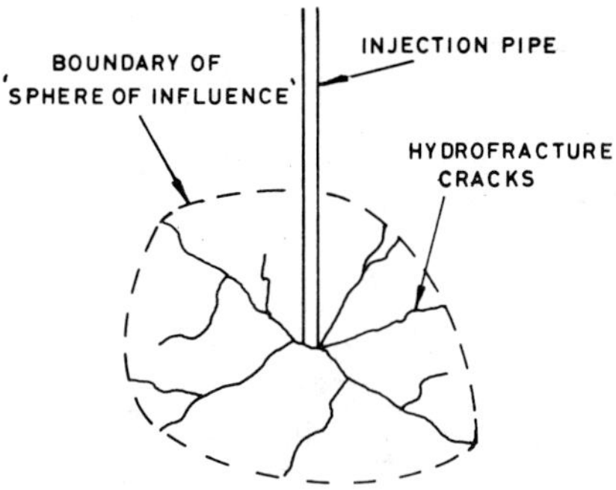

MODEL OF HYDROFRACTURE PATTERN FIG. 4.

c) No consolidation takes place outside the "sphere of influence"; rate of seepage is restricted by the permeability of the surrounding soil.

Rate of Consolidation

On these assumptions one can apply the Terzaghi "coefficient of consolidation" concept, making use of the following approximations:

a) That the permeability of the ground is reduced by a factor up to 20 during consolidation.

b) That the process is considered sufficiently complete when 75% of the excess pore water at the applied pressure has been exuded.

c) That the "radius of influence" is within the range 50 - 250 cms.

d) That the fractures are produced in a time which is short compared with the overall time of consolidation.

Briefly, the method used to quantify the process is to first calculate the pore water pressure at the boundary of the sphere for any given rate of pore water dissipation and use this as the boundary condition when solving the basic equation:

$$\frac{\delta u}{\delta t} = c_v \left(\frac{\delta^2 u}{\delta r^2} - \frac{2\,\delta u}{r\,\delta r} \right) \qquad \text{for } 0 < r \leqslant a \qquad (1)$$

Where c_v is the coefficient of consolidation and u is the pore water pressure. The solution is analogous to the problem in thermal conduction of calculating the temperature profile in a sphere when heat is radiated from its surface within an enclosure at constant temperature, which is treated by Carslaw and Jaegar (2). It gives a value of u, the pore water pressure, as a function of time and position within the sphere:

$$u = \frac{2hu_o}{r} \sum_{1}^{\infty} M_n \sin r\alpha_n$$

$$\text{where } M_n = e^{-c_v \alpha_n^2 t} \left\{ \frac{a^2\alpha_n + (ah - 1)^2}{\alpha_n^2 \, (a^2\alpha_n^2 + ah\,(ah - 1))} \right\} \sin a\alpha_n \quad (2)$$

and α_n are the roots of: $a\alpha \cot a\alpha + ah = 1$.

The volume of water expressed after time t is then given by

$$V = \int_0^a (u_o - u)\, m_v \,.4\pi r^2 dr \quad (3)$$

or, integrating,

$$V = \frac{4}{3}\pi a^3 u_o m_v \left[1 - \frac{6h}{a} \sum_1^{\infty} M_n \left\{ \frac{\sin a\alpha_n}{a^2\alpha_n^2} - \frac{\cos a\alpha_n}{a\alpha_n} \right\} \right] \quad (4)$$

where V = volume of water expressed (= volume of grout injected)

a = radius of "sphere of influence"

u_o = injection pressure

m_v = volume compressibility of the soil

$h = \frac{1}{a}\frac{k_o}{k}$ (k_o = initial permeability, k = final permeability)

Equation (4) governs the rate at which consolidation can occur. The computed relationship, over a wide range of values, is plotted in Figure 5. In the diagram the amount of water remaining is expressed as a percentage of the total water which would be exuded after infinite time, and the relationship is put in the form common to consolidation equations:

$$T = \frac{D\ a^2}{c_v} \quad (5)$$

where D = constant (see Figure 5) and

T = time (secs)

Values for D can be read off directly from Figure 5 for a range of values of permeability factors k_o/k of up to 50.

k_o/k will be typically about 10, giving a value of 0.14 for D in the above equation.

Because of the elastic behaviour to be expected of the soil outside the "sphere of influence", the consolidating effect will continue after grout injection has ceased, until the pore water pressures revert to their equilibrium levels.

For a typical clayey silt, ($c_v = 3 \times 10^{-3}$ cm^2/sec, a = 50 cms, D = 0.14) consolidation time required is about 32 hours. It is not feasible to

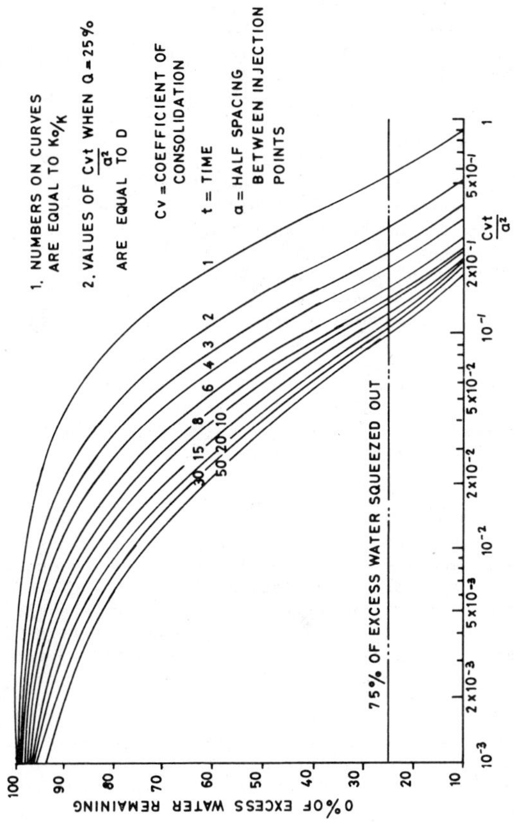

FIG. 5.

RATE OF CONSOLIDATION

inject grout continuously for this length of time without uncontrolled wasteful fissuring; repeated injections "top up" the consolidating pressure over the period, although the overall treatment must be extended over the 32 hours. The volume of grout injected must be restricted to reflect this reiterated injection technique; excess grout injected at any stage will be forced into areas outside the zone being treated, and control of the process will thereby be lost.

The "excess water remaining" after three equal time intervals is typically 47% after 1/3, 35% after 2/3 and 25% after the total injection time (ie 75% consolidation). Grout should be injected in these proportions - say 60% in the first injection and 20% in the second and third treatments, although this might well be altered on site to enable an initial tightening up of the soil to be effected by the injection of relatively small volumes of grout in the first injection stage and less likelihood of subsequent injections "running away" or coupling to other holes.

This method therefore gives a way of deducing the necessary relationship between hole spacing and overall treatment time from a knowledge of the coefficient of consolidation of the soil to be treated and the ratio of initial to final permeability of the soil.

ESTIMATION OF GROUT PRESSURES AND VOLUMES

Although obviously the higher the injection pressure used the better, consolidating pressures of greater than 10 atm above the undisturbed total stress will have little effect on the final strength of the material. The volume compressibility, m_v, falls very sharply as the pressure is increased up to about 3 atm, but changes little after 10 atm. The main advantage in using higher pressures is therefore to accelerate consolidation; the disadvantage is that grout travel will be less easily controlled.

Required grout volume can be calculated from the injection pressures and a knowledge of the soil volume compressibility m_v:

$$V = Pm_v$$

where V = volume of grout required per unit soil volume

P = injection pressure

Typically a volume 5 - 10% of the treated soil volume will be required for a loose or running silt mass. Any volume used more than this indicates either the presence of large voids or that grout is travelling out of the main treatment zone.

LIMITING SOIL PROPERTIES

Soil with a range of consolidation coefficient compatible with realistic injection times and rates and injection hole centres is defined by the equation (5). For instance, taking a four stage grout injection ground tightening programme, with an overall treatment time of 48 hours, with 1 m hole centres as being the limit for realistic treatment, then one could say that the soil must have a coefficient of consolidation given by:

$$c_v > 4 \times 10^{-3}\ cm^2/sec\ (140\ ft^2/year)$$

This represents a clayey silt, using the rough range for c_v values shown in the table below. Approximate values for m_v are also given.

Soil Category	c_v $cm^2/sec \times 10^{-3}$	m_v mm^2/N
Sandy Silt	6 - 12	10^{-1}
Medium Silt	4 - 6	1.5×10^{-1}
Clayey Silt	1 - 5	2×10^{-1}
Fine Clayey Silt	0.5 - 1.5	3×10^{-1}

From this it would appear that clayey silts should be treatable, whereas fine clayey silts would be rather more difficult, requiring more repeat grout injections and very prolonged overall treatment times.

PRACTICAL EXAMPLES OF SQUEEZE GROUTING

Seven examples of squeeze grouting are outlined in this section. Details are given in Table 1. Examples are grouped in three categories: natural soil enclosed in rock; mud runs enclosed within galleries; and extensive pressure zones traversed by tunnels.

CAVERNOUS DOLOMITES: Witwatersrand (South Africa) and Catalagzi (Turkey)

In the West Witwatersrand gold mining areas of South Africa (1) waterlogged dolomite extends over gold bearing quartzites to depths of 500 m to 1300 m, with joints widened irregularly up to 2 m to 3 m and 900 m deep. The soft insoluble residue of disolved dolomite known as "manganese wad" fills most joints. This must be stabilised if encountered during shaft sinking. It cannot be removed by flushing as more material flows in to replace it. Injections must, therefore, force it outwards or consolidate it.

Grout holes drilled down from shaft bottom at each stage of sinking are arranged in a conical cover.

For an 8 m diameter shaft, covers are typically 50 m to 60 m long with 16 holes to give a minimum nominal grout thickness of 1.4 m increasing to about 3 m at the full depth of cover.

Holes are skewed from the radial direction so that each hole overlaps its neighbour. Thus joints of any width and orientation must be intersected.

Project	Rock types and condition	Ground water pressure (atmospheres)	Injection Treatment Details					
			Description	Max. hole spacing	Injection stage length(m)	Pressure limit (atm.)	Vol. grout as % nominal vol. treated	Total time for treatment
WEST WITWATERSRAND MINE SHAFT SINKINGS S. Africa (Atherton & Garrett) 1959	Dolomites - typically leached and cavernous on joints and beds 0.1-2m wide filled with non-cohesive residual silt (manganese wad) intersected by pilot drilling.	60-70 according to depth below water level	16 no. holes per 50m deep cover x 3m annulus (max.) from shaft bottom. Holes skewed to intersect fissures. Sand:cement 3:1 injected.	Increasing 1.4m - 2.9m with depth in	Varies 15-50m according to hole intersection with joint.	Approx. 100	Cavity vol. not usually known.	Sometimes 3 months on a single hole. Most holes "tight"
CATALAGZI MINE SHAFT SINKING Turkey 1949 (Cementation Co.)	Dolomitic limestone - 2m wide joint filled with clayey silt encountered by pilot drill holes on shaft line.	27	Repetitive cement/sand injection after successive re-drilling until "take" small at pressure limit.	Not known - record incomplete	5m approx (not verified)	Approx. 100	Not known	About 2 weeks
BELLE ISLE SALT MINE ROADWAY RECOVERY Franklin LA USA 1973 (Cementation Co.)	Shaft and 7m sq section roadways in rock salt inundated by run of alluvium from surface (Fig 5). Mainly loose silty sand - but variable. Lower 1m-2m "soupy". 29,300m³ ran 300m from shaft in about 10 mins.	0 (exposed face of silt in dry salt mine)	To permit construction of concrete plug for safe shaft re-construction Phase 1: roof void filling by 1.1 cement/sand grout with 2% wt cement of bentonite, w/c 0.68. Phase 2: squeezing. Neat cement with 4% $CaCl_2$. Advancing stages when re-drill stable after injection of 0.42 tons/stage. 8 repetitions per hole.	Drilled vertically from parallel gallery above. 1.5x2.4m 1.8x1.8m	Open void 0.91m	1 48	100% 48%	6 months of which 3½ for injection.
SEIKAN TUNNEL PILOT TUNNEL RECOVERY Japan 1974 (Cementation Co.)	Plastic clay burst face of pilot tunnel advancing 65m in 24 hrs. Water broke out 30 hrs about 2000 l/min. Clay matrix with broken tuff. Parent rock broken clayey/sandy tuff at fault zone. 1000 m³ mud within arch ribs 4m x 5m high.	30 to 0 (pore pressure decreased towards exposed face of mud run)	4 no. successive bulkheads. 1st cover 40m long and 2nd 35m x 69 holes. Peripheral drainage holes. LW(cement/water + sodium silicate/water) grout. Vary water to achieve pressure limit after 2m³ per stage.	Increases to about 1m maximum in conical cover.	5 m	Just exceeds hydro-fracture pressure 20-43	51% within pilot tunnel	11 months of which 9 months grouting

Table 1 Continued . . .

Project	Rock type and condition	Ground water pressure (atmosphere)	Injection Treatment Details					
			Description	Max. hole spacing	Injection stage length(m)	Pressure limit (atm.)	Vol. grout as % nominal vol. treated	Total time for treatment
MALGOVERT TUNNEL France 1947-1953 (Rouselle)	Overthrust zone: broken schists and quartzites with pulverised ("pâté") quartzite chimneys Fig. 7. All finely fractured.	4	Successive conical covers lengths 35-55m. 36 holes per cover. Inject covers centre outwards with advancing stages Phase 1: sodium silicate/ aluminium sulphate(Francois) grout, followed immediately by sulphate resisting cement w/c 1.0 to maximum 100 kgs subject to pressure limit. Phase 2: sodium silicate/ phosphoric acid to 30% nominal ground volume.	Increasing to 2.5m 4m	About 4 m generally	25-40 after 1m^3 injection 100 10-20 usual rising to 50 max.	Varies 38% to 44%	2 years for 160 m run of tunnel out of 4 years for 15 km total length.
ROSELEND TUNNEL France 1958 (Lajeat et al)	Overthrust zone: divided roughly in two: carboniferous broken schists, triassic quartzite sands. Complex fine fracture pattern. Trias crushed. Fig. 8.	14-19	1st cover: 45m through schist/trias 92 holes. Inject blast furnace cement w/c 2 reducing to 0.5 if holes connect. As cover 2 in trias 2nd cover: 40m z 72 holes in trias. Inject: 1. sodium silicate/sodium bicarbonate followed immediately by 2. clay/cement ratio 0.25 w/c 1-2. Repetitions/ stage average 4 & 7 for covers 1 & 2.	2m 3m	5m in schists 3-4m in trias	30-70 65 80-150	Total 47% chemical 36% cement 11%	8 months - 1st cover 3 2nd cover 5 total 76m run
POLLINO NORD TUNNEL S. Italy 1970 (Cementation Co.)	Fault zone: flysch/ dolomite contact with dolomite mosaic alternating with milonite ("rock flour") chimneys about 1-4m wide. Reverting to fractured rock mosaic with clay "dusting" on fracture 2-3cm apart. Fig. 9.	4-9	Multiple covers 18-50m over 1325m. Major cover 56 holes with outer drain holes. Extra holes in milonite Phase 1: sodium silicate/ formamide thickened with $CaCl_2$: content varied to ensure pressure achieved. Phase 2: neat cement in one to 3 stages to pressure limit. w/c varied to achieve pressure.	2.5m 1.0m	20+15+15m = 3 stages 50 m max. single stage	70 100-120	chemical 3.5% cement 0.6% total 4.1%	2 years for 1325m run

Table 1 - EXAMPLES OF SUCCESSFUL SQUEEZE GROUTING

Grout consumptions are high, of order 100 tons/m of shaft. Only one or two holes per cover may intersect a joint but in extreme cases pumping through extra or re-drilled holes has continued for 3 months. Economy of material is achieved by introducing sand up to 3:1 by weight of cement with minimum water to allow pumping down the shaft.

The consolidated wad is so stiff that it can withstand blasting to remove the rock.

Similar treatment was given to clayey silt in a 2 m widened joint intersecting a shaft at a depth of about 300 m at Catalagzi, Turkey. Drill rods in pilot holes were forced out when this was encountered, followed by oozing of "toothpaste". After consolidation the shaft was sunk through the cavern whose section showed multiple layers of set grout crossing the hardened silt, roughly horizontally in thickness of 20 mm to 100 mm. No significant seepage was encountered.

RECOVERY OF MUD RUNS IN TUNNELS

Belle Isle Salt Mine, Louisiana USA (1973)

Figure (6) shows the interconnection of two shafts with roadways on two different mine levels separated by about 17 m. An accident occurred resulting in an inrush mainly of silty sand alluvium down the No 2 ventilation shaft. Some 30,000 m^3 ran about 300 m from the shaft bottom at a speed about 1 m/s, filling three roads and crosscuts at the 1216' level. It probably liquefied temporarily during flow, accounting for its extensive travel and poor and irregular compaction at rest.

To recover the shaft safely a concrete plug had to be constructed as shown in Figure (6). To facilitate this with minium danger the inrush material was consolidated by squeeze grouting.

A void at the roof where the filling had settled was first grouted with sand/cement to allow squeezing in later injections. Squeezing commenced in 0.9 m descending stages. Injections to a maximum of 0.42 tons of cement subject to a pressure limit of 48 atmospheres were repeated in each stage until it was possible to drill further without collapse of the bore. An average of eight repetitions was needed. Some inter-connections occurred between holes on a 1.8 m square grid drilled from the overlying air entry at 1163' level. The upper parts of the fill in the 7 m square road were initially moist and loose but saturated and "soupy" in the lower 2 m. After consolidation it was possible to excavate sufficient to construct the plug.

Eventually after securing the shaft, the concrete plug and consolidated fill were excavated completely. Miners reported (ref Cementation Company file) such good grout contact with the roof and walls over the last 40 ft to 50 ft of excavation that blasting was necessary to remove it. The face of grouted soil was stable, vertical and dry except for occasional small perched water pockets. Cement seams were scattered randomly in nearly horizontal layers apart from the massive block at the top.

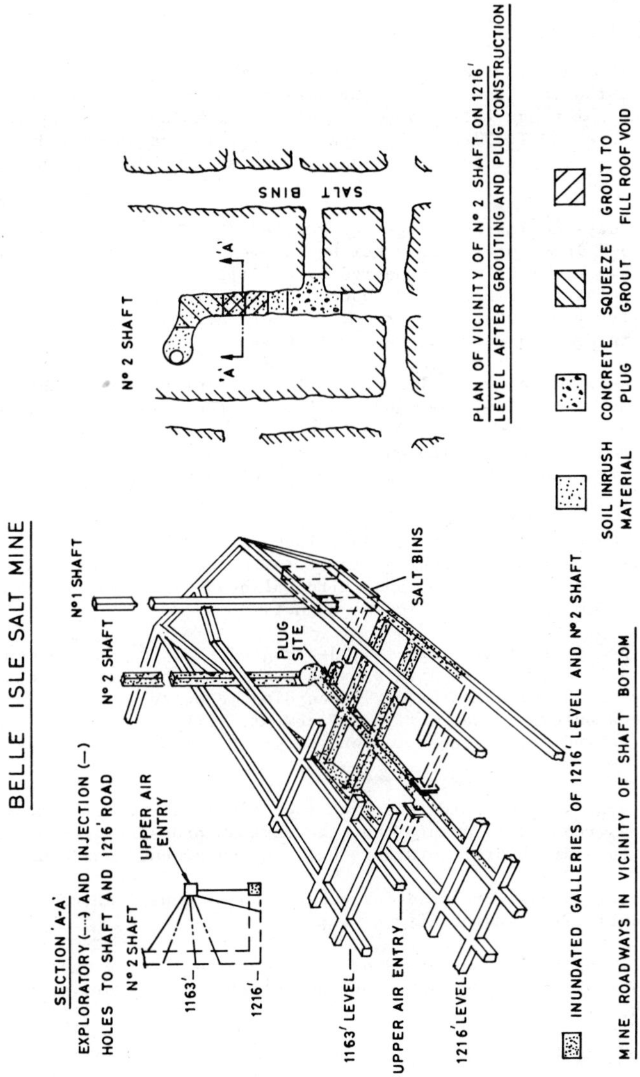
BELLE ISLE SALT MINE
SECTION 'A-A'
EXPLORATORY (----) AND INJECTION (—)
HOLES TO SHAFT AND 1216' ROAD
Nº 2 SHAFT
UPPER AIR ENTRY
1163'
1216'
1163' LEVEL
UPPER AIR ENTRY
1216' LEVEL
Nº 2 SHAFT
Nº 1 SHAFT
PLUG SITE
SALT BINS
INUNDATED GALLERIES OF 1216' LEVEL AND Nº 2 SHAFT
MINE ROADWAYS IN VICINITY OF SHAFT BOTTOM
Nº 2 SHAFT
'A'
'A'
SALT BINS
PLAN OF VICINITY OF Nº 2 SHAFT ON 1216'
LEVEL AFTER GROUTING AND PLUG CONSTRUCTION
SOIL INRUSH MATERIAL
CONCRETE PLUG
SQUEEZE GROUT
GROUT TO FILL ROOF VOID

FIG. 6.

Seikan Tunnel, Japan 1974

In the service tunnel, leading and about 100 m distant parallel to the main tunnel, the face was lost in the Hokkaido drive within a known fault zone. Competant vuggy tuff gave place with increasing frequency to bands of softer clayey tuff. Ground squeeze was experienced which induced the contractor to close the spacing of horse-shoe arch ribs 5 m high x 4 m wide.

After a few days holiday stoppage the gunited face broke when supports were removed. The squeeze could not be stemmed and the slow movement (1 m/hour) was eventually accompanied by a little water. In just over 24 hours the face had encroached about 65 m when the water flow increased to become steady at about 2000 1/min. Support arches had collapsed disturbing the ground surrounding the tunnel.

The situation was recovered with protection of four bulkheads constructed in turn 66 m, 51 m, 26 m from, and finally at, the original face. At each bulkhead grout was injected to enable advance to the next. Pits dug laterally from the tunnel allowed grouting outside as well as inside the arch line. Treatment outside was accomplished mainly from the first bulkhead.

The initial cover was effectively 40 m long with 10 longer holes of 79 total outside the tunnel reaching almost to the original face. Similar covers were used at the second and third bulkheads with length 60 m outside and 35 m inside the original tunnel.

Only one type of grout was injected, designated LW. This was a cement and sodium silicate mixture diluted with water. It produces a weak gel immediately which thickens the mix, according to concentration, and reduces bleed before slower setting of cement. Its use in these circumstances indicates a need to control grout spread as in unrestrained ground.

If 2 m^3 of grout was accepted per 5 m stage length the water content of the grout was reduced to ensure that the maximum pressure within the limit set (to avoid hydrofracture in undisturbed ground) was achieved.

Pneumatic clay spades were used for excavation of consolidated clay without trouble from instability or water.

Subsequently the fault zone ahead of the recovered service tunnel was squeeze grouted prior to crossing but no details of this are available to the authors.

ROCK PRESSURE ZONES

Malgovert and Roselend - High Pressure Water Tunnels, France

These two projects are described together as they are similar with respect to squeeze grouting (5, 6). Both are pressure tunnels for hydro-electric power aqueducts constructed in the French Alps.

They are in a region where overthrust rock masses of permo-carboniferous and carboniferous age adjoin and abut triassic "accidents". Finely fragmented carboniferous schists meet completely pulverised waterbearing triassic quartzites and limestones in over thrust folds. The crossing of these "accidents" (Figures 7 and 8) required squeeze grouting.

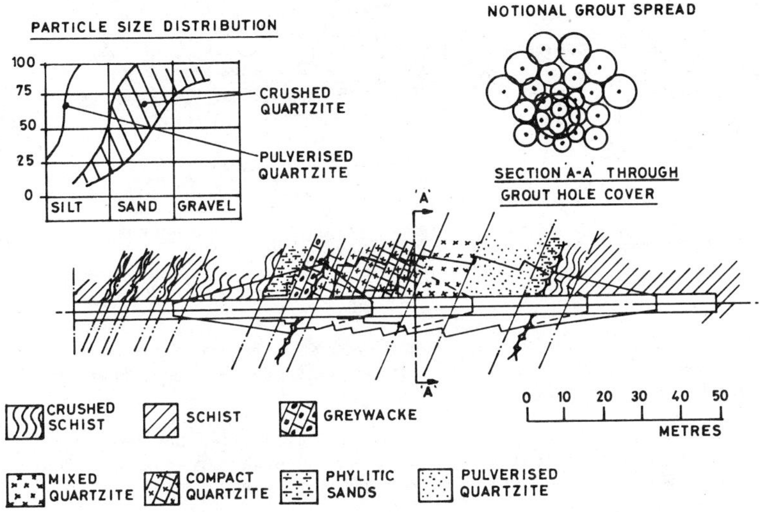

ELEVATION ON TUNNEL SHOWING STRATA AND GROUT COVERS

MALGOVERT TUNNEL (after Rousselle) FIG. 7.

At Malgovert (1947 - 53) two different adits had to be abandoned after runs of material described as "white pâté" containing huge blocks of schist flowed into the workings. In the second incident three successive bulkheads were overcome but a fourth held. The "pâté" was composed of finely pulverised virtually impervious quartzite sand. It alternated with bands in which quartzite blocks were surrounded by "pâté" (Figure 7). Permeable broken schists abutted these bands.

In the schists and mixed quartizites, the Francois gel of sodium silicate and aluminium sulphate was injected first to seal off passages too fine for subsequent cement treatment but capable of conducting away bleed water from cement grout, so restricting its entry. The quartzite sands, however, had to be squeezed by the gel and wedged by cement.

Grout hole arrays were in 5 overlapping covers of differing length in conical form for 110 m width of unstable ground (Figure 7). Maximum hole spacing was 4 m at the end of a 55 m cover.

Despite the unpromising circumstances, the treated ground was stable and reasonably watertight. Explosives were not required for excavation. Permeability was reduced by a factor of several hundreds.

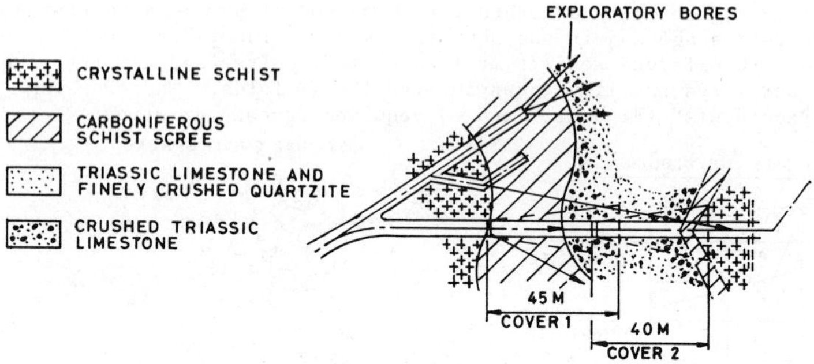

ROSELEND TUNNEL – PLAN ON EXPLORATORY ADITS AND DIVERSION FOR SHORTEST CROSSING OF PRESSURE ZONE. (AFTER Lajeat et al)

FIG. 8.

At Roselend, 1958, (Figure 8) only two covers were required for a 76 m traverse. The first ran from schists into quartzites and the second was almost wholly in quartzites.

Treatment was similar in principle to that at Malgovert. However, the initial injection of chemical gel in this instance was of sodium silicate and sodium bicarbonate and bentonite was added to the cement to further inhibit bleed.

On excavation explosives were needed for the schists but only pneumatic hand tools for the stabilised quartzites. It was noted that each cement injection produced a single fissure and these tended to be parallel in "swarms"; sometimes successive injections followed similar paths to build into thick seams whilst others formed fine tendrils through the soil. Fissures were perpendicular to the plane of the overthrust - presumably in response to least principal stress.

Pollino Road Tunnel, Italy 1970

This also was an aqueduct for hydro-electric power constructed in Calabria between Sinni and Caffaro (Figure 9). The Sinni heading driving down grade encountered a fault zone between flysch (sandstones and shales) and dolomite and was inundated. This was eventually recovered without squeeze grouting, but meantime the Caffaro drive had encountered the dolomite which became increasingly fractured with every advance, slowing progress.

Specialist grouting was adopted using conventional treatment of overlapping conical covers of holes with gradually thickening cement injections, followed in a later phase by low viscosity (2 to 3 cp) chrome-lignin chemicals in finer fractures. This was successful until, as the fault zone was approached, the rock became more finely fractured, eventually to a three dimensional mosaic of 2 to 3 cm blocks with clay partings and containing occasional milonite (rock flour) chimneys of 1 m to 2 m width. In these conditions permeation grouting proved ineffective with much grout wastage.

This was attributed to the increasing porosity and compressibility of the rock, and connection with open passages at sheared bedding planes at the fault which conducted grout uselessly away.

Subsequent treatment was by injection of a sodium silicate gel thickened with calcium chloride. The "thickness" (a combination of viscosity and shear resistance) was controlled to obtain a 70 atmosphere injection pressure. Neat cement injection followed with water content adjusted by experience to achieve 100 atmospheres pressure. This proved very successful in stabilising the ground in contrast to imperfect permeation, and simultaneously watermake was reduced about 40 times. Total grout consumption was exceptionally low at only 4% of the nominal volume treated.

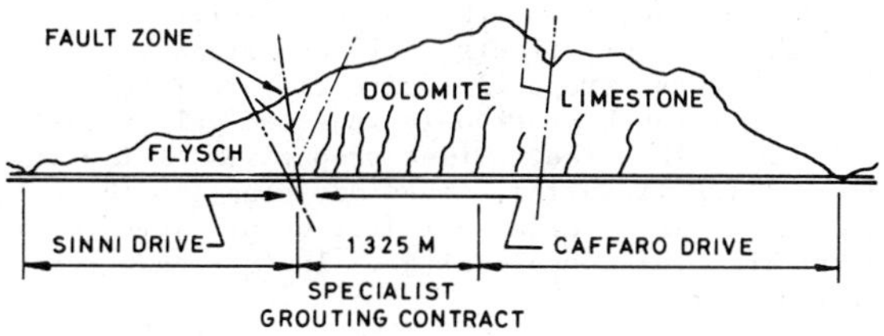

POLLINO NORD TUNNEL – GEOLOGICAL OUTLINE FIG.9.

CONTROL OF SQUEEZING PROCESS

Consolidation rate depends on the ground property c_v and pressure applied. If in attempts to speed the process, increasingly higher pressures are used, extensive fissure propagation results, often to many metres. This is wasteful outside the intended zone of treatment, and with uncertainties of ground detail can lead to connections with remote porous zones or bedding planes which absorb grout to no purpose. This has led to reservations about controllability of grouting but this depends largely on proper management of rheological properties of grouts.

Chemical grouts, like water, form hydrofractures easily. When formulated with short gel times they can form a controlled fissure network. This is because they exhibit increasing viscosity as gelation approaches and thereafter shear strength develops and quickly increases, absorbing injection pressure, leaving less for fissure propagation. Some permeation local to a fissure may occur before gelation especially in more open soil zones, at the expense of fissure propagation and consolidating pressure.

Thus in mixed ground where effective permeation is uncertain it is useless to continue with low viscosity grouts with long gel times: better to stabilise by squeezing.

The extent of hydrofracture is restricted by increasing viscosity or shear strength or both. As at Pollino Nord, calcium chloride has been added to a sodium silicate based gel with good practical effect. The extent of addition is varied until desired consolidating pressure is obtained on injection, and volumes of grout per stage controlled. Cellulose derivatives are alternative thickeners giving viscosities of 0.1 - 10 poise in the laboratory, but untried as yet on site. Both these systems give initially high viscosities available with long gel times allowing a relatively long period with effectively constant grout properties before gelation.

The original practice as in the Malgovert and Roselend tunnels (Table 1) uses grouts which give rise to weak gels instantly, but which can be pumped after initial gelation. This was the basis of the earliest processes. The LW grout used at Seikan is similar but combining cement to maintain a wedge after permanent set.

Most chemical gels are weak, but with high elasticity and compressibility. Because of the Saffman-Taylor effects, well described by Scott (7) in relation to grouting, these properties give rise to freshly mixed less viscous grout cutting through previously injected gel to form the advancing grout front. Thus the initial properties of the mix are important in controlling fissuring. They may also tip the balance of behaviour towards absorbing available injection pressure for fissure propagation rather than for consolidation. In this case resort must be made to cementitious mixes.

With cement based grouts the problem is usually that the hydrofracture pattern will be too limited. This is shown by premature pressure build-up before calculated minimum volumes have been injected. Higher water content induces more bleed and does not solve the problem.

Thus for squeeze grouting the practice is to reverse permeation grout sequences and to follow chemicals with cement injections. The chemicals line the fissures and prevent bleed from cement, which then penetrates and transmits pressure to the ground. Both sets of fissures follow the direction dictated by the least principal stress in the ground and are certainly parallel. Cement tends to follow the paths weakened by prior chemical injection. Similar behaviour is observed with hydrofracture by water injection, when even after "healing" of a hydrofracture by re-establishment of effective stresses, subsequent fissuring occurs along the same paths (4).

Re-iterated injections either from adjacent holes or through the same stage length of hole assist the squeezing process by constantly topping up pressures to maintain them near maximum during early phases of the consolidation process. This saves time for treatment and ensures a higher ultimate consolidation. In variable ground it is always good grouting practice.

CONCLUSION

Squeeze grouting has been used empirically for many years to resolve problems of stability and watermake in mines and tunnels. Some of these situations have been most unpromising having regard to the completely pulverised or shattered nature of rock coupled with high ground water pressures; yet they have been overcome. The volume of watermake on excavation may be insignificant but resulting changes in ground and hydraulic stress may induce instability. In relatively impervious an irregular ground, grouting by permeation is impracticable and consolidation by squeezing and wedging to increase effective stresses in the ground may be the only resort for stabilisation.

Where accidental mud runs have occured or in cavernous limestones the weak material is confined by rock. Here design of grouting treatment tends to concentrate on economy of method with cheap grouts, fast injection and only little restraint on injection pressures. Such disturbed materials are normally poorly compacted and absorb relatively large quantities of grout with many repetitions of injection before strength increases enough to permit excavation. The values in Table 1 of about 50% grout absorption are probably correct in ground confined by rock, but similar figures in unconfined ground are probably suggested by wastage induced by high excess grouting pressures of 30 - 50 atmospheres.

When the material to be strengthened is unconfined and extends to a large distance beyond the zone of consolidation, the squeezing process is much more difficult to control, and confronts the site engineer with many day to day decisions. The tubular geometry of tunnels leads fortuitously to the situation that a stable annulus of thickness about half that of the tunnel is sufficient to retain ground and water pressures (8). Thus with access for drilling restricted to the face, long overlapping covers of injection holes are needed to form the annulus. Drilling time becomes the critical cost factor; and hole spacing to balance the conflicting requirements of certainty of stable ground with economy of drilling is a critical decision.

Compaction grouting cannot be used in ground with relatively low porosity and its uneconomical on this scale. Thus squeezing from extensive hydrofractures becomes essential.

The theory proposed here enables the engineer to design a grouting programme and to vary it rationally on the basis of site observations and an understanding of the rheological properties of grouts.

That economy of time is necessary is reflected in the times for completion of grouting and tunnelling works, for the examples in Table 1.

ACKNOWLEDGEMENTS

The authors thank Cementation Specialist Holdings Limited for permission to publish this paper, and for access to file records of squeezing jobs, with which the Company has been associated.

REFERENCES

1. Atherton, F.G. and Garrett, W.S. "History of Cementation in Shaft Sinking", Symp. Shaft Sinking and Tunnelling, Inst. Min. Eng. London July 1959, Paper No. 12, pp 394-413.

2. Carslaw, H.S. and Jaeger, J.C. "Conduction of Heat in Solids", 2nd Ed. Oxford University Press, p 237.

3. Graf, E.D. "Compaction Grouting Technique and Observations", J. Soil Mech. and Found. Div. ASCE, 95, SM3, p 115, (Sept. 1969).

4. Jaworski ,J.W., Duncan, J.M. and Seed, H.B., "Laboratory Study of Hydraulic Fracturing", J. Soil Mech. and Found. Div. ASCE, 107, GT6, p 713 (June 1981).

5. Lajeat, J.P. and Court, M. "Rapport sur la traversee de Quartzites Fluentes par la galerie d'amenee en charge de Roselend-La Bathie", 7th Cong. Large Dams, Rome 1961, R.84.

6. Rouselle, M.J.H. "Traversee d'un Banc de Quartzites Fluentes et Sous Pression par la galerie de Malgovert", Inst. Tech. du Batiment Trav. Publics, Paris January 1955, No. 85.

7. Scott, R.A. "Fundamental Considerations Governing the Penetrability of Grouts and their Ultimate Resistance to Displacement", Symp. Grouts and Drilling Muds in Eng. Pract., I.C.E. London May 1963 pp 10-14.

8. Tan, D.Y. and Clough, G.W. "Ground Control for Shallow Tunnels by Soil Grouting", ASCE, Vol. 106, GT9, Sept. 1980, pp 1037-1057.

9. Warner, J. and Brown, D.R. "Planning and Performing Compaction Grouting", J. Soil Mech. and Found. Div., ASCE, 100, GT6, June 1974, p 653.

GROUTING METHANE DRAINAGE HOLES IN COALBEDS

by

Joseph Cervik[1]

ABSTRACT

Horizontal drainage holes drilled into gassy coalbeds are effective in reducing methane flows into mine workings during mining operations. Drainage holes were drilled 415 to 2126 ft (126 to 648 m) long. Holes were sealed with cement before interception by the mining machine to avoid formation of explosive methane-air mixtures. The cement slurry consisted of cement, flyash, and fluidifier. Excess water was used in the slurry to avoid plugging problems because of the distance the slurry was pumped. When the hole was filled with slurry, pressure was increased to squeeze excess water into the coalbed fracture system. Examination of intercepted drainage holes showed that in some cases, water had separated from the slurry and had not been squeezed into the fracture system. In no case was cement slurry forced into the fracture system. In a subsequent study, water content of the slurry was reduced to recommended quantities. Where water flow from drainage holes before grouting was less than 0.5 gal/min (1.9 L/min), holes were satisfactorily sealed. Water flow greater than 2 gal/min (7.6 L/min) before grouting diluted the slurry and water stratified in the hole. When the hole was intercepted, gas pressure forced water and then methane from the hole. The resulting methane accumulation caused the machine-mounted methane monitor to indicate the high methane accumulation, thus deenergizing electrical power to the mining machine.

INTRODUCTION

During the transformation of organic debris to coal, gaseous hydrocarbons were produced. Consequently, all coalbeds contain gas with between 80 and 99 pct methane (3)(Table 1). The heating value of the gas drained directly from coalbeds is usually more than 900 Btu/cu ft (33.5 MJ/m^3), and it is interchangeable with natural gas. Methane is a natural resource, and the Bureau encourages its recovery either by pumping it directly into commercial pipelines or by utilizing it at the mine site for power generation and heating.

---

1 Supervisory Geophysicist, Bureau of Mines, U.S. Department of the Interior, Pittsburgh, PA.

TABLE 1. - Composition and Heating Value of Coalbed Gas and Natural Gas, pct

Source (1)	CH_4 (2)	CH[1/] (3)	H_2 (4)	Inerts[2/] (5)	O_2 (6)	Btu/s cu ft (7)
Pocahontas No. 3	96.87	1.40	0.01	2.09	0.17	1059
Pittsburgh	90.75	0.29		8.84	0.20	973
Kittanning	97.32	0.01		2.44	0.24	1039
Lower Hartshorne	99.22	0.01		0.66	0.10	1058
Mary Lee	96.05	0.01		3.45	0.15	1024
Natural Gas[3/]	94.40	4.90		0.40		1068

1/ Other hydrocarbons
2/ N_2, CO_2, and He
3/ Moore, B. J., R. D. Miller, and R. D. Shrewsburg, "Analysis of Natural Gases of the United States," IC8302, Bureau of Mines, 1976.

Note: 1 Btu/cu ft. = 0.0373 MJ/m^3

Each coalbed has a characteristic gas reservoir pressure, which increases as overburden becomes thicker (2) (Table 2). Because gas pressure and gas content are directly related, deeper coalbeds are gassier and are subject to more severe methane emissions during mining than shallower ones. The high productivity necessary to a successful mining

TABLE 2. - Gas Pressure and Gas Content of Coalbeds

Coalbed (1)	Overburden, in feet (meters) (2)	Gas Pressure, in pounds per square inch gage (kilopascals) (3)	Gas Content, in cubic feet per ton (cubic centimeters/ gram) (4)
Pittsburgh	500-900 (152-274)	275 (1895)	218 (6.8)
Pocahontas No. 3	1600 (488)	650 (4478)	430 (13.4)
Mary Lee	2100 (640)	830 (5718)	512 (16)

venture can be safely maintained in deep mines if methane is drained from the coalbed prior to mining, or methane control technology is applied during mining.

Horizontal drainage holes are drilled into coalbeds to control methane during mining and to degasify coalbeds before mining. Drainage holes are routinely drilled to 1000 ft (305 m) and some have been drilled to 2500 ft (762 m) but with a great deal of difficulty (2). Every foot of horizontal hole drilled in coal is productive, and gas flow increases as hole length increases. Typical stabilized gas flows range from 40 to 250 cu ft/day/ft (3.7 to 23.0 m^3/day/m) of hole (Table 3).

TABLE 3. - Stabilized Gas Flows from Drainage Holes

Coalbed (1)	Gas Flow, in cubic feet per day per foot (cubic meters per day per meter) (2)
Pittsburgh (Fairview, WV)	125 (11.5)
Beckley (Beckley, WV)	250 (23)
Pocahantas No. 3 (Grundy, VA area)	40 to 170 (3.7 to 15.6)
Mary Lee (Warrior Basin, AL)	100 (9.2)
Lower Sunnyside (UT)	160 (14.7)

This means that a 1000-ft (305 m) drainage hole can produce from 40 000 cu ft/day (1120 m^3/day) in the Pocahontas No. 3 Coalbed to 250 000 cu ft/day (7000 m^3/day) in the Beckley Coalbed. Consequently, when a drainage hole is intercepted by mining, large quantities of methane are released directly in the cutting head of the miner. Because adequate quantities of ventilation air are not available at the interception point to dilute methane rapidly to below its explosive range (5 to 15 pct), drainage holes must be sealed before they are intercepted by mining to avoid an explosion hazard.

This paper describes a method for sealing drainage holes, problems during the sealing operation, and observations when the hole is intercepted by mining. A cement slurry, mandated by the Mine Safety and Health Administration (MSHA), was used to seal the holes.

DRAINAGE HOLE CHARACTERISTICS

The vertical trajectory of a typical drainage hole is sinusoidal

(Fig. 1). When roof strata are intercepted, the bit is pulled back about 50 ft (15 m) and a new hole is started (Fig 1-A). Early in the Bureau's drilling program, when floor strata were encountered, drilling was continued in the floor strata until the bit was brought back into the coalbed (Fig. 1-B). This practice was discontinued because the immediate floor in some cases is a clay-like material which squeezes. At present, when floor strata are intercepted, the bit is pulled back to where hole trajectory goes slightly upward, and a new hole is started.

Coalbeds in the Appalachian, Illinois, and Western coal basins are naturally fractured, and the face and butt cleats are the more prominent fracture systems (5). These fracture systems are related to regional geologic structures, with the butt cleat parallel to the axes of structures and the face cleat at right angles. Flow of methane and water from drainage holes is controlled primarily by the permeability of the fracture system.

Permeability is a difficult characteristic to measure and interpret. Coal cores for permeability measurements are difficult to obtain because coal is fractured and in many cases friable, and the coal samples relax once confining pressures are released. Also, coalbeds contain water, and gas permeability is affected by water saturation which changes with time. Stabilized water flows from drainage holes average from 0.6 gal/min (2.3 L/min) in the Pittsburgh Coalbed to 5 gal/min (19 L/min) in synclinal structures of the Beckley Coalbed.

STUDY AREA

The hole-sealing operation was conducted in the Pittsburgh Coalbed in an 18-ft (5.5-m) diameter air shaft, located 6500 ft (1982 m) from the nearest mine working (1,4) (Fig. 2). Four headings in the shaft were extended, 25 ft (7.6 m) to the north, east, south, and west (Fig 3). Five horizontal drainage holes were drilled from these headings to depths ranging from 690 to 2126 ft (204 to 648 m). These 3-1/2 to 3-5/8-in. (89-to 92-mm) diameter holes drained 940 million cu ft (26 million m^3) of methane in about 3.7 years. During this period, a set of mine entries was driven from the mine to within 500 ft (152 m) of the shaft. Degasification reduced methane emissions into these entries during mining by 70 pct.

Gas flow from the drainage holes immediately before grouting totaled 700 000 cu ft/day (19 600 m^3/day), or an average of 140 000 cu ft/day (3920 m^3/day) from each hole. Because a drainage hole is intercepted many times, the hole must be sealed along the entire length to avoid methane emissions when intercepted.

GROUTING PROCEDURES

Schedule 40 polyvinyl chloride (PVC) pipe, 1.5 in. (38 mm) in ID, was inserted to the back end of each hole, where possible. The pipe was coupled using a PVC cement and manually pushed into the hole. A grout header was attached to a 4-in. (102-mm) diameter steel pipe, which was grouted 20 ft (6 m) into the coalbed when the drainage hole was initially drilled (Fig. 4).

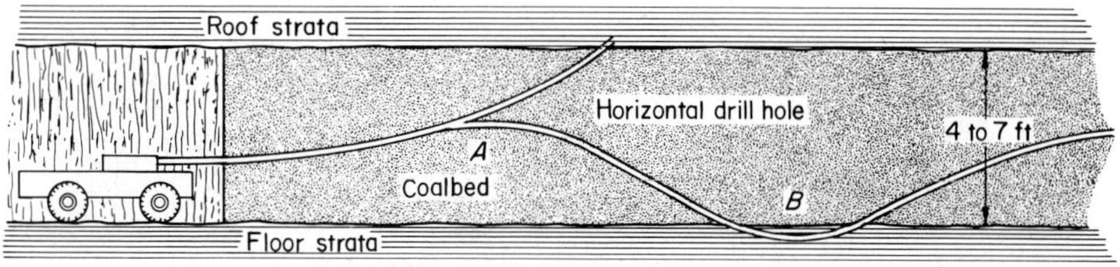

FIGURE 1. - Vertical Trajectory of Drainage Hole (1 ft. = 0.305m)

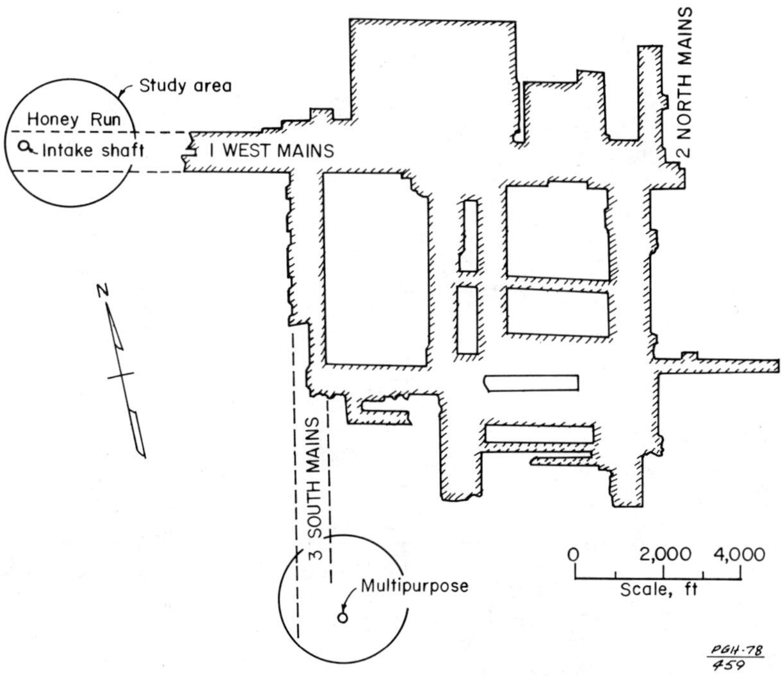

FIGURE 2. - Location of 18-ft (5.5 m) Diameter Shaft (1ft. = 0.305m)

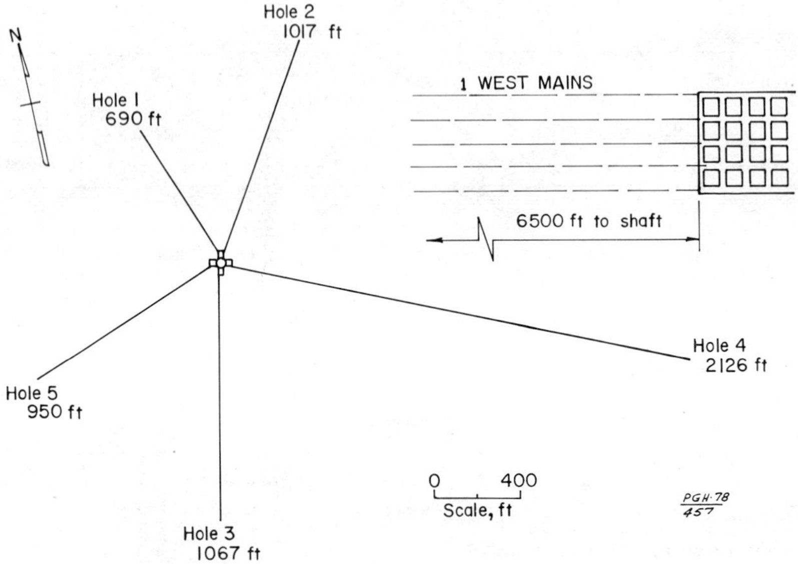

FIGURE 3. - Drainage Hole Pattern (1 ft. = 0.0.305m)

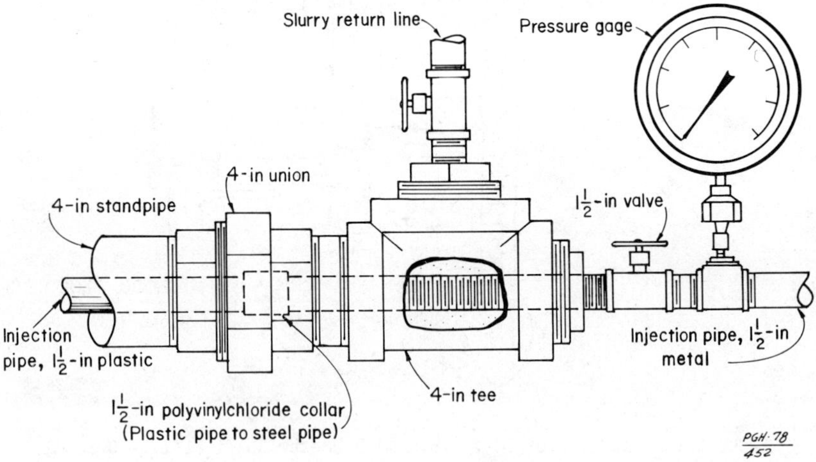

FIGURE 4. - Grout Header (1 in. = 25.4 mm)

A 3-in. (76-mm) diameter vertical pipe, located in the shaft wall, was used to transport the cement slurry from the surface to the bottom of the 700-ft (214-m) shaft (Fig. 5). A high-pressure hose connected the vertical pipe to the grout header. Initially, the vertical pipe was filled with water, and the cement slurry was poured on top of the water, which cushioned the slurry and prevented flash-setting on impact. The slurry was poured from a mixing truck through an 1/8-in. (3-mm) mesh screen and into a hopper mounted on the vertical pipe (Fig. 6). Water was drained from the vertical pipe at shaft bottom until the slurry appeared. The flexible hose was then attached to the grout header, and the slurry was forced through the PVC pipe under hydrostatic pressure. Gas and water flows from the drainage hole during the grouting operation were directed to a water trap which separated water and gas (Fig. 5). The flowing pressure at the grout header at this stage of the grouting operation was 30 to 50 psig (207 to 344 kPa).

When the slurry could no longer be forced into the hole under hydrostatic pressure, a positive displacement pump on the surface was connected to the vertical pipe (Fig. 6). Flow rates and pressure at the pump varied between 100 and 125 psig (689 and 861 kPa) and from 5 to 7 gal/min (19 to 27 L/min), respectively. Several times during the grouting operation, the flowing pressure rose rapidly to 250 psig (1722 kPa) before the slurry returned through the annulus to the collar of the hole. The rapid rise in flow pressure in each case was traced to the formation of filter cake in the flexible hose underground. When the slurry was observed at the collar of the hole, a valve in the return line was closed and the pumping was continued to force the slurry into the fracture system surrounding the drainage hole. However, we found that the pumping pressure exceeded 300 psig (2067 kPa) within 2 min, indicating that very little, if any, slurry was forced into the fracture system.

Deviations in grouting procedures occurred on holes 4 and 5 (Fig. 3). During drilling of hole 4, the clay floor was intercepted at 400 ft (122 m) and the bit followed the bottom of the coalbed to about 800 ft (244 m), where it re-entered the coalbed. PVC pipe could not be inserted beyond 400 ft (122 m) because the hole in the clay material had squeezed.

When redrilled, the hole squeezed again. Consequently, a 1-in. (25-mm) diameter metal pipe was inserted to a depth of only 1700 ft (518 m), because the metal pipe could not be negotiated through a sharp turn. The cement slurry was pumped through the metal pipe, but before the slurry returned through the annulus, either the slurry plugged in the metal pipe or the hole had squeezed around the metal pipe and prevented the slurry from reaching the collar of the hole. Consequently, the sealing operation was completed by filling the hole through the annulus.

Hole 5 had collapsed at 400 ft (122 m), and attempts to drill through the blockage failed because the bit deflected into solid coal and a new hole was started. A 1-in. (25-mm) diameter metal pipe was inserted to a depth of 400 ft (122 m), leaving 550 ft (168 m) without pipe. Although the volume of slurry pumped into the hole was close to the cal-

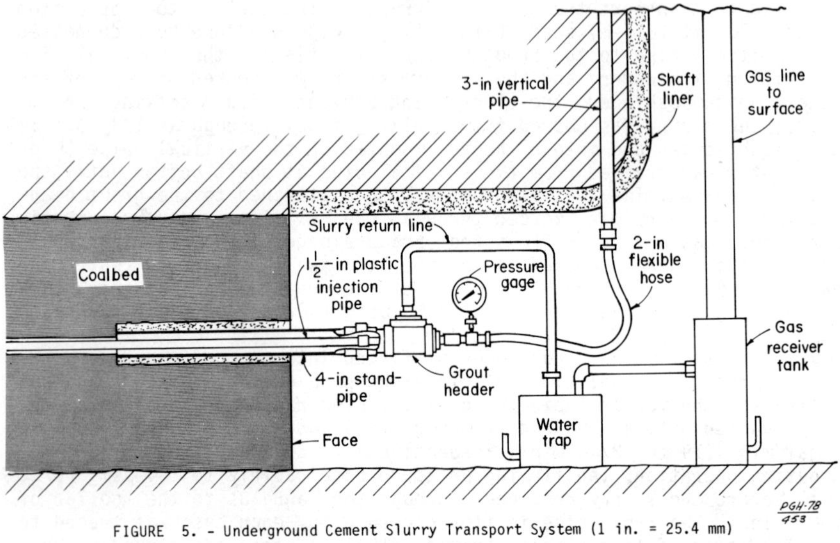

FIGURE 5. - Underground Cement Slurry Transport System (1 in. = 25.4 mm)

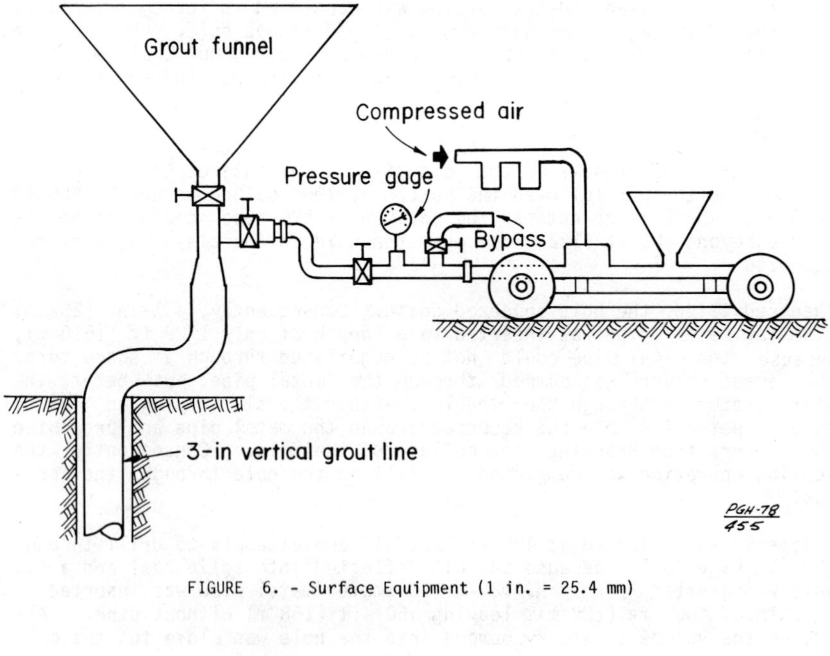

FIGURE 6. - Surface Equipment (1 in. = 25.4 mm)

culated volume of the hole, there was no assurance that the hole was filled along its entire length.

Because no guidelines were available on similar studies and to avoid plugging problems during the slurry pumping phase, the water content of the slurry was increased from the recommended level of about 10 to 15 gal (37.9 to 56.8 L) (Table 4). Consequently, a weaker cement and water separation were expected (6). A high structural strength cement is not an important consideration, but water separation could cause methane problems. When a hole is intercepted, gas pressure in the coalbed forces water from the hole, and then methane. However, we felt that by pressurizing the slurry in the hole, excess water would be squeezed from the slurry and forced into the fracture system surrounding the hole, thus avoiding stratification of the slurry.

TABLE 4 - Cement Slurry Design

Component (1)	Quantity (2)
Class A Portland Cement	1 sack (94 lb)
Fly Ash	1 sack (75 lb)
Water	15 gal
Fluidifier	2.5 lb

Note: 1 lb = 0.453 kg; 1 gal = 3.79 L

RESULTS AND DISCUSSION

Four of the five holes were intercepted by mining. No results are available for Hole 3. Hole 2 was intercepted by mining 700 ft (214 m) from the shaft (Fig. 7) and was located about in the center of the coalbed (Fig. 8). The PVC pipe was filled with cement, but the annulus was only 60 pct filled. When the hole was intercepted, gas pressure in the coalbed forced water from the hole and then methane, deenergizing electrical power to the mining machine. The cement slurry may have stratified because of excess water in the mix, or it may have been diluted further by coalbed water. The hole was subsequently intercepted six more times. In each case, the PVC pipe and annulus were filled with cement and no methane problems developed (Fig. 9). Examination of the coalbed around the hole at the interception points indicated that the cement slurry had not been forced into the fracture system. It could not be determined if excess water in the cement slurry had been squeezed into the fracture system when pressurized.

Hole 1 was intercepted next, 200 ft (61 m) from the shaft (Fig. 9), and two holes were observed in the coalbed (Fig. 10). When drilled,

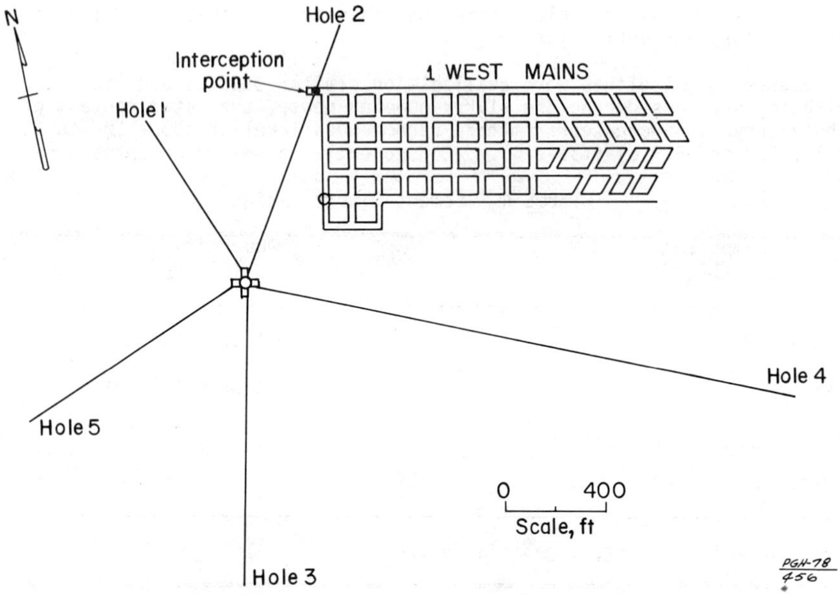

FIGURE 7. - Location of Interception of Hole 2 (1 ft. = 0.305 m)

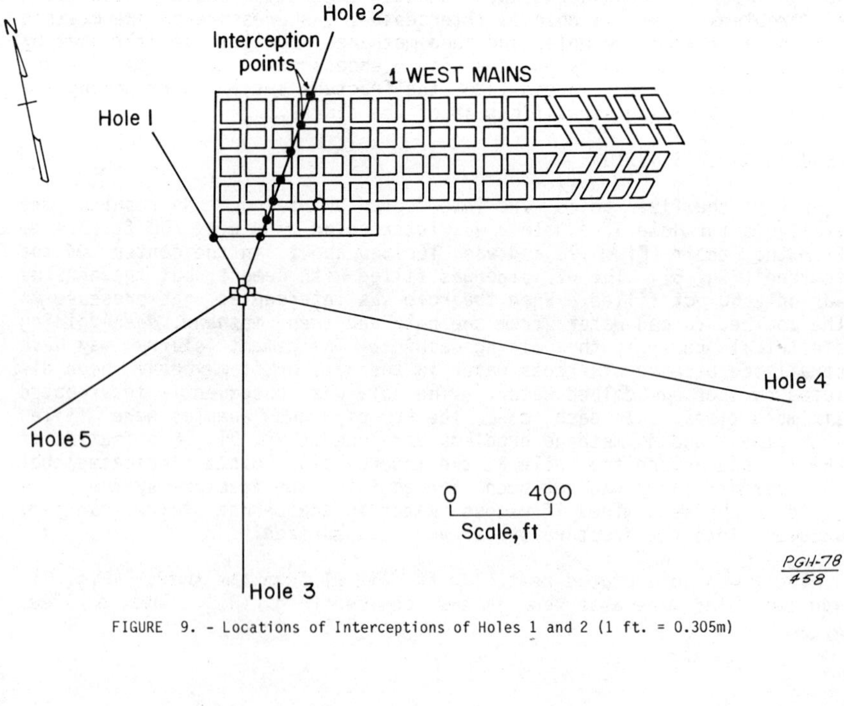

FIGURE 9. - Locations of Interceptions of Holes 1 and 2 (1 ft. = 0.305m)

FIGURE 8. - Hole 2 After Interception by Mining

FIGURE 10. - Hole 1 After Interception by Mining

the hole drifted into the roof at 275 ft (84 m). A new hole was restarted at 125 ft (38 m); at the interception point, it is about 1 ft (0.3 m) below the original hole. Both holes were filled with cement, and no methane problems developed when they were intercepted. We found no evidence of cement slurry in the coalbed surrounding the holes.

Hole 4 was intercepted 660 ft (201 m) from the shaft (Fig. 11). The hole was about 1 ft (0.3 m) below the coalbed, and the annulus did not contain grout. A small methane flow was detected, but mining was not interrupted. When the hole was intercepted in the floor a second time, 770 ft (235 m) from the shaft, methane flow from the hole deenergized electrical power to the mining machine. A quick-set cement sealed the hole and mining resumed. When intercepted a third time at 1000 ft (305 m), the hole was located in the coalbed and was filled with cement; no methane problems occurred. Apparently, the hole had squeezed around the metal pipe between 400 and 800 ft (122 and 244 m), and consequently the cement slurry could not enter this region.

Hole 5 was intercepted by mining about 3 years after grouting. Methane problems were reported, which confirmed suspicions that the cement slurry had not penetrated the blockage at 400 ft (122 m) and filled the entire hole, even though the volume of slurry pumped into the hole was close to the calculated hole volume.

Cement slurry was used to seal eight drainage holes in the Beckley Coalbed (Beckley, WV), subsequent to the studies in the Pittsburgh Coalbed. Slurry pumping rates were less than 7 gal/min (26.5 L/min), and hole lengths ranged between 415 and 910 ft (127 and 278 m). When the water content of the slurry was reduced to 9 gal (34.1 L) (Table 4), the slurry plugged the PVC pipe after about 60 gal (227 L) of slurry was pumped. A water content of 10 gal (37.9 L) proved satisfactory, and no pumping problems developed. Four of the eight holes were intercepted. No methane problems developed on two holes where water flow from the drainage holes before grouting was less than 0.5 gal/min (2 L/min). Where water flow rates were 2 and 5 gal/min (7.6 and 19 L/min), methane problems developed when intercepted by mining. Water influx during pumping apparently diluted the slurry and then water separated from the slurry.

SUMMARY AND CONCLUSIONS

Horizontal drainage holes are sealed to avoid formation of explosive methane-air mixtures when intercepted by mining. In the Pittsburgh Coalbed, five drainage holes were sealed with a cement slurry in which the water content exceeded recommended levels (Table 4). Hole 2 was intercepted seven times, and methane problems occurred only once, when cement did not fill the hole completely. No methane problems occurred when Hole 1 was intercepted. A blockage prevented the cement slurry from reaching the back part of Hole 5. Hole squeezing prevented the slurry from filling the annulus between 400 and 800 ft (122 and 244 m) in Hole 4. Methane problems developed when both there holes were intercepted by mining.

In the Beckley Coalbed, eight drainage holes were sealed with a

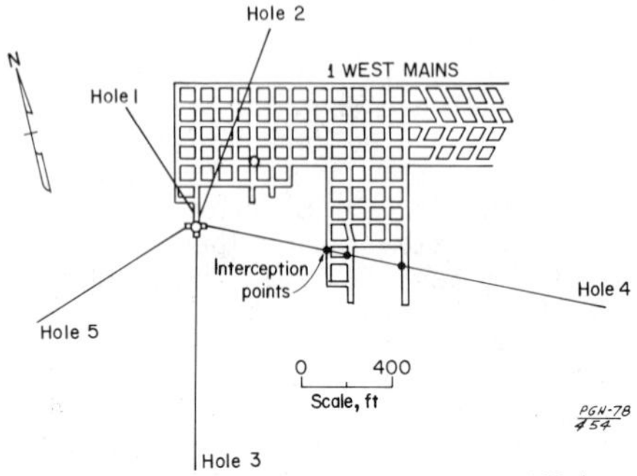

FIGURE 11. - Locations of Interceptions of Hole 4 (1 ft. = 0.305 m)

slurry in which the water content was reduced from 15 to 10 gal (56.8 to 37.9 L) (Table 4). Hole lengths were less than 1000 ft (305 m), and no pumping problems developed. Four holes were intercepted by mining. When water flow from a drainage hole was less than 0.5 gal/min (1.9 L/min), the hole was sealed satisfactorily and no methane problems occurred when it was intercepted by mining. For two drainage holes where water flow rates were 2 and 5 gal/min (7.6 and 19 L/min), methane problems developed when the holes were intercepted because the slurry had been excessively diluted and water separation occurred.

The problem of partial grouting will exist and must be dealt with after the hole is intercepted. If the intercepted drainage hole does not contain any cement, the plastic pipe must be removed and a packer inserted to seal the hole. If the intercepted drainage hole contains a partial seal, it must be recemented with a quick-set cement.

APPENDIX 1 - REFERENCES

1. Aul, G. W., J. Cervik, "Grouting Horizontal Drainage Holes in Coalbeds," RI8375, Bureau of Mines, 1979.

2. Cervik, J., "VII Methane Drainage," Second International Mine Ventilation Congress, Society of Mining Engineers of American Institute of Mining, Metallurgical, and Petroleum Engineers, Inc., New York, NY, 1980, pp. 257-264.

3. Deul, M., A. G. Kim, "Helping Finance New Mines With Revenues From Coalbed Degasification." Proceedings Third Symposiumon Underground Mining, NCA/BCR Coal Conference and Expo IV, Oct. 1977, pp. 25-41.

4. Fields, H. H., J. Cervik, T. W. Goodman, "Degasification and Production of Natural Gas From an Air Shaft in the Pittsburgh Coalbed," R18173, Bureau of Mines, 1976.

5. McCulloch, C. M., M. Deul, P. W. Jeran, "Cleat in Bituminous Coal." RI7910, Bureau of Mines, 1974.

6. Smith, D. K., ed., "Factors That Influence Cement Slurry Design," Cementing, Monograph Vol. 4, Society of Petroleum Engineers of AIME, New York, NY, 1976, p. 36.

HYDROGEOLOGICAL ASSESSMENT AND GROUTING AT SELBY

James C Black[1] C Alan Pollard[2] and Graham P Daw[3]

ABSTRACT

The paper describes grouting for water control carried out during the sinking of deep shafts at two sites within the new Selby Coalfield in Yorkshire, England. The importance of a detailed hydrogeological borehole assessment prior to shaft sinking is highlighted, with particular reference to the interpretation of test data relevant to the subsequent ground treatment.

INTRODUCTION

The National Coal Board's new Selby Coalfield Development in North Yorkshire, England, covers an area of 280 km^2 with access to the underground coal workings provided by two inclined drifts and five pairs of shafts. The strata overlying the Coal Measures contain several strong aquifers and various methods of ground treatment have been employed for ground water control during shaft and drift construction. At most of the sites grouting has played a major role. The main part of this paper describes the grouting carried out at two of the shaft locations.

In each instance a detailed hydrogeological assessment prior to shaft sinking provided a means of locating the aquifer zones and of selecting, in advance, the most appropriate method of ground treatment. Such investigations now form an important part of most shaft sinking operations and it is common practice to drill a test borehole on the centreline of the proposed shaft. The range of borehole investigative techniques is extensive and experience is required in selecting the most appropriate and cost effective test programme for each particular situation. Several authors have discussed investigation methods related to shaft sinking, (1,2). In the early part of this paper current investigation methods are reviewed with the emphasis on those most relevant to ground treatment for the control of water inflow.

Once the shaft sinking has commenced, in-shaft testing ahead of the advancing sump (shaft bottom) allows further assessment of the local aquifer characteristics, and modifications to the proposed grouting scheme as may be necessary.

1 Managing Director, Cementation Mining Limited, Doncaster, England.

2 Project Engineer (Grouting), Cementation Mining Limited, Doncaster, England.

3 Senior Project Scientist, Cementation Specialist Holdings Limited, Rickmansworth, England.

In the discussion of the grouting at the two Selby Sites the usefulness and reliability of the hydrogeological investigation data is assessed. The grouting techniques required to deal with a heavily water bearing fissured and "vuggy" limestone, are contrasted with those for a very weakly cemented waterbearing sandstone in which a combination of chemical grouting, pressure relief wells and auxiliary ground support was found to be highly effective.

REVIEW OF BOREHOLE INVESTIGATION METHODS

Requirements: To assess grouting requirements for water control the borehole investigations should provide:

(i) the location and extent of the aquifer zones,
(ii) the transmissivity and water pressure within the aquifer and hence the potential water inflow to the untreated excavation,
(iii) an indication of the nature of water flow, i.e. whether from fissures, fractures, solution passages (vughs) or rock pores,
(iv) the intrinsic permeability and hence transmissivity of rock material in each aquifer and the relative contribution of fissure and intergranular flow to the total aquifer transmissivity as this helps to determine the form of grouting, the choice of grout and the quantities required,
(v) the nature and size of fissures and rock porosity as this is an important factor determining the grouting pattern and the type of grout required,
(vi) the chemical composition of the ground water and the aquifer rock as these may place restrictions on the choice of grout.

This information is derived from four main sources: core drilling and logging, core testing, borehole water testing and borehole electrical logging. Each of these will be discussed briefly in respect of its grouting significance as it is the careful correlation and analysis of this data that forms the basis for the ground treatment design.

Core Drilling and Logging: Observations made during the drilling operation including ground water levels, drilling fluid loss, ground water inflow and drilling rates help to pin-point, at an early stage, likely problem areas.

Careful handling, logging and subsequent storage of the retrieved core is essential as frequent observations will be made of the core both at the design stage and during shaft construction. The core log should include discontinuity and core quality logs such as core loss, rock quality designation (RQD) and fracture index. A good photographic record can prove invaluable at a later date when samples have been removed and the core has deteriorated.

Essentially the main interest lies in locating zones of high permeability which may for example be evident from "vuggy" zones in limestone or coarse grained sections of sandstone obviously retaining moisture or showing signs of efflorescence. The careful inspection of apparent fractures in the core can distinguish natural partings from open and waterbearing fissures.

Core Testing: In a full hydrogeological investigation an extensive range of laboratory tests can be carried out on the core material to provide data for all aspects of shaft design. However, for grouting purposes it is the permeation properties that are most important. Permeability measurements on selected pieces of core from the aquifer zones allow an assessment of the intergranular component of flow, and an indication of whether chemical grouting is feasible. Porosity measurements are used in the analysis of the water tests, to correlate with the geophysical logs and for estimating the volume of chemical grout required. The measurement of pore size distribution is not considered routine but can provide invaluable information on the injectability requirements of the chemical grout.

Borehole Water Testing: Borehole water tests are the principal means for predicting water inflows to the proposed shaft, by providing measurements of the aquifer transmissivity and ambient piezometric head.

In recent years the drill stem test (DST), used in the petroleum industry, has been widely favoured to provide this information as the method is insensitive to the presence of residual drilling mud and to casual local constriction of passages near the borehole. It has the additional advantage over the standard aquifer pumping test of being a single hole test and of being much faster in operation.

The limitations of the DST are experienced when strong aquifers are encountered at relatively shallow depths. For these particular situations a new system has been developed which employs the same principle and method of interpretation as the DST but which uses a submersible well pump, borehole packer, and pressure transducer (3). It has additional advantages over the DST in that constant flow may be maintained throughout the flow period, and continuous monitoring of down-hole pressure is provided at a surface station.

It is our opinion that wherever possible for mineshaft groundwater assessment the DST, or at least, a technique based on the pressure recovery principle should be employed and in a later section an "in-shaft" modification of the test developed and used at Selby, is described.

Borehole Electrical Logging: A continuous record throughout the borehole length of certain physical properties allied to the nature of rock, and its contained fluids can supplement the core and water test information by:

(i) providing data where core recovery is poor and coring impracticable,
(ii) allowing correlation with the laboratory core measurements,
(iii) providing continuity where hydrological testing is either incomplete or proves unsatisfactory.

Modern processing techniques of radioactive and sonic logs, as well as providing a direct measurement of formation porosity will give indication of relative permeability and differentiate between primary and secondary porosity.

Caliper logging is invaluable for confirming the more competent areas of rock suitable for setting packers for the water testing, and is necessary to provide the borehole size correction factor used in processing the other logs.

Although not essential, logs including temperature, flowmeter and borehole T.V. surveys can often locate fairly accurately the presence of waterbearing fissures. In addition, temperature data is required to assist in the selection of chemical grout formulations.

Correlation of Data: Proper correlation and analysis of the borehole data is as important a part of the pre-assessment programme as the measurements themselves. Where possible all information should be correlated on a single chart from which the main characteristics of the various aquifer zones can be readily abstracted, (for example see Fig. 1).

Essentially the main aquifers will be identified by their high transmissivity from the water tests while within each test zone evidence of high intergranular permeability and porosity will come from core observation and laboratory tests on selected samples, and evidence of waterbearing fissures from the core examination and the various borehole electrical logs.

Water Inflow Estimation: Values of aquifer transmissivity and hydrostatic pressure obtained from the water tests together with dimensions of the shaft and knowledge of reservoir characteristics and boundary conditions where known are required to estimate potential water inflows to the shaft. These estimates can be no more reliable than the results of the in-situ water tests allow, which underlines the importance of conducting the most appropriate form of in-situ water test for a particular situation.

The basic flow phenomena are well understood and generally the geometry and boundary conditions associated with a vertical shaft provide a fairly straight forward model; the most commonly used calculation being for the case where the shaft intersects an extensive "confined" homogenous aquifer in a cylindrical reservoir.

Preliminary Grouting Predictions: Having established the location of the water bearing zones, estimated the probable level of shaft inflow and assessed its primary cause, consideration must be given to inflow control by some form of ground treatment. Where grouting is selected it is necessary to assess the type or combination of grouts that should be used in each instance.

Very simply, cement grouting is limited to relatively large interconnected cavities such as solution passages and to fissures and open joints of width greater than about 200 microns. Finer fissures and practically all intergranular porosity will require chemical grout treatment, in which case a number of factors affect the final choice, including injectability, strength, gel time, permanence, toxicity and cost.

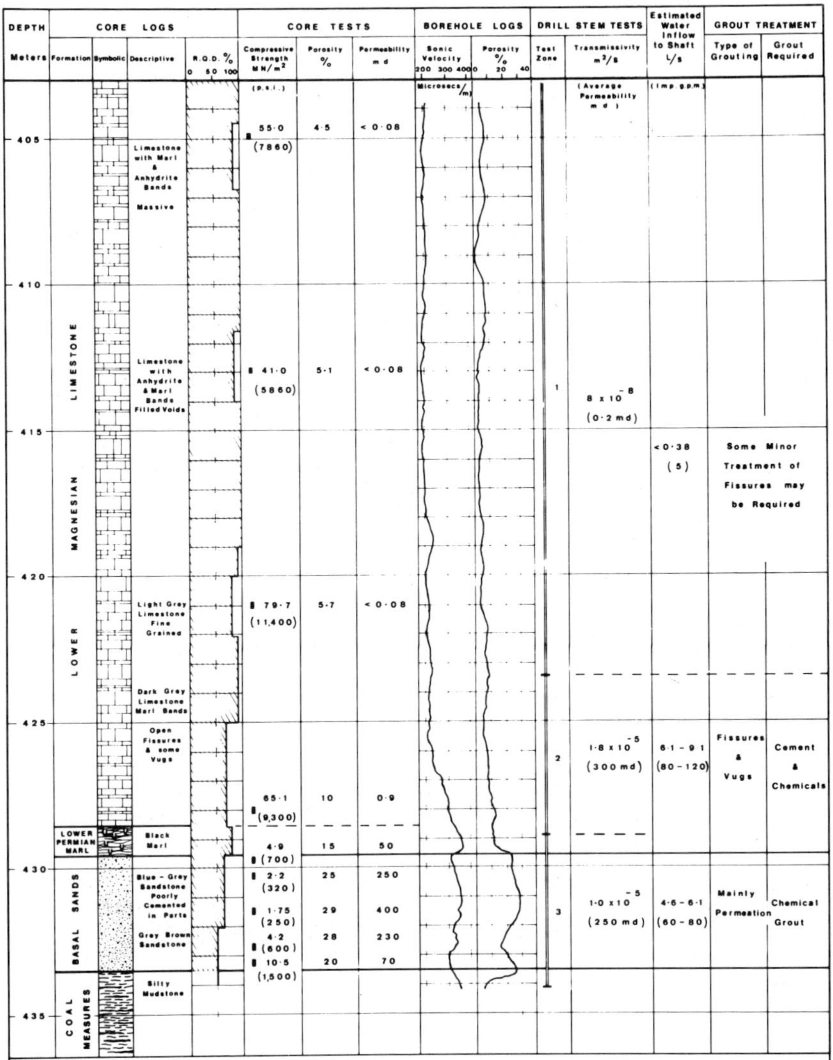

FIG 1. SECTION OF BOREHOLE INVESTIGATION DATA FOR RICCALL SHAFTS

The experienced grouting engineer will usually decide fairly early on the most likely chemical grout for a particular application and can calculate the grout hole spacing and the width of grout curtain which will be required around the shaft excavation, keeping in mind the rock strength, ground water pressure, method of excavation, and the time during which the excavation will be unlined.

IN SHAFT TESTING

Whilst primary grouting design is based on previously described borehole investigations from the surface it is essential to carry a series of exploratory boreholes ahead of the advancing shaft sump during the sinking operations to confirm predictions, and modify the design as necessary. During the Selby Project, an apparatus was developed which enables a pressure/depletion recovery test (PRT) to be carried out from such an exploratory hole. The apparatus as assembled on the shaft sump is shown in Fig 2. It is attached to the borehole standpipe valve and comprises an electronic pressure transducer whose output is fed to a digital voltmeter, a standard pressure gauge, and control valve. When the control valve is opened water flows from the aquifer producing a depletion of pressure in the aquifer. The control valve is closed after a conveniently large and measurable volume of water has been collected and the pressure recovery monitored for a period typically three times the period of flow.

The method of analysis is identical with the more conventional pressure recovery test (3) or DST where the rate of recovery of pressure after time ΔT has elapsed since cessation of flow is applied in a "Horner" plot. For a simple confined aquifer of constant transmissivity the Horner plot yields a straight line, whose slope and intercept on the pressure axis provide the transmissivity and the undisturbed aquifer water pressure.

As an illustration the results for a test in the Basal Sands at one of the Selby Shafts is included in Fig 2. In the following section the use of the test at the various sites is mentioned and comparison made where possible with the DST's.

In general terms the agreement obtained has been good and the new test has the great advantage of enabling estimates of potential water inflows to be made for single thin aquifers lying just below the completed excavation face, as opposed to the more long range aspects of tests conducted from the surface.

GROUTING AT SELBY SHAFTS

At Selby the Coal Measures are overlain by Permo-Triassic strata containing several aquifer zones requiring ground treatment for water control.

The Triassic Bunter Sandstone is the major aquifer and borehole investigations indicated potential shaft water inflows of several hundred litres/s. As it is a main water source in the Selby area, and a relatively weak sandstone it was decided to freeze this formation at all five shaft sites. Groundwater occurs to varying extents in the

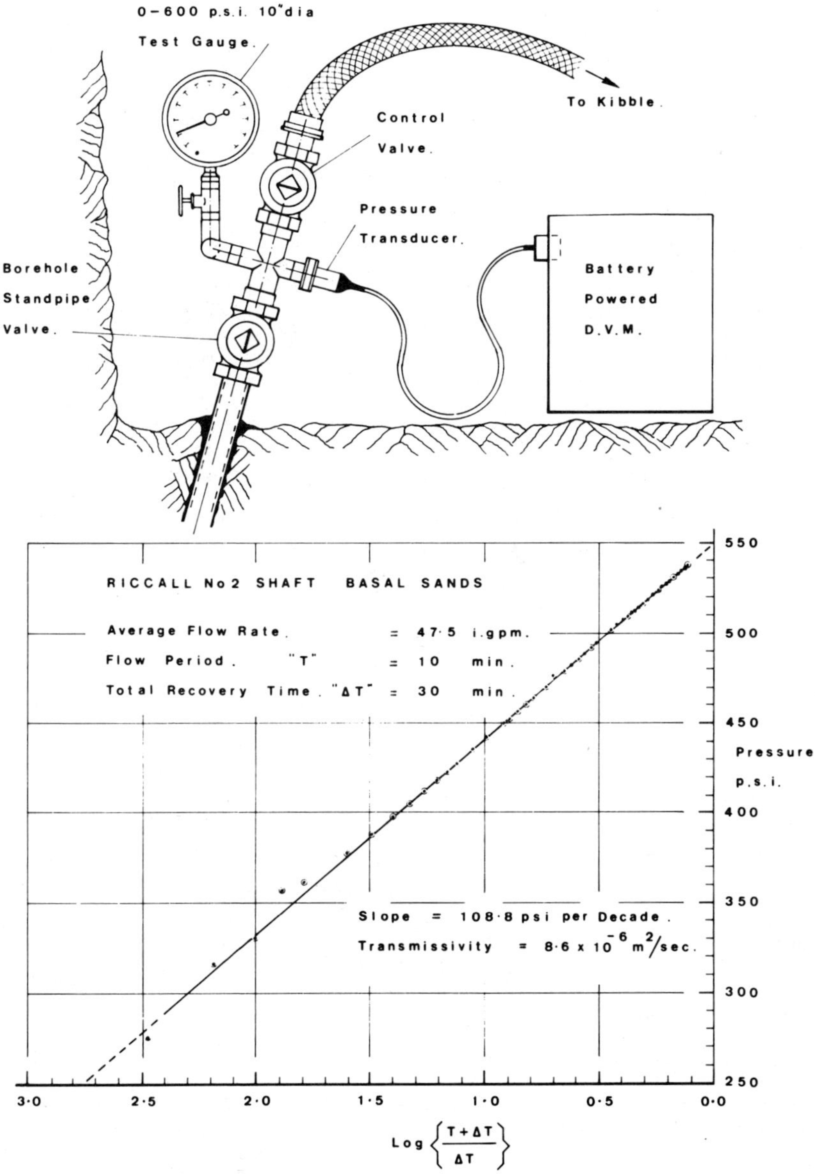

FIG 2. In-SHAFT PRESSURE RECOVERY TESTS.

Permian Magnesian Limestones, and Basal Permian Sands. Other than at drift site where the Basal Sands were frozen and at the Wistow No 1 shaft where the freeze was extended through the Limestone and Basal Sands, grouting was used as the prime means of ground treatment.

As an illustration of the variable conditions encountered and the approaches taken, the main discussion in this paper focuses on grouting of the Lower Magnesian Limestone and Basal Sands at the Riccall Shafts (Fig 3) and the No 2 Shaft at Wistow.

Lower Magnesian Limestone (L M L): The various core and borehole logs and core tests showed the L M L at the two sites to be generally fine grained, massive rock of low intrinsic permeability and porosity and relatively high strength. Variable transmissivity as evidenced by water testing and core inspection was attributed to fissuring and small open solution cavities (vughs). At Wistow these features were very pronounced over the lower 20 metres of the formation producing a high transmissivity and very high potential shaft inflow. At Riccall DST's gave a much lower transmissivity and inspection of the core indicated a finely fissured zone extending over only a few metres towards the base of the formation and minor evidence of solution cavities.

In all three shafts, grout covers approximately 30m long were employed and both cement and chemical injection were used to reduce the potential inflows. The very different conditions, and means of obtaining the reductions in potential shaft inflow are summarised in Table 1.

Wistow No 2 Shaft: A water test in the original test borehole (some 100m from shaft) had suggested a transmissivity of about 2×10^{-4} and an estimated shaft inflow of up to 75 litre/s (1000 i.gpm) from the zone 240-260m. Probe holes drilled ahead of the shaft bottom indicated that the sump for the grout cover should be located at 222m. The initial inflows into several of the primary ring of 16 holes exceeded 38 litre/s (500 i.gpm). An in-shaft pressure recovery test on one of the injection holes gave a transmissivity an order of magnitude greater than that observed at the test borehole, with a corresponding potential flow to the shaft of some 750 litre/s (10000 i.gpm).

The final grout hole pattern comprised a primary ring of 32 holes and an inner secondary ring of 16 holes. After the injection of some 185 tonnes of cement, the simultaneous inflow to four central test holes was 11.4 litre/s (150 i.gpm). This residual flow was attributed to local fine fissuring and also to narrow constrictions which prevented movement of cement grout within the network of solution cavities.

Further treatment was continued with injection by cement, followed by a sodium silicate based chemical grout, Cemex A2. This non-toxic, non-dermatitic water-stopping grout has been developed for the treatment of finely fissured rock and low permeability sandstone formations, and its use here was effective in reducing the flow through the grout cover, as monitored by the simultaneous inflow to six vertical test holes, to only 0.73 litre/s (9.6 i.gpm). In terms of grouting effectiveness, it is estimated that the grout treatment reduced the potential shaft inflow by about 1000. The shaft was then sunk a further 13m to 245m, the sump location chosen for the Basal Sands grout cover.

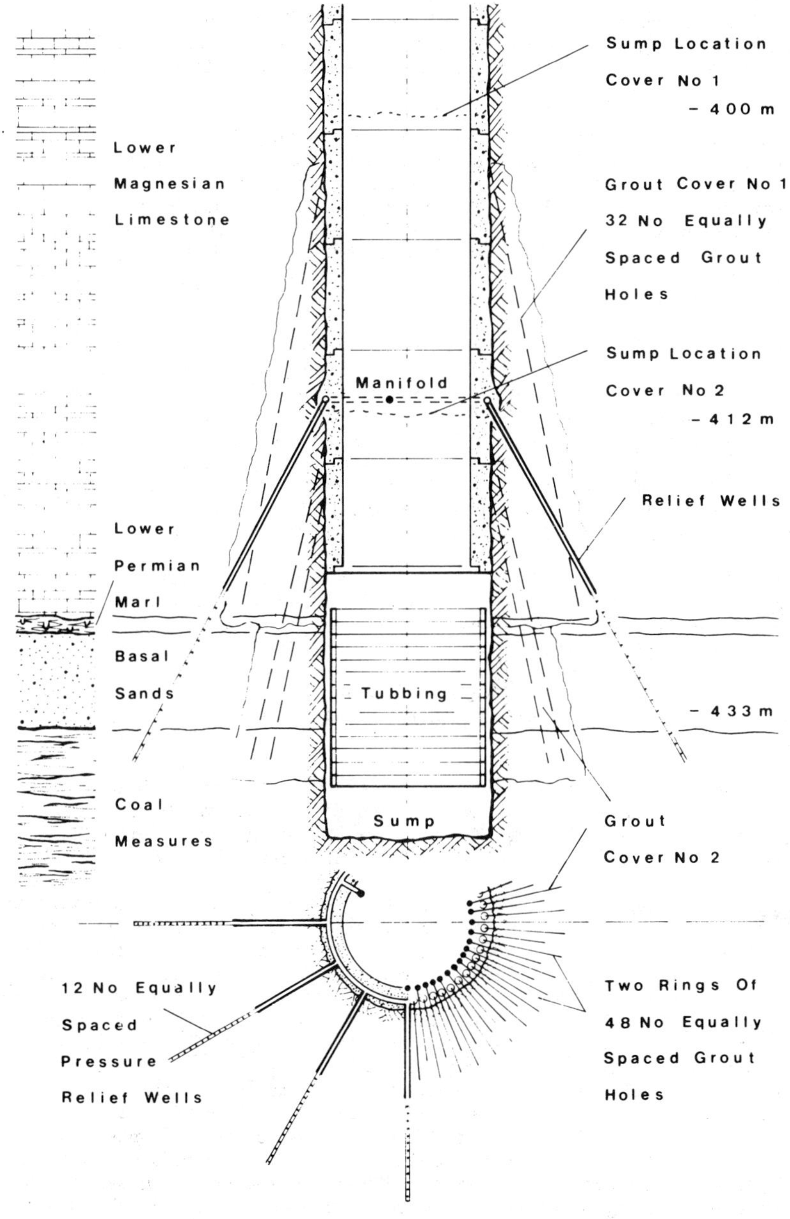

FIG 3. RICCALL No2 SHAFT GROUT COVERS

Riccall Shafts: In this instance, DST's in shaft centreline boreholes gave average transmissivities of 1.2 x 10^{-5} m^2/sec and 1.8 x 10^{-5} m^2/sec over the zones 417-426m and 423-429m in the No 1 and No 2 boreholes respectively, with corresponding potential shaft inflows of up to 7.5 litre/s (100 i.gpm).

During sinking, probe holes from the sump showed the L M L to be dry to 415m in the No 2 shaft, but one hole encountered 2.1 litre/s (28 i.gpm) inflow at 407m in the No 1 shaft. This was grouted with 13 tonnes of cement and sinking continued to the sump levels selected for the main grout covers of 395m and 400m in the No 1 and No 2 shafts respectively. A 32 hole injection pattern was used in each shaft with the geometry chosen to produce similar hole spacings of approximately 1.8m at the base of the cover.

In the No 1 shaft total flow to all injection holes was about 3.0 litre/s (40 i.gpm), with individual flows ranging from 0.08 to 0.8 litre/s (1 to 10 i.gpm). In terms of potential shaft inflow this is in reasonable agreement with the drill stem test predictions. Grouting was completed by the injection of 4.5 tonnes of cement grout and 7090 litres (1560 i.gals) of Cemex A2 chemical grout. The residual inflow to four vertical test holes within the grout cover totalled less than 0.08 litre/s (1.0 i.gpm) and the grouting can be considered to have reduced the potential shaft inflow by a factor of about 50.

In the No 2 shaft individual waterflows to the grout holes generally ranged up to 1.1 litre/s (15 i.gpm), but one hole encountered as much as 4.5 litre/s (60 i.gpm). The higher flows are in line with the higher transmissivity measured in the drill stem test for the No 2 shaft. A total of 22.2 tonnes of cement grout and 23850 litres (5250 i.gpm) of Cemex A2 chemical grout was injected. The residual flow to four central test holes was 0.25 litre/s (3.3 i.gpm) and estimated reduction in potential shaft inflow of about 35.

In retrospect it is thought that the vertical fissure encountered at 407m depth in the No 1 shaft most likely was connected to the main water-bearing zone, and the 13 tonnes of cement grout injected at that stage resulted in the fairly low subsequent cover injections.

Basal Sands: The tendency for the Basal Sands to vary considerably in thickness, texture, strength and permeability over relatively short distances was experienced initially at the Wistow site. A 2.5m thick section of generally fine grained, hard sandstone and an apparently 'dry' water test in the exploratory borehole proved to be far from the situation encountered at the actual shaft locations only some 70m distant. Here, and at Riccall the sandstone ranged from 4.5 to 6m in thickness and varied from fairly strong well cemented sections with good core recovery to soft, broken, poorly cemented sections with lost core. Indeed, two of the freeze holes for the Wistow No 1 Shaft were extended beyond the base of the Bunter Sandstone and attempts made to obtain cores through the Basal Sands. However, in both cases the core recovered showed zones of uncemented sand and the decision was taken in advance of sinking to extend the freeze through to include the Basal Sands.

TABLE 1 : LOWER MAGNESIAN LIMESTONE		WISTOW 2	RICCALL 1	RICCALL 2
Transmissivity (Borehole DST)	m^2/s	2×10^{-4}	1.2×10^{-5}	1.8×10^{-5}
Estimated Shaft Inflow	litre/s (i.gpm)	76 (1000)	7.6 (100)	9.1 (120)
Transmissivity (In-Shaft PRT)	m^2/s	3.1×10^{-3}	Not Tested	Not Tested
Estimated Shaft Inflow	litre/s (i.gpm)	760 (10000)	-	-
Aquifer Depth	m	240 - 255	417 - 426	423 - 429
	(ft)	(787 - 836)	(1368 - 1398)	(1388 - 1407)
Nature of aquifer		major vughs, fissures	few vughs, fissures	few vughs, fissures
Grout Cover extent	m	222 - 257	395 - 426	400 - 429
	(ft)	(728 - 843)	(1295 - 1398)	(1312 - 1407)
No of Rings		2	1	1
No of Injection Holes		32, 16	32	32
Hole Spacing at Base of Cover	m (ft)	1.97 (6.5)	1.74 (5.7)	1.79 (5.9)
Injection Hole Inflows				
Maximum Individual Hole	litre/s (i.gpm)	38 (500)	0.76 (10)	4.5 (60)
Quantity of Grout Injected				
Cement	tonnes (sacks)	218 (2320)	4.5 (48)	22.2 (236)
Chemical-Cemex A2	litres (i.gals)	433450 (95350)	7090 (1560)	23850 (5250)
Final Test Hole Flows				
No of Test Holes		6	4	4
All open	litre/s (i.gpm)	0.7 (9.6)	0.07 (0.9)	0.25 (3.3)
Estimated reduction in inflow		1000 : 1	50 : 1	35 : 1

TABLE 2 : BASAL SANDS		WISTOW 2	RICCALL 1	RICCALL 2
Transmissivity (Borehole DST)	m^2/s	Not Tested	2.2×10^{-5}	1.0×10^{-5}
Estimated Shaft Inflow	litre/s (i.gpm)	2.3 (30)	9.1 (120)	4.5-6.6 (60-80)
Transmissivity (In-Shaft PRT)	m^2/s	5.3×10^{-6}	2.5×10^{-5}	8.6×10^{-6}
Estimated Shaft Inflow	litre/s (i.gpm)	1.9 (25)	10.6 (140)	5.3 (70)
Aquifer Depth	m	259.2 - 263.1	426.9 - 432.5	429.6 - 433.6
	(ft)	(850 - 863)	(1400 - 1419)	(1409 - 1423)
Nature of Aquifer		Intergranular	Intergranular	Intergranular
Grout Cover extent	m	245.0 - 264.5	413.7 - 434.5	411.6 - 436.0
	(ft)	(804 - 868)	(1357 - 1426)	(1350 - 1430)
No of Rings		2	3	2
No of Injection Holes		16	48 48 48	48 48
Hole Spacing at Base of Cover	m	1.3	1.1 1.3 1.4	1.0 1.1
	(ft)	4.4	3.5 4.2 4.7	3.0 3.6
Injection Hole Inflows				
Maximum Individual Hole	litre/s (i.gpm)	1.44 (19)	38 + sand (500 + sand)	3.6 (48)
Quantity of Grout injected				
Cement	tonnes (sacks)	17.4 (185)	109.4 (116.4)	Nil
Chemical - high solids	litres (i.gals)	Nil	162750 (35800)	2410 (530)
Chemical - Cemex A2	litres (i.gals)	136900 (31020)	381900 (84000)	102300 (22500)
Final Test Hole Inflows				
No of Test Holes		1	4	4
Total Test Hole inflow	litre/s (i.gpm)	0.38 (5)	2.1 (28)	2.0 (26)
Estimated reduction in inflow		4:1	5:1	3:1
Pressure Relief				
Total Relief well inflow	litre/s (i.gpm)	2.1 (28)	7.1 (93)	3.0 (40)
Total Test Hole inflow	litre/s (i.gpm)	Not Tested	0.34 (4.5)	0.31 (4.1)
Estimated reduction in inflow		-	31:1	19:1

Laboratory tests on core samples gave in this particular shaft location high intergranular permeabilities indicating that a highly-penetrating chemical grout would be necessary to provide the injectability required to minimise the risk of "windows" in the grout cover and of erosion and instability in the less well cemented areas.

Water samples obtained during DST's indicated that the Basal Sands formation water was fairly saline. A series of tests was then conducted to ensure that selected chemical grouts were compatible with this formation water, specifically with regard to injectability, and after consideration of this and other properties such as toxicity, gel time control, strength and permanence, Cemex A2 emerged as the most suitable grout. To obtain the required injectability of this sodium silicate based grout, it was necessary to clarify the two component solutions of the grout by high speed centrifuging.

At the three sites the sump levels for the Basal Sands grouting were sited within the Lower Magnesian Limestone and the grout hole geometry was based on cover lengths varying from 20m at Wistow to 30m at Riccall. The summarised data from the three shafts is given in Table 2.

Because of the high water pressure, and the possibility that local zones of the Basal Sands were poorly cemented, and the close proximity of a major aquifer in the L M L additional precautions were taken prior to shaft sinking by (i) incorporation of a ring of 12 pressure relief wells, primarily to reduce pore water pressure on the outside of the grout curtain but also to augment its water stopping potential, (ii) severe limitation in the use of explosives through the Basal Sands, (the excavation was largely undertaken by hand-held pneumatic picks) and (iii) installation of steel tubbing to give early support to the sidewall prior to concreting the permanent lining.

Wistow No 2 Shaft: In the absence of a DST on the shaft centreline borehole, a predicted shaft inflow of some 2.3-3.0 litre/s (30-40 i.gpm) was based solely on the examination and laboratory testing of the core. Prior to grouting, a PRT from the sump level at 245m gave a transmissivity of 5.3×10^{-6} m^2/sec, and an estimated shaft inflow of 1.9 litre/s (25 i.gpm) was in good agreement with the original prediction and reasonable confirmation that intergranular permeability would be the dominant cause of flow.

Two 16 hole rings were drilled to give an average hole spacing of 1.34m at the base of the cover. Individual water inflow to the initial 16 holes ranged from 0.6-1.3 litre/s (8-16.5 i.gpm) indicative of a relatively uniformly permeable formation with no obvious fissuring.

Predetermined quantities of the clarified Cemex A2 chemical grout were injected with a total grout consumption of 137,000 litres (30120 i.gals) After grout treatment the residual inflow to a central test hole from the Basal Sands was less than 0.38 litre/s (5 i.gpm) representing a reduction by a factor of about 5. This figure of 5 represents the increase in flow resistance endowed by grouting an annular zone of about 2m in radial thickness. The permeability of the annular zone itself will have been reduced by a considerably larger factor of about 50. With the installation of the 12 pressure relief wells the inflow was further reduced and during the subsequent sinking through the Basal Sands very little water was encountered in the shaft.

Riccall No 1 Shaft: The borehole DST indicated a transmissivity of 2.2×10^{-5} m^2/sec, equivalent to an average permeability of about 300 millidarcys in an aquifer of thickness 6.6m in reasonable agreement with average measurements on core samples. The estimated shaft inflow was some 9.1 litre/s (120 i.gpm) some 4 to 5 times greater than at Wistow. Based on the collated data, the initially proposed grout cover comprised two concentric rings of 24 holes, giving hole spacings of 2.2 and 2.0 metres at the base of the cover. After the first ring of 24 standpipes had been installed and the holes drilled and injected down to the top of the Lower Permian Marl, a PRT was carried out on the first hole into the Basal Sands. This gave a transmissivity of $2.5 \times 10^{-5} m^2$/sec, in close agreement with the original DST.

Water inflows to the first four holes drilled were typically 3.8 litre/s (50 i.gpm), and these were duly injected with Cemex A2 chemical grout. During the drilling of the fifth hole unstable conditions were encountered towards the middle of the Basal Sands with the water inflow suddenly increasing from about 3.8 litre/s (50 i.gpm) to some 38 litre/s (500 i.gpm), accompanied by fine sand. The grouting schedule was then changed to one of cement injection in order to stabilise the void created, followed by chemical injection of a high solids sodium silicate based grout to help consolidate the sands, then Cemex A2 for residual water-stopping. Of the first 24 holes taken into the Basal Sands, 17 proved unstable and flowed in excess of 23 litre/s (300 i.gpm) water and as much as $4m^3$ of sand. In the second ring of 24 holes only three holes proved unstable and water inflows were generally reduced, and large quantities of high solids silicate chemical grout were injected. However due to these occurrences of instability and piping flow the completed grout cover was much modified from that originally proposed in order to provide the added security of a much thicker grout curtain. In its final form the grout cover comprised three rings of 48 holes each, and the total injection comprised 109 tonnes of cement grout; 162,750 litres (35,800 i.gals) of high solids silicate grout, and 381,900 litres (84,000 i.gals) of Cemex A2.

The simultaneous inflow to four central test holes was 2.1 litre/s (28 i.gpm), which represents a reduction in flow by grouting of some 4:1, comparable with the performance at Wistow.

The provision of a pressure relief system to augment the grouting was considered essential in order to provide the greatest security when sinking through the Basal Sands. As at Wistow, a 12 hole system was used and grouting was necessary to stabilise several of the holes. With pressure relief, the inflows to the test holes dropped to 0.34 litre/s (4.5 i.gpm), an overall reduction of 17:1. Because of the dire consequences that could occur in the event of a breach of the grout wall followed by piping, it was decided to install heavy duty steel tubbing, capable of withstanding the full hydrostatic head, as the excavation progressed. This was accomplished with only minor quantities of water encountered.

Most of the evidence available suggests that the high water inflows initially encountered could not originate from the Basal Sands but must reflect a connection to a fairly strong and local aquifer, (presumably the Lower Magnesian Limestone).

Riccall No 2 Shaft: The DST in the centreline borehole gave a transmissivity of 1.0 x 10^{-5} m^2/sec and an average permeability and estimated shaft inflow about half of that for the No 1 Shaft.

A PRT test on one of the initial grout injection holes gave a transmissivity of 8.6 x 10^{-6} m^2/sec, in good agreement with the DST. The average inflow to the first ring of 24 holes was 2.7 litre/s (35 i.gpm).

Whilst the unstable conditions which occurred at the No 1 shaft were not encountered in any grout holes at the No 2 Shaft a comprehensive pattern of grout holes comprising two rings of 48 holes were employed to provide the security of a reasonably thick grout curtain.

Cemex A2 chemical grout was used for all injections and a total quantity of 126,000 litres (27,710 i.gals) was injected into the 96 injection holes. The residual inflows to four central test holes flowing simultaneously totalled some 2 litre/s (26 i.gpm) which represents a reduction in inflow of 3:1 due to grouting and an equivalent permeability reduction factor of about 50 within the confines of the grouted annulus.

A major part of the residual inflow can probably be attributed to porous bands which even this highly clarified grout could not penetrate.

The incorporation of the pressure relief wells reduced the simultaneous inflow to the central test holes to 0.31 litre/s (4.1 i.gpm), an overall reduction of some 24:1.

Towards the completion of the sinking and tubbing installation, an inflow of 120 g.p.m. occurred on the west side of the shaft at the mid-level of the Basal Sands. Fortunately this flow was easily contained and subsequently plugged off by the injection of cement grout, and the shaft completed successfully through into the underlying Coal Measures.

CONCLUSIONS

The main prerequisite to any new shaft construction is a corehole drilled on or close to the proposed shaft centre line. The data obtained from examination of the core, water tests, core logs, and laboratory tests on selected core samples, is essential in order to properly design the shaft lining, ground treatment scheme and shaft construction method. Subsequent in-shaft testing, particularly the pressure recovery test, enables the various aquifers to be more precisely identified and grouting materials, patterns and procedures to be confirmed and finalised.

Although geologically similar, the hydrological characteristics of two formations, one a fissured and"vuggy" limestone, and the other a weakly cemented sandstone were found to vary considerably between two shaft sites - 6.8 km apart in the Selby Coalfield. Borehole investigations from the surface prior to shaft sinking gave an invaluable and acceptable indication of the ground water problems encountered during the subsequent sinking.

The combined techniques of cement and chemical grouting, pressure relief wells and the immediate installation of steel tubbing with each corresponding increment of shaft excavation, provided a successful means of sinking through a very weakly cemented sandstone.

REFERENCES

1 Adamson, J N, and Scott, R A, "Borehole Investigations and Logging Methods in Shaft Sinking", The Mining Engineer, January 1973 181-188.

2 Swaisgood, J R, and Versaw, R E "Geotechnical Investigations for Mine Shafts", Society of Mining Engineers, June 1974, 37-40.

3 Chalmers, A, Daw G P and Scott, R A, "A modified Form of Aquifer Depletion/Recovery Test for Assessing Potential Water Make into Deep Excavations", 4th Congress ISRM Montreux, September 1979, 2 67-72.

ACKNOWLEDGEMENT

The authors wish to express their appreciation to Mr M Eaton, Director of the N.C.B. North Yorkshire Area, for his permission to present this paper, and to emphasize that the views expressed are their own.

A UNIQUE UNDERPINNING OF SOIL SOLIDIFICATION UTILIZING SUPER-HIGH PRESSURE LIQUID JET

M. Shibazaki[1] and S. Ohta[2]

Abstract

In repairing and reconstructing the interiors of a structure of certain historical importance located in the center of Osaka, Japan's second largest city, a special underpinning method was developed to provide added bearing power to the foundation of the structure while preserving its appearance intact throughout the duration of the work period.

Called the column jet grouting process, the new method is by excavating the ground by force of pressurized water under an super-high pressure 400 times higher than the atmospheric pressure and replacing the excavated soil with cement grout.

The new method is characterized chiefly by two features:

1) it permits improving a foundation covering as much as 3 m^2 in area (2 m in diameter) just by drilling a hole with a diameter of only 13 cm on the floor, and 2) it is suited to indoor work as it involves the use of machines with heights up to 2 m only.

The engineering requirements necessary to maintain the structure were specified by the undertaker; that is, supporting the total weight of 28.5 GPa with 80 columns, or, in more precise terms, establishing an unconfined compressive strength of 3 MPa or more and finishing column diameters to 2 m or more.

Because the standard specifications of the column jet grouting method then available were such as to finish column diameters to 2 m, some experimentation was conducted in the field to develop a modified method of fabricating columns with a diameter of 2.5 m at least. The modified method was applied to the actual execution of the work with success.

This paper describes how the special parameters were developed from the standard ones of the column jet grouting process, with respect to the results of the execution of the work using the modified method.

[1]Senior Engineer and [2]Foundation Engineer, Chemical Grouting Company, JAPAN

I. Antecedents

(1) Development of the process

The column jet process treats the foundation with high speed water jet, and is characterized by the use of air simultaneously. Hydraulic forces have been used widely for coal mining, bedrock washing and other various purposes. When the subsoil below the ground water level is to be excavated hydraulically, hydraulic force is baffled down immediately yielding under the resistance of ground water as illustrated in Fig. 1, and fails to extend its influence far enough.

We tried all possible measures to increase the projection distance of water, and devised a method of accompanying a jet of water with air because in the open air water can be projected a long distance without losing energy. This method was refined into a practical one through various laboratory tests and field tests.

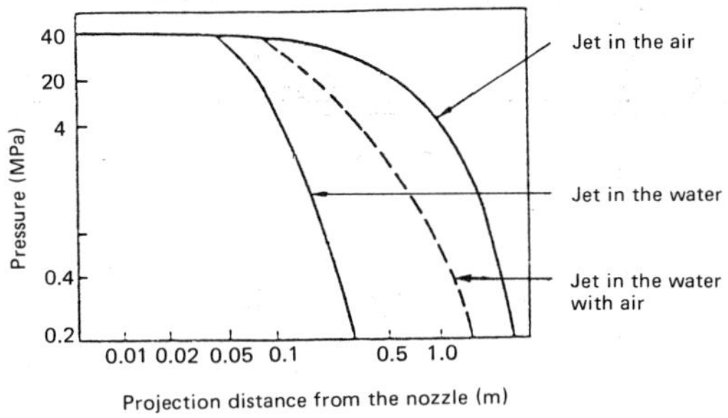

(Delivery pressure: 40 MPa; nozzle aperture: 2 mm)

Fig. 1 Relationships between the projection distance from the nozzle and pressure

(2) Application of the process

Basically, this is a displacement process in which the ground soil is broken up by a radial jet of high-pressure water and encapsulating air from an injection monitor (process pipe), discharged to the ground surface in the form of slime, and is supplanted immediately by a hardening agent, all being done in an unbroken sequence. Namely, the injection casing is lifted while being turned to perform excavation, washing away, and filling of hardening agent concurrently as illustrated in Fig. 2.

(3) Improvements and development of the jet grouting process

The jet grouting process has been developed and established viably with emphasis on the following three points.

(a) Shape and dimensions of jet nozzles

The jet nozzles are of vital importance in eliciting the maximum energy out of the jet water.

We found that the nozzle with the maximum efficiency has a rectangular cross section and has a throat with a taper angle of 0.23 radian to 0.24 and a straight mouth about 3 to 4 times as long as its aperture diameter. (See Figs. 3, 4 and 5)

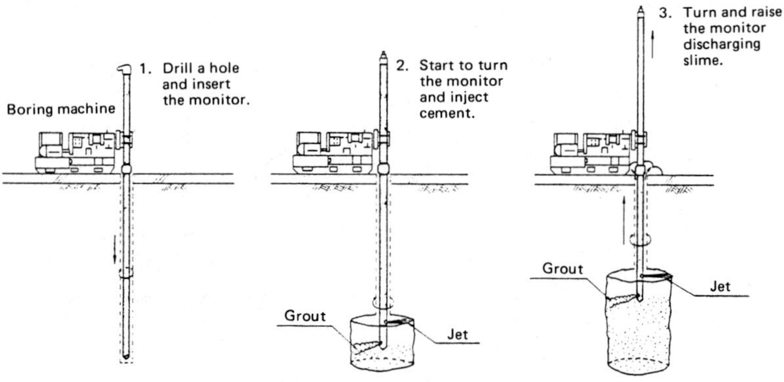

Fig. 2 Procedure

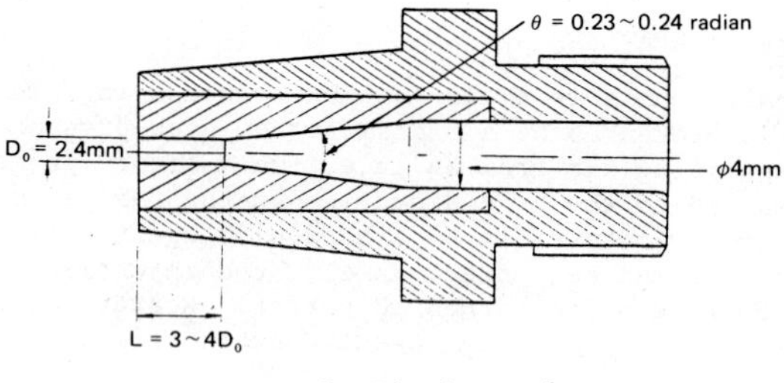

Fig. 3 Ideal nozzle

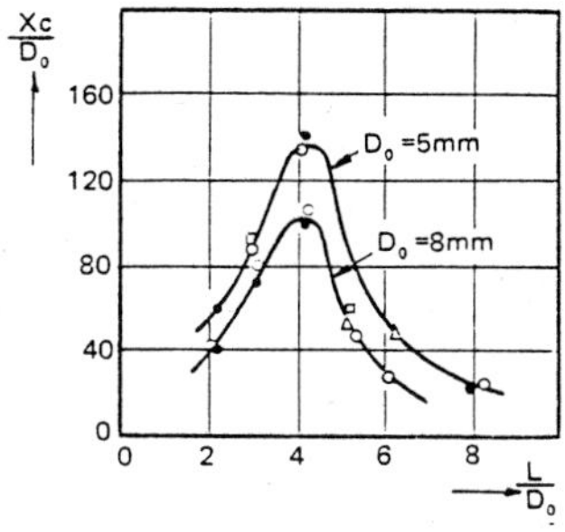

Fig. 4 Potential core length and nozzle mouth length (Shavlovsky)

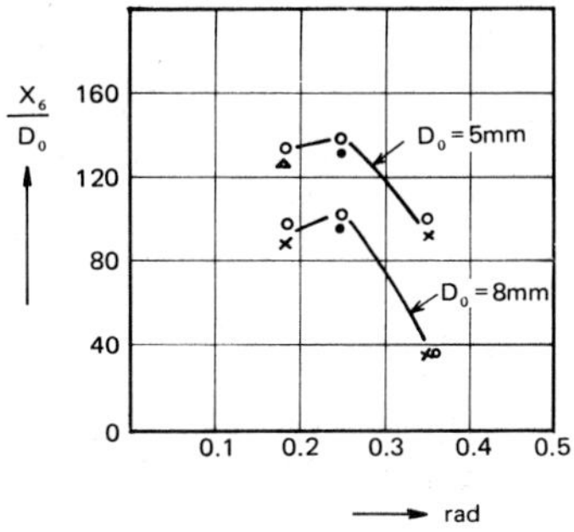

Fig. 5 Potential core length and nozzle throat taper angle (Shavlovsky)

(b) Delivery velocity of air

Unless air is blown out uniformly, the high-velocity jet of water cannot be encapsulated in the air. Thus, the nozzle is designed to let air out coaxially. In order to maximize the pneumatic effect, special provisions are made to jet the air at nearly sonic velocity.

(c) Jetting of cement suspension

The soil hydraulically excavated by high-pressure water will remain puddled, particularly if it is clay, and will defy being discharged completely to the surface. In order to prevent this, the grout is jetted in a stream-line in the lateral direction. In this way, the puddle can be completely substituted by the cement grout.

(4) Empirical data

The basic parameters governing the process were determined by laboratory tests and studies, and were corroborated by field tests which were conducted over several years. Finally, we modified a few parameters to refine the process for precise forecasting of the results from the input. The empirical data achieved are summarized in Tables 1 through 3.

Table 1. Standard parameters (fixed)

Superhigh-pressure jet water	Delivery pressure	40	MPa
	Delivery rate	1.2	ℓ/s
Hardening agent	Material	Cement	
	Mix proportion	W/C = 100%	
	Delivery pressure	2	MPa
Compressed air	Delivery pressure	700	KPa
	Delivery rate	17	ℓ/s

Table 2. Standard parameters (flexible)

Hardening agent	Delivery rate	1.2 ∿ 3.3	ℓ/s
Operating speeds	Revolving speed	5 to 10	rpm
	Lifting speed	0.83 ∿ 1.67	mm/s

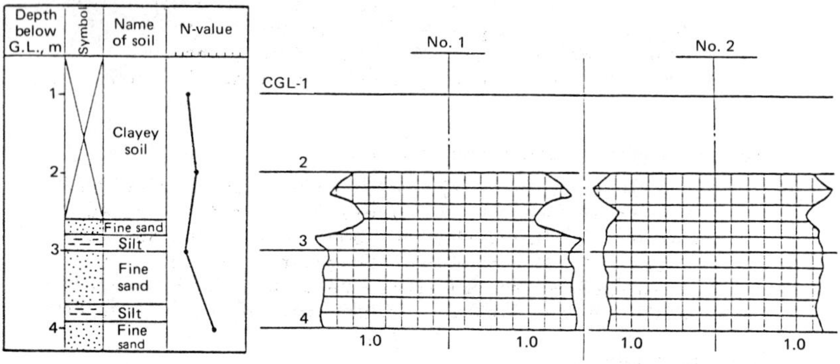

Superhigh-pressure water	Delivery pressure	40 MPa
	Delivery rate	1.2 ℓ/s
Grout	Delivery rate	3.0 ℓ/s
Compressed air	Delivery pressure	700 KPa
	Lifting speed	0.83 mm/s
	Revolving speed	5 r.p.m.

Fig. 6 Cross sections of grouted columns (sandy soil) by Test 1

Table 3. Parameters of tests

	Column No.	Process specifications							Diameter		Unconfined compression strength	Specific weight	Seismic wave		Modulus of elasticity
		Superhigh-pressure water		Hardening agent		Compressed air Pressure	Lifting speed	Revolving speed							
		Pressure MPa	Flow rate ℓ/s	Pressure MPa	Flor rate ℓ/s	KPa	mm/s	rpm	Max. cm	Average cm	MPa	kg/cm³	P km/s	S km/s	GPa
Test 1	1	40	1.2	0	1.2	700	0.8	5	340	280 ±41	3.55	1,665	2.0	0.88	2.8
Sand	2			0	1.2	700	0.8	5	293	262 ±15	2.52	1,654	1.4	0.64	1.2
	3			0	2.5	700	0.8	5	293	256 ±26	4.60	11,640	2.10	1.10	4.3
	4			0	2.5	700	0.8	10	255	223 ±21	6.78	1,650	1.80	0.85	2.5
Test 2	5	40	1.2	0	3.3	700	0.8	5	234	210 ±13	1.73	1,242	1.40	0.50	1.1
Clay	6			30	3.3	700	1.25	5	214	179 ±14	1.99	1,125	2.3	1.0	3.7
	7			30	3.3	700	1.7	5	178	161 ±14	3.51	1,074	1.60	0.69	1.0
	8			30	3.3	700	0.8	5	324	302 ±15	2.05	1,206	1.20	0.73	1.2
	9			30	3.3	0	0.8	5	134	116 ±9	6.58	1,229	1.70	1.0	3.1

Photo 1. Grouted column in sandy soil

Photo 2. Grouted column in clayey soil

II. Example of Application

(1) General

Introduced here is an example of application of the column jet process for provision of underpinning for a historic building which was built more than one hundred years ago at the center of Osaka, the second largest city in Japan. The underpinning work was conducted after three trial operations designed for respective purposes.

(2) Soil quality and environmental conditions

The site is located between two rivers. Its ground level is several meters above river water level.

As shown in Fig. 8 "First test pitting soil profile", the site is composed of sandy soil, clayey soil and then sandy soil (from top down). All these layers are alluvial. Table 4 shows the principal soil characteristics.

(3) Trial operations

For the purpose of clarifying the workmanship required on the site, three trial operations were conducted as classified below.

1st trial operation: Demonstration of standard specification

2nd trial operation: Test for ensuring a larger-diameter grouted column

3rd trial operation: Test for ensuring a tight contact between grouted column and the foundation concrete

Table 4. Soil characteristics

Depth	Soil quality	SPT	Soil characteristics
8.0	Upper sandy soil (SPu)	10 to 30	Gravel, 28%; sand, 70%; silt, 2%; Uc = 3.8 w = 20.1% $k = 4 \times 10^{-1}$ mm/s
13.0	Clayey soil (CH)	4 to 5	$q_u = 140 \sim 150$ KPa C = 60 KPa w = 40% $t = 1.8 \times 10^3$ kg/m^3
20.0	Lower sandy soil (SF)	10 to 50	Gravel, 3%; sand, 86%; others, 11% w = 18% $k = 4 \times 10^{-2}$ mm/s

(a) First trial operation

In the first trial operation, the column jet process was applied according to the specifications in Table 5 in an effort to achieve a diameter of more than 2.0 m and an unconfined compression strength of more than 0.8 MPa. The results proved successful as shown in Fig. 7 and Table 6.

Table 5. Specifications for first trial operation

Superhigh-pressure jet water	Delivery pressure	40	MPa
	Delivery rate	1.2	ℓ/s
Hardening agent	Material	Cement grout	
	Mix proportion	w/C = 100%	
	Delivery pressure	2	MPa
	Delivery rate	3.3	ℓ/s
Compressed air	Delivery pressure	700	KPa
	Delivery rate	17	ℓ/s
Operating speeds	Revolving speed	5	rpm
	Lifting speed	0.8	mm/s

Table 6. Physical test results of grouted column

Sampled depth	Dens-ity	q_u	P-wave	S-wave	Dynamic modulus of elas-ticity	Poison's ratio	Modulus of elasticity
	kg/m^3	MPa	km/s	km/s	kg/m^2		GPa
8.0 m	2,070	7.7	2.9	1.3	9.5×10^8	0.29	
12.0	1,370	2.6	1.6	0.84	2.7×10^8	0.32	
16.00	1,900	11.0					2.2
18.00							1.4

(b) Second trial operation

The second trial operation was conducted for a larger diameter with the parameters changed accordingly. By reducing the lifting speed by half, the grouted column was made larger than 3.0 m in diameter, and its unconfined compression strength was 3 MPa.

Table 7. Physical test results of grouted column

Depth	Density	q_u	P-wave	S-wave	Dynamic modulus of elasticity	Poison's ratio
m	kg/m^3	kg/m^2	m/s	m/s	kg/m^2	
9.0	1,440	3.7×10^4	1,600	930	3.3×10^8	0.27
-10.0	1,420	3.9×10^4	1,600	910	3.1×10^8	0.27

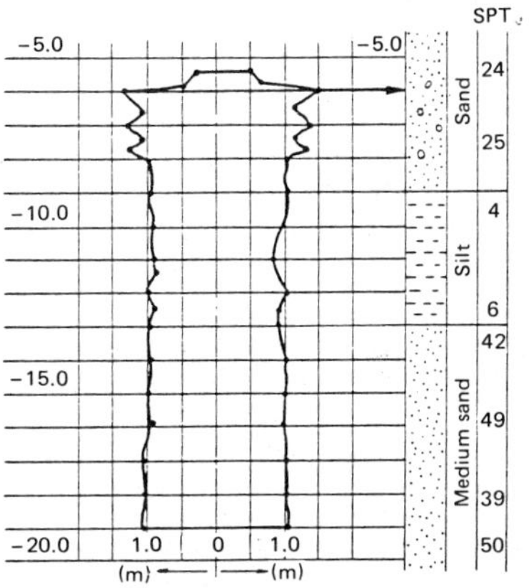

Fig. 7 Profile of column grouted in first trial operation

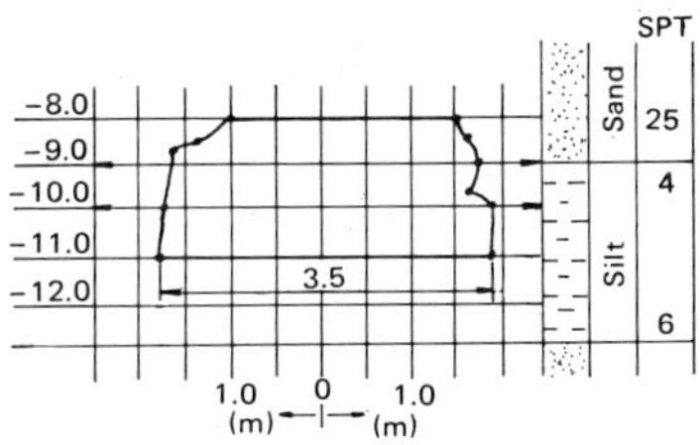

Fig. 8 Profile of column grouted in second trial operation

(c) Third trial operation

The third trial operation was designed to examine the contact between the concrete foundation bottom, piles and the grouted columns. Namely, a spread foundation of boulders and cast-in-place concrete was constructed beforehand, and columns were grouted. Core borings were then made, and it was found that the columns were integrated perfectly with the spread foundation. A cross sectional survey proved that the joints were coalescent without voids and any other defects.

(4) Final column jet grouting

(a) Selection of process

The subject building was planned to be restructured and modernized internally, and various processes were studied for the purpose of adding bearing piles, etc. The process requirements were identified as follows.

i) The process must be able to be applied as close to the wall as possible, and also be able to construct column footings right under the walls.

ii) The process must be applied through the mat foundation left intact. Namely, the holes to be made in the mat foundation must be as small as possible, and the coverage of a single grouting operation for subsoil improvement must be as wide as possible.

iii) The process must not generate vibration or shock or upheaval pressure that may crack the walls or raise the mat foundation.

The column jet grouting process is able to make piles of 2.0 m or more in diameter by making use of a pipe with a diameter as small as about 10 cm, thereby fulfilling requirements i) and ii). In addition, the in situ tests demonstrated that the process can make footings of the required bearing capacity thus satisfying requirement iii).

As a result, the column jet grouting process was employed for this project of rehabilitation of an historic building over 100 years old standing on a major thoroughfare. Most historic buildings in Japan are wooden, and stone or concrete ones are very rare. The government authorities have been addressing themselves to the preservation of the remaining historicbuildings made of stone. The rehabilitation project referred to here was to underpin the foundation of the building with the column jet grouted piles for the purpose of refurbishing the interior while leaving the external appearance intact. The bearing capacity of each pile was designed to be 1.5 GPg. The two-story building is constructed of brick with stone pillars. Its first floor has a space of about 1,600 m^2, and the total building weight is 28.5 GPg. The contact pressure is about 320 MPg. The site is subjected to uneven subsidence under the influence of subway construction work, river improvement work, etc. The building is leaning a little toward the main street.

(b) Underpinning work

The under pinning was carried out from outside for the sections right under the external walls and from inside for other parts. The working specifications are as outlined below.

- Diameter of pile : 2.5 m on the average
- Required strength : q_u = 3 MPa on the average
- Number of piles : 80
- Pile depth : 13 m (G.L. -4.0 m to -17.0 m)

The working specifications were the same as for the second trial operation.

(c) System and toolind

Figure 10 illustrates the standard system employed for the process.

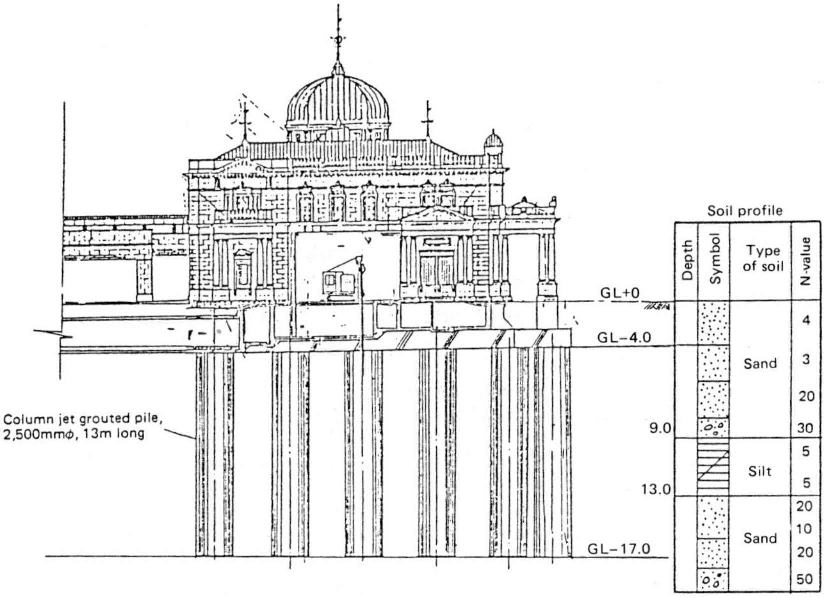

Fig. 9 Cross section showing the underpinning work

(d) Workmanship of column jet grouting, and subsequent analysis

The grouted piles were cured for more than a month, and sampling was made by core boring for the purpose of studying the workmanship. Selected samples underwent unconfined compression strength tests. The verified results are as shown in Table 8 on the next page. The finished work completely filled all requirements.

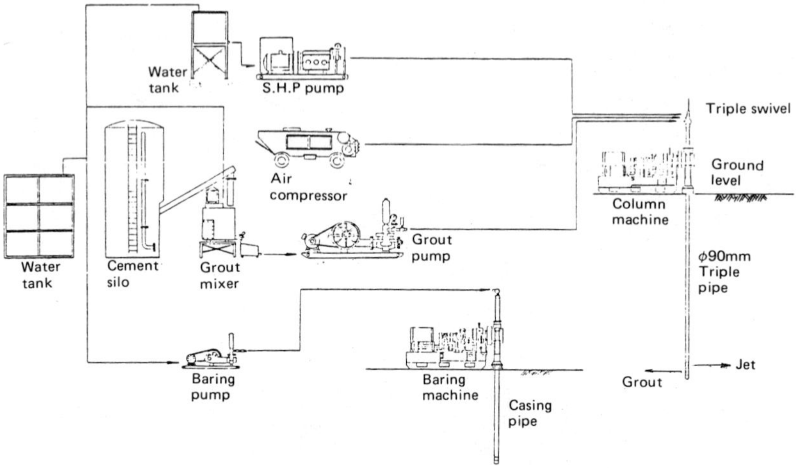

Fig. 10 A schematic view of the system

Table 8. Verified results of workmanship

Column No.	Unconfined compression strength, MPa		
	Upper sandy layer	Intermediate clayey layer	Lower sandy layer
a	4.87	0.85	12.44
b	6.27	2.50	6.55
c	9.24	3.02	6.17
d	4.74	1.30	3.38
e	6.46	1.54	1.77
f	3.72	1.80	1.20
g	6.86	0.87	4.93
h	3.24	1.54	18.94
i	2.36	1.49	20.90
j	3.40	0.91	6.33
Average	5.12	1.59	10.93

III. Conclusion

The column jet grouting process is unique in that it can improve subsoil over as wide an area as 3 to 5 m^2 at a single application by making use of water jet with pipe of a small diameter. It can be used for installing foundation piles in a very limited space as in the foregoing example. While urban civil engineering processes as represented by shield driving are liable to cause cave-in due to excavation work, etc., the column jet grouting process is quite free from such troubles, and has been applied in various ways for underpinning and for the protection of buildings from subsidence. In Japan, urban development is requiring construction deeper and deeper below ground-level, and the demand for underpinning by the column jet grouting process will increase more and more.

IV. References

Yanaida, K. On the principle and the execution of C.C.P process.

Shibazaki, M. (1973). Soil solidification utilizing super-high pressure.

Maeda, M. (1976). On the bearing capacity of piles in-situ.

Shibazaki, M (1981). Super-high pressure grouting.

Shibazaki M. Design and execution of chemical grouting.

COMPACTION GROUTING - THE FIRST THIRTY YEARS

By James Warner,[1] F. ASCE

ABSTRACT

Unlike the more conventional "penetration" type grouting procedure, Compaction Grouting involves injection of very stiff mortar-like grout material. The objective is not to fill interparticulate voids, but rather to displace and compact soil materials in place. In general, the grout remains in a homogeneous mass which grows in volume as injection is continued. Primary use is for the stabilization of fine grain soils, either as a remedial method for settlement damage or in new construction. Although the procedure has been practiced for about thirty years, its early usage was limited geographically to the southwestern portion of the United States. Presently, acceptance has become widespread, and extensive use has occurred throughout the United States. A limited amount of work has been performed in other countries, and substantial international interest has occurred within the last few years. A review of the development of the process, mechanics of injection, basic injection procedures, and current application examples are described.

INTRODUCTION

Compaction Grouting involves injection of stiff, mortar-like grout into previously drilled holes in the soil, in a closely controlled manner. As the mass of grout increases, under pressure, the soil is densified through compaction, (Fig. 1). When the diameter of the grout column or mass is relatively small, the pressures are essentially radial and therefore horizontal. However, as the size of the mass increases, considerable up-lift force develops. It is surface movement caused by such upward force that generally limits the quantity of grout placed at any given point. The upward force can be of some advantage, however, in that through judicial use thereof, settled structures can be raised to rather precise grade tolerances.

The shape of the grout mass is usually spherical or cylindrical depending upon the amount of grout hole open during any given injection sequence. In uniform soils, the shape will be quite regular whereas extensive irregularities will prevail in non-uniform soils. The size of the resulting column will be affected by the existing soil density, moisture, and mechanical properties, surface restraint conditions, and the injection pressure, rate and other factors. Masses of grout with cross sections of three feet or more in diameter are not uncommon.

One of the principal advantages of Compaction Grouting is that its maximum effect is in the weakest soil zones. It is limited to soils, and is most frequently used in materials finer than a medium sand. The

[1] Consulting Engineer, Mariposa, California

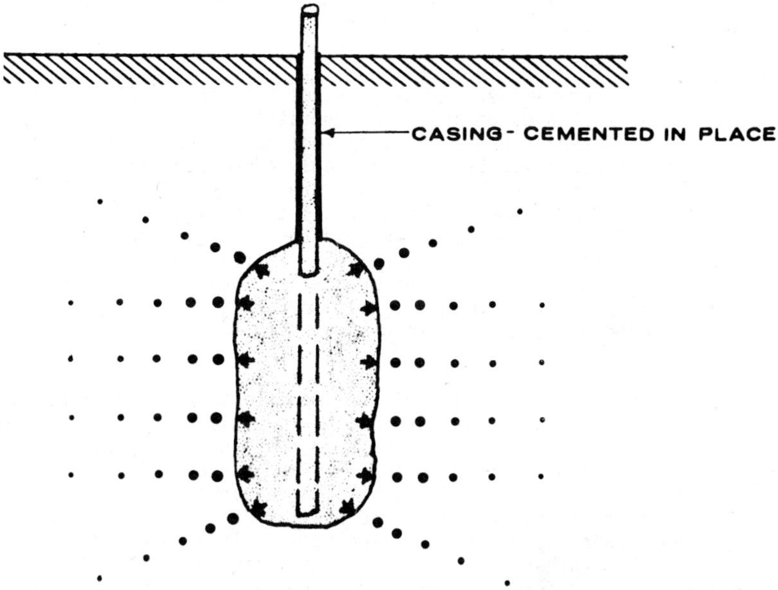

Fig. 1 - Increasing mass of grout displacing and compacting soil as it grows under injection pressure.

process may be used in clays subject to provision for adequate drainage. Where drainage is limited to the extent that high pore pressures develop, much slower pumping rates are required, and in extreme cases, the procedure is not applicable. Due to the lack of mobility of the stiff mortar-like grout employed, relatively high injection pressures are required. Accordingly, the final disposition location of the grout is reasonably controllable, as is the ability to prevent leaks into unwanted areas.

Compaction Grouting can be readily performed inside structures and other confined spaces, and its execution results in only minor interference in other operations, as large equipment is not required in the immediate injection area. Historically, its most extensive use has been in connection with settlement correction, however it has also been used for modification of in situ soils to reduce the liquefaction potential during earthquakes (1), and as a "tool of construction" to limit ground movement during soft ground tunneling (2).

The procedure was first described by Graf (3) in 1969, wherein the basic concepts were discussed. Mitchell (4) presented comparisons of the procedure with other grouting methods, in particular differentiating between the "penetration" and displacement or "compaction" mechanism, in 1970. Applicability of the procedure and the mechanics of injection including a review of original research and development leading to the then current technology were described by Brown and Warner (5) in 1973.

Criteria for planning and performing Caompaction Grouting projects were presented in 1974 by Warner and Brown (6).

MECHANICS OF INJECTION

Several researchers have performed studies in an effort to understand the actual mechanics of injection and associated soil response. Early efforts centered around evaluating soil densities, both before and after grout injection. Such efforts only added confusion, in that no uniform pattern of improvement was discernible. In many cases, it was found that the soil immediately adjacent to the injected grout mass had achieved less increase in density than the more distant surrounding soil. In more recent research, actual soil deformation has been evaluated, usually through utilization of subsurface instrumentation capable of evaluating both horizontal and vertical deformation. Such research has generally been performed in conjunction with actual field applications.

Although the exact mechanism of soil modification is not thoroughly understood, a basic understanding of the injection effect is developing. It has been fairly well established that the expanding mass of grout results in a complex system of radial and tangential stresses within the soil. Immediately adjacent to the expanding grout mass, a zone of major disruption, shearing, and plastic deformation will occur. In this zone the density of the in place soil could actually be lessened due to the disturbance. As the distance from the soil-grout interface increases, the deformation will be essentially elastic, and an appreciable increase in density can be expected.

In soils of relatively low permeability, pore pressures must be monitored during injection. In some cases, simply reducing the pumping rate will be sufficient to prevent excessive pore pressure increases. However, in extreme cases, provision for drainage must be provided. Where the normal pore pressures are appreciably raised as a result of injection, they should be allowed to relax, after which further injection will probably be required. Although Compaction Grouting is probably not the best solution to problems where relatively impermeable soils exist, applications have been made in which drainage wicks or sand drains were provided and work was performed over an extended period of time. In one case within the writer's experience, localized pore pressures in excess of the overburden pressure developed. In this instance, a number of injections were required over a period of several weeks. Additionally, a very slow injection rate (less than 0.5 cu. ft. per minute) was used.

The effective radius of a single grout hole will vary according to the soil type, moisture, relative density, as well as the grout injection rate and resulting pressure. It has been established that in most cases, significant densification will occur within a radius of one to six feet (0.3 to 1.8 m) from the interface of the soil and grout. However, it is recognized that some effect will be experienced at greater distances from the injection point. On the Bolton Hills Subway tunnel in Baltimore, Maryland (2), sub-surface instrumentation detected horizontal movements at a distance of 12 ft. (3.7 m) from the center of the grout mass. In that instance, a volume of about 40 cu. ft. (1.1 m^3) of grout was injected into a vertical zone of five ft. (1.5 m) resulting in a grout mass with a diameter of nearly three ft. (0.9 m).

A significant advantage of the compaction grouting procedure is its tendency to find and improve the weakest soil zones. Because of this factor, even though there will be some variation in the density of the soil surrounding the injected grout mass, a general improvement in the uniformity of the soil will occur. This effect will obviously be most predominant in non uniform soils. In this regard, practice has indicated that the greatest grout takes are in the vicinity of the interface of fill and natural soil in the case of apparent fill failures.

When the total volume of the injected grout and thus the horizontal cross sectional area of the mass is small, the total uplift force is relatively low. However, as the area of the mass (which increases with the square of its diameter) expands, considerable total uplift force develops. Thus, if pumping is continued indefinitely, a grout mass diameter and resultant uplift force will be developed which will cause surface upheaval. Effective compaction ceases once such uplift has occurred.

The principal controlling factor in Compaction Grouting is the grout pressure behavior. It is usually monitored at both the grout pump and also the point of injection. Pressure behavior at the point of injection should be continuously monitored and recorded. This is usually done manually with a record entry being made at each significant change of pressure, however, some work has been done in which continuous pressure recorders have been employed.

As aforementioned, injection is generally continued until a surface disturbance is noted. The pressure at which this will occur cannot be determined in advance and, in fact, it will vary widely between different holes of any given project, as well as between different project sites. The key controlling criteria for the grout injection is to not permit too rapid a pressure build-up. The pressure level is directly related to the grout injection rate and is readily controlled by adjustment thereof. On most projects the initial injection rate will be on the order of 0.7 to 1.0 cu. ft. (0.02 to 0.03 m^3) per minute. If the pressure build-up is fairly slow the rate can be slowly increased. However, as the total volume of grout injected in the hole increases, the rate is generally lowered so as not to exceed the "optimal" rate which has been determined for the specific site during the initial injection. The "optimal" injection rate will vary from project to project as dictated by varying soil properties. It is usually determined during the initial grout injections on any given project and can loosely be described as the fastest rate that can be continuously maintained without an excessively rapid pressure increase. The maximum desirable rate of pressure increase is on the order of five to seven psi (35 to 48 kN/m^2) per minute.

A great deal can be learned about the existing conditions through close observation and analysis of the pressure behavior. Where the build-up is consistent, (Fig. 2a) a relatively uniform soil is indicated. Where large fluctuations prevail, a very non-uniform condition is indicated (Fig. 2b). A sudden loss of pressure as occurs after 10 and 25 minutes of pumping on Fig. 2b is indicative of a break-through of grout into a void, such as might result from buried rocks or trash within a fill. The approximate volume of such a void can be determined by evaluation of the amount of grout injected in order to return to the pressure value at which the loss occurred. In order to facilitate such evaluations, it is important to maintain continuous records which include time, volume of grout injected, and injection pressure. A sudden loss of pressure may also be indicative of an impending surface disturbance, escape of the grout into an underground pipe or other substructure, or loss of lateral restraint such as might be provided by a retaining wall.

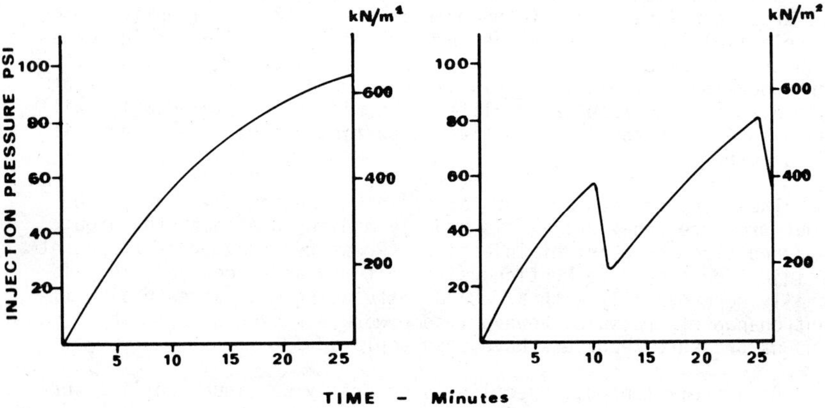

(a) (b)

Fig. 2 - Typical injection pressure behavior, uniform on the left, erratic on the right.

Because of the substantial amount of pertinent information obtained during the actual grout injection, Compaction Grouting programs are frequently altered during prosecution. In this regard it is probably best for the designer to specify the work to be done in a rather general manner. In order to obtain optimal performance, it is imperative to allow for alteration of the work as it progresses and data is retrieved.

APPLICATION PROCEDURES

A detailed description of the performance of the procedure, including material and equipment requirements was presented in Planning and Performing Compaction Grouting (6). The criteria presented therein remain appropriate, however a number of variations have developed as usage has become more widespread. Following, is a brief review of currently used application procedures. For a more complete understanding of the actual performance of the work, the reader is encouraged to refer to the various references, especially (6).

Grout Holes

Grout holes are usually two in. (50 mm) or more in diameter. The spacing varies and must take into account structural restraint as well as soil conditions. Normally, holes are placed on about eight to twelve ft. centers each way. As a rule, alternate "primary" holes are first grouted and their completion is followed by injection of the intermediate "secondary" holes. The holes are generally progressed in vertical stages of five to eight feet, working from the top down.

In practice, an oversized hole is drilled from the surface to the point at which stabilization is to begin, or a minimum of about four feet (1.2 m). A steel casing is solidly cemented into the hole. The hole is then extended, working through the casing as required, and the first stage grout injection made. At a later time, usually the next day, the hole is extended through the same casing, for the next grout stage. This sequence is then repeated until the full depth has been reached.

As the Compaction Grouting process results in densification and therefore in considerable increase in weight of the treated soil, it is crucial to extend all holes to a very competent soil or rock zone. Attainment of such, can be confirmed by extending an occasional hole an extra stage or two followed by injection. If indeed competent material has been reached, these extra stages should have very low grout takes and relatively high injection pressure.

The necessity for working from the top down has often been questioned, due to the greater expense involved by the increased drilling required. However, extensive experience and research has proven this to be a pertinent method. Because grouting of the upper stages increases their strength as well as density, their restraint capability is improved, allowing greater pressure to be used in succeeding stages. There are occasions, however, in which it might be desirable to grout in a single stage from the bottom up. Such would be particularly appropriate when working at considerable depth in a fairly thin faulty layer.

A variation of the above procedure involves drilling the holes and concurrently inserting a tight fitting steel casing full depth. A standard deformed reinforcing bar is then inserted into the casing followed by grout injection. The casing is pulled incrementally as injection continues. The pulling increments are controlled by the grout pressure, such that the grout deposition, and thus compactive effort, proceed incrementally from the bottom up. Effectiveness of this technique relies upon the maintenance of a positive seal at the bottom of the casing, a factor that is difficult if not impossible to consistently achieve according to the author's experience. Where proper performance is achieved, however, the result is a "mini-pile" that is effective in both compression and tension. Once the primary holes are thus completed, they will tend to increase the overburden restraint due to their tension reaction during injection of the secondary holes. The inclusion of the reinforcing bar also tends to maintain continuity of the pile in the event that a break in the injected grout occurs.

The reinforcing bar is sized according to the individual project requirements, but is generally on the order of one inch (2.5 cm) in diameter.

Rotary drilling is the most common method utilizing either air or water to remove the cuttings. If the holes tend to close in, rotary mud can be used; however, it should be avoided if possible as any residual mud will act as an uncontrollable lubricant when the grout is injected. In some cases the holes are pre-treated immediately prior to grout injection. Unless the natural moisture of the soil mass is substantially on the wet side of optimum, injection of water prior to grouting often will result in greater grout "takes". In especially fine-grained soils, the injection water is often treated with a wetting agent to aid its penetration into the soil. The quantity will vary, but the object is to reduce the shear strength of the soil in the vicinity of the grout hole as much as possible so that more grout can be injected. A further objective is to prevent "drying out" of the grout and subsequent premature packing during injection. In clay soils, hydrated lime slurry, or other chemical agents intended to chemically strengthen the mass, have been used for pre-treatment.

In a sense, each grout hole can be considered an exploratory hole. Therefore, careful records should be kept relative to the soils encountered during the drilling. The additional cost of carefully monitoring the drilling operation is usually small compared to the value of the data retrieved. Particular attention should be directed to the existence of organic or other unusual materials encountered. The limits of previously placed grout should also be noted when proceeding in descending stages.

Grout Material

The "Glossary of Grouting Terms" published by the ASCE Committee on Grouting in the July 1980 Journal of the Geotechnical Engineering Division defines "Compaction Grout" as follows:

> "COMPACTION GROUT - Grout injected with less than 1 in. (25 mm) slump. Normally a soil-cement with sufficient silt sizes to provide plasticity together with sufficient sand sizes to develop internal friction. The grout generally does not enter soil pores but remains in a homogeneous mass that gives controlled displacement to compact loose soils, gives controlled displacement for lifting of structures, or both."

Continuing in the "Glossary of Terms", "slump" is referred to as that consistency obtained through utilization of the standard concrete slump test, ASTM C-143.

The gradation of the sand material is probably one of the most important factors in successful application of the Compaction Grouting technique. Grout made with sand that is too coarse or harsh is subject to breakdown under high pressure wherein the water is pushed out of, and ahead of the remaining grout ingredients, resulting in blockages. Too fine a sand material will provide a grout that is overly difficult to control and may be of limited durability. An ideal sand material will be a natural, rounded material, having 100% passing a No. 8 sieve, and not more than about 20% finer than 50 microns. The inclusion of some silt size material is desirable, however, grouts containing an excessive amount of clay should be avoided. Even small clay contents can result in grout of low durability, especially when subjected to wet environments or dynamic loads. An often published figure indicating the range of preferable sand gradation is repeated herein as Fig. 3.

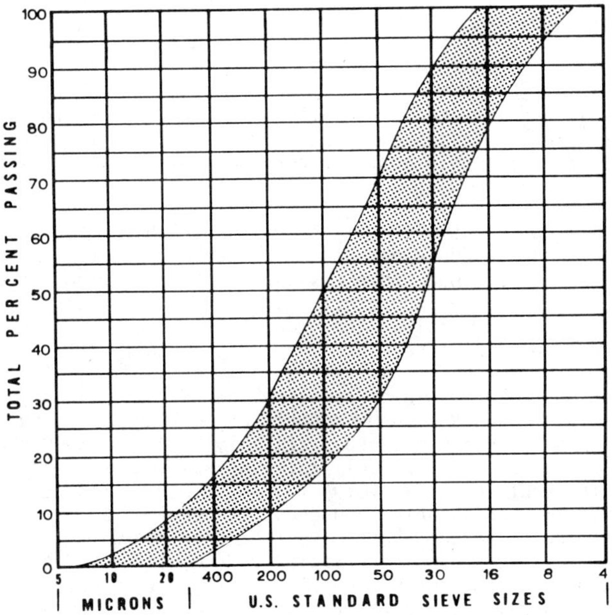

Fig. 3 - Preferred Gradation of Sand Material.

A very commonly used grout mixture contains about 12% cement, and when used with appropriate sand will provide a grout with an unconfined compressive strength of the order of four to six hundred psi (2800 to 4200 kN/m^2). It is not necessary to include cement in the grout however. Many projects have been completed in which pozzolan or hydrated lime were used in place of the cement. Such variations are pertinent to applications where low grout strengths are desired such as adjacent to a flexible culvert or tunnel lining.

In order to provide the proper control and compactive effort, it is imperative that the grout be of a very low slump, stiff plastic consistency. Although the ASCE definition of "....less than 1 in. (25 mm) slump" may be overly restrictive, experience has shown that when the slump is allowed to become greater than 1½ or 2 in. (38 or 50 mm), effective compaction is precluded and serious problems of leakage, hydraulic fracturing of the soil, and premature surface upheaval tend to occur. Additionally, control of the disposition location of the grout is lost and the amount injected is considerably lessened. Thus one of the most important single factors affecting the proper execution of Compaction Grouting is maintenance of the stiff grout consistency. Under no conditions should the grout slump be allowed to exceed about 2 in. (50 mm) as per the standard concrete slump test, ASTM C-143, where the primary purpose of the grout injection is to compact the soil. (Some exceptions might be in order where lifting structures or slabs is being performed (9). Such work would normally be done after the compaction is completed, however.)

Equipment

Because of the very stiff, relatively immobile characteristics of the grout material, special equipment is required for its handling. Conventional grouting equipment used for "pourable" mixtures is not suitable.

Mixers most often used are of the horizontal batch type with blades oriented to provide a chopping type action. Continuous pugmill type mixers can be used but where employed must be provided with a metering supply system for the grout ingredients, in order to enable positive and uniform control of the grout consistency. Continuous belt and auger type mixers have been used successfully.

The pump must be capable of handling the very low slump grout materials utilized. It should be capable of working at pressures in excess of 600 psi (4200 kN/m^2) and preferably 1000 psi (6900 kN/m^2). It must be able to work at the full range of pumping rates from virtually zero to about 2 cu. ft. (0.06 m^3) per minute. The pumping rate should be uniformly adjustable throughout its entire range while continuously pumping. The pump hopper must be provided with a force feed mechanism in order to prevent cavitation of the very stiff grout. A flow meter capable of measuring the amount of grout injected at any particular time should also be provided. Virtually all Compaction Grouting done to the present time has utilized piston type pumps.

Grout hoses are usually one and one half or two in. (38 or 50 mm) in diameter. Valves must provide a full flow opening. Standard pipe fittings should be avoided and any bends should be of the long sweep type in order to prevent major disruption of the grout flow. As with any type of grouting, suitable gauge savers must be provided for the pressure gauges which are usually located at both the pump and injection point. Because of the very stiff grout used, high pressure losses occur within the grout line, thus pressure values should be taken at the point of injection.

Controls and Records

Little effective compaction occurs once upward lift of the soil has started, and considerable damage to overlying improvements as well as the soil itself can result, should injection continue under such conditions. Therefore, pumping should be ceased immediately upon the first detection of minute surface disturbance. Accordingly, it is necessary to continuously monitor all surface improvements as well as the ground surface surrounding the point of injection for upward movement. Various instruments are used for this purpose including stringlines, surveyors' instruments, multi-station monometers, tilt meters, etc. The variety of useful control devices available is almost limitless and the engineer should make use of any he determines will be helpful for maintaining absolute control on the individual project. Extensive discussions relative to injection controls can be found in References 1, 2, 6, 8 and 9.

As with any grouting operation, continuous sequential records should be maintained. Entries are made at each significant change in injection pressure, and should include as a minimum, the time, grout pressure, amount injected to the time, and any significant events that have been observed. Obviously, any surface movements detected should be noted. It is also advisable to record the date, time, and sequence the hole is being grouted. The heading portion of a form typically used for such recording is provided in Fig. 4.

JOB NO. ______

GROUTING		MATERIAL ______ SEQUENCE ______		HOLE NO. ______ DATE GROUTED ______
Quantity	Rate	Press.	Time	REMARKS

Fig. 4 - Typical Headings for Grout Injection Record Forms.

APPLICATION EXAMPLES

Many examples and case histories are described in the various references appended hereto. Following however is a very general review of past and present usage of the technique. By far the most extensive use has been in connection with the correction of structural settlement, where both shallow and deep foundations exist. Where the foundation consists of relatively shallow spread footings, the most common solution involves grouting from the top, down in stages of about six ft. (1.8 m) as illustrated on Fig. 5. As an alternate to this method, some work has been performed wherein a relatively large grout hole, three to six in. (76 to 152 mm) in diameter, is used and the grout injected in a single stage working from the bottom, up. The holes are usually drilled utilizing equipment that permits the simultaneous installation of a tight fitting steel casing. The casing is removed incrementally as injection proceeds. A standard deformed steel reinforcing bar is often inserted in the casing prior to grout injection and remains in place, Fig. 6.

In the case of pile supported structures, the soil underlying the pile tips can be densified by grout injection, where the problem is determined to be bearing failure. In this instance, casings are usually placed to within a few feet of the pile tip elevation, and only that soil zone underlying such elevation is improved, Fig. 7. Where lack of friction is the main deficiency, grout injection is often made

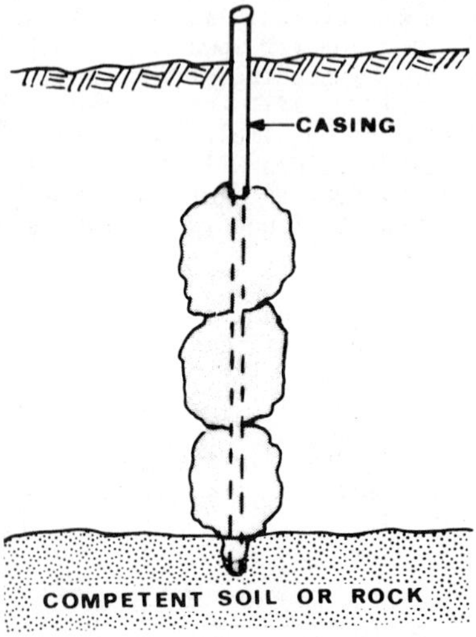

Fig. 5 - Typical stage grouting from the top down.

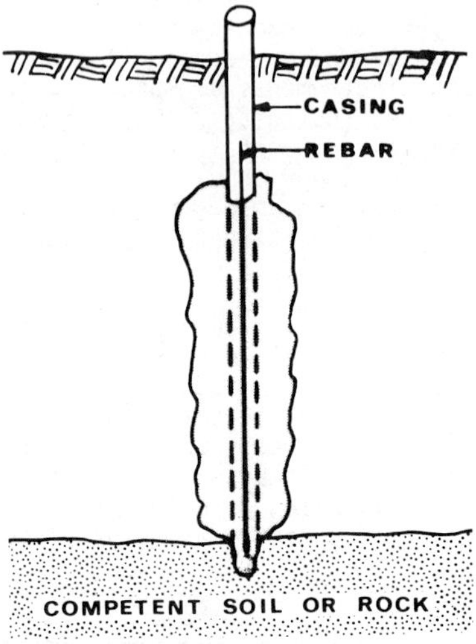

Fig. 6 - Grouting from bottom up.

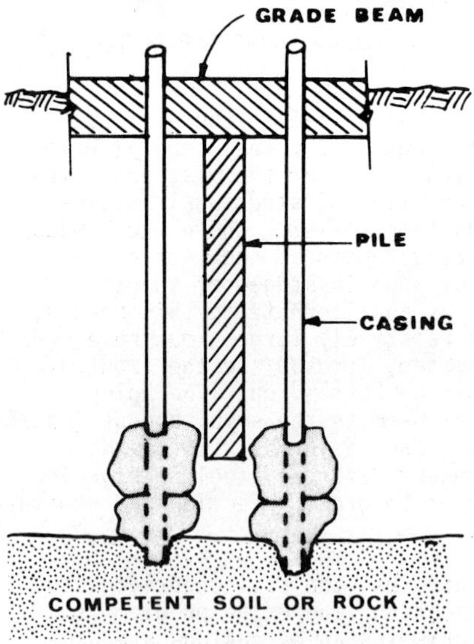

Fig. 7 - Grouting pile for increased end bearing.

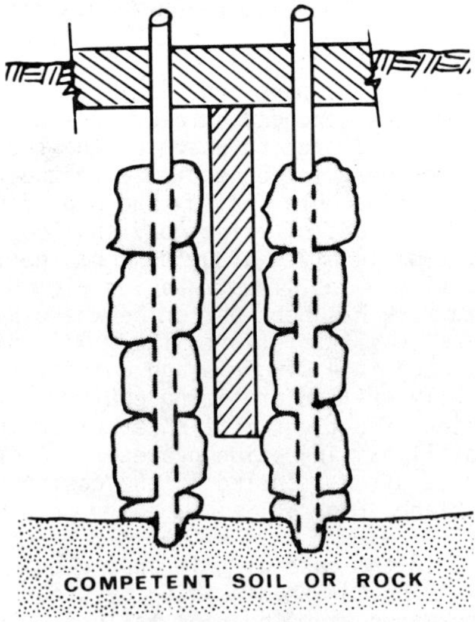

Fig. 8 - Grouting pile for increased friction.

starting at the surface, and extending to underlying competent soil or rock. This may be at an elevation greater or less than the pile tips. Such work is usually carried on from the top, down in stages as previously discussed. (Fig. 8.)

Compaction Grouting has been used to prevent or correct failure of underground pipes and culverts. This has been done either to minimize future deformation in the case of actual failures, or to prevent failure in instances where additional fill is to be placed over existing structures, increasing the overburden pressure. In such practice, a row of grout holes is usually placed longitudinally along each side of the pipe or culvert. They are cased to approximately the springline and a single injection made thereat, Fig. 9.

The technique has been used to compensate for lost ground in soft ground tunneling. In such cases, one or more lines of grout casings are placed longitudinally over the tunnel terminating at an elevation of six to eight ft. (1.8 to 2.4 m) above the roof thereof. As the tunnel shield passes, injection is made, Fig. 10, replacing any soil lost by the tunneling disturbance or into the shield tail-void. By thus replacing lost ground as it occurs, surface settlements can be virtually eliminated, and the extent of disturbed soil vastly reduced.

The procedure has been used under existing structures and pavements in order to increase their bearing capacities. Such improvement has been performed for both permanent improvement and as a "tool of construction" where temporary additional load resistance is required.

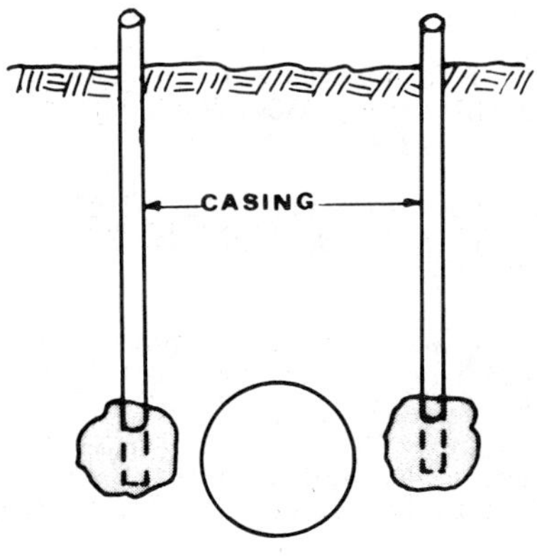

Fig. 9 - Grouting around pipe or culvert.

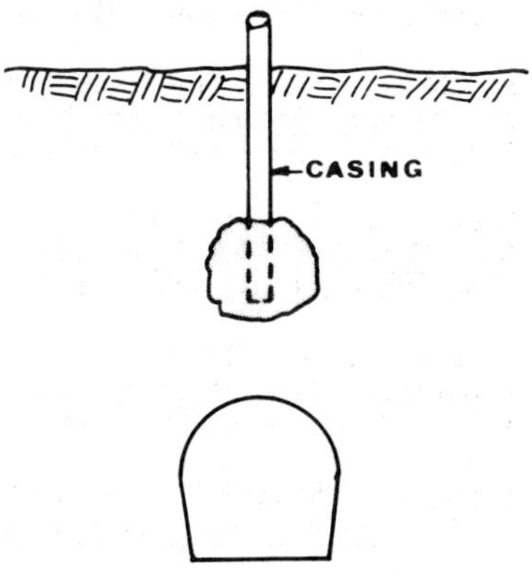

Fig. 10 - Grouting over tunnel to prevent settlement.

CONCLUSIONS

The Compaction Grouting technique has now been practiced for about thirty years. It has been used on a wide variety of applications and has proven to be an effective method for improvement of a variety of fine grain soils. The procedure differs fundamentally from the more common penetration type grouting in that relatively immobile, very stiff grout is used. Because of the relative immobility of the grout except under high pressure, close control of its final point of disposition is easily attained. Because of the very stiff grout involved, the equipment required, as well as injection procedure, varies radically from that used in penetration type grouting. The exact performance criteria for most Compaction Grouting projects are developed as the work progresses. In this regard, it is imperative to provide careful control of the injection and maintain good records, in order to insure optimal results. As experience is gained, the technique is bound to receive ever expanding usage, as well as improved understanding of the engineering mechanics of injection.

APPENDIX I. - REFERENCES

1. Warner, J., "Soil Modification to Reduce the Potential for Liquefaction," Proceedings, Second Joint Meeting, U.S./Japan Cooperative Research Program in Earthquake Engineering on Repair and Retrofit of Buildings and Lifelines, Tsukuba, Japan, May 1981.

2. Baker, W. H., MacPherson, H. H., Cording, E. J., "Compaction Grouting to Limit Ground Movements: Instrumented Case History Evaluation of the Bolton Hill Subway Tunnels, Baltimore, MD, "Technical Report U.S. Department of Transportation, October 1980.

3. Graf, E. D., "Compaction Grouting Technique," Journal of the Soil Mechanics and Foundations Division, ASCE, Vol. 95, No. SM5, Proc. Paper 6766, September 1969, pp. 1151-1158.

4. Mitchell, J. K., "In-Place Treatment of Foundation Soils," Journal of the Soil Mechanics and Foundations Division, ASCE, Vol. 96, No. SM1, Proc. Paper 7035, January 1970, pp. 73-110.

5. Brown, D. R., and Warner, J., "Compaction Grouting," Journal of the Soil Mechanics and Foundations Division, ASCE, Vol. 99, No. SM8, Proc. Paper 9908, August 1973, pp. 589-601.

6. Warner, J., and Brown, D. R., "Planning and Performing Compaction Grouting," Journal of the Geotechnical Engineering Division, ASCE Vol. 100, No. GT6, Proc. Paper 10606, June 1974, pp. 653-666.

7. Lamb, B., "Sinking Building Gets Grout Injections to Strengthen Soil," Engineering News Record, Vol. 198, No. 4, January 27, 1977.

8. Warner, J., "Compaction Grouting - A Significant Case History," Journal of the Geotechnical Engineering Division, ASCE, Vol. 104, No. GT7, Proc. Paper 13897, July 1978, pp. 837-847.

9. The Committee on Grouting of the Geotechnical Engineering Division, ASCE, "Slabjacking-State-of-the-Art," Journal of the Geotechnical Engineering Division, ASCE, Vol. 103, No. GT9, Proc. Paper 13239, September 1977, pp. 987-1005.

PRESSURE INJECTION GROUTING OF LANDFILLS USING LIME AND FLY ASH

J. R. Blacklock[1], R. C. Joshi[2], Members ASCE,
and P. J. Wright[3]

ABSTRACT

Pressure injection grouting of lime and fly ash slurry has been used successfully in a variety of geotechnical applications including ground dewatering, railroad roadbed renovation, slope stabilization, methane gas cut-off, and remedial stabilization of building foundations. A recent new application has been for stabilization of landfills to support construction of buildings and parking lots and for repair and renovation of existing buildings and parking lots. This paper presents lime/fly ash injection grouting procedures and descriptions of equipment and materials. In addition, two case histories, citing landfill stabilization projects, are presented. A unique structural characteristic of lime/fly ash slurry grout is its inherent ability to heal or recement across cracks and repair itself by a self generating mechanism. Because of this autogenous healing property, lime/fly ash mixtures are less susceptible to deterioration under repeated loading and are more resistant to attack by the elements, than other materials which do not possess this property. Because fly ash is an inexpensive by-product material and is readily available in many parts of the United States, it is an ideal material to use in this type application where large volumes are required due to the high void ratios found in landfills. The combination of lime and fly ash has proven to be extremely successful when used in the pressure injection grouting method, for strength, gas and liquid cut-off, and structural stabilization of large landfills.

INTRODUCTION

Pressure injection grouting of lime/fly ash slurry for the purpose of improving the geotechnical engineering properties of large landfills, is gaining acceptance among cost conscious land owners and developers. Pressure injection grouting of plain lime slurry has been utilized for over 20 years; however, it is only in the more recent past that the addition of fly ash to the process has made it

---

[1]Associate Professor Engineering, The University of Arkansas at Little Rock, Little Rock, Arkansas

[2]Professor, Department of Civil Engineering, The University of Calgary, Calgary, Canada

[3]Vice-President, Woodbine Corporation, Fort Worth, Texas

possible to economically deep lime/fly ash injection grout nonreactive natural and fill areas. The two principal reasons for the present favorable economics of the lime/fly ash injection grouting method of stabilization and improvement are the ready availability of an inexpensive supply of fly ash and the development of new equipment and procedures. Because the injection grouting method utilizes hydraulically pushed grout probes rather than predrilled grout holes, it is usually faster and, therefore, cheaper than stationary grout pipe methods. The anticipated future burning of large quantities of coal promises a steady supply of inexpensive fly ash for the future. The continued development of new and better equipment can, in addition, offset future cost increases due to labor and fuel. It is therefore anticipated that, in the future, lime/fly ash injection grouting projects will be more numerous, bigger, and deeper at substantial savings over competitive methods.

BACKGROUND

The development of fly ash as a readily available source of inexpensive grout solids has encouraged contractor research and development of equipment and procedures for its use in the lime/fly ash injection grouting method of improvement for landfill soil masses. This work has resulted in the application for and issuance of a United States patent for the process. Initially, the lime/fly ash injection method was an alternative method to be utilized when lime injection would not be appropriate. Gradually, however, it has become obvious that there are many uses for this new method which are not merely alternatives for lime slurry injection, but improved methods over competitive methods. The stabilization and improvement of nonreactive landfills is, perhaps, the most important utilization of lime/fly ash injection grouting to date. Considerable lime/fly ash injection grouting research is in progress at the University of Texas at Arlington and the University of Arkansas at Little Rock to generate geotechnical engineering data and to promote improved performance and economics of the methods. Other work is in progress at the University of Calgary, Calgary, Canada, Ref. 4.

EQUIPMENT

The equipment used for modern pressure injection grouting of landfill masses with lime/fly ash slurry has been designed and engineered for maximum speed reliability and precision. It was the development of this special equipment that made lime/fly ash injection economically feasible and routinely practical. The new self-contained semi-automatic injection truck (Fig. 1), equipped with a hydraulic injection system, is an essential part of the present high-production procedure. An injection fleet typically comprises a storage tank and slurry mixing unit (Fig. 2), slurry transports, injection truck, and/or an injection tractor (Fig. 3). The fleet is normally operated by two or more crewmen. The injection tractor is used for shallow applications, whereas the injection truck is used for deep applications, currently a 40-ft (12.2-m) maximum.

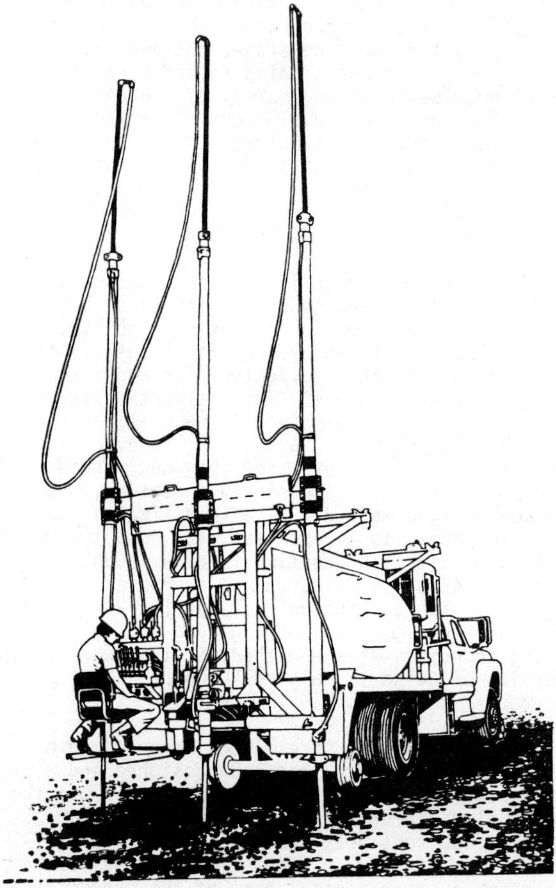

FIG. 1. - Pressure Injection Grouting Truck

FIG. 2 - Slurry Mixing Unit and Storage Tank

FIG. 3. - Tractor Injection Unit with Four Probes

GROUTING PROCEDURES

The grouting of landfills with lime/fly ash grout is best accomplished by the upstage injection method without a packer using the blanket grouting procedure of closely spaced holes grouted on a grid pattern over the entire surface area of the landfill. In this procedure, the grout can usually penetrate; filling joints, voids, and pore spaces in the landfill with the lime/fly ash grout without disturbing the formation.

This will result in the formation of a type of preplaced aggregate concrete of the grout and the solid particles and chunks of the landfill. It will strengthen the mass by filling it with a hardenable grout and, thus eliminate a major source of settlement. At the same time, it will seal off the paths for the gases and fluids and provide the necessary surface strengths for economic utilization.

A drain should be used as a means of removing excess trapped water during the injection cycle. The fluid pressure of the grout will be relieved best if the grouting progresses from one side, moving in the direction of the side chosen for draining. As the grouting progresses, closure can be achieved by splitting the hole spacing in order that there results a progressive decrease in grout take as the primary, secondary, and tertiary holes are grouted. Fracturing of the landfill by the intrusion of grout fingers, sheets and lenses along planes of weakness or between the strata of the solid particles of the landfill formation is sometimes necessary. To accomplish this, the grout must be applied at sufficient pressure to cause the strata to move. The grout take - the measured quantity of grout injected into a unit volume of the landfill - will vary as a function of the grouting pressure, grout viscosity, hole spacing, pump rate, gel time and the character of the landfill deposit.

The procedure for lime/fly ash injection grouting of landfills has many similarities to the conventional Portland cement blanket grouting procedure, in that it utilizes a closely spaced grid pattern of injections spaced uniformly over the entire surface area of the landfill. The major difference is that in the injection grouting of landfills, it is not necessary to predrill each grout hole.

The injection grout probe, being constructed of high strength steel tubing, is pushed hydraulically into the ground to the full depth prior to pumping the grout into the site. Basically, injection grouting is accomplished in the following sequence:

Step 1 - Push the injection rods to full depth.
Step 2 - Inject grout in each rod simultaneously.
Step 3 - Slowly raise injection pipes while injecting slurry.
Step 4 - Exit the ground, turn off the slurry pump and move to the next set-up.

In most instances, it is best to utilize a type of curtain grouting pattern around the perimeter of the site prior to injecting the interior. This will help contain the lime/fly ash slurry in the primary area.

MATERIALS

Lime.--In this paper the term lime refers to oxides and hydroxides of calcium. There are two types of lime commercially available: calcitic quicklime (CaO) and high-calcium hydrate, $Ca(OH)_2$. The quicklime must be slaked prior to mixing, whereas the hydrated lime, which comes in a dry powder form, is ready for immediate mixing. Laboratory testing may be used to indicate the effectiveness of any particular commercial source of lime, but it should be emphasized that the quality of the fly ash has a much greater influence on lime/fly ash pozzolanic reaction than does the lime type. It can be stated that most commercially available limes are appropriate, if a quality fly ash which meets the laboratory test series for strength and durability criteria can be economically obtained.

Fly Ash.--Fly ash is "the finely divided residue that results from the combustion of ground or powdered coal and is transported from boilers by flue gases" (ASTM Specification C 593). Fly ash is collected from the flue gases by either mechanical or electrostatic precipitation devices. Fly ash is a pozzolan and is defined as "a siliceous or siliceous and aluminous material, which in itself possesses little or no cementitious value, but which will in finely divided form and in the presence of moisture, economically react with calcium hydroxide at ordinary temperatures to form compounds possessing cementitious properties." (Ref. 6). Fly ash particles are primarily composed of silica and alumina. Secondary ingredients are carbon and oxides of iron, calcium, magnesium, and sulfur. Oxides of sodium, potassium, titanium, maganese, and phosphorus may also be present. Table 1 gives ranges of values for the chemical composition of a central Texas fly ash.

Table 1. - Composition of a Typical Central Texas Fly Ash

Principal Constituents	Percent
SiO_2	46-50
Al_2O_3	19-20
CaO	17-18
Fe_2O_3	6-7
MgO	3-4
SO_3	1-2
K_2O	0-1
Loss on Ignition	0-1

Water.--Water used in mixing lime/fly ash slurry grout shall be clean and free from injurious amounts of oils, acids, alkalis, salts, organic materials, or other substances that may be deleterious to the desired lime/fly ash-soil reactions. If nonpotable water is proposed for use, and if there is any doubt concerning compliance with the above statement, laboratory tests should be conducted to compare the lime/fly ash reaction of similar specimens incorporation potable water.

MIXTURE

Lime/fly ash grout for injection in landfills is a thick fluid mixture of lime, fly ash, and water. The ratio of lime and fly ash solids to the water is (usually) 1:2 by weight. A nonionic surfactant is always added to the grout mixture to enhance penetration of the slurry.

Lime-to-fly-ash ratios vary substantially, but the range is generally between 1:2 and 1:10 with ratios of 1:3 and 1:4 being common. Factors that tend to increase the lime requirement are (a) greater fines content in the land mass, (b) increased PI of the fines, and (c) increased pozzolanic reactivity of the fly ash (Ref. 3).

Lime/fly ash mixture proportions cannot be established based on an analysis of the separate properties of lime, fly ash, and material to be stabilized. Proportions are determined using laboratory-based mixture design procedures using component materials to be stabilized as well as samples of the actual lime and fly ash planned for injection. A typical grout mixture could be 1-lb (.454-kg) dry lime and 3-lb (1.36-kg) fly ash and 1-gal (3.8-l) H_2O. In the batch mixing method, the slurry ingredients are initially mixed and continuously agitated in 20,000-gal (75,700-l) tanks (Fig. 2). The slurry grout is then transferred to the 2000-gal (7,570-l) tanks self contained on the injection unit (Fig. 1), as the injection progresses.

CURING

The curing of lime/fly ash grout is dependent upon time, temperature and moisture factors. In general, the 60-day compressive strength for grout cured at 50°F will be obtained in 30 days at 70°F and in 10 days at 85°F (Ref. 5). This value represents approximately 50% of the one year usable strength at 75°-80°F. Under favorable temperature and conditions the mixture will gain additional strength for several years. The actual values for compressive strength of cured grout will depend upon mixture ratios, material reactivity and soil types.

AUTOGENOUS HEALING

A unique characteristic of lime/fly ash grout is its inherent ability to heal or recement across cracks by a self-generating mechanism (Ref. 1). This is called autogenous healing. Because of the autogenous healing property, lime/fly ash mixtures are less susceptible to deterioration under repeated loading and are more resistant to attach by the elements and premature overloading.

CASE HISTORIES

The following two case histories of lime/fly ash injection grouting projects are presented to illustrate the adaptability and versatility of the method. Many other geotechnical applications of lime/fly ash injection have been accomplished including dewatering projects, railroad embankment stabilization, dike stabilization and renovation of building foundations, Refs. 7 and 8.

CASE #1

Three Phase Landfill Stabilization for Existing Building
Forest Park Boulevard, Fort Worth, Texas

This 20,000-sq ft (1,858-m^2) building was constructed in 1968 in an area that was previously a sanitary landfill. Fig. 4 shows the building in the background and the condition of the city street which was also constructed over the landfill. Depth of the landfill varied from 8-10-ft (2.4-3.05-m) below the building.

The basic structure of the building was founded on piers that penetrated below the landfill into a primary formation and the outside walls were in satisfactory condition. However, the floor slab was poured on grade and had settled as much as 14-in. (355-mm) resulting in a building that was unusable and was therefore vacant except for a small nightclub operation on the north end.

Remedial stabilization of this building was accomplished in 1977 using lime/fly ash pressure injection grouting. The work was accomplished in three phases.

Phase 1.--Before any stabilization and releveling work was accomplished within the building, a curtain grouting sequence was performed around the outside of the structure. This lime/fly ash curtain was to help contain subsequent injections within the confines of the building area. Curtain grouting was performed using a self contained hydraulic injection rig (Fig. 1). A two staged injection sequence was accomplished using lime/fly ash slurry. The primary injections were made on 5-ft (1.5-m) centers to a depth of 10-ft (3-m) as close as possible to the edge of the building. Injections were accomplished upstage without packers at pressures of 50-100 psi using a slurry mixture of 4-lb (1.8-kg) solids and 1-gal (3.8-l) of water. Secondary injections were then completed between the primary injections, but set back approximately 5-ft (1.5-m) from the edge of the building. In situations where a more impermeable curtain is desired, tertiary injections would be made between the primary and secondary injections after allowing time for grout set.

Phase II.--After installation of the perimeter curtain, holes were then drilled through the floor slab on 10-ft (3-m) centers and the entire area below the buildirg was grouted downstage without packers. The lime/fly ash grout was mixed at a ratio of 5-lb (2.3-kg) dry solids per 1 gal (3.8-l) of water and injected through single grout pipes to refusal. Refusal was observed visually in the adjacent open holes that had been drilled through the floor slab. This was possible because of a 6-in. (152-mm) void that existed

FIG. 4. - Forest Park Boulevard, Fort Worth, Texas

FIG. 5. - Methane Gas Exhaust Pipes

between the slab and the subgrade. Many times grout flow could be observed as far as 20-ft (6-m) away from the grout pipe due to the high void ratio that existed in the mature landfill. During the early stages of grouting, methane gas was coming out of grout pipes and when ignited it created gas flares. As grouting progressed, the methane flare diminished until finally it was completely shut off.

Phase III.--The final phase consisted of pumping a thicker lime/fly ash grout mixture 6-lb (2.7-kg) dry solids per 1-gal (3.8-l) of water, between the floor slab and stabilized subgrade to fill the existant void and finally to raise the floor slab to its original configuration.

The floor slab has resisted further settlement for approximately four years since completion of the stabilization work and the methane gas problem has not reoccurred. The building is now fully leased.

CASE #2
Lime/Fly Ash Grouting to Relevel Floor Slab and Seal Off Methane Gas Below Existing Building
South Town East Boulevard, Mesquite, Texas

This is a 52,500-sq ft (4877-m^2) building which had been constructed over an old sanitary landfill. The building was subsequently condemned by the Fire Marshal due to excess methane gas inside the building. The building had to be vacated and had therefore become a total economic loss to the owner.

In an effort to evacuate the methane gas, the owner had installed a series of pipes throughout the building. These pipes penetrated the floor slab and were equipped with blower devices to try and push the gas out through the roof (Fig. 5). Also a collector system was used around the perimeter of the building using dry wells and vertical exhaust pipes which extended to the top of the outside walls (Fig. 6). This system did not effectively remove the gas and the owner was still not able to occupy the structure.

The work on this building was to relevel the floor in certain areas where the slab had settled approximately 2½-in. (64-mm), and to grout the landfill below the floor slab to seal off the methane gas. The work was accomplished on approximately 30,000 sq ft (2,787-m^2) of the building and progressed as follows:

1. 327 holes were drilled through the floor slab on 10-ft (3-m) centers.
2. Lime and fly ash grout was injected into the subgrade to an average depth of 9-ft (2.7-m). The procedure was started along the east and south walls of the building to form a curtain to contain subsequent grouting within the confines of the building area. Grout pipes were inserted in each hole and forced to the bottom of the landfill or until impenetratable material was encountered. Lime/fly ash grout was pumped to refusal at 50-100 psi. Grout pipes were then raised to intermediate levels and the same process repeated. The slurry consisted of 2-lb (.9-kg) lime and 2-lb (.9-kg) fly ash per 1-gal (3.8-l) of water.

FIG. 6. - Methane Exhaust Vents

3. After the landfill subgrade had been grouted to refusal, the leveling and sealing operation was initiated. A slurry mixture consisting of 2-lb (.9-kg) lime and 4-lb (1.8-kg) fly ash per 1-gal (3.8-l) of water was then injected immediately below the floor slab to relevel the distressed floor slab and finally over the entire building area tc fill any remaining voids and effectively seal off the methane gas.

The Fire Marshal made periodic checks for methane gas as the work progressed until finally no trace of gas could be found. The building was subsequently leased to a new tenant and is presently occupied. The methane gas problem has not recurred since completion of the work in July, 1980. 190-tons (172-Mg) of lime and fly ash were injected in the job.

LABORATORY TESTING

Laboratory testing for lime/fly ash injection grouting of landfills requires the following materials:

1) fly ash
2) lime
3) surfactant (wetting agent)
4) water
5) landfill fine particles.

The success of the project depends upon the correct ratio and mixture of the above materials, making certain that there is reactive compatability between each of the grout mix ingredients. It is best to use proven reactive materials from known sources in the beginning of the laboratory testing program to enhance the chances of making a successful mix during the first few trials. It is then safe to run trials with variations of unproven materials to obtain the actual mix design. A good mix should have the proper proportions of lime, fly ash and water for economics, strength and pumpability. The tests to be used should evaluate the following:

1) Durability (Fresh Water Soak Test)
2) Compressive Strength (Unconfined Compression Test)
3) Cure Rate (Time/Temperature)
4) Suspension Retention (Settlement)
5) Pumpability.

Grout testing procedures are similar to lime injection testing procedures for soil samples as discussed by Blacklock and Lawson in Ref. 2.

STRUCTURAL CONSIDERATIONS

Structural failures in landfill lime/fly ash injection grouting projects can be minimized by using properly engineered plans and specifications. An important step in this process is initial site characterization and exploration. Site characterization and exploration will include surface and subsurface exploration as well as field and laboratory testing. A grouting field test or pump test may be included in this portion of the project. This would include water and/or grout pump tests.

The close prediction of final strength and settlement rates is difficult for grouted landfills. The engineer can characterize the strength of an existing land field by using a combination of penetration and plate bearing tests. This data will be time, load, environmental and site dependent. Therefore, the proper grouting plan must allow for a large degree of latitude in the grouting specifications. The grout, no matter how strong or how fluid will not displace certain solid low strength fill materials. The grout can surround these weak solids with strong layers and walls of hardened lime/fly ash grout, and this will usually achieve the necessary increase in strength.

The new strength should be designed for the intended surface loads. In planning for a landfill grouting project, it is essential to know that the design structural loads will be within the realm of strength possibilities. From the structural standpoint the allowable strength of the lime/fly ash hardened grout is measurable. It will depend upon the mix design, time of cure and environmental conditions. These are parameters that can be predicted with reasonable accuracy. Because the grout will flow in a random pattern in and around the solid particles in the landfill, it will help to base the strength predictions upon a physical examination of the particles in the landfill mass as well as the hardened grout. The landfill can be evaluated with soil borings, trenches or pits to determine the structural character of the landfill chunks and particles. It will be possible in this manner to predict the range of the finished strength values. In landfills containing materials which defy structural characterization the engineer must proceed only with a conservative approach for the surface utilization. For instance, the initial proper use of this type of stabilized landfill might be for a park, storage lot or parking lot, utilizing a gravel, sod, or soil surface, which can be easily evaluated for more substantial use in subsequent years.

CONCLUSIONS

The pressure injection method of grouting deep landfills with lime/fly ash slurry has proven useful and economical for ground stabilization and improvement of engineering properties. The experience gained to date indicates a considerable versatility in application and improvement ranges. Lime and fly ash are both relatively inexpensive materials which serve as a useful grout when mixed with water and pressure injected into landfill masses. Strength gains are predictable in most cases, resulting in servicable stabilized land surfaces. Future applications of lime/fly ash slurry injection grouting will grow very rapidly as its potential is fully realized by the engineering profession.

REFERENCES

1. Barenberg, Ernest J., "Lime or Cement Fly Ash - Aggregate Pavements," University of Illinois, March, 1977.
2. Blacklock, J. R. and Lawson, C. H., "Handbook for Railroad Track Stabilization Using Lime Slurry Pressure Injection," Federal Railroad Administration Report No. FRA/ORD-77/30, June, 1977.
3. Goecker, W. L. et al., "Stabilization of Fine and Coarse-Grained Soils With Lime/Fly Ash Admixtures," Highway Research Board Bulletin 129, pp. 63-82, 1956, National Academy of Sciences - National Research Council, Washington, D.C.
4. Joshi, R. C. and Wright, P. J., "Soil Improvement by Lime and Lime/Fly Ash Slurry Injection Process," presented at the 1978 Symposium on Soil Reinforcing and Stabilizing Techniques, Sydney, Australia.
5. MacMurdo, F. D., and Barenberg, E. J., "Determination of Realistic Cutoff Dates for Late-Season Construction With Lime/Fly Ash and Lime-Cement-Fly Ash Mixtures." Highway Research Record No. 442 (1973), pp. 92-101.
6. "Standard Definition of Terms Relating to Hydraulic Cement," ASTMC 219-74a.
7. Wright, Paul J., "Lime/Fly Ash Injection Stabilization," presented at the October 22-27, 1978, Conference on Ash Technology and Marketing held at London, England.
8. Wright, Paul J., "Lime/Fly Ash Injection Applications," presented at the February 25-27, 1979, 5th International Ash Utilization Symposium held at Atlanta, Georgia.

LIFTING AND LEVELING HEAVY CONCRETE STRUCTURES

By John C. King[1], F. ASCE and Eduardo W. Bindhoff[2], M. ASCE

INTRODUCTION

Heavy structures that have experienced differential settlement due to foundation soil subsidence can be jacked up and righted to or near original line and grade by the use of grouting techniques. Continuous monitoring and careful measurement of movement in the structure is of paramount importance during the grouting operation. Field methods must allow a great degree of flexibility, since minute-to-minute changes in planned procedures are not uncommon. Lifting and leveling of heavy structures is, therefore, almost as much art as engineering.

A few projects that involved lifting and leveling of heavy structures will be discussed briefly followed by a detailed description of the righting of a 222 ft.-high mine headframe over a mine shaft.

GENERALIZED DESCRIPTION OF LIFTING PROCEDURE

Tall, heavy structures are best raised by grout pressure exerted in one or more of weaker soil layers at some depth below the undersurface of the foundation mat rather than at the concrete-soil interface. One of the reasons for this is that it is easier to contain the grout in deeper layers than near the surface. Another is that the grout pressure tends to squeeze water from the layers of soil that have permitted the settlement thereby increasing their strength. The grout columns, layers, and masses also increase the strength of the weak layers. And finally, the foundation slab of the structure is less subject to stress concentrations as grout pressures are raised to attain sufficient spread to lift the structure. The first step is, therefore, a determination of soil characteristics from the foundation mat to below material that is compressible under the superimposed load of the structure, taking into account dispersion of load and depth. The location and analysis of the properties of weak strata is particularly important.

To insure containment of the lifting grout, a vertical grout curtain along the perimeter of the sides of the structure where most of the lifting pressure will be exerted may have to be established. The grout curtain is not necessarily watertight but rather a combination of strings of grout bulbs and compressed soil which, backed up by in-situ soil, has sufficient strength to prevent the lifting grout from escaping laterally. The engineer will probably specify the depth of the grout curtain and that grouting be performed by the split spacing method from the bottom up with pressures limited to that which will not cause surface heaving (approximately 1 psi per foot of depth). He will

---

[1] Consulting Engineer, Cleveland, Ohio

[2] Vice-President, Intrusion-Prepakt (Intl) Inc., Cleveland, Ohio

be present during the work to make on-the-spot decisions, but generally trust a qualified superintendent and grouting crew to proceed by feel based on experience to accomplish the work.

The structure is then lifted by grouting through pipes driven at various angles to the lowest lifting area, ending in that zone at, say, 5-ft. centers.

Using a retarded cement or sand-cement grout, and employing perhaps one grout pump per 5 to 10 inserts, grout is pumped first into the area under the edge of the structure to be lifted. Pressure is brought up to approximately that computed as necessary to lift the structure, a valve closed on the end of the insert, connection changed to the next pipe, and the procedure repeated until a fluid layer of grout has been established under half or more of the area of the base of the structure. Connections are then remade to inserts near the lifting edge and pressures increased until the structure moves. How far to raise the structure in the first lift is a matter of judgment on the part of the grouting crew who will be carefully watching for signs of grout leakage between the grout curtain and the edge of the base of the structure. If it is necessary to repeat the operation to bring the structure all the way back to plumb, new pipes are driven next day to some point above the first lift, and the procedure repeated.

LEANING GRAIN ELEVATOR

To the writers' knowledge, the first major structure to be righted by a grouting procedure was a 115-ft.-high, *75,000-bu.-capacity grain elevator near Goodland, Kansas. The work was performed in 1946.

Built in 1923, the structure was soon known in the area as the "leaning tower"; a year after construction, the top of the elevator was 11 in. out of plumb. Settlement continued over the years until by April, 1946, when corrective work started, the lean was 34-5/8 in. southward from vertical. The Chicago, Rock Island, and Pacific Railroad, whose main track passed 52 ft. south of the elevator, felt threatened.

An engineering article describing the work appeared in Railway Engineering and Maintenance (now Railway Track and Structures) (Ref. 1). The opening paragraph reads:

> "BY APPLYING a heretofore untried method, a tall reinforced concrete grain elevator, which had tipped almost three feet because of uneven settlement of the soil under its foundation, was successfully restored to vertical. To effect this, the footing was extended on the low side and the elevator was lifted hydraulically back into position by means of cement grout applied at high pressures under the footing. As a further result of this work, which cost only about one-fourth of the estimated replacement expense, the foundation was materially strengthened, with further settlement considered unlikely, and the facility was returned to full operation."

The procedure selected for righting the elevator included extending the base slab a distance of 6 feet on the south side to provide the main

---

* Relevant conversion factors to SI units are given in Appendix II.

lifting area and leverage; wetting the soil under the north half of the structure to allow that side to settle as the structure rotated over the untouched material under the central area; and cutting the auxiliary structures free of the silo to permit the movement (See Figs. 1 and 2).

The slab extension, 6 ft. thick where it joined the existing structure, was heavily reinforced with radial steel welded to existing mat bars. Additional bars were grouted into horizontal holes extending as much as 20 ft. into the slab. Temporary concrete buttresses to help transfer the lift under the slab extension to the silo walls were also constructed.

Excavation for the slab extension included 4 in. for a gravel mat under the 6-ft.-thick concrete to provide a flow channel for the grout to spread under the concrete and a 2-1/2-ft.-deep trench at the perimeter to establish a barrier to contain the grout. Six 1-1/2-in.-diameter insert pipes were also placed with lower ends in the gravel stratum in advance of concreting.

The grout lifting operation was performed in five stages over a 10-day period. Lifting work started by pumping grout into the gravel layer through two inserts at a time until it appeared at adjacent pipes, then reconnecting for short periods to other pipes until the entire layer was full. Grout pressure was then raised to over 100 psi (with idle inserts valved off) until the structure began to move. At the end of the first day of grouting, the top of the silo had moved 11 in. closer to plumb.

For succeeding stages of pumping, new holes were drilled through the slab extension and hardened grout. Holes were also drilled through the foundation slab along the south side within the silo. As these holes reached the underlying soil, blow pipes were used to remove some of the material and form a pocket that would permit grout, under pressure, to penetrate and spread under the previously grouted layer. One day was normally required for this work. For the last stage of grouting some of these holes were as deep as 14 feet.

Pumping pressures to establish spread reached as high as 300 psi at which point the pressure would drop as the structure began to move.

The sequence of pumping operations was continued until the silo had been completely restored to vertical. A transit was used throughout to observe movements of the structure. No resettlement or contrary movement was noted after a raise was accomplished or after pumping was stopped, nor between pumping operations.

The basic grout mix used for the lifting consisted of two bags of portland cement, 1 bag (70 lbs.) of fly ash (Alfesil), 14 gal. water, plus 3 lbs. of Intrusion Aid, a grout fluidifier consisting of a combination of water-reducing retarder, a gas-expanding agent, and certain modifiers. Some 1500 sacks of portland cement and 1020 sacks of Alfesil plus water and admixture were used. No fine aggregate (sand) was employed in the mixes. Equipment for the work consisted of an air-operated double-tub mixer, two air-driven piston-displacement grout pumps, air-compressors, water pump, high pressure grout hoses, together with inserts, pressure gages, and plug valves.

Fig. 1. Goodland, Kansas. Constructing Footing Extension at Base of Silo. Note Grout Insert Pipes Extending Into Gravel Mat.

Fig. 2. Two Buttress Walls Were Built to Aid in Transmitting Lift Pressure from the Footing to the Elevator and Were Removed When the Work Was Completed. Note Separation of Auxiliary on Right to Permit Independent Movement of Silo.

The plan for the work, submitted by the contractor, was approved by W. E. Heimdinger, Acting Assistant Chief Engineer, Rock Island Railroad. S. T. Robinson, Division Engineer, acted in the field for the railway interests. The grouting method and admixtures used on this project were developed and (then) patented by Intrusion-Prepakt, Inc., then Chicago and now Cleveland, Ohio. General direction of the work was by Raymond Patterson, regional Vice-President of I-P, Inc., while field operations were carried out by Vic Erickson, superintendent.

SOME OTHER RECORDED RAISINGS

Settlement of a building in Farmington, New Mexico (Ref. 2), might loosely have been described as a jail break. Cracks appeared in the exterior walls of an addition to a jail building due to foundation settlement caused by unrecognized subsoil weakness. In addition to the cracks, inside partitions were sagging, openings were skewed, and doors jammed.

The situation was corrected by first grouting a subsurface curtain wall to contain subsequent grout under pressure. Then, insert pipes were driven under the foundations and carefully controlled pressure grouting applied to raise the settled portions of the jail structure to their original elevations. The work was done in the spring of 1974.

Another grain storage silo righting project for which some records are available (Ref. 3) involved four Pillsbury storage silos in Dubuque, Iowa, that had visibly settled and tilted (See Fig. 3). The problem became evident shortly after the silos were constructed when, at about 25% loading, some settlement was detected. Subsequent seasonal loadings of 35,000 bu. of corn, wheat, or oats in each of the 104-ft.-high, 25-ft.-diameter structures aggravated the problem until two of the silos tilted inward to touching although two feet apart originally. The other two silos also leaned inward to a lesser degree.

The four silos rested on a common 2-1/2-ft.-thick slab. The correcting pressure grouting was performed in the center of the silo cluster through holes drilled through the slab, Fig. 4. Soil consolidation and lifting began at 26-ft. depth, continuing in stages upward to just under the slab. The confined pressure raised the center of the slab sufficiently that the tops of the touching pair of silos spread 16 in. apart.

Foundation soil information is sketch, being based on the observations of the job superintendent of soil removed from an occasional driven grout pipe that had to be pulled out when its driving head became dislodged. His general statement was that there was a soft strata from 8 to 20 ft. below the slab. One daily job report noted "There was a lot of peat in pipe down 20 ft."

The basic grout mix used in the work consisted of 3 bags cement, 3 bags (70 lbs.) fly ash, 2 lbs. Chemix (a thickening and lubricating admixture), 9 cu. ft. grouting sand (FM probably between 1.2 and 2.0), and 26 gal. water (4.3 gal. per sack of cementing material plus moisture in sand, not recorded). A total of approximately 140 cu. yds. of grout was pumped at pressures usually between 75 to 100 psi, but occasionally at pressures up to 300 psi.

The work was done during the month of March, 1975, the last grout being pumped on March 28. On April 1, the silos were test filled to capacity; only "very little movement" was reported.

Fig. 3. Pillsbury Silos Before Grouting.

MINE HOIST HEADFRAME

The production shaft headframe (Fig. 5), over a new shaft to a copper mine at Tooele, Utah, was found a few months after construction to have suffered sufficient differential settlement and tipping to prevent the operation of skips for the removal of ore and waste from the mine. The headframe rose 224 ft. 6 in. from the subfloor elevation of 6427.50 ft. and rested on a foundation slab 77 ft. long, 52 ft. wide, by 4.5 ft. thick with an additional 8 in. topping on the inside collar subfloor. In operation, two hoists lift four skips having a capacity of 15 tons each from the loading level some 3550 ft. below.

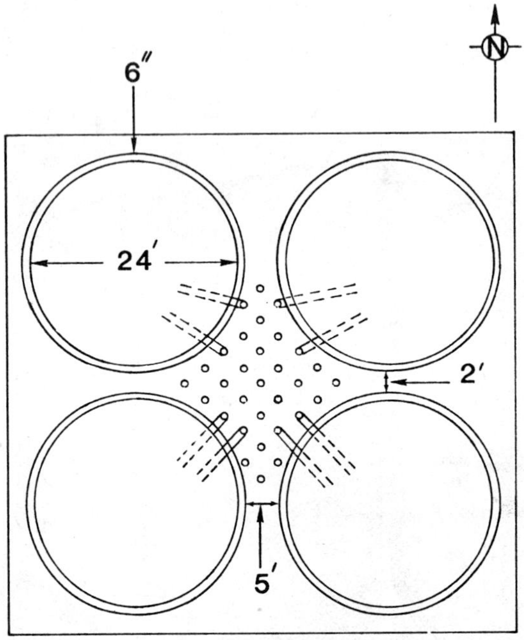

Fig. 4. Grout Insert Pattern, Pillsbury Silos

A lean of 8.9 in. south and 4.4 in. west was discovered when the shaft sinking contractor found he could not bring the shaft steel into alignment. The owner, Anaconda Copper Company, elected to use Intrusion grout and the services offered by The Prepakt Concrete Company to correct the headframe lean and prevent further settlement. The mine is located about five miles east of the town of Tooele which is approximately 35 miles southwest of Salt Lake City, Utah.

The foundation loading for the 8,000-ton headframe was 4.6 kips per square foot. On the premise that the structure, in tilting, had rotated to the north and east around axes through the mine shaft, it appeared that grout pressures on the order of 100 to 200 psi, applied at but spreading under the north and east areas of the foundation mat, should right the structure.

To contain the grout within the desired area and protect the shaft, grout curtains were established as shown in Figure 6. Lifting was accomplished by pumping through inserts placed in holes drilled in the concrete mat within the curtain, also shown on Figure 6.

Preparation for the work included removal of overburden along the north and east side to provide access to the mat and reduce the lifting load. Concurrently, additional overburden was placed on the southwest corner to load this side of the structure and hold it down.

Because of the pressures that might be used in pumping the interior curtain and in the lifting operation, twenty strain gages were placed on the shaft lining 12-1/2 ft. and 22-1/2 ft. below the subcollar level. At each level, 8 of these gages were horizontal and 2 vertical. One hundred microns in tension on any gage would call for the removal of employees at work sinking the shaft some 260 ft. below.

To monitor changes in settlement or lifting, a tiltmeter was installed on the northeast corner of the headframe at subcollar level. This instrument recorded deflections along the east-west and north-south axes to ten-thousandth of a degree in the tangent of the angle out of plumb. Other field controls included plumbing brackets on the four corners of the headframe, top targets on the east and north faces of the frame to be sighted by transit, and a series of leveling stations on the basement slab with an external benchmark for elevation control.

Fig. 5. Mine Hoist Headframe at Tooele, Utah, During Pressure Grouting.

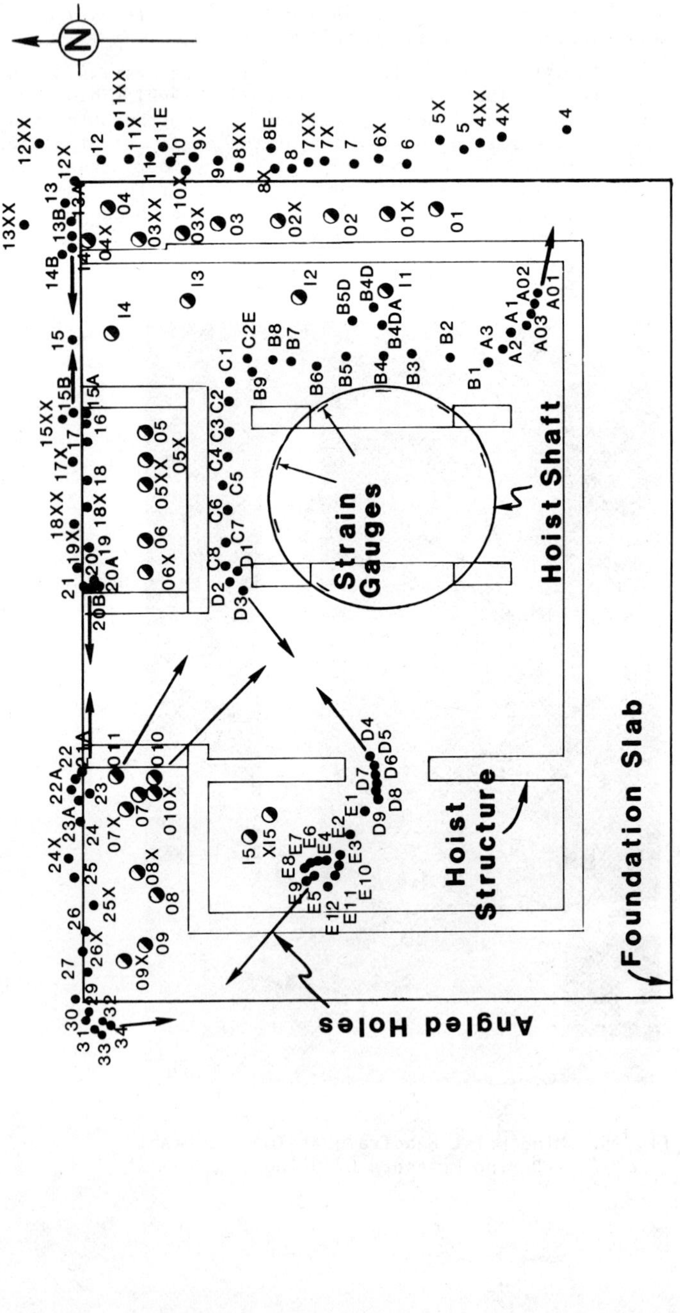

Fig. 6 Grout Hole Locations
•Curtain ◐ Lifting

Other preparatory work included cutting the conveyor shaft tunnels free from the headframe and cleaning the construction joint where the air intake service tunnel joined the headframe.

The external grout curtain was established by grouting through 1-1/2-in. flush-jointed steel pipe driven and/or drilled 5 ft. apart in two rows to a depth of 35 ft. Grouting proceeded from the bottom up at pressures generally in the range of 50 to 100 psi with quantities averaging 5 cu. ft. per vertical foot. In a soft zone, 15 to 25 ft. depth, quantities were often higher; and sometimes lower when the soil had been tightened by the previous grouting of adjacent holes. As consolidation was effected, pressures in excess of 200 psi were recorded. Some holes were angled, as indicated by arrows on the figures, to avoid or reach under obstructions. Depths of holes, total quantities of grout pumped, and average take per ft. are shown in Table 1.

The internal grout curtain was similarly established by driving and/or rotating 1-1/4-in. flush-jointed pipe through 87 holes drilled in the concrete base mat. Drill steel was used on occasion when hard zones were encountered. Depths and quantities of grout pumped are shown in Table 2.

The curtained area between the inner and outer grout curtains provided approximately 1300 sq. ft. for lifting.

In addition to the curtain wall grouting, five deep holes were drilled through the mat on the easterly side of the shaft to reach a soft area reported during excavation of the shaft. One insert (Hole 12, Fig. 6) was driven to a depth of 120 ft.; 1075 cu. ft. of grout was pumped into this area from 120 ft. depth up to 90 ft. The remaining four holes (B3, B4D, B4DA and B6) were pumped from 93 ft. up to 50 ft. and took approximately 2300 cu. ft. The grout takes for the individual holes is given in Table 3.

The lifting operation was accomplished in two stages through a pattern of holes drilled through the concrete base mat as shown in Figure 6. The first stage established a base at a depth of 32 to 35 ft. as shown in Table 4. Grout pressures were in the range of 100 to 200 psi; the volume of grout pumped into this area was 1508 c.f. A trace of movement of the tower was measured during the first day of pumping but when no further lifting occurred during establishment of the base in the next four days, grouting was stopped while the conveyor tunnels were cut free from the headframe and overburden removed from the south perimeter area.

Second stage grouting was pumped through inserts extending to just above the previously pumped base and finally at minus 20 ft. Some of the grouting was done through such of the Stage 1 inserts as could be freed and withdrawn, but generally new holes were drilled and inserts installed as needed. Pumping was alternated between the south and west sides to achieve a balanced lift of the structure. Depths and grout quantities are shown in Table 5.

Lifting progress is shown on Fig. 7. It should be explained here that all pumping of grout for the curtains, deep holes, lifting base, and start of the lifting was done with two, sometimes three, air-driven Simplex pumps. The possible need for higher pressures had been anticipated so that when further lifting could not be accomplished at the

TABLE I

VOLUME PUMPED INTO OUTSIDE CURTAIN

HOLE NUMBER	VOLUME CUBIC FEET	DEPTH FEET	CUBIC FEET PER LINEAR FOOT	ANGLE (1)	HOLE NUMBER	VOLUME CUBIC FEET	DEPTH FEET	CUBIC FEET PER LINEAR FOOT	ANGLE (1)
4	290	35	8.85		17	219	35	6.25	
4X	215	35	6.15		17X	192	35	5.48	
4XX	88	36	2.44	15	18	247	35	6.77	
5	156	35	4.44		18X	137	35	3.91	
5X	140	35	4.00		18XX	72	35	2.06	
6	222	35	6.50		19	212	35	6.05	
6X	127	35	3.68		19X	106	35	3.03	
7	196	35	5.59		20	165	35	4.71	
7X	201	29	1.39		20A	193	35	5.52	8
7XX	49	35	6.91		20B	209	35	5.96	16
8	276	35	7.89		21	177	40	4.42	23
8X	17	26-35	1.84		21A	8	10	0.80	69
8XX	111	35	3.17		22	100	15	6.63	42
8E	124	35	3.55		22A	36	20	1.82	24
9	69	31	2.21		23	120	25	4.81	12
9X	182	35	5.21		23A	147	35	4.19	6
10	218	35	6.22		24	209	35	5.97	
10X	97	20	4.85		24X	227	35	6.48	
11	74	18-35	4.34		25	260	35	7.42	
11X	214	30	7.13		25X	178	35	5.09	
11XX	59	35	1.69	10	26	276	35	7.87	
11E	98	35	2.80		26X	223	35	6.36	
12	156	32	4.87		27	287	36	7.97	
12X	367	35	10.48		28	268	35	7.65	
12XX	204	35	5.81	10	29	240	35	6.87	8
13	203	35	5.79		30	227	35	6.50	16
13A	224	35	6.41	7	31	162	35	3.77	23.5
13B	273	35	7.79	14	32	203	17	11.94	34.5
13XX	173	35	4.94	10	33	119	20	5.93	41.5
14	206	30	6.85	21	34	96	16	6.03	53
14A	31	15	2.07	48.5	Total 10,854 or 402 cu. yds.				
15	45	12	6.36						
15A	45	13	3.49	36					
15B	78	17	4.57	17					
15XX	156	35	4.52	13					
16	156	35	4.45	5					

(1) Degrees from vertical

TABLE 2

VOLUME PUMPED INTO INSIDE CURTAIN

HOLE NUMBER	VOLUME CUBIC FEET	DEPTH FEET	CUBIC FEET PER LINEAR FOOT	ANGLE (1)	HOLE NUMBER	VOLUME CUBIC FEET	DEPTH FEET	CUBIC FEET PER LINEAR FOOT	ANGLE (1)
A01	31	15	3.12	24	D1	94	25	4.87	9.5
A02	52	20	3.20	19	D2	73	20	2.43	20
A03	49	35	1.64	13.5	D3	24	15	5.10	33
A1	98	35	2.92	7	D4	51	15	5.33	44
A2	228	32	8.46		D5	160	35	5.35	35
A3	150	35	4.99		D6	161	35	2.53	21
B1	139	35	4.62		D7	126	35	4.90	14.5
B2	125	35	4.15		D8	147	35	7.69	7
B3	100	35	3.33		D9	231	35	4.79	
B4	129	35	4.28		E1	142	35	5.99	
B5	128	35	4.27		E2	180	35	6.37	
B6	152	35	5.07		E3	191	35	8.77	
B7	143	35	4.77		E4	263	35	5.33	
B8	183	36	5.23		E5	160	35	5.33	
B9	164	35	5.47		E6	148	35	4.93	6
C1	160	35	5.32		E7	179	35	5.98	12
C2	100	35	3.33		E8	4	30	0.16	18
C3	36	35	2.93		E9	133	28	6.13	23
C3E	88	35	3.48	14	E10	103	24	5.39	30
C4	105	35	5.50		E11	74	16	5.19	40
C5	165	35	4.20		E12	---	15	----	53
C6	126	35	5.05						
C7	151	35	4.56		Total 5,583 or 207 cu. yds.				
C8	137	35	4.72						

(1) Degrees from vertical

400 psi limit of the Simplex pumps, they were replaced by two high-pressure, air-driven cementation pumps. These pumps, exerting grout pressures up to 1500 psi, provided the lifting force required to move the headframe from 8.9 in. out of plumb south and 4.4 in. west to 0.9 in. south and 0.4 in. west, respectively.

Grout mixes for the curtains and the 5 deep holes varied from neat grout using 5-1/2 to 11 gal. water per sack of cement to batches containing varying amounts of mason sand up to 0.75 cu. ft. per sack of cement with 5.6 gal. water per sack. Pumping on any given hole normally started with one or more batches of thin neat grout, then progressed to thicker mixes including sand until pressures rose or the desired quantity had been placed. As areas tightened up, less and less sand was used; in some instances holes would accept only limited

TABLE 3

VOLUME PUMPED INTO DEEP HOLES

HOLE NUMBER	VOLUME CUBIC FEET	DEPTH FEET
B2	1614	90 - 120
B3	32	55 - 57
B4D	302	50 - 64
B4DA	1088	63 - 93
B6	399	58 - 80
Total	3435 or 127 cu. yds.	

TABLE 4

VOLUME PUMPED INTO LIFTING BASE

HOLE NUMBER	VOLUME CUBIC FEET	DEPTH FEET	ANGLE(1)
01	97	32 - 35	
02	94	32 - 35	
03	96	32 - 35	
04	96	32 - 35	
05	97	32 - 35	
06	99	32 - 35	
07	98	32 - 35	
08	96	32 - 35	
09	96	32 - 35	12
010	96	32 - 35	17
011	96	32 - 35	21
11	100	32 - 35	
12	96	32 - 35	
13	40	32 - 35	
14	20	32 - 35	
15	96	32 - 35	
X15	96	32 - 35	
Total	1,509 or 56 cu. yds.		

(1) Degrees from vertical

TABLE 5

VOLUME PUMPED FOR LIFT HOLES

HOLE NUMBER	VOLUME CUBIC FEET	DEPTH FEET
11	----	32
12	721	32
13	815	32
13X	346	20
14	186	32
X15	----	32
01X	32	20
01	----	32
02	818	32
02X	959	20
03	262	32
03X	1145	20
03XX	950	20
04	1847	32
04X	737	20
05	119	32
05X	1202	20
05XX	119	20
06	564	32
06X	707	20
07	89	32
07X	548	20
08	215	32
08X	308	20
09	1782	32
09X	1952	20
010	392	32
010X	319	20
011	----	32
Total	17,134 or 635 cu. yds.	

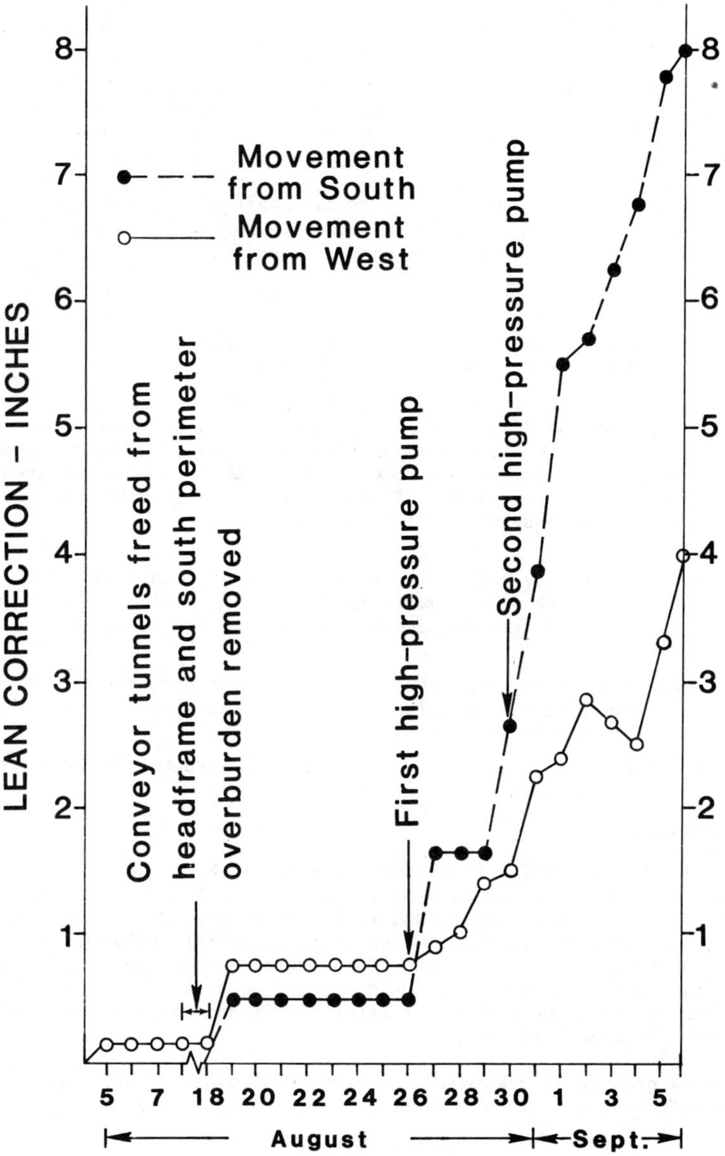

Fig. 7 Movement of Top of Headframe during Lift Grouting

quantities of neat grout. To provide additional assurance against escape of lifting grout at the northeast corner (Holes 11XX, 12XX, and 13XX, Fig. 6), 110 gal. of sodium silicate was used in proportions, per batch, 10 gal. silicate, 30 gal. water, and 1/2 sack of cement.

For lifting, pozzolan, Chemix, and Intrusion Aid were included in the mixes and sand omitted to increase penetrability and retard the setting time. Pozzolan was batched at the rate of 1 sack (70 lbs.) per 2 sacks of cement; Chemix at 4 lbs. per sack of cement; and Intrusion Aid at 2-5/8 lbs. per 3 sacks of cementing material (cement plus pozzolan). All grout was mixed in double tub, vertical shaft mixers made and furnished by The Prepakt Concrete Company except that a single tub mixer furnished grout for one of the high-pressure pumps.

CONCLUSIONS

1. Heavy, tall structures that have settled unevenly can be lifted and righted by pressure grouting procedures.
2. The procedure usually involves the establishment of a grout curtain around the portion of the structure to be lifted followed by the injection of retarded cement-base grout at fluid consistency under the structure, preferably at some depth to spread uplift stresses and generally into the weak soil zones that permitted the movement to prevent recurrance of settlement.
3. A knowledge of the soil characteristics and the horizons of types of soil is necessary. This is preferably done by soil sampling in advance of the work although, in many cases, a skilled contractor's superintendent and crew can adequately size up the situation as the inserts are driven.
4. An engineer must determine that the base of the structure is adequate for the lifting operation, increasing the size of the base if needed to provide adequate lifting area.
5. Auxiliary structures must be cut free of the structure to be raised to permit the movement.

ACKNOWLEDGMENTS

The work on the mine hoist headframe was accomplished under the general direction of Arthur H. Ditto, Sr., Project Engineer for the Anaconda Copper Company, owner, and C. O. Smith, Vice President, The Prepakt Concrete Company, grouting contractor. Cliff Kullberg was project superintendent for the grouting contractor; his daily job reports provided much of the information relied upon in the preparation of this paper. The cooperation of Peter Kiewit and Sons, shaft sinking contractor, is acknowledged.

The authors gratefully acknowledge the cooperation of Kathryn Akers McCartney for her graphic art work.

APPENDIX

APPENDIX I.--REFERENCES

1. Railway Engineering and Maintenance, May 1947, Simmons-Boardman Publishing Company.

2. Spotlight on the World of Prepakt, Summer 1974.

3. Spotlight on the World of Prepakt, Summer 1975, and Contractor's private records.

APPENDIX II.--CONVERSION FACTORS

To Convert	To	Multiply by
inches (in)	millimeters (mm)	25.40
feet (ft)	meters (m)	0.305
square feet (sq ft)	square meters (m^2)	0.093
cubic feet (cu ft)	cubic meters (m^3)	0.028
pounds (lb)	kilograms (kg)	0.454
one pound force (lbf)	neutons (N)	4.45
pounds per square inch (psi)	kiloneutons per square meter (kN/m^2)	6.9
gallons (gal)	cubic meters (m^3)	0.0038
bushel (bu)	cubic meters (m^3)	0.0036

- - - - - - - - - - - - - - - - - -

Sack, portland cement	94 lbs or 42.6 kg
Sack, fly ash (approx.)	70 lbs or 32 kg
Bushel, corn (approx.)	56 lbs or 25.4 kg

GROUTING METHOD BY USING THE FLASH-SETTING GROUT

By T. Tokoro, S. Kashima and M. Murata*

ABSTRACT: In Japan, most of big cities are located on a poor subsoil consisting of alluvium deposits complicatedly stratified and abundant in ground water. The soil stabilization by chemical grouting is indispensable for the construction of underground structures in such cities. However, any conventional grouting method using a grout with a long gel time has some inconveniences. For example, it is difficult to prevent the escape of injected grout through a stratum boundary or its dilution and diffusion caused by ground water. Thus, many problems in terms of soil stabilization effect and environmental pollution control have so far remained unsolved. Considering that the use of a grout having a very short gel time would be most effective to solve these problems once for all, we have successfully developed a new flash setting grout as well as new equipment, such as injection pipe, tip arrangement and pump, capable of injecting this new grout safely and continuously. This has made it possible to stabilize only a target soil zone almost independently of the soil conditions.

1. INTRODUCTION

In big cities in Japan, the boom of underground construction projects for line extension of national railroads, subways and the like is over. Now, instead, the construction of sewer lines and underground telephone and power lines is increasing in volume.

These underground lines are often laid in built-up areas where factories, offices, dwelling houses and other existing structures stand close together. In addition, there is a constant increase in road traffic volume. Under these circumstances, the tunneling is carried out by shield method for the most part, and the soil stabilization becomes increasingly indispensable to prevent the effect of the tunneling on the surrounding structures and roads.

Most cities in Japan, including Tokyo and Osaka, are located on alluvium deposits in river reservations, that is, poor stratified subsoil constituted by a complicated formation of clay, silt, sand, gravel and other materials. In the ground of this nature, the water table is encountered at a very high level. In tunneling in such ground, it is inevitable to wrestle with gushing water and risk of cave-in. Thus, the soil stabilization constitutes one of the most important problems for tunneling.

As a method of soil stabilization, chemical grouting has so far been widely used and appreciated in a general way. However, in such stratified ground as mentioned above, it has some inconveniences, such as return of chemical grout around the injection pipe or intensive outflow of injected grout to the ground surface through loose ground

*LAG Assoc. of JAPAN

zones. In addition, the chemical grout injected into such ground is susceptible to dilution and diffusion due to the ground water. For these reasons, it is so far been considered very difficult to ensure the positive soil stabilization over a predetermined zone in this nature of ground by the chemical grouting.

[Details of Application of the Chemical Grouting Technology in Japan]

In Japan, the chemical grouting was put to practical application toward 1955 by using a single-pipe rod provided on its top with a Y-piece. Since then, this single-pipe rod method has prevailed as the most common grouting method in Japan: two different liquids are subject to the contact mixing in the top of the single-pipe rod and the mixture thus formed is injected.

Initially, a grout consisting of water glass with sodium aluminate added as hardener was used. This grout had longer gel time (1 to 5 min) and higher strength than a grout then well-known (grout invented by Joosten) which consisted of a combination of water glass and aqueous solution of calcium chloride. Subsequently, a water-glass based grout using cement milk as hardener and other various types of water-glass based grouts were developed and put to practical use. Then, early in 1960, high-polymer type grouts, such as acrilamide type, urea type and urethane type, were successively developed. As a result of development of these typical grouts, the applications of the chemical grouting technology were diversified widely. Thus, this technology won popularity.

However, as the ground conditions became increasingly delicate and severe for grouting, the mere combination of a single-pipe rod with a grout having a gel time of several minutes often failed to give a satisfactory grouting effect in spite of injection of a very high performance grout. To cope with this situation, various measures have been tried and taken since toward 1970. These measures include use of a packer to prevent the return of the grout around the injection pipe, shortening of the gel time to prevent the grout dilution and diffusion caused by ground water, and combined injection of different types of grout into a complicated formation of soils. In general, however, all of these measures suppose very often a gel time ranging between tens of seconds and several minutes. Such gel time is sufficiently long for the grout to flow out, while it is too short for the grout to penetrate thoroughly into fine voids. A solution to this problem of gel time must be worked out.

[Appreciation of European Grouting Methods in Japan]

Though information on grouting methods put to practical application in Europe was made available, no indications of adoption or attempt of adoption of such methods were shown before the end of 1960's. This is partly because the trend at that time in Japan was to avoid the use of large equipment and complicated operational procedure, with importance attached to the simplicity.

In 1970's, as the simple grouting method calculating only upon the performance of the grout was reconsidered, European methods following closely the fundamentals and ensuring very careful work came gradually under review. In fact, efforts to put an European method to practical use in Japan were made by a certain company on the basis of the introduction of technology from France. As a result of these efforts, this European method has given many satisfactory results and has been highly appreciated.

However, there remain some problems to be solved: (1) it is still very expensive if the grouting effect with high accuracy is desired; (2) dilution and diffusion of grout may occur in a ground particularly abundant in ground water; (3) "manchette" tubes left in the ground after completion of the grouting work may impede the subsequent shield tunneling operations. Because of these problems, this method has not yet been popularized in Japan.

[Reconsideration after Occurrence of Environmental Pollution Accident]

In 1973, an accident of environmental pollution due to injection of a high-polymer type grout occurred in the Kyushu region in Japan: the grout penetrated into drinking water wells and people using these wells suffered from the contaminated water. This accident revealed two problems: harmfulness of the high-polymer material itself and thoughtless execution of the grouting work in the immediate proximity of drinking water wells.

In consequence of this accident, the Ministry of Construction published in May 1974 the "Provisional Guideline on the Execution of Construction Works using the Chemical Grouting" to lay down severe restrictions on the use of high-polymer based grouts. Thus, the use of grouts was practically limited only to water-glass based grouts.

Such restrictions on the use of these typical grouts gave naturally a heavy blow to the Japanese construction industry because they narrowed the scope of selection of grouts. On the other hand, however, they provided opportunities for proceeding seriously with the development of effective grouting methods applicable under the ground conditions so far considered too delicate for application of the chemical grouting technology and also for reviewing some unfinished studies for development of new grouting methods so far suspended.

Possible causes of pollutive effects of the chemical grouting on the surrounding environment include (1) direct inflow of the grout into the service water, (2) diffusion into the ground of the grout while being diluted by ground water and (3) dissolution of soluble constituents in the ground water after complete setting of the grout. In particular, point (1) constitutes a major cause of environmental pollution of the chemical grouting. And, no doubt, it is also a major cause of reduction in efficiency of grouting. In view of these facts, it is natural that a grouting method permitting to confine the grout in a predetermined zone has eagerly been desired from a point of view of both pollution control and economy (maximum effect with minimum material).

Backed by the above-mentioned circumstances, we actively proceeded with research and development activities on the basis of our technologies so far accumulated concerning grout, equipment and working procedure. Through various experiments, we successfully developed an epoch-making technique called "LAG method" in 1975. Outline and example of practice of this LAG method are presented in the following.

2. DEVELOPMENT OF LAG METHOD AND ITS FEATURES

[Properties and Effect of Flash-Setting Grout]

- Gel Time and Strength -

In practice, this LAG method requires the use of a grout of which gel time is as short as two or three seconds. The reasons of the use of a grout with such a short gel time are summarized by the following three points:

(1) In drilling or injecting, the injection pipe may be loosened. To fill the grout quickly around such loosened injection pipe so as to form a grout packer while continuing the injection, it is advantageous to reduce the gel time of the grout to minimum.

(2) Real ground receiving the grout is often constituted by complicated multiple beds having different coefficients of permeability. To prevent the grout from flowing out through loose beds or very porous masses, it is advantageous to reduce its gel time to minimum.

(3) The grout may be diluted by ground water. This will result in a large increase in gel time. To prevent the diffusion or outflow of the grout caused by this phenomenon, it is advantageous to reduce the gel time of the grout to minimum. (See Fig. 1).

For water-glass based grouts, many types of hardener are available: inorganic hardeners (such as phosphates, bicarbonates and bisulfates), organic hardeners (such as glyoxal and ethylene carbonate) and combinations thereof. However, such hardeners must satisfy the following conditions: (i) a short gel time must be ensured in a stable manner even under a change of temperature; (ii) even if the grout is diluted by the ground water, any extreme increase in gel time must not occur; (iii) the grout must have a sufficient strength to work as a packer. After careful studies on these conditions and others, two types of hardener have been put to practical use: combination of bisulfate and sulfate (Grout①) and a sort of bicarbonate (Grout ②). (See Fig. 3).

It proves, on the other hand, that all grouts having a short gel time are not always useful for the LAG method. In general, the gel time means a time interval between the mixing of a grout and the moment when it loses its fluidity, and not the moment when it gets its hardness. In other words, even after its gel time, the grout injected remains weak until it exhibits a certain strength, and the succeeding grout injected tends to penetrate in a concentrated manner into this weak grouted portion. This phenomenon is undesirable because it leads to a maldistribution of the grout. Therefore, in particular in a ground having loose cohesive soils intercalated or in a sand bed having a high content of fine grained soils, the quicker the grout exhibits its strength after gel time, the greater confining effect it has.

As shown in Fig. 2, the strength of flash-setting grouts now used is only less than 1 kgf/cm^2 as measured on homogeneous gels prepared for the bench tests. In the field, however, their strength after complete setting is of the order of 2 to 5 kgf/cm^2 in sandy masses. The actual results of grouting so far obtained show that this strength is generally sufficient to prevent the seepage of ground water and the cave-in of tunnels.

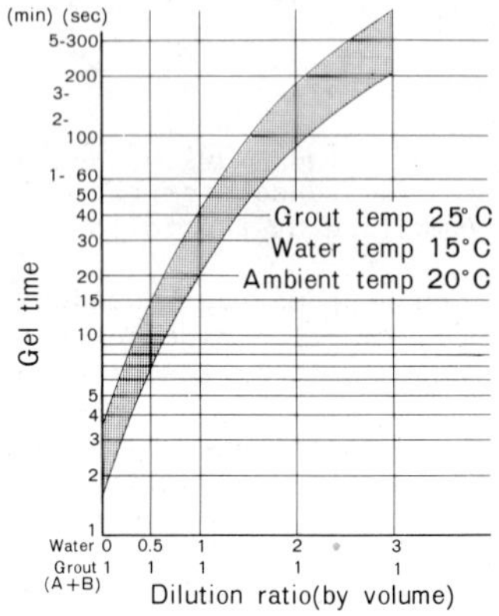

Fig. 1 Gel time of grout diluted with water

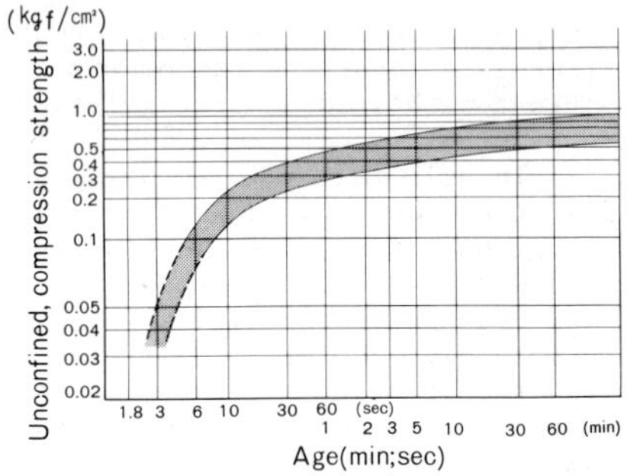

Fig. 2 Unconfined compression strength of homogeneous gel

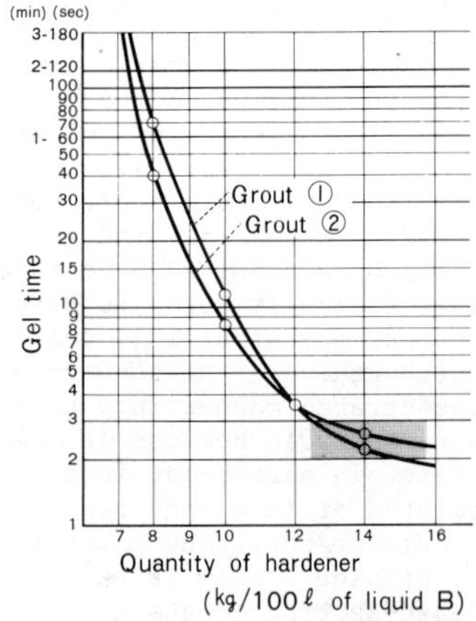

Fig. 3 Relation between content of inorganic hardener and gel time of water-glass based grouts

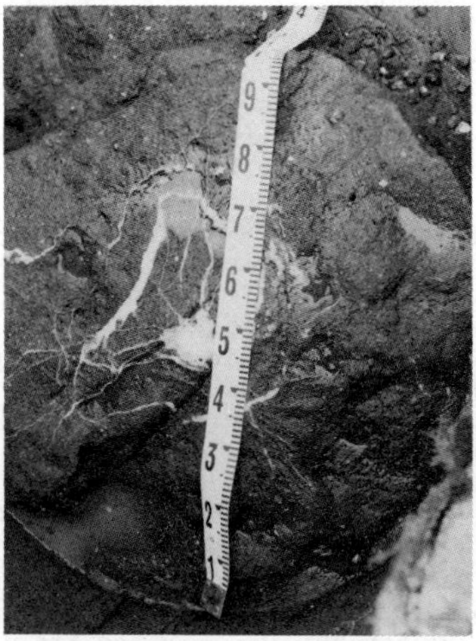

Fig. 4 Artificial soil mass stabilized by the flash-setting grout

- Penetrability of Flash-Setting Grout -

In the issues of April 1961 and January 1968 of the Journal of the Soil Mechanics and Foundations Division (ASCE), R. H. Karol et al. reported the results of experiments on gel times and penetration of grouts and derived the conclusion that a grout of shorter gel time would be more effective to obtain uniform penetration of grout into stratified deposits presenting permeability differences.

In these papers, they justify experimentally the following theory: "the grout injected first flow into more permeable zones, while hardly penetrating into less permeable zones. When these start gelling, the grout currently being injected is forced into less permeable zones. With a grout of short gel time, this procedure is smoothly repeated without loss so that the uniform penetration of the grout is obtained."

In their paper of 1968, they present in detail also the results of experiments on penetration patterns of grouts having a short gel time, giving the following description:

"As pumping begins, the grout of very short gel time is displaced radially in three dimensions from the opening in the pipe. The leading surface of the grout mass forms a sphere..... When gelation occurs, it starts at an infinitely thin shell..... As pumping continues, a finite number of channels are ruptured in this thin, incipient gel; through these channels, grout continues to flow as pumping continues..... Much of the grout trapped without the initial thin shell also gels to form a fairly thick shell containing a finite number of open channels..... The remainder of the fluid grout within the shell is forced out through the open channels as pumping continues. At the point where each of these channels come out of the initial shell into unstabilized soil, three-dimensional radial flow again begins. In this fashion, hemispheres begin to grow on the surface of the initial sphere....., and the entire process will be repeated."

The above results were obtained from the experiment conducted for a grout AM-9 (acrylamide) having a gel time of 30 seconds and at a pumping rate of 0.5 liters/min. These conditions are somewhat different from those of our water-glass based grouts, which have a gel time of 2 or 3 seconds. However, stabilized masses obtained by the LAG method in both laboratory and field show the same typical soil stabilization patterns for short gel times as those observed in the experiments of R. H. Karol et al. (See Fig. 4).

[Equipment Used and Functions thereof]

To develop equipment permitting to safely and continuously injecting into the ground without trouble such grout with a gel time of 2 or 3 seconds as described above, we carried out studies for a considerably long time. As a result of these studies, a series of devices, such as injection pipe setting machine (boring machine), agitator, injector, separate-flow pressure control device, swivel, double pipe and tip arrangement, were successfully put to practical use as special equipment having functions required for application of this LAG method.

In particular, special improvements were made on the injection pump and special functions were given to swivel, double pipe and tip arrangement. Some detailed description of these four devices is given below:

- Injection Pump -

Injection rate of grout has a close relation to the injection efficiency. If the injection rate is reduced, the soil stabilization can be made more uniform under a lower injection pressure. This means that the injection pump must satisfy especially the following conditions: (1) in a region of low injection rates, the equivalence between two liquids must be ensured and (2) even when a high pressure injection is required or when a difference in pressure between two liquids occurs, the equivalence between these liquids must be ensured.

In order to meet these conditions, a plunger pump was adopted and the following improvements were made with a view to further increasing its performance:

(1) To minimize the pulsation during operation and to ensure the equivalent discharge of two liquids, two sets of plungers were incorporated in one crankshaft.

(2) Both suction and discharge valve chests were provided with a volume substantially equal to the cylinder displacement caused by the reciprocation of plungers.

(3) The grout passage in the pump was made of an acid-proof material.

The injection pump successfully developed through these improvements and others can withstand a pressure of 50 kgf/cm^2 in normal service, and its delivery efficiency can always be kept at 93% or more, almost independently of the pressure variation (Fig. 5). In addition, it can well withstand the continuous service.

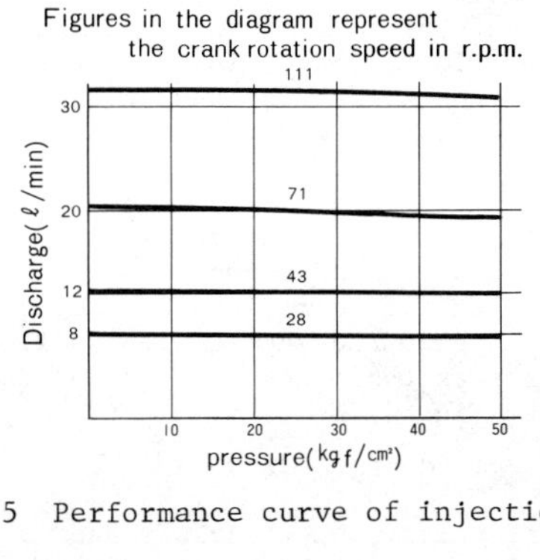

Fig. 5 Performance curve of injection pump

- Swivel and Injection Pipe -

The swivel mounted on the top of the injection pipe is a "key" device to connect two liquid feed hoses with one double pipe with both hoses completely separated from each other. In addition, this swivel is essential for permitting the injection pipe to freely rotate during boring and grouting. These functions of the swivel must effectively be filled even when the internal pressure of the pipe reaches about 40 to 50 kgf/cm^2 as a result of liquid feed.

Through studies on material and shape of packings for keeping inner and outer passages independent of each other as well as on the bearing providing the rotation, we succeeded in putting to practical use a swivel capable of filling such functions effectively.

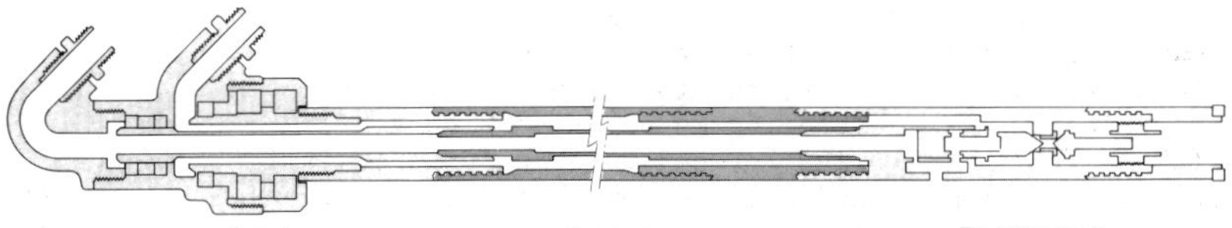

Fig. 6 Assembly drawing of injection pipe

The double pipe used for the LAG method must be enough strong to resist the pressing force applied from the surrounding ground as a result of jamming phenomenon occurring during boring or development of the injection efficiency. In addition, assembly and disassembly of double pipe sections must be easy, and two passages in the assembled double pipe must always be kept completely independent of each other. For this purpose, the double pipe was designed to permit to make a connection by positioning the inner pipe in the coupling with the aid of a snap ring on one double pipe section and by sealing the joint with two O-rings on the other double pipe section.

- Tip Arrangement -

The tip arrangement is an end device from which the grout is pumped out into the ground. In case of this LAG method, two liquids are discharged from this tip arrangement while being combined and mixed together therein. Since the LAG method uses a flash-setting grout of which gel time is only 2 or 3 seconds, the tip arrangement must meet the condition that the grout passage cannot be blocked after mixing of such flash-setting grout. In addition, it must satisfy the following four conditions:

(1) It must permit to determine and monitor the injection pressure with precision.

(2) It must permit also the horizontal injection and the upward injection with the aid of a non-return device.

(3) It must be of such construction that two liquids can be kept completely independent of each other till just before discharge.

(4) It must be of such construction that two liquids are discharged from confined ports with a view to increasing their flow velocity before their conflux and mixing.

We have so far developed four types of tip arrangement satisfying these conditions. Functions and characteristics of one of them are described below.

In grouting, the boring is first carried out by discharging water from the tip end through the outer passage in the double pipe. When the injection pipe reaches a predetermined position, the pumping of water is stopped. Then, the grouting is commenced by pumping the base component of the grout through the outer pipe passage and the hardener through the inner pipe passage. In this case, the valve placed in the inner pipe is pressed down to open the port discharging to the Point D while closing the Passage C (Fig. 7). At the same time, the port of the Passage A (base component passage) which discharges toward the Point D is opened. In this way, liquids A and B (base component and hardener) are combined together at the Point D.

In this tip arrangement, the following provisions are made to ensure that the above operation is positively done:

(1) A clearance ranging between 0.01 and 0.02 mm is left between valve outside diameter and cylinder bore of the tip arrangement so that the valve can smoothly operate and can withstand the repeated operation. In addition, a difference in hardness is provided between valve and cylinder bore.

(2) The valve is supported by a spring and the diameter of the port discharging from the Passage B to the Point D is fixed at 2.2 mm, so that when the pressure in this passage is raised to 3 kgf/cm^2 or more, the spring may positively be contracted to allow the valve to close the Passage C.

(3) The diameter of the orifice discharging from the Passage A to the Point D is fixed at 5 mm so that when the base component liquid is pumped through this passage at the rate of 6 liters/min, the pressure therein is kept at 0.5 kgf/cm^2 or less.

(4) A check valve is provided in the tip end to prevent any soil from being forced into this arrangement through the tip end during injection of grout.

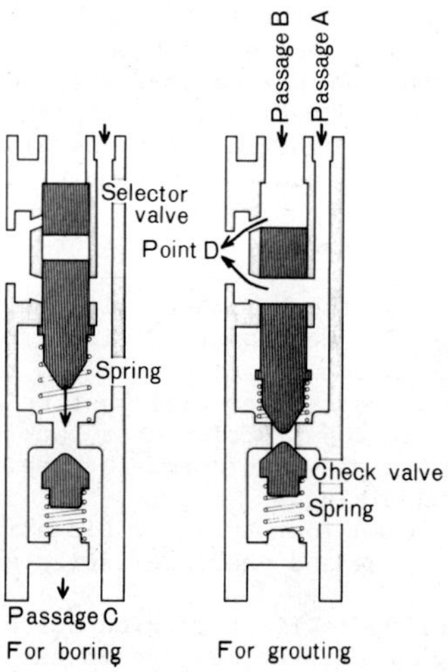

Fig. 7 Schematic of operation of tip arrangement (example)

[Grouting Procedure]

Virtually no standard grouting procedure was established heretofore. According to the conventional practice, the zone of soil mass to be stabilized, the type of grout to be injected and other essential points were carefully determined at the engineering stage, but the field grouting work was entirely given in the hands of field operators, who judged the grouting conditions and took necessary steps only by observing the injection pressure. With such practice, the effect of soil stabilization depends largely upon knowledge, skill, judgement and other conditions of field operators. This implies that, now when the grouting work is carried out at numerous sites as in Japan, this practice cannot give satisfactory results.

In view of this fact, we have established the following grouting procedure so that any field operators having no special knowledge and skill concerning the grouting can carry out the grouting work in a satisfactory manner under almost all ground conditions and working conditions only if they have a certain knowledge on equipment and grouts:

(1) Discharge rate of grout: liquid A + liquid B = about 10 lit./min.

Since late in the 1960's, the construction projects have been increasing in scale and the improvement of operation efficiency of construction works has increasingly been required. Under these circumstances, also for the grouting works, energies, have been bent on the improvement of operation efficiency. As a result, it has become popular to execute the grouting work by injecting grout at as high rate as 20 to 30 liters/min per injector unit. It is however evident that a lower injection rate of grout leads to better results of soil stabilization by chemical grouting. On this account, an injection rate equal to 1/2 to 1/3 of the conventional one has been adopted for the LAG method with stress on the soil stabilization effect and at some sacrifice of the operation efficiency.

(2) Rotation speed of injection: about 8 r.p.m.

In case of conventional grouts having a long gel time, the rotation of an injection pipe accelerates the return of the grout around it. This, combined with the inevitable pressure loss, necessarily leads to a considerable reduction in injection efficiency, or soil stabilization effect. However, it proves that, in case of a grout having a gel time of only 2 or 3 seconds, the distance of its return around the injection pipe provoked by the rotation of this latter at a low speed is only about 10 cm from the discharge port of the pipe in sand masses. With a view to form grout packers continuously and automatically by making use of this phenomenon of slight grout return and also to prevent such impossibility of withdrawal of the injection pipe as may be caused by a strong pressing force applied as a result of development of the injection efficiency, it has been decided to rotate the injection pipe at a low speed at all times during injection of grout.

(3) Sequence of injections: sequence as shown in Fig. 8.

In conventional grouting methods using grouts having a long gel time, it is normal that the outer zones of a soil mass to be stabilized are first grouted and injections into the inner zone is carried out only after completion of injections into the outer zones with a view to minimizing the deviation beyond the limits of the soil mass to be stabilized. On the other hand, for the LAG method, the sequence of injections from one side to the other side or from the central zone to the outer zones of a soil mass to be stabilized has been adopted to facilitate displacement of the ground water in that soil mass toward the outside of it. It has been proved that this sequence of injections is most effective and suitable to obtain the uniform soil stabilization throughout the predetermined zone without any omission of injection.

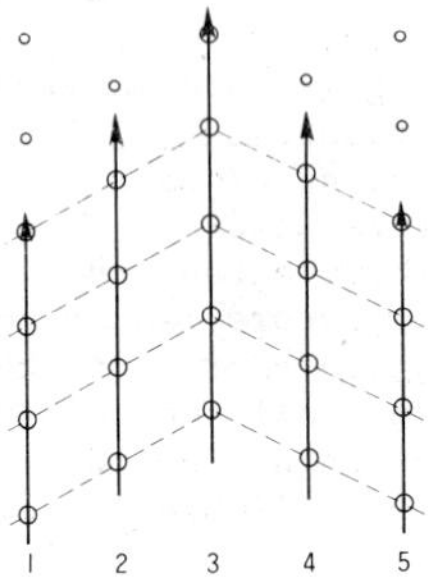

Fig. 8 Sequence of injections

3. EXAMPLE OF PRACTICAL APPLICATION OF THE LAG METHOD AND EFFECT THEREOF

The LAG method has already been practically applied to over 500 construction projects, including experimental grouting work projects and small-scale construction projects. In this section, an outline description of its practical application to the underground telephone line construction project in Tokyo is given together with a summary of such data on injection efficiency of this method as obtained from various actual field grouting works.

[Example of Practical Application]

- General Description of the Works -

The LAG method was practically applied to the works for laying telephone cables at a depth of about 27 m under the ground level. For tunneling, the shield method was adopted. As shown in Fig. 9, the ground of the site was characterized by the alternation of sand, sandy silt, silt and other materials. Such alternation extended from the ground surface down to the top of the shield. The N value of this ground was 0 to 15. In particular, the zone subject to the shield tunneling was located in a fine sand bed abundant in ground water, where the pore water pressure was as high as 2.3 kgf/cm^2. Under these conditions, it could not be expected that, in tunneling, the facing would be fully self-supporting. On the contrary, there were possibilities of water gushing and caving. It was therefore planned to stabilize the soil by chemical grouting.

Initially, the chemical grouting for soil stabilization was carried out with liquid type grout of high permeability (with gel time of several minutes) and single-pipe rod. In tunneling, however, water gushing and facing caving occurred repeatedly though the air pressure in the shield was kept at 1.8 to 1.9 kgf/cm^2. It was thus very difficult to carry out the works. Then, the grouting method was re-examined and it was decided to replace the initial method by the LAG method.

- Result of Application of the LAG Method -

Thus, the grouting was carried out by the LAG method over a tunneling length of about 300 m. The grout injection was carried out in accordance with the standard procedure for the LAG method:

- Discharge rate: 12 liters/min.
- Rotation speed of injection pipe: 8 r.p.m.
- Withdrawal speed of injection pipe: 29 to 32 min/m
- Type and gel time of grout used: grout with inorganic hardener, with gel time of 2 to 3 seconds
- Injection ratio: 40% (400 liters/m^3 of stabilized soil mass)

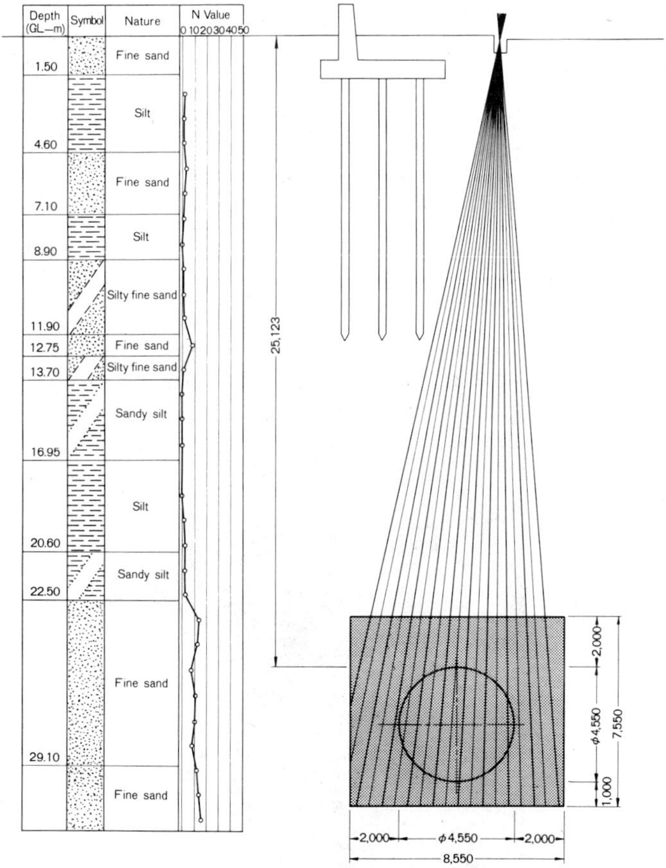

Fig. 9 Soil conditions and typical stabilized soil profile

After application of the LAG method, water gushing hardly occurred even though the air pressure in the shield was lowered to 1.3 to 1.5 kgf/cm^2. And, the facing was fully stabilized and made self-supporting. Thus, the tunneling work efficiency was remarkably increased. (See Fig. 10).

[Summary of Some Actual Data on Injection Efficiency]

Figs. 11 and 12 represent a summary of measured data of N value and coefficient of permeability before and after grouting, obtained from about 20 construction projects where the LAG method was employed.

Fig. 11 is a summary of results of standard penetration tests conducted before and after grouting. In this figure, the soils are classified into four categories. For each soil category, a frequency distribution of N'/N is represented in the form of a graph by defining the degree of soil stabilization as a ratio of the soil after grouting to that before grouting, that is N'/N, where N is the N value of the soil before grouting and N' is that after grouting.

This figure shows that the average value of N'/N is equal to 1.42 for sand ($N \geqq 30$) and 1.82 for fine sand ($N < 30$). This means that there is no remarkable difference in degree of soil stabilization between different natures of soils.

In Fig. 12, measured values of coefficient of permeability before and after grouting are plotted with values before grouting (k) and after grouting (k') taken on axis of abscissas and axis of ordinates. The solid line represents the relation between k and k' determined by the method of least squares.

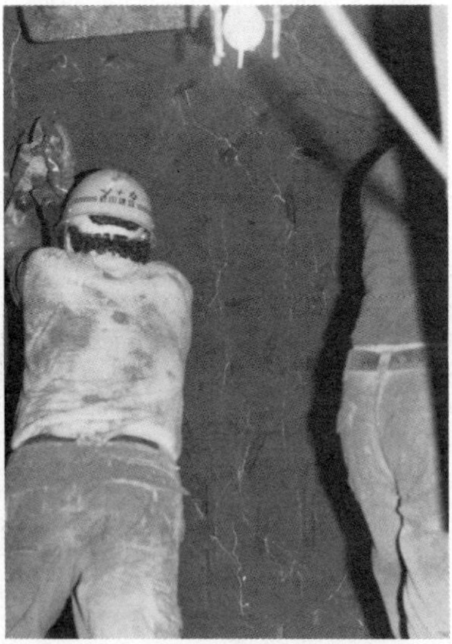

Fig. 10 Fully self-supporting facing being excavated, which has been stabilized by LAG method

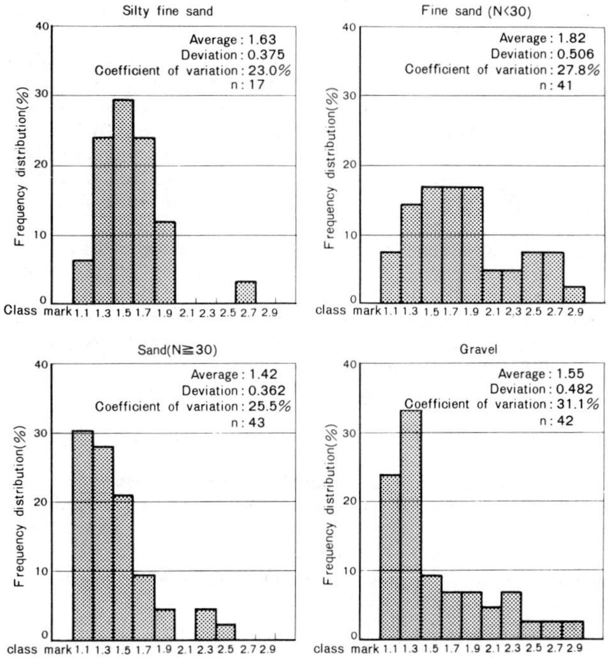

Fig. 11 Soil conditions and frequency distribution graphs of N'/N

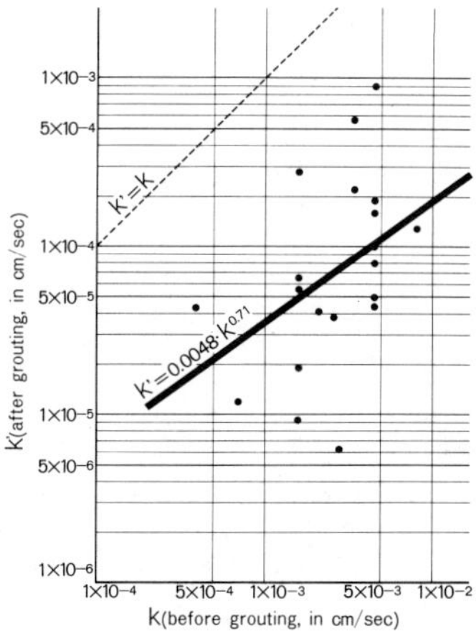

Fig. 12 Coefficient of permeability before and after grouting

4. CONCLUSION

The grouting method using a flash-setting grout is considered to be applicable under a very wide range of conditions only if equipment capable of positively pumping such grout into the ground is available. In addition, it is not too much to say that this grouting method is least influenced by various ground conditions, such as consolidation, soil formation, presence or absence of ground water and artesian condition of ground water, with most constant effect of soil stabilization being thus obtained.

Grouting methods based on this or similar concepts have been prevailing in the field of chemical grouting in Japan. Such methods have been appreciated for their ability of giving a highly reliable injection efficiency while minimizing the adverse effects on the surrounding environmental conditions. Their role in construction projects has been realized more and more profoundly.

REFERENCES:

1. R. H. Karol and A. M. Swift, "Grouting inflowing water and stratified deposits", Journal of the Soil Mechanics and Foundations Division - April 1961.

2. R. H. Karol, "Chemical grouting technology", Journal of the Soil Mechanics and Foundations Division - January 1968.

Matrix Evaluation of Structural Grouting of Rock

by Richard A. Millet, M.ASCE[1]
Richard L. Engelhardt, A.M.ASCE[2]

ABSTRACT

For an optimal grouting program, a minimum number of holes should be drilled and grouted to provide the required results. The use of the split-spacing technique does not optimize the number of grout holes necessary for a grouting program in that additional holes are drilled and grouted only for the purpose of evaluation. If the effectiveness of grouting can be evaluated prior to the final "proof" holes, such additional holes would not be needed.

By using a geologic model, which can generally be developed for any areas to be grouted, in conjunction with an evaluation matrix, the desired grouting results can be verified with a minimum of drilling and grouting of proof holes. The elimination of unnecessary proof grout holes in this manner reduces the overall cost of the grouting program by reducing both the number of holes to be drilled and grouted and the amount of time required to complete the grouting program but does not sacrifice positive engineering control of the results.

In a recent grouting program, solution cavities needed to be grouted below stable in situ rock to prevent stoping and

---

[1]Vice President, Woodward-Clyde Consultants, San Francisco, California

[2]Project Geologist, Woodward-Clyde Consultants, Wayne, NJ

further deterioration of the rock, and possible enlargement of the cavities. With a geologic model prepared for the area to be grouted and the desired engineering requirements established, an evaluation matrix was developed so that the anticipated drilling results and grout take could be compared to the actual drilling results and grout take. This approach facilitated the evaluation of whether adequate grouting had been completed and minimized unnecessary drilling and grouting.

INTRODUCTION

The purpose of this paper is to introduce a matrix evaluation approach to grouting programs. In this approach, an evaluation matrix is developed based on the desired engineering parameters (i.e., the required results) and the geology of the material being grouted. Conventional grouting evaluation uses unit-take and split-spacing as means of evaluation. These criteria are based on the same engineering parameters and geology as used for a matrix evaluation, with one important exception: split-spacing requires drilling and grouting of holes (proof holes) that are not required as part of the grouting program but are added only for the purpose of evaluation. Unit take is a part of matrix evaluation; however, evaluation is made of existing grout holes, not additional holes that would be required in a split-spacing evaluation.

To facilitate explanation of the matrix application, a specific example, by way of a case study, will be used. During a recent grouting program for the foundation of a nuclear power station, an evaluation matrix that compared both anticipated geologic conditions and grout take to actual geologic conditions and grout take was developed to determine the adequacy of the remedial grouting.

DEVELOPMENT OF THE GROUTING PLAN

For the preliminary exploration of the foundation areas for the nuclear power station, core borings were made at a spacing of approximately 100 ft center to center. These preliminary borings showed two things: (1) although the geologic formation at the site was described in the literature as an argillaceous domomite (relatively insoluble), the rock at the site contained near-surface layers of gypsiferous dolomite (relatively very soluble); and (2) cavities were found to exist in these near-surface layers in the foundation areas. To identify more precisely the location and extent of the solution activity, rock probes were drilled and geophysically logged. These rock probes were drilled at a spacing of approximately 50 ft center to center.

The results of the borings and rock probes and of a review of aerial photographs and additional information from the existing literature were combined to develop a geologic model of the site (see Figures 1 and 2). The geologic model was then evaluated with respect to the station's foundation requirement.

The results showed that some of the buildings are founded below the zone of solution activity and required no further foundation evaluation. However, some buildings were shown to be founded directly above zones of solution activity. A detailed evaluation of the location of these buildings in relation to the areal location of solution activity and the engineering parameters of the foundation was required. The results of this evaluation indicated that several buildings

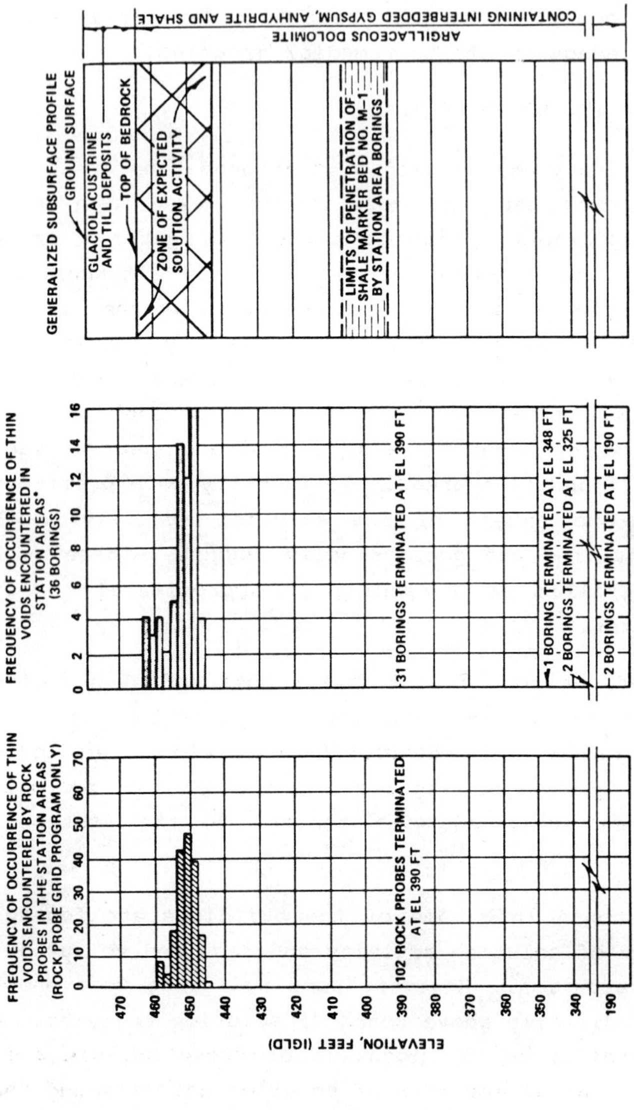

* IDENTIFICATION OF VOIDS IN STATION AREA BORINGS IS BASED ON RAPID DRILL PENETRATION RATES AND CONFIRMATION BY GEOPHYSICAL LOGGING. LOCATIONS OF RAPID DRILL PENETRATION RATES IN SOFTER SHALE OR SHALEY DOLOMITE LAYERS ARE NOT INCLUDED AS VOIDS

FIGURE 1. FREQUENCY OF OCCURRENCE OF VOIDS ENCOUNTERED VERSUS ELEVATION AND GENERALIZED SUBSURFACE PROFILE IN PROPOSED UNITS A AND B STATION AREAS

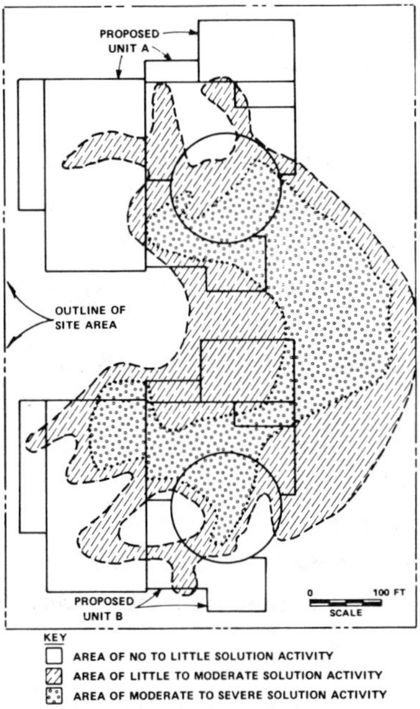

FIGURE 2. INTENSITY OF SOLUTION ACTIVITY

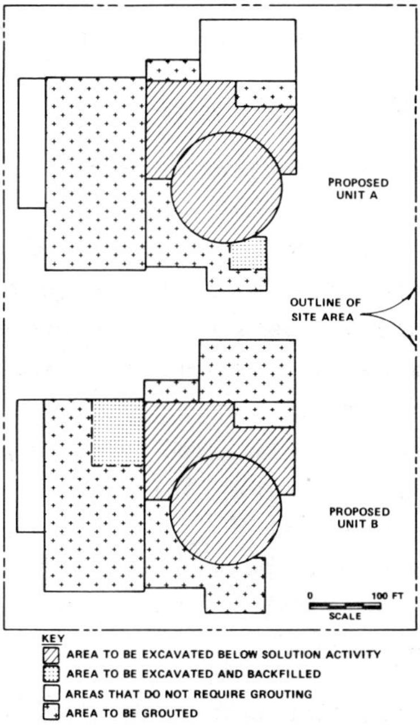

FIGURE 3. GROUTING PLAN

would require remedial treatment of the foundation rock. In areas where the top of bedrock had collapsed, the broken and weathered rock in the solution cavities would have to be excavated and the cavities backfilled with concrete (see Figure 3). In areas where the top of bedrock was intact, grouting of the solution cavities would be performed to prevent stoping and deterioration of the rock. On this basis, the conceptual grouting plan was developed.

THE GROUTING PLAN

Before grouting operations were started, a minimum number of holes (i.e., a maximum grout hole spacing) was determined. These holes were considered as primary grout holes. After the location of primary holes was determined, each hole was assigned a series number, which would indicate the sequence in which these holes would be drilled and grouted. In general, every other hole in the perimeter of a grout area was drilled and grouted and was considered as a series 1 grout hole (see Figure 4). When these holes were grouted, then the remaining primary holes in the perimeter were drilled and grouted and were considered as series 2 grout holes. Each building foundation area was considered separately.

After drilling and grouting of the perimeter holes was completed, drilling and grouting of the interior primary grout area began, and series number (e.g., series 3, 4, 5 and 6) were assigned to the grout holes (see Figure 4). Thus, no drilling was done within a distance of 15 to 20 ft from an already drilled but not grouted primary hole. This approach limited the possibility of prematurely grouting, by communication, an already drilled open hole. After all primary holes were drilled and grouted, the grout area was

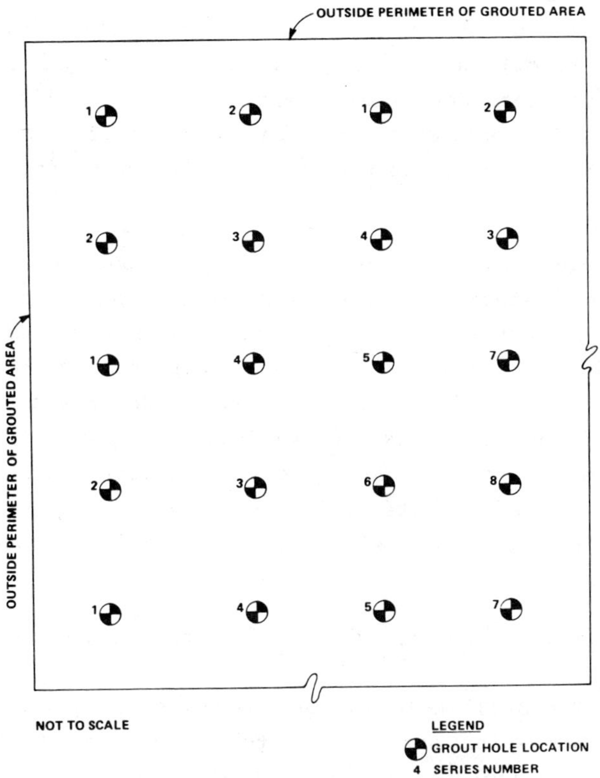

FIGURE 4. PRELIMINARY GROUT HOLE LOCATION PLAN

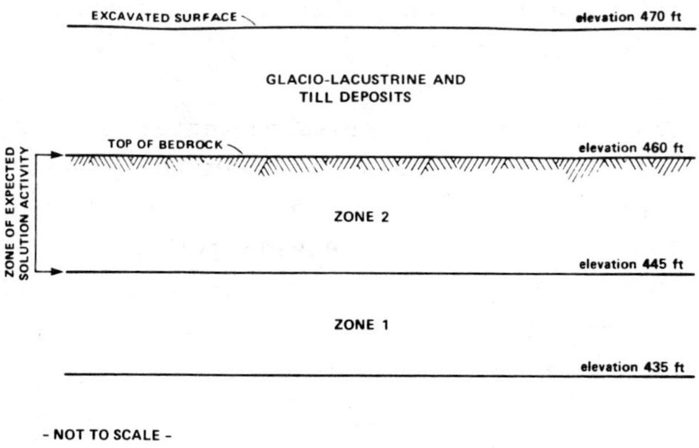

FIGURE 5. IDENTIFICATION OF ZONES OF GROUTING

evaluated based on the data collected during drilling and grouting of primary holes. If this evaluation (which is described below) determined that additional grouting was needed, then secondary holes were assigned. Following the drilling and grouting of the secondary holes, the grout area was re-evaluated and tertiary holes were assigned if necessary.

Grouting equipment included as many as four grout plants working in a station area at one time. Grout plants consisted of a double-tub arrangement with a Moyno displacement pump. Grout pressures were monitored at the grout hole using a calibrated pressure gage. Grout was injected into the foundation rock in two different zones. Inflatable pneumatic packers were used to isolate each zone in the hole. Zone 1 extended from el 435 to 445 ft, and Zone 2 extended from el 445 ft to top of rock (see Figure 5).

The following procedures were used in grouting each hole:

1. The drill hole was pressure-washed by inserting a hose-pipe arrangement to the bottom of the hole and applying compressed air and/or water. This process continued until the inspector decided that the hole was clean of all debris and rock cuttings.

2. The bedrock was pressure-tested by injecting water under simulated grouting pressure into Zone 1 and Zone 2, respectively, during a 5-minute interval. Based upon the water take, the starting grout mix to be used in each zone was selected.

3. Cement grout was injected under pressure into Zone 1 and Zone 2, respectively, until refusal was met. Refusal occurred when either a predetermined amount of grout was injected or a predetermined flow rate of injection was reached at the maximum allowable pressure.

Maxmimum allowable pressures used during grout injection were set based upon a judgment and were limited to 1 psi per foot of overburden (i.e., 25 psi and 10 psi for Zone 1 and Zone 2, respectively). Grout pressures were monitored at the injection point and by the use of in-situ heave extensometers. Grout consistency was varied during the injection process between a 5:1 to 1:1 (water:cement) mix by volume. The starting grout mix was determined by the results obtained from water pressure-testing of the grout hole. This mix was gradually thickened until refusal was met.

Grout takes were in general agreement with what was anticipated from the geologic model. The desired and anticipated trend of decrease in grout takes from primary perimeter holes to the interior primary and secondary holes respectively, was documented. The approach to the actual analysis of grouting is discussed below.

ANALYSIS OF GROUTING

As previously identified, the evaluation of grouting was to be based upon a matrix composed of the geologic model, actual grouting results, and the desired engineering results. Although considerable preconstruction subsurface data had been obtained and analyzed (see Figure 2), the drilling of the grout holes themselves was considered to be an opportunity to strengthen the understanding of the geologic model.

The most useful data gained from the grout hole drilling operations was the timing of each 1/2 ft of drill-steel penetration in bedrock. Upon review of the drill logs, zones of fast drilling were identified and defined as being a drill-steel penetration rate equal to or faster than 3 seconds per 1/2 ft. These zones of fast drilling represent the presence of solution cavities, soft shale layers, and, in some cases, variation in drilling procedures. Based upon previous studies and subsurface information at the site, zones of fast drilling below el 440 ft could generally be associated with soft shale layers and/or variations in drilling procedures. As grout hole drilling progressed, the anticipated occurrence of solution activity was compared with the actual occurrence (fast drilling) and spacial location, so that any anomalies could be quickly investigated and evaluated.

Using both the drilling information and grout take data, a matrix procedure was developed to provide a uniform basis for the evaluation of the effectiveness of grouting and for assigning additional holes for drilling and grouting. This procedure was based upon: (1) the cumulative length of fast drilling zones encountered during drilling of each grout hole; (2) the amount of grout solids injected in each hole; and (3) the sequence in which the hole was drilled and grouted. Using these factors, the evaluation matrix, presented in Figure 6, was prepared. As shown in this figure, the matrix evaluation of a grout hole for this project was set up using: (1) three categories of grout solid takes in the hole (0 to 0.5 ft^3/ft, 0.5 ft^3/ft to

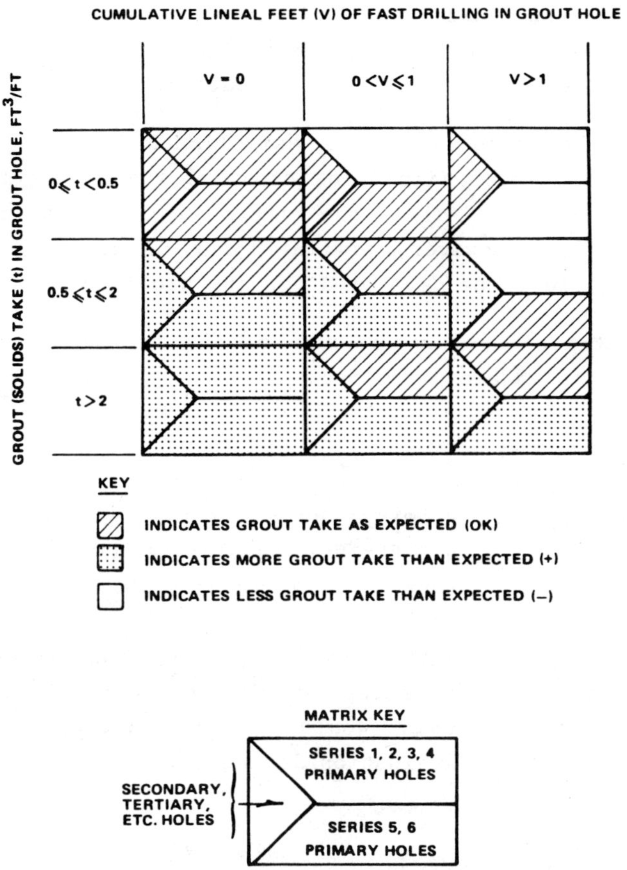

FIGURE 6. STRUCTURAL GROUTING EVALUATION MATRIX

2 ft^3/ft, and 2 ft^3/ft); (2) three categories of cumulative linear feet of fast drilling in the hole; and (3) identification of the sequence category for the hole.

These matrix categories were established based upon the site geologic model (including previous test grouting programs at the site). Thus, the matrix for a project must be site-specific. However, similar types of categorizations should be able to be developed for any grouting program.

To demonstrate the use of the matrix for this project, several examples are presented.

EXAMPLE A

Data: Primary Hole Series 1; grout take 0.2 ft^3/ft; cumulative linear feet of fast drilling in grout hole is 0.5 ft.

Results: Matrix (see Figure 6) identifies this as a case where the grout take is less than would be expected and additional evaluation is needed.

EXAMPLE B

Data: Tertiary Hole; grout take is 3 ft^3/ft; cumulative linear feet of fast drilling in grout hole is 2 ft.

Results: Matrix (see Figure 6) identifies this as a case where grout take is more than would be expected, and additional evaluation is needed.

EXAMPLE C

Data: Tertiary Hole; grout take is 0.5 ft^3/ft cumulative linear feet of fast drilling in grout hole is 0.5 ft.

Results: Matrix (see Figure 6) identifies this as a case where grout take is as expected.

The logic of the matrix is simple. As grouting continues, and if the grouting is achieving the desired results, the amount of voids (fast drilling) encountered in subsequent drilling and the amount of grout solid injected should be less. If other than an expected amount of voids or grout take was encountered, a detailed evaluation was performed. Specific review of anomalous grout holes included the evaluation of areal limits of solution-intensive areas, general performance of other near-by grout holes, the type and kind of grout refusal that was attained, and the elevation at which packers were set. In addition, information contained in drilling and grouting logs (e.g., inspector comments), previous subsurface information, reports of communication observed during drilling and grouting, and matrix evaluations of other adjacent grout holes were reviewed. During the review for this case study, three common site problems were identified:

1. Blockages in grout holes prevented the packer from being set at its proper elevation to grout Zone 1;

2. Holes were grouted by communication from another hole; and

3. Refusal criteria were unable to be attained because grout communicating with the ground surface (usually around the casing) could not be sealed off.

The first two problems were generally the result of bedrock conditions. The use of the series approach to drilling and grouting helped to minimize these two problems. The third problem was generally the result of drilling oversized holes in the overburden. This problem was minimized when holes were drilled with a smaller annulus between the casing and the hole walls.

A large-scale plan drawing of the areas being grouted was used during the evaluation to locate anomalous grout holes and to record data from the specific review (see Figure 7). This detailed evaluation plan drawing was then used in conjunction with the grout plan and the evaluation matrix to assign subsequent grout hole locations.

CONCLUSIONS AND APPLICATIONS

In the matrix evaluation of grout programs, the actual drilling and grouting information is compared to an anticipated matrix. The matrix is developed from the geologic model and the required grouting results needed to provide

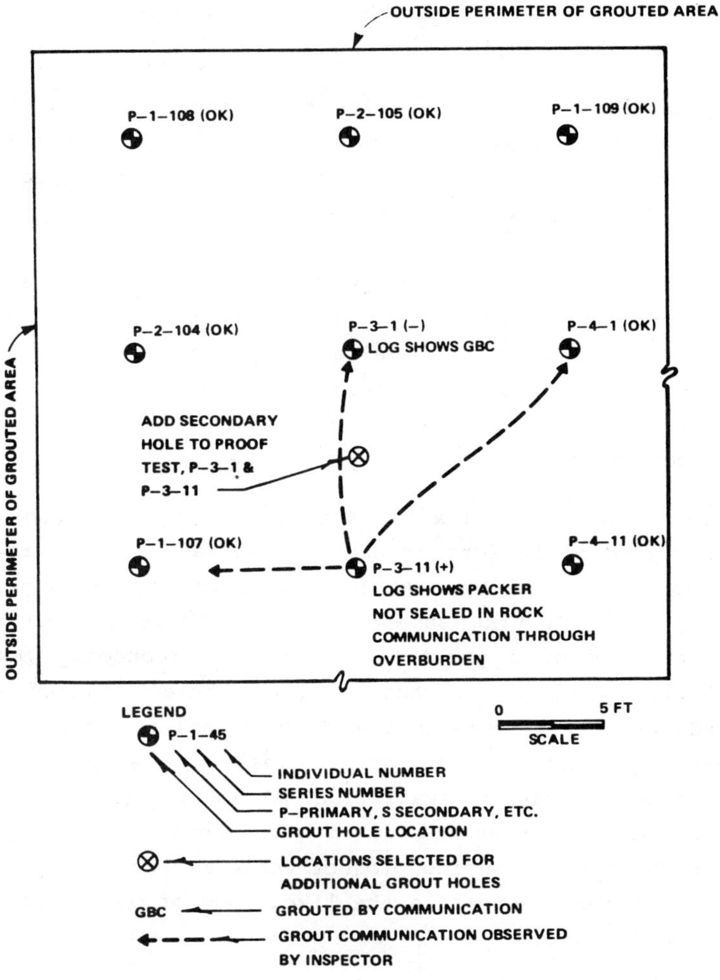

FIGURE 7. EXAMPLE OF LARGE-SCALE REVIEW DRAWING

an adequate foundation. In the case study given, over 1450 holes were evaluated in this manner.

The evaluation matrix quantifies the anticpated results of the grouting program. By using this method, anomalous grout holes are identified immediately through an objective review. Thus, anomalous holes may be given a more detailed evaluation in the same period of time that would normally be used to subjectively evaluate all grout holes. In addition, as with many major projects, client reviews and regulatory reviews are often required. By providing an objective plan in the form of the matrix, these reviews can take place in an orderly manner without having to interrupt the actual construction operations in progress.

The use of matrix evaluation should reduce the amount of time required to perform an ongoing evaluation of a grouting program. If the amount of time spent planning grout activity and selecting subsequent grout holes is reduced, indirect costs of a grouting project are reduced. Moreover, optimizing the number of grout holes drilled and grouted such that unnecessary "proof" holes are avoided reduces the direct costs of grouting. Therefore, the application of matrix evaluation to an ongoing grouting project has the potential for significantly reducing the total cost of the grouting project while increasing project control and thus increasing the likelihood of obtaining the desired end product.

ELECTRONIC MONITORING OF GROUTING

By Michael G. Jefferies[1], Brian T. Rogers[2] and
Denys W. Reades[3], A.M. ASCE.

ABSTRACT

The application of modern electronic techniques to cement grouting of dam foundations can yield a threefold advantage of lower construction cost, shorter construction time and better understanding of the reaction of the ground to the grout. The benefits of the former two advantages result in minimizing project initial cost. The third advantage offers the possibility of changing the process of grouting from an 'art' to a 'science'.

This paper outlines the application of electronic monitoring to the injection of cement grout into rock, and the advantages of this system over conventional monitoring of grouting. Equipment found satisfactory under field conditions typically encountered on a heavy civil engineering project is named to encourage other engineers to use electronics on their grouting projects.

In addition, a "first-order" means of interpreting subsurface conditions and the radius of influence of grout is presented.

ELECTRONIC TRANSDUCERS & DATA RECORDING

Conventional grout monitoring requires that the injection pressure be controlled and measured at the header assembly and that total volume injected over certain time periods be measured at the grout plant. To realize the benefits of electronic transducers, at least all of these parameters should be measured, and presented at a location such that grout injection control valves can be operated while observing both pressure and flow rate.

---

1 Snr. Civil Engr., Nova Corporation, Calgary, Alberta (formerly Golder Associates, Toronto, Ontario)

2 Geotechnical Engr., Golder Associates, Toronto, Ontario

3 Associate, Golder Associates, Toronto, Ontario

Pressure Transducer

Grout injection pressures can conveniently be monitored by attaching a pressure transducer to the injection standpipe via a gauge saver in exactly the same manner as one would attach a bourdon gauge. Two things should be borne in mind when choosing a transducer. First, small signals (such as mV) are not compatible with long cable runs required in civil engineering applications where there is a potential for signal corruption and other noise. Second, electronic components are not compatible with mud and water. Both requirements can be met by choosing a transducer which employs integral amplifiers and which is hermetically sealed. A suitable transducer for most grouting purposes is the Schaevitz P7430025-150A and one of these is illustrated on Figure 1. It should be noted that because this transducer is hermetically sealed, it measures absolute rather than gauge pressure. However, because atmospheric pressure changes are insignificant compared to the pressures and accuracy employed in grouting, this is not a limitation on the use of sealed transducers.

Flow Meter

Electronic transducers for volume measurement usually measure flow rate and then carry out the time integration to obtain volume. It turns out that having flow rate information available as well as total volume is extremely advantageous. In choosing a flow meter the same considerations of signal strength and equipment sealing as the pressure transducer applies. Ref. (1) suggests, and the authors concur, that the 'VERIFLUX' flow meter (manufactured by Kent Instruments) is suitable. This flow meter should be ordered with 'integrator card and counter' (for volume output). The flow meter is shown on Figure 2.

Processing

There are two considerations regarding processing the transducer signals. First, information must be presented to the grouting engineer in real time so as to maximize his control over the grouting. Second, the information must be recorded as a permanent record. One way of accomplishing these objectives is to bring both signals to a central location, such as adjacent to the injection control valves (as illustrated on Figure 3), and display each individual signal on a calibrated meter (simple marked up sticky labels on the face of a cheap multi-meter suffice).

A permanent record is made by recording each signal (pressure and flow rate) against time using a two channel chart recorder (Brush Gould 220) and in addition, this aids the grouting engineer as it provides a real time plot of the time rate behaviour of each parameter. However, the

FIGURE I PRESSURE TRANSDUCER FOR MONITORING OF GROUT INJECTION PRESSURES

FIGURE 2 FLOW METER

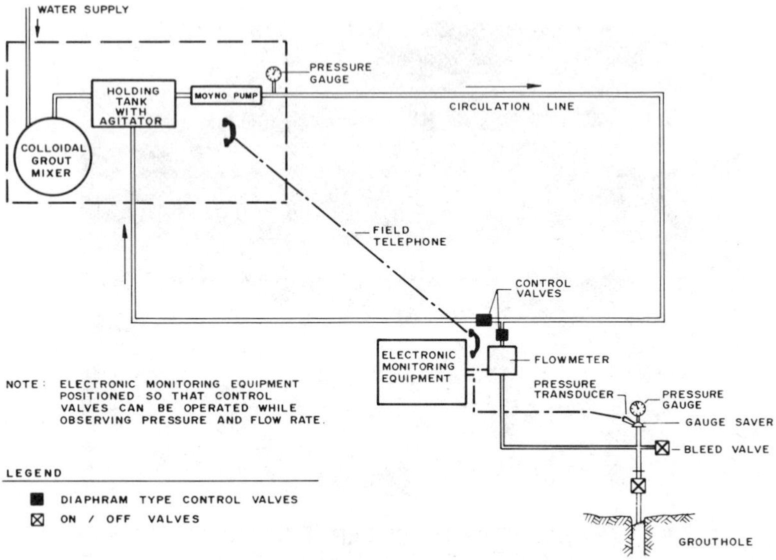

FIGURE 3 SCHEMATIC LAYOUT OF GROUT INJECTION EQUIPMENT

FIGURE 4 ELECTRONIC MONITORING EQUIPMENT

response of the ground to grout can be most clearly appreciated by automatically plotting pressure against flow rate using an X-Y plotter (Hewlett-Packard 7035B). Total volume injected is recorded on the cummulative counter provided with the flow meter and this information can be included on the chart recorder plot. Since both the 2-channel chart and the X-Y plotter utilize ink on paper, it is necessary to protect both instruments from dust and moisture. In practice this has been accomplished by having a small moveable hut which serves as a grout control centre (Figure 4).

USE OF ELECTRONICALLY ACQUIRED DATA

Water Pressure Test

One of the first tasks on a grouting job is to carry out Lugeon permeability tests. The procedure and interpretation has been described in detail by Houlsby (3) and the electronic monitoring system can be used to carry out Lugeon tests expeditiously. The test is set up in the usual way and a series of sequential ascending and descending pressures are predetermined. A 'typical' set of results for a water test is shown in Figure 5.

The test procedure differs from that described in reference (3) in that, whereas Houlsby requires each pressure to be held for 10 minutes, interpretation of the time plots printed on the 2-channel chart recorder indicates immediately whether the flow rate is stable, decreasing due to 'void filling' or increasing due to 'wash out'. In general, each pressure stage is held until the flow rate is stable before proceeding to the next pressure stage. The Lugeon value for the stage can then be calculated from the flow rate versus pressure curve recorded on the X-Y plotter, as the gradient of the plot is proportional to the Lugeon permeability, the proportion depending on the test stage length. The shape of the X-Y plot indicates the various flow patterns as described by Houlsby, for example the 'typical' plot shown in Figure 5 indicates 'laminar' flow. When the X-Y plot is curved, indicating 'turbulent' flow, the gradient of the plot at the higher pressure stages is used to calculate the Lugeon valve.

Grout Injection

A standard question faced by the grouting engineer relates to what injection pressure is appropriate for each stage of each grouthole. The real time display of grout pressure versus flow rate is of great assistance in this area because the pressure can be increased monotonically from a small value over a period of two to three minutes provided the X-Y plot remains linear or the pressure increases faster than the flow rate (turbulent flow). The

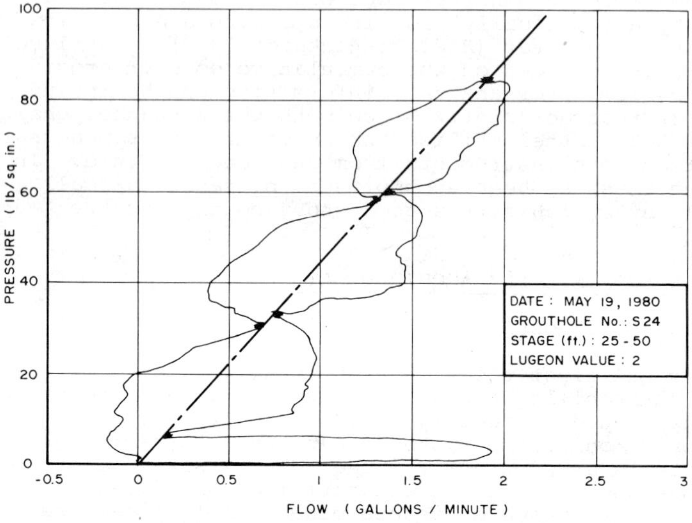

RECORD FROM X-Y PLOTTER

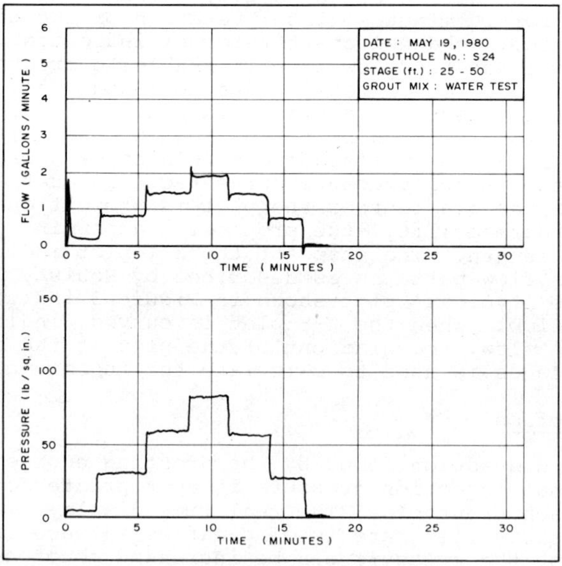

PLOT FROM CHART RECORDER

FIGURE 5 RESULTS OF LUGEON WATER TEST

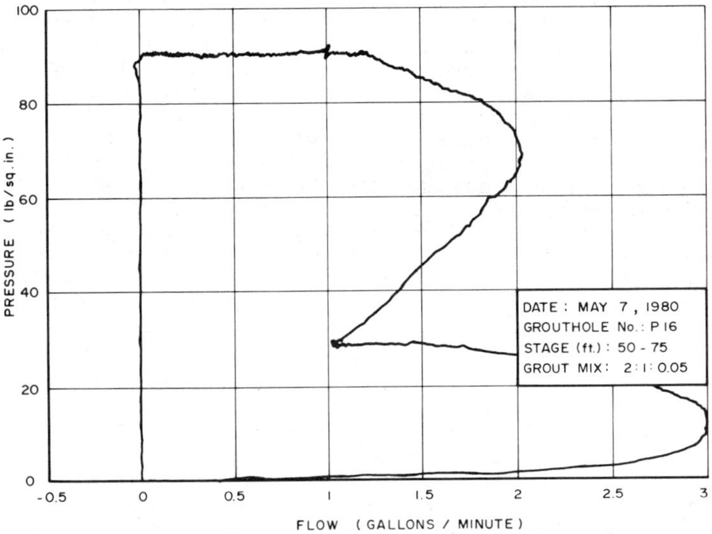

RECORD FROM X-Y PLOTTER

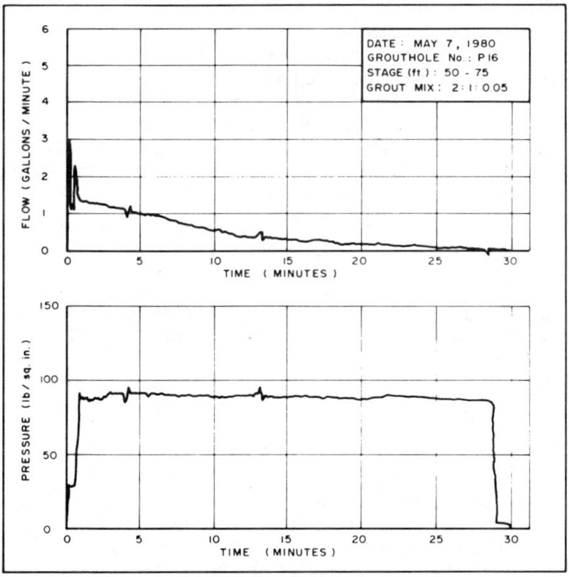

PLOT FROM CHART RECORDER

FIGURE 6 RESULTS OF GROUT INJECTION

limiting pressure desired is just prior to hydraulic fracture of the ground as this minimizes the time required to inject the desired quantity of grout. Hydraulic fracture appears on the X-Y plotter as a line near parallel to the flow axis and is easily recognized in practice because the increase in flow is extremely sudden. Remedial action when hydraulic fracture occurs is to reduce grout pressure promptly - a 10 percent reduction usually suffices to regain the monotonic pressure/flow line on the plotter. It follows from this behaviour that the grouting engineer should ensure that rapid control of pressure is possible before carrying out tests to determine the appropriate injection pressure. The authors have limited injection pressures to 80 percent of the fracture pressure for normal injection.

Another question faced by the grouting engineer concerns the appropriate mix design to be used and when to change the grout mix during grout injection. The time plot of flow rate recorded on the 2-channel chart recorder (Figure 6) gives a continuous assessment of the reaction of the ground to the grout, and interpretation of this plot allows the grouting engineer to determine the required grout mix. As an example, a high grout take with no decrease in flow rate with time, as recorded by a horizontal line of the chart recorder, would indicate that the grout should be thickened.

In addition, the flow meter can be used to determine the point of refusal to grout quite accurately. Because of limitations with the accuracy of flow measurement, the description of grout refusal has traditionally been a function of the judgement of the grouting engineer. Experience with the electronic monitoring has shown that 'traditional' refusal now appears as a distinct and measureable flow. It is helpful to have decided before starting injection what flow rate will be taken as refusal because the real time plot of declining flow rate on the 2-channel chart recorder may be quite hypnotic when first seen.

PRODUCTIVITY GAINS

The benefit of the electronic monitoring system in reducing cost and construction time stems from five factors:

a) Reduced supervision manpower: The electronic system automatically records most of the information normally required on a grouting job. In practice, the data not recorded automatically is easily handled by the grouting engineer in the normal course of his duties. There is a further gain in reduced engineering time following completion of the day's grouting as the electronic system has recorded the information graphically which eliminates much of the plotting otherwise required.

b) Reduced testing time: There are typically a large number of water tests on a grouting job and while they do not amount to a large percentage of the grouting budget, they are nevertheless a significant identifiable cost. The electronic monitoring system can reduce the time spent water testing by more than half. Since the cost of water testing is essentially only related to time, this represents a cost saving of 50 percent of this item.

c) Reduced injection time: The easy identification of maximum injection pressure with the X-Y plotter permits the largest appropriate injection pressure to be used with confidence. Since the flow rate of any viscous fluid is proportional to the differential head, increased injection pressures reduce the time required to inject a finite quantity of grout. At present, however, it must be noted that the guidelines of grouting (1), (4) have been developed over a large number of case histories using much lower injection pressures than the techniques described herein would indicate are permissible. A certain amount of care is therefore required not to inject more grout than conventional practice would indicate is required. The cost saving with this item accrues because of the reduced time per stage which permits more stages to be grouted in a shift. At present, it appears that injection time is reduced by between 10 percent and 20 percent.

d) Clearer detection of 'refusal': Conventional techniques of grout monitoring require typically about a half hour to define refusal once the grouting engineer believes he is close to it. The electronic system on the other hand, shows the flow rate in real time so that a saving of virtually a half hour per stage accumulates. This saving is particularly noticeable on grout projects with small grout takes per stage such as, for example, the grouting of relatively tight bedrock beneath a dam core where the grouting was ordered more as a matter of engineering prudence rather than because of a clear requirement for a comprehensive grouting program.

e) Reduced total construction costs: The four items above indicate how the cost of grouting itself can be reduced. A further and sometimes greater saving may accrue because of reduced delay to the general contractor. This is because grouting lies on the critical path in many heavy civil engineering projects. Harding (2) reports that this reason alone was used to justify an electronic grout monitoring system, although this was somewhat less elaborate than the one described herein.

IMPROVED UNDERSTANDING

The number and aperture of fissures being grouted during a rock grouting operation provides data which permit an assessment of how far the grout has travelled. Such an assessment of likely grout travel is helpful in the control of grouting operations. Unfortunately, few site investigations for grouting projects employ downhole impression packer techniques so that aperture information is usually not available to the grouting engineer.

Electronically acquired water test and grout injection data can be used to provide an estimate of the "effective" fissure aperture for a given grout stage. Electronic monitoring is necessary to provide the required accuracy of data if useful and consistent results are to be obtained with the technique.

The manner in which the aperture width is inferred is to idealise the flow regime of grout within the fissures in an analogous manner to the equations for spherically divergent flow in granular soils summarized by Scott (5).

The idealization is to treat grout as a viscous liquid and assume that the flow through rock fissures is governed by the Navier-Stokes equations. The simplest mathematically tractable arrangement of fissures is a set of vertically oriented fissures travelling radially away from the borehole. For such an arrangement of fissures, it can be shown that the aperture width may be estimated by:

$$a^2 = 3.46 \times \ell \times n \times \left[\frac{\mu_g - \mu_w}{P_g - P_w}\right]^{\frac{1}{2}} \times \left(Q\int_0^t Q\,dt\right)^{\frac{1}{2}} \quad \text{- Eqn (1)}$$

where: a = dominant aperture width of fissure system

ℓ = a fissure length intersecting the grout stage

n = number of fissures intersecting the grout stage

μ_g = viscosity of grout

μ_w = viscosity of water

P_g = injection pressure of grout at intersection of fissures with the grouthole

P_w = injection pressure required to establish an equal flow rate of water

Q = grout flow rate

Now both ℓ and n can be estimated from the site investigation data prior to starting grouting while μ_g can be measured with a viscometer for each type of grout in use. μ_w can be ascertained from tables of physical properties. P_w is inferred from the water test. The remaining parameters are continuously measured by the electronic system and thus Equation (1) allows the grouting engineer to assess the dominant fissure size for the particular stage being grouted. Once this has been done, the fissure dimensions can be used in the equation of volumetric compatibility to yield an "effective" radius of influence of the grout, as follows:

$$r_g = \frac{1}{\ell . n . a .} \int_0^t Q \, dt \qquad - \qquad \text{Eqn. (2)}$$

It can be noted that n can be estimated quite well from a knowledge of the characteristic fissure spacing for the site, and the stage length being grouted. On the other hand, ℓ is rather less certain; it is suggested that as a first approximation, ℓ could be taken as equal to the stage length unless other information is available. In any case 'a' depends on the square root of these two parameters which reduces the sensitivity of 'a' to error in estimating ℓ or n.

The equations presented above are for a particular simple fissure geometry which obviously cannot be expected to be applicable to every bedrock condition. Unfortunately, the complexity of the Navier-Stokes equations is such that few solutions exist so that to develop a method for generalized fissure arrangements would not appear presently possible. However, for a given particular fissure set, one could resort to numerical techniques to generate site specific correlations prior to the start of grouting operations.

DISCUSSION

The application of an electronic transducer system to cement grouting of bedrock has been described and the advantages of the system over conventional monitoring outlined. The greatest advantage in the use of the electronic system is the real time graphical display of relevant information made available to the grouting engineer. This information, when presented adjacent to the injection point, allows the grouting engineer to control the grout injection process and base his decisions on sound data. To further assist the grouting engineer, a 'first order' theory for interpreting subsurface conditions and the radius of influence of grout has been presented. This theory requires the accuracy of the electronic system to produce useful results. However, the greatest deficiency of the theory probably lies in the assumption that grout flows as a viscous fluid.

The cost benefits of an electronic monitoring system will vary from job to job, but the data provided in the paper should permit a likely saving to be estimated (on the basis of time saved) once the cost of doing the job conventionally has been estimated. Experience to date indicates that the electronic monitoring system can reduce the grouting cost to the client by at least 30 percent, with the electronic system paying for itself within the early stages of the grouting contract.

CONCLUSION

Modern electronic techniques can be used to successfully monitor cement grouting projects. The potential cost saving to the client is such that electronic techniques should be used.

While electronic monitoring is justified on cost grounds alone, the profession should not overlook the opportunity for improvements in grouting quality, reliability and marketability that result from the application of modern electronic techniques. If it can be demonstrated to the civil engineering profession at large that grout injection is an understandable process which can be monitored and which can produce predicted results reliably, then grouting will become more accepted as an engineered ground treatment process. This would benefit both owners and grouting contractors.

APPENDIX - REFERENCES

(1) "Grouting Manual"; Water Resources Commission, New South Wales, 1973.

(2) Harding, P.G.; "Multiple Grouting Techniques beat Stockholm's Groundwater", Tunnels and Tunnelling, April, 1980.

(3) Houlsby, A.C.; "Routine Interpretation of the Lugeon Water Test", Quarterly Journal of Engineering Geology, Vol. 9, The Geological Society, 1976.

(4) Houlsby, A.C.; "Engineering of Grout Curtains to Standards", Journal of Geotechnical Engineering Division, ASCE, Vol. 103, GT9, 1977.

(5) Scott, R.A.; "Fundamental Considerations Governing the Penetrability of Grouts and their Ultimate Resistance to Displacement", Grouts and Drilling Muds in Engineering Practice, Symposium, London, 1963.

QUALITY CONTROL FOR GROUTING

Peter Jay Huck[1]
Muriel Jennings Waller[2]

Key Words: Acoustic Velocity / Geophysics / Grouting Inspection / Quality Control / Radar (Borehole) / Seismic Investigations / Soil Grouting

ABSTRACT

A variety of tools are available which may be used to evaluate the results of chemical grouting in soils. Two of these tools, borehole radar and cross-hole acoustic shooting are discussed within the context of overall QC/QA programs that may be used with soil grouting.

Borehole radar may be used before and after grouting to determine the actual location of the grout, while cross-hole acoustic tests are best used to determine (qualitatively) the stiffness of the grouted zone. Although both methods have been used in the field, an extensive backlog of experience has yet to be established, so that cautious use of these systems is recommended.

1.0 INTRODUCTION

When faced with unsuitable ground conditions, the Civil Engineering designer may select soil or rock grouting as a means to improve ground conditions in-situ. In the development of his grouting design, he may consider a broad spectrum of construction control testing efforts, ranging from a program of minimum testing and overdesign on the one hand to a rigorously developed testing program coupled to an optimum grouting design on the other. Clearly, the more closely he designs the grouting effort to his project's needs, the better quality control is required, whereas if he is very conservative in his design, little effort need be spent in monitoring the actual results that may be achieved in construction. Balancing the cost of testing against the increased efficiency of the design of the grouting effort itself provides a means to minimize cost for the client. The particular mix of testing effort and degree of overdesign should be determined by the risk of grouting failure and the consequence of failure, should it occur. Under some conditions, it is appropriate to use minimal quality control, while other projects may demand intensive efforts to monitor the quality of the grouting work. Technology now exists to tailor the testing program to the project needs.

[1] Research Director, EarthTech Research Corporation, Baltimore, MD
[2] President, EarthTech Research Corporation, Baltimore, MD

2.0 GROUT MONITORING PROGRAM

We limit discussion in this paper to the use of two techniques that may be used to evaluate the results of grouting independently of the grouting operation itself. Thus they fall into the category of quality assurance tools rather than quality control tools. The techniques to be discussed are earth-probing radar and cross-hole acoustic shooting. Both of these techniques may be used in the grout pipes placed for the production grouting, so that the cost of drilling normally associated with the use of downhole instruments is in this case eliminated. Since the grout pipes must be placed anyway, it seems appropriate to use them for test purposes as well as for grouting.

The manner in which borehole radar and cross-hole acousting shooting fit into a typical grout monitoring effort is displayed in Table I. This table shows some of the steps that would be taken in both monitoring efforts of a very minor nature, and those representing a more intensive grout monitoring effort. Referring to Table I, it may be seen that the two geophysical test methods under discussion in this paper properly are used before injection begins and after its completion. Clearly, since these tests must be conducted from pipes placed for grout injection, they are tied to the grout injection schedule and the grout injection method. On large grouting efforts, drilling and injection may proceed simultaneously, but the drilling and placing of grout pipes typically has sufficient lead time so that the geophysical testing may be done without interferring with the grouting operations. It is important that these tests be conducted as early as possible, since it is known that grout may travel some distance underground. The initial surveys should be conducted before the ground is contaminated with grout or possible traces of unreacted grouting materials.

2.1 Earth-Probing Radar

Borehole radar systems are available in both transillumination mode or in transmit/receive (TDR) modes, but transillumination should be used for grouting. In transillumination mode, separate instruments are used in two adjacent boreholes, one instrument being a transmitter and the other a receiver. The radar signal passes between the two boreholes from the transmitter to the receiver. In the transmit/receive mode, a single instrument in one borehole contains both the transmitter and the receiver. The signal from the transmitter must travel out into the medium around the borehole, and be reflected by a discontinuity to bounce back to the receiver, as is done in the more familiar case of aircraft search radar. Borehole T/R systems, however, are omnidirectional, and cannot determine the azimuth of the discontinuity. The system that should be used for grout monitoring is the transillumination mode, in which the transmitter and receiver are in separate boreholes.

TABLE I

INTEGRATION OF BOREHOLE RADAR AND CROSS-HOLE ACOUSTIC TESTING INTO GROUT MONITORING PROCEDURES

GROUTING ACTIVITY	MINOR MONITORING EFFORT ONLY	INTENSE MONITORING EFFORT
Prior to Grouting	• Inspect equipment • Set elevation survey points • Establish monitoring plan and procedures	• Inspect equipment • Set elevation survey points • Establish monitoring plan and procedures
During Drilling		• Conduct pre-grout radar and cross-hole acoustic surveys
Grout Materials	• Certificates of compliance • Trial grout mixes	• Certificates of compliance • Trial grout mixes • Independent laboratory tests
During Grout	• Monitor injection pressure & flowrate • Grout samples for gel time and storage • Plot grout-take log • Heave measurements on survey points	• Monitor and record injection pressure and flowrate (strip-chart) • Grout samples for gel time and storage • Plot grout-take log • Heave measurements on survey points • In-situ deformation measurements as appropriate (inclinometer and/or in-situ strain) • Pore-pressure data • In-Situ resistivity • Acoustic emission monitoring for hydrofracturing
After Grouting	• Final heave survey • Final review and sign-off	• Final heave survey • Post-Grout radar and acoustic surveys • Final review and sign-off

The operation of transillumination radar is shown in Figure 1. The transmitter and receiver are lowered to the bottom of their respective boreholes, and then are turned on. With the transmitter continuously sending pulses to the receiver, both are raised, keeping them at the same level and recording the received signal and the depth of the two instruments continuously. The result is a profile of the medium lying between the two boreholes. In areas having grout between the two boreholes, the radar signal will be greatly attenuated, since most grouting materials are excellent absorbers of electromagnetic radiation[1]. Areas which are poorly grouted will permit more of the signal to pass through to the receiver. At the present state-of-the-art, it is possible to determine only that an area is well grouted or poorly grouted. As additional experience accumulates, it should become possible to make more quantitative statements regarding the amount of grout between the two pipes at any given level, but this is not presently possible. It is of sufficient importance to know where the grout has gone, however, that radar surveys to determine grout location are often well worth the expense of the survey.

2.1.1 Type of Radar System to be Used

A variety of earth-probing radar systems are currently available, either for purchase, or in the form of geophysical surveying contracts obtained from specialist firms. We had previously stated that borehole radar should be used. Conventional systems using surface antennas have been tested in the past and the results obtained were disappointing at best. The signal must pass through the surface overburden material, and be reflected from the top of the grout zone if a return signal is to be received back at the surface antenna. Since grout materials tend to absorb the radar signal rather than reflect it, it is difficult to see even the top of the grout zone, much less the bottom of the grout zone. Interpretation of radar signals taken from borehole instruments within or near the target grout zone is much easier than that obtained using surface antennas.

Borehole instruments currently available may use either a radar pulse, short in time but wide in frequency content, or a continuous (CW) signal in a narrow frequency band. A pulsed radar permits measuring the time of flight and the amplitude of the received signal, whereas a CW radar permits measurement of phase and signal amplitude. We have used both, basing interpretation primarily upon signal amplitude, and found both to be useful. The decision to use pulsed radar or CW instruments should be made on the availability of services in the project locale.

Future developments that will be useful include geotomography and certain new developments in radar systems that will combine the advantages of pulsed radar and CW systems. We have used geotomography, but found that the process, using instrumentation available in 1979 was slow and time consuming, and the data, while much more spectacular than a conventional borehole radar profile, was

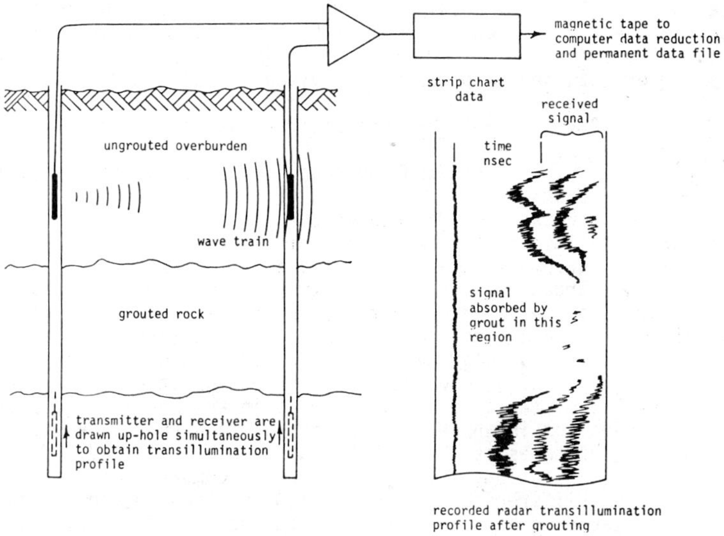

Figure 1. Use of Pulsed Borehole Radar for
Post Grouting Survey (Huck, et al, Ref. 1)

actually less informative and precise. It should be noted, however, that capabilities in geotomography are increasing rapidly(2), and that the system has certain inherent advantages over conventional borehole radar. Geotomographic services are now commercially available on a limited basis, and the increasing experience in its use and interpretation may make it the radar system of choice in some grout monitoring efforts. The user of grout monitoring services should be aware that the system exists, and is potentially quite valuable.

2.1.2 Radar Use and Interpretation

Earth-probing radar should be used at present by making surveys prior to and following grout injection. It is possible that increasing experience in the use of this tool will make it possible to eliminate the before-grouting survey, but this cannot be done with assurance at this time. The before-grout survey should be taken in the grout pipes as soon as they are placed and can be made available to the radar consultant. From the before-grout survey, it is possible to make some geotechnical interpretations, showing where layers of sand or clay may exist, or some indication of wet and dry areas. The purpose, however, of the before-grouting survey is to provide a record of the site prior to grout injection. At the completion of injection in an area, the radar profile should be repeated in the same grout holes as the initial survey was conducted. The before and after grouting profiles may then be compared on a side-by-side basis, and differences between the two ascribed to the injection of grout. Because we are looking for differences in two surveys, it is important that the two surveys be run using the same crew and instruments, and further, that the operational parameters used in the first survey be well documented so that the after grout survey may be run at the same instrument settings. If this is done, it is possible to note the changes that have occurred as a result of grouting. In areas that are well grouted, the radar signal will be absorbed in the grout zone, but not below or above it. If some volumes have been left ungrouted, then these will pass the radar signal, and be easily detected in the before and after comparison. It is helpful if the radar profile extends from below the intended grout zone to well above it, which would require some additional drilling. In these areas, the before and after surveys should be the same, since grout will (hopefully) not have intruded into areas in which it is not desired. Thus, the portions of the radar profiles that extend outside the grout zone provide calibrations of the radar survey that assures one that the system is operating properly, in addition to identifying the actual limits of the grout zone as injected.

2.2 Acoustic Cross-Hole Shooting

Acoustic cross-hole shooting(3) is a borehole survey technique based on profile of acoustic velocity of the medium between two boreholes as a function of depth. Since acoustic velocity is a function of $\sqrt{E/\rho}$, the technique is sensitive primarily to the stiffness of the grouted material. The actual surveying technique used is similar to transillumination radar, previously discussed.

2.2.1 Type of Equipment to be Used

As in transillumination radar, superior results can be obtained by using borehole instruments to profile the material between borehole pairs. Geotechnical engineers are most familiar with the use of shear, or distortional wave acoustic systems, which are useful for ranges large enough to permit separation of the shear wave and the faster compressional, or dilational (P) wave. For use in grout monitoring, short ranges are used, dictating the use of P-wave time of arrival data. The S-wave typically is buried in the noise following the P-wave at these short ranges. A number of transducers may be used to produce the desired acoustic signal, ranging from manual hammers and mechanical sounders to electrical sparkers or even sophisticated variable frequency systems. In the work described herein, a sparker was used, which produces a wide frequency spectrum. The receiver may be any transducer or hydrophone capable of the frequency response needed by the particular transmitter used. Because of the short ranges between grout pipes, a high frequency system is required.

2.2.2 Use and Interpretation of Cross-Hole Acoustic Systems

The cross-hole acoustic survey, like borehole radar, should be used in the grout pipes before and after grouting. The data should be presented in a form that displays either time of arrival or acoustic velocity between the boreholes as a function of depth. The effect of grouting is to increase the wave velocity in proportion to the improvement in structural quality of the grouted material. Figure 2 shows acoustic velocity measured across one meter (39 in.) cube grouted specimens in the laboratory, showing the rapid increase in acoustic velocity as the grout formulation in question is strengthened.

Experiments in a field grouting test(4) showed that P-wave velocities, initially in the range of 200 M/sec (600 ft/sec) before grouting, were increased to 1.5 to 2 km/sec (5000 to 6500 ft/sec) by the injection of a strong chemical grout in a dense silty sand. The post-grouting velocities are more typical of weak rock than of soil, and indeed, the grouted soil was very like a weak sandstone when the site was finally excavated. Pressuremeter tests were also conducted at this site(5). These showed increases in pressuremeter limit pressure by a factor of two, and in pressuremeter moduli by a factor of eight, indicating that the substantial increases in acoustic velocity at this site reflected actual changes in the soil properties.

The results noted above involved the use of a structural grout in a dense sand, under nearly ideal conditions. Such dramatic increases in acoustic velocity would not be seen when using a weak waterproofing grout.

2.3 Combined Acoustic and Radar Surveys

The two systems, radar and acoustic velocity, are used to

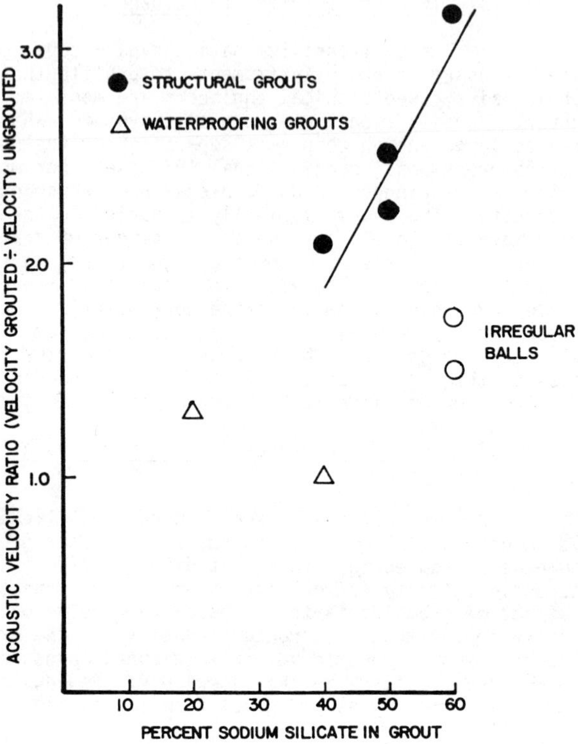

Figure 2.

INCREASE IN ACOUSTIC VELOCITY vs SILICATE PERCENTAGE

measure two different aspects of grouting quality. The radar system is used primarily to determine the location in the grout of the ground, and is useful in this regard because the electromagnetic energy travels in a relatively straight path from transmitter to receiver. The acoustic signal, on the other hand, may travel in distinctly curved paths, but its velocity is determined primarily by the stiffness of the grouted medium. Thus, cross-hole acoustic shooting provides a qualitative measure of the degree to which the material properties have been enhanced structurally. In an important or critical grouting project, it may be desirable to use both systems to obtain the best possible indication of the location and character of the grouted zone. In a less critical situation, the designer may properly elect to be more conservative in his design, and achieve confidence by overgrouting rather than by an intensive QC/QA program.

2.4 Evaluation Tools That Are Less Useful

It is probably appropriate to note that there exist a number of subsurface exploration tools whose use has been attempted, but which have not performed well in grouting evaluation systems. The first category that may be listed includes any geophysical system that relies upon measurements made from the surface. Given that a grouting project has many boreholes required for the grouting, one should consider only the use of downhole geophysical tools. Numerous attempts have been made to use electrical resistivity, radar and acoustic exploration tools from the surface, but in no case that we know of has a surface tool been shown to be as useful as a downhole system. It is more commonly the case that surface instruments may give a vague indication that the grout is somewhere down there, but little hard data can be developed, even after intensive interpretation efforts.

Having discounted surface geophysical systems, one may then take a hard look at conventional soil sampling or in-situ testing systems. The difficulty with such systems lies in the sensitive character of grouted soils. Core samples may be recovered in some cases, but the relatively rough treatment given a sample by a rotating core barrel disqualifies the sample as an undisturbed specimen. It may be examined for the presence of grout, but a strength test will not yield data in which one may place confidence. Borehole pressuremeter tests may be conducted, and useful data recovered only if extreme care is used in the preparation of boreholes. A dragbit, or fishtail bit should be used, rotating at low speed with minimal thrust in a heavy drill mud. If large pieces of grouted sand are wiped about the inner surface of the borehole, the projections formed will rupture the pressuremeter membrane with great ease. If the borehole wall is fractured by rough drilling, of course, the pressuremeter test is of little value. Probably the most effective post-grouting test, but also the most expensive, is the excavation of test pits. In a test pit, one may enter the grout zone, make personal inspection and even recover more or less undisturbed block samples by hand.

3.0 INTEGRATION OF RADAR AND ACOUSTIC TESTING INTO GROUT MONITORING EFFORTS

The techniques discussed in this paper represent only two components of a program of quality control and quality assurance tests and procedures that ideally should begin when the project is in the design phase, and run through to retention of documentation at least until successful performance has been demonstrated. The subject of QC/QA programs for grouting has been discussed in greater depth elsewhere (e.g. Ref. 6,7), and only sufficient overview will be given here to permit placing radar and acoustic velocity tests in their proper perspective.

Clearly, radar and acoustic velocities are techniques by which the grouting effort may be evaluated independently of the grouting operation iteself, whereas other tests and procedures may be used to help control the process. Both types of technique are necessary. Table I shows how the two techniques may be fit into two levels of monitoring effort, a minimum testing effort appropriate to projects in which the risks and the consequences of grouting failure are not severe, and a more comprehensive monitoring effort appropriate to projects in which grouting quality is of primary importance.

Because these two techniques both measure some element of grouting effectiveness, they should be compared with each other as well as with the process monitoring data to confirm that the various monitoring techniques are producing data that are consistent. When all systems agree, one may be confident in the results, and when they do not, then further investigation must be made to determine the unexpected conditions that may have lead to the lack of consistency.

4.0 SUMMARY

Borehole transillumination and cross-hole shooting are two geophysical techniques that may be used to evaluate the effectiveness of grouting efforts. Both require that the grout pipes be made available before and after injection. In addition, the radar requires that non-conductive grout pipes be used. Since such pipe is used with the two grout injection methods that provide the greatest control over injection (tuba-a-manchette and sleeve-pipe grouting), this restriction does not seem abhorent. While cross-hole acoustic shooting can work through steel casing, it would seem to be effective even if stage-up or stage-down grouting were used, but these systems do not make available suitable boreholes for both pre- and post-grouting surveys. Note also that the use of high-velocity steel pipe in an acoustic system will result in greater probability of undesirable signals traveling longitudinally along the pipe.

The radar system may be used to determine the limits of the grouted zone, but is not effective in measuring the structural competence of the grouted zone. Cross-hole acoustic shooting, on the other hand, provides an excellent qualitative indication of the structural properties of the grouted soil, but a poor resolution on the actual location of the grouted soil. Either system might be used alone, or they may be used in combination if the project needs demand. They both must be considered within the context of an overall monitoring program that includes process monitoring, acceptance testing of materials and tests of the neat grout produced by the grout mixing system. If the maximum benefits of any grout monitoring program are to be realized, the monitoring program should be developed at the time that the grouting effort itself is in design. If this is done, the designer may develop an overall project that incorporates a complementary mixture of monitoring effort and conservative grouting design. It is unrealistic to expect that such a monitoring program can be added to an already completed design, since much of the benefit of QC/QA efforts lie in the fact that they may be used to reduce uncertainties in the construction process.

References

1. Huck, P.J., Waller, M.J., Koerner, R.M., McCabe, W.M. and Fowler, J.C., "Monitoring and Control of Particulate Grouting in Rock", Final Report, Contract No. DOT-FH-11-9397(E), submitted to U.S. Department of Transportation, Federal Highway Administration and U.S. Department of Interior, Bureau of Reclamation, August 1980.

2. Laine, E.F., Lytle, R.J., and Okada, J.T., "Cross-Borehole Observation of Soil Grouting", Journal of Geotechnical Eng. Div., ASCE, August, 1980, pp 871-875.

3. Stokoe, K.H.,II, and Woods, (1972), "In-Situ Shear Wave Velocity by Cross-Hole Method", Proc. Amer. Soc. Civil Eng., J. Soil Mech. and Found. Eng., Vol 98, No. SM5, pp 443-460, May 1972.

4. Huck, P.J., Waller, M.J., and Fowler, J.C., "Control and Monitoring of Chemical Grout in Soils - Volume I - Technical", Final Report Contract No. DOT-FH-11-9397 to U.S. Dept. of Trans., Federal Highway Administration, August, 1980.

5. Fernandez, J.I., "In-situ Pressuremeter Testing for the Evaluation of Soil-Grout Systems", Final Report, Contract No. DOT-FH-11-9339 for U.S. Dept. of Transportation, Federal Highway Administration, June, 1980.

6. Huck, P.J. and Fowler, J.C., "Control of Rock Grouting", presented Fall Meeting, Soc. Mining Engineers of AIME, Minneapolis, Minnesota, October, 1980.

7. Huck, P.J., and Koerner, R.M., "Acoustic Emission of Monitoring of Soil and Rock Grouting", Proc. ASTM Symp. on Acoustic Emission in Geotechnical Engineering, Detroit, Michigan, June, 1981.

MULTIPLE HOLE GROUTING METHOD

Reinhard E. Mueller

ABSTRACT

Grouting, becoming more sophisticated as time passes, started as a fairly simple procedure. In the early days a grouting contractor was simply expected to produce good results without much supervision. However, experience led to the development of grouting specifications to be followed by contractors. Based on geological investigation, parameters such as maximum pressure, minimum quantity, and mix proportions, are now commonly included in grouting specifications. However, to date on-site supervision of grouting operations or pumping tests has been limited in scope, relying almost entirely on the operator's statement. The High Frequency Recorder described uses magnetic flow meters and microprocessor data manipulation to automatically record all important data during grouting operations, thus enabling the on-site engineer to evaluate the work immediately, to make the necessary adjustments for optimum performance, and to effectively oversee simultaneous grouting in multiple grout holes with a single pump. The multiple hole grouting method, thus monitored, can have an important effect on the economy of grouting operations, by increasing the speed of operations but not the manpower requirements, as well as by keeping maintenance costs and downtime low. An application case history of the MHGS is given.

MULTIPLE HOLE GROUTING (MHG)

Conventional grouting procedures involve one hole injected by one pump; faster progress requires several pumps to grout several holes. The Multiple Hole Grouting System (MHG System) enables several holes to be simultaneously injected by a single pump (Fig. 1) using a manifold to distribute the grout and to maintain equal pressure at all outlets. The number of holes that can be grouted by one pump depends on pump capacity and on the grout consumption of a ground formation.

Based on pumping tests, holes with similar characteristics are selected to ensure even penetration of the rock. The flow through each of the outlets must be monitored to guarantee the complete filling of all grout holes. In spite of careful selection of similar holes, new voids may open up during injection. With proper surveillance of the flow to

---

Product Manager, Grouting Systems, Hany & Cie. AG, Switzerland.

each grout hole, such openings can be detected immediately, blockages in holes can be traced at once, and nongrouted areas are avoided.

MULTIPLE HOLE GROUTING SYSTEM (MHGS)

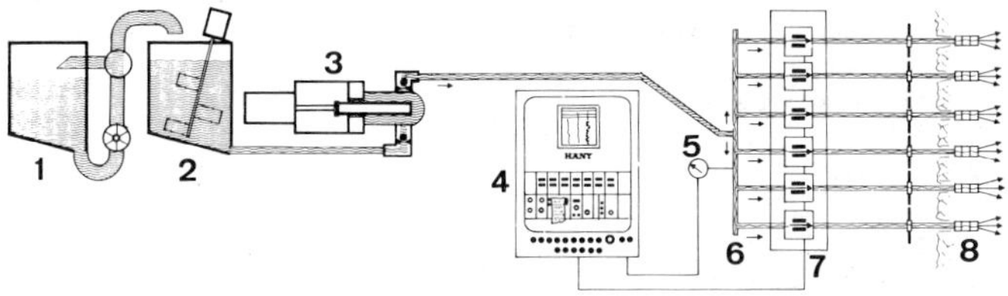

1. MIXER 2. AGITATOR 3. GROUT PUMP 4. RECORDING UNIT 5. PRESSURE TRANSDUCER
6. MANIFOLD 7. MEASURING UNIT 8. PACKER

Fig. 1. Schematic Drawing of the MHG System.

MONITORING SYSTEMS

The demand for monitoring the grouting operation and recording important data has rapidly increased during the past few years. A pump test is relatively simple to monitor requiring only a watch, water meter, pressure gauge, and, most importantly, a reliable worker. Measuring the flow of grout is far more complex. The success of the Multiple Hole Grouting (MHG) method depends on accurately monitoring the flow of grout to each grout hole.

Many grouting contractors have tried to build flow recording systems, the more successful ones using methods such as tank level gauges, stroke counters on the grout pump, etc. However, even these successful monitoring systems contain a major disadvantage in that flow rates are not directly indicated, thus complicating evaluation of results. The flow rate must be calculated from accumulated flow and elapsed time. The latest developments in electronics have made it possible to measure both pressure and flow accurately and to process the signals in such a way that diagrams and printed records are produced instantly.

High Frequency Recording System

The High Frequency Recording (HFR) System measures pressure and flow of various kinds of grout and consists of two main components, the measuring unit and the recording unit. The measuring unit, containing the flow meter and the pressure transducer, is installed anywhere in the grout line, preferably close to the pump. The recording unit is connected to the measuring unit by cables and can be placed in a remote location convenient to the operator.

Use of microprocessers allows the HFR System maximum accuracy and flexibility in data processing. The following data are recorded:

Digital Readouts:

- -Flow rate (liter/min.) at each outlet
- -Accumulated flow at each outlet, as well as the total of all outlets
- -System pressure

Chart Recorder:

- -Flow rate at each outlet
- -System pressure

Printer:

- -Date and time
- -Accumulated flow at all outlets (liters)
- -Accumulated flow at each outlet (liters)
- -Flow rate at each outlet (liters/min.)
- -Pressure (bars)
- -Sequential borehole number
- -Stage number
- -Mixing proportions
- -Total consumption of cement and additives

Recording of Flow

Flow is measured by a magnetic flow meter, which allows free passage of at least 20 mm. There are no moving parts, obstructions, or pockets which could lead to problems by cement build-up. The meter is made of materials resistant to most chemical additives in common use. Viscosity, composition of grout, and erratic flow have no influence on measurements. Flow is measured in two ranges to provide maximum accuracy even at flow rates of one liter/min. and less. The ranges (0 - 10 liters/min. and 0 - 100 liters/min.) switch automatically whenever the flow rate remains within the higher or the lower range for a certain length of time (Fig. 2). When grout consumption ceases, the flow rate digital readout starts flashing to inform the operator that the grout hole is full. Simultaneously the 0-flow is automatically printed out together with the exact time; whenever specifications require maintaining the final pressure at 0-flow over a certain period of time, both printer and chart recorder provide accurate records.

When the flow rate of 100 liters/min. is exceeded, the digital readout continues to indicate the correct flow rate but begins flashing as a warning. Should the rate exceed 120 liters/min., the system stops to process the data, which is also recorded by the printer together with date and time.

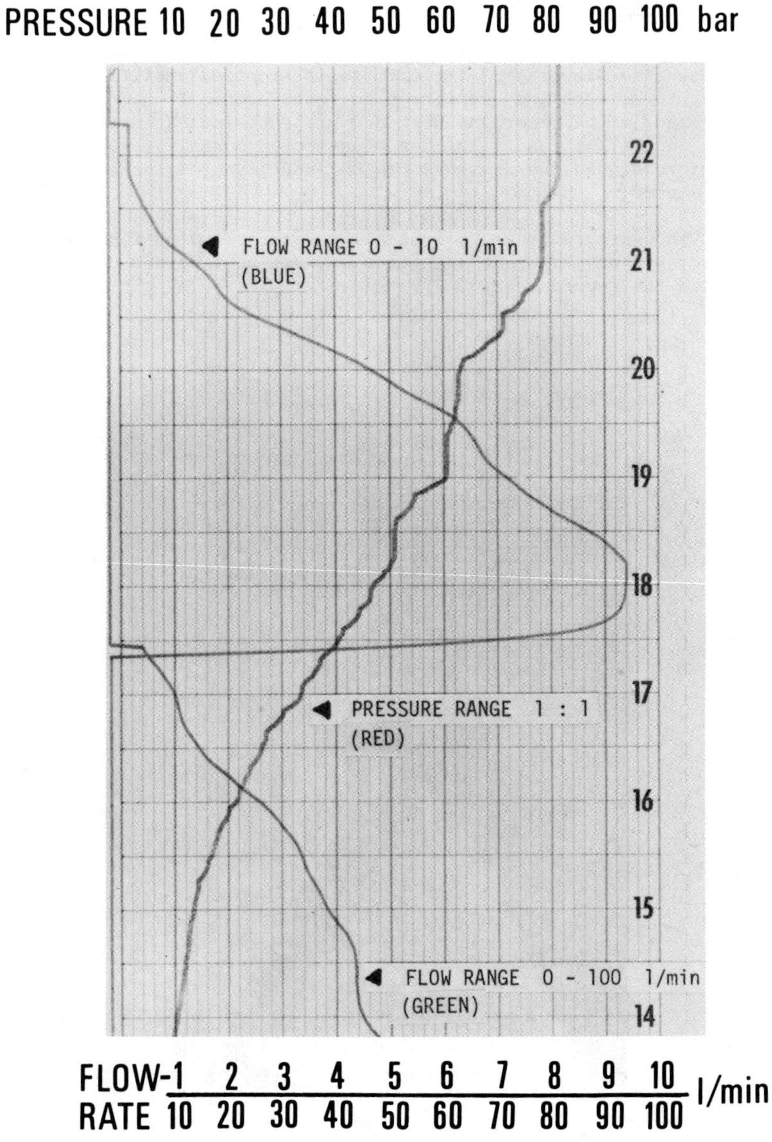

Fig. 2. Strip of recording paper showing pressure (red) and flow rate in two ranges (green and blue).

Pressure Recording

Pressure is measured by an electronic pressure transducer with a small pressure-sensing diaphragm. The electronic signals, amplified in the transducer to be much less sensitive to interference from other sources, are processed in the recording unit and pressure indicated in bars or PSI. Two ranges (1:1 and 1:10) can be selected manually. One of two methods of recording the pressure can be manually preselected, depending on requirements:

A. The chart recorder follows all pressure fluctuations created by piston pumps; actual peaks may not be recorded due to delay in the recorder (Fig. 3);

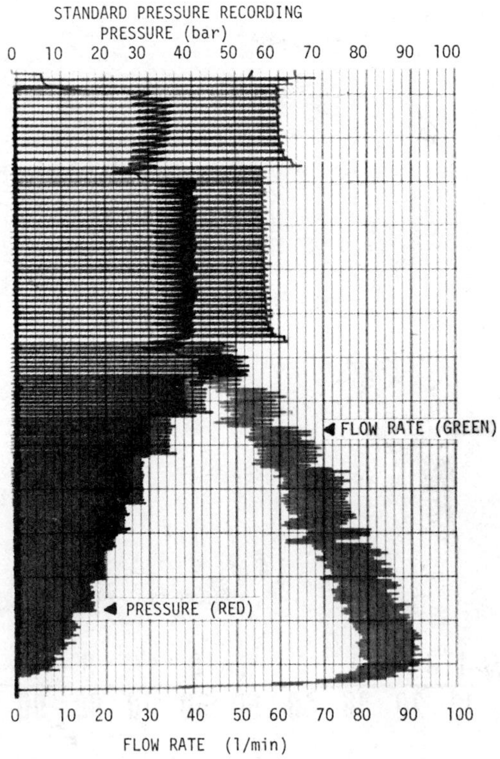

Fig. 3. Standard pressure recording with typical fluctuations of a piston pump.

B. The chart recorder records only the actual peak pressures; peak pressure is electronically sampled and held over a period of 5 sec. while the new value is sampled and held for the next 5 sec., etc. (Fig. 4).

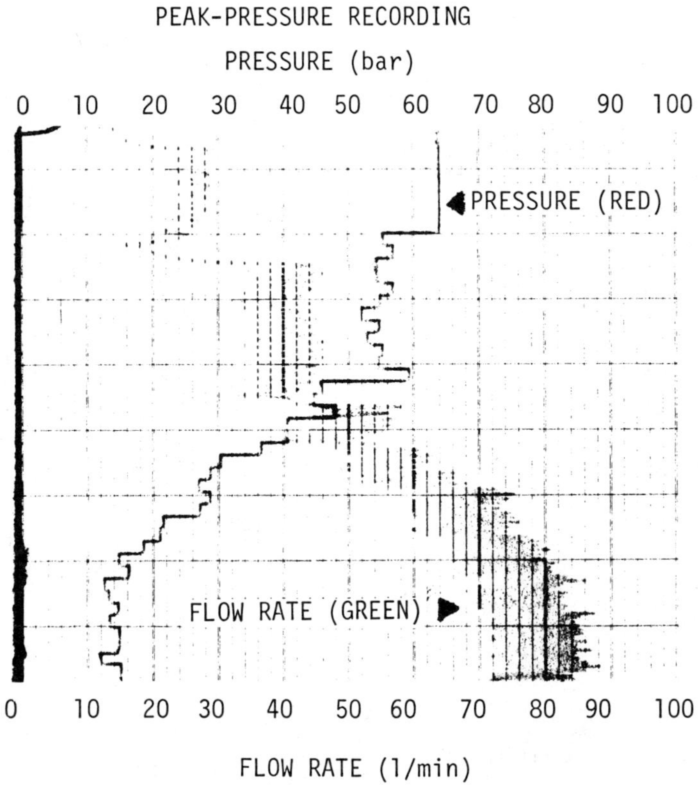

Fig. 4. Pressure recording whereby only the peak pressure is recorded.

The pressure transducer and flow meter can be installed together in the measuring unit, or the transducer can be placed at the top of the borehole or down the hole at the packer. The latter has the advantage of recording the actual water or grout pressure at any depth of the hole eliminating erroneous interpretation due to static head and friction in the grout line.

Programmable Unit

The recording unit can be supplemented by a small data terminal permitting allocation of sequential grout hole and stage numbers to the printed data. The water/cement/additive ratio can be entered so that the system automatically calculates consumption of cement and additives during the grouting operation (Fig. 5).

```
            000557
6=NUMBER    ?
            000556
5=NUMBER    ?
            000555
4=NUMBER    ?
            000554
3=NUMBER    ?
            000553
2=NUMBER    ?
            000552
1=NUMBER    ?
6=NUMBER    000551
5=NUMBER    000550
4=NUMBER    000549
3=NUMBER    000548
2=NUMBER    000547
1=NUMBER    000546
05.10.81      12:02
```

ALLOCATION OF BOREHOLE NOS. ▲

MIXING PROPORTIONS ▶

```
ADD2/C      001.3%
ADD1/C      002.0%
W/C         000.6
            1500KG/M3
ADD2        ?   KG/M3
            0002KG
ADD2        ?   KG
            2600KG/M3
ADD1        ?   KG/M3
            0003KG
ADD1        ?   KG
CEMENT      3100KG/M3
            0150KG
CEMENT      ?   KG
            0100L
WATER       ?   L
ADD1/C      002.0%
W/C         001.0
ADD2        0000KG/M3
ADD2        0000KG
ADD1        2600KG/M3
ADD1        0002KG
CEMENT      3100KG/M3
CEMENT      0100KG
WATER       0100L
05.10.81      12:03
```

PRINT-OUT OF CEMENT AND ADDITIVE CONSUMPTION ▶

```
T ADD2      000019KG
T ADD1      000029KG
T CEMENT    001509KG
T WATER     001002L
07.05.81      14:54
```

Fig. 5. Programming of borehole nos., mixing proportions, etc.

Form of Record

All data are displayed on digital readouts, recorded on a chart recorder, and printed on paper together with additional information such as date, time, grout hole number and stage, water/cement/additive ratio and consumption of grout components.

Chart Recorder

Pressure as well as flow rate in the ranges of 0 - 10 liters/ min. and 0 - 10 liters/min. are recorded on 3 different-colored channels. The pens are directly driven by linear motors and, therefore, are insensitive to vibration. Recording width is 100 mm, and paper speed is adjustable from 10 to 7200 mm/hr. The recorder is maintenance free with ink cartridges and pens that can easily be replaced.

Printer

All data are automatically printed at selected time intervals, from 1 to 99 minutes (Fig. 6), while a pushbutton allows printing at any time desired. At specific incidents, such as at 0-flow, a disconnected flow signal cable or resetting the accumulated flow to 0, the respective data are printed automatically.

PERIODIC PRINT-OUT

```
G01:08.0BAR
6*000100L  008.9L/MIN
6=NUMBER   001456
5#000072L  008.7L/MIN
5=NUMBER   001455
4*000167L  014.8L/MIN
4=NUMBER   001454
3#000087L  004.0L/MIN
3=NUMBER   001453
2*000227L  027.5L/MIN
2=NUMBER   001452
1*000134L  015.7L/MIN
1=NUMBER   001451
T*000803L
07.05.81      14:30
```

Fig. 6. Typical Periodic Print-out.

The HFR System can be equipped with a tape recorder instead of the printer. The tapes can then be processed by a microcomputer which prints out all data and plots the graphs. To secure the memorized data during power failures or equipment shutdowns, a rechargeable battery supplies power for a maximum period of 68 hours.

TECHNICAL ASPECTS OF THE HFR SYSTEM

Pump Tests

With the HFR System, no additional equipment is required to carry out pump tests. All data necessary to determine the permeability of a formation are automatically recorded: flow rate per minute, pressure, and time. For improved accuracy, the pressure transducer can be placed down the hole at the packer avoiding false information due to friction in the grout line and static head (Fig. 7).

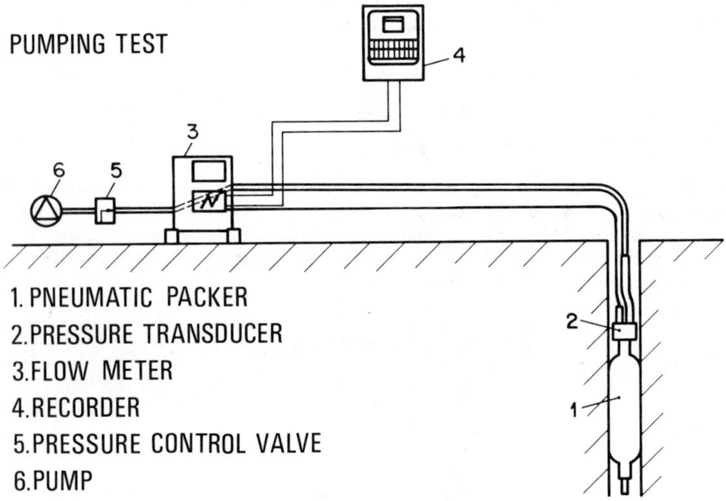

Fig. 7. Schematic of a Pump Test with Pressure Recording at the Packer.

Soil and Rock Data

Since the system produces a continuous diagram of the pressure and flow rate, the pumping test can show more than the permeability of a formation at a certain pressure; it can also indicate whether the penetration is stable or has an increasing or decreasing tendency. It may also show changes in permeability with variation of pressure.

Ground Behavior Data

The HFR System immediately signals when grout consumption ceases at a preset pressure; an immediate print-out shows the time at which consumption stopped. The chart recorder shows sudden breakthroughs into larger voids or cracks. This is important for determining the quantity and depth of the boreholes, maximum pressure and flow rate, as well as the type of grout mix.

Safety

The HFR System supplies important data to enable the engineer to take appropriate measures in every situation. Since his presence is no longer required on site at all times, supervision is simplified while useable information concerning the field grouting procedure is increased.

For delicate injections, the grout pump can be controlled electronically. Maximum allowable pressure, under constant observation at the pressure transducer, is the controlling factor. As pressure increases toward the preset maximum, the flow rate of the grout pump is gradually decreased; and, as maximum pressure is reached, the pump stops and maintains the pressure. The major advantage of this remote pressure control system is that it allows grouting within safe pressure limits, independent of friction loss in the grout line. Grouting time is thus considerably reduced since the grout pump works at maximum capacity without risking hazardous overpressure (Fig. 8).

The HFR System provides a permanent record of the actual grouting procedure which can be used to determine if specifications were met as well as for support documentation for claims.

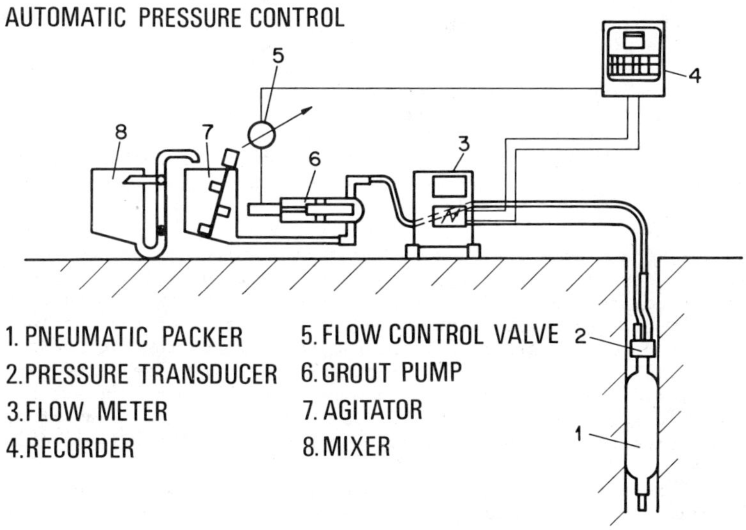

Fig. 8. Diagram of Grouting Installation with Electronic Pressure Control from the Packer.

GROUTING EQUIPMENT

Grout Mixer

The mixer should be of the high-speed colloidal type to prevent the grout from settling out of suspension. A colloidal mix is pumpable over a much longer distance than an ordinary mix and penetrates much deeper into small fissures. Water/cement ratios as low as 0.30 can be handled easily, and a high-quality mortar is still pumpable even with a low water content.

Grout Pump

A modern grout pump should be versatile; it should be able to handle low viscosity mixes at high pressure as well as stiff mixes at high capacity. The maximum pressure and flow rate should be adjustable to avoid serious damage from excess pressure or velocity. The pump should be capable of maintaining the preadjusted maximum pressure over any length of time to ensure complete filling of voids.

The choice of single- or double-action pumps can provoke argument, each type with its own advocates. Both types have advantages and disadvantages; however, recent developments have put the advantages in favor of the single-action pump. Both types allow pressure to drop when changing direction; but due to modern hydraulics, the sequence of the suction stroke on single-action pumps has become very short. On both types of pumps, the pressure measured close to the pump drops to 0 in the change-over position. However, the time sequence of the single-action pump is only a fraction of a second longer. For grouting delicate formations, the primary consideration is that a certain maximum pressure not be exceeded. A brief pressure drop within the limit does no harm. In fact, when measured after 20 to 30 m of hose, this pressure drop is almost eliminated. For minimal pressure fluctuations, an air chamber can be installed.

However, the single-action pump has many distinct advantages not available with the double-action pump; because the suction stroke is always at full speed, the velocity of grout in the suction valve is relatively high. Thus, heavy particles which tend to settle on the suction valve during the working stroke will be pulled along by the high velocity. The suction valve is always kept clear; and stiff mixes with large particles of sand, up to 5 mm and more in size, do not create problems. In double-action pumps the valve flaps or balls tend to jam after a time because of the heavy particles deposited on the valve seat.

Another advantage of single-action pumps is that, due to the much simpler transmission, piston units can easily be replaced or exchanged for other plungers with different pressure and capacity ranges.

Finally, the dismantling of a single-action pump is much simpler than for double-action pumps. This makes cleaning, maintenance, and repair work much easier especially under arduous site conditions.

Packer Systems

For grouting various formations, two main packer systems are utilized: top hole packer and deep hole packer.

A. There are two types of top hole packers:

A.1 The common recoverable packer with expandable rubber seals. This packer can be used more than once; but it must be cleaned after each use; and, when removed, the borehole must be plugged.

A.2 The nonrecoverable (one-way) packer, consisting of a sealing unit with nonreturn valve and a chucking device, is set and grouted like the conventional type. As soon as a hole is completely grouted, the chucking device can be removed while the sealing unit remains in the hole. Its nonreturn valve prevents the grout from leaking out; it is made of a material that will not damage cutting tools and complies with coal mine regulations. This is a very economical system since the packer need not be removed and cleaned, and the grouted borehole need not be plugged again. A further advantage is that the working area in tunnels is not obstructed by protruding packer ends.

B. There are various types of deep hole packers; the two most common are:

B.1 The mechanical packer which is extended by inner and outer pipes; expansion is effected mechanically by a chucking device.

B.2 The pneumatic packer, which requires only one set of extension pipes and is expanded by air. Installation time is considerably shorter compared to the double-piped mechanical packer. The pneumatic packer can be equipped with a pressure transducer which is connected to the HFR System to record the actual grout pressure at the packer.

TECHNICAL ASPECTS

Since several grout holes are injected simultaneously, even penetration of a formation is achieved. Grouting one hole at a time can allow blockages in the connections between grouted holes and neighboring holes requiring time-consuming redrilling. Grouting several holes at the same time significantly reduces the possibility of such blockages.

ECONOMIC ASPECTS

The MHG System is equipped with two manifolds, one for grout and one for clear water, allowing each grout hose and flow meter to be individually flushed. It also permits pump tests on some of the outlets parallel to the grouting operation at the other outlets.

For example, assume an average grouting time of 120 minutes with an MHGS HFR 006 using 6 outlets. The pump test can be performed simultaneously on all 6 outlets prior to grouting, or 5 outlets can be used for grouting and 1 for pump tests. It takes about 10 minutes to make the test, plus 5 minutes for changing the hose from 1 hole to another, for a total of 15 minutes per hole, or 75 minutes for 5 holes. Thus, the same equipment can perform pumping tests and grouting at the same time.

A shorter grouting period results in less downtime of other expensive construction machinery. For example, grouting ahead of a Tunnel Boring Machine (TBM) to seal or stabilize the rock results in the TBM lying idle during the grouting operation; the faster the grouting is completed, the faster the TBM can resume its operation.

The MHGS requires a smaller investment in equipment than a set-up using four or six pumps and requires less space, less power, and fewer operating personnel. Job reports are simplified for the responsible engineer since the computerized HFR System automatically supplies graphic charts and printed records of the pump test and grouting operation. The consumption of cement and additives can be automatically recorded and printed out at any time, allowing precise calculation of grouting costs, as well as inventory control of consumables.

CASE HISTORY: STOCKHOLM UNDERGROUND RAILWAY

The Precambrian geology beneath some of Sweden's principal cities has long been the source of concern for tunnellers. Drill-and-blast operations in the hard granite rock, usually overlain by surface clay, have frequently led to considerable subsurface disruption and a consequent lowering of the groundwater table resulting in surface subsidence and damage to buildings and structures.

Sweden's third largest construction contractor is using the Multiple Hole Grouting System to solve this problem in the construction of a new line for the Stockholm Metro. The contract consists of constructing 1800 m of a new 6 km long twin tube loop line for the Stockholm Metro, which starts in the suburb of Vestra Skogen, some 10 km from the city center, and runs northwest to the suburb of Hjulsta.

The contract includes the construction of two stations where the twin tube bores converge in a "trumpet" shape. The cross section of each of the twin tubes is 22 m^2 in the station areas from 70 to 150 m^2. Depth of cover over the tunnels varies from 6 to 50 m; the project is 20 m below sea level. The geology is basically granite, compressive strength 2000 kg/sq. cm, with overlying clay in some surface areas and, as is inevitable in Sweden, a significant quantity of water.

The client for the $13 million project is the Stockholm Metro Authority, although design and control of work is overseen by the Stockholm Municipal Roads Authority. Work began in August, 1979, and is expected to be complete in December, 1981, with work being performed from 6 faces.

Grouting operations are followed by excavation using 2 hydraulic 3-boom jumbos to drill 72 shotholes per 3.5 m deep round in the 22 m^2 bores.

One essential part of the job specifications is the avoidance of significant lowering of the groundwater table. Thus, along the 1800 m length of the drive, 3 grouting classifications were made regarding the amount of water ingress permissible in the tunnel.

These classifications, related to a rather complex set of regulations drawn up by municipal authorities, will be referred to here simply as Classes 1, 2, and 3. These classifications relate principally to the density of buildings above the tunnel, although in some cases the classification may reflect sound overburden (whether built on or not) where there is no danger of subsidence due to groundwater lowering. The three classes are as follows: Class 1) permissible leakage of 3 liters/min. per 100 m of tunnel, 2) 8 liters/min. per 100 m, 3) 10 liters/min. per 100 m.

Class 1 requires considerable grouting and relates to areas where structures are founded on the clay surface strata; Class 2 where buildings are founded on rock or where no buildings are present; and Class 3 to sound overburden where no grouting is necessary if no water ingress occurs.

Assignment of the particular class and the required grouting are made by a representative of the client who checks for water ingress after each set of grout holes has been drilled. Final ingress is checked after each 100 m of the drive is completed.

The work is arranged as follows: Grouting is done in 18 m lengths using an MHG System Model HFR 006. If there is no sign of water ingress after the previous blasting rounds have been completed, about seven 18 m long, 43 mm diameter, radiating grout holes are drilled in the 22 m^2 sections. If water is present in any significant amount, this figure could be trebled. To date, the average number of grout holes drilled in the 100 m^2 sections ranges from 30 to 50; the maximum number has been 80 in one face.

The automatic MHG System then delivers the grout through six hoses simultaneously. The grout consists of cement, bentonite and a flow additive, using a water/cement ratio of 3 at the start, going down to 2, then 1, and finally to 0.5 to firmly plug the hole. The amount of cement discharged into wet rock so far has varied from 8 to 6000 kg for one 18 m operation. In good conditions just enough cement grout to plug the hole is pumped in.

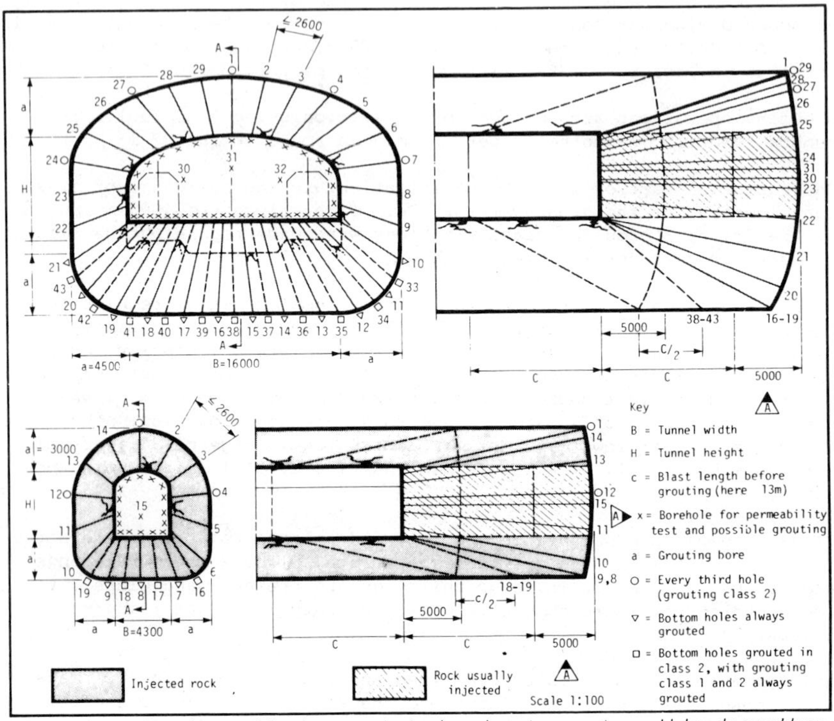

Drawings showing the application of the advance grouting in, above, the station excavations, and below, the tunnel bores.

Fig. 9. Advance Grouting in Stockholm's Subway Tunnel.

After grouting is completed, excavation takes place. In the 22 m^2 sections, full face rounds are blasted. In the station sections, two 3.5 m long pilot tunnel rounds are blasted, one 3.5 m length is then stoped, another 3.5 m pilot round is blasted, and then stoping takes place to bring the section back to full face. After blasting on both the single tube and the station sections, a grouted area about 4 m deep is retained through which the next 18 m grouting operation takes place. The whole operation (excluding mucking) is carried out by two workers drilling, two grouting and one charging.

Fig. 10. Multiple Hole Grouting System at Work in Stockholm Subway Tunnel.

In this project the MHG System's electronic monitoring equipment gave immediate digital readouts on flow and pressure for each grout outlet, recorded all data in both graph and print-out form, and permitted the client to consult the detailed records of the grouting operation at any time.

The MHG System HFR 006 has so far been in operation at this project for over 4,000 hours of grouting and pump tests, monitoring placement of over 430,000 kg of cement plus bentonite and additives for injection grouting to form an impermeable curtain around the tunnel periphery.

The saving in time and labor on such a large grouting operation is obvious. On this project an added advantage has been the minimal need for post grouting of the tunnel, which is to receive a final shotcrete and wire mesh lining (with some rockbolting).

CONCLUSION

Grouting has always been a special type of work with many unknown factors. The HFR System for automatic pressure and flow recording relieves much of this uncertainty and makes grouting a controllable and economical process. The case history discussed above shows that pregrouting with the MHG System is a viable possibility in many tunnelling situations, to reduce equipment downtime, to avoid sudden ingress of water, and to stabilize and seal the ground ahead of excavation operations. Multiple Hole Grouting may be the answer to many problems connected with impermeabilization and stabilization of rock and soil.

Reference

Harding, Piers G., "Multiple Grouting Techniques Beat Stockholm's Groundwater." Tunnels & Tunnelling, International Journal of Underground Works, April, 1980.

NEW LABORATORY TECHNIQUES TO TEST GROUTED FRACTURED ROCK

by Gérard Ballivy* and Pierre Niemants**

ABSTRATS

This paper describes new techniques for measuring the permeability of grouted fractures in rock core samples tested under different effective stress and flow conditions. The purpose of this research is to determine in which stress conditions grouted cut-offs in fractured bedrock can be washed out in a dam foundation.

A new very simple apparatus has been built to grout the fractures of NQ rock core samples; the width of natural or artificial fractures can be adjusted for each test. After curing of the grouted samples in the same environmental conditions as on the site, the permeability of grouted fractures are obtained with a new permeameter under convergent and divergent flows and under different effective stresses.

These stress and flow conditions are adjusted to those anticipated in the dam foundations with special considerations to cyclic loadings and tensile stress areas.

From these tests it is possible to determine which hydraulic gradient and external stress conditions cause washout of grout from rock fractures. These laboratory tests can be applied to assess the efficiency of different kinds of grouts and additives depending on the mineralogy of rocks and the width of fractures.

INTRODUCTION

Today in many cases the construction of new hydroelectric structures must accomodate sites presenting serious geological problems (4,6). The engineer, therefore, finds himself confronted not only by complicated geology, but also by a rocky material for which the geomechanical characteristics must be improved.

Once the grouting treatment is done and the reservoir put into operation, there are no other means of assessing the proper functioning of the grout curtain except from the measures of outflows and pressures in the foundation. A chemical analysis of the percolating water can reveal an

---

* Associate professor, Université de Sherbrooke, Sherbrooke, Quebec J1K 2R1, Canada and Technical Director, Groupe-Conseil S.M. Inc., Sherbrooke, Quebec J1L 1X8, Canada

** Research Assistant, Université de Sherbrooke, Quebec J1K 2R1, Canada eventual deterioration of the grout by leaching or chemical decay. In fact, one is obliged to recognize a deterioration of the grout curtain in certain cases without really knowing its mechanism.

The experimental device developed at the University of Sherbrooke allows simulation of the performance of a grout curtain under different stress conditions (5) and differential water pressures. The simulation of these conditions on rock samples grouted in the laboratory, leads to a better understanding of grout curtain deterioration. The possibility of bringing about the rupture of the medium gives an idea of the security factor in situ.

Completed by a classical geological and geotechnical inspection of the site, this laboratory approach brings out clarification on the type of grout to be used as well as the grouting pattern to be adopted on the terrain.

DESCRIPTION OF THE LABORATORY EQUIPMENT

The purpose of the experimental device which as been developed is to simulate the physical deterioration of an injected grout curtain resulting from the dam movements which produce tension, compression and shearing stresses in the grouted medium subjected to heavy water pressures. At the Malpasset Dam (France), it is estimated that the tensile stresses at the foundation are 400 kPa (53 psi) and those of the compression at 2.5 MPa (363 psi) (2). A dam having 120 m (396 ft) in height reaches 1.2 MPa (174 psi) in water pressure differential. Stresses resulting from leakage can provoke an hydraulic failure of the grout if the latter is simultaneously subjected to unfavorable dam stresses. The updated laboratory equipment meets the real parameters of the injection on the site and of the performance of the grout curtain in use. These parameters are as follows:

- Type of fracture: essentially there are two types of fractures i.e. the tension and shearing fractures. The latter are generally filled with loose materials (silt, clay) resulting from the shearing of the fracture. The roughness of the fracture will largely depend on the type of rock.
- Width of the fractures: an important parameter to clarify the type of grout to be used.
- The water and degree of saturation of the rock before grouting.
- The type of grout and curing conditions (temperature, relative humidity, water saturation).
- Grouting pressure.
- The state of stress in the surrounding medium after injection.
- The pressure differential to which the grouted medium is subjected.

Taking into account these different parameters has led to the development of a device able to grout fractured rock cores and an experimental apparatus capable of submitting grouted core samples to a wash-out test under conditions representative of actual field behaviour.

The rock core samples are of NQ size, with a diameter of 4.9 cm (1 15/16 in) and a height of 10.8 cm ($4\frac{1}{4}$ in), ma chined and with a central hole of 1.3 cm ($\frac{1}{2}$ in) in diameter and a depth of 9 cm ($3\frac{1}{2}$ in). Each core sample is fractured with a hammer and a chisel perpendicularly to its longitudinal axis and at its mid-height. The fracture thus obtained has the characteristics of a natural tensile fracture.

In addition to these natural fractures two other types the core sample and the argillous fractures which are natural fractures soiled with clay.

The grouting device consists of a table supporting a grout tank and three grouting celles allowing successive injections of three fractured rock core samples. An overall view of this arrangement is reproduced in figure 1. Figure 2 shows the plan of a grouting cell.

Figure 1 - Grouting device

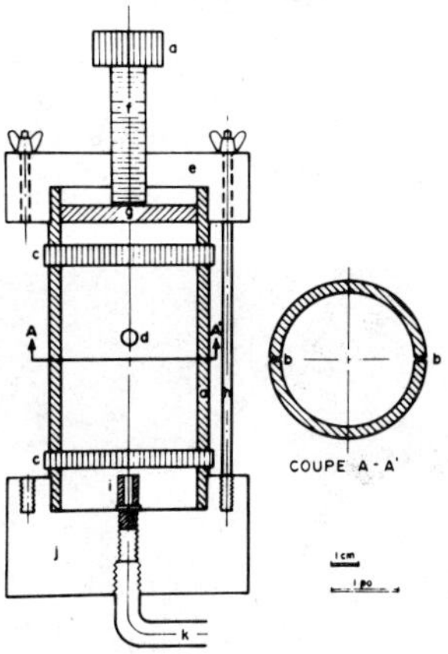

Figure 2 - Grouting cell

Figure 2 - Legend

a. P.V.C. casing
b. longitudinal rubber joint
c. clamp
d. grout outgoing orifice
e. cap
f. regulating screw
g. plate
h. stainless steel rod
i. rubber setting-coupling
j. base
k. feed pipe

The grout tank with a holding capacity of 1170 cm^3 (70 in^3) is composed of a compressed air feeding device (max. 200 kPa, 30 spi) and a spiral grout agitator connected to a variable speed rotating motor (max. 1200 rpm). The grout cell consists of a brass base with a feed pipe connected to the grout tank and a rubber setting-coupling designed to hold the rock core sample. The cap is equipped with a regulating screw for the desired fracture opening (from 0 to 3 cm, $1\frac{1}{4}$ in). A plate provides the link between the regulating screw and the upper face of the core sample. Three stainless steel rods join the base and the cap.

The two halves of the fractured rock core sample are covered with a protective rubber membrane. They are placed in a P.V.C. casing with an inside diameter of 5.1 cm (2 in) split in two lengthwise. Inside the tow parts of the casing, a cavity with a depth of 2 mm (0.079 in) is designed to received the excess grout emanating from the fracture when the grouting takes place. The position in height of this cavity is such that it always presents itself facing the fracture of the core sample. It is used to achieve homogeneous injection conditions on the whole exterior periphery of the fracture. Each half casing has an orifice for the outgoing grout into the upper part of the cavity, the outflow of which can be regulated. The tightness of the casing around the rock core sample is assured by two longitudinal rubber joints. Two clamps ensure that the core sample is perfectly set in its casing. Figure 3 shows the setting-up of a sample in the grouting casing.

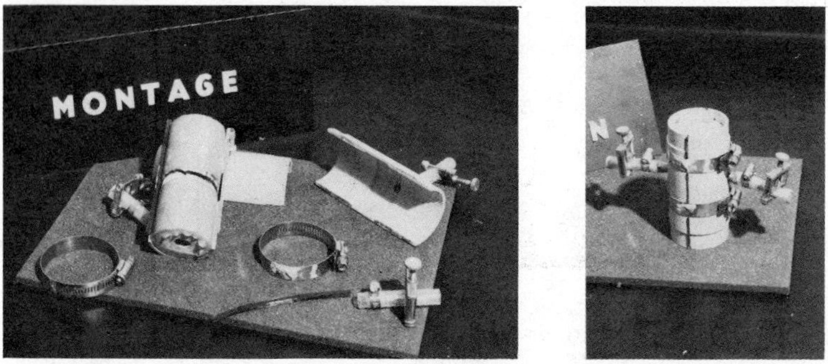

Figure 3 - Setting-up of a fissured rock core sample in the grouting casing

The adjusting of the desired fracture opening is done after having placed the casing into the base and the cap, by loosening the clamps and activating the regulating screw.

The grout under pressure (approximately 100 kPa, 15 psi) is brought successively from the tank to the three grout cells by connecting the tank outflow pipe to that of the cell in operation. The grout enters into the core sample through the central cavity and spreads into the fracture.

The injected core sample is left to rest for a few

hours (duration depends on the type of grout used). Then the central cavity of the core sample is cleaned after the casing has been removed from the grout cell. This cleaning is done by using a dampened cloth rolled up around a metal rod. This core sample manipulation right after grouting is quite simple in that the grouted core sample is held perfectly secure by the casing. The core sample, still in its casing is placed under desired curing conditions.

The stripping of the casing is done just before proceeding to the wash-out test. Then the core sample appears as illustrated in figure 4. There remains the removal of the rubber wrappings protecting the core sample and the grout ring that comes from the cavity in the casing.

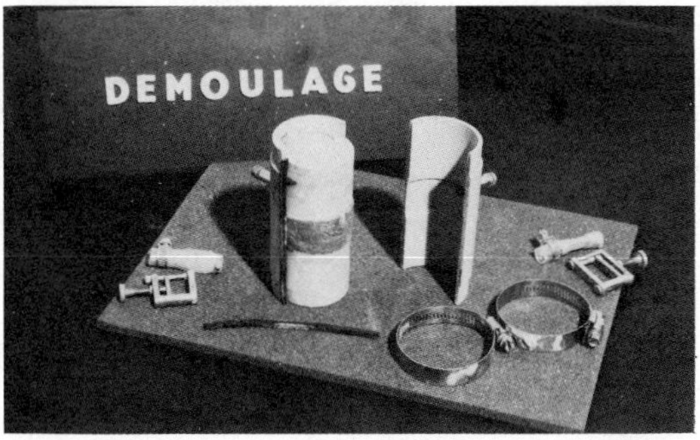

Figure 4 - Withdrawn grouted rock core sample

Figure 5 - General view of the permeameters

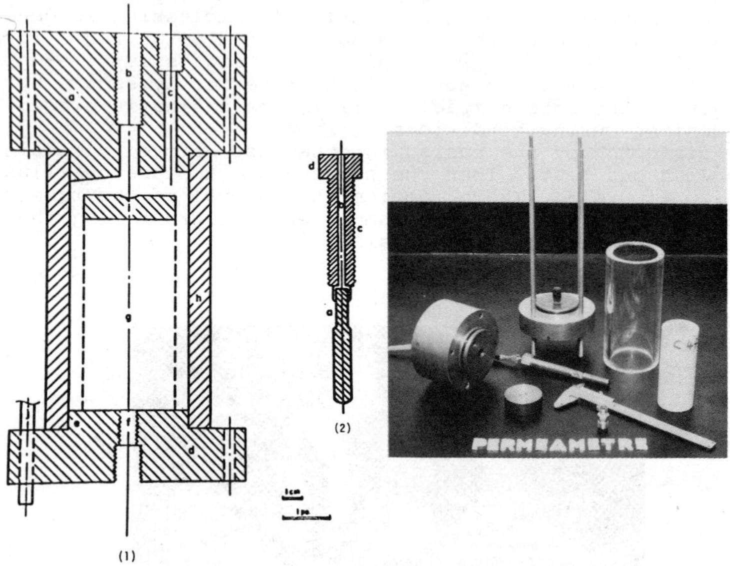

Legend:

(1) Permeameter cell

a. cap
b. central screw location
c. water circulation orifice
d. base
e. tightness "o" ring
f. rubber setting-coupling location
g. grouted rock core sample
h. cast acrylic tubing
i. pressure plate
j. stainless steel rod

(2) Central screw

a. strain gages location (square section)
b. central hole for electric wires
c. thread
d. bolt

Figure 6 - Parts of the permeameter

The cell designed for the wash-out test is, in fact, a modified radial permeameter (3) allowing the use of water pressures reaching up to 2.1 MPa (300 psi) and simultaneous axial load up to 6.7 kN (1500 lb) (external axial stress σ_a = 3.7 MPa, 540 psi). It is possible to control the imposed stresses: tangential tensile stresses in outward flow and tangential compression stresses in inward flow. The regulation of axial pressure combined with the type of flow (divergent or convergent) permits the creation of a state of compressive or tensile axial stres in the grouted fractured core sample. Figure 5 shows the three cells ready for a grout wash-out test.
Figure 6 shows the main characteristics of the cell.

The central brass screw applies the axial pressure on the sample. A section of 9.5 mm x 9.5 mm (0.375 in x 0.375 in) is provided for the mounting of four strain gages. A pressure plate (in brass) links the central screw to the core sample. The transparent cast acrylic tubing permits viewing the wash-out test in process.

The pressure source consists of two bottles of liquid nitrogen required for the application of pressure and back pressure. The percolating water is distilled and de-aired.

OBSERVATIONS ON TEST PERFORMED

Some sixty wash-out tests (1,7) have been performed, especially on ordovician fine grain limestone from St. Marc (Quebec) and also on medium grain cambro-ordovician granite from St. Gerard (Quebec) as well as on Cambrian sandstone, fine grain cemented, from Mirabel (Quebec). These three types of rocks are representative of the main foundations of dams in Quebec.

The grout used has the following characteristics: 1000 gr ordinary Portland cement (type 10) sieved (ASTM sieve no. 40), 80 gr of bentonite, 600 gr of water.

The water-cement ratio of 0.6 may appear small considering the grouts used on site, but it must be remembered that the device does not allow the excess water to run out as is the case in a natural surrounding.

The curing of the grouted core samples is done under conditions closely related to those met in practice in Quebec: 4°C (40°F) and 61% relative humidity for most of the core samples studied, or immersed in water. The curing time is from one to three days.

The opening of the fractures studied varies from 0.70 mm to 5 mm (0.03 to 0.2 in).

In a first series of tests, the grouted core sample is subjected to increasing values of the pressure differential up to 2.1 MPa (300 psi), for a constant axial load A. Increase of the differential Δp is done by stages of 100 kPa (15 psi) approximately for a flow duration of three minutes at each stage. The evolution of the permeability of the grouted core sample is followed throughout the test until the eventual wash-out. The permeability is given as follows:

$$k = \frac{Q}{2\ \pi\ \Delta p.\ L_{eff}}\ \ell n\ (R_2/R_1)$$

with Q = outflow of percolated water
Δp = pressure differential
L_{eff} = effective length on which the flow is radial
R_2 = external radius of the core sample
R_1 = internal radius of the core sample

For a given core sample the permeability can be analysed by the examination of the factor $Q/\Delta p$. The evolution of this factor through an applied pressure differential Δp allows us, therefore, to follow the process of the wash-out test. A reducing type $(Q/\Delta p)$ vs (Δp) shows a clear tendency towards an obstructing process: the grouted core sample permeability decreases even for increasing values of applied differential pressure Δp as shown in figure 7. For the series of tests carried out, an obstructing effect has been noted in 88% of the cases, either in terms of convergent or divergent flows. This obstructing process can lead either to the wash-out process, which is characterized on the graph by a more or less sudden upward movement or by a plateau indicating a constant permeability. A few tests have shown an increase as illustrated in figure 8, which indicates immediate wash-out. This type of pattern is linked with problems of premature rock-grout parting which occur at the time of the mounting or during the application of the pressure

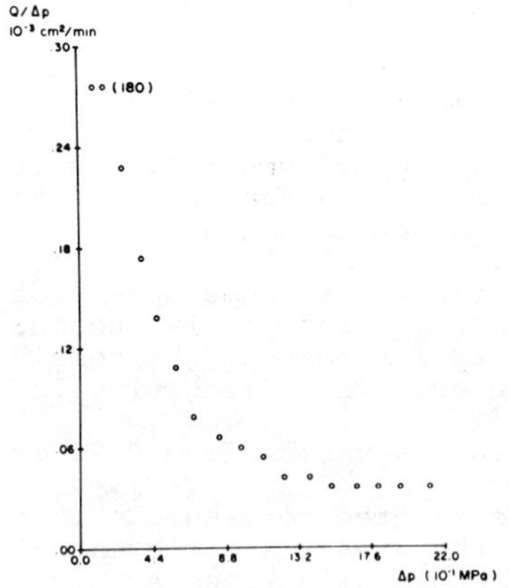

Figure 7 - Results of a test with an obstructing process
(iii) = axial load (N)

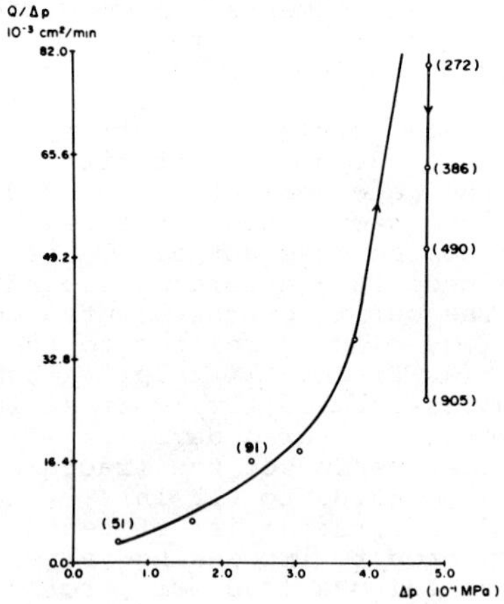

Figure 8 - Results of a test with immediate wash-out
(iii) = axial load (N)

differential Δp when the axial load A is too weak. (A < 180 N (40 lb), σ_a < 100 kPa (15 psi)). Figure 8 also shows the favorable influence of an increase of axial load A over the permeability after wash-out.

The obstructing process which was studied was more important during the convergent flow when compared to the divergent flow and was influenced by the type of grout used. The obstructing process phenomenon does not appear to be related to the rock itself: tests performed under the same conditions as those for grouted samples have shown a rock permeability practically constant with differential Δp. The permeability of an injected core sample depends on the grout whose permeability ranges from 10^{-7} cm/s (3 10^{-9} ft/sec) at three days, while that of limestone is of 10^{-11} cm/s (3 10^{-13} ft/sec).

Water tests on grout core samples have shown a permeability remarkably constant through the differential Δp until the hydraulic failure of the grout sample. This proves that the obstructing process is not caused by the grout. Hence, the obstructing process originates from a rock-grout interaction. The amount of water having percolated through the sample is likely responsible for the grout reaction and of the suspension of particles favorable to the obstructing process.

The rupture of the grouted rock core sample occurs in the following manner (figure 9):

1. DIVERGENT TYPE OF FLOW

- immediate leaching carrying out the particles: cases of young core samples (1 day) with axial load A lower than 300 N (axial stress σ_a < 170 kPa, 25 psi). An increase of axial compression stress produces grout confinement, thus improving by far its resistance to the wash-out process.
- rupture at the rock-grout interface: in the case of a bad rock-grout adhesion (smooth or argillous fracture) combined with weak axial load lower than 200 N (σ_a < 100 kPa, 15 psi) producing a state of axial tensile stress.
- rupture by exceeding tangential maximum tensile stresses, seen in the case of a core sample of two or three days given an axial load A exceeding 180 N (σ_a > 100 kPa, 15 psi).

2. CONVERGENT TYPE OF FLOW

- rupture by exceeding maximum tangential compression stresses. If after a two-day curing time this type of rupture is theoretically possible, an effect of confining of the grout occurs which renders all wash-out impossible under a differential Δp to 2.1 MPa (300 psi), even for a one-day core sample. This explains why more tests have been performed with the divergent type of flow.

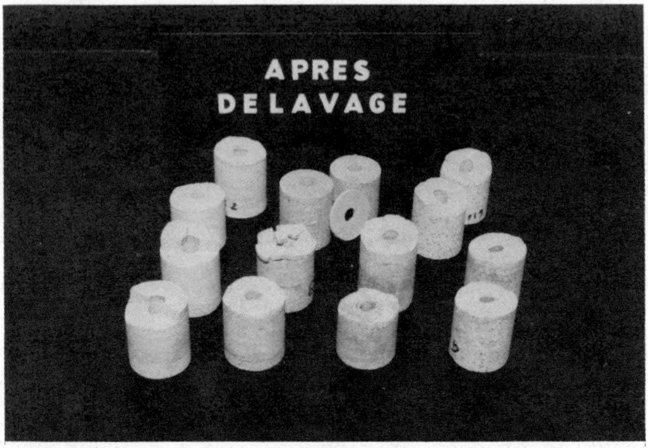

Figure 9 - Failed samples

The increase of axial load A after the wash-out process allows us to observe the post-wash-out performance of the grouted core sample. A second, even a third wash-out can then take place; figure 10 shows an example.

This increase of the axial pressure is especially profitable when the first wash-out has resulted from the parting at the rock-grout interface without bringing about an important deterioration of the injected grout. In the case of an argillous fracture, the efficacity is greatly reduced because the first wash-out has brought about a leaching of the clay covering the surface of the fracture and leaving empty spaces even for the important values of the axial load.

The overall results, discussed in their entirety in (1) and (7) show that:

- The resistance to the wash-out process is weaker the larger the fracture is.
- The resistance to the wash-out process is weaker if the injected grout is young.
- An increase of the axial load increases considerably the resistance to wash-out especially in cases of very young core samples (1 day).
- The tests on smooth or argillous fractures produce a weaker resistance to wash-out than those performed on a natural fracture, the rock-grout adhesion being of a poorer quality.
- The condition of the surface of the fracture is important. The heavy roughness obtained in the case of

granite explains its increased resistance to the wash-out process.

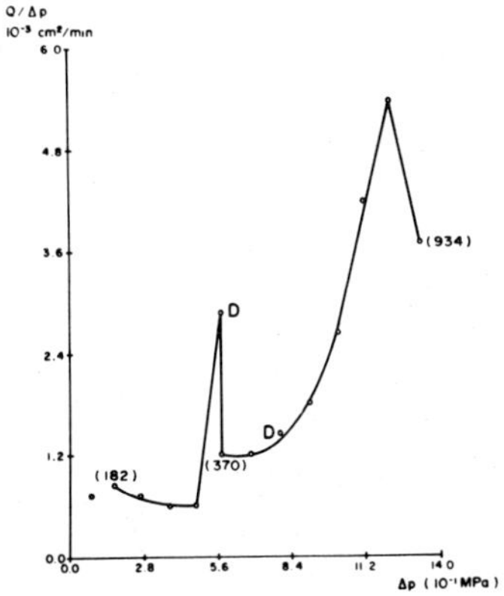

Figure 10 - Results of repeated wash-out testing
(iii) = axial load (N)
D = wash-out

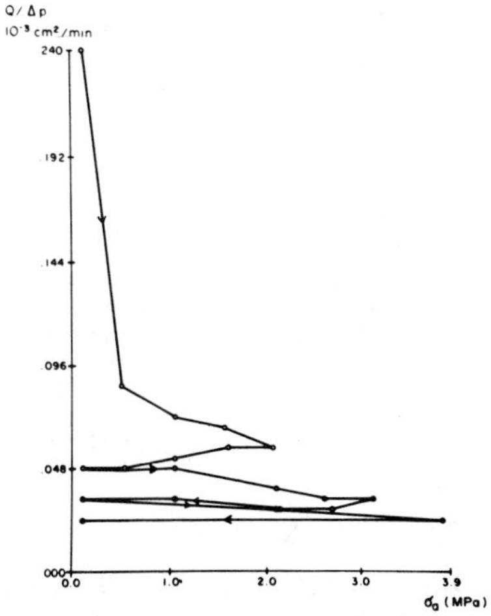

Figure 11 - Variation of the permeability with the external axial stress σ_a

A second series of tests demonstrates the influence of axial load A or axial stress σ_a over the permeability of grouted core samples subjected to a constant pressure differential of 0.7 MPa (100 psi). Three cycles of loading-unloading are applied to the sample for a maximal axial load increasing at each stage.

The general pattern of the permeability ($Q/\Delta p$) vs (σ_a) is illustrated in figure 11 which demonstrates the total irreversibility. The reduction of permeability is felt especially during the first loading stage, becoming gradually smaller in the cycles that follow. Table 1 summarizes the reduction in permeability obtained after each loading-unloading cycle for three grouted limestone core samples, the percentage being relative to the initial permeability. It can be seen that the thick fractures (3 mm, 0-12 in) have a larger reduction of permeability, which becomes more constant after a few cycles.

	e = 3.09 mm(*) natural fracture	e = 0.91 mm natural fracture	e = 2.68 mm smooth fracture
Initial permeability	100%	100%	100%
first cycle	22%	80%	21%
second cycle	17%	72%	13%
third cycle	18%	67%	10%

(*) e = Width of the fracture (1 mm = 0.04 in)

Table 1: Reduction of permeability in the cyclic test

CONCLUSION

A device has been developped to grout core samples of fractured rock in the laboratory and to subject them to a wash-out test. This gives better understanding of the type of conditions that can provoke the deterioration of a grout curtain in the rock foundation of a dam. It is possible to simulate the state of stresses and the pressure differential one can expect to find, once the dam is put into operation.

The thickness of the fracture to be grouted, its surface conditions, the relative humidity before and after grouting, also play an important role in the determination of the grout to be used in situ. These factors can also be simulated in the laboratory.

The tests performed demonstrate that there occurs an obstructing process phenomenon in the early stages of the water pressure increase. This represents a beneficial effect on which one can count in practice, if a parting of the rock-grout interface does not occur. This obstructing process has, moreover, been observed in the field on some dam sites of Hydro-Quebec.

Periodic movements of the dam can produce local tensile stresses in the grouted mass foundation and bring about a parting of the rock-grout. In placing the grouted joint under new compression, this can bring back the permeability to a value close to the initial value, unless a leaching of the filling materials (silt, clay) covering the surface of the fracture has occurred.

The first time the grouted joint is placed under compression brings about a very clear reduction in the permeability, especially when the joint is relatively thick.

The simulation in the laboratory of the wash-out phenomenon on artificially fractured core samples and field cores presenting a natural tensile or shearing joint, with a filling of argillous or silty materials and with fracture openings adapted to the results of the field geological investigation, will render possible the testing of the efficiency of various types of grout and additives. The stress and flow conditions from which grout deterioration may be expected may also been simulated. The most susceptible areas then would be treated with appropriate measures.

ACKNOWLEDGMENTS

This study was made possible through the courtesy of a grant from the Natural Science and Engineering Research Council of Canada (Grant No. A-42-28), and through the collaboration offered by the Geological and Geotechnical Service of Hydro-Quebec, specifically from Messrs. Pierre M. Crépeau, Nguyen Don and Jean Préfontaine.

REFERENCES

1. Ballivy, G., Niemants, P., "Nouvelles techniques de reconnaissance en laboratoire pour la conception des rideaux d'étanchéité dans le roc", 14e Congrès des grands barrages, Rio, Brésil, 1982, (accepté pour présentation).

2. Bellier, J., "La rupture du barrage de Malpasset", Travaux, juillet 1967, Paris.

3. Bernaix, J., "Etude géotechnique de la roche de Malpasset", Dunod, Paris, France, 1967.

4. I.C.O.L.D., "Lessons from Dan Incidents", Reduced edition, Paris, France, 1973.

5. Londe, P., Sabarly, F., "La distribution des perméabilités dans la fondation des barrages-voûtes en fonction du champ de contraintes", comptes rendus du 1er Congrès de mécanique des roches, vol. II, Lisbonne, 1966, pp. 517-521.

6. Goodman, R.E., Sundaram, P.N., "Permeability and Piping in Fractured Rocks", A.S.C.E., Journal of the Geotechnical Engineering Division, Vol. 106, GT5, May 1980, pp. 485-498.

7. Niemants, P., "Etude du comportement des voiles d'étanchéité injectés dans les fondations rocheuses de barrage", Thèse de maîtrise en Sciences appliquées, Université de Sherbrooke, 1981.

LIME SLURRY GLAZE METHOD OF SOIL TESTING

James R. Blacklock

ABSTRACT

Lime Slurry Pressure Injection (LSPI) is a specialized approach to the inplace improvement of large soil masses. It is used principally in areas where the surface must remain intact or where stabilization at large depth is required. Though the LSPI method was originally developed by FHA for use in the rehabilitation of houses constructed on expansive clay soils, it is now used for both preconstruction and remedial situations for the building and transportation fields. The exact mechanism theory by which LSPI improves or renders a soil mass stable is still being developed; however, one aspect of LSPI, which is considered to be of substantial benefit, is the formation of lime seams within the treated soil mass. Many properties are attributed to lime seams; the predominant one put forth, being that they form moisture barriers which tend to stabilize the moisture content of the soil mass. One important LSPI property which has received very little attention, is the influence of lime seams on the strength of the soil mass. The laboratory testing of lime slurry glaze samples to determine the strengthening aspect is the subject of this paper.

Strength data and material characteristics, determined from unconfined compression (U.C.) testing of laboratory manufactured lime glazed specimens, have been used in this study in an effort to determine the influence of the lime seam on soil-mass strength. During the study, the design of a new laboratory test sample and test program was developed.

Specimens treated with two types of glaze were tested in unconfined compression. These included specimens manufactured with a water glaze surface treatment and specimens manufactured with a lime slurry glaze surface treatment. Significant strength differences were determined. The water glazed specimens were less strong than the lime slurry glazed specimens after a 35 day room temperature cure. The principal means of strength gain in the treated samples was observed to be as a result of confining pressures exerted by the lime glaze jacket. Additional tests were performed on different soils both remolded and undisturbed. The positive reaction of lime slurry with the test soils resulted in a tough coating which surrounded and reinforced the samples. It is proposed that this test method be used by geotechnical engineers to help predict the success of LSPI stabilization projects.

Associate Professor Engineering, The University of Arkansas at Little Rock, Little Rock, Arkansas

---

INTRODUCTION

The reinforcement and stabilization of soils for the construction of shallow foundations for highways, railroads, airfields, parking lots, and similar facilities requires the economical improvement of large volumes of native and landfill soils. It is generally necessary to improve the soils' mechanical strength, as well as its resistance to environmental forces. The injection grouting methods of treating surface soil masses hold the best promise for the economical improvement of soils for shallow foundation construction. Materials commonly utilized for this purpose include cement, fly ash, and lime. These are mixed with water to form an injectable grout. They are usually applied for the purpose of restriction of water movement or for soil reinforcement and containment. In certain circumstances both uses are necessary; however, unfortunately some grouts are designed for only one purpose or the other. Thick sand and cement grouts might be considered double purposed; however, they have the disadvantages of requiring considerable mixing and using too much of the stabilizer material per unit of volume stabilized. Each of the materials, with differing methods of application depending on the design problem, have been used in the past in a variety of stabilization situations. Most are multi-purposed. The one material that has perhaps been used most often for soil stabilization is lime. The lime slurry pressure injection method (LSPI) of soil stabilization is one of the methods in which lime is used for soil stabilization and reinforcement. Successfully injected lime slurry develops a series of lime sheets and seams in the in-situ soil, Figure 1, which can impede moisture movement and cause the strength to be increased, Blacklock and Lawson (Ref. 2). The development of design theory and laboratory test methods for the predetermination of the liklihood of success of a particular lime injection application has been the subject of several papers (Ref. 11), Blenden (Ref. 5), Lundy and Greenfield (Ref. 12), Fohs and Kinter (Ref. 7), Higgins (Ref. 9) and Ingles and Neil (Ref. 10). There has been at least one study for the U.S. Air Force, Thompson and Robnett (Ref. 15), one for the highways, Higgins (Ref. 9), and one for the railroads, Blacklock (Ref. 3), and one for the building industry (Ref. 6).

The development of laboratory test methods specifically for lime injection was reported by Lawson (Ref. 11), Blacklock and Lawson (Ref. 2), Greeson (Ref. 8), and Blenden (Ref. 5). In Lawson's paper, a U.C. test sample which incorporated a thin perpendicular lime seam was reported. In Blacklock and Lawson's Handbook, a series of samples which utilized a method called lime inoculation was presented. In Greeson's study, a lime smeared consolidation sample was investigated. In Blenden's paper, a U.C. sample, which incorporated a drilled hole lime reinforcement method was reported. The test sample design which is discussed in this paper is one which utilizes a thin lime glaze which forms a lime surface membrane. The lime glazed samples are more nearly correct in their similitude of the lime injection process than those previously mentioned.

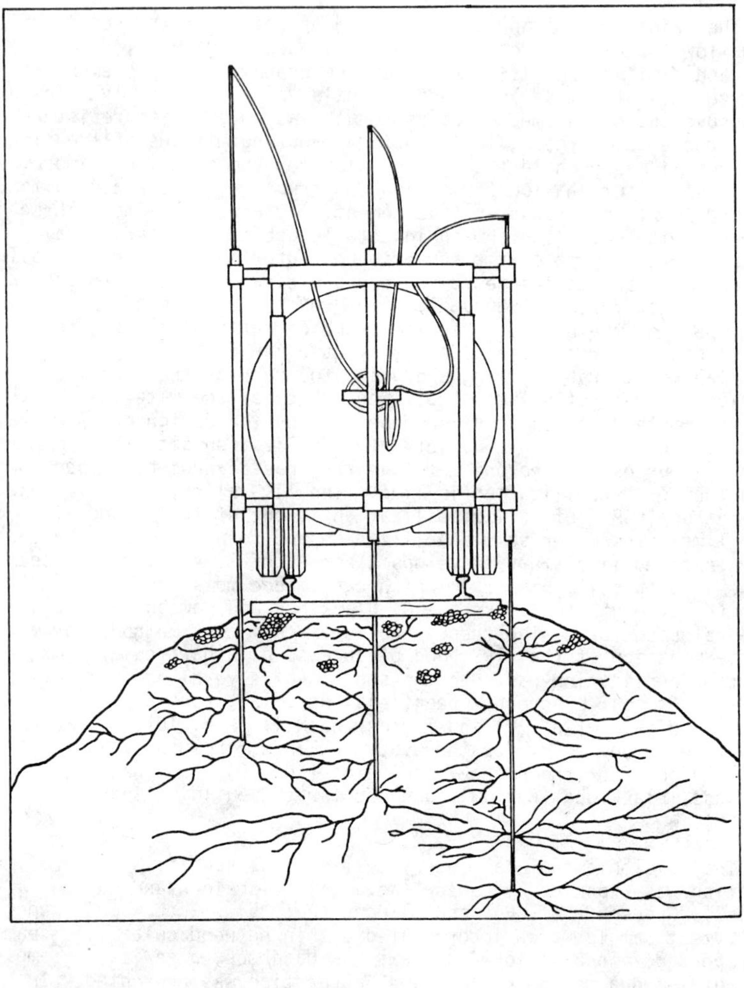

Fig. 1. Typical Railroad Subgrade Lime Seam Formation and Distribution.

BACKGROUND

The method of soil stabilization by LSPI has suffered for many years for a lack of mechanism and design theories, and representative laboratory and field test procedures. The largest setback in the development and utilization of the lime injection grouting method came about when an incomplete theory of lime stabilization by lime seam migration and growth was advanced and then subsequently investigated and renounced. The development of a complete lime injection theory is beyond the scope of this paper, however, one aspect of the mechanism of lime injection will be discussed and that is the development and utilization of lime stabilized tensile strength seams, which in effect form a composite structure of confining membranes encapsulating the soil mass being stabilized. The fact that the tensile membrane reinforcement capability of a thin lime seam and the adjacent soil has been previously ignored, is perhaps because the historic view of lime-cements has been that of a high compressive and low tensile strength material, and also, because the predominate loading environment for shallow foundation soils is one of bearing or compression. Recently a breakthrough in this area was achieved when mesh fabrics were found to increase the bearing strength of weak soil masses by supplying a much needed horizontal tensile strength component to the structural system. The fact that lime seams of injected lime slurry do not substantially migrate does in no way disqualify the lime injection grouting method as a valid method of soil stabilization. The network of lime sheets and seams, in fact, are in many cases adequate to cause the soil mass strength to be improved substantially. In addition, the soil mass is stabilized from volumetric changes brought about by environmental forces due to the sealing of the cracks by the lime slurry seam.

LIME INJECTION GROUTING

The lime injection grouting method of soil stabilization and reinforcement results in the formation of a network of thin lime sheets and seams. These lime slurry sheets and seams react with the adjacent soil to form strong relatively impervious tensile membranes locked into the soil mass. The effect of these membranes is to control the movement of moisture and to reinforce and confine the segmented portions of the soil mass.

Normal lime injection involves the mixing of 2.5-lb (1.14-kg) of hydrated lime to 1-gal (3.8-l) of water to form a uniform thin slurry and the pressure injection of approximately 7.5-gal (28.4-l) of slurry per cu yd (0.76-cu m) of soil to be treated. The surface area of lime membranes thus formed is approximately 120-sq ft (.093-sq m) per cu yd (0.76-cu m) of soil treated. The ratio of area of lime membranes and seams per unit of soil is, therefore, approximately 4.0. These numbers are based on the injection of 1.0-1.5 percent of lime per unit of weight into the subgrade. The new lime glazed soils samples discussed in this paper have a ratio of area of lime glaze membrane per unit of soil volume of 3.8.

STRESS ANALYSIS

The mechanized strength gained in a soil mass that is lime injected depends directly upon the surface area of lime membranes formed, the reactive strength of the lime soil mix, and the geometry of the seam formation. The analysis of lime injected roadbeds by the finite element method of analysis was proposed by Blacklock (Ref. 1). The development of the glazed test sample and the extension of the tensile membrane theory presented in this paper contribute significantly to the progress needed in that area. Finite element analysis of railroad subgrade soils was reported by Tarabji and Thompson (Ref. 14), and Blacklock (Ref. 4).

SPECIMEN DESIGN AND TESTING

Static Molding.--The manufacture of remolded soil specimens for strength testing is a well-established, well-documented procedure for investigation of strengths for remolded soil. Static molding has been used in the manufacture of soil-cement specimens for strength testing in highway construction for a number of years. The method of sample preparation is described in detail in ASTM D 1632-63. Briefly, it involves filling a mold of circular cross-section, Figure 2, and compressing the soil to a desired volume by closure of two pistons using a hydraulic jack, Figure 3. The procedure is illustrated in Figure 4. Since no kneading is involved, the soil has a flocculated structure. The process differs from dynamic molding, for in dynamic molding, the soil is molded predominantly by the action of hammer on soil. In static molding, the action of soil on soil results in densification

There are certain problems associated with static molding. To achieve this consistency, it is necessary to manufacture several trial specimens and make necessary adjustments to both moisture content and dry density until an acceptable specimen is obtained. The aim of manufacturing statically molded specimens is to produce specimens consistent in moisture content and dry density as well as surface texture, appearance, and constant dimensions.

With soils which have a high clay content or high plasticity, it is difficult to mix the water into the soil. This makes it virtually impossible to obtain a uniform distribution of moisture in the specimen. One way of overcoming this problem is to seal the soil in a plastic bag after mixing and allow it to equilibrate. A period of seven days will allow equilibration. Another way is to refrain from drying the soil and break it into small chunks for molding.

Obtaining a uniform density distribution is a difficult problem, more difficult to overcome than that of uniformity of moisture content. Careful study of the remolding process reveals that uniform advancement of the pistons is imperative if a uniform dry density is to be achieved. Best results are obtained if all the soil can be poured into the mold without any tamping.

Fig. 2. Moist Soil Filling Mold.

Fig. 3. Compaction of Static Molded Sample.

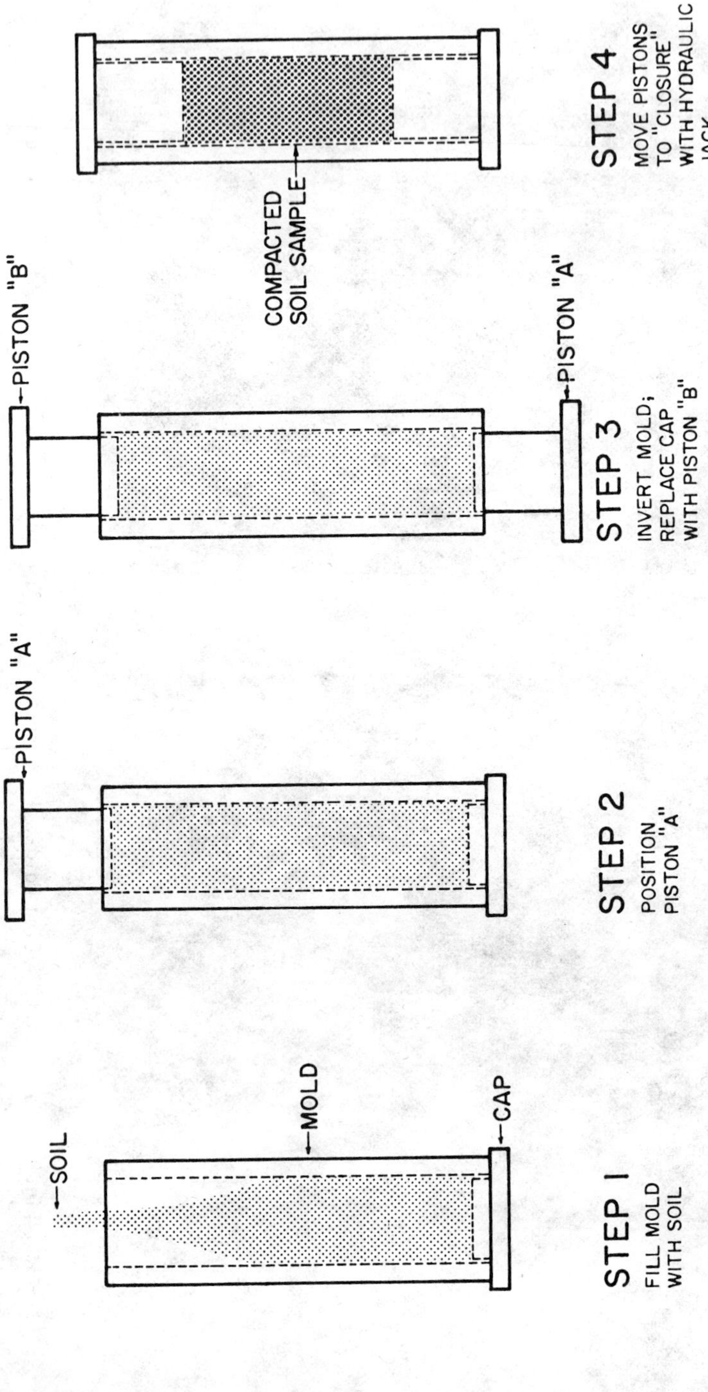

Fig. 4. Static Molding Procedure.

Specimens approximately the same size as those produced in the Harvard Compaction Test have been shown to produce consistent data. Throughout the development of the specimen-manufacturing equipment at the University of Arkansas, it has been shown that it is possible with this mold size to produce consistent specimens over long periods of time (Lawson, Ref. 11, Blendon, Ref. 5).

LIME GLAZE TREATMENT

Since this was the first study of artifically coating a sample with a lime slurry glaze in the laboratory, it was decided to keep both the specimens and the experiment simple and use equipment and materials on hand. The sample preparation and test procedure is described as follows:

1. Select a 1200 gram sample of dry pulverized soil.
2. Mix soil and water and static mold test samples using no lubricant for the mold.
3. Treat the odd numbered samples with water glaze and the even numbered samples with lime slurry glaze.
4. Seal the samples in plastic wrap and cure in moisture chamber.
5. Unwrap, weigh, and measure samples.
6. Test samples to failure in Unconfined Compression (U.C.) test machine.
7. Dry samples in oven to determine moisture content.
8. Calculate density, moisture content, and vertical stresses.
9. Plot stress curves.
10. Evaluate data.

The precise procedure for applying the water and lime glaze is necessary in order to achieve a uniform distribution of material on the sample. This is critical in the test sample, more so than in the field, since the test sample must be tested and stand alone, whereas the field soil is a portion of a huge structural system, thus being more forgiving of gaps and spaces in the lime seams. The lime seams in LSPI are often very random in nature. Two steps of the glazing procedure are shown in Figures 5 and 6. The finished product is shown in Figure 7 prior to wrapping, and in Figure 8 after wrapping prior to curing.

TEST DATA PRESENTATION

The results of one test series with an Illinois tan sandy, silty, clay obtained from a railroad problem subgrade soil are shown in Table 1. The odd numbered samples were treated with a water glaze and the even numbered samples were treated with a lime glaze. The average stress strain curves are shown in Figure 9.

CONCLUSIONS AND RECOMMENDATIONS

Conclusions.--The purpose of this study was to develop a better lime compatability laboratory strength test to be used in conjunction with LSPI soil stabilization and reinforcement investigations. Previously developed lime compatability strength tests were useful; however, it was felt that a test which more closely simulated the actual field application would be better. The initial success achieved with

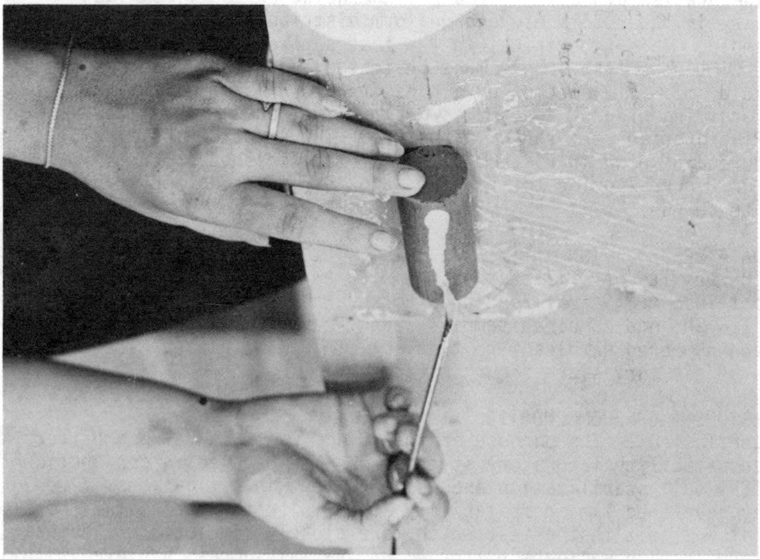

Fig. 5. Step 1 of Lime Glaze Treatment.

Fig. 6. Continuation of Lime Glaze Treatment.

Fig. 7. Glazed Sample Prior to Sealing.

Fig. 8. Glazed Sample Sealed in Plastic Wrap.

TABLE 1 - Boring 17-1 Test Data

Control Sample Data

STRAIN IN/IN	1	3	5	7	9	11	AVG STRESS
.0033	2.06	1.73	1.73	1.41	2.38	2.60	1.99
.0064	3.67	2.80	3.23	2.80	3.45	4.10	3.34
.0100	5.37	3.87	4.41	4.08	4.51	5.16	4.57
.0133	7.07	4.60	5.35	5.14	5.46	5.68	5.55
.0167	8.54	5.34	6.19	6.08	6.40	6.19	6.46
.0200	9.79	6.06	6.81	6.91	7.13	6.59	7.21
.0233	11.02	6.57	7.42	7.63	7.84	7.00	7.91
.0267	12.15	7.18	7.82	8.45	8.45	7.18	8.53
.0300	12.84	7.58	8.32	9.05	9.05	7.37	9.03
.0333	13.64	8.18	8.71	9.44	9.44	7.76	9.53
.0367	14.11	8.68	8.99	10.04	9.83	7.95	9.93
.0400	14.59	8.96	9.17	10.42	10.42	8.13	10.28
.0433	14.59	9.34	9.34	10.80	10.69	8.31	10.57
.0467	15.31	9.62	9.52	11.17	10.97	8.48	10.84
.0500	15.47	10.10	9.90	11.44	11.14	8.87	11.15
Mc =	18.70	19.95	19.78	19.72	19.74	20.41	= 19.72%
γw =	123.66	121.88	121.81	121.65	121.72	122.48	= 122.19 lb/cu.ft.
γdry=	104.18	101.61	101.69	101.61	101.65	101.72	= 102.07

Treated Sample Data

STRAIN IN/IN	2	4	6	8	10	12	AVG STRESS
.0033	2.60	2.60	1.51	1.73	2.60	2.16	2.20
.0067	3.88	4.96	3.23	3.45	5.39	4.10	4.17
.0100	5.37	6.88	4.84	5.37	7.84	6.12	6.07
.0133	6.43	8.35	6.21	7.50	9.85	7.71	7.67
.0167	7.68	9.60	7.79	8.54	11.21	8.91	8.95
.0200	9.15	10.96	9.47	9.57	12.34	10.21	10.28
.0233	9.96	11.87	11.02	10.81	13.25	11.02	11.32
.0267	10.99	12.78	12.25	11.62	13.73	11.83	12.20
.0300	11.58	13.48	13.26	12.42	14.32	12.42	12.91
.0333	12.06	13.95	14.06	13.11	14.69	12.80	13.44
.0367	12.55	14.43	14.64	13.59	15.06	13.17	13.91
.0400	13.02	14.79	15.21	14.17	15.21	13.54	14.32
.0433	13.39	15.16	15.47	14.54	15.37	13.81	14.62
.0467	13.55	15.31	15.73	14.90	15.31	14.07	14.81
.0500	13.92	15.47	15.88	15.05	15.36	14.43	15.02
Mc =	19.93	19.93	19.58	19.92	19.67	19.51	= 19.76%
γw =	122.59	122.33	122.33	122.31	122.54	122.09	= 122.36 lb/cu.ft.
γdry=	102.22	102.00	102.30	101.99	102.40	102.16	= 102.18 lb/cu.ft.

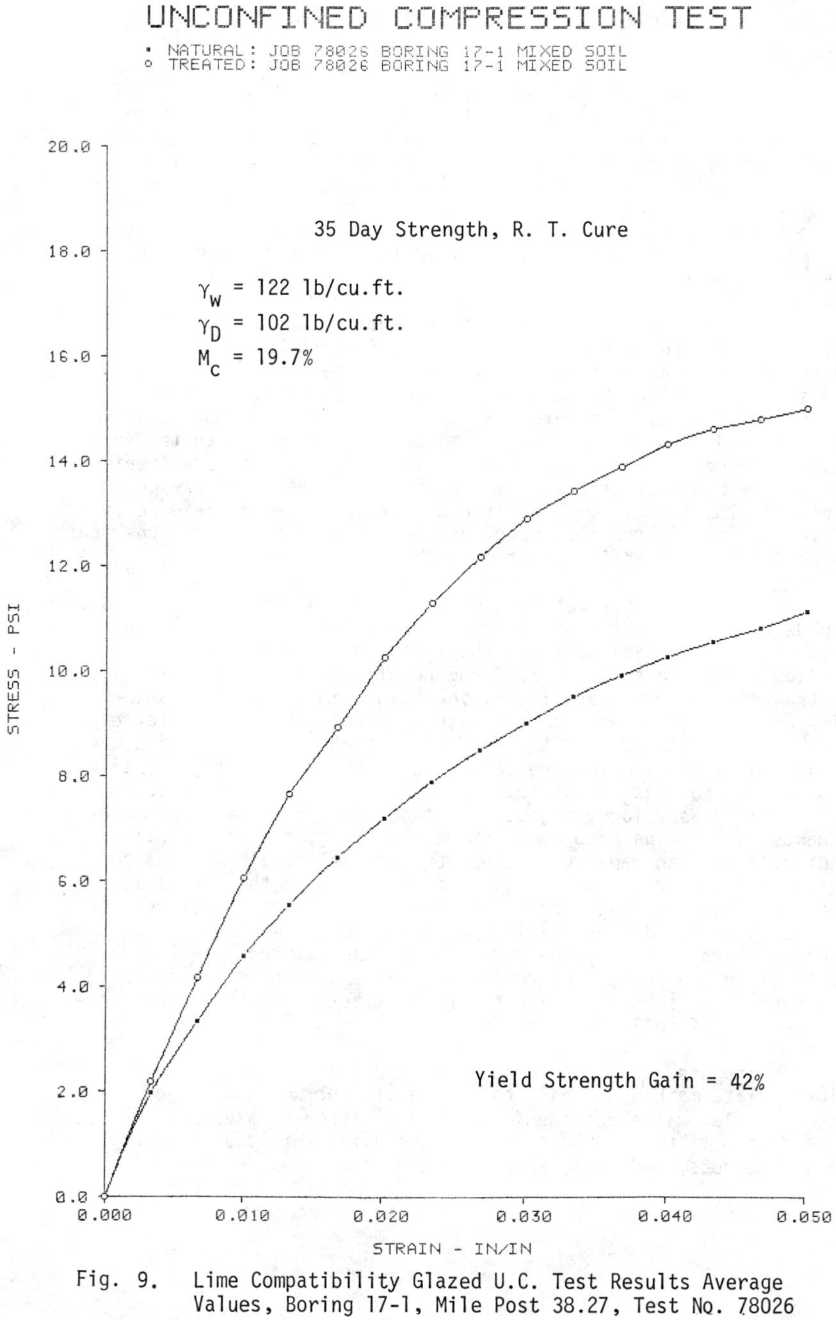

Fig. 9. Lime Compatibility Glazed U.C. Test Results Average Values, Boring 17-1, Mile Post 38.27, Test No. 78026

the first forty-four (44) remolded glazed test samples was very encouraging. The twenty-two (22) lime glazed samples showed a 35% to 86% yield strength gain with 0.5% lime after a 35 day room temperature cure. The moisture variation was well controlled having a variation of ±0.3%. Additional tests with undisturbed samples showed similar yield strength gains. In other test series, samples of sandy and silty clay mixtures were shown to register higher strength gains with smaller percentages of lime in the treatment glaze than heavier clays. The manufacture of the lime glaze treatment proved to be economical and reliable. Very little procedure experimentation was required to achieve satisfactory results.

The development of the lime seam tensile membrane soil stabilization and reinforcement notion resulted from the study and analysis, of the failure mechanism of the lime glazed samples. The strength gains registered in the test samples could not be attributed totally to compression column reinforcement. It was observed that the lime glaze formed a continuous tensile membrane which confined the soil column during the test. The failure mechanism of the lime treated samples, exhibited either a classic 45 degree shear failure or a vertical tensile split. The ratio of surface area of glaze per unit of volume soil in the test specimens was similar to that which occurs in normal lime injection grouting projects.

Recommendations.--Future highway, railroad, and airfield research and development projects, utilizing the various injection and grouting methods of soil stabilization, should include the design and test of shallow foundations, stabilized and reinforced by multiple tensile membranes of common stabilizers, including lime, lime/fly ash, and lime-cement. The study of this method in numerous full scale test applications will aid in its rapid development and utilization. An additional incentive for the development of the laboratory tensile test method relates to the future utilization of the lime slurry and lime/fly ash slurry injection methods for the repair and maintenance of existing problem subgrade soils. Highway and airfield engineers need to rapidly develop new remedial repair methods for existing failed roadbed soils. This is a problem that the railroad engineers have existed with for several years. In the future, the new construction dollar in all ground transportation areas will be greatly overshadowed by the magnitude of the maintenance-of-way dollar. In the past, many railroads have experienced the economical improvement of subgrade soils by the utilization of the lime injection method of soil stabilization.

The future development and continued improvement of the lime slurry glaze method of soil testing should include studies of different soils, different chemicals, and different glaze sample sizes. In addition, there should be a study of different loading methods, curing methods, and glaze application methods.

REFERENCES

1. Blacklock, J. R., "Analysis of Lime Injected Roadbeds," Proceedings of Roadbed Stabilization: Lime Injection Conference, Report No. FRA-OR&D-76-137, November, 1975.
2. Blacklock, J. R., and Lawson, C. H., Handbook for Railroad Track Stabilization Using Lime Slurry Pressure Injection, NTIS, 1977.
3. Blacklock, J. R., "Final Report - Evaluation of Railroad Lime Slurry Stabilization," Report No. FRA-OR&D-78-09, June, 1978.
4. Blacklock, J. R., and McAlister, B. J., "An Investigation of Stresses and Deflections," 58th Annual Meeting, Transportation Research Board, Washington, D.C., January, 1979.
5. Blenden, R. E., "A Study on the Effects of a Supernatant from Lime Slurry on the Strength of a South Dakota Clay," a report submitted in partial fulfillment of the degree of Master of Science in Engineering, Graduate Institute of Technology, Little Rock, Arkansas, 1977. Unpublished.
6. B.R.A.B., Chemical Soil Stabilization, Building Research and Advisory Board, National Academy of Sciences, Washington, D.C., 1969.
7. Fohs, D. G., and Kinter, E. B., "Migration of Lime in Compacted Soil, Public Roads, Vol. 37, No. 1, July, 1971, pp. 1-8.
8. Greeson, W. C., "Laboratory Testing of Lime Slurry Pressure Injected Soils," a report submitted in partial fulfillment of the requirements of the degree of Master of Science in Civil Engineering, University of Arkansas, 1975. Unpublished.
9. Higgins, C. M., "Lime Treatment at Depth," Final Report No. 41, Research Project No. 63-75, Louisiana Department of Highways, June, 1969.
10. Ingles, O. G., and Neil, R. C., "Lime Grout Penetration and Associated Moisture Movements in Soil."
11. Lawson, C. H., "A Study of the Effect of a Lime Seam on the Unconfined Compression Characteristics of a South Dakota Clay," a thesis submitted in partial fulfillment of the degree of Master of Science in Engineering, Graduate Institute of Technology, Little Rock, Arkansas, 1977
12. Lundy, H. L., Jr., and Greenfield, B. J., "Evaluation of Deep In-Situ Soil Stabilization by High-Pressure Lime Injection," Highway Research Record No. 235, Highway Research Board, 1968.
13. McNutt-Schneller, Inc., "Evaluation of Pressure Lime Injection by Factorial Analysis of Variance," McNutt-Schneller, Inc., Engineers, 201 Wallace Building, Little Rock, Arkansas, June 15, 1972.
14. Tarabji, S. D., and Thompson, M. R., "Finite Element Analysis of a Railway Track Support System," Report No. FRA-OR&D-76-257, July, 1976.
15. Thompson, M. R., and Robnett, O. L., "Field Evaluation of Pressure Injection Lime Treatment for Strengthening Subgrade Soils," Proceedings of Roadbed Stabilization: Lime Injection Conference.
16. Thompson, M. R., and Robnett, Q. L., "Pressure Injection Lime Treatment of Swelling Soils," 54th Annual Meeting, Transportation Research Board, Washington, D.C., January, 1975.

CEMENT GROUTS IN OFFSHORE STEEL STRUCTURES

John Colin Woodward CEng DLC FICE

ABSTRACT

This paper describes the improvements in offshore grouting techniques for deep water steel structures which have taken place in the North Sea over the past few years as a result of a wide ranging laboratory testing programme. The work covered the examination of the strength of cement-bonded pile and other tubular connections, preparation of design procedures and monitoring methods. The testing, which is still continuing, has resulted in saving material and fabrication time as well as improving quality control. Grout mixing and injection procedures which evolved in conjunction with the tests are described and particular reference is made to the development of robust underwater nuclear gauges to monitor grouting progress and quality. As all North Sea platforms have to be certified by a recognised authority for structural competence before they can be commercially operated, the work described has provided an extra degree of assurance to both designers and operators.

INTRODUCTION

The most popular form of construction for fixed offshore platforms for oil and gas production in the North Sea and most other areas is the piled steel jacket. These structures are fixed to the sea bed by tubular piles driven (or placed into predrilled holes) into the sea bed through the legs of the jacket or through tubular sleeves attached to the lower part of the jacket. The annulus between the pile and the jacket leg, sleeve or strata, is then filled with a cement grout to form the structural connection.

Axial loads in the piles for these structures are typically between 2000 and 5000 tons and penetrations range from 200 to 450 ft for driven piles. Where difficult soil conditions exist, such as calcareous sands, piles should preferably be placed and then grouted into predrilled oversize holes (so called "insert" piles) to avoid disturbance due to driving. Occasionally a combination of driven and insert piles is used in particularly unfavourable sea bed conditions. With the use of improved underwater piling hammers this dual pile system is likely to be less favoured.

Director, Wimpey Laboratories Limited,
Hayes, Middlesex UB4 OLS, UK

In order to avoid the severe surface weather conditions templates and other sea bed structures designed on the "jacket" principle are increasingly being used for drilling, production and tethering systems. Most of these structures have to use grouted connections to fix the piles to the sea bed.

In shallow water jackets, say up to 150 ft, a pile is usually driven through each leg and extends to the top of the leg so that the deck modules are welded direct onto the pile. The annular grout in the leg therefore transmits only the lateral forces between the pile and the jacket and is not normally required to support the vertical loads. Bond stresses between the steel-grout interface are low and any local debonding is unlikely to compromise the overall structural stability.

In deeper water, piles generally do not extend to the surface and are either grouped in sleeve clusters around the main legs or distributed around the base of the structure. The grouted sleeve annulus in these cases therefore forms the only structural connection between the jacket and the foundation, with the grout transmitting all the forces arising from both dead weight and live environmental loading.

Although templates are much lighter than the jackets for surface-operated drilling and production facilities, and not subjected to the same live loading, the design of the connection nevertheless needs careful investigation as the length of structures available for bonding is considerably less than for large jackets.

The trend in the past seven years for major North Sea platforms has been towards increasing pile diameters and reducing the number. This has resulted in a less radially stiff connection and further highlighted the need for high quality grout in the annulus.

In view of the importance of the grouted connection a considerable research effort has been mounted in the United Kingdom to define the principal parameters which affect this connection and establishing economic design procedures. To date, this work has concentrated on the steel-grout interface, but further work is proposed to continue investigation of the pile-soil connection.

The paper now briefly discusses this research and design work and then describes the various grout materials being used and the placement and monitoring techniques.

DESIGN OF GROUTED JACKET TO PILE CONNECTIONS

The research work undertaken by Wimpey Laboratories, the design rules which ensued, and the application to structures in the North Sea and elsewhere have been described in detail by Billington and others[1,2,3]. The basic programme consisted of four phases; static tests, fatigue tests, cyclic loading during grouting and tests on a jacket recovered from the North Sea after 15 years of service.

The tests showed that for steel-grout-steel bonding across an annulus the bond strength is proportional to

- the square root of the grout compressive strength
- a stiffness factor which covers stiffness of pile, sleeve and grout
- the height of mechanical shear keys, and inversely proportional to their spacing.

It was also shown that the surface roughness effects and the mechanical keys are additive, and the bond strength reduces slightly as the length of sleeve to diameter ratio increases.

Various limit states were considered in the tests, with the deflection state being critical, since ultimate strength may occur at deflections which would require unacceptable platform movements. A "characteristic" strength was obtained from the limit state considerations and a general bond strength design formula was evolved. References 2 and 3 describe how the various parameters for the bond strength of a connection may now be determined.

The safety factor applicable to grouted connections covers several situations which are not necessarily present for other structural components, and concerns the inherent uncertainties of offshore construction, the difficulties of inspection and quality control, and the limitation of data. Prior to the research undertaken, static tests were the only basis for design guidance and a factor of safety of 6 was mandatory for bond strength(4). The recent work has shown that this value is not unreasonable and current work is aimed at defining some of the terms in the partial factor approach to produce evidence which should lead to a reduction in this high overall figure.

DESIGN OF GROUTED INSERT PILES

It is still not possible to present a design procedure for large diameter insert piles which represents their behaviour as well as that determined for grouted steel connections. Most design methods for grouted, drilled piles therefore continue to rely on the formula for the capacity of driven piles. The American Petroleum Institute (API) codes(5) draw attention to the possibility of failure at the grout-soil interface but refer the designer to the grouted connection rules to calculate the limited grout-pile load capacity. These rules are now known to be unsafe for the large diameter, less radially stiff, connections. Tests carried out at the University of Texas(6) showed failure occurred at the pile-grout interface at bond stresses of approximately 30 lb/in^2 which is much less than the ultimate value which could be implied by the API code.

Field observations reported by a number of researchers indicate that ultimate skin friction for bored piles in clay and sand is developed at very small axial deformations whereas deflections necessary to develop ultimate end bearing are considerably higher. The implication is that the normal form of pile equation where load bearing capacity is considered as the sum of skin friction and end bearing components is not suitable for the design of insert piles.

The following installation factors all affect the capacity of an insert pile and there is wide scope for laboratory and on site investigations

- mechanical disturbance to the soil during drilling
- change of pore water conditions during drilling
- influence of drilling fluid on the grout-soil and grout-pile interfaces
- change of stress in soil below driven pile during drilling
- time between drilling and grouting
- bleed of water from grout into soil
- diffusion and mixing of soil and cement particles.

Very little is known at present about the influence which these factors have on the bond strength of a grouted connection and the designer of insert piles has to rely on his experience in interpreting soil data and writing a careful installation specification.

GROUT MATERIALS

The basic requirement for the type of grout connection being considered is a high early grout strength to ensure the stability of the structure under wave loading which can change suddenly during the offshore piling operation.

Quantities for grouting the piles on a large North Sea structure can be as high as 3000 tons. This means that a rapid injection rate is called for, and typically 1.5 tons of cement per minute are mixed to meet the schedule. It is therefore preferable to use as simple a grout mix as possible compatible with pumpability and strength. The most usual grout for the steel-pile annulus is therefore a Portland cement with a low water/cement ratio and the addition of a super-plasticiser. However several modified cements are used to provide particular grout properties such as reduced or increased densities, retardation, low heat, etc, where circumstances dictate.

Until recently, design rules took no account of grout curing temperatures when specifying grout strength, but in the North Sea the in situ grout has to achieve early strength at temperatures of 40-45°F. The low temperature has a marked effect on the 24-48 hour strength, although with ageing the curing temperature has a reducing effect on strength. Rapid hardening cement grouts can be formulated to achieve 2000 lb/in^2 unconfined compressive strength at a curing temperature of 45°F within 24 hours.

When early strengths higher than this are required it has been conclusively shown that High Alumina Cement (HAC or Ciment Fondu) grout can be used at these low temperatures enclosed in the pile annulus, without giving the "conversion" problems present in other environments, provided the water/cement ratio is lower than 0.40. For example, HAC grout cubes recovered after immersion, without protection, in the North Sea for 4 years after mixing, reached compressive strengths of

10,000 lb/in^2. More importantly, these cubes confirmed the predicted rate and degree of conversion at the 45°F temperature with an implied minimum strength of only 2% below the 24 hour strength.

The most frequently used cements for both pile grouting and making steel repair connections are API Class A and Class B oil well cements, both being mixed with sea water. The API cements have a proven reliable quality for particle size and chemical composition which is most important in the high speed operation undertaken offshore.

Additives are frequently used with these cements and the following comments on some of these are based on recent experience:-

Calcium chloride - not now generally accepted as an accelerator by certifying authorities for North Sea structures because of potential corrosion problems. The development of choride-free accelerators has been in progress for several years in Europe and formulations based on calcium formate and calcium nitrate are available. While there is still some doubt regarding the corrosion effect of these products they are much less aggressive than the chlorides and are not likely to cause problems in the steel-grout bonding for piles.

Super-plasticisers - very useful in allowing a marked reduction in water/cement ratio of grout; most common are the sodium naphthalene sulphonates which can be formulated to give an accelerating or retarding effect on grout setting time and strength development depending on the length of the polymer chain.

Granulated blast furnace slag - useful where large quantities of grout are required for, say, belled foundations at the base of insert piles; used with Portland cements to reduce hydration stresses; only moderate grout strength but cheap and easy to blend and handle.

Gilsonite - a lightweight solid hydrocarbon ground to a particle size compatible with cement for dry blending; used to reduce the density of grout; very useful when grouting insert piles to avoid fracturing weak strata but still maintaining good bond strength.

Sand - not recommended for the type of structures being considered as two of the usual advantages, cost reduction and shrinkage reduction, are not significant; also, with the long grout lines and travel up the annulus, filtration of water occurs resulting in blockages and potentially poorly grouted zones.

Bentonite/Attapulgite - not recommended for these structures due to the marked reduction in bond strength; however, useful in preventing bleed where it is necessary to use higher water/cement ratios.

Sodium silicate - not used in pile grouting although it has a useful accelerating effect; mainly used to prevent bleed in high water-cement grouts used for grouting under the base of gravity structures.

Expansion aids - gas producing agents, such as aluminium powder, are not acceptable as the expansion produced is not permanent and bond strength is reduced. Expansion of metallic aggregates in grout is satisfactory but because of high cost and practical mixing and pumping problems they are not usually recommended for annular grouting. Since it is now strongly recommended that all piles and/or sleeves have weld beads attached to ensure that mechanical bonding of the cement takes place it is not considered desirable to use any of the currently available expansion additives. There is scope for development of this technique however and various new cement formulations are being tested.

GROUTING METHODS

The most common form of annular grouting for offshore piles involves the use of a high pressure inflated packer to seal the annulus at the base of the leg or sleeve and which supports the total head of grout. This allows the grout to be injected quickly in a single lift, which can be important when grouting the initial 'pin' piles to secure the structure immediately after piling. At this early stage there can be some relative movement between the pile and sleeve, and it is useful to have a rapid setting grout available in these circumstances. Tests on HAC have shown that the oscillations of the pile and structure which can occur during grouting do not have any serious effect on bonding where shear rings are used on the pile.

Packer grouting is used for both main leg piles and skirt sleeve piles with the water being displaced up the annulus by the grout which enters just above the packer, until the grout overflows at the top of the sleeve. Packers can be placed at the top of skirt sleeves to seal the annulus in order to pump grout to the surface through a return line for sampling. Unfortunately these packers have a relatively high failure rate.

The displacement method of injection has been criticised by some designers in that grout becomes diluted by mixing with the water being displaced from the annulus and, to compensate for this, up to 50% additional material is pumped into the annulus. Dilution to this extent does not occur with the grout densities and consistencies currently being used (17.0 lb/gal). The nuclear monitoring techniques described later have demonstrated that only a very shallow zone of grout in the annulus at the water-grout interface is affected.

A method of injecting grout into a dry annulus was patented in 1974 and is now employed on the majority of platforms in the Gulf of Mexico. The method, known as "Pressure Grouting", entails sealing the top of the annulus with a plate or packer and injecting compressed air just below this seal at sufficient pressure to force out all the water past the wiper seal at the toe of the leg or sleeve. When the annulus is dewatered, the air pressure is maintained and grout is injected below the top seal, and free flows down the annulus to the base of the sleeve. As the annulus is filled with grout the air pressure is gradually released, compensating for the increase in hydrostatic head of grout. Injection continues until the head of grout in the main leg annulus balances the depth of sea water; once this grout reaches initial set the remaining part of the annulus is filled. This method requires penetration of the legs or skirt sleeves into the sea bed and difficulties can arise if either no penetration or excessive penetration occurs.

Insert piles are usually grouted through an inner centralised grout pipe connected to a grout 'shoe' with a non-return valve at the toe of the insert pile. The grout must be capable of displacing the drilling mud from the annulus. Ideally the grout should be designed so that filling can be achieved in one lift without causing fracture of the formation and hence loss of grout; monitoring of the grout level is therefore most important. Where a single lift is not possible, provision can be made for multistage injection by providing grout entry ports at the desired stage levels or by perforating the insert pile and placing a double packer to inject grout at the stage level. Care must be taken to remove any contaminated grout at the top of the first stage grout before continuing to fill the remaining annulus.

MIXING AND PUMPING METHODS

It is essential that site mixing methods produce grouts which match the trial mixes made in the laboratory. Too frequently the high shear laboratory mixers are replaced on the site by what are only grout 'stirrers'. It is also important to select quality control methods and techniques which demonstrate that the grout being produced will reach the required bond strength in the annulus.

However by far the most efficient site mixing is achieved by an automatic jet mixer in which the grout is sheared by three separate actions - through the high pressure jet, in the centrifugal pump and around the primary mixing tub. Outputs of 1.5 tons of cement per minute can be achieved by automatically controlling the rate of cement and water feed to meet a pre-set density. A nuclear density meter continuously monitors the quality of the grout passing through the jet and an electronic process controller adjusts the feed rates of cement and water. A print-out of density against time shows the high degree of quality control.

The pumps used for grouting offshore piles are usually similar to those used for oil-well cementing, ie reciprocating piston or plunger types capable of outputs of 10 barrels per min at pressures up to 5000 lb/in^2. In many cases high pressure pumps are unnecessary as the grout will syphon around the pipework and venting system and in fact high volume screw pumps have been successfully used for simple jacket pile grouting.

The correct design of grout delivery lines is an important aspect to ease pumping difficulties particularly when HAC and other viscous and high early strength grouts are used. Practical experience shows clearly that the provision of separate primary, and secondary grout lines and packer inflation lines gives the most flexibility offshore if any problems arise with packer failure or loss of grout circulation. Grout line runs of 1500 ft to each pile are not uncommon (Fig 1).

Various modifications have been tried by platform designers to reduce this amount of pipework on the large jackets; for example, a multiple packer inflation-grout line system and reach rods to operate sea bed valves from the surface, but the problems associated with 500 ft long rods can be considerable. In addition, manifolds have been used to provide grout to several lines from a main riser but again the problems with monitoring the grout flow and ensuring all the lines from the manifold fill the necessary annulus before grout starts setting can be counterproductive.

GROUT MONITORING

Having formulated and tested a grout which meets the specification in the laboratory and checked to ensure that the same grout is produced in the site mixer, the designer wants to be assured that he has this material in the annulus. Where the water depths allow, a diver can sample the grout as it overflows a sleeve, but experience shows that using a small syringe sample tube which the diver can readily handle, too often produces contaminated samples and frustration for the operator. A remote controlled TV camera can watch the flow of grout and the surface observer may be able to decide when a colour change indicates acceptable grout, but neither of these methods, in use for many years, is very satisfactory, and rarely acceptable to Certifying Authorities such as Lloyds.

Several resistivity types of grout monitor are available which indicate that the cement has begun to hydrate and set in the annulus by indicating the change in the resistance of grout. Calorimetry has also been used in an attempt to monitor grout quality but the delicate nature of the apparatus required on the structure has meant that little practical success has been achieved. It is only possible to state that cement is hydrating with these methods and no indication of strength is obtained. When employing any of these four methods it is usual to pump an extra 25% to 50% of grout to provide some surety that the annulus is adequately filled.

The most dependable method of determining the presence and potential strength of grout in the annulus (Fig 2) is to check grout density using a nuclear density meter. These gauges, developed by the UK Atomic Energy Research Establishment and Wimpey Laboratories Limited, are based on measurement of the decrease in intensity of a beam of gamma radiation after passing it through grout as it flows between a radioactive source and a detector. Each annulus usually has two gauges (one of which is a standby) each with its own source, attached to a standpipe connected to the top of the annulus. The radioactive source is usually Caesium 137 fixed in a lead shielding block, and the detector, a geiger tube housed in a waterproof container. The whole assembly has been tested to withstand shock loads of 300 G due to piling operations, hydraulic heads of over 1000 ft and vibration over

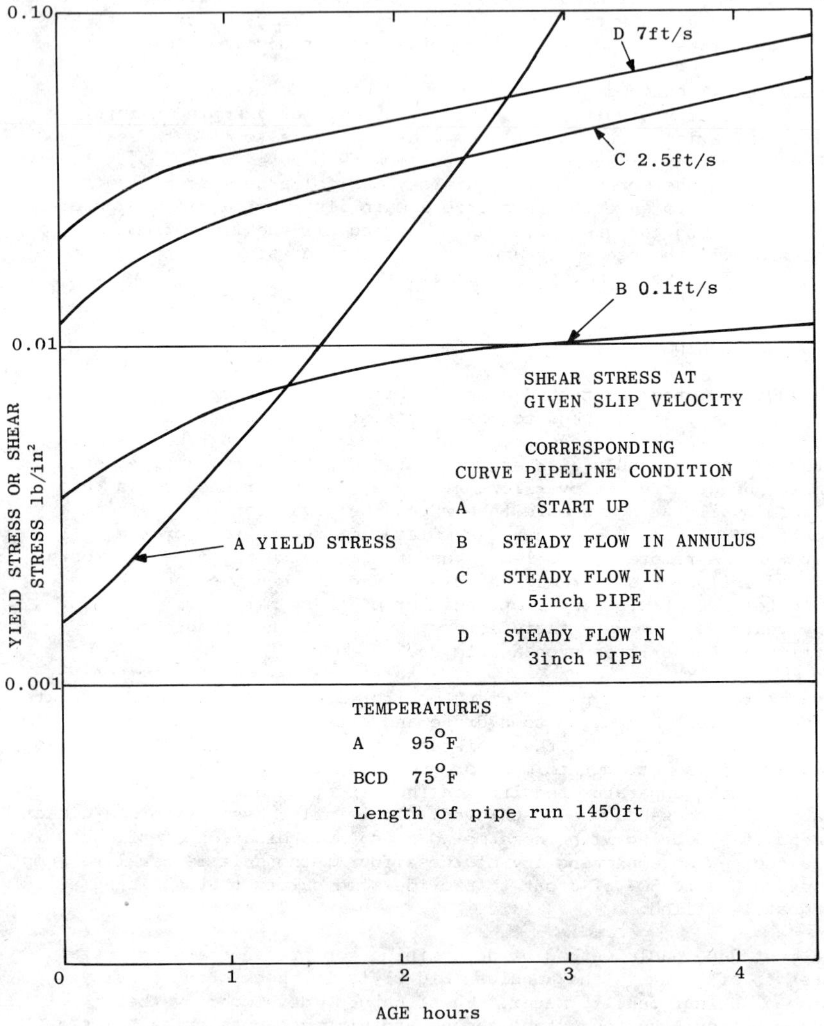

FIG.1 TYPICAL VISCOMETRIC DATA USED IN PIPELINE DESIGN CALCULATION FOR GROUT CONTAINING 17% GILSONITE WITH 0.4 WATER/CEMENT RATIO

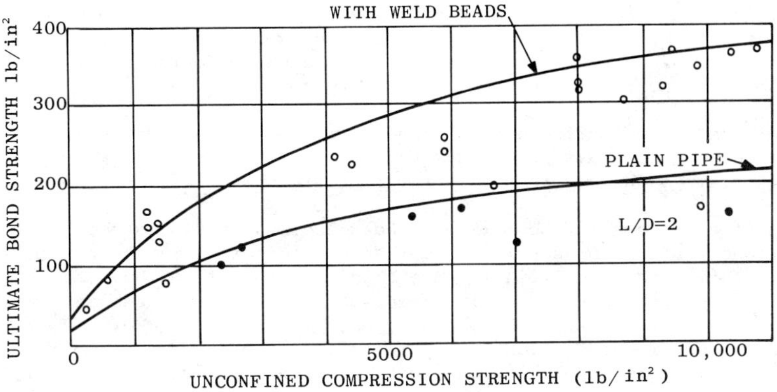

(a) RESULTS ON LABORATORY STRENGTH TESTS

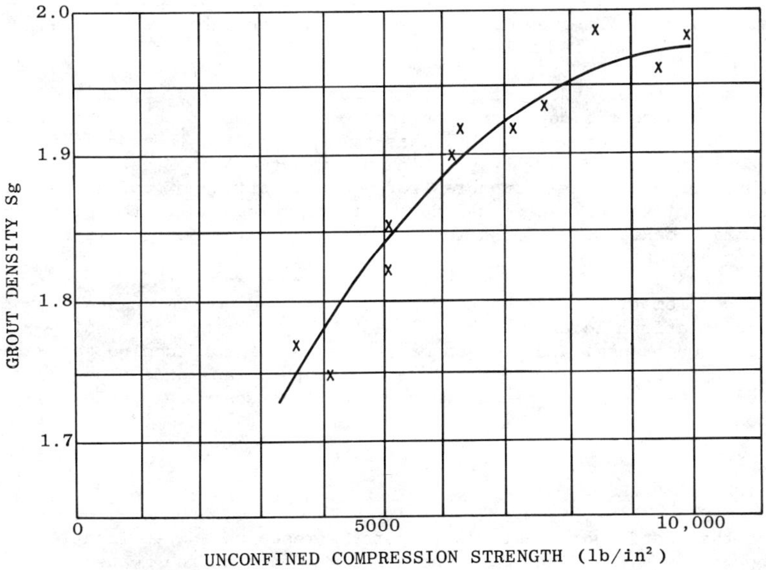

(b) RESULTS OF FIELD STRENGTH TESTS

Fig. No.2 TYPICAL RELATIONSHIP BETWEEN DENSITY AND STRENGTH OF 3inch GROUT CUBES AFTER 28 DAYS CURING AT 45°F a wide frequency range with accelerations up to 15 G. The gauges are connected to the surface through multicore coaxial cables to the monitoring equipment which can record three gauges simultaneously - accuracy of density readings achieved is approximately ± 1%.

In addition to monitoring during grouting it is possible to return to the gauges several days or weeks after grouting to check that the grout has not slumped due say to a packer failure (Fig 3).

Radioactive tracers in the form of powdered glass, sized-matched to cement and containing scandium oxide (an emitter of gamma rays capable of penetrating the steel pile), have been used to follow the rise of grout level up the annulus of insert piles. An initial batch of cement grout is seeded with the tracer and forms the leading edge of the continuous pumping operation. The tracer is detected by a scintillation detector housed in a waterproof container lowered down the inside of the pile. Any change in grout rise rate is immediately known and the necessary steps taken to deal with the problem.

The storage and use of radioactive material are subject to strict control world wide and all the above operations must be handled by classified personnel. It should be noted, however, that the levels of radioactivity are so low that even with unshielded sources the radiation level 4 ft away from the source is within the acceptable limits for unclassified workers.

These radioactive methods of grout monitoring save diver costs and therefore reduce the dependence on good weather for offshore grouting. However the multicore cable which has to be installed in conduit in the fabrication yard is costly. With the practical application of large tethered buoyant structures for oil production now under way the sea bed templates also need careful monitoring, but again the provision of a large coaxial cable in a long catenary from the surface is costly and inconvenient.

Acoustic telemetry is now being developed for a major North Sea tethered platform to enable the radioactive count from the density gauges to be transmitted via sea bed transponders to decoding software at the surface. By careful multiplexing of the transponders on the sea bed it will be possible to check several gauges within a few seconds of each other.

STRUCTURAL REPAIRS

One of the most significant results of the research and application of the work described has been the ability to design the means of strengthening and repairing tubular steel structures. Many of the early North Sea platforms are reaching the end of their anticipated design life, but as production of oil and gas is to continue at these locations, it has been essential to survey the structure and, where necessary, strengthen nodes or add additional members using the new parameters for grouted connections.

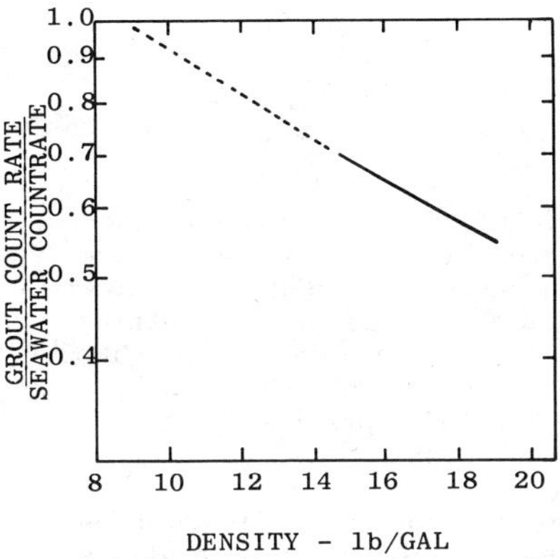

FIG.(a) CALIBRATION OF GROUT DENSITY GAUGE

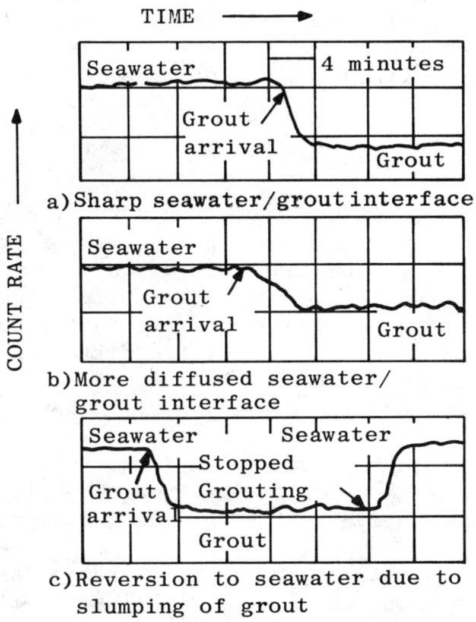

FIG.(b) RECORDED TRACES FROM GROUT DENSITY GAUGES

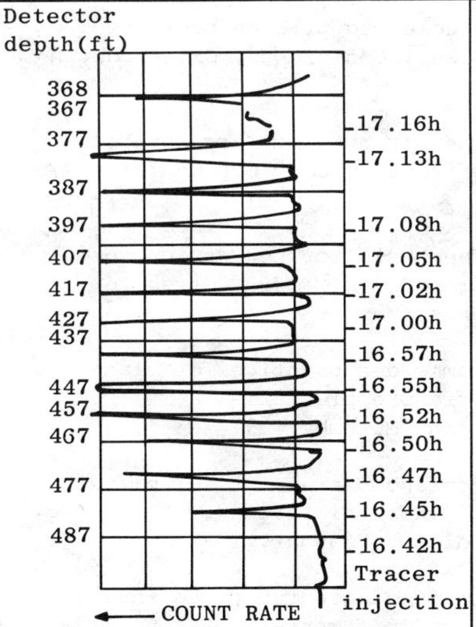

FIG.(c) RECORDED RISE OF 'LABELLED GROUT AROUND PRIMARY PILES

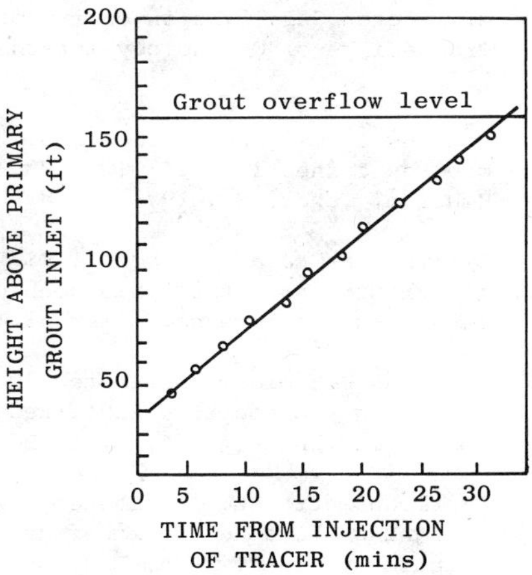

FIG.(d) MEASURED RISE OF GROUT IN PILE/SLEEVE ANNULUS

FIG. 3 - RECORDS FROM NUCLEAR DENSITY GAUGES

Similarly some relatively recently installed platforms are being required to carry heavier loads and again the grouted sleeve around a node or leg can now be properly designed and installed with confidence.

FUTURE DEVELOPMENTS

This paper has indicated several areas for future research and development in the application of grout to offshore structures. One aspect of the hardware associated with grouting is the improvement of the inflatable packer and pile grippers. For example, rapid setting grout could be used as the inflating fluid for these packers which could prevent some of the collapses which occur. This has been proposed as the sole method of gripping a pile to provide the full structural connection but, to date, has not been applied in practice.

The proposal to deform explosively the pile into the sleeve has been under test for several years but it is not considered to be a viable alternative to the annular grouting method.

It is considered that the most pressing areas for future development are those associated with insert piles and a better understanding of their modes of failure, together with improvement in high early strength grouts with low viscosity and nil shrinkage.

APPENDIX - REFERENCES

1 Billington C J: "The integrity of jacket to pile connections". Proceedings of Oceanology International Conference, OI, Brighton 1978

2 Billington C J and Lewis G H G: "The strength of large diameter grouted connections". OTC 3083, Offshore Technology Conference, Houston, Texas, May 1978

3 Billington C J and Tebbett I E: "The basis for new design formulae for grouted jacket to pile connections". OTC 3788, Offshore Technology Conference, Houston, Texas, May 1980

4 American Petroleum Institute: "Recommended practice for planning, designing and constructing fixed offshore platforms". Specification API RP2A, Ninth Edition, November 1977

5 American Petroleum Institute: "Recommended practice for planning, designing and constructing fixed offshore platforms". Specification API RP2A, Eleventh Edition, January 1979

6 Cox W R and Reese L C: "Pull out tests of grouted piles in stiff clay". OTC 2473, Offshore Technology Conference, Houston, Texas, May 1976

7 Evans G V, Parsons T V and Wallace M R G: "Nuclear grout monitoring on offshore platforms". OTC 3791, Offshore Technology Conference, Houston, Texas, May 1980

HIGHWAY GROUTING--A PERMEABLE GROUT PROVIDES DRAINAGE BELOW A CONCRETE PAVEMENT

By James M. Polatty,[1] F. ASCE

Abstract

Background for highway grouting is discussed and the details of a specialized grouting job using a permeable grout both for drainage and to restore load carrying capabilities are presented. A statistical analysis was made of the results of the daily compressive strength tests.

Introduction

Pressure grouting is recognized by most highway departments as an economical method of restoring the load carrying capabilities of portland cement concrete highways that have been in service for a number of years and are showing signs of distress. This distress or failure can be attributed to one or a combination of the following:

1. Localized consolidation of the base course or subgrade due to repetitive loading.

2. The entrance of water into the base and/or subgrade through either open joints or cracks, followed by pumping due to loading. This results in the wet soil being pumped to the surface of the pavement and the development of a void beneath the pavement.

3. Erosion of the soil from beneath the pavement by water flowing into pipes or culverts having open joints or cracks, thus creating voids below the pavement.

This distress is usually evidenced by the failure of the pavement, but an earlier indicator is shown by the presence of soil and water on the surface of the pavement. The presence of voids below the pavement can also be determined by drilling the pavement or by a close inspection of any deflection of the concrete while the pavement is under load.

Examples and Procedures

The size of the voids under a concrete highway slab, while that slab is under loading, is indicated in Fig. 1. This figure shows an interstate highway about four hours after a heavy rain. Figs. 2 and 3 show the resulting failures of concrete pavements. On Fig. 3, evidence of pumping is visible. The most economical grout mixture that can be used by the various owners usually consists of a mixture of Type I

[1] Consulting Engineer, Clinton, MS, Retired, Waterways Experiment Station.

portland cement and either a fine sand or some other filler material. The water cement ratio usually ranges between 0.6 to 0.8 depending on the fineness of the filler material and the required workability. Certain admixtures can be added to impart various special characteristics to the grout mixture such as acceleration, retardation, water reduction, and expanding aids.

FIG. 1--Highway Pumping Under load

FIG. 2--Highway Failure Area

FIG. 3--Highway Failure Area, Visible Pumping

The great majority of owners do not try to jack or lift the concrete slab but only to pump sufficient grout to fill the voids existing below the pavement. An overlay can be utilized both to level the riding surface and to increase the load carrying capabilities of the pavement. Fig. 4 is a sketch of a void below the pavement with the grouting lines in place and the grout pipe extending into the void through the drilled hole in the concrete. Fig. 5 is a sketch of a finished section with the

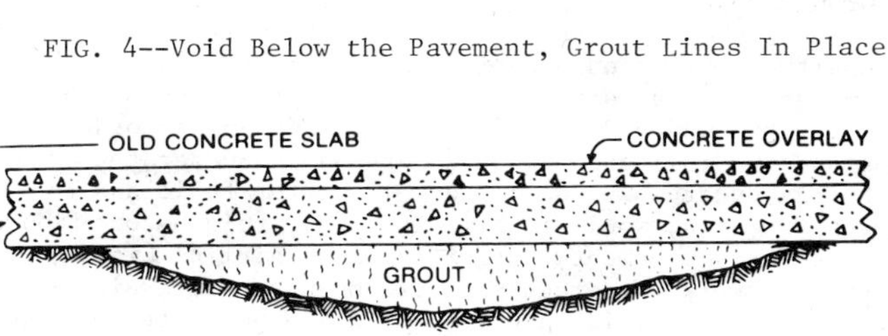

FIG. 4--Void Below the Pavement, Grout Lines In Place

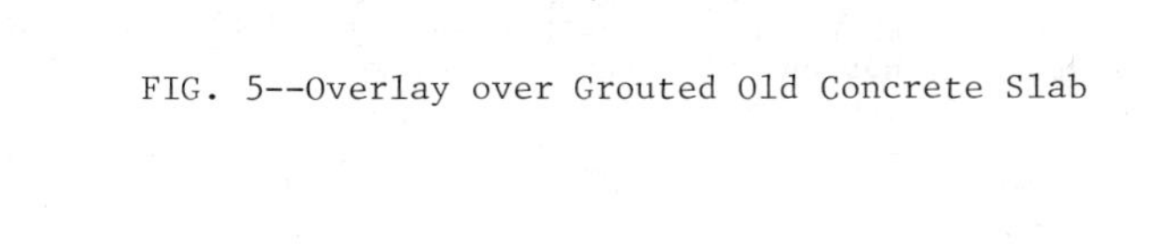

FIG. 5--Overlay over Grouted Old Concrete Slab

leveling and strengthening overlay in place. Fig. 6 and 7 show the Corps of Engineers test method CRD-C 611, "Test Method for Flow of Grout Mixtures (Flow-Cone Method)," being used in the field to control the grout mixture.

FIG. 6--Filling Flow Cone to Determine Flowability

FIG. 7--Grout Flowing Through Orifice in Flow Cone

Permeable Grout Project

The Department of Transportation and Development of the State of Louisiana in designing a section of Airline Highway (US 61) had a problem in that the highway had to be depressed (Fig. 8) in order for traffic to pass under a railroad which in turn passed under the Causeway Boulevard. This underpass, due to the high water table and the general topography of the area, would in times of heavy rain become flooded thus preventing the heavy movement of traffic along Airline Highway. To solve this problem a network of porous pipe and open grated drains were installed to take care of this water. This drainage system was connected by concrete pipe to a pump equipped sump that removed the water and allowed traffic to continue on this heavily traveled highway.

FIG. 8--Underpass on Airline Highway (US 61)

In order to maintain a low water table at all times and to insure that the drainage system would function properly, the porous pipelines were backfilled with sand. Due to the buildup of considerable sand in the collector sump, settlement of the manholes and cracking and settlement of the pavement indicated that, since the original construction, severe erosion of both the sand subgrade and the backfill around the porous pipe had occurred. A study by the Department of Transportation including a series of holes drilled through the pavement proved that there were a number of large voids below the pavement and that there was settlement of the pavement. These conditions would in all probability result in additional cracking followed by failure of the pavement due to loss of support. In addition, an elevation study of the piping system indicated that there was considerable settlement of both the concrete pipe and the manhole system.

The areas that required grouting were then indicated on the contract drawings and bids taken. The prime contractor awarded two subcontracts, one for the repair of the pipe and manholes by conventional sewer pipe technique and the second to grout through the pavement and fill the existing voids below the pavement. This paper describes the latter work. It was desired that the grout mixture be permeable, if possible, so that the water table in the area could be maintained at a low elevation, both in dry periods and in times of heavy rainfall, by using the existing drainage system.

In a project for the Defense Nuclear Agency, the Waterways Experiment Station developed and field evaluated materials that could be utilized as a shock absorber when placed around underground installations. A combination of materials that were developed and tested both in the laboratory and in field nuclear tests in Nevada was a grout mixture consisting of a portland cement slurry and a prefoamed foam. The tests indicated that the mixture in a hardened state was permeable to the flow of water and the energy absorption depended on the ratio of foam to cement slurry.

Permeable Grout Data

A grout mixture that would meet the specifications of the Department of Transportation was selected. For a 1 cu yd yield, the mixture was as follows:

Portland cement, 14 bags	6.70 cu ft
Water, 70 gallons	9.37 cu ft
Foam	10.93 cu ft
Calcium chloride, 52.6 lb	
Test data	
Density	71 lb/cu ft
Coefficient of permeability	6.1×10^{-5} cm/sec
Compressive strength	
(6 x 12 in. cylinders--7 days age)	over 200 psi

The coefficient of permeability was very close to the value for the in-place sand backfill. The test values for the cylinders was 520 psi, which would roughly translate into a California Bearing Ratio of approximately 50. The addition of the calcium chloride allowed traffic to be resumed at an early age without damage to the hardened in-place grout.

Permeable Grout Drilling and Pumping Details

The pavement in the area consisted of 12 x 20 ft slabs. The selected drilling pattern for the grout holes consisted of six holes 1½ in. in diameter drilled in each of the slabs in the areas where the State Transportation Department had indicated that voids existed. Fig. 9 shows the drilling operation with the batching, mixing, and pumping unit in the background.

FIG. 9--Drilling Grout Holes--Batching, Mixing, and Pumping Unit in Background

In general, the grouting was started at the lowest elevation and continued up grade. Fig. 10 is the header used to introduce the grout into the voids. The header was equipped with two valves in order that the grout could be returned to the wet hopper when refusal was achieved.

FIG. 10--Grout Pipe with Control Valves--Line from Pump and Return to Wet Hopper

The mixture with its high air content proved an extremely workable and flowable mixture with no bleeding. The ability of the grout mixture to flow into connecting voids below the pavement and thus fill all voids adjacent to the pumping hole was demonstrated by the excellent flow between holes. The ejection of water from cracks in the pavement is shown in Fig. 11. The flow of grout from relatively small openings, such as between the pavement and joint material, is shown in Fig. 12. When grout flowed from either a crack, joint, or adjacent hole, the pumping operation was stopped and the grout allowed to set before resuming pumping operations for that particular hole. When return was achieved from an adjacent secondary hole, that hole was plugged and grouting continued in the primary hole.

The flow of grout through openings and into the drainage system was closely monitored by light beams and observations. When evidence that the grout had entered these areas was observed, grouting in the hole in progress was stopped in order to allow the grout to set. Then, a jet of water was introduced into the drainage line in order to dilute the grout and to wash the material to the central collector sump. The recognized fact that a high air content will prevent bleeding by holding the component materials in suspension resulted in a grout mixture that had a very slight decrease in volume between its plastic and hardened state. The mixture therefore can be classed as a nonshrinkage grout. Its ability to hold the mixing water in suspension and the high air content with the increase in workability are particular desirable when

FIG. 11--Water Being Displaced from Beneath the Concrete Pavement by Pumped Grout

FIG. 12--Grout Being Extruded Through Construction Joints

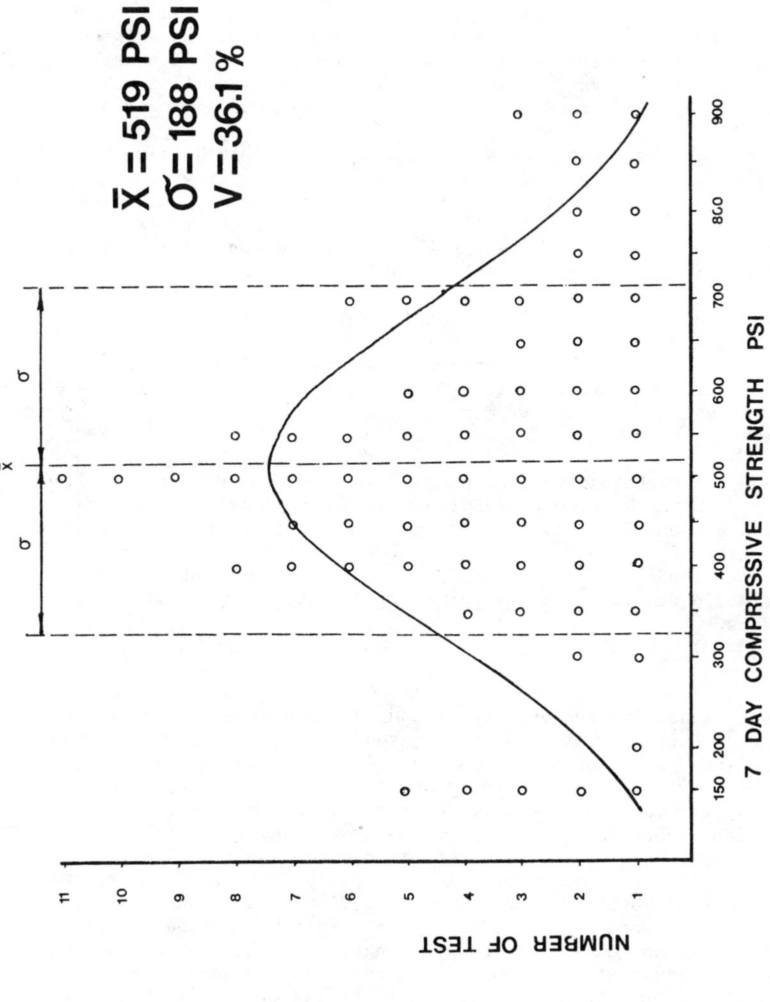

Fig 13 Normal frequency distribution-69 tests

pumping grout on a sand base or subgrade to insure flowability and prevent stoppage of the mixture.

Statistical Analysis of Test Data

Compressive test specimens (6 x 12 in.) were cast daily by the State and tested in compression at 7 days age to insure adequate strength and for control. The average strength of the 60 test specimens, tested at 7 days age, was 520 psi. In addition, field density tests were regularly conducted to insure uniform quality control.

The compressive test data are compared to a normal frequency distribution curve in Fig. 13. Also shown are the sample average strength ($\bar{X}$), the standard deviation (σ), and the coefficient of variation (V). This sample's strength variability is seen to be much greater than any evisioned in establishing standards for concrete control in ACI 214-65. This fact is not unexpected in light of the fact that:

1. The data are based on 7-day compressive strengths.

2. The mixture and the cast cylinders are of a low strength grout. Therefore, the handling, caping, and testing procedures are more critical and would have more effect on the test than for a higher strength concrete.

3. The grout mixture was proportioned and mixed in a traveling plant. By using a Kolmogorov-Smirnov goodness of fit test in analyzing the data presented in Fig. 13, it was found that if the population underlying the data was actually normally distributed with a mean of 519 psi and a standard deviation of 188 psi, then 98 percent of the time a sample would deviate from the normal distribution by more than this data does.

Conclusions

Unfortunately, the quantities and the funding as set up for the job were insufficient to thoroughly grout the entire area to refusal. Based on an inspection made this year, it can be concluded that the grout has performed in five years service, both as a permeable material and as an adequate support for the highway, in a satisfactory manner. Foamed portland cement grout appears to be a material that has a wide range of possibilities in that it can be proportioned to have a number of different characteristics. While normal frequency curves, coefficient of variation, and standard deviation are commonly used to control regular concrete production, this tool has not been used, to the author's knowledge, for any grouting work. The statistical analysis of the grout strength data indicates that the sample values do reasonably conform to a standard normal frequency distribution curve, but that the variability of group strength is much greater than that of concrete.

PRE-STRESS GROUTING OF HIGH PRESSURE WATERWAYS

Robin R.W. Harris, M.Sc., M.A.E.G., F.G.S. *

ABSTRACT

Pre-stress Grouting of waterways is a sophisticated technique which has yet to acquire widespread usage. This paper describes its application in a recently completed undertaking and deals with the preliminary preparations, the sequence of operations, the results and the conclusions derived from the work. It is hoped that this information will practically assist in the planning of similar future undertakings.

INTRODUCTION

Pre-stress, Interface or Gap grouting are terms not found in the "Preliminary Glossary of Terms Relating to Grouting". The closest definition is for Backpack grouting which is applicable in so far as it relates to the "filling of the annular space between a permanent tunnel lining and the surrounding formation". However, in Pre-stress grouting, high pressure grout is simultaneously injected through a multi-point system into a deliberately created surface of non-adhesion. The object is to induce a balanced peripheral confining stress on the concrete lining superior to the variable internal hydraulic pressures imposed during operation.

The technique is not new, having been developed by H. Lauffer and G. Seeber for the Tiroler Wasserkraftwerke AG (TIWAG) in Austria. To the author's knowledge, this system has only been used once previously outside Europe - on the Gordon Power Tunnel in Tasmania which, with a circumferential tensile working load of 285 tons per metre(87 tons per foot) was, at that time, the most highly stressed concrete-lined tunnel in the world. The Pressure Tunnels described in this paper will be subject to a circumferential tensile force of 640 tons per metre during operation (189 tons per foot), resulting from an internal pressure of 62 bars (900 p.s.i.). The state of the art has further been extended to incorporate the Pre-stressing of vertical Pressure Shafts, something not previously attempted.

The advantages of this system are primarily economic in that it is a cheaper alternative to steel lining in areas

* Contracts Manager - RODIO (SOUTH AFRICA) (PTY) LIMITED

of adequate overburden stress and where a waterproof lining is required for control of leakage.

DESCRIPTION OF THE WORKS

The Drakensberg Pumped Storage Scheme straddles the watershed between the Vaal River Catchment to the West and the Tugela River Catchment to the East. This divide also delineates the boundary between the Provinces of Natal and the Orange Free State in the Republic of South Africa. (Fig. 1).

The scheme has a dual purpose :

- to rapidly augment the generating capability of the National Grid during peak load periods via four reversible pump turbines with a continuous rating of 250MW each, and

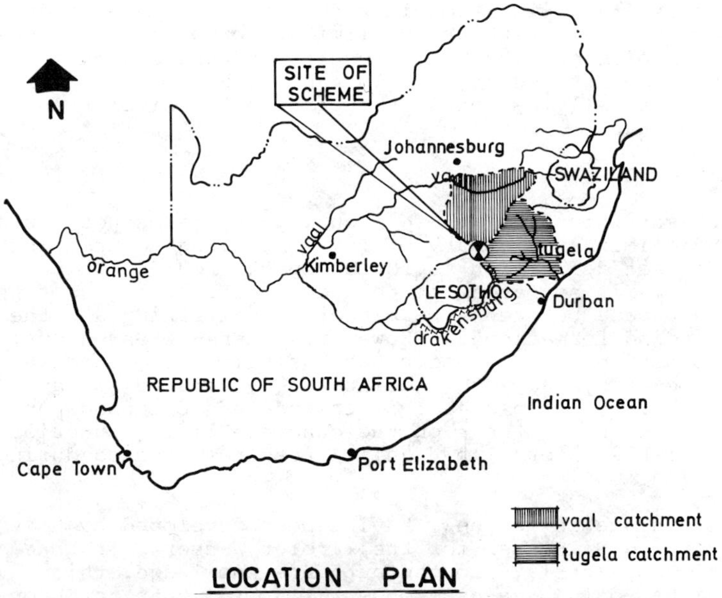

Figure 1.

- to divert water from the presently under-utilised run-off of the Tugela basin to the severely strained water resources of the Vaal Dam Catchment area. An average of 11 million cubic metres ($3{,}2 \times 10^8$ cu. feet) per week will be transferred through the scheme.

A generalised schematic layout is presented in Fig.2. The upper (Driekloof) reservoir and the lower (Kilburn) reservoir are connected by a series of tunnels, shafts and caverns excavated in the mountain side over a distance of 5 kilometres (3 miles). The live storage capacity of

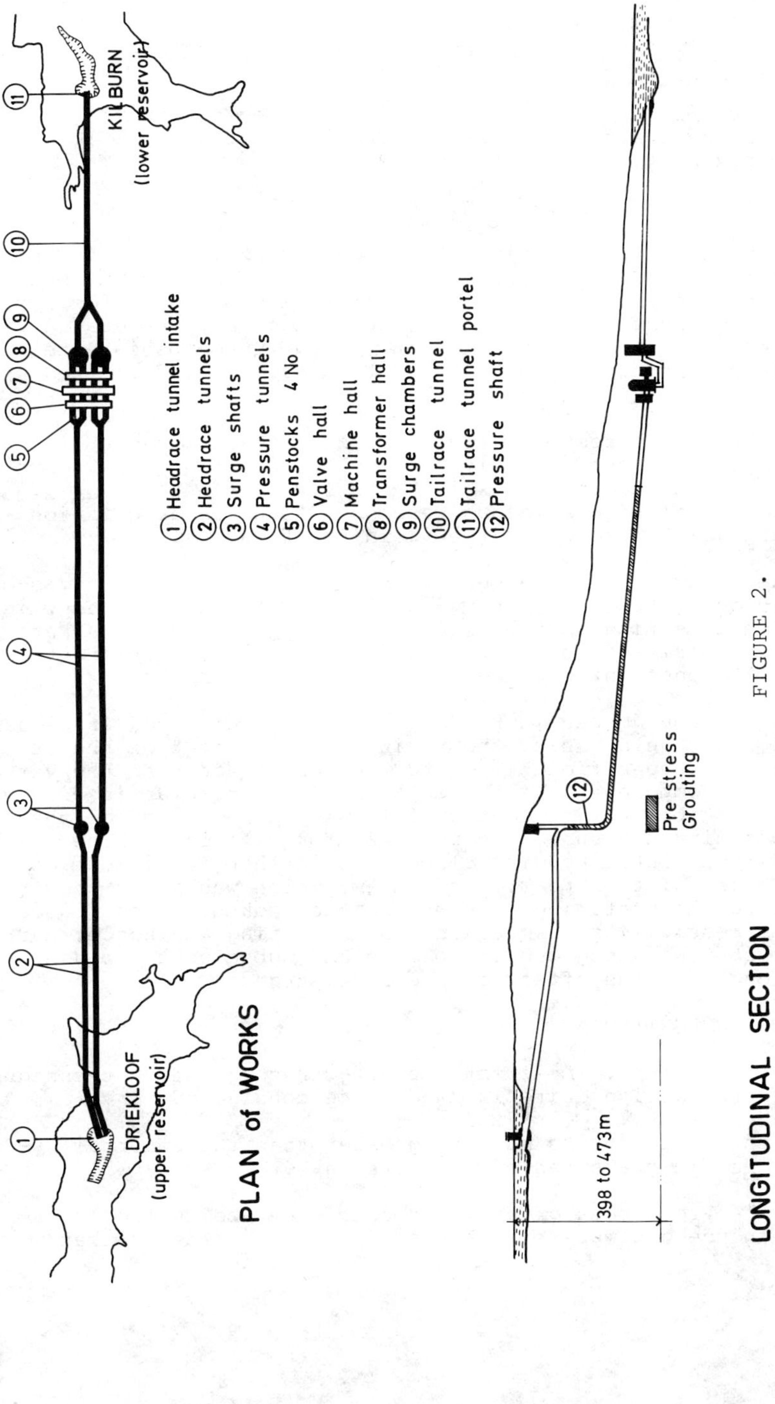
PLAN of WORKS
DRIEKLOOF
(upper reservoir)
KILBURN
(lower reservoir)
1 Headrace tunnel intake
2 Headrace tunnels
3 Surge shafts
4 Pressure tunnels
5 Penstocks 4 No
6 Valve hall
7 Machine hall
8 Transformer hall
9 Surge chambers
10 Tailrace tunnel
11 Tailrace tunnel portel
12 Pressure shaft
LONGITUDINAL SECTION
398 to 473m
Pre stress
Grouting

FIGURE 2.

these dams is 33,6 million cubic metres (1186 x 10^6 cubic feet) and the range of net head is between 398 and 463 metres (1306 and 1519 feet).

The Pre-stress grouting work was carried out on the twin Pressure Shafts and Pressure Tunnels with a total length of approximately 2100 metres (6890 feet).

GEOTECHNICAL ENVIRONMENT

The underground workings are situated exclusively in horizontally bedded Triassic sandstones, siltstones and mudstones belonging to the Beaufort series of the Karoo system. Occasional dolorite dyke intrusives, associated with the last stages of development of the Karoo basin, intersect the sedimentary strata but in this area are of limited extent and significance.

Permeability tests carried out in the early stages of the project and subsequently on Envelope Grouting work, gave values in the range 1 x 10^{-6} to 1 x 10^{-9} metres per second. Occasional higher values were only found in association with jointed zones in the vicinity of faults.

The stratagraphic sequence shows very few visible signs of tectonic stress. The principle structural features which have a bearing on the excavation and grouting works, are the clearly defined sub-horizontal bedded planes and two fault zones intersected by the Pressure Tunnels.

It was found that in keeping with the rapid headward erosion of the Tugela basin, the horizontal component of the regional stress field is 1½ to 2 times higher than the vertical component. As the mudstone in particular is highly susceptible to degradation following a reduction of the confining pressure, the regional force field combined with the horizontal bedding planes was further justification for the use of the Pre-stress system, which would not only restore the confining pressure on the rock but also reduce the possibility of cracking in the lining and hence, hydrofracturing along bedding planes and subsequent rapid erosion under the effect of pressure surges.

DESIGN CRITERIA

To effectively implement the Pre-stress grouting technique, several design criteria need to be considered :

- the field stress must be greater than the internal hydraulic pressures of the waterways.
- the Pre-stress grouting pressure must be sufficient to induce a compressive stress in the concrete so that

having taken account of the rock and concrete moduli, the creep characteristics of the concrete and the loading history to which the tunnel will be subjected, the concrete will remain in a state of net compression for the life of the scheme (50 years).

- a tunnelling method must be employed which will satisfy the above point by :

  (a) minimising damage to the integrity of the surrounding rock, and

  (b) reducing overbreak.

  It is desirable, in the first instance, to reduce the amount of injected grout lost to artificially induced fissures in the rock, as well as avoiding the potential creep (with resultant loss of stress) that could occur by inefficient filling of these fissures.

  Secondly, the excavation method should have as its objective, the formation of a circular tunnel with smooth side walls resulting in a constant concrete lining thickness. This facilitates installation of the built-in parts and furthermore, assists in the development of a uniform deformation of the lining.

- The system employed in construction of the final lining should avoid longitudinal joints wherever possible. However, should this prove impractical, the longitudinal joint should be radial.

- A surface of non-adhesion must be created between the pre-lining and final lining which will permit the introduction of grout under pressure and facilitate its desired longitudinal and circumferential propagation within the confines of the interface.

- The grout should exert a uniform pressure over the entire external circumference of the tunnel lining.

- The longitudinal migration of the grout in the direction of the tunnel axis must be controlled in order to achieve the necessary convergence over any given distance.

- The strain, once imposed, must be maintained.

EXCAVATION AND CONSTRUCTION

A machine-bored tunnel would have provided the ideal environment for Pre-stress grouting; however, for economic reasons, a drill and blast method was employed which required subsequent upgrading to conform with the design criteria.

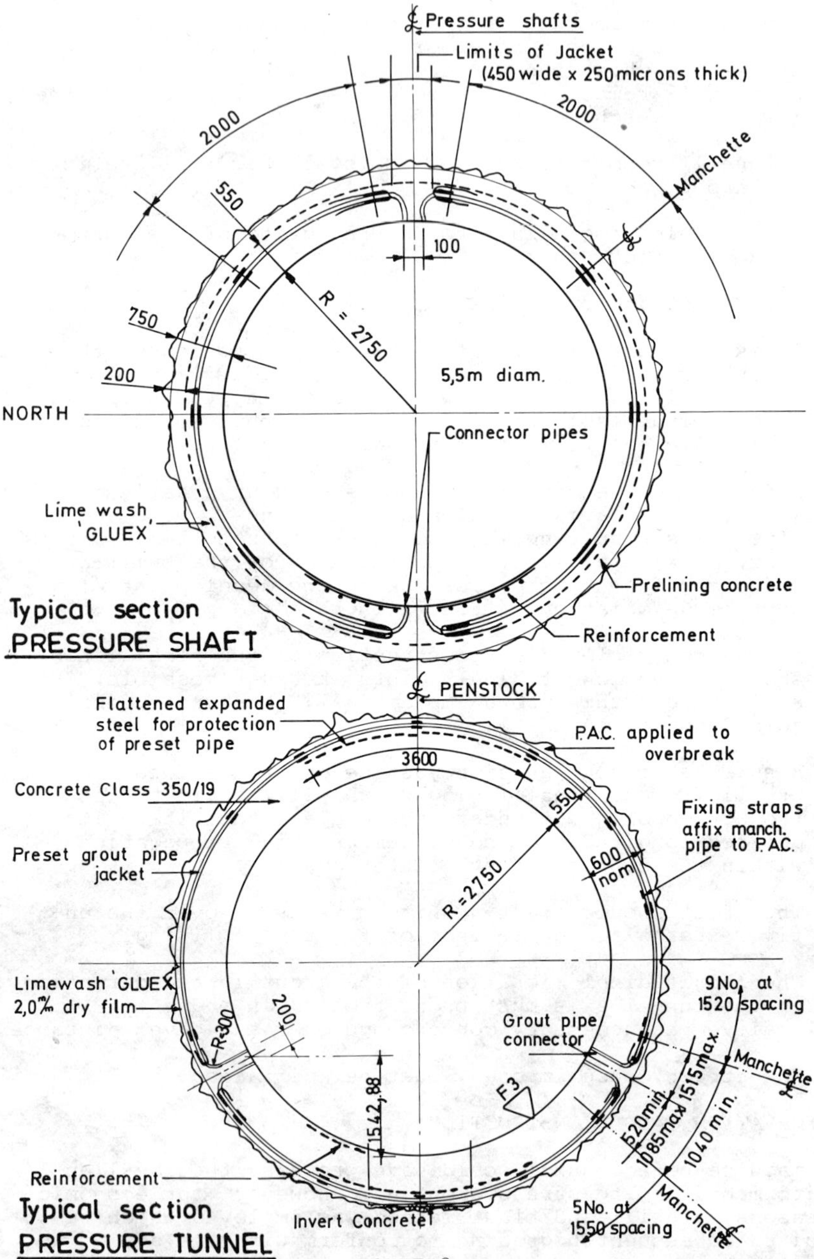

℄ Pressure shafts
Limits of Jacket
(450 wide x 250 microns thick)
2000
2000
Manchette
℄
550
100
R = 2750
750
200
5,5m diam.
NORTH
Connector pipes
Lime wash
'GLUEX'
Prelining concrete
Reinforcement
Typical section
PRESSURE SHAFT
℄ PENSTOCK
Flattened expanded
steel for protection
of preset pipe
P.A.C. applied to
overbreak
3600
Concrete Class 350/19
550
Fixing straps
affix manch.
pipe to P.A.C.
Preset grout pipe
jacket
600
nom
R=2750
Limewash 'GLUEX'
2,0% dry film
9 No. at
1520 spacing
200
R=300
Grout pipe
connector
1515 max.
Manchette
℄
1542,88
F3
520 min.
1085 max
1040 min.
Reinforcement
Manchette
℄
Typical section
PRESSURE TUNNEL
Invert Concrete
5 No. at
1550 spacing

FIGURE 3.

An interim pre-lining was constructed in the Pressure Shafts to protect the vertically intersected sedimentary strata from degradation after exposure, and to permit safe working conditions as sinking proceeded. Thus, while the pre-lining has a variable thickness, the interface between it and the final slipform is smooth and circular with a resultant uniform 550 millimetre (22 inch) thickness of the final lining. However, the pre-lining was not designed to be watertight and large quantities of grout were lost to the surrounding rock during Pre-stressing.

The excavated surface of the Pressure Tunnels required additional treatment following the decision to use the Pre-stress technique. The rock had to be trimmed back where it intruded further than a median circumference, and the over-break filled in by P.A.C. (Pnuematically Applied Concrete) behind that circumference. This work produced a smooth surface but nevertheless resulted in variations in the final lining thickness of between 650 and 850 millimetres (25 and 34 inches).

The final lining in the shafts was slipformed, while in the Pressure Tunnels a 12 metre (39 feet) long monolithic shutter was used which created circumferential construction joints at the end of each pour. In both areas a high tensile steel reinforcing lattice was placed close to the inner surface of the final lining. The reinforcement is required mainly as a crack control measure but has a secondary function of dispersing stress concentration and providing a safety control during the Pre-stressing of shafts.

PRE-STRESS GROUTING COMPONENTS

The Interface

To form a separation surface, the pre-lining surface is coated with a slaked lime product to prevent adhesion with the final lining. During the same operation any exposed fixing devices (rock bolts etc.) are cut off to prevent them acting as anchorage points in the final lining. Should the environment be moist or wet, plastic sheeting can be used, but this was not necessary on the Drakensberg Scheme.

Grout Rings

The tube-a-manchette concept as defined in the "Glossary of Terms Related to Grouting" is used. Polythene pipes of suitable flexibility and wall thickness - the former to follow pre-lining contours and the latter to withstand internal and external grout pressures - are fitted with external rubber valves positioned over holes pre-drilled at regular intervals along the pipe length. This assembly is then housed in a 450 millimetre (18 inch) wide polyethylene sleeve and fixed circumferentially to the side walls using "Hilti" nailed brackets (Fig.5). The purpose of the sleeve is to create a ready-made parting which permits an initial concentration of grout in the interface and thereby facilitate development of sufficient pressure build-up to initiate the propagation of the grout wedge.

The grout rings in the tunnels were spaced at average centres of 2,4 metres (8 feet),and in the shafts at 3,3 metres (11 feet) centres. In total, 1450 rings were installed (Fig. 4).

The injection tubes and the enclosing sleeve in the arch and invert sections of the Pressure Tunnels,were protected by expanded metal sheeting to avoid dislocation during concrete placement and as a protection from tunnel traffic.

The pipe extremities were held to the internal face of the final lining by using the connector assembly illustrated in Fig. 5. In the shafts, where a slipform was used, the connectors were strapped to a re-inforcing bar extending from the pre-lining to just behind the slipform shutter, then afterwards exposed by removal of the small cement veneer once the shutter had passed.

FIGURE 4.
ARCH RINGS INSTALLED IN A PRESSURE TUNNEL

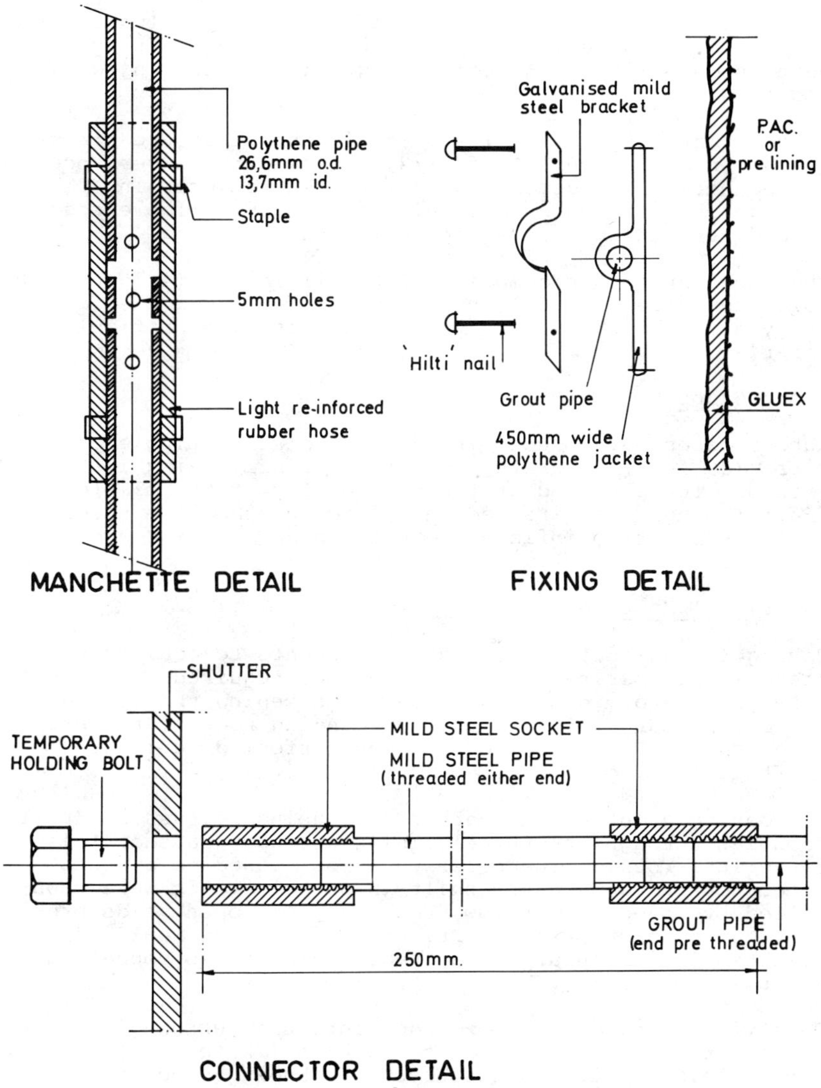
Polythene pipe
26,6mm o.d.
13,7mm id.
Staple
5mm holes
Light re-inforced
rubber hose
MANCHETTE DETAIL
Galvanised mild
steel bracket
P.A.C.
or
pre lining
'Hilti' nail
Grout pipe
GLUEX
450mm wide
polythene jacket
FIXING DETAIL
SHUTTER
TEMPORARY
HOLDING BOLT
MILD STEEL SOCKET
MILD STEEL PIPE
(threaded either end)
GROUT PIPE
(end pre threaded)
250mm.
CONNECTOR DETAIL

FIGURE 5.

Grout Arrestors

Arrestors were installed at 36 metre (118 feet) intervals on the pre-lining circumference of the Pressure Tunnels and at 20 metre (66 feet) in the second Pressure Shaft to prevent uncontrolled longitudinal migration of the injected grout.

No grout arrestors were installed in the first Pre-stress operation in Pressure Shaft 1/2. It was the experience gained here which demonstrated the need to restrain flows in order to achieve a more uniform and controllable pressure distribution over a finite length of lining.

The grout arrestors comprise a 100 millimetre (4 inch) wide polyester fleece or filter fabric ("Bidem"),which effectively allows migration of water but restrains particulate material.

Grout Stops

Rubber hoses were placed in the joints separating each 12 metre (39 feet) concrete pour of the final lining in the Pressure Tunnels, and at the extremities of the slipformed shaft linings, and wrapped in "Bidem" cloth. The hoses could be remotely inflated to staunch any leakage at these points.

Additional Measures

Having installed the necessary components to achieve the stated design criteria, it was further required to undertake preparatory grouting of cavities behind the lining and of fracture zones in the surrounding rock of the Pressure Tunnels. No pre-grouting was performed behind the shaft linings.

The Cavity grouting was carried out using percussion drilled holes through the tunnel lining extending a short distance into rock. The holes were positioned so as not to intersect and damage any built-in parts. Voids were then filled by injection at low pressure. Any open voids present during high pressure Pre-stress grouting would have resulted in irregular and unpredictable stress concentrations in the lining.

Cavity grout holes were extended into the surrounding rock and grouted at pressures up to 50 bars (725 p.s.i.) in areas where faults and fissures were known to occur. Subsequent water pressure testing in these areas indicated low final permeabilities, although grout losses during Prestressing could still be inferred for these features.

Following the pre-grouting,all drill holes were stemmed and

carefully sealed and any other observed damage to the lining repaired.

CONTROL PROCEDURES

The preparatory work described above has, as its ultimate objective, the uniform convergence of the final lining. However, the practical limitations inherent in any such system require continuous monitoring and control.

Convergence measurement was the principle system used to regulate the Pre-stress grouting. A series of convergence rods (Fig. 7) were installed at regular intervals ahead of the injection points and continually monitored. Only when a satisfactory mean convergence was obtained at the array closest to the injection point, could grouting advance.

Vibrating wire strain gauges augmented the convergence monitoring information in test sections and critical areas of the lining. Further instrumentation in the test sections included gap width monitors, radial and tangential stress cells and extensometers.

Pressure monitoring points, each consisting of a mechanical packer fitted with a valve, gauge protector and pressure gauge, were installed in holes intersecting the interface.

FIGURE 6. PRESSURE MONITORING CONSOLE.

Control measures on the actual injection equipment comprised :

- pressure relief valves to avoid exceeding safe design pressures.
- continuous clock-type pressure recorders and strategically placed pressure gauges. (Fig. 6)
- magnetic stroke counters to give an approximate measure of injected grout volume on reciprocating, positive displacement injector pumps.

THE GROUTING OPERATION

For brevity, only the grouting of the Pressure Tunnel will be described. Grouting in the Shafts would generally follow the same procedure but be subject to the access restrictions imposed by a 3 Deck stage. Pertinent data are summarised in a table at the end of this section.

The grout consisted of ordinary Portland cement and water, generally injected in the proportion of 1:1 by weight. Overall the average water/cement ratio was 1:1,33 in both the Shafts and Pressure Tunnels.

The grout was prepared in a semi-automatic surface Batch Plant, and despatched, through 25 millimetre (1 inch) pipe ranges, direct to the injection pump holding tanks. An intermediate Relay Station was used when working at the very extremities of the Pressure Tunnels. This was done to avoid excessive wastage and spilling in the Pressure Tunnel and not due to any problem arising from the 2 kilometre (1,24 mile) pumping distance. Mobility of the injection equipment was essential as the Pre-stress operation was carried out on a round the clock basis. An 8 wheel grout train was constructed which could be moved by means of an air-driven winch connected to snatch points at about 15 metre (164 feet) intervals in the Tunnel invert. The following equipment was mounted on the train :

- four double acting Rodio CMRW 70 pumps with 51 millimetre (2 inch) plungers, a 320 millimetre (12 inch) stroke length and a rated maximum pressure of 215 bars (3010 p.s.i.) using the available compressed air supply of 4 bars (56 p.s.i.).
- four agitating tanks of 300 litre (80 gallon) capacity each.
- an elevated control platform fitted with the pressure and stroke counter read-outs for each pump.

The Pre-stress grouting normally commences with the injection of water, which facilitates the opening of the separation plane. Having initiated the split, a progressively thicker water - cement mix is injected, the final thickness being dependent on the injection pressures generated as well as the "pumpability" and segregation problems associated with long delivery ranges.

The sequence of injection is governed by the need to obtain a convergence which is as uniform as possible. After attempting a number of variations, the best grouting sequence obtained involved the simultaneous injection of one lower pipe and two upper pipes, using three injector pumps.

FIGURE 7.

PRE-STRESS INJECTION AND CONVERGENCE MONITORING.

Pressures of up to 160 bars (2320 p.s.i.) were used to force open manchette rings. However, operating injection pressures of about 50 bars (725 p.s.i.) were found to produce adequate convergence and to correspond relatively well with the theoretical predictions.

As the injection points leap-frog along the Tunnel, preceded by the convergence arrays, a filtration process is taking place in the separation surface which governs the rate of advance. Grout particles accumulate in the gap as the unstable suspension segregates. This deposited cement creates the bulk required for maintaining the deformation in the final lining while the excess water is pushed ahead to develop the advancing wedge. Therefore, pumping must continue at any one point until this process of segregation is sufficiently developed.

Instrument readings indicate a rapid initial loss in strain (less than 10 percent of the original average strain) once grouting stops. This immediate loss is followed by an on-going relaxation which is related linearly to log time. It is further evident from strain gauge readings that a dendritic advance of grout takes place in the interface, causing loading on discrete sections of the circumference at any one time. The final deformation is therefore a composite of an incremental series of overlapping arches.

The relationship between the cement injected and the estimated total convergence volume obtained on the Pressure Tunnel, shows a cement take of 4 tons per cubic metre of gap volume. It is, therefore, evident that cement has been lost to the re-stressing of the rock, escape into fissures, leakage through the lining and wastage. This is confirmed by a known insitu density of 2,38 tons per cubic metre.

The only factors relating to grout loss that can be effectively controlled, are wastage and leakage through the lining. Wastage is reduced to a minimum by establishing thorough operational procedures. Leakage in the lining is more problematic.

Cracks through which leakage occurred were attributed to occasional deficiencies in the concrete quality of the lining, over-stressing of the lining and tensile stresses developed in the invert due to grout travelling preferentially in this zone. Leakage also took place through stemmed cavity grout holes and at construction joints.

In many cases leaks were eventually self-sealing. Persistent leaks were treated by either placing sealing additives in the grout or by patching the leak with "Bidem" strapped and nailed to the lining surface. Construction joints were sealed by inflating the grout stop or by deliberately bursting it with a grout containing sawdust or asbestos fibre.

ITEM	UNITS	SHAFTS		TUNNEL	
		1/2	3/4	1/2	3/4
Length of Pre-stress	metres	214	228	808	834
Internal Diametre	"	5,5	5,5	5,5	5,5
Interface " (AVE)	"	6,05	6,05	6,25	6,25
Completion Time	Days	11	11	21	18
Advance Rate (AVE)	m/hour	0,90	0,96	2,00	2,20
Injection Frequency	m^2/no	6,3	6,3	3,4	3,4
Cement - Cavity	Tons	-	-	224	280
Cement - Pre-stress	Tons	182	103	544	318

TABULATED SUMMARY.

RESULTS

The minimum long term Pre-stress requirement in the Pressure Tunnel varies from 7,4 MPa (1073 p.s.i.) at the transition to the steel lining, to 5,4 MPa (783 p.s.i.) at the base of the Pressure Shafts. The result achieved by the Pre-Stress provided an adequate factor of safety in relation to the design requirements. Even so, should tension cracks develop, the secondary control provided by the reinforcement is expected to preserve the hydraulic surface of the lining.

CONCLUSIONS

The efficacy of the Pre-stress grouting work will only be truly proven during the life of the scheme. However, certain observations made during the course of the work could facilitate the practical implementation of the Pre-stress technique in the future.

It was found that a vertical convergence generally occurred before the horizontal convergence in the tunnel. This necessitated frequent adjustments in the pattern and mode of injection in order to balance the convergence. A more equitable balance can probably be achieved by :

- avoiding a flat invert behind the lining.
- Allowing for an overlap in the manchette hoses to prevent the creation of a "dead" injection area at the connectors.
- having manchette hoses of equal length. However, in this case, access to the injection points would take precedence.

Much of the water contained in the grout mix is superfluous as it is the suspended cement which ultimately creates the bulk necessary for maintaining Pre-stress, after bleeding off the excess water. It is felt that penetration in the gap would be facilitated and the required degree of convergence attained more rapidly by stablising the grout using a suitable admixture such as Bentonite.

ACKNOWLEDGEMENTS

The author is indebted to the Electricity Supply Commission for permission to publish this paper. Special thanks are also owed to Gibb Hawkins and Partners, the Consulting Engineers responsible for the design of the tunnels and the supervision of the grouting operation, and to their Mr. Mike de Witt, for his kind advice and assistance.

USE OF GROUTING IN CAISSON CONSTRUCTION

By Clyde N. Baker, Jr.,[1] F. ASCE
Francisco A. Echevarria[2], and John P. Gnaedinger[3], F. ASCE

KEY WORDS:

Caisson Construction, Cement Grouting, Chemical Grouting, loss of ground, pumping, settlement, Soil Stabilization, voids, water flow

ABSTRACT:

The authors' experience in the use of both chemical grouting and cement grouting to facilitate caisson construction on several major projects in the Midwest is described. The cases described include: 1) grouting in advance of construction to reduce water inflow during caisson excavation, 2) grouting after caisson excavation to reduce water flow, 3) grouting to reduce soil squeeze and loss of ground, and 4) grouting to densify bearing soil loosened from "blow-ins" or excess water flow during caisson excavation. General procedures followed and results obtained are discussed.

INTRODUCTION

Background

During the writers' more than eighty years of collective experience in the fields of caisson construction and grouting, there have been numerous occasions when grouting has been proposed to solve a particular caisson construction problem but much fewer occasions when the proposal has been accepted and executed. Two major problems or conditions which can make caissons an unsuitable foundation solution, or at best, a very expensive foundation solution, are the potential for excess water inflow and soil instability. When either of these conditions is expected at a particular construction site, alternate foundation solutions usually have been used. However, sometimes because of the vagaries of nature, conditions different than anticipated are encountered during construction and what was contemplated as a routine caisson construction project can be faced with either excess water problems or soil instability, or both.

---

[1]President and Principal Engineer, Soil Testing Services, Inc., Northbrook, Illinois

[2]Vice President and Principal Grouting Engineer, Soil Testing Services, Inc., Northbrook, Illinois

[3]Chairman and Principal Engineer, Soil Testing Services, Inc., Northbrook, Illinois

Since grouting is used most often to either make soil less permeable or to improve stability, it is logical to assume that there are particular grouting procedures that could be very helpful in solving certain caisson construction problems involving excess water inflow or soil instability.

Scope

In this paper, the writers share their experience in the use of grouting in solving the following caisson problems:

1. Excess water seepage through seams and fractures in rock bearing stratum or permeable zone immediately over bearing stratum.
2. Excess water seepage around and beneath steel casing extended into rock bearing surface.
3. Caving water bearing soils in the bell zones.
4. Squeezing soils in the shaft zone.
5. Decreased bearing capacity and increased compressibility of bearing stratum caused by pumping fines out of the soil or by sudden "blow-in" during excavation.

The first problem mentioned normally is solved by grouting procedures in advance of excavation. Problems 2 through 4 involve grouting procedures during caisson excavation, and the 5th problem involves grouting procedures after the concrete has been placed in the caisson excavation and has set.

Soil Profile

The project experience that will be shared involves work in the Chicago and Detroit areas where a soil profile illustrated in Figure 1 is typical. Various caisson solutions are illustrated on this same Figure 1.

GROUTING IN ADVANCE OF CONSTRUCTION TO REDUCE WATER INFLOW DURING CAISSON EXCAVATION

History of Problem

Because of the very tall buildings built in Chicago and the resulting very high column loads involved, rock caisson foundations are an attractive foundation solution for many of the biggest projects. The rock consists of Niagaran limestone which contains fractures, joints or seams which are normally most prevalent in the surface 5 to 20 ft (1.5 to 6.1 m). The fractures and joints run vertically and horizontally and may or may not be filled with silt or clay or sand or a mixture thereof. The rock is commonly overlain by an extremely dense water bearing silt deposit containing varying amounts of sand, gravel, cobbles, boulders and broken rock. Standard practice involves open drilling through the upper clay layers to the water bearing silt deposit and then drilling under a slurry head to rock. A full length steel casing with hardened steel teeth welded to the bottom giving a saw-tooth appearance is then inserted in the hole and torqued into the rock to seal off the water bearing deposits on top of the rock.

Problems develop when the casing is dewatered and it is observed that water is entering the caisson either through seams in the rock or from the water bearing deposit on top of the rock and then down along the casing and under the casing into the hole. Because of past problems with heavy pumping of water from rock caissons in downtown Chicago pulling fines from the surrounding soil and settling adjacent structures, the City has set limits on the amount of pumping they will permit. If the pumping rate exceeds the tolerable limit [normally 5 to 15 gallons (.02 to .06 m^3) per minute], the casing must be extended further into the rock in an effort to seal off the water. This normally entails more rock removal and can be very expensive, particularly when repeated several times without success. An alternate approach to solve this problem is to grout the open fractures and seams in the rock and the permeable zone just on top of the rock at each caisson location in advance of caisson excavation. In the remainder of this section, four cases are described where grouting in advance of caisson excavation has been attempted with varying degrees of success.

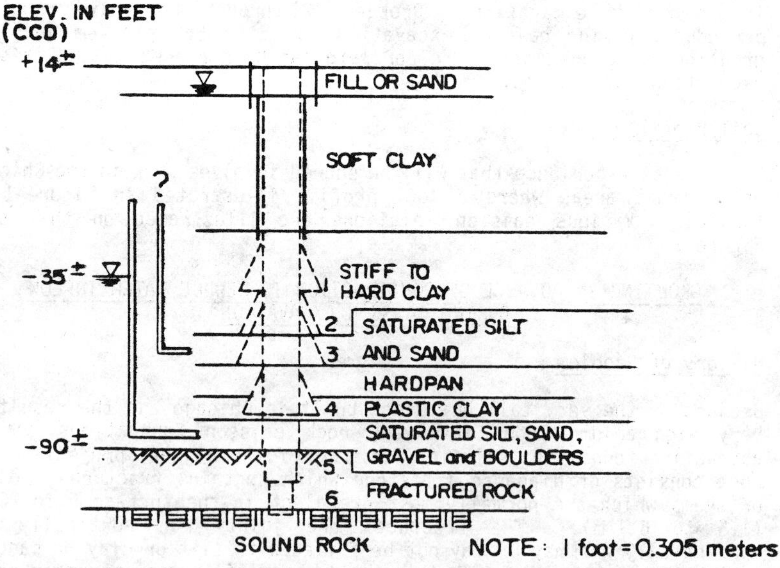

FIG. 1 TYPICAL SOIL PROFILE FOR DOWNTOWN CHICAGO

Case A

This project involved a 30-story structure adjacent to a hardpan caisson supported structure and across the street from a major rock caisson supported structure which had significant water pumping problems. The owner wanted his structure founded on rock to preclude any problems he might have if somebody built a structure adjacent to him founded on rock caissons. Borings at the site revealed conditions similar to those across the street where serious water problems had developed during construction. Initial core borings had indicated potential water problems where 100% drilling water was lost during coring in one of the borings. A major highrise office structure supported on hardpan caissons immediately adjacent to the north made any movement of soils and water beneath the hardpan more critical. In an effort to minimize the amount of water pumping that would be required during caisson excavation, pregrouting at each perimeter caisson location was proposed and accepted. It was hoped that effective grouting at the perimeter caissons would also reduce water inflow at the interior caissons. A sodium silicate grout with a proprietary name of Grout 3 was used. The procedure followed in the early portions of the work was as follows:

1. Drill NX boring to rock at center of caisson locations.
2. Drive casing to refusal on the surface of the rock.
3. Core rock a minimum 10 ft (3.05 m) or at least 8 ft (2.44 m) below major seams.
4. Water test the rock with a packer set at the bottom of the casing.
5. Pressure grout with packer set in casing.
6. Pull casing slightly and grout zone just above the rock.

Because of potential lack of back pressure after casing was pulled to permit the grouting above the rock, the procedure was changed over part of the project so that casing was driven to refusal in the material just above the rock and then this zone was grouted prior to coring the rock and grouting the rock.

Most of the grout-take was in the rock with very little grout-take in the soil above the rock. Table 1 shows a summary of the caissons, grout-take, and water seepage during caisson excavation. The results were considered excellent. Grout-take varied from 110 to 1575 gallons (0.44 to 6.3 m^3) per hole and the maximum water flow rate during caisson excavation was 5 gallons (.02 m^3) per minute and on many caissons was less than 1 gallon (.004 m^3) per minute. This compares with pumping rates in excess of 100 gallons (.4 m^3) per minute in similar rock conditions across the street. It also appeared that the perimeter grouting had been helpful in reducing water inflow rates on the interior caissons as well since the measured water inflow on the interior caissons was generally similar to the perimeter caissons.

TABLE 1

Caisson Number	Rock Elevation in feet	Caisson Level	Total Grout Take (gal)	Water Inflow (gpm)	Caisson Number	Rock Elevation in feet	Caisson Level	Total Grout Take (gal)	Water Inflow (gpm)
A4	-95.4	-103.1	210	1.0	C5	-91.1	- 95.4		2.0
F3.1	-91.3	- 95.0	420	0.5	D4	-91.0	- 94.7		1.2
F4.5	-86.6	- 94.4	175	2.0	B6	-89.5	- 96.5		1.8
F2.1	-90.5	- 97.4	110	0.5	C7	-87.8	- 92.2		0.5
F1	-93.9	-104.0	805	3.5	A6	-91.0	- 96.3	200	1.6
C4	-91.2	- 95.4		2.0	E2	-91.6	- 95.8		dry
C1	-93.6	-100.3	750	5.5	B4	-92.6	- 96.9		1.5
B1	-94.3	-109.1	300	4.7	E5	-90.0	- 93.5		2.5
C2	-95.3	-105.4		4.4	A8	-87.9	- 91.4	375	1.5
D1	-92.6	-103.1	1050	2.1	D7	-88.1	- 91.8		1.0
C3	-92.7	- 97.0		4.0	D6	-88.6	- 92.5		2.25
D3	-91.8	- 95.6		1.0	B8	-87.9	- 91.6	250	0
A2	-95.1	-106.2	225	1.8	A7	-90.4	- 95.4		1.0
E3	-90.7	- 94.7		1.5	D8	-85.5	- 89.7	1100	4.8
B3	-92.8	- 96.7		1.8	F6	-87.4	- 91.1	175	1.6
E4	-91.0	- 94.7		0.4	A5	-93.3	- 97.9		1.8
B7	-91.6	- 96.0		1.5	E8	-87.2	- 91.4	325	3.4
C6	-88.3	- 92.1		0.5	C8	-86.6	- 90.5	200	15.0
B5	-91.9	-100.1		1.5	E6	-87.3	- 93.3		3.8
D2	-92.2	- 97.5		1.2	D5	-87.9	- 91.5		1.7
A3	-95.4	-102.4	210	0.5	A1	-94.5	-101.9	325	1.0
B2	-94.6	- 98.2		2.5	F7	-86.3	- 90.3	1575	1.0

Note: 1 foot = 0.305 meters

1 gal = 0.004 cubic meters (m^3)

Case B

Because of the success achieved in pregrouting on the Case A project, on the next rock caisson foundation project in Chicago, where significant water loss was noted during the core borings, pregrouting was again proposed. In this case, the rock conditions as indicated by water loss during the core borings appeared significantly worse than at Case A.

In an effort to speed up the drilling and grouting procedures, procedures were modified so that a number of holes could be grouted at once. Procedures used were as follows:

1. Drill an NX cased borehole to rock and core approximately 20 ft (6.1 m).

2. Install a 1/2 inch (1.27 cm) PVC grout pipe with the bottom section slotted to about 3 ft (0.92 m) above the rock contact zone.

3. Backfill with pea gravel to 2 ft (0.61 m) above the slotted portion of the PVC pipe concurrently with pulling the outer casing.

4. Tremie the cement slurry into the casing on top of the pea gravel and remove the casing.

5. Allow cement slurry seal to set over night, then water test and grout with packer using flow control valves to meter materials.

The grout mix used was the modified sodium silicate grout called Grout 3, consisting of 60% liquid sodium silicate and 40% reactant.

The results of the grouting program are summarized in Table 2.

As indicated on the summary, there was 100% water loss during core drilling in the majority of the holes, indicating the extremely fractured nature of the rock. Grout-take varied from 0 to 1900 gallons (7.6 m^3) per caisson. In certain instances of high take, no back pressure built up. The total grout pumped at 22 caisson locations was 18,000 gallons (72 m^3).

During caisson excavation, it was observed that the grouting did not work. In some instances, the grout held temporarily and then blew in. In one 8 ft (2.44 m) diameter caisson, the maximum pumping rate was 1300 gallons (5.4 m^3) per minute. After supplementary grouting through three additional grout holes, the maximum pumping rate was reduced to 250 gallons (1.0 m^3) per minute. The procedure finally followed was to turn the casing further into the rock to seal off the water and if this did not work, to clean the bottom with divers and use a tremie placement method for the concrete.

TABLE 2

Caisson Number	Top of Rock Elevation (CCD) in feet	Elevation of Seams	Water Loss While Drilling	Grout Volume (gal.)	Grouting Pressure (psi)
M-9	-93.7	-98.7 (clay filled)	100% @ - 98.2	1518	45/120
M-5	-89.1	Broken rock -94.4 to -95.9 -97.1 (clay filled)	100% @ - 97.6	1289	55/120
Q-7S	-93.9	Broken rock -97.1 to -97.8 and -102.1 to -103.0 (clay filled)	100% @ -106.5	789	85/120
Q-9	-93.4	-96.9 to -97.1	100% @ - 96.9	745	90/120
P-9	-92.9	-	100% @ - 96.9	943	95/120
Q-7N	-92.9	-98.9 to -99.6 and -101.9 (clay filled)	100% @ - 98.9	798	75/120
P-7N	-93.2	-98.5 (clay filled)	50% @ -102.7	798	80/120
M-7	-89.4	Broken rock -96.4 to 101.4 -97.4 to -97.7	None	585	85/120
N-7	-92.0	Clay filled seam @ -97.5 -98.5 to -99.1	None	585	85/120
N-5	-90.1	-97.1 to -97.6	None	924	60/100
N-3	-89.5	-92.8 to -93.9, Vertical joints -95.2 to -96.2	100% @ - 92.8	618	85/120
M-3	-88.9	Clay filled joint @ -97.0	100% @ - 81.9 and @ - 94.4	1666	60/120
P-5S	-93.6	Clay filled joint @ -99.6 -101.1 to -101.5	None	966	85/120
Q-5N	-91.6	-98.4 to -98.6	50% @ - 98.4	896	80/100
P-7S	-91.2	Clay filled joint @ -99.0	None	953	60/120
Q-5S	-93.1	-99.5 (clay filled)	None	1169	80/120
Q-3	-91.6	-98.6 to -99.1	100% @ - 98.6	1855	50/100
P-5N	-92.1	-99.1 to -99.4	50% @ - 99.1	0	200/200
P-3	-91.9	-98.4 to -98.7	None	0	200/200
N-9	-95.9	Broken rock @ -97.4 to -98.4 and -100.4 and -106.9	100% @ - 97.9	0	200/200
P-7N	-91.2	Vertical joint @ -106.1 to -108.7 -97.7 to -98.0	100% @ -102.7	861	70/100
P-3	-91.4	-98.9 to -99.2 Broken rock @ -94.3 to -94.4 Clay filled joint at -98.6	None	0	200/200

Note: 1 foot = 0.305 meters

1 gal = 0.004 cubic meters (m^3)

1 psi = 6.9 kiloPascals (kPa)

Since the grouting procedure was changed between the two projects, a direct comparison is not completely possible. It may be that a one at a time, two-stage grouting procedure would have been more effective (the Case A procedure), but it is doubtful. The main problem appears to have been the much larger size cracks on the Case B project (crack widths up to 6" instead of the 1" maximum observed at Case A). This larger volume of crack required higher strength grout. In retrospect, a preferred procedure would have been to use a high strength cement grout to fill the major voids and then follow with a chemical grout to get whatever the cement grout missed.

Case C

The Case C project involved a proposed 30-story office building to be located on a site previously occupied by a structure supported on rock caissons. Initial coring and evaluation of the rock caissons to see if they might be suitable support for the new structure indicated the concrete to be porous and of poor quality, foreclosing possible re-use of the caissons. Analysis of the low quality hardpan at the site indicated belled caissons could be designed on the hardpan to carry the proposed structure. However, in some cases, the bells of the caissons came close to and actually touched the old rock caissons.

During initial caisson and bell excavation, when the bell reached full size and came close to the old rock caisson, water was observed seeping in from the direction of the rock caisson. The water seepage rate quickly increased to the point where concrete had to be placed by tremie. The water head at the bell level was approximately 30 ft (9.2 m) so that even when there was as much as 2 ft (0.61 m) distance between the old rock caisson and the bell, water sometimes broke through. The water appeared to come up through the joints in the old wood lagging that could still be seen. It could be coming up also through the porous zones in the concrete.

Because of the uncertainties of the bottom condition in the bell when water comes in quickly and because of the added costs of tremie placement and inspection by divers, it was decided to grout at the contact point zone where the bell comes closest to the old rock caisson to try to prevent future "blow-ins".

Cement grout, sodium silicate grout, and AM-9 grout were all considered, and AM-9 was finally selected in order to achieve maximum penetration of the grout into the silt, porous zones of the rock caisson, and joints in the lagging. A plan and profile showing the average soil conditions, approximate positions of the rock caissons and belled caissons, and the approximate grout hole locations, are shown on Figure 2. Different grouting procedures were used initially while trying to develop the most cost effective approach. The grout pipe backfilled with torpedo sand (later switched to pea gravel) and sealed with bentonite pellets which is shown on the Figure, was used in the first few caissons. Eventually, the fastest way was found to be to drive NX casing into the clay to a depth of about 35 ft (10.7 m) to 40 ft (12.2 m) and then to open drill below the casing down alongside the rock caisson through the hardpan to approximate Elev. -72 (-22 m).

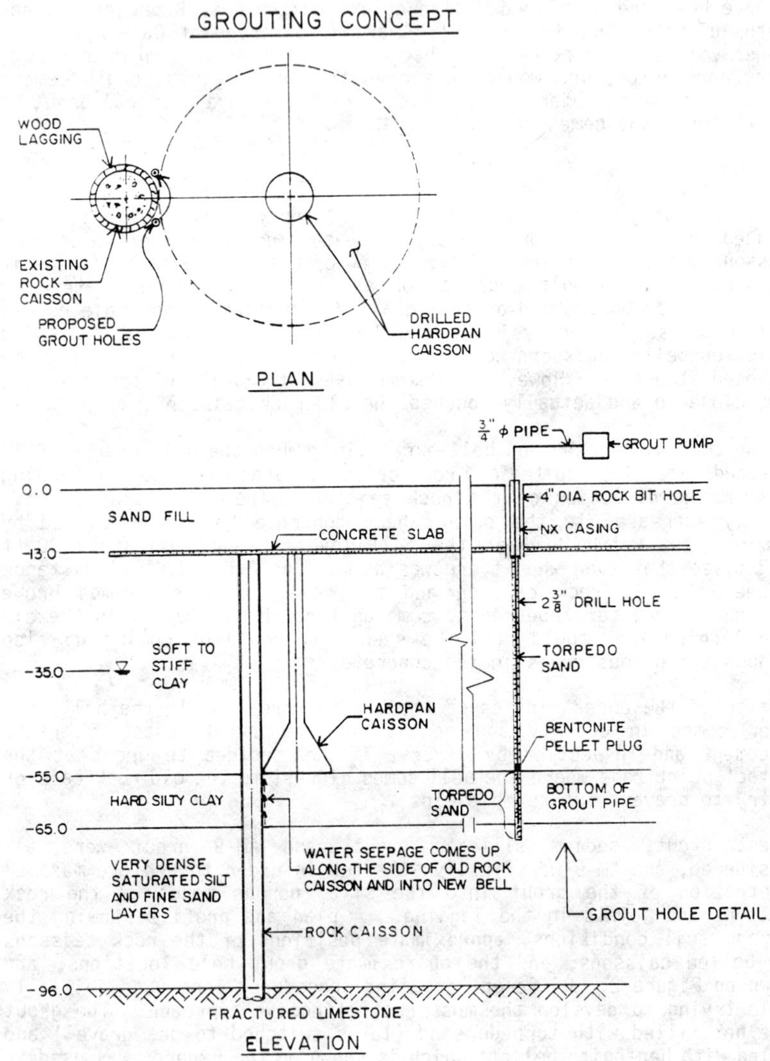
GROUTING CONCEPT
WOOD LAGGING
EXISTING ROCK CAISSON
PROPOSED GROUT HOLES
DRILLED HARDPAN CAISSON
PLAN
3/4" φ PIPE
GROUT PUMP
0.0
SAND FILL
4" DIA. ROCK BIT HOLE
NX CASING
CONCRETE SLAB
-13.0
2 3/8" DRILL HOLE
SOFT TO STIFF CLAY
-35.0
TORPEDO SAND
HARDPAN CAISSON
BENTONITE PELLET PLUG
-55.0
HARD SILTY CLAY
TORPEDO SAND
BOTTOM OF GROUT PIPE
-65.0
VERY DENSE SATURATED SILT AND FINE SAND LAYERS
WATER SEEPAGE COMES UP ALONG THE SIDE OF OLD ROCK CAISSON AND INTO NEW BELL
GROUT HOLE DETAIL
ROCK CAISSON
-96.0
FRACTURED LIMESTONE
ELEVATION

FIGURE 2

A packer was then placed at the bottom of the casing and AM-9 grout pumped using a Moyno pump and a batch mix process involving 55 gallons (0.22 m^3) at a batch. Since it would take about 9 minutes to pump the 55 gallon (0.22 m^3) batch, gel times were set for about 15 minutes to be sure to get all of the grout pumped before it gelled. Additional batches would be pumped until a back pressure of about 80 psi (552 kPa) was achieved. The amount of grout pumped generally was in the range of 2 to 4 batches, or 110 to 220 gallons (0.44 to 0.83 m^3). A total of 19 caissons were grouted, all of them successfully.

The overall effectiveness of the AM-9 grouting in stopping the seepage from the old rock caissons was repeatedly verified when the AM-9 grout was visibly exposed during bell excavation.

Case D

The Case D project involved a 35-story building at a site where poor rock conditions had been indicated by water loss during coring. The City set a 5 gallon per minute pumping limit and based on the mixed success of pregrouting at caisson locations at the previous projects just discussed, the owner agreed to experiment with pregrouting on this project. As an experiment in effectiveness, alternate perimeter caissons were grouted on the two street sides. In this way, it was believed it would be possible to see if there were any benefit in doing the pregrouting. Based on the success of the procedure used at Case A as compared to that used at Case B, the same general procedure used at Case A was followed. However, AM-9 grout was used instead of the silicate grout. This change was made based on the success achieved at Case C where the AM-9 was successful in preventing further seepage even on a caisson where a blow-in had actually occurred. The same 55 gallon (0.22 m^3) batch process used there successfully was used at Case D.

Unfortunately, the pregrouting did not prove successful. There was no discernable effect on water conditions in the pregrouted locations as compared to those not grouted. It was observed that grout-take in the holes was generally minimal; sometimes only one 55 gallon (0.22 m^3) batch and rarely over two batches. It appeared that most of the seams in the rock that eventually proved to be water bearing were initially filled with clay and silt preventing the necessary grout-take into the seams. However, upon excavation and dewatering, the seams would erode and blow into the hole and become very water active. The pregrouting did offer the advantage of providing, through the cores recovered, information regarding rock quality and approximate bearing levels which made probing for rock quality verification unnecessary at those locations.

In comparing the four projects where pregrouting was used and trying to assess why two were successful and two were unsuccessful, it appears that pregrouting with chemical grouts can be most effective in minimizing water seepage into caisson excavations when there are open seams and porous media permitting significant grout-take of at least several hundred gallons per hole. However, if the openings and channels are too large or water head too high, chemical grouts may not be strong enough to resist the water pressures or too much grout may need to be pumped before back pressure can be developed and solidifed grout created. On the other hand, if a lot of the joints, fractures, and seams are filled with fine material, the grout may not penetrate adequately and the grouting program will be ineffective. A significant reason for the AM-9 success on Case C compared to the other cases may be the more limited water head (20' vs. 55') (6.1 m vs. 16.8 m).

GROUTING DURING OR AFTER CAISSON EXCAVATION TO STOP WATER INFLOW AND CAVE-INS

History of Problem

It has been the writers' experience that grouting is seldom designed into a caisson project because of known soil and grout water conditions. More often, variations are observed in the anticipated conditions such that serious caisson construction problems develop which require use of grouting to solve. One main problem area in the typical Chicago profile illustrated in Figure 1 is the water bearing silt layer (which may or may not be under a head) which occasionally and sometimes erratically overlies the hardpan where belled caissons are normally supported. A second problem is the difficulty the caisson contractor sometimes has in effecting an adequate seal for rock caissons by the conventional technique of twisting casing into the rock. With regard to the first problem of water bearing silt over the hardpan, normal procedure calls for casing off the water bearing layer and belling in the hardpan. However, sometimes the hardpan layer is so thin that the casing can be extended only a minimum penetration into the hardpan and a successful seal is not obtained.

With regard to the second problem area, caisson contractors have sometimes "screwed" the casing as much as 18 ft (5.5 m) into the limestone bedrock (at considerable time and expense) without achieving an adequate seal. In the remainder of this section, several cases are discussed illustrating grouting solutions to these problems.

Case E

The Case E project involved a 40-story highrise to be supported on belled caissons on hardpan. The hardpan was less than 10 ft (3.05 m) thick and one of four borings indicated water bearing silt and sand on top of the hardpan. Two test caissons located in the area showing the wet silt and sand layer were successfully installed, seemingly verifying the feasiblity of the of the belled caisson design. However, during construction for the remaining caissons, conditions worse than anticipated were encountered and it was necessary to case through the sand layer into the hardpan and attempt to bell in the hardpan. Due to the limited thickness of the hardpan, a seal had to be accomplished with minimum penetration of the casing into the hardpan. The procedure used was to bottom dump a 10 bag mix cement grout in the shaft and then push casing through the grout into the hardpan. The grout outside of the casing was allowed to set up while the grout inside of the casing was stirred up and weakened with soil and water so that it could be augered out the next day. This procedure has been used successfully on many other projects, but in this case, was not successful in providing a seal. The approach that was eventually successful was to drill down on each side of the caisson with a cased bore hole and then to pressure grout with a packer set in the casing using AM-9 chemical grout. On the particular caisson in question, 385 gallons (1.54 m^3) of AM-9 grout were pumped in the two grout holes and a satisfactory seal was accomplished.

Case F

Case F involved a 50-story highrise on belled caissons on hardpan. The initial borings had indicated a variable thickness hardpan but with no special problems anticipated on top of the hardpan. However, over a portion of the site that also involved the thinnest hardpan, a wet silt layer was encountered during caisson excavation. The hardpan was not thick enough to case into the hardpan and construct a bell in the hardpan and the wet silt layer was not sufficiently stable to either bell on top of the hardpan or to rest on the wet silt. The procedure developed to permit construction to proceed involved the following:

1. Make full size bell on top of the hardpan in the wet silt letting the wet silt flow into the excavation enlarging the bell even further.

2. Measure the thickness of wet silt in the bottom of the bell after belling.

3. Bottom dump sufficient 10 bag cement grout mix for a 1 to 1 mix with the wet silt in the bottom of the bell.

4. Mix the in-place wet silt and bottom dump cement grout using the belling bucket.

5. Fill the remainder of the bell with 4 bag cement sand grout and allow to set overnight.

6. Drill through the cement grout mix extending several feet into the hardpan (depending upon thickness of the hardpan).

7. Construct bell partially in the hardpan and partially in the cement grout silt mixture on top of the hardpan.

A schematic cross-section of the completed caisson is shown in Figure 3. This procedure was used very successfully on the 9 caissons out of 52 where the procedure proved necessary. It recently has been used successfully on another major project in Chicago where a similar problem developed.

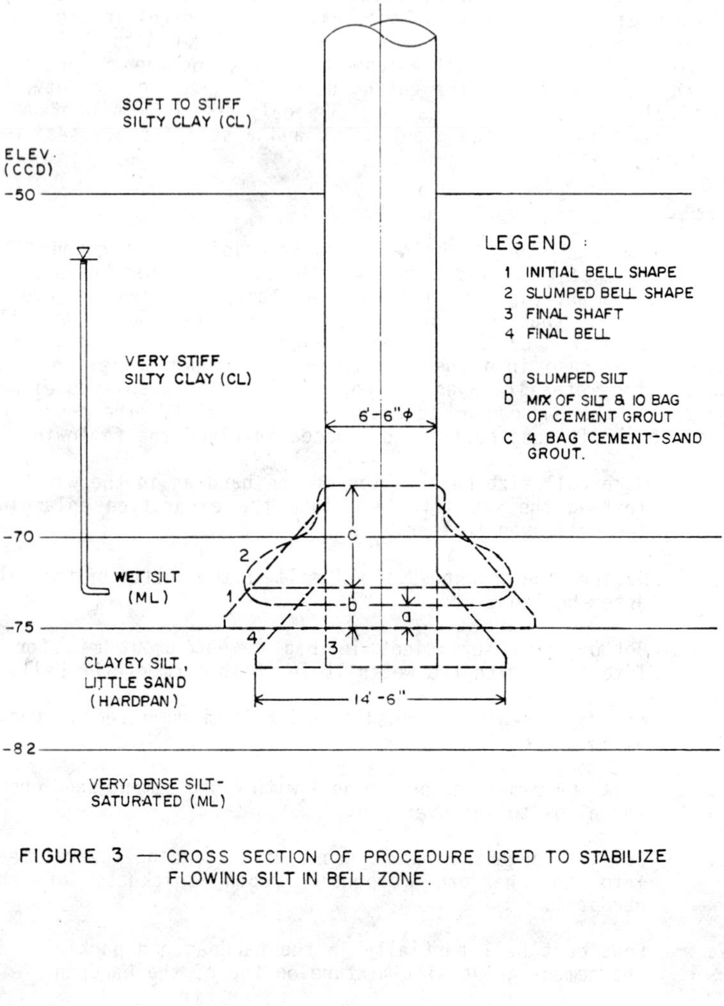

FIGURE 3 — CROSS SECTION OF PROCEDURE USED TO STABILIZE FLOWING SILT IN BELL ZONE.

Case G

This is the same project as Case D where pregrouting at the caisson locations did not prove to be significantly helpful in reducing water inflow rates on the rock caissons. During construction, it proved impractical in many cases to seal off water by extending the casing or core barrel further into rock. Two different grouting procedures were developed to solve the problem. The first procedure involved the use of AM-9 grout as follows:

1. Drill with 2-15/16 inch (7.46 cm) rock bit just outside of the core barrel (two locations on opposite sides of the core barrel) to approximately 3 ft (0.92 m) into the limestone bedrock.

2. Pump through the drill rod and bit a 55 gallon (0.22 m^3) batch of AM-9 grout (followed by just enough fresh water to clean the rod and bit) and allow the grout to set, providing some back pressure resistance for further grouting. Pump a second batch of 55 gallons (0.22 m^3) of AM-9 grout until back pressure builds up.

3. Remove the drill rod and rock bit by drilling out the grout in the bore hole.

4. Repeat the process on the opposite side of the casing.

Of the 15 caissons where this procedure was used, success was achieved in the first attempt on 10 of the 15. Three of the remaining 5 were successfully regrouted after turning the barrel deeper. One was sealed by turning the barrel deeper without grouting and one was eventually sealed with a cement grouting procedure described next.

The cement grout procedure involved the following:

1. Bottom dump 1 to 3 cubic yards (0.84 to 2.51 m^3) of a 10 bag, high slump cement grout into a flooded caisson hole.

2. Stir slowly with an auger fight to mix up any silt sediment believed to be on the bottom of the shaft.

3. Break core barrel free from rock and withdraw approximately 5 or 6 ft (1.53 to 1.83 m), allowing grout to flow against the exposed rock socket as well as into the vertical channel cut by the core barrel beneath the caisson bottom.

4. Screw barrel back down to its original position.

5. After a setting time of 5 or 6 hours, mechanically remove the excess grout inside of the barrel.

Typically, the cement grout procedure reduced the water inflows from as much as 100 gallons (0.4 m^3) per minute to less than 1 gallon (.004 m^3) per minute. Usually, some grout would remain on the bottom after the machine cleaning, but this was easily removed by hand using air spades. Use of air spades appeared to have no effect on the seal either with the cement grout or with the chemically grouted holes.

The cement grouting procedure was successful eight times in eight attempts.

In one case where the seal could not be accomplished with chemical grouting and the cement grouting procedure had to be used, it is believed that the chemical grouting was affected adversely by pumping from an adjacent caisson.

Because of the strength of the cement grout, cement grouting was performed only on those caissons where the caisson bottom was approved. The chemical grout was soft enough so that core barrels could be broken loose and turned further into the rock, but this was not possible with cement grouting.

GROUTING TO REDUCE SOIL SQUEEZE AND LOSS OF GROUND

Background of Problem

Whenever a vertical shaft extends into or through a soft clay deposit, there may be a tendency for the clay to squeeze toward the excavation, depending on the overburden pressure at a given point, the shear strength of the clay and the time that the excavation remains open. In the Chicago soil profile which includes soft clays as deep as 50 ft (15.25 m), rapidly increasing squeezing tendency has been observed when the overburden pressure to shear strength ratio exceeds 6.(1) Squeeze can also occur when the ratio is less than 6, if the time frame is long enough (days instead of hours).

On a typical hardpan project, the caisson is excavated and filled with concrete on the same day so there is a limited length of time for squeeze to occur and on many jobs, no noticeable squeeze does occur. On other projects, the overburden pressure to shear strength ratio may significantly exceed 6 and squeeze may occur in a matter of hours or even as the digging takes place.

On rock caissons, because of the difficulty in obtaining an adequate seal of core barrel into the rock and the need to quite often remove considerable amount of rock, completion of a rock caisson may take several days and, on occasion, up to several weeks. In order to insert

---

(1)"Ground Movement Associated with Drilled Pier Installations" by R. G. Lukas and C. N. Baker, Jr., presented at the ASCE Convention in Pittsburgh, PA, April, 1978 the core barrel all the way to rock [which may be as deep as 140 ft (42.3 m)] , the caisson contractors find it necessary to drill the hole slightly oversize, leaving an annular space several inches larger than the core barrel diameter. This space, if not filled artificially, eventually will squeeze shut, resulting in adjacent ground settlement.

Solutions to the Problem

In the case of the hardpan caissons, if the soft clay is soft enough that squeeze may be expected to occur during the excavation of the shaft or over a several hour period, the normal procedure is to drill the hole under a head of slurry until through the soft clay, set casing through the soft clay, remove the slurry from inside the casing and continue the shaft in the firmer materials below. Subsequently, one procedure involves placing concrete directly into the hole up into the casing, and then pulling the casing, allowing the fluid concrete pressure to displace the small amount of slurry in the annular space outside of the casing. The problem with this procedure is the potential for contamination of slurry into the concrete at some point during the casing pulling operation. While there are procedures available for assuring against this, many owners and engineers want a more fail safe system and the common procedure that has developed is to install a permanent corrugated liner inside the temporary liner, extending past the bottom of the temporary liner. The corrugated liner is filled with concrete and allowed to set overnight. The space between the corrugated liner and the temporary liner is then filled with cement grout typically with a free fall procedure and then the temporary casing is pulled. The grout then either displaces the slurry outside of the casing or mixes with it, but the integrity of the concrete within the corrugated liner is assured, as is prevention of further soft clay squeeze.

With rock caissons, the problem is different because the casing or core barrel extending all the way to rock and into rock is typically permanent. In the Chicago area, it is almost always permanent, but in other parts of the country with different soil profiles, the casing on rock caisons is often removed.

The procedure used in Chicago to minimize ground squeeze and adjacent ground settlement is to grout the annular space outside the core barrel. A grout pipe is pushed all of the way to rock outside of the core barrel or as close to rock as possible, and cement grout pumped from the bottom up displacing slurry. The success of this procedure is measured by whether the amount of grout pumped comes close to matching the theoretical volume of space to be filled. This depends in large part on when the grouting takes place. On rock caissons with few problems so that the caisson is completed within two days, grout-take approaching the theoretical volume may occur. However, on rock caissons with water problems and rock removal problems, the contractor will not want to grout around the casing until he has achieved a seal because he wants the casing free to twist further into the rock. In those cases where 5 or 6 days or more elapse from the initial digging of the hole up to the point of grouting, grout-takes of a fraction of the theoretical volume may occur.

GROUTING TO DENSIFY BEARING SOIL LOOSENED FROM BLOW-IN OR EXCESS WATER INFLOW DURING CAISSON EXCAVATION

Background of Problem

In the Chicago soil profile, and in the Detroit profile as well, if a hardpan caisson gets too close to the water bearing silt and sand stratum underlying the hardpan, there is danger of a "blow-in" occurring where soil and water quickly come up into the caisson excavation, resulting in a marked decrease in density of the underlying material and even resulting in actual voids being created beneath the hardpan as the hardpan temporarily bridges or arches over the voids. This potential problem also exists if a belled caisson comes too close to an old bore hole that extended through the hardpan (if the bore hole was not deliberately grouted). For this latter reason, it is the writers' recommendation that all bore holes extending through a potential caisson bearing stratum into a water bearing layer be grouted to prevent the bore hole from subsequently acting as a water channel.

Whether a blow-in actually occurs when penetrating close to or into the underlying dense silt stratum depends on the water head at the time, the diameter of the bell and the length of time the hole is open. Because of the typical very high density of the silt deposit, it takes time for seepage to occur and develop to the point that a pipe is created leading to the blow-in and large influx of soil and water. The high volume of water must come from a more permeable stratum such as fractures in the rock or a gravel layer sometimes overlying the rock.

If a blow-in does occur, the created voids must be filled and the loosened silt redensified if the designed foundations are to be utilized for the support of the structure. A case history is described where cement grouting was used successfully to accomplish this.

Case H - Detroit

The soil profile at Case G in Detroit generally was similar to that indicated in Figure 1, having a water bearing silt layer both on top of hardpan and below the hardpan. In addition, there were water bearing seams within the hardpan. The design called for sealing casing into the hardpan, if required, and belling in the hardpan with a design pressure of 40 ksf (1916 kPa). On caisson T-5, which was the next to last caisson being excavated in the tower portion, two water bearing sand layers were encountered within the hardpan, requiring the casing to be extended much further into the hardpan than elsewhere. The bearing stratum was tested and approved and the bell completed but before concrete could be placed into the hole, water and silt came in from below. An estimated 16 cubic yards (12.2 m^3) of silt was observed to come up into the caisson excavation as determined by measuring the bottom of the caisson prior to backfilling.

Because of the obvious lost ground, borings were made to check the capacity of the underlying silt and typical penetration resistance values of 100 had been reduced down to a range of 1 to 40. While no pressuremeter modulus values were available from the initial investigation, pressuremeter tests performed in the check borings indicated modulus values as low as 52 kg/cm^2 (5000 kPa), indicative of greatly loosened silt and much too low for the anticipated loading conditions. However, it was believed that because of the cohesionless nature of the silt, it would be possible to recompact the loose silt and fill any voids by a proper cement grouting program.

Grouting Procedure

The grouting procedure consisted of drilling nominal 4 inch (10.2 cm) diameter holes to about the lower one-third point of the underlying sand and silt zone and cementing in 1-1/4 inch (3.175 cm) diameter steel grout pipe. Explosive charges were then set off at the tip of the grout pipe and at the upper one-third point of the silt-sand zone to create an initial space for grouting. A high pressure cement grout plant capable of pumping at least 10 gallons (0.04 m^3) per minute at 600 psi (4140 kPa) was used. A cement and flyash grout (with a small amount of bentonite to keep the materials from separating) was pumped in 30 gallon (0.12 m^3) batches in each grout hole until back pressure built up to 600 psi (4140 kPa).

Results

A maximum of 74 batches of grout was pumped into one grout hole and a total of more than 30 cubic yards (23 m^3) was pumped in all 17 grout holes. Thus, the volume of grout pumped exceeded the observed amount of ground loss during the blow-in. Measurements were taken on heave of the surrounding ground and the maximum heave reported was 1/4 inch (0.64 cm).

Pressuremeter tests taken in borings performed after the grouting program indicated an increase in the pressuremeter modulus up to a range of 140 to 450 kg/cm^2 (13,750 to 44,190 kPa) which was considered a reasonable range for the original dense silt.

The tower was subsequently constructed on the caisson foundations supported on the hardpan soil over the silt without further incident.

CONCLUSIONS

1. There are a variety of soil, rock, and ground water conditions in which a grouting program may be helpful in facilitating caisson construction by reducing water seepage and increasing soil stability at critical locations.

2. There are a wide variety of grouts and grouting procedures which may prove effective for a particular condition, and some experimentation may be necessary to develop the most effective system.

3. Cement grouting under pressures of 200 to 600 psi can be effective in redensifying cohesionless, fine grain soils loosened by excess pumping or a "blow-in" during caisson excavation when adequate overburden pressure is available for confinement (80+ ft in the cases reported).

THOROUGH GROUTING CAN REDUCE LINING COSTS IN TUNNELS

By G. Dan Johnson [1]

ABSTRACT

The rock mass through which a pressure tunnel passes can provide a considerable amount of support of the internal hydraulic load on the lining if the rock is competent and thoroughly grouted to reduce voids and rock defects. Tunnel reaches which can benefit from grouting must be selected, leaving areas of weak rocks and low cover for linings designed to resist the full hydrostatic loads without rock support. The result of thorough grouting is savings in construction costs by using less steel in plate linings and reinforced concrete. Thorough grouting by contact and consolidation techniques must be achieved, and adequate results demonstrated by field testing.

Grouting specifications based on field exploration and laboratory tests should be prepared to assure the optimum in-situ strength of the rock. Contact grouting specifications should specify grout pipe patterns, grout mixes for good penetration and low shrinkage, and adequate but safe pressures and injection methods to assure thorough backfilling. The contractor usually selects mixing and transport methods for the grout. Grout may be mixed in the tunnel for small jobs or premixed outside and transported by rail, truck or pipeline.

Consolidation drilling and grouting, at much higher unit costs, is then performed to fill voids missed by contact grouting and to strengthen the surrounding rock. Thorough results are obtained by bringing "dam foundation grouting methods" into the tunnel to include water testing, thin starting mixes, stage grouting and split-spacing methods.

Planning and coordination by engineer and contractor can provide both quality and high production. During one efficient operation up to 700 tons of contact grout solids were injected per three-shift day. Final results should be verified by water testing and split-spaced check holes, and can be supplemented by sonic logging and core drilling if required.

INTRODUCTION

Pressure tunnels as typified by the one illustrated in Figure 1 rely on steel liners and, to a lesser extent, on reinforced concrete to resist internal hydrostatic pressures. Lining thicknesses of steel may vary from about one-half inch, which is considered to be about the minimum thickness for handling, up to 2-1/2 inches thick where heads are up to 2000 feet. Steel liner is expensive. For example, one ton

[1] Ebasco Services Inc., Greensboro, North Carolina

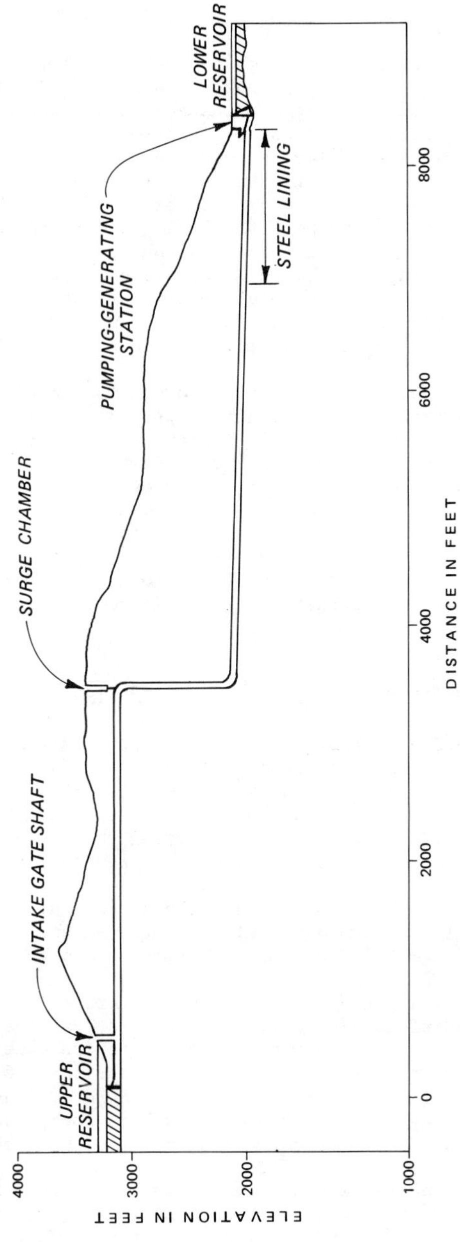

FIGURE 1. TYPICAL PUMPING-GENERATING PROJECT USING PRESSURE TUNNELS

of steel would provide only one linear foot of 1-inch-thick liner for a 16-foot-diameter tunnel at a cost in 1981 of about $3500 per linear foot. Steel costs are sufficiently high to justify varying the lining thicknesses in increments of one-sixteenth inch in accord with varying internal pressures or with variation in the amount of support provided by the surrounding concrete and rocks.

Steel costs can be substantially reduced where the rock mass is allowed to provide a maximum amount of support for the internal hydrostatic loads. However, design uncertainties due to inferred variations in the quality of the rock to support the lining and in the effectiveness of the contact (backpack) and consolidation (envelope) grouting operations often result in what may be excessive conservatism of design through excessive use of steel. Careful control of the excavation, complete filling of voids outside the lining, and proper repair of the surrounding rock by grouting can result in a more economical design by making full use of rock restraint.

During the 1974 ASCE-USCOLD Engineering Foundation Conference, there was general agreement that standard grouting procedures and specifications had changed little in the U.S. during the previous 30 years (1). The comment can also apply to tunnel grouting in the early 1980's. The aim of this paper is not to provide "all the answers" but to encourage engineers and geologists to take a new look at ways to maintain tunnel lining quality while reducing tunnel costs by using innovations as well as proven methods. Preparation of state-of-the-art specifications for excavation and grouting, followed by careful field control and testing can provide a reliable product that meets design requirements.

GROUTING TO MEET DESIGN REQUIREMENTS

The designer, according to Hartman, (2, p 114), requires the following basic information in order to design a pressure tunnel: 1) characteristics of the steel or concrete lining materials, 2) characteristics of the surrounding rocks and the exterior rock and groundwater pressures acting on the lining; and 3) ability of the lining to transmit stresses to the surrounding rock, described by Hartman as the "bedding factor".

The characteristics such as the yield strengths of the lining materials are generally uniform and easily determined and controlled during the construction. A lining can therefore be designed with confidence to support the entire internal pressures when the strength of the rock is low, the grouting results poor or more commonly when the rock and grouting conditions are unknown. Standard practice, as described by Patterson, et al. (3), has been to require steel lining where the rock cover over a pressure conduit is less than 50 percent of the maximum anticipated internal pressure head. More confidence in the rock quality and grouting results should allow less rock cover.

Assigning design values to the characteristics of the rocks outside the lining is difficult due to variations from area to area. Determining if the in-situ rock will accept the transmitted load is a separate rock mechanics problem to be resolved by mapping, boreholes or flat jacks tests, and monitoring of extensometers and other instrumentation. Methods are well described in the literature. Consolidation grouting to fill natural voids or to cement rock damaged by tunneling can usually improve the strength of the rock to support the lining.

A favorable bedding factor is obtained when grouting fills voids and repairs the rock damaged by the excavation to assure that the lining and the repaired rock will transmit stresses to the in-situ rock. Hartman states that results can be verified experimentally in test sections by measuring points on the interior of the lining for deformation using sensitive instruments before, during, and after contact grouting (2). This assures that grout has acted uniformily around the tunnel circumference.

Close instrumentation control may be difficult to achieve during production grouting outside of test sections on most projects. As an alternative to instrumentation thorough grouting can be reasonably assured by consolidation grouting in selected areas following the contact grouting, by drilling and testing of core samples of grouted rock and by sonic testing. Any voids detected by drilling, and sonic methods should be grouted by split-spacing methods and thin grouts as utilized for foundation grouting. Records of grouting and test results can assure the designer that the requirements for interaction between the lining and the surrounding rock are met and that a savings in lining costs will result from the grouting operations.

One of the best examples of utilizing rock to provide a maximum of support for steel liners is the Tehachapi Discharge Lines for the California Aqueduct which were constructed in 1968 (4). Where rock was weak, the liners were designed to carry the full hydrostatic load; however, rock mass with a calculated modulus of deformation above 5×10^6 psi was allowed to support some of the internal pressures.

Laboratory tests of intact rock cores can yield high values. For the Tehachapi project, test values for the static modulus of elasticity of rock ranged from 1.7×10^6 psi for weathered rock to 9.0×10^6 for fresh rock. However, the values for the rock mass as a whole are much smaller in that they are reduced by shears, joints and other natural defects and by the excavation. Imperfect but more representative values can be obtained in the field by in-situ testing with borehole extensometers, jacking tests, seismic and sonic methods, and by other instrumentation such as strain gages. On the Tehachapi project thorough grouting was relied upon to fill voids and to repair the rock to accept the hydrostatic loads.

Most U.S. projects, however, continue to follow a very conservative approach whereby reliance is placed on steel liners while avoiding the "unknown" conditions of grouted rock and its variables.

TUNNEL PIPING SYSTEM

Selection of the proper pipe pattern as adapted to tunnel conditions and proper pipe installation in the tunnel is a first step in careful tunnel grouting. Prior to placing the lining, grout pipes usually of 1-1/2-inch diameter must be installed, which will carry grout to the voids and vent air and water from the arch after the lining is placed. Contact grouting plans and specifications usually use some variations of a piping system such as that shown in Army Corps of Engineer manuals (5). Installation methods which have been used over the years have been modified in Figure 2 to take into account contractor innovations and selected utilization of plastic (PVC) pipe.

Pipe Pattern

Pipe spacing should be in proportion to the density of lagging because lagging will obstruct concrete placement during lining and grout flow during injecion. For example, in a tunnel with minimum obstructions by lagging and other support the pipe pattern would consist of pipes at 20-foot intervals only along the top of arch. Where there is heavy lagging, usually pipes or drill holes would be more closely spaced and installed down to the 4:30 and 7:30 clock positions, and even to the 6:00 clock position where it is necessary to fill the void below a steel liner. Any pipe for consolidation grouting of rock surrounding the tunnel should also be installed at this time if a special pattern is required. Usually many of the contact grout pipes can be redrilled for consolidation grouting during the second pass of drilling through the tunnel.

Pipe Installation

Close inspection prior to lining is important to assure that pipes are properly placed and anchored to avoid the all too common problem of pipes becoming "lost" as they are knocked askew by the concrete slick line during lining operations. If pipes become lost, then the engineer and contractor share the problem of trying to "replace pipe" by drilling. To be effective, vent pipes at the top of the arch must extend from rock to the lining. Installation is best accomplished as shown on Detail A of Figure 2 by anchoring to ribs or rock anchor pins, consisting of 2-inch-diameter pipes containing 1-1/2-inch pipe inserts. After forms are installed, the 1-1/2-inch pipes can be extended from the 2-inch pipes to make contact with the forms where the pipes can be held in place with set-screws. Pipe of PVC is adequate as inserts and is easier to work with in the confined space behind forms. However, couplings at the form contact should be of steel to accept the valves during grouting. Should the pipes be separated from the concrete form by only a fraction of an inch, the pipes will not be visible when the forms are stripped. Some contractors compress blocks of polystyrene at the pipe-form contact, which forms a blockout and aids in finding the pipe. Limited percussion drilling is required after forms are stripped to open the pipes to rock or to a void. At some tunnels, grout holes are drilled through concrete linings after lining placement in lieu of installing grout pipes. Drill holes are usually adequate below springline where there is little support to interfere with drilling and where takes are low. Generally, pipes provide the better installation for the following reasons:

1. A pipe can extend to and vent the top of an arch whereas a drill hole will extend only to the top of the concrete and not necessarily to a high point in the arch. A hole will thereby not replace a vent pipe.
2. Pipe can be extended through the maze of steel whereas drill holes must be abandoned when they encounter steel in sets, spiling or reinforcement.
3. It is difficult to retrieve the drill bits which have been drilled through wood lagging.
4. Packers instead of valves must be installed in drill holes thereby restricting high volume grout flow.

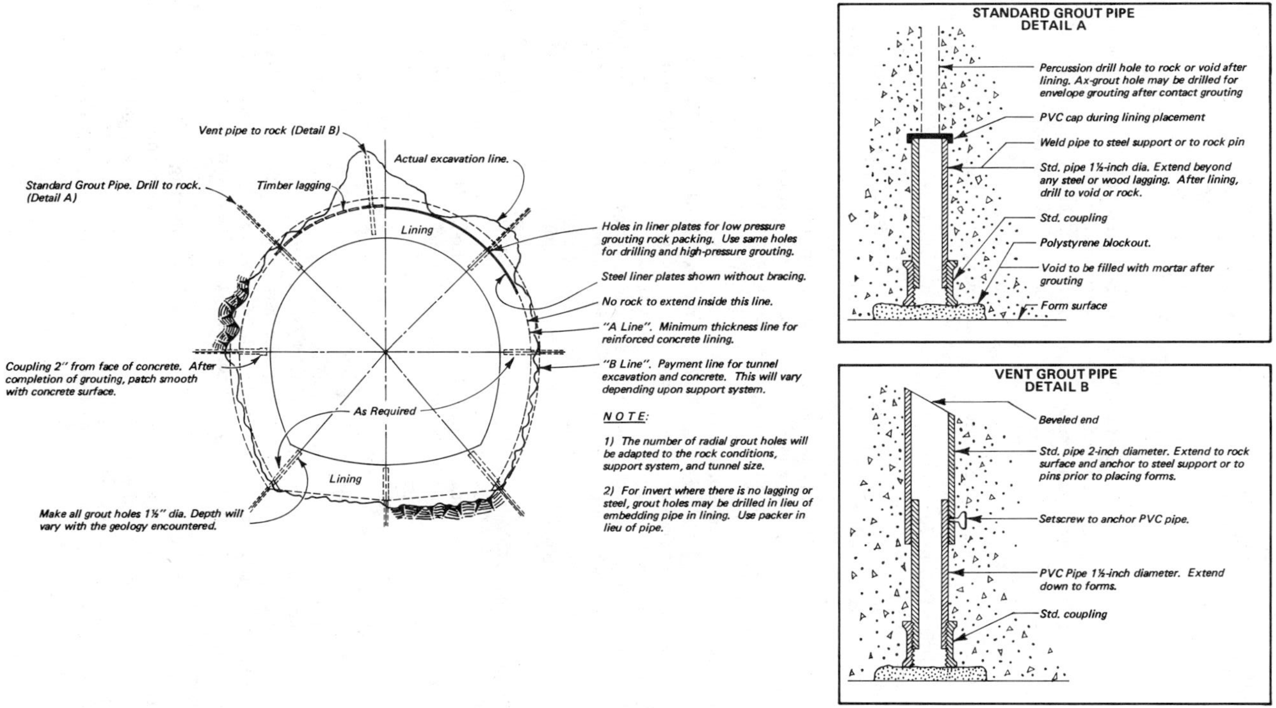

FIGURE 2. TYPICAL SECTION FOR CONCRETE LINING

5. Drilling damages the linings. Repairs to the lining require close inspection to assure that repairs are not merely surficial.

GROUT MIXES

Thorough tunnel grouting initially requires injection of a thick mix during contact grouting which will have minimum shrinkage. The thick grout will not, however, penetrate small voids in the rock. Consolidation grouting to follow the contact grouting will fill the small voids missed by the contact grout operation and penetrate into the rock.

Contact Grouting

The selection of a proper contact grout mix is determined by the objective of economically filling large voids near the lining with material which will have strength approaching that of the lining and surrounding rock. Portland cement grouts usually best meet these requirements. Most contact grouts contain sand which serves as a filler and also increases the grout strength and reduces the travel distance. Bentonite of about 2 percent by volume may be added to increase the pumpability of sanded mixes and to reduce segregation.

Techachapi Tunnels 1, 2 and 3 of the California State Water Project used a 1:1 water cement ratio plus 2 percent bentonite and up to 2 parts of decomposed granitic sand per 1 part cement. Much higher proportions of sand have been added on some jobs, especially where the sand is fine and well graded. However, the difficulty of obtaining sand of a suitable gradation (passing a #16 screen) and the wear and tear on equipment are factors to be considered. For this reason and the contractor's high price for sand, Angeles Tunnel of the California project, despite heavy lagging and large voids behind the lining, was grouted with a neat cement of 1:1 ratio plus 2 percent bentonite. Tunnels with small crown voids, due to minimal support, are usually grouted with neat cement of 1:1 or 3/4:1 water cement ratios.

Nonshrinkage grout may be required to eliminate all shrinkage voids, which otherwise would allow corrosion of steel. Such grouting is usually required below steel liner inverts. The addition of a small percentage of aluminum powder to the grout mix will nullify grout shrinkage. Following grouting in the arch with cement at Angeles Tunnel, voids under the steel liners were grouted with a mix of 7/8:1 water cement ratio plus 1-1/2 grams of aluminum powder per bag of cement. Careful control of a mix with aluminum powder is required to prevent excessive expansion and resulting damage to structures. Proper proportioning can best be achieved by laboratory mixing of the aluminum powder with a filler prior to adding the mixture to the grout batch. The contact grout mix should be adapted to the specific needs of each job with respect to pumpability, strength, fluidity and economy.

Consolidation Grouting

Consolidation grouting should utilize water testing of rock and the lining-rock contact and begin with thin mixes. Selection of mixes and other techniques are little different from foundation grouting except that confined work space, difficult access and the need to suppress dust and noise are special tunnel grouting problems to be resolved.

GROUTING PRESSURES

Grouting pressure in a tunnel, as in foundation grouting, is an important factor in thorough penetration of the grout into voids behind the lining.

Contact Grouting

Most backfill or contact grout is injected at maximum pump capability into the open voids at zero pressure except for "gage pressure" created by resistance of lines and valves. However, maximum allowable pressures of around 30 to 100 psi are usually specified to protect the linings during pressure build-up as isolated voids are filled.

In Europe high pressure contact grouting has been used to prestress the linings in compression by forcing grout along the rock/concrete or concrete/steel lining interface. Careful instrumentation control is required to maintain uniform pressures around the lining circumference during the grouting. Olive (6) describes the method as "interface grouting" at the Gordon Power Tunnel. According to Pequignot (7, p 388) at Manvoisin-Fionnay, Switzerland, a pressure of 284 psi was used in the tunnel. This pressure was 80 percent of the external pressure for which the lining was designed. Pequignot reports that pressures as high as 560 psi have been used in exceptional cases.

Consolidation Grouting

Consolidation grouting pressures are selected by the same methods as are those for foundation grouting. To avoid damage, the lowest pressures must be used in the stage nearest the tunnel lining. At Angeles Tunnel of the California State Water Project, grouting at less than the design pressure of 50 psi triggered sections of the lining at three areas to fail. The linings cracked and moved into the tunnel by separating along the invert-arch construction joint as shown in Figure 3. Subsequent studies showed that locally the joints were constructed horizontal, rather than radial to the lining surface which normally keys the arch and invert segments. The lining had been designed to withstand several times the grout pressure as applied. On this project, all damaged areas were repaired. The failure problem, although due to poor construction of the joint, reconfirmed the importance of grout pressure control.

GROUTING EQUIPMENT

Equipment for thorough grouting must be capable of mixing the required materials and of transporting them to the area where grout is injected under conditions of controlled pressure. Equipment for tunnel grouting may be specified in part by the engineer, but much of the selection will be left to the grouting contractor who must also attain high production.

Grout plants as used for injecting large volumes of thick foundation grout are adequate for contact grouting. Large jobs can benefit from bulk handling of materials. Centrifugal pump mixers are well adapted for contact grouting, especially where the thorough mixing of bentonite is required. Figure 4 shows a very automated plant of this type designed by the contractor for contact grouting a tunnel lining in California. Production from this plant reached as high as 700 tons of solids per three-shift day and averaged 472 tons. This is a plant that

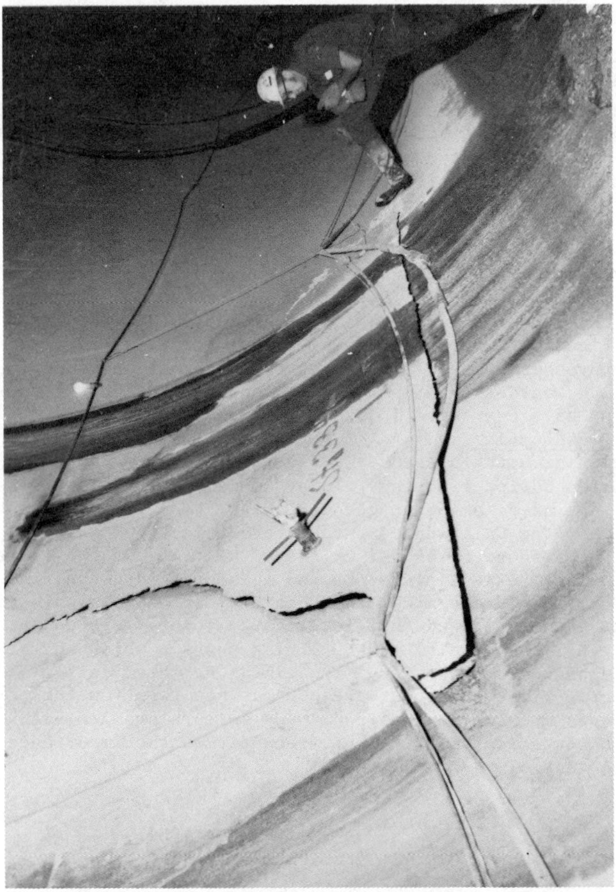

Courtesy Calif. Dept. of Water Resources

FIGURE 3. TUNNEL LINING FAILURE ALONG CONSTRUCTION JOINT.
(NOTE PACKER IN GROUT HOLE).

Courtesy Calif. Dept. of Water Resources

FIGURE 4. CONTACT GROUT MIXING PLANT.

mixed only neat cement and pozzolan; the contractor considered sand to be too difficult to handle and too abrasive to his equipment.

Grout materials for mixing or premixed grout must be transported to the injection area. This can be a difficult job in underground workings, where access is restricted and where one becomes painfully aware that two objects cannot occupy the same space at the same time. At the Angeles Tunnel job, grout was hauled to the injection area by rail in 18 cubic yard Moran Cars as shown by Figure 5. On other jobs tire-mounted transit mixers have been used to haul the grout premixed or grout has been mixed in the tunnel after the materials are hauled into the tunnel by track or rail. For larger jobs, it is usually more economical to mix outside the tunnel. Outside mixing also avoids the problem of dust and noise in the tunnel.

In the confined space of a tunnel, piping of grout is often the best solution to the transportation problem. On one job, a sand mix was pumped 5700 feet through a horizontal 1-1/2-inch line using a booster pump at the midpoint. An alternative method is to move the grout down a vertical pipe to the tunnel by gravity even if this means drilling holes especially for this purpose. A separate agitator sump and injection pump will usually be required at the end of the pipe in the injection area.

Grout pumps for transporting grout and for injecting it behind the liners must be capable of handling large volumes of thick grout at relatively low pressure and must require low maintenance. Auger pumps are most commonly used in that they are easily cleaned. However, care must be taken in selecting the proper size of pump which is designed for the grout materials. On most contact grout jobs this means a pump equipped with a short barrel of large diameter and designed for the abrasion of sand. On one job the problem was in wearing out the expensive stator on auger pumps with sanded grout. Much of the problem resulted from the use of pumps that were too small and the contractor's attempts to compensate by running the pumps at higher than design r.p.m. The contractor's final solution was to return to the old workhorse of grouting, the slush or reciprocating pump. Although this pump required extensive labor for cleaning of valves, replacement parts were relatively inexpensive. It proved helpful to periodically send a few batches of neat cement to flush out sand from the valves and pipeline.

On large operations, three mobile platforms or "highcars" must be maintained in the tunnel to provide access to workmen for the following operations:

The first operation consists of inserting valves and/or packers in the tunnel lining immediately ahead of the grouting operation and the closing of valves on holes which communicate ahead of the holes being injected. From the second car, connections are made to holes for grout injection. At this location the grouting pressure is controlled by a header man. The third operation from the last car is the recovery of valves (and packers, if used) from behind the grouting operation. Valves and packers will be cleaned, passed to the first car and used again in advance of the grouting. Unless grout is delivered by pipeline, the third car must straddle the center of the tunnel to allow grout materials to pass beneath and be delivered to the injection car.

Small items can create large problems, which can be true of valves that are installed in the lining. Improperly cleaned valves can create back pressure and give a false impression that a hole has been filled. Packers placed in drilled holes should be used in lieu of embedded pipe

Courtesy Calif. Dept. of Water Resources

FIGURE 5. TUNNEL CONTACT GROUT OPERATION.

nipples only as a last resort, because a packer provides a smaller opening than does pipe, an important consideration when high grout volume is required.

INJECTION TECHNIQUES

Grouting methods should be adopted to the desired end result and to the type of rock support system. Tunnels supported by rock bolts are much easier to contact grout than those with ribs and lagging. The ribs and lagging block the flow of concrete leaving isolated voids which must be grouted. A coating of shotcrete on the lagging adds to the problem. To minimize blockage of concrete and grout, most specifications require the removal of any lagging which can be safely removed prior to placement of the lining. Proper technique together with suitable piping, mixes, pressures, and equipment will economically fill the voids behind tunnel lining to the maximum extent practicable. The following injection procedures are recommended:

1. First, hook the lower holes on both sides of the tunnel simultaneously to move grout upward on both sides of the lining.
2. Attain as high an injection rate as possible to place the design pressure over a large tunnel area. Allow valves ahead of the grouting to remain open and vent air and water until they flow thick grout.
3. Keep communicating holes "alive" by occasionally bleeding them or by temporarily connecting them until they can be hooked at the full design pressure. Grout on a 24-hour basis, whenever possible, to avoid the sealing off of communicated holes during a shutdown. A wash pipe may be used to open holes that have sealed off but have not been hooked.
4. Establish a system of records that shows when each hole was hooked and completed or communicated. The holes can be numbered with paint on the tunnel lining and the results of grouting shown by sketch in a field book to be passed from shift to shift. Records on quantities of grout injected should be tabulated for each reach of tunnel.

Figure 6 shows a graphic presentation of grout takes in a tunnel. Such records can be useful to help assess the thoroughness of grouting for each reach and to anticipate takes of consolidation grout where that operation will follow contact grouting in selected reaches of the tunnel. Recording of overbreak during tunnel mapping will allow estimates of contact grout take and selection of areas for consolidation grouting based on the geology. An area of tunnel collapse is shown in Figure 6 where repair by grouting of the damaged rock was important. Note the high takes of both contact and consolidation grout in the area.

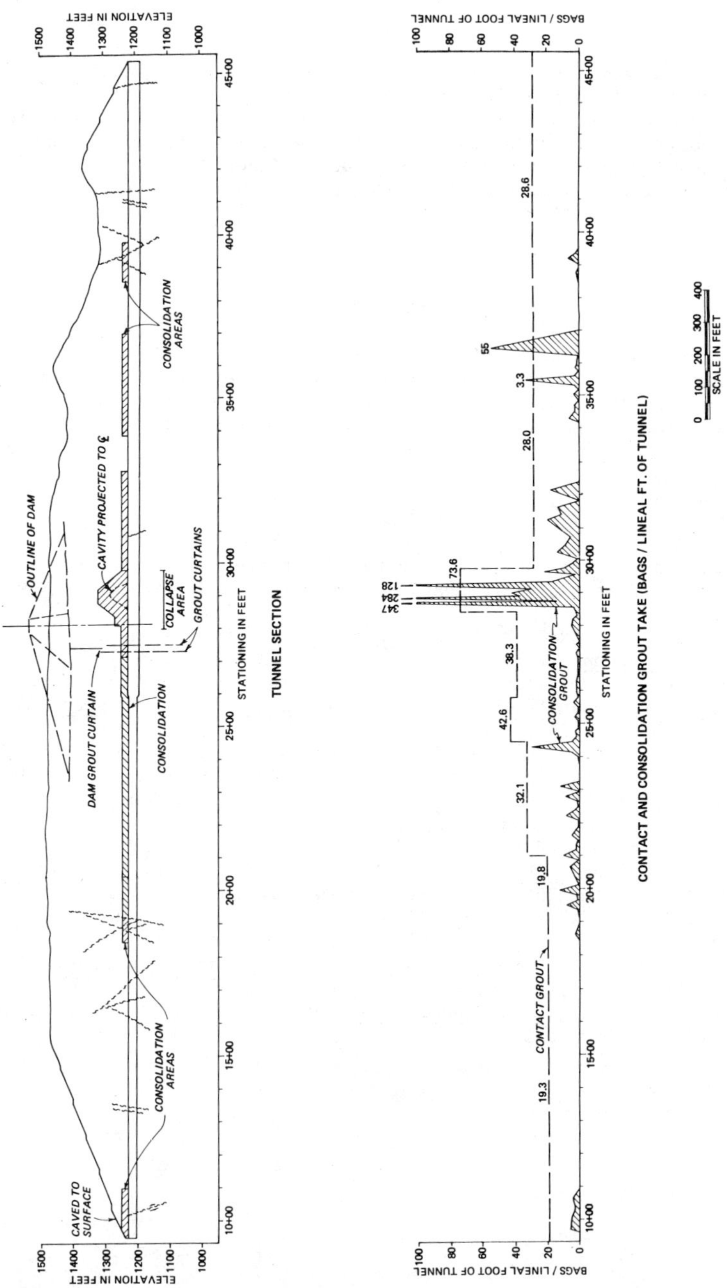

FIGURE 6. TUNNEL GROUTING SUMMARY

SUMMARY

During the last few decades, U.S. progress in tunnel grouting has lagged while the designer has put his trust in the knowns of steel and reinforced concrete linings. The grouting geologist/engineer owes it to the designer to provide thorough tunnel grouting for which satisfactory results can be demonstrated. The designer owes it to the owner to utilize rock restraint as much as possible to reduce lining costs.

ACKNOWLEDGMENT

The writer wishes to thank Raymond Sholes, now with Southern California Edison Company, for his comments on the draft. Many of the practices described herein were formulated by Mr. Sholes during grouting of tunnels for the California Department of Water Resources.

REFERENCES

1. Welsh, J.P., "Present State of the Grouting-Industry in the United States", in Foundations for Dams, ASCE, New York, 1974, p. 65-68.
2. Hartman, B.E., Rock Mechanics Instrumentation for Tunnel Construction, Terrametrics, Inc., Golden, Colorado, 1966, 153 p.
3. Patterson, F.W., Clinch, R.L., and McCaig, I.W., "Design of Large Pressure Conduits in Rock", Journal of the Power Division, ASCE, Vol. 83, No. PO6, Paper 1457, December 1957.
4. Wineland, J.A., O'Neil, A.L., and Arnold, A.B., "Tehachapi Mountains Crossing of the California Aqueduct", in Yardley, D.H., ed., Rapid Excavation -- Problems and Progress, Society of Mining Engineers, AIME, New York, 1970, p. 165-190.
5. U.S. Department of the Army, "Foundation Grouting: Planning", Engineer Manual EM 1110-2-3501, U.S. Department of the Army, Office of the Chief of Engineers, Washington, D.C., 1966 (revised 1972), 64 p.
6. Olive, R.T., "Prestressing of Gordon Power Tunnel Lining by Interface Grouting", Civil Engineering Transactions Institute of Engineers, Australia, Vol. CE18, No. 2, 1976, p. 146-154.
7. Pequignot, C.A., ed., Tunnels and Tunneling, Hutchinson and Co., London, 1963, 555 p.

SWEDISH VALVE TUBE GROUTING

By Bruce A. Lamberton[1], M. ASCE

Low density soil beneath a nuclear power facility in Florida was stabilized by injection of over 400,000 gallons of chemical grout through 17,000 ft. of 60 mm o.d. valved tubing, using the Alvik "J" drilling and injection system developed by the Stabilator Division of Skanska Cement Co. of Sweden. The Siroc process using formamide and calcium chloride reactants with sodium silicate was used for the majority of the work. Effectiveness of grout injection was demonstrated by penetration tests and permeability tests.

I. Site Description

In the summer of 1969, work on a nuclear power reactor of the Florida Power Corporation at Crystal River was interrupted by discovery of low density soil under the southeast segment of the tendon gallery. The facility included a circular reactor building adjacent to a structure containing generators and a service area. The complex is generally supported on a mat foundation imposing loads ranging from 2.5 to 7.8 ksf. Primary foundations are carried to the solutioned limerock approximately 20 ft. below the original ground surface.

Cement grouting of the limerock system is normally employed in this area in foundation preparation to fill solution channels and voids within the foundation rock. The carbonate rocks of both the Inglis and Avon formations have been subject to solution activity favoring the joint sets of the fracture system. The Piezometric Profile (Fig. 1) was gravitational subject to tidal variation with normal ground water level at +88 (plant datum). Deep well pumping, supplementing a peripheral cement grout curtain permitted excavating as deep as 35 ft. below water level (Elev. +53) although normal drawdown was to Elev. +70.

Within the plant site, the most intensive solutioning corresponds to the location of a series of fractures intersecting near the southeast edge of the Reactor building. Voids formed by the solution process were usually found to be filled with sand and silty sand transported by groundwater. These infill deposits and associated heavily leached limerock were generally found between Elev. +60 down to Elev. +25. Infill materials (Fig. 2) were identified as occasionally gravelly, coarse to fine or medium sand, containing 5% to 18% silt and clay.

---

[1] Vice-President, Intrusion-Prepakt, Inc., Cleveland, Ohio

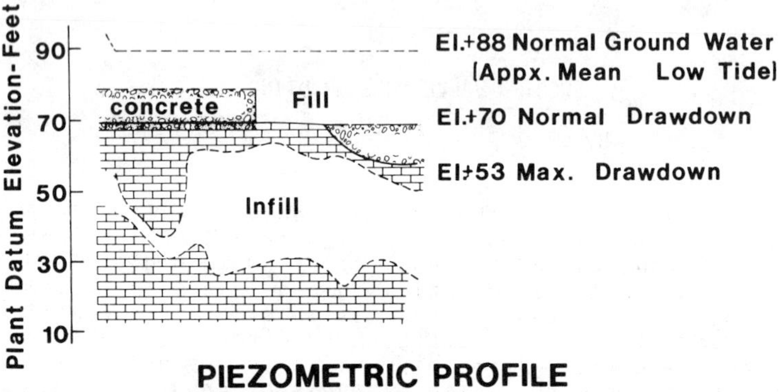

PIEZOMETRIC PROFILE
CRYSTAL RIVER UNIT NO. 3

Fig. 1

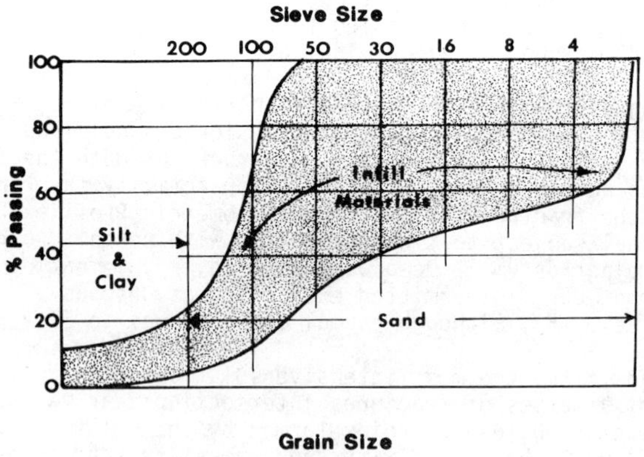

CRYSTAL RIVER UNIT NO.3

Fig. 2

Hydraulic gradients imposed by dewatering caused scour of the infill material reducing density of the sediments and increasing volume of the voids. It was in these areas (Fig. 3) that the supporting character of the foundation material was suspect, leading to the use of chemical grouting as a supplement to the scheduled cement grout foundation treatment.

II. Selection of Grout Materials

To qualify as a means for improving the structural characteristics of soil, a chemical grout must exhibit viscosity low enough to penetrate loose to medium dense silty sand containing up to 18% minus-200 mesh material. Grout stabilized sand must also exhibit an increase in strength and a decrease in compressibility, and grout stabilized soils must retain these characteristics for the useful life of the structure. This factor of permanence was the most difficult to evaluate. There was precedent to the effect that several chemical grouts in common use, including silicates, lignins, resins, and acrylamides, would permeate sand containing up to 20% minus-200 mesh material.

Of all grouts available, only the silicate types have a long record of successful use. In this reaction, insoluble silica gel is caused to percipitate from a solution of sodium silicate. Strength of the grouted soil is a function of the concentration of the silicate solution (Fig. 4). Typical job concentrations range from 40% to 60% silicate by volume. A relatively recent improvement in silicate grouting technology is known as the Siroc® process in which Formamide is the reactant and calcium chloride serves as an accelerator to control gel time. By reason of its acceptably low viscosity, adequate strength, and the long service record of silicate grouts, Siroc was chosen as the chemical grouting material.

For installation of subsurface cutoff wall, Terranier chemical grout was injected on the basis of its slightly lower viscosity. Terranier is a polyphenolic polymer pre-catalyzed with 6% formaldehyde solution. The addition of an accelerator solution of sodium dichromate causes formation of an insoluble gel. This product is no longer in commercial production, probably on largely economic grounds. However, the production and use of sodium dichromate involves a health hazard which might tend to discourage its manufacture and use in the highly safety-conscious environment of today's market.

In a few instances, where extremely rapid gel time was required, as in the immediate vicinity of deep well pumps, a modification of the Siroc system was used in which portland cement acted to accelerate the silicate/formamide reaction.

III. Drilling & Grouting Equipment

Since chemical grouting was not anticipated in the original construction schedule, it was of the utmost importance that this remedial work be performed with all possible speed. The key element in the successful timing and execution of this work was the selection of drilling and grouting equipment. Grout injection tubes were installed by means

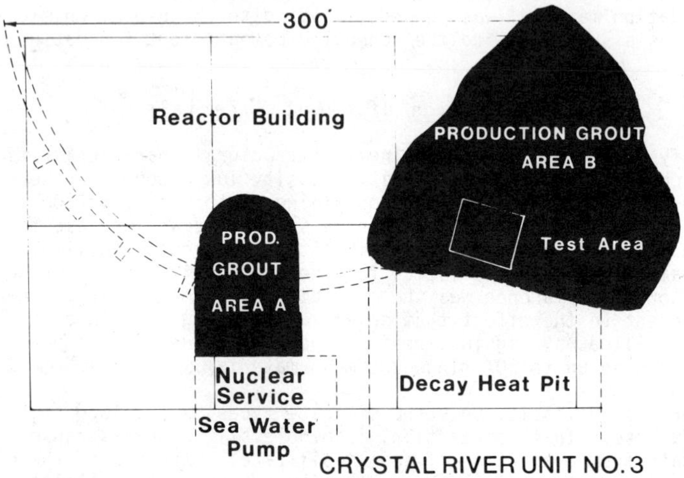
300'
Reactor Building
PRODUCTION GROUT
AREA B
Test Area
PROD.
GROUT
AREA A
Nuclear
Service
Sea Water
Pump
Decay Heat Pit
CRYSTAL RIVER UNIT NO. 3

Fig. 3

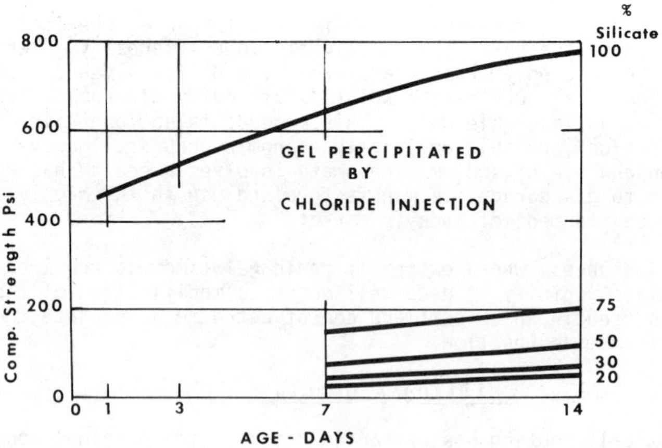
%
Silicate
100
75
50
30
20
GEL PERCIPITATED
BY
CHLORIDE INJECTION
Comp. Strength Psi
800
600
400
200
0
0
1
3
7
14
AGE - DAYS

Fig. 4

of the Swedish Alvik Drilling System, a development of the Stabilator Division of Skanska Cementjguteriet. It was the first use of this system in the United States.

A heavy percussion type drill rotates drill rod and casing down through the overburden as flushing water is forced down the drill rod, returning cuttings back through the annulus between drill rod and casing and out through a flushing assembly located below the hammer (Fig. 5). Used with an eccentric bit, it is referred to as the Alvik X System. The eccentric drilling bit is used whenever a hole must be advanced downwardly in successive stages as in a soil sampling or test grouting operation. By rotating the drill rod 180^{0} in the direction reversed to drilling, the eccentric carbide-tipped cutting bit swings back into the drill bit body permitting extraction of the rod and bit through the casing. Production with the single cutting bit is much slower than with a normal 4-point fixed cutting bit and the eccentric bit is far more expensive in first cost and somewhat more expensive in terms of bit cost per foot.

The Alvik J System employing a knock-off type bit was used almost exclusively on the Crystal River Job. Since the bit remains in the ground below the casing, redrilling and casing advancement is usually not possible. The knock-off bit, either 4- or 6-point carbide tipped, is held on the tapered shank of the central pilot bit by a friction fit.

All Alvik Systems require use of a heavy pneumatic hammer with strong positive rotation. An Atlas Copco BBE-57 was used on this job. It delivers 1950 blows per minute at a maximum torque of 253 foot pounds. It was mounted on a ROC-600 air track (Fig. 6). The drill and air track require 600 cfm air @90 psi. Flushing water is provided at a rate of 15 gpm at pressures up to 200 psi. Casing and drill rod are

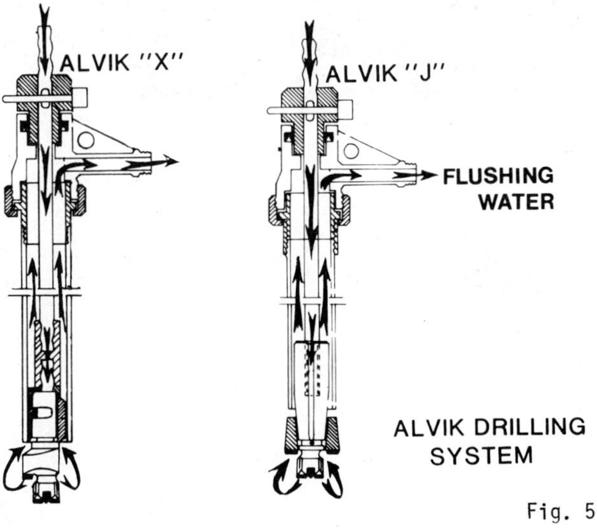

Fig. 5

(Figure 6)

Wagon Drill with Grout Tube
Handling Tripod in Foreground

maintained in precise relative position by means of a pin type locking clamp which secures rod and casing together at the driving cap.

All rotative force and hammer blows are transmitted directly to the drill rod and bit, the casing simply following down as drilling progresses. Thin rope thread couplings are shop welded to the bottom of each 10-ft. section of casing. A drill hole diameter of 70 mm provides 2.5 mm clearance for the 65 mm O.D. coupling. To remove the knock-off bit, the drill rod connection is broken immediately below the shank adapter, bit removing blocks placed over the upper end of the drill rod, and the drill reconnected and rotated slowly.

All casing used in grouting work is provided with spring leaf valves (Fig. 7), covering 10 mm diameter grouting ports inset into the casing wall on about 14" centers in diametrically opposed pairs. Casing is 60 mm outside diameter by 51 mm inside diameter. Grout is injected through a 3/4" pipe and a packer assembly, the cups of which are expanded by grout injection pressure. A pressure of approximately 30 psi is required to open the leaf springs in the absence of any restraint.

After each casing had been grouted to full depth, the casing was cleaned out and capped preparatory to regrouting if required. On this job, the relatively weak chemical gel was cleaned out with a 5000 psi water blaster and crawl nozzle. Couplings are intended to serve during casing installation only. No attempt is made to extract valved tubing which is left in place for regrouting.

Railroad tank cars delivered sodium silicate for which 10,000-gallon bulk storage was provided. Formamide was delivered in highway tankers to 4,000-gal. storage tanks. Capacity was designed to sustain two-day operation at peak production. The grout mixing plant (Fig. 8) can best be visualized as a two-step liquid handling operation -- preparation of the primary silicate and standard reactant solutions followed by metered blending of the two solutions to achieve the gel time required. Separate solutions of silicate and reagent with aniline crimson tracer dye and additional reactant if required were then prepared in 100-gal. verticle-shaft paddle-type grout mixers and delivered to small holding tanks and metering pumps adjacent to the point of injection.

For the great majority of the work on which silicate, formamide, and chloride were employed, it was possible to prepare the two solutions in such concentration that they could be metered and stream blended in a 1:1 ratio -- a particularly convenient and accurate arrangement since parallel sight gauges provide positive visual evidence of accurate proportioning. The progressive cavity, 3-stage Moyno pumps were also provided with tachometers calibrated to permit metered pumping at ratios varying from 1:1 to 8:1. Maximum pressure capability was 225 psi.

Terranier is typically pumped at ratios of basic solution to reactant ranging from 4:1 to 8:1. Pumping ratio must therefore be controlled with tachometers. From the metering pumps, the two solutions were delivered to an in-line blending manifold and pressure gauge immediately upstream of the grout injection pipe. A sampling cock was provided

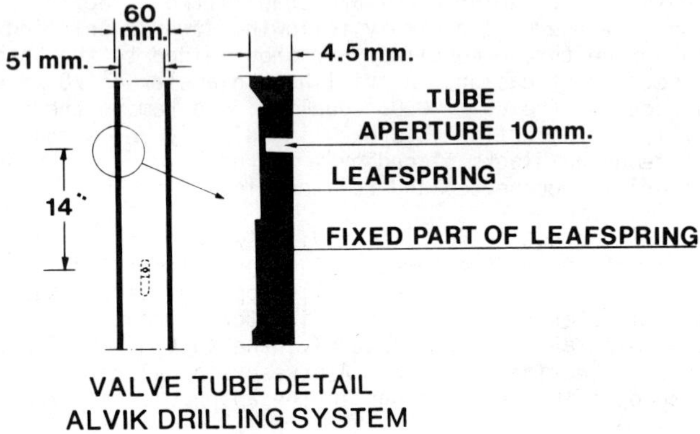
60
mm.
51 mm.
4.5 mm.
TUBE
APERTURE 10 mm.
LEAFSPRING
FIXED PART OF LEAFSPRING
14"
VALVE TUBE DETAIL
ALVIK DRILLING SYSTEM

Fig. 7

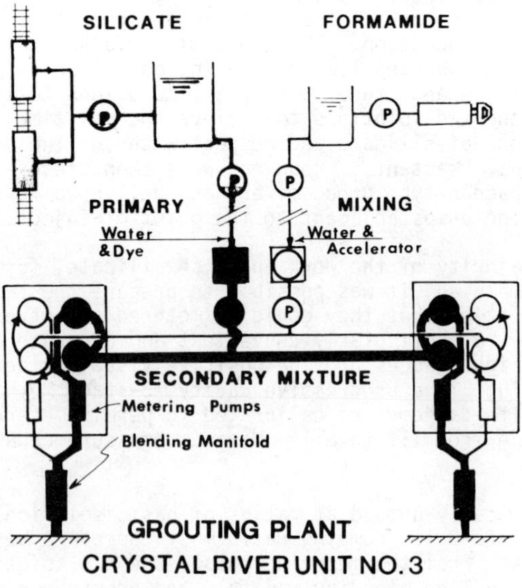
SILICATE
FORMAMIDE
P
P
P
PRIMARY
MIXING
Water
&Dye
Water &
Accelerator
P
SECONDARY MIXTURE
Metering Pumps
Blending Manifold
GROUTING PLANT
CRYSTAL RIVER UNIT NO. 3

Fig. 8

at the manifold to permit convenient check of gel time.

Each of the two chemical grout metering units had a capacity of 12 gpm although average injection rate is normally about 5 gpm per unit or about 2000 gallons per unit per 8-hour shift. Using the Alvik valve tube system, grout injection is virtually continuous with only a brief pause to release, raise, and reset the packer. Injection pipes were handled with a hand-operated, cable hoist mounted on a light-weight pipe tripod. Insert pipes were calibrated to correspond to the spacing of the valves on the casing.

In the few instances in which portland cement was used as an accelerator for silicate/formamide grout, to achieve gel times on the order of 15 s to 30 s, a double injection pipe was used and the two solutions streamblended at the bottom of the insert.

IV. Test Program

A test grouting program was conducted in two selected areas for the purpose of evaluating grouting effectiveness and establishing acceptance test criteria. Grout was injected at 45 locations in Area 1 and 39 locations in Area 2. All test grouting was conducted from Elev. +70 plant datum under drawdown conditions. Transient pressure was limited to 200 psi and sustained pumping pressure to 180 psi. Based on estimate of soil porosity and assumed radius of grout travel, grout volume was limited to 50 gal. per valve (14" vertical increments). Injection rate was in the range of 6 to 8 gpm.

In the absence of preparatory cement grouting, primary holes were injected with Siroc/cement grout to fill large voids. Primary holes in Area 1 were spaced on a 7' x 8' grid and in Area 2 on a 5' x 6' grid. The 60 remaining secondary and tertiary holes were split spaced. Terranier was injected into 9 of these holes and Siroc into the 51 remaining holes.

In all, approximately 25,000 gal. of Siroc/cement grout was injected into 24 primary holes having a total hole length of 750 ft. for an average take of 33 gal/ft. Approximately 50,000 gal. of Siroc and Terranier were injected into 60 secondary and tertiary holes totaling 1600 ft. in length for an average take of 31 gal/ft. Grout take as a percent of soil volume was 56% in the very low density soils of Area 1 and about 20% in Area 2 where much denser material was encountered between Elev. 30 and Elev. 50.

To evaluate the extent of soil improvement, N values, as measured by Standard Penetration Resitance (SPR) Tests, were measured in the test area (Fig. 9). The results of 122 tests in Area 1, considered more nearly typical of soils in the production areas, indicated that, even for the poorest infill and limerock conditions, an N value of 30 blows per foot could be achieved. Thirty-five permeability tests were conducted by the "pump-in" technique. Comparison of before and after average take values showed that chemical injection had reduced permeability by a factor of approximately 10^{-3} cm/sec.

SUMMARY OF SPR TESTS IN AREA 1
CRYSTAL RIVER UNIT NO. 3
TEST PROGRAM N VALUES

Strata	Before Grouting Avg.	After Grouting Avg.	After Regrouting Avg./Max.
Soft Limerock	15	41	58/125
Infill	0	11	25/60
Competent Limerock	25	39	88/138

Total of 122 Tests

(Figure 9)

To determine the effect of curing time and soil density on strength and compressibility of grouted soil, laboratory specimens were prepared by the puddling method in which infill sand was poured into 2" diam. by 5" high molds half filled with Siroc chemical grout while the molds were being vibrated to achieve a predetermined density.

The unconfined compressive strength of grouted specimens was greatly influenced by initial density as well as by curing time. Whereas specimens with a density of 113 pcf exhibited about 25% strength increase from 2 to 28 days, 120 pcf specimens gained over 58% strength in the same period. The modulus of elasticity, E, increases rapidly with corresponding quick decrease in axial strain at failure indicating a high rate of increase in the rigidity of the gel-soil system. The undrained shear strength of chemically stabilized soils increases almost linearly with increase in consolidation pressure.

An investigation of the effect of curing conditions clearly demonstrated that physical properties in situ would be significantly less than that of specimens subjected to standard curing or to curing in air. Stabilized soil with a relative density between 50 and 60% exhibited cohesion values in the range of 8 to 10 psi with an effective friction angle ϕ' of about 33° which was estimated to be about 2° higher than that of ungrouted soils.

In accordance with findings resulting from evaluation of the test grouting program, acceptance criteria were established for production grouting which required:

1) not less than one Standard Penetration Resistance Test per 400 ft^2 of stabilized area,

2) minimum N values of 30 blows per ft. in 75% of the tests,

3) minimum N values of 15 blows per foot in 90% of the tests,

4) no more than one N value should be less than 30 blows per ft. in any 6-ft. test hole interval,

5) at least one Permeability Test for each 800 ft^2 of stabilized area yielding K values no greater than 7 x 10^{-3} cm/sec. as averaged over the entire interval of that portion of the bore hole.

Finally, chemical grouting solutions were to be so colored that they could be readily distinguished from ground water and soil, the dye to be of such permanence that it could be detected for the duration of the work. Rhodamine crimson was selected for this use.

V. Production Grouting

As a preliminary to production grouting under the reactor area, it was necessary to install a grout curtain (Fig. 10) on the southern edge of Area B to prevent contamination of a drainage blanket underlying the Heat Decay Pit. The curtain was installed in two lifts between Elev. +45 and +70 and between +25 and +48 using principally Terranier grout with gel time controlled in the 2 to 5 minute range.

A conductivity monitoring system was installed south of the grout curtain. Ten conductivity cells on 6.7 ft. centers were installed in 2" PVC pipe perforated between elevations +50 and +65. A warning system triggered a red light in the readout system whenever conductivity reached a predetermined level. At this point, grouting nearest the warning cell was temporarily suspended and operations continued at a more remote location. The effectiveness of the curtain wall was further demonstrated by injecting dye under 50 psi pressure through a 4" casing. After 60 minutes of pumping, no trace of dye could be detected in observation pits excavated south of the curtain.

In view of the fact that the work involved stabilization of soil beneath a nuclear reactor, a strict quality control program was established. No materials were injected without the manufacturer's certification that each and every shipment met technical standards of the product. Pressure gauges were checked daily against a calibrated check gauge. Gauges showing an error in excess of 10% of full scale were replaced. Flow meters on the chemical grouting units were checked twice each shift against a calibrated container and replaced if they failed to register within 5% of true volume. Additionally, standard stroke counters on the main transfer pumps were used as a check on total quantity of material delivered to the chemical grouting units.

As each valve tube was placed, a drilling log (Fig. 11) was prepared showing graphically strata penetrated, grout observed if any, and the elevation of all valved injection points. Prior to grout injection of each tube, an "Instruction Sheet" was prepared by the Project Engineer identifying materials to be used, proportions required, gel time, and any comments pertinent to mixing or injection. During grout injection, a technician recorded for each 14" interval:

(1) depth and elevation of injection,

(2) initial and final grout injection time,

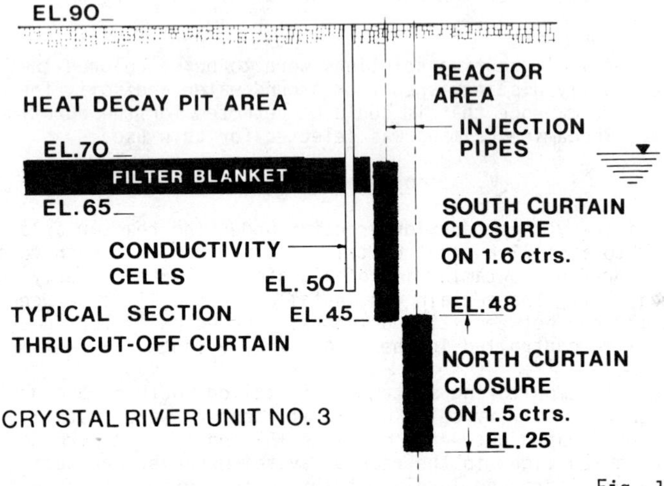

Fig. 10

FIELD DATA AQUISITION

Hole Q-15
Recorded by LA
Pump Op. LP

Date 4/14/70
Shift 2
No. Report 448

depth	elev.	time	Grout - gal. ①	②	Tot.	Ratio ①/②	°F	psi	Gel Time	Pump Rate gpm
42.43	47.59	7:20 25 5	12	12	24	1/1	74°	150/200	23	4.8
41.26	48.74	7:25 30 5	14	14	28	1/1	—	175/200	—	5.6
40.09	49.91	7:30 37 7	25	25	50	1/1	72°	170/200	26	7.1
38.92	51.08	7:37 46 9	24	24	48	1/1	—	190/200	-	5.3
37.75	52.25	7:46 53 7	25	25	50	1/1	72°	165/200	26	7.1

Remarks:

Fig. 11

(3) volume and volumetric ratio of the two solutions injected,

(4) the grout temperature during injection,

(5) the range of injection pressures observed,

(6) the average pumping rate,

(7) representative gel time of grout injected, and

(8) any remarks pertinent to the grouting operation.

Gel time determinations were made at least once each half hour and mix proportions adjusted to hold to specified limits of 8 to 12 minutes in peripheral holes and for initial injection at the bottom of each hole and 20 to 30 min. in all other areas. Injection rate ranged from 2 to 10 gpm with an optimum rate on the order of 6 to 8 gpm. Actual operating pressures ranged from 140 psi to 190 psi with a normal pressure of about 150 psi.

Grouting was conducted from the bottom up in 14" lifts corresponding to injection point spacing. A hole was considered complete when 50 gallons had been injected through each port or when injection pressure reached 200 psi, whichever came first. The only exception to this was injection of 100 gal. at the bottom holes in the cut-off curtain. If subsequent soil testing showed that soil properties were below acceptable limits, tubes in the area tested were reinjected as required. After a hole was completed, rough field data was transcribed onto a summary sheet showing graphically the quantities of grout injected and the injection pressures at each 14" interval.

Following cement grouting on an 8-ft. grid pattern, chemical grout was injected in production areas on a split spaced pattern of primary, secondary, and tertiary holes closing to a 4-ft. grid. During the period from January 9 through May 12, 1970, a total of about 52,000 gal. of grout was injected into the 86 holes forming the curtain wall and 366,000 gal. into the 371 holes in the main production areas as tabulated in Fig. 12. Grout was injected through a total of 1740 ft. of valved tubing in the curtain wall and about 15,000 ft. in the production areas. Grout take averaged 30 gal. per foot in the curtain wall and 24 gal. per foot in the production areas.

Assuming that soil volume extended 2 ft. beyond all peripheral holes, a total of 10,000 ft^3 of soil was grouted in the curtain wall for an average take of 74% and 161,000 ft^3 of soil was grouted in the production areas for a take averaging 30%. Eighty-six shifts over a 46-day period were required for grout injection in the production area for an average rate of about 2100 gal. per shift for each of the two grouting units. Drilling required only a single unit on a single shift basis to install 371 holes for an average drilling rate of one 65-ft. hole per hour.

After initial grouting, sixteen holes were drilled for acceptance testing (Fig. 13) averaging one hole for 310 ft^2 of grouted area.

CRYSTAL RIVER UNIT NO. 3

GROUT QUANTITIES INJECTED

Area	No. Holes	Grout Vol. Gal.	Grouted Footage	Take Gal/Ft.	Soil Grouted Ft^3	Take %
Cut-off	86	52,000	1,740	30	10,000	74
Production	371	366,000	15,000	24	161,000	30

GROUTING PRODUCTION

Gal. Pumped	No. Shifts	Gal. Per Shift	No. Units	Gal. Per Shift Per Unit
366,000	86	4,256	2	2,128

DRILLING PRODUCTION

Holes Drilled	Footage Drilled	No. Shifts	Holes Per Shift	Footage Per Shift
371	24,000	46	8	517

(Figure 12)

CRYSTAL RIVER UNIT NO. 3

FINAL ACCEPTANCE TEST VALUES

N Values - Blows/Ft.

	Minimum	Average	Maximum
Before Grouting	0	9.5	42
After Final Regrouting	21	80	100/1"

Permeability K - cm/sec.

	Minimum	Average	Maximum
Before Grouting	5×10^{-4}	1.0	2.0
After Final Grouting	1×10^{-4}	1×10^{-3}	3×10^{-3}

(Figure 13)

N values were determined at no more than 3-ft. intervals and often continuously through grouted zones. Fifty percent of tests were classified as practical refusal ($N \geq 50$ blows/6") and only 12% of N values were less than 30 blows per foot.

Taking full advantage of the valve tube grouting system, those tubes in zones showing marginal values were cleaned with a water blaster, regrouted, and retested before final acceptance. The 5,000 psi water blaster was equipped with a crawl nozzle in which upwardly-angled side jets forced the nozzle down through the grouting tube while forward acting jets disintegrated the chemical grout. The manner of operation is identical to that used in cleaning scale from boiler tubes. At full pressure the forward acting jet will pierce 3/8-inch plywood in a matter of seconds.

After final regrouting and retesting, the minimum N value recorded was 21 blows per foot with an average of 80 blows per foot for all material tested, an increase in the average N value before grouting of nearly 8-1/2 times. Using the "pump-in" technique, fifty permeability tests were performed at 6-ft. to 10-ft. intervals in twelve of the initial test borings, averaging one test per 420 sq. ft. of grouted area. Only one of the 50 tests showed a permeability greater than 7×10^{-3} cm/sec. The average value was 1×10^{-3} cm/sec., an improvement over initial permeability by a factor of 1000.

Refusal areas were cored with an NX double tube core barrel. Recovery from 392 feet of core drilling averaged 62%. Grout penetration into the decomposed limerock was particularly high. All samples of both infill and decomposed limerock contained samples exhibiting cohesive strength. No evidence of grout dilution was observed.

This project was performed by Intrusion-Prepakt personnel under the direction of Woodward-Gardner & Associates of Plymouth Meeting, PA. The author would particularly like to acknowledge the cooperation of Mr. Jin Kim of Woodward-Gardner who provided continuous guidance at the job site as well as much of the technical data presented in this paper.

ENGLISH / SI CONVERSION FACTORS

	Multiply	by	To obtain
Length	inches	25.4	millimeters
	feet	0.3048	meters
Area	square feet	0.0929	square meters
Volume	gallons	0.003785	cubic meters
	cubic feet	0.02832	cubic meters
Pressure	kips per square foot	0.04788	megapascals
	pounds force per square inch	0.006895	megapascals
Mass per unit volume	pounds mass per cubic foot	16.02	kilograms per cubic meter
Volume per unit length	gallons per foot	0.0124	cubic meters per meter
Torque	foot-pounds force	0.1383	meter-kilograms force
Sieve size	No. 4	=	4.75 mm
	No. 8	=	2.36 mm
	No. 16	=	1.18 mm
	No. 30	=	600 μm
	No. 50	=	200 μm
	No. 100	=	150 μm
	No. 200	=	75 μm

COMPACTION GROUTING FOR FOUNDATION STABILIZATION

by Alan N. Stilley,[1] A.M. ASCE

Abstract

Compaction grouting was used to lift and stabilize the foundations for two masonry pumphouse buildings which had experienced differential settlement. The pumphouses were part of the circulating water system for a large coal fired power plant in southern Wyoming. Exterior spread footings supporting the buildings were placed in a silty to clayey sand fill. The buildings were each about 60 by 90 feet (20 by 30 m) in plan. Settlement monitoring, started after initial structural distress of the building was noticed, indicated a maximum of about three inches (8 cm) of differential settlement had occurred. Compaction grouting, extending from immediately below the spread footings, downward to sandstone at a depth of about 15 to 30 feet (5 to 10 m), was used to lift and stabilize the structures. A low slump sand-cement grout was used. Pumping pressures at the ground surface ranged from about 50 psi (350 kN/m^2) to 800 psi (5500 kN/m^2) and about 150 to 175 cubic yards (115 to 130 m^3) of grout was placed at each building. Building foundations were lifted a maximum of about three inches (8 cm) and foundation movement appears to have stabilized. This paper presents a case history of the design of the grouting system, field control procedures, observations, and the results of the grouting.

Introduction

Compaction grouting is a relatively recent technique which has been developed in the last 10 to 15 years. It has been used primarily to lift and stabilize building foundations and floor slabs which have been placed on compressible foundation soils, or poorly compacted fill. Compaction grouting is substantially different from conventional or penetration grouting. Penetration grouting depends upon the injection of a low viscosity grout into the fractures within a rock or into voids or pores in the soil, and is usually used to limit the permeability of a soil or rock mass. Compaction grouting involves injection of a stiff, low slump grout into a soil mass. In compaction grouting, the grout normally remains as a homogeneous mass, with a distinct interface between it and the soil. Injection of the compaction grout displaces and densifies the surrounding soil mass, and can usually be used to lift foundations or

[1]Engineer, Northern Testing Laboratories, Billings, Montana.

floor slabs. In the last several years, compaction grouting has been increasingly used as an alternate to underpinning for repairing and stabilizing structures which have undergone foundation settlement. In many instances, compaction grouting has substantial advantages over underpinning or other corrective measures in terms of economics, ease of construction, and minimum disturbance to the structure. Due to the relatively recent development of the technique, and the general nature of grouting, definite design methods and procedures for estimating costs have not been well developed. The success of past projects has often depended on trial and error procedures, and the amount of experience of the engineer and contractor.

Description of Compaction Grouting

The development and history of compaction grouting is well presented in References 1 and 2. A brief description of the procedure is presented below.

Essentially, compaction grouting involves injecting a low slump grout in stages into the ground to displace and densify the soil. Normally, pipes of 1½ or 2-inch (3.8 or 5.1 cm) diameter are used. Grouting may proceed in stages from near the ground surface, downward to firm bearing, or from depth to the ground surface. It has been found that if stabilization and lifting of foundations is desired, grouting in stages from immediately below foundations, downward to sound bearing is the most successful. It is theorzied this sequence provides more overburden pressure so that higher injection pressures can be reached without allowing the grout to surface. It has been found that grout bulbs are often irregular in shape, depend upon the homogeneity of the material they are injected into, and will often take the path of least resistance in loose zones. Typically, the grout pipes are installed through the outer edge of a spread footing and cemented in place. Initial stages are injected immediately below footings and allowed to set. Drilling through the grout pipe is then performed and the next stage is injected below. This process is continued until a firm bearing material is reached. Typical stage thicknesses range from two to four feet (0.6 to 1.2 m).

Project Description and History

A large coal fired electrical generating facility was constructed in southern Wyoming starting in 1976. This unit consists of coal handling and storage facilities, water treatment and supply systems, the generation facilities, and a circulating water cooling system. Two generation units (Unit Numbers 1 and 2) of about 500 megawatts of generating capacity each were initially constructed. Each of these units depends on a circulating water cooling system. The cooling system for each unit consists of two gravity cooling towers, associated piping, and a central pumping system. Water is circulated through the generation units, allowed to cool by dropping over a series of baffles in the cooling towers, and recirculated. Pumping capacity for the system is provided by a central pump and a backup pump, located in a pumphouse. Each of these pumps is capable of pumping about 20,000 gallons per minute (1.3 m^3/sec). Circulating water lines leading in and out of the pumphouses are up to 108 inches (2.7 m) in diameter.

Each pumphouse is about 60 by 90 feet (20 by 30 m) in plan, and is constructed of masonry block bearing walls. The pumphouses are about 700 feet (210 m) apart, designed identically, and constructed at the same elevation. Pumphouse No. 1, on the east side of the site, was filled with water and in operation during most of the investigations. Pumphouse No. 2 had been structurally completed, but not filled with water.

The structures have a low portion, about 15 feet (5 m) high on the north and south sides, and a central high portion, about 35 feet (11 m) high. An overhead crane for moving and servicing the pumps, is supported on a system of steel columns and beams in the central high portion of each pumphouse. Diagonal bracing between the steel columns supporting the overhead crane creates a relatively rigid structure. Interior masonry partition walls are located in the west part of each building. A sump and storage basin extend about 35 feet (11 m) below the central portion of each pumphouse. A typical plan and photograph of one of the pumphouses is shown in Figures 1 and 2 below:

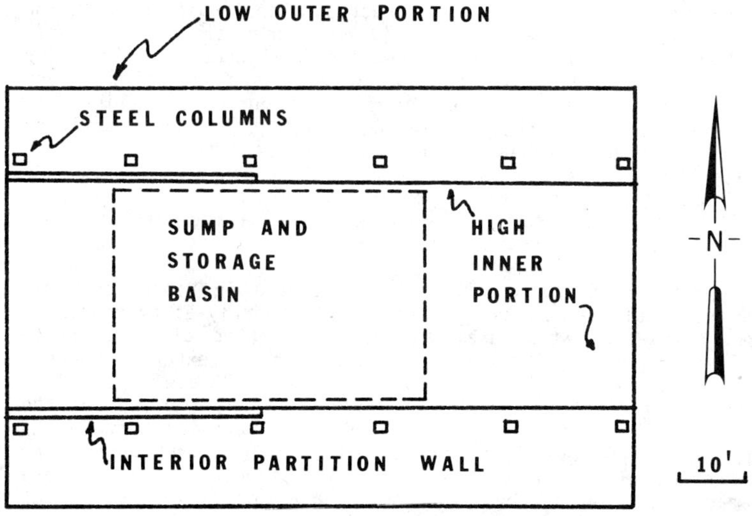

FIGURE 1. Plan View, Typical Pumphouse

FIGURE 2. Pumphouse No. 1

The exterior bearing wall portions of each pumphouse are supported on spread footings, placed about five feet (1.5 m) below ground surface (see Figure 3). Construction plans indicated exterior spread footings were three feet (1 m) wide and maximum structural loads (dead plus live) reportedly varied from about 5.5 kips per lineal foot (80 kN/m) under the low portion to about 8.7 kips per lineal foot (130 kN/m) under the central high portion. The interior portion of the pumphouses, including portions of the partition walls, and interior overhead crane columns, are supported on a mat foundation which forms the bottom of the sump and storage basin. The maximum load on this mat was reportedly about 3.9 kips per square foot (190 kN/m^2). Numerous buried water pipes and electrical conduits are associated with each structure.

Construction on the pumphouses was started in September, 1977 and was essentially complete by April, 1979. Shortly after construction was completed, structural distress was noted at both structures. This occurred primarily as cracking in the exterior masonry bearing walls, and the interior partition wall. Damage appeared to be concentrated along the west side of each structure and was the greatest at the southwest corner of each. A separation between the base and the supporting pad for the steel columns supporting the overhead crane along the west side of each structure became apparent. Damage continued to increase and by November, 1979, tearing of the connections at the base of the steel columns along the west side of the pumphouses had occurred (see Figure 6).

Site Investigations and Findings

In November, 1979, a settlement monitoring program was initiated to determine the location and rate of foundation movement. This program consisted of establishing about 20 settlement monitor points at various locations in the foundation walls and floor slabs of each pumphouse. Settlement monitors were measured by level survey to the nearest 1/100 of a foot (0.3 cm) on a weekly schedule.

In December, 1979, concern was expressed by the owner about the continued foundation movement and whether it would increase to the point operation was impaired. A geotechnical investigation was undertaken to determine the most probable cause of the damage, determine if there was potential for additional settlement, and suggest methods of stabilizing or repairing foundation movement.

Field investigations consisted of making four test borings at each structure. Test borings were made with a truck-mounted drill using hollow stem augers. The materials were carefully logged by a field engineer, standard penetration tests were performed, and thin walled tube samples were obtained. Samples from the field investigations were transported to the laboratory, where tests were conducted to determine the physical and engineering properties of the soil and rock types encountered.

The test borings revealed very similar subsurface profiles at each of the pumphouses. This generally consisted of a silty to clayey sand fill, underlain by sandstone bedrock. A review of initial site conditions, prior to construction, indicated original ground surface was about 10 to 15 feet (3 to 5 m) above the pumphouse floor elevations. Sandstone bedrock was also encountered near original ground surface so that site grading left sandstone exposed at the surface at both pumphouses.

Excavation slopes for the interior sump required backfill under other foundations in the pumphouses. The test borings indicated fill to a depth of about 17 feet (5.2 m) along the west side of each structure. Fill extended to about 10 feet (3.0 m) along the east side. Standard penetration test values indicated the consistency of the sand fill ranged from loose to medium dense, while the sandstone was very dense. Field information obtained is summarized below:

TABLE 1. Summary of Field Test Boring Information (1 foot = 0.305 m)

Boring Number	Pumphouse Number	Approximate Location	Depth of Fill, feet	Average N Value in fill, Blows per foot
5	2	S.W. Corner	17.0	17
6	2	N.W. Corner	17.0	12
7	2	S.E. Corner	10.3	7
8	2	E. Side	10.2	23
9	1	S.W. Corner	16.8	5
10	1	N. Side	17.1	15
11	1	S.E. Corner	9.7	5
12	1	E. Side	12.6	9

A typical subsurface profile is shown in Figure 3.

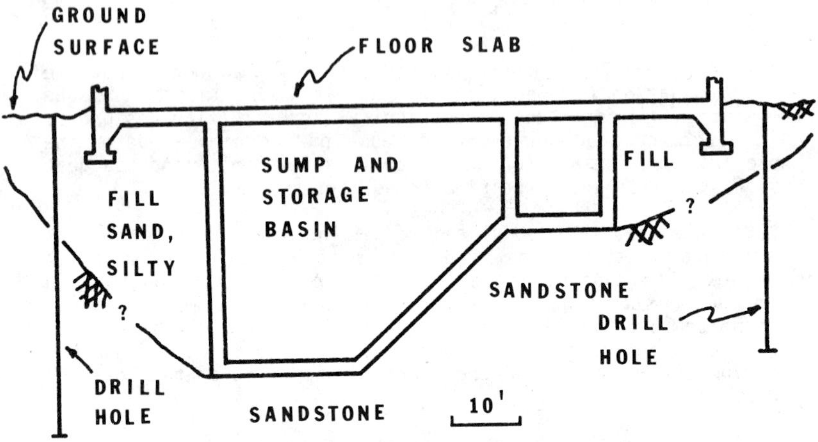

FIGURE 3. Typical Subsurface Profile, Looking North

Laboratory tests were conducted, primarily on the fill samples, to determine particle size distribution and Atterberg Limits, in-place density, and consolidation characteristics. The results of these tests are summarized on Table 2.

From the subsurface profile and building plans, it appeared that the sump and storage basin in the center portion of each pumphouse was founded in the relatively incompressible sandstone at a depth of about 35 feet (11 m), while exterior building foundations were founded on compressible fill (see Figure 3). Differential settlement between the exterior spread footings in the fill and the center portion of the building, founded in the sandstone, appeared to be the most probable cause for the structural damage. Alternate methods for repairing the damage and preventing further settlement were considered and are discussed below.

1. Monitor the structures under full loading until settlement has occurred and perform cosmetic repairs.

2. Underpin the structures through the fill to competent sandstone below.

3. Perform compaction grouting to densify the fill.

At this time, settlement and structural damage had not adversely affected operations at either of the pumphouses and it was initially decided to continue monitoring the structures until foundation settlement had ceased, and then perform repairs. The results of the settlement monitoring for Pumphouse No. 2 through July, 1980 are shown on Figure 4.

TABLE 2. Summary of Laboratory Test Results on Fill Samples
(1 foot = 0.305 m, 1 pound per cubic foot = 16.02 kg/m^3)

Drill Hole	Depth, feet	Classification	In-Place Dry Unit Weight, pcf	Percentage of Maximum Dry Density, ASTM D698, or AASHTO T99	Atterberg Limits		Gradation Percent Retained			
					Liquid Limit, %	Plasticity Index, %	Gravel	Sand	Silt	Clay
6	8.0 - 10.0	Fill, Sand, Silty	105	91	24	2	24	46	20	10
6	8.0 - 10.0	Fill, Sand, Silty	106	91	24	2	24	46	20	10
10	13.1 - 15.1	Fill, Sand, Silty	106	91	Granular		0	93	6	1
10	13.1 - 15.1	Fill, Sand, Silty	107	92	Granular		0	93	6	1
10	13.1 - 15.1	Fill, Sand, Silty	109	94	Granular		0	93	6	1
11	8.0 - 10.0	Fill, Sand, Silty	107	92	Granular		19	54	17	10
11	8.0 - 10.0	Fill, Sand, Silty	110	95	Granular		19	54	17	10
12	3.1 - 5.1	Fill, Sand, Silty	107	92	25	3	1	63	24	12

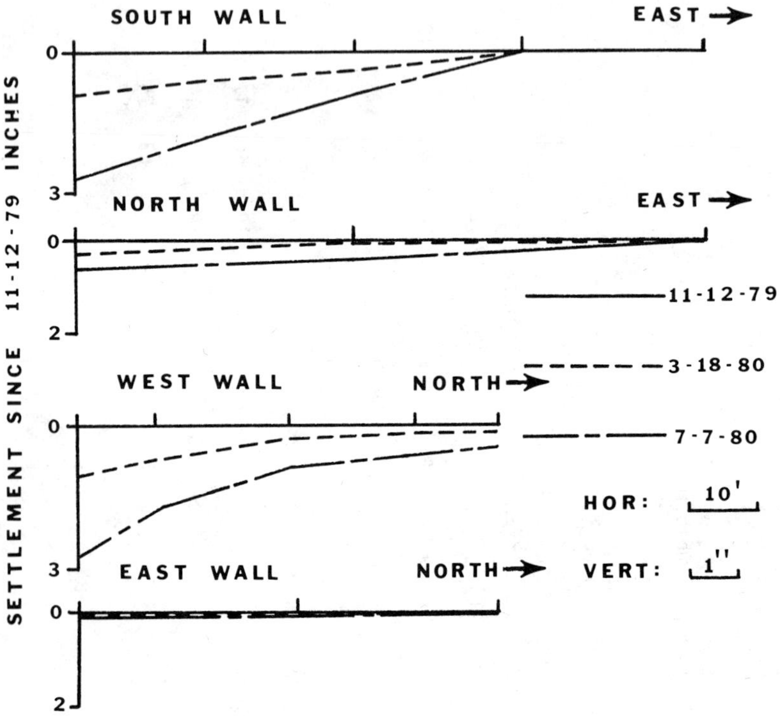

FIGURE 4. Settlement Monitoring Results, Pumphouse No. 2

Examination of the monitoring results indicated foundation settlement was continuing. The maximum settlement occurred along the west and south walls, with the southwest corner undergoing the most settlement. Between November, 1979 and July, 1980, the southwest corner of Pumphouse No. 2 had settled approximately three inches (8 cm). The rate of settlement did not appear to be slowing. At this time it was decided compaction grouting would be performed to stabilize the settlement and if possible, correct a portion of the damage.

Design of the Grouting System

An experienced grouting contractor was contacted and design and layout of the grouting system was discussed. It was decided to install the grout pipes through the outer edge of the spread footing, and inject the grout downward in stages to firm bearing. The purpose of the program was to compact the fill in a zone below the exterior spread footings, where movement was the greatest. Stabilization of the entire depth of fill below the interior floor slab would not be attempted. The interior floor slab had not undergone extensive settlement, and it was decided to core through the slab, determine if any voids were present, and fill these voids with high slump grout.

The settlement monitoring results were closely reviewed (see Figure 4) to determine the sections of foundation to be grouted. At both pumphouses, it appeared settlement of the east wall, as well as the east portions of the north and south walls, was negligible. The sections of wall that it was decided to grout are summarized on Table 3.

TABLE 3. Summary of Sections of Pumphouses to be Grouted (1 foot = 0.305 m)

Pumphouse	Wall	Length of Wall Grouted, feet
1	North	40
1	West	60 (entire length)
1	South	60
1	East	Not Grouted
2	North	20
2	West	60 (entire length)
2	South	60
2	East	Not Grouted

Based on past experience, an approximate grout hole spacing of six feet (1.8 m) was selected for the holes beneath the foundations. The number and spacing of the core holes through the floor slab would depend upon the extent of the voids encountered beneath the slab.

A preliminary attempt was made to estimate the volume of grout that would be required at each pumphouse. Based on the results of the laboratory tests, it was estimated that it would be necessary to increase the dry density of the fill by about 10 percent in order to provide sufficient support of spread footings. Calculations were made using the lengths of each pumphouse wall to be grouted (Table 3), and the depth of fill determined from the test borings (Table 1). It was also assumed that densification would occur in a zone extending about six feet (2 m) laterally beyond the edge of each spread footing. Based on these calculations, it was estimated that about 125 yd^3 (95 m^3) of grout would be required at Pumphouse No. 1 and about 100 yd^3 (75 m^3) would be required at Pumphouse No. 2.

Description of the Grouting and Field Control

Compaction grouting was started at the pumphouse locations in July, 1980. Initially, 2-inch (5 cm) pipes were installed through the outer edge of the spread footings. Installation of the pipes was accomplished by augering with a portable, compressed air powered auger to the top of the spread footing. A percussion drill was then used to penetrate the concrete footing. The steel pipes were placed immediately below the footing at a depth of about seven feet (2.1 m) below ground surface, and sealed by pouring a cement-water slurry around the pipe.

A local source of silty sand was used in the grout mix. The silty sand was run through a belt shredder prior to mixing. This unit consisted of an inclined belt which rotated at high speed. The shredder broke down any pieces of clay or silt so that proper mixing could be achieved, and by the action of the inclined belt, oversize particles, larger than about 1/2 inch (1 cm) were separated from the sand. Separation of oversize material was particularly important, as particles larger than about 1½ inches (3.8 cm) in size normally blocked the pumping system, resulting in downtime.

The grout was proportioned and mixed in a truck mounted unit. This unit consisted of adjustable feed bins to contain the sand and cement and a tank to contain water. The sand and cement were fed onto a short conveyor running under the bins. At the rear of the truck, the conveyor discharged the sand and cement into an auger feed and water was added. The auger was about six feet (2 m) long and could be rotated and lowered or lifted to control the location of discharge. The grout, which obtained the initial mixing in the feed auger, was then discharged into a storage and mixing bin on the grout pump.

The grout pump was a Sidewinder Model 4C, positive displacement pump. It consisted of a storage and mixing bin of about one-half cubic yard (0.4 m^3) capacity. The grout in the storage bin was mixed by a rotating paddle wheel mixer. The action of the paddle wheel mixer also fed the grout into the pump intake. The pumping units consisted of hydraulically operated cylinders, each having a grout volume capacity of about 0.2 $ft.^3$ (0.006 m^3). As one of the cylinders was pumping grout, the other was being filled. This resulted in relatively constant pumping rate of about 2.5 $ft.^3$/min. (0.07 m^3/min.). The pump was capable of reaching pressures of about 2500 psi (17,000 kN/m^2). Flexible hoses led from the grout pump to the steel pipes. Pressure gauges were placed in the lines at the pump outlet, and pipe near ground surface.

Prior to starting the grouting, the grout pump was calibrated. This was accomplished by pumping into a 55-gallon (0.21 m^3) drum and counting the number of pump revolutions. The volume pumped into the drum was then accurately measured, and a volume per revolution was determined.

Before the grouting started, a series of holes were cored through the interior floor slab. Occasional cracking of the floor slab had occurred and it was felt that fill settlement had probably created voids beneath the slab. Core holes revealed voids up to eight inches (20 cm) beneath the slabs. Coring was continued on a grid system until all of the voids were detected. Generally, the largest voids were located on the west end of each building.

Control of the grouting in the field consisted of (1) counting the number of pump revolutions and closely monitoring the pressure variations at the pump outlet and at the top of the grout pipe, (2) visually inspecting the ground surface and the structure in the area of the grouting, and (3) monitoring temporary benchmarks established on the building foundations, walls and floor slabs with a surveyor's level.

Generally, alternate holes along the wall were pumped, starting at the areas of the most obvious damage, and proceeding in each direction along the wall away from this point. It was found that in the upper stages in most of the holes, the pressure would build to a relatively constant value, continue at this level for most of the grouting period, and then drop off abruptly. In the upper 10 to 15 feet (3 to 5 m) of grouting, typically grout pressures would build slowly to about 100 psi (700 kN/m^2), continue at this level for some time, and then drop off. Often the drop in grout pressure was accompanied by ground fracture or heaving near the hole.

During the grouting process, the adjacent portions of the structure and the ground surface were closely observed for signs of ground heave or building movement. Temporary benchmarks were monitored during the pumping and the original set of settlement monitors was read on a daily basis at each structure, while the grouting was in progress. If fracturing or heaving of the ground occurred, the grouting at that hole location was stopped. During the initial stages for each hole, if movement of the structure was detected, grouting was stopped. If building or ground surface movement was not detected, grouting was stopped after about two cubic yards (1.5 m^3) had been placed in each stage. Occasionally, particularly at the greater depths, grout holes would block off, resulting in erratic pressure variations. These pressures occasionally reached very high levels and normally grouting was stopped at this point. This behavior was attributed to the presence of debris or large sized material in the fill.

After each stage of grout had been placed, the hole was normally left overnight. The next day the hole was drilled with a portable pneumatic powered hand drill, using water as the drilling fluid. Drilling proceeded until the last stage, placed the previous day, had been penetrated. In most instances, observation of the drilling characteristics accurately indicated when the drill had penetrated the last stage into the fill below. From the drill cuttings, it was normally possible to identify the type of fill material below the grout. The depth of each stage was accurately recorded during the drilling.

After foundation grouting had been completed, the voids beneath the floor slabs were filled. This was accomplished by placing a high slump grout through the core holes. During this operation, grout was injected through one hole until it was detected in the other holes. Very low pressures were maintained during the placement of this grout. This process was continued, moving from hole to hole, until the void had been filled.

Results of the Grouting

Grouting was started in July, 1980 and completed by September, 1980. Generally, the depth of the grouting necessary to penetrate the fill and reach the sandstone bedrock was somewhat greater than indicated by the depths to bedrock from the test borings. Since the test borings were generally 5 to 10 feet (1.5 to 3 m) from the foundation walls, whereas the grout holes were at the foundation locations, it was felt this fill depth difference was probably due to variation in the original excavation configuration. The final grout volumes placed also exceeded the original estimates by about 25 cubic yards (19 m^3) at Pumphouse No. 1, and 60 cubic yards (46 m^3) at Pumphouse No. 2. This was attributed to the greater depth of fill revealed during the grouting, as compared to the original test borings.

Summarized on Table 4 are the grout volumes and characteristics of the grouting at each structure.

TABLE 4. Summary of Grout Placed at Each Structure (1 foot = 0.305 m, 1 $yard^3$ = 0.765 m^3)

Pumphouse Number	Location	Number of Grout Holes	Depth of Final Grout Injection Below Surface, feet	Volume of Grout Placed, $yard^3$
1	South Wall	10	18 to 23	50.7
1	West Wall	7	19 to 22	50.5
1	North Wall	7	18 to 30	51.2
1	Beneath Floor Slab	17	-	4.5
			Total	156.9
2	South Wall	10	16 to 22	73.5
2	West Wall	8	17 to 21	57.3
2	North Wall	4	18 to 19	32.2
2	Beneath Floor Slab	20	-	10.1
			Total	173.1

Significant lifting or movement of the structure was normally limited to the last several grout stages after grouting had reached a depth near the sandstone bedrock. Typically, in the areas where the settlement had been the greatest, about 1/2 inch to one inch (1.3 to 2.5 cm) of lift of the building foundations was achieved during each of the last two or three grout stages that were placed. During the lifting of the structures, visible closing of cracks in the masonry exterior of the buildings could be detected. Closing of the cracks could also be audibly detected during foundation lifting. Foundation lifting was normally continued until many of the cracks had closed significantly, and in the opinion of the field engineer and contractor, further movement might increase the structural damage.

The results of the monitoring program, indicating the foundation lift achieved are shown in Figure 5.

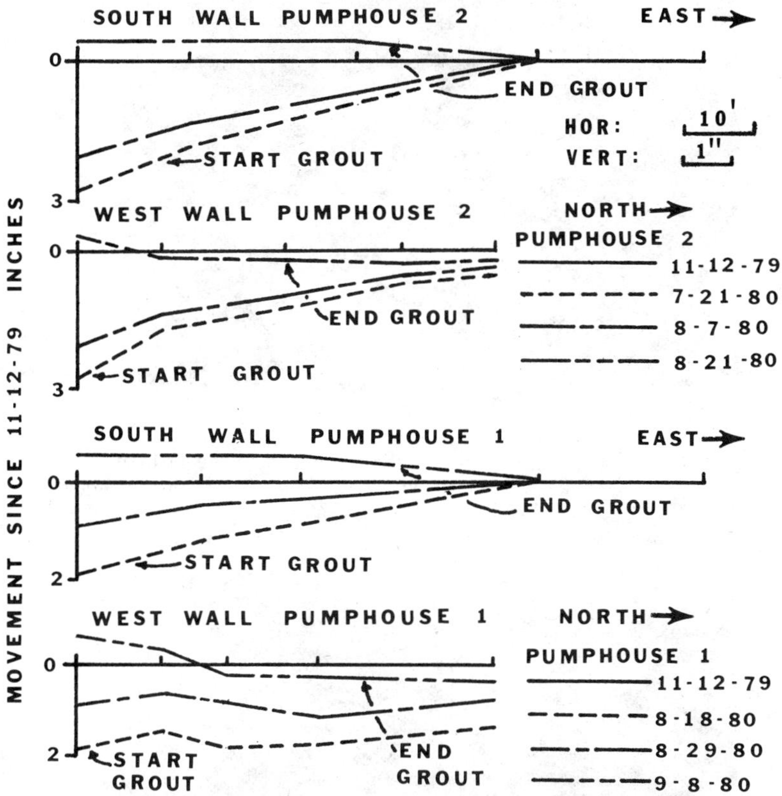

FIGURE 5. Monitoring Results Indicating Lift Achieved During Grouting

Significant structural damage had occurred at the connection at the base of a steel column supporting the overhead crane along the west wall of Pumphouse No. 2. Due to diagonal bracing between the adjacent columns, a relatively rigid frame was created. Thus, when the foundation below the column settled, damage at the connection occurred. Figures 6 and 7 illustrate the amount of foundation movement before and after the grouting had taken place.

Settlement monitoring was continued after grouting was completed. This information indicated a slight amount of foundation settlement within one to two weeks after grouting was completed. This settlement in all instances was less than 0.1 inch (0.25 cm) and may have been caused by normal shrinkage of the grout, or by

FIGURE 6. Structural Damage at Overhead Crane Column at West Wall of Pumphouse No. 2, Before Grouting

FIGURE 7. Base of Overhead Crane Column at West Wall of Pumphouse No. 2, After Grouting

consolidation of the foundation soil as pore pressures dissipated. Settlement monitoring was continued on about a monthly basis, and with both pumphouses in operation, as of about one year later, settlement has been negligible.

Conclusions

Compaction grouting was successfully used to lift and stabilize the foundations for two masonry pumphouse buildings, which had undergone settlement. Grout was generally placed in stages, to depths of 15 to 30 feet (4.5 to 9.0 m). Pumping pressures ranged from about 50 to 800 psi (350 to 5500 kN/m^2) and about 150 to 175 cubic yards (115 to 130 m^3) were placed at each structure. Foundations were lifted a maximum of about three inches (8 cm).

References

1. Brown, D. R., and Warner, J., "Planning and Performing Compaction Grouting," Journal of the Geotechnical Engineering Division, ASCE, Vol. 100, No. GT6, Proc. Paper 10606, June, 1974, pp. 653-666.

2. Graf, Edward D., "Compaction Grouting Technique and Observations," Journal of the Soil Mechanics and Foundations Division, ASCE, Vol. 95, No. SM5, Proc. Paper 6766, September, 1969, pp. 1151-1158.

SOIL MODIFICATION FOR RELOCATION OF DIE FORGING OPERATIONS

by

Antal Partos[1], M ASCE, Richard D. Woods[2], M ASCE, Joseph P. Welsh[3], M ASCE

ABSTRACT A soil modification program consisting of compaction grouting and chemical grouting was selected to improve the loose foundation soils under a vacant building to sustain the dynamic loads of forging equipment. Compacted and chemically grouted samples were laboratory tested using the resonant column technique to establish design shear moduli and vibrations at the existing facility were measured. The equipment locations were planned to suit forging operations and require lowest costs of soil modification. An approximately 330 m^2 (3600 sft) area was stabilized to a depth of approximately 9 m (29.5 ft) below floor. The increase of soil density was measured by nucleonic depth density probes and the shear moduli of the subsoil were obtained by interval crosshole tests during grouting. Floor and equipment vibrations and floor settlements were also measured approximately 2 months and 12 months after all equipment was installed. Vibration levels were below the limits for forging operations. Soil modification was a key factor in converting an existing building from a light industrial use to an application requiring die forging equipment.

INTRODUCTION

In September, 1978 K-D Tool Co., Lancaster, PA secured an option to acquire a property which included an existing building in the Walterboro Industrial Park, Colleton County, Walterboro, South Carolina. The purpose was to relocate an existing forging facility from Greenfield, Ohio. At the request of Lotz Designers and Constructors, Horsham, PA, a condition survey of the building was initiated in October, 1978 with special consideration being given to the support of forging

1. Vice President, Chief Geotechnical Engineer SITE engineers, inc. Voorhees, New Jersey.
2. Principal Consultant Stoll, Evans, Woods and Assoc. Ann Arbor, Michigan.
3. Vice President, Engineering, Marketing Hayward Baker Co. Odenton, Maryland.

equipment. This initial study was followed by detailed design and construction from October, 1979 through May, 1980. The owner favored the proposed site because moving into an existing building would minimize interruption to present forging operations. As a part of the purchase, a large piece of land adjacent to the building could also be acquired for an attractive price. The property adjoins a municipal airport and a railroad spur is also available. A good labor market, excellent housing, and inexpensive recreational facilities along with adequate electric power, water, ventilation, etc. made the proposed acquisition the first choice among many. This paper discusses the design and execution of a soil modification program for the support of four forge hammers which was completed in an existing building. Soil modification was accomplished by compaction and chemical grouting. Design criteria were checked during field work by appropriate quality control methods. Post-construction measurements of vibrations during forging operations were also obtained.

CONDITION SURVEY

The site is located within the boundaries of the Coastal Plain physiographic province. Pleistocene sands and clays are deposited over the Cooper Marl. The one-story steel frame structure is 84 x 84 m (276 x 276 ft) in size. The columns and walls are supported on spread footings; the concrete floor is a slab-on-grade. The floor appeared to be in a good condition; 12 Windsor probes indicated a concrete compressive strength of 34.47 MN/m^2 (5000 psi). Relative differences in floor elevation were generally insignificant although some small differences were observed along the north wall. Some cracks were observed on the face of exterior block walls but these cracks did not extend through the wall footings. Site drainage was poor.

Three test borings were drilled to maximum depths of approximately 22 m (72 ft) in the most probable area for the proposed forging operations. The results of two borings are shown in Fig. 1.

The soils in the upper 9 m (29.5 ft) thick stratum were grouped as Types 1 and 2 sands. Occasional thin clay layers were also encountered within the sand in the lower portions of the borings. Very loose sand underlain by loose to very loose sands was encountered from 1.5 m (5 ft) to 4.8 m (16 ft) below the floor. The ground water level was approximately 1.3 m (4.3 ft) below grade in October, 1979 but we understood that the seasonal high water level was close to the bottom of the floor.

EVALUATION OF CONDITION SURVEY

Industrial buildings with light structural and operational (floor) loads but no vibratory loads are frequently constructed with shallow spread footings and floor slabs on grade at locations where loose sands, soft silts, and high ground water levels are encountered.

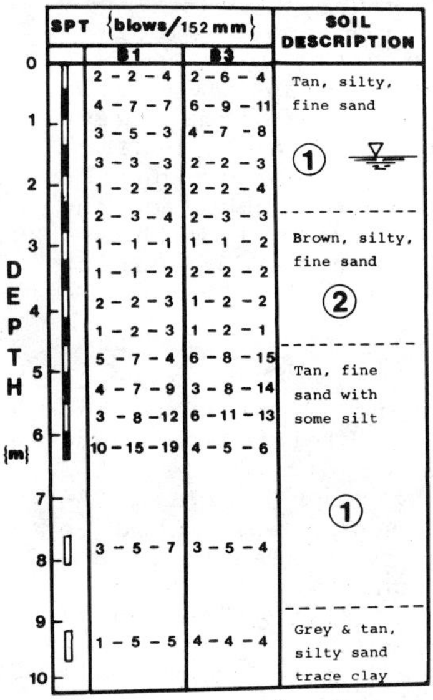

FIG. 1.-Test Boring Results

Buildings with light static service loads may not exhibit any significant structural distress, such as settlement cracks, if properly designed, constructed, and operated. In the Coastal Plain area a surface layer of cemented sands, forming a "crust", is frequently encountered as confirmed by this study. Therefore, the fact that the subsoils are relatively inadequate for many uses is not evident to the casual observer.

The dynamic (impact) loads of a forge hammer are transferred from the anvil to the foundation to the subsoil. The particles of a loose sand layer would be displaced, mainly in the vertical but also in the horizontal direction. Ground motion would not cease until densification compatible with the magnitude of impact loads is attained. Settlement of the soil under the equipment and under footings, floors, etc. would occur, resulting in serious maintenance problems and possibly an ultimate failure of equipment or structure foundations.

The buildings which were being considered for purchase did not exhibit any distress because the operations in the industrial park did not employ equipment imposing significant dynamic loads. Data for a typical forge hammer to be relocated are given in Table 1.

TABLE 1.- DESIGN DATA CHAMBERSBURG DIE FORGER #8

DEAD WEIGHT:	289 kN (32.5 tons)
RATED ENERGY:	11.7 kNm/blow (8631 ft-lbs/blow)
FALLING WEIGHT:	8.9 kN (1.0 ton)
MAX. BLOWS/MIN:	100
RAM STROKE:	.432 m (17 in.)
AIR PRESSURE:	689.5 kPa (100.0 psi)
MAX. RAM DIE WEIGHT:	2.94 kN (0.33 tons)

Based on the condition survey it was realized that serious potential problems with liquefaction and settlement of saturated fine sands existed. Therefore, deep foundations or, alternatively, soil modification to permit support of the forge hammers on footings were evaluated; a summary of conclusions follows:

1. Piles were considered for equipment support, and the damping of vibrations which would be transferred from piles to existing floors and footings was also evaluated. Piles were not selected because of great length (30 m) (98 ft), problems with damping and installation (overhead clearance), and anticipated costs of frequent maintenance of floor.

2. Spread footings supported by the existing subsoils were selected. A soil modification program was considered necessary to improve the dynamic properties of the subsoil; chemical and compaction grouting were chosen. Soil bearing capacity was of no concern. The soil modification was used to control ground motion by increasing soil stiffness. It was realized that grouting can be done with a minimum amount of destruction to existing features and with relatively no interference with planned remodeling of the building. It was also realized that the subsoils could be improved by soil modification in the future and in other areas of the same building without disruption to the forging operations.

A layout of the final soil modification plan is shown in Fig. 2 and a portion of the forging area is shown in Fig. 2A.

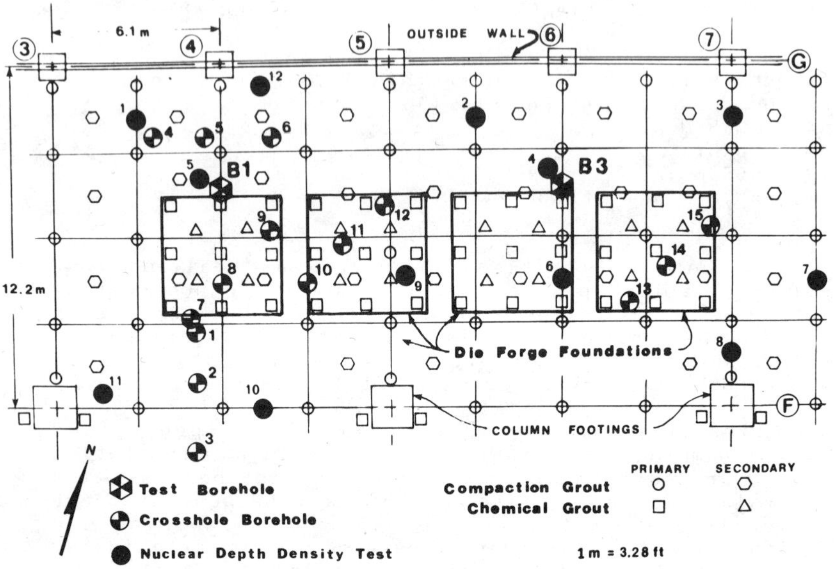

FIG. 2. - Final Soil Modification Plan

FIG. 2A. - Portion of Forging Area in Walterboro

LABORATORY TESTING

Jar samples of two types of sand obtained in Boring B-1 were initially tested in the laboratory for physical and dynamic soil properties. Particle size analyses revealed that all samples were "fine sand, trace medium sand, trace to some silt" up to 15% smaller than 0.075 mm (.00295 in.). Slight acidity was indicated by a pH of 5.6. The main purpose of the initial laboratory testing was to obtain maximum and minimum dry unit weights for Type 1 and 2 sands, so that laboratory and field tests could be correlated using values of relative density. Test results are presented in Table 2.

TABLE 2. - Maximum and Minimum Densities

Soil Type	Location Below Floor m (ft)	Description of Sand	DENSITY			
			Maximum		Minimum	
			Unit Weight kN/m^3 (pcf)	Void Ratio e	Unit Weight kN/m^3 (pcf)	Void Ratio e
1	0-2.5 (0-8) 4.8-9.0 (16-30)	Tan fine silty	17.25 (109.8)	0.51	11.67 (74.3)	1.23
2	2.5-4.8 (8-16)	Brown fine silty	15.78 (100.4)	0.64	12.45 (79.3)	1.27

Resonant column tests using procedures and equipment described by Afifi and Woods (1) were made on reconstructed dry and moist samples of Type 1 and 2 sands to determine the dynamic soil moduli which could be expected after compaction grouting. Resonant column tests were also made on grout injected samples to evaluate the maximum value of shear modulus which could be expected after completion of all grouting. Results of the initial tests on Type 1 dry sand did not confirm our expectations. Based on the Hardin equation (2) the shear wave velocity of Type 1 sand at confining pressures of 17.2 and 34.5 kN/m^2 (2.5 and 5.0 psi) should have been approximately 190 m/sec (623 ft/sec) and 200 m/sec(656 ft/sec), respectively. The measured velocities after confinement of each sample were 275 m/sec (902 ft/sec) and 330 m/sec (1083 ft/sec) at a time of 1000 min. These surprisingly high velocities suggested that some contribution to sample stiffness other than that caused by the confining pressures was occurring.

An examination of the Type 1 sand samples under a microscope showed that the dry soil grains might possibly be cemented together. Further testing revealed that the dry sand particles had a significant amount of static electric charge. Therefore, the sand particles were dispersed in water after which a slight amount of foam was noted on the water surface. After washing and drying the samples the particles did not coagulate. The filtrate was analysed and the results are presented in Table 3.

TABLE 3. - Filtrate Analysis

Constituent	Amount (%)
MBAS (Surface Active Agent)	.724
Calcium	.104
Magnesium	.0022

A significant amount of calcium and high amount of MBAS was found in the filtrate. The latter may be fatty esters or acids or alcohols and would account for the electrostatic charge (3) between the sand particles.

Since at the depth of interest the Type 1 sands are now and are expected to be in the future below the ground water level the cementation was of no further concern. Testing was discontinued on Type 1 dry sand samples. A resonant column test on a wet specimen showed a shear wave velocity of about 180 m/sec (591 ft/sec) at 1000 min at a confining pressure of 17.2 kN/m^2 (2.5 psi).

It was anticipated that the forge foundation bases would be about 2-3 m (7-10 ft) below floor level. Therefore, the principal resonant column tests were made on two sets of Type 2 sand samples. The first set of samples was remolded to a relative density of approximately 75% to simulate the anticipated optimum in-situ conditions after completion of compaction grouting. Relationships of shear wave velocity vs. time at confining pressures of 27.6 and 69 kN/m^2 (4.0 and 10.0 psi) (which are the anticipated effective confining pressures below the forge foundations) are presented in the lower portion of Fig. 3. Preliminary analysis of the dynamic response of the forge foundation soil indicated that a shear wave velocity of not less than 190 m/sec (623 ft/sec) would be necessary to avoid excessive vibrations. A dashed horizontal line in Fig. 3 shows this minimum velocity. From Fig. 3 it can be seen that compaction grouting does not achieve the minimum shear wave velocity at the low confining pressure and only slightly exceeds it at the high confining pressure.

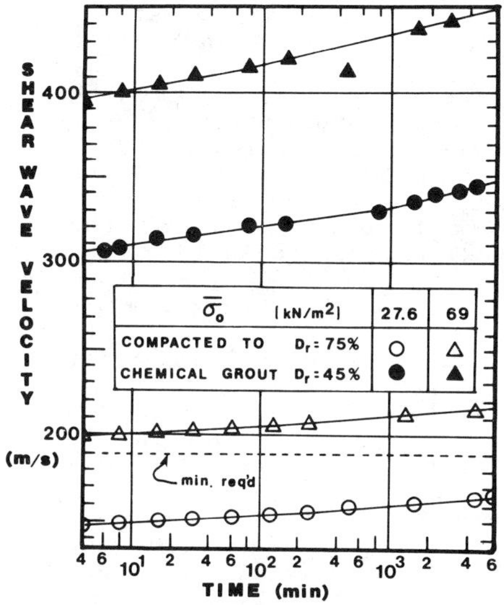

FIG. 3. - Shear Wave Velocity vs. Time

The second set of resonant column specimens was prepared by the grouting contractor, Hayward-Baker Company, by injecting specimens of Type 2 sand at a relative density of 45% with a sodium silicate grout. The grout mix consisted of 40% sodium silicate (Grade 40) with a Baume of 41.4° at 20° C (68° F), 12% a patented reactant composed of formamide and ethyl acetate, and 48% water.

This chemical grout has a gel time of about 15 minutes; it reaches about 80% strength in 20 minutes, and full strength in about 7 days. The chemical grout specimens were tested within four days of their preparation. The results of resonant column tests on one of those specimens at confining pressures of 27.6 and 69 kN/m^2 (4.0 and 10.0 psi) are shown in the upper part of Fig. 3. The shear wave velocities greatly exceed the required 190 m/sec (623 ft/sec) at each confining pressure.

It was concluded that chemical grouting alone, or in combination with compaction grouting, could improve the foundation soils to obtain the desired dynamic soil properties. On this basis a recommendation was made to the owner to exercise his option to purchase the property.

EXPLORATION AT EXISTING PLANT

To develop criteria for allowable forge vibration characteristics at the proposed Walterboro facility, vibration measurements were made at the owner's facility in Greenfield, Ohio. We understood that the new plant would have forging equipment identical or similar to that operating in Greenfield. One test boring was drilled near the foundation of a die forge and another boring was located approximately 4 m (13 ft) away from its foundation. Clayey silts of soft to medium consistency were encountered in the boreholes to a depth of 5 m (16 ft).

Vibration measurements were made in the forging room at ground floor level and in boreholes at two depths. These measurements showed that floor vibrations decayed more slowly than vibrations at the bottom of the boreholes as would be expected because the floor vibrations decay by Rayleigh wave attenuation whereas the borehole vibrations decay as body waves. Measured ground motions are shown in Fig. 4.

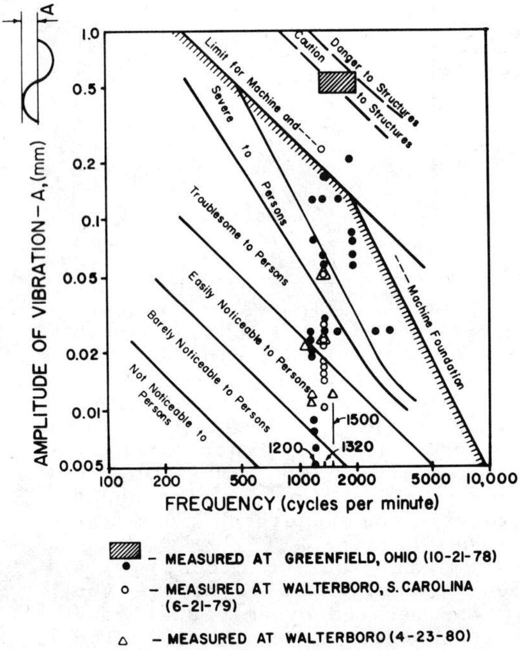

FIG. 4. - Measured Ground Motions

Maximum peak particle displacements of the forge foundations were measured as about 0.5 mm (0.0197 in.) at frequencies between 20 and 33 HZ (1200 and 1980 cycle/min). These vibration amplitudes would generally be considered excessive by many industries but, according to the plant management, they did not adversly affect operations in this situation. However, the same levels of vibration could not be tolerated at Walterboro because of the potential for settlement of the loose sands.

DESIGN OF TYPICAL FOUNDATION

Trial foundations were sized based on experience with existing forge installations and were checked for dynamic response. The geometry of the foundation selected for this installation is shown in Fig. 5. This foundation was analyzed for vertical translation and rocking using a

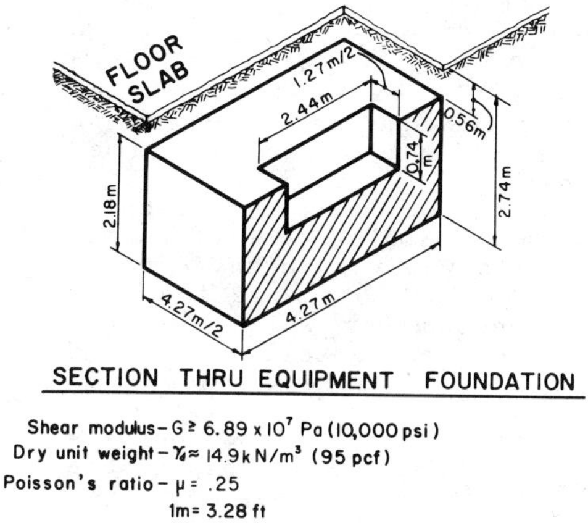

FIG. 5. - Forge Foundation

layered half-space analog suggested by Novak (4). The grouted soil under the foundation and the soil surrounding the foundation were assumed to have a shear modulus (G) of 69 and 35 MN/m^2 (10,000 and 5077 psi), respectively.

A force-time pulse based on the rated energy and operating cycle of a die forge of the type to be installed was estimated. This pulse had a peak force of 427 kN (48 tons) with a rise time of 0.005 sec, decay to zero at 0.025 sec, a negative peak of 107 kN (12 tons) at 0.03 sec, and decay to zero at 0.05 sec. Using this forcetime pulse in a phase-plane analysis (5) the maximum peak displacement was found to be 0.25 mm (0.0098 in.) at about 13 HZ (780 cycle/min).

LAYOUT OF FORGING EQUIPMENT

At least six markedly different layouts of the forging equipment were reviewed from the viewpoints of dynamic soil response and costs of soil modification. The four forge hammers were initially located in one row at approximately 6 m (20 ft) centers and approximately 4 m (13 ft) from the exterior (north) building wall. Another layout called for two rows of two forge hammers approximately 4 m (13 ft) apart and perpendicular to the exterior wall with the closest hammer being 4 m (13 ft) from the (north) wall and the furthest one located approximately 10 m (33 ft) from the north wall. The final layout is shown in Fig. 2. Each hammer was to be supported on a footing with concrete placed directly against the earth to maximize vibration damping by the surrounding soil.

SOIL MODIFICATION PLAN

Laboratory testing revealed that chemical grouting alone could provide required stiffness. However, large scale application of chemical grouting of very loose to loose uniform sands presents several problems. Because of hydraulic fracturing the sand particles can be separated and individual grout layers can be formed within the soil mass. Losing point to point contact between sand particles would pose a problem with creeping of the grout. Furthermore, the estimated cost of chemical grouting of very loose soils was very high. Therefore, compaction grouting was chosen to reduce the void ratio and improve soil stiffness after which chemical grouting was employed to keep a stabile soil matrix and provide the required stiffness. Figure 2 shows the layout of soil modification.

Compaction grouting was planned in an area of approximately 12 by 27 m (40 by 90 ft). The primary pipes were placed on 1.82 m (6.0 ft) centers with the secondary pipes in the center of the pattern. Chemical grouting was designated in the area of the four forge hammers, occupying an area of approximately 5.9 by 21.4 m (19 by 70 ft), and additionally at three interior columns. Nine primary and four secondary pipes were planned at each machine base as shown in Fig. 2. No chemical grouting was planned along the exterior wall. The limits of soil modification were established on the basis of measured and anticipated decay of Raleigh and body waves and of economic considerations. The lengths of the compaction units were 6.1 and 9.15 m (20 and 30 ft) beneath the forge foundations and at other locations, respectively. Fifty-one primary and 32 secondary compaction grout pipes of 51 mm and 76 mm (2.0 and 3.0 in.) diameter formed a 3.05 m (10 ft) grid pattern. Thirty seven primary and 16 secondary chemical grout pipes of 19 mm (0.75 in.) in diameter were arranged in the forging area with two primary pipes at each of the three column footings. The grouted lengths were 7.6 m (25 ft) at the column footings. The grouted lengths varied from 5.5 to 6.7 m (18 to 22 ft) under the foundations and along the perimeter of the foundations, respectively. The chemical grouting along the perimeter of each footing was designed to provide stability of the side walls and reduce water seepage so that the concrete for the 2.75 m (9 ft) thick footings could be placed directly against the earth.

Contract quantities were compaction grouting in 564 m (1850 ft) of pipe and 170,325 l (44,995 gal.) of chemical grouting to meet the project criteria of 70% relative density after compaction grouting and a shear wave velocity of not less than 190 m/sec (623 ft/sec) after completion of all grouting.

MONITORING PLAN

The criterion for compaction grouting was to attain 70% relative density. Therefore, initial densities and density increases had to be measured at the same location to compute relative density.

Nine m (29.5 ft) long and 48.3 mm (1.90 in.) I.D. aluminum pipes were installed at 12 locations as shown in Fig. 2. Troxler model 1351 density and model 1251 moisture gauges and a model 2651 Scaler-Ratemater (6) were used. Each gauge consists of a probe, a fitted cable and a shield and standard. The shield and standard provides automatic radiological safety during transportation and storage. It is a constant reference standard for the probe to obtain a standard one minute count rate. It also provides an efficient means to introduce the probe into the access tubing. The density probe contains a 3 millicurie source of Radium-226 and measures the total unit weight by backscatter and absorption of nuclear radiation of a spherically shaped volume approximately 13 cm (5.1 in.) in radius. The moisture probe contains a 3 millicurie source of Radium-Beryllium and detects neutrons which are thermalized by hydrogen. During testing the probes are lowered through access tubing to the desired level of testing. Generally three counts are read on the scaler and an average count rate is used. A schematic arrangement of testing and a detail of the density probe are shown in Fig. 6.

Multiple borehole crosshole tests (7) were selected to measure average subsoil shear wave velocities because it was recognized that neither compaction nor chemical grouting would produce a homogenous mass of stabilized sand.

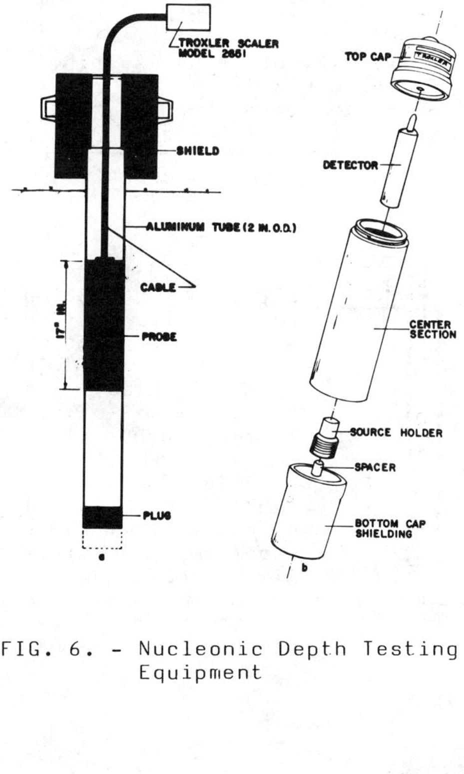

FIG. 6. - Nucleonic Depth Testing Equipment

A schematic arrangement is shown in Fig. 7. Dynamic shock waves are produced in the hammer hole and they will be received by the two geophones. The travel times are measured and velocities are computed. The equipment and instruments used are presented in Table 4, and the 15 test locations are shown in Fig. 2. Crossholes No. 1 through 7 were designated as permanent to facilitate readings of ground motion during forging operations.

Floor heaving was measured with a surveyor's level.

TABLE 4. - EQUIPMENT & INSTRUMENTS

Item Number	Common Name	Manufacturer & Model
1	Oscilloscope	Hewlett-Packard dual trace storage oscilloscope, Model 1201A
2	Oscilloscope Camera	Hewlett-Packard, Model 198A with Polaroid film holder
3	Geophone	Mandrel Industries, Electro-Tech, Model EV-8
4	Borehole Packer	University of Michigan, pneumatic geophone packer
5	Trigger Circuit	University of Michigan, capacitive circuit
6	In-Hole Hammer	University of Texas, Austin, Borehole Hammer
7	Inclinometer	Slope Indicator Company, Digitilt Model A-205

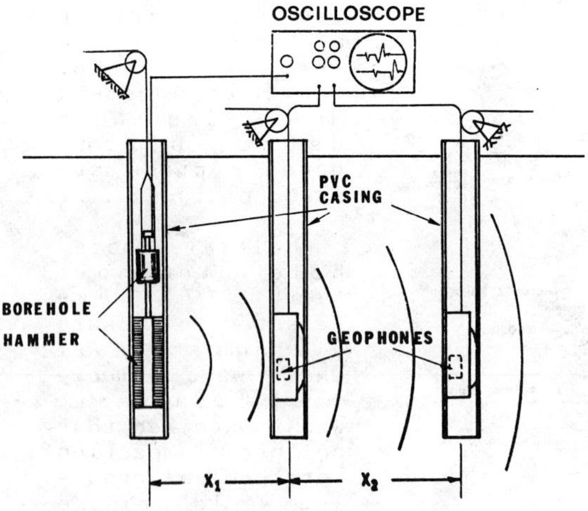

FIG. 7. - Interval Crosshole Testing

GROUTING OPERATIONS

Compaction grouting was done from February 13 to March 27, 1979. Grouting started at the bottom and advanced upwards. After initial low grout takes, 76 mm (3 in.) diameter grout pipes were installed with an Atlas Copco R 601 overburden drill working on the floor of the existing building. A batch of compaction grout mix consisted of 0.42 kN (94 lbs) Type I cement 0.33

kN (75 lbs) fly ash, 0.21 kN (47 lbs) pre-hydrated bentonite, 2.32 kN (522.5 lbs) local sand, and sufficient water to obtain less than 76 mm (3 in.) slump. The materials were mixed in a companion mixer to a Warner 1500 compaction grout pump. Field slump was measured in accordance with ASTM C 143. During grouting the pipes were raised by 0.305 m (1 ft) increments. The criteria for raising the pipe were either 30 mm (1.2 in.) heaving of floor slab or columns, or injection of 0.113 m^3 (4 cu ft) grout or pressures exceeding the maximum project pumping pressure of 5516 kPa.(800 psi). Pumping pressure fluctuated from 1379 kPa (200 psi) to 4137 kPa (600 psi). Data of compaction grouting are given in Table 5.

TABLE 5. - COMPACTION GROUTING

COMPACTION POINTS			GROUT TAKE				DIAM.
				Per Unit			Avg.
Type	No.	Length m (ft)	Total	Min. m^3 (cu ft)	Max. m^3 (cu ft)	Avg. m^3/m (cu ft/ft)	Grout Column m/(in.)
Primary	51	390.14 (1280)	79 (2,791)	0.57 (20)	2.23 (78.9)	0.20 (2.19)	0.51 (20)
Secondary	32	242.93 (797)	58.2 (2,057)	0.57 (20)	2.34 (82)	0.24 (2.59)	0.57 (23)
Total	83	633.07[a] (2077)	137.2 (4,848)	-	-	0.22 (2.39)	0.53 (21)

Note: [a] contract quantity; 564 m (1850 ft)

Assuming that the compaction grouting influenced an area of approximately 30.5 m (100 ft) by 15.2 m (50 ft) and using the average thickness of the grouted soil as 7.6 m (25 ft), the total volume of the soil is 3540 m^3 (125,000 cu ft). The total amount of compaction grout was 137.2 m^3 (4848 cft). Compaction grouting reduced the volume of the voids by an average of 3.9%.

The chemical grout pipes extended to a depth of 8.2 m (27 ft) below floor level. Chemical grouting was completed by April 10, 1979. Grout was injected through the perimeter pipes up to 0.91 m (2.8 ft) below the floor and up to 3.05 m (9.3 ft) below the floor within each foundation area. Eighteen short pipes, each 2.14 m (7 ft) in grouted length, were added to the program to improve soil stability and reduce the permeability of the footing excavation side walls.

A maximum gauge pressure of 241 kPa (35 psi) was used with flow rates varying from 1.89 liters (0.5 gal.) to 7.56 liters (2 gal.) per minute. A total of 151,216 liters (39,947 gal.) of Geloc 3 chemical grout was injected. The same mixture was used for the laboratory resonant column soil samples. The average gel time was 20 minutes. The grout was injected by means of a three unit compartmentalized tanker feeding into a three moyno pump system equipped with totalizing meters, pressure gauges, and see-through flow meters.

NUCLEONIC DEPTH DENSITY TESTS

Three hundred sixty initial readings were made at 0.305 m (1 ft) intervals at the 12 test locations. From these readings a mean average initial void ratio of 1.00 was evaluated. During nearby compaction grouting all pipes moved laterally from their initial vertical position, indicating that the lateral influence of compaction grouting exceeded approximately 1.5 to 1.8 m (4.9 to 5.9 ft). Several access tubes were also destroyed during primary and/or secondary grouting. Two to four sets of readings were made at seven density test locations during and after completion of compaction grouting. Typical void ratios at 3 test locations and project average void ratios are shown in Table 6.

TABLE 6. - VOID RATIOS (e) AT TEST LOCATIONS

Depth Below Floor m (ft)	Test Location No.					
	2		5		10	
	e_o	e_f	e_o	e_f	e_o	e_f
2.7 (8.9)	1.00	.75	.92	.69	1.00	.92
3.5 (11.5)	1.04	.85	1.04	.96	.89	.80
4.1 (13.5)	1.00	.75	.88	.82	1.00	.88
4.7 (15.4)	.96	.69	1.00	.85	1.10	.97
Average 0-4.7 (15.4)	e_{oa} = 1.00; e_{fa} = 0.82; Avg. (e_a)					

e_o = initial; e_f = after compaction grouting

An average void ratio of 0.82 was computed beneath the forging area after completion of compaction grouting; thus the initial 50% volume of voids was reduced by 5% to 45%. This improvement compares well with an approximately 4% reduction in voids for the entire area considering that the average take in the forging area was higher than the average grout take in the total area. The increase in relative density due to compaction grouting is given in Table 7.

TABLE 7. - VOID RATIOS (e) AND RELATIVE DENSITIES (D_r %)

Void Ratio (e)				Relative Density (D_r %)	
Laboratory		Field		D_r %	
e_{max}	e_{min}	e_{oa}	e_{fa}	D_{ri}	D_{rf}
1.27	0.64	1.00	0.82	40	71

D_{ri} = initial; D_{rf} = after compaction grouting

Note: For laboratory compaction grouting sample D_{rf} = 75%
For laboratory chemical grouting sample D_{ri} = 45%

Using e = 0.82, a relative density of 71.4% was computed at the end of the compaction grouting for the Type 2 sand soil which is located immediately below the equipment foundation. The volumetric stability of the foundation soils before and after grouting operation is presented in Table 8.

TABLE 8. - VOLUMETRIC STABILITY OF FOUNDATION SOILS UNDER FORGING EQUIPMENT

Phase	Void Ratio	Volume				
		Total[a]		Voids		Solids
	e	m^3 (cu ft)	%	m^3 (cu ft)	%	%
Initial	1.00	694.4 (24,522)	100	347.2 (12,261)	50	50
After Compaction Grouting	0.82	694.4 (24,522)	100	312.5 (11,035)	45	55
After Injecting 151.2 m^3 (39,947 gal) Chemical Grout	0.30	694.4 (24,522)	100	161.3 (5,696)	23	77[b]

a Total Volume;
Plan Area (21.4 x 5.9 m) x Depth (5.5 m) = 694.4 m^3

b Includes foundation soil, compaction and chemical grout materials

CROSSHOLE TESTS

To determine whether soil modification by grouting had attained a satisfactory improvement in soil modulus, the shear wave velocity of the grouted soil was measured by the multiple crosshole technique as described by Woods (7) at three stages of grouting operations. The first set of tests was made in the area of Crossholes 1 through 12 on February 25 and 26, 1979 after completion of primary compaction grouting. The second set of tests was made on March 25, 1979. At that time, secondary grouting was completed at Crossholes 1 through 12 and primary compaction grouting was completed in the remaining areas. The third set of crosshole tests was completed on April 10, 1979 at the end of all compaction and chemical grouting.

The shear wave velocity was measured at 1.5 m (4.9 ft) intervals to a maximum depth of about 10 m (32.8 ft). The locations of five sets of three Crossholes are shown in Fig. 2. Because it was desired to use the same set of boreholes at different stages of grouting, all boreholes were lined with PVC casing which was grouted in place with a weak cement grout. Inclinometer readings were made in each borehole to establish initial verticality and to identify zones which might be disturbed by the grouting operations. It turned out that the grouting operations did not disturb the boreholes by any measureable amount.

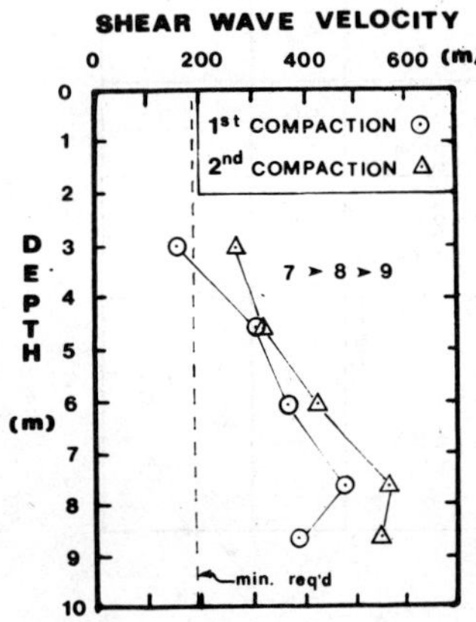

Fig. 8 Shear Wave Velocities

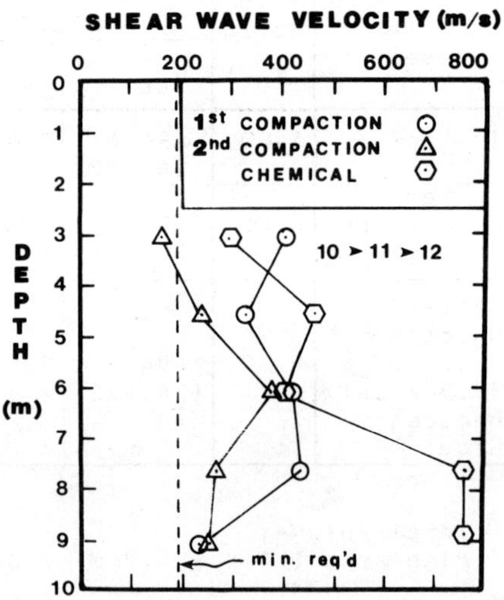

Fig. 9 Shear Wave Velocities

Initial shear wave velocities measured in February after completion of primary (1st) compaction grouting are shown with open circles in Figs. 8 and 9. The shear wave velocities were in excess of the required minimum value of 190 m/sec (623 ft/sec). After completion of secondary compaction grouting the shear wave velocities further increased at the locations of Crossholes 7 through 9 as shown with the open triangle symbols in Fig. 8. At the same time, low values, although in excess of 190 m/sec (623 ft/sec), were measured after completion of primary grouting at Crosshole locations 13 through 15 as shown with open circles in Fig. 10. At the locations of Crossholes 10 through 12 a marked reduction in velocities was measured after completion of the secondary compaction grouting as shown with open triangles in Fig. 10.

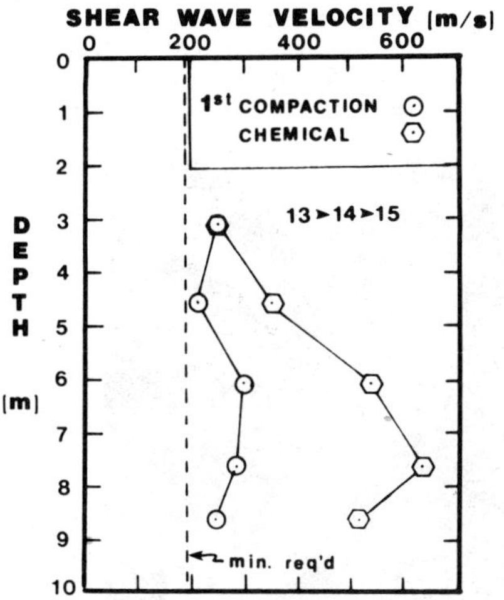

FIG. 10 - Shear Wave Velocities

Two possibilities exist to explain the decrease in velocities; either the void ratio increased or the effective confining pressure decreased. It is hard to visualize that the void ratio increased by the addition of compaction grout. However, if the first stage compaction grouting had built up very high lateral pressures then the second stage compaction grouting might have caused sufficient displacement and disturbance to eliminate or reduce the excessive lateral pressure. Indeed, evidence of this phenomenon existed since the floor heaved by an average of about 25 mm (0.98 in.) in the region of secondary compaction grouting. This floor slab heave indicated volume changes which could easily have reduced or eliminated locked-in excess stresses produced by high pressures of compaction grouting. This phenomenon was not observed at any of the other crosshole locations.

Figures 9 and 10 show open hexagonal symbols for shear wave velocities after completion of all compaction and chemical grouting. Chemical grouting was not performed at locations 1 through 6 because the locations are outside the region where forge foundations were to be built. Measured velocities increased with increasing depth (effective confinement) from 250 m/sec (820 ft/sec) up to nearly 800 m/sec (2625 ft/sec). Crosshole test results at locations 1 through 6 during and after completion of compaction grouting were similar to the data reported herein. Therefore, it was concluded that the completed soil modification was successful in attaining shear wave velocities higher than the required value of 190 m/sec (623 ft/sec), as indicated by over 450 field measurements of shear wave velocities.

FOUNDATION CONSTRUCTION

Excavation for footings started at the first foundation at the west. The second excavation was made at the third footing location measured from the west. The third excavation was made at the far east and the last excavation was made at the footing second from the west. (See Fig. 2). The center (ungrouted) portion in each area was initially excavated with a backhoe (Fig. 11). The trimming of the sides (grouted portions) was done by compressed air operated hand tools as shown in Fig. 12.

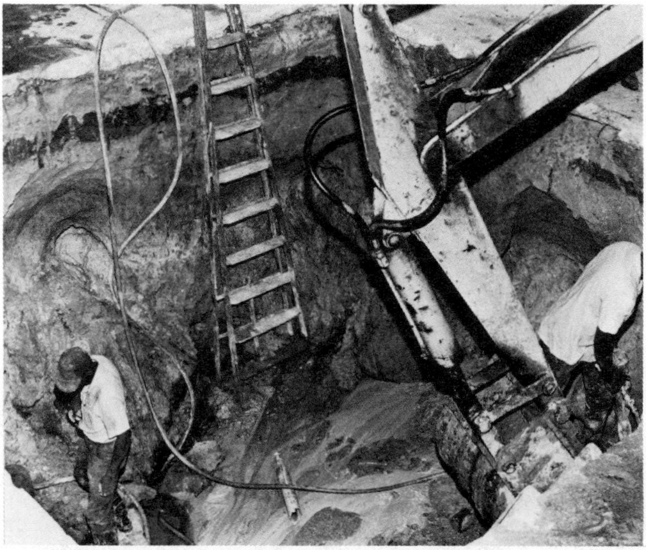

FIG. 11. - Excavation of Ungrouted Center Portion of the Foundation Area

No serious problems with water seepage were encountered, but some temporary plywood sheeting was needed to protect the sides against sloughing produced by vibration of the air hammers. Concreting was done in two steps; initially concrete was placed to the bottom of the forging equipment, then forms were placed followed by final concreting.

FIG. 12. - Trimming the Grouted Side Walls of the Excavation with Compressed Air Operated Tools

POST-CONSTRUCTION GROUND MOTIONS

The forge hammers are numbered from west (Col. line 4) to east (Col. line 7) as No. 1 to No. 4. Measurements of ground motions were made during forging operations (Hammers No. 2 and 3) on June 21, 1979; the results are shown as open circles in Fig. 4. The amplitudes were below those measured in Greenfield, Ohio and they markedly decreased to "easily" or "barely" noticeable to persons within a distance of approximately 2 to 40 m (6.6 to 131 ft) from the source of vibration, respectively.

All four hammers were operating in 1980. On April 23, 1980 measurements of ground motions were made along Column lines 3 and 5 and within a distance of approximately 3.5 to 40 m (11.5 to 131 ft) from the sources of vibrations. The amplitudes are shown as open triangles in Fig. 4, and they correlate very well with those obtained in 1979.

CONCLUSIONS

1. The compaction grouting was successful in attaining the required 70% relative density by reducing initial volume of the voids. This was the purpose of compaction grouting.

2. Chemical grouting was the essential method of soil modification to control ground motions resulting from the dynamic loads of forging operations. Shear wave velocities for chemical grout injected foundation soil greatly exceeded the design criterion of 190 m/sec (623 ft/sec).
3. Nucleonic depth density and moisture content measurements provided direct reading of in-place densities. With the aid of these readings, void ratios and relative densities were evaluated. Measurements were repeatedly made at the same locations from the start to end of compaction grouting operations. Other methods such as SPT's or cone penetration tests, etc. cannot retest the exact same locations.
4. Direct confirmation of the design criterion (shear wave velocity) of soil improvement by compaction and chemical grouting was obtained by employing interval crosshole field testing.
5. Post-construction measurements made during forging operations confirm the satisfactory performance of the grouted foundation soils in the forging area.

ACKNOWLEDGEMENTS

The authors thank Mr. J. Renninger, Vice President, K D Tool Co., Lancaster, PA for permission to publish, the late William F. Lotz, Jr., President and Gerald W. Smith, Director, Engineering, William F. Lotz, Inc., Horsham, PA for their support, and Joseph D. Slamon, SITE engineers, inc., for the quality control of this work.

REFERENCES

1. Afifi, S.S. and Woods, R.D. (1971), "Long-Term Pressure Effects on Shear Modulus of Soils," Journal of Soil Mechanics and Foundations Division Proc. ASCE, Vol. 97, No. SM 10, Oct., pp. 1445-1460.
2. Hardin, B.O. and Richart, F.E., Jr. (1963), "Elastic Wave Velocities in Granular Soils," Journal of Soil Mechanics and Foundations Division, Proc. ASCE, Vol. 89, No. SM 1, Feb., pp. 33-65.
3. Horne, R.A. Marine Chemistry (1969) John Wiley and Son, New York, pp. 322-325.
4. Novak, M. (1974), "Dynamic Stiffness and Damping of Piles," Canadian Geotechnical Journal, Vol. II, No. 4, pp. 574-578.
5. Richart, F.E., Jr., Hall, J.R., and Woods, R.D. (1970) Vibrations of Soils and Foundations, Prentice-Hall, Englewood Cliffs, New Jersey, p.414.
6. Partos, A., Koerner, R.M. (1975) "Consolidation of Soft Silts" ASTM STP 584 pp. 111-126".
7. Woods, R.D., Partos A., (1981) "Control of Soil Improvement by Crosshole Testing," Proceedings X. ICSMFE, Vol. 3, pp. 793-796.

CONSOLIDATION GROUTING AT AN EXISTING NAVIGATION LOCK

By Thom L. Neff[1], A.M. ASCE, John W. Sager[2]
and James B. Griffiths[3]

ABSTRACT

Consolidation grouting with neat cement grout was performed in the foundation rock adjacent and below one side of a very large navigation lock without a major interruption in normal operations. The grouting was part of a remedial repair program for the John Day Navigation Lock, located on the Columbia River. The foundation rock consisted of two distinct volcanic rock types with drastically different physical properties and grouting characteristics; a dense, columnar jointed basalt and a weak, altered and highly permeable flow breccia. The purpose of the grouting was to increase the rigidity of the foundation rock mass (particularly the flow breccia layer) in order to significantly reduce excessive lateral deflections of several of the large concrete wall monoliths. During the grouting period (approximately 5 months beginning about 15 March 1980) the grout, within hours following injection, was subjected to foundation strains and flowing ground water resulting from filling and emptying the lock. Environmental concerns, especially possible interruption of crucial Columbia River fish runs, required special design and construction procedures. Evaluation of the effectiveness of the grouting program utilized in situ field testing and actual pre-grout and post-grout deformation measurements.

INTRODUCTION

Project Location and Description. John Day Lock and Dam is located on the Columbia River between Oregon and Washington, 215.6 miles (347.1 km) above the river mouth and about 110 miles (177.1 km) upstream from Portland, Oregon. A photograph of the project showing the location of the navigation lock is included as Figure 1. Adjacent to the lock on the river side is the North Shore fishladder structure. The navigation lock is 675 feet (205.9 m) long and 86 feet (26.2 m) wide, and provides a maximum lift of 113 feet (34.5 m), making it one of the highest single lift locks in the world. Each of the individual concrete monoliths comprising the lock walls is separated by a contraction joint and designed as a gravity structure founded on rock. In an average month, about 150 commercial tows pass through John Day Lock, or about 4-6 tows on a

[1] Sr. Geotechnical Engr., Chas. T. Main Inc., Boston (former Prof. of Civ. Engrg., Portland State Univ., Portland, OR and Geotechnical Engr., U. S. Army Corps of Engineers, Portland
[2] Chief, Geology Section, U. S. Army Corps of Engineers, Portland
[3] Engrg Geologist, U. S. Army Corps of Engineers, Portland

Figure 1. Aerial View of John Day Project.

typical day. Figure 2 is a closer view of the lock which shows the area where the foundation grouting was performed.

Construction History. Construction of the navigation lock began in 1958. After stripping the overburden to the top of bedrock, a grout curtain was placed around the area of the future lock chamber excavation to consolidate the foundation rock, thus permitting steeper, more stable cut slopes and reducing seepage into the excavation. Following this grouting, the rock excavation was performed in two separate contracts (1959 and 1960) and the concrete monoliths were completed in 1963. Navigation lock operations began on 20 April 1968 when the project was completed and the pool first filled.

Problem Statement. Significant structural cracking with associated concrete spalling has occurred near the base of the riverside lock wall, just above the chamber floor (see Figure 3). The primary crack extends from the upper corner of the filling and emptying culvert to the face of the wall and daylights between 2 and 15 feet (0.16 m and 4.6 m) above the floor (where the spalling has occurred). The exact cause of the problem is difficult to pinpoint, but studies have revealed two most probable reasons: overstressing around the under-reinforced culvert section due to excessive foundation deformation at the lock full condition, and hydraulic surge pressures within the culvert due to the initial filling procedure. When the problem was first discovered in 1975, only two monoliths near the middle of the wall were affected. However, cyclic deflection of the lock monoliths during filling and emptying has caused the cracking and spalling to continue over the years to the point that six of the eleven riverside monoliths are now affected.

Figure 2. Aerial View of Navigation Lock.

The presence of an open, working crack in a concrete structure subject to repetitive loading (emptying and filling of the lock) represents an adverse stress condition with a relatively high probability of worsening with time. This condition required an effective repair scheme to prevent an extended, unplanned lock closure at some future date and the economic catastrophe this would produce. Field investigations and engineering analyses concluded that the cracked riverside monoliths exhibited unacceptable levels of stability and stress, and a two-part repair program was designed as a permanent solution to the problem: (a) a cement grouting program to increase the foundation rock modulus, and (b) a post-tension and epoxy grout program to restore the monoliths to their original uncracked condition. Only the foundation grouting program will be addressed in this paper.

Purpose of Grouting Program. The purpose of the consolidation grouting was to fill with grout the open joints and void spaces in the rock to make the lock foundation more rigid. A specific goal of the program was to increase the deformation modulus of a weak flow breccia layer from an assumed value of 500,000 psi (35,150 kg/cm^2) to 1,000,000 psi (70,300 kg/cm^2) (the modulus of solid flow breccia core samples tested in the lab). Finite element modeling showed that increasing the rigitity of the confining and supporting rock mass would significantly reduce the riverside monolith deflections and decrease the excessive tensile stresses around the culvert.

PRE-GROUT CONDITIONS

Site Geology. Extrusive igneous rock in the form of three stratigraphically distinct basalt lava flows belonging to the Quinton Basalt

Group comprises the local geology. The Quinton Group is part of a very thick sequence of volcanic deposits known as the Columbia River Basalt Formation, which is Mid to Late Miocene in geologic age. Within each of these flows the rock consists of vesicular to dense basalt underlying a layer of highly fragmented and partially cemented rock called flow breccia.

Foundation Rock Description and Properties. A layer of basalt forms the foundation for the gravity section of the lock wall. This rock is generally dark gray, very hard, fine-grained and dense. A characteristic pattern of prominant joints gives the appearance of "columns" when the basalt is exposed. Additional sets of fractures that formed during and after rock formation give the basalt an overall fractured and blocky appearance. Vesicles (small unconnected void spaces resulting from the rock forming process) occur at scattered locations throughout the basalt.

A layer of flow breccia underlies the upper basalt. This rock is composed in part of solidified lava and in part of ejected volcanic materials in the form of ash, lapilli (volcanic fragments) and bombs, often all welded together. The flow breccia marks the original upper surface of a lava flow. Rapid cooling, weathering, alteration, and heterogeneous nature has produced a brown colored rock that is soft, weak, crumbly, highly fractured (or brecciated), slightly cemented and rough textured. Due to the nature of deposition, the bottom flow breccia contact is very irregular. The flow breccia layer ranges from about 25 to 40 feet in thickness and contains a large number of interconnected void spaces, making the layer very permeable and particularly amenable to grout injection. The original grout curtain proved effective in reducing seepage into the excavation and preventing raveling and sloughing of the flow breccia.

Another basalt layer lies below the flow breccia and is very similar to the upper one, but exhibits less columnar structure. This unit is

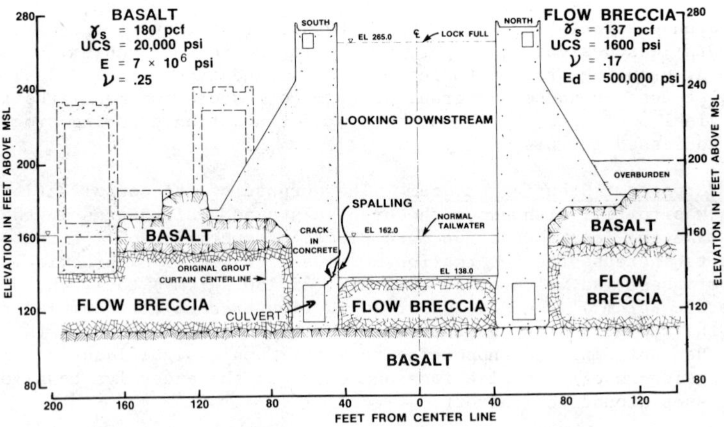

Figure 3. Lock Cross-Section and Rock Unit Properties.

the lowest layer involved in the lock foundation, and the filling and emptying culvert sections are founded on its upper surface.

Figure 3 shows the relationship of the major rock units to the lock structure. As the values in Figure 3 indicate, there is considerable difference in the typical parameters for each rock type. These rock parameters are based on laboratory and field test results as well as presumptive values taken from the literature for similar rock types.

Permeability. The basalt units, with vertical and horizontal joint sets, exhibit a permeability of about 2 x 10^{-4} ft./min. (1.02 x 10^{-4} cm/sec). The flow breccia shows a fairly wide range of permeability values. Degree of void space interconnection, percent fines within the voids, and shears within the flow breccia all affect permeability values. An average value for permeability of the flow breccia, based on calculations from pressure test data, is about 5 x 10^{-3} ft./min. (2.54 x 10^{-3} cm./sec.).

In Situ Modulus. A team from the Corps' Waterways Experiment Station carried out a program of in situ testing in January, 1980. The program sought to define the existing modulus for the flow breccia in the lateral direction by using the Goodman borehole jack, a 3-D acoustic velocity logger, and downhole and crosshole seismic techniques. Tests were conducted in four borings along an 82-foot (25 m) line parallel to the lock wall, about 129 feet (39.3 m) riverward (south) of the lock centerline. Upon completion of the grouting program, these in situ tests were repeated. The results and more complete descriptions of the details and limitations of these tests appear in the section which discusses the post-grouting data.

FIELD INSTRUMENTATION

We installed several types of instrumentation to both monitor the performance of the cracked structure and evaluate the effects of the repair. The primary systems are described below:

(a) bar extensometer - measures relative movement between adjacent monoliths, measurements taken at the crest

(b) electrolevel - measures rotation, tilt of the monoliths, at crest and toe (with assumed radius of rotation, permits calculation of horizontal and vertical deflections)

(c) electronic distance measuring (EDM) - a precise survey system that measures distances very accurately (1/16" to 1/8"), measures deflections at crest and toe of monoliths

(d) wireline - using a uniformly-tensioned, high strength piano wire, system measures total deflection between landside and riverside monoliths, and relative movement between the riverside monoliths and the fishladder.

Data from these measuring systems show a pronounced temperature effect, with maximum movements occurring in February and minimum movements occurring in August. Figure 4 is a cross-section of the river-

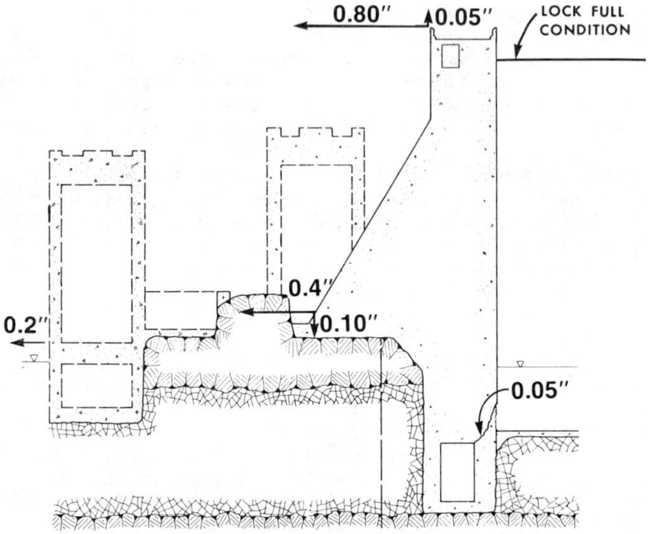

Figure 4. Typical Maximum Monolith Deflections (View looks downstream).

side wall and fishladder that shows the locations and magnitudes of a typical set of maximum monolith deflections measured prior to grouting.

GROUT PROGRAM

Design. A total of 568, 1.5 inch (3.81 cm) (EW) diameter grout holes were planned opposite nine of the structural monoliths (a distance of about 550 ft. (167.9 m)) of the riverside lock wall. Primary holes were arranged in essentially six staggered lines paralleling the lock wall. Because of space limitations, grout hole lines were close together on the surface and the holes were drilled at angles ranging from 5 to 25 degrees from vertical in order to provide coverage of the entire foundation rock mass (see Figure 5). The holes in each primary line were generally on 10 foot centers while spacing between lines was typically 6 feet. Some secondary grout holes were planned between lines of primary holes to assure complete coverage in the rock closest to the lock. A number of low-angle holes were included to intercept the high number of near-vertical joints in the basalt. Fishladder support pillars and other obstacles required altering the grout hole pattern at some locations.

In general, the grout holes were planned to extend 10 feet (3.1 m) below the contact between the flow breccia layer and underlying basalt or to the elevation of the base of the culvert portion of the structure, whichever was deeper. Typical hole depth was approximately 75 feet. The program of planned grout holes required a maximum grout travel between holes of 10 to 12 feet in the direction perpendicular to the lock in order to achieve complete coverage at the lowest elevation. The option of split spacing was provided in the contract to reduce the required grout travel if it was determined during the work that the planned pattern was not adequate.

The rock mass was divided into two zones for purposes of the grouting program. The first zone extended from the top of the hole to the upper basalt - flow breccia contact (approximately 30 feet (9.15 m)). The second zone extended through the flow breccia layer and into the underlying basalt to the bottom of the hole. Routine Corps of Engineers' stage grouting procedures were utilized. The contract permitted grouting Zone 1 in a single stage, but grout stages were limited to 5-foot (1.5 m) intervals in the flow breccia layer. Previous grouting experience in this type of material had indicated a need for short stages in the flow breccia to achieve good penetration and adequate control. Thin grout mixes were specified for the start of grout injection in order to penetrate the fine openings expected in the rock.

Some of the design details for this grouting program were based on information from a consolidation grouting program performed by the Walla Walla District of the Corps of Engineers at Little Goose Lock and Dam on the Snake River in Washington during 1978. Foundation conditions similar to those at the John Day Lock exist at the Little Goose project. Based on the results of the grouting done during original construction of the John Day lock and the consolidation grouting program at Little Goose, grout takes for the lock repair foundation grouting were expected to average about .4 sacks of cement per linear foot of hole in Zone 1 (basalt) and 1.2 sacks per linear foot in Zone 2 (primarily flow breccia). A test grouting program was not utilized because:

(1) knowledge of the groutability of the foundation, particu-

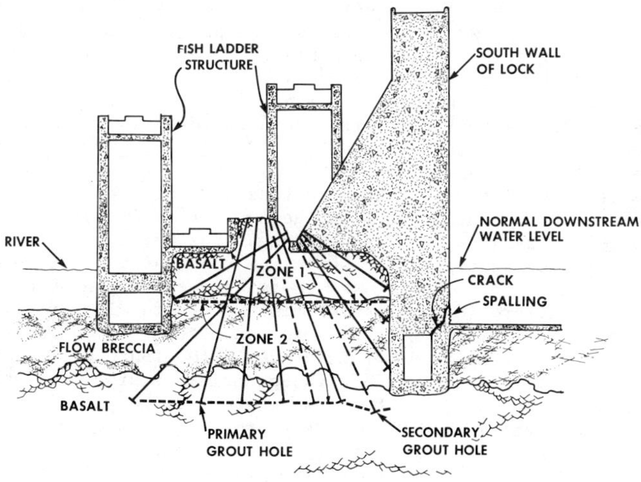

Figure 5. Typical Cross-Section Orientation of Grout Holes.

larly the flow breccia unit, was well documented in the construction history records pertaining to the original grouting of the lock area,

(2) information was available from recent grouting work done at Little Goose Dam in very similar foundation rock,

(3) evaluation of the effectiveness of a test grouting program in the foundation would require an unknown period of time that could delay the repair program, and,

(4) the degree of effectiveness resulting from a test grouting program would not likely be representative of the effectiveness of the entire grouting program.

Construction. The Contractor mobilized in early February, 1980, and soon began setting grout nipples at the assigned hole locations. Grout holes were EW size (1½-inches (3.81 cm) in diameter) and were drilled using rotary drilling equipment. As planned, the first holes drilled and grouted were the row of primary holes farthest from the lock. These holes were drilled and grouted to the full depth opposite each monolith to form a grout barrier which permitted more efficient grouting in the holes closer to the lock. Grout pressures in these barrier holes were kept low to minimize undesired grout travel. The remaining lines of primary holes, progressing toward the lock, were then drilled and grouted in Zone 1 and this procedure was repeated for Zone 2. The secondary holes were then drilled and grouted in Zones 1 and 2. In general, drilling in Zone 2 did not begin in an area until the holes had been completely grouted in Zone 1.

Prior to performing injection grouting in each stage, the hole was washed and a water pressure test performed. Grout injection was normally started with a batch of grout having a water/cement ratio of 5:1. The mix was then thickened cautiously at the discretion of the construction inspector. It was discovered during performance of the work, however, that difficulties in grout injection were encountered with mixes thicker than about 3:1; thus thick mixes were seldom used. Type I portland cement was used in the grout and a fluidifier was added to facilitage penetration into the finest fractures possible. Grout pressures were kept low (sometimes as little as 10 psi (0.7 kg/cm^2)), especially in Zone 1, to avoid damaging nearby fishladder structures and prevent surface venting. The contract specified that all grout injection be performed with the water level in the lock at tailwater elevation to avoid grouting against high seepage pressures in the rock when the lock was full and to avoid grouting while the rock was in a state of maximum strain. A period of 6 hours was set aside each day in which no lockages were allowed to provide sufficient time for performing grouting with the lock empty. The refusal criterion for each grout stage was two cubic feet of cement per hour, determined over a 10-minute period at the pressure required for that stage. Although multiple grout hookups were permitted by the contract, all holes were grouted individually.

All primary and designed secondary grout holes were completed by August, 1980. Additional secondary holes were then assigned where the rock appeared more open based on grout takes in the primary and assigned secondary holes. Locations for these additional holes were determined by the split-spacing method. A few tertiary grout holes were drilled at the end of the contract to determine the overall completeness of the grouting job. Tertiary hole locations were selected at random and grout takes were compared with those for the primary and secondary grout holes to determine if a satisfactory reduction in open voids had been achieved. Throughout the contract period, exploratory core borings were drilled to help determine the effectiveness of the grouting program in filling rock voids.

Main Problems. One of the main contract efforts focused on environmental protection. The fishladder, for passage upriver by salmon and steelhead, required careful monitoring of vibrations and possible grout leakage to assure that grouting operations had no adverse effect on fish runs. Acoustic and vibration monitoring showed the grouting operations had no significant effect on vibration and noise levels in the fishladder. A fluorescein dye compound was used to trace possible grout leakage into the fishladder or the river. Monitoring points were selected and water samples continually pumped to a fluorometer (Turner, Model No. 111) capable of sensing minute quantities of the dye. Grout leakage occurred on only a few occasions. The use of the fluorometer in this type of application was not totally successful. Instrument calibration was a significant problem and difficulties with the recorder resulted in frequent loss of data.

The duration of the contract was purposely limited to reduce the time period during which the daily lock outages were required. Also, the required 5-foot (1.53 m) maximum stages in the flow breccia reduced the rate of progress that would otherwise have been expected. These factors necessitated a round-the-clock operation to assure timely completion of the contract. Because lock filling continued during the contract period, grout was subjected to relatively high seepage gradients within hours after intrusion. Although this created the potential to flush some of the grout from the rock, no evidence was found that this actually happened to any significant degree.

Effectiveness of the grouting program was hampered somewhat by restrictions placed on grouting pressures in certain areas either to prevent damage to structures or restrict grout travel to within the area of concern. Grout takes indicated that better grout intrusion resulted where higher pressures were allowed. Limiting stage intervals to 5 feet (1.53 m) in the flow breccia resulted in many grouting stages in which little grout was intruded due to the rock being relatively "tight". However, correlations between water takes during pressure testing and grout takes during injection grouting are generally not reliable. It was considered necessary, therefore, to grout every 5-foot stage regardless of pressure test data to assure maximum grout penetration into the foundation rock.

EVALUATION OF POST-GROUT CONDITIONS

Grout Take (Actual vs. Predicted). The following shows a simple comparison of the key variables in the contract:

	Predicted	Actual
No. of Drill Holes		
primary	402	397
secondary	166	213
tertiary	27	22

	Predicted	Actual
Total Drilling Footage		
Zone 1	23,000	24,438
Zone 2	17,000	19,054
Total Grout Take (Sacks)		
Zone 1	9,200	3,540
Zone 2	20,400	29,640

The grout take in Zone 1 fell significantly below our prediction (.14 sacks/foot (0.04 sacks/m) vs. 0.4 sacks/foot (0.12 sacks/m)) and in Zone 2 the grout take exceeded our prediction (1.55 sacks/foot (0.47 sacks/m) vs. 1.2 sacks/foot (0.37 sacks/m)). The fractures in the basalt were too tight, for the most part, to accept grout. The rock directly beneath the lock wall gravity section took less grout than rock closer to the fishladder, primarily because of the existing grout curtain constructed prior to excavation for the lock chamber.

Figure 6 shows a plot of average grout take in Zone 2 for each monolith and graphically illustrates the high grout takes adjacent to the middle monoliths (15-21), where maximum monolith deflections have occurred. Cores from the exploratory borings drilled to determine effectiveness of the grouting showed the most grout adjacent the middle monoliths. The data in Figure 6 (and Table 1) emphasize the generally poor quality of the flow breccia.

In Situ Modulus. Post-grout field testing took place during the period 28-31 August 1980 to determine the effect on engineering parameters of foundation materials resulting from the cement grouting program. Obtaining accurate, repeatable data in the field with these type field tests requires careful execution. Even with the best intention, the nature of the tests presents problems of both execution and inter-

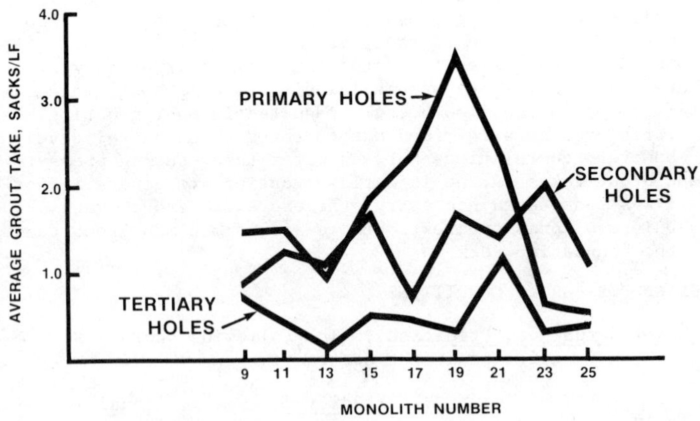

Figure 6. Average Grout Takes in Zone 2 vs. Monolith Location.

pretation. The following paragraphs summarize these limitations:

(1) Goodman jack - This test expands diametrically - opposed, curved platens against a borehole wall, using the relationship between hydraulic pressure and platen movement to calculate an in situ deflection modulus. The geometry and uniformity of the borehole wall can greatly affect the numerical test results.

(2) 3-D Acoustic Velocity - This device measures the transit time of compression (P) and shear (S) waves propagating by refraction down the borehole wall. Accurate borehole diameters at each test location provide needed data to compute dynamic Young's modulus. Large numbers of attenuating fractures and irregular hole configuration can alter signal strength and therefore affect data interpretation.

(3) Crosshole Seismic Tests - This test measures transit time of an acoustic wave as it travels from one boring through an intervening material to a second boring. Using both P-waves and S-waves permits computation of dynamic elastic parameters (averaged) for all materials on the line of travel from source to receiver. Accurate distances between holes and assumed nonrefracted ray paths are necessary to reduce the field data to represent dynamic moduli.

The following difficulties inhibited to some degree the execution of the field work, and the subsequent data reduction and interpretation:

(1) washing sand filling from original drill holes eroded material from hole walls creating irregular surfaces (especially for flow breccia),

(2) having shot and receiver at different elevations in boreholes for some crosshole tests,

(3) refraction from flow breccia to lower basalt for some crosshole tests,

(4) failure of some "zones" to allow passage of usable acoustic signals, particularly for the 6-foot (1.8 m) transmitter - to - receiver spacing (downhole), and

(5) mislocation of receiver in some crosshole work.

The data in Table 1 compare pre-grout and post-grout modulus values from the three types of field tests. The Goodman jack showed essentially no change from pre-grout to post-grout conditions. We consider this largely a result of using the same holes for both sets of tests, as the washing of sand packing from the holes resulted in irregular hole sides with loose fragments. Also, the Goodman jack stresses a relatively small volume of rock at the perimeter of the borehole and thus the washing procedure could easily mask the grout effects that took place in the rock mass.

The 3-D sonic and crosshole seismic data both show a measured increase in the in situ modulus of about 100%, from approximately 500,000

TABLE 1

COMPARISON OF PRE- AND POST-GROUT MODULI

	Pre-Grout Modulus	Post-Grout Modulus
Goodman Jack (static, localized)	psi x10^6 (kg/cm^2 x10^6)	psi x10^6 (kg/cm^2 x10^6)
Basalt	1.180[1] - 4.660[2] (0.082 - 0.33)	0.691 - 2.321 (0.048 - 0.16)
Breccia	0.085 - 0.339 (0.006 - 0.024)	0.086 - 0.329 (0.006 - 0.023)
3-D Sonic Log (dynamic, localized)		
Basalt	2.711 (0.190)	2.141 (0.150)
Breccia	0.492 (0.034)	1.015 (0.071)
Crosshole Seismic (dynamic, large volume)		
Basalt	1.980 (0.139)	2.355 (0.166)
Breccia	0.504 (0.035)	1.138 (0.08)

1 - lower number represents Goodman analysis (soft rocks)

2 - upper number represents Hustrulid analysis (hard rocks)

psi (35,150 kg/cm^2) to just over 1,000,000 psi (70,300 kg/cm^2). We have confidence in these data as good indication of an actual modulus increase because the 3-D data give reasonable localized results while the cross-hole seismic integrates data from a much larger volume of rock.

Reduction In Monolith Deflections. Using a variety of techniques, we monitored monolith deflections to both evaluate the problem and to help design, carry out and evaluate repair schemes. Post-grout measurements gave an indication of the effectiveness of the foundation grouting scheme, though such an evaluation remains somewhat complex because of the nature and magnitude of the deformations. The total outward and downward movements of the monoliths during lock filling contain aspects related to: 1) translation, 2) rigid body rotation, and 3) flexure.

The electrolevel data showed essentially no change from pre-grout to post-grout conditions. The presence of the structural crack greatly weakens the overall flexural rigidity of the lock wall/foundation system. The data emphasize both the complex nature of lock movements and the need for the two-phase repair, i.e., foundation grouting, and post-tensioned tendons and epoxy grout repair of the structural crack. The latter portion will significantly affect rotation of the monoliths.

The bar extensometer data showed a reduction in monolith deflections resulting from grouting, though only in the range of 20 to 30 percent from pre-grout movements. Since the bar extensometer measures only relative movement between monoliths, the noted reductions indicate an improvement, but do not completely explain the results since bar extensometer measurements include translation, rotation and flexure components.

The wireline system (measuring total monolith deflections) showed the most improvements, i.e., reductions of 50 percent from pre-grout measurements (Figure 7). These data, when coupled with the electrolevel and bar extensometer data, suggest that the foundation grouting had a significant effect on the lock foundation, but that the largest percentage improvement came in the translation mode.

The electronic distance measurement system (EDM) also measures total monolith deflections; however, we had some difficulty with accuracy and repeatability of the data. Currently, we can repeat measurements to within $\pm$ 0.06 to 0.12 inches (0.15 to 0.036 cm). We do, however, hope to see some further reductions using this system, after completing the tendon and epoxy portion of the repair.

Figure 7 shows a typical plot of deformation vs. time for one type of device (wireline) for the period March, 1980 through March, 1981 (the foundation grouting program took place from March 15 through September 1, 1980). We felt a complete 1-year cycle of field measurements was needed to fully evaluate the foundation grouting effectiveness under the influence of the seasonal temperature variations. Temperatures for March 1980 - March 1981 appear on Figure 7.

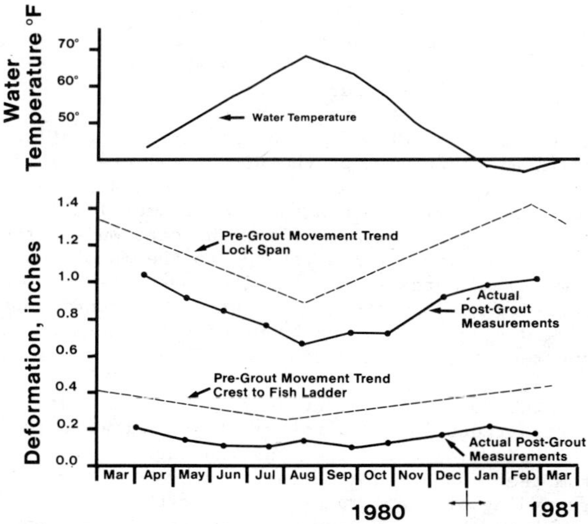

Figure 7. Pre-grout and Post-grout Wireline Data - Monolith Deflections.

SUMMARY AND CONCLUSIONS

Cost Summary. Total cost of the foundation grouting contract was \$1,452,510. The unit cost for grout in place came to about \$43.75/ft.^{3} (\$1,545.94/m^3). The amount of grout hole drilling required and quantity of grout materials used slightly exceeded our estimates, but the overall job cost was about what was predicted.

Effectiveness of Foundation Grouting Program. Considerations concerning the effectiveness of the foundation grouting program include a comparison of expected and actual deflection reductions, pre-grout and post-grout modulus measurements, along with a review of expected and actual grout takes.

Previous grouting at this site and field permeability testing were used to make grout take estimates. Actual grout take totaled approximately 33,180 cubic feet (939 m^3) (sacks) of cement, the same order of magnitude as our estimate, though variations did occur in takes within the two primary rock types and along the axis of the lock. The fact that maximum grout takes occurred in the flow breccia adjacent the middle monoliths (where maximum deformations have occurred) emphasized the need for grouting to fill voids and fractures within the foundation rock.

Pre-grout and post-grout modulus determinations showed a general increase in the in situ modulus of the flow breccia from about 500,000 psi (35,150 kg/cm^2) to about 1,000,000 psi (70,300 kg/cm^2), with good agreement between the 3-D acoustic velocity measurements and the cross-hole seismic tests.

Prediction of monolith deflection reductions as a result of foundation grouting came primarily from finite element studies of rock/structure interaction. These studies were based on rather sparse field measurements (1979 and earlier) as well as a number of assumptions regarding stress distribution, variations in rock moduli, effects of the crack, and the possible separate effects of each portion of the total repair scheme. Evaluation of this portion of the work has several inherent difficulties because of the three sources of movement, i.e., translation, rotation and flexure. Results clearly show very little effect on flexure and rotation from the foundation grouting, emphasizing the critical importance of the structural crack in affecting overall performance. The epoxy repair to the crack and installation of the post-tensioned tendons will return the cracked monoliths to monolithic structures and have a significant effect on rotation and flexure by restoring a measure of structural stiffness to the monoliths. The wireline system, which measures total monolith movements, showed deflection reductions in excess of 50 percent from pre-grout conditions. This represents very strong evidence of grout effectiveness considering that the reductions resulted almost exclusively from changes in the translation mode.

In summary, the foundation grouting achieved the objectives intended, though deflection reductions did not exactly match expectations (50 percent actual reduction vs. 70 percent expected). An additional bonus resulting from the foundation grouting program consisted of obtaining a better understanding of the complex nature of the monolith deflections, especially the contribution to monolith rotation caused by the presence of the structural crack.

RECOMMENDATIONS

Pre-Grout Investigations. Some pre-grout investigation is necessary to assist in formulating an accurate scope of work and uncovering, at least in a qualitative sense, the nature of potential problems that may be encountered during the execution of the work. Sufficient borings must be made to delineate the boundaries of the major rock types and their approximate character, especially jointing, fracturing, weathering, etc. - factors that affect rock mass permeability and grout take. Water pressure tests can provide important rock mass permeability data; however, these data can be misleading when attempting to predict grout takes. For very large jobs or important projects, field grouting tests (at different water/cement ratios and with different additives) should be performed, if time permits, to provide more accurate prediction of grout takes.

Grouting Program Design. The design of any grouting program must include all elements necessary to ensure efficient and cost-effective achievement of the primary engineering objectives of the work. Special considerations must receive proper attention, e.g., grouting with lock empty conditions for this project to avoid adverse seepage gradients during the actual grouting. The design of a grouting program should consider at least the following items:

(1) drilling method - rotary or percussion
(2) hole diameter
(3) hole spacing
(4) hole depth
(5) angle holes vs. vertical holes
(6) split-spacing option
(7) sequence of drilling and grouting
(8) washing and pressure testing
(9) grout mix and materials
(10) grout pressures
(11) grouting procedures and equipment

Grouting Construction Procedures. Close quality control and detailed record keeping by knowledgeable Field Inspectors of mix proportions, injection pressures and grout takes during grouting is very valuable. In this manner, the grouting program may be adapted to actual subsurface conditions encountered to the maximum extent possible. Our experience on this project showed mix ratios and grout pressures can significantly influence grout effectiveness. Starting with high water/cement ratios (on the order of 5:1) and gradually reducing to lower values will generally promote better penetration and void filling. Stage grouting, while adding to the cost of the work, remains the only means to achieve optimal control of the operations and permit quantitive assessment of grout takes in specific foundation zones. Post-grouting rock core drilling can provide positive evidence of grout effectiveness. Testing on such cores, as well as field tests within and between core holes, can also provide numerical assessment of changes in engineering properties brought about by the grouting.

GROUTING HALTS CHEMICALLY INDUCED SETTLEMENT

By

R. D. Hinkle, Jr.[1], M, A.S.C.E. and B. C. Yen[2], M, A.S.C.E.

ABSTRACT

Setttlement created by a chemical transformation of weathered mudstone caused extensive damage to a major electric power generating facility in northwest New Mexico. Hypotheses have been proposed regarding the cause of the problem. Most of these hypothesis involved the dissolving of gypsum found in the formation.

It was established that no change in gypsum occurred, but rather there was a change in the clays. A degradation process occurring in clay was isolated, tested and reversed in the laboratory. The process was field-tested through a lime-cement grouting program, and the settlement ceased. A calcium-magnesium replacement in the mixed-layer montmorillonite clay structure is credited with the reaction.

Two 750 megawatt turbines were raised and realigned and the grout mixture chemically reacted with the formation to arrest settlement and to achieve soil stabilization. Prior to grouting, settlement rates were .7 cm to 2 cm per year, and over three years of post-grouting monitoring shows no further settlement.

INTRODUCTION

Four Corners Generating Station (Units 4 & 5) in the Northwest corner of New Mexico has suffered rather severe settlement. Settlement rates range from 0.7 to 2 cm per year. It has been established that all settlement occurs in the upper 10 meters of the weathered rock. This paper will briefly discuss the cause and effects of the settlement and a grouting technique which was successfully used to halt the settlement.

The site is underlain by Cretaceous marine sedimentary rocks of of the Fruitland Formation and/or Kirtland Shale. The near surface deposits consist of interbedded mudstones and sandstones with shale and coal at depth. The sandstone beds are up to 0.7 m. thick while

---

1 Consulting Geotechnical Engineer, Long Beach, California

2 Professor, Department of Civil Engineering, California State University, Long Beach

Figure 1. Typical Interbedded Sandstone and Mudstone in the Upper 10m of the Project Site

Figure 2. Gypsum Crystals, Voids and Desiccation Cracks Found in Mudstones at Site

the mudstones are generally 5 to 80 mm thick with complicated interbedded sequences, Figure 1. Secondary mineralization has resulted in the deposition of soluble salts such as gypsum, epsomite and mirabilite in the bedding planes. Some voids and dessication cracks up to 6 mm thick exist where secondary salt crystallization has occurred in the bedding planes, Figure 2.

Settlement began before the total completion of the project in 1969. Maximum settlement approached 10 cm during the nine year period following construction. Differential settlement caused considerable deflection to mat foundations supporting the turbine generators. Repeated cement compaction grouting was used on the boiler structure which slowed the settlement but differential settlement continued. As the limits of adjustment of the rotating turbine shaft were reached, a more permanent corrective measure had to be performed.

MECHANISM OF SETTLEMENT

It is noted that a 1,300 acre unlined cooling water pond was constructed adjacent to the plant. Subsurface investigation showed that the water saturated all the soils below seven feet from the surface. There are many case histories about settlement induced by additional water in desert environment. For example, a case of settlement in gypsiferous clay soils was discussed by Bull(1). Large settlements occurred as a result of water inflows caused by leakage of the unlined canal of the California Water Project in the western San Joaquin Valley. The settlement was attributed to the collapsing of clay-type soils. Another case(2) of regional settlement in Texas was reported when water was introduced into an arid hypersaline environment. However, the settlement problem at the subject site became more clear when the remolded mudstone samples were subjected to long-term consolidation tests. The introduction of water during tests showed a very slow secondary compression, amounting to approximately .01 percent per day, occurs in the samples. The rate does not seem to depend on load or loading rate but, rather, is constant and linear at all loading conditions. Some tests were run for as long as 120 days with no change in this linear behavior(4). Figure 3 shows typical one-dimensional consolidation data. The clay-rich lenses of mudstone samples have an average liquid limit of 30 percent, plastic limit of 20 percent and x-ray diffraction showed that they consisted of clay materials of montmorillonite, illite and kaolinite.

Ground water tests showed that the area is super-saturated with sulfates. It is theorized that during weathering bacteria feed on the organics in the rock leaving excess sulfate as a waste. The addition of fresh water allows the cations to escape the clay structure forming the gypsum ($Ca\ So_4$). Hence the growth of gypsum crystals is believed to be from deposition of rain leached soluble cations. In the mean time, as the clay gives up its Ca^{++} and Mg^{++} ions, it shrinks in volume leaving dessication cracks and bedding plane separation cracks such as shown in Figure 2. This process is called transformation(5) that modifies the clay mineral without altering its two or three layered structural type, Figure 4. The transformed minerals are normally stable in the mudstone or shale. The

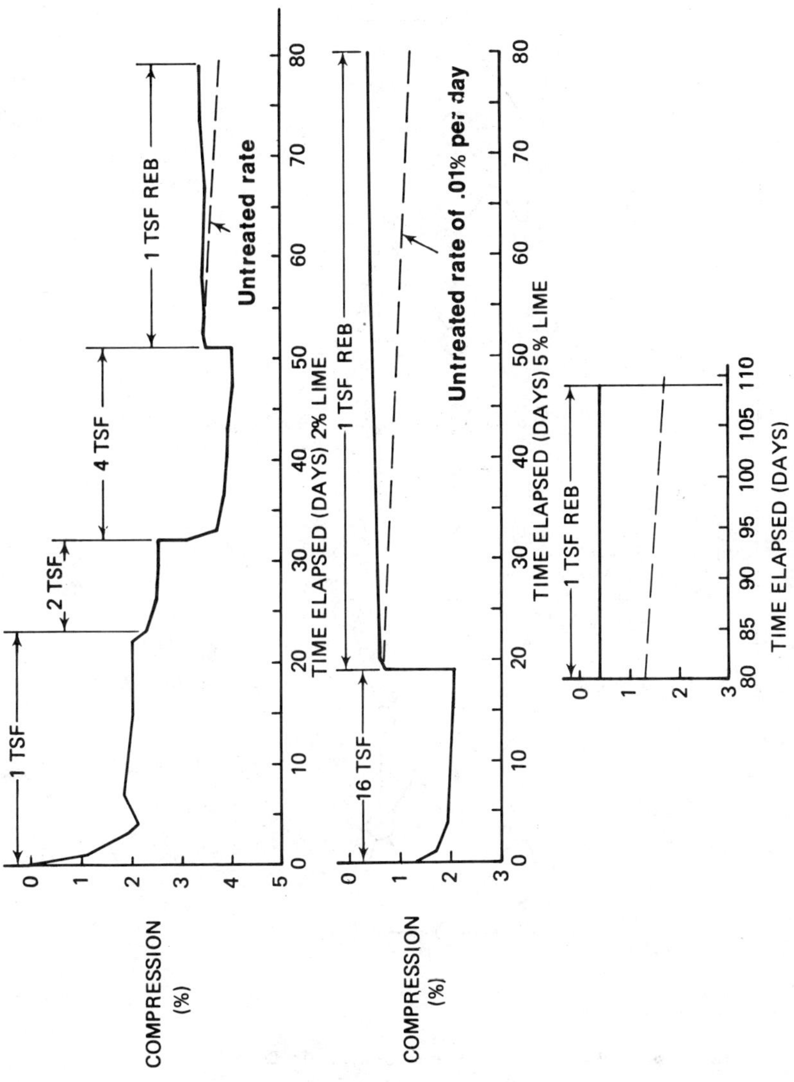

Figure 3. Consolidation Behavior of Lime Treated vs Untreated Remolded Mudstone

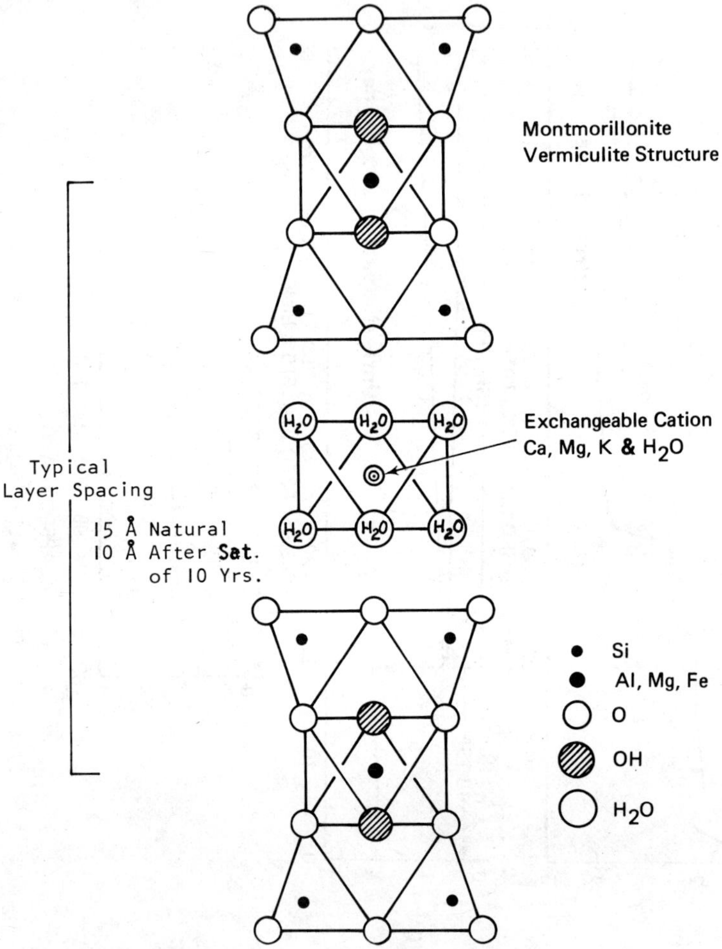

Figure 4. Clay Lattice Structure, Four Corners Area

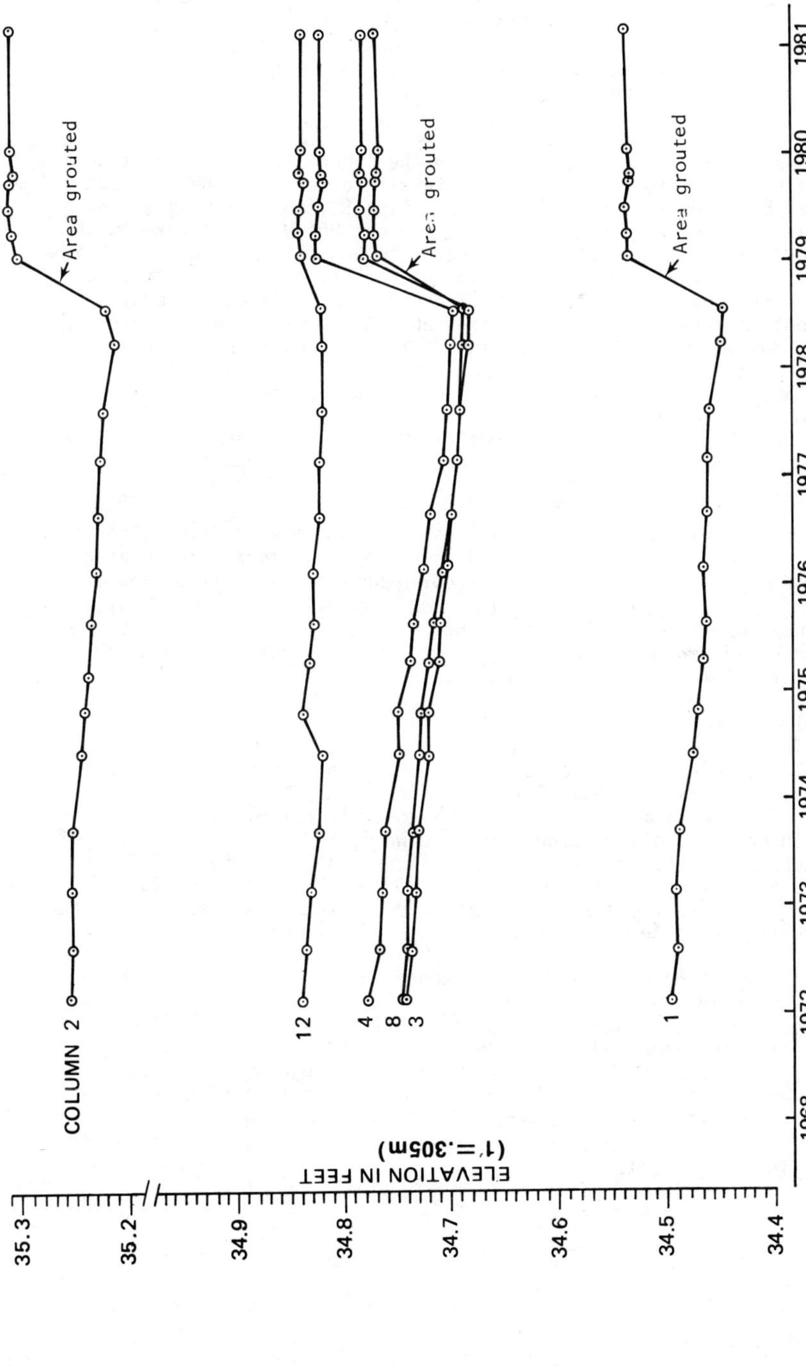

Figure 5. Typical Settlement Data, Four Corners Plant

lack of water in the desert environment allows these minerals to be stable indefinitely. Upon increased fresh water infiltration from the cooling pond, the transformation process is accelerated and the clay mineral lattice is no longer stable, and it collapses resulting in long-term settlement. Fig. 4 shows the dramatic volume change in the molecular structure.

The above scenario was proven in the laboratory by reinsertion of the cations into the clay structure with observed changes in consolidation characteristics, Figure 3. Type-S lime, which is approximately 60 percent of $Ca(OH)_2$ and 40 percent $Mg(OH)_2$, was mixed with remolded mudstone samples and subjected to consolidation tests. Note that Figure 3 showed marked improvement resulting in less settlement of lime treated samples as the Ca^{++} and Mg^{++} stabilized the clay lattice. A detailed discussion and laboratory testing were presented by Hinkle, 1978(4). The phenomenon is the same as found naturally in the weathering process where an interstitial cation is allowed to leave the clay structure leaving a smaller structure with six less chemically bound water molecules. The principal cations in this clay are Calcium (Ca^{++}) and Magnesium (Mg^{++}). These are the source of cations for the salt crystallation of gypsum ($CaSO_4$) and epsomite ($MgSO_4$).

It is noteworthy that there is an obvious zone of demarcation at the site between the weathered and unweathered bedrock. The weathered zone of approximately 10 m has a reddish-brown color while below 10 m it is a grey-blue color. Structures founded below the 10 m weathered zone have shown no settlement. Other evidence from core-recovery, electric logs and clay minerology generally substantiated the mechanism of volume changes proposed(5).

GROUTING

The turbine pedestals were reinforced concrete structures approximately 24.4 m x 35 m and 10 m high. The base mat (foundation) was approximately 2.1 m thick including a concrete circulating water line beneath one side of the pedestal as shown in Fig. 6. The top deck of the turbines contained two linked turbines which rotated at 3,600 rpm with 1,800 psi steam pressure. Figure 6 shows the differential settlement since the start-up in 1970. This settlement was too large to provide reliable operation.

The grouting operation was intended to lift and realign the turbine foundation and other structures while at the same time filling any voids beneath foundations. While lifting was occurring the grout penetrated the weakened clay layers and collapsed them to avoid future long-term settlement. It was also the objective of the grouting operation to have the lime rich grout layers spreading throughout the area to provide a source of Calcium and Magnesium. This stabilizes the clay structure thus eliminating the volume change and resulting settlement.

1. Grouting Mix Design

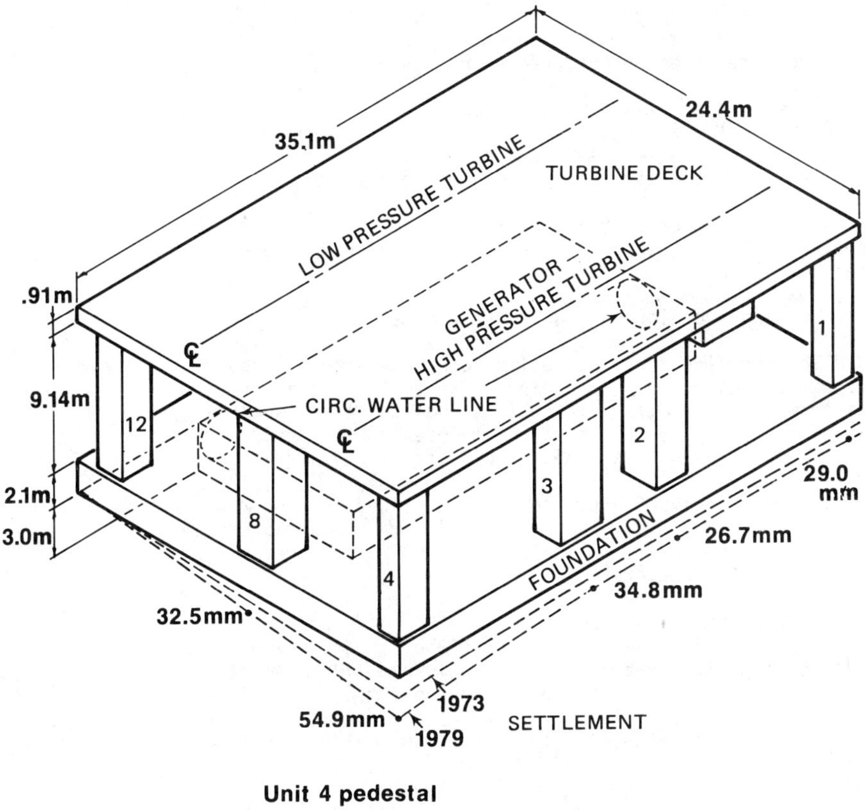

Figure 6. Schematics Showing Turbine – Generator Foundation Layout and Pre-grouting Settlements.

The settlement rates at the site showed a nearly linear behavior, Figure 5. Remolded samples of both the sandstone and mudstone were prepared and consolidated tests (Oedometer) were performed. The tests were run 120 days and the mudstone samples showed behavior similar to that observed in the field, Figure 3. Samples were prepared using the mudstones mixed with 2 percent and 5 percent (by weight) of lime and other admixtures including Portland cement and flyash. Type-S lime was found most effective, with test samples showing no settlement after 30 days curing time. It was decided to attempt to use grouting to inject the lime into the subsurface environment.

Several field tests were performed using the same mixtures yielding similar results(3). It was found that volumes of lime amounting to approximately 2 percent of the dry weight of the mudstone were adequate to halt the settlement.

Solubility tests show that the lime changed the equilibrium condition of the ground water and this stopped any solution of the crystalline salts in the formation.

2. Grouting Techniques

The grouting technique utilized was developed through several stages of trial grouting. The most critical problem was to minimize lifting of machines and structures while injecting enough grout to halt future settlement. The desired criteria allowed only 20 mils (.55 mm) of lift at any structure.

Approximately 80 grout holes were placed in and around each turbine pedestal, Figure 7. The perimeter was surrounded by holes on 1.7 m centers while the interior holes were approximately on 3.3 m centers.

Holes were drilled approximately 9.2 m deep and 6.4 cm in diameter. Drilling was done using either a percussion type drill or a CP 65 air-driven rig. A plugged 1.9 cm smooth joint steel pipe was installed to the bottom of the hole and the annualar space filled with sand to the surface. All holes were prepared in advance so that grouting could be performed quickly.

The grouting sequence consisted of grouting the perimeter holes nearest sensitive structures first, then moving in two directions across the pedestal pushing water out the two open sides.

The sequence for a hole was as follows:

1. Pull the pipe 30 to 60 cm off the bottom of the hole and wash with water to be sure it is clear.

2. Begin the hole by pumping one batch of neat grout

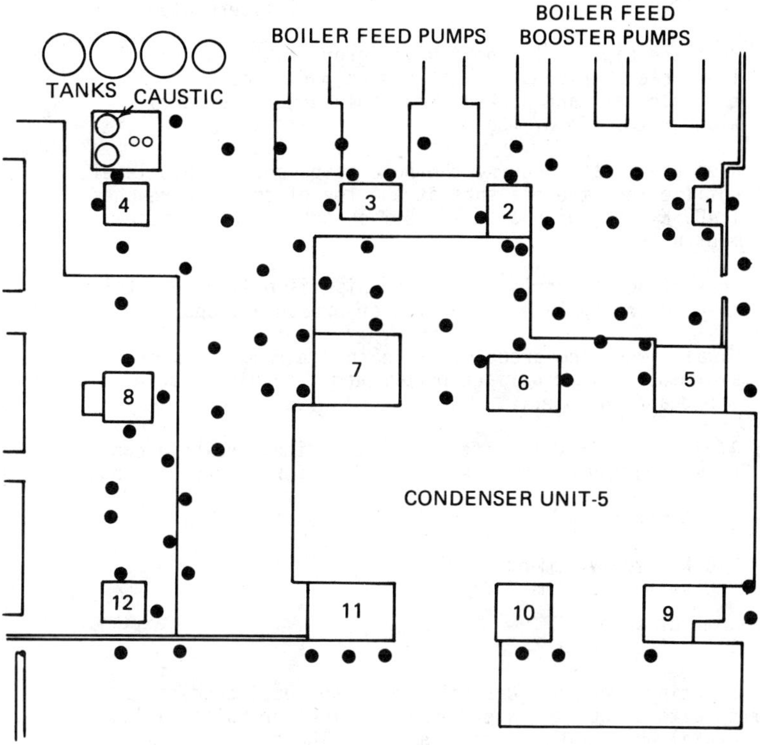

Figure 7. Arrangement of Grouting Holes for Unit 5

($1/2\ m^3$), one batch neat grout mix = 85 Kg Type-V cement
90 Kg Type-S lime
160 liters water

3. Pull the pipe up the hole until grout takes at under 1,400 KPa. This usually occurs at 5-6 m depth. Since the hole was sand packed grout only enter the formation at the tip of the pipe.

4. When grout take occurs, the hole is pumped at that level until either the pressure at the top of the hole reaches 1,400 KPa psi or uplift of .5mm occurs at any measurement point.

5. When either criteria occurs the pipe is pulled up until take occurs again with no lift or pressure drops.

6. Usually when the grout point reach 3 m from the surface a blowout at the surface occurs and the hole is abandoned and cemented.

7. If it is desired to lift at a later time, the hole can be washed and a 3 m pipe cemented in for future lifting.

The grout mix per m^3 for general areas consisted of:

170 Kg Type-V cement
180 Kg Type-S lime
$.3m^3$ of sand
320 liters water

The grouting procedure described assumed ideal conditions. It was often altered due to interference from other holes, a lens crossing several holes or very fast uplift. The grout consistency was often changed to reduce travel distance by reducing the water/cement ratio in the mix. The normal consistency would be similar to that of a thick milk shake.

The grout take was approximately 300 cubic meters per turbine pedestal; this amounts to approximately 6 percent of the total volume under the pedestal. The maximum laboratory measurement of volume change in the mudstone was 21 percent. Assuming that 30 percent of the soil column is compressible mudstone then a total possible volume change is approximately .21 x .30 = 6.3 percent.

In addition, 91000 Kg of lime was placed beneath each structure. This amounts to 91000 Kg/2,273,000 Kg mudstone or 4 percent lime by dry weight. The lab tests indicate that 2 percent lime is adequate to halt settlement.

The problems of working in an operating power plant are

many, but the most difficult ones were the time and space constraints. The grouting operations had to be performed at the start of a scheduled overhaul of the units. The turbines were cooling and moving constantly. It was attempted to stabilize the area beneath the turbine while lifting less than 2 cm and keeping the upper deck planar.

It was necessary to place approximately 300 m^3 of grout in a confined area with less than 3 m overhead clearance in less than 10 days without damaging any surrounding machinery. The adjacent units were running throughout the operation. There was evidence during test grouting that travel distance for the thin grout of up to 30 m was possible.

To complete the operation in the required time took two grouting crews (pumps and mixers) working two ten-hour shifts with a maintenance crew between shifts.

3. Monitoring Program

Due to the tight control of lifting on the project a monitoring system was required. A digital readout controlled temperature and pressure manometer system built by English Electric was installed (called Manometric System). The system consisted of a hydraulic slave measuring point at each column on the turbine pedestal (12 on each unit). The system reads and records a relative level measurement to the nearest .1 mm. It can be read at any interval desired. It was found that 1/2-hour readings during grouting were adequate.

In addition, several second order surveyor's levels were used to monitor adjacent machinery. Complete level traverses were performed approximately every three days to check all readings and establish absolute elevations.

Using these techniques, it was possible to grout the two turbine pedestals while lifting the lowest column 1.5 cm on Unit 5 and 1.3 cm on Unit 4.

Semi-annual level traverses have been performed through the plant since 1969 to identify future problems and monitor grouted structures. Fig. 5 shows some typical plotted data. It can be seen from these data that no measureable movement has occurred on Unit 5 in over three years since the grouting was performed. Previous grouting with Type-2 cement, stiff grouts show settlement rates slowed but continuing.

CONCLUSIONS

The increasing energy demand will require considerable future energy development in the Southwest United States. The weathered sedimentary clayey deposits are affected by addition of fresh water. Since the large power generators are extremely sensitive to movement, we hope that this study can help others avoid similar situations.

This investigation suggests that proper grouting can be a viable remedial method to halt chemically induced settlement. The techniques and additives can be adapted to other areas if the settlement mechanism is similar. Any formation high in crystalline gypsum in arid regions should be suspect to this reaction, especially if foundations are placed in the weathered rock-soil zone. However, it is also learned in the laboratory and field tests that the area to be grouted and the grout to be used should be thoroughly investigated. Any upset in the soil chemistry equilibrium can have drastic results. It should be noted that construction on this site using drilled pier foundations placed through the weathered zone (10 m) have performed flawlessly.

ACKNOWLEDGEMENTS

We wish to thank the Southern California Edison Company and the Arizona Public Service for the patience and understanding while these studies were being performed. The grouting procedures were developed and carried out by a project team consisting of J. Richards, D. Grime, R. Baker, D. Downs and D. Hinkle of SCE. Without the team effort this project would not have been possible. In addition, the personnel of Prepakt Concrete Company of Mesa, Arizona supplied grouting expertise during the progress of this project.

REFERENCES

1. Bull, W. B., "Alluvial Fans and Near Surface Subsidence in Western Fresno County, California", USGS Professional Paper No. 437A, 1972.

2. Degens, A. T., "Geochemistry of Sediments, A Brief Survey", Prentice-Hall, Englewood Cliffs, New Jersey, 1965.

3. Hinkle, R. D., "Report of Testing Group Program, Ash Bin Structure, Four Corners Generating Station", Southern California Edison Co., Rosemead, CA. 1978.

4. Hinkle, R. D., "Chemically Induced Settlement in the Four Corners Area of New Mexico", Thesis submitted for the Master of Science degree, Department of Civil Engineering, California State University, Long Beach, CA. 1978.

5. Hinkle, R. D., "Report of Geotechnical Conditions, Case History and Settlement Mechanism, Four Corners Generating Station, Farmington, New Mexico", Southern California Edison Company, Rosemead, CA. 1977.

6. Millott, G., "Geology of Clays", Springer-Verlag, New York, 1970.

Grouting Caverns and Soft Zones by Concrete Pumps

By C.A. Ortiz-Suarez[1], M ASCE, J.C. Agrelot[2], AM ASCE, and J.M. Cardona[3], AM ASCE

Abstract

An economical method of pressure grouting of caverns in limestone and limestone related soils is presented. Use is made of 6"Ø auger holes and concrete pumping equipment to rapidly inject portland cement grout into relatively open caverns, or very loose soils related to limestone caverns. Two case histories and a simplified analysis are presented. The analysis includes the effect of soft soil compaction and the reinforcing effect of grout columns.

Introduction

The northern third of the island of Puerto Rico is underlain by thick deposits of tropical marine limestones and other genetically related deposits. These are characteristically sandy limestones, ranging from pure and cristalline to chalky, silty, and earthy. The surface of the limestones is case hardened (recrystallized by percolating rain water) as are portions of its interior.

Chalky, silty and earthy portions of the limestone comprise most of the rocks involved. These are composed of hard calcium carbonate fragments in a calcareous ooze matrix. Extensive field and laboratory tests performed on the limestone (15) for a nuclear power station, revealed that the soft limestone behaves more as a soil than as a rock and had the following properties:

Aymamon Limestone

$$\gamma_d = 88 \text{ to } 102 \text{ pcf (ave.} = 95 \text{ pcf}$$
$$w = 17 \text{ to } 37 \text{ percent}$$
$$q_u = 2.4 \text{ ksf (ave.} = 2.1 \text{ ksf)}$$
$$C_{cu} = 1.2 \text{ ksf}$$
$$\phi_{cu} = 24°$$

Some of these rocks have developed striking karst geomorphs on a broad belt next to the northern coast of the Island and are covered by a complex of alluvial reworked blanket deposits. Some of these are nearly pure silica sand and are known as blanket sands. Various test borings at different sites of the blanket deposits show these to be of varying lithology and of different origin.

At some locations these blanket deposits are experiencing vertical erosion into enlarged joints and caverns of the underlying lime-

1 Senior Consulting Engineer, Ortiz, Agrelot & Assoc, San Juan, P.R.
2 Senior Consulting Engineer, Ortiz, Agrelot & Assoc, San Juan, P.R.
3 Geotechnical Engineer, Ortiz, Agrelot & Assoc, San Juan, P.R.

stone. This geologic process is of variable speed, depending on rainfall and/or runoff concentration.

Formerly, when the land was used primarily for agricultural purposes, the sudden formation of sinkholes in the blanket soils was of little concern or consequences, but with the growing industrial use of land (increasing runoff concentration and seepage) the problem has acquired new importance.

Previously Available Alternatives

Typically, the sudden development of sinkholes has been dealt by using one or more of the following methods:

a- Replacing lost soil with additional soil
b- Replacing lost soil with a concrete plug
c- Pressure grouting with conventional equipment

Design of buildings over sinkholes has been dealt by:

a- Using stiffened mats to bridge over sinkholes
b- Using pile foundations
c- Pressure grouting the sinkholes

If these alternatives were too expensive for the owner, the building was relocated. Whenever sinkholes were found during the construction stage, these were either pressure grouted or plugged with concrete.

The above mentioned alternatives vary significantly in cost and long term performance. Usually, replacing lost soil with additional soil is merely a temporary solution and also the least expensive of the alternatives. It is generally used in green areas and drainage ditches where, if necessary, it can be easily replaced. Occasionally driveways are also repaired using replacement soil.

Whenever structures or proposed structures are affected by the presence of sinkholes, these holes may be either plugged with concrete or pressure grouted. Concrete plugs are usually formed, after having cleaned the hole of soft soil and debris, by pouring 6 inch slump, low strength, concrete. Sometimes the concrete is brought-up to the surface of the sinkhole but generally the plug is only as high as its least lateral dimension. The sink is then backfilled with well compacted soil to the ground surface.

Pressure grouting is a more expensive solution. As practiced locally it involves drilling closely spaced 2.5"Ø groutholes, and pumping sand-cement or neat cement grout through 1.0" or 1.25"Ø pipes into the soft soils, caverns, and fissures. Aluminum powder is used to eliminate shrinkage of in place grout. Specifications developed by the U.S. Corps of Engineers, modified for local availability of materials and equipment, are used. Generally square grids are used for grouting, with hole spacings between 4 and 10 feet.

Pile and mat foundations have occasionally been used to bypass sinkholes. However, these alternatives have not been used extensively due to uncertainties of additional sinkhole development. The presence of underground piping under and adjacent to structures, normally makes engineers leery of permitting unplugged sinkholes or soft zones to remain untreated, as any leakage from broken pipes will tend to enlarge the sinkhole.

When a piled foundation is used it generally utilizes either steel H-piles or precast concrete piles (round or square - 10 inches wide), which are available locally. These are driven to refusal into the limestone below, or to a criteria developed using conservative driving formulas. However, a wide variation in pile lengths must be expected due to differential weathering of the limestone.

Case History No. 1

Pump grouting was first employed to blanket grout a large cavern revealed during the subsoil exploration for a proposed "clean" warehouse. Figure No. 1 shows the approximate location of the cavern in a plan view.

Originally the cavern was to be pressure grouted using sand-cement grout and conventional grouting equipment. The high estimated cost of the operation forced an alternative solution to be seeked.

At first, 6"Ø holes were drilled through the cavern roof using hollow stem augers on a 10 feet grid, and ready mixed grout was poured directly from the truck. This procedure left doubts as to the extent of grout penetration since the cavern was filled with muck from previous roof falls. We were especially concerned with the ability of the grout to either penetrate or pack the muck.

After the first two holes were completed, this method was discontinued and all the remaining groutholes were fitted with 4"Ø black steel pipe nipples approximately four feet long. These nipples were threaded at the top using standard threads. Nipples were then grouted in place with mortar, and to prevent the latter from falling into the hole a No. 3 segment of rebar was tacked around the bottom of the pipe. A reducing adapter was fabricated to connect a standard 3"Ø concrete pump hose to a 4"Ø standard pipe. This adapter was then screwed on the nipple and used to pump ready mixed grout into each hole. Figure No. 2 shows a typical header for this job.

A "Challenge Squeez-Crete", Model 700 truck-mounted concrete pump was used on this job. This unit is capable of developing a pumping pressure of 200 to 250 psi during continuous operation. All holes were grouted until either the nipple was jacked off the ground or the ground around it was broken and grout started to seep out.

The grout used was a locally available sand-cement grout, with a retarding agent to delay setting. It was pumped with a slump of 9 to 11 inches, containing a manufactured sand with the following properties:

Sieve Size or Number	Percent Passing
3/8"	100
4	99
8	89
16	62
30	38
50	18
100	6
200	3

Specific Gravity = 2.56
Fineness Modulus = 2.88
Absorption = 1.5%
Rodded unit wt. = 103 pcf
Soundness (Na_2So_4) =3.84% loss (weighed)

This grout contained 7.89 bags of cement per cubic yard of ready mixed grout and a water/cement ratio of 0.90. It was then thinned to a water/cement ratio of 1.5 to counteract the expected migration of water when in contact with dry soils at site.

Typically a hole was pump grouted after allowing 24 hours for the superficial nipple grout to set; and grouting was performed without pressure washing the hole.

The operator controlled the pumping rate by a hand throttle. On this first job no pressure readings were taken and the inspector controlled injection rates by the observable rise of the nipple.

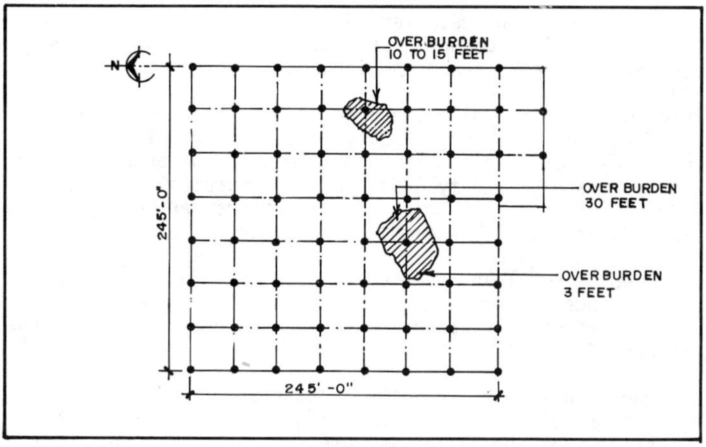

FIGURE NO. 1 PLAN VIEW OF CAVERNS UNDER "CLEAN" WAREHOUSE, CASE HISTORY NO. 1

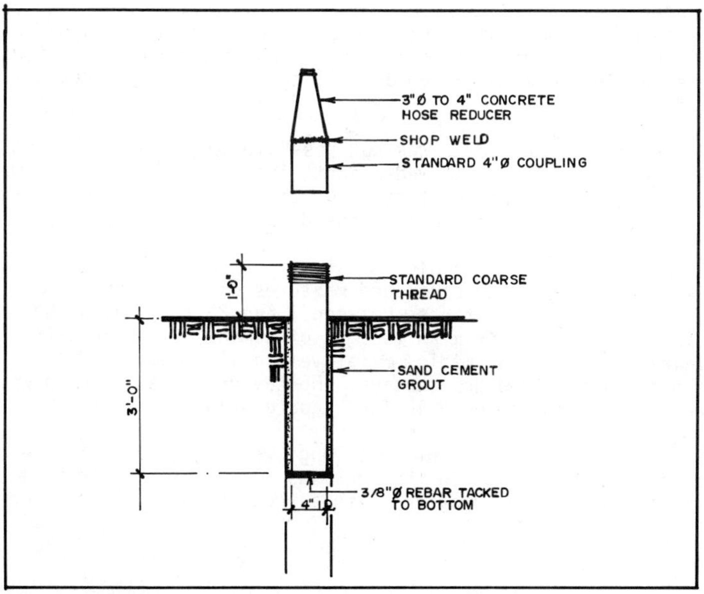

FIGURE NO. 2 TYPICAL GROUT HEADER FOR INITIAL JOB

Thus, whenever the nipple started to come out of the ground the rate was decreased. This procedure was totally based on the subjective judgement of the operator and the inspector.

Table No. 1 in the following page shows typical grout hole recap sheets for this project. During this job a total of 273 cubic yards of sand-cement grout were pumped into the caverns in 6 working days. Approximately, twenty cubic yards of surplus grout had to be dumped into a nearby waste hole. Grout intakes per hole were estimated and adjusted to the remainder of the truck load.

At present the "clean" warehouse is in satisfactory operation as it has been since construction 3 years ago. No distress is apparent anywhere in the building and no loss has been detected in the positive air pressure system.

Subsequent to this project the grout header was modified to include a pressure gage and a return line. However, the system is not well suited for the latter improvement, as will be discussed below.

Case History No. 2

Early in 1978, the exploration for a large cooling tower at an industrial plant at Barceloneta, Puerto Rico, revealed a large soft zone under the proposed site. Examination of maintenance records of the plant, showed that sinkholes had formed at least twice in the 70 by 90 feet area available for the construction of the tower. These sinkholes were corrected by filling with local blanket soils to grade.

Relocation of the tower was not possible due to plant layout, and a nearby diesel tank showed slight distress. Therefore it was decided to grout the zones stressed by the proposed cooling tower (24 ft. x 26 ft.) and the edge of the diesel tank.

Borings at the site sampled moist soils, partially saturated, red fine silty sands and silt/sand mixtures of very loose density. Penetration resistances between 5 and 25 feet below the existing grade ranged from WH (weight of hammer) to 4 blows/ft. Surficial soils were of medium relative density probably due to surface compaction and desiccation. No caverns were found in any of the borings drilled on a 10 feet square grid.

Black steel nipples, four feet long, were again used in 6"Ø auger holes as described in the previous case history. In this project the header was provided with a pressure gage at the adap-

Table No. 1 - Typical Pump Grouting Recap Sheets

Hole No.	Start of Grouting	End of Grouting	$Yards^3$ Injected	Date	Observations
301	3:22 PM	3:23 PM	0.50	2/5/78	
203	3:24 PM	3:33 PM	3	2/5/78	
222	3:33 PM	3:35 PM	0.50	2/5/78	
319	7:43 PM	12:47 PM	40.5	2/14/78	
316	12:49 PM	2:06 PM	8.5	2/14/78	
317	2:10 PM	2:15 PM	0.5	2/14/78	
315	2:18 PM	2:25 PM	1.5	2/14/78	
310	2:30 PM	3:30 PM	13	2/14/78	
310	8:05 AM	8:07 AM	0.50	2/15/78	Excessive pressure ended grouting
311	8:17 AM	8:18 AM	0.50	2/15/78	Grout, seep under excessive pressure
312	8:25 AM	8:27 AM	0.50	2/15/78	Excessive pressure forced end of grouting
313	8:32 AM	8:33 AM	0.50	2/15/78	
305	8:38 AM	8:38:30	0.50	2/15/78	
306	8:44 AM	8:53 AM	3	2/15/78	
307	8:57 AM	9:29 AM	7	2/15/78	
304	9:34 AM	10:30 AM	3	2/15/78	
220	10:54 AM	10:57 AM	0.5	2/15/78	
214	10:59 AM	11:01 AM	0.5	2/15/78	
215	11:01 AM	11:03 AM	0.5	2/15/78	
309	8:39 AM	9:05 AM	5.0	2/17/78	Pipe was jacked out of the hole

ter (3"Ø to 4"Ø reducer), and with a 3"Ø concrete hose as the return line.

A total of 44.5 cubic yards of ready mixed grout were pumped into twenty five holes at the site of the cooling tower and next to the diesel tank. At first, a series of fourteen holes were drilled and grouted for a initial grout take of 12.5 cubic yards. This phase of the project was hampered by lack of coordination and supervision. When these problems were ironed out a second series of holes were drilled and grouted, with a secondary grout take of 32.0 yards, pumped in about half a day of work.

Grout take per hole varied from approximately 0.25 cu. yds. to 5.0 cu. yds., again these volumes were subjectively calculated by the inspector. Grouting of a hole was usually stopped when grout started to seep through radial or circular cracks in the ground.

Extensive ground heave was observed with pressures over 50 psi.

As mentioned earlier the use of return valves created considerable problems. These valves were used only during the first phase of the project because they clogged very easily at a 90° bend. Thus, frequent delays were produced by disassembling this mechanism. In addition to this, using low injection rates produced long waiting periods for each 7 to 10 cu. yds. truck.

After grouting, the cooling tower was supported on a rigid mat, proportioned for an allowable bearing pressure of 1 ksf. To date the structure has performed well for 2.5 years without showing signs of distress.

Analysis

Pump grouting can help improve foundation soils by:

1. Filling Caverns
2. Compacting soft or loose soils
3. Earth Reinforcement due to frictional properties of mortar columns and seams.

These items will now be discussed:

1. Filling Caverns -

   The use of grout to fill mines has been extensively reported (1, 8, 10) and is a common practice in built-up areas suffering from subsidence due to shallow mining. In these areas the practice is to drill holes through the roof of the mine, then crushed stone is used to form a loose pillar in the mine, and finally the pillar is grouted using conventional pressure grouting techniques.

   On the other hand, pump grouting uses a thicker grout and pressurizes the entire cavern full of grout. Mine subsidence problems are very different from sinkhole problems, considering that mines have rock roofs and that pillars are accurately mapped, man-made features. Pillars can be replaced and their span reduced to avoid roof falls with a fairly high degree of confidence.

For sinkholes to endanger structures requires a soil or very weak rock roof which loses strength during periods of heavy infiltration. The span supported may also increase due to internal erosion. The conditions are very complex and almost impossible to analyse. Is for this reason that local engineers tend to fill the caverns completely, instead of building pillars in them.

2. Compacting Soft or Loose Soils -

The pump grouting procedure herein presented is analyzed as a radially expanding column of liquid with a weak cap.

Figure No. 3 shows the simplified analysis used to calculate the system's limiting pressure, which in this case is the pressure required to blow out the nipple. Local blanket deposits consist of red fine silty sand (SM), with about 10% finer than 2μ material, and with occasional lenses of red to brown sandy silts and clays. In these types of soils the ground started to heave at pressures ranging from 60 to 120 psi, which was the maximum pressure allowed by the system. However, higher pressures could be applied by using longer nipples and thus more soil cover over the grout. Nevertheless, soil or voids above the bottom of the nipple would not be compacted or grouted in that portion.

Anderson (2, 3) in 1935 and 1937 studied the injection and development of dikes, ring dikes, sills, and cone sheets of magmatic fluids into rocks using mathematical models. Results of these studies may be directly applied to our present case. If lateral soil pressures along the sides of the grouthole are less than the radial pressure of the grout then the sides will move outward. Zones which are significantly softer will move outward more rapidly, subjecting soils to lift on the upper soil-grout contact. This effect should then be able to lift the overburden creating a horizontally expanding crack under internal pressure (Griffith Crack). This crack should not extend horizontally for a long distance as the tensile stresses at its tip are greatest in slant planes at 60° to the horizontal. Therefore forming grout "sills" which will then dip either upward or downward to become a grout "dike". Based on this analysis the nipple should also rise creating tensile cracks,both radial and circular around the pipe. Both types of cracks have been observed by the writers in the field.

Pressures of the order of 50 to 100 psi result in significant plastic deformation in loose soils. As an

VOLUME OF SOIL PUSHED UPWARDS $1/3(\pi r_1^2)h_1 - 1/3(\pi r_2^2)h_2 = 11.2$ CU.FT.

WEIGHT OF SOIL PUSHED UPWARDS = 11.2 (100) = 1,120 LBS.

SURFACE AREA OF TRUNCATED CONE = $\pi r_1 \sqrt{r_1^2 + h_1^2} - \pi r_2 \sqrt{r_2^2 + h_2^2} = 21.20$ SQ. FT.

A) IF C = 250 P.S.F. AND $\phi = 0$
FORCE TO LIFT NIPPLE = WEIGHT OF SOIL + FRICTIONAL RESISTANCE
= 1120 + 21.20 (250) = 6420 LBS.

∴ HEADER PRESSURE = 6420 / .4416 = 14,539 P.S.F. OR 101 P.S.I.

B) IF C = 500 P.S.F. AND $\phi = 0°$ FORCE TO LIFT NIPPLE = 1120 + 21.20 (500) = 11720 LBS.
AND THE HEADER PRESSURE IS 11720 / (.4416) 144 = 184 P.S.I.

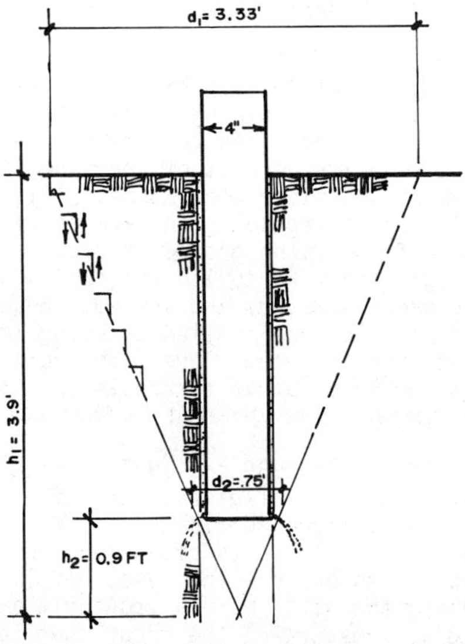

FIGURE NO. 3 SIMPLIFIED ANALYSIS OF SYSTEM'S LIMITING PRESSURE

example of this, grouthole take in Case History No, 2, where no cavern was found, was always over 0.25 cu. yds. which is about twice the initial auger hole volume of 0.14 cu. yds. All in all a decrease of porosity of over 3.5% was attained, even when ground heave is accounted for.

3. <u>Earth Reinforcement due to frictional properties of mortar columns and seams</u>.

A composite material is created which has anisotropic properties. Reinforcement of the soil mass depends on the continuity, strength and spatial arrangement of columns and seams. Usually these are injected in a 10 feet square grid, proceeding from one group of adjacent groutholes to the next so as to minimize grout hose and truck movement. In practice, this means that adjacent groutholes will be grouted while in a plastic state and squashing or break-up of grout columns may result.

These initial holes usually have much greater grout takes than the secondary holes drilled at split spacing, which should suffer less from the effect of squashing and break-up, being drilled and injected into a partially stabilized soil

For the purpose of analysis we assume that: a) secondary groutholes are unbroken, b) these retain the shape of the original auger hole, and c) the concrete has a strength (f'_c) of 1000 psi. Primary groutholes are disregarded in the analysis as is the strength gained during primary grouting.

Secondary grout columns serve as rudimentary friction piles which derive their support from the stabilized soil around it. We consider the upper five eights of each column as being embedded in a soil "pile cap" while only the lower portion, three eights, is deriving usable shear resistance. Using this approach a 25 feet long column resists an ultimate load of 3 tons.

Discussion

The method presented is an economical way to pressure grout caverns and some soft zones with ready mixed grout. The use of large displacement concrete pumps does not permit grouting where thin fissures or small passageways are open to the grout, as the pump builds up pressures very rapidly pushing the nipple out of the ground. For this reason thin fissures and small passageways will remain ungrouted.

In practice this deficiency has been reduced by drilling groutholes at spacings slightly over 5 feet. Generally the initial spacing is 10 feet with secondary holes being drilled later at split spacing. Since secondary groutholes are drilled with hollow stem augers, this permits the verification of improvement due to primary grouting. In some cases tertiary holes may be drilled where secondary grouthole takes are high.

The method has the advantage of using readily available equipment and services for which competitive bids may be obtained. Another advantage is that grouting proceeds very rapidly with injection rates of one cubic yard per minute being common.

The local cost of pump grout is in the order of $100 to $120 per cubic yard including drilling and nipple installation. The local cost of conventional pressure grout is around $250 per cubic yard. However, it should be noted than the owner is obtaining two different products for its money, when soft zones are grouted, due to the differences in the ability to penetrate.

The only difference in the way these two methods fill a cavern is that conventional pressure grout does not shrink upon setting as happens with pump grout, although the latter could be made a non-shrink grout by the use of suitable additives.

As a word of advice readers should be aware that we had a very bad experience with pump grouting of a sinkhole under an existing building. The work was being carried outside of the operating process building. This is a pre-engineered building 3-stories high with concrete footings, slabs on ground and ten feet high masonry walls, above which aluminum siding is used.

While pumping, and in a matter of 30 seconds, parts of two 20 ft. x 20 ft. slabs were raised four inches in the center, jamming partitions and doors into the hung ceiling of offices. The concrete slab failed in shear along the center of the slab creating a rough step about 2 inches high.

In this case an outer perimetral concrete apron had been angle hole drilled and the nipple was grouted to the wall, allowing 250 psi pressures to develop without noticeable movement of the nipple.

Based on this experience we now avoid pump grouting near existing structures.

REFERENCES

1. Ackenheil, A.C. and Dougherty, M.T., "Recent Developments in Grouting for Deep Mines", Journal of the Soil Mechanics and Foundations Division, ASCE, Vol. 97, No. SM 1, Proc. Paper 7042, January 1970, p. 251.

2. Anderson, E.M., "The Dynamics of the Formation of Cone-Sheets, Ring, Dykes, and Cauldron-Subsidencies", Proc. Roy. Soc. Edinburgh 56, 1935, pp. 128-157.

3. Anderson, E.M., "The Dynamics of Sheet Intrusion", Proc. Roy. Soc. Edinb., 58, 1937, pp. 242-251.

4. ASCE, Geotechnical Group, Lehigh Valley Section, "Symposium on Engineering, Construction and Maintenance Problems in Limestone Regions", August 1976.

5. Brown D.B. and Warner, J., "Compaction Grouting", Journal of the Soil Mechanics and Foundations Division, Vol. 99, No. SM 8, Proc. Paper 9908, August 1973, p. 589.

6. David, D., Sroka, E., and Golderberg, M., "Small Diameter Piles in Karstic Rock", Proc. IX International Conference on Soil Mechanics and Foundation Engineering, Vol. 1, 1977, p. 471.

7. Graf, E.D., "Compaction Grouting Technique and Observations", Journal of the Soil Mechanics and Foundations Division, Vol. 95, No. SM 5, Proc. Paper 6766, Sept. 1969, p. 1151.

8. Gray, R.E. and Meyers, J.F., "Mine Subsidence and Support Methods in Pittsburgh Area", Journal of the Soil Mechanics and Foundations Division, Vol. 96, No. SM 4, Proc. Paper 7407, July 1970, p. 1267.

9. Jennings, J.E., et al., "Sinkholes and Subsidences in the Transvaal Dolomite of South Africa", Proceedings of the Six International Conference on Soil Mechanics and Found. Eng., Vol. 1, 1965, p. 51.

10. Mansur,C.I. and Skouby, M.C., "Mine Grouting to Control Building Settlement", Journal of the Soil Mechanics and Foundations Div., Vol. 96, No. SM 2, Proc. Paper 7166, March 1970, p. 511.

11. Sowers, G.F., "Failures in Limestones in Humid Subtropics", Journ. of the Geotech. Eng. Div., Vol. 101, No. GT 8, Proc. Paper 11521, Aug. 1975, p. 771.

12. Sowers, G.F., "Settlement in Terranes of Well-Indurated Limestone", Short Course on Analysis and Design of Building Foundations, Lehigh University, Penn., Envo Publishing Co.

13. Warner, J. and Brown, D.R., "Planning and Performing Compaction Grouting", Journal of the Geotech. Eng. Div., Vol. 100, No. GT 6, Proc. Paper No. 10606, June 1974, p.653.

14. Warner, J., "Compaction Grouting", - A Significant Case History", Journal of the Geotechnical Engineering Division, Vol. 104, No. GT 7, Proc. Paper No. 13897, July 1978.

15. Yang, Z. and Hatheway, A.W., "Dynamic Response of Tropical Marine Limestone", Journal of the Geotechnical Engineering Division, Vol. 102, GT 2, Proc. Paper No. 11893, Feb. 1976, p. 123.

A DIGEST OF TYPICAL CEMENT GROUTING TAKES

by Adam Clive Houlsby, M. ASCE.

ABSTRACT

This paper has the purpose of giving actual water and cement takes in extensive grouting operations carried out at several dam sites having fairly typical foundation conditions.

The information is presented in digest form to provide a convenient reference for use by those who might wish to compare their own sites with these. However, in such comparisons, the individual vagarities and idiosyncrasies of each site must be kept in mind. Indeed, close inspection of the data presented here for any one of these dams will emphasise the vagarities commonly met with.

THE DAM SITES

The three dams in this digest were selected because they have been reported on and described elsewhere, which enables amplification of the data given here. The size of each dam, its foundation geology, the type of grouting, the w:c ratios used, the number of cases of grout applications and the performance of the completed grouting are detailed in (3) which should be read in conjunction with this paper. All three dams are of the central earth core embankment type and are located in Australia.

COMMENTS ON THE FIGURES

The data presented concerns -

(i) water taken in standard water tests carried out at 15 p.s.i. (1 bar) in each stage of each hole before the first grout application to the stage. The method is described in (4). The take is presented in lugeon units; comments on these will be found in (2).

(ii) cement taken during the grouting. This is the actual amount injected down into the holes. In some of the figs., a second amount is also shown if holes are excluded which lost cement to waste through cracks or connections. Takes are shown in cubic feet of cement per foot length of hole. The metric equivalent of this, if measured in litres of cement per centimetre, has the same numerical value.

The figures are intended to be self-explanatory. Textual comment will therefore be brief.

Pregrouting Permeabilities and Overall Cement Takes

Figs. 1, 2 and 3 show these for the three dams in terms of sections of the curtains. The pregrouting permeabilities are those measured in the primary holes before grouting. All the curtain stages are combined.

All three curtains have considerable variation in takes along their lengths. Cement takes show little relationship to permeability.

Engineer Specialising in Grouting, and Senior Surveillance Engineer, Water Resources Commission, North Sydney, N.S.W. Australia.

Effects of Successive Closures

Figs. 4, 5 and 6 compare average permeabilities before grouting of each of the closure sequences, and then they compare the changes in cement take as the closure sequences proceed. This data is given in two ways -

(a) for individual sections of the curtains;
(b) for the whole dam. The curves showing this are taken from figs. 5 and 6 of (3).

Taking averages for the whole dam can give a quite different picture to that obtained by averaging the individual sections. Whereas the former can give the classical diminishing of takes as closure proceeds, the latter can provide exceptions where some of the takes in the later steps of closure exceed earlier ones. This lack of conformity to the classical expectations becomes more apparent if the statistical sample includes even less holes until in the ultimate, where individual holes are concerned as in fig. 12, there can be marked departures from diminishing of takes as the result of closure. However, there can also be plenty of quite satisfactory cases of the classical pattern!

Variations with Depth

Figs. 7 and 8 show for two of the dams, the takes in the various depth stages of the curtains. The data is presented in much the same way as in figs. 1, 2 and 3 except for the dissection into stages. All sequences of closure are, of course, included. The 0 to 68 ft. stage for Copeton Dam includes a shallower stage than this used with some of the final closure holes. Not all sections of the curtain at this dam were taken to the third stage and hence there are gaps in the data there. Where the takes are based on fewer than 10 cases, this is indicated by an asterisk.

Permeability and cement takes sometimes reduced with depth but sometimes they did the opposite.

Takes in Multiple Applications

In figs. 9, 10 and 11, the cement takes in multiple applications are shown. Multiple applications at intervals of 24 hours minimum are given to tighten up further, following any applications with a sufficiently large take that an undesirably large quantity of water has been injected. The takes shown on these figs. for the 1st application relate to every grouted stage. Because subsequent applications were only given where the 1st application showed the need, as indicated by a take exceeding 0.25 cubic feet of cement per foot, the takes shown for 2nd applications are for a lesser number of cases than for the 1st applications. Likewise, the same criterion was used to decide when further applications were needed and the number of cases therefore reduced until by the 5th application there were very few. Applications still taking grout freely after 100 bags were arbitrarily terminated at that amount and then resumed 24 hours later.

Worthwhile quantities were injected in the multiple applications. Fig. 11 shows at 4th and 5th applications the type of runaway takes which can result from rock movements incorrectly handled. These movements were not arrested promptly. They were however effectively handled later but by then a lot of cement had been wasted.

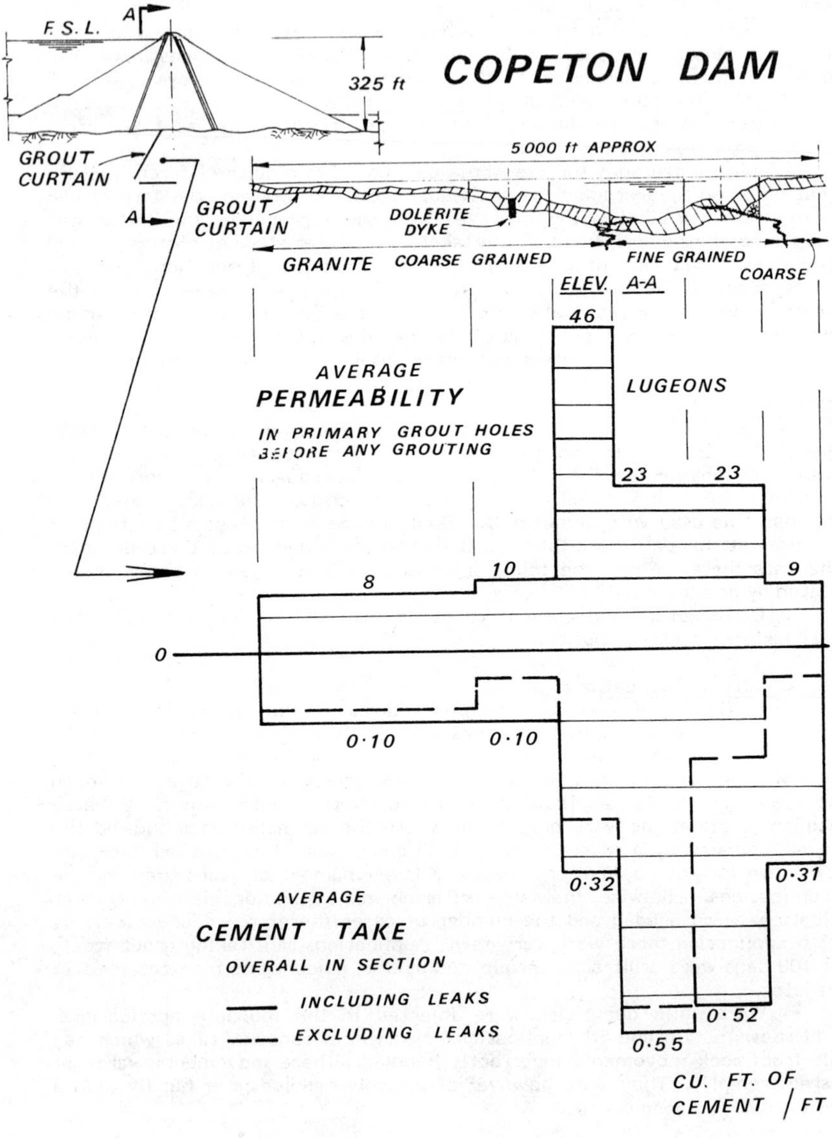

FIG 1 PRE-GROUTING PERMEABILITIES & OVERALL CEMENT TAKES

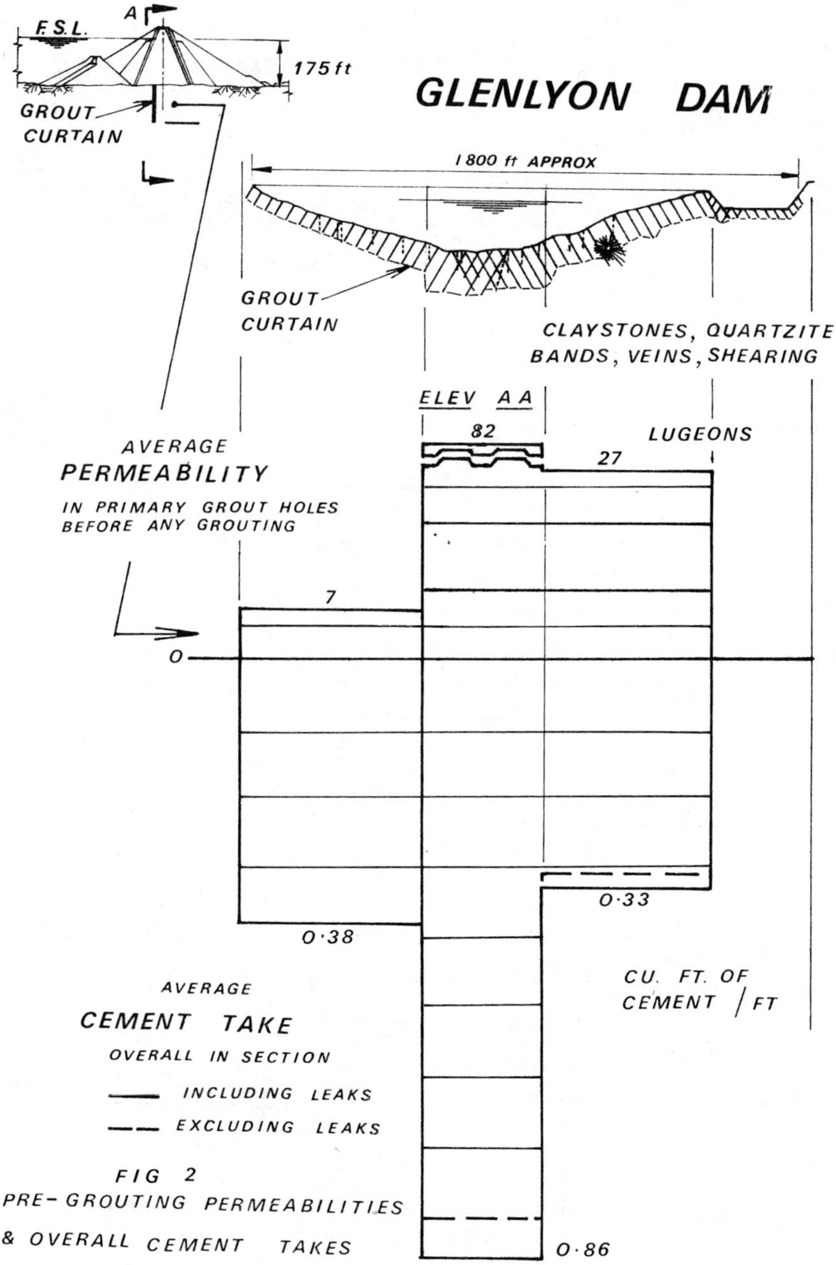

FIG 2
PRE-GROUTING PERMEABILITIES & OVERALL CEMENT TAKES

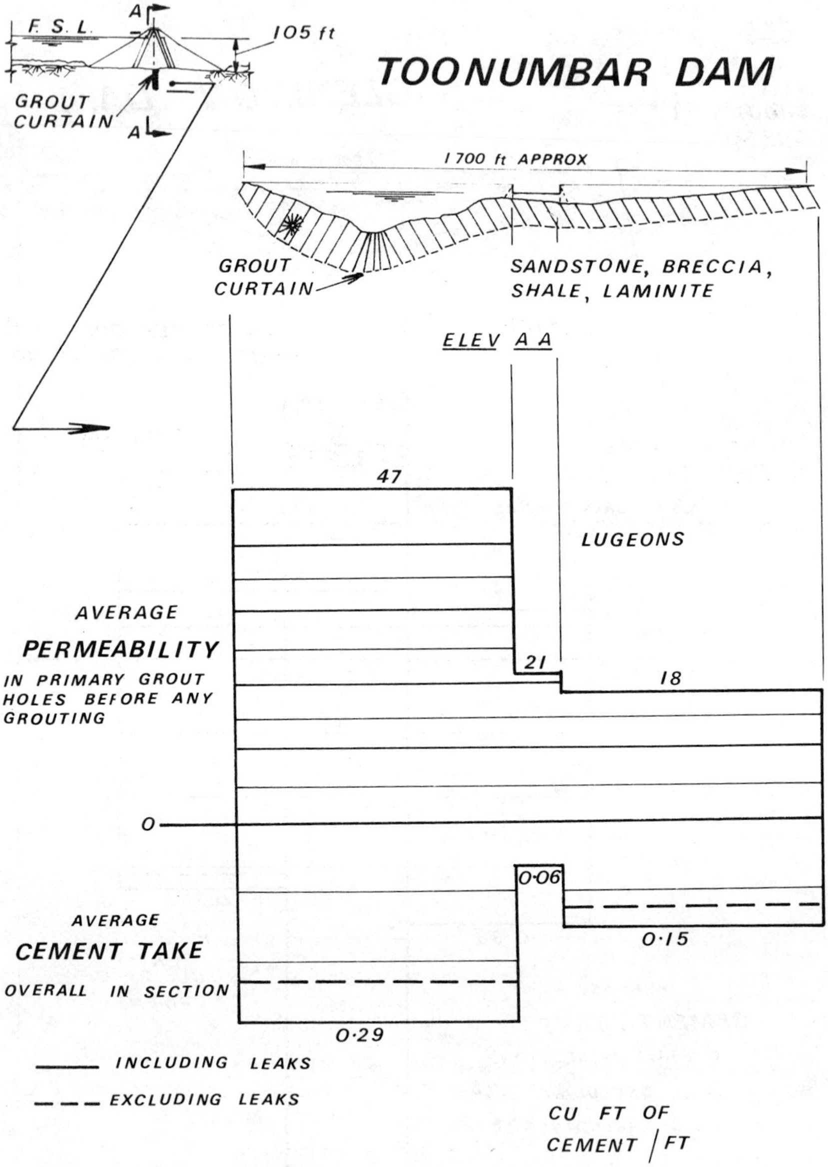

FIG 3 PRE-GROUTING PERMEABILITIES & OVERALL CEMENT TAKES

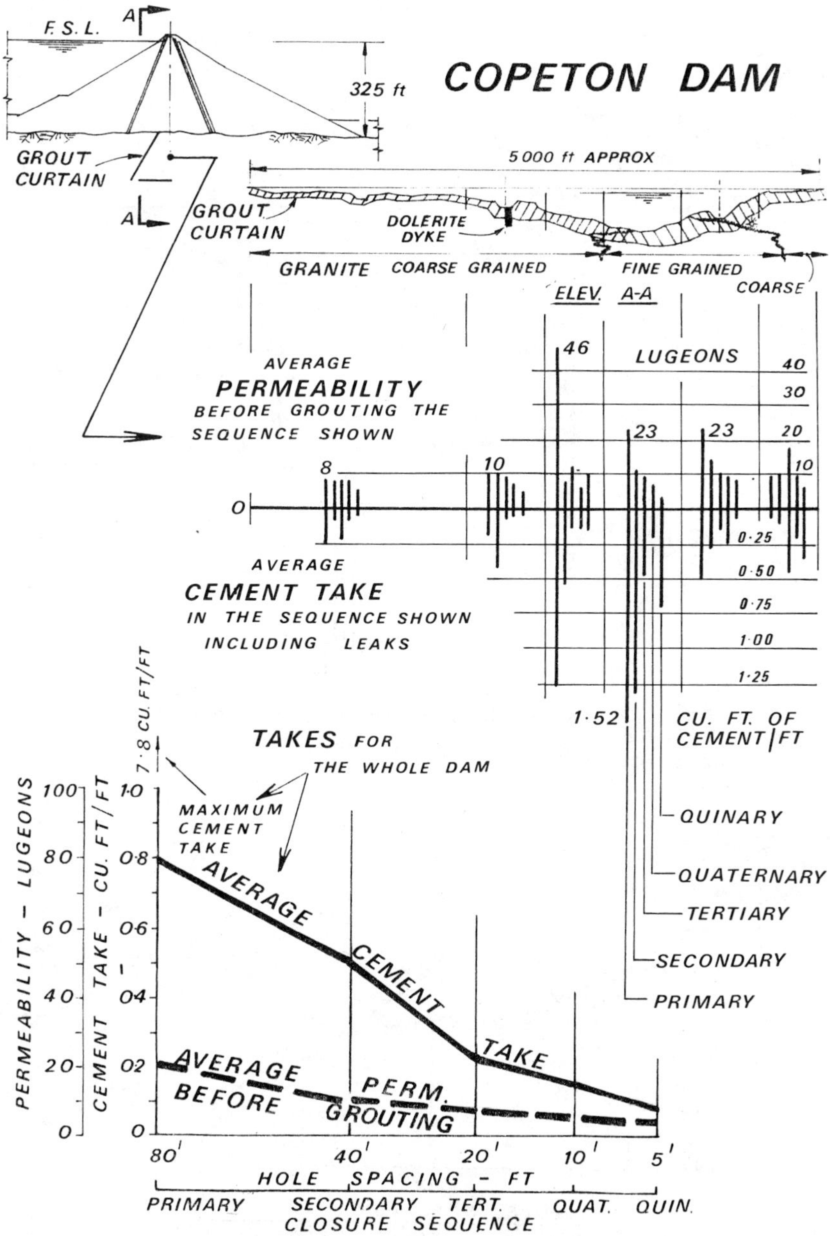

FIG 4 EFFECTS OF SUCCESSIVE CLOSURES

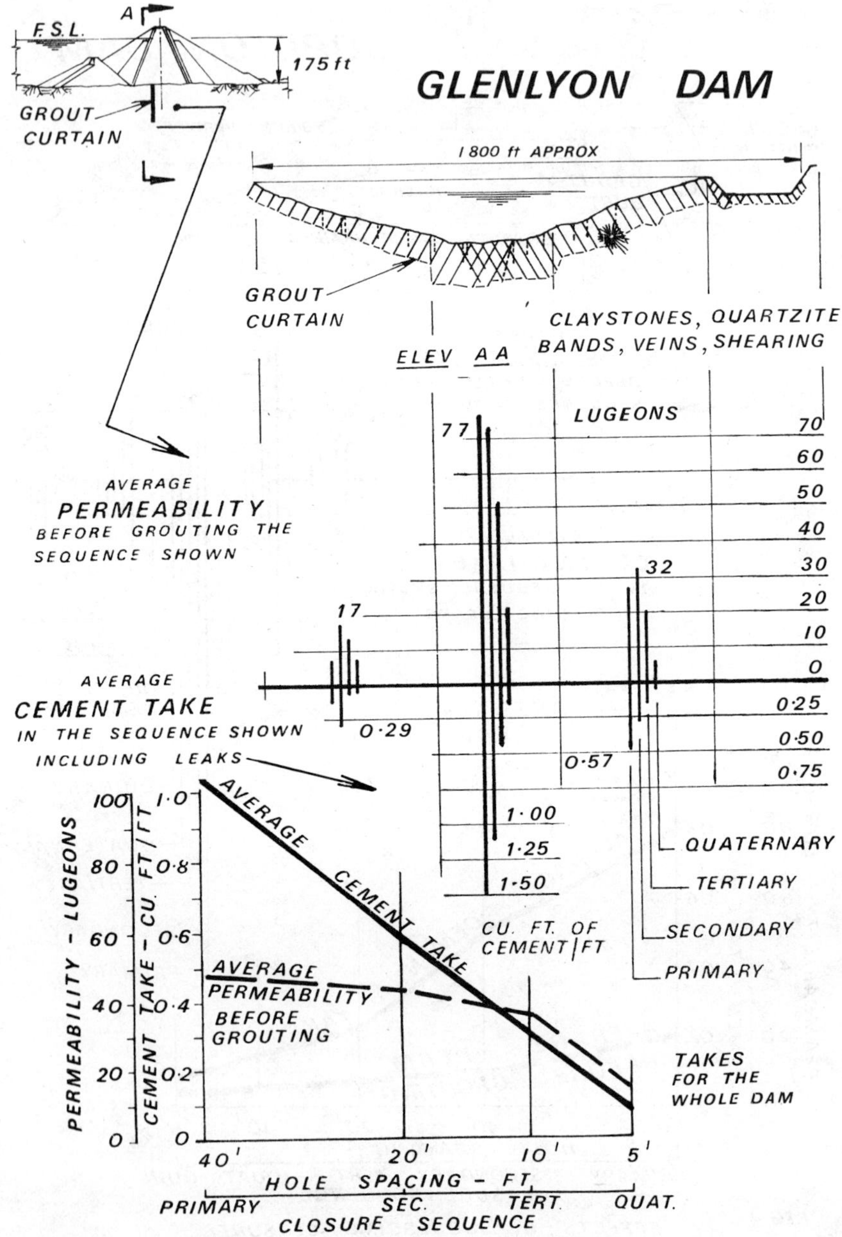

FIG 5 EFFECTS OF SUCCESSIVE CLOSURES

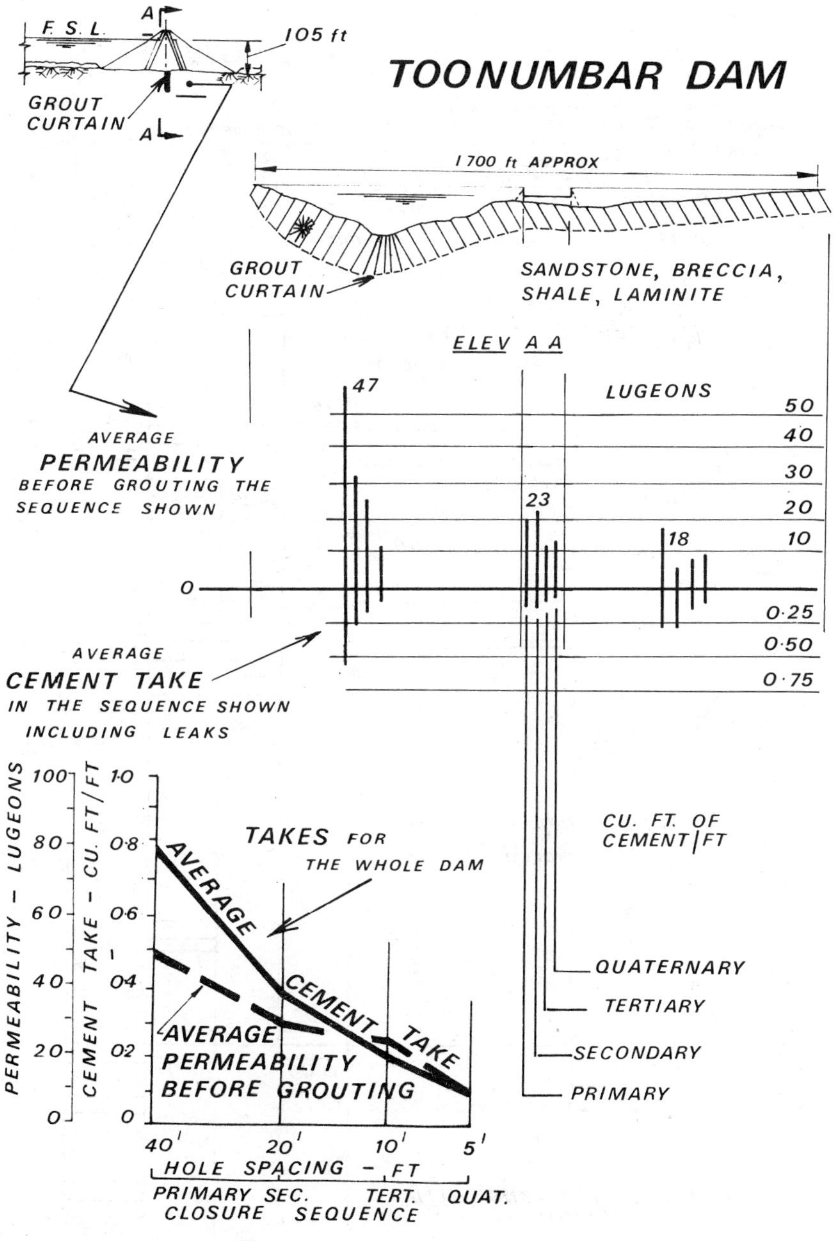

FIG 6 EFFECTS OF SUCCESSIVE CLOSURES

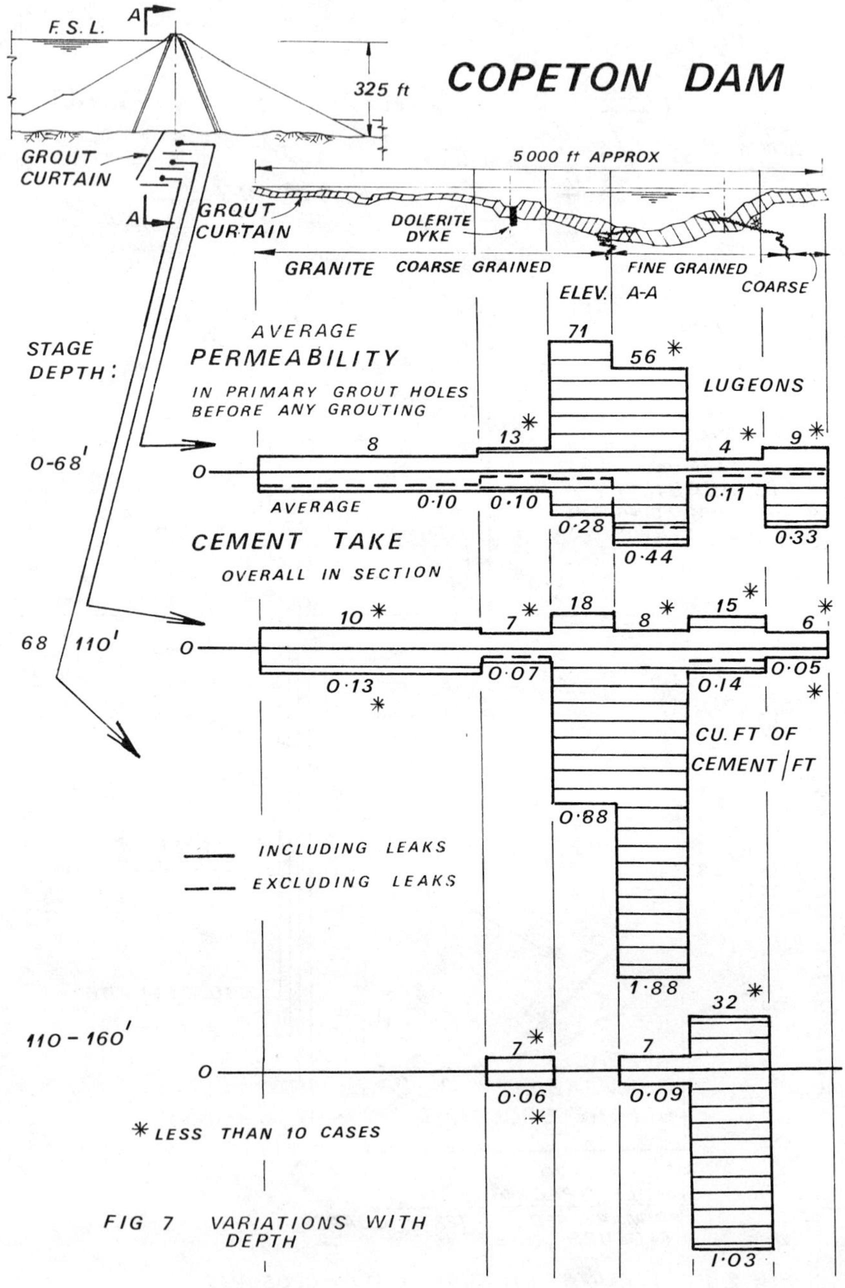

FIG 7 VARIATIONS WITH DEPTH

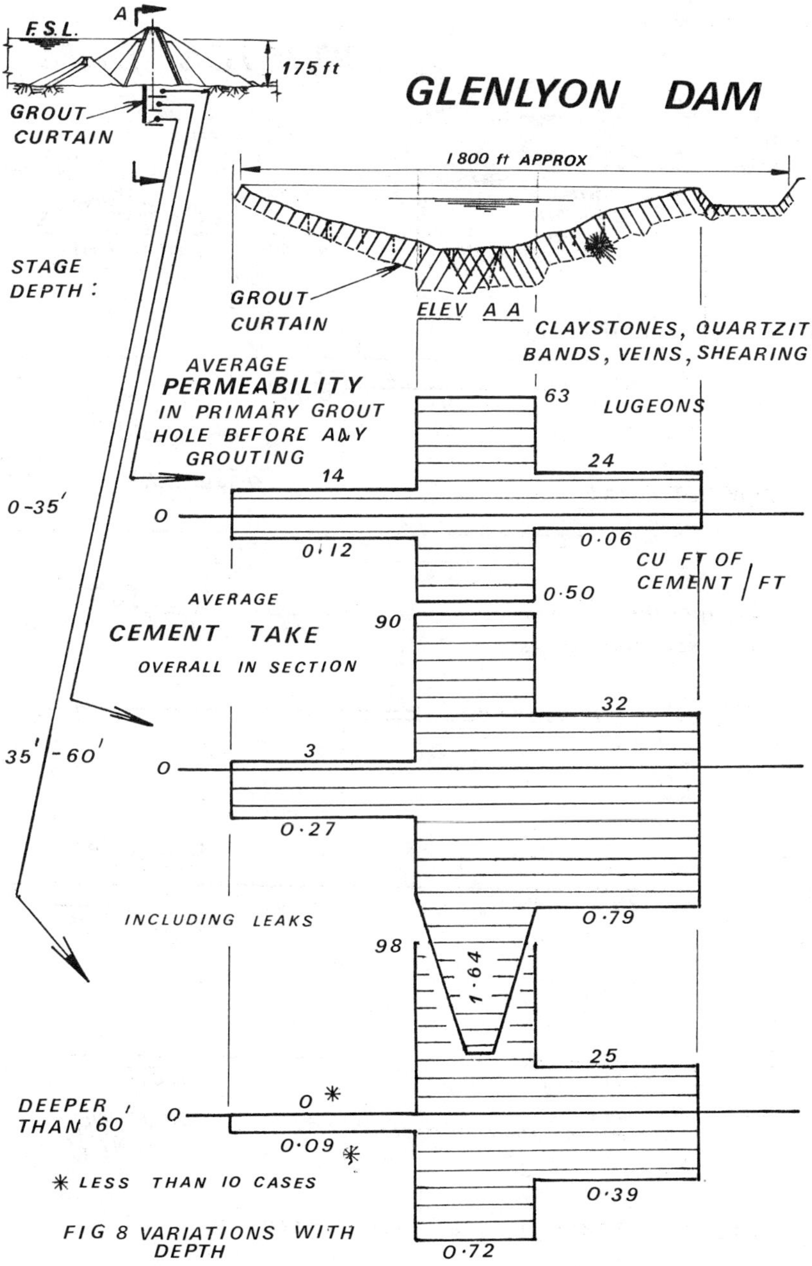

FIG 8 VARIATIONS WITH DEPTH

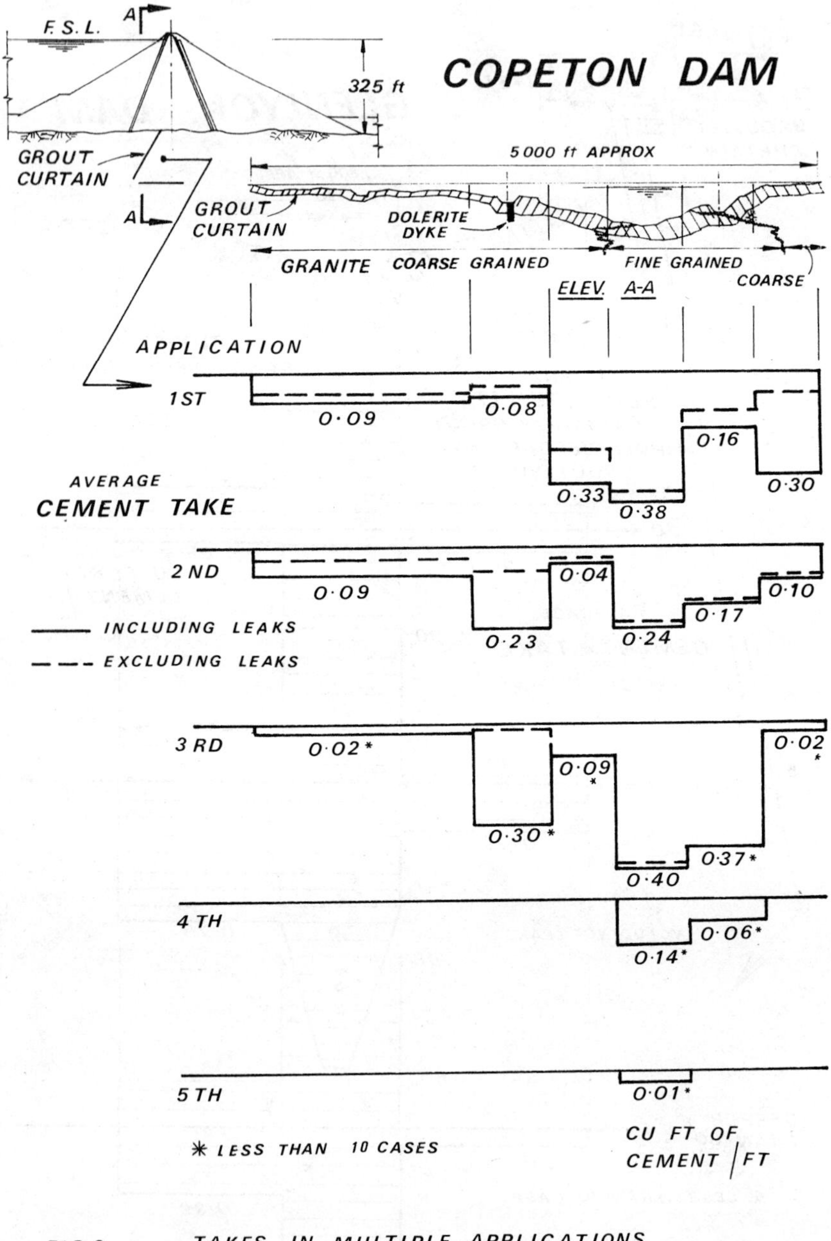

FIG 9 TAKES IN MULTIPLE APPLICATIONS

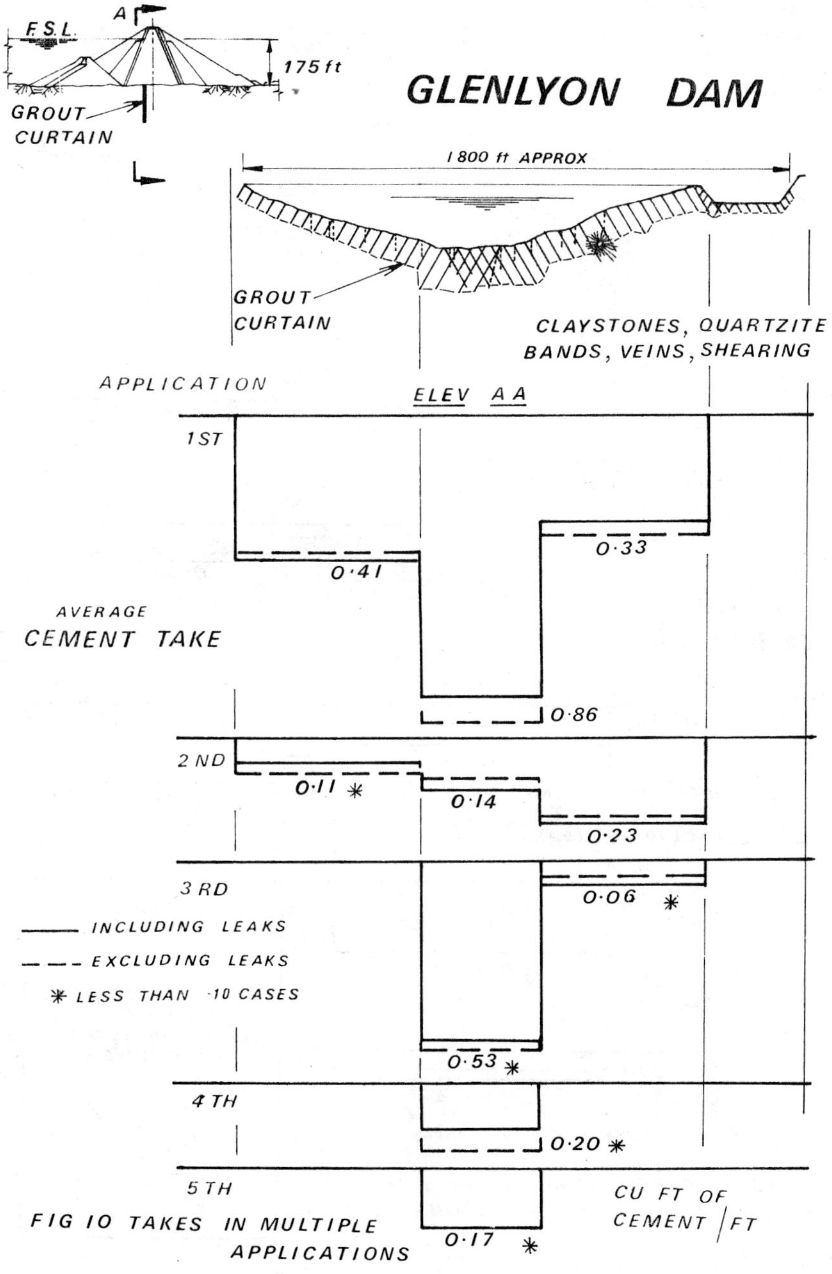

FIG 10 TAKES IN MULTIPLE APPLICATIONS

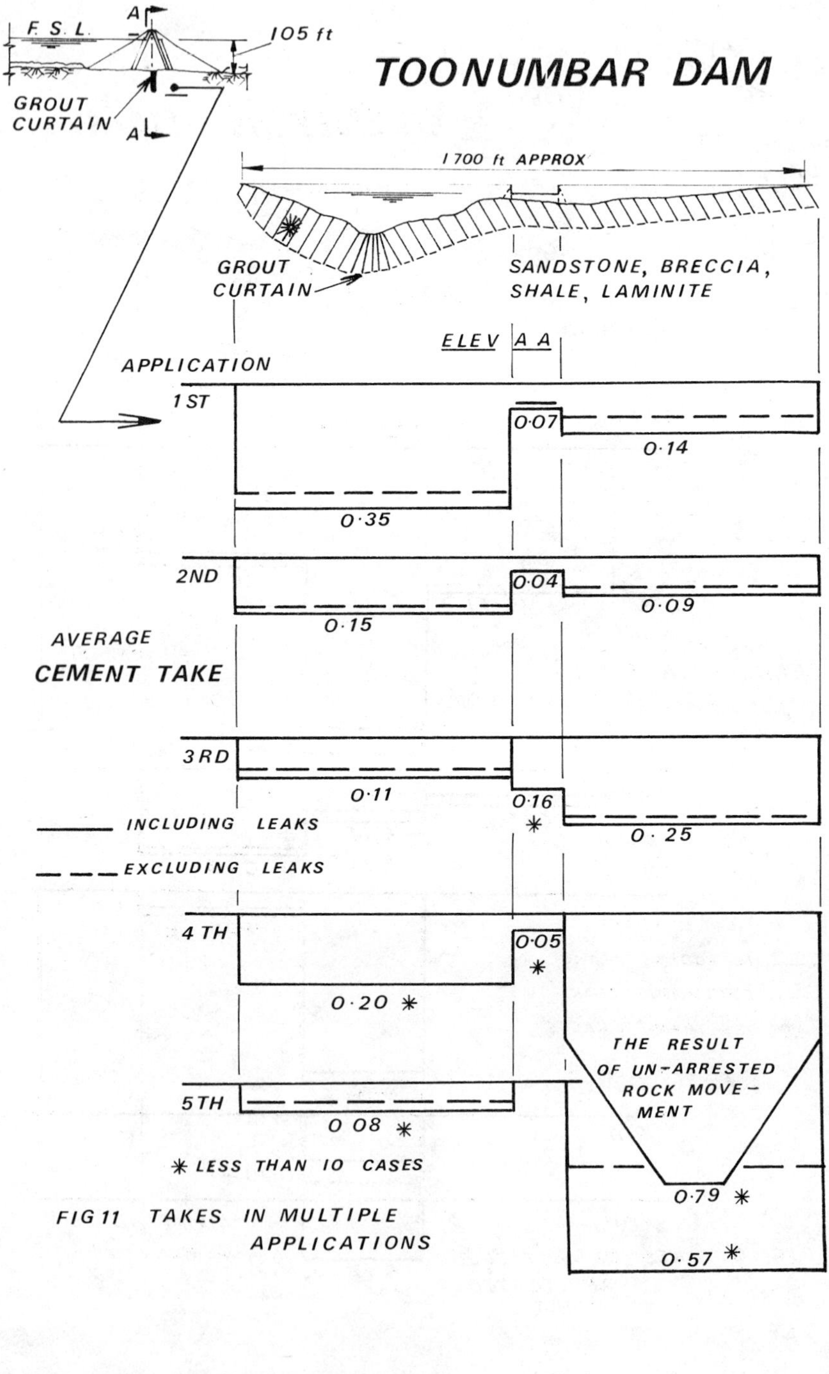

FIG 11 TAKES IN MULTIPLE APPLICATIONS

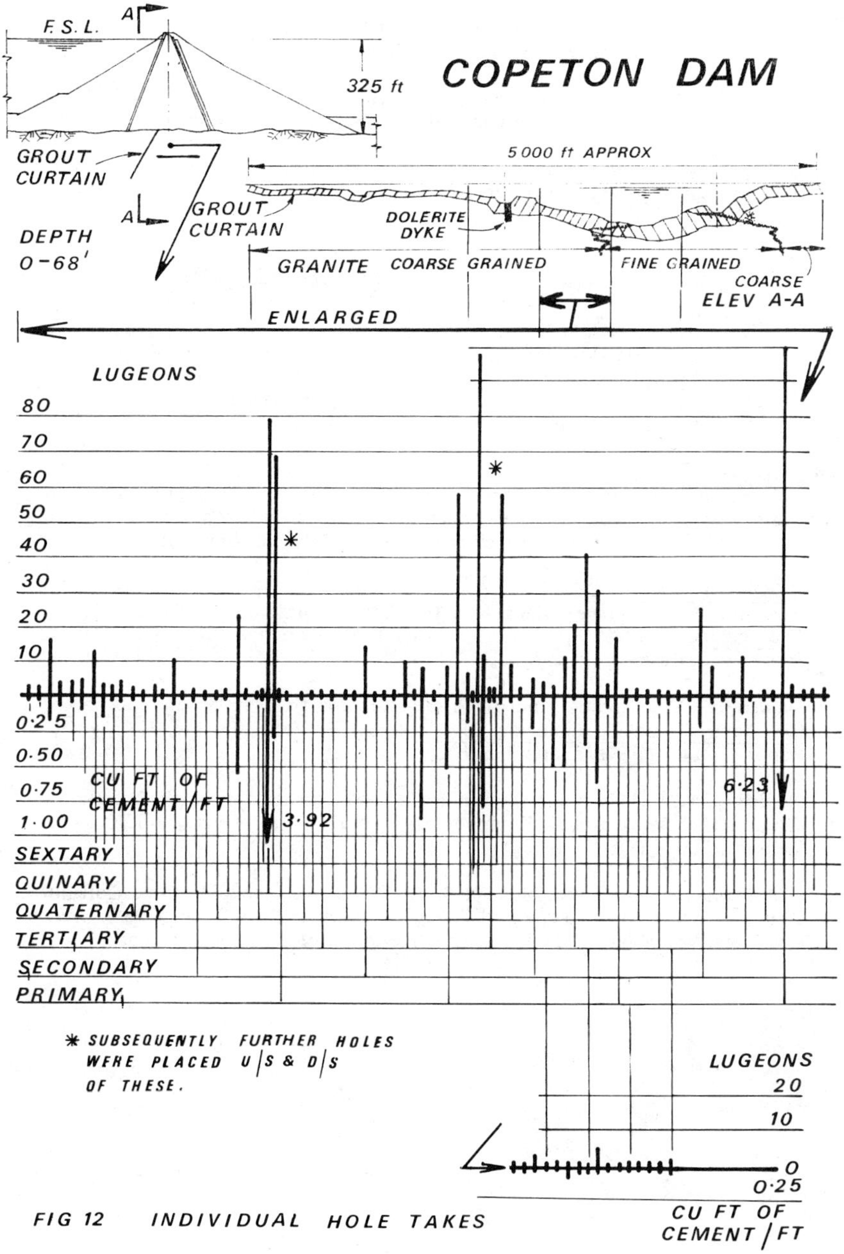

FIG 12 INDIVIDUAL HOLE TAKES

Individual Hole Takes

To conclude this presentation of data on a detailed note, fig. 12 shows the takes in every hole of a section of the curtain at Copeton Dam. Only the top stage is shown.

The highly variable nature of the takes is typical of many curtains and it emphasises the possibly misleading impressions given by averaging a number of holes. Because "windows" in grout curtains are often not permissible, there can be a need for individual examination of the takes of each hole when assessing grouting work in progress, rather than taking out averages when assessing.

SUMMARY

A summary is perhaps not really appropriate. However, it could be interesting to note -

(a) the general lack of correlation between water and cement takes;
(b) the radically different conclusions obtainable by averaging part or by the whole of a job, and the great difference possible between averages and the individual hole takes;
(c) the worthwhile quantities taken in multiple applications.

Acknowledgments

Appreciation is expressed to the Water Resources Commission, New South Wales for permission to present this data. Thanks go to Mr. U.G. Krause for conducting the detailed analysis of daunting quantities of basic data.

APPENDIX I - REFERENCES

1. Houlsby, A.C., "Towards Appropriate Metric Units for Grouting", Ground Engineering, Vol. 10, No. 5, July 1977 Foundation Publications, Brentwood, Essex, England, p. 39-42

2. Houlsby, A.C., "Engineering of Grout Curtains to Standards", Journal of the Geotechnical Engineering Division, ASCE, Vol. 103, No. GT9, Proc. Paper 13214, Sept. 1977, pp 953-970.

3. Houlsby, A.C., "Optimum Water : Cement Ratios for Rock Grouting", Grouting in Geotechnical Engineering-Specialty Conference, ASCE, New Orleans, Feb. 1982.

4. Water Resources Commission, N.S.W., "Grouting Manual", Third Edition, 1980, W.R.C., North Sydney, N.S.W. Australia, ISBN 0 7240 3786 1 (Also available from University of Missouri-Rolla, Civil Engineering Dept.).

SUBJECT INDEX

Page numbers refer to first page of paper

AUTHOR INDEX

Page numbers refer to first page of paper